KIRK-OTHMER

ENCYCLOPEDIA OF
CHEMICAL
TECHNOLOGY

FOURTH EDITION

VOLUME **17**

NICKEL AND NICKEL ALLOYS
TO
PAINT

EXECUTIVE EDITOR
Jacqueline I. Kroschwitz

EDITOR
Mary Howe-Grant

KIRK-OTHMER

ENCYCLOPEDIA OF CHEMICAL TECHNOLOGY

FOURTH EDITION

VOLUME **17**

NICKEL AND NICKEL ALLOYS
TO
PAINT

A Wiley-Interscience Publication
JOHN WILEY & SONS

New York • Chichester • Brisbane • Toronto • Singapore

This text is printed on acid-free paper.

Copyright © 1996 by John Wiley & Sons, Inc.

All rights reserved. Published simultaneously in Canada.

Library of Congress Cataloging-in-Publication Data

Encyclopedia of chemical technology/executive editor, Jacqueline
  I. Kroschwitz; editor, Mary Howe-Grant. —4th ed.
    p. cm.
  At head of title: Kirk-Othmer.
  "A Wiley-Interscience publication."
  Includes index.
  Contents: v. 17, Nickel and nickel alloys to paint
  ISBN 0-471-52686-X
  1. Chemistry, Technical—Encyclopedias. I. Kirk, Raymond E.
  (Raymond Eller), 1890–1957. II. Othmer, Donald F. (Donald
  Frederick), 1904– . III. Kroschwitz, Jacqueline I., 1942– .
  IV. Howe-Grant, Mary, 1943– . V. Title: Kirk-Othmer encyclopedia
  of chemical technology.
  TP9.E685 1992   91-16789
  660′.03 — dc20

Printed in the United States of America

10 9 8 7 6 5 4 3 2 1

CONTENTS

EDITORIAL STAFF
FOR VOLUME 17

Executive Editor: **Jacqueline I. Kroschwitz**
Editor: **Mary Howe-Grant**
Associate Managing Editor: **Lindy Humphreys**
Copy Editor: **Lawrence Altieri**
              Jonathan Lee

CONTRIBUTORS
TO VOLUME 17

Rick L. Adkins, *Bayer Corporation, New Martinsville, West Virginia*, Nitrobenzene and nitrotoluenes

Lyle F. Albright, *Purdue University, West Lafayette, Indiana*, Nitration

D. H. Antonsen, *International Nickel, Inc., Wycoff, New Jersey*, Nickel compounds

Jay A. Bardole, *Vincennes University, Vincennes, Indiana*, Paint and finish removers (under Paint)

Joseph J. Batelka, *Consultant, Savannah, Georgia*, Converting (under Packaging)

Yvette Berry, *Reckitt & Colman Inc., Wayne, New Jersey,* Odor modification

Ernst Billig, *Union Carbide Corporation, South Charleston, West Virginia,* Oxo process

Allen F. Bollmeier, Jr., *ANGUS Chemical Company, Buffalo Grove, Illinois,* Nitro alcohols; Nitroparaffins

Donald E. Bray, *Texas A&M University, College Station,* Nondestructive evaluation

David R. Bryant, *Union Carbide Corporation, South Charleston, West Virginia,* Oxo process

Daniel B. Bullen, *Iowa State University, Ames,* Nuclear fuel reserves (under Nuclear reactors)

A. R. Chapman, *Rochester Institute of Technology, Rochester, New York,* Containers for industrial materials (under Packaging)

Steven R. Childers, *Wake Forest University, Winston-Salem, North Carolina,* Opioids, endogenous

Stephen I. Clarke, *Air Products and Chemicals, Inc., Allentown, Pennsylvania,* Nitric acid

Michael J. Cruickshank, *University of Hawaii at Manoa, Honolulu,* Ocean raw materials

Larry R. Dalton, *University of Southern California, Los Angeles,* Nonlinear optical materials

Matthew Ennis, *Stanford University, Stanford, California,* Nitrogen

W. L. Godfrey, *BE Inc., Barnwell, South Carolina,* Chemical reprocessing (under Nuclear reactors)

J. C. Hall, *BE Inc., Barnwell, South Carolina,* Chemical reprocessing (under Nuclear reactors)

James G. Hansel, *Air Products and Chemicals, Inc., Allentown, Pennsylvania,* Oxygen

Thomas L. Hardenburger, *Air Liquide America Corporation, Tualatin, Oregon,* Nitrogen

Robert N. Heistand, *Consultant, Englewood, Colorado,* Oil shale

Jack H. Hicks, *Consultant, Lynchburg, Virginia,* Water chemistry of lightwater reactors (under Nuclear reactors)

Timothy E. Howson, *Wyman-Gordon, Worcester, Massachusetts,* Nickel and nickel alloys

Yury V. Kissin, *Mobil Chemical Company, Edison, New Jersey,* Introduction, High density polyethylene, Linear low density polyethylene, Polymers of higher olefins (under Olefin polymers)

G. R. Lappin, *Albemarle Corporation, Baton Rouge, Louisiana,* Olefins, higher

M. E. Leaphart II, *University of South Carolina, Columbia,* Nitrides

Arthur A. Leman, *Rohm and Haas Company, Spring House, Pennsylvania,* Architectural (under Paint)

Richard B. Lieberman, *Montell Polyolefins, Elkton, Maryland,* Polypropylene (under Olefin polymers)

Eric J. Markel, *University of South Carolina, Columbia,* Nitrides

J. Wilson Mausteller, *Consultant, Evans City, Pennsylvania,* Oxygen-generation systems

William J. Mazzafro, *Air Products and Chemicals, Inc., Allentown, Pennsylvania,* Nitric acid

Braja D. Mookherjee, *International Flavors and Fragrances, Inc., Union Beach, New Jersey,* Oils, essential

Toru Murakami, *UBE Industries Limited, Tokyo, Japan,* Oxalic acid

Raymond L. Murray, *Consultant, Raleigh, North Carolina,* Introduction, Reactor types, Waste management (under Nuclear reactors)

L. H. Nemec, *Albemarle Corporation, Baton Rouge, Louisiana,* Olefins, higher

William E. Newton, *Virginia Polytechnic Institute and State University, Blacksburg,* Nitrogen fixation

D. L. Olsson, *Rochester Institute of Technology, Rochester, New York,* Containers for industrial materials (under Packaging)

Michael Pecht, *University of Maryland, College Park,* Electronic materials (under Packaging)

Lloyd W. Pebsworth, *Polyethylene Technology, Morris, Illinois,* Low density polyethylene (under Olefin polymers)

Edwin M. Piper, *Piper Designs LLC, Littleton, Colorado,* Oil shale

Asohk Prabhu, *Nitto Denko America Inc., San Jose, California,* Electronic materials (under Packaging)

Jill Rehmann, *Fordham University, Brooklyn, New York,* Nucleic acids

Jack L. Rosette, *Forensic Packaging Concepts, Inc., Fort Mill, South Carolina,* Cosmetics and pharmaceuticals (under Packaging)

J. D. Sauer, *Albemarle Corporation, Baton Rouge, Louisiana,* Olefins, higher

Hiroyuki Sawada, *UBE Industries Limited, Tokyo, Japan,* Oxalic acid

James H. Schlewitz, *Teledyne Wah Chang, Albany, Oregon,* Niobium and niobium compounds

Dana E. Selley, *Wake Forest University, Winston-Salem, North Carolina,* Opioids, endogenous

Laura J. Sim, *Wake Forest University, Winston-Salem, North Carolina,* Opioids, endogenous

Peter A. S. Smith, *University of Michigan, Ann Arbor,* Nomenclature

Ronald Smorada, *Reemay Inc., Old Hickory, Tennessee,* Fabrics, spunbonded (under Nonwoven fabrics)

J. J. Taylor, *Electric Power Research Institute, Palo Alto, California,* Safety in nuclear power facilities (under Nuclear reactors)

John K. Tien, *Columbia University, New York, New York,* Nickel and nickel alloys

John H. Tundermann, *Inco Alloys International Inc., Huntington, West Virginia,* Nickel and nickel alloys

G. A. Townes, *BE Inc., Barnwell, South Carolina,* Chemical reprocessing (under Nuclear reactors)

E. A. Vaughn, *Clemson University, Clemson, South Carolina,* Staple fibers (under Nonwoven fabrics)

Edward Von Halle, *Consultant, Oak Ridge, Tennessee*, Isotope separation (under Nuclear reactors)

J. D. Wagner, *Albermarle Corporation, Baton Rouge, Louisiana*, Olefins, higher

David L. White, *Kwick Kleen Industrial Solvents, Inc., Vincennes, Indiana*, Paint and finish removers (under Paint)

John S. Wishnok, *Massachusetts Institute of Technology, Cambridge*, N-Nitrosamines

Richard A. Wilson, *International Flavors and Fragrances, Inc., Union Beach, New Jersey*, Oils, essential

J. A. Wojtowicz, *Consultant, Cheshire, Connecticut*, Ozone

Clyde T. Young, *North Carolina State University, Raleigh*, Nuts

NOTE ON CHEMICAL ABSTRACTS SERVICE REGISTRY NUMBERS AND NOMENCLATURE

Chemical Abstracts Service (CAS) Registry Numbers are unique numerical identifiers assigned to substances recorded in the CAS Registry System. They appear in brackets in the *Chemical Abstracts* (CA) substance and formula indexes following the names of compounds. A single compound may have synonyms in the chemical literature. A simple compound like phenethylamine can be named β-phenylethylamine or, as in *Chemical Abstracts*, benzeneethanamine. The usefulness of the *Encyclopedia* depends on accessibility through the most common correct name of a substance. Because of this diversity in nomenclature careful attention has been given to the problem in order to assist the reader as much as possible, especially in locating the systematic CA index name by means of the Registry Number. For this purpose, the reader may refer to the CAS Registry Handbook—Number Section which lists in numerical order the Registry Number with the *Chemical Abstracts* index name and the molecular formula; eg, **458-88-8**, Piperidine, 2-propyl-, (*S*)-, $C_8H_{17}N$; in the *Encyclopedia* this compound would be found under its common name, coniine [*458-88-8*]. Alternatively, this information can be retrieved electronically from CAS Online. In many cases molecular formulas have also been provided in the *Encyclopedia* text to facilitate electronic searching. The Registry Number is a valuable link for the reader in retrieving additional published information on substances and also as a point of access for on-line data bases.

In all cases, the CAS Registry Numbers have been given for title compounds in articles and for all compounds in the index. All specific substances indexed in *Chemical Abstracts* since 1965 are included in the CAS Registry System as are a large number of substances derived from a variety of reference works. The CAS Registry System identifies a substance on the basis of an unambiguous computer-language description of its molecular structure including stereochemical detail. The Registry Number is a machine-checkable number (like a Social Security number) assigned in sequential order to each substance as it enters the registry system. The value of the number lies in the fact that it is a concise and unique means of substance identification, which is independent of, and therefore bridges, many systems of chemical nomenclature. For polymers, one Registry Number may be used for the entire family; eg, polyoxyethylene (20) sorbitan monolaurate has the same number as all of its polyoxyethylene homologues.

Cross-references are inserted in the index for many common names and for some systematic names. Trademark names appear in the index. Names that are incorrect, misleading, or ambiguous are avoided. Formulas are given very frequently in the text to help in identifying compounds. The spelling and form used, even for industrial names, follow American chemical usage, but not always the usage of *Chemical Abstracts* (eg, *coniine* is used instead of *(S)-2-propylpiperidine*, *aniline* instead of *benzenamine*, and *acrylic acid* instead of *2-propenoic acid*).

There are variations in representation of rings in different disciplines. The dye industry does not designate aromaticity or double bonds in rings. All double bonds and aromaticity are shown in the *Encyclopedia* as a matter of course. For example, tetralin has an aromatic ring and a saturated ring and its structure

appears in the *Encyclopedia* with its common name, Registry Number enclosed in brackets, and parenthetical CA index name, ie, tetralin [*119-64-2*] (1,2,3,4-tetrahydronaphthalene). With names and structural formulas, and especially with CAS Registry Numbers, the aim is to help the reader have a concise means of substance identification.

CONVERSION FACTORS, ABBREVIATIONS, AND UNIT SYMBOLS

SI Units (Adopted 1960)

The International System of Units (abbreviated SI), is being implemented throughout the world. This measurement system is a modernized version of the MKSA (meter, kilogram, second, ampere) system, and its details are published and controlled by an international treaty organization (The International Bureau of Weights and Measures) (1).

SI units are divided into three classes:

BASE UNITS

length	meter[†] (m)
mass	kilogram (kg)
time	second (s)
electric current	ampere (A)
thermodynamic temperature[‡]	kelvin (K)
amount of substance	mole (mol)
luminous intensity	candela (cd)

SUPPLEMENTARY UNITS

plane angle	radian (rad)
solid angle	steradian (sr)

[†]The spellings "metre" and "litre" are preferred by ASTM; however, "-er" is used in the *Encyclopedia*.

[‡]Wide use is made of Celsius temperature (*t*) defined by

$$t = T - T_0$$

where T is the thermodynamic temperature, expressed in kelvin, and $T_0 = 273.15$ K by definition. A temperature interval may be expressed in degrees Celsius as well as in kelvin.

DERIVED UNITS AND OTHER ACCEPTABLE UNITS

These units are formed by combining base units, supplementary units, and other derived units (2–4). Those derived units having special names and symbols are marked with an asterisk in the list below.

Quantity	Unit	Symbol	Acceptable equivalent
*absorbed dose	gray	Gy	J/kg
acceleration	meter per second squared	m/s^2	
*activity (of a radionuclide)	becquerel	Bq	1/s
area	square kilometer	km^2	
	square hectometer	hm^2	ha (hectare)
	square meter	m^2	
concentration (of amount of substance)	mole per cubic meter	mol/m^3	
current density	ampere per square meter	$A//m^2$	
density, mass density	kilogram per cubic meter	kg/m^3	g/L; mg/cm^3
dipole moment (quantity)	coulomb meter	C·m	
*dose equivalent	sievert	Sv	J/kg
*electric capacitance	farad	F	C/V
*electric charge, quantity of electricity	coulomb	C	A·s
electric charge density	coulomb per cubic meter	C/m^3	
*electric conductance	siemens	S	A/V
electric field strength	volt per meter	V/m	
electric flux density	coulomb per square meter	C/m^2	
*electric potential, potential difference, electromotive force	volt	V	W/A
*electric resistance	ohm	Ω	V/A
*energy, work, quantity of heat	megajoule	MJ	
	kilojoule	kJ	
	joule	J	N·m
	electronvolt[†]	eV[†]	
	kilowatt-hour[†]	kW·h[†]	
energy density	joule per cubic meter	J/m^3	
*force	kilonewton	kN	
	newton	N	$kg·m/s^2$

[†]This non-SI unit is recognized by the CIPM as having to be retained because of practical importance or use in specialized fields (1).

Quantity	Unit	Symbol	Acceptable equivalent
*frequency	megahertz	MHz	
	hertz	Hz	1/s
heat capacity, entropy	joule per kelvin	J/K	
heat capacity (specific), specific entropy	joule per kilogram kelvin	J/(kg·K)	
heat-transfer coefficient	watt per square meter kelvin	W/(m²·K)	
*illuminance	lux	lx	lm/m²
*inductance	henry	H	Wb/A
linear density	kilogram per meter	kg/m	
luminance	candela per square meter	cd/m²	
*luminous flux	lumen	lm	cd·sr
magnetic field strength	ampere per meter	A/m	
*magnetic flux	weber	Wb	V·s
*magnetic flux density	tesla	T	Wb/m²
molar energy	joule per mole	J/mol	
molar entropy, molar heat capacity	joule per mole kelvin	J/(mol·K)	
moment of force, torque	newton meter	N·m	
momentum	kilogram meter per second	kg·m/s	
permeability	henry per meter	H/m	
permittivity	farad per meter	F/m	
*power, heat flow rate, radiant flux	kilowatt	kW	
	watt	W	J/s
power density, heat flux density, irradiance	watt per square meter	W/m²	
*pressure, stress	megapascal	MPa	
	kilopascal	kPa	
	pascal	Pa	N/m²
sound level	decibel	dB	
specific energy	joule per kilogram	J/kg	
specific volume	cubic meter per kilogram	m³/kg	
surface tension	newton per meter	N/m	
thermal conductivity	watt per meter kelvin	W/(m·K)	
velocity	meter per second	m/s	
	kilometer per hour	km/h	
viscosity, dynamic	pascal second	Pa·s	
	millipascal second	mPa·s	
viscosity, kinematic	square meter per second	m²/s	
	square millimeter per second	mm²/s	

Quantity	Unit	Symbol	Acceptable equivalent
volume	cubic meter	m^3	
	cubic diameter	dm^3	L (liter) (5)
	cubic centimeter	cm^3	mL
wave number	1 per meter	m^{-1}	
	1 per centimeter	cm^{-1}	

In addition, there are 16 prefixes used to indicate order of magnitude, as follows:

Multiplication factor	Prefix	Symbol	Note
10^{18}	exa	E	
10^{15}	peta	P	
10^{12}	tera	T	
10^{9}	giga	G	
10^{6}	mega	M	
10^{3}	kilo	k	
10^{2}	hecto	h[a]	[a]Although hecto, deka, deci, and centi
10	deka	da[a]	are SI prefixes, their use should be
10^{-1}	deci	d[a]	avoided except for SI unit-multiples
10^{-2}	centi	c[a]	for area and volume and nontech-
10^{-3}	milli	m	nical use of centimeter, as for body
10^{-6}	micro	μ	and clothing measurement.
10^{-9}	nano	n	
10^{-12}	pico	p	
10^{-15}	femto	f	
10^{-18}	atto	a	

For a complete description of SI and its use the reader is referred to ASTM E380 (4) and the article UNITS AND CONVERSION FACTORS which appears in Vol. 24.

A representative list of conversion factors from non-SI to SI units is presented herewith. Factors are given to four significant figures. Exact relationships are followed by a dagger. A more complete list is given in the latest editions of ASTM E380 (4) and ANSI Z210.1 (6).

Conversion Factors to SI Units

To convert from	To	Multiply by
acre	square meter (m^2)	4.047×10^3
angstrom	meter (m)	$1.0 \times 10^{-10\dagger}$
are	square meter (m^2)	$1.0 \times 10^{2\dagger}$

†Exact.

To convert from	To	Multiply by
astronomical unit	meter (m)	1.496×10^{11}
atmosphere, standard	pascal (Pa)	1.013×10^{5}
bar	pascal (Pa)	$1.0 \times 10^{5\dagger}$
barn	square meter (m²)	$1.0 \times 10^{-28\dagger}$
barrel (42 U.S. liquid gallons)	cubic meter (m³)	0.1590
Bohr magneton (μ_B)	J/T	9.274×10^{-24}
Btu (International Table)	joule (J)	1.055×10^{3}
Btu (mean)	joule (J)	1.056×10^{3}
Btu (thermochemical)	joule (J)	1.054×10^{3}
bushel	cubic meter (m³)	3.524×10^{-2}
calorie (International Table)	joule (J)	4.187
calorie (mean)	joule (J)	4.190
calorie (thermochemical)	joule (J)	$4.184^{\dagger}$
centipoise	pascal second (Pa·s)	$1.0 \times 10^{-3\dagger}$
centistokes	square millimeter per second (mm²/s)	$1.0^{\dagger}$
cfm (cubic foot per minute)	cubic meter per second (m³/s)	4.72×10^{-4}
cubic inch	cubic meter (m³)	1.639×10^{-5}
cubic foot	cubic meter (m³)	2.832×10^{-2}
cubic yard	cubic meter (m³)	0.7646
curie	becquerel (Bq)	$3.70 \times 10^{10\dagger}$
debye	coulomb meter (C·m)	3.336×10^{-30}
degree (angle)	radian (rad)	1.745×10^{-2}
denier (international)	kilogram per meter (kg/m)	1.111×10^{-7}
	tex‡	0.1111
dram (apothecaries')	kilogram (kg)	3.888×10^{-3}
dram (avoirdupois)	kilogram (kg)	1.772×10^{-3}
dram (U.S. fluid)	cubic meter (m³)	3.697×10^{-6}
dyne	newton (N)	$1.0 \times 10^{-5\dagger}$
dyne/cm	newton per meter (N/m)	$1.0 \times 10^{-3\dagger}$
electronvolt	joule (J)	1.602×10^{-19}
erg	joule (J)	$1.0 \times 10^{-7\dagger}$
fathom	meter (m)	1.829
fluid ounce (U.S.)	cubic meter (m³)	2.957×10^{-5}
foot	meter (m)	$0.3048^{\dagger}$
footcandle	lux (lx)	10.76
furlong	meter (m)	2.012×10^{-2}
gal	meter per second squared (m/s²)	$1.0 \times 10^{-2\dagger}$
gallon (U.S. dry)	cubic meter (m³)	4.405×10^{-3}
gallon (U.S. liquid)	cubic meter (m³)	3.785×10^{-3}
gallon per minute (gpm)	cubic meter per second (m³/s)	6.309×10^{-5}
	cubic meter per hour (m³/h)	0.2271

†Exact.
‡See footnote on p. xiii.

To convert from	To	Multiply by
gauss	tesla (T)	1.0×10^{-4}
gilbert	ampere (A)	0.7958
gill (U.S.)	cubic meter (m^3)	1.183×10^{-4}
grade	radian	1.571×10^{-2}
grain	kilogram (kg)	6.480×10^{-5}
gram force per denier	newton per tex (N/tex)	8.826×10^{-2}
hectare	square meter (m^2)	$1.0 \times 10^{4\dagger}$
horsepower (550 ft·lbf/s)	watt (W)	7.457×10^{2}
horsepower (boiler)	watt (W)	9.810×10^{3}
horsepower (electric)	watt (W)	$7.46 \times 10^{2\dagger}$
hundredweight (long)	kilogram (kg)	50.80
hundredweight (short)	kilogram (kg)	45.36
inch	meter (m)	$2.54 \times 10^{-2\dagger}$
inch of mercury (32°F)	pascal (Pa)	3.386×10^{3}
inch of water (39.2°F)	pascal (Pa)	2.491×10^{2}
kilogram-force	newton (N)	9.807
kilowatt hour	megajoule (MJ)	$3.6^{\dagger}$
kip	newton (N)	4.448×10^{3}
knot (international)	meter per second (m/S)	0.5144
lambert	candela per square meter (cd/m^3)	3.183×10^{3}
league (British nautical)	meter (m)	5.559×10^{3}
league (statute)	meter (m)	4.828×10^{3}
light year	meter (m)	9.461×10^{15}
liter (for fluids only)	cubic meter (m^3)	$1.0 \times 10^{-3\dagger}$
maxwell	weber (Wb)	$1.0 \times 10^{-8\dagger}$
micron	meter (m)	$1.0 \times 10^{-6\dagger}$
mil	meter (m)	$2.54 \times 10^{-5\dagger}$
mile (statute)	meter (m)	1.609×10^{3}
mile (U.S. nautical)	meter (m)	$1.852 \times 10^{3\dagger}$
mile per hour	meter per second (m/s)	0.4470
millibar	pascal (Pa)	1.0×10^{2}
millimeter of mercury (0°C)	pascal (Pa)	$1.333 \times 10^{2\dagger}$
minute (angular)	radian	2.909×10^{-4}
myriagram	kilogram (kg)	10
myriameter	kilometer (km)	10
oersted	ampere per meter (A/m)	79.58
ounce (avoirdupois)	kilogram (kg)	2.835×10^{-2}
ounce (troy)	kilogram (kg)	3.110×10^{-2}
ounce (U.S. fluid)	cubic meter (m^3)	2.957×10^{-5}
ounce-force	newton (N)	0.2780
peck (U.S.)	cubic meter (m^3)	8.810×10^{-3}
pennyweight	kilogram (kg)	1.555×10^{-3}
pint (U.S. dry)	cubic meter (m^3)	5.506×10^{-4}
pint (U.S. liquid)	cubic meter (m^3)	4.732×10^{-4}

†Exact.

To convert from	To	Multiply by
poise (absolute viscosity)	pascal second (Pa·s)	$0.10^{\dagger}$
pound (avoirdupois)	kilogram (kg)	0.4536
pound (troy)	kilogram (kg)	0.3732
poundal	newton (N)	0.1383
pound-force	newton (N)	4.448
pound force per square inch (psi)	pascal (Pa)	6.895×10^3
quart (U.S. dry)	cubic meter (m^3)	1.101×10^{-3}
quart (U.S. liquid)	cubic meter (m^3)	9.464×10^{-4}
quintal	kilogram (kg)	$1.0 \times 10^{2\dagger}$
rad	gray (Gy)	$1.0 \times 10^{-2\dagger}$
rod	meter (m)	5.029
roentgen	coulomb per kilogram (C/kg)	2.58×10^{-4}
second (angle)	radian (rad)	$4.848 \times 10^{-6\dagger}$
section	square meter (m^2)	2.590×10^6
slug	kilogram (kg)	14.59
spherical candle power	lumen (lm)	12.57
square inch	square meter (m^2)	6.452×10^{-4}
square foot	square meter (m^2)	9.290×10^{-2}
square mile	square meter (m^2)	2.590×10^6
square yard	square meter (m^2)	0.8361
stere	cubic meter (m^3)	$1.0^{\dagger}$
stokes (kinematic viscosity)	square meter per second (m^2/s)	$1.0 \times 10^{-4\dagger}$
tex	kilogram per meter (kg/m)	$1.0 \times 10^{-6\dagger}$
ton (long, 2240 pounds)	kilogram (kg)	1.016×10^3
ton (metric) (tonne)	kilogram (kg)	$1.0 \times 10^{3\dagger}$
ton (short, 2000 pounds)	kilogram (kg)	9.072×10^2
torr	pascal (Pa)	1.333×10^2
unit pole	weber (Wb)	1.257×10^{-7}
yard	meter (m)	$0.9144^{\dagger}$

†Exact.

Abbreviations and Unit Symbols

Following is a list of common abbreviations and unit symbols used in the *Encyclopedia*. In general they agree with those listed in *American National Standard Abbreviations for Use on Drawings and in Text (ANSI Y1.1)* (6) and *American National Standard Letter Symbols for Units in Science and Technology (ANSI Y10)* (6). Also included is a list of acronyms for a number of private and government organizations as well as common industrial solvents, polymers, and other chemicals.

Rules for Writing Unit Symbols (4):

1. Unit symbols are printed in upright letters (roman) regardless of the type style used in the surrounding text.
2. Unit symbols are unaltered in the plural.
3. Unit symbols are not followed by a period except when used at the end of a sentence.
4. Letter unit symbols are generally printed lower-case (for example, cd for candela) unless the unit name has been derived from a proper name, in which case the first letter of the symbol is capitalized (W, Pa). Prefixes and unit symbols retain their prescribed form regardless of the surrounding typography.
5. In the complete expression for a quantity, a space should be left between the numerical value and the unit symbol. For example, write 2.37 lm, *not* 2.37lm, and 35 mm, *not* 35mm. When the quantity is used in an adjectival sense, a hyphen is often used, for example, 35-mm film. *Exception:* No space is left between the numerical value and the symbols of degree, minute, and second of plane angle, degree Celsius, and the percent sign.
6. No space is used between the prefix and unit symbol (for example, kg).
7. Symbols, not abbreviations, should be used for units. For example, use "A," not "amp," for ampere.
8. When multiplying unit symbols, use a raised dot:

$$\text{N·m} \quad \text{for} \quad \text{newton meter}$$

In the case of W·h, the dot may be omitted, thus:

$$\text{Wh}$$

An exception to this practice is made for computer printouts, automatic typewriter work, etc, where the raised dot is not possible, and a dot on the line may be used.
9. When dividing unit symbols, use one of the following forms:

$$\text{m/s} \quad or \quad \text{m·s}^{-1} \quad or \quad \frac{\text{m}}{\text{s}}$$

In no case should more than one slash be used in the same expression unless parentheses are inserted to avoid ambiguity. For example, write:

$$\text{J/(mol·K)} \quad or \quad \text{J·mol}^{-1}\text{·K}^{-1} \quad or \quad \text{(J/mol)/K}$$

but *not*

$$\text{J/mol/K}$$

10. Do not mix symbols and unit names in the same expression. Write:

$$\text{joules per kilogram} \quad or \quad \text{J/kg} \quad or \quad \text{J} \cdot \text{kg}^{-1}$$

but *not*

$$\text{joules/kilogram} \quad nor \quad \text{joules/kg} \quad nor \quad \text{joules} \cdot \text{kg}^{-1}$$

ABBREVIATIONS AND UNITS

A	ampere	AOAC	Association of Official Analytical Chemists
A	anion (eg, HA)		
A	mass number	AOCS	American Oil Chemists' Society
a	atto (prefix for 10^{-18})		
AATCC	American Association of Textile Chemists and Colorists	APHA	American Public Health Association
		API	American Petroleum Institute
ABS	acrylonitrile–butadiene–styrene	aq	aqueous
abs	absolute	Ar	aryl
ac	alternating current, *n*.	*ar-*	aromatic
a-c	alternating current, *adj*.	*as-*	asymmetric(al)
ac-	alicyclic	ASHRAE	American Society of Heating, Refrigerating, and Air Conditioning Engineers
acac	acetylacetonate		
ACGIH	American Conference of Governmental Industrial Hygienists		
		ASM	American Society for Metals
ACS	American Chemical Society	ASME	American Society of Mechanical Engineers
AGA	American Gas Association		
Ah	ampere hour	ASTM	American Society for Testing and Materials
AIChE	American Institute of Chemical Engineers	at no.	atomic number
AIME	American Institute of Mining, Metallurgical, and Petroleum Engineers	at wt	atomic weight
		av(g)	average
		AWS	American Welding Society
		b	bonding orbital
AIP	American Institute of Physics	bbl	barrel
		bcc	body-centered cubic
AISI	American Iron and Steel Institute	BCT	body-centered tetragonal
		Bé	Baumé
alc	alcohol(ic)	BET	Brunauer-Emmett-Teller (adsorption equation)
Alk	alkyl		
alk	alkaline (not alkali)	bid	twice daily
amt	amount	Boc	*t*-butyloxycarbonyl
amu	atomic mass unit	BOD	biochemical (biological) oxygen demand
ANSI	American National Standards Institute		
		bp	boiling point
AO	atomic orbital	Bq	becquerel

C	coulomb	DIN	Deutsche Industrie
°C	degree Celsius		Normen
C-	denoting attachment to	*dl*-; DL-	racemic
	carbon	DMA	dimethylacetamide
c	centi (prefix for 10^{-2})	DMF	dimethylformamide
c	critical	DMG	dimethyl glyoxime
ca	circa (approximately)	DMSO	dimethyl sulfoxide
cd	candela; current density;	DOD	Department of Defense
	circular dichroism	DOE	Department of Energy
CFR	Code of Federal	DOT	Department of
	Regulations		Transportation
cgs	centimeter-gram-second	DP	degree of polymerization
CI	Color Index	dp	dew point
cis-	isomer in which	DPH	diamond pyramid
	substituted groups are		hardness
	on same side of double	dstl(d)	distill(ed)
	bond between C atoms	dta	differential thermal
cl	carload		analysis
cm	centimeter	(*E*)-	entgegen; opposed
cmil	circular mil	ϵ	dielectric constant
cmpd	compound		(unitless number)
CNS	central nervous system	*e*	electron
CoA	coenzyme A	ECU	electrochemical unit
COD	chemical oxygen demand	ed.	edited, edition, editor
coml	commercial(ly)	ED	effective dose
cp	chemically pure	EDTA	ethylenediaminetetra-
cph	close-packed hexagonal		acetic acid
CPSC	Consumer Product Safety	emf	electromotive force
	Commission	emu	electromagnetic unit
cryst	crystalline	en	ethylene diamine
cub	cubic	eng	engineering
D	debye	EPA	Environmental Protection
D-	denoting configurational		Agency
	relationship	epr	electron paramagnetic
d	differential operator		resonance
d	day; deci (prefix for 10^{-1})	eq.	equation
d	density	esca	electron spectroscopy for
d-	*dextro*-, dextrorotatory		chemical analysis
da	deka (prefix for 10^1)	esp	especially
dB	decibel	esr	electron-spin resonance
dc	direct current, *n.*	est(d)	estimate(d)
d-c	direct current, *adj.*	estn	estimation
dec	decompose	esu	electrostatic unit
detd	determined	exp	experiment, experimental
detn	determination	ext(d)	extract(ed)
Di	didymium, a mixture of all	F	farad (capacitance)
	lanthanons	*F*	faraday (96,487 C)
dia	diameter	f	femto (prefix for 10^{-15})
dil	dilute		

FAO	Food and Agriculture Organization (United Nations)	hyd	hydrated, hydrous
		hyg	hygroscopic
		Hz	hertz
fcc	face-centered cubic	i (eg, Pr^i)	iso (eg, isopropyl)
FDA	Food and Drug Administration	i-	inactive (eg, i-methionine)
		IACS	International Annealed Copper Standard
FEA	Federal Energy Administration	ibp	initial boiling point
FHSA	Federal Hazardous Substances Act	IC	integrated circuit
		ICC	Interstate Commerce Commission
fob	free on board		
fp	freezing point	ICT	International Critical Table
FPC	Federal Power Commission	ID	inside diameter; infective dose
FRB	Federal Reserve Board		
frz	freezing	ip	intraperitoneal
G	giga (prefix for 10^9)	IPS	iron pipe size
G	gravitational constant = 6.67×10^{11} N·m^2/kg^2	ir	infrared
		IRLG	Interagency Regulatory Liaison Group
g	gram		
(g)	gas, only as in H_2O(g)	ISO	International Organization Standardization
g	gravitational acceleration		
gc	gas chromatography	ITS-90	International Temperature Scale (NIST)
gem-	geminal		
glc	gas–liquid chromatography	IU	International Unit
		IUPAC	International Union of Pure and Applied Chemistry
g-mol wt; gmw	gram-molecular weight		
GNP	gross national product	IV	iodine value
gpc	gel-permeation chromatography	iv	intravenous
		J	joule
GRAS	Generally Recognized as Safe	K	kelvin
		k	kilo (prefix for 10^3)
grd	ground	kg	kilogram
Gy	gray	L	denoting configurational relationship
H	henry		
h	hour; hecto (prefix for 10^2)	L	liter (for fluids only) (5)
ha	hectare	l-	$levo$-, levorotatory
HB	Brinell hardness number	(l)	liquid, only as in NH_3(l)
Hb	hemoglobin	LC_{50}	conc lethal to 50% of the animals tested
hcp	hexagonal close-packed		
hex	hexagonal	LCAO	linear combination of atomic orbitals
HK	Knoop hardness number		
hplc	high performance liquid chromatography	lc	liquid chromatography
		LCD	liquid crystal display
HRC	Rockwell hardness (C scale)	lcl	less than carload lots
		LD_{50}	dose lethal to 50% of the animals tested
HV	Vickers hardness number		

LED	light-emitting diode	N-	denoting attachment to nitrogen
liq	liquid		
lm	lumen	n (as n_D^{20})	index of refraction (for 20°C and sodium light)
ln	logarithm (natural)		
LNG	liquefied natural gas	$^\mathrm{n}$ (as Bu$^\mathrm{n}$),	
log	logarithm (common)	n-	normal (straight-chain structure)
LOI	limiting oxygen index		
LPG	liquefied petroleum gas	n	neutron
ltl	less than truckload lots	n	nano (prefix for 10^9)
lx	lux	na	not available
M	mega (prefix for 10^6); metal (as in MA)	NAS	National Academy of Sciences
M	molar; actual mass	NASA	National Aeronautics and Space Administration
$\overline{M}_w$	weight-average mol wt		
$\overline{M}_n$	number-average mol wt	nat	natural
m	meter; milli (prefix for 10^{-3})	ndt	nondestructive testing
		neg	negative
m	molal	NF	*National Formulary*
m-	meta	NIH	National Institutes of Health
max	maximum		
MCA	Chemical Manufacturers' Association (was Manufacturing Chemists Association)	NIOSH	National Institute of Occupational Safety and Health
		NIST	National Institute of Standards and Technology (formerly National Bureau of Standards)
MEK	methyl ethyl ketone		
meq	milliequivalent		
mfd	manufactured		
mfg	manufacturing		
mfr	manufacturer	nmr	nuclear magnetic resonance
MIBC	methyl isobutyl carbinol		
MIBK	methyl isobutyl ketone	NND	New and Nonofficial Drugs (AMA)
MIC	minimum inhibiting concentration		
		no.	number
min	minute; minimum	NOI-(BN)	not otherwise indexed (by name)
mL	milliliter		
MLD	minimum lethal dose	NOS	not otherwise specified
MO	molecular orbital	nqr	nuclear quadruple resonance
mo	month		
mol	mole	NRC	Nuclear Regulatory Commission; National Research Council
mol wt	molecular weight		
mp	melting point		
MR	molar refraction	NRI	New Ring Index
ms	mass spectrometry	NSF	National Science Foundation
MSDS	material safety data sheet		
mxt	mixture	NTA	nitrilotriacetic acid
μ	micro (prefix for 10^{-6})	NTP	normal temperature and pressure (25°C and 101.3 kPa or 1 atm)
N	newton (force)		
N	normal (concentration); neutron number		

NTSB	National Transportation Safety Board	qv	quod vide (which see)
O-	denoting attachment to oxygen	R	univalent hydrocarbon radical
o-	ortho	(R)-	rectus (clockwise configuration)
OD	outside diameter	r	precision of data
OPEC	Organization of Petroleum Exporting Countries	rad	radian; radius
		RCRA	Resource Conservation and Recovery Act
o-phen	o-phenanthridine		
OSHA	Occupational Safety and Health Administration	rds	rate-determining step
		ref.	reference
owf	on weight of fiber	rf	radio frequency, n.
Ω	ohm	r-f	radio frequency, adj.
P	peta (prefix for 10^{15})	rh	relative humidity
p	pico (prefix for 10^{-12})	RI	Ring Index
p-	para	rms	root-mean square
p	proton	rpm	rotations per minute
p.	page	rps	revolutions per second
Pa	pascal (pressure)	RT	room temperature
PEL	personal exposure limit based on an 8-h exposure	RTECS	Registry of Toxic Effects of Chemical Substances
		^{s} (eg, Bu^s);	
pd	potential difference	sec-	secondary (eg, secondary butyl)
pH	negative logarithm of the effective hydrogen ion concentration		
		S	siemens
		(S)-	sinister (counterclockwise configuration)
phr	parts per hundred of resin (rubber)	S-	denoting attachment to sulfur
p-i-n	positive-intrinsic-negative		
pmr	proton magnetic resonance	s-	symmetric(al)
p-n	positive-negative	s	second
po	per os (oral)	(s)	solid, only as in $H_2O(s)$
POP	polyoxypropylene	SAE	Society of Automotive Engineers
pos	positive		
pp.	pages	SAN	styrene-acrylonitrile
ppb	parts per billion (10^9)	sat(d)	saturate(d)
ppm	parts per million (10^6)	satn	saturation
ppmv	parts per million by volume	SBS	styrene–butadiene–styrene
ppmwt	parts per million by weight	sc	subcutaneous
PPO	poly(phenyl oxide)	SCF	self-consistent field; standard cubic feet
ppt(d)	precipitate(d)		
pptn	precipitation	Sch	Schultz number
Pr (no.)	foreign prototype (number)	sem	scanning electron microscope(y)
pt	point; part		
PVC	poly(vinyl chloride)	SFs	Saybolt Furol seconds
pwd	powder	sl sol	slightly soluble
py	pyridine	sol	soluble

soln	solution	*trans-*	isomer in which substituted groups are on opposite sides of double bond between C atoms
soly	solubility		
sp	specific; species		
sp gr	specific gravity		
sr	steradian		
std	standard	TSCA	Toxic Substances Control Act
STP	standard temperature and pressure (0°C and 101.3 kPa)		
		TWA	time-weighted average
		Twad	Twaddell
sub	sublime(s)	UL	Underwriters' Laboratory
SUs	Saybolt Universal seconds	USDA	United States Department of Agriculture
syn	synthetic		
^{t} (eg, Bu^t), *t-, tert-*	tertiary (eg, tertiary butyl)	USP	*United States Pharmacopeia*
		uv	ultraviolet
T	tera (prefix for 10^{12}); tesla (magnetic flux density)	V	volt (emf)
		var	variable
t	metric ton (tonne)	*vic-*	vicinal
t	temperature	vol	volume (not volatile)
TAPPI	Technical Association of the Pulp and Paper Industry	vs	versus
		v sol	very soluble
		W	watt
TCC	Tagliabue closed cup	Wb	weber
tex	tex (linear density)	Wh	watt hour
T_g	glass-transition temperature	WHO	World Health Organization (United Nations)
tga	thermogravimetric analysis	wk	week
THF	tetrahydrofuran	yr	year
tlc	thin layer chromatography	(Z)-	zusammen; together; atomic number
TLV	threshold limit value		

Non-SI (Unacceptable and Obsolete) Units		Use
Å	angstrom	nm
at	atmosphere, technical	Pa
atm	atmosphere, standard	Pa
b	barn	cm^2
bar†	bar	Pa
bbl	barrel	m^3
bhp	brake horsepower	W
Btu	British thermal unit	J
bu	bushel	m^3; L
cal	calorie	J
cfm	cubic foot per minute	m^3/s
Ci	curie	Bq
cSt	centistokes	mm^2/s
c/s	cycle per second	Hz

†Do not use bar (10^5 Pa) or millibar (10^2 Pa) because they are not SI units, and are accepted internationally only for a limited time in special fields because of existing usage.

Non-SI (Unacceptable and Obsolete) Units		Use
cu	cubic	exponential form
D	debye	C·m
den	denier	tex
dr	dram	kg
dyn	dyne	N
dyn/cm	dyne per centimeter	mN/m
erg	erg	J
eu	entropy unit	J/K
°F	degree Fahrenheit	°C; K
fc	footcandle	lx
fl	footlambert	lx
fl oz	fluid ounce	m^3; L
ft	foot	m
ft·lbf	foot pound-force	J
gf den	gram-force per denier	N/tex
G	gauss	T
Gal	gal	m/s^2
gal	gallon	m^3; L
Gb	gilbert	A
gpm	gallon per minute	(m^3/s); (m^3/h)
gr	grain	kg
hp	horsepower	W
ihp	indicated horsepower	W
in.	inch	m
in. Hg	inch of mercury	Pa
in. H_2O	inch of water	Pa
in.-lbf	inch pound-force	J
kcal	kilo-calorie	J
kgf	kilogram-force	N
kilo	for kilogram	kg
L	lambert	lx
lb	pound	kg
lbf	pound-force	N
mho	mho	S
mi	mile	m
MM	million	M
mm Hg	millimeter of mercury	Pa
mμ	millimicron	nm
mph	miles per hour	km/h
μ	micron	μm
Oe	oersted	A/m
oz	ounce	kg
ozf	ounce-force	N
η	poise	Pa·s
P	poise	Pa·s
ph	phot	lx
psi	pounds-force per square inch	Pa
psia	pounds-force per square inch absolute	Pa
psig	pounds-force per square inch gage	Pa
qt	quart	m^3; L
°R	degree Rankine	K
rd	rad	Gy
sb	stilb	lx
SCF	standard cubic foot	m^3
sq	square	exponential form
thm	therm	J
yd	yard	m

BIBLIOGRAPHY

1. The International Bureau of Weights and Measures, BIPM (Parc de Saint-Cloud, France) is described in Appendix X2 of Ref. 4. This bureau operates under the exclusive supervision of the International Committee for Weights and Measures (CIPM).
2. *Metric Editorial Guide (ANMC-78-1)*, latest ed., American National Metric Council, 5410 Grosvenor Lane, Bethesda, Md. 20814, 1981.
3. *SI Units and Recommendations for the Use of Their Multiples and of Certain Other Units (ISO 1000-1981)*, American National Standards Institute, 1430 Broadway, New York, 10018, 1981.
4. Based on *ASTM E380-89a (Standard Practice for Use of the International System of Units (SI))*, American Society for Testing and Materials, 1916 Race Street, Philadelphia, Pa. 19103, 1989.
5. *Fed. Reg.*, Dec. 10, 1976 (41 FR 36414).
6. For ANSI address, see Ref. 3.

R. P. LUKENS
ASTM Committee E-43 on SI Practice

Continued

NEUTRON ACTIVATION.   See ANALYTICAL METHODS; RADIOACTIVE TRACERS.

NICKEL AND NICKEL ALLOYS

The first reported use of nickel [7440-02-0], Ni, was in a nickel–copper–zinc alloy produced in China in the Middle Ages and perhaps earlier. Alloys of nickel may have been used in prehistoric times. The metal was first isolated for analytical study in the mid-1700s by Axel Cronstedt, who named it nickel, which derives from the German word *kupfernickel*, or false copper.

Nickel occurs in the first transition row in Group 10 (VIIIB) of the Periodic Table. Some physical properties are given in Table 1 (1–4). Nickel is a high melting point element having a ductile crystal structure. Its chemical properties allow it to be combined with other elements to form many alloys.

In the United States in 1992, 57% of the nickel consumed was used in stainless steels and alloy steels (see STEEL), 28% in nonferrous and high temperature alloys (qv), 9% in electroplating (qv), and the remaining 6% consumed primarily as catalysts (see CATALYSIS) in ceramics (qv), in magnets (see MAGNETIC MATERIALS), and as nickel salts (5,6) (see NICKEL COMPOUNDS). The U.S. markets for nickel alloys (wrought nickel, nickel alloys, and superalloys) in 1992 were ca 40% in the transportation and aircraft industries, 16% in the chemical–petrochemical industry, 18% in the electrical equipment industry, 11% in construction, machinery, and fabricated metal products, and 15% in other uses. In the 1990s, these proportions remained quite constant, with transportation and aircraft usage decreasing slightly. The total U.S. 1992 consumption of nickel was ca 156,500 metric tons. Of this amount, about 121,000 t was primary nickel and 35,500 t was recycled nickel. The world primary nickel consumption in 1992 was 813,900 t. Hence, the United States consumes about 15% of the world's primary nickel production.

1

Table 1. Physical Constants of Nickel[a]

Property	Value
atomic weight	58.71
crystal structure	fcc
lattice constant 25°C, nm	0.35238
melting point, °C	1453
boiling point (by extrapolation), °C	2732
density at 20°C, g/cm^3	8.908
specific heat at 20°C, kJ/(kg·K)[b]	0.44
avg. coefficient of thermal expansion × 10^{-6}, °C^{-1}	
at 20–100°C	13.3
20–300°C	14.4
20–500°C	15.2
thermal conductivity, W/(m·K)	
at 100°C	82.8
300°C	63.6
500°C	61.9
electrical resistivity at 20°C, $\mu\Omega$·cm	6.97
temperature coefficient of resistivity at 0–100°C, ($\mu\Omega$·cm)/°C	0.0071
Curie temperature, °C	353
saturation magnetization, T[c]	0.617
residual magnetization, T[c]	0.300
coercive force, A/m[d]	239
initial permeability, mH/m[e]	0.251
max permeability, mH/m[e]	2.51–3.77
modulus of elasticity × 10^3, MPa[f]	
tension	206.0
shear	73.6
Poisson's ratio	0.30
reflectivity, %	
at 0.30 μm	41
0.55 μm	64
3.0 μm	87
total emissivity, μW/m^2 [g]	
at 20°C	45
100°C	600
500°C	120
1000°C	190
thermal neutron cross section, neutron velocity of 2200 m/s, m^2 [h]	
absorption	4.5×10^{-28}
reaction cross section	17.5×10^{-28}

[a]Refs. 1–4.
[b]To convert J to cal, divide by 4.184.
[c]To convert T to G, multiply by 1.0×10^4.
[d]To convert, A/m to Oe, divide by 79.58.
[e]To convert mH/m to G/Oe, multiply by 795.8.
[f]To convert MPa to psi, multiply by 145.
[g]To convert μW/m^2 to erg/(s·cm^2), multiply by 10^{-3}.
[h]To convert m^2 to barn, divide by 1.0×10^{-28}.

Reserves and Resources

Nickel comprises ca 3% of the earth's composition and is exceeded in abundance by iron (qv), oxygen (qv), silicon (see SILICON AND SILICON ALLOYS), and magnesium (see MAGNESIUM AND MAGNESIUM ALLOYS). However, although nickel comprises ca 7% of the earth's core, it comprises only about 0.009% of the earth's crust, ranking 24th in order of abundance in the crust. Fortunately, ore forms amenable to economic mining exist. The 1992 economic reserve and reserve base quantities, the countries of occurrence, the ore grades, and the mine production rates are listed in Table 2.

Canada, Cuba, and Russia have the largest economic reserves, whereas the United States has less than 0.1% of the world's estimated reserves. Russia produced about 25%, Canada 22%, and New Caledonia 12% of the world's nickel in 1992. The United States produced less than 1% of the world's nickel in the same year. The net import reliance of the United States as a percent of apparent consumption remained about 75% over the period 1987–1992.

The trends in total world mine production rates from 1987 to 1992 are evident in Table 3. An 8-yr averaging shows ca 2% growth in annual consumption. The average price of nickel has varied from year to year; the actual price more than doubled from 1985 to 1988. However, third quarter 1993 prices dropped below mid-1980 prices to <$4.50/kg. Based on the 1992 world nickel consumption level of 813,900 t and the average annual London Metal Exchange (LME) nickel price, the 1992 monetary value for the nickel mining and refining industry would be approximately 6×10^9.

The world economic (proven) nickel reserves are estimated at 47.0×10^6 t. At the 1992 world rate of mine production, these reserves would be expected to last at least until the year 2050. If, however, annual mine production increases at a rate that reflects a predicted increase in the world primary nickel consumption of 2% annually, these reserves would be depleted before 2030 (6,8,9).

Table 2. World Nickel Reserves and Resources,[a] t × 10³

Location	Reserve base[b]	Economic reserves Total	Economic reserves Ore-grade fraction, %	Mine capacity	Refinery and smelter capacity	Mine production
Canada	14,000	6,200	1.5–3.0	205	155	192.1
New Caledonia	15,000	4,500	1.0–3.0	91	45	106.9
United States	2,500	23	0.8–1.5	5	50	6.7
Cuba	23,000	18,000	1.3–2.0	54	30	32.0
Japan					112	
Russia	7,300	6,600	0.9–4.0	300	300	220.0
Australia	6,800	2,200	1.5–5.0	75	54	57.7
Indonesia	13,000	3,200	0.5–2.5	64	5	77.6
other[c]	28,400	6,277	0.8–4.0	316	422	200.5
World total	*110,000*	*47,000*		*1,110*	*1,173*	*893.5*

[a]Refs. 5–7.
[b]Includes demonstrated resources that are economic (reserves), marginally economic (marginal resources), and some that are subeconomic (subeconomic resources) as of 1992.
[c]Primarily China, Papua (New Guinea), South Africa, and the Dominican Republic.

Table 3. World Nickel Mine Production and Prices[a]

Year	World mine production, nickel content, $t \times 10^3$	Average annual price, $/kg	Average constant price,[b] $/kg
1987	890.5	4.82	4.82
1988	917.7	13.78	13.34
1989	970.5	13.34	12.39
1990	937.2	8.86	7.91
1991	922.9	8.16	6.97
1992	893.5	7.00	5.78

[a] Refs. 5 and 6.
[b] Price based on the 1987 U.S. dollar.

In addition to the reported economic reserves, there are substantial nickel resources which could be amenable to mining and refining once appropriate technology becomes available. The single largest such resource is seabed nodules which contain ca 1% nickel and which could represent up to 800×10^6 t of nickel (see OCEAN RAW MATERIALS).

Nickel Ores. The two types of nickel ore which can be mined economically are classified as sulfide and lateritic (10). As of 1992, the sulfide deposits accounted for just over 50% of the nickel produced worldwide. The most common nickel sulfide is pentlandite [53809-86-2], $(Ni,Fe)_9S_{16}$, which is almost always found in association with chalcopyrite [1308-56-1], $CuFeS_2$, and large amounts of pyrrhotite [12063-67-1], Fe_7S_8. Other, much rarer nickel sulfides include millerite [1314-04-1], NiS, heazlewoodite [12035-71-1], Ni_3S_2, and the sulfides of the linnaeite series, $(Fe,Co,Ni)_3S_4$. The nickel sulfides were formed thousands of meters below the surface of the earth by the reaction of sulfur with nickel-bearing rocks. These sulfides generally are found in northern regions where glacial action has planed away much of the overlying weathered surface rock. Important sulfide deposits are found in Canada, Russia, and Finland.

In contrast to the sulfide ores, the lateritic ores were formed over long periods of time as a result of weathering of exposed nickel-containing rocks. The lateritic weathering process resulted in nickel solutions that were redeposited elsewhere in the form of oxides or silicates. One type of laterite is nickeliferous limonitic iron laterite $(Fe,Ni)O(OH)\cdot nH_2O$ which consists primarily of hydrated iron oxide in which the nickel is dispersed in solid solution. The other type of laterite is nickel silicate in which nickel is contained in solid solution in hydrated magnesium–iron minerals, eg, garnierite [12178-41-5], $(Ni,Mg)_6Si_4O_{10}(OH)_8$. Lateritic ores occur primarily in tropical regions, eg, New Caledonia, or in regions which were once at least subtropical for extended periods, eg, Oregon. These deposits are distributed widely and constitute the largest nickel reserves.

Extraction and Refining

The treatments used to recover nickel from its sulfide and lateritic ores differ considerably because of the differing physical characteristics of the two ore types. The sulfide ores, in which the nickel, iron, and copper occur in a physical mixture as distinct minerals, are amenable to initial concentration by mechanical methods, eg, flotation (qv) and magnetic separation (see SEPARATION, MAGNETIC). The lateritic ores are not susceptible to these physical processes of beneficiation, and chemical means must be used to extract the nickel. The nickel concentration processes that have been developed are not as effective for the lateritic ores as for the sulfide ores (see also METALLURGY, EXTRACTIVE; MINERALS RECOVERY AND PROCESSING).

Sulfide Ores. *Pyrometallurgical Processes.* Sulfide ores first undergo crushing and milling operations to reduce the material to the necessary degree of fineness for separation. Froth flotation or magnetic separation processes separate the sulfides from the gangue. Most sulfide ores then undergo a series of pyrometallurgical processes consisting of roasting, smelting, and converting. Roasting, in which much of the iron is oxidized and a large portion of the sulfur is removed as sulfur dioxide, is carried out in multihearth furnaces, fluidized-bed roasters, or rotary kilns. The material then is smelted in reverberatory furnaces or blast furnaces or by flash smelting or arc-furnace smelting. During smelting, a siliceous slag containing iron oxide and other oxide compounds is removed and the sulfur content is further reduced, yielding an impure copper–nickel–iron–sulfur matte. In the converting or Bessemerizing stage, the matte is charged into a horizontal-type converter and the molten matte with silica added is blown with air. This procedure removes virtually all the remaining iron in a slag as well as more of the sulfur as sulfur dioxide, yielding a sulfur-deficient copper–nickel matte.

The matte can be treated in different ways, depending on the copper content and on the desired product. In some cases, the copper content of the Bessemer matte is low enough to allow the material to be cast directly into sulfide anodes for electrolytic refining. Usually it is necessary first to separate the nickel and copper sulfides. The copper–nickel matte is cooled slowly for ca 4 d to facilitate grain growth of mineral crystals of copper sulfide, nickel–sulfide, and a nickel–copper alloy. This matte is pulverized, the nickel and copper sulfides isolated by flotation, and the alloy extracted magnetically and refined electrolytically. The nickel sulfide is cast into anodes for electrolysis or, more commonly, is roasted to nickel oxide and further reduced to metal for refining by electrolysis or by the carbonyl method. Alternatively, the nickel sulfide may be roasted to provide a nickel oxide sinter that is suitable for direct use by the steel industry.

Electrolytic Refining. The electrolytic refining process generally is carried out in a divided cell using anodes which are cast from the crude metal or from nickel sulfides. The electrolyte is pumped continuously through the cell, and impure anolyte, which forms by solution of the anode, is pumped out of the electrolyzing tank and through a purification train to remove soluble impurities. The impure anolyte is prevented from coming into direct contact with the cathode by the use of a porous diaphragm. The nickel cathode starting sheets are made by deposition onto stainless steel blanks from which the nickel sheets are stripped after 2 d and then the cathode is built up in ca 10 d. The final nickel metal that is obtained has a purity exceeding 99.9 wt %. The electrorefining process also facilitates the recovery of precious metals and other metals of value, eg, cobalt, that remain in the insoluble anode residues which are collected during the nickel refining process.

Carbonyl Process. Crude nickel also can be refined to very pure nickel by the carbonyl process. The crude nickel and carbon monoxide (qv) react at ca 100°C to form nickel carbonyl [13463-39-3], $Ni(CO)_4$, which upon further heating to ca 200–300°C, decomposes to nickel metal and carbon monoxide. The process is highly selective because, under the operating conditions of temperature and atmospheric pressure, carbonyls of other elements that are present, eg, iron and cobalt, are not readily formed.

In the carbonyl process, the liquid is purified, vaporized, and rapidly heated to ca 300°C which results in the decomposition of the vapor to carbon monoxide and a fine high purity nickel powder of particle sizes <10 μm. This product is useful for powder metallurgical applications (see METALLURGY, POWDER). Nickel carbonyl can also be decomposed in the presence of nickel powder, upon which the nickel is deposited. This process yields nickel pellets, typically about 0.8 cm dia and of >99.9 wt % purity.

Hydrometallurgical Processes. Hydrometallurgical refining (see SUPPLE- MENT) also is used to extract nickel from sulfide ores. Sulfide concentrates can be leached with ammonia (qv) to dissolve the nickel, copper, and cobalt sulfides as amines. The solution is heated to precipitate copper, and the nickel and cobalt solution is oxidized to sulfate and reduced, using hydrogen at a high temperature and pressure to precipitate the nickel and cobalt. The nickel is deposited as a 99 wt % pure powder.

Lateritic Ores. *Pyrometallurgical Processes.* Nickel oxide ores are processed by pyrometallurgical or hydrometallurgical methods. In the former, oxide ores are smelted with a sulfiding material, eg, gypsum, to produce an iron–nickel matte that can be treated similarly to the matte obtained from sulfide ores. The iron–nickel matte may be processed in a converter to eliminate iron. The nickel matte then can be cast into anodes and refined electrolytically.

A different type of nickel product is obtained by roasting the nickel matte to the oxide, grinding and compacting the oxide, and reducing the oxide to metal using charcoal in a muffle furnace. The metal sinters to form rondelles that contain ca 99.3 wt % nickel. Alternatively, the nickel oxide ore may be smelted without a sulfiding agent and reduced using coke in an electric furnace to produce ferronickel. Ferronickel generally contains 20–30 wt % nickel, but the nickel content may be higher.

Hydrometallurgical Processes. The hydrometallurgical treatments of oxide ores involve leaching with ammonia or with sulfuric acid. In the ammoniacal leaching process, the nickel oxide component of the ore first is reduced selectively. Then the ore is leached with ammonia which removes the nickel into solution, from which it is precipitated as nickel carbonate by heating. A nickel oxide product used in making steel is produced by roasting the carbonate.

In the acid-leaching process, the oxide ore is leached with sulfuric acid at elevated temperature and pressure, which causes nickel, but not iron, to enter into solution. The leach solution is purified, followed by reaction with hydrogen sulfide and subsequent precipitation of nickel and cobalt sulfides. The nickel sulfide is refined by conversion to a sulfate solution and reduction with hydrogen to produce a high purity nickel powder.

Commercial Forms of Nickel

Nickel is available in many commercial forms. The main forms marketed are those listed in Table 4. The very pure unwrought nickels, primarily the electrolytic cathodes and carbonyl pellets and, to a lesser degree, the briquettes and rondelles, are used for production of alloys in which contamination by undesired elements must be minimized in order to obtain desired properties, eg, in nickel-base superalloys and magnetic materials. Pure nickel powder is utilized in the production of porous plates for batteries (qv) and in the production of powder metallurgy parts. A large amount of nickel is available as nickel oxide sinter and ferronickel, which is widely used in the steel and foundry industries as an economical nickel source.

Table 4. Commercial Forms of Nickel

Type	Nickel content, wt %[a]	Uses
electrolytic (cathode)	>99.9	alloy production, electroplating
electrolytic rounds	>99.9	electroplating
carbonyl pellets	>99.7	alloy production, electroplating
briquettes	99.9	alloy production
rondelles	99.3	alloy production
powder	99.74	sintered parts, battery electrodes
nickel oxide sinter	76.0	steel and ferrous alloy production
ferronickel[b]	20–50	steel and ferrous alloy production
nickel salts[c]		electroplating, catalysts
nickel chloride	24.70	
nickel nitrate	20.19	
nickel sulfate	20.90	

[a]Values are approximate.
[b]Different grades of ferronickel are produced, and the nickel content denoted includes 1–2 wt % Co.
[c]Nickel content is theoretical.

Alloys

Properties. Selected chemistries and properties of commercially available nickel and typical cast and wrought nickel alloys are given in Tables 5 and 6 (1–3,8,9,11–32). Nickel-base alloys provide excellent mechanical properties from cryogenic temperatures through temperatures in excess of 1000°C. The development paths of typical wrought nickel alloys is depicted in Figure 1 (10). Nickel alloys are strengthened by solid solution hardening, carbide strengthening, and precipitation hardening.

Nickel. Nickel metal is available in many wrought forms and usually is designated as Nickel 200 or Nickel 201 [39369-16-9] and according to the Unified Numbering System (UNS) as UNS N02201, 205 (UNS N02205), and 270 [128355-61-3] (UNS N02270). Nickel 200 is the general-purpose nickel used in ambient-temperature applications in food processing (qv) equipment, chemical containers, caustic-handling equipment and plumbing, electromagnetic parts,

Table 5. Nominal Chemical Composition of Nickel Alloys,[a] wt %

Alloy	CAS Registry Number	Ni	Fe	Cr	Mo	Mn	Si	C	Al	Ti	Other
Nickel 200	[12671-92-0]	99.5	0.1			0.25	0.05	0.06			0.05 Cu
MONEL alloy 400	[11105-19-4]	65.5	1.5			1.0	0.25	0.15			31.5 Cu
MONEL alloy K-500	[11105-28-5]	65.0	1.0			0.5	0.15	0.15	3.0	0.5	29.5 Cu
NIMONIC alloy 75	[11068-69-2]	76.0	2.5	20.0		0.5	0.5	0.1		0.5	
INCONEL alloy 600	[12606-02-9]	75.5	8.0	15.5		0.5	0.2	0.08			
INCONEL alloy 625	[12682-01-8]	62.0	2.5	22.0	9.0	0.2	0.2	0.05	0.2	0.2	3.5 Nb
INCOLOY alloy 800	[11121-96-3]	31.0	46.0	21.0		0.8	0.5	0.05	0.4	0.4	
HASTELLOY alloy B-2	[61608-60-4]	65.5	2.0	1.0	28.0	1.0	0.1	0.02			2.5 Co
HASTELLOY alloy C-22	[98686-65-8]	54.5	4.0	21.5	13.5	0.5	0.05	0.01			2.5 Co, 3.0 W, 0.3 V
HASTELLOY alloy C-276	[12604-59-0]	55.5	5.0	16.0	16.0	1.0	0.05	0.02			2.5 Co, 4.0 W
HASTELLOY alloy G-3	[77644-65-6]	40.5	19.5	22.0	7.0	1.0	1.0	0.01			5.0 Co, 0.5 (Cb+Ta), 1.5 W, 2.0 Cu
INCONEL alloy 718	[12606-10-9]	52.5	18.5	19.0	3.0	0.2	0.2	0.04	0.5	0.9	5.0 Nb
INCOLOY alloy 909	[95569-75-8]	38.0	42.0				0.4	0.01	0.1	1.5	13.0 Co, 4.7 Nb
B-1900	[12773-54-5]	64.5		8.0	6.0			0.1	6.0	1.0	10.0 Co, 4.0 Ta, 0.015 B, 0.1 Zr
MAR-M247	[97265-31-1]	59.5		8.5	0.5			0.15	5.5	1.0	10.0 Co, 10.0 W, 0.015 B, 0.1 Zr, 1.5 Hf, 3 Ta
RENÉ 80	[12612-07-6]	60.0		14.0	4.0			0.17	3.0	5.0	9.5 Co, 4.0 W, 0.015 B, 0.03 Zr
WASPALOY	[11068-93-2]	58.0		19.5	4.5			0.08	1.3	3.0	13.5 Co, 0.006 B, 0.06 Zr
UDIMET 500	[11068-87-4]	53.5		18.0	4.0			0.08	2.9	2.9	18.5 Co, 0.006 B, 0.05 Zr
UDIMET 700	[11068-91-0]	53.0		15.0	5.0			0.08	4.5	3.5	18.5 Co, 0.03 B
NIMONIC alloy 80A	[11068-71-6]	76.0		19.5		0.2	0.2	0.05	1.5	2.5	0.003 B, 0.06 Zr
NIMONIC alloy 115	[51204-21-8]	59.0		14.5	3.5			0.15	5.0	4.0	13.5 Co, 0.16 B, 0.04 Zr
INCONEL alloy MA754	[62112-97-4]	78.5		20.0				0.05	0.3	0.5	0.6 Y_2O_3

[a]MONEL, DURANICKEL, BRIGHTRAY, INCONEL, INCOLOY, and NIMONIC are trademarks of the Inco family of companies; HASTELLOY and C-22 are trademarks of Haynes International; ILLIUM is a trademark of Stainless Foundry and Engineering Co.; UDIMET is a trademark of Special Metals Corp.; MAR-M is a trademark of Martin Marietta Corp.; RENÉ is a trademark of General Electric Co.; and WASPALOY is a trademark of United Technologies Corp.

Table 6. Properties of Nickel Alloys[a]

Alloy	UNS Number	Melting range, °C	Yield strength,[b] MPa[c]				100-h rupture strength, MPa[c]		
			20°C	538°C	760°C	982°C	649°C	812°C	982°C
Nickel 200	N02200	1435–1446	103–931	139[d]					
MONEL alloy 400	N04400	1299–1349	172–1173	179[d]					
MONEL alloy K-500	N05500	1316–1349	241–1380	648[d]					
NIMONIC alloy 75	N06075	1340–1380	275	210	172	70	255	39	10
INCONEL alloy 600	N06600	1355–1415	285	220	180	41	160	55	19
INCONEL alloy 625	N06625	1290–1350	490	415	415	140	440	125	32
INCOLOY alloy 800	N08800	1355–1385	250	180	150		240	63	21
HASTELLOY alloy B-2[e]	N10665	1320–1350	412						
HASTELLOY alloy C-22	N06022	1357–1399	373–1391	234	214				
HASTELLOY alloy C-276	N10276	1323–1371	356	233					
HASTELLOY alloy G-3	N06985	1260–1343	311	186	165	94			
INCONEL alloy 718	N07718	1260–1335	1125	1020	800		725		
INCOLOY alloy 909	N19909	1395–1430	975	850	440		510		
B-1900[f]		1275–1300	825	870	808	415		505	170
MAR-M247		1221–1357	958	875	841	380		572	193
RENÉ 80								350[g]	165
WASPALOY	N07001	1330–1355	795	725	675	140	760	275	45
UDIMET 500	N07500	1300–1395	840	795	730	230	930	305	83
UDIMET 700		1205–1400	965	895	830	305	828	400	110
NIMONIC alloy 80A	N07080	1360–1390	620	530	505	62	595	195	14
NIMONIC alloy 115		1260–1315	865	795	800	240		400	110
INCONEL alloy MA754	N07754	1320–1390	662	504	262	166			131[h]

[a]Refs. 1–3, 8,9,11–32.
[b]Where two numbers appear, the first refers to the annealed or solution heat-treated condition, the second to the condition when maximum strength is achieved by cold-working or aging. Otherwise the number refers to the alloy heat-treated for optimum strength.
[c]To convert MPa to psi, multiply by 145.
[d]Value is at 316°C.
[e]3.18-mm sheet.
[f]As cast.
[g]Value is at 871°C.
[h]Value is at 1093°C.

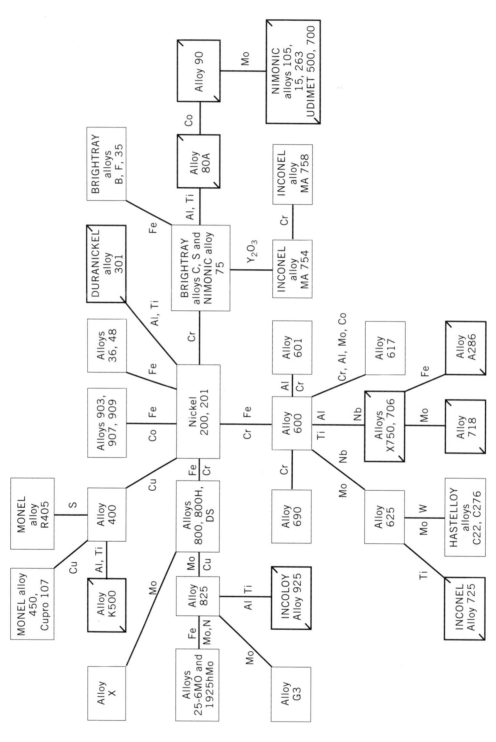

Fig. 1. Development of wrought nickel alloys (see Tables 5 and 6), where (☐) represent solid solution material and (◹) represent precipitation-hardenable material. Alloys 25-6MO and 1925hMo are superaustenitic stainless steels.

and aerospace and missile components. Nickel 201 has a much lower trace carbon content than the 200 and is thus more suitable for elevated temperature applications. The lower carbon content prevents elevated temperature stress–corrosion cracking. Nickel 205 is low in carbon but contains trace amounts of magnesium; Nickel 270 is one of the purest, ie, 99.98 wt %, commercial nickels. DU-RANICKEL alloy 301 (UNS N03301), which contains about 4.5 wt % aluminum and 0.5 wt % titanium, can be aged to form very fine γ'-$Ni_3(Al,Ti)$ [12003-81-5] precipitates. This type of alloy combines high strength and hardness with the excellent corrosion resistance that is characteristic of Nickel 200. Various of these nickel metals also are used as welding (qv) electrodes for joining ferritic or austenitic steels to high nickel-containing alloys and for welding the clad side of nickel-clad steels.

Nickel has excellent corrosion-resistance properties. Nickel and nickel alloys are useful in reducing environments and under some oxidizing conditions in which a passive oxide film is developed. In general, nickel is very resistant to corrosion in marine and industrial atmospheres, in distilled and natural waters, and in flowing seawater. Nickel has excellent resistance to corrosion by caustic soda and other alkalies. In nonoxidizing acids, nickel does not readily discharge hydrogen. Hence, nickel has fairly good resistance to sulfuric acid, hydrochloricacid, organic acids, and other acids, but has poor resistance to strongly oxidizing acids such as nitric acid. Nickel has excellent resistance to neutral and alkaline salt solutions. Nonoxidizing acid salts are moderately corrosive, and oxidizing acid salts and oxidizing alkaline salts generally are corrosive to nickel. Nickel also is resistant to corrosion by chlorine, hydrogen chloride, fluorine, and molten salts.

Wrought and cast nickel anodes and sulfur-activated electrodeposited rounds are used widely for nickel electrodeposition onto many base metals. Nickel also can be plated by an electroless process (see ELECTROLESS PLATING). Nickel plating provides resistance to corrosion for many commonly used articles, eg, pins, paper clips, scissors, keys, fasteners, etc, as well as for materials used in food processing (qv), the paper (qv) and pulp (qv) industries, and the chemical industry, each of which is often characterized by severely corrosive environments. Nickel plating is used in conjunction with chromium plating to provide decorative finishes and corrosion resistance to numerous articles. Nickel plating is used to salvage worn, corroded, or incorrectly machined parts. Nickel electroforming, in which nickel is electrodeposited onto a mold which subsequently is separated from the deposit, is used to form complex shapes, eg, printing plates, tubing, nozzles, screens, and grids.

Porous nickel electrodes made from nickel powder are used in storage batteries (qv) and fuel cells (qv). Nickel–cadmium batteries have attractive properties including long operating and storage lives, high rate discharge capability, high rate charge acceptance, and high and low temperature capability.

Nickel also is an important industrial catalyst. The most extensive use of nickel as a catalyst is in the food industry in connection with the hydrogenation or dehydrogenation of organic compounds to produce edible fats and oils (see FATS AND FATTY OILS).

Nickel Alloying.   Nickel is alloyed into low alloy steels, ferritic alloy steels, and austenitic stainless steels through the conventional steelmaking processes, eg, open hearth, basic oxygen conversion, and the argon–oxygen decarburization (AOD) processes. The AOD process is used to produce a substantial quantity of the stainless steels in the world. It is a highly productive process that yields cleaner products at lower operating and materials costs as compared to the older conventional electric-arc-furnace (EAF) steelmaking practice. EAF or AOD melting and air-induction melting (AIM) are used for some nickel-base alloys. Electroslag remelt (ESR) processing also is used to further refine these steels and nickel alloys.

Nickel alloys that are heavily alloyed with other elements including the nickel-base and iron-base superalloys, also are produced by vacuum-induction melting (VIM). In the VIM process, the melting, alloying, melt treatments, and ingot casting are carried out under vacuum. Industrial VIM furnaces generally can process up to 20-t batches. For further alloy refinement, VIM castings are used as electrodes and are ESR- or vacuum-arc remelted (VAR). Investment castings of the chemically complex nickel-base alloys, especially those containing the reactive elements aluminum and titanium, also are carried out under vacuum. More recently, directional solidification techniques, in which the heat is extracted directionally through a controlled solidification rate and temperature gradient, are used to produce either monocrystalline nickel-base superalloys or polycrystalline structures having long columnar grains. Gas powder-atomizing techniques, which involve VIM master melts, also are used routinely to produce fine nickel-base powders for subsequent powder metallurgical consolidation of near-net-shape components. Melting technologies involving electron-beam and plasma melting are also being used to melt nickel alloys (see PLASMA TECHNOLOGY).

Nickel–Copper.   In the solid state, nickel and copper form a continuous solid solution. The nickel-rich, nickel–copper alloys are characterized by a good compromise of strength and ductility and are resistant to corrosion and stress corrosion in many environments, in particular water and seawater, nonoxidizing acids, neutral and alkaline salts, and alkalies. These alloys are weldable and are characterized by elevated and high temperature mechanical properties for certain applications. The copper content in these alloys also ensure improved thermal conductivity for heat exchange. MONEL alloy 400 is a typical nickel-rich, nickel–copper alloy in which the nickel content is ca 66 wt %. MONEL alloy K-500 is essentially alloy 400 with small additions of aluminum and titanium. Aging of alloy K-500 results in very fine γ'-precipitates and increased strength (see also COPPER ALLOYS).

Typical applications for the nickel–copper alloys are in industrial plumbing and valves, marine equipment, petrochemical equipment, and feedwater heat exchangers (see PIPING SYSTEMS). The age-hardened alloys are used as pump shafts and impellers, valves, drill parts, and fasteners (see PUMPS). Nickel–copper alloys also are used as coated electrodes or filler alloys for welding purposes. Coinage is typically an alloy of 75 wt % Cu and 25 wt % Ni.

Copper and nickel can be alloyed with zinc to form nickel silvers. Nickel silvers are ductile, easily formed and machined, have good corrosion resistance, can be worked to provide a range of mechanical properties, and have an attractive white color. These alloys are used for ornamental purposes, as silverplated and uncoated tableware and flatware; in the electrical industry as contacts, connections, and springs; and as many formed and machined parts (see ELECTRICAL CONNECTORS).

Nickel–Chromium. Nickel and chromium form a solid solution up to 30 wt % chromium. Chromium is added to nickel to enhance strength, corrosion resistance, oxidation, hot corrosion resistance, and electrical resistivity. In combination, these properties result in the nichrome-type alloys used as electrical furnace heating elements. The same alloys also provide the base for alloys and castings which can withstand hot corrosion in sulfur and oxidative environments, including those containing vanadium pentoxides which are byproducts of petroleum combustion in fossil-fuel electric power plants and in aircraft jet engines. Alloy additions to nickel–chrome usually are ca 4 wt % aluminum and ca ≤1 wt % yttrium. Without these additions, the nichrome-type alloys provide hot oxidation or hot corrosion resistance through the formation of surface nickel–chromium oxides. Aluminum provides for surface Al_2O_3 formation and the yttrium or other rare-earth additions improve the adherence of the protective oxide scales to the nickel–chromium–aluminum substrates.

Nickel–Iron. A large amount of nickel is used in alloy and stainless steels and in cast irons. Nickel is added to ferritic alloy steels to increase the hardenability and to modify ferrite and cementite properties and morphologies, and thus to improve the strength, toughness, and ductility of the steel. In austenitic stainless steels, the nickel content is 7–35 wt %. Its primary roles are to stabilize the ductile austenite structure and to provide, in conjunction with chromium, good corrosion resistance. Nickel is added to cast irons to improve strength and toughness.

Many nickel–iron alloys have useful magnetic characteristics and are used in a wide range of devices in the electronics and telecommunication fields. Some nickel–iron alloys are magnetically soft and have attractive properties of high initial permeability, high maximum magnetization and low residual magnetization, low coercive force, and low hysteresis and eddy-current losses. These properties are sensitive to alloying and to precipitate and grain morphologies. Important soft magnetic alloys are based on compositions of 78 wt % Ni–22 wt % Fe, 65 wt % Ni–35 wt % Fe, and 50 wt % Ni–50 wt % Fe, which often include a few weight percent of molybdenum, copper, or chromium.

The majority of permanent magnets are made from magnetically hard alloys of nickel and iron that are characterized by high values of residual magnetization and coercive force. The many Alnico alloys, consisting of (14–28) wt % Ni–(5–35) wt % Co–(6–12) wt % Al–(0–6) wt % Cu–(0–8) wt % Ti–balance iron, are precipitation-strengthened, hard, brittle alloys in which the magnetic properties are sensitive to heat treatments which determine precipitate and grain morphologies.

Some nickel–iron alloys have anomalously low thermal-expansion coefficients within certain temperature ranges. This behavior results from a balance between the normal thermal expansion and a contraction caused by magnetostriction. These alloys, eg, nickel–iron alloys having 36 wt %, 42 wt %, or 50 wt % nickel, and a 29 wt % Ni–17 wt % Co–54 wt % Fe alloy, are used as glass-to-metal joints and in metrology equipment, thermostats and thermometers, cryogenic structures and devices (see CRYOGENICS), and many other electrical and engineering applications.

Demands for improved efficiency in aircraft gas turbines led to the use of a family of age hardenable, controlled expansion superalloys for engine seals and casings. INCOLOY alloys 903 [*61107-16-2*] (UNS N19903), 907 [*107652-*

23-3] (UNS N19907), and 909 evolved from a continuing effort to improve the environmental resistance of this Cr-free, Fe–Ni–Co based system.

Another anomalous property of some nickel–iron alloys, which are called constant-modulus alloys, is a positive thermoelastic coefficient which occurs in alloys having 27–43 wt % nickel. The elastic moduli in these alloys increase with temperature. Usually, and with additions of chromium, molybdenum, titanium, or aluminum, the constant-modulus alloys are used in precision weighing machines, measuring devices, and oscillating mechanisms (see WEIGHING AND PROPORTIONING).

Nickel–Molybdenum. Molybdenum in solid solution with nickel strengthens the latter metal and improves its corrosion resistance, eg, in the HASTELLOY alloys. HASTELLOY alloy B-2 is noted for its superior resistance to corrosion by hydrochloric acid at all concentrations up to the boiling point; by other nonoxidizing acids, such as sulfuric and phosphoric; and by hot hydrogen chloride gas. Other nickel–molybdenum alloys contain chromium, which improves the resistance to corrosion and, especially, to oxidation. The Ni–Cr–Mo HASTELLOY alloy C-22, which also contains cobalt and tungsten, is resistant to a wide range of chemical process environments, including strong oxidizing acids, organic and inorganic media, chlorine, and brine. HASTELLOY alloy C-276 also has excellent resistance to corrosion by oxidizing environments, oxidizing acids, chloride solutions, and other acids and salts.

Another set of nickel alloys, which have a high chromium content, a moderate molybdenum content, and some copper, are the ILLIUM alloys. These cast alloys are wear and erosion resistant and highly resistant to corrosion by acids and alkalies under both oxidizing and reducing conditions.

Nickel–Iron–Chromium. A large number of industrially important materials are derived from nickel–iron–chromium alloys. These alloys are within the broad austenitic, gamma-phase field of the ternary Ni–Fe–Cr phase diagram and are noted for good resistance to corrosion and oxidation and good elevated temperature strength (see HIGH TEMPERATURE ALLOYS). Examples are the INCONEL alloys, which are based on the INCONEL alloy 600 composition. Alloy 600 is a solid solution alloy with good strength and toughness from cryogenic to elevated temperatures and good oxidation and corrosion resistance in many media. In addition, the alloy is easily fabricated and joined. Many modifications of alloy 600 have been made to produce other alloys with different characteristics. For example, INCONEL alloy 601 [*12631-43-5*] (UNS N06601) contains aluminum for improved high temperature oxidation resistance, INCONEL alloy 625 contains molybdenum and niobium in solid solution for better strength, and INCONEL alloy 690 [*54385-90-9*] (UNS N06690) with further additions of chromium was developed for use in the nuclear industry and is particularly noted for its resistance to corrosion by high purity water (see NUCLEAR REACTORS).

Other alloys have been developed for use in particular corrosive environments at high temperatures. Several of these are age-hardenable alloys which contain additions of aluminum and titanium. For example, INCONEL alloys 718 and X-750 [*11145-80-5*] (UNS N07750) have higher strength and better creep and stress rupture properties than alloy 600 and maintain the same good corrosion and oxidation resistance. Alloy 718 exhibits excellent stress rupture properties up to 705°C as well as good oxidation resistance up to 980°C and is widely used in gas turbines and other aerospace applications, and for pumps, nuclear reactor parts, and tooling.

The INCOLOY alloys exemplify another class of nickel–iron–chromium alloys. INCOLOY alloy 800 is resistant to hot corrosion, oxidation, and carburization and has good elevated-temperature strength. Modifications of alloy 800 impart different strength or corrosion-resistance characteristics. For example, INCOLOY alloy 801 [12605-97-9] (UNS N08801) contains more titanium, which, with appropriate heat treatments, can age-harden the alloy and provide increased resistance to intergranular corrosion; INCOLOY alloy 802 [51836-04-5] (UNS N08802) contains more carbon which provides improved high temperature strength through carbide strengthening. INCOLOY alloy 825 [12766-43-7] (UNS N08825) and HASTELLOY alloy G-3 contain molybdenum, copper, and other additions and are exceptionally resistant to attack by aggressive corrosive environments.

The corrosion- and heat-resistant alloys, eg, alloys 600 and 800, are used extensively in heat-treating equipment, nuclear and fossil-fuel steam generators, heater-element sheathing and thermocouple tubes, and in chemical and food-processing equipment. Alloys 625 and 825 are used in chemical processing, pollution control, marine and pickling equipment, ash-pit seals, aircraft turbines and thrust reversers, and radiation waste-handling systems. The age-hardened INCONEL and INCOLOY alloys are used in gas turbines, high temperature springs and bolts, nuclear reactors, rocket motors, spacecraft, and hot-forming tools. There are also nickel–iron–chromium alloys used as welding electrode and filler metals.

Nickel-Base Superalloys. Superalloys, which are critical to gas-turbine engines because of their high temperature strength and superior creep and stress rupture-resistance, basically are nickel–chromium alloyed with a host of other elements. The alloying elements include the refractory metals tungsten, molybdenum, or niobium for additional solid-solution strengthening, especially at higher temperatures and aluminum in appropriate amounts for the precipitation of γ' for coherent particle strengthening (see REFRACTORIES). Titanium is added to provide stronger γ', and niobium reacts with nickel in the solid state to precipitate the γ''-phase; γ'' is the main strengthening precipitate in the 718-type alloys. Cobalt, generally present in many superalloys in large (≥ 10 wt %) amounts, enhances strength, oxidation, and hot-corrosion resistance which is also provided by the chromium in the alloy. Small excess amounts of carbon usually are present in superalloys for intentional carbide precipitation at grain-boundaries which, as discrete and equiaxed particles, can provide obstacles for grain-boundary sliding and motion, thus suppressing creep at high temperatures. Small or trace amounts of elements, eg, zirconium, boron, and hafnium, may be present and these enhance grain-boundary strength and improve ductility. The strength and elevated-temperature properties of a superalloy are dependent on the volume fraction of the fine γ'-precipitates, which can be increased to ca 60 wt %, depending on the aluminum and titanium content. Besides precipitation control at the grain boundaries, improved heat resistance can result from either the elimination of grain boundaries or through the growth of aligned grains with minimum grain boundaries perpendicular to the principal applied stress direction, eg, in turbine-blade applications.

Because of constitutional complexity, the exact chemistries of nickel-base superalloys must be controlled carefully in order to avoid the precipitation of deleterious topologically close-packed (TCP) phases and extraneous carbides after long-term high temperature exposure. Heat-treatment schedules and thermomechanical treatments in the case of wrought alloys also are important to provide optimum strength and performance.

Oxide-Dispersion-Strengthened Alloys. Through mechanical alloying and other powder-metallurgical techniques, highly hot-oxidation and corrosion-resistant nickel–chromium matrices are strengthened by very fine dispersions of somewhat chemically inert oxide particles to produce alloys such as INCONEL alloy MA754. These oxide dispersions replace γ' as the main strengthening agent and provide strength benefits close to the melting temperature. Gamma-prime precipitation strengthening usually begins to decline above 800°C. The oxide-dispersion-strengthened (ODS) nickel-, iron-, and cobalt-base alloys are used mainly in bar and sheet forms in gas turbine vanes in combustion chambers and as exhaust hardware in very high temperature applications.

Nickel Aluminide Intermetallics. Cast and wrought nickel aluminides are being developed for high temperature applications. Using additions of 0.1% Zr to β-NiAl, and alloys containing $\geq 40\%$ Al, an adherent, protective oxide is formed that improves cyclic oxidation performance. Additions of B, Cr, Co, and titanium diboride, TiB_2, reportedly enhance ductility and the strength of nickel aluminides. Bars, wires, and continuously cast strip have been produced that have excellent strength and oxidation resistance (33) (see also GLASSY METALS).

Nickel-Matrix Composites. Nickel and nickel–titanium compositions are used as powder binders for carbide cutting tools (see CARBIDES; TOOL MATERIALS). Gas turbine engine engineers have begun development of ceramic-filled honeycomb structures using nickel-base oxide dispersion-strengthened alloys. Additional studies are underway to develop nickel-base superalloy composites using tungsten, molybdenum, silicon carbide, and sapphire fibers (34) (see also ABLATIVE MATERIALS; METAL-MATRIX COMPOSITES).

BIBLIOGRAPHY

"Nickel and Nickel Alloys" in *ECT* 1st ed., Vol. 9, pp. 271–288, by W. A. Mudge and W. Z. Friend, The International Nickel Co., Inc.; in *ECT* 2nd ed., Vol. 13, pp. 735–753, by J. B. Adamec and T. E. Kihlgren, The International Nickel Co., Inc.; in *ECT* 3rd ed., Vol. 15, pp. 787–801 by J. K. Tien and T. E. Howson, Columbia University.

1. S. J. Rosenberg, *Natl. Bur. Stand. Monogr.* **106** (1968).
2. W. L. Mankins and S. Lamb, *Nickel and Nickel Alloys* in *ASM Handbook*, 10th ed., Vol. 2, ASM International, Materials Park, Ohio, 1990, pp. 428–445.
3. W. Betteridge, *Nickel and Its Alloys*, Industrial Metals Series, MacDonald and Evans, Ltd., London, 1977.
4. R. C. Weast and D. R. Lide, *Handbook of Chemistry and Physics*, 70th ed., Chemical Rubber Publishing Co., Boca Raton, Fla., 1989–1990.
5. P. H. Kuch, *Nickel 1991*, U.S. Bureau of Mines, Washington, D.C., Apr. 1993.
6. *World Nickel Statistics*, Vol. III, No. 5, International Nickel Study Group, May 1993.
7. P. H. Kuch, *Nickel*, U.S. Bureau of Mines, Washington, D.C., Jan. 1993.
8. J. K. Tien, R. M. Arons, and R. W. Clark, *J. Metals* **28**(12), 26 (1976).

9. J. K. Tien and R. M. Arons, *AIChE Symp. Ser.* **73**(170) (1977).
10. J. R. Boldt, Jr., and P. Queneau, *The Winning of Nickel*, D. Van Nostrand Co., Inc., New York, 1967.
11. F. B. White-Howard, *Nickel, An Historical Review*, D. Van Nostrand Co., Inc., New York, 1963.
12. J. L. Everhart, *Engineering Properties of Nickel and Nickel Alloys*, Plenum Press, New York, 1971.
13. *Nickel 200, 201 & DURANICKEL alloy 301*, Inco Alloys International, Huntington, W. Va., 1988.
14. *MONEL alloys*, Inco Alloys International, Huntington, W. Va., 1989.
15. *INCONEL alloy 600*, Inco Alloys International, Huntington, W. Va., 1987.
16. *INCOLOY alloys 800 and 800HT*, Inco Alloys International, Huntington, W. Va., 1989.
17. *Resistance to Corrosion*, Inco Alloys International, Huntington, W. Va., 1985.
18. *High Temperature, High Strength Nickel Base Alloys*, Nickel Development Institute, Toronto, Canada, 1987.
19. *Properties of Some Metals and Alloys*, Inco Limited, New York, 1981.
20. *Oxide Dispersion Strengthened Superalloys*, Inco Alloys International, Huntington, W. Va., 1990.
21. *1983 Databook, Metal Progress*, American Society of Metals, Metals Park, Ohio, June 1983.
22. *Alloy Performance Data*, Special Metals Corp., New Hartford, N.Y., 1986.
23. *HASTELLOY alloy B-2*, Haynes International, Kokomo, Ind., 1977.
24. *HASTELLOY alloy C-22*, Haynes International, Kokomo, Ind., 1980.
25. *HASTELLOY alloy C-276*, Haynes International, Kokomo, Ind., 1978.
26. *HASTELLOY alloy G-3*, Haynes International, Kokomo, Ind., 1976.
27. S. Purushothaman and J. K. Tien, in *Properties of High Temperature Alloys*, Electrochemical Society, Inc., Princeton, N.J., 1976, pp. 3–41.
28. S. D. Antolovich and co-workers, eds, *Superalloys 1992*, TMS Publication, Warrendale, Pa., 1992.
29. K. C. Russell and D. F. Smith, *Physical Metallurgy of Controlled Expansion Invar-Type Alloys*, TMS Publication, Warrendale, Pa., 1990.
30. E. A. Loria, *Superalloys 718, 625 and Various Derivatives*, TMS Publication, Warrensdale, Pa., 1991.
31. J. M. Svaboda, in *Metals Handbook*, 9th ed., Vol. 15, ASM International, Materials Park, Ohio, 1988, pp. 815–823.
32. E. F. Bradley, *Superalloys, A Technical Guide*, ASM International, Materials Park, Ohio, 1988.
33. V. R. Parameswaran, *J. Minerals Met. Mater.*, 41–43 (June 1992).
34. J. Doychak, *J. Minerals Met. Mater.*, 46–51 (June 1992).

JOHN H. TUNDERMANN
Inco Alloys International

JOHN K. TIEN
Columbia University

TIMOTHY E. HOWSON
Wyman-Gordon

NICKEL COMPOUNDS

Nickel [*7440-02-0*], Ni, recognized as an element as early as 1754 (1), was not isolated until 1820 (2). It was mined from arsenic sulfide mineral deposits (3) and first used in an alloy called German Silver (4). Soon after, nickel was used as an anode in solutions of nickel sulfate [*7786-81-4*], $NiSO_4$, and nickel chloride [*7718-54-9*], $NiCl_2$, to electroplate jewelry. Nickel carbonyl [*13463-39-3*], $Ni(CO_2)_4$, was discovered in 1890 (see CARBONYLS). This material, distilled as a liquid, decomposes into carbon monoxide and pure nickel powder, a method used in nickel refining (5) (see NICKEL AND NICKEL ALLOYS).

Nickel has a [Ar] $3d^8 4s^2$ electronic configuration and forms compounds in which the nickel atom has oxidation states of -1 through $+4$. Whereas reagents yield an array of compounds in a variety of nickel oxidation states (6–8), Ni(II) represents the bulk of all known compounds. As of this writing, >237,000 (9) compounds of nickel have been reported. The primary uses for nickel compounds, aside from nickel refining and electroplating (10), are in steel (qv) making, catalysis (qv), storage batteries (qv), specialty chemicals, and specialty ceramics (qv).

Simple nickel salts form ammine and other coordination complexes (see COORDINATION COMPOUNDS). The octahedral configuration, in which nickel has a coordination number (CN) of 6, is the most common structural form. The square-planar and tetrahedral configurations (11), in which nickel has a coordination number of 4, are less common. Generally, the latter group tends to be reddish brown. The 5-coordinate square pyramid configuration is also quite common. These materials tend to be darker in color and mostly green (12).

Examples of stable crystalline derivatives of Ni(III) and Ni(IV) are the hexafluoride anions [*32698-29-6*], NiF_6^{3-} and [*23712-86-9*], NiF_6^{2-} (13,14). A review of dicarbamates stabilizing Ni(III) and Ni(IV) is available (15). Examples of the binuclear and diamagnetic species of Ni(I) are the cyanonicklates $K_4[Ni_2(CN)_6]$ [*40810-33-1*] and $K_6[Ni_2(CN)_8]$ (16,17). Dihydrohexacarbonyldinickel($-$I) [*12549-35-8*], $H_2Ni_2(CO)_6$, an example of nickel in the -1 oxidation state, decomposes above $-33°C$ (18). Many Ni(0) compounds have been prepared. Nickel carbonyl, $Ni(CO)_4$, is the most common (19). Other types of Ni(0) compounds are nickel dicyclooctadiene [*33221-58-8*], $(C_8H_{12})_2Ni$ (20), $(C_6H_5NC)_4Ni$ [*23411-45-2*] (21), $Ni(P(OCH_3)_3)_4$ [*14881-35-7*] (22); $((C_5H_4N)_2)_2Ni$ [*15186-68-2*] known as $Ni(bipy)_2$ (23); and the anion $[Ni(CN)_4]^{4-}$ [*15453-80-2*] (24).

Inorganic Compounds

Nickel Oxides. *Properties.* Nickel oxide [*1313-99-1*], NiO, is a green cubic crystalline compound, mp 2090°C, density 7.45 g/cm^3, the properties of which are related to its method of preparation. Green nickel oxide is prepared by firing a mixture of water and pure nickel powder in air at 1000°C or by firing a mixture of high purity nickel powder, nickel oxide, and water in air (25,26). Whereas this temperature is required for full development of the crystal, the temperature is high enough that an equilibrium leading to dissociation back to the elements is established. Consequently, it is virtually impossible to obtain green nickel oxide made by high temperature firing that does not have traces of nickel metal. Single whiskers of green nickel oxide have been made by the closed-tube transport method from oxide powder formed by the decomposition of nickel sulfate using HCl as the transport gas (27). Green nickel oxide, free of nickel metal, also is formed by thermal decomposition of nickel carbonate [3333-67-3], $NiCO_3$, or nickel nitrate [13138-45-9], $Ni(NO_3)_2$. Green nickel oxide is a refractory material that is inert in most aqueous systems. At high temperatures it reacts with other ceramic materials in the presence of various fluxes (28).

Black nickel oxide, NiO, a microcrystalline form, results from calcination of the carbonate or nitrate at 600°C. This incompletely annealed product typically has more oxygen than its formula indicates, ie, 76–77 wt % nickel compared to the green form which has 78.5% nickel content. This results from chemisorption of oxygen on the surface of the crystal defects. Black nickel oxide compositions are chemically reactive and form simple nickel(II) salts when heated with mineral acids. Both black and green nickel oxide can be converted to the metal by heating with carbon, carbon monoxide, or hydrogen. Both green and black nickel oxide fuse with potassium hydroxide at 700°C to form potassium nickelate [50811-97-7], K_2NiO_2 (27). Other nickel oxides, eg, Ni_2O_3 [1314-06-3], density 4.84 g/cm³, NiO_2 [12035-36-8], and Ni_3O_4 [12137-09-6], have been reported. Although detailed characterization evidence is lacking, many claims have been made for the electrochemical and photochemical properties (29).

Manufacture.   Several nickel oxides are manufactured commercially. A sintered form of green nickel oxide is made by smelting a purified nickel matte at 1000°C (30); a powder form is made by the desulfurization of nickel matte. Black nickel oxide is made by the calcination of nickel carbonate at 600°C (31). The carbonate results from an extraction process whereby pure nickel metal powder is oxidized with air in the presence of ammonia (qv) and carbon dioxide (qv) to hexaamminenickel(II) carbonate [67806-76-2], [Ni(NH₃)₆]CO₃ (32). Nickel oxides also are made by the calcination of nickel carbonate or nickel nitrate that were made from a pure form of nickel. A high purity, green nickel oxide is made by firing a mixture of nickel powder and water in air (25).

Uses.   The sinter oxide form is used as charge nickel in the manufacture of alloy steels and stainless steels (see STEEL). The oxide furnishes oxygen to the melt for decarburization and slagging. In 1993, >100,000 metric tons of nickel contained in sinter oxide was shipped to the world's steel industry. Nickel oxide sinter is charged as a granular material to an electric furnace with steel scrap and ferrochrome; the mixture is melted and blown with air to remove carbon as CO_2. The melt is slagged, poured into a ladle, the composition is adjusted, and the melt is cast into appropriate shapes. A modification of the use of sinter oxide is its injection directly into the molten metal (33).

Green nickel oxide powder, used in the refining of nickel, is agglomerated to a particular shape and then reduced to metal in a furnace. Green and black nickel oxides are used in the ceramic industry for making frit, ferrites (qv), and inorganic colors (see CERAMICS; COLORANTS FOR CERAMICS). Black nickel oxide is used for the manufacture of nickel salts and specialty ceramics. Nickel dioxide [12035-36-8], NiO_2, also known as Nickel Black, has been reported to have a solar absorbance coefficient of >0.92 (29). Green and black nickel oxides are used for nickel catalyst manufacture by admixing, usually when wet, with a powdered ceramic support material. The mixture is formed into a suitable shape and then reduced with hydrogen to form the finished catalyst. Green nickel oxide displays a negative temperature coefficient (NTC) and is used in the formation of thermisters, heat-sensitive electric switches that are extremely sensitive to changes in temperature and are used in process control (qv) in chemical manufacturing. NTC devices are also used in computer circuitry, in air conditioners, and fire detection equipment (34). Green nickel oxide is used in a mixture with other high purity metal oxides in varistors or voltage surge arrestors in lightning strike devices or as in-line varistors in electronics (35). The equilibrium of nickel oxide with its elements is important commercially: nickel oxide is used along with aluminum azide as the propellent for automotive air bags. The mixture is rapidly heated by resistance wire to 1400°C and the reaction of dissociated oxygen and aluminum azide to release nitrogen gas to fill the bag is complete in microseconds (36).

Nickel oxide is used commercially to make nickel fibers in a process whereby a water slurry containing nickel oxide and a cellulose (qv) type binder is forced through tiny orifices to form green fibers. Subsequent steps include drying and reduction with hydrogen. The resulting nickel fibers are matted together and used for the filtration (qv) of gases (37,38).

Nickel Sulfate. *Properties.* Nickel sulfate hexahydrate [*10101-97-0*], $NiSO_4 \cdot 6H_2O$, is a monoclinic emerald-green crystalline salt that dissolves easily in water and in ethanol. When heated, it loses water and above 800°C decomposes into nickel oxide and SO_3. Its density is 2.03 g/cm^3.

Manufacture. The preferred method for making nickel sulfate is adding nickel powder to hot dilute sulfuric acid. Adding sulfuric acid to nickel powder in hot water enhances the formation of H_2S. Hydrogen sulfide always forms as a by-product upon reaction of metallic nickel and sulfuric acid. The liberated hydrogen is absorbed by the metal and then reduces the sulfate anion to H_2S.

Nickel sulfate also is made by the reaction of black nickel oxide and hot dilute sulfuric acid, or of dilute sulfuric acid and nickel carbonate. The reaction of nickel oxide and sulfuric acid has been studied and a reaction induction temperature of 49°C determined (39). High purity nickel sulfate is made from the reaction of nickel carbonyl, sulfur dioxide, and oxygen in the gas phase at 100°C (40). Another method for the continuous manufacture of nickel sulfate is the gas-phase reaction of nickel carbonyl and nitric acid, recovering the solid product in sulfuric acid, and continuously removing the solid nickel sulfate from the acid mixture (41). In this last method, nickel carbonyl and sulfuric acid are fed into a closed-loop reactor. Nickel sulfate and carbon monoxide are produced; the CO is thus recycled to form nickel carbonyl.

Uses. The principal use for nickel sulfate is as an electrolyte for the metal-finishing application of nickel electroplating (qv). Nickel sulfate also is used as the electrolyte for nickel electrorefining. High purity nickel sulfate is used in electroless plating (qv) (42), where nickel sulfate and a reducing agent, eg, sodium hypophosphite, are brought together in hot water in the presence of the workpiece to be plated (43). Another application for nickel sulfate is as a nickel strike solution, which is used for replacement coatings (qv) or nickel flashing on steel that is to be porcelain-enameled (see ENAMELS, PORCELAIN AND VITREOUS). Nickel sulfate is also used as an intermediate in the manufacture of other nickel chemicals and as a catalyst intermediate.

Nickel Nitrate. *Properties and Preparation.* Nickel nitrate hexahydrate [13478-00-7], $Ni(NO_3)_2 \cdot 6H_2O$, is a green monoclinic deliquescent crystal, mp 56°C, density 2.05 g/cm^3, that is extremely soluble in water. Nickel nitrate hexahydrate loses water on heating and eventually decomposes, forming nickel oxide. The loss of the individual waters of hydration upon heating the hexahydrate has been studied, and the existence of the anhydrous covalent compound $(Ni(NO_3)_2$ [13138-45-9], before it decomposes, can be observed using thermal analysis techniques. The latter compound is prepared by the addition of methyl glyme to nickel nitrate hexahydrate followed by vacuum distillation and drying (44). Nickel nitrate hexahydrate can be prepared by the reaction of dilute nitric acid and nickel carbonate.

Manufacture. Nickel nitrate is made commercially by several methods. Nickel metal reacts vigorously with nitric acid and, if the reaction is not closely controlled, excess heating occurs and causes breakdown of the nitric acid. Nickel ammonium nitrate [22026-79-5], $H_3N \cdot xHNO_3 \cdot xNi$, also forms in the commercial methods that use nitric acid and metallic nickel because nickel absorbs the released hydrogen and catalytically reduces the nitrate anion to ammonia. The methods vary as to the amount of ammonia formed and the relative concentrations of acid and metal control the ammonia formation. The use of solid nickel such as electrolytic nickel or nickel briquettes enhances ammonia formation. Nickel powder, added slowly to a stirred mixture of nitric acid and water, yields nickel nitrate containing the least ammonia. Adding nitric acid to nickel powder in water results in the formation of considerable quantities of ammonium nitrate. A method to eliminate the ammonia formation employs the addition of nitric acid to a mixture of black nickel oxide powder and hot water. The reaction is controlled by using a cooling coil or cold water condenser because the reaction is highly exothermic (39).

Uses. Nickel nitrate is an intermediate in the manufacture of nickel catalysts, especially those that are sensitive to sulfur and therefore preclude the use of the less expensive nickel sulfate. Nickel nitrate also is an intermediate in loading active mass in nickel–alkaline batteries of the sintered plate type (see BATTERIES, SECONDARY CELLS). Typically, hot nickel nitrate syrup is impregnated in the porous sintered nickel positive plates. Subsequently, the plates are soaked in potassium hydroxide solution, whereupon nickel hydroxide [12054-48-7] precipitates within the pores of the plate.

Nickel Halides. *Properties.* Nickel forms anhydrous as well as hydrated halides. The properties of the anhydrous salts are given in Table 1.

Table 1. Properties of Anhydrous Nickel Halides[a]

Compound	CAS Registry Number	Mp, °C	Density, g/cm^3	Color	Solubility, 0°C, g/100 mL H$_2$O
nickel difluoride	[10028-18-9]	1000[b]	4.63	light green	4
nickel dichloride	[7718-54-9]	1001	3.56	yellow	64
nickel dibromide	[13462-88-9]	963	5.10	orange	113
nickel diiodide	[13462-90-3]	797	5.83	black	124

[a]Refs. 45 and 46.
[b]Sublimes.

Nickel chloride hexahydrate [7791-20-0] is formed by the reaction of nickel powder or nickel oxide with a hot mixture of water and HCl. Nickel fluoride [13940-83-5], $NiF_2 \cdot 4H_2O$, is prepared by the reaction of hydrofluoric acid on nickel carbonate. Nickel bromide [18721-96-5], $NiBr_2 \cdot 6H_2O$, is made by the reaction of black nickel oxide and HBr. The reaction of hydriodic acid with nickel carbonate yields nickel iodide [7790-34-3], $NiI_2 \cdot 6H_2O$.

Uses. Nickel chloride hexahydrate is an important material in nickel electroplating. It is used with nickel sulfate in the conventional Watts plating bath (47). Nickel chloride is an intermediate in the manufacture of certain nickel catalysts, and it is used to absorb ammonia in industrial gas masks. Anhydrous nickel chloride [7718-54-9] is formed from a mixture of nickel powder and sodium chloride at the anode during recharging of sodium nickel chloride batteries, which have possible use as the power source in electric vehicles (48). Nickel bromide has limited use in nickel electroplating. The reaction of nickel chloride or nickel bromide with dimethoxyethane yields ether-soluble $NiX_2 \cdot 2C_2H_4(OCH_3)_2$ compounds which are useful as nickel-containing reagents for a variety of reactions used to form coordination compounds of nickel (49). Nickel fluoride is used for cold sealing of anodic coatings on aluminum (50,51).

Nickel Carbonate. Nickel carbonate [3333-67-3], $NiCO_3$, is a light-green, rhombic crystalline salt, density 2.6 g/cm^3, that is very slightly soluble in water. The addition of sodium carbonate to a solution of a nickel salt precipitates an impure basic nickel carbonate. The commercial material is the basic salt $2NiCO_3 \cdot 3Ni(OH)_2 \cdot 4H_2O$ [29863-10-3]. Nickel carbonate is prepared best by the oxidation of nickel powder in ammonia and CO_2. Boiling away the ammonia causes precipitation of pure nickel carbonate (32).

Nickel carbonate is used in the manufacture of catalysts, in the preparation of colored glass (qv), in the manufacture of certain nickel pigments, and as a neutralizing compound in nickel electroplating solutions. It also is used in the preparation of many specialty nickel compounds.

Nickel Hydroxides. Nickel hydroxide [12054-48-7], $Ni(OH)_2$, is a light-green, microcrystalline powder, density 4.15 g/cm^3. It decomposes into nickel oxide and water when heated at 230°C, and is extremely insoluble in water. A solution of nickel sulfate which is treated with sodium hydroxide yields the gelatinous nickel hydroxide which, when neutralized, forms a fine precipitate that can be filtered. Another industrial route for the manufacture of nickel hydroxide is by electrodeposition at an inert cathode using metallic nickel as the anode and nickel nitrate as the electrolyte. High purity crystalline nickel hydroxide can be made from nickel nitrate solution and potassium hydroxide by subsequently extracting the gelatinous precipitate with hot alcohol. Nickel hydroxide is an intermediate in the manufacture of nickel catalysts. The principal use for nickel hydroxide is in the manufacture of nickel–cadmium batteries. Product morphology, particle size, and method of synthesis all bear important roles in the utility of nickel hydroxide in nickel alkaline storage batteries (52). Nickel hydroxide can be formed electrochemically within the sintered positive-electrode plaque if the porous plaque is used as the cathode (53). A variation of this method involves two porous electrodes and a d-c pole-reversing technique (54).

When nickel hydroxide is oxidized at the nickel electrode in alkaline storage batteries the black trivalent gelatinous nickel hydroxide oxide [12026-04-9],

Ni(OH)O, is formed. In nickel battery technology, nickel hydroxide oxide is known as the nickel active mass (see BATTERIES, SECONDARY CELLS). Nickel hydroxide nitrate [56171-41-6], Ni(OH)NO$_3$, and nickel chloride hydroxide [25965-88-2], NiCl(OH), are frequently mentioned as intermediates for the production of nickel powder in aqueous solution. The binding energies for these compounds have been studied (55).

Nickel Fluoroborate. Fluoroboric acid and nickel carbonate form nickel fluoroborate [14708-14-6], Ni(BF$_4$)$_2$·6H$_2$O. Upon crystallization, the high purity product is obtained (47). Nickel fluoroborate is used as the electrolyte in specialty high speed nickel plating. It is available commercially as a concentrated solution.

Nickel Cyanide. Nickel cyanide tetrahydrate [20427-77-4], Ni(CN)$_2$·4H$_2$O, forms apple-green plates which are, like other metal cyanides, highly poisonous. When the tetrahydrate is heated to 200°C, anhydrous nickel cyanide [557-19-7], Ni(CN)$_2$, forms. Further heating causes decomposition. Nickel cyanide is made by the reaction of potassium cyanide and nickel sulfate. Nickel cyanide, highly insoluble in water, precipitates from the reaction medium. Nickel cyanide is soluble in aqueous alkali cyanides as well as in other bases, including ammonium hydroxide and alkali metal hydroxides. The latter yield the stable, water-soluble orange tetracyanonickelate(II). An aqueous solution of K$_2$[Ni(CN)$_4$] [14220-17-8] does not yield a nickel sulfide precipitate in the presence of hydrogen sulfide. Nickel cyanide has been used in the Reppe process for the conversion of acetylene to butadiene (qv) and other products (56) (see ACETYLENE-DERIVED CHEMICALS; CYANIDES).

Nickel Sulfamate. Nickel sulfamate [13770-89-3], Ni(SO$_3$NH$_2$)$_2$·4H$_2$O, commonly is used as an electrolyte in nickel electroforming systems, where low stress deposits are required. As a crystalline entity for commercial purposes, nickel sulfamate never is isolated from its reaction mixture. It is prepared by the reaction of fine nickel powder or black nickel oxide with sulfamic acid in hot water solution. Care must be exercised in its preparation, and the reaction should be completed rapidly because sulfamic acid hydrolyzes readily to form sulfuric acid (57).

Nickel Sulfide. Nickel, like iron and cobalt, forms monosulfides which may show considerable deviation from stoichiometry without exhibiting heterogeneity. Nickel sulfide [1314-04-1], NiS, occurs naturally as the mineral millerite, and has a trigonal crystalline form and a yellow metallic luster; density 5.65 g/cm^3, mp 797°C. It is insoluble in water. Nickel sulfides often are thought of as binary alloys of sulfur and nickel because the metallic appearance of the sulfides resembles alloys more than chemical compounds. Other nickel sulfides include two subsulfides, Ni$_2$S [12137-08-5] and Ni$_3$S$_2$ [12035-72-2]. The latter is found as the mineral heazlewoodite. Another naturally occurring sulfide is polydymite [12137-12-1], Ni$_3$S$_4$ (58).

Nickel sulfide, NiS, can be prepared by the fusion of nickel powder with molten sulfur or by precipitation using hydrogen sulfide treatment of a buffered solution of a nickel(II) salt. The behavior of nickel sulfides in the pure state and in mixtures with other sulfides is of interest in the recovery of nickel from ores, in the high temperature sulfide corrosion of nickel alloys, and in the behavior of nickel-containing catalysts.

Other Nickel Salts. *Nickel Arsenate.* Nickel arsenate [7784-48-7], $Ni_3(AsO_4)_2 \cdot 8H_2O$, is a yellowish green powder, density 4.98 g/cm³. It is highly insoluble in water but is soluble in acids, and decomposes on heating to form As_2O_5 and nickel oxide. Nickel arsenate is formed by the reaction of a water solution of arsenic anhydride and nickel carbonate. Nickel arsenate is a selective hydrogenation catalyst for inedible fats and oils (59).

Nickel Phosphate. Trinickel orthophosphate [14396-43-1], $Ni_3(PO_4)_2 \cdot 7H_2O$, exists as apple-green plates which decompose upon heating. It is prepared by the reaction of nickel carbonate and hot dilute phosphoric acid. Nickel phosphate is an additive to control the crystal size of zinc phosphate in conversion coatings which are applied to steel prior to its being painted (see METAL SURFACE TREATMENTS).

Nickel Double Salts. Nickel ammonium chloride [16122-03-5], $NiCl_2 \cdot NH_4Cl \cdot 6H_2O$, nickel ammonium sulfate [15699-18-0], $NiSO_4 \cdot (NH_4)_2SO_4 \cdot 6H_2O$, and nickel potassium sulfate [10294-65-2], $NiSO_4 \cdot K_2SO_4 \cdot 6H_2O$, are prepared by crystallizing the individual salts from a water solution. These have limited use as dye mordants and are used in metal-finishing compositions (59).

Nickel Amine Complexes. The thermal stability of the hexaamminenickel(II) halides increases with the size of the halide (7). Decomposition temperatures at 13 kPa (100 mm Hg) for $Ni(NH_3)_6Cl_2$ [10534-88-0], $Ni(NH_3)_6Br_2$ [13601-55-3], and $Ni(NH_3)_6I_2$ [13859-68-2] are 398, 433, and 450°C, respectively. The thermal decomposition of tetrakispyridinenickel(II) dichloride [14076-99-4] (60) and the bidendate ethylenediamine (en) complex [Ni(en)₃]Cl₂ [13408-70-3] (61) have also been studied.

Organic Compounds

Nickel plays a role in the Reppe polymerization of acetylene where nickel salts act as catalysts to form cyclooctatetraene (62); the reduction of nickel halides by sodium cyclopentadienide to form nickelocene [1271-28-9] (63); the synthesis of cyclododecatrienenickel [39330-67-1] (64); and formation from elemental nickel powder and other reagents of nickel(0) complexes that serve as catalysts for oligomerization and hydrocyanation reactions (65) (see also ORGANOMETALLICS).

Nickel Carbonyl. *Properties.* Nickel carbonyl, $Ni(CO)_4$, is a colorless liquid having a high vapor pressure. The vapor density is ca four times that of air. As a liquid it is miscible in all proportions with most organic solvents and is practically insoluble in water. Nickel carbonyl reacts slowly with hydrochloric or sulfuric acid but reacts vigorously with nitric acid or the halogens, forming the corresponding nickel salts upon the liberation of CO. Bromine water is a useful reagent for the controlled decomposition of nickel carbonyl and for destroying residual amounts of it in a chemical apparatus. The properties of nickel carbonyl are given in Table 2 (see also CARBONYLS).

Thermodynamic properties (71,72), force constants (73), and infrared absorption characteristics (74) are documented. The coordinatively unsaturated species, $Ni(CO)_3$ and $Ni(CO)_2$, also exist and the bonding and geometry data have been subjected to molecular orbital treatments (75,76).

Manufacture. Nickel carbonyl can be prepared by the direct combination of carbon monoxide and metallic nickel (77). The presence of sulfur, the surface

Table 2. Physical Properties of Nickel Carbonyl[a]

Property	Value
melting point, °C	−17
crystallization point, °C	−25
density at 20.0°C, g/mL	1.3103
electronic absorption, nm	240[b]
molecular rotation at 578 nm, μrad	2.768
bond distances, pm	
Ni — C	183.8
C — O	114.1
vapor pressure data, kPa[c]	
at 0°C	17.1
16.1°C	36.2
21.1°C	44.3
28.5°C	60.0
35.1°C	77.7
42.2°C	101
critical temperature, °C	200

[a]Refs. 66–70.
[b]Intense.
[c]To convert kPa to psi, multiply by 0.145.

area, and the surface activity of the nickel affect the formation of nickel carbonyl (78). The thermodynamics of formation and reaction are documented (79). Two commercial processes are used for large-scale production (80). An atmospheric method, whereby carbon monoxide is passed over nickel sulfide and freshly reduced nickel metal, is used in the United Kingdom to produce pure nickel carbonyl (81). The second method, used in Canada, involves high pressure CO in the formation of iron and nickel carbonyls; the two are separated by distillation (81). Very high pressure CO is required for the formation of cobalt carbonyl and a method has been described where the mixed carbonyls are scrubbed with ammonia or an amine and the cobalt is extracted as the ammine carbonyl (82). A discontinued commercial process in the United States involved the reaction of carbon monoxide with nickel sulfate solution.

Substituted Nickel Carbonyl Complexes. The reaction of trimethyl phosphite and nickel carbonyl yields the monosubstituted colorless oil, $(CO)_3NiP(OCH_3)_3$ [17099-58-0], the disubstituted colorless oil, $(CO)_2Ni[P(OCH_3)_3]_2$ [16787-28-3], and the trisubstituted white crystalline solid, $(CO)Ni[P(OCH_3)_3]_3$ [17084-87-6] (mp 98°C). Liquid complexes result from the reaction of trifluorophosphine with nickel carbonyl yielding $(CO)_3Ni(PF_3)$ [14264-32-5], $(CO)_2Ni(PF_3)_2$ [13859-78-4], and $(CO)Ni(PF_3)_3$ [14219-40-0]. A bidendate substituted nickel carbonyl, $(CO)_2Ni(o\text{-}C_6H_4[P(CH_3)_2]_2)$ [76404-14-3], a white crystalline compound (mp 123°C), is known. Substituted arsine complexes and a substituted stibine derivative have been isolated. Reviews of N-, P-, As-, and Sb- donor nickel carbonyl complexes are available (83,84). The electrochemical reduction of bisbipryridinenickel(II) with CO_2 in DMF solvent gives the nickel zero complex, (2,2′-bipyridine-N,N')-dicarbonylnickel(0)

[*14917-14-7*] (T-4) (85). An example of the C-ylide nickel carbene complex is (1,3-diethyl-2-imidazolidene-ylidene)nickel(0) tricarbonyl (86).

π-Cyclopentadienyl Nickel Complexes.   Nickel bromide dimethoxyethane [*29823-39-9*] forms bis(cyclopentadienyl)nickel [*1271-28-9*] upon reaction with sodium cyclopentadienide (63). This complex, known as nickelocene, π-$(C_5H_5)_2Ni$, is an emerald-green crystalline sandwich compound, mp 173°C, density 1.47 g/cm^3. It is paramagnetic and slowly oxidizes in air. A number of derivatives of nickelocene are known, eg, methylnickelocene [*1292-95-4*], which is green and has mp 37°C, and bis(π-indenyl)nickel [*52409-46-8*], which is red, mp 150°C (87,88).

Substituted derivatives of nickelocene, where one ring has been replaced, include the complex cyclopentadienyl nitrosyl nickel [*12071-73-7*], (π-C_5H_5)NiNO, a red liquid, mp −41°C. A review of nitrosyl complexes with nickel is available (89). The dimer complex di-μ-carbonyl-bis(η^5-cyclopentadienyldinickel) [*12170-92-2*], (π-C_5H_5NiCO)$_2$, is made in reversible reaction from nickel carbonyl and nickelocene (63). The complex is a red-violet solid and is diamagnetic, and spectroscopic studies show the presence only of bridging carbonyl groups.

Tetrakisligand Nickel(0) Complexes.   Tetrakisligand nickel(0) complexes are made by several methods. One procedure is the substitution of CO in nickel carbonyl. Tetrakistrichlorophosphinenickel(0) [*36421-86-0*], Ni(PCl$_3$)$_4$, yellow, mp 120°C (dec) is a product of CO substitution synthesis (22). Another method of preparation involves substitution of more powerful donor ligands, such as triphenylphosphine, in other tetrakisligand nickel(0) complexes. The red solid tetrakistriphenylphosphinenickel(0) [*15133-82-1*], mp 125°C, and the yellow solid tetrakistrimethylphosphinenickel(0) [*28069-69-4*], mp 185°C (dec), are examples of the ligand-substitution method of preparation (90). A third method of preparation involves the direct reaction of nickel powder and ligands where halogens are on the donor atom. Examples of complexes prepared by this method are Ni(PF$_3$)$_4$ [*13858-65-9*] and Ni(CH$_3$PCl$_2$)$_4$ [*76404-15-4*] (91). A fourth method of preparation of NiL$_4$ complexes involves reaction of nickel halides and ligands in the presence of a reducing agent, eg, zinc metal powder (92). The mixed ligand complex Ni(P(o-CH$_3$C$_6$H$_4$)$_3$)$_3$ (NCCH(CH$_3$)CHCH$_2$) [*41686-95-7*] has been prepared from anhydrous nickel chloride [*7718-54-9*], 3-pentenenitrile, and tri-o-tolylphosphine (93).

Tetrakisligand nickel(0) complexes have tetrahedral structures. Electronic structures have been studied and conformational analysis performed. Quantitative equilibria measurements of the ligands in these complexes imply a dominant role for ligand steric effects when the complexes are employed as catalysts (94).

Tetrakisligand nickel(0) complexes catalyze the reaction of ethylene and butadiene to give 1,4-hexadiene (95), the isomerization of 1-butene to 2-butene (96), and hydrocyanation of butadiene to form adiponitrile (97,98). The thermal decomposition of tetrakis(triorganophosphite)nickel(0) complexes in high boiling solvents is a method for depositing a high purity coating of nickel on steel (99).

Other Complexes.   Several other classes of organonickel complexes are known. Allyl bromide and nickel carbonyl react to give a member of the π-allyl system [*12012-90-7*], [π-C_3H_5NiBr]$_2$ (100). Tris(η^2-ethene)nickel [*50696-82-7*] reacts with acetylene and 1,2-bis(diisopropylphosphino)ethane to form the ethyne complex (η^2-ethyne)(1,2-bis(diisopropylphosphino)ethane)nickel and related species (101). A review of nickel η-bonded complexes containing unsaturated organic molecules is available (102).

1,2,3,4-Tetramethylcyclobutadiene dichloride [76404-16-5] can be prepared by reaction of nickel carbonyl and 3,4-dichlorotetramethylcyclobutene (CBD) in polar solvents (103). The complex is black-violet, mp 185°C (dec).

The reaction of a mixture of 1,5,9-cyclododecatriene (CDT), nickel acetylacetonate [3264-82-2], and diethylethoxyaluminum in ether gives red, air-sensitive, needle crystals of (CDT)Ni [12126-69-1] (66). Crystallographic studies indicate that the nickel atom is located in the center of the 12-membered ring of (CDT)Ni (104). The latter reacts readily with 1,5-cyclooctadiene (COD) to yield bis(COD) nickel [1295-35-8] which has yellow crystals and is fairly air stable, mp 142°C (dec) (20). Bis(COD)nickel also can be prepared by the reaction of 1,5-COD, triethylaluminum, and nickel acetylacetonate.

In another class of compounds a nickel complex serves as anion. Two examples are tetraethylammonium tetrachloronickelate [5964-71-6], ((C_2H_5)$_4$ N)$_2$[NiCl$_4$] (105), and tetraethylammonium triphenylphosphinetribromonickelate [41828-60-8], [(C_2H_5)$_4$N][(C_6H_5)$_3$PNiBr$_3$] (106).

Nickel salts form coordination compounds with many ligands. Dibromobis(tri-n-butylphosphine)nickel(II) [15242-92-9], [(n-C_4H_9)$_3$P]$_2$NiBr$_2$, dicyanoammineaquanickel(II), Ni(NH$_3$)(H$_2$O)(CN)$_2$, and bromonitrosobis(triphenylphosphine)nickel(II) [14586-72-2], are complexes used for syntheses in preparative organonickel chemistry.

Reduction of compounds of the type LNiX$_2$ and L$_2$NiX$_2$ where L is a ligand yields hydride complexes, ie, LNiHX and L$_2$NiHX (see HYDRIDES). The LNiHX are generally stable only at low temperatures; the L$_2$NiHX are more stable. A high degree of stability can result when bulky ligands are employed, eg, chlorohydridobis(tricyclohexylphosphine)nickel [25703-57-5], HNi(P(cyclo-C_6H_{11})$_3$)$_2$Cl, has mp 150°C (dec) (107).

The presence of strongly electron-donating ligands has a large effect on the synthesis of alkyl and aryl nickel compounds. Whereas dimethylnickel [54836-89-4], Ni(CH$_3$)$_2$, cannot be isolated, even at -130°C, bis(tricyclohexylphosphine)dimethylnickel [36427-03-9] [(cyclo-C_6H_{11})$_3$P]$_2$Ni(CH$_3$)$_2$, is stable at ambient temperature (108). Bis(triphenylmethyl)nickel [7544-48-1], [(C_6H_5)$_3$C]$_2$Ni, has violet crystals and mp 120°C (dec). This compound, considered to be a ligand-free nickel aryl complex (109), can be prepared by the reduction of nickel salts in the presence of hexaphenylethane. Chlorobis(triphenylphosphine)phenylnickel [38415-93-3], [(C_6H_5)$_3$P]$_2$NiC$_6$H$_5$Cl, mp 122°C, is a bisligand aryl nickel halide (110). The complex ((CH_3)$_3$P)$_2$Ni(CH$_3$)$_2$ [60802-48-4] reacts further to yield the orange-red solid, mp 49–51°C ((CH_3)$_3$P)$_3$Ni(CH$_3$)$_2$ [42725-08-4], containing three phosphines (111).

π-Complexes of alkylnickel and arylnickel also have been prepared. The stronger electronic donating compounds impart greater stability. Methyl Grignard reagent and π-allylnickel [12077-85-9] yield the dimer of π-allylmethylnickel, a complex which cannot be isolated above -78°C. However, tricyclohexylphosphine π-allylmethylnickel [76422-11-2], π-$CH_2CHCH_2NiCH_3$(P(cyclo-C_6H_{11})$_3$), is a stable yellow solid, mp 50°C (dec) (112). Tri-n-butylphosphine-π-cyclopentadienylmethylnickel [7298-70-0], P(C_4H_9)$_3$-

(π-C_5H_5)NiCH$_3$, can be prepared by the reaction of a cyclopentadienyl ligand nickel halide and methyl Grignard reagent. It is a greenish brown solid, mp 29–30°C (113).

Nickel(0) compounds containing CO_2 ligands are of interest for environmental studies. When dry CO_2 is bubbled into Ni(PR$_3$)$_4$ solutions, red-orange diamagnetic complexes, [Ni(PR$_3$)$_2$CO$_2$], form (114). Nickel complexes also form with macroligands such as tetraazamacrocycles (115). Macro ligand complexes of nickel have been studied in the redox process of CO_2 in water–CH$_3$CN solutions (116,117). Cationic nickel ligand complexes [Ni(C$_6$Cl$_5$)L(PR$_3$)]$^+$ have been reported (118). There is much interest in this field because of possible biochemical implications.

Nickel Salts and Chelates. Nickel salts of simple organic acids can be prepared by reaction of the organic acid and nickel carbonate of nickel hydroxide; reaction of the acid and a water solution of a simple nickel salt; and, in some cases, reaction of the acid and fine nickel powder or black nickel oxide.

Nickel acetate tetrahydrate [6018-89-9], Ni(C$_2$H$_3$O$_2$)·4H$_2$O, is a green powder which has an acetic acid odor, density 1.74 g/cm^3. When heated, it loses its water of crystallization and then decomposes to form nickel oxide. Nickel acetate is used as a catalyst intermediate, as an intermediate in the formation of other nickel compounds, as a dye mordant, as a sealer for anodized aluminum, and in nickel electroplating (59).

Nickel formate dihydrate [15694-70-9], Ni(HCOO)$_2$·2H$_2$O, is a green monoclinic crystalline compound which melts with decomposition to nickel oxide at 180°C; density 2.15 g/cm^3. Nickel formate is used in the preparation of fat-hardening nickel hydrogenation catalysts (119).

Other simple nickel salts of organic acids include the oxalate [20543-06-0], oleate [68538-38-5], and stearate [2223-95-2]. The latter two have been used as oil-soluble nickel forms in the dyeing of synthetic polyolefin fibers (see DRIERS AND METALLIC SOAPS). Nickel oxalate has been used as a catalyst intermediate (59).

Nickel acetylacetonate [3264-82-2], Ni(C$_5$H$_7$O$_2$)$_2$, is a green powder which can be made by the aqueous reaction of a soluble nickel salt with 2,4-pentanedione. It is the simplest of the bidentate coordination compounds of nickel. Its use is primarily in preparative organonickel chemistry. Nickel acetylacetonate complexes with aluminoxanes have been reported to be active copolymerization catalysts for styrene and 2-norbornene (120). Other well-known chelates include ethylenediaminebisacetylacetonatenickel(II) [42948-35-6], nickel phthalocyanine [14045-02-8], and nickel dimethylglyoxime [13478-93-8] (121,122). The last two compounds have been studied as pigments. A review of polydentate nickel complexes is available (123). Nickel also forms derivatives with organic sulfur compounds, eg, trithiocarbonato triamminenickel [39282-88-7], (NH$_3$)$_3$NiCS$_3$, dimethyldithiocarbamatenickel [15521-65-0], Ni[(CH$_3$)$_2$NCS$_2$]$_2$, and nickel ethyl xanthate [52139-56-7], Ni(C$_2$H$_5$OCS$_2$)$_2$ (124).

Economic Aspects

Estimated 1994 worldwide usage of nickel compounds exclusive of use in nickel refining is shown in Table 3. Nickel compound prices are also given.

Table 3. Annual 1994 Worldwide Usage and Prices of Nickel Compounds

Compound	Formula	Quantity used,[a] t $\times$ 10^3	Price,[b] \$/kg
nickel oxide sinter	NiO	100	4.00
nickel sulfate	$NiSO_4 \cdot 6H_2O$	15	2.50
green nickel oxide	NiO	10	4.50
nickel nitrate	$Ni(NO_3)_2 \cdot 6H_2O$	8	3.50
black nickel oxide	NiO	6	6.00
nickel carbonate	$NiCO_3$	5	4.00
nickel chloride	$NiCl_2 \cdot 6H_2O$	3	4.00
nickel acetate	$Ni(C_2H_3O_2)_2 \cdot 4H_2O$	c	
others		5	5.00

[a]Estimated for 1994.
[b]As of January 1994.
[c]Usage is included in others.

The approximate worldwide annual usage of nickel chemicals at 10^3 t, other than for steel and nickel refining, in 1994 was, for plating salts, 12–15; catalysts, 10–12; specialty ceramics, 3–4; specialty chemicals, 2–3; and other specialties, 1–2.

Analytical Methods

Analytical determination of nickel in solution is usually made by atomic absorption spectrophotometry and, often, by x-ray fluorescence spectroscopy.

Nitric acid can be used for the dissolution of nickel from many inorganic substances. In some cases perchloric acid is used in combination with nitric acid. Simple organic forms of nickel also can be dissolved in nitric acid. In the case of complicated structural organic forms of nickel, oxidation calorimetry must be used to decompose the substances.

Nickel also is determined by a volumetric method employing ethylenediaminetetraacetic acid as a titrant. Inductively coupled plasma (ICP) is preferred to determine very low nickel values (see TRACE AND RESIDUE ANALYSIS). The classical gravimetric method employing dimethylglyoxime to precipitate nickel as a red complex is used as a precise analytical technique (122). A colorimetric method employing dimethylglyoxime also is available. The classical method of electrodeposition is a commonly employed technique to separate nickel in the presence of other metals, notably copper (qv). It is also used to establish calibration criteria for the spectrophotometric methods. X-ray diffraction often is used to identify nickel in crystalline form.

Health and Safety Factors

Eye and Skin Contact. Some nickel salts and aqueous solutions of these salts, eg, the sulfate and chloride, may cause a primary irritant reaction of the eye and skin. The most common effect of dermal exposure to nickel is allergic contact dermatitis. Nickel dermatitis may occur in sensitized individuals following close and prolonged contact with nickel-containing solutions or metallic objects such as jewelry, particularly pierced earrings. It is estimated that 8–15%

of the female human population and 0.2–2% of the male human population is nickel-sensitized (125).

Although most nickel sensitization results from nonoccupational exposures, nickel dermatitis was historically a problem in workplaces where there was a high risk of continuous contact with soluble nickel, eg, in electroplating (qv) shops. Improved personal and industrial hygiene has largely eliminated this problem. However, there are a few occupations involving wet nickel work, particularly where detergents facilitate the penetration of skin by nickel, where hand eczema may occur (126).

Protective equipment and clothing such as face shields and gloves should be worn and safety showers should be available wherever there is a possibility of being splashed or otherwise contacted by nickel-containing solutions. If dermatitis should occur, the possibility that it is nickel-related should be brought to the attention of a physician.

Inhalation. *Nickel Carbonyl.* Nickel carbonyl is an extremely toxic gas. The permissible exposure limit (PEL) in the United States is 1 part per billion (ppb) in air (127). The American Conference of Governmental Industrial Hygienists (ACGIH) threshold limit value (TLV) for an 8-h, time-weighted average concentration is 50 ppb (128). Nickel carbonyl may form wherever carbon monoxide and finely divided nickel are brought together. Its occurrence has been suspected but never demonstrated in some industrial operations, eg, welding of nickel alloys.

Concentrations of nickel carbonyl as low as 30 ppm in air for 30 min may be lethal for humans. Individuals exposed to these high concentrations show immediate symptoms of dizziness, headache, shortness of breath, and vomiting. These early symptoms generally disappear in fresh air, but delayed symptoms may develop 12–36 h later. These latter symptoms include shortness of breath, cyanosis, chest pain, chills, and fever. In severe exposure cases, death results from pneumonitis.

Nickel carbonyl should be used in totally enclosed systems or under good local exhaust. Plants and laboratories where nickel carbonyl is used should make use of air-monitoring devices, alarms should be present in case of accidental leakage, and appropriate personal respiratory protective devices should be readily available for emergency uses. Monitoring of urinary nickel levels is useful to help determine the severity of exposure and identify appropriate treatment measures. Some large-scale users of nickel carbonyl maintain a supply of sodium diethyldithiocarbamate, or Antabuse, a therapeutic agent, on hand for use in case of overexposure.

Other Nickel Compounds. With the exception of nickel carbonyl, nickel compounds are not known to be acutely toxic by inhalation. The potential chronic toxicity of nickel compounds is likely to be of greater concern. In particular, the incidence of cancer of the nose and lungs has been found to be significantly increased in a number of obsolete nickel refinery workers. Based on epidemiological and experimental results, the International Agency for Research on Cancer (IARC) (129) has concluded that all nickel compounds are Category 1, ie, known human carcinogens. There has generally been a lack of epidemiological evidence of a carcinogenic risk associated with exposure to metallic nickel and nickel alloys (130–132). The IARC has classified metallic nickel as a Category 2B carcinogen, ie, possibly carcinogenic to humans. Cancer risk in nickel refineries has been related primarily to exposure to soluble nickel at concentrations >1 mg/m^3 and to less soluble forms of nickel, principally nickel subsulfide and nickel oxides, at concentrations >10 mg/m^3 (131).

Epidemiological studies of nickel-producing and nickel-using workers seldom indicate excess mortality from nonmalignant respiratory disease. Evidence for such effects exists mainly as a few reports of isolated incidents of asthma, pulmonary fibrosis, chronic bronchitis, and emphysema in nickel workers. Nickel may or may not play a causal role in these incidents (131).

Some nonmalignant respiratory effects have been observed in experimental animals during acute or subchronic exposures. Soluble and moderately soluble compounds were more toxic than were insoluble compounds and produced different effects. Sulfate and subsulfide produced fibrosis whereas nickel oxide did not.

It is good practice to keep concentrations of airborne nickel in any chemical form as low as possible and certainly below the relevant standard. Local exhaust ventilation is the preferred method, particularly for powders, but personal respirator protection may be employed where necessary. In the United States, the Occupational Safety and Health Administration (OSHA) personal exposure limit (PEL) for all forms of nickel except nickel carbonyl is 1 mg/m^3. The ACGIH TLVs are respectively 1 mg/m^3 for Ni metal, insoluble compounds, and fume and dust from nickel sulfide roasting, and 0.1 mg/m^3 for soluble nickel compounds. The ACGIH is considering whether to lower the TLVs for all forms of nickel to 0.05 mg/m^3, based on nonmalignant respiratory effects in experimental animals.

Uses

Catalysts. Nickel is an important hydrogenation catalyst because of its ability to chemisorb hydrogen. One important nickel catalyst is Raney nickel. Raney nickel catalyst is used widely in laboratory and industrial hydrogenation processes. It is the most active, least specific of the nickel catalysts. Raney nickel catalyst has been used in a continuous hydrogenation process by filling a tube with chunks of the nickel–aluminum alloy and activating the surface by passing a solution of caustic over it, thereby removing some of the surface aluminum. Periodic flushes with caustic enable the catalyst to be reactivated in place. A review is available (133).

A number of variations of the nickel–aluminum catalyst have been developed. One involves rolling nickel and aluminum foil at 630°C followed by leaching with caustic (134). Nickel–aluminum alloy has been flame-sprayed on the inside of steel tubes, followed by leaching (135), and has been suggested for use as a continuous reactor for the conversion of synthesis gas to methane (136) (see FUELS, SYNTHETIC). Another method is the electroplating of nickel on the inside wall of stainless steel tubing, followed by aluminizing the nickel surface and activating with a caustic leach (137). Other alloying compositions include a nickel–iron–aluminum alloy which, upon caustic activation, is used as a catalyst for the selective hydrogenation of organic nitro compounds (see AMINES BY REDUCTION) (138). A composition of nickel and silicon yields the nickel silicides [12035-57-3], NiSi, and [12201-89-7], NiSi$_2$, which, upon caustic leaching, activate a nickel surface (139). Nickel–boron alloy [12007-00-0], when activated with caustic, has been claimed to be a more reactive hydrogenation catalyst than the nickel–aluminum catalyst (140).

Supported nickel catalysts of the precipitated and impregnated types are used for methanation, steam-hydrocarbon reforming, petrochemical hydrogenation, and fat hardening (see CATALYSTS, SUPPORTED). The nickel compound and the ceramic carrier in a dry reduction technique are heated in a stream of inert gas to decompose the nickel compound to nickel oxide, which then is reduced with hydrogen to nickel metal. Precipitated catalysts generally are made from nickel carbonate and nickel hydroxide. Nickel nitrate, nickel chloride, hexammine nickel carbonate, and nickel acetate solutions are used for impregnation (141). Supported catalysts of nickel on alumina or nickel on zirconia have been suggested as a heterogeneous methanation system for the hydrogenation of CO. The latter catalyst has been reported to be quite active in the presence of H_2S (142).

Nickel is used with other elements for special types of hydrogenation catalysts. Nickel sulfide [*16812-54-7*] and nickel tungsten sulfide catalysts are used when high concentrations of sulfur compounds are present in the hydrogenation of petroleum distillates. Nickel–molybdenum catalysts are used to denitrogenate petroleum fractions that are high in nitrogen-containing components (see PETROLEUM).

Black nickel oxide is used as an oxygen donor in three-way catalysts containing rhodium, platinum, and palladium (143). Three-way catalysts, used in automobiles, oxidize hydrocarbons and CO, and reduce NO_x. The donor quality, ie, the ability to provide oxygen for the oxidation, results from the capability of nickel oxide to chemisorb oxygen (see EXHAUST CONTROL, AUTOMOTIVE).

Nickel and other transition metals function as solvent-catalysts for the transformation of carbon species into the diamond allotrope. At temperatures high enough to melt the metal or metal–carbon mixture and at pressures high enough for diamond to be stable, diamond forms by what is probably an electronic mechanism (see CARBON, DIAMOND–SYNTHETIC).

Important organic chemical intermediates are manufactured using nickel organometallic compounds as catalysts. One example is Shell's Higher Olefins Process (SHOP), a commercial method for the oligerimerization of ethylene (qv) to detergent-range alpha-olefins. The catalyst is a ligand-stabilized nickel chelate dissolved in a polar solvent, into which ethylene is pressurized. The resulting alpha-olefins are insoluble in the solvent and thus easily separated from the reaction mixture (144–147).

Another example is the du Pont process for the production of adiponitrile. Tetrakisarylphosphitenickel(0) compounds are used to affect the hydrocyanation of butadiene. A multistage reaction results in the synthesis of dinitrile, which is ultimately used in the commercial manufacture of nylon-6,6 (144–149).

Nickel Carbonyl.   Nickel carbonyl serves as an intermediate in nickel refining, the capacity of which at Inco Limited is > 100,000 t/yr. High purity nickel pellets for melting and dissolving are a product of the carbonyl-refining process. The nickel powders useful in nickel chemical synthesis and for making nickel alkaline-battery electrodes and powder metallurgical parts are derived from the carbonyl-refining process (150–155). Nickel carbonyl also is used in a carbonylation reaction in the synthesis of acrylic and methacrylic esters from acetylene and alcohols (156) (see ACRYLIC ACID AND DERIVATIVES; METHACRYLIC ACID AND

DERIVATIVES). Nickel carbonyl has been proposed as a catalyst of as an addition agent for a variety of organic reactions including catalysis, polymerization, and other carbonylation reactions.

The surface of the decomposition of nickel carbonyl is important in numerous commercial applications. Graphite, always present in the nickel powder formed from nickel carbonyl, results from the disproportionation of CO. The quantity of graphite formed can be related to a few ppm of iron carbonyl present in the nickel carbonyl gas. A smooth coating of nickel forms when graphite powder is used as a substrate for nickel carbonyl decomposition. An autocatalytic process involving graphite may thus occur during nickel carbonyl decomposition. The facile decomposition of nickel carbonyl on the surfaces of aluminum and alumina powders leads to smoothly nickel-coated substrate particles (157–159). For most metallic and oxide particles in the presence of nickel carbonyl, the primary nucleation species reverts to the newly formed graphite interface. Scanning electron micrographs particles show the growth of nickel warts, suggesting that the population of nucleating species on these substrates is very low.

Nickel-coated powder products are used as conductive pigments for application in surface coatings (qv), adhesives (qv), injection molding powders, and for sealants (160). Nickel-coated aluminum powder is also used in plasma spray applications for hard-facing metals (see PLASMA TECHNOLOGY). Nickel-coated alumina is useful in grinding-wheel applications (161) and as an interface material for the cementation of nickel to alumina in electronic applications. Nickel carbonyl gas is used commercially to coat graphite fiber tows which are useful in lightning-strike devices, eletromagnetic interference (EMI) shielding in plastic parts, and a variety of epoxy composites requiring conductivity (162,163). The polyacrylonitrile (PAN) graphite fiber, however, does not form a nucleation surface for nickel carbonyl decomposition. A promotor such as a Lewis acid must be added to the gas stream and the nickel film rapidly forms on the heated graphite fiber surface. The nickel, however, is not bonded to the fiber as it is to the powder substrates. The commercial value for nickel-coated particles and fibers in 1993 was estimated to be $15 million (see CARBON AND GRAPHITE FIBERS).

Electroplating. The second-largest application for nickel chemicals is as electrolytes in nickel electroplating (qv). In ordinary plating systems, nickel present in the electrolyte never forms on the finished workpiece; the latter results from dissolution and transfer from nickel anodes. Decorative nickel plating is used for automobile bumpers and trim, appliances, wire products, flatware, jewelry, and many other consumer items. A comprehensive review of nickel electroplating has been compiled (164).

Specialty Ceramics. Black or green nickel oxide and nickel carbonate are used extensively in the ceramic industry (see CERAMICS). Nickel oxide is added to glass frit compositions which are used for porcelain-enameling of steel. Nickel enhances the adhesion of glass (qv) to steel through the formation of spinel structures of the mixed silicates of iron and nickel at the interface (165). Nickel oxides also are used in the manufacture of the magnetic nickel–zinc ferrite powders, which are made into parts for use in electric motors, antennas, and cathode ray tube (CRT) yokes (see FERRITES). They are also used for shielding and filtering electromagnetic radiation interference (EMI) in electronics equipment (166–168).

The nickel silicides, ie, Ni_3Si [12059-22-2], Ni_5Si_2 [12059-27-7], Ni_2Si [12059-14-2], and NiSi [39467-10-2], are electroconductive materials that are useful in resistors and resistance heating elements (169). Nickel silicides are also used to discharge static electricity from glass in automobile windshields. Nickel boride [12007-00-0], NiB, made from the exothermic reaction of boron with molten nickel, is used as a getter for the removal of oxygen from the nickel film that is manufactured for use as the screen backing in mercury-vapor displays.

Nickel niobium [59913-35-8], NiNb, is a glassy-type compound formed by reaction of niobium oxide and nickel powder. Adding nickel niobium to molten nickel–iron alloys is an important and convenient way of introducing this refractory metal, niobium, to the structure and therefore reducing the oxidation potential for corrosion. Nickel aluminide [12003-81-5], Ni_3Al, an intermetallic compound formed by the reaction of molten aluminum and nickel pellets, is a high strength, light, ductile material. It can be used in powder form to make compacted products that can be sintered to high strength, such as gears (170).

Nickel phosphorus compounds also have been studied for their electroconductive nature (see ELECTROLESS PLATING). Nickel selenite [15060-62-5], nickel phosphate, nickel tungstate [14177-51-6], nickel chromite [12018-18-7], potassium nickel molybdate [59228-72-7], nickel oxide, and nickel nitrate have been used as glass colorants (see COLORANTS FOR CERAMICS). Nickel oxide also is employed as a colorant in ceramic body stains used in ceramic tile, dishes, pottery, and sanitary ware. Nickel oxide imparts avocado green and gray colors in ceramic glazes. When fired, nickel oxide, antimony oxide, and titanium dioxide produce the yellow chalking pigment, nickel (antimony) titanate [11118-07-3]. This pigment is used extensively in exterior house paint and in vinyl house siding because of its good weatherability. Nickel titanate has been recommended as a replacement for heavy-metal pigments, such as lead chromate and cadmium yellow (171). Nickel cobalt aluminate is a useful fade-resistant blue pigment for exterior paint application (172) (see PIGMENTS).

Plastics Additives. Many claims have been made for the use of nickel chemicals as additives to various resin systems. By far the most important application is as uv-quenchers in polyolefins (173,174). Among the useful nickel complexes in these systems are dibutyldithiocarbamate nickel [13927-77-0], nickel thiobisphenolates, and nickel amide complexes of bisphenol sulfides (175). The nickel complex of O,O-dimethylcyclohexyldithiophosphate increases the uv-stability of high density polyethylene (HDPE) (176). Several classes of nickel compounds are effective as light stabilizers in poly(vinyl chloride) (PVC), although none have become commercially important, because of objectionable coloring of the plastic. Nickel aminothiobisphenolates have been claimed as light stabilizers in ABS graft copolymers (177). Nickel dialkylhydroxyphenylalkylphosphonate imparts light stability to ABS terpolymers and polyurethanes (178) (see ANTIOXIDANTS; UV STABILIZERS).

Nickel dialkyldithiocarbamates stabilize vulcanizates of epichlorhydrinethylene oxide against heat aging (178). Nickel dibutyldithiocarbamate [56377-13-0] is used as an oxidation inhibitor in synthetic elastomers. Nickel chelates of substituted acetylacetonates are flame retardants for epoxy resins

(179). Nickel dicycloalkyldithiophosphinates have been proposed as flame-retardant additives for polystyrene (180–182) (see FLAME RETARDANTS; HEAT STABILIZERS).

Organic Dyes and Pigments. A number of nickel pigments have been reported, eg, the nickel disazomethine complex [61312-95-6] which is prepared from 2-hydroxy-1-naphthaldehyde (183), the water-soluble nickel azo–azomethine complex (184), and nickel chelates of azines and disazines (185). Nickel Azo Yellow [51931-46-5] is a commercially important pigment with excellent lightfastness and good heat stability and bleed resistance (186,187). Other nickel azo pigments which have had some success include Nickel Azo Gold and Nickel Azo Red (188) (see AZO DYES). The lightfastness in other pigment systems, eg, the quinacridones and iron blue, have been improved by the addition of 1–2-wt % nickel in soluble salt form.

Nickel also has been used as a dye site in polyolefin polymers, particularly fibers. When a nickel compound, eg, the stearate or bis(p-alkylphenol) monosulfide, is incorporated in the polyolefin melt which is subsequently extruded and processed as a fiber, it complexes with certain dyes upon solution treatment to yield bright fast-colored fibers which are useful in carpeting and other applications (189). Nickel stearate complexing of disperse mordant dyes has been studied (190).

Agricultural Chemicals. Many claims exist for the use of nickel chemicals as nematocides, miticides, and other pesticides (qv) (191). However, extensive testing of many classes of nickel complexes in insecticide, fungicide, nematocide, and herbicide programs leaves little doubt that, except in the case of a few selected fungus organisms, nickel chemicals afford little more efficacy than the nonnickel-containing derivatives (see HERBICIDES; INSECT CONTROL TECHNOLOGY). The application of nickel ion, particularly the halides, as a commercial fungicide to control blister blight of tea (qv) was formerly practiced in the Far East (192). Tea is one of a few botanical species that naturally contains nickel (193), and in India the nickel content of certain plants grown in lateritic soils has been found to be up to 100 ppm (194). Nickel sulfate is used to control rust in bluegrass seed crops in the northwestern United States (195). The use of nickel sulfate has been thoroughly explored for the control of cereal rusts, but no commercial application has resulted (196). Nickel sulfate has been proposed as an additive to wood (qv) chips against fungus attack in long-term chip storage piles (197). Evidence exists for the possible requirement of nickel ion in chicks, rats, and pigs (198), and studies show that nickel ion activates certain enzymes and may have a metabolic role (see FEEDS AND FEED ADDITIVES; MINERAL NUTRIENTS).

Other Specialty Chemicals. In fuel-cell technology, nickel oxide cathodes have been demonstrated for the conversion of synthesis gas and the generation of electricity (199) (see FUEL CELLS). Nickel salts have been proposed as additions to water-flood tertiary crude-oil recovery systems (see PETROLEUM, ENHANCED OIL RECOVERY). The salt forms nickel sulfide, which is an oxidation catalyst for H_2S, and provides corrosion protection for downwell equipment. Sulfur-containing nickel complexes have been used to limit the oxidative deterioration of solvent-refined mineral oils (200).

Nickel salts and soaps have been used in electrosensitive copy paper for image development. Nickel bis-(3,5 di-*tert*-butylsalicylate) [68569-24-4] has been studied in pressure-sensitive color developer sheets (201). It has also been used for color stabilization of color copy paper (see ELECTROPLATING).

Nickel phosphate complexes with ammonia have been used for high speed photographic image amplification (202), and nickel chelated quenching compounds, which stabilize image dyes in photographic film, also have been used (see PHOTOGRAPHY) (203). Nickel (2-hydroxy-4-methoxybenzophenone-5-sulfonate) [130543-71-4] reduces ultraviolet photofading of the dyes Rose Crystal Violet and Erioglaucine (204).

Nickel Chemical Waste Reduction

The obvious destination for nickel waste is in the manufacture of stainless steel, which consumes 65% of new refined nickel production. Stainless steel is produced in a series of roasting and smelting operations. These can be hospitable to the various forms of nickel chemical waste. In 1993, 3×10^3 t of nickel from nickel-containing wastes were processed into 30×10^3 t of stainless steel remelt alloy (205,206) (see RECYCLING, NONFERROUS METALS). This quantity is expected to increase dramatically as development of the technology of waste recycle collection improves.

Recovery from waste begins with pulverization of brittle forms, shredding of storage batteries and mixing with other powdery forms such as spent catalysts, and using spent nickel plating and other nickel-containing chemical pickle solutions to form a slurry. The slurry is then dehydrated into pellet form and charged into reverberatory furnaces where it meets up with scrap and other alloy-forming ingredients in the molten state for smelting. The finished premelt product is cast into metal pigs and shipped to the steel industry. There it is used with other forms of feed to produce finished stainless steel products.

BIBLIOGRAPHY

"Nickel Compounds" in *ECT* 1st ed., Vol. 9, pp. 289–304, by J. G. Dean, The International Nickel Co., Inc.; in *ECT* 2nd ed., Vol. 13, pp. 753–765, by D. H. Antonsen and D. B. Springer, The International Nickel Co., Inc.; in *ECT* 3rd ed., Vol. 15, pp. 801–819, by D. H. Antonsen, The International Nickel Co., Inc.

1. A. F. Cronstedt, *Mineralogie* (*Stockholm*) **218** (1758).
2. A. Berthier, *Ann. Chem. Phys.* **14**(2), 52 (1820); **25**, 94 (1824).
3. F. B. Howard-White, *Nickel, An Historical Review*, Methuen & Company, London, 1963, pp. 69–71.
4. J. R. Bolt, Jr., and P. Queneau, *The Winning of Nickel*, Longmans Canada Ltd., Toronto, 1967, p. 83.
5. J. F. Thompson and N. Beasley, *For The Years to Come*, G. P. Putnam's Sons, New York, 1960, p. 104.
6. G. Wilkinson, F. G. A. Stone, and E. W. Abel, eds., *Comprehensive Organometallic Chemistry*, Pergamon Press, Oxford, U.K., 1982, Chapt. 37.1–37.9.
7. G. Wilkinson, R. D. Gillard, and J. A. McCleverty, eds., *Comprehensive Coordination Chemistry*, Pergamon Press, Oxford, U.K., 1987, Chapt. 50.
8. Gmelin's *Nickel*, 8th ed., Vol. 57, Part B.2, 1965–1967.
9. Personal communication, *Chemical Abstracts Structure File and Dictionary File Printout for Nickel Compounds*, Washington, D.C., Oct. 6, 1993.

10. G. A. DiBari, *Inco Nickel Currents* **1**(2), 6–14 (1993).

11. W. Streurer and W. Adlhart, *Acta Crystallogr. Sect. A*, **39**, 44 (1983).

12. R. Morassi, I. Bertini, and L. Sacconi, *Coord. Chem. Rev.* **11**, 343 (1973).

13. H. Bode and E. Voss, *Ann. Anorg. Allgem. Chem.* **269**, 165 (1952).

14. R. Hoppe, *Rec. Trav. Chim. Pays-Bas* **75**, 569 (1956).

15. J. Willemse, J. A. Cras, J. J. Steggerda, and C. P. Keijzers, *Struct. Bonding* **28**, 83 (1976).

16. Y. Xie, S. Huang, X. Xin, and A. Dai (original in Chinese), *Chem. Abstr.* **113**, 143956d (1990).

17. R. Nast and H. Kasperl, *Chem. Ber.* **92**, 2135 (1959).

18. G. Piacchioni and V. Valenti, *J. Organomet. Chem.* **224**(1), 89 (1982).

19. U.S. Pat. 455,230 (June 30, 1891), L. Mond.

20. R. A. Schunn, *Inorg. Synth.* **15**, 5 (1974).

21. D. L. Cronin, J. R. Wilkinson, and L. J. Todd, *J. Magn. Resonance* **17**(3), 353, (1975).

22. W. A. Levason and C. A. McAuliffe, *Accounts Chem. Res.* **11**, 363 (1978).

23. B. J. Henne and D. E. Bartak, *Inorg. Chem.* **23**, 369 (1984).

24. R. Del Rosario and L. S. Stuhl, *J. Amer. Chem. Soc.* **106**(4), 1160 (1984).

25. U.S. Pat. 4,053,578 (Oct. 11, 1977), B. Hill and W. H. Elwood, Jr. (to The International Nickel Co., Inc.).

26. Can. Pat. 1,035,545 (Aug. 1, 1978), R. Sridhar and H. Davies (to Inco Ltd., Canada).

27. S. Saito, K. Kurosawa, and S. Takemoto, *Bull. Univ. Osaka Prefect Serv. A.* **24**(1), 123 (1975).

28. M. A. Hagan, *The Encyclopedia of Chemistry*, 2nd ed., Reinhold Publishing Corp., New York, 1967, p. 445.

29. M. Koltun, G. Gukhman, and A. Gavrilina, *Proc. SPIE-Int. Soc. Opt. Eng.* 1727, 1992.

30. *Nickel Oxide Sinter 75*, Product Data Sheet A-789, International Nickel Co., Inc., New York, 1977.

31. *Inco Nickel Oxide*, Product Data Sheet A-1083, International Nickel Co., Inc., New York, 1975.

32. Can. Pat. 828,670 (Dec. 2, 1969), A. Illis, H. J. Koehler, and B. J. Brandt (to The International Nickel Co. of Canada, Ltd.).

33. C. R. Quail, Jr., and Q. R. Skrabec, *Metal. Producing* **33**, 54 (Apr. 1979).

34. *An Introduction to General Characteristics of Thermisters*, Product Brochure, Dale Electronics Inc., El Paso, Tex., 1985.

35. S. R. Korn, *Surface Mount Technol.*, 11 (Aug. 1986).

36. H. J. Schmidt, *Textileveredlung* **26**(7–8), 293 (1991).

37. U.S. Pat. 4,312,678 (Jan. 26, 1982), S. L. Colucci (to National Standard Co.).

38. *Fibrex Nickel Fibers*, Product Brochure, National Standard Co., Mishawaka, Ind., 1987.

39. R. N. Rhoda, D. H. Antonsen, and B. Hill, *Ind. Eng. Chem. Prod. Res. Dev.* **20**, 398 (1981).

40. U.S. Pat. 3,869,257 (Mar. 4, 1975), H. P. Beutner and G. Flick (to The International Nickel Co., Inc.).

41. U.S. Pat. 3,857,926 (Dec. 31, 1974), H. P. Beutner and C. E. O'Neill (to The International Nickel Co., Inc.).

42. M. Yonemitsu, S. Hirano, and H. Ikeda, *Sumitomo Keikinzoku Giho (Jpn.)*, **32**(3), 198, 1991.

43. K. L. Lin and C. S. Jong, *Mater. Chem. Phys.* **35**(1), 53 (1993).

44. Jpn. Pat. 61 254 216 (Nov. 12, 1986), S. Kitatsume (to Mitsubishi Yuka Fine Chemicals Ltd.).

45. R. Weist, ed., *Handbook of Chemistry and Physics*, 67th ed., Chemical Rubber Co., Boca Raton, Fla., 1986–1987.

46. S. Budavari, ed., *Merck Index*, 11th ed., Merck & Co., Inc., Rahway, N.J., 1989, Nos. 6409 and 6418.

47. G. A. DiBari, *Metal Finishing Guidebook Directory*, 47th annual ed., Metal and Plastics Publication, Inc., Hackensack, N.J., 1979, p. 270.

48. R. J. Bones, D. A. Teagle, S. D. Booker and F. Cullen, *J. Electrochem. Soc.* **136**(5), 26 (May 1989).

49. L. G. L. Ward, D. P. Jordan, and D. H. Antonsen, *157th American Chemical Society Meeting*, Minneapolis, Minn., Apr. 1969.

50. Jpn. Pat. 63 30 326 (Feb. 1988), N. Myamoto and T. Maeda.

51. J. Swanepool and A. M. Heyns, *Spectrochim. Acta, Part A*, **47**a, 243 (1991).

52. V. A. Ettel, *Nickel–Cadmium Battery Update-92 Conference, Munich, Germany, Oct. 1992*, Cadmium Association, London, 1993.

53. U.S. Pat. 3,827,911 (Aug. 6, 1974), D. F. Pickett (to United States of America, Air Force Materials Laboratory).

54. Ger. Offen 2,653,984 (June 1, 1978), G. Crespey, P. Matthey, and M. Gutjahr (to Daimler-Benz A. G.).

55. A. I. Eremin, I. D. Kovalev, and A. M. Potapov, *Zh. Neorg. Khim.* **37**(6), 1341 (1992).

56. J. W. Copenhaver and M. H. Bigelow, *Acetylene and Carbon Monoxide Chemistry*, Reinhold, Princeton, N.J., 1949.

57. O. Z. Standritchuk, V. I. Maksin and A. K. Zapol'skii, *Zh. Fiz. Khim.* **63**(9), 2332 (1989).

58. Ref. 4, p. 7.

59. G. G. Hawley, ed, *The Condensed Chemical Dictionary*, 10th ed., Van Nostrand Co., New York, 1981, p. 724.

60. A. K. Majumdar and co-workers, *J. Inorg. Nucl. Chem.* **26**, 2177 (1964).

61. T. D. George and W. W. Wendladt, *J. Inorg. Nucl. Chem.* **25**, 395 (1963).

62. W. Reppe and co-workers, *Justus Liebigs Ann. Chem.* **560**, 1 (1948).

63. K. W. Barrett, *J. Chem. Soc.* **51**, 422 (1974).

64. D. J. Bauer and C. Kurger, *J. Organomet. Chem.* **44**, 397 (1972).

65. Ref. 6, Chapt. 56.

66. K. A. Walsh, *U. S. Atomic Energy Commission Report No. LA-1649*, Washington, D.C., 1953.

67. W. Hieber and O. Vohler, *Z. Anorg. Allg. Chem.* **294**, 219 (1958).

68. L. Hedberg, T. Iijima, and K. Hedberg, *J. Chem. Phys.* **70**, 3224 (1979).

69. J. W. Cederberg, C. H. Anderson and N. F. Ramsey, *Phys. Rev.* **136**, A960 (1964).

70. F. Gallais and J. M. Savarialt, *Rev. Chim. Miner.* **11**, 592 (1974).

71. J. A. Conner, *J. Organomet. Chem.* **94**, 195 (1975).

72. J. A. Conner, *Top. Curr. Chem.* **71**, 71 (1977).

73. L. H. Jones, R. S. McDonnell, and M. Goldblatt, *J. Chem. Phys.* **48**, 2663 (1968).

74. A. D. Cornier, J. D. Brown, and K. Nakamoto, *Inorg. Chem.* **12**, 3011 (1973).

75. J. K. Burdett, *Inorg. Chem.* **14**, 1058 (1975).

76. M. Elian and R. Hoffmann, *Inorg. Chem.* **14**, 1058 (1975).

77. L. Mond, C. Langer, and F. Quincke, *J. Chem. Soc.* **57**, 749 (1890).

78. L. Mond and R. Nasini, *Z. Phys. Chem.* **8**, 150 (1891).

79. D. H. Stedman, *Science* **208**, 1029 (1980).

80. D. H. Antonsen and J. H. Tundermann, *Metals Handbook*, 9th ed., Vol. 7, American Society of Metals, Metals Park, Ohio, 1984, pp. 134–138.

81. Ref. 4, p. 383.

82. U.S. Pat. 3,252,791 (May 24, 1966), G. R. Frysinger and H. P. Buetner (to The International Nickel Co., Inc.).

83. P. W. Jolly and G. Wilke, *The Organic Chemistry of Nickel*, Vol. I, Academic Press, New York, 1974, Chapt. 8.

84. Ref. 6, Chapt. 37.3.4.1.2.
85. L. Garnier, Y. Rollin, and J. Perichon, *J. Organometal. Chem.* **367**(3), 347 (1989).
86. M. F. Lappert and P. L. Pye, *J. Chem. Soc. Dalton Trans.*, 2172 (1977).
87. F. H. Kohler, K. H. Doll, and W. Prosdorf, *Angew. Chem. Int. Ed. Engl.* **19**, 479 (1980).
88. F. H. Kohler, *J. Organometal. Chem.* **110**, 235 (1976).
89. R. Eisenberg and C. D. Meyer, *Accounts Chem. Res.* **8**, 26 (1975).
90. Ref. 6, Chapt. 37.3.4.2.
91. N. J. Taylor, *J. Chem. Soc. Commun.* **8**, 476 (1985).
92. T. M. Bolthaze and R. C. Grabiek, *J. Org. Chem.* **45**, 5425 (1980).
93. W. Aresta, C. F. Nobile, and A. Sacco, *Inorg. Chim. Acta* **12**, 167 (1975).
94. C. A. Tolman, *J. Am. Chem. Soc.* **92**, 2956 (1970).
95. C. A. Tolman, *J. Am. Chem. Soc.* **92**, 6785 (1970).
96. C. A. Tolman, *J. Am. Chem. Soc.* **92**, 2994 (1970).
97. U.S. Pat. 3,631,191 (Dec. 28, 1971), N. J. Kane (to E. I. du Pont de Nemours & Co., Inc.).
98. C. W. Westan, *Inorg. Chem.* **16**(6), 1313 (1977).
99. D. P. Jordan and D. H. Antonsen, *Ind. Eng. Chem. Prod. Res. Develop.* **8**, 208 (June 1969).
100. B. Henc, *J. Organometal. Chem.* **191**, 425 (1980).
101. K. R. Poeschke, *Angew. Chem. Int. Ed. Engl.* **26**, 1288 (1987).
102. L. Malatesta and S. Cenimi, *Zerovalent Compounds of Metals*, Academic Press, London, 1974.
103. H. Hobart and H. J. Reigel, *J. Organometal. Chem.* **229**(1), 85 (1982).
104. G. N. Schrauzer and H. Thyret, *J. Amer. Chem. Soc.* **82**, 6420 (1960).
105. N. S. Gill and R. S. Nyholm, *J. Chem. Soc.*, 3997 (1959).
106. F. A. Cotton and D. M. L. Goodgame, *J. Am. Chem. Soc.* **82**, 2967 (1960).
107. Ref. 83, Chapt. 4.
108. S. Otsuka, K. Tani, and T. Yamagata, *J. Chem. Soc. Dalton Trans.*, 2491 (1973).
109. A. Schott and co-workers, *Justus Liebigs Ann. Chem.* **3**, 508 (1973).
110. M. Trouple, Y. Rollin, S. Sibille, F. Fauvarque, and J. Perichon, *J. Chem. Soc.* (S), 24 (1980).
111. H. H. Karsch, H. F. Klein, and H. Schmidbaur, *Chem. Ber.* **107**, 93 (1974).
112. B. Bogdanovic, H. Bonnemann, and G. Wilke, *Angew. Chem.* **78**, 591 (1966).
113. M. D. Rausch, Y. F. Chang, and H. B. Gordon, *Inorg. Chem.* **8**, 1355 (1969).
114. D. A. Palmer and R. Van Eldik, *Chem. Rev.* **83**, 651 (1983).
115. A. M. Tait and H. D. Busch, *Inorg. Synth.* **18**, (1978).
116. F. V. Lovecchio, E. S. Gore and H. D. Busch, *J. Am. Chem. Soc.* **96**, 3109 (1974).
117. B. Fiser and R. Eisenberg, *J. Am. Chem. Soc.* **102**, 7636 (1980).
118. M. Wada, *Inorg. Chem.* **14**, 1415 (1975).
119. L. L. Bircumshaw and J. Edwards, *J. Chem. Soc.* **1800** (1950).
120. Jpn. Pat. 04 45 113 (Feb. 1992), H. Maezawa, J. Matsumoto, and S. Asahi.
121. R. K. Panda, S. Acharya, G. Nedgi, and D. Ramaswany, *J. Chem. Soc. Dalton Trans.*, 1471 (1984).
122. Ref. 46, No. 3235.
123. F. Mani and C. Mealli, *Inorg. Chim. Acta* **63**, 97 (1982).
124. Z. L. Tesic, T. J. Janjic, and M. B. Celap, *J. Chromatogr.* **628**(1), 148 (1993).
125. T. Menné and E. Nieboer, *Endeavor* **13**, 117–122 (1989).
126. T. Fisher, in H. I. Maibach, T. Menné, eds., *Nickel and the Skin: Immunology and Toxicology*, CRC Press, Boca Raton, Fla., 1989, pp. 117–132.
127. *Code of Federal Regulations, OSHA General Industry*, Title 29, Chapt. XVII, Part 19.10, Washington, D.C.

128. *TLV for Chemical Substances and Physical Agents in the Workroom Environment, with Intended Changes for 1980*, ACGIH, Cincinnati, Ohio, 1985, Table Z.1.

129. *IARC Monographs on the Evaluation of Carcinogenic Risks to Humans of Chromium, Nickel and Welding*, Vol. 49, International Agency for Research on Cancer (IARC), World Health Organization, Geneva, Switzerland, 1990, pp. 257–446.

130. D. L. Cragle, D. R. Hollis, C. M. Shy, and T. H. Newpot, in F. W. Sunderman and co-workers, eds., *Nickel in the Human Environment*, IARC Scientific Publications No. 53, World Health Organization, Geneva, Switzerland, 1984, pp. 57–63.

131. International Committee on Nickel Carcinogenesis in Man (ICNCM), *J. Scand. Work Environ. Health* **16**(1), 1–84 (1990).

132. C. K. Redmond and co-workers, *High Nickel Alloys Workers Study Update*. Final Report to the Nickel Producers Environmental Research Association, Durham, N.C., Nov. 1993.

133. Ref. 46, No. 8124.

134. J. Yasumura and T. Yoshino, *I&EC Prod. Res. Dev.* **7**, 252 (1968).

135. U.S. Pat. 3,271,326 (Sept. 6, 1966), A. J. Forney and J. J. Demeter (to United States of America).

136. J. H. Field and co-workers, *I&EC Prod. Res. Dev.* **3**, 150 (June 1964).

137. U.S. Pat. 3,846,344 (Nov. 5, 1974), F. E. Larsen and E. Snape (to The International Nickel Co., Inc.).

138. Ger. Offen. 2,713,374 (Sept. 28, 1978), H. J. Becker and W. Schmidt (to Bayer AG).

139. A. B. Fasman, B. K. Almashev, and B. Usenov, *Kinet. Katal.* **17**, 1353 (1976).

140. Y. Nitta, T. Imanaka, and S. Teranishi, *Nypan Kagaku Kaishi* **9**, 1362 (1976).

141. P. H. Emmett, ed., *Catalysis*, Vol. 1, Reinhold Book Corp., New York, 1954, Chapt. 7, p. 342.

142. R. A. Della Betta, A. G. Piken, and M. Shelef, *J. Catal.* **40**, 173 (1975).

143. B. J. Copper and L. Keck, *SAE Technical Paper series, No. 800461*, Society of Automotive Engineers, Inc., Feb. 1980.

144. Ref. 6, Chapt. 56.2.

145. U.S. Pat. 4,020,121 (1977), A. T. Kister and E. F. Lutz (to Shell Oil Co.).

146. E. R. Freitas and C. R. Gum, *Chem. Eng. Prog.* **75**(1), 73 (Jan. 1979).

147. W. Keim, R. Appel, S. Gruppa and F. Knoch, *Angew. Chem. Int. Ed. Engl.* **26**, 1012 (1987).

148. G. W. Parshall, *Homogeneous Catalysis*, John Wiley & Sons, Inc., New York, 1980, p. 70.

149. C. A. Tolman, *J. Am. Chem. Soc.* **94**, 2994 (1972).

150. E. J. Hayes and D. H. Antonsen, *Metal Powd. Rep.* **41**(1) (1986).

151. V. A. Tracey, *I&EC Prod. Res. Dev.* **25**, 582 (1986).

152. A. J. Catotti, *Nickel Cadmium Battery Application Engineering Handbook*, General Electric Co., Gainesville, Fla., 1975, p. 2-2.

153. V. A. Tracey, *I&EC Prod. Res. Dev.* **18**, 234 (Sept. 1979).

154. *Sintering Data for Inco Nickel Powder A-1293*, The International Nickel Co., Inc., New York, 1979.

155. D. H. Antonsen, in P. A. Lewis, ed., *Pigment Handbook*, 2nd ed., Vol. 1, Wiley-Interscience, New York, 1988, pp. 823–827.

156. M. Salkind, E. H. Riddle, and R. W. Keefer, *Ind. Eng. Chem.* **51**, 1232 (1959).

157. E. L. Rees, F. W. Heck, and G. A. DiBari, *Modern Developments in Powder Metallurgy*, Vols. 18–21, Metal Powder Industries Federation, Princeton, N.J., 1988.

158. *Nickel-Coated Graphite-60*, Product Data Sheet, Novamet Specialty Products Corp., Wyckoff, N.J., 1988.

159. *Nickel-Coated Aluminum*, Product Data Sheet, Novamet Specialty Products Corp., Wyckoff, N.J., 1988.

160. A. C. Hart, *Mater. Edge*, 14 (Nov.-Dec. 1989).
161. *Nickel-Coated Alumina-25*, Product Data Sheet, Novamet Specialty Products Corp., Wyckoff, N.J., 1988.
162. J. A. E. Bell and G. Hansen, "Properties of Nickel-Coated Carbon and Kevlar Fibers Produced by the Decomposition of Nickel Carbonyl," the *23rd International SAMPE Technical Conference, Anaheim, Calif.*, Society for the Advancement of Material and Process Engineering, Covina, Calif., Oct. 1991.
163. J. A. E. Bell and G. Hansen, "Nickel Coated Fibers for Aerospace Applications," the *24th International SAMPE Technical Conference, Toronto, Canada*, Society for the Advancement of Material and Process Engineering, Covina, Calif., Oct. 1992.
164. *Inco Guide to Nickel Plating*, International Nickel Inc., New York, Rev. 1989.
165. A. I. Nedeljkovic, *Vitr. Enameller* **27**(3), 45 (1976).
166. W. R. Buessem, O. V. Gigliotti, S. H. Linwood, and J. W. Proske, *Powder Met. Int.* **6**(3), 120 (1974).
167. *Bulletin No. 2100*, Ferrite Division, Ferronics Inc., Fairport, N.Y., 1979.
168. C. U. Parker, *EMC Test Design*, 26 (Jan. 1994).
169. Jpn. Pat. 78, 537 98 (May 16, 1978), M. Hattori and co-workers (to Matushita Electrical Industrial Co., Ltd.).
170. V. K. Sikka, *Modern Developments in Powder Metallurgy*, Vols. 18–21, Metal Powder Industries Federation, Princeton, N.J., 1988, p. 543.
171. *Plastics Compounding, 1994/95 Redbook*, Advanstar Comm. Inc., Cleveland, Ohio, 1994, p. 50.
172. U.S. Pat. 3,748,165 (July 24, 1973), B. Hill (to The International Nickel Co., Inc.).
173. Ref. 171, p. 41.
174. A. Zweig and W. A. Hendersen, Jr., *J. Polym. Sci. Polym. Chem. Ed.* **13**, 717 (1975).
175. Ger. Offen. 2,519,594 (Nov. 27, 1975), L. Avar and K. Hofer (to Sandoz).
176. USSR Pat. 1,257,078 (Sept. 15, 1986).
177. U.S. Pat. 3,624,116 (Nov. 30, 1971), L. G. L. Ward (to The International Nickel Co., Inc.).
178. A. Zweig and W. A. Henderson, Jr., *J. Polym. Sci. Poly. Chem. Ed.* **13**, 993 (1975).
179. U.S. Pat. 3,310,575 (Mar. 21, 1967), J. D. Spivak (to Geigy Chemical Corp.).
180. U.S. Pat. 4,006,119 (Feb. 1, 1977), H. C. Beadle and I. Gibbs (to R. T. Vanderbilt Co., Inc.).
181. Ger. Offen. 2,631,475 (Jan. 19, 1978), N. Buhl, W. Kleeberg, and R. Wiedenmann (to Siemans AG).
182. U.S. Pat. 3,812,080 (May 21, 1974), A. M. Feldman (to American Cyanamid Co.).
183. Ger. Offen, 2,610,308 (Sept. 15, 1977), T. Papenfuhs and H. Volk (to Hoechst AG).
184. Ger. Offen, 2,533,958 (Feb. 17, 1977), T. Papenfuhs (to Hoechst AG).
185. U.S. Pat. 2,877,252 (Sept. 15, 1961), D. Hein and co-workers (to American Cyanamid Co.).
186. Ref. 171, p. 46.
187. P. A. Lewis, ed. *Pigment Handbook*, 2nd ed., Vol. 1, Wiley-Interscience, New York, 1988, p. 513.
188. U.S. Pat. 3,338,938 (Aug. 29, 1967), A. S. Matlack (to Hercules, Inc.).
189. U.S. Pat. 4,066,388 (Jan. 3, 1978), R. Botros (to American Color & Chemicals).
190. G. I. Belova and I. A. Ledneva, *Izhevsk*, 66–70 (1987).
191. *Nickel Compounds as Fungicides, 5351*, International Nickel Co., Inc., New York, 1971.
192. C. S. V. Ram, *Curr. Sci. (India)* **30**, 57 (1961).
193. K. E. Burke and C. H. Albright, *J. Assoc. Offic. Anal. Chem.* **53**, 531 (1970).
194. A. K. Shyam, *Indian Miner* **43**(1), 73 (1989).
195. J. R. Hardisen, *Phytopathology* **53**, 209 (Feb. 1963).

196. U.S. Pat. 2,971,880 (Feb. 14, 1961), H. L. Keil and H. P. Frohlich (to Rohm & Haas Co.).
197. A. Assarsson, I. Croan, and E. Frisk, *Sv. Papperstidn.* **73**, 493 (1970).
198. F. H. Neilson and H. E. Sauberlich, *Proc. Soc. Exp. Biol. Med.* **134**, 845 (1970).
199. J. M. King, Jr., *Advanced Technology Fuel Cell Programs*, EPRI EE-335, Research Project 114-1, Final Report, Power Systems Division, United Technologies Corp., South Windsor, Conn., Oct. 1976.
200. U.S. Pat. 4,090,970 (May 23, 1978), M. Braid (to Mobil Oil Co.).
201. Jpn. Pat. 10 72 889 (Mar. 17, 1989), M. Anzai, A. Utsunomya, and M. Yamaguchi (to Hodogaya Chemical Co., Ltd.).
202. H. E. Spencer and J. E. Hill, *Photogr. Sci. Eng.* **20**, 260 (1976).
203. U.S. Pat. 4,050,938 (Sept. 27, 1977), W. F. Smith, Jr. And G. A. Reynolds (to Eastman Kodak Co.).
204. H. Oda, *Nippon Kasei Gakkaishi* **39**(6), 583 (1988).
205. R. H. Hanewald, M. E. Schweers, and J. C. Onuska, *Recycling Metal-Bearing Wastes through Pyrometallurgical Technology*, INMETCO (INCO Co.), Ellwood City, Pa., 1992.
206. *Making Metals a Reusable Resource*, INMETCO (INCO Co.), Ellwood City, Pa., May 1993.

General References

Gmelin's *Nickel*, 8th ed., Vol. 57, 1965–1967.
J. G. Dean, *Ind. Eng. Chem.* **51**, 48 (Oct. 1959).
P. W. Jolly and G. Wilke, *The Organic Chemistry of Nickel*, Vol. I, Organometallic Complexes, Academic Press, Inc., New York, 1974.
P. W. Jolly and G. Wilke, *The Organic Chemistry of Nickel*, Vol. II, Organic Synthesis, Academic Press, Inc., New York, 1975.
R. B. Pannell, K. S. Chung, and C. H. Bartholomew, *J. Catal.* **46**, 340 (1977).
R. B. King, *Transition-Metal Organometallic Chemistry, An Introduction*, Academic Press, New York, 1969, Chapt. VII.
P. A. Lewis, ed., *Pigment Handbook*, Wiley-Interscience, New York, 1988.
Metals Handbook, 9th ed. Vol. 7, *Powder Metallurgy: Production of Metal Powders*, American Society of Metals, Metals Park, Ohio, 1984.
G. P. Tyroler and C. A. Landolt, *Extractive Metallurgy of Nickel and Cobalt*, The Metallurgical Society, Inc., Warrendale, Pa., 1988.
H. Topsoe, B. S. Clausen, N. Topsoe, and J. Hyldtoft, *Symposium on the Mechanism of HDS/HDN Reactions*, Vol. 38, No. 3, Division of Petroleum Chemistry, Preprints, American Chemical Society, Chicago, Ill., July 1993.

D. H. ANTONSEN
International Nickel, Inc.

NICKEL SILVER.   See COPPER ALLOYS.

NICOTINAMIDE.   See VITAMINS.

NICOTINE.   See ALKALOIDS; INSECT CONTROL TECHNOLOGY.

NIELSBOHRIUM.   See ACTINIDES AND TRANSACTINIDES.

NIOBIUM AND NIOBIUM COMPOUNDS

Niobium

Niobium, discovered by Hatchett in 1801, was first named columbium. In 1844, Rosed thought he had found a new element associated with tantalum (see TANTALUM AND TANTALUM COMPOUNDS). He called the new element niobium, for Niobe, daughter of Tantalus of Greek mythology. In 1949, the Union of Pure and Applied Chemistry settled on the name niobium, but in the United States this metal is still known also as columbium. Sometimes called a rare metal, niobium is actually more abundant in the earth's crust than lead.

Niobium is important as an alloy addition in steels (see STEEL). This use consumes over 90% of the niobium produced. Niobium is also vital as an alloying element in superalloys for aircraft turbine engines. Other uses, mainly in aerospace applications, take advantage of its heat resistance when alloyed singly or with groups of elements such as titanium, zirconium, hafnium, or tungsten. Niobium alloyed with titanium or with tin is also important in the superconductor industry (see HIGH TEMPERATURE ALLOYS; REFRACTORIES; SUPERCONDUCTING MATERIALS).

Properties. Elemental niobium [7440-03-1], Nb, has a cosmic abundance of 0.9 relative to silicon $\equiv$ 106 (1), an average value of 24 ppm in the earth's crust (2), and a comparable value on the lunar surface (3). Niobium is a monoisotopic element, although a search for residual radionuclides from the formation of the solar system has established the natural abundance of ^{92}Nb [13982-37-1], having a half-life, $t_{1/2}$, of 1.7×10^8 yr, to be $1.2 \times 10^{-10}\%$ (4). In addition, minute amounts of ^{94}Nb [14681-63-1], $t_{1/2} = 2.03 \times 10^4$ yr, and ^{95}Nb [13967-76-5], $t_{1/2} = 35$ d, occur in nature; the former from neutron capture by the stable isotope, and the latter as the daughter of ^{95}Zr in the fission products of ^{235}U. Niobium-93 has a nuclear spin of 9/2 and a thermal neutron-capture cross section of $1.1 \pm 0.1 \times 10^{-28}$ m^2 (1.1 barns) which makes it of much interest to the nuclear industry (see NUCLEAR REACTORS).

Niobium, like vanadium, undergoes no phase transitions from room temperature to the melting point. It is a steel-grey, ductile, refractory metal having a higher melting point than molybdenum and a lower electron work function than tantalum, tungsten, or molybdenum. Niobium closely resembles tantalum in its properties; the former is only slightly more chemically reactive. The metal is resistant to most gases below 200°C, but is air oxidized at 350°C, developing an oxide film of increasing thickness which changes from pale yellow to blue to black at 400°C. Absorption of hydrogen at 250°C and nitrogen at 300°C occurs to form interstitial solid solutions which greatly affect the mechanical properties. Niobium is attacked by fluorine and gaseous hydrogen fluoride and is embrittled by nascent hydrogen at room temperature. It is unaffected by aqua regia and mineral acids at ordinary temperatures, except hydrofluoric acid in which it dissolves. Niobium is attacked by hot concentrated hydrochloric and sulfuric acids, dissolving at 170°C in concentrated sulfuric acid, and by hot alkali carbonates and hydroxides, which cause embrittling.

The most common oxidation state of niobium is +5, although many anhydrous compounds have been made with lower oxidation states, notably +4 and +3, and Nb^{5+} can be reduced in aqueous solution to Nb^{4+} by zinc. The aqueous chemistry primarily involves halo- and organic acid anionic complexes. Virtually no cationic chemistry exists because of the irreversible hydrolysis of the cation in dilute solutions. Metal–metal bonding is common. Extensive polymeric anions form. Niobium resembles tantalum and titanium in its chemistry, and separation from these elements is difficult. In the solid state, niobium has the same atomic radius as tantalum and essentially the same ionic radius as well, ie, $Nb^{5+} \approx Ta^{5+}$ = 68 pm. This is the same size as Ti^{4+} (68 pm) and Li^+ (69 pm). Some properties of niobium are listed in Table 1; corrosion data are presented in Table 2.

Table 1. Properties of Niobium

Property	Value	Reference
atomic number	41	
atomic weight	92.906	
atomic volume, cm^3/mol	10.8	
atomic radius, nm	0.147	
electronic configuration	$[Kr]4d^45s^1$	
ionization potential, eV	6.77	
crystal structure	bcc	
lattice constant at 0°C, pm	330.04	
density at 20°C, g/cm^3	8.66	
mp, °C	2468 ± 10	
bp, °C	5127	
latent heat of fusion, kJ/mol^a	26.8	5
latent heat of vaporization, kJ/mol^a	697	5
heat of combustion, kJ/mol^a	949	6
heat capacity, $J/(mol·K)^a$		5
at 298 K	24.7	
1500 K	29.7	
3000 K	33.5	
entropy, $J/(mol·K)^a$		5
at 298 K	36.5	
1500 K	79.6	
3000 K	111.6	
vapor pressure at 2573 K, mPa^b	22	7
evaporation rate at 2573 K, $\mu g/(cm^2·s)$	1.9	7
thermal conductivity at 298 K, W/(m·K)	52.3	
coefficient of linear thermal expansion, 291–373 K, °C^{-1}	7.1×10^6	8
volume electrical conductivity, % IACS^c	13.3	9
electrical resistivity, $\Omega·m$	$13–16 \times 10^{-6}$	
temperature coefficient of resistivity, °C^{-1}	3.95×10^{-3}	
work function, eV	4.01	
secondary emission (primary δ_{max} = 400 V), eV	1.18	
positive ion emission, eV	5.52	

^{a}To convert J to cal, divide by 4.184.
^{b}To convert mPa to μm Hg, divide by 133.3.
^{c}IACS = International Annealed Copper Standard. Pure copper = 100%.

Table 2. Corrosion of Niobium Metal[a]

Medium	Temperature, °C	Concentration, wt %	Corrosion rate, μm/yr
sulfuric acid[b]	23	96	0.5
	50	40	5
	100	20	0.5
	145	96	5,000
	bp	70	dissolves
	190[c]	1–10	slight
	250[c]	1–10	slight
	250[c]	20	250
	250[c]	30	1,300
hydrochloric acid	bp	1	
	bp	5	
	bp	10	100
	bp	15	450
	bp	20	1,000
	190[c]	5	30
	190[c]	10	500
	190[c]	15	14,000
hydrofluoric acid	23	all	very high
nitric acid	bp	70	
	190[c]	70	
	250[c]	70	
phosphoric acid	23	85	0.5
	100	85	80
aqua regia	23		0.5
	55		20
sodium hydroxide	23	10	20
	23	40	30
potassium hydroxide	23	40	90
zinc chloride	bp	40	
ferric chloride	23	10	
formic acid	bp	10	
acetic acid	bp	≤99.7	
oxalic acid	bp	10	20
citric acid	bp	10	20
lactic acid	bp	10	10
	bp	85	2.5
trichloroacetic acid	bp	50	
trichloroethylene	bp	99[d]	

[a]Refs. 10–13.
[b]Hydrogen embrittlement at higher temperatures.
[c]In sealed tubes.
[d]1% water present.

Occurrence. Niobium and tantalum usually occur together. Niobium never occurs as the metal, ie, in the free state. Sometimes it occurs as a hydroxide, silicate, or borate; most often it is combined with oxygen and another metal, forming a niobate or tantalate in which the niobium and tantalum isomorphously replace one another with little change in physical properties except density. Ore concentrations of niobium usually occur as carbonatites and are associated with tantalum in pegmatites and alluvial deposits. Principal niobium-bearing minerals can be divided into two groups, the titano- and tantalo-niobates.

Titano-niobates consist of the salts of niobic and titanic acids. The important minerals of this group are pyrochlore [12174-36-6], loparite [12173-83-0], koppite [12198-49-1], and others. Pyrochlore, the most important, is complex and of varying composition. The general formula for a typical Canadian pyrochlore is $(Na,Ca)_2(Nb,Ti)_2O_6[F,OH]$; for a typical Brazilian pyrochlore, $(Ba,Ca)_2(Nb,Ti,Ce)_2O_6[O,OH]$. The color of pyrochlore ranges from dark grey to brown to orange-brown. Pyrochlore occurs in carbonatite complexes primarily in Brazil and Canada as well as in Kenya, Uganda, Nigeria, Zaire, Norway, and the United States (14). It also occurs with calcite, dolomite, apatite, magnetite, and some silicates. The density of pyrochlore is ca 4.0–4.4 g/cm^3. The tantalum content usually is low, ca 0.1–0.3% on a metal basis.

Tantalo-niobates consist of the salts of niobic and tantalic acids. The general formula for this group is $(Fe,Mn)(Nb,Ta)_2O_6$. The minerals consist of isomorphic mixtures of the four possible salts. The compound is a columbite if niobium is predominant and a tantalite if tantalum is predominant. These minerals are brown to black and usually contain titanium, tin, tungsten, and other impurities. The density of columbite is ca 5 g/cm^3 and that of tantalite is ca 8 g/cm^3. A regular gradation exists between these limits in proportion to the tantalum content. Columbite–tantalite minerals usually are finely disseminated in granitic rocks and associated pegmatites or occur as enriched concentrations in alluvial (placer) deposits. The main sources of columbites are Nigeria, Zaire, and Malaysia, and a number of smaller occurrences in other parts of the world. Columbites frequently are associated with cassiterite [1317-45-9], SnO_2, deposits and niobium occurs in tin slags during processing. These slags from Thailand and Malaysia are processed for their niobium and tantalum content.

Extraction, Refining, and Metallurgy. The stability of niobium ores is reflected in their occurrence as the residues of advanced weathering processes. Thus, stringent conditions are required to render the niobium extractable. The process of extracting and refining niobium consists of a series of consecutive operations. Frequently, several steps are combined: the upgrading of ores by preconcentration (15); an ore-opening procedure to disrupt the niobium-containing matrix; preparation of a pure niobium compound; reduction to metallic niobium; and refining, consolidation, and fabrication of the metal.

The most straightforward process is the direct conversion, ie, reduction, of the niobium concentrate to metallic niobium. The primary method is the aluminothermic reduction of a pyrochlore and iron–iron oxide mixture (16). A single batch usually contains from 1–10 metric tons of pyrochlore together with the necessary aluminum powder, iron scrap, and/or iron oxide. Frequently small amounts of lime or fluorspar are used as fluxing agents. Sometimes a small quantity of a powerful oxidizer, eg, sodium chlorate, which provides additional reaction heat is added. A typical reactor consists of a refractory-lined steel shell. Sometimes a floor of slag from previous reduction reactions is used. Typically, a small amount of a starter mixture, eg, aluminum powder and sodium chlorate, is ignited electrically to start the reaction. In another variation, aluminum powder and barium peroxide are used, and a small water spray initiates the aluminum–peroxide combustion. In some cases, the reaction is started using a small quantity of reactants, and additional pyrochlore–aluminum–iron mixture is fed into the reaction mixture by means of a chute until the reactor vessel is filled with molten metal and slag.

In the batch reactions, the reaction time varies from 2–3 min to ca 25 min, primarily depending on the size of the reaction. At the completion of the reaction, the molten ferroniobium is at the bottom of the reactor and the slag floats on top. Most of the impurities go into the slag. Some of the more easily reduced metals go into the ferroniobium. Typical commercial-grade ferroniobium has the following wt % composition: Nb, 62–67; Fe, 28–32; Si, 1–2.5; Al, 0.5–2.0; Ti, 0.1–0.4; P, 0.05–0.15; S, 0.05–0.1; and C, 0.05–0.1.

Minor amounts of tantalum, tin, lead, bismuth, and other elements also occur in the ferroniobium. After cooling for 12–30 h, the metal is separated from the slag and crushed and sized for shipment. The recovery of niobium in the aluminothermic reaction is 87–93%. Larger reactions generally give better recoveries.

Ferroniobium also is produced in an electric furnace procedure. Essentially the same reactants are used as in the aluminothermic method. Because the electric furnace provides additional energy input, the quantity of aluminum can be substantially reduced and partly substituted by other reducing agents, eg, ferrosilicon. The total heat input can be better controlled using the electric furnace procedure. Recovery of niobium is therefore generally better than for the aluminothermic method. The metal and slag which are produced can be tapped or skimmed using standard electric furnace procedures. Generally, the production volume of ferroniobium is not sufficient to take advantage of the control and semicontinuous operation offered by the electric furnace (see FURNACES; ELECTRIC).

In addition to the standard ferroniobium, there is a lesser but significant demand for high purity niobium alloys, mainly high purity ferroniobium and nickel–niobium. These high purity alloys are used in the fabrication of nickel- and cobalt-based superalloys which are used primarily in jet engine and aerospace applications (see HIGH TEMPERATURE ALLOYS). The alloys are produced by reducing a pure niobium oxide in the presence of iron or nickel in an aluminothermic reaction using carefully controlled conditions and raw materials (17). In some cases, the reactions are carried out in water-cooled copper reactors to avoid contamination by refractory materials. The key raw material for these alloys is a pure (99 wt %) niobium oxide, which can be produced only by chemical procedures.

Direct attack by hot 70–80 wt % hydrofluoric acid (qv), sometimes with nitric acid (qv), is effective for processing columbites and tantalo-columbites. Yields are >90 wt %. This method, used in the first commercial separation of tantalum and niobium, is used commercially as a lead-in to solvent extraction procedures. The method is not suited to direct processing of pyrochlores because of the large alkali and alkaline-earth oxide content therein, ie, ca 30 wt %, and the corresponding high consumption of acid.

Concentrated sulfuric acid (97 wt %) at 300–400°C has been used to solubilize niobium from columbite and pyrochlore (18,19). The exothermic reaction is performed in iron or silicon–iron crucibles to yield a stable sulfato complex. The complex is filtered free of residue and is hydrolyzed by dilution with water and boiling to yield niobic acid which is removed by filtration as a white colloidal precipitate.

Fusion with caustic soda at 500–800°C in an iron crucible is an effective method for opening pyrochlores and columbites (20). The reaction mixture is flaked and leached with water to yield an insoluble niobate which can be converted to niobic acid in yields >90 wt % by washing with hydrochloric acid.

The reaction of chlorine gas with a mixture of ore and carbon at 500–1000°C yields volatile chlorides of niobium and other metals. These can be separated by fractional condensation (21–23). This method, used on columbites, is less suited to the chlorination of pyrochlore because of the formation of nonvolatile alkali and alkaline-earth chlorides which remain in the reaction zone as a residue. The chlorination of ferroniobium, however, is used commercially. The product mixture of niobium pentachloride, iron chlorides, and chlorides of other impurities is passed through a heated column of sodium chloride pellets at 400°C to remove iron and aluminum by formation of a low melting eutectic compound which drains from the bottom of the column. The niobium pentachloride passes through the column and is selectively condensed; the more volatile chlorides pass through the condenser in the off-gas. The niobium pentachloride then can be processed further.

The reaction of finely ground ores and an excess of carbon at high temperatures produces a mixture of metal carbides. The reaction of pyrochlore and carbon starts at 950°C and proceeds vigorously. After being heated to 1800–2000°C, the cooled friable mixture is acid-leached leaving an insoluble residue of carbides of niobium, tantalum, and titanium. These may be dissolved in HF or may be chlorinated or burned to oxides for further processing.

Once the niobium ore has been opened, the niobium must be separated from the tantalum and/or impurities. The classical method of doing this is the addition of an excess of potassium fluoride to hydrofluoric acid solutions of niobium ores to precipitate the complex fluorides and oxyfluorides of niobium, tantalum, and titanium. These are redissolved in dilute hydrofluoric acid, a 3 wt % HF solution containing $K_2NbOF_5 \cdot H_2O$ [19200-74-9], $K_2TiF_6 \cdot H_2O$, and K_2TaF_7 with respective solubilities at 15°C of ca 77 g/L, 12 g/L, and 5 g/L. The hydrate of K_2TiF_6 is isomorphous with $K_2NbOF_5 \cdot H_2O$, but K_2TaF_7 is not. Dipotassium tantalum heptafluoride is stable in 3 wt % HF. Its solubility increases with HF concentration and temperature, and it is 50 times more soluble at 85°C than at 15°C. The species changes to $KTaF_6$ at 45% HF. The solubility of K_2NbOF_5 is high in 3 wt % HF and increases with temperature. At 6–40 wt % HF, the occurring species are dipotassium niobium heptafluoride [36354-32-2], K_2NbF_7, and $KNbF_6$ [16919-14-5] at >45 wt % HF. Increasing the concentration of KF depresses the solubility of all the complex salts. Repeated recrystallization can produce K_2TaF_7 with less than 0.01 wt % niobium content, but the niobium remains in the mother liquor and, hence, still contains the titanium impurity. Thus, the compound's usefulness is limited in the economical production of pure niobium.

The recrystallization of complex fluoride salts has been replaced completely by solvent extraction techniques which are used extensively. Many acid–solvent combinations have been reported in the literature, eg, HF–methyl isobutyl ketone (MIBK) (24), $HF–HNO_3$–MIBK (25), HF–HCl–MIBK (26), and HF–H_2SO_4–MIBK (27). Commercial processes involve the use of various acids in combination with HF and either MIBK or tributyl phosphate (TBP), generally starting with an upgraded feed material. The dissolution in HF is performed in rubber or polyethylene-lined tanks to form the fluoride feed solution. The separation then may be effected by two methods: either the tantalum is extracted first by contacting a low acidity aqueous feed with the organic phase and subsequently raising the acidity of the aqueous phase and extracting the niobium with fresh organic, or both niobium and tantalum are extracted into the organic phase from a strongly acidic medium and the niobium is back-extracted form the organic with dilute acid. The metal values are back-extracted from the organic phase with a low acidity aqueous phase from which they are precipitated as oxides by addition of ammonia or as double fluorides by addition of potassium fluoride.

Another solvent extraction scheme uses the mixed anhydrous chlorides from a chlorination process as the feed (28). The chlorides, which are mostly of niobium, tantalum, and iron, are dissolved in an organic phase and are extracted with 12 N hydrochloric acid. The best separation occurs from a mixture of MIBK and diisobutyl ketone (DIBK). The tantalum transfers to the hydrochloric acid leaving the niobium and iron, the DIBK enhancing the separation factor in the organic phase. Niobium and iron are stripped with hot 14–20 wt % H_2SO_4 which is boiled to precipitate niobic acid, leaving the iron in solution.

Another method of purifying niobium is by distillation of the anhydrous mixed chlorides (29). Niobium and tantalum pentachlorides boil within about 15°C of one another which makes control of the process difficult. Additionally, process materials must withstand the corrosion effects of the chloride. The system must be kept meticulously anhydrous and air-free to avoid plugging resulting from the formation of niobium oxide trichloride, $NbOCl_3$. Distillation has been used commercially in the past.

A process has been developed to recover niobium from ferroniobium (30). The need for this process came about when Brazil would only export niobium in the form of ferroniobium. The process starts with a hydriding step, so as to be able to crush the alloy. Screening precedes a nitriding step, followed by an acid leach of the iron nitrides. This leaves the niobium nitride for further processing to the pure niobium metal.

Once purification of the niobium has been effected, the niobium can be reduced to the metallic form. The double fluoride salt with potassium, K_2NbF_7, can be reduced using sodium metal. The reaction is carried out in a cylindrical iron vessel filled with alternating layers of K_2NbF_7 and oxygen-free sodium:

$$K_2NbF_7 + 5\ Na \longrightarrow Nb + 2\ KF + 5\ NaF$$

Use of excess sodium drives the reaction, usually done under an argon or helium blanket, to completion. After cooling, the excess sodium is leached with alcohol and the sodium and potassium fluorides are extracted with water, leaving a mass of metal powder. The metal powder is leached with hydrochloric acid to remove iron contamination from the crucible.

Fused-salt electrolysis of K_2NbF_7 is not an economically feasible process because of the low current efficiency (31). However, electrowinning has been used to obtain niobium from molten alkali halide electrolytes (32). The oxide is dissolved in molten alkali halide and is deposited in a molten metal cathode, either cadmium or zinc. The reaction is carried out in a ceramic or glass container using a carbon anode; the niobium alloys with the cathode metal, from which it is freed by vacuum distillation, and the niobium powder is left behind.

Niobium pentoxide can be reduced with carbon in a two-step process, called the Balke process. Formation of the carbide is the first step. The oxide is mixed with the stoichiometric amount of lamp black, placed in a carbon crucible, and heated in vacuum to 1800°C:

$$Nb_2O_5 + 7\ C \longrightarrow 2\ NbC + 5\ CO$$

The carbide then is remixed with a stoichiometric amount of oxide, compacted into chunks, and refired to >2000°C under reduced pressure:

$$5\ NbC + Nb_2O_5 \longrightarrow 7\ Nb + 5\ CO$$

The product chunks are hydrided, crushed, and dehydrided. The resultant powder is blended and pressed into bars which are purified by high temperature sintering. The sintering removes all of the carbon and most of the oxygen and is followed by consolidation by either arc or electron-beam melting.

Niobium pentoxide also is reduced to metal commercially by the aluminothermic process. The finely ground powder is mixed with atomized aluminum and an accelerator compound which gives extra heat during reaction, then is ignited. The reaction is completed quickly and, after cooling, the slag is broken loose to free the metal derby which is purified by electron-beam melting.

The pentachloride $NbCl_5$ can be reduced with hydrogen to yield a metal powder of high purity comparable to electron-beam-melted metal from other reduction processes. However, the large excess of hydrogen which is required and the attending safety problems make this route undesirable. Niobium pentachloride also can be reduced by the Kroll process, ie, decomposition of the halide using magnesium, and by reduction with oxygen-free sodium to yield niobium sponge, which must be consolidated.

Powder and sponge may be compacted at 0.7 GPa (6900 atm) into bars which are presintered at 1400–1500°C in vacuum. The bars then are resistance-heated in high vacuum to slightly below the melting point. After cooling, the bars are rolled to consolidate the pores and are resintered at 2300°C to yield a fabricable metal product of 98% theoretical density.

Arc melting also can be used to consolidate bars of metal that are pressed from powder or sponge and used as consumable electrodes in a low voltage, high current arc. The bar is suspended vertically and the molten metal falls from the bottom of the bar onto a water-cooled copper crucible, from which it is removed as an ingot.

The most common method in commercial use is electron-beam melting. The furnace essentially is a large thermionic vacuum tube. A high current, ie, >10 A, beam of electrons at >20 kV is focused magnetically on the bottom of a suspended metal compact bar. The molten metal globules fall into a pool on top of the ingot which is contained in a water-cooled copper cylinder. The ingot is retracted at the rate that it is formed. Impurities are boiled out of the molten pool and either are pumped away or deposited on the furnace walls as slag. The final ingot is 20–30 cm in diameter and 1.5–2 m long. Scrap niobium can be recycled by the hydride–dehydride process. Heating niobium under hydrogen pressure results in the formation of niobium hydride, which is brittle enough to crush and size. The crushed hydride can be reheated in a vacuum to form niobium metal powder.

Niobium metal is available as ingot, sheet, rod, and wire and can be fabricated and formed by most metallurgical and engineering techniques (33). Cold working is necessary to avoid embrittlement which results from absorption of oxygen and nitrogen. Niobium can be reduced by 50–90% without intermediate annealing. When work hardening occurs, annealing may be done in an inert atmosphere or at <10 mPa (7.5×10^{-5} mm Hg) at 1300–1400°C. The recrystallization temperature is 1050°C.

Economic Aspects. Information on the economic aspects of niobium and its compounds is available (34–36). As can be seen in Table 3, Brazil has the most niobium-bearing ore reserves, and is also the leading producer of niobium mineral concentrates (36). Canada is a distant second. United States has some small reserves, but as of this writing is not doing any mining of niobium ores. In Brazil, pyrochlore is mainly mined by mechanized open pit, whereas Canada's pyrochlore comes from underground mining. Columbite, which is the chief ore being mined in most of the other countries, is mined by simple hand operations to hydraulic monitors and dredges at placer deposits. The principal concentrate producers are the Araxa (Companhia Brasilira de Metallurgica e Mineracao) and Catala mines (Mineracao Catalao de Goias Ltda) in Brazil, plus Niobec (a Teck/Cambior 50/50 joint venture) in Canada.

Table 3. World Niobium Reserves and Reserves Base, t $\times$ 10^{3} [a,b]

Country	Reserves	Reserve base[c]	1993 Production
Brazil	3311	3629	9.756
Canada	136	408	2.393
United States		[d]	
Africa			
Nigeria	64	91	0.017
Zaire	32	91	0.455
Rwanda			0.030
Zimbabwe			0.014
other[e]			<0.001
Australia			0.050
Thailand			0.001
other market economy countries	6	9	
World total	*3549*	*4228*	

[a]Ref. 36.
[b]Niobium content.
[c]Reserve base includes demonstrated resources that are economic (reserves), marginally economic (marginal reserves), and some that are subeconomic (subeconomic resources).
[d]Negligible.
[e]Namibia and South Africa each had <0.5 t.

The average price of contained niobium in concentrates for the period of 1960 to 1993 is shown in Figure 1. The peak price of almost $22/kg in 1977 resulted from increasing demand, inflation, and higher labor costs. The gradual decrease of prices throughout the 1980s was because of the large quantities of pyrochlore produced in Brazil and Canada. Prices for principal niobium products are shown in Table 4. Surplus supplies of niobium coming from the former USSR in the early 1990s kept prices low throughout the niobium industry.

The value of contained metal in annual production of niobium in the Western World was $90 million at the average 1993 price (34). U.S. consumption of niobium has been in the range of 3300 to 3500 metric tons from 1989–1993 of which over 90% is used as alloy additions in steels. Of the five domestic U.S. producers of niobium products in 1993, only two integrated niobium from raw

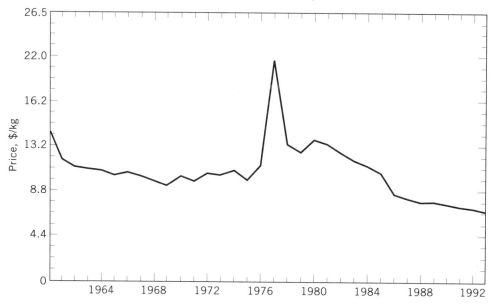

Fig. 1. Average annual price of contained niobium in concentrates, based on constant 1987 dollars (36).

Table 4. 1995 Prices of Niobium Products

Product	Price, $/kg
columbite ore, contained Nb	6.17–7.05
ferroniobium	
regular grade	14.51
vacuum grade	39.68
niobium metal	
sheet or foil	150–220
powder finer mesh	165–190
powder coarser mesh	100–130
niobium oxide	17.64
niobium–titanium alloys, billet, and rod	100–150
Nb alloys (C-103, Nb–1%Zr) as bar, rod, or sheet	165–285

material to end products: Shieldalloy Metallurgical Corp. and Cabot Corp., which process both niobium and tantalum products. Reading Alloys, Inc. and Teledyne Wah Chang Corp. are principal producers of high purity niobium products; Kennametal Inc. is a principal producer of niobium carbides.

Analytical Methods. The lack of stable niobium compounds resulting from a pronounced tendency for hydrolysis to colloidal suspensions of the hydrated oxide has given rise to an extensive body of literature (37–40). The analysis of niobium by classical methods involves solubilization of the sample; separation of gross interferences by selective precipitation, solvent extraction, ion exchange (qv), or another chromatographic technique; and determination by colorimetric, spectrometric, or gravimetric methods. Three of the most popular methods for dissolution of niobium metal, alloys, oxide, ores, and minerals include the following. (*1*) Addition of nitric acid to hydrofluoric acid aids in dissolving metal and alloy samples. Hydrofluoric acid solutions are used in anion-exchange separations, colorimetric determinations, and solvent-extraction separations. (*2*) Fusion of niobium ores with potassium hydroxide or potassium carbonate followed by water leaching of the cooled melts dissolves the niobium and tantalum as niobates and tantalates, respectively. Addition of sodium chloride to saturate the alkaline solution precipitates almost quantitatively the sparingly soluble sodium salts. Acidification of a suspension of the sodium salts precipitates the mixed hydrated oxides for further processing. (*3*) Fusion with potassium bisulfate, $KHSO_4$, or pyrosulfate, $K_2S_2O_7$, is the most widely used route to solubilize the ignited oxides or ores. The sample is fused in a quartz crucible with an eight- to tenfold excess of reagent to a clear melt at 650–800°C. The cooled melt is dissolved in a saturated solution of ammonium oxalate or 20 wt % tartaric acid or citric acid. A large number of precipitation reactions for the determination of niobium and tantalum are based on these acid solutions of oxalato or tartrato complexes. It is necessary to maintain a large excess of oxalic or tartaric acid for stability of the complexes. The oxalic acid solutions are more stable but the tartrato complexes can be made alkaline without precipitation of niobic and tantalic acids.

Of the gravimetric procedures, precipitation with tannin from a slightly acidic oxalate solution is probably the best known. The hot oxalate extract of a potassium pyrosulfate fusion is adjusted with ammonia to pH 3.7–4.0 using Bromothymol Blue. Dropwise addition of a 2 wt % solution of tannin precipitates the lemon-yellow complex of tantalum. Several grams of ammonium chloride are added to prevent peptization, and the solution is digested and filtered. Further addition of tannin and ammonia to the clear filtrate precipitates the vermilion-colored complex of niobium. Many other organic precipitants have been used, including Cupferron, *N*-benzoyl-*N*-phenylhydroxylamine, 8-quinolinol (Oxine), phenylarsonic acid, pyrogallol, and others.

Solvent extraction techniques are useful in the quantitative analysis of niobium. The fluoro complexes are amenable to extraction by a wide variety of ketones. Some of the water-insoluble complexes with organic precipitants are extractable by organic solvents and colorimetry is performed on the extract. An example is the extraction of the niobium–oxine complex with chloroform (41). The extraction of the niobium–pyrocatechol violet complex with tridodecylethylammonium bromide and the extraction of niobium–pyrocatechol–sparteine complex with chloroform are examples of extractions of water-soluble complexes.

Colorimetry is performed on the extract (42,43). Colorimetry may also be performed directly on the water-soluble complex, eg, using ascorbic acid and 5-nitrosalicylic acid (44,45).

Chromatographic methods play a prominent role in the clean separation of niobium from tantalum and other metals (see CHROMATOGRAPHY). The use of methyl ethyl ketone as the eluent for the cellulose column separation of niobium and tantalum in dilute hydrofluoric acid–ammonium fluoride solutions represented a significant advance over the tannin precipitations. The anionic-exchange separation of niobium from tantalum and other metals from a hydrochloric–hydrofluoric acid medium is accomplished by elution with ammonium chloride–hydrofluoric acid (46). This same type of ion-exchange separation is used to determine Nb along with Ta in various matrices, such as Inconel (47), FeNb alloys (48), and titanium (49). Once niobium is isolated, it may be determined gravimetrically by precipitation as an insoluble complex or as the earth acid, by firing to the oxide at 800–1000°C, and by weighing. Alternatively, the precipitate may be redissolved and determined colorimetrically in aqueous solution or in an organic extract.

The determination of the presence of the interstitial gases hydrogen, nitrogen and oxygen, and carbon are important because of the affects they have on the physical properties of niobium and its alloys. Hydrogen, nitrogen, and oxygen are most often determined by inert gas fusion. The gas of interest is captured and separated through a chromatographic procedure where detection is by infrared spectroscopy or thermal conductivity (50,51). Carbon determinations are almost entirely done by automated instrumentation. Combustion of the sample in a stream of oxygen converts the carbon to carbon dioxide which is then detected by infrared spectroscopy (52).

Spectroscopic methods for the determination of impurities in niobium include the older arc and spark emission procedures (53) along with newer inductively coupled plasma source optical emission methods (54). Some work has been done using inductively coupled mass spectroscopy to determine impurities in niobium (55,56). X-ray fluorescence analysis, a widely used method for niobium analysis, is used for routine work by niobium concentrates producers (57,58). Paying careful attention to matrix effects, precision and accuracy of x-ray fluorescence analyses are at least equal to those of the gravimetric and ion-exchange methods.

Health and Safety Factors.   Toxicity data on niobium and its compounds are sparse. The most common materials, eg, niobium concentrates, ferroniobium, niobium metal, and niobium alloys, appear to be relatively inert biologically. Limited animal experiments show high toxicity for some salts which are related to disturbance of enzyme action. Niobium hydride has moderate fibrogenic and general toxic action. Recommended maximum allowable concentrations are 6 mg/m^3 (59). Recommended maximum permissible concentration of Nb in reservoir water is 0.01 mg/L. Rats receiving 0.005 mg/kg daily over nine months show changes in cholinesterase activity but no effects are observed at doses of 0.0005 mg/kg. The threshold for affecting clarity and biological oxygen demand (BOD) is 0.1 mg/L (60).

Unstable niobium isotopes that are produced in nuclear reactors or similar fission reactions have typical radiation hazards (see RADIOISOTOPES). The metastable ^{93}Nb, $t_{1/2} = 14$ yr, decays by 0.03 MeV gamma emission to stable ^{93}Nb; ^{95}Nb, $t_{1/2} = 35$ d, a fission product of ^{235}U, decays to stable ^{95}Mo by emission of 0.16 MeV β- and 0.77 MeV γ-radiation; and ^{97}Nb, $t_{1/2} = 72$ min, decays to stable ^{97}Mo by emission of 1.27 MeV β- and 0.66 MeV γ-radiation. Inhalation experiments on mice have been performed using aerosols of radioniobium ^{95}Nb, prepared at 100, 250, 600, and 1100°C to give particles of different chemical compositions (61). The activity was restricted to the lungs for the 600 and 1100°C aerosols, but was translocated partially to the skeleton for the 100 and 250°C aerosols. At the lower temperatures, the highest radiation dose was to the lungs, skeleton, and liver; whereas the radiation dose was delivered exclusively to the lungs at the higher temperatures.

Coarse metal niobium and niobium alloy powders are difficult to ignite, even at 300°C, but high surface area material such as 5 micrometer powder may autoignite at room temperature. Fire fighting procedures for niobium and niobium hydride powder suggest letting the fire burn itself out. Small fires can be controlled by smothering with dry table salt or using Type D dry powder fire-extinguishing material. Under no circumstances should water be used, as a violent explosion may result. Carbon dioxide is not effective in extinguishing burning metal (62).

Uses. Niobium, as ferroniobium, is used extensively in the steel industry as an additive in the manufacture of high strength, low alloy (HSLA) and carbon steels. The ferroniobium acts as a grain refiner to increase yield and tensile strength at additions as low as 0.02 wt %. Normal usage is 0.03–0.1 wt %. The most important application of niobium HSLA or microalloyed steels is in oil- and gas-pipeline steels, particularly those which may experience operating conditions below −25°C (see PIPELINES). The niobium microalloyed steels also are widely used in automobiles, buildings, bridges, ships, towers, concrete reinforcing bars, etc, where the strength-to weight and cost-per-unit-strength ratios are particularly advantageous. The distribution of the uses of iron and nickel alloys of niobium is given in Table 5.

Addition of niobium to austenitic stainless steels inhibits intergranular corrosion by forming niobium carbide with the carbon that is present in the steel. Without the niobium addition, chromium precipitates as a chromium carbide film at the grain boundaries and thus depletes the adjacent areas of chromium and

Table 5. 1992 U.S. Consumption of Ferroniobium and Nickel Niobium, kg of Nb[a]

Material	Quantity
steel	
carbon	837,541
stainless and heat resisting	346,936
high strength, low alloy	931,096
tool and unspecified	8,044
superalloys	411,620
alloys and miscellaneous	20,076
Total consumption	*2,555,313*

[a]Ref. 36.

reduces the corrosion resistance. An amount of niobium equal to 10 times the carbon content is necessary to prevent precipitation of the chromium carbide.

Niobium is also important in nonferrous metallurgy. Addition of niobium to zirconium reduces the corrosion resistance somewhat but increases the mechanical strength. Because niobium has a low thermal-neutron cross section, it can be alloyed with zirconium for use in the cladding of nuclear fuel rods. A Zr–1%Nb [11107-78-1] alloy has been used as primary cladding in the countries of the former USSR and in Canada. A Zr–2.5 wt % Nb alloy has been used to replace Zircaloy-2 as the cladding in Candu-PHW (pressurized hot water) reactors and has resulted in a 20% reduction in wall thickness of cladding (63) (see NUCLEAR REACTORS).

Niobium is a common additive to the nickel- and cobalt-based superalloys. Addition levels are typically 1–2.5 wt % in the cobalt-based alloys and 2–5 wt % in the nickel-based alloys. Niobium-based alloys with tungsten, titanium, and zirconium have superior strength and corrosion resistance up to 1200°C; eg, C-103, a Nb–Hf–Ti alloy having Ta, Ti, and Zr as low impurities, retains high yield strength at higher temperatures. It is used in rocket nozzles and afterburners along with turbine blades by the aerospace and aircraft industries. Sodium vapor lamps take advantage of the Nb–1%Zr alloy's resistance to metallic sodium heated to 871–927°C (64). The alloy is also used in the SP 100 reactor program for the space station.

A use small in total quantity consumed is as anodized niobium for jewelry. Niobium oxide is used as a substitute for the higher priced tantalum oxide in optical glass (65). A niobium alloy, Ti–45% Nb, originally developed as a superconducting material, has found another application in ventlines of gold mining autoclaves where these autoclaves better resist corrosion and erosion while resisting the danger of ignition (66).

The corrosion resistance of niobium and its high electrical conductivity and ductility make it a valuable structural material for chemical and metallurgical applications. The heat-transfer coefficient of niobium is more than twice that of titanium and three times higher than zirconium and stainless steels. Niobium is corrosion-resistant to most media, with the exception of hydrofluoric acid and hot concentrated hydrochloric and sulfuric acids. The pickling solution for removal of normal surface oxides is one part nitric acid, one part sulfuric acid, two parts hydrofluoric acid, and four parts water (by volume) (see METAL SURFACE TREATMENTS). Niobium also shows good corrosion resistance to sulfidizing atmospheres of low oxygen potential, which may be used in the production of substitute natural gas from sulfur-containing materials (67). Liquid sodium, potassium, sodium–potassium alloys, or lithium have little effect on niobium up to 1000°C, and its resistance to many other liquid metals is good.

Niobium is used as a substrate for platinum in impressed-current cathodic protection anodes because of its high anodic breakdown potential (100 V in seawater), good mechanical properties, good electrical conductivity, and the formation of an adherent passive oxide film when it is anodized. Other uses for niobium metal are in vacuum tubes, high pressure sodium vapor lamps, and in the manufacture of catalysts.

Niobium carbide is used as a component of hard metals, eg, mixtures of metal carbides that are cemented with cobalt, iron, and nickel. Along with tantalum carbide, niobium carbide is added to impart toughness and shock and erosion resistance. The spiraling rise in the price of tantalum has spurred the development of a hafnium carbide–niobium carbide substitute for tantalum carbide (68). These cemented carbides are used for tool bits, drill bits, shovel teeth, and other wear-resistant components; turbine blades; and as dies in high pressure apparatus (see CARBIDES).

Niobium and many of its alloys exhibit superconductivity, ie, the lack of electrical resistance at very low temperatures, thus they are of great interest for power generation (qv), propulsion devices, fusion energy (qv) research, electronic devices, particle accelerators, and other applications. Niobium becomes superconducting at 9.15 K. Other niobium compounds and their transitional temperatures are NbTi [12384-42-8], 9.5 K; Nb$_3$In [12030-07-8], 9 K; Nb$_3$Sn [12035-04-0], 18 K; Nb$_3$Al [12003-75-7], 18.8 K; Nb$_3$Ga [12024-05-4], 20 K; and Nb$_3$Ge [12025-22-8], 23 K. Most superconducting devices use niobium–titanium because of ease of its fabrication into magnet wire, which is its most common application. Where very high magnetic fields are necessary, niobium–tin, Nb$_3$Sn, conductors are used, even though the intermetallic nature of Nb$_3$Sn makes fabrication difficult. Improved methods of fabrication should lead to wider use of Nb$_3$Sn and to commercial application of niobium–aluminum and niobium–aluminum–germanium superconductors (see SUPERCONDUCTING MATERIALS). Niobium oxide has started replacing the more expensive tantalum oxide as a raw material in lanthanum borate glasses. It has also found wide applications in ultrasonics, acoustooptics, electrooptics, and nonlinear optics.

Niobium Compounds

Niobium Boride. A number of niobium boride phases have been described in the literature, ie, Nb$_2$B [12344-74-0], Nb$_3$B [56450-58-9], Nb$_3$B4 [12045-89-5], NbB, and NbB$_2$. Only the monoboride and the diboride melt congruently; NbB$_2$ decomposes at the melting point to NbB and boron (69). Some of the properties of these niobium borides together with other niobium compounds are listed in Table 6.

The most common methods of preparation have been hot-pressing, sintering, or remelting powdered mixtures of elemental boron with niobium or niobium hydride (103,104). Other methods are the reduction of a mixture of Nb$_2$O$_5$ and B$_2$O$_3$ with aluminum, silicon, or magnesium (105); carbon reduction at 2000°C of Nb$_2$O$_5$ and B$_2$O$_3$ (106); reaction of carbon with B$_4$C and Nb$_2$O$_5$ (107); electrolysis of molten mixtures of Nb$_2$O$_5$ with alkali metal and alkaline-earth metal borates and fluorides to produce NbB$_2$ (108,109); chemical vapor deposition (CVD) onto a hot substrate by a mixture of niobium halide, boron halide, hydrogen, and argon (77–110); and CVD of boron on niobium accompanied or followed by diffusion into the substrate (111) (see THIN FILMS, FILM FORMATION). Niobium diboride generally is a gray powder; is unattacked by hydrochloric acid, nitric acid, or aqua regia; is attacked slowly by hot sulfuric or hydrofluoric acid; and is dissolved rapidly by molten alkali, hydroxides, carbonates, bisulfates, and sodium peroxide. It is oxidized in air at red heat.

Table 6. Properties of Niobium Compounds

Compounds	CAS Registry Number	Molecular formula	Lattice	Lattice constant, pm	Density, g/cm^3	Mp, °C	Bp, °C	Specific resistivity, $\mu\Omega\cdot$cm	Refs.
niobium boride	[12045-19-1]	NbB	orthorhombic	$a = 329.8$ $b = 872.4$ $c = 316.6$	7.5	2000		64.5^a	70,71
niobium diboride	[12007-29-3]	NbB$_2$	hexagonal	$a = 308.9$ $c = 330.3$	6.9^b	3050		65^{a,c}	70–76
diniobium carbide	[12011-99-3]	Nb$_2$C	hcp	$a = 312.7$ $c = 497.2$	7.8	3090			77
niobium carbide	[12069-94-2]	NbC	fcc	$a = 447.1$	7.788^d	3600	4300	180 max^e	75,78–83
niobium pentafluoride	[7783-68-8]	NbF$_5$	monoclinic	$a = 963$ $b = 1443$ $c = 512$ $\beta = 96.1^f$	3.54	79	234		84,85
niobium fluorodioxide	[15195-33-2]	NbO$_2$F	cubic	$a = 390.2$					86
niobium pentachloride	[10026-12-7]	NbCl$_5$	monoclinic	$a = 183.0$ $b = 1798$ $c = 588.8$ $\beta = 90.6^f$	2.74^g	208.3	248.2		87,88
niobium trichloro-monoxide	[13597-20-1]	NbOCl$_3$	tetragonal	$a = 1087$ $c = 396$	3.72	vacuum sublimes at ca 200			89,90
niobium pentabromide	[13478-45-0]	NbBr$_5$	orthorhombic	$a = 612.7$ $b = 658$ $c = 1855$	4.36	254	365		91,92
niobium tribromo-monoxide	[14459-75-7]	NbOBr$_3$				vacuum sublimes at 180	ca 320 dec		

Name	CAS Registry Number	Formula	Crystal system	Lattice parameters	Density	mp	bp	References
niobium pentaiodide	[13981-86-7]	NbI$_5$	monoclinic	$a = 1058$ $b = 658$ $c = 1388$ $\beta = 109.1^f$		ca 200 dec		93
niobium hydride	[13981-86-7]	NbH	bcc		6–6.6			
diniobium nitride	[12033-43-1]	Nb$_2$N	hcp	$a = 305.6$–304.8 $c = 495.6$	8.08	2050		94,95
niobium nitride	[24621-21-4]	NbN	fcc	$a = 438.2$–439.2	8.4^a		200^a, 450 (at mp)	95–97
niobium oxide	[12034-57-0]	NbO	cubic	$a = 421.08$	7.30			98
niobium dioxide	[12034-59-2]	NbO$_2$	tetragonal	$a = 1371$	5.90			99
α-niobiumpentoxide	[1313-96-8]	α-Nb$_2$O$_5$	monoclinic	$a = 2116$	4.55	1491±2		100–102

^{a}At 25°C.
^{b}Has a Mohs' hardness of 8+.
^{c}Thermal conductivity value is 17 W/(m·k) at 23°C.
^{d}Has a Mohs' hardness of 9+.
^{e}Thermal conductivity value is 14 W/(m·k) at 23°C.
^{f}Units are degrees.
^{g}Has a hardness of 208.3.

Niobium Carbide. Apparently three solid single-phase regions exist in the niobium–carbon system, ie, a solid solution of carbon in niobium (bcc), Nb_2C (hexagonal), and NbC (fcc) (77). The compositional range of Nb_2C is very limited, whereas NbC varies from $NbC_{0.7}$ to nearly stoichiometric NbC. Thermodynamic values for these phases have been reported (81,112,113). Industrial preparation utilizes Nb_2O_5 and carbon as starting materials. The reaction starts at ca 675°C but temperatures of 1800–2000°C are needed for completion of the reaction. Heating the elemental powders also produces NbC if a sufficiently high final temperature is used. Chemical vapor deposition (CVD) can be used to deposit NbC on a hot surface by reaction of $NbCl_5$ and hydrogen and hydrocarbons. Niobium carbide powder has a gray metallic color up to a composition of $NbC_{0.9}$; the color changes to lavender upon addition of carbon up to $NbC_{0.99}$. NbC is unreactive and resists boiling in aqua regia; a mixture of HNO_3 and HF is needed for dissolution. NbC burns on heating in air to >1100°C and can be converted to the nitride by heating in nitrogen or ammonia.

Niobium Halides and Oxyhalides. All possible halides of pentavalent niobium are known and preparations of lower valent halides generally start with the pentahalide. Ease of reduction decreases from iodide to fluoride.

Niobium Pentafluoride. Niobium pentafluoride is prepared best by direct fluorination of the metal with either fluorine or anhydrous hydrofluoric acid at 250–300°C. The volatile NbF_5 is condensed in a pyrex or quartz cold trap, from which it can be vacuum-sublimed at 120°C to yield colorless monoclinic crystals. It is very hygroscopic and reacts vigorously with water to give a clear solution of hydrofluoric acid and H_2NbOF_5 [12062-01-0]. This acid also is formed by dissolving niobium metal or niobic acid in hydrofluoric acid and, at high acid concentrations, it is converted to H_2NbF_7. Addition of potassium fluoride to a solution of H_2NbOF_5 precipitates $K_2NbOF_5 \cdot H_2O$, which is soluble in hot water and can be recrystallized from a saturated solution to give large monoclinic platelets. The high solubility of $K_2NbOF_5 \cdot H_2O$ in water was the basis for early separations of tantalum and niobium because the corresponding tantalum salt K_2TaF_7 is 12 times less soluble in 1 wt % HF at 20°C.

Niobium Dioxide Fluoride. Niobium dioxide fluoride, NbO_2F, is formed on dissolution of niobium pentoxide in 48 wt % aqueous hydrofluoric acid, evaporation of the solution to dryness, and heating to 250°C.

Niobium Pentachloride. Niobium pentachloride can be prepared in a variety of ways but most easily by direct chlorination of niobium metal. The reaction takes place at 300–350°C. Chlorination of a niobium pentoxide–carbon mixture also yields the pentachloride; however, generally the latter is contaminated with niobium oxide trichloride. The pentachloride is a lemon-yellow crystalline solid that melts to a red-orange liquid and hydrolyzes readily to hydrochloric acid and niobic acid. It is soluble in concentrated hydrochloric and sulfuric acids, sulfur monochloride, and many organic solvents.

Niobium Oxide Trichloride. Niobium oxide trichloride, $NbOCl_3$, also can be prepared in a variety of ways, ie, oxidation of the pentachloride by air, reaction of the pentoxide with the pentachloride, reaction of carbon tetrachloride or HCl gas with the pentoxide at 400–700°C, and as an intractable impurity in most preparations of the pentachloride. It is a white solid that sublimes at ca 200°C, is thermally unstable, and forms the pentoxide and the pentachloride at higher temperatures.

Niobium Pentabromide. Niobium pentabromide is most conveniently prepared by reaction of bromine with niobium metal at ca 500°C. It is a fairly volatile yellow-red compound that is hygroscopic and readily hydrolyzes. It is soluble in water, alcohol, and ethyl bromide.

Niobium Oxide Tribromide. Niobium oxide tribromide, NbOBr, is a yellow-brown solid which is readily hydrolyzed by moist air. It is prepared by reaction of bromine with a mixture of niobium pentoxide and carbon at 550°C. It decomposes in vacuum to the pentabromide and pentoxide at 320°C.

Niobium Pentaiodide. Brass-yellow crystals of niobium pentaiodide are formed by direct reaction of excess iodine with niobium metal in a sealed tube (114). It is thermally unstable and decomposes to the tetraiodide [*13870-21-8*] at 206–270°C in vacuum (115).

Niobium Hydride. Hydrogen reacts exothermically with niobium to form a stable interstitial solid solution. In a gas-phase reaction at 300–1500°C and hydrogen pressures of 0–101 kPa (0–1 atm), the lattice parameter and heat of solution increase with hydrogen content up to the composition $NbH_{0.85}$ (116). X-ray studies show a linear relation between atomic volume and the hydrogen content (117). The absorption of hydrogen is proportional to the square root of the hydrogen pressure, which indicates dissociation of molecular hydrogen at the metal surface and diffusion of hydrogen atoms (118). Although the hydride is stable at room temperature, heating to 500°C at 0.67 kPa (5 mm Hg) decomposes the hydride to hydrogen and niobium metal. The expansion on absorption and contraction on desorption of hydrogen by the metal lattice usually leaves the metal in powder form after a hydride–dehydride operation, a process of commercial value for production of metal powder and recovery of metal scrap. An unstable dihydride, NbH_2 [*13981-96-9*], has been prepared by the cathodic hydrogenation of niobium foil in 6 N sulfuric acid (119). It is very unstable and decomposes in vacuum or air to the monohydride (see HYDRIDES).

Niobium Nitrides. The uptake of nitrogen by niobium metal proceeds by the exothermic formation of an interstitial solid solution of nitrogen atoms in the bcc lattice of the metal. The solubility of nitrogen in the metal is proportional to the square root of the nitrogen partial pressure until the formation of the nitride phase Nb_2N. This relation holds from 1200–2400°C and over almost 10 orders of magnitude of the pressure (120). At the solubility limit of the solid solution α-phase, the hcp β-phase appears, which has a composition of $NbN_{0.4}$ to $NbN_{0.5}$. Further absorption of nitrogen leads to the formation of a fcc δ-phase with a homogeneity of $NbN_{0.88}$ to $NbN_{0.98}$. This δ-phase is stable only above 1230°C. Niobium nitride can be prepared by heating the metal in nitrogen or ammonia to 700–1100°C (83), by heating the pentoxide and carbon to 1250°C in the presence of nitrogen (121), and by CVD using $NbCl_5$, H_2, and N_2 (111,122). The nitride is a light gray powder with a yellow cast; it is insoluble in HCl, HNO_3, and H_2SO_4; it is attacked by hot caustic, lime, or strong alkalies with the evolution of ammonia. It reacts when heated in air to form the pentoxide and to liberate nitrogen. Nb_2N is resistant to acids by reacts with strong alkali to liberate nitrogen rather than ammonia (see NITRIDES).

Niobium Oxides. The solubility of oxygen in niobium obeys Henry's law to the solubility limit of the first oxide phase of $850–1300°C$ (123). The amount of oxygen in solution in niobium is 1.3 at. % at $850°C$ and nearly 2 at. % at $1000°C$ (124). Only three clearly defined anhydrous oxides of niobium have been obtained in bulk, ie, NbO, NbO_2, and Nb_2O_5. Niobium monoxide, NbO, is obtained by hydrogen reduction of the pentoxide at $1300–1700°C$ or by heating a compressed mixture of the metal powder with NbO_2 in argon at $1700°C$. It has a gray metallic appearance. Niobium dioxide, NbO_2, also can be obtained by hydrogen reduction of the pentoxide at $800–1300°C$, by heating a properly proportioned mixture of the pentoxide and the metal, or by thermal dissociation of the pentoxide at $1150°C$ in an argon sweep. Niobium dioxide is black with a bluish cast, is a strong reducing agent in the dry state, and is converted to the pentoxide on ignition in air.

The considerable confusion existing in the literature regarding the polymorphism of Nb_2O_5 seems to have been resolved. Three distinct phases have been identified: a low temperature phase T, a middle temperature phase M, and a high temperature phase H (125). With regard to the amorphous oxide produced from the hydrolysis of niobic acid, conversion to the crystalline T form occurs at $500°C$ and is followed by transformation to the M form at $1000°C$ and to the H form at $1100°C$. These phases were renamed γ (T), β (M), and α (H). The transition of amorphous to γ-phase was found to occur at $440°C$ (126). Heating at $830°C$ irreversibly converts the γ-phase to a mixture of β- and α-forms and further heating to $1095°C$ irreversibly transforms this mixture to the pure α-form. It was concluded that the β-form is an imperfectly crystallized α-phase. Subsequent work has demonstrated the existence of another metastable phase, ϵ (127). It appears that three crystalline forms of Nb_2O_5 are detectable at atmospheric pressure and that the α-form is the only stable structure and is monoclinic with 14 formula units in the unit cell (99,128). The α-phase can be prepared by heating the metal carbide, nitride, or niobic acid at $>1100°C$ and has been prepared in the form of large single crystals (129). It generally is an eggshell-white powder, and turns yellowish on heating because of the formation of oxygen vacancies in the lattice. It is insoluble in acids, except hydrofluoric, and can be dissolved by fused alkali pyrosulfates, carbonates, or hydroxides.

Niobic Acid and Salts. *Niobic Acid.* Niobic acid, $Nb_2O_5 \cdot xH_2O$, includes all hydrated forms of niobium pentoxide, where the degree of hydration depends on the method of preparation, age, etc. It is a white insoluble precipitate formed by acid hydrolysis of niobates that are prepared by alkali pyrosulfate, carbonate, or hydroxide fusion; base hydrolysis of niobium fluoride solutions; or aqueous hydrolysis of chlorides or bromides. When it is formed in the presence of tannin, a voluminous red complex forms. Freshly precipitated niobic acid usually is colloidal and is peptized by water washing, thus it is difficult to free from traces of electrolyte. Its properties vary with age and reactivity is noticeably diminished on standing for even a few days. It is soluble in concentrated hydrochloric and sulfuric acids but is reprecipitated on dilution and boiling and can be complexed when it is freshly made with oxalic or tartaric acid. It is soluble in hydrofluoric acid of any concentration.

Niobates. Niobic acid is amphoteric and can act as an acid radical in several series of compounds, which are referred to as niobates. Niobic acid is soluble in solutions of the hydroxides of alkali metals to form niobates. Fusion of the anhydrous pentoxide with alkali metal hydroxides or carbonates also yields niobates. Most niobates are insoluble in water with the exception of those alkali metal niobates having a base-to-acid ratio greater than one. The most well-known water-soluble niobates are the 4:3 ad the 7:6 salts (base:acid), having empirical formulas $M_8Nb_6O_{19}$ (aq) and $M_{14}Nb_{12}O_{37}$ (aq), respectively. The hexaniobate is hydrolyzed in aqueous solution according to the pH-dependent reversible equilibria (130), when the pH is ca 9.

$$Nb_6O_{19}^{8-} \text{ (aq)} \underset{}{\overset{H_2O}{\rightleftharpoons}} HNb_6O_{19}^{7-} \text{ (aq)} + OH^-$$

$$Nb_6O_{19}^{7-} \text{ (aq)} \underset{H_2O}{\overset{}{\rightleftharpoons}} HNb_6O_{18}^{6-} \text{ (aq)} + OH^-$$

The 7:6 salts are the acid salts of the normal 4:3 hexaniobates. The formulas can be written as $M_7H(Nb_6O_{19})$ (aq). Further hydrolysis can take place. At pH ca 4.5, the irreversible precipitation of niobic acid occurs.

$$H_6(Nb_6O_{18}) \longrightarrow 3\,Nb_2O_5 + 3\,H_2O$$

The potassium salts are the most soluble and other salts usually are precipitated by addition of the appropriate metal chloride to a solution of the corresponding potassium salt. The metaniobates, $MNbO_3$, and orthoniobates, $MNbO_4$, generally are prepared by fusion of the anhydrous mixed oxides. The metaniobates crystallize with the perovskite structure and are ferroelectric (131) (see FERROELECTRICS). The orthoniobates are narrow band-gap semiconductors (qv) (132).

Sodium metaniobate(1:1) [67211-31-8], $Na_2O \cdot Nb_2O_5 \cdot 7H_2O$ or $Na_2Nb_2O_6 \cdot 7H_2O$, separates as colorless triclinic crystals as a result of concentrating the mother liquor from the preparation of the 7:6 sodium niobate by spontaneous evaporation. It also can be obtained by fusion of the anhydrous pentoxide in sodium hydroxide or carbonate.

Potassium niobate(4:3) [12502-31-7], $4K_2O \cdot 3Nb_2O_5 \cdot 16H_2O$ or $K_8Nb_6O_{19} \cdot 16H_2O$, is obtained by dissolving niobic acid in a concentrated solution of potassium hydroxide. The large monoclinic crystals are separated by concentrating the solution. The salt is very soluble; at room temperature a saturated solution contains 425 g/100 g of water. It is much more soluble in hot water and prone to form supersaturated solutions.

Sodium niobate(7:6) [12201-59-1], $7Na_2O \cdot 6Nb_2O_5 \cdot 31H_2O$ or $Na_{14}Nb_{12}O_{37} \cdot 31H_2O$, forms a crystalline precipitate when a hot solution of a soluble niobium compound is added to a hot concentrated sodium hydroxide solution. It is insoluble in the presence of excess sodium hydroxide but is sparingly soluble in pure water. It also can be formed by addition of sodium hydroxide or chloride to a solution of the 4:3 potassium niobate.

Lithium niobate [12031-63-9], $Li_2O \cdot Nb_2O_5$ or $LiNbO_3$, is prepared by the solid-state reaction of lithium carbonate with niobium pentoxide. After being separately predried at 150–200°C, the stoichiometric amounts of the oxides are carefully mixed and heated to 600°C in a platinum crucible. The temperature is increased slowly for 12 h from 600 to 800°C and is maintained at 800°C for another 12 h. The mixture is cooled, crushed, and reheated to 900°C for 12 h. The product may not be completely homogeneous.

BIBLIOGRAPHY

"Columbium" in *ECT* 1st ed., Vol. 4, pp. 314–324, by C. W. Balke, Fansteel Metallurgical Corp.; "Niobium and Niobium Compounds" in *ECT* 2nd ed., Vol. 13, pp. 766–784, by P. A. Butters, Murex Ltd.; in *ECT* 3rd ed., Vol. 15, pp. 820–840, by P. H. Payton, Teledyne Wah Chang Albany.

1. G. G. Goles, "Cosmic Abundances," in *Handbook of Geochemistry*, Vol. 1, Springer-Verlag, Berlin, New York, 1969.
2. M. R. Krishnadev and A. Galibois, *Proceedings of the 3rd Interamerican Conference on Material Technology*, 1972, p. 581.
3. W. Von Engelhardt and R. Stengelin, *Earth Planet. Scl. Lett.* **42**, 213 (1979).
4. K. E. Apt, J. D. Knight, D. C. Camp, and R. W. Perkins, *Geochim. Cosmochim. Acta* **38**, 1485 (1974).
5. D. R. Stull and G. C. Sinke, *Thermodynamic Properties of the Metals*, American Chemical Society, Washington, D.C., 1956.
6. G. L. Humphrey, *J. Am. Chem. Soc.* **76**, 978 (1954).
7. R. Speiser, P. Blackburn, and H. L. Johnston, *J. Electrochem. Soc.* **106**, 52 (1959).
8. C. R. Tottle, *Nucl. Eng.* **3**, 212 (1958).
9. J. R. Darnell and L. F. Yntema, in B. W. Gonser and E. M. Sherwood, eds., *The Technology of Columbium (Niobium)*, John Wiley & Sons, Inc., New York, 1958.
10. D. F. Taylor, *Ind. Eng. Chem.*, 639 (Apr. 1950).
11. C. R. Bishop, *Corrosion* **19**(9), 308t (1963).
12. D. L. Macleary, *Corrosion* **18**, 67t (1962).
13. *ACS Monogr.* **158** (1963).
14. D. P. Gold, M. Vallee, and J. P. Charette, *Can. Mining Metall. Bull.* **60**, 1131 (1967). M. Vallee and F. Dubuc, *Can. Mining Metall. Bull.* **63**, 1384 (1970); C. Carbonneau and J. C. Caron, *Can. Mining Metall. Bull.* **58**, 281 (1965).
15. M. Robert, *Can. Min. J.* **38** (Mar. 1978).
16. H. Stuart, O. de Souza Paraiso, and R. de Fuccio, *Iron Steel*, 11 (May 1980).
17. F. Perfect, *Trans. Metall. Soc. AIME* **239**, 1282 (1967).
18. F. J. Kelly and W. A. Gow, *Can. Mining Metall. Bull.* **58**, 843 (1965).
19. U.S. Pat. 3,607,006 (Sept. 21, 1971), E. P. Stambaugh (to Molybdenum Corp. of America).
20. U.S. Pat. 4,182,744 (Jan. 8, 1980), R. H. Nielsen and P. H. Payton (to Teledyne Industries, Inc.).
21. S. L. May and G. T. Engel, *U.S. Bur. Mines Rep. Invest.* **6635**, (1965).
22. F. Habashi and I. Malinsky, *CIM Bull.* **68**(761), 85 (1975).
23. U.S. Pat. 3,153,572 (Oct. 20, 1964), W. E. Dunn, Jr. (to E. I. du Pont de Nemours & Co., Inc.).
24. D. J. Soissan, J. J. McLafferty, and J. A. Pierret, *Ind. Eng. Chem.* **53**(11), 861 (1961).
25. C. H. Faye and W. R. Inman, *Research Report MD210*, Dept. Mines and Technical Surveys, Ottawa, Canada, 1956.
26. J. R. Werning, K. B. Higbie, J. T. Grace, B. F. Speece, and H. L. Gilbert, *Ind. Eng. Chem.* **46**, 644 (1954).
27. C. W. Carlson and R. H. Nielson, *J. Met.* **12**, 472 (June 1960).
28. J. R. Werning and K. B. Higbie, *Ind. Eng. Chem.* **46**, 2491 (1954).
29. B. R. Steele and D. Geldart, "Extraction and Refining of the Rarer Metals," *Proceedings of the Symposium on the Institute of Mining and Metallurgy*, London, 1956, pp. 287–309.
30. U.S. Pat. 5,322,548 (June 21, 1994), B. F. Kieffer, J. R. Peterson, T. R. McQueary, M. A. Rossback, and L. J. Fenwick (to Teledyne Industries, Inc.).
31. T. K. Mukherjee and C. K. Gupta, *Trans. SAEST* **11**(1), 127 (1976).

32. U.S. Pat. 3,271,277 (Sept. 6, 1966), L. F. Yntema.

33. *Tool. Prod.* **45**(2), 76 (1979).

34. *Minerals Handbook 1994–1995*, Phillip Corwson, Stocton Press, New York, pp. 189–194.

35. *Niobium—Survey of World, Production, Consumption and Prices*, 3rd ed., Roskill Information Services Ltd., London, Oct. 1992.

36. *Columbium and Tantalum*, U.S. Bureau of Mines, Washington, D.C., Nov. 1994.

37. W. R. Schoeller, *The Analytical Chemistry of Niobium and Tantalum*, Chapman and Hall, London, 1937.

38. W. R. Schoeller and A. R. Powell, *The Analysis of Minerals and Ores of the Rarer Elements*, 3rd ed., Griffin, London, 1955.

39. R. W. Moshier, *Analytical Chemistry of Niobium and Tantalum*, Pergamon, New York, 1964.

40. I. M. Gibalo, transl. J. Schmorak, *Analytical Chemistry of Niobium and Tantalum*, Ann Arbor-Humphrey Science Publishers, Ann Arbor, Mich., 1970.

41. J. L. Kassner, A. G. Porrata, and E. L. Grove, *Anal. Chem.* **27**, 493 (1955).

42. Y. Shijo, *Bull. Chem. Soc. Jpn.* **50**, 1011 (1977).

43. A. G. Ward and O. Borgen, *Talanta* **24**, 65 (1977).

44. R. N. Gupta and B. K. Sen, *J. Inorg. Nucl. Chem.* **37**, 1548 (1975).

45. G. C. Shivahare and D. S. Parmar, *Ind. J. Chem.* **13**, 627 (1975).

46. S. Kallman, H. Oberthin, and R. Liu, *Anal. Chem.* **34**, 609 (1962).

47. ASTM E-1473, *Standard Methods for the Chemical Analysis of Nickel, Cobalt, and High Temperature Alloys*, American Society for Testing and Materials, Philadelphia, Pa., 1994.

48. ASTM E-367, *Standard Method for the Chemical Analysis of FeCb*, American Society for Testing and Materials, Philadelphia, Pa., 1994.

49. ASTM E-120, *Standard Test Methods for the Chemical Analysis of Titanium and Titanium Alloys*, American Society for Testing and Materials, Philadelphia, Pa., 1994.

50. B. Langford, I. H. Stice, L. Latimer, and R. E. Walsh, *Determination of Hydrogen by Impulse Furnace Fusion Employing Chromatographic Detection*, Teledyne Wah Chang Corp., Albany, Oreg., 1974.

51. B. Langford and I. H. Stice, *The Determination of Nitrogen and Oxygen by Inert Gas Fusion*, Teledyne Wah Chang Corp., Albany, Oreg., 1994.

52. B. Langford, L. Latimer, and K. Ash, *Determination of Carbon Utilizing a LECO IR--12 Analyzer*, Teledyne Wah Chang Corp., Albany, Oreg., 1977.

53. Ref. 38, p. 260.

54. J. Fraley, *Elemental Analysis by Argon Plasma Emission Spectroscopy*, Teledyne Wah Chang Corp., Albany, Oreg., 1988.

55. G. Beck, *Elemental Analysis by Argon Plasma Mass Spectroscopy*, Teledyne Wah Chang Corp., Albany, Oreg., 1988.

56. R. Herzog and F. Dietz, *Chem. Biol. Lab.* **73**, 67–109.

57. K. Fujimori and F. DiGiorgi, *Metal. ABM* **30**(204), 751 (1974).

58. H. L. Giles and G. M. Holmes, *X-Ray Spectrom.* **7**(1), 2 (1978).

59. G. A. Shkurko, *Gig. Tr.* **9**, 74 (1973).

60. L. A. Sazhina and L. N. Elnichnyky, *Gig Sanit.* **6**, 8 (1975).

61. R. G. Thomas, S. A. Walker, and R. O. McClellan, *Proc. Soc. Exp. Biol. Med.* **138**(1), 228 (1971).

62. *Niobium Base Alloy Powders Material Safety Data Sheet*, No. 516, Teledyne Wah Chang Corp., Albany, Oreg., 1992.

63. B. A. Cheadle, W. J. Langford, and R. I. Coote, *Nucl. Eng. Int.* **24**(289), 50 (1979).

64. *Outlook*, Vol. 9, No. 2, Teledyne Wah Chang Corp., Albany, Oreg., 1988.

65. T. Ichimura, in H. Stuart, ed., *Niobium Proceedings of the International Symposium*, The Metallurgical Society of AIME, Warrendale, Pa., 1981, p. 603.

66. *Outlook*, Vol. 13, No. 3, Teledyne Wah Chang Corp., Albany, Oreg., 1992.

67. K. N. Straffod and J. R. Bird, *J. Less Common Met.* **68**, 223 (1969).

68. P. H. Booker and R. E. Curtis, *Cutting Tool Eng.* **18** (Sept.–Oct. 1978).

69. F. Fairbrother, *The Chemistry of Niobium and Tantalum*, Elsevier Publishing Co., New York, 1967.

70. L. H. Anderson and R. Kiessling, *Acta Chem. Scand.* **4**, 160 (1950).

71. *Tech. Data Sheet No. 4-B*, Borax Consolidated Ltd., London.

72. G. V. Samsonov and L. Ya. Markovski, *Usepkhi. Khim.* **25**(2), 190 (1958).

73. F. W. Glaser, *J. Met.* **4**, 391 (1952).

74. B. Post, F. W. Glaser, and D. Moskowitz, *Acta Metall.* **2**, 20 (1954).

75. S. J. Sindeband and P. Schwartzkopf, "The Metallic Nature of Metal Borides," *97th Meeting of the Electrochemical Society*, Cleveland, Ohio, 1950; *Powder Metall. Bull.* **5/3**, 42 (1950).

76. K. Moers, *Z. Anorg. Chem.* **198**, 243 (1931).

77. E. K. Storms and N. H. Krikorian, *J. Phys. Chem.* **64**, 1461 (1960).

78. E. K. Storms, N. H. Krikorian, and C. P. Kempter, *Anal. Chem.* **32**, 1722 (1960).

79. E. K. Storms, *The Refractory Carbides*, Academic Press, Inc., New York, 1967, p. 70.

80. R. Kieffer and F. Kölbl, *Powder Met. Bull.* **4**, 4 (1949).

81. R. Hultgren, P. D. Desai, D. T. Hawkins, M. Gleiser, and K. K. Kelley, *Selection Values of the Thermodynamic Properties of Binary Alloys*, American Society for Metals, Metals Park, Ohio, 1973, p. 500.

82. I. E. Campbell, ed., *High Temperature Technology*, John Wiley & Sons, Inc., New York, 1956.

83. H. Takeshita, M. Miyaki, and T. Sano, *J. Nucl. Mat.* **78**, 77 (1978).

84. Ref. 53, p. 76.

85. A. J. Edwards, *J. Chem. Soc.*, 3714 (1964).

86. L. K. Frevel and H. W. Rinn, *Acta Cryst.* **9**, 626 (1956).

87. J. H. Canterford and R. Colton, *Halides of the Second and Third Row Transition Elements*, John Wiley & Sons, Inc., New York, 1968.

88. D. R. Sadoway and S. W. Flengas, *Can. J. Chem.* **54**(11), 1692 (1976).

89. D. E. Sands, A. Zalkin, and R. E. Elson, *Acta Cryst.* **12**, 21 (1959).

90. Ref. 53, p. 103.

91. S. S. Berdonosova, A. V. Lapitskii, D. G. Berdonosova, and L. G. Vlasov, *Russ. J. Inorg. Chem.* **8**, 1315 (1963).

92. S. S. Berdonosova, A. V. Lapitskii, and E. K. Bakov, *Russ. J. Inorg. Chem.* **10**, 173 (1965).

93. W. Littke and G. Brauer, *Z. Anorg. Allg. Chem.* **325**, 122 (1963).

94. G. Brauer, *Z. Elektrochem.* **46**, 39 (1949).

95. N. Schönberg, *Acta Chem. Scand.* **8**(2), 208 (1954).

96. Ref. 53, p. 190.

97. R. Kieffer and P. Schwartzkopf, *Hartstoffe and Hartmetalle*, Springer, Vienna, Austria, 1953.

98. Ref. 53, p. 23.

99. Ref. 53, p. 24.

100. Ref. 53, p. 26.

101. A. Reisman and F. Holtzberg, in A. M. Alper, ed., *High Temperature Oxides*, Part II, Academic Press, Inc., New York, 1970, p. 220.

102. B. M. Gatehouse and A. D. Wadsley, *Acta Cryst.* **17**, 1545 (1964).

103. R. Kiessling, *J. Electrochem. Soc.* **98**(4), 166 (1957).

104. L. Brewer, D. L. Sawyer, D. H. Templeton, and C. H. Dauben, *J. Am. Ceram. Soc.* **34**, 173 (1951).

105. U.S. Pat. 2,678,870 (1950), H. S. Cooper.
106. P. M. McKenna, *Ind. Eng. Chem.* **28**, 767 (1936).
107. G. A. Meerson and G. V. Samsonov, *Zh. Prikl. Khim.* **27**, 1115 (1954).
108. A. Andrieux, *Compt. Rend.* **189**, 1279 (1929).
109. J. T. Norton, H. Blumenthal, and S. J. Sindeband, *J. Met.* **185**, 749 (1949).
110. S. Motojima, K. Sugiyama, and U. Takahashi, *J. Cryst. Growth* **30**, 233 (1975).
111. C. F. Powell, J. H. Oxley, and J. M. Blocher, Jr., *Vapor Deposition*, John Wiley & Sons, Inc., New York, 1966, p. 346.
112. Ref. 64, p. 72.
113. JANAF Thermochemical Tables, 1975 Suppl., *J. Phys. Chem. Ref. Data* **4**(1), 51 (1975).
114. R. F. Rolsten, *J. Am. Chem. Soc.* **79**, 5409 (1957).
115. J. D. Corbett and P. W. Seabaugh, *J. Inorg. Nucl. Chem.* **6**, 207 (1958).
116. S. Komjathy, *J. Less Common Met.* **2**, 466 (1960).
117. H. Wenzl and J. Physique, *Colloque C7* **38**(Suppl. 12), C7 221 (1977).
118. A. Sieverts and H. Moritz, *Z. Anorg. Chem.* **247**, 124 (1941).
119. G. Brauer and H. Muller, *J. Inorg. Nucl. Chem.* **17**, 102 (1961).
120. G. Horz, *Electrochem. Soc. Proc. Symp. Prop. High Temp. Alloys* (77-1), 753 (1976).
121. E. Friederich and L. Sittig, *Z. Inorg. Chem.* **143**, 293 (1925).
122. T. Takehashi, H. Itoh, and T. Yamaguchi, *J. Crystal Growth* **46**, 69 (1979).
123. W. Nickerson and C. J. Altstetter, *Scr. Metall.* **7**, 229 (1973).
124. R. Lauf and C. Altstetter, *Scr. Metall.* **11**, 983 (1977).
125. G. Brauer, *Z. Anorg. Allg. Chem.* **248**, 1 (1941).
126. F. Holtzberg, A. Reisman, M. Berry, and M. Berkenblit, *J. Am. Chem. Soc.* **79**, 2039 (1957).
127. A. Reisman and F. Holtzberg, *J. Am. Chem. Soc.* **81**, 3182 (1959).
128. Ref. 86, p. 222.
129. I. Shindo and H. Komatsu, *J. Cryst. Growth* **34**(1), 152 (1976).
130. G. Jander and D. Ertel, *J. Inorg. Nucl. Chem.* **14**, 77 (1960).
131. A. Rauber, in K. Kaldis, ed., *Current Topics in Materials Science*, Vol. 1, North Holland Publishing Co., Amsterdam, the Netherlands, 1978, p. 481.
132. G. G. Kasimov, E. G. Vovkotrub, E. I. Krylov, and I. G. Rozanov, *Inorg. Mat.* **11**(6), 891 (1975).

General References

G. L. Miller, *Metallurgy of the Rarer Metals—6 Tantalum and Niobium*, Butterworths Scientific Publications, London, 1959.
R. J. H. Clark and D. Brown, *The Chemistry of Vanadium, Niobium and Tantalum*, Pergamon Press, Elmsford, N. Y., 1975.
D. L. Douglass and F. W. Kunz, eds., "Columbium Metallurgy," *Proceedings Symposium, New York, June 9–10, 1960*, Interscience Publishers, New York, 1961.
B. W. Gonser and E. M. Sherwood, eds., *The Technology of Columbium (Niobium)*, John Wiley & Sons, Inc., New York, 1958.
F. T. Sisco and E. Epremian, *Columbium and Tantalum*, John Wiley & Sons, Inc., New York, 1963.
S. Gerardi, "Niobium," in *Metals Handbook*, 10th ed., American Society for Metals, Metals Park, Ohio, 1990, pp. 565–571.
T.I.C. Bulletin, Tantalum–Niobium International Study Center, Brussels, Belgium.

JAMES H. SCHLEWITZ
Teledyne Wah Chang

NITRATION

Nitration is defined in this article as the reaction between a nitration agent and an organic compound that results in one or more nitro (—NO$_2$) groups becoming chemically bonded to an atom in this compound. Nitric acid is used as the nitrating agent to represent C-, O-, and N-nitrations. O-nitrations result in esters. N-nitrations result in nitramines.

$$C-H + HNO_3 \longrightarrow C-NO_2 + H_2O$$
$$C-OH + HNO_3 \longrightarrow C-O-NO_2 + H_2O$$
$$N-H + HNO_3 \longrightarrow N-NO_2 + H_2O$$

In the examples, a nitro group is substituted for a hydrogen atom, and water is a by-product. Nitro groups may, however, be substituted for other atoms or groups of atoms. In Victor Meyer reactions which use silver nitrite, the nitro group replaces a halide atom, eg, I or Br. In a modification of this method, sodium nitrite dissolved in dimethyl formamide or other suitable solvent is used instead of silver nitrite (1). Nitro compounds can also be produced by addition reactions, eg, the reaction of nitric acid or nitrogen dioxide with unsaturated compounds such as olefins or acetylenes.

Nitrations are highly exothermic, ie, ca 126 kJ/mol (30 kcal/mol). However, the heat of reaction varies with the hydrocarbon that is nitrated. The mechanism of a nitration depends on the reactants and the operating conditions. The reactions usually are either ionic or free-radical. Ionic nitrations are commonly used for aromatics; many heterocyclics; hydroxyl compounds, eg, simple alcohols, glycols, glycerol, and cellulose; and amines. Nitration of paraffins, cycloparaffins, and olefins frequently involves a free-radical reaction. Aromatic compounds and other hydrocarbons sometimes can be nitrated by free-radical reactions, but generally such reactions are less successful.

Ionic Nitration Reactions

Acid mixtures containing nitric acid and a strong acid, eg, sulfuric acid, perchloric acid, selenic acid, hydrofluoric acid, boron trifluoride, or an ion-exchange resin containing sulfonic acid groups, can be used as the nitrating feedstock for ionic nitrations. These strong acids are catalysts that result in the formation of nitronium ions, NO$_2^+$. Sulfuric acid is almost always used industrially since it is both effective and relatively inexpensive.

Most ionic nitrations are performed at 0–120°C. For nitrations of most aromatics, there are two liquid phases: an organic and an acid phase. Sufficient pressure, usually slightly above atmospheric, is provided to maintain the liquid phases. A large interfacial area between the two phases is needed to expedite transfer of the reactants to the interface and of the products from the interface. The site of the main reactions is often at or close to the interface (2). To provide large interfacial areas, a mechanical agitator is frequently used.

Mechanism. The NO$_2^+$ mechanism has been accepted since about 1950 for the nitration of most aromatic hydrocarbons, glycerol, glycols, and numerous other hydrocarbons in which mixed acids or highly concentrated nitric acid are used. The mechanism has been discussed in detail and critically analyzed (1). NO_2^+ attacks an aromatic compound (ArH) as follows:

$$ArH + NO_2^+ \longrightarrow \left[Ar \begin{array}{c} H \\ \diagup \\ \diagdown \\ NO_2 \end{array} \right]^+ \longrightarrow ArNO_2 + H^+$$

For an alcohol, glycol, or glycerol, or for amines, the reaction may be represented:

$$ROH + NO_2^+ \longrightarrow \left[RO \begin{array}{c} H \\ \diagup \\ \diagdown \\ NO_2 \end{array} \right]^+ \longrightarrow RONO_2 + H^+$$

$$RNHR' + NO_2^+ \longrightarrow \left[\begin{array}{c} NO_2 \\ | \\ RNH \\ | \\ R' \end{array} \right] \longrightarrow RN(NO_2)R' + H^+$$

When sulfuric acid is present in the mixed acids, the following ionization reactions occur. These ionic reactions are rapid, and equilibrium concentrations of NO_2^+ are likely to be present at all times in the acid phase. NO_2^+ concentrations depend mainly on the composition of the mixed acids but decrease to some extent as the temperature increases (3).

$$H_2SO_4 + HNO_3 \rightleftharpoons NO_2^+ + HSO_4^- + H_2O \tag{1}$$

$$H_2SO_4 + H_2O \rightleftharpoons HSO_4^- + H_3O^+ \tag{2}$$

$$2\,HNO_3 \rightleftharpoons NO_2^+ + NO_3^- + H_2O \tag{3}$$

$$HNO_3 + H_2O \rightleftharpoons NO_3^- + H_3O^+ \tag{4}$$

Figure 1 indicates how the NO_2^+ concentrations vary at 20°C as a function of the molar composition of the acid mixture; these results were determined using Raman spectra readings.

To model previously researched NO_2^+ data (3), equilibrium constants were first calculated for equations 1–3. At 20°C, $K_1 = 0.0622$, $K_2 = 211.7$, and $K_3 = 154 \times 10^{-6}$. The experimental data are predicted in general within the bounds of experimental accuracy. Because equation 4 is not an independent equation in this set, K_4 can be calculated at 20°C as 0.524.

Benzene, toluene, and other aromatics that are easily nitrated can sometimes be nitrated using acids having zero NO_2^+ concentrations (Fig. 1). Two explanations for this are (1) NO_2^+ is actually present but in concentrations too low

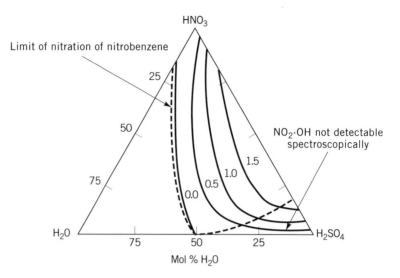

Fig. 1. Concentration of NO_2^+ in mol/1000 g solution. NO_2^+ concentrations increase with decreased amounts of water. The highest NO_2^+ concentrations occur at approximately 2:1 molar ratios of H_2SO_4 to HNO_3. At high concentrations of H_2SO_4, almost all of the HNO_3 is ionized to form NO_2^+. About 3% of pure HNO_3 ionizes to produce NO_2^+.

to be measured by Raman spectra, and (2) NO_2^+ is hydrated to form $H_2NO_3^+$, which is also a nitrating agent.

A considerable body of experimental results have been reported when N_2O_5–HNO_3 and N_2O_5–solvent mixtures were used as nitrating agents (4,5). NO_2^+ is produced in such mixtures as follows:

$$N_2O_5 \rightleftharpoons NO_2^+ + NO_3^-$$

The overall nitration reaction between an aromatic and N_2O_5 mixtures is as follows:

$$ArH + N_2O_5 \longrightarrow ArNO_2 + HNO_3$$

Mixtures of HNO_3, H_2SO_4, and SO_3 also result in high concentrations of NO_2^+, and toluene can be readily nitrated at -40 to $-10°C$ as a result (6). At these low temperatures, the formation of the meta-isomer of mononitrotoluene (MNT) is greatly reduced. Such a reduction is highly desired in the production both of dinitrotoluenes (DNTs) employed to produce intermediates for polyurethane production and of trinitrotoluene (TNT), which is a high explosive. m-MNT results in the production of undesired DNT and TNT isomers (see NITROBENZENE AND NITROTOLUENES).

Substituted aromatics, eg, alkylbenzenes, sometimes experience attack at the substituent position by NO_2^+ (7). A cyclohexadienyl cation is formed; it is unstable and the nitro group migrates on the ring to a carbon atom that is attached to a hydrogen. Loss of the proton results in a stable nitroaromatic.

In addition to the conventional mixed acids commonly used to produce DNT, a mixture of NO_2 and H_2SO_4 (8), a mixture of NO_2 and oxygen (9), and just HNO_3 (10) can also be used. Terephthalic acid and certain substituted aromatics are more amenable to nitrations using HNO_3, as compared to those using mixed acids. For compounds that are easily nitratable, acetic acid and acetic anhydride are sometimes added to nitric acid (qv). Acetyl nitrate, which is a nitrating agent, is produced as an intermediate as follows:

$$(CH_3CO)_2O + HNO_3 \longrightarrow CH_3CONO_3 + CH_3COOH$$

Acetic anhydride and acetic acid increase the solubility of the two phases in each other, and they are employed for the commercial N-nitration of hexamethylene-tetramine [100-97-0] (11) to form cyclotrimethylenetrinitramine [121-82-4] (RDX), $(CH_2)_3(NNO_2)_3$. Renewed consideration has been given to replacing H_2SO_4 with an improved solid catalyst to reduce the environmental problems of disposal or reconcentration of the waste acid and to increase production of desired nitrated isomers. For example, a catalyst with suitable pore size might increase the production of 4-MNT and reduce that of 3-MNT when toluene is nitrated.

The nitronium-ion mechanism predicts reasonably well most aspects of aromatic nitrations including isomer distribution, kinetics, and relative rates in mixtures (1). Isomer distribution, eg, for mononitration of toluene, is predicted well. However, the distribution is difficult to explain when the mixed acids contain ≥90 wt % H_2SO_4. In general, the rates of nitration are essentially proportional to the NO_2^+ concentrations (Fig. 1). The relative rates of nitration when mixtures of aromatics are used are in accord with substituents on the aromatic ring affecting the reactivity of the compound because of polar and steric factors. Aromatics with attached alkyl groups, eg, toluene, ethylbenzene, or cumene, are more reactive than benzene. An alkyl group increases the electron density of the ring at the ortho and para positions. As a result, NO_2^+ reacts predominately at these positions of the ring. Mass-transfer effects, and hence agitation, however, affect the relative importance of isomers produced (12). Aromatics with attached nitro groups, eg, mononitrobenzene (MNT), are always nitrated with considerably more difficulty than the unnitrated aromatics. Attached nitro groups reduce the electron density of the ring and are meta-directing. Generally, an aromatic compound is almost completely mononitrated before being di- or trinitrated, since more stringent conditions are required for the subsequent nitration steps than for the initial one.

Nitrous acid or nitrite salts may be used to catalyze the nitration of easily nitratable aromatic hydrocarbons, eg, phenol or phenolic ethers. It has been suggested that a nitrosonium ion (NO^+) attacks the aromatic, resulting initially in the formation of a nitrosoaromatic compound (13). Oxidation of the nitrosoaromatic then occurs:

$$ArNO + HNO_3 \longrightarrow ArNO_2 + HNO_2$$

The nitrosonium ion is produced from nitrous acid and nitric acid.

$$HNO_2 + HNO_3 \rightleftharpoons NO^+ + NO_3^- + H_2O$$

Some investigators, however, believe NO_2^+ is the nitrating agent for these easily nitrated aromatics.

Numerous by-products have been detected during aromatic nitrations (14,15). Two types of oxidations involving HNO_3 occur during toluene nitrations: (1) oxidation of the methyl group and (2) ring oxidations or decompositions yielding gaseous by-products. Ring decompositions are reported (16) to occur primarily in the acid phase and are the principal method of oxidation during the manufacture of TNTs. However, numerous oxidations also occur in the hydrocarbon phase owing to the solubility of HNO_3 in nitroaromatics. HNO_3 solubility in general increases as the amounts of DNTs and especially TNTs increase in the organic phase, and as the concentrations of HNO_3 increase in the acid phase. H_2SO_4 and water are, however, much less soluble in the organic phase. Dissolved HNO_3, which is essentially anhydrous, is an effective oxidizing agent especially at higher temperatures. Oxidation-type by-products include benzaldehyde, nitrobenzaldehydes, nitrophenols, nitrocresols, nitrobenzoic acid, tetranitromethane, nitrous acid, NO_x, and water. Condensation by-products are also often produced in significant quantities. The white compound produced in TNT processes is a solid and is an example of condensation.

Zeolites have recently been employed as solid catalysts for the vapor-phase nitration of aromatics with nitric acid. Additional research is required to improve yields and to minimize loss of catalytic activity as the nitration progresses (see MOLECULAR SIEVES).

Kinetics of Aromatic Nitrations. The kinetics of aromatic nitrations are functions of temperature, which affects the kinetic rate constant, and of the compositions of both the acid and hydrocarbon phases. In addition, a larger interfacial area between the two phases increases the rates of nitration since the main reactions occur at or near the interface. Larger interfacial areas are obtained by increased agitation and by the proper choice of the volumetric % acid in the liquid–liquid dispersion. At high % acid, the dispersions are acid-continuous, but at low % acid, they are hydrocarbon-continuous. Unfortunately no quantitative evidence is available on how the interfacial area varies as the % acid in the dispersion changes. Research indicates, however, that at least for some dispersions, a maximum interfacial area occurs in both the acid-continuous and the hydrocarbon-continuous regions, but not at the boundary between the two (17). The viscosities and densities of the two phases and the interfacial tension between the phases are important physical properties affecting the interfacial area. Such properties are, of course, dependent on both temperature and the respective compositions of the phases. Temperature also changes the solubilities of various compounds in either the acid or hydrocarbon phase (18,19). Such dissolved compounds often result in by-product formation.

Rates of nitration determined over a range of temperatures in two-phase dispersions have been used to calculate energies of activation from 59–75 kJ/mol (14–18 kcal/mol). Such energies of activation must be considered as only apparent, since the true kinetic rate constants, NO_2^+ concentrations, and interfacial area all change as temperature is increased.

Increased agitation of a given acid–hydrocarbon dispersion results in an increase in interfacial areas owing to a decrease in the average diameter of the dispersed droplets. In addition, the diameters of the droplets also decrease to relatively low and nearly constant values as the volume % acid in the dispersions approaches either 0 or 100%. As the droplets decrease in size, the ease of separation of the two phases, following completion of nitration, also decreases.

Industrial Applications. Significant process changes have occurred in many nitration plants. Batch nitrations were used almost exclusively in the 1940s, but continuous-flow units are widely used in the 1990s, especially in plants having relatively large production capacities. A well-designed continuous-flow plant often offers all of the following advantages, per unit weight of product, as compared to batch units: increased safety, decreased energy requirements, reduced amounts of undesired by-products, fewer environmental problems, reduced labor requirements, and lower operating expenses.

Many nitrated products are explosives, including DNT, TNT, and nitroglycerine (NG). At least some mononitroaromatics can also be exploded under certain conditions (20). Because of the high heats of nitration, runaway reactions followed by severe explosions have occurred in industrial batch nitrators. To minimize these potential hazards, the compositions of the feed acids and reaction conditions are currently better controlled than formerly. The objective is to so operate that most of the HNO_3 reacts within the reactor, and the resulting used acid is mainly a mixture of H_2SO_4 and water. In some processes, 99% or more of the feed HNO_3 reacts. Dispersions (or mixtures) of such a waste acid and the nitration product are relatively safe to handle.

Considerable effort has been made to minimize energy requirements in the nitration plants. For example, both Chemetics, Inc. (21) and Josef Meissner GmbH and Co. (22) have developed adiabatic processes for the production of nitrobenzene (NB). The heat of nitration is used to provide energy for preheating the feed streams and for flashing of the used acid phase. In the Chemetics process, mixed acids and a 10% excess of benzene are fed to the bottom of a series of well-agitated reactors. The temperature increases to a considerable extent because of the nitration. Over 99.5% of the nitric acid reacts, and about 90% of the benzene is converted to NB. The two phases are separated, and the hydrocarbon phase is distilled to separate NB and benzene. The hot acid phase containing mainly H_2SO_4 and water is flashed at subatmospheric pressure to obtain an acid containing up to 70% H_2SO_4. Only a small amount of heat needs to be added to reconcentrate the acid. The reconcentrated acid is combined with feed HNO_3 and is then recycled. Six such plants have been built, including the world's largest NB plant. In Meissner units, benzene and mixed acids (a mixture of 98% HNO_3 and 98% H_2SO_4) are contacted in a special mixing device and the dispersion then flows to a separation device. The nitration in this unit is apparently completed within, at most, several seconds.

Centrifugal separators are used in many modern processes to rapidly separate the hydrocarbon and used acid phases. Rapid separation greatly reduces the amounts of nitrated materials in the plant at any given time. After an explosion in a TNT plant (16), decanters (or gravity separators) were replaced with centrifugal separators. In addition, rapid separation allows the hydrocarbon phase to be quickly processed for removal of the dissolved nitric acid, NO_x, etc. These dissolved materials lead to undesired side reactions. The organic phase generally contains some unreacted hydrocarbons in addition to the nitrated product.

A significant concern in all nitration plants using mixed acids centers on the disposal method or use for the waste acids. They are sometimes employed for production of superphosphate fertilizers. Processes have also been developed to reconcentrate and recycle the acid. The waste acid is frequently first stripped with steam to remove unreacted HNO_3 and NO_x. Water is then removed by low pressure evaporization or vacuum distillation. The resulting acid can often be concentrated to 93–96% H_2SO_4 at the nitration plant. When the waste acid contains large amounts of dissolved organics, as occurs in a few nitration plants, the acid is then often converted at high temperatures, in the presence of air, to mainly SO_2 plus carbon oxides. The SO_2 is then oxidized to SO_3 and converted back to concentrated H_2SO_4.

DNT processes have been developed by Meissner (22), Chematur Engineering (23), and Biazzi (24). In all of the processes, one reactor (or a series of reactors) is used for the production of MNT and another for DNT production. In each process, highly concentrated H_2SO_4 and HNO_3 are used to prepare the mixed acids for the nitration of MNT, to produce DNTs. The acid mixture present after the dinitration step contains mainly H_2SO_4 and water. It is mixed with concentrated HNO_3, and the resulting mixed acids are used for the nitration of toluene to produce MNTs. The used acid mixture after the mononitration step generally contains 70–72% H_2SO_4, 0.1–0.5% HNO_3, and NO_x; small amounts of dissolved organics, including MNTs and DNTs; and the remainder water. In the Meissner process, a single mixing device with low residence time is used for the mononitration reactor, and a second one for the dinitration reactions. In the Chematur process, a single pump reactor is used for each nitrating step. This reactor provides intense mixing of the two phases so that the nitrations are completed in less than one second. Chematur claims that its pump reactor provides much better mixing of the two phases, compared to an agitated autoclave reactor. In the Biazzi process, continuous-flow agitated reactors are provided, and heat-transfer surfaces are used to remove the heats of nitration. More than one reactor in series is employed by Biazzi for each nitration step.

Nitroglycerin (NG) processes of Meissner (22) and Chematur (23,25) employ an injection nozzle for the intimate mixing and hence rapid nitration of the glycerol and mixed acids. In its NG process, Biazzi (24) uses a reactor of similar design to that used in its DNT process. All three companies, however, employ just one reactor for the nitration of glycerol. The feed acids employed for these NG processes contain at most only a small amount of water. The used acids produced contain typically 70–72% H_2SO_4, 8–12% HNO_3, 0.5–3.0% NG, small amounts of NO_x, and 12–18% water. In these processes, as well as the DNT processes, the amount of explosive materials in the reactor is low, especially in those reactors which require only a small residence time to complete the nitrations. These processes are highly automated, which reduces labor costs and also serves to minimize the dangers to plant operators.

Flow processes are also available (11,22,23,26) for the production of cellulose nitrates (see CELLULOSE ESTERS, INORGANIC). The cellulose repeating unit

($C_6H_{10}O_5$) contains three hydroxyl groups, all of which can be nitrated. The number of groups which are nitrated (or esterified) determines whether the product is used as a plastic, as lacquer, or as gunpowder. Operating conditions are selected that reduce the molecular weight of the cellulose to only a limited extent during nitration. When the cellulose linters (or fibers) are contacted with concentrated mixed acids, diffusion of the acids into the fibers and of water out of the fibers occurs. NO_2^+ concentrations expectably vary as a function of residence time and radial position in the fibrous material.

Nitrations Using N_2O_5.   Considerable worldwide interest has occurred in the late 1980s and the early and mid-1990s for nitrations using N_2O_5, which itself is generated by two processes. First, an electrolysis process is used with a solution of HNO_3 and N_2O_4 to produce a solution of HNO_3 and N_2O_5 (27–29). Second, a solution of an inert organic solvent such as CH_2Cl_2 and N_2O_4 is contacted with ozone to produce a solution containing N_2O_5 (5). A large pilot plant has been used to investigate nitrations using N_2O_5–HNO_3 solutions. Production of nitramines (or N-nitrations) is particularly promising, since these compounds are more stable in the presence of N_2O_5–HNO_3 solutions as compared to mixed acids containing H_2SO_4 (5). Good results have been obtained for the production of the high explosives, cyclotetramethylenetetranitramine [2691-41-0] or (HMX) (30) and DADN. Another high explosive, polynitrofluorene, has been produced via C-nitrations, for the first time. The overall exothermicities of the reactions are less when N_2O_5 is used, as compared to mixed acids (5). When N_2O_5 is employed, part of N_2O_5 is generally converted to HNO_3. With mixed acids, the water produced is dissolved with high exothermicity in H_2SO_4.

Solutions of CH_2Cl_2 and N_2O_5 have only mild nitrating power. Yet some nitrations are rapid: the N_2O_5 reacts on an almost stoichiometric basis, and only minimal residual acid is present upon completion of the nitration. Some nitrations having unique characteristics can be accomplished (5). For example, organic compounds with three- or four-atom rings which contain either oxygen or nitrogen heteroatoms nitrate readily. The rings are opened, and a nitrate or nitramine group is formed at each end of the resulting molecules. Essentially no water is produced as a by-product. For ethylene oxide, $NO_2OCH_2CH_2ONO_2$ is formed. Certain plasticizers and melt-castable explosives have also been produced, some for the first time.

Free-Radical Nitrations of Paraffins

Both vapor-phase and liquid-phase processes are employed to nitrate paraffins, using either HNO_3 or NO_2. The nitrations occur by means of free-radical steps, and sufficiently high temperatures are required to produce free radicals to initiate the reaction steps. For liquid-phase nitrations, temperatures of about 150–200°C are usually required, whereas gas-phase nitrations fall in the 200–440°C range. Sufficient pressures are needed for the liquid-phase processes to maintain the reactants and products as liquids. Residence times of several minutes are commonly required to obtain acceptable conversions. Gas-phase nitrations occur at atmospheric pressure, but pressures of 0.8–1.2 MPa (8–12 atm) are frequently employed in industrial units. The higher pressures expedite the condensation and recovery of the nitroparaffin products when cooling water is employed to cool the product gas stream leaving the reactor (see NITROPARAFFINS).

Chemistry. Free-radical nitrations consist of rather complicated nitration and oxidation reactions (31). When nitric acid is used in vapor-phase nitrations, the reaction of equation 5 is the main initiating step where NO_2 is a free radical, either $\cdot NO_2$ or $\cdot ONO$. Temperatures of $>$ca 350°C are required to obtain a significant amount of initiation, and equation 5 is the rate-controlling step for the overall reaction. Reactions 6 and 7 are chain-propagating steps.

$$HNO_3 \longrightarrow \cdot OH + NO_2 \tag{5}$$

$$RH + \cdot OH \longrightarrow R\cdot + H_2O \tag{6}$$

$$R\cdot + HNO_3 \longrightarrow RNO_2 + \cdot OH \tag{7}$$

When nitrogen dioxide is used, the main reaction steps are as in equations 8 and 9.

$$RH + NO_2 \longrightarrow R\cdot + HNO_2 \longrightarrow R\cdot + \cdot OH + NO \tag{8}$$

$$R\cdot + \cdot NO_2 \longrightarrow RNO_2 \tag{9}$$

These reactions occur as low as 200°C. The exact temperature depends on the specific hydrocarbon that is nitrated, and reaction 8 is presumably the rate-controlling step. Reaction 9 is of minor importance in nitration with nitric acid, as indicated by kinetic information (32).

An important side reaction in all free-radical nitrations is reaction 10, in which unstable alkyl nitrites are formed (eq. 10). They decompose to form nitric oxide and alkoxy radicals (eq. 11) which form oxygenated compounds and low molecular weight alkyl radicals which can form low molecular weight nitroparaffins by reactions 7 or 9. The oxygenated hydrocarbons often react further to produce even lighter oxygenated products, carbon oxides, and water.

$$R\cdot + \cdot ONO \longrightarrow RONO \tag{10}$$

$$RONO \longrightarrow NO + RO\cdot \tag{11}$$

Reactions 8 and 9 are important steps for the liquid-phase nitration of paraffins. The nitric oxide which is produced is oxidized with nitric acid to reform nitrogen dioxide, which continues the reaction. The process is complicated by the presence of two liquid phases; consequently, the nitrogen oxides must transfer from one phase to another. A large interfacial area is needed between the two phases.

Dinitroparaffins, such as 2,2-dinitropropane, can be produced during liquid-phase nitrations which are at relatively low temperatures. Dinitroparaffins have apparently never been produced, however, with gas-phase nitrations which are at higher temperatures.

Nitroparaffins are produced that contain either the same or fewer number of C atoms as compared to the feed n-paraffin. When propane is nitrated, 1-nitropropane, 2-nitropropane, nitroethane, and nitromethane are all produced, in addition to a significant amount of oxygenated hydrocarbons (containing one or two carbon atoms) and carbon oxides. When nitric acid is the nitrating agent at 410–420°C, the nitroparaffin mixture contains approximately 25, 40, 10, and 25 wt %, respectively, of these four nitroparaffins (33). For nitrations of propane with NO_2 at 280°C, the relative amounts are about 14, 52, 9, and 25 wt %, respectively (34). When n-butane is nitrated, both 1-nitro- and 2-nitrobutane are produced, in addition to the above four nitroparaffins. When the nitrating temperature is lowered, the average molecular weight of the nitroparaffin mixture increases, ie, fewer C–C bonds are broken. Side reactions which occur in the reactor, but have not been shown in equations 5–11, include the attack by free radicals and subsequent reaction of nitroparaffins initially produced. At lower temperatures such as those used for NO_2 nitrations, some higher molecular weight nitroparaffins are converted to lower molecular weight ones. At higher temperatures as used for HNO_3 nitrations, the side reactions lead mainly to the destruction of nitroparaffins.

Only 20–40% of the HNO_3 is converted in the reactor to nitroparaffins. The remaining HNO_3 produces mainly nitrogen oxides (and mainly NO) and acts primarily as an oxidizing agent. Conversions of HNO_3 to nitroparaffins are up to about 20% when methane is nitrated. Conversions are, however, often in the 36–40% range for nitrations of propane and n-butane. These differences in HNO_3 conversions are explained by the types of C–H bonds in the paraffins. Only primary C–H bonds exist in methane and ethane. In propane and n-butane, both primary and secondary C–H bonds exist. Secondary C–H bonds are considerably weaker than primary C–H bonds. The kinetics of reaction 6 (a desired reaction for production of nitroparaffins) are hence considerably higher for both propane and n-butane as compared to methane and ethane. Experimental results also indicate for propane nitration that more 2-nitropropane [79-46-9] is produced than 1-nitropropane [108-03-2]. Obviously the hydroxyl radical attacks the secondary bonds preferentially even though there are more primary bonds than secondary bonds.

Conversions per pass of NO_2 to nitroparaffins tend to be significantly less than when HNO_3 is used. When propane is nitrated with NO_2, conversions are as high as 27%, but they are much less for nitrations of methane and ethane. The remaining NO_2 reacts mainly to produce NO, and a considerable number of oxidation steps occur. The theoretically maximum conversion of NO_2 to nitroparaffins is 66.7%.

$$2\,RH + 3\,NO_2 \longrightarrow 2\,RNO_2 + H_2O + NO$$

HNO_3 conversions to nitroparaffins pass through a maximum at paraffin:HNO_3 molar ratios of approximately 4:1 to 6:1 (32). At higher ratios, a high fraction of the HNO_3 reacts to form alkyl free radicals. At lower ratios, a large fraction of HNO_3 decomposes, as in reaction 5.

Attempts have been made to increase the conversions of both HNO_3 and NO_2 to nitroparaffins by adding oxygen, ozone, or halogens such as chlorine or bromine to the feed mixture (35–38). Significantly higher conversions, often in the 50–70% range, have been obtained. Such additives promote the formation of free radicals, but increase the production of oxygenated or halogenated compounds. These additives have apparently never been employed commercially.

When cyclohexane is nitrated to produce nitrocyclohexane [1122-60-7], the following techniques minimize undesired C–C breakage in the ring: low temperature nitrations with NO_2 (31), careful control of the reactor temperatures, and use of halogen additives. Oxygen increases the level of C–C bond breakage.

Processes for Paraffin Nitrations.   Propane is thought to be the only paraffin that is commercially nitrated by vapor-phase processes. Temperature control is a primary factor in designing the reactor, and several approaches have been investigated (32). Excess amounts of propane are used to help provide a heat sink and help moderate the temperature. As already indicated, excess but controlled amounts of propane also increase the HNO_3 conversions to nitroparaffins. A significant amount of steam is also present in the reactor as a heat sink whenever HNO_3 is used as the nitrating agent. Between 60–70% HNO_3 (and the remainder water) is commonly used as the feed acid. Reactors in laboratory units have often been designed for good heat transfer. Relatively small-diameter tubular reactors, fluidized-bed reactors, and molten salt reactors have all been successfully used.

In a commercial unit, a spray nitrator (39) is operated adiabatically. The liquid HNO_3 feed is sprayed directly into the hot and preheated propane feed. The heat of nitration provides the heat to vaporize the HNO_3 and to preheat it to the desired temperature for nitration. At one time, several spray nitrators were operated in series, with additional HNO_3 being sprayed into each nitrator (32). In such an arrangement, the optimum propane:HNO_3 ratios did not occur, and considerable amounts of nitroparaffins degraded.

For vapor-phase processes, the product stream from the nitrator must be separated. The nitroparaffins, excess propane, and NO plus NO_2 (which are converted back to HNO_3), are recovered. The oxygenated products are removed, but there are generally insufficient amounts for economic recovery.

The vapor-phase process of Société Chemique de la Grande Paroisse for production of nitroparaffins employs propane, nitrogen dioxide, and air as feedstocks (34). The yields of nitroparaffins based on both propane and nitrogen dioxide are relatively high. Nitric oxide produced during nitration is oxidized to nitrogen dioxide, which is adsorbed in nitric acid. Next, the nitric dioxide is stripped from the acid and recirculated.

Nitromethane [75-52-5] is produced in China. Presumably a modified Victor Meyer method is being employed. Nitromethane is transported in drums or smaller containers. Two tank cars of nitromethane exploded in separate incidents in the 1950s. Both explosions occurred in the switching yard of a railroad station. In both cases, essentially adiabatic vapor compression of the nitromethane–air mixture in the gas space of the tank car resulted in the detonation of the liquid nitromethane. Other nitroparaffins do not, however, detonate in this manner.

Health and Safety Factors

The danger of an explosion of a nitrated product generally increases as the degree of nitration increases, eg, trinitroaromatics are more hazardous as compared to dinitroaromatics or especially mononitroaromatics. Nitroaromatics and some polynitrated paraffins are highly toxic when inhaled or when contacted with the skin. All nitrated compounds tend to be highly flammable.

BIBLIOGRAPHY

"Nitration" in *ECT* 1st ed., Vol. 9, pp. 314–330, by Willard deC. Crater, Hercules Powder Co.; in *ECT* 2nd ed., Vol. 12, pp. 784–796, by Lyle F. Albright, Purdue University; in *ECT* 3rd ed., Vol. 15, pp. 841–853, by Lyle F. Albright, Purdue University.

1. G. S. Olah, R. Malhotra, and S. C. Narang, *Nitration: Methods and Mechanism*, VCH Publishers, New York, 1989.
2. C. Hanson and H. A. M. Ismail, *Chem. Eng. Sci.* **32**, 775 (1977).
3. M. B. Zaman, *Nitronium Ions in Nitrating Acid Mixtures*, Ph.D. dissertation University of Bradford, U.K., 1972.
4. J. F. Fischer, in H. Feuer and A. T. Nielsen, eds., *Nitro Compounds: Recent Advances in Synthesis and Chemistry*, VCH Publishers, New York, 1990, Chapt. 3.
5. G. E. Bagg, personal communication, Defense Research Agency, Fort Halstead, Kent, U.K., 1993.
6. M. E. Hill and co-workers, in *ACS Symposium Series No. 22*, American Chemical Society, Washington, D.C., 1976, Chapt. 17, pp. 253–271.
7. P. C. Myhre, in Ref. 6, Chapt. 4, pp. 87–94.
8. U.S. Pat. 4,123,466 (Oct. 31, 1978), C. Y. Lin, F. A. Stuber, and H. Ulrich (to the Upjohn Co.).
9. U.S. Pat. 4,028,425 (June 7, 1977), E. Gilbert (to United States of America).
10. U.S. Pat. 4,918,250 (Apr. 17, 1990) and 5,001,272 (Mar. 19, 1991), R. W. Mason, P. C. Imm, and K. J. Bordelon (to Olin Corp.).
11. T. Urbanski, *Chemistry and Technology of Explosives*, Vols. 1–3, The Macmillan Co., New York, 1964, 1965; Vol. 4, Permagon Press, Elmsford, N.Y., 1983.
12. B. Milligan, *IEC Fundamentals* **25**, 83 (1986).
13. C. A. Bunton and co-workers, *J. Chem. Soc.*, 2628 (1950).
14. L. F. Albright, D. F. Schiefferle, and C. Hanson, *J. Appl. Chem. Biotechnol.* **26**, 522 (1976).
15. H. Suzuki, *Synthesis*, 217 (Apr. 1977).
16. E. Gilbert, in S. M. Kaye, ed., *Encyclopedia of Explosives and Related Items*, Vol. 9, U.S. Army Armament Research and Development Command, Dover, N.J., 1980, T235–286.
17. D. J. am Ende, Ph.D. dissertation, Purdue University, Ind., 1993.
18. C. Hanson and H. A. M. Ismail, *J. Appl. Chem. Biotechnol.* **25**, 319 (1975).
19. D. F. Schiefferle, C. Hanson, and L. F. Albright, in Ref. 6, Chapt. 11, pp. 176–189.
20. R. C. Dartnell and T. A. Ventrone, *Chem. Eng. Prog.* **67**(6), 58 (1971).
21. A. Tisan, personal communication, Chemetics International Co. Ltd., Vancouver, B.C., Canada, 1993.
22. G. Pelster and G. Langecker, personal communication, Josef Meissner GmbH and Co., Köln, Germany, 1993.
23. D. G. Klima, personal communication, Chematur Engineering, Karlskoga, Sweden, 1993.
24. Technical data, Biazzi, SA, Chailly/Montreux, Switzerland, 1993.
25. B. Brunnberg, in Ref. 6, Chapt. 24, pp. 341–343.
26. Ref. 16, N62–68.
27. U.S. Pat. 4,443,308 (Apr. 17, 1984), C. C. Coon, J. E. Harrar, R. K. Pearson, and R. R. McGuire (to United States of America).

28. U.S. Pat. 5,120,408 (June 9, 1992), J. R. Marshall, D. J. Schriffin, F. C. Walsh, and G. E. Bagg (to United Kingdom).

29. U.S. Pat. 5,181,996 (Jan. 26, 1993), G. E. Bagg (to United Kingdom).

30. U.S. Pat. 4,432,902 (Feb. 21, 1984), R. R. McGuire, C. C. Coon, J. E. Harrar, and R. K. Pearson (to United States of America).

31. R. Lee and L. F. Albright, *Ind. Eng. Chem. Proc. Des. Dev.* **4**, 411 (1965).

32. L. F. Albright, *Chem. Eng.* **73**(12), 149 (1966).

33. G. B. Bachman, L. M. Addison, J. V. Hewett, L. Kohn, and A. Millikan, *J. Org. Chem.* **17**, 906 (1952).

34. U.S. Pat. 3,780,115 (Dec. 18, 1973), P. L'Honore, G. Cohen, and J. Jacquinot (to Société Chemique de la Grande Paroisse).

35. G. G. Bachman, H. B. Hass, and L. M. Addison, *J. Org. Chem.* **17**, 914 (1952).

36. G. B. Bachman, H. B. Hass, and J. V. Hewett, *J. Org. Chem.* **17**, 928 (1952).

37. G. B. Bachman, J. V. Hewett, and A. Millikan, *J. Org. Chem.* **17**, 935 (1952).

38. G. B. Bachman and L. Kohn, *J. Org. Chem.* **17**, 942 (1952).

39. U.S. Pat. 2,418,241 (Apr. 1947), L. A. Stengel and R. G. Egly (to Commercial Solvents Corp.).

Lyle F. Albright
Purdue University

NITRIC ACID

Nitric acid [7697-37-2], HNO_3, also known as *aqua fortis*, azotic acid, hydrogen nitrate, or nitryl hydroxide, is a chemical of major industrial importance. As of 1993, it ranked thirteenth in production volume for chemicals made in the United States (ca 7 million t/yr for 1989–93); production in western Europe is estimated to be about 19 million t/yr. Because of its properties as a very strong acid and a powerful oxidizing agent, as well as its ability to nitrate organics, nitric acid is essential in the production of many chemicals (eg, pharmaceuticals, dyes, synthetic fibers, insecticides, and fungicides), but is used mostly in the production of ammonium nitrate for the fertilizer industry (see FERTILIZERS). By the end of the nineteenth century, its industrial importance had already become established in the production of explosives and dyestuffs. After World War II nitric acid production grew rapidly with the expanding use of synthetic fertilizers. Because of the increased popularity of urea as a fertilizer, production has leveled off in the 1990s. Most growth in demand has come from the production of polyurethanes, fibers, and ammonium nitrate-based explosives. Other uses for nitric acid are in the manufacture of explosives (trinitrotoluene, nitroglycerin, etc), metal nitrates, nitrocellulose, and nitrochlorobenzene, the treatment of metals (eg, the pickling of stainless steels and metal etching), as a rocket propellant, and for nuclear fuel processing.

The first reports of nitric acid have been credited to Arab alchemists of the eighth century. By the Middle Ages it was referred to as *aqua fortis* (strong water) or *aqua valens* (powerful water). From that time onward, nitric acid was produced primarily from saltpeter [7757-79-1] (potassium nitrate) and sulfuric acid. In the nineteenth century, Chilean saltpeter [7631-99-4] (sodium nitrate) from South America largely replaced potassium nitrate. However, at the beginning of the twentieth century newer manufacturing technologies were introduced. In Norway, where electricity was inexpensive, electric arc furnaces were used to make nitrogen oxides, and subsequently nitric acid, directly from air. The commercial life of these furnaces was relatively brief and most were shut down by 1930. At about the same time, a different production method was being developed. In 1908, at Bochum, Germany, Ostwald piloted a 3-t per day nitric acid process based on the catalytic oxidation of ammonia with air. In 1913 the synthesis of ammonia from coal, air, and water was successfully demonstrated using the Haber-Bosch process. With a secure and economical supply of ammonia, ammonia oxidation became firmly established as an industrial route to nitric acid manufacture. Process developments continued and plant scale increased to commercial quantities in both Europe and the United States. The first full-size plant to be built in the United States was installed in 1917 by Chemical Construction Co. (Muscle Shoals, Alabama). The process operated at atmospheric pressure and used multiple ammonia oxidation converters. Since those early days, ammonia oxidation has become the basis of all commercial nitric acid production. There have been many advances in plant design leading to improved process performance and higher production capacities at increased operating pressures. The plants of the 1990s have single-train capacities up to 2000 t/d and operate at pressures up to 1.5 MPa (14.8 atm). More details on the history of nitric acid and development of the manufacturing process are available (1–4).

In the modern ammonia oxidation process, most nitric acid is produced as a weak acid (50–65 wt %). High monopressure processes minimize capital investment, whereas split- or dual-pressure processes optimize ammonia conversion efficiency and catalyst use. Weak acid is suitable for use in the production of fertilizers, but stronger acid (up to 99 wt % acid) is required for many organic reactions of industrial importance. Direct strong nitric (DSN) processes make the acid directly from nitrogen oxides obtained by the oxidation of ammonia. Nitric acid concentration (NAC) processes use extractive distillation to concentrate the weak acid. A dehydrating agent such as sulfuric acid or magnesium nitrate is used to enhance the volatility of HNO_3 so that distillation methods can surpass the maximum boiling azeotrope of nitric acid. Several DSN processes can produce weak and strong acids simultaneously.

Physical Properties

Crystals of pure nitric acid are colorless and quite stable. Above the melting point of −41.6°C, nitric acid is a colorless liquid that fumes in moist air and has a tendency to decompose, forming oxides of nitrogen. The rate of decomposition is accelerated by exposure to light and increases in temperature. Depending on the concentration of dissolved nitrogen dioxide, the color may range from yellow to red. The normal boiling point of nitric acid is 83.4°C, but when heated the liquid gradually decomposes to form a maximum boiling azeotrope at 120°C and 69 wt % HNO_3 (5).

Nitric acid is completely miscible with water. The freezing point curve for aqueous solutions of nitric acid has two maxima corresponding to melting points for the two hydrates of nitric acid: the monohydrate (77.77 wt % acid) at -37.62°C and the trihydrate (53.83 wt % acid) at -18.47°C (6). Local minima occur at about 32, 71, and 91 wt % acid (7). There is some variability in the data available on partial pressures of acid and water vapor over solutions of nitric acid. This is most evident in the vapor pressure measurements of one component when the solution concentration of that component is very low. The decomposition of nitric acid at high temperatures and concentrations can also lead to some doubt about the accuracy of vapor pressure measurements. The vapor pressure and density of such acid-containing nitrogen dioxide (ie, fuming nitric acid) increase with the percentage of dioxide present. Data that have been subjected to thermodynamic tests and corroborated by the measurements of other studies are therefore the most reliable (5,8–11). The density, viscosity, and thermal conductivity of nitric acid solutions are given in Table 1. A detailed tabulation of densities is available (12). For all temperatures, density increases with acid concentration; viscosity reaches a maximum at ca 60–70 wt % acid (13).

Thermodynamic data for nitric acid are given in Table 2. Properties for the ternary systems sulfuric acid–nitric acid–water (5,14) and magnesium nitrate–nitric acid–water (11,15–17) used in processes for concentrating nitric acid are available.

Table 1. Physical Properties of Nitric Acid Solutions at 20°C

HNO_3, wt %	Bp^a at 101 kPa,^{b} °C	Partial pressures,^{c} kPa^b HNO_3	Partial pressures,^{c} kPa^b H_2O	Density,^{d} g/mL	Specific heat,^{e} J/(g·K)^f	Viscosity,^{e} mPa·s(=cP)	Thermal conductivity,^{e} W/(m·K)
0.0	100.0		2.27	0.9982	4.19	1.0	0.60
10.0	102.2		2.12	1.0543	3.73	1.1	0.57
20.0	104.4		1.91	1.1150	3.39	1.2	0.53
30.0	107.3		1.61	1.1800	3.18	1.4	0.50
40.0	110.8	0.01	1.28	1.2463	3.01	1.6	0.47
50.0	114.7	0.04	0.94	1.3100	2.85	1.9	0.43
60.0	118.2	0.12	0.61	1.3667	2.64	2.0	0.40
70.0	119.3	0.37	0.33	1.4134	2.43	2.0	0.37
80.0	112.1	1.22	0.12	1.4521	2.22	1.9	0.34
90.0	96.0	3.57	0.02	1.4826	1.97	1.4	0.31
100.0	83.4	6.20	0	1.5129	1.76	0.9	0.28

^{a}Ref. 5.
^{b}To convert kPa to mm Hg, multiply by 7.5.
^{c}Refs. 8 and 9.
^{d}Ref. 12.
^{e}Ref. 13.
^{f}To convert J to cal, divide by 4.184.

Table 2. Thermodynamic Properties of Nitric Acid and Its Hydrates

Property[a]	HNO_3	$HNO_3 \cdot H_2O$	$HNO_3 \cdot 3H_2O$
nitric acid, wt %	100.0	77.77	53.83
freezing point, °C[b]	−41.59	−37.62	−18.47
heat of fusion, kJ/mol[b]	10.48	17.5	29.12
heat of formation at 25°C, kJ/mol[c]	−174.10	−473.46	−1056.04
free energy of formation at 25°C, kJ/mol[c]	−80.71	−328.77	−811.09
entropy at 25°C, J/(mol·K)[c]	155.60	216.90	346.98
heat of vaporization at 25°C, kJ/mol[c]	39.04		

[a]To convert kJ to kcal, divide by 4.184.
[b]Ref. 6.
[c]Ref. 18.

Chemical Properties

Nitric acid is a strong monobasic acid, a powerful oxidizing agent, and nitrates many organic compounds. Until the end of the nineteenth century, it was made by heating a metallic nitrate salt with less volatile concentrated sulfuric acid. Removal of the volatile nitric acid permits the reaction to go to completion. This method is still used for laboratory preparation of the acid.

Acidic Properties. As a typical acid, it reacts readily with alkalies, basic oxides, and carbonates to form salts. The largest industrial application of nitric acid is the reaction with ammonia to produce ammonium nitrate. However, because of its oxidizing nature, nitric acid does not always behave as a typical acid. Bases having metallic radicals in a reduced state (eg, ferrous and stannous hydroxide becoming ferric and stannic salts) are oxidized by nitric acid. Except for magnesium and manganese in very dilute acid, nitric acid does not liberate hydrogen upon reaction with metals.

Oxidizing Properties. Nitric acid is a powerful oxidizing agent (electron acceptor) that reacts violently with many organic materials (eg, turpentine, charcoal, and charred sawdust) (19,20). The concentrated acid may react explosively with ethanol (qv). Such oxidizing properties have had military application: nitric acid is used with certain organics, eg, furfuryl alcohol and aniline, as rocket propellant (see EXPLOSIVES AND PROPELLANTS).

Depending on acid concentration, temperature, and the reducing agent involved, any of the following oxidations may occur:

$$4\ HNO_3 + 2\ e^- \longrightarrow 2\ NO_3^- + 2\ H_2O + 2\ NO_2 \tag{1}$$

$$8\ HNO_3 + 6\ e^- \longrightarrow 6\ NO_3^- + 4\ H_2O + 2\ NO \tag{2}$$

$$10\ HNO_3 + 8\ e^- \longrightarrow 8\ NO_3^- + 5\ H_2O + N_2O \tag{3}$$

$$10\ HNO_3 + 8\ e^- \longrightarrow 9\ NO_3^- + 3\ H_2O + NH_4^+ \tag{4}$$

$$16\ HNO_3 + 12\ e^- \longrightarrow 14\ NO_3^- + 4\ H_2O + 2\ NH_3OH^+ \tag{5}$$

Concentrated nitric acid favors the formation of nitrogen peroxide, whereas low strength acid favors the generation of nitric oxide. Concentrated acid converts the oxides, sulfides, etc, of most elements in a low oxidation state to a higher level, eg, sulfur dioxide is oxidized to sulfuric acid. As a general rule for metals, those below hydrogen in the electrochemical series yield nitrogen peroxide and nitric oxide. Those above hydrogen react to produce nitrogen, ammonia, hydroxyl amine, or nitric oxide when treated with nitric acid.

Most nonmetallic elements (except nitrogen, oxygen, chlorine, and bromine) are oxidized to their highest state as acids. Heated with concentrated acid, sometimes in the presence of a catalyst, sulfur, phosphorus, arsenic, and iodine form sulfuric, orthophosphoric, orthoarsenic, and iodic acid, respectively. Silicon and carbon react to produce their dioxides.

Nitric acid reacts with all metals except gold, iridium, platinum, rhodium, tantalum, titanium, and certain alloys. It reacts violently with sodium and potassium to produce nitrogen. Most metals are converted into nitrates; arsenic, antimony, and tin form oxides. Chrome, iron, and aluminum readily dissolve in dilute nitric acid but with concentrated acid form a metal oxide layer that passivates the metal, ie, prevents further reaction.

Organic Reactions. Nitric acid is used extensively in industry to nitrate aliphatic and aromatic compounds (21). In many instances nitration requires the use of sulfuric acid as a dehydrating agent or catalyst; the extent of nitration achieved depends on the concentration of nitric and sulfuric acids used. This is of industrial importance in the manufacture of nitrobenzene and dinitrotoluene, which are intermediates in the manufacture of polyurethanes. Trinitrotoluene (TNT) is an explosive. Various isomers of mononitrotoluene are used to make optical brighteners, herbicides (qv), and insecticides. Such nitrations are generally attributed to the presence of the nitronium ion, NO_2^+, the concentration of which increases with acid strength (see NITRATION).

$$HNO_3 + H^+ \rightleftharpoons NO_2^+ + H_2O \tag{6}$$

$$ArH + NO_2^+ \longrightarrow ArNO_2 + H^+ \tag{7}$$

Alcohols and glycerols are nitrated by esterification in a mixture of concentrated nitric and sulfuric acids. This reaction is of importance in the production of nitroglycerin from glycerol and nitrocellulose from cellulose.

$$ROH + HONO_2 \longrightarrow RONO_2 + H_2O \tag{8}$$

Dilute nitric acid can be used to oxidize an aliphatic hydrocarbon. For example, a significant use for nitric acid is the oxidation of cyclohexanol and cyclohexanone (qv) to produce adipic acid (qv). Most adipic acid is used for the production of nylon-6,6.

$$3\,C_6H_{11}OH + 3\,C_6H_{10}O + 14\,HNO_3 \longrightarrow 6\,HOOC(CH_2)_4COOH + 14\,NO + 10\,H_2O \tag{9}$$

Manufacture and Processing

Almost all commercial quantities of nitric acid are manufactured by the oxidation of ammonia with air to form nitrogen oxides that are absorbed in water to form nitric acid. Because nitric acid has a maximum boiling azeotrope at 69 wt %, the processes are usually categorized as either weak (subazeotropic) or direct strong (superazeotropic). Typically, weak processes make 50–65 wt % acid and direct strong processes make up to 99 wt % acid. Strong acid may also be made indirectly from the weak acid by using extractive distillation with a dehydrating agent. Nitric acid concentration processes use a dehydrating agent such as sulfuric acid or magnesium nitrate to enhance the volatility of HNO_3 so that distillation methods can surpass the azeotropic concentration of nitric acid.

WEAK ACID PROCESS

Historically, different design philosophies between the United States and Europe have led to the development of two basic types of weak acid plant: the high monopressure and the dual-pressure processes. The high monopressure process has been favored in the United States because of its lower capital cost and traditionally lower energy and ammonia prices. In Europe, where allowable capital payback periods and energy costs have traditionally been higher, the dual-pressure process evolved. In the 1990s, these processes continue to advance in design and are competitive on a worldwide basis (22). Many licensors offer both designs. Figures 1 and 2 show examples of the two kinds of processes. The different types of NO_x abatement systems illustrated in these diagrams are interchangeable for either process.

Monopressure Process. Monopressure processes are either medium pressure, 0.3–0.6 MPa (3–6 atm), or high pressure, 0.7–1.2 MPa (7–12 atm). The high pressure process has been the most prevalent design. Higher operating pressures reduce equipment size and capital cost. The capital cost of the high monopressure process is about 10–14% lower than that of the dual-pressure process. Higher gauze temperatures and operating pressures accommodate a more efficient recovery of process energy, either as steam or as reheated tail gas, providing power for air compression. One licensor (23) has found additional cost advantages by going to a vertical equipment layout, resulting in a smaller process footprint, less piping, and the elimination of weak acid pumps.

Dual-Pressure Process. Dual-pressure processes have a medium pressure (ca 0.3–0.6 MPa) front end for ammonia oxidation and a high pressure (1.1–1.5 MPa) tail end for absorption. Some older plants still use atmospheric pressure for ammonia conversion. Compared to high monopressure plants, the lower oxidation pressure improves ammonia yield and catalyst performance. Platinum losses are significantly lower and production runs are extended by a longer catalyst life. Reduced pressure also results in weaker nitric acid condensate from the cooler condenser, which helps to improve absorber performance. Due to the split in operating conditions, the dual-pressure process requires a specialized stainless steel NO_x compressor.

Process Description. Air is supplied to the process from a compressor that is powered by an expander and a makeup driver, all having a common shaft. The expander is a turbine that recovers energy from spent process gases as they are reduced to atmospheric pressure. The makeup driver, usually a steam turbine or electric motor, meets the balance of any power requirement for air compression. Ammonia and air are mixed such that there is an excess of oxygen and are passed over a platinum catalyst to produce nitric oxide, NO, water vapor, and

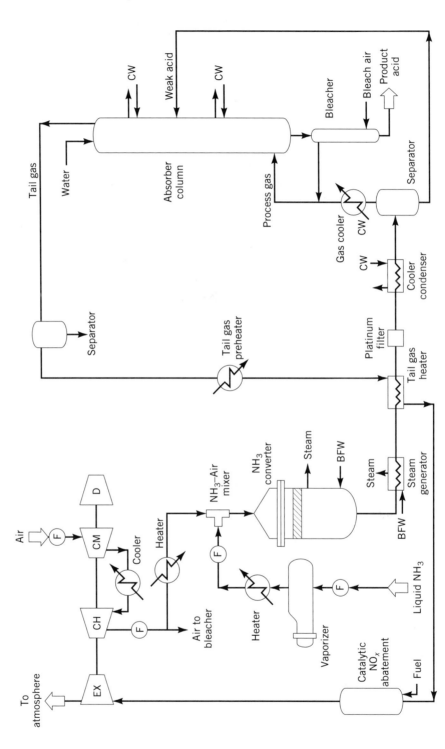

Fig. 1. Monopressure process using catalytic NO$_x$ abatement, where BFW = boiler feed water, CH = high level compression, CM = medium level compression, CW = cooling water, D = makeup driver, EX = expander, and F = filter.

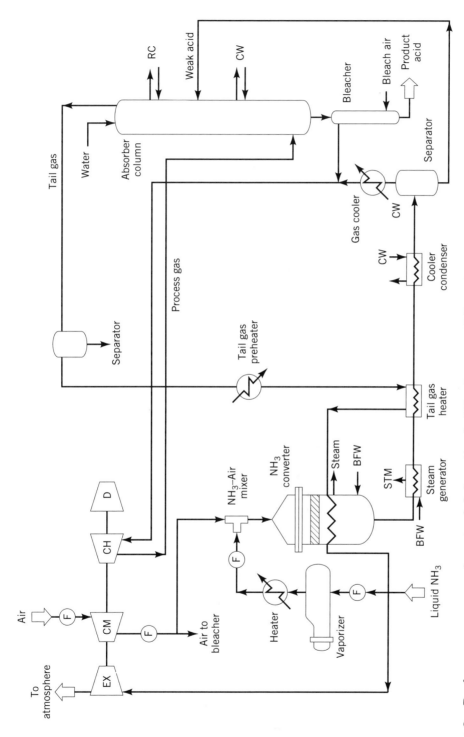

Fig. 2. Dual-pressure process using extended absorption for NO_x abatement. RC = refrigerated cooling; see Figure 1 for other definitions.

much heat. The resulting gases are cooled, thus generating steam that can be exported or used to power the steam turbine. As the process gases cool, nitric oxide is further oxidized to form nitrogen dioxide, NO_2, in equilibrium with its dimer, dinitrogen tetroxide, N_2O_4. Because hot liquid nitric acid is corrosive, the extent to which heat can be usefully recovered from the hot process gas in the heat train is limited by a need to remain above the dew point for HNO_3. After leaving the heat train, the process gases are further cooled in a cooler–condenser to remove condensate. This exchanger uses expensive materials of construction that are resistant to corrosion by hot acid. The process gases then enter a column where the equilibrium mix of NO_2 and N_2O_4, known as nitrogen peroxide, is absorbed into water, producing nitric acid. Nitric oxide, released by formation of the nitric acid, must be oxidized to complete the conversion of nitrogen oxides to nitric acid. Spent gases from absorption contain residual levels of nitrogen oxides, NO_x, which for environmental reasons have to be removed before discharge to the atmosphere.

Compression and Expansion.   For many reasons the compressor–expander set may be considered the heart of a nitric acid process. To a large extent it sets the energy efficiency and operating flexibility of the process and represents about 35% of total equipment cost. Air compressors are typically multistage centrifugal units with intercoolers, while for large-capacity plants axial flow compressors are often used for the initial stages of compression. A high pressure process may use as much as 470 kWh/t of HNO_3 for air compression, with over 80% of this energy provided by power recovery in the expander. For a given flow, the amount of energy available for recovery is set by its absolute temperature. The fraction of available energy actually recovered by an expander is a function of its polytropic or adiabatic efficiency and the pressure expansion ratio. The upper limit for temperature is primarily determined by an expander's mechanical design. Hot expanders are of a multistage gas turbine design and construction, permitting operation at temperatures up to 680°C, The less expensive cold expanders operate below ca 540°C and are of simpler design, eg, steam turbine designs modified for operation with nitric plant tail gas. Hot expansion maximizes the power provided by an expander for air compression. Being more energy efficient than a typical steam turbine makeup driver, a hot versus a cold expander can increase steam for export by as much as 0.5 t/t of HNO_3. Steam export can also be enhanced by replacing the steam makeup driver with an electric motor. Every ton of steam thus saved increases electrical consumption by ca 200 kWh.

Both the compressor and expander are carefully matched to the process and its design capacity. The rate of oxidation of nitric oxide dictates that production capacity is proportional to pressure cubed. The expander, however, varies plant operating pressure linearly with gas flow rate. Consequently, the range of energy efficient operation is limited and most nitric plants can turn down to only about 70% of design capacity. Reduced rates often require back-pressuring of the process to maintain a suitable operating pressure. This results in a loss of expander energy recovery. Maintaining pressure at reduced flow rate brings operation of the compressor closer to its surge point. A portion of compressed air is sometimes vented to atmosphere to increase flow rate and avoid surge conditions with a resulting loss in process energy efficiency. Various schemes have been used to increase the range of efficient turndown. Flow rate can be increased by bypassing some air around part of the process and into the expander (24). Selection of compressors and expanders with adjustable stators and nozzle sizes can also help extend the range of efficient turndown (25).

In the split- or dual-pressure process, low to medium pressure gases (ca 0.3–0.6 MPa) containing nitrogen oxides are compressed to ca 1.1–1.5 MPa for efficient absorption in water to make nitric acid. Stainless steel is used for construction in this corrosive environment and, because of the potential for unreacted ammonia from the converter combining with nitric acid to form ammonium nitrate, there can be additional operating and safety considerations caused by the accumulation of ammonium nitrate solids. Successful operation of dual-pressure processes has shown that these problems have been overcome. One reported solution to the buildup of ammonium nitrate solids is a periodic injection of steam to parts of the compressor (26).

Conversion of Ammonia.   Ammonia [7664-41-7], mixed with air and having an excess of oxygen, is passed over a platinum catalyst to form nitric oxide and water (eq. 10). The $\Delta H_{298\ K} = -226$ kJ/mol of NH_3 consumed (-54 kcal/mol). Heats of reaction have been derived from heats of formation at 25°C (18).

$$NH_3 + 1.25\ O_2 \longrightarrow NO + 1.5\ H_2O \tag{10}$$

Various metal oxides can and have been used as catalysts for this reaction, but none have found wide use in industry. Interest in finding a less expensive material than platinum continues, with most recent activity centering on cobalt catalysts (26,27). In the 1990s, platinum remains the catalyst of choice. Used as multiple layers of a fine wire mesh or gauze, it typically contains 5–10 wt % of rhodium for increased strength and improved reaction yield. Use of up to 5 wt % palladium reduces the overall cost. Using such platinum-based catalysts, the conversion of ammonia to nitric oxide is high (93–98%). The optimum reaction temperature for conversion efficiency increases with operating pressure. Overall, the combination of higher operation pressure and temperature results in decreased conversion efficiency. At typical gauze temperatures of 800–940°C, nitric oxide is thermodynamically unstable and slowly decomposes into nitrogen and oxygen. Decomposition losses are minimized by avoiding excessive catalyst contact time and rapidly cooling the gases as they exit the converter. Fortunately, the conversion of ammonia to NO is very rapid and catalyst contact time is ca $10^{-3}–10^{-4}$ s. The formation of nitrogen by direct combustion of ammonia usually represents the most significant yield loss (eq. 11), where $\Delta H_{298\ K} = -317$ kJ/mol of NH_3 consumed (75.8 kcal/mol). Small amounts of nitrous oxide N_2O are also formed but much lower reaction temperatures of around 500°C are required for N_2O to become a significant yield loss.

$$NH_3 + 0.75\ O_2 \longrightarrow 0.5\ N_2 + 1.5\ H_2O \tag{11}$$

Besides the direct oxidation of ammonia, various theories have been postulated for the reaction mechanism, including the formation of an intermediate imide (NH), nitroxyl (HNO), or hydroxylamine (28). Parameters affecting the conversion efficiency of ammonia to nitric oxide have also been studied (28–34). Gas velocity (linear, space, and mass), catalyst (composition, surface area, and age), operating conditions (pressure and temperature), and converter design (mixing and flow distribution) all have a significant effect on conversion efficiency. Reduced ammonia feed concentration and operation at low pressure are reported to improve nitric oxide yield. The greater selectivity toward nitric oxide formation under such conditions is probably caused by operating under gas-phase, mass-transfer limiting conditions for ammonia, which results in low concentrations of ammonia at the catalyst surface. Reactions involving the formation of N_2 and N_2O at the catalyst surface are thought to be one order higher in ammonia concentration than the desired reaction for NO. Decomposition of NO at the catalyst also appears to be enhanced by pressure, because the reaction is second order in the partial pressure of NO (32). Operation at atmospheric pressure is generally credited with a 3% yield benefit over operation at high pressure, ca 1.0 MPa (10 atm). Because low ammonia feed concentration improves yield, air and ammonia are often preheated so that an optimum feed concentration and temperature can be attained for the desired reaction. Most commercial processes operate with ammonia feed concentrations of between 10.5 and 12.0 vol %; the lower explosive limit for ammonia in air is 13.8 vol % at 0.1 MPa, diminishing to 12.4 vol % at 0.8 MPa (33,34). Industrial experience has shown that significant yield losses can result from impurities such as oil and rust present in feedstocks, so various methods of impurity removal, including filtration (ie, of the air, ammonia, and ammonia–air mixture) are commonly employed (35–37). Particles of rust and equipment made of mild steel promote the decomposition of ammonia even before it reaches the gauze.

The performance of a platinum gauze catalyst varies with its age. Changes in efficiency are affected by structural and compositional changes to the catalyst. A new gauze has smooth rounded wires of catalyst. After a short period of operation, conversion efficiency increases as the surface roughens and metal migrates to the surface to form small metal growths that increase surface area of the catalyst. As these growths increase, the catalyst structure is weakened and catalyst is lost. The catalyst surface becomes enriched with inactive rhodium oxide, Rh_2O_3, and ammonia conversion efficiency begins a gradual decline. Platinum is lost preferentially to palladium or rhodium. The rate of metal loss is a strong function of temperature; as much as a tenfold increase has been reported for a temperature change from 820 to 920°C (34). The amount of precious metal loss is financially significant and various methods, such as downstream glass fiber filters, are used to recover lost particles of catalyst.

Volatilized platinum oxides may be recovered on noble metal catchment gauzes that are placed immediately below the platinum oxidation gauze. Catchment gauzes are made from palladium or palladium combined with a base metal (38,39). Platinum recovery rates, reported to be as high as 70–85%, are to some extent offset by losses of palladium. Newer specialized designs of gauze have become available that reach optimum conversion efficiency in a matter of hours rather than days of plant operation and, overall, provide a higher conversion efficiency (38). Such gauzes replace only a portion of the standard converter gauze sheets and provide an easy light-off because they have a lower ignition temperature. This approach is similar to the well-known practice of placing a few used catalyst sheets on top of a new gauze pack to facilitate easy light-off and improved initial catalyst performance. Other gauze designs use knitted wire construction rather than a weave (39). The increased surface area per unit mass of catalyst is reported to improve ammonia conversion and possibly increase catalyst life because of reduced migration of rhodium to the catalyst surface. Knitted gauzes are approximately 75% lighter than conventional gauzes (four knitted sheets are equivalent in weight to one woven sheet).

Actual converter design and performance vary with the operating conditions. Single-train capacities of up to 2500 t/d HNO_3 for a high pressure process and 1500 t/d for a medium pressure process are claimed, but most large-capacity processes have used multiple converters, each with capacities of ca 600–1000 t/d HNO_3 and diameters of up to 6 or 7 m. Catalyst loading for most plants is reported (22) to be in the range from 1.5 to 1.9 kg per t/h of nitric production and probably lowest for high pressure operation (34). In approximate terms, gauze diameter varies inversely with pressure and the number of layers of gauze varies from four at atmospheric pressure to 24–45 layers for high pressure operation. Ammonia conversion efficiency and optimum gauze temperature vary with operating pressure. Typical values for atmospheric operation are a 97–98% conversion efficiency at 790–850°C. High pressure operation, ie, at 1.0 MPa (10 atm), provides a 94–95% conversion efficiency at about 940°C. Gross catalyst losses increase from around 0.05–0.10 g/t for atmospheric to medium pressure plants up to 0.2–0.4 g/t HNO_3 for high pressure processes. The greater catalyst loss for high pressure operation has encouraged the development and use of recovery catchment gauzes and mechanical filters. High efficiency mechanical filters recover as much as 50% of such losses. The combined use of catchment gauzes and filters is credited with 70% or greater recovery when used in a high pressure process (23). A gradual deterioration in gauze efficiency limits plant operating time. Operating time is shortest for high pressure operation at 2–3 months. This increases to 4–6 months for medium pressure plants and 8–12 months for low pressure combustion. The advantages of low pressure operation are offset by the capital cost for larger process equipment.

Oxidation of Nitric Oxide. Nitric oxide [*10102-43-9*] reacts slowly with oxygen to yield nitrogen dioxide [*10102-44-0*] according to the reversible reaction (eq. 12) for which $\Delta H_{298 \text{ K}} = -57$ kJ/mol of NO consumed (13.6 kcal/mol).

$$2 \text{ NO} + \text{O}_2 \rightleftharpoons 2 \text{ NO}_2 \tag{12}$$

Low temperatures strongly favor the formation of nitrogen dioxide. Below 150°C equilibrium is almost totally in favor of NO_2 formation. This is a slow reaction, but the rate constant for NO_2 formation rapidly increases with reductions in temperature. Process temperatures are typically low enough to neglect the reverse reaction and determine changes in NO partial pressure by the rate expression (40–42) (eq. 13). The rate of reaction, and therefore the time required to achieve a given extent of oxidation, is proportional to pressure cubed. This is of great significance in plant design and economics. The volume or size of equipment needed to oxidize the nitric oxide is inversely proportional to pressure cubed.

$$d(P_{NO})/dt = -k(P_{NO})^2(P_{O_2}) \qquad (13)$$

Nitrogen dioxide rapidly forms an equilibrium mixture with its dimer, dinitrogen tetroxide ($\Delta H_{298\ K} = -28.6$ kJ/mol of NO_2 consumed). The formation of tetroxide is favored by low temperature and high pressure.

Absorption of Nitrogen Oxides. There have been numerous studies and reports on the reaction mechanisms and rate-controlling steps for the absorption of nitrogen oxides into water (43–46). The overall reaction to form nitric acid may be represented by equation 14, where $\Delta H_{298\ K} = -46$ kJ/mol of NO_2 consumed.

$$3\ NO_2(g) + H_2O(l) \rightleftharpoons 2\ HNO_3(aq_\infty) + NO(g) \qquad (14)$$

Equilibrium for this reaction, expressed in terms of the constituent partial pressures, is a function of temperature, with low temperature favoring the formation of nitric acid (43). The equilibrium for reaction 14 has been more commonly reported in terms of the partial pressure of nitrogen oxides as a strong function of acid strength. Replacing the partial pressure of NO with that of N_2O_4 by use of the equilibrium expression for NO_2-dimer formation eliminates temperature as a significant variable for typical absorption applications (44). Besides equilibrium, models used to simulate or design the absorption process (45–51) must also account for large changes in absorption efficiency, ie, the approach to equilibrium, with acid strength. The mechanisms and rate-controlling steps for acid formation change with concentration and extent of oxidation to nitrogen peroxide. At high concentrations of nitrogen peroxide (>40 wt %), ie, for nitric acid concentrations, N_2O_4 is the principal route to acid formation. At lower concentrations of nitrogen peroxide, a combination of mechanisms involving NO_2, N_2O_3, and HNO_2 become more important (45,50).

Before entering the absorption column, the process gases are cooled in a cooler–condenser to remove condensate. Because most of the water for absorption comes from the combustion of ammonia, it is important to remove the condensate with a minimum content of nitric acid. This optimizes the height at which condensate may be fed to the column. If the condensate contains too much nitric acid, the column may not be able to provide the desired product acid strength. It is therefore important not to oversize the cooler–condenser. For condensation at atmospheric pressure, it is possible to remove water containing less than 10 wt % acid. At higher pressures of up to 0.9–1.4 MPa (9–14 atm), the condensate typically contains 45–50 wt % acid. Formation of nitric acid in the condensate generates nitric oxide, which lowers the extent of oxidation to NO_2. To make product acid of high strength, the nitrogen oxides must be fully oxidized before entering the absorption column. Air, used to strip dissolved NO_x in the product acid, is mixed with the process gas to enhance the oxidation process. As the gases are cooled and oxidized, further conditioning of the gas stream is caused by staged contacting with the acid condensate. Nitric oxide reacts with nitric acid per equation 14 to restore equilibrium forming the dioxide and water.

The design of nitric absorption columns is a specialized process requiring the individual tailoring of each tray for its cooling requirements and the gas volumes needed for oxidation of the nitric oxide. A good summary of absorber design aspects is available (52). Modern processes use columns having bubble cap or sieve trays. Capacities of around 1800 t/d HNO_3 have been achieved in a single column. Columns are up to 6 m in diameter, 80 m high, and contain as many as 30–50 trays. Discounting NO_x abatement requirements, the economic optimum for absorber efficiency results in tail gas NO_x concentrations of ca 1500–2500 ppmv. For abatement reasons single absorption columns reach tail gas NO_x compositions of ca 200 ppmv. The need to make processes more efficient and to meet environmental regulations has led to new advances in understanding absorption chemistry. One such development is the High Efficiency Absorption (HEA) technology developed by Rhône Poulenc. The specifics of this technology are not well known, but it can reduce the size of an absorber column by up to 35% (53). HEA is reported (54) to be applicable in regions in which the gas-phase concentration of NO_x is low (<8000 ppm) and the rate-limiting step for nitric acid formation is typically the gas-phase oxidation of NO. By directly oxidizing nitrous acid in the liquid phase to nitric acid, the usual decomposition of nitrous acid into nitric oxide and acid is circumvented and the large oxidation volumes needed to regenerate NO_2 from NO are greatly reduced.

NO_x Abatement. Source performance standards for nitric acid plants in the United States were introduced by the U.S. EPA in 1971 (55). These imposed a discharge limit of 1.5 kg of NO_x as equivalent nitrogen dioxide per 1000 kg of contained nitric acid, which corresponds to about 200–230 ppmv of nitrogen oxides in vented tail gas, whereas concentrations after absorption may contain as much as 2000–3000 ppmv of nitrogen oxides. Regulations and a review of abatement methods used in the EC are available (56). Various methods have been used to reduce tail gas NO_x concentrations to an acceptable level for discharge to the atmosphere (57). The most commonly employed methods are extended absorption, selective catalytic reduction (SCR), and nonselective catalytic reduction (NSCR).

Extended absorption uses an additional column to oxidize and react the nitrogen oxides with water to form acid. Bubble cap trays hold their liquid seal on shutdown and are therefore preferred for minimizing NO_x abatement problems during start-up and shutdown. Low partial pressures of nitrogen oxides lead to low rates of oxidation and a need for large amounts of gas-phase holdup. Extended absorption requires relatively few trays but large oxidation volumes, and often employs refrigeration to promote the oxidation process and minimize column size. This method of abatement is most effective for high pressure absorption in which abatement to less than 200 ppmv NO_x can be achieved in a single column. In general, medium pressure plants use two absorption columns to achieve concentrations of ca 500 ppmv NO_x. Cold tail gases are reheated by heat exchange with hot process gas to increase power recovery in the expander.

Selective catalytic abatement uses a catalyst and ammonia fuel to selectively reduce nitrogen oxides in preference to combustion with the much higher levels of oxygen in the tail gas. This method can operate over a wide range of pressures but requires temperatures to be 210–410°C. For effective abatement, a slight excess of ammonia is used, leaving 5–20 ppmv in the treated tail gas. Higher concentrations of ammonia present a potential safety hazard that has to be avoided by the use of good process controls. Ammonia at low temperatures can form ammonium nitrate and nitrite, which might accumulate in downstream equipment and pose an explosion hazard upon being heated. Base metal oxides (titanium and vanadium) are commonly used as the catalysts, although platinum, palladium, or zeolites may also be used. The catalysts are usually shaped as honeycombs or flat parallel plates. The treated tail gases are also reheated for improved expander energy recovery.

Nonselective abatement uses a catalyst and fuel (gaseous hydrocarbon such as propane or natural gas) to reduce nitrogen oxides to nitrogen and, in the process, combust any remaining free oxygen in the tail gas. This consumes significantly more fuel than a selective reduction system, but the energy thus generated is mostly retrieved as power in the expander. Platinum, palladium, and rhodium, either in the form of pellets or as a honeycomb, are typically used as catalysts in these systems. The minimum temperature for inlet tail gas depends on the fuel in use and its ignition temperature. Hydrogen, which has the lowest ignition temperature, requires about 150–200°C. A large temperature rise due to combustion, about 130°C for every volume percent of oxygen, places a limit on total oxygen content for tail gas exiting the absorber. Free oxygen tends to be consumed preferentially and fuel must be fed in slight stoichiometric excess to the oxygen. Leftover hydrocarbons are discharged to the atmosphere, including small amounts of carbon monoxide and hydrogen cyanide. Relatively few nitric acid plants use this method of abatement, which was one of the first available when source performance standards were introduced in 1971.

In an economic comparison of these three common abatement systems, a 1991 EPA study (58) indicates extended absorption to be the most cost-effective method for NO_x removal, with selective reduction only matching its performance for small-capacity plants of about 200–250 t/d. Nonselective abatement systems were indicated to be the least cost-effective method of abatement. The results of any comparison depend on the cost of capital versus variable operating costs. A low capital cost for SCR is offset by the ammonia required to remove the NO_x. Higher tail gas NO_x concentrations make this method less attractive. The investment required for extended abatement is partially recovered by increased yield of nitric acid product. Other methods of abatement have been proposed but for various reasons have not found general commercial acceptance. They include scrubbing (eg, with ammonia, urea, and caustic chemicals) as well as molecular sieve adsorption (57,59). One chemical absorption method (60) uses weak acid (25–30 wt %) to remove nitrogen oxides from the tail gas to form nitrous acid, HNO_2. Steam stripping regenerates the clean acid for recycle. The removed NO_x is compressed and fed to the plant's main absorption column. Environmental regulations may reasonably be expected to become stricter in the future. This could lead to plant designs using a combination of NO_x abatement methods, eg, extended absorption with SCR.

DIRECT STRONG NITRIC PROCESSES

Direct strong nitric (DSN) processes use many of the production steps described for weak acid production. Ammonia and air are combusted over platinum and the process gas stream is cooled to the point at which weak acid condensate is

formed. Tail gases are reheated by hot process gas and provide energy recovery
in a gas expander. DSN processes differ from weak acid plants in the additional
processing steps required to achieve superazeotropic strengths of nitric acid.
The combustion of ammonia and cooling of process gas are often carried out
at atmospheric pressure. Excess water can then be directly removed as process
condensate containing only 1–3 wt % nitric acid. Processes using medium or high
pressure ammonia combustion have a 45–50 wt % nitric condensate, requiring
either a rectification step to remove excess water from the plant or the coproduction of weak acid. The dehydrated process gases are usually compressed,
mixed with air and nitrogen peroxide obtained from the bleaching of product
acid, and then cooled to promote the full oxidation of nitric oxide to nitrogen
peroxide. The gases are sometimes contacted with weak to azeotropic strength
acid so that the nitric oxide can react with nitric acid per equation 14 to restore
chemical equilibrium by forming the dioxide and water.

The key to producing strong acid is obtaining high partial pressures of
nitrogen peroxide. In older processes such as HOKO (61), nitrogen peroxide is
absorbed in chilled 98 wt % acid, distilled off, and condensed. The peroxide is
mixed with oxygen and weak nitric acid in an autoclave at 5 MPa to produce
strong acid. Processes such as SABAR generate strong acid by azeotropic rectification (62). Nitrogen peroxide in the process gas is chemically absorbed into
azeotropic or greater strength acid. Air stripping and rectification of the resulting acid stream produces strong acid of 98–99 wt % strength. Various process
schemes (63–65) fall between these two examples, some of which co-produce both
strong and weak acid. Flow schemes for many of the DSN processes, as well as
nitric concentration (NAC) plants, are available (66,67).

DSN plants have found little application in the United States, but several
have been built in Europe. Some older-style HOKO processes, although more
capital-intensive and less economical than later DSN processes, continue to
operate in the 1990s.

NITRIC ACID CONCENTRATION

Nitric acid has a maximum boiling azeotrope at 69 wt %, so it is not possible
to obtain greater concentrations by the simple distillation of the weak acid. Nitric acid concentration (NAC) processes use extractive distillation to concentrate
weak acid up to 99 wt %. A dehydrating agent, such as sulfuric acid or magnesium nitrate, is used to enhance the volatility of HNO_3 so that distillation
methods can surpass the azeotropic concentration of nitric acid. Weak acid and
dehydrating agent are fed to a distillation column. Water removed from the acid
dilutes the dehydrating agent, which is removed as a bottom stream and later
concentrated for reuse in the process. Superazeotropic acid vapors pass to the
bottom of a rectification section in which the acid is concentrated up to 99 wt %.
The strong nitric vapors are condensed overhead and a portion of the acid is
returned to the column as reflux.

One of the most commonly practiced NAC technologies uses sulfuric acid
of 82–96 wt % as the dehydrating agent (66,67). The spent dilute sulfuric acid,
typically ca 65–70 wt %, often requires multiple vacuum-stage evaporation for
reconcentration and use in the NAC process. Sulfuric acid is sufficiently strong

dehydrating agent that at high feed concentrations of ca 93–96 wt % H_2SO_4 the rectification section can be eliminated. The resulting higher operating temperatures can lead to increased decomposition of the nitric acid product. One licensor reduces sulfuric acid use by preconcentrating the weak nitric acid (eg, to ca 67 wt %) by simple distillation without a dehydrating agent. Some processes use live steam injected into the base of the column to remove any residual nitric acid from the spent sulfuric acid.

Magnesium nitrate has become popular as a dehydrating agent in commercial NAC processes (17,68). At high concentrations, magnesium nitrate solutions are quite viscous and have high melting points, so feed concentration is typically restricted to around 72 wt %. Recovered nitrate, ca 68 wt %, is concentrated for reuse in a single-stage, steam-heated vacuum flash drum. The ease of concentrating aqueous magnesium nitrate solution and the use of less expensive materials of construction make magnesium nitrate NAC technology less capital intensive than the sulfuric acid method. Magnesium nitrate at 72 wt % is a less effective dehydrating agent than sulfuric acid, which has a lower melting point and viscosity and so can be used in higher concentrations. Therefore, processes using magnesium nitrate use rectification to increase the concentration of nitric acid up to 99 wt %.

Because of restrictions in equipment size, magnesium nitrate processes were initially limited to small plants. Improvements in the materials of construction have led to increased capacities and a lower capital cost. Sulfuric acid processes are usually preferred when reconcentration of the sulfuric acid is not required, ie, when the dilute sulfuric can be used to make another product.

Other Processes

Most other routes for making nitric acid involve the formation of nitrogen oxide directly from the air using various energy sources, such as electric arc furnaces and shock waves (4). Air heated to extremely high temperatures forms nitric oxide (ca 8 vol % at 4000 K), which must then be rapidly quenched to avoid decomposition of the nitric oxide on cooling. In Norway an electric arc process found successful application until ca 1930. The Wisconsin process, developed during World War II, heated air to >2000°C by contact with a bed of fuel-heated magnesia pebbles. Rapid quenching by contact with a second cold bed yielded 1–2 wt % nitric oxide. Later, a method using three beds of silica gel in series was developed to recover the nitric oxide. The first bed dehumidified the gas, while the second catalytically oxidized the nitric oxide to dioxide and the third adsorbed the nitrogen dioxide, which then could be stripped in a highly concentrated form (69,70). A small plant was built in 1953 for the U.S. government to demonstrate the technology. The plant operated successfully for 18 months, but the process did not prove its economic viability (71).

Alternative approaches to nitric oxide formation include irradiation of air in a nuclear reactor (72) and the oxidation of ammonia to nitric oxide in a fuel cell generating energy (73). Both methods indicate some potential for commercial application but require further study and development.

Materials of Construction

Early nitric acid plants were severely restricted in design by the available materials of construction. Until high chromium content iron was developed in the 1920s, commonly used construction materials were acid brick, earthenware, and glass (qv). As better materials of construction came along, nitric acid process design could be improved. Higher operating pressures resulted in lower plant capital costs. The following is an overview of the materials of construction used in modern processes. More detailed information concerning the application and corrosion resistance of specific materials can be found (74–77).

Weak Acid. Stainless steels (SS) have excellent corrosion resistance to weak nitric acid and are the primary materials of construction for a weak acid process. Low carbon stainless steels are preferred because of their resistance to corrosion at weld points. However, higher grade materials of construction are required for certain sections of the weak acid process. These are limited to high temperature areas around the gauze (ca 900°C) and to places in which contact with hot liquid nitric acid is likely to be experienced (the cooler condenser and tail gas preheater).

Catalyst baskets and gauze supports must be resistant to oxidation, nitriding, and distortion from high temperatures. Typical materials of construction are high strength alloys made of iron–nickel–chromium and nickel–chromium. Alloys made of nickel–chromium–tungsten–molybdenum are also finding application in such service.

High chromium (20–27%) SS provides better corrosion resistance to nitric acid at elevated temperatures than low carbon SS. It has been used to replace low carbon steel in cooler condenser and tail gas preheater service, providing sufficiently extended life to justify its additional cost.

Zirconium has seen increasing use with nitric acid. The excellent corrosion resistance of zirconium is a result of the formation of a strongly adhering inert oxide film. The film is diffusion-bonded to the metal and has the ability to repair itself should it be damaged. Zirconium rivals tantalum in its corrosion resistance to nitric acid (76), having low corrosion rates at all concentrations up to the boiling point. Its resistance extends up to 230°C and 65 wt %. It is susceptible to stress corrosion cracking, which can be prevented by avoiding high, sustained tensile stresses (78). Zirconium is used in nitric acid service for cooler condensers, tail gas preheaters, and reboilers.

Duplex stainless steels (ca 4% nickel, 23% chrome) have been identified as having potential application to nitric acid service (75). Because they have a lower nickel and higher chromium content than typical austenitic steels, they provide the ductability of austenitic SS and the stress–corrosion cracking resistance of ferritic SS. The higher strength and corrosion resistance of duplex steel offer potential cost advantages as a material of construction for absorption columns (see CORROSION AND CORROSION CONTROL).

Strong Nitric Acid. Materials of construction commonly used in the production of strong acid (98–99 wt %) are aluminum, tantalum, borosilicate glass, glass-lined steel, high silica cast iron, and high silica stainless steels. Stainless steel is often used for the storage and shipment of up to 95 wt % nitric acid. At higher concentrations, corrosion rates for stainless steel become excessive and aluminum is the commonly accepted material of construction. The use of aluminum alloys is restricted to high acid concentrations of 80–100 wt % and moderate temperatures of less than 38°C. High silica stainless steels (ca 4–7 wt % silicon) have demonstrated good corrosion resistance to concentrated nitric acid. However, corrosion resistance does not significantly extend to weaker strengths of acid. These steels are commonly used in direct strong acid processes and apparently played a key role in making such processes successful. High silica cast iron also displays good corrosion resistance to strong acid, but is more commonly used in columns and pump casings.

Titanium is resistant to nitric acid from 65 to 90 wt % and dilute acid below 10 wt %. It is subject to stress–corrosion cracking for concentrations above 90 wt % and, because of the potential for a pyrophoric reaction, is not used in red fuming acid service. Tantalum exhibits good corrosion resistance to nitric acid over a wide range of concentrations and temperatures. It is expensive and typically not used in conditions where other materials provide acceptable service. Tantalum is most commonly used in applications where the nitric acid is close to or above its normal boiling point.

Glass offers good resistance to strong acid at high temperatures. However, it is subject to thermal shock and a gradual loss in integrity as materials such as iron and silica are leached out into the acid. Nonmetallic materials such as PTFE, PVDC, PVDF, and furan can be used for nitric acid to a limited degree, but are mainly restricted to weak acid service at ambient to moderate temperatures.

Process Performance and Economics

Unit consumption of the key raw material, ammonia (qv), is typically in the range of 0.28–0.29 t/t HNO$_3$. Nitric acid plant designs are adjusted to meet the economic requirements associated with specific sites. There are many different equipment and process design variations that result in different process performance. Therefore, it is not practical to report typical performance figures for each type of nitric acid process. Table 3 lists references for design and performance information on some of the processes offered by nitric technology licensors. Each

Table 3. References for Design and Performance Information

Licensor	Weak acid	Direct strong nitric (DSN)	Nitric acid concentration (NAC)
Espindesa, Spain	79	66,79	
Grande Paroisse, France	80		
Hercules, USA			66
Humphreys & Glasgow, Ltd.	60	65	
Rhône Poulenc, France	81	79	79,81
Sumitomo Chem. Co., Japan		61,66	
Uhde GmbH, Germany	81,82,83		
Weatherly, USA	81,82		
Zimmer AG, Germany[a]		59,66	

[a]A subsidiary of Lurgi AG, formerly Davy McKee AG (Bamag).

design variation affects not only process efficiency but also capital cost. Production economics are therefore a combination of operating costs (determined by process performance and local material costs) and fixed costs that are related to capital investment.

Comparison of the economics of the high-monopressure process and the dual-pressure process has led to extensive debate. A 1979 comparative study (22) of these two processes indicated that both are competitive and suggested that the monopressure was probably most favored for small to intermediate-size plants, the dual-pressure process is preferred for very large capacities. Although the economic analysis was not conclusive, it did highlight some important economic parameters. Lower platinum losses were found to be the most important operating cost, which compensates for the higher capital cost of the dual-pressure process. Changes in the cost of ammonia and energy were found to have less effect. The lower ammonia yield of a high monopressure process was mostly offset by increased steam generation.

Similar economic comparisons have been made for the production of strong nitric acid using the various direct strong processes and weak acid plants in combination with a concentration (NAC) process (84,85). These comparisons indicate that the newer DSN technologies such as SABAR are competitive with the older, more established processes such as HOKO and extractive distillation. Process economics can be affected to a large degree by site specifics. For example, if there is a use for dilute sulfuric acid, then an NAC process using sulfuric acid as the dehydrating agent may be more favored. An inexpensive source of oxygen resulting from on-site coproduction of nitrogen may be advantageous for a HOKO-style process.

Process Licensors. Some of the well-known nitric acid technology licensors are listed in Table 3. Espindesa, Grande Paroisse, Humphreys and Glasgow, Rhône Poulenc, Uhde, and Weatherly are all reported to be licensors of weak acid technology. Most weak acid plant licensors offer extended absorption for NO_x abatement. Espindesa, Rhône Poulenc, Weatherly, and Uhde are also reported (53,57) to offer selective catalytic reduction (SCR) technology.

Espindesa, Rhône Poulenc, Sumitomo, Uhde, and Davy McKee are licensors of direct strong acid processes. Davy McKee (formerly Bamag) is now Zimmer AG, a subsidiary of Lurgi AG. Chemetics, Hercules, Jenaer Glaswerk Schott (Germany), and Plinke (Germany) are all reported to license nitric acid concentration plants (see LICENSING).

Production. The rapid post-war growth of nitric acid production in the United States came to an end in 1970, by which time production had reached 7 million t/yr. Except for market cycles and a production peak of 9 million t/yr in 1980–1981, the demand for nitric acid has remained fairly flat at ca 7 million t/yr for 1991–93. Table 4 lists U.S. production figures for 1965 to 1992. Since the early 1980s, urea has been displacing ammonium nitrate as a fertilizer. The resulting reduction in demand for nitric acid has to some extent been offset by the increased use of ammonium nitrate in explosives and by the growth in production of polyurethane foams and nylon-6,6. Predictions show only a marginal growth in demand (ca 1%) as the century ends. Existing commercial production capacity exceeds demand by ca 1–2 million t/yr; an additional capacity of 1.5 million t/yr exists in U.S. Army facilities. Production in Europe was estimated at 18.5 million

Table 4. U.S. Production of Nitric Acid,[a] 10³ t/yr, 100% Basis

Year	Production	Year	Production	Year	Production
1965	4443	1977	7256	1985	6935
1970	6897	1978	7198	1986	6111
1971	6929	1979	8088	1987	6554
1972	7204	1080	8375	1988	7249
1973	7619	1981	8249	1989	7574
1974	7366	1982	6704	1990	7030
1975	6828	1983	6321	1991	7189[b]
1976	7068	1984	7074	1992	7295[b]

[a]Ref. 86 unless otherwise noted.
[b]Ref. 87.

tons for 1989, and demand is expected to remain at this level for several years. The 1989 production levels in Canada, Mexico, and Japan are reported to be 1.0, 0.3, and 0.6 million t/yr, respectively. References 86–89 were used for production figures and market trend information. Reference 86 provided the most extensive source of market information on nitric acid. It contains historical production data for the United States, Canada, Mexico, western Europe, and Japan, producer lists indicating plant location, size, product concentration, and data on nitric acid use.

Prices. Most nitric acid produced in the United States is for captive consumption for which merchant market pricing does not apply. Published mid-year merchant market prices on a railroad tankcar works basis for 1983–1993 are as follows (Table 5). The price of nitric acid remained constant during 1983–1988 and then changed abruptly by a reduction of $15/t (100% basis) for weak acid and an increase of $40/t (100% basis) for strong acid. This change may indicate a relative undersupply of strong acid, which is used in the growing production of nitrated organics (polyurethanes) and adipic acid (nylon-6,6). The actual versus published price of nitric acid is affected by many factors. Shipping costs are prohibitive, so geographic location and regional supply availability are important. Price is also affected by the supply and demand of ammonia, the principal raw material for nitric acid manufacture.

Table 5. Merchant Market Prices,[a] $/t Nitric Acid, 100% Basis

Nitric acid, wt %	1983–1988	1989–1993
94.5–98	280	320
52–67	195	175–185

[a]Ref. 89.

Specifications and Standards

Shipment. The Department of Transportation (DOT) classifies nitric acid as a hazardous material requiring proper packaging, labeling, and shipping documentation for transportation. The DOT defines three categories of nitric acid: (*1*) nonfuming, more than 70 wt % acid; (*2*) nonfuming, less than 70 wt % acid; and (*3*) red fuming nitric acid. All are in Hazards Class 8, meaning corrosive material, and must be labeled "corrosive." Red fuming nitric acid must also be labeled "oxidizer" and "poison." Each category of nitric acid has its own packaging authorization number (ie, packaging requirements) for both bulk and nonbulk shipping. Depending on these packaging requirements, nitric acid may be shipped in either stainless steel or aluminum. Bulk nitric acid can be shipped by railcar, tank truck, or portable tank. Nonbulk packaging includes drums. For other than red fuming acid, a variety of smaller containers are permitted, all of which require various forms of individual packaging, depending on acid strength and mode of transportation. Packages are mostly glass or earthenware containers ranging in size from 0.5 to 2.5 l. Transportation on passenger aircraft or railcar is forbidden for all categories of nitric acid. Red fuming acid is forbidden on cargo aircraft. The full DOT requirements for packaging and shipping nitric acid are available (90).

Standards. The ACS defines two grades of reagent acid (91): nitric acid, having a concentration of 69.0–71.0 wt % HNO_3, and nitric acid, 90%, having a concentration of ≥ 90 wt % HNO_3. Both have maximum allowable levels of chlorides, sulfates, arsenic, heavy metals, iron, and residue after burning. Impurity limits defined for the 90% grade are the least stringent. Nitric acid must be colorless and free from suspended matter, whereas the 90% grade has a maximum content for dissolved oxides as N_2O_3. The USP (92) has definitions that refer to those of the ACS.

Commercially, nitric acid concentrations are graded in terms of degrees Baume as follows:

Degrees Baumé	Approximate wt % HNO_3
36	52.3
38	56.5
40	61.4
42	67.2

Analytical and Test Methods

Qualitative Analysis. Nitric acid may be detected by the classical brown-ring test, the copper-turnings test, the reduction of nitrate to ammonia by active metal or alloy, or the nitrogen precipitation test. Nitrous acid or nitrites interfere with most of these tests, but such interference may be eliminated by acidifying with sulfuric acid, adding ammonium sulfate crystals, and evaporating to a low volume.

In the brown-ring test, concentrated sulfuric acid is carefully poured down the side of an inclined test tube to form a separate layer beneath the solution to be tested. After the two layers have been cooled without mixing, a few drops of a ferrous sulfate solution are placed inside the inclined tube and drained down to the interface. The appearance within a few minutes of a brown layer of ferronitrosulfate, $(Fe(NO))SO_4$, at the junction of the two liquids shows the presence of nitrates. Nitric acid or nitrates may be detected after the removal of nitrous acid by warming a mixture of the material to be tested, a 1:1 solution of sulfuric acid, and copper turnings in a test tube. The evolution of brown fumes indicates the presence of nitrates. Nitrates are easily reduced by active metals or alloys in alkaline solution to give ammonia. Suitable reducing agents are aluminum, zinc, or Devarda's alloy (50 wt % Cu, 45 wt % Al, 5 wt % Zn).

Nitric acid may be precipitated by nitron [2218-94-2] (4,5-dihydro-1,4-diphenyl-3,5-phenylimino-1,2,4-triazole). The yellow precipitate may be seen at dilutions as low as 1:60,000 at 25°C or 1:80,000 at 0°C. To prevent nitrous acid from interfering with the test results, it may be removed by treating the solution with hydrazine sulfate, sodium azide, or sulfamic acid.

Quantitative Analysis.   The total acidity of nitric acid solution may be determined by conventional titration using phenolphthalein as the indicator.

Other Acidic Impurities.   Sulfuric Acid.   The sample is evaporated to dryness on a steam bath. The residue is removed with water, and evaporation is repeated until the sample is free from nitric acid fumes. The residue is diluted with water and titrated with standardized sodium hydroxide solution using phenolphthalein as an indicator.

Hydrochloric acid is determined gravimetrically as silver chloride.

Nitrous Acid.   Lower oxides of nitrogen and nitrous acid [7782-77-6] generally are determined and reported as NO_2. In the absence of organic matter or other reducing agents, the determination may be made by titrating with potassium permanganate solution. A sharp end point is achieved by rapidly adding the permanganate solution to the solution containing the oxide of nitrogen or nitrous acid. If the addition of permanganate solution is low, some oxidation of the sample by dissolved air occurs resulting from the dilution of the solution with water.

Other Methods.   Ion chromatography using conductance detection can be used to measure low (< 1%) levels of nitrite, chloride, sulfate, and other ions in nitric acid. Techniques for ion chromatographic analysis are available (93).

Ferrous Sulfate Titration.   For determination of nitric acid in mixed acid or for nitrates that are free from interferences, ferrous sulfate titration, the nitrometer method, and Devarda's method give excellent results. The determination of nitric acid and nitrates in mixed acid is based on the oxidation of ferrous sulfate [7720-78-7] by nitric acid and may be subject to interference by other materials that reduce nitric acid or oxidize ferrous sulfate. Small amounts of sodium chloride, potassium bromide, or potassium iodide may be tolerated without serious interference, as can nitrous acid up to 50% of the total amount of nitric acid present. Strong oxidizing agents, eg, chlorates, iodates, and bromates, interfere by oxidizing the standardized ferrous sulfate.

Possible interferences and variation of results from modified techniques can be avoided by titrating the sample in exactly the same way and by employing approximately the same amounts of materials as in the initial standardization of the ferrous sulfate against a known quantity of nitric acid. The ferrous sulfate solution is added in a thin stream until the initially yellowish solution turns brown. The titration is complete when the faint brownish-tinged end point is reached.

Nitrometer Method.   The nitrometer method also is used to determine nitric acid or nitrates in mixed acid or oleum. It involves the measurement of the volume of NO gas that is liberated when mercury is oxidized by nitric acid. The method is based on the following reaction:

$$2\,HNO_3 + 3\,H_2SO_4 + 3\,Hg \longrightarrow 3\,HgSO_4 + 4\,H_2O + 2\,NO \tag{15}$$

Devarda's Method. Nitrogen in nitrates or nitric acid also may be determined by the Kjeldahl method or by Devarda's method. The latter is both convenient and accurate when no organic nitrogen is present. The nitrate is reduced by Devarda's alloy to ammonia in an alkaline solution. The ammonia is distilled and titrated with standard acid.

Health and Safety Factors

Nitric acid and the oxides of nitrogen found in its fumes are highly toxic and capable of causing severe injury and death. It is corrosive and can destroy human tissue. Nitric acid is regulated by OSHA, which lists it as a Process Safety Hazardous Chemical and Air Contaminant. Under SARAH, the EPA lists it as an Extremely Hazardous Substance and Toxic Chemical. Per OSHA, the 1991 permissible exposure limits for nitric acid are 2 ppm (5 mg/m^3) for an 8-h time-weighted average and 4 ppm (10 mg/m^3) for a 15-min short-term exposure. Exposure limits may vary according to local and national regulations. Inhalation symptoms may take several hours to appear. They include irritation of the throat and nose, coughing, chest pain, difficulty in breathing, giddiness, nausea, ulceration of the nasal mucous membranes, pulmonary edema, and chemical pneumonia. The symptoms resulting from skin contact vary from moderate irritation to severe burn, depending on contact time and strength of the nitric acid. Signs of contact may include a yellow discoloration of the skin; severe burns may penetrate deeply causing ulceration and the scarring of tissue.

First aid practices for the treatment of exposure to nitric acid should be obtained from a current version of the Material Safety Data Sheet or other appropriate safety literature. Acid in contact with the skin must be removed immediately by washing with large amounts of water for a period of at least 15 min. A 5% solution of triethanolamine should be applied to affected areas of the skin. Continuous flushing with large quantities of water is required for acid burns to the eye; the eyelids should be gently lifted to ensure adequate washing. Medical attention should be sought immediately. For cases of ingestion, a poison control center should be contacted for advice. A conscious and alert victim who is not convulsing should drink several glasses of water for dilution and follow with lime milk or milk of magnesia. Vomiting should not be induced, nor should attempts be made to neutralize the acid with sodium bicarbonate. If exposure occurs by inhalation, the person should be moved to fresh air and given support in breathing as needed. The effects of inhalation can be fatal and may be delayed for several hours. It is important to consult a physician immediately after the administration of first aid. Additional health and safety information is available (19,20,94,95).

Uses

The largest use of nitric acid (ca 74–78% of total U.S. production) is for the manufacture of ammonium nitrate. About 75% of ammonium nitrate [6484-52-2] is used in fertilizers (qv); the remainder is used for chemicals, explosives, and miscellaneous other uses. Partly because of the increased popularity of urea, the use of ammonium nitrate as a fertilizer has declined. This has been offset to some extent by the growth in the use of ammonium nitrate for explosives and other chemical uses. Overall, the production of ammonium nitrate in the United States is expected to remain flat for several years. The next three largest uses for nitric acid are in the manufacture of cyclohexanone (ca 8–9%), dinitrotoluene (ca 4%), and nitrobenzene (ca 3–4%). Cyclohexanone [108-94-1] is a raw material for manufacture of adipic acid, which reacts with hexamethylenediamine to make nylon-6,6. Dinitrotoluene [25321-14-6] is hydrogenated to toluenediamine [26764-44-3], which is used to make toluenediisocyanate [1321-38-6] (TDI). Nitrobenzene [98-95-3] (qv) is hydrogenated to make aniline, which is a raw material for the manufacture of methylene diphenyl diisocyanate [101-68-8] (MDI). TDI is used to make flexible polyurethane foams, elastomers, and coatings, whereas MDI is used for rigid foams (see AMINES, AROMATIC–METHYLENEDIANILINE). Other uses of nitric acid are in the production of explosives, metal nitrates, nitrocellulose [9004-70-0], nitrochlorobenzene, metal treatments (eg, the pickling of stainless steels and metal etching), rocket propellants, and nuclear fuel processing (86,88).

BIBLIOGRAPHY

"Nitric Acid" in *ECT* 3rd ed., Vol. 15, pp. 853–871, by D. J. Newman, Barnard and Burk, Inc.

Overall and historical

1. J. W. Mellor, *Comprehensive Treatise on Inorganic and Theoretical Chemistry*, Vol. VIII, Longmans, Green and Co., Ltd., London, 1928, pp. 555–562.
2. T. H. Chilton, *Strong Water*, The M.I.T. Press, Cambridge, Mass., 1968.
3. J. K. Bradley and G. Drake, *Chem. Age India* **33**(1) (Jan. 1982).
4. I. Brunborg and Per B. Holmesland, in C. Keleti, ed., *Nitric Acid and Fertilizer Nitrates*, Vol. 4, Marcel Dekker, Inc., New York, 1985, pp. 1–17.

Physical properties

5. S. R. M. Ellis and J. M. Thwaites, *J. Appl. Chem.* **7**, 152 (1957).
6. W. R. Forsythe and W. F. Giauque, *J. Am. Chem. Soc.* **64**, 48 (1942).
7. G. Smith, *Nitrogen*, (168), 21–28 (Nov.–Dec. 1990).
8. R. Vandoni and M. Laudy, *J. Phys. Chim.* **49**, 99–102 (1952).
9. G. Aunis, *J. Phys. Chim.* **49**, 103–108 (1952).
10. R. Flatt and F. Benguerel, *Helv. Chem. Acta* **45**, 1765 (1962).
11. S. Takeshi and co-workers, *Kagaku Kogaku, Ronbunshu* **11**(3) (1985).
12. R. H. Perry and C. H. Chilton, eds., *Chemical Engineers' Handbook*, sixth ed., McGraw-Hill Book Co., Inc., New York, 1984.
13. T. R. Bump and W. L. Sibbitt, *Ind. Eng. Chem.* **47**, 1665 (1955).

14. C. McKinley and G. G. Brown, *Chem. Metal. Eng.* **49**, 142–144 (1942).
15. B. E. Thompson and co-workers, *Vapor Liquid Equilibrium of the Mg(NO₃)₂-HNO₃-H₂O System*, ORNL/MIT-360, Oak Ridge National Laboratory, Oak Ridge, Tenn., Dec. 1983.
16. R. T. Jublin, J. L. Marley, and R. M. Counce, *J. Chem. Eng. Data* **31**, 86–88 (1986).
17. J. G. Sloan, *The Thermodynamic Behaviour of Electrolytic and Mixed Solvents, Symp.*, Advanced Chemistry Series, 155, 1975, pp. 128–142.
18. D. D. Wagman and co-workers, *J. Phys. Chem. Ref. Data* **11** (Suppl. 2) (1982).

Chemical properties

19. N. I. Sax and R. J. Lewis, Sr., *Dangerous Properties of Industrial Materials*, Vol. III, 7th ed., Van Nostrand Reinhold, New York, 1989.
20. L. Bretherick, *Handbook of Reactive Chemical Hazards*, 4th ed., Butterworth Publishers, Stoneham, Mass., 1990, pp. 1147–1178.
21. L. F. Albright and C. Hanson, eds., *Industrial and Laboratory Nitrations*, ACS Symposium Series 22, Washington, D.C., 1976.

Weak acid process

22. R. L. Harvin, D. G. Leray, and L. R. Roudier, in A. I. More, ed., *Fertilizer Acids, Proceedings of the Br. Sulphur Corp. International Conference on Fertilizer*, 3d ed., Vol. 1, Paper 1, 1979.
23. G. A. Smith, *Nitrogen 88, Br. Sulphur Corp. 12th International Conference*, Geneva, Mar. 27–29, 1988, pp. 191–200.

Compressor–expander machinery

24. J. S. Cichowski, *Chem. Eng.*, 149–150 (Nov. 15, 1982).
25. W. Hanggeli, *Turbomachinery International* (Mar. 1982).

Ammonia conversion

26. O. Jurovack and V. Berezny, *Chemicky Prumysl (Czech.)*, **40/65**(2), 59–61 (1989).
27. M. M. Karavaev and co-workers, *Khim. Prom. (Russian)* **1**, 32–35 (1991).
28. H. Holtzmann, *Chem. Ing. Technol.* **39**, 89–95 (1967).
29. R. M. Heck, J. C. Bonacci, W. R. Hatfield, and T. H. Hsiung, *Ind. Eng. Chem. Proc. Des. Dev.* **21**, 73–79 (1982).
30. L. Handforth and J. N. Tilley, *Ind. Eng. Chem.* **26**(12) (1934).
31. F. Sperner and W. Hohmann, *Platinum Metals Rev.* (Jan. 1976).
32. S. P. S. Andrew, Ref. 4, pp. 31–40.
33. H. Connor, *Platinum Metals Rev.* (Jan. 1967).
34. H. Connor, *Platinum Metals Rev.* (Apr. 1967).
35. *Chem. Proc.* (May 1977).
36. G. R. Gillespie and D. Goodfellow, *Chem. Eng. Prog.* **70**(3) (Mar. 1974).
37. E. Baur and F. K. Pethick, *Nitrogen* (192), 19–22 (July–Aug. 1991).
38. *Nitrogen*, (183), 27–32 (Jan.–Feb. 1990).
39. B. T. Horner, *Nitrogen* (193), 32–36 (Sept.–Oct. 1991).

NO oxidation

40. M. Bodenstein, *Z. Elektrochem.* **24**, 183 (1918).
41. M. Bodenstein, *Z. Phys. Chem.* **100**, 68 (1922).
42. J. D. Grieg and P. G. Hall, *Trans. Faraday Soc.* **63**, 665 (1967).

NO$_x$ absorption mechanisms

43. P. J. Hoftyzer and F. J. G. Kwanten, in G. Nonhebel, ed., *Gas Purification Processes for Air Pollution Control*, Newnes–Butterworths, London, 1972, pp. 164–187.
44. J. J. Carberry, *Chem. Eng. Sci.* **14**, 189 (1959).
45. S. P. S. Andrew and D. Hanson, *Chem. Eng. Sci.* **14**, 105 (1961).
46. D. N. Miller, *AIChE J.* **33**, 1351 (1987).

Models

47. M. Koukolik and J. Marek, *Proceedings of the 4th European Symposium on Chemical Reaction Engineering*, Brussels, 1968, pp. 347–359.
48. H. Holma and J. Sohlo, *Comput. Chem. Eng.* **3**, 135–141 (1979).
49. U. Hoffmann and G. Emig, *Ger. Chem. Eng.* **2**, 282–293 (1979).
50. R. M. Counce and J. J. Perona, *AIChE J.* **29**, 26–32 (Jan. 1983).
51. K. W. Wiegand, E. Scheibler, and M. Thiemann, *Chem. Eng. Technol.* **13**, 289–297 (1990).
52. S. P. S. Andrew, Ref. 4, pp. 41–59.
53. "Latest Nitric Acid Process Design," *Nitrogen* (165) (Jan.–Feb. 1987).
54. "Absorber Design in Nitric Acid Plants," *Nitrogen* (188), 21–28 (Nov.–Dec. 1990).

NO$_x$ abatement

55. *A Review of Standards of Performance for New Stationary Sources*, EPA-450/8-84-011, EPA, Research Triangle Park, N.C., 1984.
56. *Tech. Notes on Best Avail. Tech. Not Entailing Excessive Cost for Nitric Acid Prod.*, Comm. Eur. Communities, Report EUR 13004 EN, Luxembourg, 1990.
57. *Nitrogen* (171), 25–32 (Jan.–Feb. 1988).
58. *Alternative Control Techniques Document: Nitric and Adipic Acid Manufacturing Plants*, EPA-450/3-91-026, EPA, Research Triangle Park, N.C., Dec. 1991.
59. L. Hellmer, *Chem. Eng.* **82**, 98–99 (1975).
60. A. Horton, in A. I. More, ed., *Fert. Acids, Proc. Br. Sulphur Corp. Int. Conf. Fert.*, 3rd ed., Vol. 1, Paper 5, 1979.

Strong nitric and concentration processes

61. G. Fauser, *Chem. Met. Eng.* **39**, 430 (1932).
62. L. Hellmer, *Chem. Eng. Prog.* **68**, 67–71 (Apr. 1972).
63. T. Ohrul, M. Okubo, and O. Imai, *Hydrocarbon Proc.* **57**, 163 (Nov. 1978).
64. L. M. Marzo and J. M. Marzo, *Chem. Eng.* **87**, 54–55 (Nov. 1980).
65. R. J. H. Hanbury, *Chem. Eng.*, 50–51 (Dec. 1972).
66. *Nitrogen* (129) (Jan.–Feb. 1981).
67. G. D. Honti, Ref. 4, pp. 99–129.
68. *Chem. Eng. News* **36**, 40–41 (June 5, 1958).

Other processes

69. N. Cilbert and F. Daniels, *Ind. Eng. Chem.* **40**, 1719 (1948).
70. E. G. Foster and F. Daniels, *Ind. Eng. Chem.* **43**, 986, 992 (1951).
71. E. D. Ermenc, *Chem. Eng. Prog.* **42**, 149, 488 (1956).
72. P. Harteck and S. Dondes, *Science* **146**, 30 (1964).
73. C. G. Vayenas and R. D. Farr, *Science* **208**, 593 (May 9, 1980).

Materials of construction

74. J. Eimers, Ref. 4, pp. 149–157.
75. "Materials for Nitric Acid," *Nitrogen*, (203) (May–June 1993).

76. R. D. Cooks, *Materials of Construction for Nitric Acid*, Process Industries Corrosion, National Association of Corrosion Engineers, 1986, pp. 259–263.
77. D. W. McDowell, Jr., "Handling Mineral Acids," *Chem. Eng.* (Nov. 11, 1974).
78. T.-L. Yau, *NACE* **39**(5) (May 1983).

Process performance and economics

79. "Opinion," *Fertilizer Focus*, 38–44 (Mar. 1990).
80. P. Lesur and M. Pottier, in A. I. More, ed., *Fert. Acids, Proc. Br. Sulphur Corp. Int. Conf. Fert.*, 3rd ed., Vol. 1, Paper 3, 1979.
81. "Opinion," *Fertilizer Focus*, 13–21 (Mar. 1988).
82. "Opinion," *Fertilizer Focus*, 6–11 (Mar. 1992).
83. K. Hopfer and I. Dayasagar, *Recent Adv. Inorg. Acids. Ind., Proc. Lect. Ser.*, Indian Chemical Manufacturers Association, Bombay, India, 1980.
84. L. M. Marzo and J. M. Marzo, *Fert. Acids, Proc. Br. Sulphur Corp., 3d Int. Conf. Fert.*, London, Nov. 1979.
85. S. M. Mokashi, *Recent Adv. Inorg. Acids. Ind., Proc. Lect. Ser.*, Indian Chemical Manufacturers Association, Bombay, India, 1980.

Production and pricing statistics

86. L. Fujise, D. H. Lauriente, M. Jaeckel, and Y. Sakuma, *CEH Mrk. Res. Rpt. Nitric Acid*, SRI International, Menlo Park, Calif., 1991.
87. M. Reisch, *Chem. Eng. News* (Apr. 12, 1993).
88. *Chem. Prod. Synopsis*, Mannsville Chemical Products Corp., Ashbury, N.J., Apr. 1990.
89. *Chem. Mark. Rep.*, Schnell Publishing Co. Inc., New York, July 1, 1989–1993.

Packaging and shipping requirements

90. *DOT Hazardous Materials Code of Federal Regulations*, Title 49, Washington, D.C.

Specifications and standards

91. *Reagent Chemicals*, 8th ed., ACS Specifications, ACS, Washington, D.C., 1993.
92. *U.S. Pharmacopeia XXII and National Formulary XVII*, U.S. Pharmacopeial Conventions, Inc., Washington, D.C., 1990.

Analytical and test methods

93. W. R. Jones and P. Jandik, "New Methods for Chromatographic Separations of Anions," *Am. Lab.* (June 1990).

Safety and health

94. L. Parmeggiani, ed., *Encyclopedia of Occupational Health and Safety*, Vol. 2, 3rd ed., International Labor Office, Geneva, Switzerland, 1983.
95. *Dangerous Properties of Industrial Materials Report*, May–June 1985, pp. 64–67.

STEPHEN I. CLARKE
WILLIAM J. MAZZAFRO
Air Products and Chemicals, Inc.

NITRIDES

At elevated temperatures and pressures, nitrogen combines with most elements to form nitrogen compounds. In the presence of metals and semimetals, it forms nitrides where nitrogen has a nominal valence of -3. Atomic nitrogen, which reacts much more readily with the elements than does molecular nitrogen, forms nitrides with elements that do not react with molecular nitrogen even at very high pressures. The binary compounds of nitrogen are shown in Figure 1 (1). These compounds may be classified, according to their chemical and physical properties, into four groups: salt-like, metallic, nonmetallic or diamond-like, and volatile nitrides. The nitrides of the high melting transition metals, eg, TiN, ZrN, and TaN, are characterized by high melting points, hardness (qv), and resistance to corrosion and are referred to as refractory hard metals (see REFRACTORIES). The nonmetallic compounds, eg, BN, Si_3N_4, and AlN, are corrosion- and heat-resistant ceramic-like industrial materials having semiconductor properties (see ABRASIVES; SEMICONDUCTORS).

An alphabetical list of nitrides together with CAS Registry Numbers is given in Table 1.

Properties

Salt-Like Nitrides. The nitrides of the electropositive metals of Groups 1 (IA), 2 (IIA), and 3 (IIIB) form salt-like nitrides having predominantly heteropolar (ionic) bonding and are regarded as derivatives of ammonia. The composition of these nitrides is determined by the valency of the metal, eg, Li_3N, Ca_3N_2, and ScN. The thermodynamic stability of the salt-like nitrides increases with increasing group number. For example, the nitrides of the alkali metals are only marginally or not at all stable, whereas the rare-earth metals are effective nitrogen scavengers in metals and alloys (see LANTHANIDES). The salt-like nitrides generally are electrical insulators or ionic conductors, eg, Li_3N. The nitrides of the Group 3 (IIIB) metals are metallic conductors or at least semiconductors and thus represent a transition to the metallic nitrides. The salt-like nitrides are characterized by sensitivity to hydrolysis. These compounds react readily with water or moisture to give ammonia and the metal oxides or hydroxides.

Lithium nitride can be prepared by the reaction of lithium metal and nitrogen. Lithium nitride crystals are dark red and melt at 845°C. The electrical ionic conductivity of the bulk material is 6.6×10^{-4} S/cm at 25°C and 8.3×10^{-2} S/cm at 450°C. The ionic conductivity of single crystals of Li_3N is extremely anisotropic. The value parallel to the c-axis is 1×10^{-5} S/cm at 20°C; the value perpendicular to the c-axis is 1.2×10^{-3} S/cm.

Metallic Nitrides.   Properties of metallic nitrides are listed in Table 2. The nitrides of the transition metals of Groups 6 and 7 (IVB–VIIB) generally are termed metallic nitrides because of metallic conductivity, luster, and general metallic behavior. These compounds, characterized by a wide range of homogeneity, high hardness, high melting points, and good corrosion resistance, are grouped with the carbides (qv), borides (see BORON COMPOUNDS), and silicides (see SILICON COMPOUNDS) as refractory hard metals. They crystallize in highly

1 (IA)	2 (IIA)	3 (IIIB)	4 (IVB)	5 (VB)	6 (VIB)	7 (VIIB)	8 (VIIIB)	9 (VIIIB)	10 (VIIIB)	11 (IB)	12 (IIB)	13 (IIIA)	14 (IVA)	15 (VA)	16 (VIA)	17 (VIIA)	18 (VIIIA)
H_3N																	He
Li_3N	Be_3N_2											BN	$(CN)_2$ CN_2	N_2	O_2N ON_2 ON	F_3N	Ne
Na_3N	Mg_3N_2											AlN	Si_3N_4	PN	SN	Cl_3N	Ar
K_3N	Ca_3N_2	ScN	Ti_2N $TiN_{0.9}$ TiN	V_2N VN	Cr_2N CrN	Mn_4N Mn_2N Mn_3N_2	Fe_4N Fe_2N	Co_3N Co_2N	Ni_3N	Cu_3N	Zn_3N_2	GaN	Ge_3N_4	AsN	SeN	Br_3N	Kr
Rb_3N	Sr_3N_2 Sr_2N	YN	ZrN	Nb_4N_3 Nb_2N NbN $NbN_{0.95}$	Mo_2N MoN	$TcN_{0.75}$	Ru	Rh	Pd	Ag_3N	Cd_3N_2	InN	Sn_3N_2 Sn_3N_4	SbN	TeN	I_3N	Xe
Cs_3N	Ba_3N_2	LaN	Hf_3N_2 HfN	Ta_3N_5 Ta_2N TaN $TaN_{0.8}$ $TaN_{0.1}$	W_2N WN	Re_2N	Os	Ir	Pt	Au_3N	Hg_3N_2	TlN	Pb_3N_4 Pb_3N_2	BiN	Po?	At?	Rn
Fr?	Ra?	Ac?	Rf?	Ha?	106?												

CeN	PrN	NdN	Pm?	SmN	EuN	GdN	TbN	DyN	HoN	ErN	TmN	YbN	LuN
Th_3N_4 ThN	PaN_2	U_2N_3 UN	NpN	PuN	AmN	CmN	BkN	Cf?	Es?	Fm?	Md?	No?	Lr?

Fig. 1. Binary compounds of nitrogen. See Table 1.

109

Table 1. Alphabetical List of Nitrides

Compound	CAS Registry Number	Formula
aluminum nitride	[24304-00-5]	AlN
americium nitride	[12296-96-1]	AnN
ammonia	[7664-41-7]	NH_3
antimony nitride	[12333-57-2]	SbN
arsenic nitride	[26754-98-3]	AsN
barium nitride	[12047-79-9]	Ba_3N_2
berkelium nitride	[56509-31-0]	BkN
berylium nitride	[1304-54-7]	Be_3N_2
bismuth nitride	[12232-97-2]	BiN
boron nitride	[10043-11-5]	BN
bromine nitride	[15162-90-0]	Br_3N
cadmium nitride	[12380-95-9]	Cd_3N_2
californium nitride	[70420-43-8]	CfN
calcium nitride	[12013-82-0]	Ca_3N_2
carbon nitride	[12069-92-0]	CN_2
cerium nitride	[25764-08-3]	CeN
chlorine nitride	[10025-85-1]	Cl_3N
chromium nitride	[24094-93-7]	CrN
chromium nitride (2:1)	[12053-27-9]	Cr_2N
cobalt nitride (2:1)	[12259-10-8]	Co_2N
cobalt nitride (3:1)	[12432-98-3]	Co_3N
copper nitride	[1308-80-1]	Cu_3N
curium nitride	[56509-28-5]	CmN
cyanogen	[2074-87-5]	$(CN)_2$
dinitrogen tetraoxide	[10544-72-6]	N_2O_4
dysprosium nitride	[12019-88-4]	DyN
erbium nitride	[12020-21-2]	ErN
europium nitride	[12020-58-5]	EuN
fluorine nitride	[13967-06-1]	F_3N
gadolinium nitride	[25764-15-2]	GdN
gallium nitride	[25617-97-4]	GaN
germanium nitride	[12065-36-0]	Ge_3N_4
gold nitride	[13783-74-9]	Au_3N
hafnium nitride (1:1)	[25817-87-2]	HfN
hafnium nitride (3:2)	[12508-69-9]	Hf_3N_2
holmium nitride	[12029-81-1]	HoN
indium nitride	[25617-98-5]	InN
iodine nitride	[21297-03-1]	I_3N
iron nitride (2:1)	[12023-20-0]	Fe_2N
iron nitride (4:1)	[12023-64-2]	Fe_4N
lanthanum nitride	[25764-10-7]	LaN
lead nitride (3:2)	[58572-21-7]	Pb_3N_2
lead nitride (3:4)	[75790-62-4]	Pb_3N_4
lithium nitride	[26134-62-4]	Li_3N
lutetium nitride	[12125-25-6]	LuN
magnesium nitride	[12057-71-5]	Mg_3N_2
manganese nitride (2:1)	[12163-53-0]	Mn_2N
manganese nitride (3:2)	[12033-03-3]	Mn_3N_2

Table 1. *(Continued)*

Compound	CAS Registry Number	Formula
manganese nitride (4:1)	[12033-07-7]	Mn_4N
mercury nitride	[12136-15-1]	Hg_3N_2
molybdenum nitride	[12033-19-1]	MoN
molybdenum nitride (2:1)	[12033-31-7]	Mo_2N
neodymium nitride	[25764-11-8]	NdN
neptunium nitride	[12058-90-1]	NpN
nickel nitride	[12033-45-3]	Ni_3N
niobium nitride	[11092-17-4]	NbN
niobium nitride (2:1)	[12033-63-5]	Nb_2N
niobium nitride (4:3)	[12163-98-3]	Nb_4N_3
nitrogen	[7727-37-9]	N_2
nitrous oxide	[10024-97-2]	N_2O
phosphorus nitride	[17739-47-8]	PN
plutonium nitride	[12033-54-4]	PuN
potassium nitride	[29285-24-3]	K_3N
praseodymium nitride	[25764-09-4]	PrN
protactinium nitride	[75733-54-9]	PaN_2
rhenium nitride	[12033-55-5]	Re_2N
rubidium nitride	[12136-85-5]	Rb_3N
samarium nitride	[25764-14-1]	SmN
scandium nitride	[25764-12-9]	ScN
selenium nitride	[12033-59-9]	SeN
silver nitride	[20737-02-4]	Ag_3N
α-silicon nitride	[12033-89-5]	Si_3N_4
sodium nitride	[12136-83-3]	Na_3N
strontium nitride	[12033-82-8]	Sr_3N_2
sulfur nitride	[28950-34-7]	SN
tantalum nitride	[12033-62-4]	TaN
tantalum nitride (2:1)	[12033-63-5]	Ta_2N
tantalum nitride (3:5)	[12033-94-2]	Ta_3N_5
tellerium nitride	[59641-84-8]	TeN
terbium nitride	[12033-64-6]	TbN
thallium nitride	[12033-67-9]	TlN
thorium nitride	[12033-65-7]	ThN
thorium nitride (3:4)	[12033-90-8]	Th_3N_4
thulium nitride	[12033-68-0]	TmN
tin nitride (3:2)	[75790-61-3]	Sn_3N_2
tin nitride (3:4)	[75790-62-4]	Sn_3N_4
titanium nitride (1:1)	[25583-20-4]	TiN
titanium nitride (2:1)	[12169-08-3]	Ti_2N
tungsten nitride	[12058-38-7]	WN
tungsten nitride (2:1)	[12033-72-6]	W_2N
uranium nitride	[25658-43-9]	UN
vanadium nitride	[24646-85-3]	VN
vanadium nitride (2:1)	[12209-81-3]	V_2N
ytterbium nitride	[24600-77-9]	YbN
yttrium nitride	[25764-13-0]	YN
zinc nitride	[1313-49-1]	Zn_3N_2
zirconium nitride	[25658-42-8]	ZrN

Table 2. Properties of Metallic Nitrides[a]

Nitride	Color	Lattice parameter,[b,c] nm	Density, g/cm³	Hardness[d]	Mp, °C	Heat conductivity, W/(m·K)	Coefficient of thermal expansion, K⁻¹ × 10⁻⁶	Electrical resistivity, μΩ·cm	Transition temperature, K
TiN	golden yellow	0.4246	5.43	2000	2950	29.1	9.35	25	4.8
ZrN	pale yellow	0.4577	7.3	1520	2980	10.9	7.24	21	9
HfN	greenish yellow	0.4518	14.0	1640	3330	11.1	6.9	33	
VN	brown	0.4139	6.10	1500	2350	11.3	8.1	85	7.5
NbN	dark gray	0.4388	8.47	1400	2630 dec	3.8	10.1	78	15.2
ε-TaN	dark gray	[e]	14.3	1100	2950 dec	9.54		128	1.8
δ-TaN	yellowish gray	0.4336	15.6	3200	2950 dec				17.8
CrN	gray	0.4150	6.14	1090	1080 dec[f]	11.7		640	not superconductive
Mo₂N	gray	0.416[g]	9.46	1700	790 dec[h]		6.7		5.0
W₂N	gray	0.412[g]	17.7		dec				
ThN	gray	0.5159	11.9	600	2820			20	
UN	dark gray	0.4890	14.4	580[i]	2800	15.5	8.0	176	
PuN	dark gray	0.4907	14.4		2550				

[a]Ref. 1.
[b]Structures are fcc of the NaCl type, unless otherwise noted.
[c]Lattice parameters are room temperature values.
[d]Value is microhardness unless otherwise noted.
[e]Structure is hexagonal B 35. Lattice parameters are α: 0.5191 nm; c: 0.2906.
[f]At 0.1 MPa (14.5 psi).
[g]Structures are fcc.
[h]At 0.7 MPa (101.5 psi).
[i]Value is Knoop hardness.

112

symmetrical, metal-like lattices. The small nitrogen atoms occupy the interstitial voids within the metallic host lattice forming interstitial alloys similar to the generally isotypic carbides. Metallic nitrides can be alloyed with other nitrides and carbides of the transition metals to give solid solutions. Complete solid solubility has been demonstrated for a great number of combinations (2). Similarly, oxynitride and oxycarbide interstitial alloys form over wide O–N and C–N composition ranges (3). At high temperatures, all pseudobinary systems between cubic mononitrides and monocarbides of the 4 (IVB) and 5 (VB) metals show complete miscibility, with the exception of the pairs ZrN–VN, HfN–VN, ZrN–VC, HfN–VC, and HfC–VN. TaN has a cubic high temperature modification that is completely miscible with all other cubic monocarbides and mononitrides (4,5) (Table 3).

Although there are several hundred binary nitrides, only a relative few ternary bimetallic metal nitrides are known (6). A group of ternaries of the composition $M_x M'_y N_z$, where M is an alkali, alkaline-earth, or a rare-earth metal and M' is a transition or post-transition metal, have been synthesized (6). Most of these compounds react readily with available oxygen, including H_2O. These compounds have simple stoichiometries, integral values of x,y, and exhibit conductivities ranging from insulating to metal-like.

Metallic nitrides are wetted and dissolved by many liquid metals and can be precipitated from metal baths. The stoichiometry is determined not by the valency of the metal, but by the number of interstitial voids per host atom. The metallic nitrides are stable against water and all nonoxidizing acids except hydrofluoric acid. The thermodynamic stability decreases with increasing group number from the nitrides of the 4 (IVB) metals. The nitrides of Mo and W can

Table 3. Solubilities in the Nitride–Nitride and Nitride–Carbide Systems[a]

Compound	ZrN	HfN	VN	NbN	TaN Cubic	TaN Hexagonal
			Nitride–nitride system			
TiN	○	○	○	○	○	◎
ZrN		○	●	○	○	◎
HfN			●	○	○	◎
VN				○	○	◎
NbN					○	◎
TaN (cubic)						◎
			Nitride–carbide system			
TiC	○	○	○	○	○	◎
ZrC	○	○	●	○	○	◎
HfC	○	○	●	○	○	◎
VC	●	●	○	○	○	◎
NbC	○	○	○	○	○	◎
TaC	○	○	○	○	○	◎

[a] Completely miscible (○), partially miscible in the cubic phase (◎), and not at all or very slightly miscible (●).

be prepared only by the action of nitrogen under high pressure or reaction with atomic nitrogen or dissociating ammonia. The same is true for the nitrides of the iron group metals (7,8). No nitrides of the platinum-group noble metals are known.

Nonmetallic (Diamond-Like) Nitrides. Some properties of nonmetallic nitrides are listed in Table 4. The nitrides of some elements of Groups 13 (IIIA) and 14 (IVA) eg, BN, Si_3N_4, AlN, GaN, and InN, are characterized by predominantly covalent bonding. These are stable chemically, have high degrees of hardness (eg, cubic BN) and high melting points, and are nonconductive or semiconductive. The structural elements of diamond-like nitrides are tetrahedral, M_4N, which are structurally related to diamond. Although the most common graphite-like form of BN does not contain these structural elements, boron nitride is considered a diamond-like nitride for two reasons: the existence of a diamond-like form at high pressures and the chemical and physical behavior of BN. Diamond-like nitrides have stoichiometric compositions having no homogeneity range and, as a rule, do not form solid solutions with each other. The preparation and properties of hexagonal BN are discussed elsewhere (see BORON COMPOUNDS, REFRACTORY).

Silicon nitride can be heated in air up to 1450–1550°C. In nitrogen, inert gas, or reducing atmosphere, Si_3N_4 can be heated up to 1750°C. Above 1750°C,

Table 4. Properties of Nonmetallic (Diamond-Like) Nitrides

Nitride	Structure	Lattice parameter,[a] nm	Density, g/cm^3	Micro-hardness	Maximum stability temperature, °C	Heat conductivity, W/(m·K)	Coefficient of thermal expansion, $\beta \times 10^{-6}$
BN	hexagonal		2.3	[b]	3000	15	7.51
	a	0.2504					
	c	0.6661					
	fcc, Zn blende	0.3615	3.4	[c]			
AlN	hexagonal wurtzite		3.05	1230	2200	30	4.03
	a	0.311					
	c	0.4975					
GaN	hexagonal wurtzite		5.0		600[d]		
	a	0.319					
	c	0.518					
Si_3N_4	hexagonal		3.2	3340	1900	17	2.75
	a	0.7748					
	c	0.5618					
	hexagonal						
	a	0.7608					
	c	0.2911					

[a]Values are at room temperature.
[b]Like graphite.
[c]Approaching diamond.
[d]In vacuo.

decomposition and sublimating evaporation become severe. When in the presence of carbon, however, Si_3N_4 stability depends on temperature and pressure. The equilibrium temperature for the reaction

$$Si_3N_4 + 3\ C \longrightarrow 3\ SiC + 2\ N_2$$

at normal pressure is 1700°C. Under reduced pressure of 130 mPa (1 μm Hg), the decomposition temperature is <1100°C and at 3.0 MPa (435 psi), the decomposition temperature is >1900°C (9).

Volatile Nitrides. The nitrogen compounds of the nonmetallic elements generally are not very stable. These nitrides decompose at elevated temperatures. Some are explosive and decompose upon shock. They form distinct molecules similar to organic compounds, and at low temperatures are gaseous, liquid, or easily volatilized solids. Exceptions are $(SN)_x$, which is polymeric, chemically stable, and has semimetallic properties; and $(PNCl_2)_x$, which has attracted some scientific interest as inorganic rubber (see INORGANIC HIGH POLYMERS). None of the volatile nitrides has obtained any substantial industrial application except ammonia (hydrogen nitride) and nitrogen oxide (oxygen nitride). Gaseous nitrogen fluorides are explosive; Cl_3N, a dark-yellow liquid, evaporates somewhat on heating and explodes. I_3NNH_3 [15823-38-8] detonates at the slightest touch.

Preparation

Nitriding Metals or Metal Hydrides. Metals or metal hydrides may be nitrided using nitrogen or ammonia. Pure metal powders or pure metal hydride powders yield nitride products that are nearly as pure as the precursors.

The nitrides of Groups 4(IVB) and 5(VB) elements form at ca 1200°C. The nitrides of magnesium and aluminum form at 800°C. Aluminum nitride, obtained by heating aluminum powder in the presence of ammonia or nitrogen at 800–1000°C, is formed as a white to grayish blue powder. A grade of especially high purity results from the decomposition of $AlCl_3$–NH_3 vapor mixtures.

The nitrides of the alkali earth metals form at 300–400°C and lithium nitride can form at room temperature. Raising the temperature shortens the reaction time and promotes a complete reaction. Nitrides of metals that do not react or react slowly with molecular nitrogen at normal pressure may require pressures up to 100 MPa (14,500 psi) or more (7). Even these high pressures do not suffice for the preparation of thermodynamically unstable nitrides, eg, the nitrides of the iron group metals (Fe, Co, Ni), rhenium nitride, Re_2N, and the high nitrides of Mo and W (8). In these cases, nitriding in a stream of purified ammonia at 600–1000°C leads to the formation of the desired nitrides. Atomic nitrogen is an even more powerful nitriding agent and can be produced by ionizing molecular nitrogen by the action of electrical discharges (10). These transition-metal nitrides can also be formed at high temperatures in a self-propagating, high temperature combustion reaction of loosely packed metal exposed to a gas.

A combustion front separates the products from reactants and the procedure can be made self-sustaining by gas recycle–restock and careful control of the combustion regime (11).

Metal Oxides. A process based on the reaction of metal oxides and nitrogen or ammonia in the presence of carbon is economical and has possibilities for large-scale production because less expensive metal oxides can be used in place of metal powders. However, the products, which contain oxygen and carbon, are not very pure. Removal of residual amounts of oxygen and carbon is difficult, especially in cases where carbon and oxygen atoms are taken into solution within the nitride lattice, which is usually the case for nitrides of the transition metals. Low reaction temperatures generally favor the formation of nitrides and high temperatures promote stable carbides, leading to carbon contamination.

Metal Compounds. Many nitrides, eg, BN, AlN, TiN, ZrN, HfN, CrN, Re_2N, Fe_2N, Fe_4N, and Cu_3N, may be prepared by the reaction of the corresponding metal halide and ammonia. An intermediate step in this method is the thermal decomposition of the ammonia–halide complex that forms. Nitrides also may be obtained by the reaction of ammonia and oxygen-containing compounds, eg, oxyhalides such as $VOCl_3$ and CrO_2Cl_2; ammonium–oxo complexes, eg, NH_4VO_3 and NH_4ReO_4; or oxides such as GeO_2, B_2O_3, and V_2O_3, and ferrous metal oxides. These nitrides, however, are not very pure and may contain residual oxygen and halogen. On nitriding pure carbides at 3–30 MPa (435–4350 psi) and 1100–1700°C, the carbides of the 4(IVB) and 5(VB) metals are transformed into carbonitrides and free carbon, with the exception of TaC, which is stable. The carbides of the 6(VIB) metals react to form $Cr_3(C,N)_2$ and $Mo(C,N)$. Like TaC, WC is stable under these conditions (12).

Precipitation from the Gas Phase. The van Arkel gas decomposition process gives especially pure nitrides and nitride films, which under certain conditions may precipitate as single crystals. The nitrides include TiN, ZrN, HfN, VN, NbN, BN, and AlN. In this process, a gaseous reaction mixture consisting of a volatile metal halide, nitrogen, and hydrogen is conducted over a hot substrate, eg, tungsten wire. The metal halide decomposes and the resulting nitride deposits on the wire. The deposition temperature is from 1000 to 1500°C, and this procedure is limited by the thermal stability of the nitride.

Other Methods of Preparation. The nitrides, Si_3N_4, Ge_3N_4, Zn_3N_2, Cd_3N_2, and Ni_3N, also may be produced by thermal decomposition of the corresponding metal amide or imide. Rb_3N and Cs_3N are obtained by azide decomposition. AlN and Si_3N_4 can be produced by the carbothermal reduction of intercalation compounds, magadiite– and montmorillonite–polyacrylonitrile (13). Nitrides low in nitrogen also can be synthesized from nitrides having a higher nitrogen content by decomposition in a vacuum or by reduction with hydrogen. For example, UN can be produced from U_2N_3, Co_3N from Co_2N, and Ta_2N from TaN. Nitrides and complex nitrides of ferrous metals and of high melting transition metals present in steels and superalloys can be isolated by electrolysis (14).

Silicon nitride occurs in two forms, α-Si_3N_4 and β-Si_3N_4. Pure Si_3N_4 is white, but the colors of commercial materials may be tan, gray, or black because of residual silicon or impurities. Si_3N_4 may be prepared by nitriding silicon powder at 1200–1400°C or, for extremely fine-grained Si_3N_4, by the reaction of $SiCl_4$ or SiH_4 and N_2 or NH_3 (see also ADVANCED CERAMICS).

The formation of nitrides from gaseous halides, ammonia, and nitrogen (atomic and molecular) in a plasma torch is possible (see PLASMA TECHNOLOGY). A specific type of plasma processing called cathodic arc plasma deposition (CAPD) has been used successfully to produce films of nitrides and carbonitrides (15). Material is evaporated by vacuum arcing. The source of the material is the cathode in the circuit and is ignited from this surface to create the arc. The voltages necessary range from 15 to 50 V and the arc spots generate the plasma supply. Plasma nitriding offers several advantages over other processes. It is nonpolluting and energy efficient, provides flexible deposition conditions without sacrifice of quality, minimizes distortion, and is easily applicable to compound film deposition.

Ion implantation (qv) directly inserts nitrogen into metal surfaces. A carefully polished and cleaned metal surface at room temperature in a vacuum ($\sim$0.133 mPa (1-μm Hg)) can be directly implanted with 80-keV nitrogen ions (10) (see METAL SURFACE TREATMENTS, CASE HARDENING). In an alternative synthesis, argon ions (Ar^+) of 8 keV can be used to ionize gas-phase nitrogen to obtain the same results (17).

Nitride-Containing Layers. Besides case hardening, the hardening, ie, increase in nitrogen content, achieved by nitriding special alloy steels is technologically significant in the heat treatment of high quality parts, such as gears. Hardness (qv) properties are imparted by the resulting coatings of needle-shaped precipitates of the nitrides and carbonitrides of iron, aluminum, chromium, molybdenum, etc. The nitriding steels (nitroalloy steels) that are developed especially for this process contain ca 0.4% C, 1% Al, 1.5% Cr, 0.2% Mo, 1% Ni, and trace amounts of other elements. In gas nitriding, the parts made from such steels are annealed in ammonia at ca 500°C for up to 100 h, whereby 0.1–1-mm thick coatings form. The hardness of these coatings exceeds that of the precipitation-hardened parts by ca 30%.

In nitriding or carbonitriding of condensed materials, molten cyanides (qv) are used at ca 570°C; this method produces fairly thick coatings of nitrides or carbonitrides after ca 1 h. Another process involves the use of an atmosphere containing activated nitrogen atoms or ions that are formed by the action of an electrical glow discharge; the temperature is ca 560°C and the time required for completion of the ion nitriding is 10–12 h (10). The advantage of all of these methods is the lack of distortion during surface hardening, unlike quench hardening, which usually results in at least small changes in dimensions and, at worst, in distortions.

Wear-resistant layers can be deposited on the surface of nearly every kind of material (eg, steel, cast iron, and cemented carbides) by a chemical vapor deposition (CVD) process (18–20) (see THIN FILMS, FILM FORMATION TECHNIQUES). Passing a stream of a mixture of gases containing, for example, $TiCl_4$ vapor, H_2, and N_2, over the surface of heated metallic or nonmetallic bodies results in the deposition of a thin uniform film of TiN. Optimum temperature for this process is 900–1100°C. The film is extremely hard and wear resistant and improves the cutting performance of carbide tips for machining steel and long chipping materials. Carbonitride Ti(C,N) can be deposited analogous to TiN by feeding N_2, $TiCl_4$, H_2, and CH_4 into the gas stream. Whereas the color of TiN is pure golden yellow, the color of $Ti(C_xN_{1-x})$ can be varied from red golden (ca $Ti(C_{0.1}N_{0.9})$) to a deep purple (ca $Ti(C_{0.4}N_{0.6})$). These films can be polished to a good finish with consequent excellent luster.

One disadvantage of CVD is the necessary high temperature for adherent and pore-free coatings. Deposition of TiN by a sputtering process obviates the use of high temperatures and results in substrate temperatures below 200°C; but consequently deposition rates generally are low (<1 mm/h). Ion plating is a process in which the following occur simultaneously: evaporation of titanium atoms, reaction with very dilute nitrogen gas in the gas phase, and deposition of these highly active atomic clusters onto a solid metallic surface. Thus higher deposition rates occur without a substrate temperature increase when ion plating is used (21). Cemented carbides also can be treated by this process, eg, $TiC-Mo_2C(Mo)-Ni$ or $WC-TiC-TaC-Co$. If the latter is high in TiC, it can be nitrided at 1150–1350°C in an autoclave. After ca 20 h at 4–8 MPa N_2, a 10-mm thick film that is high in Ti(C,N) is produced (22).

Manufacture and Processing

Nitride Coatings. Carbide tips coated with titanium nitride or titanium carbonitride usually are manufactured by a CVD process using $TiCl_4$, H_2, and N_2 in a hot-wall reactor. Most of the large carbide producers, eg, Metallwerk Plansee (Austria), Sandvik AB (Sweden), Carboloy, Division of General Electric, and Teledyne Firth Stirling (U.S.), and Krupp Widia (Germany), manufacture nitride-coated carbide tips. The portion of tips that are coated is as high as 50–70%. Most are nitride coated. Teledyne Firth Stirling manufactures HfN-coated tips as well as TiN- and Ti(C,N)-coated tips (22).

To obtain an adherent, uniform nitride coating, the lapped tips are positioned on grids made of heat-resistant wire inside the hot-wall reactor. The reactor is thoroughly flushed with nitrogen and hydrogen, the temperature is brought to ca 950–1100°C by moving the preheated furnace over the reactor, and $TiCl_4$ vapor is fed to the gas stream. The pressure within the furnace is usually kept below atmospheric, but it can be maintained at ambient pressure with equally good results. After completion of the coating cycle, the reactor is cooled by removing the furnace. The tips are removed after cooling and may be conditioned by sand blasting or tumbling to round the cutting edges.

Union Carbide (now Advanced Ceramics Technologies) developed a process similar to CVD to produce pyrolytic boron nitride-shaped bodies. The process employs graphite mandrels in a high temperature, low pressure reaction chamber. Vapors deposit on the mandrel to produce a thick, high purity, anisotropic, impervious BN layer. In many cases, the desired product is simply slipped off the cooled graphite mandrel. Increasingly, the desired product is not the free-standing pyrolytic BN object but a BN coating on a shaped graphite body. Graphite shapes having adherent BN coatings are used routinely for r-f susceptors, resistance heaters, heat shields, and nozzles (see ABLATIVE MATERIALS; REFRACTORY COATINGS).

Silicon Nitride. Silicon nitride is manufactured either as a powder as a precursor for the production of hot-pressed parts or as self-bonded, reaction-sintered, silicon nitride parts. α-Silicon nitride, used in the manufacture of Si_3N_4 intended for hot pressing, can be obtained by nitriding Si powder in an atmosphere of H_2, N_2, and NH_3. Reaction conditions, eg, temperature, time, and atmosphere, have to be controlled closely. Special additions, such as Fe_2O_3 to the precursor material, act as catalysts for the formation of predominately α-Si_3N_4. Silicon nitride is ball-milled to a very fine powder and is purified by acid leaching. Silicon nitride can be hot pressed to full density by adding 1–5% MgO.

Self-bonded reaction-sintered Si_3N_4 is manufactured according to the flow sheet in Figure 2. Silicon powder, ball-milled to <63 μm, eventually is purified by acid leaching to remove abraded iron particles, and is conditioned by adding an organic agent as a plasticizer or deflocculant. The mixture can be cold pressed,

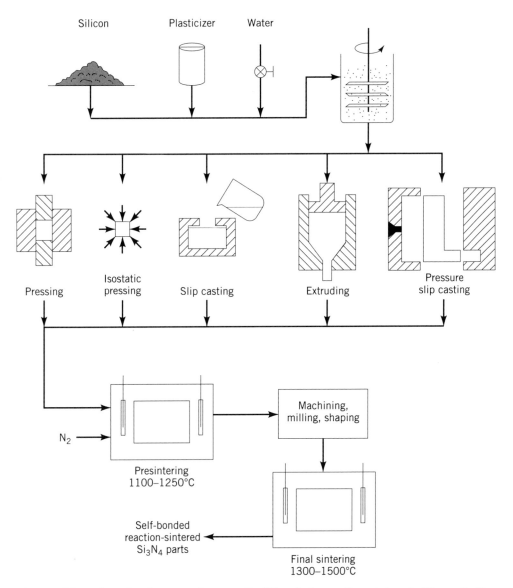

Fig. 2. Flow sheet for the manufacture of self-bonded, reaction-sintered silicon nitride. Courtesy of Annawerk, Ceranox (Roedental, Germany).

isostatically pressed, or extruded. Slip casting or pressure slip casting has been applied successfully to obtain complex preforms.

The organic bonder or the deflocculant is removed by heating the preforms and prenitriding them at 1000–1250°C. The prenitriding treatment imparts a strength level to the preforms that is sufficient to allow them to be handled, machined, milled, or drilled to the required tolerances using conventional tools. During the final nitriding stage at 1300–1500°C, all the silicon particles are converted to Si_3N_4. No shrinkage takes place either during the presintering process or during the final sintering stage.

Economic Aspects

Small amounts of TiN, HfN, and other metallic nitrides are produced on a pilot-plant scale. Titanium nitride is sold for $40–100/kg, depending on purity and grain size. Prices for HfN are ca $400/kg.

Annual production of powdered BN is ca 180–200 metric tons per year and its cost is $50–250/kg, depending on purity and density. The price of cubic boron nitride is similar to that of synthetic diamond bort. Hot-pressed, dense BN parts are 3–10 times more expensive than reaction-sintered parts.

Annual production of aluminum nitride is 50–100 t and it is sold for ca $40/kg. Extra high purity, ie, high heat conductive aluminum nitride, is sold for $50–70/kg.

Annual production of silicon nitride is ca 100–200 t. Utility-grade silicon nitride costs $4–5/kg in 100- to 500-kg quantities. The reaction-sintered parts are sold for $120 to $300/kg, depending on complexity of shape. Hot-pressed, fully dense Si_3N_4 parts are priced 5–10 times higher than reaction-sintered parts.

Analytical and Test Methods

Procedures for the quantitative analysis of nitrogen in nitrides are similar to those used for carbides (qv). The same is true for chemical assay methods, although the stability of many nitrides to acids creates difficulties, especially in the Kjeldahl determination of nitrogen content. Consequently, modifications of the Dumas combustion method have been developed (23,24). The Dumas method entails heating the nitride sample in a CO_2 stream in the presence of an oxidant. The nitrogen in the sample is given off and detected, usually by a gas chromatograph. Cu_2O, CuO, and V_2O_5 are commonly used as oxidants (25). Generally good results are obtained for most transition-metal nitrides, but a small amount of the nitrogen typically is not evolved and must be accounted for by hot extraction. Some of the commercially important nitrides cannot be analyzed by this modified method. AlN requires a V_2O_5 oxidant to obtain good results; BN and Si_3N_4 do not react with any of these oxidants. Boron nitride and silicon nitride can be analyzed by a method that makes use of the reaction of nitrides with molten alkalies (26). Nearly all of the nitrides, including the very stable ones, can be dissolved in polytetrafluoroethylene (PTFE) bombs using a mixture of hydrofluoric acid and perchloric acid at 190°C under pressure (27). The resulting solution contains all of the nitrogen in the form of NH_4^+ ions, unlike the Dumas method. Reproducible nitrogen values also are obtained by direct gas determination according to modified high temperature extraction processes (28).

Numerous spectroscopic methods for the qualitative and quantitative analysis of nitrides are available. Methods include x-ray diffraction, which is commonly used for identification of nitrides and the determination of crystalline structure. X-ray diffraction has also been used to characterize residual stress distributions in coated steels (see NONDESTRUCTIVE EVALUATION) (29). Electron microscopies and electron diffraction are commonly used to characterize high strength nitrides and alloys. Auger electron and Rutherford back-scattering spectroscopy are useful for determining the spatial distribution of atoms in nitride films (30). Auger electron spectroscopy is a useful method to analyze interfacial regions and in-depth profiling when combined with ion sputtering (31). X-ray photoelectron spectroscopy (xps) characterizes the nature of nitride chemical bonding by the determination of oxidation states of metals and defect structures (32). High resolution solid-state nmr has been utilized to describe nonstoichiometric nitrides and oxynitrides (33).

Health and Safety Factors

Toxicology.   As a chemical group, toxicity of nitrides generally stems from the possible reactions with water to form toxic fumes (especially ammonia) rather than from the nitride. There are, of course, exceptions. The salt-like nitrides decompose when in contact with water or moisture to form ammonia, which can irritate the respiratory organs and mucous membranes. The metallic nitrides are very stable chemically, but fine powder or dust of the nitrides of the transition metals can be pyrophoric, especially the nitrides of the actinide metals, UN, ThN, PuN, and those having high surface area, eg, Mo_2N and W_2N. These can ignite in air and during communitive operations. Moreover, nitrides of the actinide metals are carcinogenic.

The diamond-like nitrides, especially as dust, can irritate the lungs or cause scratching of the eyes owing to mechanical means. BN is particularly troublesome because of a lack of solubility in many compounds. Nitrides of the 11(IB) and 12(IIB) metals and especially the volatile nitrides have to be handled with extreme care because of their instability and high degree of toxicity. Hazards associated with the nitrides of the Group 11(IB) and 12(IIB) metals are determined by reactivity with other substances. For example, the addition of concentrated HNO_3 to copper(I) nitride creates a violent explosion. Mild heating of Hg_3N_2 evolves highly toxic mercury vapors and an eventual explosion (34).

Uses

High Strength and Hardness.   Most applications of metallic nitrides stem from their hardness. For example, aluminum nitride can be used to form structural parts that exhibit high degrees of hardness (see ADVANCED CERAMICS, STRUCTURAL). AlN can be prepared in the same way as ceramics (qv), ie, by powder metallurgy from nitride powders or by sintering mixtures of Al_2O_3 and C in the presence of nitrogen. These parts are stable to oxidizing and carbon-containing furnace gases up to 800°C. Above 800°C, they are stable only under a protective gas or in a vacuum. Because of good chemical and thermal stability, sintered parts made of AlN have been proposed for use in nozzles, thermocouple-protecting tubes, crucibles, and boats, eg, in the manufacture of semiconductors (35).

More recently, a liquid-phase metal sintering process was successfully applied to nitrides and carbonitrides, which normally have poor wetting properties. The oxygen content of the prematerial is kept below 0.05% and suitable bonding alloys based on Ni–Mo and Ni–Mo–C are introduced (36,37). This binder composition is especially effective in wetting the nitride or carbonitride particles during sintering and in removing the adhering oxide layers. Cemented titanium carbonitrides having a nickel–molybdenum binder compare well with cemented carbides of this ISO-P series corresponding to C-5 to C-8 grades of the U.S. industry code for cemented carbides (see CARBIDES, CEMENTED CARBIDES). TiN is scarcely soluble in solid iron and is much less so than TiC, hence TiN has favorable frictional qualities and little tendency toward local welding and seizing during cutting operations. Sealing rings of cemented carbonitride have proved satisfactory in difficult chemical environments.

Quaternary carbonitride alloys based on (Ti,Mo)(C,N) tend to separate into two isotypic phases: one (eg, Ti(C,N)) is rich in nitrogen and poor in molybdenum, and the other (eg, (Ti,Mo)C) contains nearly all of the molybdenum but little nitrogen. This spinodal decomposition reaction has been applied to the production of cemented carbonitride alloys, ie, spinodal alloys, that show superior cutting performance and very little wear, especially when used to machine cast iron and chilled cast iron (38). More recent work has been carried out with regard to TaN(HfN)–TaC(HfN) alloys and alloys containing the cubic TaN (39,40). Hot-pressed cemented hard metals based on TaN–ZrB$_2$ are far more suitable for the machining of high melting metals and superalloys than are cemented carbides (41).

The nitride coating of cemented carbide tips, ie, throwaway tips, has increased greatly. Layers of TiN, Ti(C,N), and HfN that are 3–6 mm thick are applied using CVD (18–20). HfN-containing layers also are produced. In a U.S. process, HfN coatings are manufactured by treating Hf sponge with elemental chlorine, thus introducing HfCl$_4$ *in situ* (23). An important European development features multilayers on carbide tips. The carbide surface first is coated with TiC, then with Ti(C,N), and finally by TiN layers; thus the excellent flank wear of TiC coatings is combined with the excellent crater wear of TiN coatings (42). The golden-colored TiN layers are used in the jewelry industry, mainly as the coating of scratch-free watch cases. In Japan, Ti(C,N,O) layers with up to 1.60% O$_2$ are used instead of carbide, nitride, or carbonitride layers (43).

Within the system Mo–C–N, a molybdenum–carbonitride-phase Mo(C,N) having the WC-type structure can be prepared (12). This phase is completely miscible with WC. Solid solutions (Mo,W)(C,N) between Mo(C,N) and WC behave similarly to WC when sintered with a cobalt or nickel binder and give cemented carbonitrides with properties that are similar to those that are based on WC–Co. In these cemented carbonitrides, a substantial part of tungsten can be replaced with molybdenum, with two parts of tungsten equivalent to one part of molybdenum. In situations of tungsten shortage, such as occurred in the mid-1970s, this replacement is expected to gain technical and economical significance (44).

Transition-metal nitrides improve hardness properties. Nitrogen-alloyed chromium steels, as compared to nitrogen-free austenitic steels, are characterized by increased strength without loss of toughness (45). The corrosion resistance does not decrease if the chromium content is increased to compensate for the amount of chromium that is bound by nitrogen. The industrial preparation of these steels takes place in pressure-melt equipment at ca 2.0 MPa (290 psi) nitrogen, whereby as much as 1.8% N may go into solution. As an alloying constituent, nitrogen can be applied not only by melting under pressure but by employing ferrochromium. This is obtained by treating 50–70% ferrochromium with nitrogen at 950–1150°C in calcium cyanamide pusher furnaces. From 3.5 to 4.5% N is taken up. Research on the ternary system of Group 4 (IVB), 5 (VB), and 6 (VIB) metals as bases for future nitride applications has been reported (36,37). Cubic NbN and Nb(C,N) are superconductors having high transition temperatures (NbN, 17.3 K; Nb(C,N), 17.8 K). Pseudobinary systems containing nitrides have topped these values (NbN–TiC and NbN–NbC, ~18.0 K). There have not been technical applications because of the extreme brittleness of these compounds (see SUPERCONDUCTING MATERIALS).

Nuclear Applications. Use of the nitrides of uranium-235 and thorium as fuels and breeders in high temperature reactors has been proposed (see NUCLEAR REACTORS). However, the compounds most frequently used for this purpose are the oxides and carbides. Nitrides could be useful in high temperature breeder reactors because of their stability when in contact with molybdenum as a cladding material. Only those mononitrides that are prepared by powder metallurgy, eg, by hot pressing into sintered parts, and those that are high melting, chemically stable, and metallic, as well as the solid solutions of such nitrides and monocarbides, possibly in the form of dispersions in stainless steels, may be suitable for use as nuclear fuel materials (46). The properties of interest for use in reactors have been studied in detail (see Table 2). The higher nitrides of uranium, ie, β-U_2N_3 and α-UN_{2-x}, and of thorium, ie, Th_3N_4, which might be formed during the reaction of the metals with ammonia or pressurized nitrogen, can be decomposed easily by treating the compounds in vacuum at high temperatures.

The rare-earth nitrides do not have any technical applications. These are high melting compounds but are hydrolyzed easily by moisture and are not stable under normal atmospheric conditions.

Solid Electrolytes. Of the salt-like nitrides, only Li_3N has attracted technical interest. Lithium nitride has an uncommonly high ionic conductivity in the solid state and hence is considered a suitable electrolyte for lithium–sulfur batteries, conferring a favorable capacity as compared to conventional batteries (qv). Lithium nitride is used as a solvent catalyst in the production of cubic BN.

Refractories. Hexagonal boron nitride is a soft white powder and resembles graphite in crystal structure, in texture, and in many other properties, except that it is an electrical insulator. It is used in the refractories industry as a mold-facing and release agent. Structural parts made of BN are manufactured by hot pressing. Characteristics include low density, easy workability, good heat resistance, and especially good thermal conductivity and stability to thermal shock, excellent corrosion resistance, and the ability to provide electrical insulation. Boron nitride-based structural parts are used as wall liners in plasma-arc devices, such as gas heaters, arc-jet thrusters, and high temperature magneto-hydrodynamic devices (47) (see MAGNETOHYDRODYNAMICS). BN also is used as a crucible material for reactive metal melts because of its nonwetting properties. Boron nitride composites, eg, $BN-Si_3N_4$ and $BN-SiC$, are stable to molten zinc and to covering salts, such as borax. As an additive, BN adjusts the electrical conductivity of sintered TiB–TiC boats that are used in vacuum evaporation of aluminum. Boron nitride also is used as a solid dopant source for semiconductors and would be a good material for use in electronic devices operating at high temperature.

Sintered parts of Si_3N_4 may be manufactured from Si_3N_4 powder or from silicon powder by means of cold pressing or slip casting with subsequent sintering in the presence of nitrogen. Silicon preforms that have been sintered under nitrogen at 1200°C have sufficient strength to be machined using conventional carbide tools to close dimensional tolerances. Thus even complicated forms (holes, grooves, threads, etc) can be prepared. During final sintering at 1200–1400°C, no change in dimensions occurs but hardness is increased to a level where only grinding with diamond wheels is effective. Parts with good strength may be obtained, however residual porosities are 15–30%. Such parts are characterized by low thermal expansion, good thermal shock resistance, resistance to creep, and high electrical resistivity. They exhibit good resistance to corrosion by acids but are less resistant to hydrofluoric acid and alkali–hydroxide solutions. They are also stable toward reactive gases and nonferrous metals (48,49).

Sintered parts of Si_3N_4 can be made into tubes, crucibles, boats, nozzles, etc. The tubes can be used for protecting thermocouples or for measuring the temperature in nonferrous melts or gas temperatures in the steel industry (see TEMPERATURE MEASUREMENTS). Other parts are suitable as linings of cooling and purification towers for SO_2 roasting gas, cyclone dust separators, spray nozzles, and combustion nozzles, or they may serve as parts in pumps for melting of light metals and linings for Wankel engines. An extensive research effort is being made to develop reaction-sintered or hot-pressed silicon nitride turbine blades as a high temperature material for use in automotive engines (50).

Silicon nitride is one of the few nonmetallic nitrides that is able to form alloys with other refractory compounds. Numerous solid solutions of β-Si_3N_4 and Al_2O_3 have gained technical interest. Many companies have begun to mass produce reaction-sintered and hot-pressed Si_3N_4 parts.

Abrasives. The graphitic form of BN can be converted into a high pressure diamond-like form by applying pressures of 5–9 GPa (50–90 kbar) and temperatures of 1500–2000°C in the presence of catalysts, such as alkali metals, alkaline-earth metals, antimony, tin, and lead (qv). Lithium nitride seems to be the best catalyst. Pure cubic BN is colorless, but the technical product always contains either excess boron and is brown or black, or excess lithium nitride and is yellow. The cubic boron nitride is stable in air at atmospheric pressure to 2000°C. Cubic boron nitride approaches the hardness of diamond and has been introduced successfully as an abrasive for steel, especially high alloy steel and superalloys (see ABRASIVES). Cubic BN is not used for grinding cemented carbides. The commercial use of cubic BN grit for grinding wheels has grown steadily over the years. The material competes with diamond bort (51) especially for grinding operations involving high alloy steels.

Coatings and Lubrication. Union Carbide has developed proprietary water-based and aerosol boron nitride coatings for use in plastics, metals, ceramics (qv), glass (qv), and composite material production. Typical applications include use as a release agent, high temperature lubricant, and protective coating for casting, forging, molding, heat treating, welding, brazing, and laser cutting. The coatings are inert to 800°C in oxidizing environments and 2000–3000°C in inert environments or vacuum.

Catalysis. The development of high surface area porous Mo_2N and W_2N has spurred interest in nitrides for catalytic applications (52–58). High surface area powders are produced by slow (0.6 K/min) heating of the respective oxides in rapidly flowing ammonia or nitrogen–hydrogen mixtures. Surface areas as high as 220 m^2/g have been produced. The freshly prepared nitrides are pyrophoric, oxidizing vigorously if exposed directly to air, and must be passivated by slow, controlled oxidation. It is possible to manipulate the solid-phase chemistry, replacing nitrogen in the metal lattice to form high surface area molybdenum carbides, borides, and metal. High specific surface area γ-Mo_2N has been studied as a catalyst for NH_3 synthesis, ethane hydrolysis, quinoline hydrodenitrogenation, CO hydrogenation, and hydrodesulfurization.

Electronic and Optoelectronic Applications. Aluminum nitride, characterized by covalent bonding, is a semiconductor having a relatively wide band gap. Owing to the high thermal conductivity and resistivity and reasonably constant dielectric constant of aluminum nitride, commercial applications exist as substrates for semiconductor chips and in manufacturing sintered polycrystalline ceramic heat sinks (59). Primarily because AlN has the highest known surface acoustic wave velocity, usage as an excellent surface acoustic wave device, even under adverse conditions, could be developed. The only problem is the difficulty in obtaining good single crystals (59). Optoelectronically, AlN emits in the uv range (200 nm). However, AlN is hydrolyzed slowly by water vapor and moisture. A p–n junction diode utilizing cubic BN with a part of the emission spectrum in the uv range has been created (59).

The mononitrides of Group 13(IIIA) metals Ga and In also have desirable electronic and optoelectronic properties (see also Table 4). GaN has potential usage as blue light-emitting diodes, in color displays, and in solid-state lasers. The same holds for InN, except the emission wavelength is around 600 nm (red). These compounds can be prepared by the solid-state reactions of Ga_2O_3 or In_2O_3 and ammonia. The utilization of the luminescence properties of GaN and InN is, however, limited by the thermal decomposition of these compounds, which occurs above 600°C (see LASERS; LIGHT-GENERATION, LIGHT-EMITTING DIODES).

BIBLIOGRAPHY

"Nitrides" in *ECT* 1st ed., Vol. 9, pp. 345–352, by L. S. Foster, Watertown Arsenal; in *ECT* 2nd ed., Vol. 13, pp. 814–825, by F. Benesovsky, Metallwerk Plansee A. G., Reutte, and R. Kieffer, Technische Hochschule, Vienna.; in *ECT* 3rd ed., Vol. 15, pp. 871–887, by F. Benesovsky, Metallwerk Plansee A. G., Reutte, and R. Kieffer and P. Ettmayer, Technische Hochschule, Vienna.

1. R. Kieffer and P. Ettmayer, *High Temperature, High Pressures* **6**, 253 (1974).
2. P. Duwez and F. Odell, *J. Electrochem. Soc.* **97**, 299 (1950).

3. N. E. Brese and M. O'Keefe, *Struct. Bonding (Berlin)*, **79**, 307 (1992).
4. R. Kieffer, H. Nowotny, P. Ettmayer, and G. Dufek, *Metallurg (Moscow)* **26**, 701 (1972).
5. F. Gatterer, G. Dufek, P. Ettmayer, and R. Kieffer, *Mh. Chem.* **106**, 1137 (1975).
6. F. J. Disalvo, *Science* **247**, 649 (1990).
7. P. Ettmayer, H. Priemer, and R. Kieffer, *Metallurg (Moscow)* **23**, 307 (1969).
8. D. J. Jack and K. H. Jack, *Mater. Sci. Eng.* **11**, 1 (1973).
9. H. Rassaerts and A. Schmidt, *Planseeber. Pulvermetall.* **14**, 110 (1966).
10. J. Klausler, *Fachber. Oberflachentech.* **6**, 201 (1968).
11. H. W. Dandekar, C. C. Agrafiotis, J. A. Puzynski, and V. Hlavacek, *Chem. Eng. Sci.* **45**, 2499 (1990).
12. R. Kieffer, H. Nomotny, P. Ettmayer, and M. Freudhofmeier, *Metallurg (Moscow)* **25**, 1335 (1971).
13. K. Kuroda, Y. Sugahara, and C. Kato, *Br. Ceram. Proc.* **37**, 15 (1986).
14. W. Koch, *Metallkundliche Analyse*, Stahleisen, Dusseldorf, 1965.
15. H. Randawa and P. C. Johnson, *Surf. Coat. Technol.* **31**, 303 (1987).
16. B. X. Lui, X. Zhou, and H. D. Li, *Phys. Status Solidi A* **113**, 11 (1989).
17. Y. Baba and T. A. Sasoki, *J. Vac. Sci. Technol. A* **6**, 2945 (1988).
18. R. Kieffer, D. Fister, H. Schoof, and K. Mauer, *Powder Metall. Int.* **4**, 1 (1973).
19. U.S. Pat. 3,717,496 (1970) (to Deutsche Edelstahlwerke Krefeld Germany).
20. W. Schintlmeister, O. Pacher, K. Pfaffinger, and T. Raine, *J. Electrochem. Soc.* **123**, 924 (1976).
21. H. K. Pulker, R. Buhl, and E. Moll, "Ion Plating," paper presented at *The 9th Plansee Seminar*, 1977.
22. O. Rudiger, H. Grewe, and J. Kolaska, *Wear* **48**, 267 (1978).
23. E. Rudy, B. F. Kieffer, and E. Baroch, *Planseeber. Pulvermetall.* **26**, 105 (1978).
24. J. Rottmann and H. Nickel, *Fresenius Anal. Chem.* **247**, 208 (1969).
25. W. Lengauer, *Talanta* **38**, 659 (1991).
26. H. Puxbaum and A. Vendl, *Fresenius 2. Anal. Chem.* **287**, 134 (1977).
27. W. Werner and G. Tolg, *Fresenius 2, Anal. Chem.* **276**, 103 (1975).
28. G. Paesold, K. Muller, and R. Kieffer, *Fresenius 2. Anal. Chem.* **232**, 31 (1967).
29. T. Hirsch and P. Mayr, *Surf. Coat. Technol.* **36**, 729 (1988).
30. K. Takahashi and M. Iwaki, *Nucl. Instrum. Methods Phys. Res., Sect. B* **45**, 669 (1990).
31. S. Hofmann, *J. Vac. Sci. Technol. A* **4**, 2789 (1986).
32. C. N. R. Rao and J. Gopalakrishnan, *New Directions in Solid State Chemistry*, Cambridge University Press, New York, 1986.
33. A. I. Gusev, *Phys. Status Solidi B* **156**, 11 (1989).
34. N. I. Sax, *Dangerous Properties of Industrial Materials*, Van Nostrand Reinhold Co., New York, 1979.
35. G. Long and L. M. Forster, *J. Electrochem. Soc.* **109**, 1176 (1962).
36. R. Kieffer and P. Ettmayer, *High Temperature, High Pressures* **6**, 253 (1974).
37. R. Kieffer, P. Ettmayer, and M. Freudhofmeier, *Metallurg (Moscow)* **25**, 1335 (1971); R. Kieffer, P. Ettmayer, and M. Freudhofmeier, in H. H. Hausner, ed., *Modern Developments in Powder Metallurgy*, Vol. 5, Plenum Press, New York, pp. 201–214, 1971.
38. E. Rudy, *J. Less Common Met.* **33**, 43 (1973).
39. R. Kieffer, G. Dufek, P. Ettmayer, and R. Ducreux, papers 5–7, presented at *The IV European Symposium for Powder Metallurgy*, Grenoble, France, 1975.
40. M. Komac, T. Kosmac, and F. Thummler, *Planseeber. Pulvermetall.* **25**, 101 (1977).
41. V. Murata and E. D. Whitney, *Am. Ceram. Soc. Bull.* **46**, 643 (1967); *Am. Ceram. Soc. Bull.* **48** 698 (1969); *Am. Ceram. Soc. Bull.* **47**, 617 (1968).
42. W. Schintlmeister and O. Pacher, *Metallurg (Moscow)* **28**, 690 (1974); *Planseeber. Pulvermetall.* **23**, 260 (1975).

43. T. Sadahiro, S. Yamaya, K. Shibuki, and N. Ujiie, *Wear* **48**, 291 (1978).
44. R. Kieffer, P. Ettmayer, and B. Lux, paper 33 presented at *The Recent Advances in Hardmetal Production Conference*, Loughborough, U.K., 1979.
45. J. Frehser and C. Kubisch, *Berg-u. Huettenmaenn. Montash.* **108**, 369 (1963).
46. S. J. Paprocki and co-workers, *Rep. Battelle Mem. Inst. BMI*, 1365 (1959).
47. J. Fredrickson and W. H. Redanz, *Met. Prog.* **87**(2), 97 (1965).
48. J. F. Collins and R. W. Gerby, *J. Met.* **7**, 612 (1955).
49. A. M. Sage and J. H. Histed, *Powder Met.* **8**, 196 (1961).
50. E. Gugel and G. Liemer, *Ber. Dtsch. Keram. Ges.* **50**, 151 (1973).
51. L. Coes, Jr., "Abrasives," in *Applied Mineralogy*, Vol. 1, Springer-Verlag, Vienna, 1971.
52. L. Volpe and M. Boudart, *J. Solid State Chem.* **59**, 332 (1985).
53. E. J. Markel and J. W. Van Zee, *J. Catal.* **126**, 643 (1990).
54. S. T. Oyama, J. C. Schlatter, J. E. Metcalf, and J. M. Lambert, *Ind. Eng. Chem. Res.* **27**, 1639 (1988).
55. L. Volpe and M. Boudart, *J. Phys. Chem.* **90**, 4878 (1986).
56. G. S. Ranhotra, A. T. Bell, and J. A. Reimer, *J. Catal.* **108**, 40 (1987).
57. J. C. Schlatter, S. T. Oyama, J. E. Metcalf, and J. M. Lambert, *Ind. Eng. Chem. Res.* **27**, 1648 (1988).
58. S. T. Oyama and D. J. Sajkowski, *Prepr. Am. Chem. Soc. Div. Pet. Chem.* **35**(2), 233 (1990).
59. R. F. Davis, *Proc. IEEE* **79**, 702 (1991).

General References

R. Marchand, Y. Laurent, J. Guyador, P. L'Haridon, and P. Verdier, *J. Eur. Cer. Soc.* **8**, 197 (1991).
W. Langauer and P. Ettmayer, *High Temp.–High Pressures* **22**, 13 (1990).
R. Freer, ed., NATO ASI Series: *The Physics and Chemistry of Carbides, Nitrides, and Borides*, Vol. 185, Kluwer Academic Publishers, Boston, Mass., 1990.
A. Rabenau, *Solid State Ionics* **6**, 277 (1982).
L. E. Toth, *Transition Metal Carbides and Nitrides*, Academic Press, Inc., New York, 1971.
G. V. Samsonov, *Nitridij*, Naukova Dumka, Kiev, USSR, 1969.
H. Goldschmidt, *Interstitial Alloys*, Butterworths, London, 1967.
R. Kieffer and F. Benesovsky, *Hartstoffe*, Springer-Verlag, Vienna 1963.

Eric J. Markel
M. E. Leaphart II
University of South Carolina

NITRIDING.   See Metal-surface treatments.

NITRILE RUBBER.   See Elastomers, synthetic.

NITRILES.   See Supplement.

NITRO ALCOHOLS

A nitro alcohol is formed when an aliphatic nitro compound with a hydrogen atom on the nitro-bearing carbon atom reacts with an aldehyde in the presence of a base. Many such compounds have been synthesized, but only those formed by the condensation of formaldehyde (qv) and the lower nitroparaffins (qv) are marketed commercially. The condensation may occur one to three times, depending on the number of hydrogen atoms on the nitro-substituted carbon (R and R' = H or alkyl), and yield nitro alcohols with one to three hydroxyl groups.

$$RR'CHNO_2 + CH_2O \overset{OH^-}{\rightleftharpoons} RR'C(CH_2OH)NO_2$$

$$RCH(CH_2OH)NO_2 + CH_2O \overset{OH^-}{\rightleftharpoons} RC(CH_2OH)_2NO_2$$

$$HC(CH_2OH)_2NO_2 + CH_2O \overset{OH^-}{\rightleftharpoons} C(CH_2OH)_3NO_2$$

In addition to the mononitro compounds, monohydric and dihydric dinitro alcohols have been prepared but are not available commercially. The formation, properties, and reactions of nitro alcohols have been reviewed (1,2).

Physical Properties

The physical properties of the commercially available nitro alcohols are given in Table 1. Except for nitrobutanol, these nitro alcohols are white crystalline solids when pure. They are thermally unstable above 100°C and purification by distillation is a hazardous procedure.

The nitro alcohols generally are soluble in water and in oxygenated solvents, eg, alcohols. The monohydric nitro alcohols are soluble in aromatic hydrocarbons; the diols are only moderately soluble even at 50°C; at 50°C the triol is insoluble.

Chemical Properties

The nitro alcohols can be reduced to the corresponding alkanolamines (qv). Commercially, reduction is accomplished by hydrogenation of the nitro alcohol in methanol in the presence of Raney nickel. Convenient operating conditions are 30°C and 6900 kPa (1000 psi). Production of alkanolamines constitutes the largest single use of nitro alcohols.

Nitro alcohols form salts upon mild treatment with alkalies. Acidification causes separation of the nitro group as N_2O from the parent compound, and results in the formation of carbonyl alcohols, ie, hydroxy aldehydes, from primary nitro alcohols and ketols from secondary nitro alcohols.

$$2\ RCHOHCH{=}\overset{+}{N}\overset{O^-}{\underset{O^-}{\diagup}}\ M^+ \overset{H^+}{\longrightarrow} 2\ RCHOHCH{=}\overset{+}{N}\overset{O^-}{\underset{O^-}{\diagup}}\ H^+ \longrightarrow 2\ RCHOHCHO\ +\ N_2O{\uparrow}$$

Table 1. Physical Properties and Toxicity Data of Nitro Alcohols[a]

Compound	CAS Registry Number	Structural formula	Mol wt	Mp, °C	Bp, °C	Solubility in water at 20°C g/100 mL	LD$_{50}$, g/kg
2-nitro-1-butanol (NB)	[609-31-4]	CH$_3$CH$_2$CCH$_2$OH, (H top, NO$_2$ bottom)	119.12	−47 to −48	105[b]	54	1.2
2-methyl-2-nitro-1-propanol (NMP)	[76-39-1]	CH$_3$CCH$_2$OH, (CH$_3$ top, NO$_2$ bottom)	119.12	90	94[c]	350	1.0
2-methyl-2-nitro-1,3-propanediol (NMPD)	[77-49-6]	HOCH$_2$CCH$_2$OH, (CH$_3$ top, NO$_2$ bottom)	135.12	ca 160	dec	80	4.0
2-ethyl-2-nitro-1,3-propanediol (NEPD)	[597-09-1]	HOCH$_2$CCH$_2$OH, (C$_2$H$_5$ top, NO$_2$ bottom)	149.15	56	dec	400	2.8
2-hydroxymethyl-2-nitro-1,3-propanediol (TRIS NITRO)	[126-11-4]	HOCH$_2$CCH$_2$OH, (CH$_2$OH top, NO$_2$ bottom)	151.12	175–176	dec	220	1.9

[a]Ref. 3.
[b]At 1.3 kPa (10 mm Hg).
[c]At 1.95 kPa (15 mm Hg).

Nitro alcohols react with amines to form nitro amines. Such a reaction can be carried out with a wide variety of primary and secondary amines, both aliphatic and aromatic; a basic catalyst is required if aromatic amines are involved. The products of reactions between dihydric nitro alcohols and amines are nitrodiamines, many of which are good fungicides (qv). Dihydric nitro alcohols, primary amines, and formaldehyde react to yield nitrohexahydropyrimidines (4). Nitrohexahydropyrimidines can be reduced to the corresponding amines, some of which are good fungicides or bactericides, eg, hexetidine [141-94-6] (5-amino-1,3–bis(2-ethylhexyl)-5-methylhexahydropyrimidine).

Esters of nitro alcohols with primary alcohol groups can be prepared from the nitro alcohol and an organic acid, but nitro alcohols with secondary alcohol groups can be esterified only through the use of an acid chloride or anhydride. The nitrate esters of the nitro alcohols are obtained easily by treatment with nitric acid (qv). The resulting products have explosive properties but are not used commercially.

On dehydration, nitro alcohols yield nitro-olefins. The ester of the nitro alcohol is treated with caustic or is refluxed with a reagent, eg, phthalic anhydride or phosphorus pentoxide. A milder method involves the use of methane sulfonyl chloride to transform the hydroxyl into a better leaving group. Yields up to 80% after a reaction time of 15 min at 0°C have been reported (5). In aqueous solution, nitro alcohols decompose at pH 7.0 with the formation of formaldehyde. One mole of formaldehyde is released per mole of monohydric nitro alcohol, and two moles of formaldehyde are released by the nitrodiols. However, 2-hydroxymethyl-2-nitro-1,3-propanediol gives only two moles of formaldehyde instead of the expected three moles. The rate of release of formaldehyde increases with the pH or the temperature or both.

Manufacture and Processing

The nitro alcohols available in commercial quantities are manufactured by the condensation of nitroparaffins with formaldehyde [50-00-0]. These condensations are equilibrium reactions, and potential exists for the formation of polymeric materials. Therefore, reaction conditions, eg, reaction time, temperature, mole ratio of the reactants, catalyst level, and catalyst removal, must be carefully controlled in order to obtain the desired nitro alcohol in good yield (6). Paraformaldehyde can be used in place of aqueous formaldehyde. A wide variety of basic catalysts, including amines, quaternary ammonium hydroxides, and inorganic hydroxides and carbonates, can be used. After completion of the reaction, the reaction mixture must be made acidic, either by addition of mineral acid or by removal of base by an ion-exchange resin in order to prevent reversal of the reaction during the isolation of the nitro alcohol (see ION EXCHANGE).

The purification of liquid nitro alcohols by distillation should be avoided because violent decompositions and detonation have occurred when distillation was attempted. However, if the distillation of a nitro alcohol cannot be avoided, the utmost caution should be exercised. Reduced pressure should be utilized, ie, ca 0.1 kPa (<1 mm Hg). The temperature of the liquid should not exceed 100°C; hot water should be used as the heating bath. A suitable explosion-proof shield should be placed in front of the apparatus. At any rise in pressure, the distillation should be stopped immediately. The only commercially produced liquid nitro alcohol, 2-nitro-1-butanol, is not distilled because of the danger of decomposition. Instead, it is isolated as a residue after the low boiling impurities have been removed by vacuum treatment at a relatively low temperature.

Economic Aspects

The nitro alcohols in Table 1 are manufactured in commercial quantities; however, three of the five of them are used only for the production of the corresponding amino alcohols. 2-Methyl-2-nitro-1-propanol (NMP) is available as the crystalline solid or as a mixture with silicon dioxide. 2-Hydroxymethyl-2-nitro-1,3-propanediol is available as the solid ($9.15/kg), a 50% solution in water ($2.33/kg), a 25% solution in water ($1.41/kg), or as 1-oz (28.3-g) tablets ($76.00/case of 144).

Health, Safety, and Environmental Factors

Acute oral LD_{50} data for nitro alcohols in mice are given in Table 1. Because of their low volatility, the nitro alcohols present no vapor inhalation hazard. They are nonirritating to the skin and, except for 2-nitro-1-butanol, are nonirritating when introduced as a 1 wt % aqueous solution in the eye of a rabbit. When 0.1 mL of 1 wt % commercial-grade 2-nitro-1-butanol in water is introduced into the eyes of rabbits, severe and permanent corneal scarring results. This anomalous behavior may be caused by the presence of a nitro-olefin impurity in the unpurified commercial product.

   Because it is the nitro alcohol with greatest potential for human exposure, additional testing of 2-hydroxymethyl-2-nitro-1,3-propanediol has been conducted. In a 90-day dermal study in rats, no effects were observed with animals exposed to 1000 mg/kg/d except for a slight yellow staining at the site of application. No teratogenic effects were noted in either rats or rabbits when tested by gavage. Additionally, 2-hydroxymethyl-2-nitro-1,3-propanediol was not found to be mutagenic in *in vitro* tests, ie, the salmonella reverse-mutation test, a chromosome aberration study with Chinese hamster ovary cells, and in an unscheduled DNA synthesis study with primary rat hepatacytes.

   In a battery of tests, which determine the tendency of chemicals to inhibit aquatic organisms, accumulate in such organisms, and degrade in the environment, 2-hydroxy-2-nitro-1,3-propanediol was found to have low potential for harm in the environment (7).

Uses

The nitro alcohols are useful as intermediates for chemical synthesis. In particular, they are used to introduce a nitro functionality and, by reduction of the resultant intermediate, an amino functionality.

   Antimicrobials.   In slightly alkaline aqueous solutions, nitro alcohols are useful for the control of microorganisms, eg, in cutting fluids, cooling towers, oil-field flooding, drilling muds, etc (8–15) (see INDUSTRIAL ANTIMICROBIAL AGENTS;

PETROLEUM). However, only 2-hydroxymethyl-2-nitro-1,3-propanediol (TRIS NITRO) is registered by the U.S. Environmental Protection Agency (EPA). Under the provisions of the 1988 amendments to the Federal Insecticide, Fungicide and Rodenticide Act (FIFRA), EPA has published a "reregistration eligibility document" which allows for the continued registration of this useful biocide (16).

Polymers. All nitro alcohols are sources of formaldehyde for cross-linking in polymers of urea, melamine, phenols, resorcinol, etc (see AMINO RESINS). Nitrodiols and 2-hydroxymethyl-2-nitro-1,3-propanediol can be used as polyols to form polyester or polyurethane products (see POLYESTERS; URETHANE POLYMERS). 2-Methyl-2-nitro-1-propanol is used in tires to promote the adhesion of rubber to tire cord (qv). Nitro alcohols are used as hardening agents in photographic processes, and 2-hydroxymethyl-2-nitro-1,3-propanediol is a cross-linking agent for starch adhesives, polyamides, urea resins, or wool, and in tanning operations (17–25). Wrinkle-resistant fabric with reduced free formaldehyde content is obtained by treatment with 2-methyl-2-nitro-1-propanol as the cross-linker (26).

Stabilizers. Nitro alcohols can be used to prevent the decomposition of p-phenylenediamine color-developing agents (27). 2-Hydroxymethyl-2-nitro-1,3-propanediol and 2-nitro-1-butanol have been used as additives for the stabilization of 1,1,1-trichloroethane.

Other. 2-Nitro-1-butanol is an excellent solvent for many polyamide resins, cellulose acetate butyrate, and ethylcellulose. It can be utilized in paint removers for epoxy-based coatings. 2-Hydroxymethyl-2-nitro-1,3-propanediol is useful for control of odors in chemical toilets. Its slow release of formaldehyde ensures prolonged action to control odor, and there is no reodorant problem which sometimes is associated with the use of free formaldehyde. 2-Hydroxymethyl-2-nitro-1,3-propanediol solutions are effective preservative and embalming fluids. The slow liberation of formaldehyde permits thorough penetration of the tissues before hardening.

BIBLIOGRAPHY

"Nitro Alcohols" in *ECT* 1st ed., Vol. 7, pp. 375–381, by E. B. Hodge, Commercial Solvents Corp.; in *ECT* 2nd ed., Vol. 13, pp. 826–834 by R. H. Dewey, Commercial Solvents Corp.; in *ECT* 3rd ed., Vol. 15, pp. 910–916, by A. F. Bollmeier.

1. H. B. Hass and E. F. Riley, *Chem. Rev.* **32**, 373 (1943).
2. B. M. Vanderbilt and H. B. Hass, *Ind. Eng. Chem.* **32**, 34 (1940).
3. *NP Series, Technical Data Sheet No. 15*, ANGUS Chemical Co., Jan. 1989.
4. M. Senkus, *J. Am. Chem. Soc.* **68**, 10 (1946).
5. J. Melton and J. E. McMurry, *J. Org. Chem.* **40**, 2138 (1975).
6. W. E. Noland, in H. E. Baumgarten, ed., *Organic Syntheses*, Coll. Vol. 5, John Wiley & Sons, Inc., New York, 1973, pp. 833–838.
7. D. Freitag and co-workers, *GSF-Ber.*, 73–83 (Apr. 1992).
8. H. O. Wheeler and E. O. Bennett, *Appl. Microbiol.* **4**, 122 (1956).
9. E. O. Bennett and H. N. Futch, *Lubr. Eng.* **16**, 228 (1960).
10. U.S. Pat. 3,001,936 (Sept. 26, 1961), E. O. Bennett and E. B. Hodge (to Commercial Solvents Corp.).
11. U.S. Pat. 3,789,008 (Jan. 29,1974), D. W. Young (to Elco Chemicals Inc.).
12. Ger. Pat. 2,530,522 (Jan. 27, 1977), P. Voegele (to Henkel & Cie GmbH).

13. U.S. Pat. 3,542,533 (Nov. 25, 1970), M. S. Beach (to Eastman Kodak Co.).
14. U.S. Pat. 4,113,444 (Sept. 12, 1978), P. M. Bunting and co-workers (to Gulf Research & Development Co.).
15. U.S. Pat. 3,592,893 (July 13, 1971), H. G. Nosler and co-workers (to Henkel & Cie GmbH).
16. "Tris(hydroxymethyl)nitromethane," *Reregistration Eligibility Decision*, U.S. EPA, Washington, D.C., 1993.
17. U.S. Pat. 3,897,583 (July 29, 1975), C. Bellamy (to Uniroyal, SA).
18. M. L. Happich and co-workers, *J. Am. Leather Chem. Assoc.* **65**(3), 135 (1970).
19. U.S. Pat. 3,809,585 (May 7, 1974), H. L. Greenberg (to U.S. Dept. of the Navy).
20. Jpn. Pat. 74 94,748 (Sept. 9, 1974), Y. Hori and co-workers (to Dai-ichi Kogyo Seiyaku Co., Ltd.).
21. U.S. Pat. 3,982,993 (Sept. 28, 1976), R. L. Fife (to Georgia-Pacific Corp.).
22. Ger. Pat. 1,958,914 (Aug. 6, 1970), J. Delmenico and co-workers (to Commonwealth Scientific and Ind. Research Org.).
23. U.S. Pat. 3,475,383 (Oct. 28, 1969), F. D. Stewart (to B. F. Goodrich Co.).
24. U.S. Pat. 4,039,495 (Aug. 2, 1977), J. H. Hunsucker (to IMC Chemical Group, Inc.).
25. U.S. Pat. 4,298,638 (Nov. 3, 1981), J. H. Hunsucker (to ANGUS Chemical Co.).
26. U.S. Pat. 4,431,699 (Feb. 14, 1984), J. H. Hunsucker (to ANGUS Chemical Co.); U.S. Pat. 4,478,597 (Oct. 23, 1984), J. H. Hunsucker (to ANGUS Chemical Co.).
27. Brit. Pat. 1,468,015 (Mar. 23, 1977), R. Cowell and co-workers (to May and Baker Ltd.).

ALLEN F. BOLLMEIER, JR.
ANGUS Chemical Company

NITROBENZENE AND NITROTOLUENES

Nitrobenzene

Nitrobenzene [98-95-3] (oil of mirbane), $C_6H_5NO_2$, is a pale yellow liquid with an odor that resembles bitter almonds. Depending on the purity, its color varies from pale yellow to yellowish brown.

Nitrobenzene was first synthesized in 1834 by treating benzene with fuming nitric acid (1), and was first produced commercially in England in 1856 (2). The relative ease of aromatic nitration has contributed significantly to the large and varied industrial applications of nitrobenzene, other aromatic nitro compounds, and their derivatives.

Physical Properties. Nitrobenzene is readily soluble in most organic solvents and is completely miscible with diethyl ether and benzene. Nitrobenzene is only slightly soluble in water with a solubility of 0.19 parts per 100 parts of water at 20°C and 0.8 pph at 80°C. Nitrobenzene is a good organic solvent. For example, it is used in Friedel-Crafts reactions because aluminum chloride is soluble in nitrobenzene. The physical properties of nitrobenzene are summarized in Table 1.

Chemical Properties. Nitrobenzene reactions involve substitution on the aromatic ring and reactions involving the nitro group. Under electrophilic conditions, the substitution occurs at a slower rate than for benzene, and the nitro group promotes meta substitution. Nitrobenzene can undergo halogenation, sulfonaton, and nitration, but it does not undergo Friedel-Crafts reactions. Under nucleophilic conditions, the nitro group promotes ortho and para substitution.

The reduction of the nitro group to yield aniline is the most commercially important reaction of nitrobenzene. Usually the reaction is carried out by the catalytic hydrogenation of nitrobenzene, either in the gas phase or in solution, or by using iron borings and dilute hydrochloric acid (the Bechamp process). Depending on the conditions, the reduction of nitrobenzene can lead to a variety of products. The series of reduction products is shown in Figure 1 (see AMINES BY REDUCTION). Nitrosobenzene, N-phenylhydroxylamine, and aniline are primary reduction products. Azoxybenzene is formed by the condensation of nitrosobenzene and N-phenylhydroxylamine in alkaline solutions, and azoxybenzene can be reduced to form azobenzene and hydrazobenzene. The reduction products of nitrobenzene under various conditions are given in Table 2.

Manufacturing and Processing. Nitrobenzene is manufactured commercially by the direct nitration of benzene using a mixture of nitric and sulfuric

Table 1. Physical Properties of Nitrobenzene

Property	Value	Reference
mp, °C	5.85	3
bp, °C		3
at 101 kPa[a]	210.9	
13 kPa[a]	139.9	
0.13 kPa[a]	53.1	
density, g/cm^3		
d_4^{0b}	1.223	3
d_4^{10}	1.213	3
d_4^{25}	1.199	4
refractive index, n_D^{20}	1.55296	3
viscosity at 15°C, mPa·s(=cP)	2.17	5
surface tension at 20°C, mN/m(=dyn/cm)	46.34	5
dielectric constant at 25°C	34.82	5
specific heat at 30°C, J/g^c	1.509	5
latent heat of vaporization, J/g^c	331	5
latent heat of fusion, J/g^c	94.2	5
heat of combustion at constant volume, MJ/mol^c	3.074	3
flash point (closed cup), °C	88	4
autoignition temperature, °C	482	6
explosive limit at 93°C, vol % in air	1.8	6
vapor density (air = 1)	4.1	4

[a]To convert kPa to mm Hg, multiply by 7.5.
[b]Supercooled liquid.
[c]To convert J to cal, divide by 4.184.

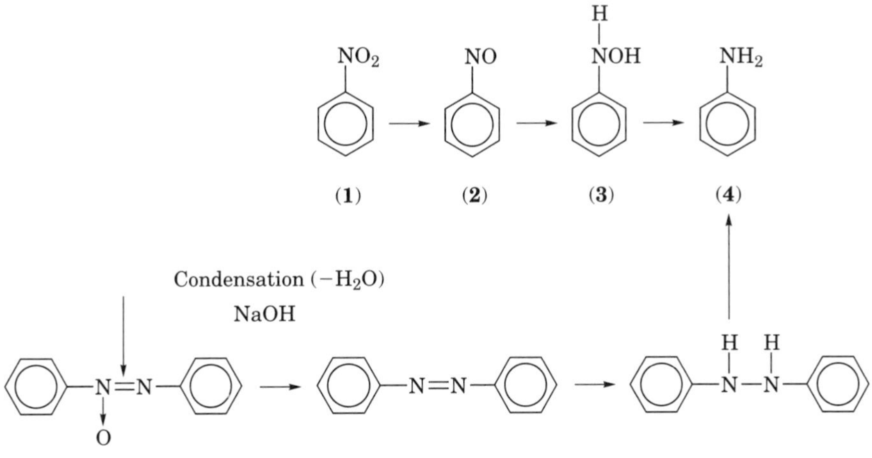

Fig. 1. Reduction products of nitrobenzene (**1**): nitrosobenzene [*98-95-3*] (**2**); *N*-phenylhydroxylamine [*100-65-2*] (**3**); aniline [*62-53-3*] (**4**); azoxybenzene [*495-48-7*] (**5**); azobenzene [*103-33-3*] (**6**); and hydrazobenzene [*122-66-7*] (**7**).

Table 2. Reduction Products of Nitrobenzene

Reagent	Product
Fe, Zn, or Sn + HCl	aniline
H_2 + metal catalyst + heat (gas phase or solution)	aniline
$SnCl_2$ + acetic acid	aniline
Zn + NaOH	hydrazobenzene, azobenzene
Zn + H_2O	*N*-phenylhydroxylamine
Na_3AsO_3	azoxybenzene
$LiAlH_4$	azobenzene
$Na_2S_2O_3$ + Na_3PO_4	sodium phenylsulfamate, $C_6H_5NHSO_3Na$

acids, which commonly is referred to as mixed acid or nitrating acid. Because two phases are formed in the reaction mixture and the reactants are distributed between them, the rate of nitration is controlled by mass transfer between the phases as well as by chemical kinetics (7). The reaction vessels are acid-resistant, glass-lined steel vessels equipped with efficient agitators. By vigorous agitation, the interfacial area of the heterogeneous reaction mixture is maintained as high as possible, thereby enhancing the mass transfer of reactants. The reactors contain internal cooling coils which control the temperature of the highly exothermic reaction (see EXPLOSIVES AND PROPELLANTS; NITRATION). In addition, batch reactors can be equipped with circulating external heat-exchange loops. This allows for efficient temperature control during the beginning of the reaction, when the reactor is not full enough to touch the internal cooling coils.

Nitrobenzene can be produced by either a batch or a continuous process. With a typical batch process, the reactor is charged with benzene, then the nitrating acid (56–60 wt % H_2SO_4, 27–32 wt % HNO_3, and 8–17 wt % H_2O) is added slowly below the surface of the benzene. The temperature of the mixture is maintained at 50–55°C by adjusting the feed rate of the mixed acid and the amount of cooling. The temperature can be raised to about 90°C toward the end of the reaction to promote completion of reaction. The reaction mixture is fed into a separator where the spent acid settles to the bottom and is drawn off to be refortified. The crude nitrobenzene is drawn from the top of the separator and washed in several steps with a dilute base, such as sodium carbonate, sodium hydroxide, magnesium hydroxide, etc, and then water. Depending on the desired purity of the nitrobenzene, the product can be distilled. Usually a slight excess of benzene is used to ensure that little or no nitric acid remains in the spent acid. The batch reaction time generally is 2–4 hours, and typical yields are 95–98 wt % based on benzene charged.

Because a continuous nitration process generally offers lower capital costs and more efficient labor usage than a batch process, most if not all of the nitrobenzene producers use continuous processes. A typical continuous process for the production of nitrobenzene is given in Figure 2. Benzene and the nitrating acid (56–65 wt % H_2SO_4, 20–26 wt % HNO_3, and 15–18 wt % water) are fed into the nitrator, which can be a stirred cylindrical reactor with internal cooling coils and external heat exchangers or a cascade of such reactors. The nitrator also can be designed as a tubular reactor, eg, a tube-and-shell heat exchanger with appropriate cooling, involving turbulent flow (8). Generally, with a tubular reactor, the reaction mixture is pumped through the reactor in a recycle loop and a portion of the mixture is withdrawn and fed into the separator. A slight excess of benzene usually is fed into the nitrator to ensure that the nitric acid in the nitrating acid is consumed to the maximum possible extent and to minimize the formation of dinitrobenzene. The temperature of the nitrator is maintained at 50–100°C by varying the amount of cooling. The reaction mixture flows from the nitrator into a separator or centrifuge where it is separated into two phases. The aqueous phase or spent acid is drawn from the bottom and concentrated in a sulfuric acid reconcentration step or is recycled to the nitrator where it is mixed with nitric and sulfuric acid immediately prior to being fed into the nitrator. The crude nitrobenzene flows through a series of washer–separators where residual acid is removed by washing with dilute base followed by final washing with water. The product then is distilled to remove water and benzene, and if required, the nitrobenzene can be refined by vacuum distillation. Reaction times of 10–30 minutes are typical, and theoretical yields are 96–99%. The nitration process is unavoidably associated with the disposal of wastewater from the washing steps. This water can contain nitrobenzene, mono- and polynitrated phenolics, carboxylic acids, other organic by-products, residual base, and inorganic salts from the neutralized spent acid which was present in the product. Treatment of this wastewater represents a significant expense for the producer. Generally wastewater is extracted with benzene to remove the nitrobenzene, and benzene that is dissolved in the water is stripped from the water prior to the final waste treatment.

In the last few years several modifications to the traditional mixed acid nitration procedure have been reported. An adiabatic nitration process was developed for the production of nitrobenzene (9). This method eliminated the

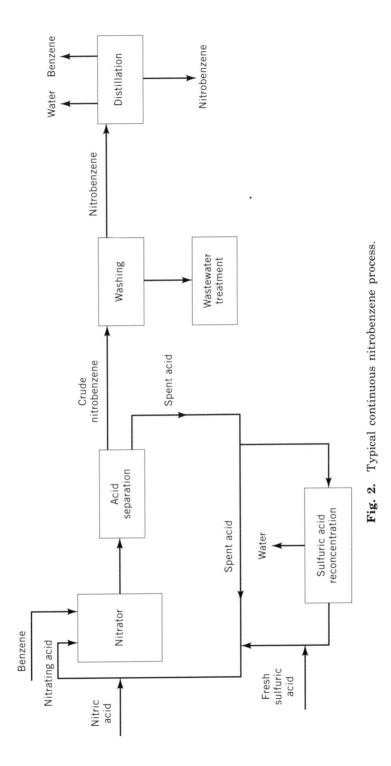

Fig. 2. Typical continuous nitrobenzene process.

need to remove the heat of reaction by excessive cooling. The excess heat can be used in the sulfuric acid reconcentration step. An additional advantage of this method is the reduction in reaction times to 0.5–7.5 minutes.

For the process step involving handling of spent sulfuric acid, several patents have been issued in which improvements in this step were a main claim. The azeotropic nitration of benzene essentially eliminates the need to reconcentrate sulfuric acid (10,11). The nitration step is carried out at higher than usual temperatures (120–160°C). Because excess benzene is used, the higher temperature allows water to be removed as a water–benzene azeotrope. The water is separated and the benzene phase, containing approximately 8% nitrobenzene, is recycled back into the reactor. The dry sulfuric acid is then reused continuously.

Another concentration method involves passing an inert gas such as N_2 or CO_2 through the reaction medium (12). As the gas passes through, it becomes humidified and carries captured water with it. Most of the energy required for the gas humidification comes from the heat of reaction. An advantage is that expensive drying equipment is not needed. Also, the sulfuric acid mist formed in typical concentrators is minimized. Du Pont uses a similar process in its nitrobenzene production facility.

When the spent sulfuric acid must be reconcentrated, it has been found that reconcentration to 100% sulfuric acid is not necessary (13). An energy savings can be realized if the sulfuric acid is concentrated under a vacuum to only 75–92%. If the process is carried out at 130–195°C, these medium concentration ranges are sufficient to destroy trace organics, thus preventing any loss in the efficiency or capacity of the nitration process. This process is adaptable to existing manufacturing installations.

Because the highest possible interfacial area is desired for the heterogeneous reaction mixture, advances have also been made in the techniques used for mixing the two reaction phases. Several jet impingement reactors have been developed that are especially suited for nitration reactions (14). The process boosts reaction rates and yields. It also reduces the formation of by-products such as mono-, di-, and trinitrophenol by 50%. First Chemical (Pascagoula, Mississippi) uses this process at its plant. Another technique is to atomize the reactant layers by pressure injection through an orifice nozzle into a reaction chamber (15). The technique uses pressures of typically 0.21–0.93 MPa (30–135 psi) and consistently produces droplets less than 1 μm in size. The process is economical to build and operate, is safe, and leads to a substantially pure product.

Due to increasingly strict environmental regulations, effort has been put into reducing the amount of contaminants in the waste stream of the nitration process. For instance, residual nitrobenzene can be removed from wastewater with a multistage extraction process in which the organic and aqueous phases are run countercurrent to each other (16). In the first step the organic phase comes into contact with a countercurrent stream of alkali water which neutralizes residual acid. In the second step, the organic phase is extracted with water to remove residual alkali. Finally, the combined water layers are extracted with a stream of 0.5 parts of benzene per 1 part water. The benzene extracts only nitrobenzene, leaving any nitrophenols or picric acids in the water. This method can extract up to 99.44% of nitrobenzene from the wastewater.

The need for neutralization of the organic phase with alkali can be reduced by extracting the acidic contaminants using molten salts, preferably a mixture of zinc nitrate and magnesium nitrate, at 55 to 70°C (17). Since the two phases are not miscible, the organic layer can be removed, leaving behind the acidic contaminants in the molten salt. The salts are then regenerated by flashing off nitric acid. If necessary, the organic phase can undergo a polishing neutralization.

The acidic contaminants can also be removed by employing a system that utilizes extractions, precipitation, distillation, and other treatments for rendering the waste stream acceptable for current disposal standards (18–20). First Chemical Corp. uses such a system. Residual nitric acid can be removed by a multistage countercurrent liquid–liquid extraction. The nitric acid (ca 25%) is then reconcentrated by distillation for further use.

Environmental aspects, as well as the requirement of efficient mixing in the mixed acid process, have led to the development of single-phase nitrations. These can be divided into liquid- and vapor-phase nitrations. One liquid-phase technique involves the use of ≥ 98% by weight nitric acid, with temperatures of 20–60°C and atmospheric pressure (21). The molar ratios of nitric acid:benzene are 2:1 to 4:1. After the reaction is complete, excess nitric acid is vacuum distilled and recycled. An analogous process is used to simultaneously produce a nitrobenzene and dinitrotoluene mixture (22). A conversion of 100% is obtained without the formation of nitrophenols or nitrocresols. The nitrobenzene and dinitrotoluene are separated by distillation.

The use of vapor-phase nitrations has been one of the most active areas in aromatic nitration chemistry since the 1970s. Although several approaches have been reported, most of the patents issued have one technique in common: the use of solid nitration catalysts (qv). The nitric acid–benzene mixture flows through the solid catalyst in a continuous process using the appropriate molar ratios, temperatures, and pressures. Obviously this technique eliminates the need for neutralizing or washing the product to remove acidic contaminants. The catalysts range from silica–alumina types to transitional metal catalysts. The percent conversion varies depending on the catalyst used. Table 3 summarizes some of the reported results.

Economic Aspects. The two main areas affecting the economic aspects for nitrobenzene production are process related costs, including raw material costs, energy requirements, waste treatment, etc, and the U.S. and world demand for products made from nitrobenzene. The most significant costs related to the production process are the raw material costs. These typically are at least 85% of the production costs for nitrobenzene. Annual statistics for the U.S. nitrobenzene

Table 3. Vapor-Phase Nitration of Benzene

Catalyst	Conversion, %	Reference
alumina–silica–metal oxide	99	23
acidic sheet clay, acidic composite oxides	91	24
solid supported sulfuric catalysts	96	25
silica–alumina zeolites	95	26
molecular sieves, ≥ 5×10^{-4} μm	92	27
nitrate salts (KNO_3, $NaNO_3$, $LiNO_3$)	40	28

market are shown in Table 4 (29,30). As of mid-1995, the principal U.S. producers of nitrobenzene were Rubicon Inc., First Chemical Corp., Du Pont, BASF Corp., and Bayer Corp. (formerly Miles, Inc).

Specifications and Test Methods.   Specifications for double-distilled nitrobenzene are listed in Table 5. Several qualitative spot tests are applicable to nitrobenzene and depend on a characteristic color developed by its reaction with certain reagents (31). However, these tests are not specific to nitrobenzene because other aromatic nitro compounds yield colored products that are similar or only slightly different in color. One example of such a test is the heating or fusing of nitrobenzene with diphenylamine, which yields a reddish yellow color.

Table 4. U.S. Production, Sales, and Prices of Nitrobenzene[a]

Year	Production, t	Sales, t	Price,[b] $/kg
1960	73,600	2,800	0.24
1962	90,500	4,100	0.24
1964	108,500	4,300	0.21
1966	148,300	6,200	0.21
1968	180,500	5,200	0.19
1970	248,400	7,600	0.19
1972	250,000	5,700	0.19
1974	229,800	8,800	0.21
1976	185,500	8,800	0.51
1978	378,750	9,163	0.51
1980	412,770	na	na
1982	351,444	8,482	0.73
1984	464,933	12,746	0.73
1986	434,497	na	0.73
1988	557,919	na	0.73
1990	532,972	na	0.73
1992	612,350	na	0.73
1994	739,356	na	0.73

[a]Ref. 29.
[b]Ref. 30.

Table 5. Specifications of Double-Distilled Nitrobenzene[a]

Property	Value
purity, %	≥99.8
color	clear, light yellow to brown
freezing point, °C	≥5.13
distillation range[a] (first drop), °C	≥207
dry point,[b] °C	212
moisture, %	<0.1
acidity as HNO_3, %	<0.001

[a]95% boiling at 207–210°C.
[b]Temperature at which no liquid remains.

In general, calorimetric methods also are subject to interferences from aromatic nitro compounds. Certain colorimetric methods are based on the nitration of nitrobenzene to *m*-nitrobenzene and subsequent determination by the generation of a red-violet color with acetone and alkali. A general titrimetric method for the determination of aromatic nitro compounds is based on reduction with titanium(III) sulfate or chloride in acidic solution followed by back-titration of excess titanium(III) ions with a standard ferric alum solution (32). Although the freezing and distillation ranges of nitrobenzene are commonly used indicators of purity, most modern techniques use instrumental methods such as gas chromatography and high pressure liquid chromatography.

Health and Safety Factors. Nitrobenzene is a very toxic substance; the maximum allowable concentration for nitrobenzene is 1 ppm or 5 mg/m^3 (6). It is readily absorbed by contact with skin and by inhalation of vapor. If a worker was exposed for eight hours to 1 ppm nitrobenzene in the working atmosphere, about 25 mg of nitrobenzene would be absorbed, of which about one-third would be by skin absorption and the remainder by inhalation. The primary effect of nitrobenzene is the conversion of hemoglobin to methemoglobin; thus the conversion eliminates hemoglobin from the oxygen-transport cycle. Exposure to nitrobenzene may irritate the skin and eyes. Nitrobenzene affects the central nervous system and produces fatigue, headache, vertigo, vomiting, general weakness, and in some cases unconsciousness and coma. There generally is a latent period of 1–4 hours before signs or symptoms appear. Nitrobenzene is a powerful methemoglobin former, and cyanosis appears when the methemoglobin level reaches 15%. Chronic exposure can lead to spleen and liver damage, jaundice, and anemia. Alcohol ingestion tends to increase the toxic effects of nitrobenzene; thus alcohol in any form should not be ingested by the victim of nitrobenzene poisoning for several days after the nitrobenzene poisoning or exposure. Impervious protective clothing should be worn in areas where risk of splash exists. Ordinary work clothes that have been splashed should be removed immediately, and the skin washed thoroughly with soap and warm water. In areas of high vapor concentrations (>1 ppm), full face masks with organic-vapor canisters or air-supplied respirators should be used. Clean work clothing should be worn daily, and showering after each shift should be mandatory.

With respect to the hazards of fire and explosion, nitrobenzene is classified as a moderate hazard when exposed to heat or flame. Nitrobenzene is classified by the ICC as a Class-B poisonous liquid.

Uses. The largest end use for nitrobenzene is in the production of aniline (see AMINES, AROMATIC). Approximately 95–98% of nitrobenzene is converted to aniline; the demand for nitrobenzene fluctuates with the demand for aniline. Production grew at an average annual rate of almost 5% from 1984 to 1988, but dropped by over 4% during the 1989–1990 economic downturn (29). For 1990, 96% of the 532,972 metric tons of nitrobenzene produced was used for aniline. The 22,680 metric tons of nitrobenzene left were used to produce a variety of other products, such as *para*-aminophenol [123-30-8] (PAP) and nigrosine dyes (Table 6). The U.S. producers of PAP are Mallinckrodt, Inc., Rhône-Poulenc, and Hoechst Celanese, with combined production capacities >35,000 metric tons (as of May 1995). Mallinckrodt is the largest producer, with over 50% of capacity. PAP primarily is used as an intermediate for acetaminophen [103-90-2].

Table 6. U.S. Consumption of Nitrobenzene,[a] t

Year	Aniline	Other[b]	Total
1976	333,391	9072	342,463
1978	369,678	9072	378,750
1980	403,698	9072	412,770
1982	331,123	9072	340,195
1984	451,325	13,608	464,933
1986	419,573	15,876	435,449
1988	537,508	20,412	557,920
1990	510,292	22,680	532,972
1992	600,103	12,247	612,350
1993	657,892[c]	13,426	671,318

[a] Ref. 29.
[b] For example, p-aminophenol, dyes, pigments, chemical intermediates, and solvents.
[c] Projected for 1993.

A smaller volume use of PAP is the production of dyestuffs and resins. Nigrosine dyes are widely used as black colorants in plastics, inks (qv), textiles, and shoe polishes.

DERIVATIVES

Mononitrochlorobenzenes. *Properties.* The physical properties of the ortho, meta, and para isomers of nitrochlorobenzene are summarized in Table 7.

o-Nitrochlorobenzene crystallizes in light yellow, monoclinic needles. It is insoluble in water and very soluble in benzene, diethyl ether, and hot ethanol. o-Nitrochlorobenzene reactions involve the nitro group, chlorine atom, and aromatic ring. The nitro group can be partially reduced to the corresponding intermediate or fully to the amino group. The aromatic ring can be nitrated, leading to the formation of 2,4-dinitrochlorobenzene [97-00-7] and 2,6-dinitrochlorobenzene

Table 7. Physical Properties of Mononitrochlorobenzenes

Property	o-Nitrochloro-benzene	m-Nitrochloro-benzene	p-Nitrochloro-benzene
CAS Registry Number	[88-73-3]	[121-73-3]	[100-00-5]
melting point, °C	32.5[a]	46 (stable)[b]	83[c]
		24 (labile)[b]	
boiling point, °C$_{kPa}$[d]	246$_{100}$[a]	236$_{101}$[a]	242$_{101}$[a]
	119$_{1.1}$[b]		113$_{1.1}$[b]
density,[e] g/mL	1.368	1.534	1.520
flash point,[c] closed cup, °C	123	103	110

[a] Ref. 5.
[b] Ref. 3.
[c] Ref. 6.
[d] To convert kPa to mm Hg, multiply by 7.5.
[e] d_4^{22}.

[606-21-3], or it can be sulfonated, yielding 3-nitro-4-chlorobenzenesulfonic acid [121-18-6]. The chlorine atom can be replaced easily by OH, OCH_3, OC_6H_5, NH_2, etc by nucleophilic attack. Treatment of o-nitrochlorobenzene with aqueous sodium hydroxide at 130°C results in the formation of o-nitrophenol [88-75-5], and with aqueous methanolic potassium hydroxide at high temperature and pressure, o-nitroanisole [91-23-6] is formed. When o-nitrochlorobenzene is treated with aqueous ammonia under high temperature and pressure, o-nitroaniline [88-74-4] is formed. o-Nitrochlorobenzene condenses with aniline to form 2-nitrodiphenylamine [119-75-5].

m-Nitrochlorobenzene is a pale yellow crystalline solid which can exist as a stable or labile form in the solid state. It is insoluble in water, very soluble in benzene and diethyl ether, and soluble in acetone, chloroform, and hot ethanol. Unlike the ortho and para isomers, the chlorine atom of m-nitrochlorobenzene is not activated for nucleophilic substitution.

p-Nitrochlorobenzene crystallizes in light yellow monoclinic prisms. It is insoluble in water and very soluble in benzene, diethyl ether, and hot ethanol. p-Nitrochlorobenzene undergoes the same reactions described for the ortho isomer to yield the analogous para derivatives. Tin(II) chloride and hydrochloric acid convert p-nitrochlorobenzene to p-chloroaniline. The aromatic ring of the para isomer can undergo additional substitution by nitration to yield 2,4-dinitrochlorobenzene, by chlorination to yield 3,4-dichloronitrobenzene [99-54-7], or by sulfonation to yield 2-chloro-5-nitrobenzenesulfonic acid [96-73-1]. The chlorine atom is activated and, as with the ortho isomer, can be easily displaced by nucleophilic attack. Treatment with aqueous ammonia at elevated temperature and pressure results in the formation of p-nitroaniline [100-01-6]. Treatment with aqueous sodium hydroxide under pressure gives p-nitrophenol [100-02-7]. p-Nitrochlorobenzene reacts with sodium disulfide to form 4,4'-dinitrodiphenyl disulfide [100-32-3] which is an intermediate in the preparation of sulfanilamide derivatives.

Manufacture and Processing. Chlorobenzene can be nitrated at 40–70°C with a nitrating acid consisting of 52.5 wt % H_2SO_4, 35.5 wt % HNO_3, and 12 wt % H_2O. The technique and equipment are similar to that described for the nitration of benzene. The resulting product is a mixture of isomers containing about 34 wt % o-nitrochlorobenzene, 65 wt % p-nitrochlorobenzene, and 1 wt % m-nitrochlorobenzene. The mixture is cooled to a temperature slightly above its freezing point (~15°C) and a large portion of the para isomer slowly crystallizes and is separated from the mother liquor. The liquid mixture of isomers is separated by a combination of fractional distillation and crystallization. Other methods of preparing the mononitrochlorobenzenes are chlorination of nitrobenzene, the diazotization of nitro anilines and replacement by chlorine (Sandmeyer reaction), and the reaction of phosphorus pentachloride with the nitrophenols. These reactions are used on a laboratory scale but are not of commercial interest.

Several patents have been issued that offer improvements in chloronitrobenzene production. Most are vapor-phase nitrations using solid catalysts. Table 8 lists some of these results.

Economic Aspects. U.S. production of chloronitrobenzenes in 1993 was 54,431 metric tons per year of which 19,099 metric tons were the ortho isomer and 35,332 metric tons the para isomer. The meta isomer is not isolated in U.S.

Table 8. Vapor-Phase Nitration of Chlorobenzene

Catalysts	Conversion, %	Reference
alumina—silica—metal oxide	100	23
group 4b–3b mixed oxide	80	33
acidic sheet clay, acidic composite oxides	90	24
solid-supported sulfuric catalysts	93	25
HNO_3/phosphoric acid mixture (anhydrous)	100	34

production. The bulk, fob prices of o- and p-chloronitrobenzene were $1.72/kg and $2.01/kg, respectively. Chloronitrobenzenes are manufactured by Du Pont and Monsanto Co.

Health and Safety Factors. The mononitrochlorobenzenes are toxic substances which may be absorbed through the skin and lungs giving rise to methemoglobin. Their toxicity is about the same as or greater than that of nitrobenzene. The para isomer is less toxic than the ortho isomer, and the maximum allowable concentration that has been adopted for p-nitrochlorobenzene is 1 mg/m^3 (0.1 ppm) (6). The mononitrochlorobenzenes are moderate fire hazards when exposed to heat or flame. They are classified by the ICC as Class-B poisons. The same handling precautions should be used for these compounds as are used for nitrobenzene.

Uses. o-Nitrochlorobenzene is used in the synthesis of azo dye intermediates such as o-chloroaniline (Fast Yellow G Base), o-nitroaniline (Fast Orange GR Base), o-anisidine (Fast Red BB Base), o-phenetidine, and o-aminophenol (see AZO DYES). It also is used in corrosion inhibitors, pigments, and agriculture chemicals. p-Nitrochlorobenzene is used principally in the production of intermediates for azo and sulfur dyes. Other uses include pharmaceuticals (qv), photochemicals, rubber chemicals (qv), and insecticides (see INSECT CONTROL TECHNOLOGY). Typical intermediates manufactured from the para isomer are p-nitroaniline (Fast Red GC Base), p-anisidine, p-aminophenol, p-nitrophenol, p-phenylenediamine, 2-chloro-p-anisidine (Fast Red R Base), 2,4-dinitrochlorobenzene, and 1,2-dichloro-4-nitrobenzene.

Other Nitrochlorobenzenes. *2,4-Dinitrochlorobenzene.* This compound is a yellow solid that can exist in three forms, one stable and two labile. The stable α-form crystallizes in yellow rhombic crystals from diethyl ether; mp α is 53.4°C, β 43°C, and γ 27°C; bp at 101 kPa (= 1 atm) 315°C with slight decomposition; $d_4^{22}\alpha$ 1.697 g/cm^3, $d_4^{20}\beta$ 1.680 g/cm^3; flash point (closed cup) 194°C; vapor density (air = 1) 6.98 (5,6). 2,4-Dinitrochlorobenzene [97-00-7] is insoluble in water and soluble in benzene, hot ethanol, diethyl ether, and carbon disulfide.

2,4-Dinitrochlorobenzene can be manufactured by either dinitration of chlorobenzene in fuming sulfuric acid or nitration of p-nitrochlorobenzene with mixed acids. Further substitution on the aromatic ring is difficult because of the deactivating effect of the chlorine atom, but the chlorine is very reactive and is displaced even more readily than in the mononitrochlorobenzenes.

2,4-Dinitrochlorobenzene is used primarily in the manufacture of azo dyes; other areas include the manufacture of fungicides, rubber chemicals, and explosives. It is produced by Eastman Kodak Co. and Sandoz Chemical Corp. and its bulk fob price is $2.73/kg. 2,4-Dinitrochlorobenzene is more toxic than nitrobenzene. It is an extremely powerful skin irritant and must be handled with great care.

3,4-Dichloronitrobenzene. This compound crystallizes in needles from ethanol and has both a stable and labile form. The stable α-form has a melting point of 42–43°C, and the β-form is a liquid which changes to the α-form at 15°C (6); bp at 101 kPa (1 atm) 225–256°C; d_4^{75} 1.4558 g/cm^3; vapor density (air = 1), 6.6 (3,5,6). It is insoluble in water and soluble in diethyl ether, hot ethanol, and benzene. It can be prepared by the chlorination of *p*-nitrochlorobenzene or by the nitration of *o*-dichlorobenzene. The isomers are separated by fractional distillation and crystallization. 3,4-Dichloronitrobenzene [99-54-7] is produced by Du Pont.

2,5-Dichloronitrobenzene. 2,5-Dichloronitrobenzene [89-61-2] crystallizes in prisms or plates from ethanol; mp 56°C; bp at 101 kPa (1 atm) 267°C; d_4^{75} 1.4390 g/cm^3; vapor density (air = 1), 6.6 (3,6). It is insoluble in water and soluble in hot ethanol, diethyl either, and benzene. It is prepared by the nitration of *p*-dichlorobenzene and is used extensively in the manufacture of dyestuff intermediates. 2,5-Dichloronitrobenzene is produced by Du Pont.

Nitrotoluenes

MONONITROTOLUENES

The mononitration of toluene results in the formation of a mixture of the ortho, meta, and para isomers of nitrotoluene. The presence of the methyl group on the aromatic ring facilitates the nitration, but it also increases the ease of oxidation.

Properties. *o*-Nitrotoluene [88-72-2] is a clear yellow liquid. The solid is dimorphous and the melting points of the α- and β-forms are −9.55 and −3.85°C, respectively. *o*-Nitrotoluene is infinitely soluble in benzene, diethyl ether, and ethanol. It is soluble in most organic solvents and only slightly soluble in water (0.065 g in 100 g of water at 30°C). The physical properties of *o*-nitrotoluene are listed in Table 9.

The strong electron-acceptor action of the nitro group in *o*-nitrotoluene confers increased reactivity on the methyl group; thus the methyl group is easily oxidized. Oxidation with potassium permanganate or potassium dichromate causes the formation of *o*-nitrobenzoic acid [552-16-9]. When boiled with a sodium hydroxide solution, *o*-nitrotoluene exhibits the phenomena of autoxidation and reduction and yields anthranilic acid. When the oxidation is carried out with manganese dioxide and sulfuric acid, *o*-nitrobenzoic acid or *o*-nitrobenzaldehyde

Table 9. Physical Properties of *o*-Nitrotoluene

Property	Value	Reference
melting point, °C		
α	−9.55	3
β (stable form)	−3.85	3
boiling point, °C		
at 101 kPa[a]	221.7	35
0.13 kPa[a]	50.0	6
density, g/mL		
d_{15}^{19}	1.1622	6
d_4^{20}	1.163	4
d_4^{60}	1.124	5
refractive index, n_D^{20}	1.5474	3
surface tension at 20°C, mN/m(= dyn/cm)	44.1	5
viscosity at 20°C, mPa·s(=cP)	2.37	5
heat of combustion at constant volume, MJ/mol[b]	3.75	5
vapor density (air = 1)	4.72	6
flash point (closed cup), °C	106	6

[a]To convert kPa to mm Hg, multiply by 7.5.
[b]To convert J to cal, divide by 4.184.

[552-89-6] is formed, depending on the reaction conditions. One method of reducing *o*-nitrotoluene to *o*-toluidine is by iron powder and hydrochloric acid. Alkaline reduction with iron or zinc leads in a stepwise fashion to azoxy, azo, and hydrazo compounds, depending on the reaction conditions. Nitration of *o*-nitrotoluene gives 2,5-dinitrotoluene [121-14-2] and 2,6-dinitrotoluene [606-20-2]. Chlorination of *o*-nitrotoluene in the absence of iron yields *o*-nitrobenzyl chloride [612-23-7], *o*-chlorotoluene, or *o*-chlorobenzyl chloride, depending on the reaction conditions. In the presence of iron, chlorination results in the formation of 2-nitro-6-chlorotoluene [83-42-1] and 2-nitro-4-chlorotoluene [89-59-8].

m-Nitrotoluene [99-08-1] is a clear yellow liquid that freezes at 16.1°C. It is soluble in most organic solvents, such as ethanol, benzene, and diethyl ether, and is only sparingly soluble in water, 0.05 g/100 g of water at 30°C. The physical properties of *m*-nitrotoluene are given in Table 10.

m-Nitrotoluene does not have an active methyl group as do the ortho and para isomers. It is oxidized readily to *m*-nitrobenzoic acid [121-92-6] by chromic acid and more slowly by potassium hexacyanoferrate(III) in alkaline solution. *m*-Nitrobenzaldehyde [99-61-6] is the chief product of the electrolytic oxidation of *m*-nitrotoluene. Acid, neutral, or catalytic reduction of *m*-nitrotoluene yields *m*-toluidine. Nitration of *m*-nitrotoluene yields primarily 3,4-dinitrotoluene [610-39-9] and small amounts of 2,3-dinitrotoluene [602-01-7] and 2,5-dinitrotoluene [619-15-8].

p-Nitrotoluene [99-99-0] crystallizes in colorless rhombic crystals. It is only slightly soluble in water, 0.044 g/100 g of water at 30°C; moderately soluble in methanol and ethanol; and readily soluble in acetone, diethyl ether, and benzene. The physical properties of *p*-nitrotoluene are listed in Table 11.

The methyl group of *p*-nitrotoluene is activated by the para nitro group. *p*-Nitrotoluene is oxidized to *p*-nitrobenzoic acid [62-23-7] by potassium hexa-

Table 10. Physical Properties of *m*-Nitrotoluene

Property	Value	Reference
melting point, °C	16.1	5
boiling point, °C		
at 101 kPa[a]	231.9	5
13.3 kPa[a]	156.9	35
0.13 kPa[a]	50.2	35
density, g/mL, d_4^{20}	1.1581	5
refractive index, n_D^{20}	1.5470	3
surface tension at 25°C, mN/m(=dyn/cm)	43.5	5
viscosity at 20°C, mPa·s(=cP)	2.33	5
heat of combustion at constant volume, MJ/mol[b]	3.732	37
vapor density (air = 1)	4.72	6
flash point (closed cup), °C	106	6

[a]To convert kPa to mm Hg, multiply by 7.5.
[b]To convert J to cal, divide by 4.184.

Table 11. Physical Properties of *p*-Nitrotoluene

Property	Value	Reference
melting point, °C	53.5	5
boiling point, °C		
at 101 kPa[a]	238.5	5
1.2 kPa[a]	104.5	3
0.13 kPa[a]	53.7	6
density, g/mL		
d_4^{20}	1.286	35
d_4^{73}	1.1038	3
refractive index, $n_D^{62.6}$	1.5346	4
surface tension at 60°C, mN/m(=dyn/cm)	42.3	5
viscosity at 60°C, mPa·s(=cP)	1.2	5
heat of combustion at constant volume, MJ/mol[b]	3.718	5
vapor density (air = 1)	4.72	6
flash point (closed cup), °C	106	6

[a]To convert kPa to mm Hg, multiply by 7.5.
[b]To convert J to cal, divide by 4.184.

cyanoferrate(III) in alkaline solution, potassium permanganate, or potassium dichromate. *p*-Nitrotoluene is converted to *p*-nitrobenzaldehyde [*555-16-8*] by electrolytic oxidation in an acetic acid–sulfuric acid mixture or by treatment with lead(IV) oxide in concentrated sulfuric acid. *p*-Nitrotoluene is reduced by iron and hydrochloric acid to *p*-toluidine. Alkaline reduction with iron leads to the formation of a mixture of azoxy, azo, and hydrazo compounds, depending on the reaction conditions. Nitration of *p*-nitrotoluene gives 2,4-dinitrotoluene. Chlorination can occur on either the aromatic ring or the methyl group, and the resulting product depends on the catalyst and reaction conditions used. Under free-radical reaction conditions, *p*-nitrobenzyl chloride [*100-14-1*] is formed, and in the presence of iron or antimony(III) chloride, 4-nitro-2-chlorotoluene [*121-86-8*] is obtained. *p*-Nitrotoluene undergoes sulfonation, yielding 2-methyl-5-nitrobenzenesulfonic acid [32784-87-5]. Heating p-nitrotoluene with an alcoholic potassium hydroxide solution results in the formation of 4,4'-dinitrostilbene [2501-02-2].

Manufacture and Processing. Mononitrotoluenes are produced by the nitration of toluene in a manner similar to that described for nitrobenzene. The presence of the methyl group on the aromatic ring facilitates the nitration of toluene, as compared to that of benzene, and increases the ease of oxidation which results in undesirable by-products. Thus the nitration of toluene generally is carried out at lower temperatures than the nitration of benzene to minimize oxidative side reactions. Because toluene nitrates at a faster rate than benzene, the milder conditions also reduce the formation of dinitrotoluenes. Toluene is less soluble than benzene in the acid phase, thus vigorous agitation of the reaction mixture is necessary to maximize the interfacial area of the two phases and the mass transfer of the reactants. The rate of a typical industrial nitration can be modeled in terms of a fast reaction taking place in a zone in the aqueous phase adjacent to the interface where the reaction is diffusion controlled.

Mononitrotoluenes can be produced by either a batch or continuous process. With a typical batch process, the toluene is fed into the nitrator and cooled to about 25°C. The nitrating acid (52–56 wt % H_2SO_4, 28–32 wt % HNO_3, and 12–20 wt % H_2O) is added slowly below the surface of the toluene and the temperature of the reaction mixture is maintained at 25°C by adjusting the feed rate of the nitrating acid and the amount of cooling. After all of the acid is added, the temperature is raised slowly to 35–40°C. After completion of the reaction, the reaction mixture is put into a separator where the spent acid is withdrawn from the bottom and is reconcentrated. The crude product is washed in several steps with dilute caustic and then water. The product is steam distilled to remove excess toluene and then dried by distilling the remaining traces of water. The resulting product contains 55–60 wt % o-nitrotoluene, 3–4 wt % m-nitrotoluene, and 35–40 wt % p-nitrotoluene. The yield of mononitrotoluenes is ca 96%.

Although several patents pertain to the production of nitrotoluene, most of these processes report yields of only 50% or lower (15,23–25,27,28). The main thrust for these patents has been the nitration of benzene, with toluene being included to strengthen the patent estate. A few higher yield processes, however, have been reported. Toluene reacts with nitronium salts, such as NO_2BF_4, in the presence of catalytic amounts of crown ethers or polyethers to give nitrotoluene in 70% yield (38). The reaction occurs at room temperature and gives lower yields of m-nitrotoluene, which is desirable. Another process reacts toluene with nitric acid (qv) in the presence of solid silica–alumina catalysts in the gas phase to form nitrotoluene in 95% yield (26). Both of these methods have the advantage of eliminating the use of sulfuric acid.

The separation of the isomers is carried out by a combination of fractional distillation and crystallization. In a fractional vacuum distillation step, the distillate, obtained at a head temperature of 96–97°C at 1.6 kPa (12 mm Hg), is fairly pure o-nitrotoluene and can be purified further by crystallization. The meta isomer is distilled from a mixture of m- and p-nitrotoluene and can be purified further by additional distillation and crystallization steps. The bottoms product from the distillation steps is cooled in a crystallizer to obtain p-nitrotoluene.

Effort has focused on increasing the amount of the para isomer formed in the mononitration of toluene, because it generally is in the greatest demand of the three isomers. In a typical nitration with mixed nitric and sulfuric acids, the ratio of p-nitrotoluene:o-nitrotoluene (para:ortho) usually is 0.6. Nitration of toluene with nitric acid in the presence of phosphoric acid leads to an increase in formation of the para isomer with a para:ortho ratio of 1.11 (7,39,40). Nitration with nitric acid in the presence of various aromatic sulfonic acids either in solution or on a support, eg, diatomaceous earth, results in increased selectivity toward para substitution with para:ortho ratios of 0.8–1.5 (40–42). Normally with this technique, a large excess of toluene and a relatively large amount of catalyst must be used with highly concentrated nitric acid. Another approach to increase the selectivity for para nitration has been the addition of anhydrous calcium sulfate, where the amount added is based on moles of water formed in the nitration reaction; para:ortho ratios of 1.20–1.35 are obtained (43). Gas-phase nitration of toluene with nitric acid in the presence of a carrier substance, based on SiO_2 and/or Al_2O_3 and impregnated with a high boiling inorganic acid, has been described (44). This technique is carried out with a large excess of toluene at 0.7–6.7 kPa (5.3–50.3 mm Hg) and 100–140°C; para:ortho ratios of 1.2–2.0 are obtained. The effect of the different catalysts on the isomer distribution of the mononitrotoluenes is summarized in Table 12.

If pure isomers are required, the ortho and meta compounds can be prepared by indirect methods. o-Nitrotoluene can be obtained by treating 2,4-dinitrotoluene with ammonium sulfide followed by diazotization and boiling with ethanol. m-Nitrotoluene can be prepared from p-toluidine by acetylation, nitration deacetylation, diazotization, and boiling with ethanol. A fairly pure p-nitrotoluene, which has been isolated from the isomeric mixture, can be purified further by repeated crystallization.

Economic Aspects. Annual 1993 U.S. production of the mononitrotoluenes is 26,000 metric tons, with about 16,120 metric tons of the ortho isomer, 780 metric tons of the meta isomer, and 9,100 metric tons of the para isomer. The prices of o-, m-, and p-nitrotoluene in bulk fob are \$1.15/kg, \$2.54/kg, and \$3.64/kg, respectively. The mononitrotoluenes are manufactured by Du Pont and First Chemical Corp.

Analytical and Test Methods. o-Nitrotoluene can be analyzed for purity and isomer content by infrared spectroscopy with an accuracy of about 1%.

Table 12. Isomer Ratio of Mononitrotoluenes with Different Catalysts

Catalyst	Para, wt %	Ortho, wt %	Meta, wt %	p/o ratio	Conversion, %	Reference
normal mixed acid	38	58	4	0.6	98	
H_3PO_4	50.2	45.1	4.3	1.11	99.6	39
m-benzenedisulfonic acid (BDA)	43.6	53.8	2.9	0.81	92.8	41
BDA on Celite 545				1.53	92	42
anhydrous $CaSO_4$	54.4	43.2	2.3	1.26	89	43
5% H_2SO_4 on Al_2O_3	62.5	34.0	3.5	1.84	61.1	44
NO_2BF_4/polyether				0.91	70	38
silica–alumina zeolite	55	40	5	1.4	95	26
alumina–silica–metal oxide	52	48	0	1.09	48.7	23

p-Nitrotoluene content can be estimated by the decomposition of the isomeric toluene diazonium chlorides because the ortho and meta isomers decompose more readily than the para isomer. A colorimetric method for determining the content of the various isomers is based on the color which forms when the mononitrotoluenes are dissolved in sulfuric acid (45). From the absorption of the sulfuric acid solution at 436 and 305 nm, the ortho and para isomer content can be determined, and the meta isomer can be obtained by difference. However, this and other colorimetric methods are subject to possible interferences from other aromatic nitro compounds. A titrimetric method, based on the reduction of the nitro group with titanium(III) sulfate or chloride, can be used to determine mononitrotoluenes (32). Chromatographic methods, eg, gas chromatography or high pressure liquid chromatography, are well suited for the determination of mononitrotoluenes as well as its individual isomers. Freezing points are used commonly as indicators of purity of the various isomers.

Health and Safety Factors. The toxic effects of the mononitrotoluenes are similar to but less pronounced than those described for nitrobenzene. The maximum allowable concentration for the mononitrotoluenes is 2 ppm (11 mg/m^3) (6). Mononitrotoluenes are low grade methemoglobin formers (4) and may be absorbed through the skin and respiratory tract. The toxicity of alkyl nitrobenzenes decreases with an increasing number of alkyl groups and increases with an increasing number of nitro groups. The mononitrotoluenes represent moderate fire hazards when exposed to heat or flame. The same precautions used in handling nitrobenzene should be used for these compounds.

Uses. *o*-Nitrotoluene is used in the synthesis of intermediates for azo dyes, sulfur dyes, rubber chemicals, and agriculture chemicals. Typical intermediates are *o*-toluidine, *o*-nitrobenzaldehyde, 2-nitro-4-chlorotoluene, 2-nitro-6-chlorotoluene, 2-amino-4-chlorotoluene (Fast Scarlet TR Base), and 2-amino-6-chlorotoluene (Fast Red KB Base). *p*-Nitrotoluene is used principally in the production of intermediates for azo and sulfur dyes. Typical intermediates are *p*-toluidine, *p*-nitrobenzaldehyde, and 4-nitro-2-chlorotoluene.

DINITROTOLUENES

Dinitration of toluene results in the formation of a number of isomeric products, and with a typical sulfuric–nitric acid nitrating mixture the following mixture of isomers is obtained: 75 wt % 2,4-dinitrotoluene [*121-14-2*], 19 wt % 2,6-dinitrotoluene [*606-20-2*], 2.5 wt % 3,4-dinitrotoluene [*610-39-9*], 1 wt % 2,3-dinitrotoluene [*602-01-7*], and 0.5 wt % 2,5-dinitrotoluene [*619-15-8*]. The dinitrotoluenes are a moderate fire and explosion hazard when exposed to heat or flame. The maximum allowable concentration in air is 1.5 mg/m^3 (0.2 ppm). Dinitrotoluenes are used as intermediates for the production of toluene diisocyanate and dyestuffs. They also are used as explosives.

2,4-Dinitrotoluene crystallizes in yellow needles from carbon disulfide and is soluble in a number of organic solvents. It is only slightly soluble in water, 0.03 g/100 g of water at 22°C. Its physical properties are listed in Table 13.

2,4-Dinitrotoluene can be prepared by the nitration of *p*-nitrotoluene with yields of ca 96% or it can be obtained from the direct nitration of toluene. 2,4-Dinitrotoluene is oxidized to 2,4-dinitrobenzoic acid [*610-30-0*] by potassium

Table 13. Physical Properties of 2,4-Dinitrotoluene

Property	Value	Reference
melting point, °C	64–66	5
boiling point at 101 kPa (= 1 atm), °C	300^a	5
density, g/mL		
d_4^{15}	1.521	6
d_4^{71}	1.321	5
vapor density (air = 1)	6.27	6
heat of combustion, MJ/mol^b	3.568	5
flash point, °C	207	6

^{a}Slight decomposition.
^{b}To convert J to cal, divide by 4.184.

permanganate or chromic acid, and is reduced to 2,4-diaminotoluene by iron and acetic acid. It is reduced partially by zinc chloride and hydrochloric acid to 2-amino-4-nitrotoluene [99-55-8] and by ammonium sulfide to 4-amino-2-nitrotoluene [119-32-4].

BIBLIOGRAPHY

"Nitrobenzene and Nitrotoluenes" in *ECT* 1st ed., Vol. 9, pp. 388–401, by H. H. Bieber and A. G. Hill, American Cyanamid Co.; in *ECT* 2nd ed., Vol. 13, pp. 834–853, by H. J. Matsuguma, Dept. of the Army; in *ECT* 3rd ed., Vol. 15, pp. 916–932, by K. L. Dunlap, Mobay Chemical Corp.

1. E. Mitscherlich, *Ann. Phys. Chem.* **31**, 625 (1834).
2. A. Gero, *Textbook of Organic Chemistry*, John Wiley & Sons, Inc., New York, 1963.
3. J. R. A. Pollock and R. Stevens, eds., *Dictionary of Organic Compounds*, 4th ed., Eyre & Spottiswoode Publishers Ltd., London, 1965.
4. G. D. Clayton and F. E. Clayton, eds., *Patty's Industrial Hygiene and Toxicology*, 3rd rev. ed., John Wiley & Sons, Inc., New York, 1981.
5. J. A. Dean, *Lange's Handbook of Chemistry*, 14th ed., McGraw-Hill Book Co., New York, 1992.
6. N. R. Sax, *Dangerous Properties of Industrial Materials*, 8th ed., Van Nostrand Reinhold, New York, 1992.
7. L. F. Albright and C. Hanson, eds., *Industrial and Laboratory Nitrations*, American Chemical Society, Washington, D.C., 1976.
8. U.S. Pat. 3,092,671 (June 4, 1963), S. B. Humphrey and D. R. Smoak (to United States Rubber Co.).
9. U.S. Pat. 4,021,498 (May 3, 1977), V. Alexanderson, J. B. Trecek, and C. M. Vanderwaart (to American Cyanamid Co.).
10. U.S. Pat. 3,928,475 (Dec. 23, 1975), M. W. Dassel (to Du Pont).
11. U.S. Pat. 3,981,935 (Sept. 21, 1976), R. McCall (to Du Pont).
12. U.S. Pat. 4,331,819 (May 25, 1982), R. McCall (to Du Pont).
13. U.S. Pat. 4,772,757 (Sept. 20, 1988), G. Lailach and co-workers (to Bayer Aktiengesellschaft).
14. U.S. Pat. 4,994,242 (Feb. 19, 1991), J. B. Rae and E. G. Hauptmann (to Noram Engineering and Constructors Ltd.).
15. U.S. Pat. 4,973,770 (Nov. 27, 1990), C. M. Evans (to C-I-L, Inc.).
16. U.S. Pat. 4,241,229 (Dec. 23, 1980), J. W. Priegnitz (to UOP Inc.).

17. U.S. Pat. 5,099,079 (Mar. 24, 1992), A. B. Quakenbush (to Olin Corp.).
18. U.S. Pat. 4,925,565 (May 15, 1990), E. G. Adams and R. B. Barker (to First Chemical Corp.).
19. U.S. Pat. 4,986,917 (Jan. 22, 1991), E. G. Adams, A. C. Bayer, A. D. Farmer, and B. J. Hook (to First Chemical Corp.).
20. U.S. Pat. 4,986,920 (Jan. 22, 1991), E. G. Adams, A. C. Bayer, A. D. Farmer, and B. J. Hook (to First Chemical Corp.).
21. WO Appl. 89/12620 (28 Dec. 89), R. W. Mason (to Olin Corp.).
22. U.S. Pat. 4,935,557 (June 19, 1990), R. V. Carr and B. A. Toseland (to Air Products and Chemicals, Inc.).
23. U.S. Pat. 4,415,744 (Nov. 15, 1983), I. Schumacher and K. B. Wang (to Monsanto Co.).
24. U.S. Pat. 5,004,846 (Apr. 2, 1991), H. Sato, K. Hirose, K. Nagai, H. Yoshioka, and Y. Nagaoka (to Sumitomo Chemical Co., Ltd.).
25. U.S. Pat. 5,030,776 (July 9, 1991), H. Sato, K. Nagai, H. Yoshioka, and Y. Nagaoka (to Sumitomo Chemical Co., Ltd.).
26. U.S. Pat. 4,418,230 (Nov. 29, 1983), J. Bakke and J. Liaskar (to Aktiebolaget Bofors).
27. U.S. Pat. 4,426,543 (Jan. 17, 1984), I. Schumacher and K. B. Wang (to Monsanto Co.).
28. U.S. Pat. 4,804,792 (Feb. 14, 1989), R. W. Mason and K. Steely (to Olin Corp.).
29. *Chemical Economic Handbook*, SRI International, Menlo Park, Calif., Oct. 1991.
30. *Chem. Mark. Rep.* (July 1960–1993).
31. F. Feigl, *Spot Tests in Organic Analysis*, 5th ed., Elsevier Publishing Co., New York, 1956.
32. Y. A. Gawargious, *The Determination of Nitro and Related Functions*, Academic Press, Inc., New York, 1973.
33. U.S. Pat. 4,618,733 (Oct. 21, 1986), I. Schumacher (to Monsanto Co.).
34. U.S. Pat. 4,476,335 (Oct. 9, 1984), S. Takenaka, T. Nishida, and J. Kanemoto (to Mitsue Toatsu Chemicals, Inc.).
35. S. Budavari, M. J. O'Neil, A. Smith, and P. E. Heckelman, eds., *Merck Index*, 11th ed., Merck & Co., Inc., Rahway, N.J., 1989.
36. J. Timmermans, *Physico-Chemical Constants of Pure Organic Compounds*, Elsevier Publishing Co., Inc., New York, 1950.
37. T. Urbanski, *Chemistry and Technology of Explosives*, Vol. 1, MacMillan Co., New York, 1964.
38. U.S. Pat. 4,392,978 (July 12, 1983), R. L. Elsenbaumer and E. W. Wasserman (to Allied Corp.).
39. Fr. Pat. 1,541,376 (Oct. 4, 1968), C. L. Hakansson, A. A. Arvidsson, and I. Loken (to Aktiebolag Bofors).
40. T. Kameo and O. Manabe, *Nippon Kagaku Daishi*, 1543 (1973).
41. U.S. Pat. 3,196,186 (July 20, 1965), A. W. Sogn and J. G. Natoli (to Allied Chemical Corp.).
42. T. Kameo, S. Nishimura, and O. Manabe, *Nippon Kagaku Daishi*, 122 (1974).
43. U.S. Pat. 3,957,889 (May 18, 1976), B. Milligan and D. G. Miller (to Air Products and Chemicals, Inc.).
44. U.S. Pat. 4,112,006 (Sept. 5, 1978), H. Schubert and F. Wunder (to Hoescht Atkiengesellschaft).
45. F. D. Snell and C. T. Snell, *Colorimetric Methods of Analysis*, 3rd ed., Vols. 3 and 4, D. van Nostrand Co., Inc., Princeton, N.J., 1954.

RICK L. ADKINS
Bayer Corporation

NITROFURANS.   See ANTIBACTERIAL AGENTS, SYNTHETIC.

NITROGEN

Nitrogen [7727-37-9], atomic number 7, is a nonmetallic element situated between carbon and oxygen in the Periodic Table. It is the most abundant element accessible to human beings which exists uncombined with any other elements (1). Most of the nitrogen in the atmosphere occurs as a diatomic gas N_2, sometimes referred to as dinitrogen, and comprises 78.03% by volume and 75.45% by weight of the earth's atmosphere. Industry annually isolates millions of tons of nitrogen from air. Nitrogen-containing compounds are essential to all life.

Nitrogen was discovered in 1772 by Rutherford (2), and about that same time, Scheele, Priestly, and Cavendish were also working with this "burnt" or "dephlogisticated" air. The name nitrogen from the Latin *nitrum*, or the Greek combination, *nitron* "native soda" and *gene* "forming" was suggested by Chaptal in 1790 when it was learned that niter and nitric acid were formed from this constituent (3). In the 1780s nitrogen oxides were produced by combining nitrogen and oxygen using an electrical discharge. Nitrogen was first liquified by Cailletet in 1877. In the early 1900s atmospheric nitrogen was first used for large-scale industrial purposes. Calcium cyanamide was first produced in 1895 by the Frank-Caro process. In 1900, Birkeland-Eyde developed the first industrial oxidation of nitrogen. Ostwald was awarded the 1909 Nobel Prize for work that led to the industrial-scale catalytic oxidation of NH_3 to HNO_3. Two more Nobel Prizes were granted for the Haber-Bosch process for the catalytic synthesis of ammonia (qv) from N_2 and H_2. This process reached industrial scale by 1913. In the 1990s, five of the 15 largest volume industrial chemicals produced in the United States contain nitrogen: ammonia, nitrogen (gaseous and liquified), ammonium nitrate, nitric acid (qv), and urea (qv) (1,4).

Physical Properties

Nitrogen is the lightest of the Group 15 elements and has an atomic weight of 14.008. There are two naturally occurring stable isotopes: ^{14}N (relative atomic mass 14.003, natural abundance 99.634%) and ^{15}N (relative atomic mass 15.000, natural abundance 0.366%). Both of these isotopes have a nuclear spin and are used in nmr experiments (5). The ground state electronic configuration of the nitrogen atom is $1s^2 2s^2 2p_x{}^1 2p_y{}^1 2p_z{}^1$ with three unpaired electrons (^4S). Nitrogen can have oxidation states ranging from +5 to −3. The most common oxidation states are +5 (eg, nitric acid), +3 (eg, nitrous acid), 0 (eg, molecular nitrogen), and −3 (eg, ammonia). The electronegativity of N (~3 on the Pauling scale) is only exceeded by oxygen and fluorine. It has a single-bond covalent radius of ~70 pm. Unlike the heavier elements of Group 15, nitrogen readily forms multiple bonds with itself and other atoms (6,7). The nitrogen atom typically has a maximum coordination number of four, and can be found in the tetrahedral, pyramidal, planar, angular, and linear geometries. Pentacoordinate nitrogen bonding has been demonstrated in gold(I) complexes with trigonal bipyramid geometry (8). Theoretical calculations indicate that sixfold nitrogen coordination may be possible (9).

Molecular nitrogen, N_2, is a colorless, odorless, diamagnetic, noncombustible gas at standard pressure (101.3 kPa) and temperature (0°C). The gas condenses to a colorless liquid at $-195.8°C$ at atmospheric pressure. Depending on the temperature, solid molecular nitrogen exists in one of two forms at atmospheric pressure, α and β, both of which are white. At extreme pressures and at room temperature, nitrogen solidifies into two additional phases, δ and ϵ. A stable stoichiometric solid compound of nitrogen and helium of composition $He(N_2)_{11}$ at a pressure of 9 GPa (1,300,000 psig) has been reported (10). Other properties of molecular nitrogen are shown in Tables 1 and 2 (6,7,11). Density is represented by Q.

The deviation from the perfect gas law is not great at ordinary pressures and temperatures. At the highest pressure normally encountered commercially, 41 MPa (6000 psig), the compressibility factor of nitrogen is 1.3629 at 25°C (12).

Table 1. Physical Properties of Molecular Nitrogen

Property	Value
molecular weight	28.0134
N–N bond length, pm	109.8
first ionization energy, eV	15.58
critical point	
$\quad T_{crit}$, K	126.2
$\quad P_{crit}$, MPa[a]	3.39908
$\quad Q_{crit}$, g/L	314.03
boiling point,[b] K	77.35
weight of liquid at boiling point,[b] kg/m³	808.5
heat of vaporization, kJ/kg[c]	199
triple point	
$\quad T$, K	63.15
$\quad P$, kPa[a]	12.463
$\quad Q_g, Q_l, Q_s$, g/L	0.6803, 867.7, 947
heat of fusion, kJ/kg[c]	25.8
Q,[d] g/L	1.2505
relative density (air)[d]	0.967
specific heat capacity,[d] J/(g·K)	1.039
dynamic viscosity,[d] mPa·s(=cP)	15.9×10^{-3}
thermal conductivity, mW/(m·K)	23.86
compressibility factor (PV/RT)	0.9995

[a]To convert MPa to psi, multiply by 145.
[b]At 101.3 kPa = 1 atm.
[c]To convert kJ to kcal, divide by 4.184.
[d]At 273.15 K and 101.3 kPa = 1 atm.

Table 2. Physical Properties of Solid Molecular Nitrogen

Form[a]	Temperature range, K	Crystal structure	Q, g/L	Vapor pressure, kPa[b]
β	35.6–63.15	hexagonal	948[c]	6.3[c]
α	0–35.6	cubic	1027[d]	1.47×10^{-11d}

[a]Heat of transformation $(\alpha - \beta)$: 228.9 kJ/mol (54.71 cal/mol).
[b]To convert kPa to mm Hg, multiply by 7.5.
[c]Measured at 60 K.
[d]Measured at 21.2 K.

Chemical Properties

The chemistry of molecular nitrogen is marked by its relative inertness (13). The electronic configuration of the N_2 molecule in terms of molecular orbital theory is $1\sigma_g^2 1\sigma_u^2 2\sigma_g^2 2\sigma_u^2 1\pi_u^4 3\sigma_g^2 1\pi_g^0$. There is a large (8.3 eV) energy gap between the highest occupied molecular orbital (HOMO), $3\sigma_g$, and the lowest unoccupied molecular orbital (LUMO), $1\pi_g$. The LUMO is high enough that molecular nitrogen only takes electrons from alkali metals or other strongly electron yielding metals. The HOMO of dinitrogen is so low that the ionization energy (15.6 eV) approaches that of argon (15.75 eV). N_2 has a short (109.76 pm) interatomic distance and high (945 kJ/mol (226 kcal/mol)) dissociation energy. At ordinary pressures, even at temperatures approaching 3000°C, there is no appreciable dissociation (6). The triple bond of dinitrogen is nonpolar and has a significantly lower reactivity than other comparable triple bond systems, such as the isoelectronic compound carbon monoxide. Because of the stability of the nitrogen–nitrogen bond, and highly symmetrical electron distribution of the molecule, the intermolecular forces are generally very small.

Despite the stability of molecular nitrogen, the chemistry of nitrogen is extremely important. There is essentially a limitless supply of nitrogen in the atmosphere. However, most biological species can use only nitrogen which has been chemically combined with some other element or compound, so-called fixed nitrogen, and the abundance of this type of nitrogen is relatively limited. Agriculture and industry depend on an abundance of nitrogen in forms other than gaseous N_2. There is much research interest in elucidating inexpensive methods of converting molecular nitrogen to some form of fixed nitrogen (see NITROGEN FIXATION).

In most solvents, N_2 has a relatively low solubility. The solubility in water at 0°C and 101.3 kPa (1 atm) of N_2 is only 23.5 ppt (by volume) (14). Nitrogen is slightly soluble in iron and steel through the reaction N_2 (gas) $\rightleftharpoons$ 2 N (dissolved), where N (dissolved) represents atomic nitrogen dissolved in iron. The solubility of nitrogen at one atmosphere pressure in pure liquid iron at 1600°C is about 0.045% by weight, and in α-phase iron at 900°C is about 0.0045% by weight (15). Nitrogen has a significant alloying effect in steels. The presence of nitrogen can strengthen low carbon steels at the expense of increased strain aging properties. Nitrogen in amounts up to 0.02% has been used to strengthen high strength low alloy steels (16). Dissolved nitrogen strengthens and increases the pitting resistance of austenitic stainless steels but lowers the mechanical properties of ferritic stainless steels. The quantity of dissolved nitrogen in carbon steels can be controlled by deoxidizing the melt with aluminum, which combines with the nitrogen to form nitrides, producing aluminum-killed steel; by inerting the ladle with argon; by vacuum arc melting; by vacuum degassing; or by electroslag remelting processes. Through these methods, dissolved nitrogen levels can be maintained as low as 40–50 ppm.

Because of the inherent stability of the N_2 molecule, high temperatures are often used to coax its reactivity. Temperature elevation promotes N_2 reactions with chromium, silicon, titanium, aluminum, boron, calcium, strontium, beryllium, magnesium, and lithium to form nitrides; at 400°C, nitrogen reacts with oxygen and chlorine to form nitrosyl chloride; at 900°C, graphite and sodium carbonate react with nitrogen to form carbon monoxide and sodium cyanide; at 1500°C, nitrogen reacts with acetylene to form hydrogen cyanide. Also at high temperatures, N_2 and O_2 react to give NO. Arguably, the most important industrial process which utilizes N_2 as a constituent is ammonia formation at high pressures from nitrogen and hydrogen in the presence of heat and a catalyst, known as the Haber-Bosch process (1,11).

When molecular nitrogen is subjected to the actions of a condensed electrode discharge or to a high frequency, it is changed to an activated, unstable condition, which returns to its ground state with the emission of a greenish yellow glow. In its excited state, the gas is chemically active; it combines in the cold with mercury, sulfur, and phosphorus. This activity is the result of the presence of excited N_2 molecules at various energy levels, as well as the presence of atomic nitrogen. An example of this excited state chemistry is when activated nitrogen causes the dissociation of carbon dioxide that is normally impervious to the actions of ground state N_2 (17).

Dinitrogen complexes are molecules that contain dinitrogen bound to a metal. The first dinitrogen complex, $[Ru(N_2)(NH_3)_5]^{2+}$, was reported in 1965 as a product of the interaction of hydrazine and $RuCl_3$ (aq) (18). There are hundreds of complexes in the 1990s with dinitrogen as a ligand (19,20). The bonding of a neutral, dipole-free and not easily polarized N_2 molecule to a metal is generally quite labile. N_2-metal binding can exhibit a variety of geometries, eg, end-on (η^1) and side-on (η^2) or bridging between two metals. End-on η^1 binding is the most common and is a combination of a coordinate bond between the electron pair on one of the nitrogen atoms and an acceptor orbital of the metal, and back-bonding of loosely held metal electrons into the π^* orbital of the N_2 ligand (21).

Manufacture and Processing

Atmospheric air is the feedstock for all commercial nitrogen production processes. Nitrogen is separated from air commercially by cryogenic distillation, pressure swing adsorption, membrane permeation, or hydrocarbon combustion processes. Cryogenic distillation is the most cost-effective technology for production of large quantities of relatively pure nitrogen and is the most commonly used. Pressure swing adsorption and membrane permeation are the most economical processes for production of lower purity nitrogen in low to moderate volume ranges (25–500 m^3/h (1000–20,000 SCFH)). All statements of volume (m^3) are at normal conditions ($t = 25°C$, pressure $= 101.3$ kPa). Both are rapidly growing technologies. Industry estimates indicate that noncryogenic separation will eventually account for greater than 30% of all commercial nitrogen production (22). Combustion-based processes are in decline in most applications due to displacement by noncryogenic processes but are still widely used in heat treatment where residual contaminants play an active process role. The choice of the most economic technology is principally driven by required nitrogen purity and flow rate as shown in Figure 1. Liquid nitrogen is produced exclusively from cryogenic processes.

Cryogenic Air Separation. This method has been in commercial use for the production of nitrogen since the beginning of this century. Most nitrogen is produced in large tonnage cryogenic distillation plants with oxygen and argon

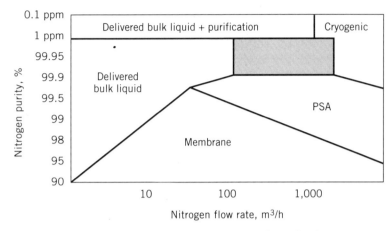

Fig. 1. Approximate economic range of nitrogen supply technologies (at median site conditions). Shaded area represents bulk liquid or PSA membrane plus deoxo or liquid assist cryogenic. To convert m^3/h to SCFH, multiply by 40.

as coproducts. The nitrogen and oxygen are either utilized directly in gaseous form at adjacent industrial facilities with distribution by pipeline, or some or all is liquified to enable distribution and storage in vacuum-insulated vessels. The delivered liquid nitrogen is then used directly or vaporized as needed or is vaporized and stored under pressure in cylinders.

If the required gaseous nitrogen flow rate is relatively constant and is greater than about 400 m^3/h (15,000 SCFH), an on-site cryogenic production plant may be the most economical mode of supply. Figure 2 shows a typical cryogenic nitrogen separation process. Air is compressed, cooled to remove excess water vapor, and purified to remove residual carbon dioxide, water vapor, and other contaminants which could freeze in the process by either reversing heat exchangers or molecular sieve pressure swing adsorption (Fig. 3). In the reversing heat-exchanger process, the incoming compressed air is cooled by countercurrent heat exchange with cold oxygen-rich waste gas and nitrogen product exiting the process. Residual contaminants in the air are frozen out onto the heat-exchanger surfaces. At periodic intervals, the flow through the heat exchangers is reversed to remove the contaminants. In the molecular sieve process, two beds of activated alumina and activated carbon are used to remove the contaminants by reversible adsorption, with the incoming air flow alternating between beds to allow regeneration of the depleted bed. The molecular sieve process is advantageous in nitrogen plants that may have insufficient waste gas available to clean out the reversing heat exchangers or that must be started up and shut down frequently. Most nitrogen plants use the molecular sieve process due to increased reliability and lower cost (see MOLECULAR SIEVES).

The clean compressed air is then further cooled to near the dew point of air (ca $-176°C$ at 450 kPa) through countercurrent heat exchange with outflowing streams of waste gas and cold nitrogen product and is separated into

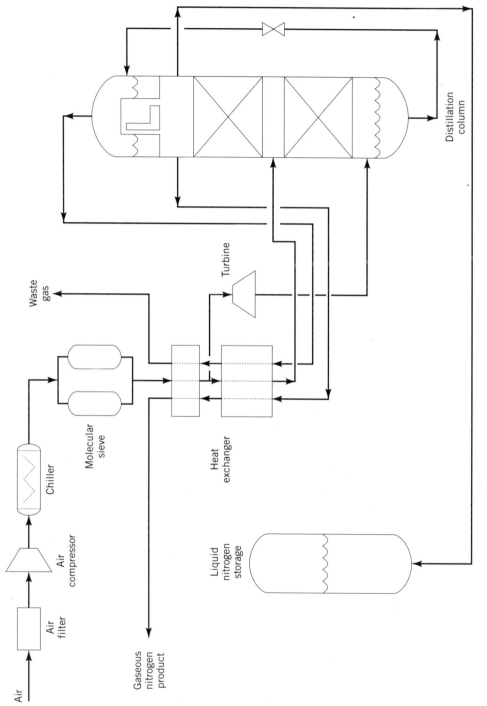

Fig. 2. Cryogenic nitrogen generation process.

its constituents in a single distillation column. Nitrogen product emerges from the top of the column and is available at ca 140–450 kPa (20–65 psig). Additional refrigeration is required to accommodate heat losses in the system from process efficiencies and imperfect thermal insulation. This is provided by either expanding a portion of the incoming compressed air through an expansion turbine coupled to an external brake to provide isoenthalpic cooling, or through the vaporization of a small quantity of delivered liquid nitrogen. In larger plants with an expansion turbine, liquid nitrogen in volumes between 1 to 5% of the capacity of the plant can be produced for use as backup. If more liquid nitrogen production ($\leq 10\%$) or higher nitrogen product pressures (ca 450–1000 kPa (65–145 psig)) are required, an alternative cycle can be used which expands a portion of the oxygen-rich waste gas rather than the incoming air through the expansion turbine.

Capacities of on-site cryogenic nitrogen plants range from 250 to over 50,000 m^3/h (10,000 to over 2,000,000 SCFH). A level of 1 ppm residual O_2 is readily achievable with present technology. Through the addition of more separation trays and at the expense of reduced nitrogen recovery, levels of less than 1 ppb O_2 are possible for demanding semiconductor manufacturing applications (see SEMICONDUCTORS). In 1993, the approximate fully installed capital cost of a standard 1300 m^3/h (50,000 SCFH) plant was $2 million, and that of a 5250 m^3/h (200,000 SCFH) plant was $4.6 million. Electrical power requirements range from 0.20–0.30 kWh/m^3 (0.53–0.79 kWh/100 SCF).

Pressure Swing Adsorption. PSA systems operate on the principle of reversible selective adsorption of oxygen on a carbon molecular sieve (CMS). This process was invented in 1965 and commercialized in the 1970s by the German research institute on coal (qv), Bergbau-Forschung GmbH. Modern CMS is manufactured from bituminous coal which is oxidized in air at a temperature below the ignition point and mixed with an organic binder to form pellets which are carbonized and processed. These pellets have pores with diameters similar in size to oxygen and nitrogen molecules. Oxygen is adsorbed by the CMS at a faster rate than nitrogen due to polarity and molecular size effects, the oxygen molecule being smaller than nitrogen.

A typical nitrogen PSA system is shown in Figure 3. Air is compressed, filtered, cooled to remove excess moisture, and passed alternately through two beds of CMS. Oxygen, carbon dioxide, and water vapor are selectively adsorbed in the CMS matrix. When one bed is saturated with oxygen, the air flow is diverted to the second. The first bed is depressurized to atmosphere, releasing the adsorbed oxygen, carbon dioxide, and water vapor and the process is repeated. Product nitrogen is collected in a nitrogen receiver and is available at ca 690 kPa (100 psig). Atmospheric argon is concentrated in the product nitrogen.

Nitrogen PSA systems are economical for producing gaseous nitrogen on-site at flow rates ranging from 25 to 800 m^3/h (1,000 to 30,000 SCFH) at purities up to 99.8% (~0.2% residual oxygen). More production can be obtained from a given system at lower purities, provided sufficient air is available. For higher purities, the residual oxygen can be combined with hydrogen (qv) in a supplemental deoxo system. The water reaction product is removed in a regenerative dryer. However, residual hydrogen may be present in the product nitrogen and there are significant additional costs of hydrogen supply and the deoxo system.

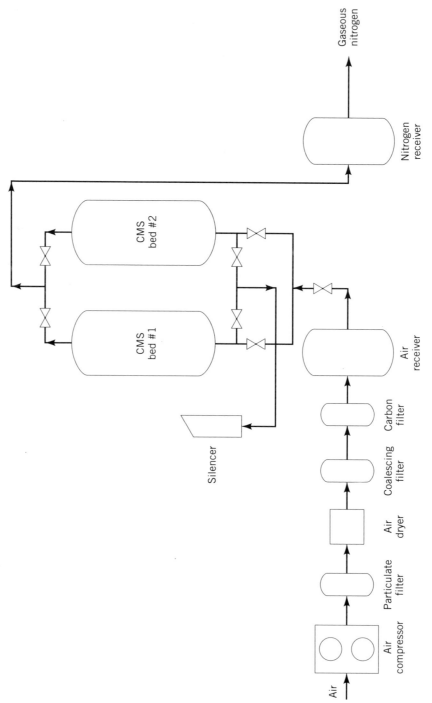

Fig. 3. Pressure swing adsorption nitrogen generation system. CMS = carbon molecular sieve.

In 1993, the approximate installed capital cost for a 4000-SCFH PSA plant was $150,000, and $350,000 for a 20,000-SCFH system, both producing nitrogen at 99.5% purity. Electrical power requirements ranged from 0.40 to 0.48 kWh/m^3 (1.05–1.25 kWh/100 SCF) for 99.5% nitrogen.

Membrane Permeation. The use of hollow-fiber polymeric membranes for air separation is a newer technology, commercially viable since the early 1980s. Membrane systems are displacing PSA systems in the lower purity, lower flow rate range. Membrane nitrogen generation systems operate on the principle of selective gaseous permeation through a membrane. Commercially available systems use hollow-fiber membranes (qv) fabricated from polymers such as polysulfones, polyimides, and polycarbonates which permeate oxygen faster than nitrogen through a solution–diffusion mechanism. A typical membrane fiber outer diameter is 100–200 μm with a wall thickness of 30–50 μm. In the membrane process shown in Figure 4, air is compressed and passed through a series of filters to remove any residual oil, which can be detrimental to membrane longevity, from the compressor and excess water vapor. The air is then heated to the optimum process temperature for the given polymer (usually 40–60°C) and is fed axially into the center of thousands of the hollow fibers packed in a tube-and-shell configuration. The nitrogen is concentrated during its passage down the fibers and is collected as the nitrogen product. Atmospheric argon is also concentrated in the product stream (see MEMBRANE TECHNOLOGY).

Nitrogen membrane systems are economical for producing gaseous nitrogen at flow rates ranging from 3 to 3000 m^3/h (100–100,000 SCFH), depending on purity produced. Purities up to 99.5% are economical, depending on flow rate. As with PSA systems, the residual oxygen can be removed by use of a supplemental deoxo system. Membrane technology has proven versatile because of the simplicity of the process and the light weight of the membranes.

In 1993, the approximate installed capital cost of a 26 m^3/h (1000 SCFH) membrane system was $25,000, and $100,000 for a 260 m^3/h (10,000 SCFH) system at 98% purity (balance oxygen). Electrical power requirements ranged from 0.30 to 0.60 kWh/m^3 (0.80–1.50 kWh/100 SCF).

Inert Gas Generators. Inert gas generators remove atmospheric oxygen by combustion with natural gas or propane. These systems find wide application in metal heat treating but have been largely displaced by other more cost-effective and reliable nitrogen generation technologies in other applications. In metal heat treating, inert gas generators are commonly called exothermic generators. Combustion products include controllable amounts of CO, H_2, CO_2, and water vapor, depending on the specific air–fuel gas input mixture. Excess water vapor and CO_2 can be removed through a refrigerant dryer and a molecular sieve adsorber, respectively. Two types of exothermic atmospheres, rich and lean, are commonly specified in heat treating applications. Rich exothermic gas contains 10–21% CO and H_2, and 5% CO_2, produced with a 6.5 to 1 air–fuel gas input ratio. Lean exothermic gas contains 1–4% CO and H_2, and 11% CO_2, produced with a 9.0 to 1 air–fuel gas input ratio (23). The balance of both atmospheres is nitrogen with a dew point of 40°C. These gas mixtures are commonly used for annealing ferrous and nonferrous materials where decarburization and brightness are not factors. Exothermic generators produce surplus heat which can be used to generate steam.

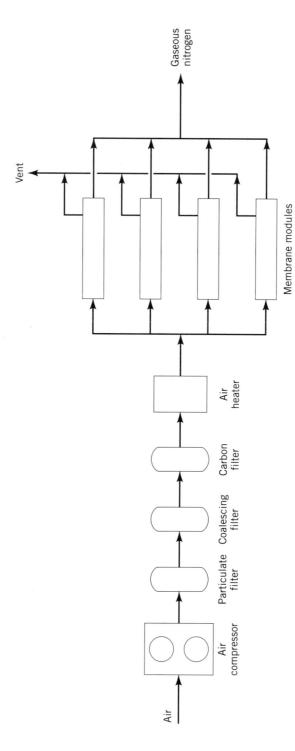

Fig. 4. Membrane permeation nitrogen generation system.

Shipment

Nitrogen is shipped and stored in gaseous form in steel cylinders under high pressure and in liquid form in vacuum-insulated containers. Because of the weight of high pressure storage vessels and the volatile nature of liquid nitrogen, most nitrogen is produced and consumed locally within a radius of less than 250 km. An exception is the large-scale distribution of nitrogen at moderate pressures (4000 kPa (600 psig)) in extensive pipeline networks serving the petrochemical and petroleum refining industries along the U.S. Gulf Coast of Texas and Louisiana, and serving the steel and chemical industries in northern France, the Benelux countries, and in the Chicago area in the United States. Nitrogen with oxygen is produced by numerous cryogenic air separation plants located along the pipeline networks (see CRYOGENICS). The U.S. Gulf Coast network is one of the world's longest with a total length of over 1200 km with 25 air separation plants supplying nitrogen and oxygen with a total capacity of over 20,000 t/d to over 70 companies.

Moderate quantities of nitrogen are shipped in liquid form in vacuum-insulated tanker trucks with typical capacities of 19 t liquid nitrogen (6200 gal) or 15,300 m³ (580,000 SCF) when vaporized. The nitrogen is delivered to vacuum-insulated storage vessels at user sites with capacities ranging from 5,700 to 50,000 L (1,500 to 13,000 gal), containing from 3,700 to 31,500 m³ (140,000 to 1,200,000 SCF) gaseous nitrogen upon vaporization. Liquid nitrogen is consumed directly from the vessels or gaseous nitrogen is produced as needed by vaporization in heat exchangers heated by free convection or forced ambient air, steam, or electricity. Heat leakage through the vessel insulation evaporates up to 0.5% of the vessel volume per day. Storage vessels are typically maintained at pressures up to 1700 kPa (250 psig) and pressure relief valves divert any excess pressure into the user's pipeline, eliminating loss of nitrogen to the atmosphere. Higher pressure liquid storage vessels (up to 4000 kPa (600 psig)) are available. Figure 5 shows a typical liquid nitrogen storage vessel and vaporizer.

Smaller quantities of liquid nitrogen are shipped in liquid cylinders, which are pressurized stainless steel vacuum-insulated containers with capacities of 160–180 L of liquid nitrogen, producing 100–120 m³ (3800–4600 SCF) gaseous nitrogen when vaporized. Moderate quantities of high pressure gaseous nitrogen are shipped in tube trailers, which contain a number of horizontal steel cylinders the length of a truck trailer. Tube trailers contain from 1160 to 3860 m³ (44,000–147,000 SCF) gaseous nitrogen at a pressure of 18,200 kPa (2640 psig). Very small quantities of gaseous nitrogen are shipped in individual steel cylinders ranging in capacity from 0.5 to 13 m³ (20 to 490 SCF) at pressures from 18,200 to 41,400 kPa (2640–6000 psig).

Economic Aspects

The principal international producers and distributors of nitrogen are L'Air Liquide S.A. (France), The BOC Group Plc (U.K.), Air Products and Chemicals, Inc. (U.S.), and Praxair, Inc. (U.S.). There are many other smaller regional producers.

A total of 21.3×10^9 m³ (810×10^9 SCF) of nitrogen was produced in the United States in 1992, of which 70% was distributed by pipeline and 30% in

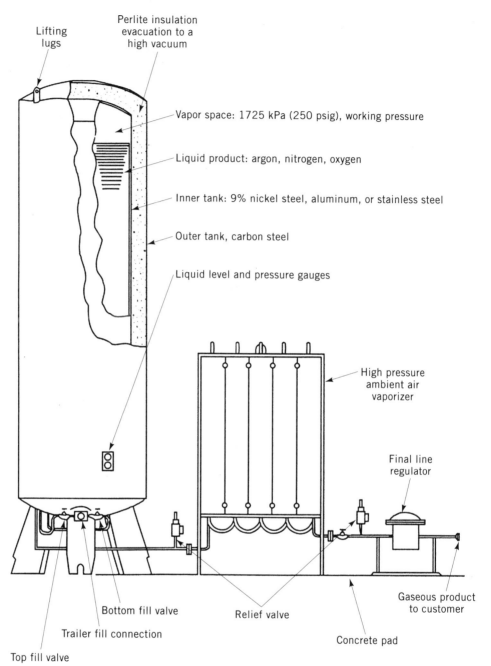

Fig. 5. Liquid nitrogen storage vessel and vaporizer.

either liquid or gaseous form (24). The total value of U.S. nitrogen shipments was $840 million in 1991. In 1991, total worldwide production was approximately 45×10^9 m^3 (1.7×10^{12} SCF). International trade is negligible due to distribution economics. In the United States, nitrogen is produced in the largest quantity of any industrial gas, followed by oxygen. In Japan and western Europe, nitrogen is the second leading industrial gas after oxygen.

Nitrogen production in the United States has grown substantially because of growth in applications. Production has increased an average of 7.1% per year since 1970 (24). In the 1990s there has been a slowing of the growth rate but the increase in noncryogenic production should spur continued production increases in the 5% per year range through the rest of the twentieth century.

Specifications, Standards, and Quality Control

In the United States, the Compressed Gas Association lists nine grades of nitrogen, differentiated by oxygen content, dew point, total hydrocarbon content, and other contaminant levels (25). These grades, more often specified in government than commercial contracts, are shown in Table 3. Commercial cryogenically produced liquid nitrogen usually meets or exceeds Type II, Grade L. Higher purity Class II liquid nitrogen meets Type II, Grade M specifications. Commercially available compressed gaseous nitrogen, always produced from vaporized liquid nitrogen, usually meets or exceeds Type II, Grade L requirements. There are several specialty grades of compressed nitrogen typically available, eg, zero-grade which contains <0.2 ppm total hydrocarbons and a minimum nitrogen purity of 99.995%, ultra high purity (UHP) which contains <1 ppm water and a minimum nitrogen purity of 99.999%, oxygen-free which contains <0.5 ppm oxygen and a minimum nitrogen purity of 99.995%, and pre-purified which contains a minimum nitrogen purity of 99.998%.

Table 3. Grades of Industrial Nitrogen

Gaseous and liquid nitrogen	Nitrogen, min % (mol/mol)	Oxygen[a]	Water, ppm (v/v)	Dew point, °C	Total hydrocarbon content[a]
B[b]	99.0				
E[c]	99.5	0.5%	26	−63	58
F	99.9	0.1%	32	−60	
G	99.95	500	26	−63	
H[c]	99.99	50	11	−75	5
K	99.995	20	16	−70	
L	99.998	10	4	−90	
M	99.999	5	2	−100	5
Q[d]	99.999	1	2	−100	1

[a]Units in ppm (mol/mol) unless otherwise noted.
[b]10 ppm CO.
[c]Type II <1.0 mg/L permanent particulates.
[d]5 ppm each of Ar, Ne, He, and CO.

Health and Safety Factors

Gaseous nitrogen is nontoxic and nonflammable but does not support life. Nitrogen should be stored and used only in well-ventilated areas. Special care must be taken entering an enclosed area which may be enriched in nitrogen. Deaths occur every year when personnel enter confined spaces in which nitrogen has displaced oxygen. Portable air packs or an air-fed mask must be used when working in an area with oxygen content below 19.5% by volume. Oxygen level analyzers with low level alarms should be used whenever the possibility of an oxygen-deficient atmosphere exists.

Liquid nitrogen and its vapor are extremely cold and can rapidly freeze human tissue. Liquid nitrogen spills should be flushed with water to accelerate evaporation. When exposed to liquid nitrogen, carbon steel, rubber, and plastic become embrittled and may fracture under stress. Copper, brass, bronze, Monel, aluminum, and 300 series austenitic stainless steels remain ductile and are acceptable for cryogenic service. Liquid nitrogen in poorly insulated containers can concentrate and condense atmospheric oxygen on the exterior surfaces which may cause a serious fire hazard. One volume of liquid nitrogen produces 696.5 volumes of gaseous nitrogen at atmospheric pressure and room temperature so extreme pressures can be generated if liquid nitrogen vaporizes in a confined space. Storage vessels or handling equipment should be provided with multiple pressure relief devices to prevent the buildup of high pressure. A pressure relief valve for primary protection and a frangible disk for secondary protection are commonly provided on commercial liquid nitrogen storage vessels (26).

Uses

Applications for nitrogen are widespread in both its gaseous and liquid phases. Gaseous nitrogen is usually used as an inert blanketing or carrier gas. Liquid nitrogen is used as an expendable nonreactive, nontoxic refrigerant. Very few applications, excepting the large-scale synthesis of ammonia from atmospheric nitrogen, use nitrogen as a reactant.

Chemical Process Industry.   Gaseous nitrogen is used extensively in the chemical process industry as a blanketing, inerting, or purging agent to prevent oxidation of sensitive materials or to prevent the formation of an explosive mixture. Chemical storage tanks are frequently maintained under a slight positive pressure of nitrogen. During withdrawal of product from the storage tank, nitrogen is injected into the tank to maintain headspace pressure; during filling of the storage tank, nitrogen is vented from the tank to atmosphere, a flare system, or a solvent recovery system. Process vessels, piping, and storage vessels are purged with nitrogen to remove oxygen or control reactant levels. Purging is performed by one of three methods, ie, displacement, pressurization, or dilution. Displacement purging is often used to purge piping systems. The volume of nitrogen required corresponds to the volume of the pipe. Pressurization purging involves the repeated pressurization with nitrogen and venting of a vessel to remove the contaminant. Dilution purging assumes complete mixing of the vessel contents and the injected nitrogen, and the contaminant concentration is slowly reduced through time.

For blanketing flammable liquids or solids, considerable oxygen concentrations can be tolerated in the blanketing nitrogen without risk of formation of an explosive mixture. Table 4 shows the maximum permissible oxygen levels in nitrogen that can be used to blanket a number of common substances (27). Noncryogenically produced nitrogen is often used in such applications because of its economy and residual oxygen content (28).

Table 4. Maximum Permissible Oxygen
Concentrations for Various Substances[a]

Gas or vapor	Oxygen, %
methane	12
ethane	11
propane	11.5
ethylene	10
benzene	11.4
toluene	9.5
styrene	9.0
gasoline	12
acetone	11.5
carbon disulfide	5
hydrogen	5
methanol	10
dust	
soy flour	15
charcoal	17
aluminum	2
magnesium	0
sulfur	12
butadiene–styrene	13

[a] Percent of oxygen in nitrogen above which combustion can take place.

Much nitrogen is utilized during polymerization operations where oxygen is a polymerization inhibitor. Nitrogen is often used for the pressure transfer of hazardous or flammable liquids and as a drying agent for oxidation-sensitive materials such as nylon resins and fibers (see DRYING). Nitrogen is used to agitate and deoxygenate liquids and solid–liquid mixtures. It is also used as a carrier gas for other reactants such as oxygen in the regeneration of spent reforming catalyst in petroleum (qv) refining. Nitrogen is used as a blowing agent (see BLOWING AGENTS) in foamed plastic production and in blow molding of plastic containers. It can be used as a stripping agent to remove volatile organic compounds from process wastewater.

Liquid nitrogen is used in cold traps to remove and recover solvents or volatile organic compounds from gas streams to reduce atmospheric emissions. Liquid nitrogen can be used to accelerate the cooldown time for process reactors (29).

Food Industry. Large volumes of liquid nitrogen are used in the food industry for cryogenic freezing, where it competes with liquid carbon dioxide and mechanical freezing technologies. Cryogenic freezing of frozen foods reduces cell damage which results in reduced thawed product quality. By freezing the food very rapidly, inter- and intracellular ice crystal growth is minimized, avoiding rupture and damage of the cell walls of the food product. Concentration of water-soluble components between ice crystals during the freezing process is also reduced. Both effects reduce syneresis or drip loss due to release of cellular fluid, an effect which adversely impacts the texture, flavor, aroma, and nutritional value of the thawed product. Cryogenic freezing also reduces dehydration during the freezing process.

Types of cryogenic freezers include immersion freezers, in which the product is fed by conveyor belt directly through a bath of liquid nitrogen; in-line freezers, in which liquid nitrogen is sprayed onto the product traveling on a belt; and spiral freezers, in which an annular liquid nitrogen spray freezes the food product traveling on a space-conserving spiral conveyor. Immersion freezers provide the fastest freezing rates, at greater than 1000°C/min, but only utilize the latent heat of vaporization of liquid nitrogen. Efficiencies are less than 50% versus in-line freezers which use the additional cooling effect of vaporized nitrogen and have efficiencies approaching 90%. However, the capital cost of an immersion freezer is much less than that for in-line or spiral freezers (30). Liquid nitrogen consumption in food freezing applications is in the range of 1.0 to 2.0 kg/kg of product (31). Actual consumption is influenced by product and process variables.

Liquid nitrogen also finds application in pressurizing aluminum and plastic food and noncarbonated beverage cans and bottles. Several drops of liquid nitrogen are injected into the can or bottle prior to sealing. After sealing, the nitrogen vaporizes to provide internal pressurization which provides sidewall strength and allows the use of very thin sidewall materials. Nitrogen consumption is in the range of 0.12 to 0.25 g per can or bottle (32,33).

Other growing uses for liquid nitrogen include batch chilling of meat products and cookie dough, and cooling flour and sugar during warm season processing.

Gaseous nitrogen is used in controlled and modified atmosphere food storage and packaging to maintain the quality of fresh fruits and vegetables by retarding oxygen-dependent ripening and decay processes. Controlled atmospheres are extensively used in the storage of apples and pears to allow extended storage prior to sale without significant quality degradation. Controlled atmosphere storage rooms for apples and pears are typically maintained at 98–99% nitrogen concentrations with balance controlled amounts of oxygen and carbon dioxide. Nitrogen can be provided from vaporized liquid nitrogen, noncryogenic generators, or combustion-based generators.

Modified atmospheres that predominantly contain nitrogen are increasingly used in the packaging of fresh pasta, nuts (qv), potato chips, dry goods (such as coffee), breads, and prepared salads to extend shelf-life.

Primary Metallurgy and Heat Treatment.   In ferrous metallurgy (qv), nitrogen is typically provided by pipeline as a coproduct with oxygen to steel producing facilities for use as a scavenging, stirring, and inert blanketing gas during steel processing and casting operations. Nitrogen can be used to replace more expensive argon in the argon–oxygen decarburization (AOD) process during the manufacture of austenitic stainless steels. In nonferrous metallurgy, nitrogen finds application in degassing of aluminum melts to remove hydrogen and reduce oxide formation. Nitrogen is used to shroud and cool aluminum extrusion dies to reduce surface oxidation and increase die life. The use of PSA nitrogen generators in this application has been reported (34).

In most heat treatment processes for metal fabricated parts, nitrogen serves as a nonreactive, passive constituent of the furnace atmosphere. It can be used alone for the annealing of aluminum, copper, and some low carbon steels. Most often it is used in combination with other gases such as H_2 and CO to produce reactive atmospheres for sintering, carburizing, and carbonitriding ferrous parts. Nitrogen reacts with some stainless steels and cannot be used in their heat treatment. At high temperatures, atomic nitrogen combines with iron to form finely divided nitrides, producing a hardened nitrided or carbonitrided surface layer (35).

Electronics Manufacturing.   In the manufacture of semiconductors, gaseous nitrogen is used in the largest quantity of any gas as an inert carrier in epitaxy and diffusion processes, ion implantation, chemical vapor deposition, annealing, and plasma etching. Nitrogen purity requirements are extremely rigorous, reaching 1 ppb contaminant levels. On-site cryogenic nitrogen plants are often used as the supply mode. If delivered liquid nitrogen is used, careful purging of transfer lines is required to maintain purity. Facility piping is usually electropolished stainless steel.

In electronic component assembly operations, standard purity nitrogen finds uses in atmospheres for wave soldering and infrared reflow soldering machines. This application has grown because manufacturers are converting soldering operations from older technologies, that required post-cleaning with chlorofluorocarbons, to no-clean technologies, that require nitrogen atmospheres. Nitrogen is also used extensively in various utility applications such as dry box storage of work-in-process components, blanketing of process chemicals, burn-in oven atmospheres, and nitrogen-driven pneumatic tools.

Oil and Natural Gas Production.   Nitrogen is used in oil and natural gas well completion and stimulation applications as well as oil and natural gas enhanced recovery. Applications in well completion and stimulation include use as a foaming agent for fracturing, acidizing, cementing, and gravel packing fluids. Nitrogen up to 95% by volume is used in these fluids to reduce density and hydrostatic gradient in the wells, thereby reducing the amount of energy required to return the fluid to the surface after the operation (36). Nitrogen for these applications historically has been delivered in liquid form by tanker truck and vaporized on-site. In the 1990s, portable on-site membrane generators have begun to find application.

Nitrogen is used for pressure maintenance in oil and gas reservoirs for enhanced recovery. It is sometimes used as a miscible agent to reduce oil viscosity and increase recovery in deep reservoirs. Other applications include recovery of oil in attic formations, gas cap displacement, and a sweep gas for miscible CO_2 slugs. Nitrogen competes with CO_2, a more miscible gas with hydrocarbons (qv), in most of these applications. The production mode is typically by on-site cryogenic separation plants. In 1990, nitrogen production in enhanced recovery operations was 20×10^6 m^3/d (750 million SCF/d) (37).

Two applications well suited to noncryogenic nitrogen production technologies include the replacement of cushion gas with nitrogen in underground natural gas storage reservoirs, pioneered by Gaz de France (38,39), and the use of nitrogen to displace natural gas in underground coal seams (40).

Other Applications.   Other applications for gaseous nitrogen include use as a blanketing agent for float glass manufacture, pressurizing agent in aerosols, inflation agent in aircraft tires and landing struts, purging agent for electrical cabinets, and pressurization agent for autoclaves used in the production of composite materials (41).

A significant application for liquid nitrogen is the cryogenic grinding of plastic or heat-sensitive materials. Plastic materials become brittle at cryogenic temperatures enabling easier grinding or deflashing operations. Cryogenic grinding is performed on both thermoplastic and thermosetting resins, old rubber tires for recycling, spices, coffee, coloring agents and pigments (qv), and wax. In the United States, 40×10^6 m^3/yr (1.5 billion SCF/yr) of liquid nitrogen is consumed in these applications (42).

Other liquid nitrogen applications include freezing biological specimens such as livestock semen and whole blood, cryosurgery, shrink fitting metal parts, and paint removal. Liquid nitrogen can be used for ground freezing to allow excavation in unstable wet soils (43), freeze plugging pipe sections to allow repairs while the rest of the pipeline remains pressurized (44), and cooling concrete in hot weather (45).

BIBLIOGRAPHY

"Nitrogen" in *ECT* 1st ed., Vol. 9, pp. 404–406, by E. S. Gould, Polytechnic Institute of Brooklyn; in *ECT* 2nd ed., Vol. 13, pp. 857–863, by J. W. Hall, Union Carbide Corp.; in *ECT* 3rd ed., Vol. 15, pp. 932–941, by R. W. Schroeder, Union Carbide Corp.

1. N. N. Greenwood and A. Earnshaw, *Chemistry of the Elements*, 1st ed., Pergamon Press, New York, 1984.
2. M. E. Weeks, *Discovery of the Elements*, 7th ed., updated by H. M. Leicester, *J. Chem. Educ.* (1968).
3. S. C. Bevan, S. J. Gregg, and A. Rosseinsky, *Concise Etymological Dictionary of Chemistry*, Applied Science Publishers Ltd., London, 1976.
4. *Chem. Eng. News*, 10–13 (Apr. 12, 1993).
5. J. Mason, ed., *Multinuclear NMR*, Plenum Press, New York, 1987.
6. F. A. Cotton and G. Wilkinson, *Advanced Inorganic Chemistry*, 5th ed., Wiley-Interscience, New York, 1988.
7. K. Jones, *Comprehensive Inorganic Chemistry*, Vol. 2, Pergamon Press, New York, 1991, pp. 147–199.
8. A. Grohmann, J. Riede, and H. Schmidbaur, *Nature* **345**, 140 (1990).
9. M. Mingos and R. Kankers, *J. Organometallic Chem.* **384**, 405 (1990).
10. W. L. Vos and co-workers, *Nature* **358**, 46 (1992).
11. D. R. Lide, ed., *Handbook of Chemistry and Physics*, 72nd ed., CRC Press, Cleveland, Ohio, 1991.
12. R. H. Perry and C. H. Chilton, eds., *Chemical Engineers' Handbook*, 5th ed., McGraw-Hill Book Publishing Co., Inc., New York, 1973, pp. 3–107.
13. G. Henrici-Olive, *Coord. Catal.* **9**, 289–305 (1976).
14. R. Battino, ed., *Nitrogen and Air*, Solubility Data Series, Vol. 10, Pergamon Press, New York, 1982.
15. *Handbook of Iron and Steelmaking*, United States Steel Corp., Pittsburgh, Pa., 1985, pp. 404–405.
16. *Metals Handbook*, 10th ed., Vol. 1, ASM International, Materials Park, Ohio, 1990.
17. A. N. Wright and C. A. Winkler, *Active Nitrogen*, Academic Press, Inc., New York, 1968.
18. A. D. Allen and C. V. Senoff, *Chem. Comm.*, 621–622 (1965).
19. P. Pelikan and R. Boca, *Coord. Chem. Rev.* **55**, 55 (1984).
20. G. J. Leigh, *Sci. Prog.* **291**, 389–412 (1989).
21. T. Yamabe, K. Hori, T. Minato, and K. Fukui, *Inorg. Chem.* **19**, 2154–2159 (1980).

22. *Chem. Eng.* **97**, 37 (1990).
23. *Plant Eng.* **46**, A4–A7 (1992).
24. *Current Industrial Reports, Industrial Gases*, Series M28C, U.S. Department of Commerce, Bureau of the Census, Washington, D.C., 1992.
25. *Handbook of Compressed Gases*, 3rd ed., Compressed Gas Association, Inc., Arlington, Va., 1990.
26. *Safe Handling of Cryogenic Liquids*, 2nd ed., CGA P-12, Compressed Gas Association, Inc., Arlington, Va., 1987.
27. *Standard on Explosion Prevention Systems*, NFPA 69, National Fire Protection Association, Quincy, Mass., 1986.
28. T. L. Hardenburger, *Chem. Eng.*, 136–146 (Oct. 1992).
29. J. Hoose, *Hydrocarbon Proc.* **63**, 71–72 (Oct. 1984).
30. R. Macrae, ed., *Encyclopedia of Food Science, Food Technology and Nutrition*, Vol. 3, Academic Press, Inc., San Diego, Calif., 1993, pp. 2060–2065.
31. Y. H. Hui, *Encyclopedia of Food Science and Technology*, Vol. 2, John Wiley & Sons, Inc., New York, 1992, p. 1254.
32. *Food Eng.* **60**, 94 (Jan. 1988).
33. *Food Eng.* **59**, 79 (July 1987).
34. D. C. Miller, Z. S. Pukanecz, R. D. Ritter, and R. H. Maynard, *Light Metal Age* **48**, 44–46 (Feb. 1990).
35. *Metals Handbook*, 10th ed., Vol. 4, ASM International, Materials Park, Ohio, 1990, p. 543.
36. V. L. Ward, *SPE Prod. Eng.* **1**, 275–278 (July 1986).
37. B. Evison and R. E. Gilchrist, *Petrol. Eng. Intl.* **64**, 69–71 (Mar. 1992).
38. C. Bellier, *Innovation Technologie: Industries et Techniques* **662** (Sept. 1, 1989).
39. S. E. Foh, Y. A. Shikari, R. M. Berry, and F. LaBaune, paper presented at the *SPE Gas Technology Symposium*, Dallas, Tex., June 13, 1988.
40. M. D. Stevenson, W. V. Pinczewski, and R. A. Downey, paper presented at the *SPE Gas Technology Symposium*, Calgary, Alberta, Canada, June 28, 1993.
41. P. Laut, *Adv. Mater. Proc.* **143:5** (May 1993).
42. R. G. Scurlock, ed., *History and Origins of Cryogenics*, Oxford University Press, New York, 1992, p. 243.
43. *Cryogenics* **26**, 572 (Oct. 1986).
44. C. Schelling, *Plant Eng.* **40**, 55–57 (July 24, 1986).
45. C. Ibbetson, *Civil Eng.*, 24–26 (June 1987).

General References

K. D. Timmerhaus and T. M. Flynn, *Cryogenic Process Engineering*, Plenum Press, New York, 1989.
R. G. Scurlock, ed., *History and Origins of Cryogenics*, Oxford University Press, New York, 1992.
Handbook of Compressed Gases, 3rd ed., Compressed Gas Association, Arlington, Va., 1990.

THOMAS L. HARDENBURGER
Air Liquide America Corporation

MATTHEW ENNIS
Stanford University

NITROGEN FIXATION

More than 99.9% of the nitrogen (qv) on earth is present as the dinitrogen molecule, N_2, of which somewhat more than 97% is trapped in primary and sedimentary rocks (2×10^{17} and 4×10^{14} metric tons, respectively) and about 2% (4×10^{15} t) is free in the atmosphere (1,2). In comparison, all living plants and animals together contain only 1×10^{10} t of nitrogen, and there is an additional 1.2×10^{10} t of nitrogen in the form of dead organic matter distributed about equally among land and sea and 6×10^{11} t of soluble inorganic forms of nitrogen (excluding N_2) in the sea (1–5). Thus only a very small proportion of the nitrogen present on earth is, at any one time, in a usable or fixed form. Various transformations in the nitrogen cycle (Fig. 1) allow nitrogen to move between the atmospheric inert pool and the fixed, usable terrestrial pools.

Nitrogen fixation, also called dinitrogen fixation, is a pivotal process in the global cycling of nitrogen. Nitrogen fixation is involved with the atmosphere-to-terrestrial direction of the cycling. Nitrification and denitrification convert ammonia to nitrate and then, via nitrogen oxides, to dinitrogen which is lost to the atmosphere. Leaching and erosion of soils result in the movement of fixed nitrogen between land and sea. The biological world stays just ahead of a nitrogen deficiency because the fixation rate slightly exceeds the denitrification rate (6). Only about one-third of the available nitrate is assimilated by plants, one-third is leached away, and one-third is denitrified and lost to the atmosphere (7).

Dinitrogen is fixed either by natural processes or by industrial ammonia (qv) production (1,8,9). The estimates for the annual biological contribution range around $100–200 \times 10^6$ t. Industrial fixation contributes about 50×10^6 t/yr for fertilizer uses (see FERTILIZERS). Other processes, eg, lightning and combustion, are estimated to fix about 30×10^6 t/yr. Thus the biological process represents

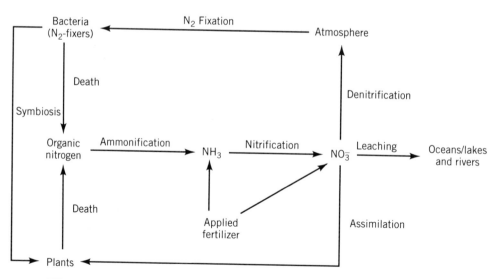

Fig. 1. Biological pathways and processes involved in the nitrogen cycle.

the majority (ca 65%) of the total annual fixation rate, contributing about three times as much as the commercial production of fertilizer.

Plants depend on the availability of nitrogenous compounds produced from atmospheric N_2 either commercially or biologically. The availability of fertilizer nitrogen is almost always the limiting factor in crop productivity. In nature, however, only a relatively few species of bacteria have the capability of converting N_2 into ammonia, which can then be incorporated into amino acids (qv) and the precursors of nucleic acids (qv). These microbes usually do so for their own benefit. In certain crops, such as the legumes, ie, peas, beans, alfalfa, etc, nature has provided a mechanism for biological interaction between the plant and nitrogen-fixing bacteria. The plant receives fixed nitrogen directly from the bacteria which are harbored in nodules on its roots. The most important food crops, however, such as the cereal grains, ie, rice, wheat, and corn, and root and tuber crops, do not harbor symbiotic partners. Hence, for crop productivity to reach commercially acceptable levels, extensive augmentation by commercially fixed nitrogen is necessary.

Considerable progress in the understanding of biological nitrogen fixation has been made since the 1970s. For example, the x-ray crystal structures of both component proteins of nitrogenase, the enzyme responsible for catalyzing nitrogen fixation, have been published (10), providing the biological information necessary for building effective synthetic model systems for N_2 reduction. Although a well-defined mechanism for biological nitrogen reduction is still lacking, a useful numerical model has been developed (11). Moreover, purely chemical processes have been devised that bind N_2 and in some cases activate the nitrogen sufficiently so that reduced nitrogen compounds, ie, ammonia and/or hydrazine, are produced on protonation.

Industrial Processes

Until the early nineteenth century, the fixed usable nitrogen stockpiled over millions of years by various natural processes was enough to sustain the needs of the earth's population. Then the dramatic growth of cities and populations led to the beginnings of the nitrogenous fertilizer industry. Guano, hardened bird droppings, was imported into Europe from Peru, as was saltpeter (sodium nitrate) from Chile. These fertilizer forms were supplemented in the industrialized nations by the ammoniacal by-products from coal gas. Further increases in demand led to the invention of several nitrogen fixation processes. Some were exploited commercially.

The first such process was the Birkeland-Eyde process for N_2 oxidation, implemented in 1905 (12). In this process, air is passed through an electric arc at temperatures above 3000°C to generate nitric oxide [10102-43-9], NO.

$$N_2 + O_2 \rightleftharpoons 2\,NO$$

On cooling the air stream, further oxidation gives nitrogen dioxide [10102-44-0], NO_2, which on absorption into water gives a mixture of nitric, HNO_3, and nitrous, HNO_2, acids (see NITRIC ACID).

$$2\,NO + O_2 \longrightarrow 2\,NO_2$$

$$2\,NO_2 + H_2O \longrightarrow HNO_3 + HNO_2$$

The low (ca 2%) yield of NO, the tendency to revert to N_2 and O_2 if the product stream is not quenched rapidly, the consumption of large (ca 60,000 kWh/t N_2 fixed) amounts of electricity, and the concomitant expense to sustain the arc all led to the demise of this process. The related Wisconsin process for oxidizing N_2 at high temperatures in a pebble-bed furnace was developed in the 1950s (13). Although a plant that produced over 40 t/d of nitric acid was built, the product recovery costs were not economically competitive.

At about the same time that the Birkeland-Eyde process was developed, the Frank-Caro cyanamide process was commercialized (14). In this process limestone is heated to produce lime, which then reacts with carbon in a highly energy-demanding reaction to give calcium carbide. Reaction with N_2 gives calcium cyanamide [150-62-7], which hydrolyzes to ammonia and calcium carbonate (see CYANAMIDES).

$$CaCO_3 \longrightarrow CaO + CO_2$$

$$CaO + 3\,C \longrightarrow CaC_2 + CO$$

$$CaC_2 + N_2 \longrightarrow CaCN_2 + C$$

$$CaCN_2 + 3\,H_2O \longrightarrow CaCO_3 + 2\,NH_3$$

Even though the overall energy requirement of this process is only ca 20–25% of the arc process, the Haber-Bosch process, also developed in the early 1900s, proved to be more economical. A fourth process, the Serpak process for the catalytic nitriding of aluminum, was never commercially exploited to any significant degree, mainly because of large energy requirements (12).

The synthetic ammonia industry of the latter part of the twentieth century employs only the Haber-Bosch process (12–15), developed in Germany just before World War I. Development of this process was aided by the concurrent development of a simple catalyzed process for the oxidation of ammonia to nitrate, needed at that time for the explosives industry. N_2 and H_2 are combined directly and equilibrium is reached under appropriate operating conditions. The resultant gas stream contains ca 20% ammonia.

$$N_2 + 3\,H_2 \rightleftharpoons 2\,NH_3$$

When this reaction was first discovered, a considerably higher (ca 1300°C) temperature was required than that used in the 1990s. Thus, until Haber discovered the appropriate catalyst, this process was not commercially attractive. As of this writing (ca 1995), the process suffers from the requirement for significant quantities of nonrenewable fossil fuels. Although ammonia itself is commonly used as a fertilizer in the United States, elsewhere the ammonia is often converted into solid or liquid fertilizers, such as urea (qv), ammonium nitrate or sulfate, and various solutions (see AMMONIUM COMPOUNDS).

The Haber-Bosch Process. A modern ammonia plant performs two distinct functions. The more energy-demanding and complex function is the preparation and purification, from various feedstocks, of synthesis gas, known as syngas, which contains N_2 and H_2 in a 1:3 ratio. The second function is the catalytic conversion of syngas to ammonia (Fig. 2). In the years since commercial introduction in 1913, many process changes have been made in syngas production to lower costs and to give greater efficiencies.

Synthesis Gas Production. Through World War II, coal (qv) was the primary raw material for ammonia synthesis, either directly as in the original water gas plant or indirectly via coke-oven gas. As of the 1990s petroleum-based products are preferred and represent ca 88% of all feedstocks. Dihydrogen for synthesis gas is produced either by steam reforming of natural gas and other lighter hydrocarbons (qv), such as naphtha, or by the partial oxidation of heavy oils and coal. Other processes include coal carbonization, oil refining, and the electrolysis of water (14) (see COAL CONVERSION PROCESSES; GAS, NATURAL; HYDROGEN).

In the catalytic steam reforming of natural gas (see Fig. 2), the hydrocarbon stream, principally methane, is desulfurized and, through the use of superheated steam (qv), contacts a nickel catalyst in the primary reformer at ca 3.04 MPa (30 atm) pressure and 800°C to convert methane to H_2.

$$CH_4 + H_2O \longrightarrow 3\ H_2 + CO$$

Reforming is completed in a secondary reformer, where air is added both to elevate the temperature by partial combustion of the gas stream and to produce the $3:1::H_2:N_2$ ratio downstream of the shift converter as is required for ammonia synthesis. The water gas shift converter then produces more H_2 from carbon monoxide and water. A low temperature shift process using a zinc–chromium–copper oxide catalyst has replaced the earlier iron oxide-catalyzed high temperature system. The majority of the CO_2 is then removed.

$$CO + H_2O \longrightarrow H_2 + CO_2$$

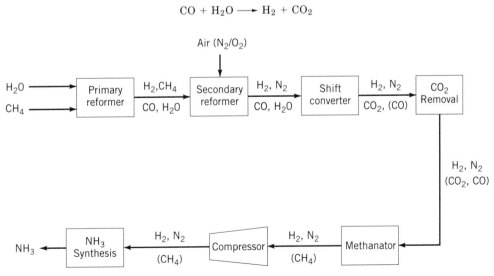

Fig. 2. The Haber-Bosch process. Gases in parentheses are minor constituents of the mixture.

The partial-oxidation process differs only in the initial stages before the water gas shift converter. Because it is a noncatalyzed process, desulfurization can be carried out further downstream. The proportions of a mixture of heavy oil or coal, etc, O_2, and steam, at very high temperature, are so adjusted that the exit gases contain a substantial proportion of H_2 and carbon monoxide.

$$3\ C + H_2O + O_2 \longrightarrow H_2 + 3\ CO$$

These gases are then fed to the water gas converter as in the steam-reforming process, after which they are compressed to ca 20.3 MPa (ca 200 atm) for processing in the catalytic ammonia converter.

A breakthrough in this area was the development of centrifugal turbine compressors, which gave a significant energy savings compared to the reciprocating compressors, but only at ammonia outputs of greater than 600 t/d. This economy means that all newer plants operate at 1000–2000 t/d. Energy can also be saved by increasing the operating pressure for the reformers in synthesis gas production because this lowers compression costs at the catalytic ammonia converter and improves heat recovery from the unreacted steam. The general trend is for ammonia plants to become independent of external electricity sources. Many recovery systems are employed in such a way that waste heat is recaptured in a reusable form.

Catalytic Conversion to Ammonia.   A large number of catalysts have been tested in the ammonia synthesis reaction. Haber originally proposed osmium as the catalyst and achieved an equilibrium yield of ammonia of 8% at 550°C and 15.1 MPa (150 atm). However, on an industrial scale, only iron, cobalt, molybdenum, and tungsten are practical. The best and most economical catalyst is metallic iron, using alumina and potassium oxide as promoters. The alumina acts as a structural promoter, preventing catalyst sintering. K_2O provides an electronic promoter, acting to increase the synthesis rate per unit area of the catalyst (16) and also aiding in desorption of ammonia from the surface. Of the three known iron oxides, only magnetite, Fe_3O_4, yields an efficient catalyst after reduction with H_2 to spongy iron (14) (see CATALYSIS).

Because the ammonia synthesis reaction is an equilibrium, the quantity of ammonia depends on temperature, pressure, and the H_2-to-N_2 ratio. At 500°C and 20.3 MPa (200 atm), the equilibrium mixture contains 17.6% ammonia. The ammonia formed is removed from the exit gases by condensation at about -20°C, and the gases are recirculated with fresh synthesis gas into the reactor. The ammonia must be removed continually as its presence decreases both the equilibrium yield and the reaction rate by reducing the partial pressure of the N_2–H_2 mixture.

The mechanism of the synthesis reaction remains unclear. Both a molecular mechanism and an atomic mechanism have been proposed. Strong support has been gathered for the atomic mechanism through measurements of adsorbed nitrogen atom concentrations on the surface of model working catalysts where dissociative N_2 chemisorption is the rate-determining step (17). The likely mechanism, where (ad) indicates surface-adsorbed species, is as follows:

$$H_2 \rightleftharpoons 2\ H\ (ad)$$

$$N_2 \rightleftharpoons 2 \, N \, (ad)$$

$$N \, (ad) + 3 \, H \, (ad) \rightleftharpoons NH_3 \, (ad)$$

$$NH_3 \, (ad) \rightleftharpoons NH_3$$

Operational Constraints and Problems. Synthetic ammonia manufacture is a mature technology and all fundamental technical problems have been solved. However, extensive know-how in the construction and operation of the facilities is required. Although apparently simple in concept, these facilities are complex in practice. Some of the myriad operational parameters, such as feedstock source or quality, change frequently and the plant operator has to adjust accordingly. Most modern facilities rely on computers to monitor and optimize performance on a continual basis. This situation can produce problems where industrial expertise is lacking.

Biological Systems

In contrast to the large industrial facilities required to produce ammonia economically, some microorganisms are capable of diazotrophy, ie, the ability to use N_2 gas as the sole source of nitrogen for growth. Only prokaryotes, ie, those living things without an organized nucleus (eubacteria, cyanobacteria, archebacteria, and actinomycetes) can perform biological nitrogen fixation, the result of which is the reduction of N_2 to ammonia. Such bacteria can be either free-living, such as *Azotobacter* and *Clostridium*, or symbiotic, like the rhizobia. The latter group, in tight associations with higher leguminous plants, are much more important agriculturally. In exchange for the fixed nitrogen supplied by the bacterium, the legume supplies a protective environment in the form of the root nodule and energy in the form of carbohydrate generated by photosynthesis. Thus renewable solar energy (qv) powers this fertilizer production system. Ammonia fertilizer from the Haber process involves energy costs in production, in transportation to the user, and in storage for what is usually a seasonal industry. As food demands increase and fossil fuel reserves deplete, the exploitation of biological nitrogen fixation becomes more and more attractive as an alternative to commercial fertilizer production. Research in this area ranges from employing molecular genetic techniques to engineer nonlegume cash crops such as corn and wheat (see WHEAT AND OTHER CEREAL GRAINS) to fix enough N_2 for its own requirements (see GENETIC ENGINEERING), through the increased use of associative symbioses to the development of catalysts based on nitrogenase for N_2-reducing processes.

    The ability to fix N_2 is widely spread among bacterial genera (18,19). It occurs in free-living obligate aerobes, microaerobes, facultative anaerobes, and strict anaerobes, with examples in each group capable of photosynthesis, as well as in a variety of associations with higher plants, liverworts, and fungi. Although farmers recognized the benefits of crop rotations utilizing legumes centuries ago, the source of that benefit was unknown to them. The first report of nitrogen fixation, in 1838, involved a comparison of the growth and nitrogen content of cereals to that of leguminous plants in both greenhouse and field experiments (20,21). Mainly clover was used in rotations with wheat and tuber crops. When it was speculated that nitrogen entered plants directly, there was much skepticism. It was not until the 1880s that this work was convincingly confirmed (22). The perplexing question of the source of the fixed nitrogen was solved by localizing the activity to the bacteria-filled nodules on the roots of pea plants (21,23).

Plant–Bacterial Associations. It is known that N_2-fixing bacteria enter into associations with a variety of hosts (4,18,24–27). In addition to the *Rhizobium*–leguminous plant associations, first recognized in the 1880s (20–25), these involve actinorhizal symbioses where *Frankia* is harbored in root nodules on nonleguminous plants, such as alder; cyanobacteria intimately associated with fungi, ferns, cycads, and even the angiosperm, *Gunnera*; and the less formalized associative symbioses of grasses with *Azotobacter* or *Azosprillum*, for example.

Rhizobium–Legume Associations. The legumes (family Leguminoseae) include temperate and tropical flowering plants, ranging from small plants, like clover, to bushes and large trees, such as acacia. Only about 3,500 of the known 17,000 species of legumes have been tested for root nodulation. About 90% of the species tested in the subfamilies, Papilionoideae and Mimosoideae, are nodulated by rhizobia; less than 25% of the Caesalpinioideae species are nodulated. All nodules on legumes are caused by colonization by rhizobia, which are grouped into three genera: *Rhizobium*, which are usually fast-growing, unable to fix N_2 *ex planta*, and have plasmid-borne nitrogen fixation (*nif*) genes and a narrow host range; *Bradyrhizobium*, which are usually slow growers, some of which can fix N_2 *ex planta*, and have chromosomal *nif* genes and often a broad host range; and *Azorhizobium*, of which only one species is known as a fast grower capable of fixing N_2 *ex planta*. Two other genera, *Sinorhizobium* and *Photorhizobium*, have been proposed but are not yet generally accepted (18). There is only one known case of rhizobia nodulating a nonlegume. *Parasponia* (28), a woody member of the elm family that grows in southeast Asia and the Pacific islands, has nodules.

The best-studied associations are those for important crops, such as the pulses, ie, peas and beans, including soybeans (see SOYBEANS AND OTHER OILSEEDS), clovers, and alfalfa (1,18,29). These associations show a degree of specificity. Certain rhizobia only infect certain host plants. However, some legumes may be nodulated by several rhizobial species, and a single rhizobial species may nodulate more than one legume host. The rhizobia are divided into species and biovars, based on the range of acceptable hosts. This situation gives the appearance of promiscuity. However, recognition among plant and microbe is a highly refined, but still incompletely understood, process (30–33).

The first visibly discernible stage of the developing symbiosis is root hair deformation, branching, and curling. The rhizobia promote invagination of the external surface of the root at these deformed root hairs to form an infection thread through which they enter the plant. As the infection thread elongates, and in some symbioses even before it is formed, root inner cortical cells are induced to divide. These cells become the nodule primordium. The thread enters these cells and releases the rhizobia, which remain confined within vesicles bound by the plant-derived peribacteroid membrane. Within the infected plant cells, the bacteria cease growing and enlarge and often take on unusual shapes to become bacteroids (29). Infection occurs as early as the appearance of the first leaves and fixation continues usually until pod filling.

Certain strains of rhizobia are relatively ineffective as N_2 fixers. Thus seeds are coated with inoculum, ie, a dried effective rhizobial culture in peat, before planting so that indigenous strains of rhizobia in the soil are out-competed and good fixation rates ensured. This practice is important because, once a plant is infected by one strain, the ability of other strains to invade the plant is greatly decreased. Infection is also inhibited if the soil contains significant amounts of fixed nitrogen (18,34).

During the early rhizobial root-colonization stage, various components of root exudates, especially flavonoids and isoflavonoids, act both in rhizobial chemotaxis and to stimulate the expression of the rhizobial nodulation (*nod*) genes (35–38). These inducing compounds are active at low concentrations and appear to be exuded specifically in the area of the root most responsive to nodulation (39). The first of the compounds to be identified was luteolin [*491-70-3*] (36), which, because it is also produced by seeds and even by nonleguminous plants, cannot be host-specific as expected. It appears to be the range of compounds produced by the plant that determines host-specificity (38). Every stage in nodule formation is accompanied by the expression of nodule-specific plant genes, each encoding a plant protein called a nodulin (40). In these early stages of nodulation, the rhizobial *nod* genes (41–43) are stimulated by the flavonoids, which, together with the product of the rhizobial regulatory *nodD* gene, induce the expression of the other *nod* genes. This concerted action results in the synthesis of the bacterial signal molecules called Nod factors. Nod factors consist of three to five $\beta1$–4-linked N-acetylglucosamine residues as shown in Figure 3. The nonreducing sugar is acylated with a variety of either saturated or unsaturated fatty acids and may also be modified by either O-acetylation or carbamylation. Substitution at the reducing sugar, for example, with sulfate in *Rhizobium meliloti* (32) or with 2-*O*-methylfucose in *Bradyrhizobium japonicum* (45), is essential for biological activity. The common *nodABC* genes are required for the core oligosaccharide, and the other host-specific *nod* genes are responsible for the appended side chains, which likely confer host specificity on the individual Nod factors (43).

Rhizobia were once thought to fix N_2 only after becoming bacteroids within the plant nodule. It is now known, however, that some of them can fix *ex planta* (18), but only when the O_2 concentration is below 0.5%. This low O_2 requirement mimics in nodules the role of leghemoglobin, which is to supply a high flux of O_2 at a low concentration to the bacteroids for metabolism without causing damage to the O_2-sensitive nitrogenase (46,47). The occurrence of leghemoglobin, however, is not restricted to legume nodules. The genes for leghemoglobin are widespread in plants and are even expressed at low levels in some of them (48). Leghemoglobin may act generally to signal an O_2 deficit and, thus the need for the plant to shift from oxidative to fermentative metabolic processing (49). Leghemoglobin is the protein that gives cut nodules their red color and is plant-produced (50).

Nitrogenase is of bacterial origin and the rhizobial nitrogenase has similar properties to the enzyme from the free-living organisms (51,52). In some symbioses, nitrogenase-produced H_2 is evolved directly from the nodule (53), whereas other symbioses recapture this H_2 through a bacterial uptake hydrogenase and so recycle this otherwise lost energy. This type of recapture has been suggested

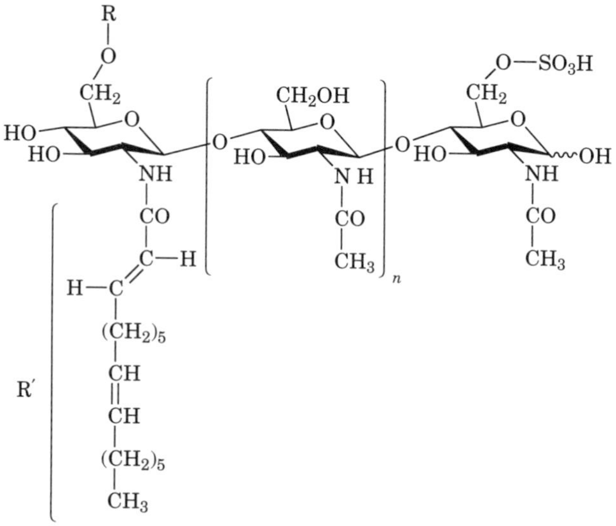

Fig. 3. The structure of the nodulation (Nod) factors of *Rhizobium meliloti* 2011 (44), where n is 2 or 3, R is —H or —$COCH_3$, and R' is $C_{16}{:}2$ as shown, $C_{16}{:}1$, or $C_{16}{:}3$, ie, a C_{16} fatty acid chain having from 1 to 3 double bonds. The *N*-acetyl glucosamine residues and an acyl moiety, R', are present in all known Nod factors, but the number of residues, n, the length and composition of R', and the substituent at R can all vary. The SO_4^{2-} is peculiar to *R. meliloti*. Other organisms have other substituents in this position. Courtesy of Kluwer Academic Publishers.

as both an index of efficiency and a criterion for the selection of rhizobia for agricultural use (54).

Although root nodules are the most common sites of N_2-fixing symbioses, some tropical legumes like *Sesbania* produce stem nodules in association with *Azorhizobium caulinodans* (55). In contrast to root nodules, some stem nodules are photosynthetic and contain, in the case of *Aeschynomene indica*, rhizobia themselves capable of photosynthesis (56). This close relationship of photosynthesis to fixation may ease the energy supply demand of nodules.

Nodulated Nonleguminous Angiosperms. The wide variety of host plants nodulated by actinomycetes appear to be unrelated taxonomically, but almost all are woody trees and shrubs. The best-known example involves *Alnus* (alder), which has root nodules called actinorhizae harboring actinomycetes of the genus *Frankia* (18,57,58). The plants have a wide geographical distribution and often are the first plant types to colonize poor or devastated soils. These have, therefore, an important ecological role, and some may be of great significance in biomass production. Alder is increasing in importance in timber production in the northwestern United States, because the growth of Doublas fir is apparently stimulated by intercropping with alder (59).

Distinct *Frankia* species are apparent from DNA-hybridization studies. There are several *Frankia* host-specificity groups based on the ability to infect the same group of plants (60). Unlike other N_2-fixing microsymbionts, this organism is multicellular and differentiated. Similarly to the process with legumes, the hyphae-like filaments penetrate the plant tissue and end in club-shaped vesicles, which are the site of N_2 fixation (61). Cell-free, partially purified preparations of nitrogenase from these vesicles have properties similar to those of the purified enzyme from free-living bacteria (62). Most actinorhizal nodules (61) have little or none of a hemoglobin-like protein present, and the only barrier to O_2 diffusion, and for nitrogenase protection, appears to be the vesicle envelope. Thus most actinorhizae show maximum rates of N_2 fixation at atmospheric O_2 levels. However, some actinorhizae both produce a hemoglobin-like protein, with about 50% amino acid sequence similarity to that in soybean, and have a physical barrier. In these last, the hemoglobin-like protein probably plays a similar role in O_2 diffusion to the one it has in legume root nodules.

Cyanobacterial Associations. Symbiotic associations range from lichens, involving a fungus, through liverworts and ferns, to gymnosperms and an angiosperm (18,63–66). Most are less formalized than the legume–*Rhizobium* symbiosis. Almost all the involved cyanobacteria are capable of growth and N_2 fixation without the host. In the symbiosis, the cyanobacterium's primary function is to provide fixed nitrogen for both partners, as shown by the increased number of heterocysts, the specialized N_2-fixing cells. No specialized structure, like the nodule, is developed to accommodate the symbiosis. The microsymbiont usually invades normal host structures, like the leaf cavity in *Azolla*, although modifications of these structures may subsequently occur.

Lichens are exceptional in that the symbiotic state is classified as a separate organism. All N_2-fixing lichens have a fungal and cyanobacterial symbiont (most often a *Nostoc* species), but some may have a green alga as an additional partner. Each lichen genus accommodates only one cyanobacterial genus, either in layers just below the surface or restricted to spherical bodies called cephalodia. In bryophytes, ie, mosses and liverworts, the cyanobacteria, again often a *Nostoc* species, are enclosed in cavities in the ventral side of the thallus. Bryophytes also form casual epiphytic associations with cyanobacteria.

The water fern *Azolla* is unique in being the only known fern genus to associate symbiotically with a N_2-fixing cyanobacterium, *Anabaena azollae*. It is globally distributed and is variously considered as either a waterway-blocking weed or as an important contributor of fixed nitrogen to rice culture. Either as a green manure or grown in dual culture, estimates indicate that *Azolla* can supply fixed nitrogen to rice in amounts comparable to those supplied by the rhizobial symbiont to the legume (65,67,68). In nature, the fern is always associated with the cyanobacterium, but it can be freed of the microsymbiont and, when provided with a fixed-nitrogen source, grown alone. The alga from this association is difficult to grow alone and the evidence for reinfection of an algal-free plant is controversial. The microsymbiont lives in a cavity in the dorsal leaf lobe of the fern, and both are photosynthetic. Transmission from fern to fern is ensured by retention of some *Anabaena* filaments within the sexual reproductive organs, so that as the young fern develops through these filaments, its leaves become rapidly infected (68).

Only the Cycads, which commonly grow in Africa, South America, and Australia, among the gymnosperms form N_2-fixing associations. These plants produce modified lateral roots called coralloid roots, which often extend above the ground. Infection of these roots is common and almost always involves *Nostoc* (18,63). The site of entry of the microsymbiont is unknown, but they are usually located intercellularly. In the angiosperms, only *Gunnera* forms an association with *Nostoc*, through invasion of its secretory glands just behind the shoot apex (66). Although located above the ground and within the stem, these sites of nitrogen fixation are still often referred to as nodules. Unusual aspects of this cyanobacterial association and indications of its advanced nature are the intracellular location of the bacteria, the well-developed vascular system surrounding the microsymbiont, and the high rates of fixation and efficient transport of fixed nitrogen such that all the needs of the host are met.

Associative Symbioses. There are some rather informal associations in which a measure of interdependence exists among some grasses (family Gramineae) and bacteria, but where no specialized structure is developed. The best-characterized examples are the association of the tropical grass *Paspalum* with the bacterium *Azotobacter paspali*, and that of the grass *Digitaria* with *Azospirillum brasilense* (69). In the former, a mucilaginous sheath forms around the root, within which the bacteria live and fix N_2. The bacteria do not invade the plant tissue. In the *Digitaria–Azospirillum* example, however, the roots are invaded but no nodule develops. The extent to which the plants benefit from the association is uncertain. *Azospirillum lipoferum*, which occurs in temperate zones, associates with certain corn and sorghum cultivars, but the effect on the plants appears to be small (70,71). More formalized endophytic associations have been discovered involving *Acetobacter diazotrophicus* and *Herbaspirillum* spp. with sugar cane (72), and *Azoarcus* spp. with Kallar grass and possibly rice (73). It appears that some of these associations can supply up to 60% of the fixed nitrogen for the host plant's growth, indicating a significant agronomic and economic potential.

Other Associations. It is often difficult to separate fixation arising from free-living species on the surface from that which is intimately associated with plant organs. Further, whether invasion is directed or adventitious is not always clear. For example, *Pseudomonas* spp. and *Azospirillum* spp. are often found inside the roots of grasses, but how these entered and what function they perform is unclear. Other associations occur with leaf surfaces, and the N_2-fixing bacteria that multiply in the phyllosphere could upgrade the local soil conditions after rain washes their products onto the ground. A more formal association, involving the leaves of certain tropical angiosperms like *Psychotria* and *Klebsiella* bacteria exists. Here, the bacteria occupy nodules on the leaf surface but, even though they are known to fix N_2 as a free-living organism, the *Klebsiella* cease to do so in the nodule. Although some benefit is likely to be derived by each of the partners, it is not from nitrogen fixation (74). Similarly, mycorrhizae are symbiotic fungi which were once thought to fix N_2 in association with the roots of the gymnosperm *Podocarpus*. In fact, they only create an environment conducive to fixation by soil bacteria (27,75).

Finally, associations with animals exist (76). For higher animals, such as humans with *Klebsiella* and ruminants with *Clostridium*, it is unlikely that much is contributed to the nitrogen status of the animal because sufficient fixed nitrogen likely exists in the diet to repress any N_2-fixing activity by the bacteria. However, for termites and shipworms, the associations are significant. *Citrobacter* infects the intestinal tract of termites and can, if necessary, fix N_2 there. The amount of N_2 fixed and the benefit gained by the insect appears to depend on its diet (77,78). The N_2-fixing, cellulose-decomposing bacterium that inhabits the Deshayes gland in wood-boring shipworms has yet to be identified. It contributes significantly to the mollusc's well-being by providing up to 35% of its fixed nitrogen requirement, thus forming a significant association (79). This whole area of animal symbioses has been only sparsely researched.

Free-Living Microorganisms. Except for the cyanobacteria, free-living bacteria are generally not agriculturally important, contributing only about 0.1% of the fixed nitrogen of a leguminous association. The free-living cyanobacteria, by comparison, contribute about 2–5% as much as a leguminous association. The difference probably lies in the cyanobacteria's ability to photosynthesize, which relieves their dependence on the often limiting carbon, and therefore energy-yielding, substrates in the soil.

Free-living bacteria are, however, used as the source of the enzyme nitrogenase, responsible for N_2 fixation (1,4,26,80), for research purposes because these are easier to culture. The enzyme is virtually identical to that from the agriculturally important rhizobia. These free-living N_2-fixers can be simply classified into aerobes, anaerobes, facultative anaerobes, photosynthetic bacteria, and cyanobacteria.

Aerobes and Microaerophiles. The best-studied examples of aerobes are the azotobacters. All are obligate aerobes, which efficiently fix N_2 only in air. Many other genera fix N_2 but are usually more sensitive to O_2, eg, *Corynebacterium* and *Azospirillum*, and require microaerophilic conditions for fixation. A similar situation arises for the remarkable N_2 fixer, *Thiobacillus ferrooxidans*, which grows at pH 2 by oxidizing ferrous iron to ferric. All these organisms have systems, some more efficient than others, to protect nitrogenase from damage by O_2 while they are growing aerobically. Similarly, *Rhizobium* fixes only when the microaerophilic conditions existing in the nodule are simulated in a plant-free culture. Many other genera oxidize gases, such as H_2 or methane, to derive the energy necessary for growth and fixation.

Anaerobes. These bacteria, typified by *Clostridium pasteurianum*, are found in soil and water and, as obligate anaerobes, cannot use O_2. The clostridia metabolize glucose and related compounds to butyrate, carbon dioxide, and H_2, and many species fix N_2. A second group of N_2-fixing anaerobes are the sulfur bacteria, such as *Desulfovibrio*. Again, not all fix N_2, but these grow by reducing sulfate or other oxidized sulfur compounds to sulfide, which is in part responsible for the smell of polluted environments. *Desulfovibrio*, which occurs naturally in the sea, is ecologically important as the principal nonphotosynthetic N_2-fixing contributor to the formation of nitrogen-containing marine sediments. A third group of anaerobic N_2 fixers are the methanogens, which are members of the Archeae, the third Kingdom of living things.

Facultative Anaerobes. These bacteria can grow with or without O_2, but they fix N_2 only under anaerobic conditions. The well-studied genus *Klebsiella* occurs in soil, water, and animal intestines and contains a number of species that fix N_2. They are related to *Escherichia coli*, which, although not a naturally occurring N_2 fixer itself, can be genetically modified to fix N_2. Other genera, including *Citrobacter* and *Enterobacter*, are also of this type. Another N_2 fixer is *Bacillus*, which, like *Clostridium*, has the ability to form spores to survive unfavorable conditions.

Photosynthetic Bacteria and Cyanobacteria. The photosynthetic bacteria are anaerobes that do not produce O_2 from photosynthesis and fix N_2 only in the light. Like the cyanobacteria, they can use CO_2 as their sole carbon source for growth. The photosynthetic bacteria are usually colored and include both sulfur and nonsulfur bacteria. *Chromatium* is red and oxidizes sulfur or sulfides to sulfate during photosynthesis to produce the reductant, and possibly the energy, necessary for N_2-fixation. *Rhodospirillum* is a purple, nonsulfur, N_2-fixing genus. The cyanobacteria differ from these because cyanobacteria grow aerobically and like plants also produce O_2 from photosynthesis. To protect nitrogenase from the evolved O_2, some, like *Anabaena*, produce heterocysts, which are incapable of evolving O_2, where nitrogenase is located. Others, like the filamentous *Plectonema*, do not have heterocysts, but fix N_2 only under lower O_2 pressures and low light intensity. However, the unicellular *Gloethece* can fix N_2 in air.

Nitrogenase. The biological catalyst that reduces atmospheric N_2 to ammonia is the metalloenzyme nitrogenase which exists in three genetically distinct forms: the conventional Mo-based system, Mo-nitrogenase, and two more recently discovered alternative systems, V-nitrogenase and nitrogenase-3. As early as 1930 (81) molybdenum, Mo, was believed to be absolutely essential for nitrogen fixation, even though vanadium, V, was found to be almost as stimulatory to bacterial growth on N_2 (82). The requirement for Mo was solidified both by the isolation of the larger of the two component proteins of nitrogenase, the molybdenum–iron, MoFe, protein (83) having a FeMo-cofactor (84), and later by the discovery of nitrogen-fixation specific (*nif*) genes involved in Mo-specific functions. Biochemical–genetic studies have also shown that biological nitrogen fixation can occur in the absence of Mo (85–87). The V-based system was demonstrated first when added vanadium stimulated diazotrophic growth of strains of both *A. vinelandii* and *A. chroococcum* from which the Mo-nitrogenase structural genes had been deleted and later by the isolation of V-nitrogenase (88,89). *A. vinelandii* has an additional nitrogenase, nitrogenase-3, which may operate with Fe only (90).

In 1960 Mo-nitrogenase was first isolated as a cell-free extract from *Clostridium pasteurianum* (91). Since then, many species have yielded nitrogenase. All of the enzymes are highly O_2 sensitive. In general, the enzymes from all N_2-fixing bacteria are very similar (52). Thus Mo-nitrogenase has been isolated as a ca 300,000 mol wt nitrogenase complex (92,93), which is more stable to O_2 than its individual component proteins. The individual components, the MoFe protein and the Fe protein, have also been separately purified (83,94). Neither has activity alone. All N_2-fixing species examined have Mo-nitrogenase. In some species, the Mo system occurs alone, whereas in others it is found in all permutations with the V-based and third systems. It is not understood why each species has its own combination of these nitrogenases (95). The biosynthesis of each enzyme is transcriptionally regulated by the availability of the appropriate metal ion in the medium. If Mo is present, only Mo-nitrogenase is synthesized. When Mo is absent and V is plentiful, V-nitrogenase is synthesized. When both are absent, nitrogenase-3 is produced. However, vanadium only represses transcription of the nitrogenase-3 structural genes if the V-nitrogenase structural genes are present (96).

All three *Azotobacter* nitrogenases comprise two separately purifiable component proteins. Each has a specific homodimeric Fe protein. The Fe protein-1 is transcribed from *nifH*; Fe protein-2 from *vnfH*; and Fe protein-3 from *anfH*. Each is about 60,000 mol wt, having a single [4 Fe—4 S] cluster bridging the two identical subunits. There is about 90% sequence identity among Fe protein-1 and Fe protein-2, but only about 60% identity when Fe protein-3 is compared with either Fe protein-1 or Fe protein-2. The second component is of about 200,000 mol wt and takes the form of a tetrameric $\alpha_2\beta_2$-MoFe protein encoded by *nifDK*: a hexameric VFe protein, encoded by *vnfDGK*, or a hexameric FeFe protein encoded by *anfDGK*. The VFe and FeFe proteins have an additional 13,000-mol wt δ-subunit, encoded by *vnfG* (89,97,98) and *anfG* (99), respectively. Both the MoFe protein and VFe protein contain two types of prosthetic groups: the FeMo (FeV)-cofactors and the P-cluster pairs. This is probably the case for the FeFe protein also. The best-characterized of the alternative nitrogenases are the V-nitrogenases from *A. vinelandii* (97) and *A. chroococcum* (88,98,100) and the Fe-nitrogenases from *A. vinelandii* (100), *Rhodobacter capsulatus* (101), and *Rhodospirillum rubrum* (102,103). In heterologous crosses, Fe protein-1 and Fe protein-2 cross-complement well with the MoFe protein and VFe protein, but not with the FeFe protein (98). The Fe protein-3 is virtually ineffective with either the MoFe protein or the VFe protein (99).

In addition, Fe protein-1 has at least two noncatalytic roles related to nitrogen fixation, and presumably Fe protein-2 and Fe protein-3 do also. The first function is somehow involved in the early stages of FeMo-cofactor biosynthesis because mutant strains having a deletion in *nifH* produce neither Fe protein nor FeMo-cofactor, rather only a FeMo-cofactor-deficient apo-MoFe protein (104,105). The second role involves insertion of preformed FeMo-cofactor into these apo-MoFe proteins (106).

Comparative Genetics of the Three Nitrogenase Systems of Azotobacter vinelandii. Of the 20 contiguous *nif* genes found for the Mo-nitrogenase in *K. pneumoniae* (107), 19 have been identified, cloned, and sequenced in *A. vinelandii*. These are, however, distributed among two unlinked clusters (108,109). The principal cluster includes the structural genes *nifHDK*; four of six FeMo-cofactor biosynthetic genes, *nifENVH*; plus *nifM*, which is involved in Fe-protein processing; *nifS*, which is involved in providing S^{2-} for cluster biosynthesis (110); and others (108). The minor cluster encompasses the other two FeMo-cofactor-associated genes, *nifBQ*, and the regulatory genes, *nifLA* (109).

A number of nitrogen-fixation genes associated with both of the genetically distinct Mo-independent enzymes have also been identified. These include the structural genes for both the V-nitrogenase *vnfHDGK* and the Fe-nitrogenase *anfHDGK* (111,112), two *nifA*-like genes (113) and one *nifEN*-like sequence (114). Apparently, each system has its own specific NifA-like protein (VnfA or AnfA) as a positive regulator. In contrast, the second *nifEN*-like (called *vnfEN*) set of genes appears to have a common function in both the *vnf* and *anf* systems. The *nifEN*-gene products likely form a template on which FeMo-cofactor is biosynthesized before insertion into the separately biosynthesized apo-MoFe protein (115). The *vnfEN*-gene products probably work similarly for both the FeV-cofactor and the putative Anf, ie, FeFe, cofactor (114).

The situation is complicated by *nifMBVUS* being essential for all three systems (109,116,117). The *nifM* requirement indicates that all three Fe proteins are so similar that a single NifM protein can process them all. Because *nifBV* are involved with the biosynthesis of the FeMo-cofactor of Mo-nitrogenase, their common requirement therefore suggests similar cofactors in all three systems. Further, the absolute *nifB* requirement indicates that its function cannot be Mo-specific. The requirement for *nifB*, but not *nifEN*, is unexpected because *nifB* and *nifN* are fused in *Clostridium pasteurianum* (118), which expresses a Mo-independent nitrogenase (119). For *nifV*, which provides homocitrate, an organic component of FeMo-cofactor (120), the obvious interpretation is that homocitrate is a constituent of all three cofactors, although its role in nitrogen fixation remains unclear. The products of *nifUS* have been reported to be required for full activation of both components, particularly the Fe protein, of the Mo-nitrogenase (121) with *nifS* being identified as the provider of S^{2-} for cluster synthesis (110).

Organization of the nif Genes. In early studies, *K. pneumoniae* was the organism of choice for identifying the genes the products of which were concerned with nitrogen fixation. Using genetic complementation, biochemical reconstitution and *in vivo* DNA-directed gene expression, 20 *nif* genes were identified (122). Later, using the cloned *K. pneumoniae nifHDK* genes as a heterologous probe, similar structural genes were identified in a variety of N_2-fixing organisms. Using DNA-sequence analysis, the *nif*-gene clusters of many organisms are being mapped. Some differences in organization occur and not all *nif* genes have been found in all organisms (123).

Requirements for Catalysis and Substrate Reduction. All three nitrogenases have the same requirements for catalytic activity: magnesium adenosine triphosphate (MgATP), a low potential reductant, and an anaerobic environment (124,125). Under appropriate conditions, a variety of substrates in addition to the physiological substrate, N_2, can be reduced. Figure 4 shows the Mo-based system. During catalysis, N_2 is reduced to form NH_3 plus magnesium adenosine diphosphate (MgADP) and phosphate. Because MgADP is an inhibitor of nitrogenase catalysis by competing for the MgATP-binding sites on the Fe protein, this inhibition is circumvented in *in vitro* assays by using the creatine phosphate-creatine phosphokinase system to reconvert MgADP to MgATP. Reductants capable of supporting nitrogenase turnover have redox potentials more negative than -300 mV and include naturally occurring ferredoxins and flavodoxins *in vivo* and sodium dithionite, $Na_2S_2O_4$, *in vitro* (125). The optimal stoichiometry for nitrogenase function is four moles MgATP hydrolyzed for each pair of electrons transferred to substrate; this ratio is independent of the substrate reduced.

$$N_2 + 16\ MgATP^{2-} + 4\ S_2O_4^{2-} + 24\ H_2O \xrightarrow{nitrogenase} 2\ NH_3\ (aq) + H_2 + 16\ MgADP^-$$
$$+16\ HPO_4^{2-} + 8\ HSO_3^- + 16\ H^+$$

In contrast to the situation with the alternative nitrogenases, but with the notable exception of the *C. pasteurianum* proteins, the component proteins from all Mo-based nitrogenases interact as heterologous crosses to form catalytically active enzymes (52). Carbon monoxide, CO, is a potent inhibitor of all nitrogenase-catalyzed substrate reductions, with the exception of H^+ reduction

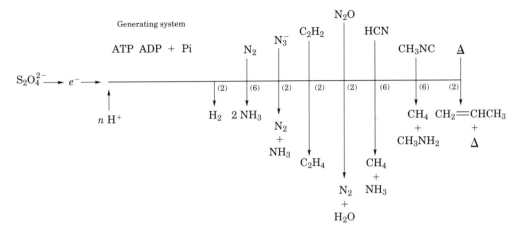

Fig. 4. Requirements, substrates, and products of Mo-nitrogenase catalysis, where I is the MoFe protein; II the Fe protein; and Pi is inorganic phosphate. The generating system is composed of creatine phosphate and creatine phosphokinase to recycle the inhibitory MgADP produced during catalysis to MgATP. The numbers in parenthesis represent the number of electrons required for the reaction shown. ⊿ is cyclopropene; Δ cyclopropane.

(126). Molecular hydrogen has a unique involvement with Mo-nitrogenase and with N_2 reduction in particular. There are four well-documented roles both *in vivo* and *in vitro*. Via hydrogenase, H_2 can act as a reductant for nitrogenase; it is the product in the absence of other reducible substrates in an ATP-dependent, CO-insensitive reaction (127); H_2 is a specific competitive inhibitor of N_2 reduction, neither affecting reduction of any other substrate nor its own evolution (128); and under an N_2-D_2 atmosphere, HD is formed in a CO-sensitive, MgATP-requiring reaction (129–132).

Substrate reduction is accomplished by a series of sequential associations and dissociations of the two proteins, and during each cycle, two molecules of MgATP are hydrolyzed and a single electron is transferred from the Fe protein to the MoFe protein (11,133), with the dissociation step being rate-limiting at about 6 s^{-1} (11). Although the kinetics of all the partial reactions have been measured, little is known about the physical details of the mechanism, including how the two proteins interact, the pathway of electron flow within the MoFe protein, the nature of possible reaction intermediates, etc. The recently reported structures of the *Azotobacter* Fe protein (134) and MoFe protein (10,135–137) have, however, allowed speculative, computer-generated docking models for the possible protein–protein contacts to be suggested (134,138–140). Both mutagenesis at Arg-101 (138) and chemical cross-linking experiments involving Glu-113 (139) of Fe protein-1 have implicated these residues in Fe protein–MoFe protein interactions. Similar studies have also implicated residues Lys-399 of the β-subunit (139) and Asp-161 of the α-subunit (141) of the MoFe protein in docking.

Stopped-flow calorimetric studies at 5°C suggest that MgATP hydrolysis precedes electron transfer within the Fe protein–MoFe protein complex; however, stopped-flow spectrophotometry at 20°C indicates that the protons, detected by a pH-sensitive dye and presumably produced as a result of MgATP hydrolysis, are released slower than, and therefore after, electron transfer (142). In all probability the 20°C studies better reflect the actual physiological situation. The MgATP binding causes an uncharacterized conformational change in Fe protein-1 (and -2) that affects both the redox potential (143), the shape of the electron paramagnetic resonance (epr) signal (100,144,145), and the accessibility of the iron (146) in the [4 Fe–4 S] cluster of the Fe protein (see IRON COMPOUNDS). Further, MgATP hydrolysis occurs only when the Fe protein and MoFe protein are complexed. These observations, together with details from the Fe-protein crystal structure, which shows well-separated binding sites for the [4 Fe–4 S] cluster and the nucleotide, are the basis of a hypothesis by which MgATP binding and hydrolysis coordinates and regulates the unidirectional flow of electrons to the MoFe protein (147).

The strict requirement for ATP makes biological N_2-fixation an energy-consuming process, and organisms preferentially use fixed nitrogen when it is available. Despite energy demand, biological nitrogen fixation is an attractive alternative to the Haber process, particularly if the apparently wasteful evolution of H_2 can be eliminated. It has been argued that Mo cannot be the site at which substrate binds and is reduced because of the discovery of the two alternative nitrogenases, which are Mo-independent. However, the observed differences in catalytic rates and preferences among these three nitrogenases reflect real differences in affinity for the various substrates, which in turn may reflect direct involvement of this heterometal atom with the substrate (95).

The deduced amino acid sequences of the α- and β-subunits of the VFe and FeFe proteins show significant (55%) identity with one another, but less (32%) with those of the MoFe protein (112). Even so, the conservation of both the domain structures around the eight strictly conserved cysteine residues and the spacing between them remain constant (95,141,148,149) and indicates that all three protein types have the same general structural features. Subtle but important differences are, however, likely to exist. For example, when the FeV-cofactor center is extracted from the VFe protein and used to reconstitute an apo-MoFe protein, the resulting hybrid protein, unlike both parents, can no longer fix N_2 even though it continues to catalyze the reduction of other substrates (150).

Structure of the MoFe Protein.   Extensive spectroscopic studies of the MoFe protein, the application of cluster extrusion techniques (84,151), x-ray anomalous scattering, and x-ray diffraction (10,135–137,152) have shown that the MoFe protein contains two types of prosthetic groups, ie, protein-bound metal clusters, each of which contains about 50% of the Fe and S^{2-} content. Sixteen of the 30 Fe atoms and 14–16 of the 32–34 S^{2-} constitute one type of prosthetic group; the two P-cluster pairs, and the remaining 14 Fe atoms and 18 S^{2-}, together with both Mo atoms, comprise the other type, the two FeMo-cofactors. The distribution of these cluster types within the protein, as determined by x-ray techniques, is shown in Figure 5. These clusters are distributed in pairs, separated by about 7.0 nm, ie, at opposite ends of the protein. Each pair consists of one FeMo-cofactor and, about 1.9 nm away, one P-cluster pair.

The two P-cluster pairs in the MoFe protein were originally thought (154) to be four discrete, but spectroscopically unusual [4 Fe–4 S] clusters (155). In the dithionite-reduced state of the MoFe protein, these exist in the biologically unique, all-ferrous state, that is as $[4 Fe–4 S]^0$. Reformulation as an [8 Fe–8 S]

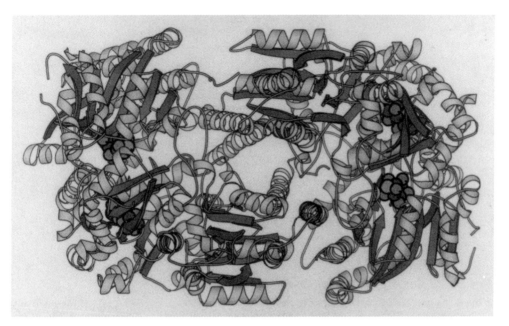

Fig. 5. Ribbons diagram of the MoFe protein $\alpha_2\beta_2$-tetramer from *A. vinelandii* viewed down the pseudo-twofold symmetry axis. α-Helical regions are shown as coils; β-sheet regions are shown as arrows; the remaining regions are shown as threads. Each half of the molecule, which may be visualized by drawing a line through the center from the 1 o'clock to the 7 o'clock position, represents an α,β-dimer, each encompassing one P-cluster pair and one FeMo-cofactor. The β-subunits make all contacts among the α,β-dimers and are arranged symmetrically around the central channel and below the α-subunits. The α-subunits, which do not contact one another, are located on the left and right sides of the molecule with their FeMo-cofactors at the lower left (cluster of spheres) and upper right, whereas the P-cluster pairs are located upper left and lower right (153).

cluster, based on electron paramagnetic spectroscopic (epr) studies of the solid-thionine-oxidized MoFe protein (156), was confirmed by solution of the crystal structure by x-ray diffraction methods. The P-cluster pair is modeled as two [4 Fe–4 S] cubes bound to each other through two cysteinyl bridges, from cysteinyl-88 of the α-subunit, α-Cys-88, and the β-subunit cysteinyl-95 residue, β-Cys-95, and a disulfide bond with only two terminal Cys–Fe interactions each (Fig. 6) (10,135). This structure also confirms the predictions from primary sequence conservation and extrusion characteristics that the P-cluster pairs would occur at the subunit interfaces and that α-Cys-62, α-Cys-88, α-Cys-154, β-Cys-70, β-Cys-95 and β-Cys-153 would provide all the necessary thiolate ligands (157). Mutagenesis studies have shown that some single amino-acid substitutions at α-Cys-88, providing a bridging thiolate, and β-Cys-153, providing a terminal thiolate, do not completely interrupt the ability to fix N_2. All substitutions at the other four conserved Cys residues result in loss of the ability to fix N_2 (157,158).

The VFe protein also has the equivalent of P-cluster pairs which have similar properties to those found in the MoFe protein (159). No information is available on whether P-cluster pairs exist in the FeFe protein, but because of the relatively high sequence identity and the similar genetic basis of its biosynthesis,

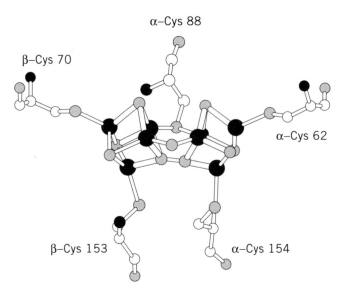

Fig. 6. View of the nitrogenase MoFe protein P-cluster pair where (●) represents Fe, (◉) S, and (○) C as modeled (153). The side chain of one of the bridging cysteinyl (Cys) residues, β-Cys-95, has been omitted for the sake of clarity.

the occurrence seems highly likely. The catalytic role assigned to the P-cluster pair involves accepting electrons from the Fe protein for storage and future delivery to the substrate via the FeMo-cofactor centers. As of this writing (ca early 1995), this role has yet to be proved.

Each of the two FeMo-cofactor centers consists of one Mo atom, seven Fe atoms, nine sulfides, and a homocitrate molecule (10,135–137). This prosthetic group is the origin of the biologically unique, characteristic $S = 3/2$ epr spectrum of the MoFe protein. The group can be extracted into an organic solvent, usually N-methylformamide (NMF), in which it is stable, and can reconstitute all activity and spectroscopic properties of an apo-MoFe protein when added to an appropriate extract (84). Using either the unique epr signal or x-ray absorption spectroscopy (xas) as a monitor of ligand binding, only one thiolate was found to bind to Fe per isolated FeMo-cofactor, suggesting that only one thiolate bond is involved in binding FeMo-cofactor to its protein matrix (160–161). The dithionite-reduced VFe protein from both *A. vinelandii* and *A. chroococcum* exhibits spectroscopy consistent with the presence of a FeV-cofactor, which can also be extracted into NMF (150). Similarly, a FeFe-cofactor likely exists because a $S = 3/2$ epr signal has been elicited from the purified FeFe protein from *R. capsulatus* (162).

Sequence analysis and mutagenesis experiments indicated that the thiolate ligand is provided by α-Cys-275 (163–165). These predictions have been confirmed by the MoFe-protein crystal structure, in which the FeMo-cofactor is shown to be bound to the α-polypeptide only and by only two direct bonds, through the α-Cys-275 and α-His-442 residues (Fig. 7). In addition, indirect (hydrogen-bonded) interactions occur through α-Gln-191 and α-Gln-440, both of which hydrogen bond to the homocitrate, α-Arg-359, α-Arg-96, and α-His-

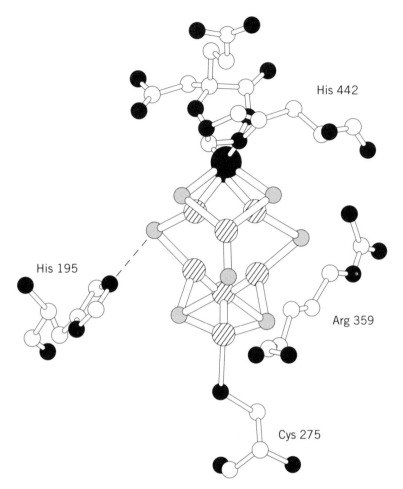

Fig. 7. View of the FeMo-cofactor prosthetic group of the nitrogenase MoFe protein with some of the surrounding amino acid residues where (●) represents the molybdenum coordinated to α-His-442 and homocitrate (at the top), (◎) represents the iron, interspersed with the sulfur (**O**) and carbon (○) atoms (153).

195. Many of these interactions were predicted through mutagenesis studies (166–168). Site-directed mutagenesis of the *A. vinelandii nifD* gene has produced evidence to indicate that the FeMo-cofactor is the site of substrate reduction because a single amino acid substitution at either α-Gln-191, by Lys or Glu, or α-His-195 by Asn results in simultaneous changes in both substrate specificity, eg, both C_2H_4 and C_2H_6 are produced from C_2H_2, and epr spectrum, ie, the spectral signature of FeMo-cofactor (166).

Although FeMo-cofactor is clearly implicated in substrate reduction catalyzed by the Mo-nitrogenase, efforts to reduce substrates using the isolated FeMo-cofactor have been mostly equivocal. Thus the FeMo-cofactor's polypeptide environment must play a critical role in substrate binding and reduction. Also, the different spectroscopic features of protein-bound vs isolated FeMo-cofactor clearly indicate a role for the polypeptide in electronically fine-tuning the substrate-reduction site. Site-directed amino acid substitution studies have been used to probe the possible effects of FeMo-cofactor's polypeptide environment on substrate reduction (163–169). Catalytic and spectroscopic consequences of such substitutions should provide information concerning the specific functions of individual amino acids located within the FeMo-cofactor environment (95,122,149).

Regulation of Nitrogen Fixation.   Both the synthesis and activity of nitrogenase are under tight genetic control in nitrogen-fixing cells, mainly because nitrogenase is a principal cellular component having a high energy demand. Regulation responds to three environmental factors: nitrogen status, O_2 tension, and metal-ion availability. The presence of fixed nitrogen, eg, nitrate, is a significant regulatory factor, and bacteria utilize the fixed nitrogen until it is totally depleted rather than synthesize nitrogenase to fix their own (170). This control mechanism is important for free-living N_2 fixers, such as *A. vinelandii*, but less so for symbiotic organisms, eg, *B. japonicum*, which are adapted to export fixed nitrogen to their host. Of primary importance is O_2 tension, because both nitrogenase proteins are extremely sensitive to oxidative damage (171). Finally, because nitrogenase consists of metal-containing proteins, the availability of certain metal ions becomes regulatory, particularly in those organisms that biosynthesize the alternative nitrogenases (116,148,172,173). Regulation of N_2 fixation can be exerted at several different levels in different organisms. For example, NH_3 causes a reversible post-translational modification in photosynthetic bacteria, in which the Fe protein is ADP-ribosylated and consequently inactivated (174). In strict aerobes, the enzyme can undergo conformational protection to guard against O_2 damage by forming a protein aggregate, which dissociates to resume fixation when the O_2 stress is relieved. In most filamentous cyanobacteria, *nif*-gene expression is developmentally regulated, such that nitrogenase is only expressed in the specialized, nonphotosynthetic heterocysts (175). However, the most uniform, although by no means all-encompassing, mechanism is one of transcriptional regulation of the *nif* genes themselves by the specific regulatory genes, *nifLA*, which encode the positive regulatory protein, NifA, and the negative regulatory, NifL, respectively (176).

NifA proteins are of one of two types. One group consists of O_2-sensitive NifA proteins, which are found in the rhizobia *Rhodobacter* and *Azospirillum*. These proteins have a proposed iron-binding motif on a linker between the central and C-terminal domains that may form a redox-sensitive site to sense changes in O_2 tension. The *A. vinelandii* and *K. pneumoniae* NifA proteins are not O_2-sensitive and do not have this motif. The *nifL* gene product is the negative regulator, which also modulates NifA activity in response to changes in O_2 and fixed nitrogen status, but the mechanism by which it does so remains unclear (177). Regulation of *nifA* expression is varied among organisms and depends on preferred physiology and ecology. For example, fixed nitrogen status is usually more important for free-living N_2-fixers; O_2 tension is an overriding concern of the organisms involved in symbiotic associations with plants.

In enteric bacteria, expression of the *nif* genes is under general nitrogen regulation by the *ntr* system. This system consists of essentially a sensor–activator pair of proteins, encoded by *ntrB* and *ntrC*. When fixed nitrogen is limiting, the sensor, NtrB, phosphorylates (and so activates) the activator, NtrC, which then interacts with, and allows expression from, all *ntr*-regulated promoters, including that of the *nifLA* transcription unit. When nitrogen is sufficient, NtrB dephosphorylates (and so deactivates) NtrC, which, in turn, switches off the *nifLA* promoter and so stops *nif* transcription. The activity of the sensor, NtrB, is modulated through the activity of the gene products of the *gln* (glutamine-producing) system. Regulation of nitrogen fixation is obviously a complex, demanding process.

Chemical Approaches

Dinitrogen has a dissociation energy of 941 kJ/mol (225 kcal/mol) and an ionization potential of 15.6 eV. Both values indicate that it is difficult to either cleave or oxidize N_2. For reduction, electrons must be added to the lowest unoccupied molecular orbital of N_2 at -7 eV. This occurs only in the presence of highly electropositive metals such as lithium. However, lithium also reacts with water. Thus, such highly energetic interactions are unlikely to occur in the aqueous environment of the natural enzymic system. Even so, highly reducing systems have achieved some success in N_2 reduction even in aqueous solvents.

Dinitrogen-Reducing Systems. The binding of N_2 to a metal center is the first step in activating molecular nitrogen toward reduction. Since the first compound of this type, [Ru(NH$_3$)$_5$(N$_2$)]Br$_2$ [*15246-25-0*] was synthesized (178), most transition metals have been found to form similar compounds (179,180). Many dinitrogen compounds are so stable that they are unreactive toward reduction and so have little chance to form the basis of a catalytic system.

Nonaqueous Systems. The first nonbiological N_2-reducing system was reported in 1964 (181) at about the same time that the first metal–N_2 complexes were prepared and when the first active N_2-fixing cell-free bacterial extracts were being produced. It used titanium tetrachloride [*7550-45-0*], TiCl$_4$, or dichlorobis(η^5-cyclopentadienyl)titanium(IV) [*1271-19-8*], (η^5-C$_5$H$_5$)$_2$TiCl$_2$, with ethylmagnesium bromide or lithium naphthalenide as reductant in ethyl ether (181–183). Although the mechanism of reduction remains unclear, the conversion of N_2 into nitride is likely. These systems are not catalytic, however, because the solvolysis needed to liberate NH$_3$ also destroys the active species. A truly catalytic effect was demonstrated using a mixture of TiCl$_4$, metallic Al, and AlBr$_3$ at 50°C, when NH$_3$ was obtained at 200 mol/g·atom Ti via the catalytic nitriding of aluminum (184).

A number of reaction products have been isolated from the (η^5-C$_5$H$_5$)$_2$TiCl$_n$–N$_2$–reductant system, where $n = 1, 2$, all of which assume an intense blue color in solution. Spectroscopic absorption occurs at a maximum, λ_{max}, of ca 600 nm. The relationship among these products is unclear (185,186), but the lability of the C$_5$H$_5$ ring may be an important complicating factor. When (η^5-C$_5$R$_5$)$_2$TiCl$_2$ [*11136-36-0*], R = CH$_3$, is used, two distinct interconvertible N$_2$ complexes are formed (187). X-ray crystallography shows one to be a mono-N$_2$ complex, [(η^5-C$_5$R$_5$)$_2$Ti]$_2$(N$_2$) [*11136-46-2*] having a linear Ti–N $=$ N–Ti bridge and an N–N bond length of 0.116 nm (188). The second is a tri-N$_2$ complex of Ti that is unstable above -80°C. However, both zirconium analogues are thermally stable and the structure of [(η^5-C$_5$R$_5$)$_2$Zr(N$_2$)]$_2$(N$_2$) [*54387-50-7*], R = CH$_3$, shows a similar linear Zr–N $=$ N–Zr bridge where the N–N bond is 0.118 nm plus one end-on terminal N_2 ligand, the N–N bond is 0.1115 nm, on each zirconium atom (Fig. 8) (189).

The mono-N_2 Ti complexes apparently cannot produce hydrazine or ammonia directly, but in the presence of excess reductant do give NH_3. In contrast, the tri-N_2 complexes of both Zr and Ti react directly with HCl to liberate two N_2 molecules and produce hydrazine at a ratio of 0.9 mol/mol of complex from the third (188). The structure of the tri-N_2–Zr complex gives no clue to the basis for this reactivity. No significant differences are evident either among the terminal and bridging N_2 ligands or when compared to the structure of other metal–N_2 complexes that do not produce hydrazine or ammonia on protonation. Related Ti(III)–N_2 complexes also show this variation in requirements for the production of hydrazine or ammonia (190,191).

Early on, it appeared that no direct connection existed between the highly reducing systems that produce ammonia or hydrazine from N_2 and the well-defined metal–N_2 complexes. However, just as metal–N_2 compounds have been isolated from the reducing systems, so too have a number of metal–N_2 compounds been degraded to ammonia or hydrazine. The mononuclear, tertiary phosphine complexes of Mo and W are the best-studied examples. The reaction of trans-$M(N_2)_2(dppe)_2$, where dppe = 1,2-bis(diphenylphosphino)ethane (192) and M = Mo or W, [25145-64-6] and [41700-58-7], respectively, with excess acid does not proceed past the hydrazido(2-), ie =N—NH_2, stage (193).

$$trans\text{-}Mo(N_2)_2(dppe)_2 + 2\,HBr \longrightarrow [Mo(N_2H_2)Br(dppe)_2]Br + N_2$$

However, when either $P(C_6H_5)(CH_3)_2$ or $P(C_6H_5)_2(CH_3)$ is used to form cis- or trans-$M(N_2)_2(PR_3)_4$, M = Mo or W, respectively, followed by treatment with acid, ammonia yields of about 2 mol or 0.7 mol per mole of complex for M = W and Mo, respectively, are produced (193,194). These and related data have been used to suggest a possible stepwise sequence for the reduction and protonation of

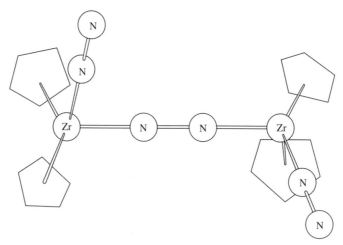

Fig. 8. Molecular structure of $\{[\eta^5\text{-}C_5(CH_3)_5]_2Zr(N_2)\}_2(N_2)$ (after Ref. 189). Each pentagon represents a pentamethyl cyclopentadienyl ring system.

N_2 on a single molybdenum atom in nitrogenase (194). However, acidification leads to complete destruction of the complex. Using both the stabilizing effect of the chelating phosphine triphos, $(C_6H_5)P[CH_2CH_2P(C_6H_5)_2]_2$, and the lability of a simple phosphine, a cyclic system has been devised in which the product, $MoBr_3$(triphos), can be used to prepare the starting $Mo-N_2$ complex again (195).

$$2\ Mo(N_2)_2[P(C_6H_5)_3](triphos) + 8\ HBr \longrightarrow 2\ MoBr_3(triphos) + 2\ NH_4Br + 3\ N_2 + 2\ P(C_6H_5)_3$$

A further measure of improvement has resulted in a cyclic process for the electrochemical synthesis of NH_3 (196). The catalyst is $trans$-$W(N_2)_2$(dppe)$_2$ or less effectively its Mo analogue. The choice of acid is important so as to provide a ligand during part of the cycle, yet also an effective leaving group when the catalyst is reformed. A sulfonic acid is used.

Production of NH_3 in these Mo and W systems has certain features in common with the Ti and Zr systems. Each has more than one bound N_2 ligand in the metal's coordination sphere, but only one is reduced in a process initiated by acid addition. Protonation of N_2 occurs as the phosphine or other N_2 ligands, which stabilize the lower oxidation states of the metals, are successively replaced by acid counterions. These latter favor the higher oxidation states, and this ligand exchange encourages the transfer of more metal electron density to N_2, thus promoting further protonation. Because the chelating phosphines, eg, dppe, are much less easily replaced, $M(N_2)_2$(dppe)$_2$ does not produce ammonia readily, whereas using the more easily displaced, simple phosphines, all six metal electrons are used to produce ammonia. In the mixed-phosphine system, ammonia is formed but only three electrons per atom of molybdenum are used. In the binuclear zirconium(II) system (Fig. 8), only four electrons are available and so only hydrazine is formed. Here also, electron flow from metal to N_2 could be encouraged by loss of N_2 and coordination of the appropriate acid anion. Similarly, only hydrazine is produced from other dinuclear Ta and Nb compounds, which have bridging N_2 (197).

These ammonia- and hydrazine-forming reactions appear to be conducted under mild conditions. However, the reducing power, eg, of magnesium metal, which has $E^0 = -2.4$ V, is built into these systems during the preparation of the metal$-N_2$ complexes. The complete degradation of the simple phosphine complexes of Mo and W during ammonia formation does not favor catalysis. An advantage is offered by the electrochemical and the mixed-phosphine systems and by the zirconocene system, where the product of acid degradation is the starting material for preparation of the metal$-N_2$ complex. It remains to be seen if significantly milder reductants can effect the synthesis of these or similarly reactive N_2-containing species (198,199) and if these can be operated cyclically or catalytically for long periods (see also ORGANOMETALLICS).

Aqueous Systems. It has been reported that many strong reducing agents produce minute amounts of ammonia from N_2 in aqueous solution in the presence of derivatives of transition metals. However, spurious results are easily obtained because (*1*) low metal concentrations are used and contaminating species may occur; (*2*) the Nessler test for ammonia is not specific; (*3*) the system may scavenge traces of ammonia or nitrogen oxides, which are subsequently reduced, from the

N_2 gas; and (4) nitrogen-containing substances added to the reaction mixture may be degraded to ammonia. Only those systems that have been substantiated through the use of $^{15}N_2$ are discussed herein.

Aqueous systems that reduce N_2 to either hydrazine or ammonia are known. Aqueous or aqueous–alcoholic solutions of sodium molybdate(VI) [7631-95-0], Na_2MoO_4, or oxotrichloromolybdenum(V) [13814-74-9], $MoOCl_3$, with $TiCl_3$ as reductant and Mg^{2+} at pH 10–14, produce some N_2H_4 at 25°C and 0.1 MPa (1 atm) N_2. However, at 50–100°C and 5.1–15.2 MPa N_2, yields of hydrazine reach 100 mol/mol Mo. At the higher temperature, some NH_3 is produced also. Vanadium(II) or Cr(II) are equally effective as reductants. The reaction mixture is heterogeneous, and hydroxide-bonded polynuclear entities may furnish the reducing capacity of Ti(III) through Mo(III) to N_2. This system's efficiency is about 1% of the biological systems. Replacement of both Mo and Ti by V(II) gives, at alkaline pH, lower temperatures and 10.1 MPa (100 atm) N_2 pressure, quantitative reduction, 0.22 mol N_2H_4 per mol of V within minutes (200). A four-electron reaction is proposed, with a tetramer of V(II) ions oxidizing to V^{3+}. At room temperature or higher, NH_3 and H_2 are produced because the hydrazine is further reduced by more V(II) (200).

$$(V^{4+})_4 + N_2 + 4\,H_2O \longrightarrow 4\,V^{3+} + N_2H_4 + 4\,OH^-$$

A related homogenous aqueous–alcoholic system, composed of V(II) complexes of catechol and its derivatives, reduces N_2 to ammonia and H_2. Only catecholates are active in this system, which is sensitive to pH. This system has been likened to nitrogenase by suggesting that both use a sequence of two four-electron reductions to evolve one H_2 for every N_2 reduced (201).

A third aqueous N_2-reducing system has been developed based on the knowledge that nitrogenase contains iron, molybdenum, sulfide, and thiol groups. It was reported that 3–5 μmol of NH_3 are produced from about 5 mmol Na_2MoO_4, 2.5 mmol thioglycerol, 0.1 mmol $FeSO_4 \cdot 5H_2O$, and 0.25 g $NaBH_4$ in 50 mL of borate buffer (pH 9.6) under 13.7 MPa (135 atm) N_2 (202). In the absence of molybdate, no NH_3 is obtained. Yields up to ca 0.04 mol NH_3/mol Mo are obtained with a molybdenum–cysteine complex under 0.1 MPa (1 atm) N_2. Specific stimulation of activity by ATP is reported but acids may produce the same effect (203). These molybdothiol systems are suggested to produce diazene, N_2H_2, that disproportionates to hydrazine, N_2H_4, which is then reduced to NH_3 (202). Rather more successful are both the $[MoO(CN)_4(H_2O)]^{2-}$ [55493-45-3] and $NaBH_4$ systems, which give up to 0.3 mol NH_3 per mol complex (204), and the 6:1 MoO_4^{2-}–insulin mixture with $NaBH_4$, which produces 65 mol ammonia/mol Mo in 30 min at 23°C under 0.1 MPa N_2 (205).

All the chemical N_2-reducing systems require further development to become important as N_2-reduction methods, but insight into the binding and activation of N_2 toward reduction is gained through them as well as into the induction of internal redox reactions. The first step in enzymic N_2 reduction undoubtedly involves the binding of N_2 to a transition-metal site. There is nothing in the known enzymology that precludes either changes in the metal coordination sphere or protonation of bound N_2, or both, as initiators of the redox process.

Even so, these inorganic systems for N_2 reduction are not expected to provide a useful means of ammonia production for the near future.

Nitrogenase Structural Models. Structural modeling has concentrated on the nitrogenase substrate-binding site, ie, the FeMo-cofactor. Two principal activities have been undertaken: the study of FeMo-cofactor itself after isolation from the protein (206,207) and the chemical synthesis of mixed molybdenum−iron−sulfur cluster compounds. Inspiration for the latter was a Mo x-ray absorption spectroscopic (xas) study, particularly the analysis of the entended x-ray absorption fine structure (exafs) region, of the nitrogenase MoFe protein and isolated FeMo-cofactor (208,209). These studies showed that the Mo atom was surrounded by three to four bound sulfur atoms at 0.237 nm, ca three iron atoms at a distance of 0.270 nm, and two to three oxygen or nitrogen atoms at a distance of 0.212 nm (209). The existence of these features has since been substantiated by the x-ray structure data, which provides a complete set of structural features with which any synthetic model must be compatible.

The early prototypes of Mo−Fe−S clusters (210−211) consisted of two $[MoS_4(FeSR)_3]$ cube-like structures triply bridged via the two Mo atoms by sulfide or thiolate, RS^-, groups. These complexes have Mo−Fe and Mo−S distances of ca 0.273 and 0.237 nm, respectively, within each cube. These distances are very close to those occurring in the MoFe protein, but neither the atomic ratio of molybdenum-to-iron nor the spectroscopy of FeMo-cofactor is duplicated (210). Many other variations on the Mo−Fe−S-cluster theme have been synthesized (212,213). Of note are the single-cubane clusters, which are prevented from bridging to one another by a catecholate ligand bound to the Mo atom. These clusters bind, among other ligands, CN^-, a substrate of nitrogenase, but do not bind N_2, and although they also exhibit a S = 3/2 epr signal, the clusters do not duplicate the metal ratio of FeMo-cofactor. Also of interest are the compounds with two bridged $(MoFe_3S_4)$ cores, which do bind N_2H_4 as a second bridge between the Mo atoms. None of these Mo−Fe−S cluster compounds bind N_2 to form a stable complex, although certain of them have been used as catalysts in systems that produce NH_3 from N_2 (214).

Outlook

The mature Haber-Bosch technology is unlikely to change substantially in the foreseeable future. The centers for commercial ammonia production may, however, relocate to sites where large quantities of natural gas are flared from crude oil production, eg, Saudi Arabia or Venezuela. Relocation would not offset the problems for agriculture of high transportation and storage costs for ammonia production and distribution. Whereas the development of improved lower temperature and pressure catalysts is feasible, none is on the horizon as of this writing.

The discovery of chemical N_2 fixation under ambient conditions is more compatible with a simple, complementary, low temperature and low pressure system, possibly operated electrochemically and driven by a renewable energy resource (qv), such as solar, wind, or water power, or other off-peak electrical power, located near or in irrigation streams. Such systems might produce and apply ammonia continuously, eg, directly in the rice paddy, or store it as an increasingly concentrated ammoniacal solution for later application. In fact, the Birkeland-Eyde process of N_2 oxidation in an electric arc has been reconsidered in just such a context (215) for areas where fertilizer production capacity of a few t/yr can make a significant impact on agricultural production. Thus simple, inexpensive, small-scale systems may have a place in areas where power is cheap and where the high capital investment of an ammonia plant cannot be justified. The developing countries, some of which have very limited fossil-fuel supplies or fertilizer demands, are the likely initial target locations for such systems. These same areas might also be the beneficiaries of any process that can take advantage of the simultaneous reductive and oxidative photocatalytic N_2 fixation that occurs in hydrous iron(III) oxide-loaded Nafion films (216), on zinc-doped gallium phosphide semiconductors (217), or even the natural photofixation that occurs on minerals like rutile in desert sands (218).

Other important contributions could be made by improving the utilization of applied nitrogen fertilizer. Less than 50% of the nitrogen applied is actually assimilated by plants. To this end, slow-release fertilizer can make an impact as can development of nitrification and denitrification inhibitors (see CONTROLLED RELEASE TECHNOLOGY, AGRICULTURAL). All such strategies would prevent ammonia losses to the atmosphere and ground water. However, the effects of such inhibitors on the nitrogen cycle are unclear.

The exploitation of the benefits of biological fixation is expected to demand, in the shorter term, an increased use of legumes and other symbiotic systems in agriculture, taking care to match the most effective rhizobial strains with the appropriate cultivar. If these associations could be manipulated to start fixation earlier or to continue it later into the plant's growth, a substantial benefit could accrue. Another significant benefit would result if the ability to fix N_2 in the presence of fixed-nitrogen sources were to be conferred on bacteria. Cyanobacteria are remarkably self-sufficient, and their presence in rice paddies should continue to be exploited, together with the *Azolla–Anabaena* association, to enhance the vitally important production of rice. The continuing discovery of new nonleguminous (associative) symbioses indicates avenues through which both N_2 fixation and the delivery of fixed nitrogen to crop plants may be enhanced. The development of associations, possibly with the principal food crops that do not enter into N_2-fixing symbioses, could have dramatic effects on both fertilizer usage and food production.

Genetic manipulation of N_2 fixation appears to be the ultimate solution both for reducing fossil-fuel energy inputs to fertilizer production and for increasing food supplies (219). Understanding of the recognition and infection processes followed by genetic manipulation could result in new or enhanced symbiotic associations, because many plants appear already to have most of the genes necessary to produce a nodule-like structure. Perhaps the best possibility is to transfer the ability to fix N_2 to those bacteria that already reside within the cells of plants. The success of transferring the *nif* genes from one bacterial genus to another has opened up the possibility of transfer to a crop plant. However, genetic transfer is not enough. Possibly nitrogenase could be relocated to the chloroplasts of leaves where, if properly protected from the O_2 evolved by photosynthesis, it could take advantage of directly available reducing equivalents produced from sunlight (220).

BIBLIOGRAPHY

"Nitrogen Fixation" in *ECT* 2nd ed., Suppl. Vol., pp. 604–623, by R. H. Stanley, Titanium Intermediates Ltd., in *ECT* 3rd ed., Vol. 15, pp. 942–968, by W. E. Newton, Charles F. Kettering Research Laboratory.

1. R. C. Burns and R. W. F. Hardy, *Nitrogen Fixation in Bacteria and Higher Plants*, Springer-Verlag, Berlin, 1975, p. 43.
2. C. C. Delwiche, Ambio **6**, 106 (1977).
3. R. M. Garrels, F. T. Mackenzie, and C. Hunt, *Chemical Cycles and the Global Environment: Assessing Human Influences*, Kaufman, Inc., Los Altos, Calif, 1975; A. J. M. Schoot Uiterkamp, in P. M. Gresshoff, L. E. Roth, G. Stacey, and W. E. Newton, eds., *Nitrogen Fixation; Achievements and Objectives*, Chapman and Hall, London, 1990, p. 55.
4. W. E. Newton and W. H. Orme-Johnson, eds., *Nitrogen Fixation*, University Park Press, Baltimore, Md., 1980.
5. R. H. Burris, in Ref. 4, Vol. 1, p. 7.
6. W. J. Payne, J. J. Rowe, and B. F. Sherr in Ref. 4, Vol. 1, p. 29; F. E. Clark and T. Rosswell, eds., *Terrestrial Nitrogen Cycles*, Ecological Bulletin #33, Swedish National Research Council, Stockholm, 1981.
7. S. H. Wittwer in A. Hollaender, ed., *Genetic Engineering for Nitrogen Fixation*, Plenum Press, New York, 1977, p. 515.
8. C. C. Delwiche, *Sci. Amer.* **223**, 136 (1970).
9. E. W. Paul, *Ecological Bulletin* #26, Swedish National Research Council, Stockholm, 1978.
10. C. Kim and D. C. Rees, *Science* **257**, 1677 (1992).
11. R. N. F. Thorneley and D. J. Lowe, *Biochem. J.* **224**, 887 (1984).
12. F. A. Ernst, *Fixation of Atmospheric Nitrogen*, van Nostrand Co., New York, 1928.
13. F. Daniels, *Chem. Eng. News* **33**, 2370 (June 6, 1955).
14. C. A. Vancini, *Synthesis of Ammonia*, CRC Press, Cleveland, Ohio, 1971.
15. F. Haber, *Z. Angew. Chem.* **27**, 473 (1914).
16. G. Ertl, D. Prigge, R. Schloegl, and M. Weiss, *J. Catal.* **79**, 359 (1983).
17. G. Ertl, *Catal. Rev.-Sci. Eng.* **21**, 201 (1980).
18. J. I. Sprent and P. Sprent, *Nitrogen Fixing Organisms: Pure and Applied Aspects*, Chapman and Hall, London, 1990.
19. J. P. W. Young, in G. Stacey, R. H. Burris, and H. J. Evans, eds., *Biological Nitrogen Fixation*, Chapman and Hall, London, 1992, p. 43.
20. J. B. Boussingault, *Ann. Chim. Phys., 2nd Ser.* **67**, 5 (1838).
21. P. W. Wilson, *The Biochemistry of Symbiotic Nitrogen Fixation*, University of Wisconsin Press, Madison, Wis., 1940.
22. H. Hellriegel and H. Wilfarth, *Z. Ver. Rubenzucker-Ind. Deutsch. Reichs*, 1888.
23. A. Quispel, in H. Bothe, F. J. de Bruijn, and W. E. Newton, eds., *Nitrogen Fixation: Hundred Years After*, Gustav Fischer, Stuttgart, Germany, 1988, p. 3.
24. E. B. Fred, I. L. Baldwin, and E. McCoy, *Root Nodule Bacteria and Leguminous Plants, Studies in Science No. 5*, University of Wisconsin Press, Madison, Wis., 1932.
25. P. W. Wilson, in J. R. Postgate, ed., *Chemistry and Biochemistry of Nitrogen Fixation*, Plenum Press, London, 1971, p. 1.
26. R. Palacios, J. Mora, and W. E. Newton, eds., *New Horizons in Nitrogen Fixation*, Kluwer Academic Publishers, Dordrecht, the Netherlands, 1993.
27. G. Bond, *Ann. Rev. Plant Physiol.* **18**, 107 (1967).
28. M. J. Trinick, *Nature (London)* **244**, 459 (1973).
29. J. M. Vincent, in Ref. 4, Vol. II, p. 103.

30. S. R. Long, in G. Stacey, R. H. Burris, and H. J. Evans, eds., *Biological Nitrogen Fixation*, Chapman and Hall, New York, 1992, p. 560.
31. W. M. Barbour, S.-P. Wang and G. Stacey, in Ref. 19, p. 648.
32. P. Lerouge and co-workers, *Nature (London)* **344**, 781 (1990).
33. I. Vijn, L. das Neves, A. van Kammen, H. Franssen, and T. Bisseling, *Science* **260**, 1764 (1993).
34. G. E. Ham in Ref. 4, Vol. II, p. 131.
35. J. L. Firman, K. E. Wilson, L. Rossen, and A. W. B. Johnson, *Nature* **324**, 90 (1986).
36. N. K. Peters, J. W. Frost, and S. R. Long, *Science* **233**, 977 (1986).
37. J. W. Redmond and co-workers, *Nature (London)* **323**, 632 (1986).
38. B. G. Rolfe, *BioFactors* **1**, 3 (1988).
39. W. D. Bauer, *Ann. Rev. Plant Physiol.* **32**, 407 (1981).
40. J.-P. Nap and T. Bisseling, *Science* **250**, 948 (1990).
41. R. F. Fisher and S. R. Long, *Nature (London)* **357**, 655 (1992).
42. H. Spaink, *Plant Mol. Biol.* **20**, 977 (1992).
43. J. Denarie, F. Debelle, and C. Rosenberg, *Ann. Rev. Microbiol.* **46**, 497 (1992).
44. Ref. 26, p. 24.
45. J. Sanjuan and co-workers, *Proc. Nat. Acad. Sci. USA* **89**, 8789 (1992).
46. W. E. Newton and C. J. Nyman, eds., *Proceedings of the First Symposium on Nitrogen Fixation*, Washington State University Press, Pullman, Wash., 1976.
47. C. A. Appleby, in Ref. 45, Vol. 1, p. 274.
48. D. Bogusz and co-workers, *Nature (London)* **331**, 178 (1988).
49. C. A. Appleby, D. Bogusz, E. S. Dennis, and W. J. Peacock, *Plant Cell Environ.* **11**, 359 (1988).
50. M. J. Dilworth, *Biochim. Biophys. Acta* **184**, 432 (1969).
51. D. W. Israel, R. L. Howard, H. J. Evans, and S. A. Russell, *J. Biol. Chem.* **249**, 500 (1974).
52. D. W. Emerich and R. H. Burris, *J. Bacteriol.* **134**, 936 (1978).
53. K. R. Schubert and H. J. Evans, *Proc. Nat. Acad. Sci. USA* **73**, 1201 (1976).
54. D. J. Arp, in Ref. 19, p. 432.
55. B. Dreyfus and Y. R. Dommergues, *Appl. Environ. Microbiol.* **41**, 97 (1981).
56. A. R. J. Eaglesham and co-workers, in P. M. Gresshoff, L. E. Roth, G. Stacey, and W. E. Newton, eds., *Nitrogen Fixation: Achievements and Objectives*, Chapman and Hall, London, 1990, p. 805.
57. W. Newcomb and S. M. Wood, *Int. Cytol. Rev.* **109**, 1 (1987).
58. D. D. Baker and B. C. Mullin, in Ref. 19, p. 259.
59. J. O. Dawson, *Outlook Agric.* **15**, 202 (1986).
60. D. Callaham, P. Del Tredici, and J. G. Torrey, *Science* **199**, 899 (1978).
61. J. G. Torrey, *Bioscience* **28**, 586 (1978).
62. D. R. Benson, D. J. Arp, and R. H. Burris, *Science* **205**, 688 (1979).
63. J. W. Millbank, in A. Quispel, ed., *The Biology of Nitrogen Fixation*, Elsevier Science Publishing Co., Inc., New York, 1974, p. 238.
64. J. W. Millbank, in R. W. F. Hardy, and W. S. Silver, eds., *A Treatise on Nitrogen Fixation*, Sect. III, John Wiley & Sons, Inc., New York, 1977, p. 125.
65. G. A. Peters, *Bioscience* **28**, 580 (1978).
66. W. B. Silvester, in P. S. Nutman, ed., *Symbiotic Nitrogen Fixation in Plants*, Cambridge University Press, London, 1976, p. 521.
67. T. A. Lumpkin and D. L. Plucknett, *Econ. Bot.* **34**, 111 (1980).
68. G. A. Peters and J. C. Meeks, *Ann. Rev. Plant Physiol. Plant Mol. Biol.* **40**, 193 (1989).
69. J. Döbereiner, J. M. Day, and P. J. Dart, *J. Gen. Microbiol.* **71**, 103 (1972).
70. J. F. W. von Bulow and J. Döbereiner, *Proc. Nat. Acad. Sci. USA* **72**, 2389 (1974).
71. J. Döbereiner, V. M. Reis, and A. C. Lazarini, in Ref. 23, p. 717.

72. J. Döbereiner, V. M. Reis, M. A. Paula, and F. Olivares, in Ref. 26, p. 671.

73. B. Reinhold-Hurek and T. Hurek, in Ref. 26, p. 691.

74. W. S. Silver, Y. M. Centifanto, and D. J. D. Nicholas, *Nature (London)* **199**, 396 (1963).

75. B. N. Richards and G. K. Voight, *Nature (London)* **201**, 310 (1964).

76. G. J. Waughman, J. R. S. French, and K. Jones, in W. J. Broughton, ed., *Nitrogen Fixation*, Vol. 1, Clarendon Press, Oxford, U.K., 1981, p. 135.

77. J. R. Benemann, *Science* **181**, 164 (1973).

78. G. D. Prestwich and B. L. Bentley, *Oecologia* **49**, 249 (1981).

79. J. B. Waterbury, C. B. Calloway, and R. D. Turner, *Science* **221**, 1401 (1983).

80. G. Stacey, R. H. Burris, and H. J. Evans, eds., *Biological Nitrogen Fixation*, Chapman and Hall, London, 1992.

81. H. Bortels, *Arch. Mikrobiol.* **1**, 333 (1930).

82. H. Bortels, *Zentral. Bakter. Parasitenk. Abt. II* **87**, 476 (1933).

83. W. A. Bulen and J. R. LeComte, *Proc. Natl. Acad. Sci. USA* **56**, 979 (1966).

84. V. K. Shah and W. J. Brill, *Proc. Natl. Acad. Sci. USA* **74**, 3249 (1977).

85. P. E. Bishop, D. M. L. Jarlenski, and D. R. Hetherington, *Proc. Nat. Acad. Sci. USA* **77**, 7342 (1980); P. E. Bishop and co-workers, Science **232**, 92 (1986).

86. P. E. Bishop, M. E. Hawkins, and R. R. Eady, *Biochem. J.* **238**, 437 (1986).

87. T. Kentemich, G. Danneberg, B. Hundeshagen, and H. Bothe, *FEMS Microbiol. Lett.* **51**, 19 (1988).

88. R. L. Robson and co-workers, *Nature (London)* **322**, 388 (1986).

89. R. L. Robson, P. R. Woodley, R. N. Pau, and R. R. Eady, *EMBO J.* **8**, 1217 (1989).

90. R. N. Pau, L. A. Mitchenall, and R. L. Robson, *J. Bacteriol.* **171**, 124 (1989).

91. J. E. Carnahan, L. E. Mortenson, H. F. Mower, and J. E. Castle, *Biochim. Biophys. Acta* **44**, 520 (1960).

92. W. A. Bulen and J. R. LeComte, *Methods Enzymol.* **24**, 456 (1972).

93. H. Haaker and C. Veeger, *Eur. J. Biochem.* **77**, 1 (1977).

94. L. E. Mortenson, *Biochim. Biophys. Acta* **127**, 18 (1966).

95. W. E. Newton, in Ref. 26, p. 5.

96. F. Luque and R. N. Pau, *Mol. Gen. Genet.* **227**, 481 (1991).

97. B. J. Hales and co-workers, *Biochemistry* **25**, 7251 (1986).

98. R. R. Eady and co-workers, *Biochem. J.* **244**, 197 (1987).

99. J. R. Chisnell, R. Premakumar, and P. E. Bishop, *J. Bacteriol.* **170**, 27 (1988).

100. R. R. Eady, T. H. Richardson, R. W. Miller, M. Hawkins, and D. J. Lowe, *Biochem. J.* **256**, 189 (1988).

101. K. Schneider, A. Müller, U. Schramm, and W. Klipp, *Eur. J. Biochem.* **195**, 653 (1991).

102. L. J. Lehman and G. P. Roberts, *J. Bacteriol.* **173**, 5705 (1991).

103. P. W. Ludden, R. Davis, R. Petrovich, and G. P. Roberts in Ref. 26, p. 143.

104. W. A. Filler and co-workers, *Eur. J. Biochem.* **160**, 371 (1986).

105. A. C. Robinson, B. K. Burgess, and D. R. Dean, *J. Bacteriol.* **166**, 180 (1986).

106. A. C. Robinson, T. W. Chun, J.-G. Li, and B. K. Burgess, *J. Biol. Chem.* **264**, 10088 (1989).

107. W. Arnold and co-workers, *J. Mol. Biol.* **203**, 715 (1988).

108. M. R. Jacobson and co-workers, *J. Bacteriol.* **171**, 1017 (1989).

109. R. D. Joerger and P. E. Bishop, *J. Bacteriol.* **170**, 1475 (1988).

110. L. Zheng, R. H. White, V. L. Cash, R. F. Jack, and D. R. Dean, *Proc. Nat. Acad. Sci. USA* **90**, 2754 (1993).

111. R. D. Joerger and co-workers, *J. Bacteriol.* **171**, 1075 (1989).

112. R. D. Joerger and co-workers, *J. Bacteriol.* **172**, 3400 (1990).

113. R. D. Joerger, M. R. Jacobson, and P. E. Bishop, *J. Bacteriol.* **171**, 3258 (1989).

114. E. D. Wolfinger and P. E. Bishop, *J. Bacteriol.* **173**, 7565 (1991).
115. K. E. Brigle, M. C. Weiss, W. E. Newton, and D. R. Dean, *J. Bacteriol.* **269**, 1547 (1987).
116. P. E. Bishop and R. Premakumar, in Ref. 19, p. 736.
117. C. Kennedy and D. R. Dean, *Mol. Gen. Genet.* **231**, 494 (1992).
118. J.-S. Chen, S.-Z. Wang, and J. L. Johnson, in Ref. 56, p. 483.
119. M. J. Dilworth, R. R. Eady, R. L. Robson, and R. W. Miller, *Nature (London)* **327**, 167 (1987).
120. T. R. Hoover and co-workers, *Nature (London)* **329**, 855 (1987).
121. M. R. Jacobson and co-workers, *Mol. Gen. Genet.* **219**, 49 (1989).
122. D. R. Dean and M. R. Jacobson in Ref. 19, p. 763.
123. Ref. 26, p. 44.
124. J. E. Carnahan and J. E. Castle, *Ann. Rev. Plant Physiol.* **14**, 125 (1963).
125. W. A. Bulen, R. C. Burns, and J. R. LeComte, *Proc. Nat. Acad. Sci. USA* **53**, 532 (1965).
126. R. W. F. Hardy, E. Knight, Jr., and A. J. D'Eustachio, *Biochem. Biophys. Res. Commun.* **20**, 539 (1965).
127. W. A. Bulen, J. R. LeComte, R. C. Burns, and J. Hinkson, in A. San Pietro, ed., *Non-Heme Iron Proteins: Role in Energy Conversion*, Antioch Press, Yellow Springs, Ohio, 1965, p. 261.
128. J. L. Hwang, C. H. Chen, and R. H. Burris, *Biochem. Biophys. Acta* **292**, 256 (1973).
129. G. E. Hoch, K. C. Schneider, and R. H. Burris, *Biochim. Biophys. Acta* **37**, 273 (1960).
130. B. K. Burgess, S. Wherland, W. E. Newton, and E. I. Stiefel, *Biochemistry* **20**, 5140 (1981).
131. J. H. Guth and R. H. Burris, *Biochemistry* **22**, 5111 (1983).
132. B. B. Jensen and R. H. Burris, *Biochemistry* **24**, 1141 (1985).
133. R. V. Hageman and R. H. Burris, *Biochemistry* **17**, 4117 (1978).
134. M. M. Georgiadis and co-workers, *Science* **257**, 1653 (1992).
135. M. K. Chan, J. Kim, and D. C. Rees, *Science* **260**, 792 (1993).
136. J. Kim and D. C. Rees, *Biochemistry* **33**, 389 (1994).
137. J. T. Bolin and co-workers, in E. I. Stiefel, D. Coucouvanis, and W. E. Newton, eds., *Molybdenum Enzymes, Cofactors and Model Systems*, American Chemical Society, Washington, D.C., 1993, p. 186.
138. R. G. Lowery and co-workers, *Biochemistry* **28**, 1206 (1989).
139. A. Willing and J. B. Howard, *J. Biol. Chem.* **265**, 6596 (1990).
140. D. Wolle, C.-H. Kim, D. R. Dean, and J. B. Howard, *J. Biol. Chem.* **267**, 3667 (1991).
141. C.-H. Kim, L. Zheng, W. E. Newton, and D. R. Dean in Ref. 26, p. 105.
142. R. N. F. Thorneley and co-workers, *Biochem. J.* **264**, 657 (1989); R. E. Mensink, H. Wassink, and H. Haaker, *Eur. J. Biochem.* **208**, 289 (1992).
143. J. Bergstrom, R. R. Eady, and R. N. F. Thorneley, *Biochem. J.* **251**, 165 (1988).
144. W. G. Zumft, G. Palmer, and L. E. Mortenson, *Biochim. Biophys. Acta* **292**, 413 (1973).
145. B. J. Hales, D. J. Langosh, and E. E. Case, *J. Biol. Chem.* **261**, 15301 (1986).
146. G. A. Walker and L. E. Mortenson, *Biochemistry* **13**, 2382 (1974).
147. D. Wolle, D. R. Dean, and J. B. Howard, *Science* **258**, 992 (1992).
148. R. N. Pau, *Trends in Biochemical Sciences*, **14**, 183 (1989); M. J. Dilworth and R. R. Eady, *Biochem. J.* **277**, 465 (1991).
149. W. E. Newton and D. R. Dean, in Ref. 137, p. 216.
150. B. E. Smith, R. R. Eady, D. J. Lowe, and C. Gormal, *Biochem. J.* **250**, 299 (1988).
151. D. M. Kurtz, Jr., and co-workers, *Proc. Nat. Acad. Sci. USA* **76**, 4986 (1979).
152. J. T. Bolin and co-workers, in Ref. 56, p. 117.
153. J. T. Bolin, Purdue University, personal communication, 1993.

154. R. Zimmermann and co-workers, *Biochim. Biophys. Acta* **537**, 185 (1978).
155. P. J. Stephens, in T. G. Spiro, ed., *Molybdenum Enzymes*, Wiley-Interscience, New York, 1985, p. 117.
156. W. R. Hagen, H. Wassink, R. R. Eady, B. E. Smith, and H. Haaker, *Eur. J. Biochem.* **169**, 457 (1987).
157. D. R. Dean and co-workers, *Mol. Microbiol.* **4**, 1505 (1990).
158. H. D. May, D. R. Dean, and W. E. Newton, *Biochem. J.* **277**, 457 (1991).
159. J. E. Morningstar, M. K. Johnson, E. E. Case, and B. J. Hales, *Biochemistry* **26**, 1795 (1987).
160. B. K. Burgess, E. I. Stiefel, and W. E. Newton, *J. Biol. Chem.* **255**, 353 (1980).
161. S. D. Conradson and co-workers, *Proc. Nat. Acad. Sci. USA* **91**, 1290 (1994).
162. A. Müller, K. Schneider, K. Knittel, and W. R. Hagen, *FEBS Lett.* **303**, 36 (1992).
163. K. E. Brigle, W. E. Newton, and D. R. Dean, *Gene* **37**, 37 (1985).
164. K. E. Brigle and co-workers, *Proc. Nat. Acad. Sci. USA* **84**, 7066 (1987).
165. H. M. Kent, M. Bainea, C. Gormal, B. E. Smith, and M. Buck, *Mol. Microbiol.* **4**, 1497 (1990).
166. D. J. Scott and co-workers, *Nature (London)* **343**, 188 (1990).
167. D. J. Scott, D. R. Dean, and W. E. Newton, *J. Biol. Chem.* **267**, 20002 (1992).
168. H. Thomann, M. Bernardo, W. E. Newton, and D. R. Dean, *Proc. Nat. Acad. Sci. USA* **88**, 6620 (1991).
169. H. M. Kent and co-workers, *Biochem. J.* **264**, 257 (1989).
170. R. M. Pengra and P. W. Wilson, *J. Bacteriol.* **75**, 251 (1958).
171. S. Hill, *FEMS Microbiol. Rev.* **54**, 111 (1988).
172. P. E. Bishop and R. D. Joerger, *Ann. Rev. Plant Physiol. Plant Mol. Biol.* **41**, 109 (1990).
173. R. N. Pau, in Ref. 26, p. 117.
174. G. P. Roberts and P. W. Ludden, in Ref. 19, p. 135.
175. R. Haselkorn and W. J. Buikema, in Ref. 19, p. 166.
176. M. J. Merrick, in Ref. 19, p. 835; M. Buck and W. Cannon, *Mol. Microbiol.* **6**, 1625 (1992); T. R. Hoover, E. Santero, S. Porter, and S. Kustu, *Cell* **63**, 11 (1990).
177. A. Contreras and M. Drummond, *Gene* **103**, 83 (1991).
178. A. D. Allen and C. V. Senoff, *Chem. Commun.* 621 (1965).
179. J. Chatt, J. R. Dilworth, and R. L. Richards, *Chem. Rev.* **78**, 589 (1978).
180. R. L. Richards, in M. J. Dilworth and A. R. Glenn, eds., *Biology and Biochemistry of Nitrogen Fixation*, Elsevier, Amsterdam, the Netherlands, 1991, p. 58.
181. M. E. Vol'pin and V. B. Shur, *Dokl. Akad. Nauk SSSR* **156**, 1102 (1964); M. E. Vol'pin, V. B. Shur, and M. A. Ilatovskaya, *Izv. Akad. Nauk SSSR Ser. Khim.* **19**, 1728 (1964).
182. G. Henrici-Olive and S. Olive, *Angew. Chem. Int. Ed. Engl.* **6**, 873 (1967).
183. E. E. van Tamelen, G. Boche, and R. Greeley, *J. Amer. Chem. Soc.* **90**, 1677 (1968).
184. M. E. Vol'pin, M. A. Ilatovskaya, L. V. Kosyakova, and V. B. Shur, *J. Chem. Soc. Chem. Comm.*, 1074 (1968).
185. Yu. G. Borod'ko and co-workers, *J. Chem. Soc. Chem. Comm.*, 1178 (1972).
186. R. H. Marvich and H. H. Brintzinger, *J. Amer. Chem. Soc.* **93**, 2046 (1971); G. P. Pez and S. C. Kwan, *J. Amer. Chem. Soc.* **98**, 8079 (1976).
187. J. M. Manriquez and co-workers, *J. Amer. Chem. Soc.* **100**, 3078 (1978).
188. R. D. Sanner and co-workers, *J. Amer. Chem. Soc.* **98**, 8358 (1976).
189. R. D. Sanner, J. M. Manriquez, R. E. Marsh, and J. E. Bercaw, *J. Amer. Chem. Soc.* **98**, 8351 (1976).
190. J. H. Teuben, *J. Organometal. Chem.* **57**, 159 (1973).
191. M. J. S. Gynane, J. Jeffery, and M. F. Lappert, *J. Chem. Soc. Chem. Comm.* 34, (1978).

192. M. Hidai, K. Tominari, and Y. Uchida, *J. Amer. Chem. Soc.* **94**, 1010 (1972).
193. J. Chatt, G. A. Heath, and R. L. Richards, *J. Chem. Soc. Dalton Trans.* 2074 (1974).
194. J. Chatt, A. J. Pearman, and R. L. Richards, *J. Chem. Soc. Dalton Trans.* 1852 (1977); J. Chatt, *Chemtech.* **11**(3), 162 (1981).
195. J. A. Baumann and T. A. George, *J. Amer. Chem. Soc.* **102**, 6153 (1980).
196. C. J. Pickett and J. Talarmin, *Nature (London)* **317**, 652 (1985).
197. S. M. Rocklage, H. W. Turner, J. D. Fellmann, and R. R. Schrock, *Organometallics* **1**, 703 (1982).
198. P. Sobota and B. Jezowska-Trzebiatowska, *Coord. Chem. Rev.* **26**, 71 (1978).
199. L. P. Didenko, A. G. Ovcharenko, A. E. Shilov, and A. G. Shilova, *Kinet. Katal.* **18**, 1078 (1977).
200. A. Shilov and co-workers, *Nature (London)* **231**, 460 (1971).
201. L. A. Nikonova and co-workers, *J. Mol. Catal.* **1**, 367 (1975–1976).
202. G. N. Schrauzer, *Angew. Chem. Int. Ed. Engl.* **14**, 514 (1975).
203. A. P. Khrushch, A. E. Shilov, and T. A. Vorontsova, *J. Amer. Chem. Soc.* **96**, 4987 (1974).
204. E. L. Moorehead, P. R. Robinson, T. M. Vickrey, and G. N. Schrauzer, *J. Amer. Chem. Soc.* **98**, 6555 (1976).
205. B. J. Weathers, J. H. Grate, N. A. Strampach, and G. N. Schrauzer, *J. Amer. Chem. Soc.* **101**, 925 (1979).
206. W. E. Newton in Ref. 19, p. 877.
207. B. K. Burgess, *Chem. Revs.* **90**, 1377 (1990).
208. S. P. Cramer and co-workers, *J. Am. Chem. Soc.* **100**, 3814 (1978).
209. S. D. Conradson and co-workers, *J. Am. Chem. Soc.* **109**, 7507 (1987).
210. T. E. Wolff and co-workers, *J. Am. Chem. Soc.* **100**, 4630 (1978).
211. G. Christou, C. D. Garner, F. E. Mabbs, and T. J. King, *J. Chem. Soc. Chem. Comm.* 740, (1978).
212. D. Coucouvanis, in E. I. Stiefel, D. Coucouvanis, and W. E. Newton, eds., *Molybdenum Enzymes, Cofactors and Model Systems*, American Chemical Society, Washington, D.C., 1993, p. 304.
213. D. Coucouvanis and co-workers, *Inorg. Chem.* **28**, 4181 (1989).
214. K. Tanaka, Y. Hozumi, and T. Tanaka, *Chemistry Lett. (Japan)*, 1203 (1982).
215. R. W. Treharne, D. R. Moles, M. R. Bruce, and C. K. McKibben, in S. Ahmed and H. P. M. Gunasena, eds., *Proceedings of the Second Review Meeting of I.N.P.U.T.S. Project*, East-West Center, Honolulu, Hawaii, 1978, p. 279.
216. K. Tennakone and co-workers, *J. Chem. Soc., Chem. Commun.*, 579 (1991).
217. C. R. Dickson and A. J. Nozik, *J. Amer. Chem. Soc.* **100**, 8007 (1978).
218. G. N. Schrauzer and co-workers, *Proc. Nat. Acad. Sci. USA* **80**, 3873 (1983).
219. R. Dixon, J. Vandeleyden, and D. Romero, in Ref. 26, p. 765.
220. M. Merrick and R. A. Dixon, *Trends Biotechnol.* **2**, 162 (1984).

WILLIAM E. NEWTON
Virginia Polytechnic Institute and State University

NITROGEN OXIDES. See AIR POLLUTION; EXHAUST CONTROL; SUPPLEMENT.

NITROGEN SULFIDE. See EXPLOSIVES AND PROPELLANTS.

NITROGEN TRICHLORIDE. See CHLORAMINES AND BROMAMINES.

NITROPARAFFINS

Nitroparaffins (or nitroalkanes) are derivatives of the alkanes in which one hydrogen or more is replaced by the electronegative nitro group, which is attached to carbon through nitrogen. The nitroparaffins are isomeric with alkyl nitrites, RONO, which are esters of nitrous acid. The nitro group in a nitroparaffin has been shown to be symmetrical about the R–N bond axis, and may be represented as a resonance hybrid:

$$R\text{—}\overset{+}{N}\!\!\begin{array}{c}\diagup O^-\\ \diagdown\!\!\!_{\displaystyle O}\end{array} \quad\longleftrightarrow\quad R\text{—}\overset{+}{N}\!\!\begin{array}{c}\diagup O\\ \diagdown\!\!\!_{\displaystyle O^-}\end{array}$$

Nitroparaffins are classed as primary, RCH_2NO_2, secondary, R_2CHNO_2, and tertiary, R_3CNO_2, by the same convention used for alcohols. Primary and secondary nitroparaffins exist in tautomeric equilibrium with the enolic or aci forms.

The nitroparaffins are named as derivatives of the corresponding hydrocarbons by using the prefix "nitro" to designate the NO_2 group (1), eg, 1,1-dinitroethane, $CH_3CH(NO_2)_2$. The salts obtained from nitroparaffins and the so-called nitronic acids are identical and may be named as derivatives of either, eg, sodium salt of aci-nitromethane, or sodium methanenitronate [25854-38-0].

Nitromethane, nitroethane, 1-nitropropane, and 2-nitropropane are produced by a vapor-phase process developed in the 1930s (2).

Physical Properties

The physical constants of the lower mononitroparaffins and of a number of polynitroparaffins are listed in Tables 1–3. Most polynitroparaffins are colorless crystalline or wax-like solids at or near room temperature. They are insoluble in water and alkanes but soluble in most other organic solvents. The lower mononitroparaffins are colorless, dense liquids with mild odors. The boiling points of the mononitroparaffins are much higher than those of the isomeric nitrates; for example, the normal boiling point of nitromethane is 101.2°C, whereas that of methyl nitrate is −12°C. This phenomenon may be attributed in large part to intermolecular hydrogen bonding. Accurate vapor-pressure determinations (3) for the lower nitroparaffins have been made and adapted to an Antoine equation (see Tables 1 and 2). A nomograph was constructed from these data (4). The properties of azeotropes of nitroparaffins with water or with organic liquids (5) and critical solution temperature data of nitroparaffins in binary (6) and ternary (7) systems have been described.

The molecular configuration of the $C\text{–}NO_2$ group in nitromethane has been determined by electron-diffraction methods (8). These data indicate that the nitrogen atom and atoms attached to it lie in the same plane, and that the O–N–O bond angle (127 ± 3°) is greater than the C–N–O bond angles (116 ± 3°). The spreading of the O–N–O bond angle beyond 120° is attributed to the repulsion of the negatively charged oxygen atoms of the highly polar system. Infrared (9) and mass spectra (10) for this group have also been discussed.

Table 1. Physical Properties of the Lower Mononitroparaffins

Property	Nitromethane	Nitroethane	1-Nitropropane	2-Nitropropane
CAS Registry Number	[75-52-5]	[79-24-3]	[108-03-2]	[79-46-9]
molecular weight	61.041	75.068	89.095	89.095
boiling point at 101.3 kPa,[a] °C	101.20	114.07	131.18	120.25
vapor pressure,[b] kPa[a]	3.64	2.11	1.01	1.73
freezing point, °C	−28.55	−89.52	−103.99	−91.32
density,[b] g/mL	1.138	1.051	1.001	0.988
coefficient of expansion per °C	0.00122	0.00112	0.00101	0.00104
refractive index, n_D^{20}	1.38188	1.39193	1.40160	1.39439
surface tension,[b] mN/m(=dyn/cm)	37.48	32.66	30.64	29.87
viscosity,[b] mPa·s(=cP)	0.647	0.677	0.844	0.770
heat of combustion (liq) at 25°C kJ/mol[c]	−708.4	−1362	−2016	−2000
heat of vaporization, kJ/mol[c]				
(liq) at 25°C	38.27	41.6	43.39	41.34
at bp	34.4	38.0	38.5	36.8
heat of formation (liq) at 25°C, kJ/mol[c]	−113.1	−141.8	−168.0	−180.7
specific heat at 25°C				
J/(mol·°C)[c]	106.0	138.5	175.6	175.2
J/(g·°C)[c]	1.74	1.85	1.97	1.97
dielectric constant at 30°C	35.87	28.06	23.24	25.52
dipole moment, $\times 10^{-30}$ °C·m[d]				
gas	11.68	11.94	12.41	12.44
liquid	10.58	10.64		
aqueous azeotrope				
bp, °C	83.59	87.22	91.63	88.55
wt % nitroparaffin	76.4	71.0	63.5	70.6
pH of 0.01 M aqueous solution of 25°C	6.4	6.0	6.0	6.2
solubility in water, wt %				
at 20°C	10.5	4.6	1.5	1.7
at 70°C	19.3	6.6	2.2	2.3
solubility[b] of water in nitroparaffin, wt %	1.8	0.9	0.6	0.5
at 70°C	7.6	3.0	1.7	1.6
Antoine's constant[e]				
A	6.399073	6.300057	6.252442	6.208143
B	1441.610	1435.402	1474.299	1422.898
C	226.939	220.184	215.986	218.341
critical temperature, °C	315	388	402	344

[a]To convert kPa to mm Hg, multiply by 7.5. [b]At 20°C. [c]To convert J to cal, divide by 4.184. [d]To convert C·m to debye, divide by 3.336×10^{-30}. [e]$\log p_{kPa} = A - B/(t + C)$.

Table 2. Physical Constants of C-4 and Higher Mononitroparaffins

Property	1-Nitro-butane	2-Nitro-butane	1-Nitro-2-methyl-propane	2-Nitro-2-methyl-propane	Nitro-cyclohexane
CAS Registry Number	[627-05-4]	[600-24-8]	[625-74-1]	[594-70-7]	[1122-60-7]
freezing point, °C	−81.33	glass	−76.85	26.23	−34
boiling point, °C	152.77	139.50	141.72	127.16	205.5−206
Antoine's constant[a]					
A	6.220403	6.202795	6.199044	6.112625	
B	1523.797	1494.318	1483.643	1396.948	
C	208.778	216.542	212.905	212.989	
vapor pressure at 20°C, kPa[b]	0.36	0.77	0.64	solid	
density at 25°C, g/mL	0.96848	0.96036	0.95848	solid	1.0680_4^{19}
refractive index, n_D^{20}	1.41019	1.40407	1.40642	1.39175^{30}	1.4608

[a] $\log\ p_{kPa} = A - B/(t + C)$.
[b] To convert kPa to mm Hg, multiply by 7.5.

Table 3. Physical Constants of Polynitro Compounds

Compound	CAS Registry Number	Mp, °C	Boiling point °C	at kPa[a]	Sp gr	Refractive index, n_D^t	Water solubility[b]
dinitro-methane	[625-76-3]		39−40	0.266	1.524	1.4480^{20}	
trinitro-methane	[517-25-9]	14.3 (dec)	45−47	2.933	1.5967_4^{24}	1.445511_{He}^{24}	sol
tetranitro-methane	[509-14-8]	13.8	125.7		1.6377_4^{21}	1.43416^{21}	insol
1,1-dinitro-ethane	[600-40-8]		185−186		1.3503_{23}^{23}	1.4346^{20}	sl sol
1,2-dinitro-ethane	[7570-26-5]	39−40	135	0.800	1.4597_4^{20}	1.4488^{20}	sl sol
1,1,1-trinitroethane	[595-86-8]	57	68	2.266	1.4223_4^{77}	1.4171_α^{77}	insol
2,2-dinitro-propane	[595-49-3]	54	185				insol
1,1-dinitro-cyclohexane	[4028-15-3]	36	142−143	4.666	1.2452_4^{21}	1.4732^{21}	insol

[a] At 101.3 kPa (=1 atm) if not indicated. To convert kPa to mm Hg, multiply by 7.5.
[b] All named compounds are soluble in ethanol and in ethyl ether.

Most organic compounds, including aromatic hydrocarbons, alcohols, esters, ketones, ethers, and carboxylic acids are miscible with nitroparaffins, whereas alkanes and cycloalkanes have limited solubility. The lower nitroparaffins are excellent solvents for coating materials, waxes, resins, gums, and dyes.

The thermal characteristics of higher nitroparaffins are quite different from those of nitromethane. The nitropropanes provide nearly twice as much heat as does nitromethane when burned in air or oxygen. When the only source of oxygen is that contained within the molecule, nitropropanes yield only 20% as much energy as nitromethane on burning.

Chemical Properties

The chemical reactions of the nitroparaffins have been discussed in depth (11–19), and their utility for the synthesis of heterocyclic and other compounds has been noted (20,21).

Tautomerism.   Primary and secondary mononitroparaffins are acidic substances which exist in tautomeric equilibria with their nitronic acids.

$$RCH_2NO_2 \rightleftharpoons RCH{=}NO_2H \quad \text{and} \quad RR'CHNO_2 \rightleftharpoons RR'C{=}NO_2H$$

The nitro isomer is weakly acidic; the nitronic acid isomer (aci form) is much more acidic. A comparison of the ionization constants of the two forms in water at 25°C is given in Table 4.

An equilibrium mixture of the isomers usually contains a much higher proportion of the true nitro compound. The equilibrium for each isomeric system is influenced by the dielectric strength and the hydrogen-acceptor characteristics of the solvent medium. The aci form is dissolved and neutralized rapidly by strong bases, and gives characteristic color reactions with ferric chloride.

Polynitroparaffins are stronger acids than the corresponding mononitroparaffins. Thus 1,1-dinitroethane has an ionization constant of 5.6×10^{-6} in water at 20°C; trinitromethane is a typical strong acid with an ionization constant in the range of 10^{-2} to 10^{-3}. Neutralization of these substances occurs rapidly, and they may be titrated readily.

In addition to neutralization, prolonged action of alkaline reagents can effect oxidation–reduction and extensive decomposition. 1,1-Dinitroparaffins and trinitromethane are more stable than are mononitro compounds during neutralization and subsequent regeneration, and therefore more rigorous experimental conditions are permissible.

Table 4. Ionization Constants of Lower Mononitroparaffins

Compound	K_{nitro}	K_{aci}
nitromethane	6.1×10^{-11}	5.6×10^{-4}
nitroethane	3.5×10^{-9}	3.9×10^{-5}
2-nitropropane	2.1×10^{-8}	7.7×10^{-6}
1-nitropropane		2.0×10^{-5}

Acidification of mononitroparaffin salts immediately gives the nitronic acid. Many nitronic acids have been isolated and stored as crystalline solids. In solution, a nitronic acid either isomerizes slowly into its more stable nitro form or undergoes some irreversible transformation. The isomerization occurs sufficiently slowly that it can be measured by conductometric or halometric methods.

Salts. Nitroparaffins dissociate to form ambidentate anions, which are capable of alkylation at either the carbon or oxygen atom (22).

$$R-\overset{\overset{\displaystyle R'}{|}}{C}=\overset{+}{N}\overset{\nearrow O^-}{\underset{\searrow O^-}{}} \longleftrightarrow R-\overset{\overset{\displaystyle R'}{|}}{\overset{-}{C}}-\overset{+}{N}\overset{\nearrow O^-}{\underset{\searrow O}{}}$$

Reactions of alkyl, allylic, and benzylic halides with these salts usually give carbonyl compounds, presumably through the nitronic ester (O-alkylation) as an intermediate. With certain benzyl halides substituted in the para or ortho position with nitro groups, the reaction gives almost exclusively C-alkylation. For example, sodium 2-propanenitronate [12384-98-4] reacts with p-nitrobenzyl chloride (23) or with p-nitrobenzyltrimethylammonium iodide (24) to form 2-methyl-2-nitro-1-(p-nitrophenyl)propane [5440-67-5]. This reaction depends on the nature of the leaving group (22) and is inhibited by powerful electron acceptors (25). It has been considered a radical-anion (25) or chain process (26).

Alkali salts of primary nitroparaffins, but not of secondary nitroparaffins, react with acyl cyanides to yield α-nitroketones by C-acylation (27).

$$RCH{=}NO_2M \; + \; R'\overset{\overset{\displaystyle O}{||}}{C}{-}CN \; \longrightarrow \; R'\overset{\overset{\displaystyle O}{||}}{C}CH(NO_2)R \; + \; MCN$$

Most other acylating agents act on salts of either primary or secondary nitroparaffins by O-acylation, giving first the nitronic anhydrides which rearrange to give, eg, nitrosoacyloxy compounds (28).

Alkaline solutions of mononitroparaffins undergo many different reactions when stored for long periods, acidified, or heated. Acidification of solutions of mononitro salts is best effected slowly at 0°C or lower with weak acids or buffered acidic mixtures, such as acetic acid–urea, carbon dioxide, or hydroxylammonium chloride. If mineral acids are used under mild conditions, eg, dilute HCl at 0°C, decomposition yields a carbonyl compound and nitrous oxide (Nef reaction).

$$CH_3\overset{\overset{\displaystyle NO_2Na}{||}}{CH} \; \xrightarrow[\text{0°C}]{\text{HCl}} \; CH_3\overset{\overset{\displaystyle O}{||}}{CH} \; + \; N_2O$$

Reaction with nitrous acid can be used to differentiate primary, secondary, and tertiary mononitroparaffins. Primary nitroparaffins give nitrolic acids, which dissolve in alkali to form bright red salts.

$$RCH{=}NO_2H + HNO_2 \xrightarrow{-H_2O} [RCH(NO)NO_2] \longrightarrow RC({=}NOH)NO_2 \xrightarrow{NaOH}$$

$$RC(NO){=}NO_2Na + H_2O$$

Secondary nitroparaffins give alkali-insoluble nitroso derivatives known as pseudonitroles. As monomers in the liquid state, pseudonitroles have a characteristic blue color; as solids they exist as white crystalline dimers. Tertiary nitroparaffins do not react with nitrous acid and no color develops.

$$R_2C{=}NO_2H + HNO_2 \longrightarrow R_2C(NO)NO_2 + H_2O$$

With sodium azide, salts of secondary nitroparaffins rearrange to N-substituted amides (29). With SO_2, primary or secondary nitroparaffins give imidodisulfonic acid salts (30). Potassium nitroform reacts quantitatively with nitryl chloride in ether to form tetranitromethane (31).

Sodium methanenitronate reacts with phenyl isocyanate in benzene to give the readily separable sodium salts of nitroacetanilide and nitromalondianilide. Except as the salt, nitromethane is unreactive with phenyl isocyanate at temperatures up to 100°C; the higher homologues do not give condensation products that can be isolated.

Acid Hydrolysis. With hot concentrated mineral acids, primary nitroparaffins yield a fatty acid and a hydroxylamine salt. If anhydrous acid and lower temperatures are used, the intermediate hydroxamic acid can be recovered.

$$CH_3CH_2CH_2NO_2 + H_2O + H_2SO_4 \longrightarrow CH_3CH_2COOH + H_3NOH{\cdot}HSO_4$$

$$CH_3CH_2NO_2 \xrightarrow[HCl]{anhydrous} CH_3C({=}NOH)OH$$

Halogenation. In the presence of alkali, chlorine replaces the hydrogen atoms on the carbon atom holding the nitro group. If more than one hydrogen atom is present, the hydrogen atoms can be replaced in stages; exhaustive chlorination of nitromethane yields chloropicrin, ie, trichloronitromethane [76-06-2]. The chlorination can be stopped at intermediate stages. Bromination or iodination takes a similar course, but bromopicrin [464-10-8] and iodopicrin [39247-25-1] tend to be less stable.

Halonitroparaffins can be prepared in which the halogen and nitro groups are not on the same carbon atom. The direct chlorination of nitroparaffins to give nongeminal substitution is promoted by irradiation in anhydrous media (32,33). For example, nitroethane yields 2-chloro-1-nitroethane [625-47-8], and 1-nitropropane yields both 2- [503-76-4] and 3-chloro-1-nitropropane [13021-02-8] on treatment with alkalies (34). Replacement of the hydroxy group in nitro alcohols (qv) with halogen yields vicinal halonitroparaffins. Action of phosphorus tribromide on nitro alcohols in dimethylformamide gives *vic*-bromonitroparaffins (35).

The acid chloride of *aci*-nitromethane, $CH_2=N(Cl)O$ (mp $-43°C$, bp $2-3°C$), is formed by fusion of nitromethane and picrylpyridinium chloride (36). It is hydrolyzed to nitrosomethane, reduces potassium permanganate strongly, and exhibits no reactions characteristic of hydroxamic acids.

Reaction With Carbonyl Compounds. Primary and secondary nitroparaffins undergo aldol-type reactions with a variety of aldehydes and ketones to give nitro alcohols (11). Those derived from the lower nitroparaffins and formaldehyde are available commercially (see NITRO ALCOHOLS). Nitro alcohols can be reduced to the corresponding amino alcohols (see ALKANOLAMINES).

These reversible reactions are catalyzed by bases or acids, such as zinc chloride and aluminum isopropoxide, or by anion-exchange resins. Ultrasonic vibrations improve the reaction rate and yield (see ULTRASONICS). Reaction of aromatic aldehydes or ketones with nitroparaffins yields either the nitro alcohol or the nitro olefin, depending on the catalyst. Conjugated unsaturated aldehydes or ketones and nitroparaffins (Michael addition) yield nitro-substituted carbonyl compounds rather than nitro alcohols. Condensation with keto esters gives the substituted nitro alcohols (37); keto aldehydes react preferentially at the aldehyde function.

Most nitroparaffins do not react with ketones, but in the presence of alkoxide catalysts, nitromethane and lower aliphatic ketones give nitro alcohols; in the presence of amine catalysts dinitro compounds are obtained.

$$2\ CH_3NO_2\ +\ CH_3\overset{\overset{\displaystyle O}{\|}}{C}CH_3\ \xrightarrow{R_2NH}\ O_2NCH_2\underset{\underset{\displaystyle CH_3}{|}}{\overset{\overset{\displaystyle CH_3}{|}}{C}}CH_2NO_2\ +\ H_2O$$

Nitro olefins can be made in some cases by dehydration of the aromatic nitrohydroxy derivatives. Subsequent reduction yields the aromatic amine. The following three-step reaction yielding 2-amino-1-phenylbutane illustrates the synthesis of this class of valuable pharmaceutical compounds.

$$CH_3CH_2CH_2NO_2\ +\ C_6H_5CHO\ \longrightarrow\ C_6H_5CHOH\underset{\underset{\displaystyle NO_2}{|}}{C}HCH_2CH_3\ \xrightarrow{-H_2O}\ C_6H_5CH=C\overset{\diagup CH_2CH_3}{\diagdown NO_2}$$

$$\xrightarrow{+H_2}\ C_6H_5CH_2\underset{\underset{\displaystyle NH_2}{|}}{C}HCH_2CH_2$$

In the presence of amine salts of weak acids, the nitro olefin is formed directly.

$$C_6H_5CHO + CH_3CH_2NO_2 \xrightarrow[\text{HOOCCH}_3]{\text{RNH}_2} C_6H_5CH{=}C\overset{CH_3}{\underset{NO_2}{\diagdown}} + H_2O$$

Mannich-Type Reactions. Secondary nitroparaffins, formaldehyde (qv), and primary or secondary amines can react in one step to yield Mannich bases.

$$R_2CHNO_2 + CH_2O + R_2'NH \longrightarrow R_2\underset{NO_2}{\overset{|}{C}}CH_2NR_2' + H_2O$$

Primary nitroparaffins react with two moles of formaldehyde and two moles of amines to yield 2-nitro-1,3-propanediamines. With excess formaldehyde, Mannich bases from primary nitroparaffins and primary amines can react further to give nitro-substituted cyclic derivatives, such as tetrahydro-1,3-oxazines or hexahydropyrimidines (38,39). Pyrolysis of salts of Mannich bases, particularly of the boron trifluoride complex (40), yields nitro olefins by loss of the amine moiety. Closely related to the Mannich reaction is the formation of sodium 2-nitrobutane-1-sulfonate [76794-27-9] by warming 1-nitropropane with formaldehyde and sodium sulfite (41).

Reduction. The lower nitroparaffins are reduced readily to the corresponding primary amines with a number of reducing agents. Partial reduction yields aldoximes, ketoximes, or N-substituted hydroxylamines. Suitable reduction methods range from iron and hydrochloric acid to high pressure hydrogenation over Raney nickel or noble-metal catalysts. Reaction conditions have been developed also for the reduction of olefinic or carbonyl groups with or without reduction of the nitro group present, and for the reduction of the nitro group leaving the other groups untouched (42). Some of the products obtained are useful as pharmaceuticals.

Oxidation. Nitroparaffins are resistant to oxidation. At ordinary temperatures, they are attacked only very slowly by strong oxidizing agents such as potassium permanganate, manganese dioxide, or lead peroxide. Nitronate salts, however, are oxidized more easily. The salt of 2-nitropropane is converted to 2,3-dimethyl-2,3-dinitrobutane [3964-18-9], acetone, and nitrite ion by persulfates or electrolytic oxidation. With potassium permanganate, only acetone is recovered.

$$3 \; \overset{CH_3}{\underset{CH_3}{\diagup}}C{=}\overset{+}{N}O_2Na \xrightarrow{[O]} CH_3\overset{CH_3}{\underset{NO_2}{\overset{|}{C}}}{-}\overset{CH_3}{\underset{NO_2}{\overset{|}{C}}}CH_3 + CH_3\overset{O}{\overset{||}{C}}CH_3 + NaNO_2$$

α-Nitroacetaldehyde [5007-21-6] is formed in low yield by the oxidation of nitroethane with selenium dioxide. This product is easily oxidized in air to nitroacetic acid [625-75-2], which spontaneously decarboxylates to nitromethane.

Addition to Multiple Bonds. Mono- or polynitroparaffins with a hydrogen on the carbon atom carrying the nitro group add to activated double bonds under the influence of basic catalysts (43–45). Thus nitromethane forms tris-(β-cyanoethyl)nitromethane [1466-48-4] with acrylonitrile and 2-nitropropane yields 4-methyl-4-nitrovaleronitrile [16507-00-9]. These Michael-type condensations with acrylic compounds take place in liquid ammonia without catalyst (46). In the presence of dehydrating agents, such as phenyl isocyanate or phosphorus oxychloride, nitroparaffins add to activated olefins, such as methyl acrylate, to give oxazines (47).

Nitroparaffins add 1,4 to conjugated systems; methyl vinyl ketones, for example, yield the corresponding γ-nitro ketone, which can be reduced to a γ-nitro alcohol (48). More than one vinyl group may react with primary nitroparaffins (49).

Conjugated nitro olefins can function as the acceptor to which nitroparaffins may add to form polynitro derivatives (50–51). Nitro olefins also undergo Diels-Alder condensations with compounds such as maleic anhydride or anthracene (52).

Nitroparaffins and HBF_4 react with alkynes through conjugate addition of a proton and the nitronate ion at the triple bond as illustrated by nitromethane and 5-decyne (53).

$$C_4H_9C{\equiv}CC_4H_9 + CH_3NO_2 \xrightarrow{HBF_4} \underset{\underset{N=CH_2}{|}}{\overset{\overset{O}{\|}}{C_4H_9\underset{|}{C}CHC_4H_9}}$$

Other Reactions. α-Nitroalkanoic acids or their esters can be prepared (54–56) by treating nitroparaffins with magnesium methyl carbonate, or with triisopropylaluminum and carbon dioxide. These products are reduced readily to α-amino acids.

$$CH_3NO_2 + (CH_3OCOO)_2Mg \longrightarrow \underset{N-O}{\overset{O}{\diagdown}}\kern-1em\underset{\downarrow}{}\ Mg + 2\,CH_3OH + CO_2$$

1,1,1-Trinitroparaffins can be prepared from 1,1-dinitroparaffins by electrolytic nitration, ie, electrolysis in aqueous caustic sodium nitrate solution (57). Secondary nitroparaffins dimerize on electrolytic oxidation (58); for example, 2-nitropropane yields 2,3-dimethyl-2,3-dinitrobutane, as well as some 2,2-dinitropropane. Addition of sodium nitrate to the anolyte favors formation of the former. The oxidation of salts of *aci*-2-nitropropane with either cationic or anionic oxidants generally gives both 2,2-dinitropropane and acetone (59); with ammonium peroxysulfate, for example, these products are formed in 53 and 14%

yields, respectively. Ozone oxidation of nitroso groups gives nitro compounds; 2-nitroso-2-nitropropane [5275-46-7] (propylpseudonitrole), for example, yields 2,2-dinitropropane (60).

O-Acylation of 2-nitropropane occurs on reaction with either ketene or acetic anhydride (61) in the presence of dry sodium acetate at 70–80°C. Ketovinylation of 2-nitropropane at the 1-position occurs on treatment of sodium 2-propanenitronate with a chlorovinyl ketone (62).

Furoxans are formed by the dehydration of two moles of a nitroparaffin (63).

$$2\ RCH_2NO_2 \xrightarrow[\text{in CHCl}_3]{POCl_3,\ (C_2H_5)_3N} \left[2\ RC\equiv N\longrightarrow O \right] \longrightarrow$$

Preparation and Manufacture

Synthetic methods suitable for preparation of a wide variety of nitroparaffins have been reviewed (64–67).

A general one-step method for preparation of primary and secondary nitroparaffins from amines by oxidation with m-chloroperbenzoic acid in 1,2-dichloroethane has been reported (68). This method is particularly useful for laboratory quantities of a wide variety of nitroparaffins because a large number of amines are readily available from ketones by oxime reduction and because the reaction is highly specific for nitroparaffins.

Higher nitroalkanes are prepared from lower primary nitroalkanes by a one-pot synthesis (69). Successive condensations with aldehydes and acylating agents are followed by reduction with sodium borohydride. Overall conversions in the 75–80% range are reported.

The only method utilized commercially is vapor-phase nitration of propane, although methane (70), ethane, and butane also can be nitrated quite readily. The data in Table 5 show the typical distribution of nitroparaffins obtained from the nitration of propane with nitric acid at different temperatures (71). Nitrogen dioxide can be used for nitration, but its low boiling point (21°C) limits its effectiveness, except at increased pressure. Nitrogen pentoxide is a

Table 5. Effect of Temperature on the Nitration of Propane with Nitric Acid

Product	Nitration temperature		Type of substitution
	505–510°C	790–795°C	
nitromethane, mol %	22.0	32.3	alkyl cleavage
nitroethane, mol %	16.6	24.2	alkyl cleavage
1-nitropropane, mol %	13.2	24.2	primary
2-nitropropane, mol %	48.2	19.3	secondary
ratio of cleavage products to substitution products	0.628	1.30	

powerful nitrating agent for n-alkanes; however, it is expensive and often gives polynitrated products.

Nitromethane, nitroethane, 1-nitropropane, and 2-nitropropane are made on a large scale at the Sterlington, Louisiana, plant of ANGUS Chemical Co. In the manufacturing process (11,72–74), nitric acid (qv) reacts with excess propane at 370–450°C and 0.81–1.2 MPa (8–12 atm). Stainless steel is the preferred material of construction. The reaction products are cooled and the nitroparaffins and associated by-products, such as aldehydes and ketones, condense. Propane and nitric oxide remain in the gas stream. The propane is separated and recycled to the reactor. The nitric oxide is converted to nitric acid, which, mixed with fresh acid, is fed to the reactor. The crude nitroparaffins are washed to remove oxygenated impurities, and then are fractionated to commercial-grade nitroparaffins (Table 6). The ANGUS Chemical plant has a production capacity of 41,000 t of nitroparaffins. The single-drum prices for 1994 were as follows: nitromethane, \$6.05/kg; nitroethane, \$6.16/kg; 1-nitropropane, \$5.94/kg; and 2-nitropropane, \$5.65.

Table 6. Typical Properties of Commercial-Grade Nitroparaffins

Property	Nitromethane	Nitroethane[a]	1-Nitropropane[a]	2-Nitropropane
distillation range at 101.3 kPa[b] (90% min), °C	100–103	112–116	129–133	119–122
vapor density (air = 1)	2.11	2.58	3.06	3.06
change of density with temperature, 0–50°C, g/mL per °C	0.0014	0.0012	0.0011	0.0011
weight per L at 20°C, kg	1.13	1.05	1.00	0.9
flash point, °C				
Tag open cup	44.4	41.1	48.9	37.8
Tag closed cup	35.6	30.6	35.6	27.8
lower limit of flammability, vol %	7.3[c]	3.4[d]	2.2[e]	2.5[f]
ignition temperature, °C	418	414	420	428
evaporation rate[g]	139	121	88	110
evaporation number[h]	9	11	16	10
hydrogen bonding parameter, γ	2.5	2.5	2.5	2.5
solubility parameter, δ	12.7	11.1	10.7	10.7

[a]A mixture of 1-nitropropane and nitroethane is marketed as NiPar 640.
[b]To convert kPa to mm Hg, multiply by 7.5.
[c]At 33°C.
[d]At 30°C.
[e]At 34°C.
[f]At 27°C.
[g]n-Butyl acetate = 100.
[h]Diethyl ether = 1.

The only other nitroparaffin manufactured on a large scale was nitrocyclohexane [1122-60-7], made by liquid-phase nitration of cyclohexane. Nitrocyclohexane was the starting material for ϵ-caprolactam via reduction to cyclohexanone oxime. This process has been superseded by other, more efficient processes (see CAPROLACTAM). Nitrocyclohexane is not being produced in large quantities for either captive use or sale.

The preparation of polynitroparaffins has been reviewed (75). 2,2-Dinitropropane has been produced in pilot-plant quantities by liquid-phase nitration starting from either propane or 2-nitropropane (76,77) (see NITRATION).

gem-Dinitroparaffins are made conveniently from primary or secondary mononitroparaffins by the oxidative nitration of mononitroparaffins in alkaline solution using silver ion as the oxidizing agent (Shechter-Kaplan reaction) (78–80). This reaction has been used to prepare 1,1-dinitroethane on a tonnage scale.

$$R_2C{=}NO_2^- + NO_2^- + 2\,Ag^+ \longrightarrow R_2C(NO_2)_2 + 2\,Ag$$

Shipment and Storage

The four commercial nitroparaffins are available in drums; they are also available in bulk except for nitromethane, for which shipment in tank cars or trucks is prohibited.

Safety factors have been of prime consideration in the development of recommendations (81,82) for the storage and safe handling of nitromethane during recovery operations and transfer in piping systems. Bulk shipments of specified mixtures containing nitromethane and a diluent are permitted under a Department of Transportation (DOT) exemption. Nitromethane preferably should be stored in the 208-L (55-gal) drums in which it is shipped. These containers are of lightweight construction and there is little possibility that they might develop sufficiently high internal pressure either to ignite the nitromethane or to allow it to burn as a monopropellant. Bulk-storage tanks should be isolated, buried, or barricaded to protect them from projectile impacts should an explosion occur in nearby equipment or facilities. It is recommended that nitromethane be diluted with another suitable liquid prior to bulk storage. Despite the fact that nitromethane may be detonated under certain conditions, it is not classified as an explosive for shipping purposes.

Commercial-grade nitroparaffins are shipped and stored in ordinary carbon steel. However, wet nitroparaffins containing more than 0.1–0.2% water may become discolored when stored in steel for long periods, even though corrosion is not excessive. Aluminum and stainless steel are completely resistant to corrosion by wet nitroparaffins. Storage in contact with lead (qv), copper, or alloys containing these metals should be avoided. Polymeric materials for gaskets, hoses, and connections should be tested for their suitability before exposure to nitroparaffins.

Because of their flash points, nitroparaffins are classified as flammable liquids under DOT regulations (hazard class 3, PG III). Nitromethane and nitroethane fires can be extinguished with water, CO_2, foam, or class ABC dry chemical extinguishers. Nitroparaffins should not be exposed to dry caustic soda, lye, or similar alkaline materials.

Specifications

The specifications of the four commercial nitroparaffins are given in Table 7.

Table 7. Specifications of Commercially Available Nitroparaffins

Assay	Nitromethane	Nitroethane	1-Nitropropane	2-Nitropropane
purity,[a] min wt %	98.0	98.0	98.5	96.0
total nitroparaffins, min wt %	99.0	99.0	99.0	99.0
specific gravity at 25/25°C	1.124–1.135			
acidity as acetic acid, max wt %	0.1		0.2	0.1
water, max wt %	0.1	0.2	0.1	0.1
color, max APHA	20	20	20	20

[a]Determined by gas chromatography.

Analytical Methods

The nitroparaffins have been determined by procedures such as fractionation, titration, colorimetry, infrared spectroscopy, mass spectrometry, and gas chromatography. The early analytical methods and uses of polynitroparaffins as analytical reagents have been reviewed (11). More recent qualitative and quantitative methods have also been reviewed (83).

A titration method for primary and secondary nitroparaffins using hypochlorite gives good accuracy ($\pm 0.1\%$) (84). It is based on the following equation:

$$CH_3CH_2NO_2 + 2\ NaClO \longrightarrow CH_3CCl_2NO_2 + 2\ NaOH$$

A number of colorimetric methods are available (83); however, spectroscopic methods are generally preferred.

Data on infrared curves for many nitroparaffins and their sodium salts have been reported (10,85–89). References 87, 90 and 91 give uv spectra. Accurate analysis and positive identification of the components of a mixture of several nitroparaffins can be obtained by mass spectrometry (qv) (92).

Gas chromatography is probably the most versatile method for analyzing nitroparaffins (93–95), eg, in the presence of nitric acid esters (96).

High performance liquid chromatography (hplc) may be used to determine nitroparaffins by utilizing a standard uv detector at 254 nm. This method is particularly applicable to small amounts of nitroparaffins present, eg, in nitro alcohols (qv), which cannot be analyzed easily by gas chromatography. Suitable methods for monitoring and determination of airborne nitromethane, nitroethane, and 2-nitropropane have been published by the National Institute of Occupational Safety and Health (NIOSH) (97). Ordinary sorbent tubes containing charcoal are unsatisfactory, because the nitroparaffins decompose on it unless the tubes are held in dry ice and analyzed as soon after collection as possible.

Health and Safety Factors

Toxicology. The nitroparaffins have minimal effects by way of actual contact. There were neither systemic effects nor irritation in dermal studies in rabbits. Human exposure of a prolonged or often-repeated nature has led to low grade irritation attributable to removal of oil from the skin, an effect produced by most organic solvents. Eye irritation potential of all four nitroparaffins has been determined in rabbits. Other than a transient slight redness and some lachrymation, no effects were noted. The average Draize score was 0.0. The acute oral toxicity, LD_{50}, of all four nitroparaffins has been determined in the rat (Table 8).

Inhalation is the chief route of worker exposure. Comparative data from acute or subchronic inhalation exposures with rats (98) indicate that nitromethane and nitroethane are the least toxic of the nitroparaffins by this route and do not induce methemoglobin formation. The nitropropanes are less well tolerated; 2-nitropropane is more toxic than 1-nitropropane and is more likely to cause methemoglobinemia.

The 1994–1995 threshold limit values as recommended by the American Conference of Governmental Industrial Hygienists (ACGIH) are given in Table 8. These time-weighted average values are those levels to which nearly all workers may be exposed for an 8-h workday and a 40-h work week without adverse effect (99).

A comprehensive study of the tolerance of laboratory animals to vapors of 2-nitropropane was reported in 1952 (100). In a study published in 1979, rabbits and rats survived exposure to nitromethane for six months at 750 and 100 ppm, respectively, with no unexpected findings (101). Similarly, no compound-related effects were found for rabbits exposed to 2-nitropropane at 200 ppm or for rabbits or rats exposed at 27 ppm. Liver damage was extensive in male rats exposed at 207 ppm for six months, and hepatocellular carcinomas were observed. Subsequently, the International Agency for Research on Cancer (IARC) found that there is "sufficient evidence" to conclude that 2-nitropropane causes cancer in rats but that epidemiologic data are inadequate to reinforce the conclusion in humans (102). The National Toxicology Program also concluded that it "may reasonably be anticipated to be a carcinogen" (103).

Table 8. Oral Toxicity[a] and Threshold Limit Values[b] of Nitroparaffins

Nitroparaffin	LD_{50}, mg/kg	TLV, ppm	TLV, mg/m^3
nitromethane	1210 ± 322	20	50
nitroethane	1625 ± 193	100	307
1-nitropropane	455 ± 75	25	91
2-nitropropane	725 ± 160	10	36

[a] In rat.
[b] Ref. 99.

Because of these findings with 2-nitropropane, lifetime inhalation studies have been conducted in rats at 200 ppm with nitromethane and nitroethane (104), and at 100 ppm with 1-nitropropane (105). In no instance was cancer found in any exposed rats. This difference was confirmed by an independent study in which oral administration of 2-nitropropane caused liver tumors, whereas 1-nitropropane administration did not (106). Also, mutagenicity testing has provided clear evidence that the primary nitroparaffins are not genotoxic or mutagenic. 2-Nitropropane, in contrast, is mutagenic in the Ames test. Many studies have compared the genotoxicity of the primary nitroparaffins to that of 2-nitropropane (107); possible mechanisms of carcinogenicity and the fact that tumors developed only in rats at 2-nitropropane levels and lengths of exposure sufficient to cause hepatotoxicity also have been discussed. An excellent review of the toxicity of 2-nitropropane has been published by the World Health Organization (108).

Safe Handling. Any work area where nitroparaffins are present should be ventilated adequately so as to maintain concentration levels below the accepted exposure limit. Fresh-air masks should be supplied to workers entering confined spaces, eg, storage tanks, containing a high concentration of nitroparaffin vapors. Nitroparaffins have high heats of adsorption on respirator canisters containing Hopcalite, a mixture of copper, cobalt, manganese, and silver present in some canisters for converting carbon monoxide to carbon dioxide; the use of respirator masks containing these substances may lead to fire in the presence high concentrations of the nitroparaffins.

The ignition temperature of the lower homologues are relatively high for organic solvents. When ignited, nitromethane burns with a lazy flame that often dies spontaneously, and in any case is extinguished readily with water which floats on the heavier nitromethane. Nitropropanes burn more vigorously, but less so than gasoline.

Some dry-chemical fire extinguishers contain sodium or potassium bicarbonate; these should not be used on nitromethane or nitroethane fires. Dry chemical extinguishers can be used on nitropropane fires.

Three conditions have been identified under which nitromethane can be detonated: (1) nitromethane can explode if subjected to a severe shock such as that of a high explosive with more power than a No. 8 blasting cap; (2) it can be initiated by a rapid compression under adiabatic conditions; and (3) nitromethane can be detonated by heating it under confinement to near the critical temperature (315°C). These conditions combining high pressure with high temperature are the same as those under which nitromethane burns as a monopropellant. Certain compounds, eg, amines or strong oxidizing agents, when present in admixture with nitromethane, can sensitize it to decomposition by strong shock. The addition of such sensitizers should be avoided unless the nitromethane is intended for use as an explosive.

In designing facilities for handling and processing nitromethane, it is recommended that nitromethane not be processed in high pressure equipment. All vessels for nitromethane service should be protected to prevent adiabatic compression. Detonation traps should be installed at each end of transfer lines and in every 61 m (200 feet) of continuous line. Nitromethane lines should be located underground or in channels wherever possible. Pressure relief devices (rated ~ 690 kPa = 100 psig) should be installed between closed valves (81).

The insensitivity of nitromethane to detonation by shock under normal conditions of handling has been demonstrated by a number of full-scale tests. Sensitivity to shock increases with temperature; at $\cong 60°C$, nitromethane can be detonated by a No. 8 blasting cap. Nitroethane can be initiated only when heated near its boiling point under heavy confinement; neither it or the nitropropanes can be detonated in unconfined conditions.

Environmental Concerns. Few data on the environmental effects of the nitroparaffins are available. However, they are known to be of low toxicity to the fathead minnow (109). Based on their uv spectra, the nitroparaffins would be expected to undergo photolysis in the atmosphere. The estimated half-life of 2-nitropropane in the atmosphere is 3.36 h (110). Various values have been determined for the half-life of nitromethane, but it is similar to 2-nitropropane in persistence (111). Reviews of the available data on the environmental effects of nitromethane and 2-nitropropane have been published by the U.S. Environmental Protection Agency (112,113).

When disposed of, all the nitroparaffins are considered to be hazardous waste. All have the characteristic of ignitability; however, 2-nitropropane also is a listed hazardous waste (U171) because of its toxicity. The preferred method of disposal is by incineration. Generally, the nitroparaffins must be blended with other flammable liquids to ensure compliance with NO_x emission standards.

Uses

The nitroparaffins have been utilized for many applications (114). Some of these uses have been discontinued because of economic and environmental considerations. For instance, significant quantities of 1-nitropropane once were used for the production of hydroxylammonium sulfate and propionic acid by hydrolysis. The need to dispose of an acid waste stream from this process made it uneconomical, so it was discontinued.

This article limits discussion to the current uses of each of the nitroparaffins and omits uses that are either no longer of commercial interest or that have not yet become of commercial importance.

Nitromethane. The nitroparaffins are used widely as raw materials for synthesis. Nitromethane is used to produce the nitro alcohol (qv) 2-(hydroxymethyl)-2-nitro-1,3-propanediol, which is a registered biocide useful for control of bacteria in a number of industrial processes. This nitro alcohol also serves as the raw material for the production of the alkanolamine (qv) 2-amino-2-(hydroxymethyl)-1,3-propanediol, which is an important buffering agent useful in a number of pharmaceutical applications.

Halogenation of nitromethane is utilized to produce two economically important pesticides, chloropicrin [76-06-2], a soil fumigant, and bronopol, a biocide useful for control of microbial growth in cosmetics and industrial applications.

Nitromethane also is used in the synthesis of the antiulcer drug, ranitidine [66357-35-5]. A two-step process utilizing nitromethane, carbon disulfide, potassium hydroxide, and dimethyl sulfate yields 1,1-bis(methylthio)-2-nitroethene [13623-94-4], which reacts further to produce ranitidine.

Significant quantities of nitromethane find use as fuel in drag racing and as a hobby fuel. In addition, nitromethane is used in explosive applications, particularly in shaped charges for specialized applications. It has the advantage of not in itself being classified as an explosive, but it can be made cap-sensitive on-site, thus simplifying the transport of the explosive system to remote locations.

Only small amounts of nitromethane are used as solvent, but it is used in specialized applications such as the solublization of α-cyanoacrylate glue and acrylic polymers. Also, nitromethane is useful as solvent for single-phase Friedel-Crafts reactions (115).

Finally, nitromethane has been used in large quantities as a stabilizer for 1,1,1-trichloroethane. The use of this degreasing solvent is expected to decline and disappear under the provisions of the Montreal Protocol (116), which bans ozone-depleting substances, of which this is one.

Nitroethane. The principal use of nitroethane is as a raw material for synthesis in two applications. It is used to manufacture α-methyl dopa, a hypertensive agent. Also, the insecticide S-methyl-N-[(methylcarbamoyl)-oxy]thioacetimidate [16752-77-5], can be produced by a synthesis route using nitroethane as a raw material. The first step of this process involves the reaction of the potassium salt of nitroethane, methyl mercaptan, and methanol to form methyl methylacetohydroxamate. Solvent use of nitroethane is limited but significant. Generally, it is used in a blend with 1-nitropropane.

1-Nitropropane. The alkanolamines (qv), 2-amino-2-ethyl-1,3-propanediol and 2-amino-1-butanol, are produced by the two-step process described previously.

Though less important economically, solvent usage consumes a larger portion of the 1-nitropropane production than is consumed of the other nitroparaffins for this use. In particular, a blend of nitroethane and 1-nitropropane has been developed which is useful as an additive for improved coatings performance (116).

2-Nitropropane. As much as 9100 t of 2-nitropropane once were consumed for use in coatings annually. Concern about toxicity and a general movement to low volatile organic compound (VOC) coatings have resulted in almost the complete disappearance of this use for 2-nitropropane. However, derivatives such as 2-methyl-2-nitro-1-propanol (used in tire cord adhesive) and 2-amino-2-methyl-1-propanol (a pigment dispersant and buffer), have served as an outlet for 2-nitropropane production.

BIBLIOGRAPHY

"Nitroparaffins" in *ECT* 1st ed., Vol. 9, pp. 428–455, by H. Shechter, Ohio State University, and R. B. Kaplan, E. I. du Pont de Nemours & Co., Inc.; in *ECT* 2nd ed., Vol. 13, pp. 864–888, by J. L. Martin and P. J. Baker, Jr., Commercial Solvents Corp.; in *ECT* 3rd ed., Vol. 15, pp. 969–987, P. J. Baker, Jr., and A. F. Bollmeier, Jr., International Minerals & Chemical Corp.

1. *IUPAC Nomenclature of Organic Chemistry*, Pergamon Press, New York, 1979, p. 275.
2. U.S. Pat. 1,967,667 (July 24, 1934), H. B. Hass, E. B. Hodge, and B. M. Vanderbilt (to Purdue Research Foundation); *Ind. Eng. Chem.* **28**, 339 (1936).

3. E. E. Toops, Jr., *J. Phys. Chem.* **60**, 304 (1956).

4. B. Fader, *Chem. Process. (Chicago)* **19**(8), 174 (1956).

5. L. H. Horsley, *Adv. Chem. Ser.* **6**, (1952); L. H. Horsley and W. S. Tamplin, *Adv. Chem. Ser.* **35**, (1962); L. H. Horsley, *Adv. Chem. Ser.* **116**, (1973).

6. A. W. Francis, *Adv. Chem. Ser.* **31**, (1961).

7. A. W. Francis, *Liquid–Liquid Equilibriums*, John Wiley & Sons, Inc., New York, 1963, pp. 174, 214; A. W. Francis, *J. Chem. Eng. Data* **11**, 234 (1966).

8. L. O. Brockway, J. Y. Beach, and L. Pauling, *J. Am. Chem. Soc.* **57**, 2693 (1935).

9. G. Geiseler and H. Kessler, in T. Urbanski, ed., *Nitro Compounds: Proceedings of an International Symposium, Warsaw, Poland, 1963*, Pergamon Press Ltd., Oxford, U.K., 1964, pp. 187–194.

10. R. T. Aplin, M. Fischer, D. Becher, H. Budzikiewicz, and C. Djerassi, *J. Amer. Chem. Soc.* **87**, 4888 (1965); N. M. M. Nibbering, Th. J. de Boer, and H. J. Hofman, *Rec. Trav. Chim.* **84**, 481 (1965); A. V. Iogansen and G. D. Litovchenko, *Zh. Prikl. Spektroskopii, Akad. Nauk Belorussk. SSR* **21**(3), 243 (1965); *Ibid.* **3**(6), 538 (1965).

11. H. B. Hass and E. F. Riley, *Chem. Rev.* **32**, 373 (1943).

12. N. Levy and J. D. Rose, *Q. Rev. (London)* **1**, 358 (1947).

13. P. A. S. Smith, *The Chemistry of the Open-Chain Organic Nitrogen Compounds*, Vol. II, W. A. Benjamin, Inc., New York, 1966, pp. 391–454.

14. H. H. Baer and L. Urbas, in H. Feuer, ed., *The Chemistry of the Nitro and Nitroso Groups*, Pt. 2, Wiley-Interscience, New York, 1970, pp. 75–200.

15. D. Seebach, E. W. Colvin, F. Lehr, and T. Weller, *Chimia* **33**, 1 (1979).

16. H. Fuer, ed., *Tetrahedron* **19**(Suppl. 1), (1963).

17. T. Urbanski, ed., *Nitro Compounds: Proceedings of an International Symposium, Warsaw, Poland, 1963*, Pergamon Press Ltd., Oxford, U.K., 1964.

18. N. Ono and A. Kaji, *Synth. Org. Chem. Japan* **38**, 115 (1980).

19. H. Feuer and A. T. Nielsen, eds., *Nitro Compounds, Recent Advances in Synthesis and Chemistry*, VCH Publishers, Inc., New York, 1990.

20. T. Urbanski, *Synthesis*, 613 (1974).

21. M. T. Shipchandler, *Synthesis*, 666 (1979).

22. N. Kornblum and P. Pink, in Ref. 16, pp. 17–22.

23. H. B. Hass, E. J. Berry, and M. L. Bender, *J. Am. Chem. Soc.* **71**, 2290 (1949).

24. H. Shechter and R. B. Kaplan, *J. Am. Chem. Soc.* **73**, 1883 (1951).

25. R. C. Kerber, G. W. Urrey, and N. Kornblum, *J. Am. Chem. Soc.* **87**, 4520 (1965).

26. G. A. Russell and W. C. Danen, *J. Am. Chem. Soc.* **88**, 5663 (1966).

27. G. B. Bachman and T. Hokama, *J. Am. Chem. Soc.* **81**, 4882 (1959).

28. E. H. White and W. J. Considine, *J. Am. Chem. Soc.* **80**, 626 (1958).

29. L. G. Donaruma and M. L. Huber, *J. Org. Chem.* **21**, 965 (1958).

30. H. L. Wehrmeister, *J. Org. Chem.* **25**, 2132 (1960).

31. T. Urbanski, Z. Novak, and E. Morag, *Bull. Acad. Polon. Sci., Ser. Sci. Chim.* **11**(2), 77 (1963).

32. U.S. Pat. 2,337,912 (Dec. 28, 1943), E. T. McBee and E. F. Riley (to Purdue Research Foundation).

33. U.S. Pat. 3,099,612 (July 30, 1963), L. A. Wilson (to Commercial Solvents Corp.).

34. U.S. Pat. 3,100,806 (Aug. 13, 1963), P. Bay (to Abbott Laboratories).

35. U.S. Pat. 3,054,829 (Sept. 18, 1962), G. B. Bachman and R. O. Downs (to Purdue Research Foundation).

36. K. Okon and G. Aluchna, *Bull. Acad. Polon. Sci., Ser. Sci. Chim. Geol. Geograph.* **7**, 83 (1959).

37. N. J. Leonard and A. B. Simon, *J. Org. Chem.* **17**, 1262 (1952).

38. M. Senkus, *J. Am. Chem. Soc.* **68**, 1611 (1946); **72**, 2967 (1950).

39. T. Urbanski and co-workers, in Ref. 17, pp. 195–218.

40. W. D. Emmons, W. N. Cannon, J. W. Dawson, and R. M. Ross, *J. Am. Chem. Soc.* **75**, 1993 (1953).
41. U.S. Pat. 2,477,870 (Aug. 2, 1949), M. H. Gold and L. J. Draker (to The Visking Corp.).
42. S. L. Ioffe, V. A. Tartakovskii, and S. S. Novikov, *Russ. Chem. Rev.* **35**, 19 (1966).
43. S. S. Novikov, I. S. Korsakova, and K. K. Babievskii, *Usp. Khim.* **26**, 1109 (1957).
44. E. E. Hamel in Ref. 16, pp. 85–95.
45. M. B. Frankel in Ref. 16, pp. 213–217.
46. S. Wakamatsu and K. Shimo, *J. Org. Chem.* **27**, 1609 (1962).
47. E. Profft, *Chem. Tech. (Berlin)* **8**, 705 (1956).
48. H. Shechter, D. L. Ley, and L. Zeldin, *J. Am. Chem. Soc.* **74**, 3664 (1952).
49. H. Feuer and R. Harmetz, *J. Org. Chem.* **26**, 1061 (1961).
50. V. V. Perekalin, in Ref. 17, pp. 135–157.
51. V. V. Perekalin, *Unsaturated Nitro Compounds*, Gosudarst. Nauch.-Tekh. Izdatel. Khim. Lit., Leningrad, USSR, 1961; in Israel Program for Scientific Translation, Jerusalem, Israel, 1963.
52. M. H. Gold and K. Klager, in Ref. 16, pp. 77–84.
53. G. V. Roitburd and co-workers, *Tetrahedron Lett.* **48**, 4935 (1972).
54. U.S. Pat. 3,055,936 (Sept. 25, 1962), M. Stiles and H. L. Finkbeiner (to Research Corp.).
55. H. L. Finkbeiner and M. Stiles, *J. Amer. Chem. Soc.* **85**, 616 (1963).
56. H. L. Finkbeiner and G. W. Wagner, *J. Org. Chem.* **28**, 215 (1963).
57. A. P. Hardt, F. G. Borgardt, W. L. Reed, and P. Noble, Jr., *Electrochem. Technol.* **1**, 375 (1963).
58. C. T. Bahner, *Ind. Eng. Chem.* **44**, 317 (1952).
59. H. Shechter and R. B. Kaplan, *J. Amer. Chem. Soc.* **75**, 3980 (1953).
60. U.S. Pat. 3,267,158 (Aug. 16, 1966), A. J. Havlik (to Aerojet-General Corp.).
61. T. Urbanski and W. Gurzynska, *Rocz. Chem.* **25**, 213 (1951).
62. V. F. Belyaev and R. I. Shamanovskaya, *Zh. Org. Khim.* **1**, 1388 (1965).
63. T. Mukaiyama and T. Hoshino, *J. Amer. Chem. Soc.* **82**, 5339 (1960).
64. A. V. Topchiev, *Nitration of Hydrocarbons and Other Organic Compounds*, Pergamon Press Ltd., Oxford, U.K., 1959.
65. N. Kornblum, in A. C. Cope, ed., *Organic Reactions*, Vol. 12, John Wiley & Sons, Inc., New York, 1962, pp. 101–156.
66. H. O. Larson, in Ref. 14, Part 1, 1969, pp. 301–348.
67. R. F. Purcell, in J. J. McKetta ed., *Encyclopedia of Chemical Processes and Design*, Vol. 31, Marcel Dekker, Inc., New York, 1990, pp. 267–281.
68. K. E. Gilbert and W. T. Borden, *J. Org. Chem.* **44**, 659 (1979).
69. G. B. Bachman and R. J. Maleski, *J. Org. Chem.* **37**, 2810 (1972).
70. U.S. Pat. 4,329,523 (May 11, 1982), R. James and R. Egly (to International Minerals & Chemical Corp.).
71. H. B. Hass and H. Shechter, *Ind. Eng. Chem.* **39**, 817 (1947).
72. F. A. Lowenheim and M. K. Moran, eds., *Faith, Keyes, and Clark's Industrial Chemicals*, 4th ed., John Wiley & Sons, Inc., New York, 1975.
73. R. N. Shreve, *The Chemical Process Industries*, McGraw-Hill Book Co., Inc., New York, 1956, p. 933.
74. J. C. Reidel, *Oil & Gas J.* **54**(36), 110 (1956).
75. P. Noble, Jr., F. G. Borgardt, and W. L. Reed, *Chem. Rev.* **64**, 7 (1964).
76. U.S. Pat. 2,489,320 (Nov. 29, 1949), E. M. Nygaard and W. I. Denton (to Socony-Vacuum Oil Co.).
77. W. I. Denton, R. B. Bishop, E. M. Nygaard, and T. T. Noland, *Ind. Eng. Chem.* **40**, 381 (1948).

78. R. B. Kaplan and H. Shechter, *J. Amer. Chem. Soc.* **83**, 3535 (1961).
79. U.S. Pat. 2,997,504 (Aug. 22, 1961), H. Shechter and R. B. Kaplan (to Ohio State Research Foundation).
80. U.S. Pat. 3,000,966 (Sept. 19, 1961), K. Klager (to Aerojet-General Corp.).
81. *Nitromethane: Storage and Handling Guidelines, Technical Data Sheet No. 2*, ANGUS Chemical Company, Buffalo Grove, Ill., Aug. 1993.
82. *Nitroparaffins and Their Hazards*, Research Report No. 12, Committee on Fire Prevention and Engineering Standards, National Board of Fire Underwriters, New York, 1959.
83. C. J. Wassink and J. T. Allen, in F. D. Snell and L. S. Ettre, eds., *Encyclopedia of Industrial Chemical Analysis*, Vol. 16, John Wiley & Sons, Inc., 1972, pp. 412–448.
84. L. R. Jones and J. A. Riddick, *Anal. Chem.* **28**, 1137 (1956).
85. C. Frejacques and M. Leclercq, *Mem. Poudres* **39**, 57 (1957).
86. J. R. Nielsen and D. C. Smith, *Ind. Eng. Chem., Anal. Ed.* **15**, 609 (1943).
87. R. N. Hazeldine, *J. Chem. Soc.*, 2525 (1953).
88. N. Kornblum, H. E. Ungnade, and R. A. Smiley, *J. Org. Chem.* **21**, 377 (1956).
89. Z. Buczkowski and T. Urbanski, *Spectrochim. Acta* **18**, 1187 (1962).
90. R. L. Foley, W. M. Lee, and B. Musulin, *Anal. Chem.* **36**, 1100 (1964).
91. M. J. Kamlet and D. J. Glover, *J. Org. Chem.* **27**, 537 (1962).
92. J. C. Neerman and O. S. Knight, *Chem. Eng.* **56**(11), 125 (1949).
93. R. M. Bethea and T. D. Wheelock, *Anal. Chem.* **31**, 1834 (1959).
94. W. Biernacki and T. Urbanski, *Bull. Acad. Polon. Sci., Ser. Sci. Chem.* **10**, 601 (1962).
95. R. M. Bethea and F. S. Adams, *Anal. Chem.* **33**, 832 (1961); *J. Chromatogr.* **8**, 532 (1962).
96. E. Camera, D. Pravisani, and W. Ohman, *Explosivstoffe* **13**(9), 237 (1965).
97. NIOSH, *Manual of Analytic Methods*, 3rd ed., 2nd Suppl. NIOSH, Cincinnati, Ohio, 1988.
98. J. Dequidt, P. Vasseur, and J. Potencier, *Bull. Soc. Pharm. Lille* 83, 131, 137 (1972); 29 (1973).
99. *1994–1995 Threshold Limit Values and Biological Exposure Indices*, American Conference of Governmental Industrial Hygienists, Cincinnati, Ohio, 1994.
100. J. F. Treon and F. R. Dutra, *AMA Arch. Ind. Hyg. Occup. Med.* **5**, 52 (1952).
101. T. R. Lewis, C. E. Ulrich, and W. M. Busey, *J. Environ. Pathol. Toxicol.* **2**, 233 (1979).
102. *IARC Monographs on the Evaluation of the Carcinogenic Risk of Chemicals to Humans*, Vol. 29, International Agency for Research on Cancer, Lyon, France, 1982, pp. 331–343.
103. *Fifth Annual Report on Carcinogens, Summary*, National Toxicology Program, Research Triangle Park, N.C., 1989, pp. 199–201.
104. T. B. Griffin and co-workers, *Ecotoxicol. Environ. Safety* **16**, 11 (1988).
105. T. B. Griffin and co-workers, *Ecotoxicol. Environ. Safety* **6**, 268 (1982).
106. E. S. Fiala and co-workers, *Carcinogenisis* **8**(12), 1947 (1987).
107. R. A. Davis, in G. D. Clayton and F. E. Clayton, eds., *Patty's Industrial Hygiene and Toxicology*, 4th ed., Vol. II, Pt. A, John Wiley & Sons, Inc., New York, 1993, pp. 599–662.
108. *Environmental Health Criteria 138: 2-Nitropropane*, World Health Organization, Geneva, Switzerland, 1992.
109. M. W. Curtis and C. H. Ward, *J. Hydrol.* **51**(1–4), 359, (1981).
110. L. T. Cupitt, *Project Summary: Fate of Toxic and Hazardous Materials in the Air Environment*, U.S. Environmental Protection Agency, Research Triangle Park, N.C., 1980.
111. W. D. Taylor and co-workers, *Int. J. Chem. Kinet.* **12**, 231 (1980).
112. *Health and Environmental Effects Profile for Nitromethane*, U.S. Environmental Protection Agency, Cincinnati, Ohio, 1985.

113. *Health and Environmental Effects Profile for 2-Nitropropane*, U.S. Environmental Protection Agency, Cincinnati, Ohio, 1985.

114. R. E. Kass, *Specialty Chem.* (Aug. 1983).

115. L. Schmerling, *Ind. Eng. Chem.* **40**, 2072 (1948).

116. "Montreal Protocol on Substances that Deplete the Ozone Layer," Sept. 16, 1987; *Code Fed. Reg.*, Title 40, Part 82, U.S. EPA, Washington, D. C., 1988.

117. G. N. Robinson, *Amer. Paint J.* **77**(28), 42–47 (Dec. 14, 1992).

ALLEN F. BOLLMEIER, JR.
ANGUS Chemical Company

N-NITROSAMINES

N-Nitrosodialkylamines (N-nitrosamines) were first characterized in the late nineteenth century (1,2). Although they have been used as synthetic intermediates and solvents, and possess interesting structural and spectroscopic properties, their potential uses have been eclipsed by their toxicity and especially their genotoxicity (3–7). Most of the N-nitrosamines tested are carcinogenic in animals. Nitrosamines have induced tumors in every animal species tested (8). Although there is no direct causal evidence, humans are probably susceptible to cancers induced by nitrosamines, and several specific nitrosamines are strongly suspected to be human carcinogens (9–12). The primary focus of research on these compounds in the 1990s is therefore on biochemistry, utility as experimental mutagens and carcinogens, and on potential human exposure to them (13–28).

Properties

Many of the chemical, physical, and biological properties of more than 20 selected N-nitrosamines have been summarized (29). N-Nitrosamines (**1**) encompass a wide range of structural types because the single feature common to them is the NNO functionality, and there are few restrictions on the groups that can be attached to the remaining two valences on the amine nitrogen, where R_1 or R_2 can be alkyl, aryl, or mixed. When R_1 or $R_2 = CH_2X$, X may be H, alkyl, aryl, halogen, or alkoxy. When either R_1 or $R_2 = $ aryl, various substituents may be attached to the rings. When R_1 or $R_2 = $ H, or when X = OH, the resulting primary N-nitrosamines or α-hydroxy-N-nitrosamines are generally unstable (30–34). Examples of other structural types, including derivatives of cyclic amines such as N-nitrosomorpholine [59-89-2] (**2**) or N-nitrosopyrrolidine [930-55-2] (**3**), have also been characterized (7,8,29,35–42).

(1)   (2)   (3)

N-Nitrosamines typically are light yellow volatile solids or oils. The electron delocalization in the NNO functionality sufficiently restricts rotation around the N–N bond that the *E* (4) and *Z* (5) isomers of unsymmetrically substituted examples can often be separated (43).

(4)   (5)

The spectroscopic properties of the *N*-nitrosamines, especially the nmr and mass spectra, vary widely depending on the substituents on the amine nitrogen (44–47). The nmr spectra are affected by the *E–Z* isomerism around the N–N partial double bond and by the axial–equatorial geometry resulting from conformational isomerism in the heterocycles (44,45). Some general spectral characteristics for typical dialkylnitrosamines and simple heterocyclic nitrosamines are given in Table 1.

Synthesis. The classic laboratory synthesis of *N*-nitrosamines is the reaction of a secondary amine with acidic nitrite [*14797-65-0*] at ca pH 3. The primary nitrosating intermediate is N_2O_3 arising from nitrous acid [*7782-77-6*] (48).

Table 1. Characteristic Spectral Properties of *N*-Nitrosamines

Property	Functionality involved	Value
uv–vis λ_{max}, nm	NNO	230–235, 330–375
ir absorption, cm^{-1}	N–N stretch	1040–1160
	N–O stretch	1430–1500
nmr absorption, ppm	α-CH (*E*)	5.8–6.9 CCl$_4$)
		6.1–6.32 (C$_6$H$_6$)
	α-CH (*Z*)	6.3–6.6 (CCl$_4$)
		6.6–6.8 (C$_6$H$_6$)
mass spectral fragmentation, m/z	entire molecule	M, M-17, M-30, M-31

Primary and tertiary amines also form *N*-nitrosamines (41,49,50). Although these nitrosations are generally slower and give lower yields than is the case with secondary amines, there are exceptions including some drugs such as aminopyrine [*58-15-1*] (49,51,52). There are a number of known catalysts for nitrosation reactions including halides, pseudohalides, sodium thiocyanate [*534-18-9*], or formaldehyde [*50-00-0*] (53,54). Nitrosations of higher amines such as dihexylamine [*143-16-8*] are enhanced by micelle formation (55). Secondary amines can be nitrosated at nonacidic pH by N_2O_3 or N_2O_4, or by other NO donors such as nitrosyl chloride [*2696-92-6*] (56,57). *N*-Nitrosamines can be formed under some conditions by transnitrosation which is the transfer of the NO group from a nitrosamine to an amine (58,59).

Efficient nitrosations of amines with inorganic nitrosyl compounds also have been reported (60).

$$R_1 R_2 NH + Fe(CN)_5 NO^{2-} \longrightarrow R_1 R_2 NNO$$

Inhibition of nitrosation is generally accomplished by substances that compete effectively for the active nitrosating intermediate. *N*-Nitrosamine formation *in vitro* can be inhibited by ascorbic acid [*50-81-7*] (vitamin C) and α-tocopherol [*59-02-9*] (vitamin E) (61,62), as well as by several other classes of compounds including pyrroles, phenols, and aziridines (63–65). Inhibition of intragastric nitrosation in humans by ascorbic acid and by foods such as fruit and vegetable juices or food extracts has been reported in several instances (26,66,67).

Reactions.   The chemistry of the *N*-nitrosamines is extensive and will be only summarized here (8,35,42). Most of the reactions of the nitrosamines, with respect to their biological or environmental behavior, involve one of two main reactive centers, either the nitroso group itself or the C–H bonds adjacent (α) to the amine nitrogen. The nitroso group can be removed readily by a reaction which is essentially the reverse of the nitrosation reaction, or by oxidation or reduction (68,69).

$$\begin{array}{c} R_1 \\ \diagdown \\ R_2 \diagup \end{array} N\!-\!NO \ + \ CF_3COOH \quad \xrightarrow{\text{oxidation}} \quad \begin{array}{c} R_1 \\ \diagdown \\ R_2 \diagup \end{array} N\!-\!NO_2$$

$$\begin{array}{c} R_1 \\ \diagdown \\ R_2 \diagup \end{array} N\!-\!NO \ + \ Zn/CH_3COOH \quad \xrightarrow{\text{reduction}} \quad \begin{array}{c} R_1 \\ \diagdown \\ R_2 \diagup \end{array} N\!-\!NH_2$$

The effects of uv radiation on *N*-nitroso compounds depend on the pH and the medium. Under neutral conditions and in the absence of radical scavengers, these compounds often appear chemically stable, although the *E–Z* equilibrium, with respect to rotation around the N–N bond, can be affected (70). This apparent stability is due to rapid recombination of aminyl radicals and nitric oxide [10102-43-9] formed during photolysis. In the presence of radical scavengers nitrosamines decay rapidly (71). At lower pH, a variety of photoproducts are formed, including compounds attributed to photoelimination, photoreduction, and photo-oxidation (69). Low concentrations of most nitrosamines, even at neutral pH, can be eliminated by prolonged irradiation at 366 nm. This technique is used in the identification of *N*-nitrosamines that are present in low concentrations in complex mixtures (72).

Reactions at the α-carbons have been of considerable interest because it is at these positions that enzymatic oxidation, which is believed to initiate the events leading to carcinogenic metabolites, generally occurs (5,7,8,73). The α-hydrogens exchange readily as shown in the following where D represents ^{2}H. This exchange apparently results from stabilization of an anionic intermediate by electron delocalization (74,75).

$$HC \diagdown N\!-\!NO \quad \underset{H_2O,\ H^+}{\overset{D_2O,\ D^+}{\rightleftharpoons}} \quad DC \diagdown N\!-\!NO$$

This property has been exploited in syntheses of *N*-nitrosamine derivatives by the reaction of electrophiles (E) with α-lithiated intermediates. These intermediates are prepared by hydrogen–lithium exchange using lithium diisopropylamide [4111-54-0] (LDA) (76,77).

$$HC \diagdown N\!-\!NO \quad \xrightarrow{\text{LDA}} \quad LiC \diagdown N\!-\!NO \quad \xrightarrow{\text{E}} \quad EC \diagdown N\!-\!NO$$

Handling and Disposal

N-Nitrosamines are potentially hazardous and should be handled in designated hoods and with protective clothing. Nitrosamines can be destroyed by treatment with aluminum–nickel alloy under basic conditions (78,79).

Analytical and Test Methods

The potential exposure of humans to *N*-nitrosamines through diet, occupational or other sources, or their possible formation *in vivo*, has created a continuing need for the detection and confirmation of low levels of *N*-nitrosamines in complex mixtures of organic chemicals or biological matrices. Many of the *N*-nitrosamines that are recognized as environmentally significant are sufficiently volatile and stable for analysis by gas chromatography (gc). The less volatile *N*-nitrosamines can be analyzed by high performance liquid chromatography (hplc) (35). A variety of detection techniques such as polarography, spectrophotometric cleavage of the NNO bond followed by detection of the resulting nitrite, alkali flame detection, electrolyte conductivity (Coulson detector), and mass spectrometry, have been used, but most of these techniques are not sufficiently sensitive, selective, or accessible for routine analysis of low levels of nitrosamines present in complex mixtures (80). The most significant contribution to *N*-nitrosamine analysis has been the thermal energy analyzer. This detector, which responds to nitric oxide released from the *N*-nitroso compounds by pyrolysis, is sensitive and highly selective for the NNO functionality and can be used for both gc and hplc, although it is somewhat limited as an hplc detector (81,82). Some classes of compounds, eg, alkyl nitrites and C-nitroso compounds, give false-positive responses, but various screening methods, such as the destruction of nitrosamines by uv radiation, can usually distinguish these types of molecules from *N*-nitrosamines (70,83,84). Photohydrolytic systems based on release of nitric oxide by uv irradiation are used as alternative hplc detectors (82,85). A method which is potentially useful for nonvolatile *N*-nitroso compounds and based on acidic denitrosation has been described (86).

Confirmation of the identities of nitrosamines generally is accomplished by gas chromatography–mass spectrometry (gc/ms) (46,87). High resolution gc/ms, as well as gc/ms in various single-ion modes, can be used as specific detectors, especially when screening for particular nitrosamines (87) (see ANALYTICAL METHODS; TRACE AND RESIDUE ANALYSIS).

Health and Safety Factors

Toxicity. Many *N*-nitrosamines are toxic to animals and cells in culture (4,6–8,88). *N*-Nitrosodimethylamine [62-75-9] (NDMA) is known to be acutely toxic to the liver in humans, and exposure can result in death (89). Liver damage, diffuse bleeding, edema, and inflammation are toxic effects observed in humans as a result of acute and subacute exposure to NDMA. These effects closely resemble those observed in animals dosed with NDMA (89,90).

Carcinogenicity. Some of the toxicological properties of a selected group of nitrosamines are listed in Table 2. The number of nitrosamines that have

Table 2. Toxicological Properties of Some Representative *N*-Nitrosamines in the BD Rat[a]

Compound	CAS Registry Number	LD$_{50}$	log$(1/D_{50})^b$	Principal target organ
N-nitrosodimethylamine (NDMA)	[62-75-9]	40	2.3	liver
N-nitrosodiethylamine (NDEA)	[55-18-5]	280	3.2	liver, esophagus
N-nitrosodiethanolamine (NDELA)	[1116-54-7]	7500	0.005	liver
N-nitrosodipropylamine	[621-64-7]	480	0.05	liver, esophagus
N-nitrosodiisopropylamine	[601-77-4]	850	2.1	liver
N-nitrosopyrrolidine (NPYR)	[930-55-2]	900	1.0	liver
N-nitrosomorpholine (NMOR)	[59-89-2]	320	1.9	liver
N-nitrosodicyclohexylamine	[947-92-2]		c	
N-nitrosoproline (NPRO)	[7519-36-0]		c	
N-nitrosomethyl(benzyl)amine	[937-40-6]	18	3.1	esophagus
N-nitrosopiperidine	[100-75-4]	200	1.9	liver, esophagus
N-nitrosonornicotine	[16543-55-8]		d	

[a]BD rat represents a particular strain used to test carcinogenicity of some *N*-nitroso compounds. Refs. 7 and 62.
[b]D$_{50}$: dose causing tumors in 50% of the test animals; increasing values for log$(1/D_{50})$ represent higher carcinogenicity (51).
[c]Not carcinogenic to the BD rat.
[d]Not investigated in the BD rat. Suspected human carcinogen (12).

been tested for carcinogenicity exceeds 200. If related *N*-nitroso compounds such as nitrosamides or nitrosoureas are included, the number of tested compounds is well over 300 (5,7,8,29,35,38,91). Most are carcinogenic, although the potency varies dramatically within the series (7,75). The mean dose for the formation of tumors by *N*-nitrosodiethylamine (NDEA), for example, is only ca 0.0006 mol/kg body weight, whereas *N*-nitrosoproline is considered noncarcinogenic (7,92). Carcinogenicity has been observed both with single, relatively large doses and with long-term chronic exposure to lower doses (7,93). The *N*-nitrosamines generally are organ selective (7,8,94). *N*-Nitrosodimethylamine, for example, is primarily a liver carcinogen, *N*-nitrosomethylbenzylamine is primarily an esophageal carcinogen, and *N*-nitrosobutyl(4-hydroxybutyl)amine is primarily a bladder carcinogen. Other target organs include the nose, bladder, pancreas, lungs, and kidneys (7,8,95). Most nitrosamines are not direct-acting carcinogens, but require metabolic activation in order to exert their carcinogenic effect. The potency and the organ selectivity of the *N*-nitrosamines are therefore determined by complex interactions involving molecular structures and the spectrum of metabolizing enzymes in the test animal (96,97). There are consequently species- and sex-related differences in both potency and organ selectivity (7,8,98,99). *N*-Nitroso-*N*-methyl(2-oxopropyl)amine, for example, is an esophageal carcinogen in the rat and a pancreatic carcinogen in the hamster (100,101). The principal enzymes involved in nitrosamine metabolism are the cytochrome P450 enzymes, with P450 2E1 being the most widely studied (102). Other known or suspected nitrosamine-metabolizing enzyme systems include sulfotransferases, alcohol dehydrogenases, and peroxidases (61,102–104). The mechanisms involved in nitrosamine carcinogenesis are not completely understood in detail, especially with respect to organ selectivity or for nitrosamines that have no α-hydrogens (*N*-nitrosodiphenylamine [*86-30-6*]) or are oxidized at other positions (*N*-nitrosodiethanolamine or *N*-nitroso-*N*-methyl(2-oxopropyl)amine)). The following sequence, however, is generally accepted for the metabolic activation of most simple dialkyl *N*-nitrosamines to electrophilic intermediates (5,7,8,31,32,87,105–108). Other metabolic pathways include denitrosation, oxidation at sites other than the α-hydrogens, and chain-shortening (99,109–112).

$$
\underset{(6)}{\overset{\displaystyle \mathrm{RCH_2}}{\underset{\displaystyle \mathrm{R'CH_2}}{>}}\!\!\mathrm{N-NO}} \xrightarrow{\text{enzyme}} \underset{(6)}{\overset{\displaystyle \mathrm{RCHOH}}{\underset{\displaystyle \mathrm{R'CH_2}}{>}}\!\!\mathrm{N-NO}} \longrightarrow \underset{(7)}{\mathrm{RCHO} + [\mathrm{R'\,CH_2\,NHN{=}O}]}
$$

$$
\underset{(7)}{[\mathrm{R'CH_2\,NHN{=}O}]} \rightleftharpoons \underset{(8)}{[\mathrm{R'CH_2N{=}NOH}]} \xrightarrow{\mathrm{H^+}} \underset{(9)}{[\mathrm{R'CH_2N{=}NOH_2^+}]} \xrightarrow{?} \underset{(10)}{[\mathrm{R'CH_2N_2^+}]}
$$

There is substantial evidence for the initial enzymatic α-oxidation to (**6**) and for the alkylation of nucleic acids by the resulting electrophiles (5,7,31,32,108,113–116). The intervening intermediates (**7–10**) are hypothesized largely by analogy with the known behavior of primary nitrosamines (30,116) and from reactive synthetic α-oxidized nitrosamines (33,34). Oxidation of nitrosamines in microsomes may lead to intermediates (**9**) and (**10**) *in vitro*, and these may also be responsible for alkylation of nucleic acids *in vivo* (106,107). Microsomes are cell particles of the smallest size with a piece of endoplasmic reticulum attached, and contain the predominant biotransformation systems, including the cytochrome P450 enzymes. Tumors are thought to arise from mispairing during subsequent replications of the alkylated DNA. Most alkylation appears to occur at the N-7 position of DNA guanine (**11**), but alkylation at the O-6 position of guanine is more closely correlated with tumor formation than is alkylation at N-7 (114).

(**11**)

Mutagenicity. The *N*-nitrosamines, in general, induce mutations in standard bacterial-tester strains (117). As with carcinogenicity, enzymatic activation, typically with liver microsomal preparations, is required. Certain substituted *N*-nitrosamine derivatives (**12**) induce mutations without microsomal activation (31,33,34). Because the α-acetoxy derivatives can hydrolyze to the corresponding α-hydroxy compounds, this is consistent with the hypothesis that enzymatic oxidation leads to the formation of such unstable α-hydroxy intermediates (**13**) (118). However, for simple *N*-nitrosamines, no systematic relationship has been found between carcinogenicity and mutagenicity (117,119–123).

Human Exposure. *N*-Nitrosamines have been reported in pesticide preparations (124), corrosion inhibitors (125), lubricating fluids and cosmetics (qv), ie, *N*-nitrosodiethanolamine (126,127), sunscreens (128), rubber products including baby-bottle nipples and pacifiers (115,129), foods including cheese, processed meats, beer (qv), cooked bacon, and powdered milk (42,130–133), and tobacco products (52,134–136) (see INSECT CONTROL TECHNOLOGY; CORROSION AND CORROSION CONTROL; LUBRICATION AND LUBRICANTS; FOOD TOXICANTS, NATURALLY OCCURRING). In addition to exposure by preformed *N*-nitroso compounds there is exposure by nitrosation of amines in the mouth, stomach, or other sites in the body. Formation of nitrosamines inside the body has been demonstrated unambiguously in humans and in experimental animals. Amine- and nitrite-fed animals develop tumors that are identical with those expected from the corresponding *N*-nitrosamines (137). The formation *in vivo* of *N*-nitrosamines from amines and nitrite or nitrite precursors can be directly observed for nonmetabolized nitrosamines such as *N*-nitrosoproline (138) and when nitrosamine metabolism is blocked (53,139,140). Nitrosation in humans can be monitored by urinary levels of *N*-nitrosoproline or *N*-nitrosothiazolidine carboxylic acid (nitrosothioproline) (92,141–144). In the case of *N*-nitrosoproline, the precursors (proline and nitrate) can be administered systematically (92). These results have stimulated a large number of studies concerning possible relationships between nitrosation in the body and elevated cancer risk (138,145–149), and whether this nitrosation can be blocked or inhibited, especially by dietary components (26,66,67,138,150).

Nitrite is also formed in the body. Saliva, for example, contains efficient nitrate-reducing bacteria and consequently provides a constant low (ca 7 pg/mL) level of nitrite, that increases rapidly following ingestion of nitrate (151,152). Nitrate synthesis involves production of nitric oxide by a large number of cell types including endothelial cells, which line cavities and vessels, macrophages, which engulf and digest cells and microorganisms, neurons, and liver cells (153–156). The nitric oxide can react with oxygen to form N_2O_3, which is a nitrosating agent at neutral or near-neutral pH (48,57,157,158). There is thus the possibility of nitrosamine formation at sites in the body other than the stomach.

　There is insufficient evidence to unequivocally link nitrosamine exposure to elevated risk for human cancer (159). There are, however, a number of specific cases, especially with respect to the tobacco-related nitrosamines, in which exposure to *N*-nitroso compounds is of concern. The strongest evidence in this context is probably that relating to oral cancer rates among habitual users of smokeless tobacco (snuff). Oral cancer rates among this group are significantly elevated over those of nonusers, and *N*-nitrosonornicotine, and 4-(methylnitrosamino)-1-(3-pyridinyl)-1-butanone [*64091-91-4*], both of which are potent animal carcinogens, are present along with less potent *N*-nitroso compounds in smokeless tobacco (52,160–163). Urinary metabolites of these compounds can be detected in users. Both DNA and protein adducts, which are evidence for the presence of the active electrophilic metabolites of the nitrosamines, have been detected in human samples (12,164–166). This evidence, although remaining circumstantial, is nonetheless becoming increasingly persuasive (149,150,164,166).

BIBLIOGRAPHY

"*N*-Nitrosamines" in *ECT* 3rd ed., Vol. 15, pp. 988–996, by J. S. Wishnok, Massachusetts Institute of Technology.

1. A. Geuther, *Lieb. Ann.* **128**, 151 (1863).
2. A. Geuther and E. Schiele, *J. Prakt. Chem.* **4**, 485 (1871).
3. H. A. Freund, *Ann. Intern. Med.* **10**, 1144 (1937).
4. J. M. Barnes and P. N. Magee, *Brit. J. Indust. Med.* **11**, 167 (1954).
5. P. N. Magee and J. M. Barnes, *Adv. Cancer Res.* **10**, 163 (1967).
6. P. N. Magee and J. M. Barnes, *Brit. J. Cancer* **10**, 114 (1956).
7. H. Druckrey, R. Preussmann, S. Ivankovic, and D. Schmähl, *Z. Krebsforsch.* **69**, 103 (1967).
8. R. Preussmann and B. W. Stewart, in C. E. Searle, ed., *Chemical Carcinogens* (*ACS Monograph 182*), 2nd ed., American Chemical Society, Washington, D.C., 1984, pp. 643–828.
9. H. Bartsch, H. Ohshima, and B. Pignatelli, *Mut. Res.* **202**, 307 (1988).
10. H. Bartsch and R. Montesano, *Carcinogenesis* **5**, 1381 (1984).
11. S. Preston-Martin, *CRC Crit. Rev. Toxicol.* **21**, 295 (1991).
12. P. G. Foiles and co-workers, *Chem. Res. Toxicol.* **4**, 364 (1991).
13. I. K. O'Neill, J. Chen, and H. Bartsch, eds., *Relevance to Human Cancer of N-Nitroso Compounds, Tobacco Smoke and Mycotoxins,* IARC Scientific Publication No. 105, International Agency for Research on Cancer, Lyon, France, 1991.
14. H. Bartsch, I. K. O'Neill, and R. Schulte-Hermann, eds., *Relevance of N-Nitroso Compounds to Human Cancer: Exposures and Mechanisms*, IARC Scientific Publications No. 84, International Agency for Research on Cancer, Lyon, France, 1987.
15. I. K. O'Neill and co-workers, eds., *N-Nitroso Compounds: Occurrence, Biological Effects and Relevance to Human Cancer*, International Agency for Research on Cancer, Lyon, France, 1984.
16. J.-P. Anselme, ed., *N-Nitrosamines*, American Chemical Society, Washington, D.C., 1979.
17. R. A. Scanlan and S. R. Tannenbaum, eds., *N-Nitroso Compounds*, American Chemical Society, Washington, D.C., 1981.

18. R. N. Loeppky and C. J. Michejda, eds., *Nitrosamines and Related N-Nitroso Compounds*, American Chemical Society, Washington, D.C., 1994.

19. C. E. Searle, ed., *Chemical Carcinogens*, 2nd ed., American Chemical Society, Washington, D.C., 1984.

20. W. Lijinsky, *Chemistry and Biology of N-Nitroso Compounds*, Cambridge University Press, Cambridge, U.K., 1992.

21. W. Lijinsky, *Carcinogenesis* **14**, 2373 (1993).

22. Y. Liu and co-workers, *Carcinogenesis* **14**, 2383 (1993).

23. J. Kanno, R. R. Maronpot, M. Takahashi, T. Kasuga, and Y. Hayashi, *Carcinogenesis* **14**, 2389 (1993).

24. W. Thamavit and co-workers, *Carcinogenesis* **14**, 2415 (1993).

25. Z. Guo, T. J. Smith, H. Ishizaki, and C. S. Yang, *Carcinogenesis* **12**, 2277 (1991).

26. M. A. Helser, J. H. Hotchkiss, and D. A. Roe, *Carcinogenesis* **13**, 2277 (1992).

27. K. D. Brunnemann and D. Hoffmann, *Carcinogenesis* **13**, 2407 (1992).

28. S. Y. Brendler, A. Tompa, K. F. Hutter, R. Preussmann, and B. L. Pool-Sobel, *Carcinogenesis* **13**, 2435 (1992).

29. *IARC Working Group on the Evaluation of Carcinogenic Risk of Chemicals to Humans, Some N-Nitroso Compounds*, IARC Monograph No. 17, International Agency for Research on Cancer, Lyon, France, 1978.

30. J. March, *Advanced Organic Chemistry: Reactions, Mechanisms, and Structure*, 3rd ed., Wiley-Interscience, New York, 1985, 313 pp.

31. J. E. Baldwin and co-workers, *Tetrahed. Lett.* **5**, 333 (1976).

32. B. Gold and W. B. Linder, *Am. Chem. Soc.* **101**, 6772 (1979).

33. A. Maekawa and co-workers, in H. Bartsch and I. K. O'Neill, M. Castegnaro, and M. Okada, eds., *N-Nitroso Compounds: Occurrence and Biological Effects*, IARC Scientific Publication No. 41, International Agency for Research on Cancer, Lyon, France, 1982, pp. 379–396.

34. M. Mochizuki, M. Osabe, T. Anjo, E. Suzuki, and M. Okada, *J. Cancer Res. Clin. Oncol.* **108**, 290 (1985).

35. E. A. Walker and co-workers, eds., *Environmental Aspects of N-Nitroso Compounds*, IARC Scientific Publication No. 19, International Agency for Research on Cancer, Lyon, France, 1978.

36. J.-P. Anselme, ed., in Ref. 16.

37. H. H. Hiatt, J. D. Watson, and J. A. Winstein, eds., *Origins of Human Cancer, Cold Spring Harbon Conference on Cell Proliferation*, Cold Spring Harbor Laboratory, Cold Spring Harbor, N.Y., 1977.

38. D. Schmähl, *Oncology* **37**, 193 (1980).

39. M. C. Archer, S. R. Tannenbaum, and J. S. Wishnok, in E. A. Walker, P. Bogovski, and L. Griciute, eds., *Environmental N-Nitroso Compounds: Analysis and Formation*, IARC Scientific Publication No. 14, Lyon, France, 1976, pp. 141–145.

40. R. A. Scanlan, *CRC Crit. Rev. Food Technol.* **5**, 357 (1975).

41. A. L. Fridman, F. M. Mukhametshin, and S. S. Novikov, *Russ. Chem. Rev.* **40**, 34 (1971).

42. M. L. Douglas, B. L. Kabacoff, G. A. Anderson, and M. C. Cheng, *J. Soc. Cosmetic Chem.* **29**, 581 (1978).

43. W. T. Iwaoka, T. Hansen, S.-T. Hsieh, and M. C. Archer, *J. Chromatog.* **103**, 349 (1975).

44. Y. L. Chow and C. J. Colon, *Can. J. Chem.* **46**, 2827 (1968).

45. R. R. Fraser and L. K. Eng, *J. Am. Chem. Soc.* **98**, 5895 (1976).

46. W. T. Rainey, W. H. Christie, and W. Lijinsky, *Biomed. Mass Spectrom.* **5**, 395 (1978).

47. J. W. Pensabene, W. Fiddler, C. J. Dooley, R. C. Doerr, and A. E. Wasserman, *J. Ag. Food Chem.* **204**, 274 (1972).

48. S. S. Mirvish, *Toxicol. Appl. Pharmacol.* **31**, 325 (1975).
49. W. Lijinsky, L. Keefer, E. Conrad, and R. Van de Bogart, *J. Natl. Cancer Inst.* **49**, 1239 (1972).
50. B.-G. Osterdahl and S. Slorach, *Food Add. Contam.* **5**, 581 (1988).
51. B. Spiegelhalder and R. Preussmann, *Carcinogenesis* **6**, 545 (1985).
52. G. Eisenbrand and D. Schmähl, in D. D. Breimer and P. Speiser, eds., *Topics in Pharmaceutical Sciences*, Elsevier, Amsterdam, the Netherlands, 1981, pp. 291–305.
53. T. Y. Fan and S. R. Tannenbaum, *J. Ag. Food Chem.* **21**(2), 237 (1973).
54. L. K. Keefer and P. P. Roller, *Science* **181**, 1245 (1973).
55. J. D. Okun and M. C. Archer, *J. Natl. Cancer Inst.* **58**, 409 (1977).
56. B. C. Challis and co-workers, in E. A. Walker, M. Castegnaro, L. Griciute, and R. E. Lyle, eds., *Environmental Aspects of N-Nitroso Compounds*, IARC Scientific Publication No. 19, International Agency for Research on Cancer, Lyon, France, 1978, p. 127.
57. B. C. Challis and S. A. Kyrtopoulos, *J. Chem. Soc. Perkin. I*, 299 (1979).
58. S. S. Singer, *J. Organic Chem.* **43**, 4612 (1978).
59. R. N. Loeppky, W. Tomasik, and B. E. Kerrick, *Carcinogenesis* **8**, 941 (1987).
60. H. Maltz, M. A. Grant, and M. C. Navaroli, *J. Organ. Chem.* **36**, 363 (1971).
61. S. S. Mirvish, L. Wallcave, M. Eagan, and P. Shubik, *Science* **177**, 65 (1972).
62. W. J. Mergens and co-workers, in Ref. 56, p. 199.
63. Y. T. Bao and R. N. Loeppky, *Chem. Res. Toxicol.* **4**, 382 (1991).
64. A. L. Wilcox, Y. T. Bao, and R. N. Loeppky, *Chem. Res. Toxicol.* **4**, 373 (1991).
65. R. N. Loeppky and Y. T. Bao, in I. K. O'Neill and H. Bartsch, eds., *Nitroso Compounds: Biological Mechanisms, Exposures and Cancer Etiology*, International Agency for Research on Cancer, Lyon, France, 1992, pp. P16a–b.
66. S. S. Mirvish, *Cancer Res. (7 Suppl).* **54**, 1948S (1994).
67. M. A. Helser and J. H. Hotchkiss, *J. Ag. Food Chem.* **42**, 129 (1994).
68. G. Eisenbrand and R. Preussmann, *Arzneim. Forsch.* **207**, 1513 (1970).
69. Y. L. Chow, *Accts. Chem. Res.* **6**, 354 (1973).
70. C. J. Michejda, N. E. Davidson, and L. K. Keefer, *Chem. Comm.*, 633 (1976).
71. C. J. Michejda and T. Rydstrom, in Ref. 15, p. 365.
72. W. Fiddler, R. C. Doerr, and E. G. Piotrowski, in Ref. 56, p. 33.
73. J. S. Wishnok, M. C. Archer, A. S. Edelman, and W. M. Rand, *Chem.-Biol. Interac.* **20**, 43 (1978).
74. L. K. Keefer and C. H. Fodor, *J. Am. Chem. Soc.* **92**, 5747 (1970).
75. D. Seebach and D. Enders, *Angew. Chem. Int. Ed. Eng.* **14**, 15 (1975).
76. B. Renger, H.-O. Kalinowski, and D. Seebach, *Chem. Berich.* **110**, 1866 (1977).
77. D. Seebach, D. Enders, and B. Renger, *Chem. Berich.* **110**, 1852 (1977).
78. G. Lunn, E. B. Sansone, A. W. Andrews, and L. K. Keefer, *Cancer Res.* **48**, 522 (1988).
79. G. Lunn, E. B. Sansone, and L. K. Keefer, *Carcinogenesis* **4**, 315 (1983).
80. A. E. Wassermann, in P. Bogovski, R. Preussmann, and E. A. Walker, eds., *N-Nitroso Compounds, Analysis and Formation*, IARC Scientific Publication No. 3, International Agency for Research on Cancer, Lyon, France, 1972.
81. D. H. Fine, F. Rufeh, and B. Gunther, *Analyt. Lett.* **6**, 731 (1973).
82. D. E. G. Shuker and S. R. Tannenbaum, *Analyt. Chem.* **55**, 2152 (1983).
83. T. Y. Fan and co-workers, in Ref. 56, p. 3.
84. T. J. Hansen, M. C. Archer, and S. R. Tannenbaum, *Analyt. Chem.* **51**, 1526 (1979).
85. J. J. Conboy and J. H. Hotchkiss, *Analyst.* **114**, 155 (1989).
86. B. Pignatelli and co-workers, in R. N. Loeppky and C. J. Michejda, eds., *Nitrosamines and Related N-Nitroso Compounds*, American Chemical Society, Washington, D.C., 1994, pp. 102–118.
87. T. Gough, *Analyst.* **103**, 785 (1978).
88. L. Catz-Biro, W. Chin, M. C. Archer, M. S. Pollanen, and M. A. Hayes, *Toxicol. Appl. Pharmacol.* **102**, 191 (1990).

89. R. D. Kimbrough, in P. N. Magee, ed., *Banbury Report 12: Nitrosamines and Human Cancer*, Cold Spring Harbon Laboratory, Cold Spring Harbor, 1982, pp. 25–34.
90. W. Chin, V. M. Lee, and M. C. Archer, *Chem. Res. Toxicol.* **6**, 372 (1993).
91. S. S. Mirvish, *J. Tox. Environ. Health.* **2**, 1267 (1977).
92. H. Ohshima and H. Bartsch, *Cancer Res.* **41**, 3658 (1981).
93. P. N. Magee and J. M. Barnes, *J. Pathol. Bacteriol.* **84**, 19 (1962).
94. R. Preussmann and M. Wiessler, *Trends Pharm. Sci.* **8**, 185 (1987).
95. H. Druckrey, R. Preussmann, and S. Ivankovic, *Ann. NY Acad. Sci.* **163**, 676 (1969).
96. J. S. Wishnok, in J.-P. Anselme, ed., *N-Nitrosamines*, ACS Symposium Series 101, American Chemical Society, Washington, D.C., 1979.
97. A. S. Edelman, P. L. Kraft, W. M. Rand, and J. S. Wishnok, *Chem.-Biol. Interact.* **31**, 81 (1980).
98. W. Lijinsky, in Ref. 96, p. 165.
99. W. Lijinsky, in Ref. 86, pp. 250–266.
100. W. Lijinsky, M. D. Reuber, J. E. Saavedra, and G. M. Singer, *J. Natl. Cancer Inst.* **70**, 959 (1983).
101. P. Pour and co-workers, *Cancer Res.* **40**, 3585 (1980).
102. C. S. Yang and co-workers, in Ref. 86, pp. 168–178.
103. G. Eisenbrand and C. Janzowski, in Ref. 86, pp. 179–194.
104. M. Stiborova, E. Frei, H. H. Schmeiser, M. Wiessler, and P. Anzenbacher, *Cancer Lett.* **63**, 53 (1992).
105. P. D. Lawley and co-workers, eds., *Topics in Chemical Carcinogenesis*, University of Tokyo Press, Tokyo, Japan, 1972, 272 pp.
106. K. K. Park, J. S. Wishnok, and M. C. Archer, *Chem.-Biol. Interact.* **18**, 349 (1977).
107. K. K. Park, M. C. Archer, and J. S. Wishnok, *Chem.-Biol. Interact.* **29**, 139 (1980).
108. D. F. Heath, *Biochem. J.* **85**, 72 (1962).
109. B. Gold, J. Farber, and E. Rogan, *Chem.-Biol. Interact.* **61**, 215 (1987).
110. E. Suzuki and M. Okada, *Gann.* **72**, 547 (1981).
111. M. Bonfanti, C. Magagnotti, M. Bonati, R. Fanelli, and L. Airoldi, *Cancer Res.* **48**, 3666 (1988).
112. M. Bonfanti, C. Magagnotti, R. Fanelli, and L. Airoldi, *Chem.-Biol. Interact.* **59**, 203 (1986).
113. O. G. Fahmy, M. J. Fahmy, and M. Weissler, *Biochem. Pharmacol.* **24**, 2009 (1975).
114. A. E. Pegg, *Adv. Cancer Res.* **25**, 195 (1977).
115. D. C. Havery and T. Fazio, *J. Off. Analyt. Chem.* **66**, 1500 (1983).
116. C. J. Collins, *Accts. Chem. Res.* **4**, 315 (1971).
117. J. McCann, E. Choi, E. Yamasaki, and B. N. Ames, *Proc. Natl. Acad. Sci. USA* **72**, 5135 (1975).
118. J. B. Guttenplan, F. Hutterer, and A. J. Garro, *Mut. Res.* **35**, 415 (1976).
119. L. B. Kier and L. H. Hall, *J. Pharm. Sci.* **67**, 725 (1978).
120. T. K. Rao, J. A. Young, and W. Lijinsky, *Mutat. Res.* **66**, 1 (1979).
121. W. J. Dunn, III, and S. Wold, *Biorgan. Chem.* **10**, 29 (1981).
122. J. B. Guttenplan, in H. Bartsch, I. K. O'Neill, and R. Schulte-Hermann, eds., *The Relevance of N-Nitroso Compounds to Human Cancers: Exposures and Mechanisms*, IARC Scientific Publication No. 84, International Agency for Research on Cancer, Lyon, France, 1987, pp. 129–131.
123. J. B. Guttenplan, *Carcinogenesis* **14**, 1013 (1993).
124. S. Z. Cohen and co-workers, in *IARC Working Group on the Evaluation of Carcinogenic Risk of Chemicals to Humans*, IARC Monograph No. 17, International Agency for Research on Cancer, Lyon, France, 1978, p. 333.
125. M. C. Archer and J. S. Wishnok, *J. Environ. Sci. Health.* **AII**, 587 (1976).

126. T. Y. Fan and co-workers, *Science* **196**, 70 (1977).
127. T. Y. Fan and co-workers, *Fd. Cosmet. Toxicol.* **15**, 423 (1977).
128. D. C. Havery and H. J. Chou, in Ref. 86, pp. 20–33.
129. N. P. Sen, S. W. Seaman, and S. C. Kushwaha, *J. Chromatogr.* **463**, 419 (1989).
130. A. R. Tricker and S. J. Kubacki, *Food Add. Contam.* **9**, 39 (1992).
131. B. J. Canas, D. C. Havery, F. L. Joe, and T. Fazio, *J. Assoc. Analyt. Chem.* **69**, 1020 (1986).
132. D. Forman, *Cancer Surv.* **6**, 719 (1987).
133. R. A. Scanlan and J. F. Barbour, in I. K. O'Neill, J. Chen, and H. Bartsch, eds., *Relevance to Human Cancer of N-Nitroso Compounds, Tobacco Smoke and Mycotoxins*, IARC Scientific Publication No. 105, International Agency for Research on Cancer, Lyon, France, 1991, pp. 242–243.
134. K. D. Brunnemann, L. Yu, and D. Hoffmann, *Cancer Res.* **37**, 3218 (1977).
135. D. Hoffmann and J. D. Adams, *Cancer Res.* **41**, 4305 (1981).
136. K. D. Brunnemann, J. C. Scott, and D. Hoffmann, *Carcinogenesis* **3**, 6936 (1982).
137. J. Sander and G. Buerkle, *Z. Krebsforsch.* **73**, 54 (1969).
138. S.-H. Lu and co-workers, *Cancer Res.* **46**, 1485 (1986).
139. J. S. Wishnok and co-workers, in Ref. 122, pp. 135–137.
140. R. H. Liu, B. Baldwin, B. C. Tennant, and J. H. Hotchkiss, *Cancer Res.* **51**, 3925 (1991).
141. H. Ohshima, M. Friesen, I. K. O'Neill, and H. Bartsch, *Cancer Lett.* **20**, 183 (1983).
142. H. Ohshima, I. K. O'Neill, M. Friesen, J.-C. Bereziat, and H. Bartsch, *J. Cancer Res. Clin. Oncol.* **108**, 121 (1984).
143. T. Tahira, M. Tsuda, K. Wakabayashi, M. Nagao, and T. Sugimura, *Gann.* **75**, 889 (1984).
144. M. Tsuda, N. Frank, S. Shigeaki, and T. Sugimura, *Cancer Res.* **48**, 4049 (1988).
145. W. G. Stillwell and co-workers, *Cancer Res.* **51**, 190 (1991).
146. Y. Wu and co-workers, *Int. J. Cancer* **54**, 713 (1993).
147. T. Knight and co-workers, *Int. J. Cancer.* **50**, 736 (1992).
148. P. Srivatanakul and co-workers, *Int. J. Cancer.* **48**, 821 (1991).
149. T. M. Knight and co-workers, *Europ. J. Cancer.* **27**, 456 (1991).
150. C. D. Leaf, A. J. Vecchio, and J. H. Hotchkiss, *Carcinogenesis* **8**, 791 (1987).
151. S. R. Tannenbaum, A. J. Sinskey, M. Weisman, and W. Bishop, *J. Natl. Cancer Inst.* **53**(1), 79 (1974).
152. S. R. Tannenbaum, M. Weisman, and D. Fett, *Fd. Cosmet. Toxicol.* **14**, 549 (1976).
153. D. J. Stuehr and M. A. Marletta, *Cancer Res.* **47**, 5590 (1987).
154. M. A. Marletta, P. S. Yoon, R. Iyengar, C. D. Leaf, and J. S. Wishnok, *Biochemistry* **27**, 8706 (1988).
155. R. Iyengar, D. J. Stuehr, and M. A. Marletta, *Proc. Natl. Acad. Sci. USA.* **84**, 6369 (1987).
156. M. A. Marletta, *Chem. Res. Toxicol.* **1**, 249 (1988).
157. B. C. Challis and co-workers, in Ref. 33, pp. 11–20.
158. B. C. Challis and S. A. Kyrtopoulos, *J. Chem. Soc., Perkin Trans. 2* **12**, 1296 (1978).
159. H. Bartsch, in Ref. 13, pp. 1–10.
160. B. Prokopczyk, M. Wu, J. E. Cox, and D. Hoffmann, *Carcinogenesis* **13**, 863 (1992).
161. K. D. Brunnemann, L. Genoble, and D. Hoffmann, *J. Agric. Fd. Chem.* **33**, 1178 (1985).
162. S. S. Hecht and D. Hoffmann, *Carcinogenesis* **9**, 875 (1988).
163. S. S. Hecht and D. Hoffmann, *Cancer Surv.* **8**, 273 (1989).
164. S. S. Hecht and co-workers, *New Engl. J. Med.* **329**, 1543 (1993).
165. S. S. Hecht, S. G. Carmella, P. G. Foiles, and S. E. Murphy, *Cancer Res. (Suppl.)* **54**, 1912S (1994).
166. S. G. Carmella and co-workers, *Cancer Res.* **50**, 5438 (1990).

General References

References 7, 8, 13–20, and 29.

JOHN S. WISHNOK
Massachusetts Institute of Technology

NOBELIUM.   See ACTINIDES AND TRANSACTINIDES.

NOISE POLLUTION AND ABATEMENT.   See SUPPLEMENT.

NOMENCLATURE

Chemical nomenclature embraces several subcategories: names for chemical elements and compounds; names for classes of compounds and substances; names for other substances, such as mixtures and composites; names for particles, processes and transformations, properties, effects, units of measurement, techniques, instruments and apparatus, and even for theories and concepts. Only the first three are considered to be the heart of chemical nomenclature; the others are generally thought of as terminology, although it is not in all cases easy to draw a sharp boundary between the two. This article emphasizes the first group but also gives some information about the various compendia of terminology.

The largest part of the subject is the nomenclature of organic compounds, simply because there are so many of them, and of such diverse nature. The types of compounds and structures differ considerably among organic, inorganic, and biochemical substances, and each of their respective nomenclatures has developed somewhat differently, although not independently. Macromolecular nomenclature and pharmaceutical nomenclature have practical requirements of their own. It is therefore appropriate to treat each of these several areas separately.

Concern with chemical nomenclature has grown on a broad international scale as the importance of consistent, uniform nomenclature is increasingly recognized. When one compound is known by more than one name and further, when one name may refer to more than one compound, serious confusion can result. The effect of such confusion is especially acute in indexes and compilations, and it is therefore natural that much of the work in systematizing names of chemical substances has been done in connection with such works. For example, although the Beilstein Institute uses the International Union of Pure and Applied Chemistry (IUPAC) names with only slight modification (1), more far-reaching variations have been developed by Chemical Abstracts Service, which is principally concerned with indexing (2). The usage of the Royal Society of Chemistry (London) represents a further variant (3). Various committees, both national and international, are working toward a consistent, systematic nomenclature.

The greatly increased interest of industrial chemists and government agencies is an encouraging development. Whereas these groups formerly seemed willing to adopt any name without considering its general acceptability, they now realize that a uniform system of naming is essential to effective handling of information in relation to the many thousands of substances with which they and other groups have to be concerned. Among the areas in which nomenclature plays a key role are patent law, trade and customs regulations, identification of controlled substances, pharmaceutical and health information, and studies of the environment and pollution.

Fortunately, the basic scheme of most chemical nomenclature in all languages is derived from a common source, and there has always been substantial uniformity in practice. Nevertheless, in some instances there have been significant differences beyond those inherent in the languages that have caused inconvenience. For example, some elements are or have been known by completely different names in different countries, eg, beryllium vs glucinum, nitrogen vs azote, niobium vs columbium, tungsten vs wolfram, and potassium vs kalium. The elements that have been historically familiar owing to existence in the free state (iron, lead, sulfur, carbon, copper, silver, and gold) have widely different names because they belong to the commonly spoken language, and their use is not at all confined to science and technology. The names for some common compounds also differ widely; "water" is an example.

Partly because of the increasing percentage of the chemical literature that is published in English (4) and the increased worldwide attention paid to international recommendations, these differences have decreased somewhat. For example, recommendations for the adaptation of IUPAC inorganic nomenclature to the German language (5) include such changes as those from Wasserstoff to Hydrogen and from the spellings Jod and Aethyl to Iod and Ethyl, respectively. There are even changes within the English-speaking world; eg, the Royal Society of Chemistry uses the spelling "sulfur" instead of "sulphur," so that British and American usage is the same.

As early as 1886, an American Chemical Society Committee on Nomenclature and Notation recognized the value of uniform practice among English-speaking chemists by recommending that there be no serious departure from the system then in use by the British Chemical Society without extremely compelling reasons. Further progress in British and American agreement was made in 1923, when the nomenclature committees of the two chemical societies adopted 10 rules covering the more commonly disputed points. Interest in British–American cooperation was reaffirmed by the two societies in 1951.

The first international gathering was the Geneva Congress of 1892, but no further advance by international conference was made for many years. In 1913, the Council of the International Association of Chemical Societies, in its third session in Brussels, appointed commissions on inorganic and organic nomenclature, but this initiative ended abruptly with the outbreak of World War I. Work was not resumed until 1921, when the IUPAC appointed commissions on the nomenclature of inorganic, organic, and biological chemistry at its second conference, held in Brussels. These commissions subsequently produced many valuable reports and comprehensive recommendations. Their work is ongoing, and involves collaboration with national committees throughout the world.

In the United States, the Committee on Nomenclature of the American Chemical Society is the clearinghouse for nomenclature recommendations and adoptions, aided by various divisional nomenclature committees of the Society. Close liaison is maintained with the various nomenclature bodies of IUPAC. Progress is being made not only in improved nomenclature, but also in the extension of nomenclature recommendations to newly developing areas of chemistry.

Important as names are, they cannot serve all purposes (6). There are other, complementary means of identifying chemical compounds, eg, structural formulas, notation systems, and registry numbers. None of these are nomenclature, however. Although structural formulas are sometimes easier to recognize than names, the reverse may also be true. Structural formulas can be space-consuming, they may be troublesome to reproduce, and they are difficult to arrange in an order useful for retrieval of information. Systems for representing chemical structures by fragmentation coding, topological coding, and linear notation have been developed (7); they have particular advantages for electronic handling. The notation systems are more compact than language-based nomenclature, but they are less familiar to most chemists, and many of them suffer from the disadvantage of difficulty of recognition.

Of the various systems, the one originated by Dyson (8) was once adopted by IUPAC, but has been largely abandoned. The system originated by Wiswesser (9) is still widely applied and has catalyzed the development of significant innovations in computerized handling of information about chemical structures. The registry numbers generated by Chemical Abstracts Service (10) also provide an unambiguous identification for chemical compounds, but they are in general assigned arbitrarily, and therefore have no structural information built into them.

Symbols for the Elements

Although symbols are not a part of nomenclature, the two are closely related, and the former have played an extremely important role in chemistry. Ancient Egyptian inscriptions include hieroglyphics for gold, silver, copper, iron, and lead (11). In the time of alchemy, the known metals were associated with the seven classical planets, and the same symbols (which were not letters) were used for both a metal and the corresponding planet. Dalton devised a somewhat similar system based on the circle as a conventional representation of the atom. He and others used symbols to represent composition. Berzelius (12) contributed the modern, alphabetic symbols, which have become completely international. Interesting accounts of the evolution of chemical symbols have been given (13,14). Because of the difficulty of establishing priority of discovery for most of the elements of atomic number above 100, and because of the need to refer to hypothetical elements with higher atomic numbers, IUPAC has developed interim symbols and names for such elements (15). For example, element 104 would have the interim symbol Unq, and the name unnilquadium.

Inorganic Nomenclature

Perhaps no subject in chemistry has undergone less change over the twentieth century than inorganic nomenclature. This longevity attests to the fundamental soundness of the original proposals of Guyton de Morveau (16,17) that established it, but it also suggests why inconsistencies and confusions have remained as well, which have continued to disconcert chemists. The development of inorganic nomenclature has again accelerated, however, and the inconsistencies are being eliminated.

In the days of alchemy and the phlogiston theory, no system of nomenclature that would be considered logical in the 1990s was possible. Names were not based on composition, but on historical association, eg, Glauber's salt for sodium sulfate decahydrate and Epsom salt for magnesium sulfate; physical characteristics, eg, spirit of wine for ethanol, oil of vitriol for sulfuric acid, butter of antimony for antimony trichloride, liver of sulfur for potassium sulfide, and cream of tartar for potassium hydrogen tartrate; or physiological behavior, eg, caustic soda for sodium hydroxide. Some of these common or trivial names persist, especially in the nonchemical literature. Such names were a necessity at the time they were introduced because the concept of molecular structure had not been developed, and even elemental composition was incomplete or indeterminate for many substances.

The System of Guyton de Morveau, Lavoisier, and Co-Workers. Although Bergman (18) proposed a new chemical nomenclature in 1784 that had some suggestion of system, credit for making the first attempt toward a convenient chemical nomenclature belongs to Guyton de Morveau (16,17). His pioneer work led to publication in 1787 of *Methode de Nomenclature Chimique*, written in collaboration with Lavoisier, Berthollet, and Fourcroy, which proved to be a landmark in the development of chemistry. The pioneer work of this committee on nomenclature was widely publicized by the use of its nomenclature in Lavoisier's *Traité Elementaire de Chimie* (1789). The earlier, highly unsystematized practices were replaced quickly by this new nomenclature, the general plan of which continues in use in the 1990s.

Boyle stated clearly the concept of an element in 1661, but Lavoisier was the first to regard a substance as an element until it is shown to be otherwise, and also the first to compile a list of elements. The fundamental principle of the new nomenclature was that the name of a compound should exhibit the elements involved and their relative proportions, if known. The combinations of oxygen with other elements played a dominant role. Thus the product of the union of a simple nonmetallic substance with oxygen was called an acid, whereas that of the union of a metal with oxygen was called an oxide. The union of an acid and an oxide produced a salt. The acids or oxides were given names in which the generic part was the word "acid" or "oxide" and the specific part was an adjective derived from the name of the other element, eg, *acide sulfurique* and *oxide plombique* (sulfuric acid and lead oxide). The same principle supplied names for sulfides and phosphides. The resemblance to the binomial nomenclature system for classifying plants and animals originated by Linnaeus is obvious.

In cases where a substance combines with oxygen to produce more than one acid or oxide, a clear distinction was drawn by altering the termination or by adding a prefix derived from Greek, eg, *acide sulfureux* and *acide sulfurique* for sulfurous and sulfuric acids; *oxide de plomb blanc* and *oxide plombique* for lead monoxide and lead dioxide; and dinitrogen oxide, nitrogen dioxide, and dinitrogen pentoxide.

The names adopted for salts consisted of a generic part derived from the acid and a specific part from the metallic base: *l'oxide de plomb* + *l'acide sulfurique* ⟶ *le sulfate de plomb*. The names for salts of acids containing an element in different degrees of oxidation were given different terminations: *sulfite de soude* and *sulfate de soude* for sodium sulfite and sulfate, and *nitrite de baryte* and *nitrate de baryte* for barium nitrite and nitrate.

Berzelius (19) further applied and amplified the nomenclature introduced by Guyton de Morveau and Lavoisier. It was he who divided the elements into metalloids (nonmetals) and metals according to their electrochemical character, and the compounds of oxygen with positive elements (metals) into suboxides, oxides, and peroxides. His division of the acids according to degree of oxidation has been little altered. He introduced the terms anhydride and amphoteric and designated the chlorides in a manner similar to that used for the oxides.

This system of nomenclature has withstood the impact of later experimental discoveries and theoretical developments that have since the time of Guyton de Morveau and Lavoisier greatly altered the character of chemical thought, eg, atomic theory (Dalton, 1802), the hydrogen theory of acids (Davy, 1809), the dualistic theory (Berzelius, 1811), polybasic acids (Liebig, 1834), Periodic Table (Mendeleev and Meyer, 1869), electrolytic dissociation theory (Arrhenius, 1887), and electronic theory and modern knowledge of molecular structure.

Established Practice in the English Language. The nearly literal translation of the French terms into English, Russian, and other languages resulted in the system whose use has become standard practice in English-speaking as well as other countries. The system has been molded by the fact that elemental composition and valence (or oxidation number) are the principal variables for most inorganic compounds other than the most complex, whereas connectivity and the possibility for isomers have been of little concern. Binary compounds are systematically designated by two words, the first referring to the more electropositive constituent and the second, ending in -ide, referring to the more electronegative constituent, eg, sodium chloride. In cases where the metal exhibits two oxidation states, the lower is indicated by the termination -ous and the higher by -ic, as in cuprous oxide and cupric oxide. Many ternary compounds that contain well-established groups are named as though they were binary compounds, eg, sodium hydroxide, calcium cyanide, and ammonium nitrate.

Ternary compounds are also named by citing the more electropositive constituent first. The various oxidation states of the more electropositive element are designated by a system of prefixes and terminations added to a stem characteristic of the element, except in the case of coordination compounds (qv). Examples are as follows (see CHLORINE OXYGEN ACIDS AND SALTS):

Acids		Salts	
$HClO$	hypochlorous acid	$NaClO$	sodium hypochlorite
$HClO_2$	chlorous acid	$NaClO_2$	sodium chlorite
$HClO_3$	chloric acid	$NaClO_3$	sodium chlorate
$HClO_4$	perchloric acid	$NaClO_4$	sodium perchlorate

The name cyanates indicates that the compounds contain an oxygen atom, in contrast to cyanides.

Modified Forms in Common Use. There are numerous situations in which the foregoing system does not meet all requirements. In the formation of binary compounds, several elements exhibit more than two states of oxidation. One method, recommended by IUPAC, of handling these situations is the use of prefixes derived from Greek to indicate stoichiometric composition (20), eg, titanium dichloride, $TiCl_2$; titanium trichloride, $TiCl_3$; titanium tetrachloride, $TiCl_4$; dinitrogen oxide (nitrous oxide), N_2O; (mono)nitrogen (mon)oxide (nitric oxide), NO; dinitrogen trioxide, N_2O_3; dinitrogen tetraoxide, N_2O_4; and dinitrogen pentaoxide, N_2O_5. Other accepted methods of indicating proportions of constituents are the Stock system (oxidation number) and the Ewens-Bassett (charge number) system.

Some elements form acids with more than four oxidation states, requiring other combinations of prefixes and suffixes: $H_4P_2O_6$, intermediate between H_3PO_3 and H_3PO_4, is known as hypophosphoric acid, and the salts M_2FeO_3, intermediate between M_2FeO_4 and $MFeO_2$, are sometimes called perferrites. Here again the oxidation number and charge number systems offer advantages. Ortho-, meta-, and pyro- prefixes or numerical prefixes to denote stages of hydroxylation of acids also find use (see PHOSPHORIC ACID AND THE PHOSPHATES). In many instances, special names have been created to deal with unusual situations, eg, the thionic acids, $H_2S_xO_6$ ($x > 2$); dithionous acid, $H_2S_2O_4$, an alternative to hydrosulfurous acid; and nitroxylic acid, H_2NO_2, in place of hydronitrous acid.

Systems of Compounds. The nomenclature system of Guyton de Morveau and co-workers was designed specifically for oxygen compounds. As early as 1826 (21,22), it became evident that the halogens could play much the same role in many other compounds as oxygen does in the familiar oxygen salts. By 1840, Hare was writing of chloro acids and chloro bases, and recognized several classes of salts: oxy-, sulfo- (now called thio-), seleni-, telluri-, chloro-, fluoro-, cyano-, etc (23). Remsen (24) was a proponent of this system of nomenclature, but it received its fullest treatment from Franklin (25,26) in connection with his concept of systems of compounds (27).

The analogies are shown by the following reactions:

$$K_2O + B_2O_3 \longrightarrow K_2O \cdot B_2O_3 \text{ or } 2\,KBO_2$$

$$K_2S + B_2S_3 \longrightarrow K_2S \cdot B_2S_3 \text{ or } 2\,KBS_2$$

$$KF + BF_3 \longrightarrow KF \cdot BF_3 \text{ or } KBF_4$$

$$K_3N + BN \longrightarrow K_3N \cdot BN \text{ or } K_3BN_2$$

The products resulting from such reactions should, therefore, have analogous names. If KBO_2 is a borate, KBS_2 is a thioborate and KBF_4 is a fluoroborate. Similarly, the replacement of an oxygen atom by a sulfur atom or two fluorine atoms is understandable. However, the relationship of K_3BN_2 is less obvious, until one considers the dehydration and deammoniation schemes:

$$B(OH)_3 \xrightarrow{-H_2O} OBOH$$

$$B(NH_2)_3 \xrightarrow{-NH_3} HNBNH_2$$

Formally, both of the resulting compounds are borates. Franklin encountered difficulty in choosing a suitable prefix to indicate that the substance K_3BN_2 was a salt of a boric acid containing nitrogen in the same sense that KBS_2 was a salt of a boric acid containing sulfur. The prefix nitro- had been in use for a long time, with a quite different meaning. Because these compounds are best studied in liquid-ammonia solution, Franklin called them ammono salts and the corresponding oxygen salts aquo salts. Although this practice probably prevented general adoption of his scheme of nomenclature for nitrogen compounds, the names thio-, chloro-, etc, did become widespread, especially for sulfur and halogen compounds. On this basis, there is no inconsistency in calling KPF_6 a phosphate even though the analogous oxygen compound, K_3PO_4, has a formula of different appearance, and one cannot be derived from the other by simple substitution.

Although the foregoing pattern of nomenclature is useful, it does lead to some difficulties. Many quaternary compounds contain oxygen and another electronegative element. In the series M_2CO_3, M_2CO_2S, M_2COS_2, and M_2CS_3, the names are carbonates, (mono)thiocarbonates, dithiocarbonates, and trithiocarbonates, respectively. However, in practice both the prefixes mono- and tri- are often omitted, and it is uncertain whether the omission signifies the mono- or the completely substituted compound; for example, consider thiosulfate, $S_2O_3^{2-}$, and chloroplatinate, $PtCl_6^{2-}$. The situation is somewhat more complicated when oxygen and fluorine are present in the same compound, because one is bivalent and the other univalent, and the coordination number toward fluorine is different from that toward oxygen: H_3PO_4, H_2PO_3F, HPO_2F_2, and HPF_6. Furthermore, investigators have not always been consistent in choosing the same reference state for the names of the oxygen salts and the halogen salts. Thus, for rhenium(IV), the salts M_2ReO_3 are known as rhenites, whereas the chloro salts, M_2ReCl_6 are known as chlororhenates. A further variant is represented by chlorosulfonic acid, a name preferred by organic chemists for $ClSO_3H$, and chlorosulfuric acid, the name for the same coupound preferred by inorganic chemists. Here the name is derived by the principle of substitution; chlorine replaces hydrogen in the hypothetical compound HSO_3H (compare the organic analogue, benzenesulfonic acid, $C_6H_5SO_3H$), or it replaces OH in $(OH)_2SO_2$ (H_2SO_4). Only a few other inorganic compounds have been named in such ways; the examples chlorochromic acid, $HCrO_3Cl$, and fluorophosphoric acid are analogous to the latter type of name. Names such as ferrocyanide and ferricyanide are archaic remnants of a time when all such species were regarded as double salts, but they are in common use owing to convenience.

Coordination Compounds. The approach of Werner (28) to the problem of naming ternary and higher order compounds is based on an entirely different point of view. By considering all such substances as complex or coordination compounds, he succeeded in naming a wide variety of them according to a single general pattern. To designate the oxidation state of the element serving as the center of coordination, Werner chose the characteristic endings suggested by Brauner (29), but these have been totally superseded by the oxidation-number and charge-number systems.

The main essentials of the Werner scheme are given here by example (30–32).

Formula	Name
$Cr(NH_3)_6^{3+}(NO_3)_3$	hexamminechromium(III) nitrate
$Pt(H_2O)_2(NH_3)_4^{4+}(Cl^-)_4$	diaquotetraammineplatinum(IV) chloride
$CoCl(NO_2)(en)_2Br$	chloronitrobis(ethylenediamine)cobalt(III) bromide
$HAgCl_2$	hydrogen dichloroargentate(I) or dichloroargentic acid
$Na_2(SSO_3)^{2-}$	sodium thiotrioxosulfate(VI)

The Stock Oxidation-Number System.   Stock sought to correct many nomenclature difficulties by introducing Roman numerals in parentheses to indicate the state(s) of oxidation, eg, titanium(II) chloride for $TiCl_2$, iron(II) oxide for FeO, titanium(III) chloride for $TiCl_3$, iron(III) oxide for Fe_2O_3, titanium(IV) chloride for $TiCl_4$, and iron(II,III) oxide for Fe_3O_4. In this system, only the termination -ate is used for anions, followed by Roman numerals in parentheses. Examples are potassium manganate(IV) for K_2MnO_2, potassium tetrachloroplatinate(II) for K_2PtCl_4, and sodium hexacyanoferrate(III) for $Na_3Fe(CN)_6$. Thus a set of prefixes and terminations becomes unnecessary.

The oxidation-number system is easily extended to include other coordination compounds. Even the interesting substances represented by the formulas $Na_4Ni(CN)_4$ and $K_4Pd(CN)_4$ create no nomenclature problem; they become sodium tetracyanonickelate(0) and potassium tetracyanopalladate(0), respectively.

The Charge-Number (Ewens-Bassett) System.   The oxidation state of an atom as expressed by the oxidation number is a formal concept for partitioning the electric charge between atoms in a molecule or chemical structure. For many chemical structures, this formal procedure may lead to representations of charge distribution that are inconsistent with experiment. Therefore, Ewens and Bassett (33) proposed to express only the total charge on an ion without representing valence and its associated arbitrariness of assigning electronic distribution within a given structure. Examples include titanium(2+) chloride for $TiCl_2$; titanium(3+) chloride for $TiCl_3$; potassium tetrachloroplatinate(2−) for K_2PtCl_4; and sodium tetracyanonickelate(4−) for $Na_4Ni(CN)_4$.

Negatively coordinated groups are given before neutral coordinated groups in the examples of Werner scheme names above. Ewens and Bassett presented good reasons why that order should be reversed. The IUPAC rules (20) recommend that ligands be cited in alphabetical order regardless of their anionic, cationic, or neutral nature. Other applications of the oxidation-number and charge-number systems are specified in the rules.

Coordination nomenclature is also extended to neutral, ie, uncharged, compounds. The ligands are cited in alphabetical order, followed by the name of the central atom (generally a metal), without spaces. Examples are tetraethyllead for $Pb(C_2H_5)_4$, (dimethylamido)pentafluorotungsten for $WF_5[N(CH_3)_2]$, and dichlorodiethyltin for $Cl_2Sn(C_2H_5)_2$. By virtue of having devised additional methods for indicating metal-to-metal attachments and distribution of ligands in unsymmetrical compounds, this type of coordination nomenclature is especially useful with organometallic compounds.

International Agreement. The first report of the Commission for the Reform of the Nomenclature of Inorganic Chemistry was written in 1926 by Delépine (34). The work of the Commission continued with the issuance of several reports, followed in 1940 by a comprehensive set of rules (35,36) which were well received. These led to the 1959 rules (37), which in turn underwent further evolution (20). These later rules were expanded and improved in 1990 to provide the basis for naming inorganic compounds (20). They retain most of the well-established names for binary and pseudobinary compounds and for the oxoacids of the nonmetals and their derivatives. Examples are potassium chloride, KCl; manganese dioxide, MnO_2; hydrogen sulfide, H_2S; phosphorus trichloride, PCl_3; phosphoryl chloride, $POCl_3$; and perchlorate, ClO_4^-. Alternatives are offered based on coordination nomenclature, but they are not meant to supersede the customary names. Examples of these are hydrogen trioxonitrate, HNO_3; dioxonitrate, NO_2^-; and tetraoxomanganic(VI) acid, $HMnO_4$. Such alternatives are useful in naming less common and more complex substances. For some of these another innovation is offered, in which all groups attached to the central atom are indicated by coordination nomenclature, eg, hydroxotrioxorhenium(VII), $HOReO_3$, otherwise known as perrhenic acid. In coordination names for oxoacids, O is always designated oxo, and HO hydroxo. Trivial names are retained for certain hydrides, eg, hydrazine, N_2H_4; silane, SiH_4; diborane, B_2H_6; and phosphine, PH_3. Traditional names for some groups or ions are also retained, eg, nitrosyl, NO; uranyl, UO_2; and chromyl, CrO_2. Any other charged and neutral chemical entities, especially those containing a central metal surrounded by groups that may be ions, radicals, or molecules, are named according to the rules for coordination compounds.

Elemental composition, ionic charge, and oxidation state are the dominant considerations in inorganic nomenclature. Connectivity, ie, which atoms are linked by bonds to which other atoms, has not generally been considered to be important, and indeed, in some types of compounds, such as cluster compounds, it cannot be applied unambiguously. However, when it is necessary to indicate connectivity, italicized symbols for the connected atoms are used, as in trioxodinitrate(N,N), O_2N-NO^{2-}. The nomenclature that has been presented applies to isolated molecules (or ions). For substances in the solid state, which may have more than one crystal structure, with individual connectivities, two devices are used. The name of a mineral that exemplifies a particular crystal structure, eg, rutile or perovskite, may be appended. Alternatively, the crystal structure symmetry, eg, rhombic or triclinic, may be cited, or the structure may be stated in a phrase, eg, face-centered cubic.

The IUPAC Commission on Nomenclature of Inorganic Chemistry continues its work, which is effectively open-ended. Guidance in the use of IUPAC rules (38) as well as explanations of their formulation (39) are available. A second volume on nomenclature of inorganic chemistry is in preparation; it will be devoted to specialized areas. Some of the contents have had preliminary publication in the journal *Pure and Applied Chemistry*, eg, "Names and Symbols of Transfermium Elements" in 1944.

Organic Nomenclature

Modern organic nomenclature is such that it can be better understood by first tracing how it developed. Organic substances played a minor role in the *Methode de Nomenclature Chimique*, published by Guyton de Morveau, Lavoisier, and co-workers in 1787, which became the basis for subsequent inorganic nomenclature. Eighteen organic acids were given their present names (succinic, malic, etc), and several other substances were mentioned, such as alcohol, ether (including esters as well as true ethers), starch, gluten, and camphor. Gaseous hydrocarbons, the only ones included, were lumped together as carbonated hydrogen gas. Thus a few common names were incorporated into the new method, but no systematic organic names were possible because of lack of knowledge. Little else could be done, for the basis for determining elemental composition, as in empirical formulas, did not yet exist.

The practice of assigning ad hoc names to organic compounds was neither avoidable, nor burdensome when only a small number of compounds were recognized. Such ad hoc names are termed "trivial" or "traditional," to indicate that they contain no encoded structural information. They are useful for common compounds, and many of them are retained to this day, but they are not helpful in understanding chemical relationships. As they proliferated, the number and variety of them became unmanageable. The development of systematic nomenclature was driven by this circumstance, and was made possible by advances in understanding and determining the structure of molecules.

Systematic nomenclature is in essence a scheme for encoding structural information in a name. For organic chemistry, it probably began in 1832, when Justus Liebig's journal, *Annalen der Chemie*, was born and when Liebig and Woehler published their memorable article on the radical of benzoic acid (40). This radical (C_6H_5CO in the modern formula) they termed "benzoyl," thus coining -yl (from the Greek *hyle*, meaning stuff or material), one of the most useful suffixes in chemistry. By radical or compound radical, they meant a group of atoms that remains unaltered in chemical transformations. The word group is used for almost any portion of a molecule considered as a unit for convenience in naming or otherwise. The name ethyl soon followed. These two names, one of an acid group (radical) and the other of a hydrocarbon group (radical), may be regarded as the progenitors of the host of group names used in the 1990s. From them it was an easy step to the combinations of benzoyl chloride, ethyl iodide, ethyl oxide, etc, many of which still survive. These binary names are analogous to the binary inorganic names introduced in 1787.

It was many years before organic nomenclature shook off the influence of electrochemical theory and its binary names. Gradually, as facts accumulated, it became clear that this theory must give way to a unitary conception of the molecule. At the same time, the phenomenon of substitution, or replacement of one atom or group of atoms by another, was recognized to be of central importance. Some binary names are still used, either as a true expression, as for salts, or for convenience, as in ethyl sulfide or acetyl chloride, but for the most part the principle of substitution is used without regard to whether such replacement can actually be effected experimentally. Usually the atom replaced is hydrogen, and the replacement may be indicated by either a prefix or a suffix. Thus, in naming CH_3Cl chloromethane rather than methyl chloride, the replacement of one atom of hydrogen in methane, CH_4, by chlorine is indicated. The name methanol for CH_3OH indicates that one hydrogen atom of methane has been replaced by hydroxyl, the characteristic group of alcohols, denoted by the suffix -ol. A third group of names is formed by combining a class name with a specific word, as in ethyl alcohol or benzophenone oxime. Whatever the method or combination of methods used, there must be a name for a parent compound to form a basis for it. This may be a trivial, ie, traditional or common as distinguished from systematic, name such as camphor or naphthalene, or it may be partly systematic, such as butane (from butyric), or fully systematic, such as pentane. Such methods show how, because of its inherent problems, organic chemical nomenclature is different from that of botany and zoology or mineralogy, and even from inorganic chemistry.

By 1866, based on earlier proposals, it was possible for Hofmann (41) to arrange hydrocarbons in series by their empirical formulas, ie, methane, CH_4, and methene, CH_2; ethane, C_2H_6, ethene, C_2H_4, and ethine, C_2H_2; propane, C_3H_8, propene, C_3H_6, and propine, C_3H_4; and quartane, C_4H_{10}, quartene, C_4H_8, and quartine, C_4H_6. These are known as the Hofmann-Gerhardt names.

Hofmann's scheme has been modified by replacing quartane with butane and continuing the homologous series with the Greek forms pentane, hexane, etc, which are still used. For the C_nH_{2n} series, ie, the olefins, the names methylene, ethylene, propylene, etc, came into use instead of Hofmann's terms, but the names propene, butene, pentene, etc were revived in the Geneva system and are the preferred terms. For C_2H_2, ethine has never replaced the older term acetylene, but propine, butine, etc, reappeared in the Geneva system. The ending -yne is used, as in propyne, etc, to avoid confusion with the ending -ine of organic bases such as aniline.

The Hofmann-Gerhardt names did not distinguish between isomers. This defect proved decidedly negative for their future use, for the realization was rapidly developing that a central requirement of nomenclature for organic compounds is to express connectivity and the consequences of isomerism. The requirements of organic nomenclature thus differ fundamentally from those of inorganic nomenclature for simple compounds. Different methods of distinguishing isomers arose: $CH_3CH_2CH_2CH_3$ became normal butane (abbreviated to n-butane), and $CH_3CH(CH_3)_2$ isobutane or trimethylmethane. Of olefins, $CH_2{=}CHCH_2CH_3$ became α-butylene or ethylethylene; $CH_3CH{=}CHCH_3$, β-butylene or symmetrical dimethylethylene; and $CH_2{=}C(CH_3)_2$, isobutylene or unsymmetrical dimethylethylene. It thus becomes evident that as the number of carbon atoms and therefore the number of isomers increases, the coining of such names meets with insuperable difficulties. The situation with regard to hydrocarbons had its parallel in the nomenclature of alcohols and other types of compounds.

The Geneva System.   Until nearly the end of the nineteenth century, only limited attempts were made to meet the needs systematically; for the most part, authors followed their own tastes. The state of affairs finally became so confused that an international congress, held in connection with the Paris Exposition of 1889, appointed an International Commission for the Reform of Chemical Nomenclature, leading to a historic meeting in Geneva in 1892 at which the resolutions afterward known as the basis of the Geneva system (42–44) were adopted. Mention of only a few of the thirty-four participants from nine countries suffices to indicate the Commission's authority: Armstrong and Gladstone from Britain, Friedel and Bouveault from France, and von Baeyer and E. Fischer from Germany.

The Geneva Conference was strongly influenced by the need for names that would be suitable for systematic indexing of organic compounds. The groundwork was laid by a French subcommission. One of the chief principles of the system was the selection of the longest straight chain of carbon atoms in the molecule as a parent structure. Thus the names butane, pentane, etc would refer to the normal (unbranched) isomers only. The parent hydrocarbon could then be modified by attaching to its name one or more prefixes or suffixes to specify chemically characteristic features (commonly termed functional groups). A representative selection is given in Table 1. For small compounds, this was all that was necessary, as in the name ethanol (ethane plus the suffix -ol) for CH_3CH_2OH, because only one isomer was possible. When two or more different positions of attachment of a prefixed or suffixed group exist, a position designator, called a locant, is necessary. These are arabic numerals, set off by hyphens, starting with 1 at an end of the chain. Accordingly, $CH_3CH=CHCH_3$ became 2-butene, and $CH_3CH_2CHOHCH_2CH_3$ became 3-pentanol. The position of the locants was permissive, providing that there was no ambiguity, and pentanol-3 was considered equally acceptable. In the 1990s, however, the official IUPAC recommendation is to place the locant immediately before the feature that it locates, as in but-2-tene and pentan-3-ol for the foregoing examples. Branching of a carbon chain is handled in an analogous manner, by citing and locating the branch, as in $CH_2=CHCH_2CH(CH_3)_2$, 4-methylpent-1-ene. In special cases, a letter locant, the italicized symbol for an element, is used to indicate attachment to an atom other than carbon, as in *N,N*-dimethylacetamide, $CH_3CON(CH_3)_2$.

This was a great advance, as complex hydrocarbons of various kinds could be clearly named if their structures were known. The Geneva system, modified and expanded by subsequent Commissions, is used systematically in the fourth and fifth editions of *Beilstein's Handbuch der Organischen Chemie* (1), and as the basis of IUPAC nomenclature. In 1992 a commemorative symposium on the centennial anniversary of the Geneva Conference was held in Geneva (45).

The International Union and the Definitive Report. The original Geneva rules found only partial acceptance, for they were far from complete, related chiefly to aliphatic compounds, and did not meet problems raised by subsequent discoveries. The next important step was the *Definitive Report of the Commission on the Reform of the Nomenclature of Organic Chemistry*, adopted by the Commission and by the Council of the International Union of Chemistry in 1930 at a meeting in Liège (46–48). This report used the Geneva rules as a basis for modification, and many of the 68 Liège rules deal with topics not touched in the original Geneva report.

Table 1. Prefixes and Suffixes for Some Principal Functional Groups[a]

Formula	Class name	Prefix	Suffix	Radicofunctional form
—COOH	carboxylic acids	carboxy	-carboxylic acid[b] or -oic acid	
—SO$_3$H	sulfonic acids	sulfo	-sulfonic acid	
—COOR	esters	alkoxy carbonyl	alkyl -oate or carboxylate[b]	
—COX	acid halides	halocarbonyl	-oyl halide or carbonyl halide[b]	
—CONH$_2$	amides	carbamoyl	-amide or carboxamide[b]	
		cyano		
—CN	nitriles		-nitrile or -carbonitrile[b]	alkyl[c] cyanide
—CH=O	aldehydes	formyl	-al or -carbaldehyde[b]	
>C=O	ketones[d]	oxo	-one	dialkyl[c] ketone
—OH	alcohols	hydroxy	-ol	alkyl[c] alcohol
—SH	thiols (or mercaptans[e])	sulfanyl (mercapto[e])	-thiol	alkyl mercaptan
—NH$_2$	amines	amino	-amine	
=NH	imines	imino	-imine	
—OR	ethers	alkoxy or aryloxy		(di)alkyl ether
—Cl	chlorides	chloro		alkyl[c] chloride
—NO$_2$	nitro compounds	nitro		
>SO$_2$	sulfones	alkyl[c] sulfonyl		(di)alkyl[c] sulfone

[a]In order of precedence (see text).

[b]The shorter form implies no additional carbon atoms, and is used when the group is part of a chain. The long form implies one more carbon atom than the parent structure, and is used when the group is attached to a ring, or for other reasons is not conveniently named as part of a carbon skeleton.

[c]Or aryl.

[d]Both bonds from the carbonyl group must be to a carbon atom.

[e]Has been widely used, but is no longer officially recommended.

One provision of the *Definitive Report* is that only one kind of function (the principal function) should be expressed by the ending of the name, ie, by a suffix. The others, if there are any, are designated by prefixes (Rule 51). Thus CH$_3$CHOHCOCH$_3$ is called not butan-3-ol-2-one, but 3-hydroxy-2-butanone. An important modification of the Geneva system is that the fundamental chain used as a basis in an aliphatic compound is not necessarily the longest chain in the molecule, but must be the longest chain of those containing the maximum number of occurrences of the principal functional group (Rule 18) (49). This shifts the importance for naming from side chains such as methyl and ethyl to functional groups such as —COOH and —OH (50). Rules 14–16 deal with parent heterocyclic compounds, and Rule 16 gives recognition to the "a" (or oxa–aza) method for replacement nomenclature, based on hydrocarbon names, which has extensive use in the naming of heterocycles and long-chain acyclic compounds containing heteroatom interruptions. An example of the latter is $CH_3CH_2O(CH_2CH_2O)_3CH_2CH_2OH$, 3,6,9,12-tetraoxatetradecan-1-ol, as an alternative to the conventional type of name (51). Sections B and C of the current IUPAC Rules (52–54) extend this type of nomenclature.

The concept of the principal function raises the question of how priority is determined when two or more different functional groups are present. No arbitrary rule can be entirely satisfactory, but an order has been codified in IUPAC recommendations (52–54), and an essentially similar order is used by Chemical Abstracts Service. In general, a higher state of oxidation takes precedence over a lower one, as shown in Table 1.

The use of prefixes and suffixes for distinguishing the various radicals, groups, and functions has caused some problems, because some groups, eg, HS —, have borne more than one name, and some names, eg, anisyl, have had more than one meaning. The *Definitive Report* included only a limited number of prefixes and suffixes. Chemical Abstracts Service publishes its own lists (2). A formally approved list of prefixes and corresponding suffixes, together with rules for their formation, has appeared in IUPAC publications (52–55); the most recent version can be found in the comprehensive *Guide to the Use of IUPAC Nomenclature of Organic Compounds* (54). An important departure from earlier recommendations is that the systematic names of acyl groups derived from carboxylic acids must end in -oyl, common traditional or trivial names, such as acetyl and oxalyl, excepted. The purpose of this rule is to distinguish unambiguously between hydrocarbyl groups and acyl groups. Thus anisyl can only mean methoxyphenyl, whereas anisoyl refers only to methoxybenzoyl.

The ending -yl (or -oyl) is standard for univalent groups (with certain traditional exceptions, such as succinoyl). It may be combined with a sign for unsaturation, as in propenyl, $CH_3CH=CH-$, or ethynyl, $CH\equiv C-$. The ending -ylene is one device for denoting a bivalent group in which the two free valences are on different atoms, but, with the exception of methylene, $-CH_2-$, and ethylene, $-CH_2CH_2-$, the ending -diyl, with locants as appropriate, is preferred, as in propane-1,3-diyl, $-CH_2CH_2CH_2-$. When the two free valences of a bivalent group form a double bond, the ending is -ylidene, as in ethylidene, $CH_3CH=$. For a trivalent group forming a triple bond, the ending is -ylidyne, as in ethylidyne, $CH_3C\equiv$.

For indicating the number of groups of the same kind, the prefixes di-, tri-, tetra-, etc are used when the expressions are simple, and bis-, tris-, tetrakis-, etc when they are complex; for example, "dichloro," but "bis(dimethylamino)." The prefix bi- is used to denote the joining of two groups of the same kind together, as in biphenyl, $C_6H_5 - C_6H_5$, or the doubling of a compound with loss of two hydrogen atoms, as in biarsine, H_2AsAsH_2.

A historical account of the development of organic nomenclature from the time preceding the Geneva Conference to fairly recent times is available (56).

The IUPAC Commission on Nomenclature of Organic Chemistry has continuing responsibility for revising and expanding the rules that appeared in the *Definitive Report*. The several sections of revisions that had been first published in the journal *Pure and Applied Chemistry* have been collected in book form (52,54); subsequent revised sections are similarly published from time to time

(57–59). These revisions are often based on other significant contributions, such as for organosilicon compounds (60), organophosphorus compounds (61), and stereochemical configuration (62,63). The Hantzsch-Widman system for naming heterocyclic rings containing one heteroatom (64) has been revised, and the nomenclature of ions and radicals has been dealt with comprehensively (49).

A considerable number of trivial or semitrivial (traditional) names have been retained by IUPAC for compelling practical reasons; the approved ones are available (54). A very brief selection is shown in Table 2. Other contributions to organic nomenclature are available (2,50,51). A first catalog of all known ring systems (65) was first replaced in 1976 (66) and again in 1984 (67). The latter source provides the names and numbering schemes of ring systems and cage parents. Diagrams for all known cyclic and acyclic stereoparent structures are also available (68). Recent developments include the use of computerized approaches to nomenclature of organic compounds. A comprehensive computer program called AUTONOM (69) has been developed at the Beilstein Institute and has been made available commercially. Structures may be drawn on a computer monitor, and in a few seconds the complete IUPAC name is displayed. For the increasingly small number of structures that the program cannot handle (currently about 25%), failure is explicitly indicated. The reverse problem, deriving a structure from a systematic name, has been addressed by a different group, at the University of Hull (70). These two groups are the most prominent in the area of computerized handling of organic nomenclature, but there are others engaged in alternative approaches.

Table 2. Compounds Used as Parent Structures with Trivial Names

Chemical formula	Name	Systematic equivalents
C_6H_6	benzene	
$C_{10}H_6$	naphthalene	
C_5H_5N	pyridine	
C_4H_4S	thiophene	
C_4H_4O	furan	
C_6H_5OH	phenol	benzenol
$C_6H_5NH_2$	aniline	benzenamine
$H_2N\mathrm{-\!-}NH_2$	hydrazine	diazane
NH_3	ammonia or amine	azane
$HO\mathrm{-\!-}OH$	hydrogen peroxide	dioxidane
CH_3COCH_3	acetone	propan-2-one
H_2NCONH_2	urea	carbamide

Biochemical Nomenclature

The nomenclature of biochemical compounds is in large measure a part of organic nomenclature. However, it has its own special problems, arising partly from the fact that many biochemical compounds must be given names before their chemical structures have been fully determined, and partly from the interest in grouping them according to biological function as much as to chemical class.

The IUPAC Commission on Nomenclature of Biological Chemistry was established in 1921, along with the organic and inorganic commissions. It worked actively and closely with the organic commission. Early subjects of concern were carbohydrates, proteins, enzymes, and fats (71). More recently, this Commission shared its work with a corresponding Commission of the International Union of Biochemistry, which is now the International Union of Biochemistry and Molecular Biology (IUBMB); this led to the establishment of the Joint Commission on Biochemical Nomenclature (JCBN) in 1964.

The Joint IUPAC/IUBMB Commission has published many recommendations dealing with the nomenclature of natural products (72), steroids (73), carbohydrates (74), cyclitols (75), tetrapyrroles (76), corrinoids (77), quinones with isoprenoid side-chains (78), carotenoids (79), vitamins (80–82), amino acids and peptides (83,84), lipids (85), etc, a large number of which are collected in book form (72).

The IUBMB Commission on Nomenclature has issued a number of recommendations dealing with areas of a more biochemical nature (72), such as peptide hormones (86), conformation of polypeptide chains (87), abbreviations for nucleic acids and polynucleotides (88), iron–sulfur proteins (89), enzyme units (90), etc. The Commission has also produced rules and recommendations for naming enzymes (91,92).

The presence of many chiral centers in compounds of biochemical significance or natural-product interest has led to the use of stereoparents (70). These are parent structures having trivial names that imply (without explicitly expressing) a particular steric configuration. Common examples are the names of the simple sugars, exemplified by glucose. This simplifying nomenclature device is indispensable in carbohydrate and steroid chemistry, among others. The following structure represents 6-O-methyl-D-glucose, a specific configuration of 2,3,4,5-tetrahydroxy-6-methoxyhexanal.

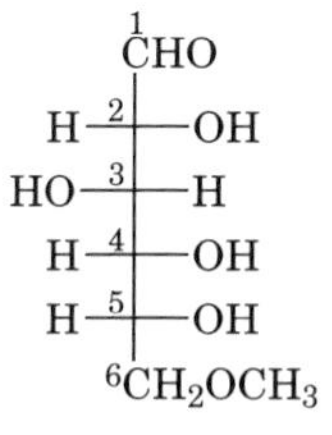

$$
\begin{array}{c}
{}^{1}\mathrm{CHO} \\
\mathrm{H}{-}{}^{2}{-}\mathrm{OH} \\
\mathrm{HO}{-}{}^{3}{-}\mathrm{H} \\
\mathrm{H}{-}{}^{4}{-}\mathrm{OH} \\
\mathrm{H}{-}{}^{5}{-}\mathrm{OH} \\
{}^{6}\mathrm{CH_2OCH_3}
\end{array}
$$

Although it is not strictly within the subject of biochemical nomenclature, it is appropriate to mention the existence of standardized generic names for pharmaceutical drugs. Such names are essentially coined, or trivial, names, but they often include syllables from the systematic organic names, and endings that reflect a structural class, eg, -cillin (from penicillin), or an important area of medical application, eg, -vir (antiviral). Glossaries of these generic names are published periodically (93), and a glossary of United States Approved Names (USAN) is published annually (94).

Macromolecular Nomenclature

The nomenclature of macromolecules can be complicated when there is little or no regularity in the molecules; for such molecules, the structural details may also be uncertain. In cases where the macromolecule is a polymeric chain with some uncertainties about regularity in its structure, a simple expedient is to name the polymer after the monomer that gave rise to it. Thus there are source-based names such as poly(vinyl chloride).

The first attempt to formulate a systematic nomenclature for polymers was based on the smallest repeating structural unit; it was published in 1952 by a Subcommission on Nomenclature of the IUPAC Commission on Macromolecules (95). The report covered not only the naming of polymers, but also symbology and definitions of terms. However, these nomenclature recommendations did not receive widespread acceptance. Further progress was slow, with a report on steric regularity in high polymers published in 1962 and updated in 1966 (96).

In 1967, the Polymer Nomenclature Committee of the American Chemical Society published proposals for naming linear polymers on the basis of their chemical structure (97), which were then introduced into *Chemical Abstracts* (CA) *Indexes* and published in their final form in 1968 (98).

A Macromolecular Division of IUPAC was created in 1967, and it created a permanent Commission on Macromolecular Nomenclature, parallel to the other nomenclature commissions. The Commission over the years has issued recommendations on basic definitions, stereochemical definitions and notations, structure-based nomenclature for regular single-strand organic polymers and regular single-strand and quasisingle-strand inorganic and coordination polymers, source-based nomenclature for copolymers, and abbreviations for polymers. All of these are collected in a compendium referred to as the IUPAC Purple Book (99).

Recommendations on additional aspects of macromolecular nomenclature such as that of regular double-strand (ladder and spiro) and irregular single-strand organic polymers continue to be published in *Pure and Applied Chemistry* (100,101). Recommendations on naming nonlinear polymers and polymer assemblies (networks, blends, complexes, etc) are expected to be issued in the near future.

Examples of the two macromolecular nomenclature systems are as follows. For source-based names for homopolymers and copolymers: polyacrylonitrile, poly(methyl methacrylate), poly(acrylamide-*co*-vinylpyrrolidinone), polybutadiene-*block*-polystyrene, and poly(propyl methacrylate)-*graft*-poly(1-vinylnaphthalene). Structure-based examples are as follows: poly(oxy-1,4-phenylene) (**1**), poly(oxyethyleneoxyterephthaloyl) (**2**), and poly[imino(1-oxo-1,6-hexanediyl)] (**3**).

(1)   (2)   (3)

The more familiar source-based names for these polymers are poly(phenylene oxide) (**1**), poly(ethylene terephthalate) (**2**), and polycaprolactam (**3**).

Nomenclature in Other Areas of Chemistry

A number of glossaries of terms and symbols used in the several branches of chemistry have been published. They include physical chemistry (102), physical–organic chemistry (103), and chemical terminology (other than nomenclature) treated in its entirety (104). IUPAC has also issued recommendations in the fields of analytical chemistry (105), colloid and surface chemistry (106), ion exchange (107), and spectroscopy (108), among others.

BIBLIOGRAPHY

"Nomenclature" in *ECT* 1st ed., Vol. 9, pp. 473–485, by W. C. Fernelius (Inorganic), The Pennsylvania State College, and A. M. Patterson (Organic), Antioch College; in *ECT* 2nd ed., Vol. 14, pp. 1–13, by K. L. Loening, Chemical Abstracts Service; in *ECT* 3rd ed., Vol. 16, pp. 28–46, by K. L. Loening, Chemical Abstracts Service.

1. *Beilstein's Handbuch der Organischen Chemie*, 4th ed., Springer-Verlag, Berlin and Heidelberg, 1972, 4th and 5th Supplements, 1973–1995.
2. *Chemical Abstracts Index Guide*, Appendix IV, Chemical Abstracts Service, Columbus, Ohio, 1994, 125I–272I.
3. R. S. Cahn, ed., *Handbook for Chemical Society Authors* (Special Publications No. 14), The Chemical Society, London, 1961.
4. D. B. Baker, *Chem. Eng. News* **54**(20), 23 (May 10, 1976).
5. E. Hayek, *Oesterr. Chem. Z.* **76**, 2 (1975).
6. O. C. Dermer, G. Gorin, and K. L. Loening, *Int. J. Sociol. Lang.* **11**, 61 (1976).
7. J. E. Rush, *J. Chem. Inform. Comput. Sci.* **16**, 202 (1976).
8. G. M. Dyson, *A New Notation and Enumeration System for Organic Compounds*, 2nd ed., Longmans, Green & Co., Inc., New York, 1949. *Rules for International Union of Pure and Applied Chemistry Notation for Organic Compounds*, John Wiley & Sons, Inc., New York, 1961; G. M. Dyson, M. F. Lynch, and H. L. Morgan, *Inform. Stor. Retr.* **4**, 27 (1968).
9. W. J. Wiswesser, *A Line-Formula Chemical Notation*, Thomas Y. Crowell Co., New York, 1954. E. G. Smith and P. A. Baker, *The Wiswesser Line-Formula Chemical Notation (WLN)*, 3rd ed., Chemical Information Management, Inc. (CIMI), Cherry Hill, N.J., 1975.
10. P. G. Dittmar, R. E. Stobaugh, and C. E. Watson, *J. Chem. Inform. Comput. Sci.* **16**, 111 (1976).
11. M. P. Crosland, *Historical Studies in the Language of Chemistry*, Harvard University Press, Cambridge, Mass., 1962.
12. J. J. Berzelius, *Thompson's Annals of Philosophy* **2**, 443 (1813).

13. I. W. D. Hackh, *J. Amer. Pharmacol. Ass.* **7**, 1038 (1918).

14. G. Rasch, *Z. Chem.* **6**, 297 (1966).

15. International Union of Pure and Applied Chemistry, *Pure Appl. Chem.* **51**, 381 (1979).

16. L. B. Guyton de Morveau, *J. Phys.* **19**, 370 (1782).

17. L. B. Guyton de Morveau, *Ann. Chim. Phys.* **25**, 205 (1798).

18. T. Bergman, *Meditationes de Systemate Fossilium Naturali*, 1784.

19. J. J. Berzelius, *J. Phys.* **73**, 253 (1811).

20. International Union of Pure and Applied Chemistry, *Nomenclature of Inorganic Chemistry, Recommendations 1990*, G. J. Leigh, ed., Blackwell, Oxford, U.K., 1990.

21. P. A. von Bonsdorff, *Ann. Chim. Phys.* **34**, 142 (1826).

22. P. Boullay, *Ann. Chim. Phys.* **34**, 337 (1826).

23. R. Hare, *Compendium of the Course of Chemical Instruction*, 4th ed., Philadelphia, Pa., 1840.

24. I. Remsen, *Amer. Chem. J.* **11**, 298 (1889).

25. E. C. Franklin, *Amer. Chem. J.* **47**, 285 (1912).

26. E. C. Franklin, *J. Amer. Chem. Soc.* **46**, 2137 (1924).

27. E. C. Franklin, *The Nitrogen System of Compounds*, Reinhold Publishing Corp., New York, 1935.

28. A. Werner, *Neuere Anschauungen auf dem Gebiete der Anorganischen Chemie*, 3rd ed., Vieweg, Braunschweig, 1913, pp. 92–95.

29. B. Brauner, *Z. Anorg. Chem.* **32**, 10 (1902).

30. O. Ohman, *Z. Angew. Chem.* **33**, 326 (1920).

31. A. Rosenheim, *Z. Angew. Chem.* **33**, 78 (1920).

32. A. Stock, *Z. Angew. Chem.* **32**, 373 (1919); A. Stock, *Z. Angew Chem.* **33**, 79 (1920).

33. R. V. G. Ewens and H. Bassett, *Chem. Ind.* **27**, 131 (1949).

34. M. Delépine, *Bull. Soc. Chim.* **43**, 289 (1928).

35. W. P. Jorissen, H. Bassett, A. Damiens, F. Fichter, and H. Rémy, *Ber.* **A73**, 53 (1940); *J. Chem. Soc.* 1404 (1940); *J. Amer. Chem. Soc.* **63**, 889 (1941).

36. R. J. Meyer, *Chem. Weekblad* **33**, 722 (1936).

37. International Union of Pure and Applied Chemistry, *Nomenclature of Inorganic Chemistry*, Butterworths, London, Eng., 1959; *J. Amer. Chem. Soc.* **82**, 5523 (1960).

38. International Union of Pure and Applied Chemistry, *How to Name an Inorganic Substance*, Pergamon Press, Oxford, UK, 1977.

39. W. C. Fernelius, K. L. Loening, and R. Adams, "Notes on Nomenclature." See issues of *J. Chem. Educ.* beginning **48**, 433 (1971) through **55**, 30 (1978).

40. J. Liebig and F. Wöhler, *Ann. Pharm.* **3**, 249 (1832).

41. A. W. von Hofmann, *Proc. Roy. Soc. (London)* **15**, 57 (1866/7).

42. A. Pictet, *Arch. Sci. Phys. Nat.* **27**, 485 (1892).

43. F. Tiemann, *Ber.* **26**, 1595 (1893).

44. H. E. Armstrong, *Proc. Roy. Soc. (London)* **8**, 127 (1892); H. E. Armstrong, *Nature* **46**(1177) 56 (1892).

45. M. V. Kisakürek, ed., *Organic Chemistry: Its Language and Its State of the Art*, Verlag Helvetica Chimica Acta, Basel, Switzerland, and VCH, Weinheim, Germany, 1993.

46. International Union of Chemistry Commission on the Reform of Nomenclature in Organic Chemistry, *Ber.* **A65**, 11 (1932); *Bull. Soc. Chim.* **45**, 973 (1929); *Recl. Trav. Chim.* **48**, 641 (1929); *Helv. Chim. Acta* **14**, 868 (1931); *J. Chem. Soc.* 1607 (1931).

47. A. M. Patterson, *J. Amer. Chem. Soc.* **55**, 3905 (1933).

48. P. E. Verkade, *Recl. Trav. Chim.* **51**, 185 (1932).

49. W. H. Powell, *Revised Nomenclature for Radicals, Ions, Radical Ions, and Related Species*, IUPAC Recommendations, 1993; *Pure Appl. Chem.* **65**, 1357 (1993).

50. J. H. Fletcher, O. C. Dermer, and R. B. Fox, *Advan. Chem. Ser.* No. 126 (1974).
51. R. S. Cahn and O. C. Dermer, *Introduction to Chemical Nomenclature*, 5th ed., Butterworth, London, 1979.
52. International Union of Pure and Applied Chemistry, *Nomenclature of Organic Chemistry*, Sections A, B, C, D, E, F, and H, Pergamon Press, Oxford, U.K., 1979.
53. International Union of Pure and Applied Chemistry, *Nomenclature of Organic Chemistry, Definitive Rules for Section C. Characteristic Groups Containing Carbon, Hydrogen, Oxygen, Nitrogen, Halogen, Sulfur, Selenium and /or Tellurium*, Butterworth, London, 1965.
54. J.-C. Richer, R. Panico, and W. H. Powell, *Guide to the Use of IUPAC Nomenclature of Organic Compounds*, Blackwell, Oxford and London, 1994.
55. *C. R. Conf. Union Intern. Chim. Pure Appl., 15th*, 127–185 (1949).
56. P. E. Verkade, *A History of the Nomenclature of Organic Chemistry*, D. Reidel Publishing Co., Dordrecht, Boston, and Lancaster, 1985.
57. International Union of Pure and Applied Chemistry, *Pure Appl. Chem.* **45**, 11 (1976).
58. International Union of Pure and Applied Chemistry, *Eur. J. Biochem.* **86**, 1 (1978).
59. International Union of Pure and Applied Chemistry, *Pure Appl. Chem.* **51**, 353 (1979).
60. Ref. 55, pp. 127–132.
61. *Chem. Eng. News* **30**, 4515 (1952).
62. R. S. Cahn, C. Ingold, and V. Prelog, *Angew. Chem. Int. Ed. Engl.* **5**, 385 (1966).
63. R. S. Cahn, *J. Chem. Educ.* **41**, 116 (1964).
64. *Pure Appl. Chem.* **51**, 1995 (1979).
65. A. M. Patterson, L. T. Capell, and D. F. Walker, *The Ring Index*, 2nd ed., American Chemical Society, Washington, D.C., 1960, and Suppl. I–III, 1963, 1964, 1965.
66. *Parent Compound Handbook*, Chemical Abstracts Service, Columbus, Ohio, 1976, and supplements. *Ring Systems Handbook*, Chemical Abstracts Service, Columbus, Ohio, 1988 and supplements; 1993 and supplements.
67. *Ring Systems Handbook*, Chemical Abstracts Service, Columbus, Ohio, 1993, and supplements.
68. *Chemical Abstracts Index Guide*, Chemical Abstracts Service, Columbus, Ohio, 1992.
69. J. L. Wisniewski, in W. A. Warr, ed., *Chemical Structures 2*, Springer-Verlag, Heidelberg and New York, 1993, pp. 55–63. D. O'Sullivan, *Chem. Eng. News* 31 (Aug. 13, 1990). L. Goebels, A. J. Lawson, and J. L. Wisniewski, *J. Chem. Inform. Comput. Sci.* **31**, 216 (1991).
70. G. H. Kirby, M. R. Lord, and J. D. Rayner, in W. A. Warr, ed., Chemical Structures 2, Springer-Verlag, Heidelberg and New York, 1993, pp. 43–53. D. I. Cooke-Fox, G. H. Kirby, and J. D. Rayner, *J. Chem. Inform. Comput. Sci.* **29**, 112 (1989) and preceding papers.
71. J. E. Courtois, *Advan. Chem. Ser.* **8**, 83 (1953).
72. International Union of Biochemistry and Molecular Biology, *Biochemical Nomenclature and Related Documents*, 2nd ed., Portland Press, London, 1992.
73. International Union of Pure and Applied Chemistry and IUPAC-IUB Commission on Biochemical Nomenclature, *Pure Appl. Chem.* **31**, 285 (1972); *Biochemistry* **8**, 2227 (1969), **10**, 4994 (1971); *Pure Appl. Chem.* **61**, 1783 (1989).
74. IUPAC-IUB Commission on Biochemical Nomenclature, *Biochemistry* **10**, 3983 (1971), 4995 (1971); *Pure Appl. Chem.* **53**, 1901 (1981), **54**, 207, 211, 1517, 1523 (1982), **55**, 1269 (1983).
75. IUPAC-IUB Commission on Biochemical Nomenclature, *Eur. J. Biochem.* **57**, 1 (1975).
76. IUPAC-IUB Joint Commission on Biochemical Nomenclature, *Pure Appl. Chem.* **59**, 779 (1987).

77. IUPAC-IUB Commission on Biochemical Nomenclature, *Biochemistry* **13**, 1555 (1974).

78. IUPAC-IUB Commission on Biochemical Nomenclature, *Eur. J. Biochem.* **53**, 15 (1975).

79. IUPAC-IUB Commission on Biochemical Nomenclature, *Pure Appl. Chem.* **41**, 405 (1975); *Biochemistry* **10**, 4827 (1971); *Biochemistry* **14**, 1803 (1975).

80. IUPAC-IUB Commission on Biochemical Nomenclature, *Biochemistry* **13**, 1056 (1974).

81. IUPAC-IUB Commission on Biochemical Nomenclature, *Eur. J. Biochem.* **46**, 217 (1974).

82. *Ibid.*, **2**, 5 (1967).

83. IUPAC-IUB Commission on Biochemical Nomenclature, *Biochemistry* **14**, 449 (1975).

84. *Ibid.*, **11**, 1726 (1972).

85. IUPAC-IUB Commission on Biochemical Nomenclature, *Eur. J. Biochem.* **79**, 11 (1977).

86. IUPAC-IUB Commission on Biochemical Nomenclature, *Biochemistry* **14**, 2559 (1975).

87. IUPAC-IUB Commission on Biochemical Nomenclature, *Pure Appl. Chem.* **40**, 291 (1974); *Biochemistry* **9**, 3471 (1970).

88. IUPAC-IUB Commission on Biochemical Nomenclature, *Pure Appl. Chem.* **40**, 277 (1974); *Biochemistry* **9**, 4022 (1970).

89. IUPAC-IUB Commission on Biochemical Nomenclature, *Biochemistry* **12**, 3582 (1973).

90. International Union of Biochemistry, *Eur. J. Biochem.* **97**, 319 (1979).

91. International Union of Biochemistry, *Enzyme Nomenclature Recommendations 1964*, Elsevier, Amsterdam, The Netherlands, 1965.

92. International Union of Biochemistry, *Enzyme Nomenclature 1978*, Academic Press, New York, 1979, Suppl. 1; *Eur. J. Biochem.* **104**, 1 (1980).

93. *International Nonproprietary Names for Pharmaceutical Substances, Cumulative List No. 8*, World Health Organization, Geneva, Switzerland, 1988; *WHO Drug Inform.* **6**(3) and Special Section (1992).

94. *USAN and USP Dictionary of Drug Names*, 30th ed., USP Convention, Inc., Rockville, Md., 1993.

95. International Union of Pure and Applied Chemistry, *J. Polym. Sci.* **8**, 257 (1952).

96. International Union of Pure and Applied Chemistry, *J. Polym. Sci.* **56**, 153 (1962); *Pure Appl. Chem.* **12**, 645 (1966).

97. R. B. Fox, *Polym. Prepr. (Amer. Chem. Soc., Div. Polym. Chem.)* **8**(2), e–r (1967); *J. Chem. Doc.* **7**, 74 (1967).

98. American Chemical Society, *Macromolecules* **1**, 193 (1968).

99. International Union of Pure and Applied Chemistry, *Compendium of Macromolecular Nomenclature*, Blackwell, Oxford, U.K., 1991.

100. International Union of Pure and Applied Chemistry, *Pure Appl. Chem.* **65**, 1561 (1993).

101. International Union of Pure and Applied Chemistry, *Pure Appl. Chem.* **66**, 873 (1994).

102. International Union of Pure and Applied Chemistry, *Manual of Symbols and Terminology for Physicochemical Quantities and Units*, Pergamon Press, Oxford, UK, 1979; *Pure Appl. Chem.* **51**, 1 (1979).

103. V. Gold, compiler, *Glossary of Terms Used in Physical–Organic Chemistry: Recommendations 1982*, Blackwell, Oxford and London, 1982.

104. V. Gold, A. D. McNaught, and P. Sehmi, compilers, *Compendium of Chemical Terminology*, Blackwell, Oxford and London, 1987.

105. International Union of Pure and Applied Chemistry, *Compendium of Analytical Nomenclature*, Peramon Press, Oxford, U.K., 1977.

106. International Union of Pure and Applied Chemistry, *Pure Appl. Chem.* **31**, 577 (1972); *Pure Appl. Chem.* **46**, 71 (1976).

107. International Union of Pure and Applied Chemistry, *Pure Appl. Chem.* **29**, 619 (1972).

108. International Union of Pure and Applied Chemistry, *Pure Appl. Chem.* **30**, 651 (1972); *Pure Appl. Chem.* **45**, 99 (1976); *Pure Appl. Chem.* **45**, 105 (1976); *IUPAC Inf. Bull. Append. Provis. Nomencl.* **54**, 1 (Dec. 1976).

General References

P. Fresenius, *Organic Chemical Nomenclature: Introduction to the Basic Principles*, John Wiley & Sons, Inc., New York, 1989 (trans. from German ed., 1983).

E. W. Godly, *Naming Organic Compounds: A Systematic Instruction Manual*, Ellis Horwood Ltd., Chichester, U.K., 1989.

B. P. Block, W. H. Powell, and W. C. Fernelius, *Inorganic Chemical Nomenclature: Principles and Practices*, American Chemical Society, Washington, D.C., 1990.

D. Hellwinkel, *Die Systematische Nomenklatur der Organischen Chemie*, 3rd ed., Springer-Verlag, Berlin and Heidelberg, 1982.

W. Liebscher, *Handbuch zur Anwendung der Nomenclatur Organisch-Chemischer Verbindungen*, Akademie-Verlag, Berlin, 1979.

H. A. Favre, *La Nomenclature pour la Chimie Organique*, Ordre des Chimistes du Quebec, Montréal, Canada, 1992.

F. G. Alcaraz, *Nomenclatura de Química Orgánica*, Universidád de Murcia, Murcia, Spain, 1991.

Peter A. S. Smith
University of Michigan

NONDESTRUCTIVE EVALUATION

The technology of nondestructive evaluation (NDE) includes all nondamaging or nonintrusive methods for determining material identity; for evaluating material properties, composition, structure, or serviceability; and for detecting discontinuities and defects in materials. Nondestructive tests commonly are used for quality control (qv), process control (qv), and for reliability assurance of materials, protective coatings (qv), components, welds, assemblies, structures, and operating systems. The use of these tests helps prevent premature failure of materials during processing, manufacturing, or assembly, and during service under anticipated operating stresses and environments. Proper testing can lower manufacturing and operating costs, minimize insurance risks, and help prevent interruptions of service and potential disasters that might cause loss of life or usefulness of costly facilities.

Nondestructive tests differ from methods of laboratory analysis and testing where specimens are generally sectioned, broken, damaged, or destroyed. Nondestructive tests can be performed on materials, components, and structures or systems that actually are to be used. Thus, effective use of NDE requires engineering knowledge of the structure, the performance characteristics, and service environment, as well as the test method. More complete information on all of the topics discussed herein are available (1–6).

Operating companies and regulatory agencies (qv) have a vital responsibility for specifying and managing the use of nondestructive testing during erection and service of costly facilities and systems such as chemical plants, petroleum refineries, off-shore drilling platforms, nuclear power plants, and transport systems where failures during services are potentially disastrous. Effective in-service nondestructive tests of materials and additional nondestructive tests made during maintenance (qv) shutdown periods can permit early detection and subsequent correction of hidden damage or deterioration (see also MATERIALS RELIABILITY; MATERIALS STANDARDS AND SPECIFICATIONS). Such test programs provide protection for the public, as well as for corporations and management from failure costs and penalties, which can include high litigation and insurance costs (7–10).

Nondestructive tests applied to consumer products ensure safety and proper operation for the purchaser. Tests might also be used periodically by the consumer during service to determine whether deterioration or conditions that might lead to premature failure have developed as a result of improper handling or storage, or misuse or abuse by operators. Inspection is especially important in cases where cracking or failure has already begun but has not progressed to the point of propagating toward a disastrous sudden failure. Use of NDE can add to customer satisfaction, eliminate the need for recalls and repairs, and help to protect both manufacturers and service industries from damage suits and litigation. The proper use of nondestructive tests can be a factor of economic survival for organizations manufacturing and servicing consumer products (7–10).

Program Requirements

Education and Certification. Nondestructive tests only provide indications which must be interpreted. An effective NDE program requires educated and experienced personnel at all levels. Management personnel who have technical degrees are essential. Decisions on inspection frequency, method, and the critical nature of the findings require engineering knowledge. Further, NDE operators must be adequately trained and supervised for evaluation to function properly. A feeling of trust and support must also exist within the NDE program. Associates degrees provide excellent entry level education for NDE technicians. A list of educational institutions offering NDE education is regularly published by the American Society of Nondestructive Testing (11,12). Government, technical society, and industry codes and specifications often control the certification and qualification of nondestructive test personnel and their supervision.

Test operators and engineers must evaluate indications from test results and make decisions concerning the suitability of the material for further processing or use. For the NDE team, this decision involves interpretation of the test indications based on prior knowledge of the nature, composition, and structure of the material and the effects of prior process handling. Items to be considered include the environment in which the material is to be used, the stresses and operating conditions to which it is to be subjected, the material properties and discontinuities that could contribute to failure in its intended service, and the probabilities of failure under various combinations of loading or environmental conditions. If failure is thought to be possible, the dangerous material should be replaced or repaired, and further NDE made to confirm adequate repair or replacement.

An effective NDE program relies heavily on periodic certification of the competence of its personnel (13,14). Certification programs designate levels of competence for all levels of personnel. Level I technicians are able to carry out instructions in an NDE; Level III supervisors are qualified to evaluate the needs of the test and devise a scheme that assures the desired level of quality or safety.

Risk-Based Inspection.   Inspection programs developed using risk analysis methods are becoming increasingly popular (15,16) (see HAZARD ANALYSIS AND RISK ASSESSMENT). In this approach, the frequency and type of in-service inspection (ISI) is determined by the probabilistic risk assessment (PRA) of the inspection results. Here, the results might be a false acceptance of a part that will fail as well as the false rejection of a part that will not fail. Whether a plant or a consumer product, false acceptance of a defective part could lead to catastrophic failure and considerable cost. Also, the false rejection of parts may lead to unjustified, and sometimes exorbitant, costs of operation (2). Risk is defined as follows:

$$\text{risk} = P(F) \times C(F)$$

where $P(F)$ = probability of failure and $C(F)$ = consequences of failure. The consequences of failure might be a direct cost as well as a loss of good will or injury or death of an employee or customer.

In the simplest terms, a fault-tree for risk analysis requires the following information: probability of detection of a particular anomaly for an NDE system, repair or replacement decision for an item judged defective, probability of failure of the anomaly, cost of failure, cost of inspection, and cost of repair. Implementation of a risk-based inspection system should lead to an overall improvement in the inspection costs as well as in the safety in operation for a plant, component, or a system. Unless the database is well established, however, costs may fluctuate considerably.

Human Factors.   Several nontechnical factors can significantly affect the results of a nondestructive inspection. Many of these are classified as human factors (1,2,17). Operator experience affects the probability of detection of most flaws. Typically, an inexperienced operator has more false rejects, known as Type II errors, than an experienced operator. A poor operator has few false rejects but is more likely to miss a defect in the inspection, known as a Type I error. Operator fatigue, boredom, or an unfavorable environment such as lighting, cold, or rain may further affect performance. Thus it usually is a good investment for the inspection company to assure that the operator environment is most amenable to inspection, that the equipment is suitable for the task, and that the operator is alert and well rested.

Visual Inspection. Visual inspection should always be regarded as the first defense against failure (1). Without scientific proof, it is estimated that 80% of defects are found by visual inspection. A pilot walking around an aircraft, or a mechanic observing a machine in operation often finds defects very quickly. The cost of this inspection is minimal. Human factor considerations are particularly important for the visual inspection process. Although the visual inspection is perhaps the most inexpensive and finds the most defects, the 20% of the defects remaining after the visual inspection must also be found; thus the more costly and technically elaborate NDE methods are needed.

Automation and Control of Tests. Increased use of automated systems for controlling and analyzing inspection systems, processes, and results has led to improved efficiency and confidence in NDE (18–22). Equipment must often operate under severe conditions of test material and instrumentation contamination, vibration, temperature, corrosive exposure, and inaccessibility. In a large chemical plant or petroleum refinery, for example, the structures themselves often make access difficult or hazardous, both for test equipment and for test personnel (23). In such cases, remote locations of probes and sensitive test signal detectors may require transmission of initial test signals to electronic analysis equipment and data analyzers in protected locations. This introduces the possibilities of damage not only to test equipment, but also to the transmission and communication systems that provide these interconnections. In large installations, surveillance systems designed to detect human or equipment intrusions into critical locations may become a vital part of nondestructive test systems designed to monitor plant or transportation equipment and prevent disastrous failure during its operating life.

Principles

Most nondestructive tests require use of suitable probing media to cause the test objects to emit signals that can be detected and interpreted in terms of material properties or defects. These resultant test signals may be converted to visible or audible indications, analogue electrical signals, meter or digital display readings, computer data, or images of many different types. Output signals, images, or data must then be evaluated by the human test operator or by automated test systems.

Each type of probe used in active nondestructive test systems has specific capabilities and limitations. The probe type, ie, a coil, slender probe, flat transducer, etc, affects the response from the discontinuity or feature to be evaluated. For example, surface, near-surface, surface-connected, interior, back-surface, flat surface, curved surface, rough or smooth surface anomalies and discontinuities such as cracks, porosity, inclusions, lamination, segregation, leaks, bonding, corrosion pinholes or attack, erosion, wall thinning, missing components, etc, all can be detected using NDE. No single type of inspection method, however, is adequate to detect all of the material conditions and discontinuities that can influence the serviceability of materials or systems. Thus, for reliable testing, two or more basically different types of test methods are usually necessary to detect and confirm conditions that may affect performance. Different materials or test-object geometries may require totally different tests or combinations of tests to ensure reliability. In many cases, the test methods appropriate for materials identification and selection prior to processing or welding are quire different from those that should be used after manufacture, erection, or during maintenance shutdown periods of facilities. In particular, the test method should be selected with consideration of materials, shapes, accessibility, and conditions or defects that affect the integrity and serviceability of the material, component, structure, or system to be tested.

External Energy Tests.   Material response to an external energy or probing medium input is the nondestructive test output signal. Many test systems involve measuring the reflection or scattering of the probing medium by the material of the test object. For example, reflected signals are used in ultrasonic testing for thickness or defects (see ULTRASONICS). Scattering is used in x-ray diffraction and topological analyses where images are reconstructed from a single diffraction line, in optical holographic nondestructive tests of opaque test objects, where interference patterns are formed and recorded on fine-grain photographic film, and in various light-scattering tests for analysis of fine threads or surface conditions (see X-RAY TECHNOLOGY). Still other active systems involve the conversion of incident probing medium energy into different forms of emitted energy, as when neutron beams stimulate emission of gamma rays from test materials. This is used in some forms of oil-well logging or neutron radiography or in examination of some fine arts objects. In some cases, the test object serves simply to obstruct or attenuate the incident beam or field of the probing medium, creating a shadow of varying intensity as a result of interfering effects of the material upon the propagation of the exploring medium, as in the case of optical comparators.

Passive Tests.   Passive nondestructive tests are those for which no specific probing medium, other than environmental or service loading conditions, need be applied to test objects to obtain test signals. Signals are emitted when changes in internal stress distributions, phase or structure, temperature, corrosion attack, external loading, or other operating conditions produce them. One dramatic example of a passive signal is the loud bang that accompanies the failure of a weld or beam when it is overloaded or fails by virtue of defects. Smart materials, containing embedded sensors (qv), may be effective for passive tests (see SUPPLEMENT).

Surface and Near-Surface Defect Detection

Inspecting external surfaces or walls for cracks and leaks, or gauging the dimensions and surface geometry of test objects, requires techniques having special capabilities. Surface inspections typically involve techniques using gases, vapors, fluids, and particles. Motion may be imparted to such active, probing media by mechanical, hydraulic, or electromagnetic forces, or it may result from gas or vapor pressure, diffusion, permeation, osmosis solubility, evacuation, pumping, or other factors. In other cases, capillary action is all that is required. Additionally, devices such as calipers, mechanical gauges, and micrometers may be useful for surface inspections. Subsurface discontinuities and defects that have no openings to these externally accessible surfaces cannot, as a general rule, be detected by tests based on the active media techniques (see also SURFACE AND INTERFACE ANALYSIS).

Visual Inspection and Optical Tests.   Direct visual inspection and examination using optical aids and gauges is the most common of all forms of nondestructive tests. Microscopic, telescopic, and electronic aids to vision include borescopes and magnifiers. These devices are applied for inspecting inaccessible areas within turbine engines and along the internal length of long tubes. Corrosion, distortion, and general weld quality may be detected using these techniques. Additionally, typical go–no go gauges may be used for assuring dimensional quality in welds (3). Visual examination of welds also may locate common defects such as unfilled craters, undercutting, overlap, and incomplete penetration. Quantitative measurements utilize geometrical optics, stroboscopic tests, spectrographic comparators, densitometers, refractometers, interferometers, tests for thickness of thin layers, phase-contrast techniques, and tests employing polarized light and measurements of birefringence. Advanced techniques include light section, schlieren (shadow), and diffraction methods of optical imaging and measurements. The effects of light scattering are used in the Eberhardt fine thread test, the dual pinhole fine structure test, and wave front reconstruction techniques for extreme magnification, ie, up to 10^7 times enlargement reconstructed from an x-ray diffraction pattern (see MICROSCOPY).

Liquid Penetrant Inspection.   Liquid penetrants provide unique sensitivity to surface cracks and other discontinuities open to the test surface (1–6,24). Advances in formulation of carrier liquids and of color or fluorescent dye systems containing developers have permitted the detection of discontinuities having dimensions smaller than a half wavelength of light, which are not visible using the light microscope or metallograph in the absence of the liquid tracer. Some high resolution liquid penetrant and developer systems have shown reproducible test indications of crack-like discontinuities too fine to be replicated for imaging with the electron microscope. Precleaning of test surfaces to remove grease, oil, rust, scale, or other contamination coatings is needed prior to use of liquid penetrant tests (see METAL SURFACE TREATMENTS). Contaminants can fill or cover discontinuity openings and inhibit entry of the liquid tracer.

Liquid penetrant applied to test surfaces by various means enters cracks and defects by capillary action during the specified penetrant dwell period. Excess penetrant is removed from the surface by water spray, emulsifiers, or solvent wiping. Developers then are applied and form a surface coating into which entrapped penetrant is drawn from the opening to form visible color or fluorescent images. When fluorescent penetrants and black light (uv) illumination are used, these become high contrasting bright indicators. The vehicle and dyes are specially formulated for rapid entry and entrapment in discontinuities and easy removal of excess penetrant without disturbing the entrapped tracer. Depending on the developer used, the image formed can be greatly enlarged and highly visible.

As for other tests that involve application of liquids or chemical tracers, such test materials must be removed from the test surfaces and discontinuities after use to avoid later corrosion or damage to the base materials. In aerospace and nuclear test applications involving sensitive materials such as austenitic stainless steels, the tracers and removers must be formulated with a minimum of halogen or sulfur impurities so that residues that might be retained within discontinuities or crevices in assembled components do not contribute to later stress corrosion cracking or other deterioration. This precaution is essential in the case of various high temperature and exotic alloys such as nickel-base alloys, austenitic stainless steels, and titanium and its alloys (see HIGH TEMPERATURE ALLOYS; NICKEL AND NICKEL ALLOYS; STEEL). Oil-base tracers and other liquids that might react with liquid or gaseous oxygen (qv) should be avoided when testing structures to be used for oxygen storage.

Other test media and techniques include post-emulsification penetrants, penetrants that form gels resistant to easy removal from entrapments, penetrants that concentrate dye constituents as their carrier liquids evaporate during test processing, and penetrants that form strippable coatings in the developers. Still other penetrant systems are formulated for use at abnormally low or high temperatures for special test applications.

Filtered-Particle Inspection. Solids containing extensive interconnected porosity, eg, sintered metallic or fired ceramic bodies formed of particles that are typically of 0.15-mm (100-mesh) screen size, are not inspectable by normal liquid penetrant methods. The preferred test medium consists of a suspension of dyed solid particles, which may be contained in a liquid vehicle dyed with a different color. Test indications can form wherever suspensions can enter cracks and other discontinuities open to the surface and be absorbed in porous material along interior crack walls. The solid particles that form test indications are removed by filtration along the line of the crack at the surface where they form color or fluorescent indications visible under near-ultraviolet light (1,3).

Magnetic Flux Leakage. Static magnetic fields are used widely to test ferromagnetic materials for surface and near-surface defects (1–6). Steels and other ferromagnetic materials are strongly magnetized using external coils, yokes, or magnets (see MAGNETIC MATERIALS). The relationship of current and magnetic field directions is described by the right-hand rule, with the thumb of the right hand in the current direction and the fingers in the direction of the field lines. Circular magnetization of tubes or round bars is done by passing a high d-c or a-c magnetizing current through a central conductor or through the test object itself. Longitudinal fields are generated in round parts such as bars and tubes by inserting them in an energized circular coil. Additionally, magnetic yokes having movable legs are used to excite longitudinal fields in both flat and irregularly shaped parts. Flux lines in a saturated ferromagnetic material are diverted by cracks, loss of wall thickness, or other anomalies, so that these flux lines leak into the surrounding medium, as shown in Figure 1.

Ferromagnetic materials are differentiated from those that are nonferromagnetic by the permeability μ as defined by the following:

$$\mu = \mu_r \mu_0 = \frac{B}{H}$$

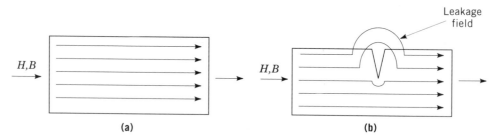

Fig. 1. Schematic of magnetic flux leakage tests where H,B represents an external applied magnetic field. See text. (**a**) Saturated material having no crack, and (**b**) leakage field caused by crack.

where μ_0, the permeability in a vacuum, $= 4\pi(10^{-7})\,H/m$, and μ_r = relative permeability, B = flux density $= \Phi/A$ (in T), H = field strength (in A/m), Φ = total flux (in Wb), and A = area (in m). Other parameters useful in defining the magnetic response of a material are the saturation flux density, B_s; remanence or retentivity, B_r; and the coercive force, H_c. Typical low and high strength steels may have the values shown in Table 1.

The magnetic leakage field may be detected in several ways, most frequently by a moving conducting coil or sensor, very fine ferromagnetic particles, or recording tape. A moving probe coil or a semiconductor magnetic field detector passing through a leakage field causes a voltage to be generated. The voltage may be analyzed to determine the presence and characteristics of typical defects. Finely divided, low remanence magnetic particles are another common technique for indicating flaws. When these particles are applied to the surfaces of the suitably magnetized test object, they collect where leakage magnetic flux exists at the material surface. These accumulations usually define cracks, porosity, or near-surface submerged cavities. The particles may be coated with lubricants and color dyes or fluorescent materials to aid visibility. Parts tested using visible dye coatings are inspected under white light; those tested using fluorescent magnetic particles are inspected under near-ultraviolet radiation (365 nm) from a filtered mercury-arc lamp. Typically, wet fluorescent magnetic particles define small discontinuities better than dry particles in visible light. Tests of these types are often used on piping, tubing, and weld root beads and completed fillet welds on steel structures. They also are used widely on jet engine turbine blades made of ferromagnetic materials, including some cobalt-base alloys. In other test systems, replication is made of the magnetic field patterns on flexible magnetic recording media. Magnetic recording techniques permit electronic imaging of defects, and provide records that can be stored, transmitted, or reproduced

Table 1. Magnetic Parameters for Steel[a]

Steel	μ_r	B_r, T[b]	B_s, T[b]	H_c, A/m
low strength	450	0.94	1.5	880
high strength	450	1.49	1.7	1150

[a] Ref. 2.
[b] To convert T to G, multiply by 10^4.

(see INFORMATION STORAGE MATERIALS, MAGNETIC). Magnetic leakage flux tests have wide application in industrial inspection. Typical are tests of longitudinal welds in pipe during fabrication, and oil-field equipment, including oil-well drill pipe, tubing, line pipe, and structures. Pipelines (qv) are regularly inspected for thin wall using the magnetic flux leakage principle. Internal inspection devices called pigs are used for inspecting buried pipelines. Similar tests can be made on structures and piping in chemical plants and petroleum (qv) refineries.

Eddy-Current and Magnetic Induction Tests. Time-varying magnetic fields excited by alternating current are used both in eddy-current and magneto-induction nondestructive tests (1–6,25,26). Figure 2 shows a typical eddy-current excitation probe. The varying input voltage acting on the round iron core causes the magnetic field exiting the core ends. When held near an electrically conducting material, the alternating magnetic field excites eddy currents at the surface of the part, just beneath the probe. As for a-c electric power circuits, the impedance (Z) to the flow of current results from two effects: resistive heat-energy loss resulting from current through resistance, R, and reactance created by magnetic field momentum effects which tends to cause the current to lag the voltage that induces it. This reactance, X_L, is related to energy storage in the magnetic field of the test coils. The resistance R is related to the rate of power dissipation and production of heat, both by the current in the magnetizing coil and by the eddy currents in the test material. The total coil impedance, in the presence of the test object, can be represented by the following relation:

$$Z = R + jX_L$$

Both resistance and reactance are measured in ohms. The impedance may be represented in quadrant 1 on a complex plane diagram, where R and X_L

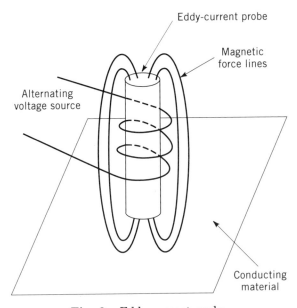

Fig. 2. Eddy-current probe.

are plotted along the horizontal and vertical axes, respectively (Fig. 3). On the complex impedance plane, the empty coil point, A, corresponding to the coil impedance when no test material lies in its field, is on the vertical, reactive axis. Introduction of a test material into the coil field changes the impedance to a point B in quadrant 1 having both resistive and reactive components. As the test frequency, material conductivity, or material thickness increase, the combined impedance, Z, tends to follow a curved locus, moving downward and to the right as eddy-current effects increase. By calibration of the eddy-current test instrument and test coil on sheets or bars of known materials, each such change can be quantitatively related to test material conductivity, relative magnetic permeability, thickness, nonconducting coating thickness, or presence of discontinuities that interrupt the normal circular flow paths of the eddy currents.

Eddy-current reactions are different for ferromagnetic and nonferromagnetic materials. Significant material variables, test object geometry, and changes in composition or temperature are measurable using eddy current in nonferromagnetic metals and alloys (1). For steel and other ferromagnetic materials, property determination is more difficult, but detection and location of discontinuities is feasible and can be performed at high speeds. Many steel producers use eddy-current tests extensively to control quality of bars, billets, sheets, and plate, and welded or seamless pipes and tubes. Aerospace manufacturers use eddy-current tests to detect material fatigue and other cracking conditions, corrosion effects, and damage from fires or overheating of aluminum alloys, for example. Other tests are made to measure and control the thickness and continuity of protective metallic coatings (qv), or the thickness of nonconducting coatings on conducting base materials.

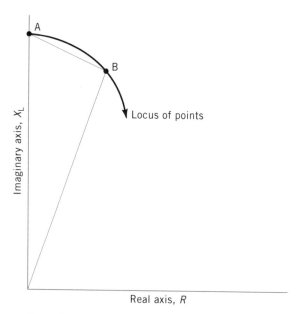

Fig. 3. Impedance plane diagram for eddy-current inspections, where A is the empty coil point and B is impedance in a coil field. See text.

Eddy-Current Penetration Depth and Defect Sensitivity. Eddy currents are most sensitive to discontinuities at the excited surface which interrupt or distort the induced flow paths. Magnetic fields of varying intensity can be produced by coils that carry time-varying electric currents, or by permanent magnets moving with respect to the test object surfaces. The magnetizing coils act like the primary windings of electric power transformers and induce voltages in conductors within their magnetic field of influence by Faraday's law of induction. If the test material is an electric conductor, these induced voltages set up eddy currents in accordance with Ohm's law. The eddy currents, in turn, create their own magnetic fields which react upon the primary magnetizing coil fields by opposing them in accordance with Lenz's law. In metallic conducting and ferromagnetic test materials, time-varying magnetic fields can induce varying states of magnetization and eddy currents at and below the surface. In general, the greatest magnetization and eddy-current density are found in the surface layer adjacent to the exciting magnetic field. The field intensity and eddy current density diminish exponentially beneath this excited surface. The standard depth of penetration, d, for eddy currents in the test material is the depth at which the current density drops by 37% of the value at the surface. This depth (in mm) is given by the following:

$$d = (\pi f \mu \sigma)^{1/2}$$

where f is frequency in Hz, μ is permeability, and σ is electrical conductivity in mho/m.

Frequencies from 1 kHz to 50 MHz are used for various applications (3). Ferromagnetic materials have a skin-effect response to eddy currents which restricts the penetration depth. Nonferromagnetic materials on the other hand can be inspected to greater depth. In 6061-T6 aluminum, for example, a coil having a 1-kHz frequency effectively penetrates the surface to a depth of 3.2 mm (1). The same probe in steel penetrates to a depth of 0.5 mm. Higher frequency probes penetrate to lesser depths, thus these are sensitive to smaller defects on the surface. The eddy currents themselves create magnetizing forces that tend to oppose the magnetizing field of the exciting test coil. This generally results in a reduction in the net magnetic field, and a time lag, within the test coil and throughout the portion of the test object under the influence of the magnetizing field. By measuring the magnetizing coil impedance or the magnetic field intensity just above the excited surface in the vicinity of the magnetizing coil, the inspector can obtain test signals that are related to the test coil diameter, magnetizing current magnitude and frequency, the test material conductivity, its magnetic permeability, and interruptions or distortions of eddy-current flow caused by discontinuities or defects.

Radioactive Krypton-85 Gas Penetrant Inspection. In a highly sensitive surface inspection technique, test objects are first placed in an evacuated chamber that is then filled with radioactive krypton-85 [*13983-27-2*], [85]Kr. The gas, inert to chemical reactions, is preferentially adsorbed in microcracks, porosity, shrinkage, and related surface connected discontinuities (1). After exposure, the krypton is pumped out of the chamber, returned to a storage vessel, and air is readmitted to the chamber. Krypton-85 is more readily purged from the free surfaces of the test object. The radioactive gas trapped within surface discontinuities is not purged. Krypton molecules, having diameters of only 0.3 nm, are able to diffuse into anomalies as small as 1 nm in width. Flaws are then visualized by placing a radiation sensitive film or emulsion dispersion on the test part surface, exposing it via the penetrating beta-ray electron emission (0.662 Mev) from [85]Kr entrapments. Photographic processing of the x-ray film or surface coating provides permanent detailed images of surface connected discontinuities (see RADIOISOTOPES).

Electron Surface-Transit Inspection with Solid Probe.   Conducting metallic surfaces of test objects can be inspected for geometric discontinuities, oxides, or surface contamination conditions by means of a solid conducting probe, preferably a gold-plated electrical contact, drawn across the surface in a manner similar to drawing a line with a pencil. Contact interruptions can occur when the probe crosses a surface crack or groove, or when metal-to-metal contact is interrupted by a layer of insulating oxide, giving rise to triboelectric effects, which induce electrical signals both in the probe and in the conducting test object. The resultant wide-band pulse signals can be amplified and detected by audio, am radio, fm radio, or other electronic signal detectors and amplifiers over a wide frequency range from 10 Hz to >100 MHz. When the probe is moved continuously and smoothly, triboelectric signals are generated at each contact interruption. If the tip of the probe is caused to vibrate vertically so as to make and break electrical contact with the test surface at a known frequency, a continuous sequence of pulse signals can be generated. Changes in contact potentials caused by varying contamination states or intermittent surface coatings, as in the case of oxide layers, rust, or temper colors or by surface cracks or grooves, also are indicated by changes in the pulse amplitude signal.

Electrified Particle Tests.   Static electric fields serve to test dielectric materials such as glass (qv), porcelain, enamel (see ENAMELS, PORCELAIN AND VITREOUS), plastics, or organic resin products, and coatings. Electrified particle tests use calcium carbonate [471-34-1], $CaCO_3$, particles blown through rubber or other nozzles to produce high static charge, and sprayed with dry air dielectric surfaces or coatings over conducting base materials, which are usually grounded (1). The cloud of charged particles provides the electric field which serves as the probing medium; the particles attracted and held electrostatically on the dielectric surfaces serve as indicators of location and geometry of surface cracks and discontinuities. This test is capable of reproducibly detecting surface cracks too narrow to admit visible light or invisible light via a light microscope or metallograph.

Electron Imaging Tests.   Analytical instruments such as the electron microscope and the scanning electron microscope (sem) employ high energy electron beams to image thin test objects, replicas of surfaces of thick objects, and other applications. These surface imaging tests are made using electrons decelerated to energies of 0 to ~ 4–5 eV to produce positive images when the electrons enter the test object and pass to ground through a resistor, and the test object is electrically connected through a capacitor to the video input of a television picture tube or video recorder. If the electron energy exceeds the crossover voltage so that the secondary electron emission ratio exceeds unity, each minute element of the electron image reverses polarity at its characteristic energy level. This provides a quantitative measure of the nature of the surface layer of contamination which strongly influences the work function of the metal.

Inspection for Internal Defects

Energy transmitted for use in nondestructive testing includes acoustic, static electric, magnetic, electromagnetic, gravitational fields, dynamic electromagnetic fields, and photon beams. Depending on wavelengths relative to the typical spacing between atoms in the test object, such probing media may be limited to testing exposed surfaces, eg, light irradiation of opaque materials, or may be capable of penetrating to great depths, as can high energy x- or γ-rays. Among the advantages of the energy transmission methods is the possibility of visualizing test object surfaces or interior volumes. This also makes possible the detection of objects at considerable distances, precise dimensional or displacement measurement, and the inspection of objects that are in motion, as in flash photography or radiography, or in motion picture or television imaging and recording. When imaging is feasible, there is a further psychological advantage. Visual information often is more convincing to management and workers. Digital imaging also allows tests results to be transmitted to remote points, stored, and retrieved for use by others at later times. Further, image enhancement may be useful for special interpretation purposes.

Ultrasonic Tests. Ultrasonic nondestructive tests are made by introducing beams of high frequency mechanical stress waves into test materials (see ULTRASONICS) (1–6,27). The velocity of sound propagation in a test material is directly proportional to the square root of the elastic moduli and inversely proportional to the square root of density of material. Sonic waves travel in bulk solids both as longitudinal and shear (transverse) waves. In longitudinal waves the particle displacement is in the direction of the wave movement, and for shear waves the particle displacement is perpendicular to the direction of wave movement. Typically, shear wave velocity (C_2) is about one-half to two-thirds of the longitudinal wave velocity (C_1) in the same material. Shear waves do not propagate in fluids. Wave speed for some common engineering materials are given in Table 2.

A variety of other wave modes are useful in nondestructive evaluation. For example, Rayleigh, or surface waves are useful for detecting very small surface breaking defects. These waves, traveling exclusively on the material surface, are the same type as seen on the top surface of a pond when a stone is dropped onto it. Plate waves, also called Lamb modes, are useful for inspecting plates, tubing,

Table 2. Wave Speeds for Common Materials, m/s[a]

Material	C_1	C_2
steel	5900	3230
aluminum	6320	3130
clear acrylic resin (PMMA)	2670	1120
water	1483	

[a] Refs. 1 and 2.

and layered media such as composite materials (qv). In each of these cases, the path of travel may curve with the surface or shape allowing the ultrasonic energy to reach normally inaccessible surfaces.

Ultrasonic inspection decisions generally are based on several parameters, notably amplitudes, duration, and arrival times of transmitted, refracted, and reflected waves. Most tests are made by pulsed ultrasonic beams generated by piezoelectric transducers excited by high voltage spikes occurring at regular intervals, ie, the pulse repetition rate (see PIEZOELECTRICS). The initial pulse, which occurs when the high voltage spike strikes the probe, is commonly referred to as the main bang. The front surface echo occurs when the sound pulse initially enters the test piece. In contact testing the transducer is in contact with the test object, and the main bang and the front surface echo occur simultaneously. For immersion testing, where the probe is separated from the object being inspected by a fluid, the front surface echo appears after the main bang.

Frequently, a single ultrasonic transducer serves both as the sender of the ultrasonic pulse and as a receiver for the sound waves reflected from surfaces and interior discontinuities. The receiver transforms the stress pulse back into electrical oscillations. All of the signals are displayed on an oscilloscope screen for interpretation. For a material of length L having a wave speed C_1, the anomaly shown in Figure 4 would reflect a signal back to the send–receive probe. This is a pulse echo test arrangement. The arrival time of the reflected signal, compared to the arrival time of the back-surface echo, would indicate the depth below the inspection surface of the hidden flaw. Large area reflectors (anomalies) would be expected to reflect larger amplitude signals. This is not always the case, however, because the shape and orientation of the anomaly considerably influences the character of the reflected pulse. In many cases, other information such as frequency content and arrival times at multiple probe locations is required for a more knowledgeable decision on the characteristics of the unknown reflector.

For contact testing a couplant normally is used between the probe and the test piece. This material may be oil, water, or some gel or other liquid or paste. Compatibility with the test object is important, so that no unexpected chemical attack occurs, causing a crack to initiate. Whereas the frequency range

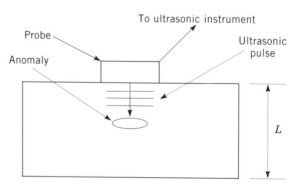

Fig. 4.   Schematic of an ultrasonic nondestructive testing set-up of material of length, L.

for ultrasonic tests may extend from approximately 50 kHz to 50 MHz, the range most commonly used for metallic test materials is 0.5–25 MHz.

Immersion Testing. In immersion testing, the test object typically is submerged in a tank of water, and the sound beam from the search tube is coupled through several centimeters of water to the front surface of the test object. If the incident beam is aligned normal to the surface of the test object, part of the energy of the incident beam strikes the surface of the test object and is reflected back through the water to the face of the transducer element. The reflected pulse causes a sequence of signals known as water multiples. The initial pulse striking the water and the test piece interface also excites an inspection pulse in the specimen which behaves exactly as it would for the contact method. In immersion testing the initial pulse moves if the distance from the probe to the object is changed, and the main bang always stays stationary at time zero. It is good practice to lengthen the water path so that the first water multiple signal returns after the back-surface echo from the test object has been received. This avoids confusion in interpreting between water multiples and echoes from significant discontinuities within the test material.

In immersion testing, the search tube which supports the transducer can be changed as needed to obtain any desired angle of incidence. The search tube can be moved in regular scanning patterns to cover the entire test object area for detailed inspection. Large tanks or vessels containing fluids may be inspected using a remote, submerged manipulator located internally to scan the walls with either normal or angled probes. Many large immersion inspection systems provide automatic scanning under computer control, computer data analysis, storage of ultrasonic test data, and recording of plan and side views of the ultrasonic response of the test object. The B (side view) and C (plan view) scans may be displayed in color or multiple gray scale indicating the depths of the indications.

Angle Beam Testing. Flaws oriented perpendicular to the test surface are more easily detected with an angle beam where the inspection beam enters the test piece at an angle away from the normal. In welds, defects hidden under T-joints and lack of fusion discontinuities are more easily detected using angle beams. The relationship of the incident, reflected, and refracted waves is governed by Snell's law which states that for all waves the ratio of sine of the incident angle to the wave speed is constant. In Figure 5, for a solid–solid interface, the incident longitudinal wave traveling at speed C_1 and at angle θ_1 excites the reflected longitudinal and shear waves at θ_1'' and θ_2'', respectively. Also, the refracted longitudinal and shear waves occur at θ_1' and θ_2', respectively, in the lower material. For most applications of angle beam inspection, the incident beam is inclined so that θ_1' is greater than 90°, thus eliminating the longitudinal wave beam and leaving only the transverse wave beam to be used for inspection. Using a still greater angle of incidence, the transverse wave can also be eliminated, leaving a strong Rayleigh wave traveling on the surface. In fluids, there are no shear waves excited. The surface wave travels slightly slower than the shear wave, and depth penetration is about one wavelength into the material. The wave can be reflected from surface connected cracks, sharp ridges, and even from liquid drops on the surface of a dry material. The oblique angle for angle beam inspection is achieved in contact testing through the use of

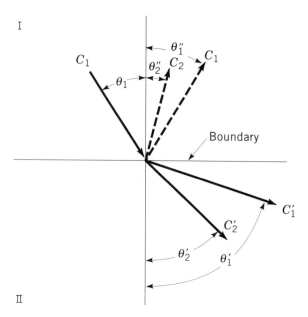

Fig. 5. Reflection and refraction at a solid–solid interface at an oblique angle. See text (2).

an intermediary wedge material having a lower longitudinal wave speed than the material being inspected. Clear acrylic resin, typically polymethyl methacrylate (PMMA), is commonly used for wedge material.

Special contact transducers having wedges providing incidence angles for specific applications are used widely in industry. For example, normal incidence is used in tests for laminations within sheets, and for sheet or plate thickness where the back surface of the test material parallels, to within perhaps 10°, the front surface. Shear wave transducers typically used for weld examination have 45°, 60°, or 70° inspection (refracted) angles. To locate discontinuities, the transducers may be moved back and forth over selected surface areas until the angled search beam approaches normal incidence on the flaw surface, maximizing the reflection signal. In immersion ultrasonic tests, the angle of incidence of the sound beam can be readily changed.

In most ultrasonic tests, the significant echo signal often is the one having the maximum amplitude. This amplitude is affected by the selection of the beam angle, and the position and direction from which it interrogates the flaw. The depth of flaws is often determined to considerable precision by the transit time of the pulses within the test material. The relative reflecting power of discontinuities is determined by comparison of the test signal with echoes from artificial discontinuities such as flat-bottomed holes, side-drilled holes, and notches in reference test blocks. This technique provides some standardized tests for sound beam attenuation and ultrasonic equipment beam spread.

Acoustic Emission Tests. Acoustic emission tests are based on the detection and analysis of relatively low level mechanical vibrations, often at ultrasonic frequencies, created within test materials by environmental or internal stress, or external loading conditions (1–3). The lowest intensity acoustic signals are often ascribed to dislocation movements in which a few atoms shift within the solid material. Other signals, usually of higher intensity, may be caused by internal inclusions which crack or break free of their bonds with the metallic base material under certain conditions of stress. Still greater signals result from small sudden extensions of cracks which propagate within the material, as a result of cyclic stress, stress corrosion, fatigue, creep, or thermal cycling. These signals are usually considered to be highly significant because they are direct indications of local failures within the test material. Of course, signals of greater magnitude, which are often quite audible to human observers, accompany significant fractures.

Large-scale instrumentation and analysis systems have been applied in experiments designed to locate sources of acoustic emission, discriminate against insignificant noises, ie, those noises coming from outside the system or material under test, and to record trends and predict potential failures before these occur. Systems such as pressure vessels, chemical and nuclear power reactor pressure systems, bridges, dams, aircraft, crane booms, and space vehicles have been instrumented so that a database could be developed of in-service acoustic emission characteristics. As these databases increase and more knowledge is gained about each particular structure, the utilization of acoustic emission in industrial applications is expected to increase. The emergence of more advanced data analysis equipment could facilitate this growth.

Penetrating Eddy-Current Tests in Nonmagnetic Material. Conducting materials having a low permeability, μ, may be inspected to considerable depth using eddy current. This is because of the absence of the skin effect for these materials. Low frequency probe coils, larger than the conventional ones used for inspecting ferromagnetic materials, are used for these inspections. Penetration depths of greater than 12 mm for 304 stainless steel are reported (1). In an electron-beam weld in 6–4 titanium, lack of fusion in the root area was found for a thickness of 25 mm (25). Penetration depths of approximately 4 mm are possible for low frequency tests in aluminum.

Microwave Tests. Microwave tests may be considered to be a high frequency type of electromagnetic induction or eddy-current test when used with conducting test materials (1). Wavelengths are in the range of a few centimeters and microwaves penetrate only to shallow depths in highly conducting or ferromagnetic materials. However, microwaves often penetrate to considerable depth in dielectric materials, suffering phase shift and attenuation effects somewhat like those observed for eddy currents. Microwaves can be generated by electronic circuits, and can be beamed or propagated from horns, dish antennas, and many special types of antenna systems (see MICROWAVE TECHNOLOGY).

Internal and surface flaws may be located and material property investigations may be conducted using microwaves (1). For use on industrial and consumer products, microwave test equipment is miniaturized to probe small regions of test materials by reflection or transmission measurements from a short range. Microwave tests are used for measurements of highway pavement thickness. Thickness tests of metals employ radar transmitters and detectors on both sides of moving strips. Microwaves can respond to various surface coatings. Surface cracks in conducting materials may respond like slotted antenna arrays to scanning fields or beams of microwaves. Microwaves are used to detect and evaluate faults in microwave plumbing or transmission systems, where impedance mismatches cause reflection, with or without phase reversal, of signal pulses.

X-Ray and γ-Ray Tests. Radiographic imaging tests (1–6,28) are widely used to examine interior regions of metal castings, fusion weldments, composite structures and brazed honeycomb mechanisms, and many other metallic and nonmetallic components. Radiographic energy, in the form of photons, may be excited using either an x-ray tube or a radioisotope, such as iridium-192 [14694-69-0], ^{192}Ir, or cobalt-60 [10198-40-0], ^{60}Co. Higher energy levels penetrate greater thickness of the same material. X-ray energies in the 8–160 kV range penetrate up to 76 mm of light alloys and 6 mm of steel. Radioisotopes such as iridium-192 are used for thinner steel pipe walls or lower density metals. Cobalt-60 is used for steel thickness in the range of 25 to 190 mm. Penetrating radiation, introduced at intensity I_0 on the side of the part facing the source or tube, is recorded by film, real-time imaging systems, or by detection gauges on the opposite side (Fig. 6). Anomalies in the path are imaged on the recorded plane, provided that the alignment and exposure are correct. Maximum detection sensitivity occurs for radiography when the significant dimension of the anomaly is oriented parallel to the x-ray beam. Radiographic tests are made on pipeline welds, pressure vessels, nuclear fuel rods, and other critical materials and components that may contain three-dimensional voids, inclusions, gaps, or cracks aligned so that portions are parallel to the radiation beam. Because penetrating radiation tests measure the mass of material per unit area in the beam path, these can respond to changes in thickness or material density, and to inclusions of density different from that of the material in which the inclusions are embedded. The typical film radiographic test sensitivity used in industrial inspection can detect thickness or density differences equal to about 2% of material thickness. For the most critical nuclear and aerospace applications, however, specifications may call for demonstration of better contrast sensitivity in such penetrating radiation tests. When cracks follow irregular paths, some parts of which are aligned with the path of the radiation beam for 2% or more of the total beam path length in the test material, it is possible to visualize the cracks with great clarity if the cracks effectively reduce the thickness of material through which the probing beams must pass. However, x-ray and γ-ray film imaging tests are insensitive to cracks, disbonds, and laminar discontinuities that lie in planes perpendicular

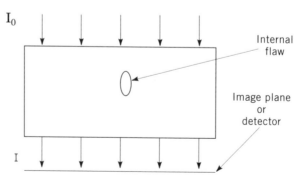

Fig. 6. Radiographic inspection. See text.

to the radiation beams or to defects representing little change in thickness. In extrusions and hot forgings, many discontinuities, which may originally have been three-dimensional voids or discontinuities, are flattened and greatly elongated in the deformation process. These discontinuities are usually parallel to the surface of the product and penetrating radiation tests rarely detect such laminar flaws. In fusion-welded pipe and tubing, however, often both the longitudinal welds made in the mill and the circumferential or girth welds made during tubing erection or during the laying of pipelines in the field provide ideal subjects for radiography inspections. In large-diameter pipelines, x-ray or γ-ray sources are often propelled into the open end of the pipeline on crawlers having detectors of girth weld locations for distances of several kilometers. The isotope source traveling on the pipe centerline emits radiation circumferentially, producing an interpretable weld image and a permanent record for further reference.

For pipelines in service in chemical plants, it is not usually convenient to place a radiation source inside the pipe and position it to irradiate each welded joint. The radioisotope source container may be placed on the outer surface of the pipe. The radiation beams then pass through two pipe wall thicknesses to expose films placed diametrically opposite the radiation source, also on the outside of the pipe wall. Other methods, such as magnetic particle inspection of welds in steel pipe, or ultrasonic inspection of welds in pipes of all materials, supplement x-rays in many critical applications. The ultrasonic tests can often detect the thin, laminar discontinuities parallel to the pipe surface or the incomplete fusion discontinuities along the weld preparation surface. These flaws often cannot be revealed by radiography.

Neutron Imaging Test. Beams and fields of energetic particles, such as neutrons, electrons, and other charged particles, produce useful images during transmission through or reflection from test materials (1–6,28). Neutron beams, which can pass through much greater thicknesses and masses of many materials than can x-rays or γ-rays, are, however, highly attenuated by hydrocarbons (qv) and other materials having high neutron scattering cross sections. Many metals are basically transparent to neutrons, thus neutron beams have proved useful in locating hidden corrosion in panels where the corrosion product is hydrogenous, and for evaluating the placement of polymeric parts in metal valves. Additionally, real-time systems have been used for imaging the flow of polymer and metal material into casting molds. Neutron activation is also used for the examination of oil paintings such as those attributed to Old Masters. Techniques for neutron imaging are similar to those used for x-ray imaging (see FINE ART EXAMINATION AND CONSERVATION).

Positron Annihilation. Positrons, positive beta-rays emitted by an atomic nucleus, have found significant application in investigating interior conditions of materials (29,30). Positrons may be viewed as antimatter. Their path through a material is significantly affected by imperfections. In metals, alloys, and polymers, positron emissions enable detection of deformation, creep, and fatigue damage. Use of positrons involves knowledge of their release quantity and expected travel path through a material. Determination of a deviation from the expected quantity and path indicates abnormal conditions in the material.

Infrared and Thermal Conduction Tests. Infrared or thermal radiation involves much shorter wavelengths than do microwave tests, but the former provides similar capabilities for nondestructive testing (1–6). Infrared waves transmit heat into surface layers of test materials. These waves are reflected from thick layers and partially transmitted through very thin layers, such as thin coatings on test materials. Infrared photographic films, infrared television camera tubes, and special night vision devices used in military and surveillance operations serve as sensitive remote detectors of surface temperature pattern. Infrared heat lamps have been used to excite test material surfaces, causing selective heating dependent on reflectivity of surface-to-infrared wavelengths, and conversion of incident heat into surface temperature patterns that depend on the thermal conductivity and heat capacity of subsurface materials.

Infrared patterns can lead to detection of lack of bonding between surface platings, between coatings and base metal, or within brazed honeycombs and other composite materials where bonding is vital. More recent developments involve infrared television or image tubes similar to the types used for night vision in military and other applications.

Optical Holography Tests. Optical holography (qv) utilizes coherent monchromatic radiation in the visible wavelength region (1). Three-dimensional images are formed on the holographic photographic film or plate when the film or plate is irradiated with the same type of radiation used to form it. A stable platform is required both for the optic system and for supporting test objects in fixed positions during exposure and viewing. Monochromatic photon beams typically generated using helium–neon or carbon dioxide lasers having emitted radiation wavelengths in the optical region, measure displacement or vibration of surfaces exposed to the radiation beam. As in the Michaelson-Morely experiment, the monochromatic (single-frequency) photon beam emitted from the laser is split using an angled glass partial reflector. The reference beam passes through a spatial filter (pinhole and lens) and is reflected by mirrors onto a sensitive photographic plate in a relatively uniform field intensity. The main beam passes through a similar pinhole and lens system to reduce higher order effects, and illuminates the surface of the test object. Light reflected from the test surface also illuminates the sensitive photographic plate, causing an interference pattern at that film plate surface where the pattern interacts with the reference beam. Differences in the arrival time of the waveforms create interference patterns. After exposure and development, the image produced on the photographic plate contains an ultrafine pattern which, viewed while the plate is illuminated by the reference beam only, reproduces the image of the test object, apparently in three dimensions. To reveal anomalies or changes in the part, the test object is again illuminated by means of the main beam where no part of the optical system or test object position is changed. Inspection through the previously exposed hologram on the photographic plate yields interference patterns having optical fringes surrounding anomalies where the test object surface has moved by small distances, eg, one or more half wavelengths of the laser radiation.

Other techniques are also used to produce holographic images. All holographic procedures have applications in industrial optical holography for inspection of materials, particularly brazed or adhesively bonded honeycomb sandwich structures, or adhesion of coatings on metal and other surfaces. Optical holographic tests are also used for inspection of tires for lack of bond between plies, shoulder anomalies, and other defects. These tests also serve to indicate vibration modes and amplitudes, including nodes and regions of intense local vibration, for machine parts and structures, vibrating diaphragms, and sonic or ultrasonic transducers.

High Voltage Probe and Corona Tests. Static electric fields and high voltage probes are used to detect internal voids in dielectric materials (1). An insulated high voltage probe, typically having a spherical conducting tip, is moved over the accessible surfaces to create a static field in the test material. Changes in the minute current used to charge the probe or radio detection of corona created within the internal voids can be used to locate and detect internal discontinuities in many cases. Tests involving excitation with a-c electric fields at low (60 Hz) frequencies are used to detect voids in the dielectric materials used on encased transformers, capacitors, and other electrical components (see EMBEDDING). Similar tests using probes at high voltages can be used to detect cracks and pinholes in dielectric coatings on conducting base metals by electric sparks from probe to base metal.

Materials Characterization and Identification

Nondestructive evaluation plays an important role in material property investigations. Among the many possible applications, the various techniques are able to determine the elastic constants and the texture as well as the chemical content necessary for alloy identification. Microwave tests are used most widely for measurements of moisture content of grains, foods, paper (qv), and ceramics, and in thickness tests of highway pavements. Thickness tests of metals employ radar transmitters and detectors on both sides of moving strips. Microwaves can respond to various surface coatings. Microwave phase and amplitude changes during transmission or reflection have been used for materials evaluation of ceramics, organic resin materials and composites, and rock samples from the earth. Microwaves are used to detect and evaluate faults in microwave plumbing or transmission systems, where impedance mismatches cause reflection, with or without phase reversal or signal pulses.

Probing media that use chemical reactions to identify materials, evaluate surface coatings, detect leaks and surface connected discontinuities, or measure corrosion resistance are typically limited to surfaces and regions accessible to fluid chemical probing (1,31). Chemical spot tests and modifications involving electrochemical reactions commonly are used to identify materials or to sort accidentally mixed metals and alloys. A metal-sensing liquid penetrant test system that responds only when cracks in nonmetallic composite materials extend to internal base metal or metallic reinforcements has been described (24). Eddy-current testing is able to sort several materials because of differences in conductivity (1,3,25,26,32). By calibration of the eddy-current test instrument and test coil on sheets or bars of known materials, observed differences can be quantitatively related to test material conductivity, relative magnetic permeability, or changes in composition or temperature for nonmagnetic metals and alloys. Applications for superconducting materials have been discussed (32).

For steel and other ferromagnetic materials, property determination is more difficult. Other tests are made to measure the continuity of protective metallic coatings. Residual stresses induced in welded structures and in components in service owing to chemical attack may contribute to early failure. Both ultrasonic and radiographic techniques have shown applications which are useful in determining residual stresses (27,28,33,34). Ultrasonic techniques use the acoustoelastic effect where the ultrasonic wave velocity changes with stress. The x-ray diffraction (xrd) method uses Bragg's law of diffraction of crystallographic planes to experimentally determine the strain in a material. The result is used to calculate the stress. As of this writing, whereas xrd equipment has been developed to where the technique may be conveniently applied in the field, convenient ultrasonic stress measurement equipment has not. This latter technique has shown an ability to differentiate between stress relieved and nonstress relieved welds in laboratory experiments.

Inspections

Leak Detection and Measurement. Leak testing utilizes fluid probing media for detection and location of leaks and for measurements of leakage rates through individual complex systems (31). Leak detection and prevention is vital where systems contain poisonous or hazardous chemicals, fuels, oxidants, radioactive materials, high pressures, or vacuum. Capillary of osmosis effects may permit leakage even in the absence of a pressure differential across the pressure boundary of a system. The leakage rate for gases and vapors is typically measured in terms of the mass flow rate, ie, the equivalent of the product of the pressure at a specific point in the leak path and the volume flow rate, in $Pa \cdot m^3/s$ (atm at $STP \cdot cm^3/s$). The leakage rate thus describes the number of molecules of gas that pass through the leak in unit time. Leakage can be measured by the drop in pressure within closed pressurized systems, or by the rise in pressure in closed vacuum systems, after correction for effects of temperature changes, outgassing, condensation, or other effects that may change the state of vapors and gases enclosed within the system (see also PRESSURE MEASUREMENT). Alternatively, leakage rates may be determined dynamically by pumping fluids into or from the volume under test and measuring pressures and flow rates. In more elaborate leak tests, tracer gases, often diluted or pressured up using sensitive detectors, are used to measure leakage rates quantitatively and to compare these leakage rates with those of standard reference leaks of the same tracer gas or gas mixtures.

Significantly higher leak testing sensitivity is attainable using gaseous tracers rather than liquid tracers (1). Smaller leaks can become clogged by liquids held in narrow passageways by strong capillary forces so that they are not detected until the liquid blockages are removed by exposure to vacuum and possibly heating. For this reason pressure vessels, pipelines, and other closed systems preferably should not be exposed to liquid contamination prior to leak testing with gaseous tracers. The choice of leak testing tracer gas and detection method depends on whether the system under test is to be pressurized, evacuated, or operated at atmospheric pressure, and whether it is to be operated within a hood or test chamber that is pressurized, evacuated, or at atmospheric pressure. Details of many different methods and procedures for leak testing are given (1,3,31).

Although liquid tracers are generally far less sensitive than gaseous ones, liquid tracers have unique advantages for many applications. For example for systems already filled with water, gasoline, lubricating oils, or liquid chemicals, it is often convenient to add a miscible liquid tracer to the contained fluid, and permit its distribution throughout the entire system in that fluid. Such liquid tracers typically pass through any existing leaks and form visible colors, stains, fluorescence, or other readily detectable indications. Some types of liquid tracers tend to evaporate at exit points of leaks, concentrating their dyes to form vivid leak indications. Others require application of a developer to enhance the indications formed at leak exits.

Chemical detection of leaks using gaseous tracers has considerable usage for closed systems and other equipment into which it is not feasible to introduce liquid leak tracers. When ammonia (qv) gas, which is usually considered dangerous, is the leak tracer, points of leakage can be found with strips of pH indicator paper moistened with water or by brushing bromocresol purple dye on the outside of pipes and vessels and allowing it to dry. Leaks are indicated by a change in the powdered dye color from light yellowish green to vivid purple. The ammonia gas concentration detected can be as low as 50 ppm NH_3 in air. Indicator dyes sensitive to pH can be used with tracer gases that change the pH of liquids they contact. Acids or bases can be added to the dye solution to shift the pH as close as feasible to the color change-over point to add sensitivity to leak detection. Chemical smoke or fumes are also used as leak indicators with tracer gases such as ammonia, hydrogen sulfide, and carbon dioxide (qv). Ammonia is usually prohibited on brass parts because of its strong corrosiveness. Hydrogen chloride vapor reacts with ammonia to form a white chemical fog or mist of ammonium chloride. Sulfur dioxide from a sulfur candle can also be used with ammonia tracer gas to form a white mist of ammonium sulfide. Carbon dioxide also serves as a reactant for ammonia. Precautions must be employed using any of these chemical leak detectors.

Another leak indicator consisting of agar–agar solution containing sodium carbonate and phenolphthalein can be sprayed onto surfaces where leakage is suspected. Leaking carbon dioxide forms visible white spots in the red film formed by this indicating solution. Smoke bombs and smoke candles ignited within test volumes are also used to permit visible or odorous smoke to escape through leaks. Many other chemical leak-indicator systems have been described (31).

High voltage probe and corona tests can be used to detect cracks and pinholes in dielectric coatings on conducting base metals by electric sparks from probe to base metal. Analogous tests are used to detect leaks into evacuated systems, where a high frequency spark coil may be used to create the electric field. If leaks exist in glass or other transparent dielectrics containing intermediate vacuum pressures, a white spark to the glass, or a visible electrical glow discharge within the evacuated system serves as a common leak detector. The Krypton-85 gas technique is a highly sensitive surface inspection technique applied by first filling a chamber with the gas, which is subsequently removed. Because the molecules are able to diffuse into crevasses as small as 1 nm in width, traces of the gas left after the evacuation of the chamber indicate possible locations of small cracks and leaks.

Flaw Sizing. Correct flaw sizing is one of the most crucial aspects of NDE. Given that load characteristics and fracture mechanics may be reasonably determined, it then remains that the flaw size be accurately estimated. Using a confident assessment of the flaw size, the risks associated with continued operation may be determined. Found flaws range from the obvious surface breaking flaws that may be detected visually to the deeply hidden, very small defects that are crucial in high strength materials. For surface breaking flaws, the length may be reasonably estimated unless the ends are tightly closed. In the latter case, the end location may need to be located using other techniques. Whereas liquid penetrant and wet magnetics reveal surface defects, the length of tightly closed crack ends may be underestimated. Eddy-current techniques may be more sensitive to correctly locating crack ends. Wet magnetics or eddy current usually are more sensitive than dry magnetics or penetrant. Depth of cracks is best determined using ultrasonics, even though this method gives rise to uncertainties. The electric current conduction method also may be used for depth determination, but care must be exercised to avoid seriously faulty readings. For submerged, crack-like flaws, the technique through which the flaw was found is often best for size estimation. For example, a flaw discovered by ultrasonics may well be sized with the same technique. Complimentary methods are also useful. Length can often be confirmed using radiography, if that is possible.

Where ultrasonics is the best choice of technique, tip diffraction, where the probes are located at several positions, is helpful. Loss of metal wall, including pits and grooves, in piping and pressure vessels normally can be sized well using ultrasonics and/or radiography. Hydrogen and methane blisters usually are incorrectly sized from the visible surface because the crack tip may extend for some distance beyond the bulge. The ends of the crack may be located using ultrasonics. Typically, the true flaw size is always larger than the estimated size. Hydrogen attack is one of the most difficult flaws to size correctly because there is no distinct planar growth pattern. Often ultrasonic attenuation or back-scatter techniques are best for this type of flaw.

Ultrasonic flaw sizing using tip diffraction methods is gaining favor. Previously, the amplitude-down technique was used. The amplitude-down technique assumed that the edge of a crack could be located by moving the probe so that half of the energy would bypass the crack when the center of the beam was directly over the flaw end. Owing to the irregular nature of the crack end, this technique may lead to size estimates of a flaw to be one-half or less than the true size (35). Tip diffraction techniques were shown in the same study to yield estimates within 15% of the true size. Using tip diffraction, two echoes are expected from a surface crack, the main echo from the root of surface breaking cracks, for example, and the separate echo from the crack tip. The difference in arrival times of the root echo and the tip echo are used to estimate the crack depth. Critically refracted or high angle longitudinal, sometimes called creeping waves, perform quite well in sizing subsurface defects (36).

Electric current conduction tests, in which current is introduced into test materials by direct electric contacts and potential drops are measured across specific zones of the test surface using additional contact electrodes, are used to measure effects of corrosion wall-thinning, as in petroleum product storage tank roofs, or to detect cracks in welds, metallic sheets and plate, and railroad rails (1,6). Modifications of such systems measure the resultant magnetic field distributions created by the current conduction, as for Hall effect semiconductor devices, moving pick-up coils, and even by thermal effects resulting from resistance heating within the test materials (1). Difficulties in applying direct electric contact testing include the possibility of sparks or burns created by passing high current through poor electrode contact areas, and the thermoelectric and triboelectric effects that may produce false signal voltages owing to short-circuit paths that the current may find through touching crack surfaces. Alternatively, thermoelectric or triboelectric voltage signals can be used to identify and sort metals and alloys where surface contamination effects do not interfere.

Imaging Techniques. Imaging techniques are useful aids in NDE. Obviously, the more that is known about the size and orientation of an anomaly, as well as other shape characteristics, the more informed the decision on its disposition. Some surface techniques, such as penetrant and magnetic particle, are inherently imaging techniques because the indications are visible. Little may be known about the shape below the surface, however. Basic radiographic methods are, in some situations, able to produce good images showing an anomaly's characteristics. Depth resolution, again, may be lacking. Advancements in the imaging capabilities of radiography, ultrasound, and magnetic testing have occurred throughout the 1990s (see also IMAGING TECHNOLOGY).

Image reconstruction techniques generally require that several shots be made of the suspect area. This involves the placement of the probe, source, sensor, etc, at several locations around the part and the storage in a computer of the locations of the source and sensor and the observed response. For example, in both ultrasound and radiography, it is easy to perceive that where some information about the defect is obtained using one source or probe location, a second view of the same defect from another location gives more information on its characteristics. Multiple views can often give enough information to yield a high resolution reconstruction of the defect and the surrounding region. An additional discussion on image reconstruction using ultrasound and radiography, as well as other methods, is available (1–3,6).

Flaw imaging using ultrasound includes the holography, speckle interferometry, tomography, scanning acoustic microscopy, and pitch-catch methods (1–3,6). Whereas all of these have shown unique abilities in imaging hidden defects, holography, speckle interferometry, tomography, and scanning acoustic microscopy have been most useful as research tools. Application of the pitch-catch method, which as emerged as a more useful industrial tool, involves multiple probe placements as well as recognition of arrivals from several different wave paths and modes. The principles used are the same as those for ultrasonic flaw sizing. The data set consisting of the arrival times of the various modes and the probe placement is stored in the computer and used for image reconstruction. These methods have been quite successful in generating detailed images in either color or gray scale that reveal considerable information about the area in question. These images, may then be sent over data transmission lines for evaluation at remote locations.

Computed tomography (CT) is perhaps the most well-known imaging technique used in research and industrial NDE (1–3,6). Typically, a source and receiver are arranged in a circular fashion where the item being tested is located near the center of the circle. In the simplest case, a single source location is used and the scattering patterns are obtained at several receiving positions. The scattering patterns are then used to reconstruct the image of the hidden defect. In more modern units, both the source and receiver are manipulated to different locations allowing more data to be accumulated for each flaw. The detail and resolution of the reconstructed image are also greatly increased. These techniques are widely used in medical applications, and have found increased use in many industries where the ability to generate quick images of internal anomalies is important. In addition to the CT methods, magnetic resonance imaging (MRI) has seen growth, although industrial applications trail those of the other techniques.

Magnetic flux leakage (MFL) imaging appears to have originated with the use of high speed computer processing of MFL detected by coil sensors rotating around the outer surface of oil-field tubes (37) for the detection of new tube flaws. The results are presented on a fold-out two-dimensional map of the pipe, so that the flaws are shown in their true locations. Whereas the severity of the flaws is not assessed by this technique, the length and angle to the pipe axis is recorded so that the inspector may more easily perform prove-up. The imaging of the flaw locations, along with use of noise reduction algorithms, also permits clearer detection of inner surface imperfections, and thus indicates that the sensitivity of the technique should be raised. Typically 5% deep electrical discharge machined (EDM) test notches can be detected on the internal diameter surface in 9.5 mm (0.375 in.) wall thickness.

Fast computers are also permitting the MFL from much larger flaws in pipe lines to be detected and imaged. Here, the pipe is magnetized longitudinally using a permanent or electromagnet, and signal amplitudes from flaws are used on a color scale to predict flaw severity, where maps of the corroded regions are drawn. Principal pipeline pig manufacturers have indicated that in using computer enhancement, the condition of many kilometers of pipe can be mapped. Attempts are also being made to implement similar signal processing on equipment designed for tank floor inspection. In both cases, the system detects corrosion on the far surface (see PIPELINES; TANKS AND PRESSURE VESSELS).

Thickness Measurement. Accurate thickness measurement is important in determining the remaining life in many piping and pressure vessel installations. Ultrasonics provide good thickness determination capability, where the time of flight of the pulse in the section in question is compared to the time of flight in a sample of identical material having known thickness and geometry. Typical accuracies are 1% or 0.01 mm for a 1-mm thick wall (1–3,6,24). Whereas this method can be used quite reliably, errors can arise because of signal interpretation problems and because of the overwhelming amount of data to be collected for a test. Well trained operators are, however, able to extract useful information from the signal, and modern data storage and analysis equipment facilitate the data presentation in order to minimize the likelihood of an error.

Radiographic imaging of corrosion wall thinning at tube bends and other large corrosion pitting regions may be applied to piping, storage tanks, and pressure vessels used in the chemical and petroleum industries. However, fine corrosion pits that penetrate from the inside diameter pipe surface only part way through the pipe wall thickness are hard to detect by any available nondestructive test. Ultrasonic tests are more widely used for thickness measurements of tanks and storage structures, where access to both sides of the wall is usually difficult or impossible while the tank is in service and contains fluids. γ-Ray scanning of process vessels is used to determine the density of material within the vessel during operation. Additionally, the process is used to check for flooding and proper process operations in various locations.

Supplies of Equipment and Services. A successful NDE program depends on qualified suppliers of services, training, and equipment. Potential suppliers of these are listed in yearly update sources (11,38).

Economic Aspects

Nondestructive evaluation is a service industry, and NDE economic activity is directly related to the economy of the basic industries, ie, aerospace, utilities, petrochemical, automotive, metals, or other (39). Overall, the NDE equipment market is small, totaling only $1 billion worldwide in 1990. The distribution of equipment sales among the various industrial groups is relatively equal. Aerospace is at the top with 26% of the market; petrochemical is at the bottom with 10%. Ultrasonic equipment dominates worldwide equipment sales; x-ray film and equipment also show strong markets. Eddy-current, magnetic particle, dye penetrant, and other techniques represent small portions of total sales. There are mixed forecasts for the future growth of NDE equipment sales. Growth in equipment sales may be tied to growth in the primary industries and the future of these industries is uncertain. Growth areas in NDE appear to be in equipment improvement and the development of new engineering systems for more efficient and reliable NDE. It is likely that the growth in NDE is shifting from a marketing oriented industry to one emphasizing engineering and education. NDE is seen to be moving to the role of a tool used by industry for economic benefit related to minimizing losses and maximizing income. Growth in the demand for NDE in engineering education is likely to continue (12).

Nondestructive evaluation requires specialized equipment, trained operators, and is an intrusion in operating or manufacturing schedule, which may involve considerable costs. The benefits may also be significant. Although the direct accountable benefit of NDE may be difficult to establish, a method for estimating benefit based on costs averted has been described (40). A return on investment case study for a power generating station reported significant savings a few years after the implementation of a nondestructive evaluation plan (41). Systematic NDE, used with maintenance procedures, decreased the energy lost at the power plant from cracked and leaking components. In implementing the NDE plan, the first effort was to attack the areas of greatest energy losses. It was only after several overhaul cycles that savings began to show. For this particular plant, an average cost of inspection for each boiler was $143,000 in 1989, a 23% increase over the 1979 inflation adjusted rate. Benefit, however, was realized when maintenance costs decreased from $10.97/MWh in 1979 to $4.34/MWh in 1989, the figures adjusted for inflation. Besides direct savings in reduced energy losses, savings also were realized in better schedules for repairs and maintenance as a result of the NDE plan.

BIBLIOGRAPHY

"Nondestructive Testing," in *ECT* 3rd ed., Vol. 16, pp. 47–72, by R. C. McMaster, The Ohio State University.

1. D. E. Bray and D. McBride, *Nondestructive Testing Techniques*, John Wiley & Sons, Inc., New York, 1992.
2. D. E. Bray and R. K. Stanley, *Nondestructive Evaluation*, McGraw-Hill Book Co., Inc., New York, 1989.
3. ASM International, *Metals Handbook*, Vol. 17, 9th ed., ASM International, Materials Park, Ohio, 1989.
4. R. C. McMaster, ed., *Nondestructive Testing Handbook*, The Ronald Press Co., New York, 1959, 1963, The American Society for Nondestructive Testing, Columbus, Ohio, 1977, 1979.
5. W. J. McGonnagle, *Nondestructive Testing*, 2nd ed., The American Society for Nondestructive Testing, Columbus, Ohio, 1969.
6. R. Halmshaw, *Non-destructive Testing*, Edward Arnold, London, 1987.
7. R. N. Pangborn, C. E. Bakis, and A. E. Holt, *J. Pressure Vessel Technol.* **113**(2), 163–169 (1991).
8. R. D. Barer and B. F. Peters, *Why Metals Fail*, Gordon and Breach, Science Publishers, New York, 1970.
9. *Source Book in Failure Analysis*, American Society for Metals, Metals Park, Ohio, 1974.
10. F. R. Hutchins and P. M. Unterweiser, *Failure Analysis: The British Engine Technical Reports*, American Society for Metals, Metals Park, Ohio 1981.
11. C. Lopez, *Mater. Eval.* **53**(2) (1995).
12. D. E. Bray, *Mater. Eval.* **51**(6), 651–652, 654–655 (June 1993).
13. *ASNT Recommended Practice No. SNT-TC-1A, Personnel Qualification and Certification in Nondestructive Testing*, American Society for Nondestructive Testing, Columbus, Ohio, 1992, periodically revised.
14. *Supplements to ASNT Recommended Practice No. SNT-TC-1A, Questions and Answers for Qualifying NDT Level III Personnel*, American Society for Nondestructive Testing, Columbus, Ohio, 1977, periodically revised.
15. K. R. Balkey and co-workers, *Mech. Eng.* **112**(3), 68–74 (1990).
16. ASME Research Task Force on Risk-Based Inspection Guidelines, *Risk-Based Inspection—Development of Guidelines*, CRTD Vol. 20-1, The American Society of Mechanical Engineers, New York, 1991.
17. J. R. Dickens and D. E. Bray, *Mater. Eval.* **52**(9), 1033–1034, 1036–1038, 1040–1041 (Sept. 1994).
18. C. D. Cowfer, ed., *Use of Computers in NDE Engineering and Data Acquisition Systems*, NDE Vol. 2, The American Society of Mechanical Engineers, New York, 1986.
19. J. L. Rose and G. H. Thomas, *Mater. Eval.* **38**(1), 69 (1980).
20. W. E. Woodmansee, *Mater. Eval.* **38**(3), 33 (1980).
21. R. D. Sachs, J. D. Elkins, and J. H. Smith, *Mater. Eval.* **30**, 121, 135 (June 1972).
22. A. J. Rogovsky and J. L. Rose, *Mater. Eval.* **37**, 47 (Mar. 1979).
23. *API Guide for Inspection of Refinery Equipment*, 2nd ed., American Petroleum Institute, Division of Refining, New York, 1976, periodically revised.
24. R. C. McMaster, ed., "Liquid Penetrant Inspection," *Nondestructive Testing Handbook*, Vol. 2, 2nd ed., American Society for Nondestructive Testing, Columbus, Ohio, and the American Society for Metals, Cleveland, Ohio, 1982.
25. D. J. Hagemaier, *Fundamentals of Eddy-Current Testing*, American Society for Nondestructive Testing, Columbus, Ohio, 1990.

26. R. C. McMaster, P. McIntire, and M. Mester, ed., "Electromagnetic Testing," *Nondestructive Testing Handbook*, Vol. 4, 2nd ed., American Society for Nondestructive Testing, Columbus, Ohio, 1986.

27. P. McIntire, ed., "Ultrasonic Testing," *Nondestructive Testing Handbook*, Vol. 7, 2nd ed., The American Society for Nondestructive Testing, Columbus, Ohio, 1991.

28. P. McIntire, ed., "Radiography and Radiation Testing," *Nondestructive Testing Handbook*, Vol. 3, 2nd ed., The American Society for Nondestructive Testing, Columbus, Ohio, 1985.

29. C. F. Coleman and A. E. Hughes, "Positron Annihilation," in R. S. Sharpe, ed., *Research Techniques in Nondestructive Testing*, Vol. 3, Academic Press, London, 1977, pp. 355–391.

30. A. J. Hill, D. R. Overberg, and B. W. Cherry, *Proceedings of the 1993 ASME Pressure Vessels and Piping Conference*, PVP Vol. 257, Denver, Colo., July 25–29, 1993, pp. 93–95.

31. R. C. McMaster, ed., "Leak Testing," *Nondestructive Testing Handbook*, Vol. 1, 2nd ed., The American Society for Nondestructive Testing, Columbus, Ohio, and The American Society for Metals, Metals Park, Ohio, 1982.

32. K. T. Hartwig, "An Eddy-Current Decay Technique for Low Temperature Resistivity Measurements," in G. Birnbaum and G. Free, eds., *ASTM STP 722*, American Society for Testing and Materials, Philadelphia, Pa., 1981, pp. 157–172.

33. T. Leon-Salamanca and D. E. Bray, "Ultrasonic Measurement Of Residual Stress In Steels Using Critically Refracted Longitudinal Waves (L_{CR})," *Res. NDE* **6**(4) 1995 (in press).

34. J. C. Spanner, Jr., ed., *Proceedings of the 1994 ASME Pressure Vessels and Piping Conference*, PVP Vol. 276, NDE Vol. 12, Minneapolis, Minn., June 19–23, 1994, 185 pp.

35. C. D. Cowfer and O. F. Hedden, *Pressure Vessel Technol.* **113**(2), 170–173 (1991).

36. J. M. Davis, *Proceedings of the 1991 ASME Pressure Vessels and Piping Conference*, PVP Vol. 216, NDE Vol. 9, San Diego, Calif., June 23–27, 1991, pp. 35–37.

37. R. K. Stanley, "Imaging of Magnetic Flux Leakage Signals for High Quality Assessment of Oilfield Tubular Products," *Proceedings of the 13th World Conference on NDT*, Sao Paulo, Brazil, Oct. 1992.

38. C. Lopez, ed., *Buyer's Guide, Mater. Eval.* **52**(6) (1993), (revised annually).

39. C. D. Wells, *Insight*, **36**(5), 331–341 (1994).

40. E. P. Papadakis, *Mater. Eval.* **50**(6), 774–776 (1994).

41. D. Pruitt and S. Lebsack, *Mater. Eval.* **52**(11), 1279–1281 (1994).

Donald E. Bray
Texas A&M University

NONLINEAR OPTICAL MATERIALS

Nonlinear optical (NLO) materials are the building blocks of the emerging technology of photonics, ie, the acquisition, transmission, processing, and storage of information using photons (light quanta). The interactions between photons are weak compared to interactions between electrons, thus long distance transmission of information over fiber optic lines (see FIBER OPTICS) with minimal signal degradation is facilitated. Moreover, compared to electronics, photonics offer advantages in terms of speed (bandwidth) of information processing.

Analogous to electronic materials, photonic materials can be divided into two classes: linear (passive) and nonlinear (active). Linear materials are employed in the construction of transmission lines (waveguides), passive directional couplers, gratings, etc; nonlinear materials are used in the construction of switches, modulators, frequency doublers and triplers, active beam steering elements, limiters, amplifiers, rectifiers, and transducers. Nonlinear optical phenomena arise when applied external fields, ie, either light or low frequency electrical fields, are sufficiently strong to compete with internal electrostatic interactions. Thus, nonlinear optical materials are typically those containing weakly bound (highly polarizable) electrons. Nonlinear optical phenomena include harmonic generation, sum- and difference-frequency generation, optical parametric oscillation, rectification, intensity-dependent refraction, phase conjugation, multiple wave mixing, Raman scattering, Brillouin scattering, induced opacity and reflectivity, and multiple photon absorption.

Further subclassification of nonlinear optical materials can be explained by the following two equations of microscopic, ie, atomic or molecular, polarization, p, and macroscopic polarization, P, as power series in the applied electric field, E (disregarding quadrupolar terms which are unimportant for device applications):

$$p = \alpha E + \beta EE + \gamma EEE + \ldots \tag{1}$$

where α is the linear polarizability, β and γ are the first and second (atomic or molecular) hyperpolarizabilities, respectively; and

$$P = \chi^{(1)} E + \chi^{(2)} EE + \chi^{(3)} EEE + \ldots \tag{2}$$

where $\chi^{(1)}$ is the linear optical susceptibility, $\chi^{(2)}$ the second-order NLO susceptibility, and $\chi^{(3)}$ the third-order NLO susceptibility. The second-order coefficients are zero for centrosymmetric symmetries. Typical of power series expressions, the magnitudes of the coefficients decrease for higher order terms. Because practical applications of third-order optical nonlinearity are rare, terms beyond third order can be ignored. The small magnitudes of optical nonlinearities often require long interaction lengths, eg, as found in waveguides, for reasonable device performance. Each coefficient is complex with a real component corresponding to index of refraction phenomena and an imaginary component corresponding to absorption phenomena. Real and imaginary components are related by the Kramers-Kronig relationship (1,2). When speaking of NLO devices, however, it is common to speak of exploiting either a nonlinear index change only or a nonlinear absorption change. Nonlinear index of refraction changes are typically exploited in regions of transparency, ie, regions far removed from one-photon resonances, but nonlinear absorption phenomena are usually exploited near resonance.

Materials are also classified according to a particular phenomenon being considered. Applications exploiting off-resonance optical nonlinearities include electrooptic modulation, frequency generation, optical parametric oscillation, and optical self-focusing. Applications exploiting resonant optical nonlinearities include sensor protection and optical limiting, optical memory applications, etc. Because different applications have different transparency requirements, distinction between resonant and off-resonance phenomena are thus application specific and somewhat arbitrary.

Nonlinear optical materials can also be classified according to atomic composition into organic and inorganic materials. Inorganic materials range from crystalline lithium niobate [12031-63-9], LiNbO$_3$, to amorphous semiconductor materials such as gallium arsenide [1303-00-0], GaAs (see ELECTRONIC MATERIALS, SEMICONDUCTORS). Organic materials can be either crystalline such as polydiacetylene, or polymeric materials incorporating a variety of organic chromophores such as disperse red 1 [2872-52-8], 4-(N-ethyl-N-2-hydroxyethyl)amino-4'-nitroazobenzene. For most organic materials, however, the optical nonlinearity can be associated with weakly bound π-electrons. As of this writing (ca mid-1995), no single material has been proven suitable for significant commercial applications, thus discussion herein focuses on the material requirements for various applications.

Second-Order Nonlinear Optical Materials

Many classes of second-order material applications can be envisioned by noting the sinusoidal nature of electromagnetic radiation and rewriting equation 2 as

$$P = \chi^{(1)}E_0 \cos(\omega t - kz) + (1/2)\chi^{(2)}(E_0)^2[1 + \cos(2\omega t - 2kz)] + \dots \quad (3)$$

where ω is the frequency of the elctromagnetic radiation, t the time, and k and z are the wave vector and spatial coordinate (position), respectively.

Two applications can be identified for second-order NLO materials: frequency doubling, ie, the 2ω term; and electrooptic modulation, ie, the first term in brackets. Frequency doubling results in the generation of a new light beam at twice the frequency (or half the wavelength) of the original light beam. For example, the generation of light at 532-nm wavelength can be realized by doubling the 1064-nm light from a yttrium aluminum garnet (YAG) laser (see LASERS). The low frequency term is associated with electrooptic (EO) modulation, ie, the Pockel's effect. This term, $(1/2)\chi^{(2)}(E_0)^2$, can be characterized by frequencies ranging from 0 Hz (dc) to 100 GHz; that is, frequencies that span the radio frequency, microwave, and millimeter wave regions.

Materials for Electrooptic Modulation. The fundamental phenomenon of Pockel's effect is a phase change, $\Delta\phi$, of a light beam in response to a low frequency electric field of voltage, V. Relevant relationships for collinear electrical and optical field propagation are as follows (1–6):

$$\Delta\phi = \pi n^3 r_{\text{eff}} VL/\lambda h, \qquad r_{\text{eff}} = -8\pi\chi^{(2)}/n^4 \quad \text{and} \quad V_\pi L = \lambda h/n^3 r_{\text{eff}} \quad (4)$$

where n is the index of refraction, r_{eff} the effective electrooptic coefficient, L the modulation length, h the gap distance between the electrodes (film thickness), λ

the optical wavelength, and V_π the voltage required to realize a phase shift of π. Sometimes a parameter Γ is added to the denominator of the $V_\pi L$ equation to take into account the radio-frequency field coupling efficiency, ie, losses owing to the resistivity of metal drive electrodes. Consequently, the $V_\pi L$ product is a measure of the modulation efficiency of the device. The Pockel's effect is used to transduce electrical signals into optical signals, as, for instance, in the cable television industry, for photonic detection of radar, for voltage sensing in the electric power industry, and to effect switching in local area networks. The most commonly encountered device configurations are shown in Figure 1. For the Mach-Zehnder configuration, the phase shift induced in one branch of the modulator by Pockel's effect results in amplitude modulation when the modulated and unmodulated beams are recombined at the output.

$$I_o = I_i \, \sin^2[(\phi_{ba} + \Delta\phi)/2] \quad \text{and} \quad (V_\pi L)_{MZ} = \lambda h/n^3 r_{33} \tag{5}$$

where I_i and I_o are, respectively, the input and output intensities, ϕ_{ba} is the phase difference between the two arms, $\Delta\phi$ the phase modulation produced by the EO effect, and the subscript MZ denotes Mach-Zehnder.

The birefringent (BR) modulator makes use of polarized light and tensorial nature of the electrooptic coefficient. For example, poled organic polymer films are characterized by two nonzero components for the electrooptic tensor: r_{33} and

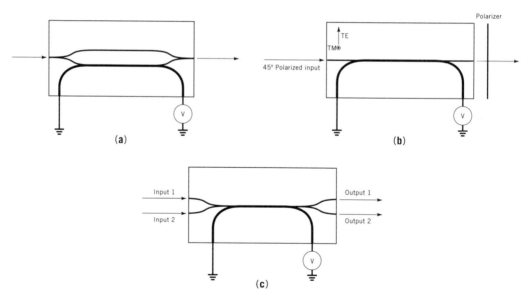

Fig. 1. Representative device configurations exploiting electrooptic second-order nonlinear optical materials are shown. Schematic representations are given for (**a**) a Mach-Zehnder interferometer, (**b**) a birefringent modulator, and (**c**) a directional coupler. In (**b**) the optical input to the birefringent modulator is polarized at 45 degrees and excites both transverse electric (TE) and transverse magnetic (TM) modes. The applied voltage modulates the output polarization. Intensity modulation is achieved using polarizing components at the output.

r_{13}, parallel and orthogonal to the poling direction, respectively. In order to obtain amplitude (intensity) modulation, the input beam is 45-degree polarized to excite both transverse electric (TE) and transverse magnetic (TM) modes. An analyzer then converts the polarization modulation into an amplitude modulation:

$$\Delta n(\text{TE}) = n^3 r_{13} V/2h \qquad \Delta n(\text{TM}) = n^3 r_{33} V/2h \qquad (6)$$

$$\Delta \phi(\text{TE/TM})_{\text{sp}} = 2\pi \Delta n L/\lambda = \pi V L(n^3 r_{13} - n^3 r_{33})/\lambda h \qquad (7)$$

$$I_o = I_i \, \sin^2[(\phi_{\text{sp}} + \Delta \phi_{\text{sp}})/2] \qquad (V_\pi L)_{\text{BR}} = 1.5(V_\pi L)_{\text{MZ}} \qquad (8)$$

Compared to the Mach-Zehnder modulator, the modulation efficiency of the birefringent modulator is lower because $V_\pi L$ is 1.5 times higher, ie, $r_{33} = 3r_{13}$ for dipolar chromophores. Directional couplers (Fig. 1c) consist of two side-by-side waveguides separated by a few micrometers. The overlap of the guided waves in the two waveguides couples energy back and forth between the waveguides. When a low frequency voltage is applied, the intensities at the output ports are determined by either modulation of the phase mismatch, $\Delta\beta$, or the coupling coefficient, κ. The smallest voltage required for switching is 1.7 times larger than that of a Mach-Zehnder device.

Two parameters that characterize the performance of electrooptic modulators are $V_\pi L$ and the bandwidth given by the following:

$$L\Delta\omega = c/4\left[n(\omega) - \epsilon_{\text{eff}}^{1/2}\right] \qquad (9)$$

where ϵ_{eff} is the dielectric constant. The bandwidth is defined by the velocity mismatch of the optical and the low (radio, microwave, or millimeter wave) frequency waves propagating in the material. In Table 1, drive voltage and bandwidth characteristics for three of the most promising and diverse classes of electrooptic materials are compared. Although single-crystal lithium niobate permits realization of digital-level drive voltages, the bandwidth is limited to a few gigahertz unless sophisticated device architectures are employed, which in turn require higher drive voltages. Lithium niobate also exhibits a relatively low (0.1 GW/cm^2) optical damage threshold owing to its large photorefractive effect.

Table 1. Comparison of Electrooptic Modulator Performance Parameters of NLO Materials[a]

Parameter	LiNbO$_3$	NLO polymer	GaAs
r_{eff}, pm/V	30	45	1.5
$n^3 r_{\text{eff}}^b$	300	220	64
ϵ	30	2.8	10
$n^3 r_{\text{eff}}/\epsilon^b$	10	130	6
space-bandwidth product, GHz·cm	7	120	10
$V_\pi L$, V·cm	5	5	0.3
optical loss, dB/cm	0.2	0.2–1.0	2

[a]At optical wavelength of 1.3 μm.
[b]Values given are figures of merit.

For organic EO materials such as polymers, both n and ϵ are determined by the π-electrons of the NLO chromophore and $n^2 \approx \epsilon$ leading to bandwidths > 100 GHz for a 1-cm device length. Operation of organic polymer electrooptic modulators has been demonstrated for frequencies as high as 60 GHz (6). Unlike crystalline inorganic NLO materials, organic chromophores seldom assemble naturally in noncentrosymmetric space groups. Moreover, when organic NLO crystals are obtained, these are usually characterized by growth anisotropy that renders the crystals unusable for device applications. To obtain noncentrosymmetric macroscopic lattices of organic NLO chromophores, extensive use has been made of sequential synthesis and electric field poling methods.

Sequential synthesis film fabrication approaches include Langmuir–Blodgett methods, Merrifield-type covalent coupling reactions, sequential synthesis exploiting ionic interactions, and molecular beam epitaxy methods (4,7). Unfortunately, to prepare films on the order of 1 μm, required for device fabrication, a high degree of perfection is required for the deposition of each layer. As of this writing (mid-1995), sequential syntheses are still not competitive for the fabrication of electrooptic modulators, even though these methods are of great interest for frequency doubling applications because of opportunities afforded for phase-matched second harmonic generation (see THIN FILMS, FILM FORMATION TECHNIQUES).

The method of choice for the fabrication of organic EO materials appears to be the electric field poling of NLO chromophores in polymer matrices near the glass-transition temperature, T_g, of the polymer matrix. A deficiency of this approach is the relaxation of poling-induced order when the poling field is removed. To overcome this problem, a variety of lattice-hardening reactions have been developed. These have generally been successful in permitting the fabrication of NLO-active polymer lattices that can withstand temperatures on the order of 100°C for long (thousands of hours) periods of time and even higher temperatures for such short periods of time as associated with the deposition of metal electrodes (4,7). The final glass-transition temperature appears to be the critical parameter in defining the long-term stability of poling-induced optical nonlinearity both for chromophore–polymer composites and for chromophores covalently attached to the polymer host lattices (8). Certain advantages have also been observed for the covalent incorporation of chromophores instead of chromophore–polymer composites. These include the realization of higher chromophore number densities, prevention of phase separation and chromophore aggregation, avoidance of chromophore sublimation at high processing temperatures, reduced plastization of the polymer lattices for a finite chromophore loading, elimination of poling-induced light scattering arising from an inhomgeneous distribution of chromophores, and avoidance of chromophore extraction with application of cladding layers in the fabrication of EO modulator waveguide structures (see COMPOSITE MATERIALS, POLYMER-MATRIX).

A putative advantage of organic NLO materials over crystalline inorganic materials is that active chromophores can be systematically varied in the former to increase optical nonlinearity (Table 2). An even more important feature of polymeric materials is that they can be readily integrated with semiconductor electronics and fiber optic transmission lines. Buried channel nonlinear optical waveguides can be fabricated by a variety of techniques, including reactive ion

Table 2. NLO Chromophores and Corresponding $\mu\beta$ Values[a]

Chromophore base	R	Nonlinearity, $\mu\beta$, esu $\times 10^{-48}$
H_3C—N(—H_3C)—⟨benzene⟩—R	NO_2	140[b,c]
	—⟨benzene⟩—NO_2	275[d]
	—(CH=CH)$_n$—⟨benzene⟩—NO_2	580[e], 482[f] ($n = 1$) 813[f] ($n = 2$) 1074[f] ($n = 3$) 1700[f] ($n = 4$)
	—N=N—⟨benzene⟩—NO_2	800[g,h]
	—(CH=CH)$_n$—CH= (thiobarbituric ring, N—CH_2CH_3, N—CH_2CH_3, S)	370[f] ($n = 0$) 1457[f] ($n = 1$) 3945[f] ($n = 2$) 9831[f] ($n = 3$)
	—(CH=CH)$_n$—CH= (isoxazolone with phenyl)	312[f] ($n = 0$) 1202[f] ($n = 1$) 3156[f] ($n = 2$) 8171[f] ($n = 3$)
H_3C—CH_2—N(—H_3C—CH_2)—⟨benzene⟩—R	—N=N—⟨benzene⟩—C(=C(CN)$_2$)—CN	4100[g,h]
	—CH=CH—⟨thiophene (X)⟩—R′	560[d] ($X = O$, $R' = NO_2$) 700[d] ($X = S$, $R' = NO_2$) 2400[i] ($X = S$, $R' = $ thiobarbituric derivative with H_3C—CH_2)
julolidine—(CH=CH)$_2$—C(CH_3)(=CH—CH)—CH=CH—C(=CH—CH)(CH_3)—CH= isoxazolone with phenyl		28,500[f]

[a]Measured at 1.907 μm unless otherwise indicated. μ is the dipole moment and β is the molecular first hyperpolarizability defined in equation 1. [b]Measured at 1.3 μm. [c]Ref. 9. [d]Ref. 10. [e]Ref. 11. [f]Ref. 12. [g]Measured at 1.58 μm. [h]Ref. 13. [i]Ref. 14.

etching and electron cyclotron resonance etching, photochemical lithography, and spatially selective poling (7,15). Mechanically stable coupling to silica fibers can be accomplished by silicon V-groove techniques (7). As the name implies, silicon V-groove technology involves etching a V groove in a silicon substrate, positioning a silica fiber in that groove against the electrooptic waveguide, and overcoating the fiber with electrooptic polymer.

Lithium niobate exists as single crystals and cannot be readily grown on other substrates. There have, however, been attempts at growth on semiconductor substrates. Channel waveguides have been obtained through ion or proton exchange involving lithium niobate single crystals, and lithium niobate films have been grown heteroepitaxially on lithium tantalate, $LiTaO_3$, single crystalline substrates (16). Because of the difficulty entailed in obtaining lithium niobate waveguides on semiconductor substrates, lithium niobate modulators are typically flip-chip bonded to a substrate; or alternatively, the electronics, fibers, and modulators are interconnected via cables and fibers (17). An effort has also been made to obtain all-semiconductor systems that rely primarily on GaAs or related compounds for both optical and electronic components (see PACKAGING, ELECTRONIC MATERIALS) (18).

A market for electrooptic modulators does not exist as of this writing; much of the motivation for developing prototype devices derives from anticipated needs of the information superhighway and the cable television industry. Various corporate research laboratories have produced prototype modulators that can operate to 60 GHz (7,19–28). Nontechnical issues such as the early implementation of direct (satellite transmission) television may also affect the market success of these electrooptic modulators.

Materials for Frequency Doubling. Second-order NLO materials can be used to generate new frequencies through second harmonic generation (SHG), sum and difference frequency mixing, and optical parametric oscillation (OPO). The first, SHG, is given in equation 3.

Second harmonic generation has some of the same material requirements as electrooptic modulation; for instance, a significant magnitude of $\chi^{(2)}$ that translates by a linear relationship into an adequate SHG coefficient, d. However, unlike the electrooptic modulation, SHG has a more stringent transparency requirement because transparency is needed at both the fundamental and the second harmonic wavelengths. A more challenging requirement for SHG is to avoid destructive interference by phase matching the fundamental and second harmonic. For a collinear propagation of the fundamental and second harmonic beams, the wave vector difference between the two waves is given by the following, where $k = n\omega/c$:

$$\Delta k = k_{2\omega} - 2k_{\omega} \tag{10}$$

This equation leads to a variation of the second harmonic power, given by

$$P(2\omega) = \left\{ 128\pi^5 [P(\omega)]^2 / WN(2\omega)_{\text{eff}} N^2(\omega)_{\text{eff}} \lambda^2 c \right\} \cdot$$

$$(F_{112})^2 \int |\chi^{(2)}(x) \exp(i\Delta kx) \, dx|^2 \tag{11}$$

where N is the mode index, W the spatial extent of the fundamental in the y direction, and F_{112} the normalized overlap integral. If a periodic variation of $\chi^{(2)}$ exists along x such that

$$\chi^{(2)}(x) = \chi^{(2)}(a) + \chi^{(2)}(b) \, \cos(2\pi x/\Lambda) \qquad (12)$$

where Λ is the grating period, then the integral of equation 11 has a nonvanishing value of $\chi^{(2)}(b)/2$. This method of phase matching is called quasi phase matching (3,4,29) and has been applied to both inorganic and organic materials. For example, quasi phase-matched SHG has been observed for LiNbO$_3$, LiTaO$_3$, potassium titanyl phosphate (KTP) [109657-81-0], KTiOPO$_4$, crystals as well as for NLO chromophore-containing polymers. The required periodic domain structures were obtained by proton exchange (30,31), titanium diffusion (32,33), asymmetric temperature growth (34), laser-heat pedestal growth (35,36), quick heat treatment (37), electron injection (38), and electric field poling (39). Quasi phase matching in organic polymeric thin films has been accomplished by photochemical processing (40) and electric field poling (41).

Phase matching in crystalline materials is typically achieved by using birefringence associated with different propagations to offset the natural dispersion in the medium (3–6). This can be accomplished either through angle or by temperature tuning. Other methods of achieving phase matching of the fundamental and second harmonic waves co-propagating in the material include the use of anomalous dispersion, Cerenkov radiation, and counter-directed waves (3–5). A variety of material processing schemes have been utilized to achieve the particular geometries required for many phase-matching schemes.

As for electrooptic materials, materials for frequency doubling can also be divided into first generation inorganic crystalline materials and second generation organic materials. Such division, however, does not imply that inorganic materials are obsolete. Indeed, only inorganic materials have achieved commercial success as frequency doublers for infrared lasers. Representative inorganic materials are listed in Table 3 (3–6). Owing to the relatively low nonlinear coefficients, d, of most inorganic materials, long interaction lengths are usually required; these, in turn, necessitate large, high quality single crystals. Efforts have also been made to fabricate nonlinear optical waveguides. Single crystal films are often used to avoid light scattering problems. For organic materials, the trade-off between optical nonlinearity and optical transparency continues to be a problem. Materials having large d (eg, 50–200 pm/V) values typically exhibit finite absorption at the second harmonic. This is especially problematic in attempting to develop frequency doublers for diode lasers operating at approximately 820 nm. Research, however, has focused on extending the window of transparency of organic frequency doubling materials into the blue (42,43). A commonly quoted (3) figure of merit (FOM) for frequency doubling materials is d^2/n^3, where d is the second harmonic generation coefficient and n the index of refraction. If this relationship is applied to the materials in Table 1, a value of approximately 4 is obtained for lithium niobate and a value of approximately 500 for the NLO polymer. Obviously, such a FOM is useful only if the materials in question can satisfy the transparency requirements of a particular application.

Table 3. Second Harmonic Generation Coefficients of Inorganic Materials

Material	Second harmonic generation coefficients, pm/V	Transparency window, nm
lithium niobate	$d_{15} = 6.3$ and $d_{22} = 3.3$	400–2,500
potassium dihydrogen phosphate	$d_{36} = d_{14} = d_{25} = 0.5$	200–1,500
ammonium dihydrogen phosphate	$d_{36} = d_{14} = d_{25} = 0.6$	200–1,200
potassium titanyl phosphate	$d_{33} = 15$, $d_{13} = 7$, $d_{32} = 5$, $d_{24} = 8$, $d_{15} = 7$	350–4,500
lithium formate monohydrate	$d_{31} = d_{15} = 0.107$ and $d_{32} = d_{24} = 1.25$	230–1,200
urea	$d_{36} = d_{14} = d_{25} = 1.42$	200–1,430
gallium arsenide	$d_{36} = d_{14} = d_{25} = 90$	900–17,000

Commercial frequency doublers have relied on inorganic materials. The commercial future of doublers depends on not only the improvement in second-order materials but also the development of diode lasers capable of operating in the visible frequency domain.

Materials for Parametric Oscillation and Amplification. The need for frequency-agile laser sources has driven the development of optical parametric oscillators (OPOs) and optical parametric amplifiers (OPAs). A variety of these devices has become available from commercial laser component suppliers such as Continuum and Spectra Physics. Both OPOs and OPAs rely on difference frequency generation. For operation in the visible frequency domain, inorganic materials such as β-barium borate [13701-59-2], BaB_2O_4, or KTP are required (44). Because organic chromophores have band edges that extend through much of the visible spectral region, use is limited to the near infrared (45). OPOs and OPAs for use with pulses as short as a nanosecond are under development and commercial devices are available from a number of suppliers. Those devices having picosecond- and femtosecond-pulse capability are at the stage of specialty instrumentation requiring considerable expertise for successful operation.

Materials for Spatial Light Modulation. In a spatial light modulator, light impinges on a photoconductor, eg, amorphous silicon, that is reverse-biased with an applied voltage. Charge carriers are formed in the photoreceptor in an image-wise fashion. The pattern of charge then drifts, under the influence of the field, to the interface between the light blocking layer and the photoreceptor where it is trapped. Electric field gradients are generated in an adjacent electrooptic film which, in turn, generates patternwise variations in the electric field-dependent refractive index in the electrooptic layer. A readout beam, impinging on the electrooptic film side of the modulator, is then used to decipher the information stored in the electrooptic film. The electrooptic layer can be either a liquid crystalline material or a second-order NLO material (Pockel's effect). Several types of electrooptic materials have been utilized in the construction of spatial light modulators; for example, potassium dihydrogen phosphate has been used in conjunction with high resistivity silicon to construct a device having spatial resolution approaching 10 line pairs/mm and frame-grabbing rates of 1 kHz (46). Although second-order nonlinear optical materials permit high data rates, their small electrooptic coefficients do not permit them to compete effectively with liquid crystalline materials (qv).

Third-Order Nonlinear Optical Materials

Third-order processes include third harmonic generation (THG), self-focusing, self-defocusing, self-phase modulation, saturable absorption, and reverse saturable absorption. Because of the weak magnitude of third-order optical nonlinearities, practical applications are rare; the few applications that do exist typically exploit long interaction lengths. A wide range of materials have been studied in the search for materials containing usable optical nonlinearities; such materials include inorganic crystals and glasses; doped glasses; simple gases and liquids (atomic and molecular as well as organic and inorganic); semiconductors; quantum dots, wells and wires; conductive particle composites; organic and polymer crystals; amorphous organic and inorganic polymers; and biological complexes. Mechanisms of optical nonlinearity can be found in the literature for many types of optical nonlinearity and various classes of materials contributing to these nonlinear phenomena.

    Applications Involving Nonlinear Index Phenomena. The index of refraction, n, can be expressed for nonlinear optical materials as

$$n = n_0 + n_2 I \tag{13}$$

where I is the intensity of light and n_2, the nonlinear index, can be related to the third-order susceptibility by

$$n_2 = 16\pi^2 \chi^{(3)}/c(n_0)^2 \tag{14}$$

where n_0 is the linear index of refraction and c is the speed of light. Index of refraction phenomena are typically associated with the bending of a light beam or a phase shift of light passing through a material.

    Nonlinear refraction phenomena, involving high intensity femtosecond pulses of light traveling in a rod of Ti:sapphire, represent one of the most important commercial exploitations of third-order optical nonlinearity. This is the realization of mode-locking in femtosecond Ti:sapphire lasers (qv). High intensity femtosecond pulses are focused on an output port by the third-order Kerr effect while the lower intensity continuous wave (CW) beam remains unfocused and thus is not effectively coupled out of the laser.

    The dynamic Kerr effect (DKE) is the third-order analogue of the second-order Pockel's effect. DKE also gives rise to phase shifts given by

$$\Delta\phi = 2\pi\Delta nFL/\lambda \tag{15}$$

where $\Delta n = n_2 I$ and F accounts for any cavity resonance enhancement. However, in the Kerr effect, the field controlling the nonlinear phase shift is an optical field leading to an all-optical control of phenomena such as local area network switching. Indeed, prototype devices have been produced that are competitive in performance with electrooptic switches.

A crucial advantage of switching exploiting the DKE over switching based on the Pockel's effect is that the switching speed for the latter is limited at high speeds by electrode resistivity. In contrast, switching exploiting the DKE is typically limited by phase relaxation times that are on the order of a few hundred femtoseconds. The disadvantage of DKE switches, however, is the long interaction length required. The dynamic Kerr effect has been utilized in nonlinear optical loop mirrors (NOLMs) and tetrahertz optical asymmetric demultiplexers (TOADs) to produce ultrafast (eg, 100 Gbits/s) all-optical demultiplexing. Both silica fiber and semiconductor materials have been used in device performance demonstrations (47–50).

In an effort to identify materials appropriate for the application of third-order optical nonlinearity, several figures of merit (FOM) have been defined (1–5,51–53). Parallel all-optical (Kerr effect) switching and processing involve the focusing of many images onto a nonlinear slab where the transmissive characteristics are a function of the total light intensity. The interaction length and focal spot size are limited by diffraction. For a phase shift greater than π to be achieved, n_2 must be $> n_0 \lambda^2/(4FP_s)$, where P_s is the laser power. For $P_s = 10$ mW, $F = 1$, and $\lambda = 1$ μm, n_2 must then be $>10^{-11}$ m^2/W. This requires $\chi^{(3)}$ to be $>10^{-5}$ esu, which is a number very difficult to achieve. Because in serial configurations light is confined in a waveguide, a smaller optical nonlinearity can be compensated for by a longer interaction length. The practical interaction length, L, is limited by optical losses; it is described $L = A/\alpha$, where A is the acceptable loss and α the attenuation coefficient. The optically induced phase shift is then given by

$$\Delta\phi = 2\pi\Delta n F A/\lambda\alpha \tag{16}$$

Both nonlinear and linear absorption can contribute to losses (3,53), ie,

$$\alpha = \alpha_1 + \alpha_2 I + \alpha_3 I^2 + \dots \tag{17}$$

where α_1, α_2, and α_3 are the one photon (linear) absorption, two-photon (nonlinear) absorption, and three-photon (nonlinear) absorption coefficients, respectively. One FOM definition for serial processing is

$$W = n_2 I/\lambda\alpha_1 \tag{18}$$

for one-photon absorption, and

$$T^{-1} = n_2/2\lambda\alpha_2 \tag{19}$$

for two-photon absorption. The switching power of channeled devices can be estimated from

$$P_s = \Delta\phi/(n_2 L/\sigma_{\text{eff}}) \tag{20}$$

where σ_{eff} is the effective waveguide cross-sectional area.

Besides W, there are two other definitions of FOM for all-optical processing:

$$W' = n_2/\alpha \quad \text{and} \quad W'' = n_2/\alpha\tau \tag{21}$$

where τ is the switching speed. Any nonlinearity, even a thermal effect, can turn on instantaneously. In order to avoid cross-talk for series pulses, the relaxation time of the optical nonlinearity, τ, must be short. The first definition (W) emphasizes the maximum nonlinear phase shift obtainable in conjunction with the lowest loss. The second definition (W') focuses on the trade-off between nonlinear response and optical loss. The third definition (W'') also takes relaxation time into account. Comparing materials using these different definitions is important because FOMs always vary with offset from one- and two-photon resonances. A rough comparison of four examples representing four classes of materials is presented in Table 4. To provide the most meaningful comparison, optical nonlinearities have been measured far from resonance. High FOM for silica glasses is not particularly meaningful because unrealistically long interaction lengths would be required. It is usually impractical to build devices limited by such space requirements.

For measurements carried out closer to resonances, thermal contributions to the optical nonlinearity, given by the following formula, can compete with electronic contributions (54,55):

$$n_2(\text{thermal}) = (\alpha\tau_t/\rho C_p)(dn/dT) \tag{22}$$

where dn/dT is the thermooptic coefficient, ρ the density of the material, C_p the specific heat capacity at constant pressure, and τ_t the thermal relaxation time.

Table 4. Comparison of Kerr Effect Materials

Material	n_2, m^2/W	τ, s	Δn	α, cm^{-1}	W^a	T^b	W', m^3/W	W'', m^3/(W·s)
GaAlAs	10^{-12}	10^{-8}	2×10^{-3}	30	1	3	3×10^{-15}	3×10^{-7}
doped CdS glass	10^{-14}	10^{-11}	5×10^{-5}	3	0.3		3×10^{-17}	3×10^{-6}
silica glass	10^{-20}	10^{-14}	10^{-6}	10^{-5}	$>10^3$	<1	$>10^{-17}$	$>10^{-3}$
polydi-acetylene (PTS) crystal	10^{-16}	10^{-14}	$>10^{-3}$	<0.8	>10	<0.1	$>10^{-17}$	$>10^{-3}$
diamino-nitro-stilbene polymer (DANS)	10^{-17}			<0.2	>5	0.2	5×10^{-16}	

^{a}See equation 18.
^{b}See equation 19.

Degenerate Four-Wave Mixing.   Degenerate four-wave mixing (DFWM) is also called phase conjugation or real-time holography (qv). It is a Kerr-type phenomenon where irradiation of a sample using three beams results in a fourth beam that satisfies wave vector conservation. In other words, two beams are used to write a grating that is read by a third beam, which generate a signal that is the phase conjugate to the reading beam. DFWM can be employed as an optical signal processing technique. When an object beam is scrambled by propagating through an aberrating medium, the conjugate beam propagates back through the medium along the same path as followed by the object wave. The scrambling is thus undone. Such devices are useful when high power laser beams must be brought into tight focus, or transmitted through the atmosphere, or used to produce a pattern.

Applications Involving Nonlinear Absorption Phenomena.   Saturable absorption (hole-burning) is a change (typically a decrease) in absorption coefficient which is proportional to pump intensity. For a simple two level system, this can be expressed as

$$\alpha(I) = \alpha_0/(I + I/I_s) \tag{23}$$

where I_s, the saturation light intensity is equal to $\Delta E/(\sigma \cdot \tau)$, in which ΔE is the energy difference between the ground and excited state, σ the absorption cross section, and τ the excited state to ground state relaxation time. One application of saturable absorption is in volatile computer memory, dynamic random access memory (DRAM), where the refresh time depends on excited state relaxation time. A tunable laser is used to write many spectral Bennet holes, ie, bits of information, within an inhomogeneously broadened spectrum at the same spatial location. The storage density, which can reach 10^7, is given by the ratio of the inhomogeneous line width to the homogeneous packet width (56). Such a density is literally more than a thousand times better than the best video disk.

Reverse saturable absorption is an increase in the absorption coefficient of a material that is proportional to pump intensity. This phenomenon typically involves the population of a strongly absorbing excited state and is the basis of optical limiters or sensor protection elements. A variety of electronic and molecular reorientation processes can give rise to reverse saturable absorption; many materials exhibit this phenomenon, including fullerenes, phthalocyanine compounds (qv), and organometallic complexes (see ORGANOMETALLICS; SUPPLEMENT).

Photorefractive Materials

Photorefractivity can be thought of as a four part process. Initially, pairs of spatial frequency modes of a single input beam interfere with the photorefractive material to produce a periodic intensity distribution. The input light causes impurities (inorganics) or electron donors/acceptors (organics) to release charges (electrons or holes) that migrate through the dark regions of the material and become trapped. This results in a periodic charge distribution in the material which, in turn, yields a periodic electrical field. If no external electric field is present, this periodic electrical field is phase shifted by $\pi/2$ from the original light interference pattern. Finally, the induced space-charge field alters the refractive index periodically through the linear electrooptic effect. The index of refraction grating created in this manner combines with the intensities of the spatial frequency modes and introduces a nonlinear phase for each mode. The overall effect is a Kerr-like phenomenon realized with a second-order nonlinear optical chromophore. Photorefractive materials range from crystalline inorganic materials such as lithium niobate to organic polymer composites and homopolymer materials.

Economic Aspects

As of the mid-1990s, large-scale commercial applications have yet to be found for nonlinear optical materials. This situation may change dramatically before the year 2000 (57). Among the most promising classes of materials for extensive commercial application are second-order materials for use as signal transducers in cable television (CATV) and for real-time intercomputer communication applications; switches in local area optical networks; ultrafast digital-to-analogue converters; and alternatives to electronic switches for ultrafast switching operations, for radio frequency and microwave power distribution, and for remote voltage sensing. A number of commercial suppliers of lithium niobate switches exists, such as United Technologies. Of the two companies that offer polymeric electrooptic modulator materials for sale, IBM Corp. has poly(disperse red 1 methacrylate-*co*-methyl methacrylate [*11989-05-8*], and AdTech Systems Research, Inc. has LD-3 NLO-polymer, which is a methylmethyacrylate polymer containing the 4-(bis(2-hydroxyethyl)amino-4′-(6-methacryloyl-hexyl) sulfonyl) azobenzene chromophore. A small market also exists for frequency doublers based on inorganic crystalline materials. Similarly, a niche market application has been found for BBO and KTP materials for optical parametric oscillators and amplifiers. These devices are available through laser suppliers such as Coherent, Continium, and Spectral Physics, commercial applications of third-order materials are even less common than they are for second-order materials. The most significant application is probably the exploitation of the dynamic Kerr effect to achieve mode locking in Coherent's Ti:sapphire laser.

BIBLIOGRAPHY

1. P. N. Prasad and D. J. Williams, *Introduction to Nonlinear Optical Effects in Molecules and Polymers*, Wiley-Interscience, New York, 1990.
2. R. W. Boyd, *Nonlinear Optics*, Academic Press, Inc., New York, 1992.
3. G. I. Stegeman and W. Torruellas, *Electrical, Optical, and Magnetic Properties of Organic Solid State Materials*, Materials Research Society, Pittsburgh, 1994, pp. 397–412.
4. L. R. Dalton and co-workers, *Molecular Electronics and Molecular Electronic Devices*, CRC Press, Inc., Boca Raton, Fla., 1993, pp. 125–207.
5. W. Nie, *Adv. Mat.* **5**, 520 (1993).
6. J. F. Reintjes, *Encyclopedia of Modern Physics*, Academic Press, Inc., New York, 1990, pp. 361–414.
7. L. R. Dalton and co-workers, *Adv. Mat.* **7**, 519 (1995).
8. D. M. Burland, R. D. Miller, and C. A. Walsh, *Chem. Rev.* **94**, 31 (1994).
9. H. E. Katz and co-workers, *J. Am. Chem. Soc.* **109**, 6561 (1987).

10. L.-T. Cheng, *J. Phys. Chem.* **95**, 10643 (1991).
11. V. P. Rao and co-workers, *Proc. SPIE* **1775**, 32 (1992).
12. S. R. Marder and co-workers, *Science* **263**, 511 (1994).
13. C. W. Dirk, *Chem. Mater.* **2**, 700 (1990).
14. A. K. Jen and co-workers, *Mat. Res. Soc. Symp. Proc.* **328**, 413 (1994).
15. W. Wang and co-workers, *Appl. Phys. Lett.* **65**, 929 (1994).
16. A. Baudrant, H. Vial, and J. Duval, *J. Cryst. Growth* **43**, 197 (1978).
17. C. Burke and co-workers, *J. Lightwave Tech.* **10**, 610 (1992).
18. L. Eldada and co-workers, *J. Lightwave Tech.* **10**, 1610 (1992).
19. D. Girton and co-workers, *Appl. Phys. Lett.* **58**, 1730 (1991).
20. C. C. Teng, *Appl. Phys. Lett.* **58**, 1538 (1992).
21. W. H. G. Horthuis and co-workers, *Proc. SPIE* **2025**, 516 (1993).
22. G. R. Mohlmann and co-workers, *Proc. SPIE* **2285**, 355 (1994).
23. A. J. Ticknor, G. F. Lipscomb, and R. Lytel, *Proc. SPIE* **2285**, 386 (1994).
24. B. A. Smith and co-workers, *Proc. SPIE* **2025**, 499 (1993).
25. J. I. Thackara and co-workers, *Proc. SPIE* **2025**, 564 (1993).
26. J. C. Chon and co-workers, *Proc. SPIE* **2285**, 340 (1994).
27. K. W. Beeson and co-workers, *Proc. SPIE* **2025**, 488 (1993).
28. J. C. Claude, P. Robin, and V. Dentan, *Proc. SPIE* **2025**, 467 (1993).
29. J. A. Armstrong and co-workers, *Phys. Rev.* **127**, 1918 (1962).
30. K. Shinozaki and co-workers, *Appl. Phys. Lett.* **59**, 510 (1991).
31. K. Mizuuchi, K. Yamamoto, and T. Taniuchi, *Appl. Phys. Lett.* **59**, 1538 (1991).
32. K. Shinozaki and co-workers, *Appl. Phys. Lett.* **58**, 1934 (1991).
33. X. Cao, R. Strivastava, and R. V. Ramaswamy, *Opt. Lett.* **17**, 592 (1992).
34. Y. Lin, L. Mao, and S. Cheng, *Appl. Phys. Lett.* **59**, 516 (1991).
35. E. J. Lim and co-workers, *Appl. Phys. Lett.* **59**, 2207 (1991).
36. D. H. Jundt and co-workers, *Appl. Phys. Lett.* **59**, 2657 (1991).
37. K. Mizuuchi and K. Yomamoto, *Appl. Phys. Lett.* **60**, 1283 (1992).
38. W. Hsu and M. C. Gupta, *Appl. Phys. Lett.* **60**, 1 (1992).
39. F. Ahmed, *J. Bangladesh Acad. Sci.* **13**, 175 (1989).
40. L. R. Dalton and co-workers, in P. N. Prasad, ed., *Frontiers of Polymer Research*, Plenum Press, New York, 1992, pp. 115–123.
41. R. A. Norwood and G. Khanarian, *Electron. Lett.* **26**, 2105 (1990).
42. G. H. Cross and co-workers, *Proc. SPIE* **2285**, 11 (1994).
43. D. Hissink and co-workers, *Proc. SPIE* **2025**, 37 (1993).
44. *J. Opt. Soc. Amer. B* **10**(9), **10**(11) (1993).
45. D. Josse and co-workers, **64**, 3655 (1994).
46. D. Armitage, W. W. Anderson, and T. J. Karr, *IEEE J. Quant. Elec.* **QE21**, 1241 (1985).
47. D. M. Patrick and A. D. Ellis, *Electron. Lett.* **29**, 227 (1993).
48. D. M. Patrick, A. D. Ellis, and D. M. Spirit, *Electron. Lett.* **29**, 702 (1993).
49. T. Morioka and co-workers, *Electron. Lett.* **30**, 591 (1994).
50. A. D. Ellis and D. M. Spirit, *Electron. Lett.* **29**, 2115 (1994).
51. J. P. Hermann and J. Ducuing, *J. Appl. Phys.* **45**, 1500 (1974).
52. S. R. Friberg and P. W. Smith, *IEEE J. Quant. Elec.* **QE23**, 2089 (1987).
53. G. I. Stegeman and R. H. Stolen, *J. Opt. Soc. Amer. B* **6**, 652 (1989).
54. X. F. Cao and co-workers, *J. Appl. Phys.* **65**, 5012 (1989).
55. A. N. Bain and co-workers, *Spectros. Int. J.* **8**, 71 (1990).
56. U. P. Wild and co-workers, *Appl. Opt.* **29**, 4329 (1990).
57. R. S. Service, *Science* **267**, 1918 (1995).

LARRY R. DALTON
University of Southern California

NONWOVEN FABRICS

STAPLE FIBERS

Traditional textile fabrics are made by weaving or knitting. Nonwoven fabrics are similar to woven and knitted fabrics in that both are planar, inherently flexible, porous structures composed of polymer-based materials. The main difference between the two is the manner in which the fabric is made.

A woven fabric is assembled by interlacing two or more sets of yarns at right angles to one another in a designated order. A knitted fabric is assembled by intermeshing loops from one or more yarn strands into a two-dimensional array. In both weaving and knitting, yarn is the transporting media for the polymer-based material which is generally in the form of continuous filaments or twisted discontinuous fibers. The shortest path from fiber to fabric is found by placing a predetermined number of fibers or filaments into a two-dimensional array and locking them together. The processes involved in making this transformation are the essence of nonwoven manufacturing technology. The fabrics that result are technically sophisticated engineered structures that can be made to resemble in appearance, and exceed in properties, many woven or knitted fabrics.

A nonwoven fabric can be assembled by mechanically, chemically, or thermally interlocking layers or networks of fibers, filaments, or yarns. Fabrics made from textile fibers in this manner have been classified as dry-laid nonwovens.

Commodity paper sheets are made by discharging a mixture of fibers suspended in a dilute slurry of water onto a fine mesh screen, which allows the water to pass through and the fibers to be deposited (se PAPER). The fibers are then removed from the screen in the form of a continuous mat, composed of about 80% water and 20% fiber, pressed between rolls to mechanically remove more water, subjected to heat to dry out the remaining water, and wound into rolls. Fibers most often used in papermaking are cellulose (qv), and either wood pulp or chopped plant fibers such as hemp, cotton (qv), sisal, or flax. When wetted and mechanically treated, cellulosic fibers split into a network of fine fibrils, which are drawn into intimate contact with other fibers by surface tension forces during drying, and interlock by the mechanism of hydrogen bonding. A nonwoven fabric can also be made by suspending fibers in water or some other fluid (including air), controlling the way by which the fibers and suspending media are separated, and then mechanically, chemically, or thermally interlocking the fibers together. Fabrics made in this manner have been classified as wet-laid nonwovens.

Plastic films are made by extruding molten polymer through a narrow slit. Upon cooling, the polymer solidifies in the form of a flat film, is stretched to align the polymer molecules, and wound into rolls. Thermoplastic fibers such as nylon, polyester, and polypropylene are also made by extruding molten polymer through a narrow hole, solidifying the polymer by cooling, stretching to align polymer molecules, and winding onto a cylindrical package. Nonwoven fabrics can also be made by extending the fiber extrusion process to include interlocking fibers or filaments concurrent with their extrusion, modifying the porosity of a film by perforating it, or modifying the film manufacturing process in order to form porous films concurrent with their extrusion. Fabrics made this way are called polymer-laid nonwovens.

History

Documentation of the early history of fabrics made by nonwoven methods is limited but fascinating. A method used since the eighth century BC to make the felt used in nomadic dwellings consists of placing loose wool on an old dampened felt, soaking the fibers with water, layering the old felt, new wool, and a skin from a freshly killed yak around a pole which serves as an axle, and dragging the bundle by horseback around in circles for several hours (1). Nonwoven felts are one of the oldest, if not indeed the oldest, forms of fabric. Mongolian tribes of Asia were early users, and reference to felt materials is contained in the writings of ancient Greeks and Romans (2).

The feltmaking process may have been rediscovered by a Benedictine monk during an eighth century pilgrimage from Caen to the shrine at Mont-Saint-Michel (3). During the journey, the monk placed some greased wool, plucked from a wandering sheep, in his sandals to relieve his feet. At the end of the day, he examined the wool and noted that the fibers had interlocked into a matted fabric due to the working action of his feet against the sandal and the presence of heat and moisture. The discovery is said to mark the birth of felting in the Western world, and resulted in the monk Fuetre (St. Clement) being named the patron saint of the felt industry.

Although the existence of some form of nonwoven fabric can be traced to antiquity (4,5), the emergence of nonwoven fabric manufacturing as an industry has been relatively recent. Finding a better way to make gauze resulted in Johnson & Johnson's involvement in nonwovens in the 1930s (6). After 10 years of experimentation, the nonwovens department and its Masslinn fabrics became a division of Chicopee Manufacturing Corp. Rayon replaced cotton in most commercial products, and the fabrics were put to use in bed pads, dental napkins, surgical towels, disposable diapers, sanitary napkins, hand towels, casket linings, pattern markers, ribbons, and wiping cloths. In the 1970s, production facilities using jets of water to bond fibers together to produce nonwoven surgical gauze were completed. In the 1990s, most woven gauze applications have been replaced by nonwoven fabrics.

Other companies with early involvement in developing nonwovens as textile replacements include Avondale Mills, Kimberly-Clark, The Kendall Co., and the West Point Manufacturing Co. Freudenburg of Germany, a worldwide producer of nonwoven interlinings (another woven fabric replacement), began efforts in the 1930s to find a substitute for leather (qv) (7).

Considering the success demonstrated by textile and paper producers (8) in making fabrics directly from fibers, a logical next step would be to make fabrics directly from the materials used to make the fibers themselves, cutting out the fiber-production step altogether. Therefore, following the successful introduction of synthetic fibers into traditional woven and knitted fabrics, fiber producers devoted some focus on nonwovens. The quest by synthetic fiber producers for new applications and markets resulted in the introduction of synthetic staple into nonwoven products in the mid-1950s. The knowledge gained in adapting these new fibers to the various commercial nonwoven processes available at the time resulted in an increased understanding of how these processes work. These discoveries, in turn, brought about improvements to those processes and expanded into the development of new technologies for making fabrics directly from fibers, and then expanded further into the development of methods for making fabric directly from the fiber-forming polymer itself. Two closely related technologies illustrate methods for making fabric directly from the fiber-forming polymers: spunbonds and meltblowns (see NONWOVEN FABRICS, SPUNBONDED).

The spunbond process transforms polymer directly to fabric by extruding filaments, orienting them as bundles or groupings, layering them on a conveying screen in a patterned array, and interlocking them by thermal fusion, mechanical entanglement, chemical adhesives, or combinations of these methods. Technology for commercially producing spunbonds was developed in the 1950s in Germany by Freudenberg and in the United States by Du Pont. These early systems, as well as others developed later in the Netherlands, England, and France, were kept in-house. A spunbond technology system developed by Lurgi (Germany) was sold on a licensing basis during the early 1970s (9).

Meltblown fabrics, like most spunbonds, are manufactured directly from thermoplastic resins. The resin-in-chip form is heated, hence the term melt, to the liquid state, and as it passes through an extrusion orifice is injected (blown) with sonic velocity air at about 250–500°C. The fast-moving airstreams effectively stretch or attenuate the molten polymer and solidify it into a random array of very fine-diameter fibers. The fibers are then condensed (separated from the airstream) as a randomly entangled web and compressed between heated rolls. The combination of fine-diameter fibers, random entanglement, and close packing brings about a fabric structure with a large surface area and many small pores. Meltblown technology was developed at the U.S. Naval Research Laboratories (10). Researchers at Esso (11) and others independently at 3M improved the process during the mid-1960s and commercialized products in the late 1960s. In the early 1970s, Exxon began a worldwide licensing program, which attracted firms such as Kimberly-Clark, Riegel Paper Co., Pall, and Electrolux (see LICENSING). Since the 1970s, developments in meltblown technology have focused on improvements in die design, web handling, resin requirements, energy utilization, and other means to enhance productivity and fabric strength.

Definitions

A number of definitions have been drafted to help distinguish nonwovens from other fabrics. The Textile Institute defines nonwovens, in general, as ..."textile structures made directly from fibre rather than yarn. Fabrics are normally made from continuous filaments or from fibre webs or batts strengthened by bonding using various techniques: these include adhesive bonding, mechanical interlocking by needling or fluid jet entanglement, thermal bonding and stitch bonding." The American Society for Testing and Materials (12) defines a nonwoven fabric as "A textile structure produced by bonding or interlocking of fibers, or both, accomplished by mechanical, chemical, or solvent means and combinations thereof."

Lengthy definitions and specifications have also been made by the International Standards Organization (13) and the INDA Association of the Nonwoven Fabrics Industry.

As items of commerce, nonwovens are recognized internationally by government agencies, trade associations, and business enterprises. In the United States, statistical data on establishments engaged in manufacturing nonwoven fabrics are reported under Standard Industrial Classification 2297. Nonwoven fabrics are produced in weights ranging from less than 10 g/m^2 (0.3 oz/yd^2) to as much as 3500 g/m^2 (100 oz/yd^2) and are sold in the form of rolls cut to lengths and widths to accommodate specific application requirements.

As engineered structures, nonwovens can be designed to have appearances, textures, and other aesthetic properties comparable to traditional wovens and knits, and performance and functional properties superior to traditional wovens and knits. Nonwovens are, indeed, a distinct class of fiber-based materials with the characteristics of fabric and many of its useful properties.

Nonwoven Processes

The basic concept employed in producing a nonwoven fabric is to transform fiber-based materials into two-dimensional sheet structures with fabric-like qualities, primarily flexibility, porosity, and mechanical integrity. In practice, this concept is carried out through the use of one of a number of different technologies, depending on the fiber material used and/or the fabric characteristics desired. For the most part, nonwoven fabric manufacturing processes are based on primary technologies transferred from three basic manufacturing industries, textiles, paper, and plastics, and various combinations of established processes from one or more of these industries (14). Accordingly, processes for manufacturing nonwoven fabric can be grouped into one of four general technology bases: textile, paper, extrusion, and hybrid, meaning any combination of these technologies.

Textile technology is used to mechanically or aerodynamically arrange textile fibers into preferentially oriented webs. Fabrics produced by these systems are referred to as dry-laid nonwovens. Dry-laid nonwovens are manufactured with machinery associated with staple fiber processing, such as cards and garnetts, which are designed to manipulate preformed fibers in the dry state. Also included in this category are nonwovens made from filaments in the form of tow, and fabrics composed of staple fibers and stitching filaments or yarns, ie, stitchbonded nonwovens.

Papermaking technology is used to process wood pulp fibers, synthetic fibers longer than wood pulp, and other fibers that differ in other ways from pulps. Included in this category are methods for producing dry-laid pulp and wet-laid nonwovens. These fabrics are manufactured with machinery associated with pulp fiberizing, such as hammer mills, and paperforming, ie, slurry pumping onto continuous screens, designed to manipulate short fibers suspended in a fluid.

Extrusion technology is used to produce spunbond, meltblown, and porous-film nonwovens. Fabrics produced by these systems are referred to individually as spunbonded, meltblown, and textured- or apertured-film nonwovens, or generically as polymer-laid nonwovens. These fabrics are produced with machinery associated with such polymer extrusion methods as melt-spinning, film casting, and extrusion coating. In polymer-laid systems, fiber structures are simultaneously formed and manipulated.

Nonwoven technologies that employ machinery and processing principles traditionally used to manufacture textile, paper, or extruded materials, when viewed collectively, form what may be termed the primary or basic nonwoven fabric manufacturing systems. These systems are or can be continuous processes. Common to each of these systems are four sequential phases: fiber selection and preparation, web formation, bonding, and finishing.

Fiber selection and preparation involves choosing the fiber or fiber material for a specific application and purifying it so that the process yields a fabric with properties sufficient to perform its intended function. Web formation is the process by which individual fibers or fibrous materials are arranged in order to bring about the physical properties desired in the fabric structure. Bonding is the process by which the fibers or fibrous materials are interlocked in order to provide the integrity or strength desired in the fabric structure. Finishing includes slitting the fabric to the width desired, winding the fabric in roll form or cutting it to the length desired, and treating the fabric surface chemically or mechanically to bring about enhanced functional or aesthetic properties.

An outline of the three basic nonwoven manufacturing systems according to parent technology and the four manufacturing phases common to each is given in Table 1.

Nonwoven hybrid technology includes (1) methods to combine two or more nonwoven fabrics made by any of the primary nonwoven manufacturing systems, (2) methods to provide a combination of fabric properties, and (3) methods to produce true composite nonwoven structures. Combining systems employs lamination technology, or at least one basic nonwoven web formation or consolidation method, to join two or more fabric layers. Combination systems utilize at least one basic nonwoven web formation, bonding, or finishing element to enhance the properties of one or more fabric substrates. Composite systems integrate two or more basic nonwoven web formation technologies to produce nonwoven structures.

The various nonwoven processes and the fabrics made from each have a number of common characteristics. In general, textile technology-based processes provide maximum product versatility, because most all textile fibers and bonding systems can be utilized and conventional textile fiber processing equipment readily adapted at minimal cost. Extrusion technology-based processes provide somewhat less versatility in product properties, but yield fabric structures with exceptional strength-to-weight ratios, as is the case with spunbonds; high surface area-to-weight characteristics, a benefit of using meltblown technology; or high property uniformities per unit weight, as is the case with textured films, at modest cost.

Paper technology-based nonwoven processes provide the least product versatility and require a high investment at the outset, but yield outstandingly uniform products at exceptional speeds. Hybrid processes provide combined technological advantages for specific applications.

The arrangement of information provided in Table 1 can be used to trace the routes for producing most nonwoven fabrics. Also, several fundamental aspects relevant to the general nature of nonwoven fabrics can be inferred,

Table 1. Basic Nonwoven Fabric Manufacturing Systems

System	Textile			Paper		Extrusion		
	Garnetting	Carding	Air-laid fiber	Air-laid pulp	Wet-laid	Spunbond	Meltblown	Film
fiber selection and preparation	Natural and manufactured textile fibers			Natural and manufactured fiber/pulp		Fiberforming polymer chips		
	mechanical opening and volumetric blending			mechanical opening gravimetric feeding	wet slurry	mechanical, electrostatic, aerodynamic filament orientation	aerodynamic fiber orientation and shattering	perforate, cast; cast and aperture
web formation	Mechanical			Fluid				
	parallel fiber layers randomized batts cross-lapped layers		isotropic fiber layers	random fiber mattes	controlled fiber layers	pattern layering on conveyor screen	collection on conveyor screen or shape	heat, heat stretch, perforate, heat, stretch
							Cooling	
web consolidation (bonding)	Mechanical			Mechanical		Mechanical		
	stitchbonding, needle-punching, hydroentangling			hydroentangling		needle-punching		
	Chemical							
	sprayed latex or powder; saturated, printed, or frothed latex; solvent							
	Thermal							
	thermal calender, radiant or convection oven, vacuum drum or mold, laminating, sonic welding							
finishing	Slitting and winding							
	other application-dependent physical or chemical surface treatments							

namely that (1) the properties of nonwoven fabrics are fiber material dependent; (2) the orientation of individual fibers or fibrous materials in a nonwoven fabric can be selectively arranged in order to efficiently utilize their property advantages; (3) the fiber arrangements are interlocked by bonding to yield flat, flexible, porous sheet structures in the form of rolls; and (4) specific fabric properties can be enhanced by incorporating finishing processes into the fabric manufacturing sequence.

The four nonwoven fabric manufacturing phases are further interrelated in that production efficiencies and optimum end use product performance are dependent on the mutual compatibility of these manufacturing phases. A key factor in the evolution of the various nonwoven manufacturing technologies has been the development of methods to process a wide variety of fibers and polymers, and to maintain control of individual fiber placement, fiber-to-fiber bonding, and fabric surface energy through the application of basic principles known to textile, paper, and extrusion scientists, technologists, and practitioners.

From a practical standpoint, the fiber or polymer must interact or process freely with the dynamics of web formation, and the resulting fiber network must be in register with the interlocking arrangement or media, in order for the fabric structure to transmit the maximum potential inherent in the properties of individual fibers. Ultimately, if a nonwoven fabric is to be totally effective and its properties fully utilized, it must be appropriately configured to meet its end use application or appropriately placed in the end use item in such a way that the performance of the product reflects the position and characteristics of individual fibers.

Fibers for Nonwovens

Nonwoven fabrics made directly from polymers are discussed elsewhere (see NONWOVEN FABRICS, SPUNBONDED). Emphasis here is on nonwovens made from fibers. The properties of nonwoven fabrics are highly influenced by the properties of their constituent fibers (see FIBERS, SURVEY).

The terms fiber and textile fiber, in general, refer to materials that can be spun into yarn for use in strand products such as threads, cords, cables, and ropes, or as intermediate materials for fabric-making processes such as weaving, knitting, or tufting. Fibers can also be transformed directly into fabrics by placing them on a surface and interlocking or bonding them by mechanical, chemical, thermal, or solvent means. The way in which fibers respond to being placed on a flat surface and being mechanically interlocked by repeated stroking of metallic needles determines fiber processing performance in needle-punching operations. The properties of a needle-punched fabric are determined by the properties of the fiber, the fiber processing performance, and the nature and intensity of the needling operation. Fiber properties are, perhaps, the most important ingredient in a successfully engineered needle-punched fabric.

Technically, a fiber is a unit of matter characterized by a high ratio of length to width. Textile fibers also exhibit sufficient strength and extensibility, elasticity, flexibility, and temperature stability to endure the environments they are expected to be used in. Textile filaments are fibers with lengths which are, for all practical purposes, continuous. Spunbonds used for geotextiles (qv), roofing, and filtration are examples of nonwoven fabrics composed of continuous filaments. Discontinuous fibers, such as wool, cotton, and acrylic, used in non-woven fabrics designed to serve as vibration felts, furniture batting, and blankets, respectively, are referred to as staple fibers. Staple fibers range in length from about 2 to 20 mm. Fibers with thicknesses greater than about 100 μm (500 mils) are generally considered coarse bristles; fibers with lengths less than a centimeter are generally not processed on textile-based processing machinery into commodity products.

Typical textile fibers used, for example, in a needle-punched filter fabric, are a blend of 3.3- and 6.6-dtex (3- and 6-denier) polyester staple. These fibers are ~5 cm long, have diameters ranging from 18 to 25 μm, mass-per-unit-length or linear density values ranging from ~350 to 650 mg per 1000 m, and length-to-width ratios in the order of 1000 to 1.

Virtually all fibers are composed of long-chain molecules or polymers arranged along the fiber axis. Essential requirements for fiber formation include long-chain molecules with no bulky side-groups, strong main-chain bonding, parallel arrangement of polymer chains, and chain-to-chain attraction or bonding. Basic phases in the fiber formation process are obtaining a suitable polymeric material, converting the material to liquid form, solidifying the material into fiber dimensions, and treating the fiber to bring about desired properties. These four phases are present in the formation of natural as well as manufactured synthetic fibers, the principal differences being the amount of time and energy required.

In the production of manufactured fibers, suitable material is obtained from naturally occurring polymers such as cellulose or protein, or is synthesized from simple chemical compounds. Conversion to liquid form is achieved by chemical reaction, heat, pressure, or combinations thereof. Longitudinal fiber dimensions are achieved by passing the liquid polymer through spinnerettes and coagulating the material in a chemical bath (wet spinning), hot air stream (dry or solution spinning), or cold air stream (melt spinning). Several physical properties are obtained by controlled stretching, or drawing, of the fiber by a factor of 4 to 10, which aligns the molecular chains along the fiber axis. Alternatively, chemical and/or physical fiber variants can be derived through the use of additives or modification of the polymer's chemistry or fiber surface.

A selection of fiber property data is given in Table 2 as an illustration of the range of fiber properties available commercially for use in manufacturing nonwoven fabrics. In general, fiber diameters range from 5 to >40 μm for natural fibers, and from less than 10 μm (microdenier) to as high as needed for manufactured fibers.

Trends in staple fiber shipments to nonwoven producers are illustrated in Figure 1 and represent nearly 20% of all U.S. shipments of these fibers. Usage of polyester staple consumed in fiberfill are not included in these data. In 1992, some 182,000 t of virgin staple polyester were consumed in this application.

Web Formation

Web formation, the second phase in manufacturing nonwoven fabrics, transforms fibers or filaments from linear elements into planar arrays in the form

Table 2. Properties of Some Commercially Available Textile Fibers

Fiber	Density, g/mL	Modulus, N/tex[a]	Tenacity, N/tex[a]	Elongation, %	Regain, %	T_m, °C
cotton	1.52	4.85	0.26–0.44	7	7	
jute	1.52	17.2	0.44–0.52	2	12	
wool	1.31	2.38	0.08–0.17	40	13	
rayon	1.54	4.85–7.5	0.08–0.44	8–20	11	177
acetate	1.32	3.53	0.11	25–45	6.5	260
nylon	1.14	2.65	0.44–0.79	15–50	3–5	260
polyester	1.38	4.41–8.38	0.35–0.71	15–50	0.4	254
acrylic	1.16	6.44	0.17–0.26	20–30	1–1.5	
polypropylene	0.91	7.76	0.26–2.64	20	0.01–0.1	177
Nomex[b]	1.38	8.83	0.35–0.44	20–30	4	371
Kevlar[b]	1.44	42.34	0.79–1.14	1.5–4	5	482
sulfur	1.37	2.65–3.53	0.26–0.35	25–35	0.6	285
glass	2.56	30.89	0.79–1.76	2–5	0	1482

[a]To convert N/tex to g/den, multiply by 11.33.
[b]Polyamides.

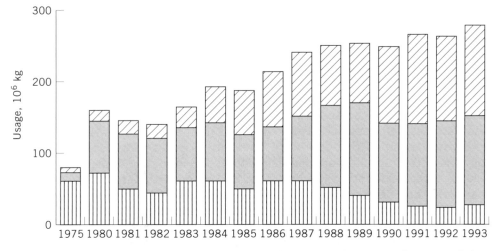

Fig. 1. Trends in staple fiber usage in nonwoven fabric manufacturing where ▨ is olefin, ▦ is PET, and ▥ is rayon.

of preferentially arranged layers of lofty and loosely held fiber networks termed webs, batts, mattes, or sheets. As indicated in Table 1, mechanical and fluid means are used to achieve the fiber arrangement desired by adapting fiber processing machinery designed to manufacture textile, paper, or extrusion products. Basic fabric parameters established at web formation, in addition to fiber orientation, are unfinished product weight and manufactured width. In all nonwoven manufacturing systems, the fiber material is deposited or laid on a forming or conveying surface. The physical environment at this phase is dry when textile technology is used, wet when papermaking technology is used, and molten when extrusion technology is used; hence the terms dry-laid, wet-laid, and fluid polymer-laid.

Textile Carding.   Textile fiber-processing machine design is based on fiber length and diameter. In the preparation of discontinuous fibers such as cotton and wool for conversion into yarns, the carding process transforms entangled fiber mats weighing about 500 g/m^2 into parallel strands or slivers weighing about 5 g/m. In carding, the fiber mats are decreased in mass per unit length, and individual fibers are provided a parallel orientation. The two primary actions involved are termed carding and stripping.

The carding action is the combing or working of fibers between fine surfaces or points oriented in opposing directions. Actual carding or parallelization of fibers occurs when one of the surfaces moves at a speed greater than the other. The stripping action occurs when the points are arranged in the same direction and the more quickly moving surface removes or transfers the fibers from the more slowly moving surface. On carding machines, the combing or working points, termed clothing, are mounted in the form of flexible wire pins (fillet clothing) or spirally wrapped, sawtoothed wires (metallic clothing) on cylinders, rolls, or stationary curvilinear surfaces. Clothing geometry and cylinder and roll/surface arrangement and dimensions are principal parameters in the design of a carding machine configuration.

Woolen cards, for example, were designed to process a rather wide range of fiber lengths (<1–20 cm) and diameters (<20–50 μm) with additional objectives of removing contaminants, mixing fibers, preserving fiber length, extracting as few fibers as possible, and delivering as many as 100 slivers. Conventional woolen cards, consequently, consist of a series of relatively wide and large cylinders to achieve productivity and accommodate fiber length requirements; multiple rolls to work and mix fibers on the large cylinders; and smaller cylinders and rolls to take fibers to and transfer them from each working area.

Carding sets of this type were frequently used in U.S. woolen mills during the 1950s and 1960s and consist of three 1.5 m main cylinders with seven sets of workers and strippers and a breast section with three sets of workers and strippers. This configuration can be considered as three cards in tandem. A Scotch Feed placed between the first and second card reorients the fibers 90° and provides a longitudinally blended feed mat for the intermediate. Each carding section is typically fillet wire-clothed with increasing wire density and decreasing working and stripping zone settings in order to achieve optimum carding effectiveness. Machines were produced in widths of 1.5, 2.1, and 2.5 m (60, 84, and 100 in.). The size and complexity of woolen cards was considered necessary to accomplish the objectives of processing pure wool mixes at production rates of about 114 kg/h.

When processing carpet wools and wool–synthetic blends, less carding action is required. In these cases, the carding set can be shortened, or, to avoid overworking the fiber and to increase productivity, metallic wire can be used. When processing synthetic fibers on shortened woolen card sets, metallic wire usage is most common. Newer generation cards, as illustrated in Figure 2, designed for processing worsted and/or synthetic fibers, are termed compact cards. Machines of this type typically process fiber at comparable or higher rates than those of traditional woolen cards of similar machine width.

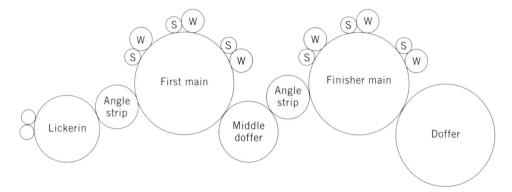

Fig. 2. Compact card for processing worsted (W) and/or synthetic (S) fibers.

Cotton cards, on the other hand, were designed to process shorter fibers (15–30 mm) and a more narrow range of fiber diameters (15–30 μm). Additional requirements include eliminating very short fiber segments and extracting non-fibrous material such as seed coat particles, leaf and stalk fragments, and dirt. A traditional cotton card consists of a roll-to-plate mat feeding assembly, a fiber-from-mat separating roll (lickerin), one large (1.1 m) cylinder and several curvilinear surfaces (revolving flats) between which the carding action takes place, a smaller (68 cm) cylinder which removes fibers (doffs) from the carding cylinder, and a web-condensing and sliver-coiling assembly. Cotton cards are generally one meter wide and typically operate at production rates in the 45–50 kg/h range.

When short (35–50 mm) synthetic fibers are processed on cotton cards, the flats are often replaced with stationary granular surfaces in order to minimize fiber extraction and damage. Fibers up to 150 mm in length are processed on cotton cards with workers and strippers, as illustrated in Figure 3. In this instance, the roller top configuration is used as a means to preserve fiber length and provide a more gentle carding action.

Garnetts were designed to thoroughly disentangle textile fibers which were reclaimed from various fiber or textile manufacturing operations or regenerated from textile threads and rags. Usage requirements include the ability to handle a very wide range of fiber lengths and diameters, preservation of fiber length,

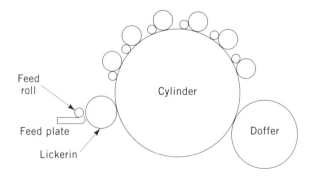

Fig. 3. Cotton card with workers and strippers (doffers).

maximum mixing, minimum extraction, and maximum throughput or productivity. Fibers processed on traditional garnetts are often placed in bales for processing at later times or in other locations. Garnetts are compact, versatile, highly productive machines. Common to most garnetts are a feeding section, a gentle opening section, a working section consisting of from one to four (depending on the degree of disentanglement required) 50–75 cm cylinders with or without a worker and stripper rolls, and one to four 30–75 cm doffers. Garnetts are built in 1.5- to 2.5-m widths and can process fiber at rates as high as 600 kg/h.

Historically, the choice of textile fiber-processing machinery to form nonwoven fabric webs involved considerations of fiber length and thickness, the extent of mixing and individual fiber alignment desired, and the degree of nonfibrous matter extraction and preservation or selective establishment fiber length distribution required for efficient manufacturing. In general, to make a web of parallel-oriented, long, coarse fibers, woolen cards were used, finer and shorter fibers were processed on cotton cards, and fibers with a wide range of dimensions were processed on garnetts. The use of these types of machinery to produce cotton and wool felts, and battings and waddings for padding and stuffing applications, provided the foundation for many dry-laid nonwoven manufacturing enterprises.

Nonwoven Cards.   Modern, high speed cards designed to produce nonwoven webs show evidence of either a cotton or wool fiber-processing heritage and have processing rate capabilities comparable to those of garnetts. Contemporary nonwoven cards are available in widths up to 5 m and are configured with one or two main cylinders, roller or stationary tops, one or two doffers, or various combinations of these principal components.

Single-cylinder cards are usually used for products requiring machine-direction or parallel-fiber orientation. Double-cylinder cards, frequently termed tandem cards, are basically two single-cylinder cards linked together by a section of stripper and feed rolls to transport and feed the web from the first working area to the second. The coupling of two carding units in tandem distributes the working area and permits greater fiber throughput at web quality levels comparable to slower single-cylinder machines.

Roller-top cards have five to seven sets of workers and strippers to mix and card the fibers carried on the cylinder. The multiple transferring action and re-introduction of new groupings of fibers to the carding zones provides a doubling effect which enhances web uniformity. Stationary-top cards have strips of metallic clothing mounted on plates positioned concavely around the upper periphery of the cylinder. The additional carding surfaces thus established provide expanded fiber alignment with minimum fiber extraction.

Double-doffer cards are generally used to conserve manufacturing space or optimize throughput while maintaining web quality. The double-doffer configuration splits the web, which essentially doubles the output of lightweight structures or yields an additional doubling action for heavier ones. A contemporary nonwoven card configuration consisting of roller-topped cylinder, stationary-topped cylinder, and single doffer is illustrated in Figure 4.

Web Layering.   Forming fibers into a web on carding or garnetting machinery is a mechanical transfer operation and takes place at the doffer. The web is formed as the doffer strips and accumulates fibers from the cylinder. The

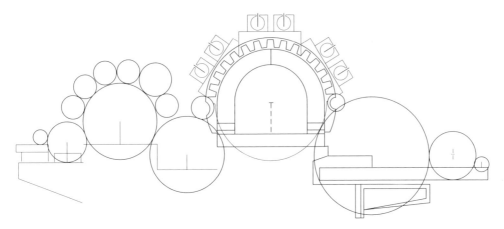

Fig. 4. Contemporary nonwoven card configuration.

number of fibers accumulated and the mass of each fiber determine the weight of the web. For a given fiber orientation, web weight per unit area is limited by the ratio of the surface speed of the cylinder to the surface speed of the doffer. For most mechanical web-formation devices, this ratio ranges from 10 to 15. Thus for cotton cards processing cotton-like dimension fibers, the practical maximum web weight per doffer is about 10 g/m^2. Similarly, practical maximums for woolen card and garnett webs are 20–25 and 50 g/m^2, respectively.

Nonwoven fabrics are produced in weights ranging from less than 10 to several hundred grams per square meter, and fiber orientations ranging from parallel (a ratio of 6 to 10 times more fibers aligned in the machine direction as opposed to the cross-machine direction) to biaxial to isotropic or random. Web building or web-layering to achieve a desired weight can be accomplished by folding from one forming machine, collection from multiple forming machines, or cross-lapping.

Web folders or straight plaiters are used with cotton cards to produce items such as surgical waddings, and with woolen cards and garnetts to produce padding and cushion filler. The resulting batt is limited in width to the width of the forming machine. Delivery is in the form of individual stacks of parallel fiber layers. Layering of webs from two or more cards or garnetts arranged in tandem onto a conveying apron or screen provides continuous delivery. Tandemly arranged, roller-top cotton carding lines are used to form webs for products such as diaper and feminine pad facings, interlinings, and wipes. In this instance, web weight is controlled by the number of cards included and finite adjustment of cylinder or doffer speed ratios. Density gradient and multifiber laminate webs can be formed by processing fibers of different sizes and chemical types on individual cards in the line. Web characteristics include a high degree of fiber parallelization and increased uniformity due to the doubling effect of layering.

Cross-lapping is essentially the plaiting or folding of a fiber web onto a conveying device placed at an angle of 90° to the forming unit. Delivery is continuous and fiber orientation is biaxial. In addition to being a means of determining a range of product weights, cross-lapping is also a means of determining a range of product widths. Additives, such as binder and particulate matter, can also be deposited onto individual web layers at the lapping stage.

Cross-lapping can be achieved by doffing webs onto reciprocating floor aprons (blamires), inclined aprons (camel back) reciprocating onto stationary floor aprons or conveyers, or runout (horizontal) apron folders reciprocating onto stationary floor aprons or conveyors. Cross-lapped web layers may be formed from one or several cards or garnetts in a row. Layered web widths may range from several centimeters to several meters. Cross-lapped webs are used in the production of highloft and needled structures.

Web Spreading and Web Drafting.   Spreading layers of parallel fiber webs is a means of simultaneously increasing web width, decreasing web weight, and altering fiber orientation. Controlled stretching or drafting web layers is a means of simultaneously increasing web throughput, decreasing web weight, and altering fiber orientation.

Spreading devices typically consist of modules of bowed rolls of increasingly wider widths operating at speeds slightly greater than the conveying speed of the input web. Fibers move longitudinally but mostly horizontally past one another, resulting in a lateral stretching or drafting of the web and overall repositioning of individual fibers. Width increases of 50–250% are common. Web-drafting devices consist of a series of top-and-bottom roll sets of the same width operating at successively increasing speeds. When heavy cross-lapped layers are drafted, predominantly cross-oriented webs are provided a more isotropic arrangement. Draft ratios of six and higher are practiced in some nonwoven operations. Web drafters are also used as a means for on-line weight control.

Continuous filament bundles (tow) can be formed into webs by spreading the tow bands over a series of threaded rolls. Tow spreaders operate in principle in a manner similar to web spreaders, with the addition of air-assisted spreading bars between adjacent expander rolls. Each spreader module effectively doubles material width. Spread filament webs can be layered with other webs, used in longitudinal form, or cross-lapped to provide a wide range of widths and weights. Weights from 25 to several hundred grams per square meter, and widths from 50 to several meters, are used for various nonwoven products. Throughput speeds exceeding 500 m/min and capacities in the 200–300 kg/h range are possible. Continuous filament webs can also be formed by layering, spreading, or cross-lapping filament yarns from beams or creels commonly used in weaving and knitting operations.

Random Cards.   Fiber orientation ratios as low as 3:1 can be achieved on cards by expanding the condensing action at doffing through the addition of scrambling or randomizing rolls operating at successively slower surface speeds. Proper selection of clothing wire and speed ratios can yield webs with increased z-direction fiber orientation, resulting in increased thickness and loft; throughput speed, however, is decreased. Cards specifically designed to produce random webs at contemporary throughputs are configured with several small cylinders that hurl the fibers onto adjacent doffers or cylinders which in turn transfer the fibers centrifugally onto subsequent cylinders. The roll arrangement for a random card for producing webs for nonwovens in the 10–50 g/m^2 range is illustrated in Figure 5**a**. A random card designed to produce highloft structures in the 40–200 g/m^2 is shown in Figure 5**b**.

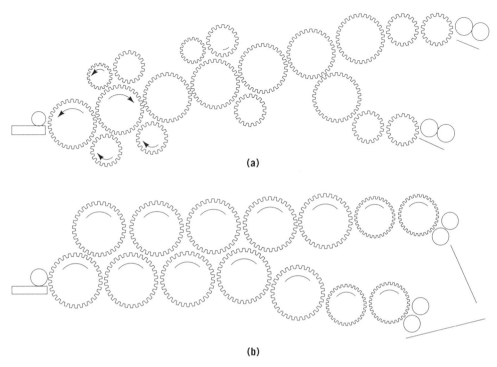

Fig. 5. Random card roll arrangements designed for (**a**) lightweight nonwovens, and (**b**) highloft nonwovens.

Aerodynamic Web Formation. Air-laid nonwovens can be grouped into two categories: those formed from natural or synthetic textile fibers and those formed from natural or synthetic pulps. An aerodynamic web formation machine designed to process textile fibers is shown in Figure 6. The basic elements are a preformed feed mat, a feeding arrangement, a fiber separation device, an air-generating means, an air-regulation means, and a fiber collection or condensing means.

As fibers in the feed mat pass between the feed roll and feed plate, they are separated by metallic wire teeth on the lickerin roll and carried to an air venturi where they are stripped and tumbled until they strike a moving, perforated collection surface. At the collection surface, the airborne fibers follow paths of least resistance and accumulate in a self-leveling manner while the air passes through perforations. Fiber orientation in the web is isotropic in layers corresponding to the number of fibers transferred from the wire teeth to the air-transportation zone, the intensity of the air, and the speed of the collection surface.

Three-dimensional webs can be made on air-forming machines, provided the fibers used are relatively short and stiff and the webs made are of relatively low density. Air-forming machines allow for production of web thicknesses up to several centimeters, and weights ranging from 30 to 3000 g/m^2 at widths from one to several meters. Production rates, depending on web weight, range from 5 to 150 m/min.

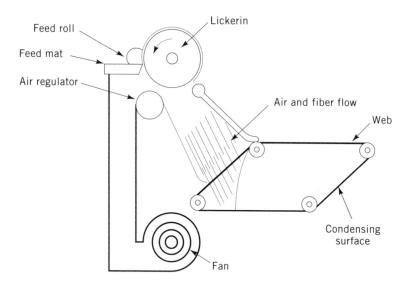

Fig. 6.   Aerodynamic web formation.

Textile fibers can be air-formed directly into end use configuration by including a shaped condensing surface or, as in the production of pillows, an air-permeable collection package. Aerodynamic web formation is a suitable means for processing brittle fibers such as glasses and ceramics, and stiff fibers such as metallics and wood.

Fibers of different diameters, lengths, shapes, and densities fractionate, or break up, when processed together in airstreams. This fractionation results in the formation of webs with different top and bottom surface characteristics, as well as varying density and porosity gradients. Such structures are well suited for many filtration applications.

Short Fiber Systems.   The web formation phase of the papermaking process occurs between the headbox and the forming wire. In this area, the fibers, suspended in a dilute water slurry, are deposited on a moving screen which permits the water to pass and the fibers to collect. Traditional papers utilize a variety of wood pulps or other short (1–4 mm) cellulosic fibers which pack together to form relatively dense, nonporous, self-adhered sheets. The use of textile fibers, instead of cellulose-based materials, with papermaking machinery distinguishes wet-laid nonwoven manufacturing from traditional paper manufacturing. Both manufacturing methods, however, transport the fibers in a water slurry. The use of papermaking fibers on air-laid nonwoven machinery bridges a gap between textile and paper systems. In both technologies, the transport medium is a fluid; water in wet-laid nonwovens, and air in dry-laid pulps.

Dry-Laid Pulp.   A principal objective of using air to form webs from natural and synthetic fiber pulps is to produce relatively lofty, porous structures from short fibers, without using water. Early technical developments in air-laid pulp processing were made by Kroyer in Denmark.

Air- or dry-laid pulp machinery can be envisioned as a series of forming-unit modules. Each module consists of two to four perforated drums through which airborne fibers are circulated and further agitated by mechanical beaters placed in close proximity to the inner drum surfaces. As the fibers circulate and separate by the force of the air and the sweeping action of the beaters, they are pulled through the drum perforations by a vacuum onto a condensing conveyor.

Air-laid pulp-forming lines generally consist of three or more forming heads in tandem. Line widths range from 1 to 3 m and operate at speeds of some 400 m/min. Web weights range from 70 to 2000 g/m^2 at throughputs of about 1000 kg/h. Air-laid pulp lines can be modified to process mixtures of textile and pulp fibers and to accommodate the addition of particulate matter.

Wet-Laid Web Formation. In the wet-lay or wet-forming process, fibers are suspended in water, brought to a forming unit where the water is drained off through a forming screen, and the fibers deposited on the screen wire. A principal objective of wet-laid nonwoven manufacturing is to produce structures with textile fabric characteristics, primarily flexibility and strength, at papermaking speeds. This can be done by incorporating textile fibers at web formation.

As a general rule, however, textile fibers do not wet out readily, are difficult to disperse, and tend to tangle with one another. Consequently, large amounts of water are necessary to keep the fibers suspended. Further, if the slurry is not handled properly, the fibers tangle and cause poor sheet formation. Two approaches to resolving these difficulties are increasing slurry–dilution ratio and controlling fiber orientation.

Forming machine designs that have been commercially successful include the inclined-wire fourdriner and the cylinder former. Inclining the forming wire and suction boxes on a fourdriner machine to an angle of 5–30° expands the forming area, which in turn decreases the flow requirements for web formation, increases drainage, and aligns fibers along the machine direction. The cylinder former configuration also provides an expanded forming area. Another benefit of this design is that higher vacuum pressures can be used, which results in the ability to produce both heavy and dense as well as light and relatively impermeable structures.

Web Consolidation

Nonwoven bonding processes interlock webs or layers of fibers, filaments, or yarns by mechanical, chemical, or thermal means. The extent of bonding is a significant factor in determining fabric strength, flexibility, porosity, density, loft, and thickness. Bonding is normally a sequential operation performed in tandem with web formation, but it is also carried out as a separate and distinct operation.

In some fabric constructions, more than one bonding process is used as a means to enhance the physical or chemical properties of the fabric. In mechanical bonding, fibers are interlocked by friction; methods include needle-felting, stitchbonding, and hydroentangling. Chemical bonding methods involve applying adhesive binders to webs by saturating, spraying, printing, or foaming. In thermal bonding, heat is used to fuse or weld fibers together.

Needle-Punching. In this method, sometimes called needle-felting, fiber webs are mechanically interlocked by physically repositioning some of the fibers from a horizontal to a vertical position. Fiber repositioning is achieved by intermittently passing a barbed needle into the web to move groups of fibers from one layer to another, and then withdrawing the needle without disturbing the newly oriented fibers. The degree of interlocking depends mainly on the extent to which the needle penetrates the web (depth of penetration), the needling density (penetrations per unit area of fabric), and the number of groups of fibers repositioned per penetration, a function of the needle design.

The basic elements of a needle-punch machine, or needle loom, are illustrated in Figure 7 and consist of a web-feeding mechanism, a needle beam with a needle board and needles (ranging in number from 500 to 7500 per meter of machine width), a stripper plate, a bed plate, and a fabric take-up mechanism.

The fiber web, sometimes carried or reinforced by a scrim or other fabric, is guided between the metal bed and stripper plates, which have openings corresponding to the arrangement of needles in the needle board. During the downstroke of the needle beam, each barb carries groups of fibers, corresponding in number to the number of needles and number of barbs (up to 36) per needle, into subsequent web layers to a distance corresponding to the penetration depth. During the upstroke of the needle beam, the fibers are released from the barbs, and interlocking is accomplished. At the end of the upstroke, the fabric is advanced by the take-up, and the cycle is repeated. Needling density is determined by both the distance advanced and the number of penetrations per stroke.

The development of a mechanical process for producing felt is dated to 1820, and has been attributed to J. R. Williams (15). The transition from interlocking fibers by working the scales on adjacent fiber surfaces against one another, to working the fibers by a scaled external member in the form of a barbed penetrating device took place during the last quarter of the nineteenth century. This transition was made possible by the development of mechanisms and machinery to produce needled nonwovens in a factory environment.

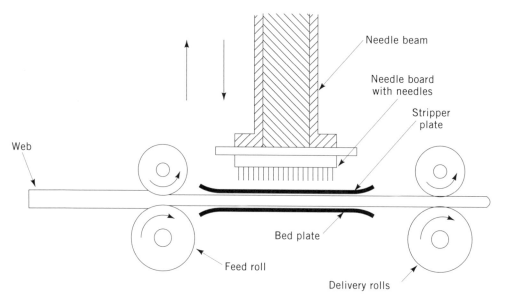

Fig. 7. Basic elements of a needle-punch machine.

Needle looms are produced in widths ranging from several centimeters to several meters. Virtually all needle looms employ reciprocating motion to provide the penetration action. The most common needle loom configuration is the single upper-board, downstroke arrangement. Other arrangements include double upper-board, single upper- and lower-board, and double upper- and lower-board. To achieve high penetration densities on both sides of a fabric, needle looms of differing configurations are often placed in tandem.

Needle looms with low density boards are used to lightly consolidate webs and are termed pre-needlers or tackers. Machines with multiple or high density needle capabilities are referred to as consolidation or finishing needle looms. Machines designed to produce patterned or raised surfaces are termed structuring looms and are used as a mechanical finishing process.

Fabric weights range from 50 to 5000 g/m^2, and needling densities range from fewer than a hundred to several thousand penetrations per square centimeter.

Most needled fabric is made in flat form; however, tubular fabric, ranging in diameter from a few millimeters to papermakers' felt dimensions, can be made on some machines.

Needled nonwovens are sometimes mistaken for fabrics which have been felted or fabrics made directly from fibers which have been interlocked by a combination of mechanical work, chemical action, moisture, and heat. Fabrics which have been felted are generally composed of yarns spun from wool fibers and have undergone a controlled shrinkage by subjection to the fulling process, a mechanical beating in the presence of lubricating agents. Fabrics made directly from fibers which have been interlocked by a combination of mechanical work, chemical action, moisture, and heat are felts. Felts are generally composed of wool or fur fibers and are physically held together by the interlocking of scales on individual fibers. Fiber interlocking in a felt is achieved by a process called hardening, which consists of passing fiber webs between oscillating and vibrating plates in the presence of steam. Following hardening, the felt is subjected to a fulling process. Felt density, stiffness, and tenacity are dependent on web weight and extent of hardening and fulling.

A needled felt, on the other hand, is a fabric composed of natural, synthetic, or a combination of natural and synthetic fibers physically interlocked by the action of a needle loom with or without combination of other textile fabrics and with or without suitable combination of mechanical work, chemical action, moisture, and heat, but without weaving, knitting, stitching, thermal bonding, or adhesives (16).

Early needle-punched nonwovens were made from coarse animal hair and vegetable fibers, and were used as carpet underlays and spring padding for mattress and furniture. The 1920s and 1930s saw the development of better needle-punch machinery, and during the late 1950s, needled synthetic fiber products were introduced to the home furnishings and apparel markets. Several efforts were launched in the 1960s to produce simulated leather with needled fabric as a substrate (see LEATHER-LIKE MATERIALS).

Principle applications of needle-punched nonwovens for the 1990s include automotive, apparel components, blankets, carpeting, carpet padding, coating substrates, filtration, furniture, geotextiles, insulation, roofing substrates, and wall coverings. In 1990, the production of needle-punched fabric was estimated to approach 91,000 t and 606 million square meters (15).

Stitchbonding. This is a mechanical bonding method that uses knitting elements, with or without yarn, to interlock fiber webs. Sometimes called stitchthrough or web knitting, this technology was developed in eastern Europe during the late 1940s and produces fabric structures which most closely resemble textiles. Maliwatt and Arachne machines employ yarn; Malivlies and Arabeva machines use modified knitting needles to interlock the fibers. Both families of machines operate essentially on the same principle, but differ in the positioning of the knitting elements, direction of web passage, and type of needles used.

The sequence of operations for a web-consolidation cycle on an Arachne machine is as follows. The web is guided upward and positioned between the web-holder table and the knock-over table and penetrated by the needle. After passing through the web, the hook of the needle is provided with a yarn properly placed by the guide and lapping motion. As the needle is withdrawn, the yarn is secured in the hook of the needle by the closing motion of the yarn guide. When the needle reaches the end of the upward stroke, the yarn is pulled through the previously formed loop, the loop is cast off, the fabric is advanced, and the cycle is repeated. Similar functions are served by the Arachne web-holder table and the needle loom bed plate, the Arachne knock-over table and the needle loom stripper plate, and the Arachne knitting needle and the needle loom needle. Thus, when yarn is eliminated, stitchbonding and needle-felting methods interlock fibers similarly.

Stitchbonded fabrics are used in home furnishings, footwear, filtration, packaging, and coating. Machine widths are in the 2-m range; fabric weights range from about 75–250 g/m². A variation of stitchbonding is used to make multiaxial-layered yarn and yarn-and-sheet structures for composite material reinforcement substrates.

Hydroentanglement. This is a generic term for a nonwoven process that can be used for either web consolidation, fabric surface-texturing purposes, or both. The mechanism is one of fiber rearrangement within a preformed web by means of fluid forces. When used for bonding, hydroentanglement repositions individual fibers into configurations that result in frictional interlocking. When used as a surface-texturing means, hydroentanglement repositions fibers into open-patterned arrangements.

Also termed spunlaced or jet-laced nonwovens, fabrics of this type have been sold commercially since the early 1970s and have been successfully used in applications such as interlinings, bedding, wound dressings, coating substrates, roofing, wipes, and surgical gowns. General characteristics of hydroentangled nonwovens include textile-like tacticity (softness, drape, and bulk), mechanical properties (strength, elongation, and abrasion), and appearance; good absorbency and wetting; and the absence of nonfibrous components such as binder chemicals or processing residues. Hydroentangled nonwovens are produced in weights ranging from about 25 to 175 g/m².

Early development of hydroentanglement was carried out independently at the Chicopee division of Johnson & Johnson and at Du Pont during the late 1950s and early 1960s. Chicopee interests focused on improving a process called Keybak, which was used to produce apertured nonwoven webs by spraying water on staple fiber webs sandwiched between an apertured plate and a fine-mesh screen. Du Pont researchers were looking for a way to mechanically interlock fiber webs without the use of felting needles.

The hydroentanglement process, as illustrated by Du Pont patent drawings (17), involves subjecting the web and its conveying device to increasingly higher pressure jets of water. When the water jet strikes the web, it moves individual fibers away from the high points of the conveying means and is deflected by the conveying surface. As a result, voids are created in the web, and fibers intermingle. Whether the fabric surface is visibly smooth or openly patterned depends on the wire design or surface geometry. When highly interlocked, mechanically bonded (spunlace) structures are desired, high water pressure and plain or three-harness fine mesh (70–100) wire or screen is used. The resulting fabric surface is comparatively smooth and the overall structure is relatively strong because of a large amount of individual fiber entanglement. When open-surface (apertured) structures are desired, lower water pressure and conveying wire combinations or surfaces with preferred patterning configurations and depths are used, and a fabric surface with an overall aperture geometry reflective of wire or surface contour is established. A wide variety of aperture shapes (circles, ovals, squares, and rectangles) and lines (straight, diagonal, and chevron) are possible through appropriate wire or embossing pattern selection. Individual aperture shape or hole clarity is a function of fiber dimensions, jetting pressure, and wire interlacing or embossment shape and height.

Chemical Bonding.   Sometimes called resin bonding, chemical bonding is a general term describing the technologies employed to interlock fibers by the application and curing of a chemical binder. The chemical binder most frequently used to bond nonwovens is waterborne latex (see LATEX TECHNOLOGY). Most latex binders are made from vinyl materials, such as vinyl acetate, vinyl chloride, styrene, butadiene, acrylic, or combinations thereof. The monomer is polymerized in water, and the polymeric material takes the form of suspended (emulsified) particles. Thus the emulsion polymerization of vinyl acetate yields a vinyl acetate polymer binder and the copolymerization of styrene and butadiene yields a styrene–butadiene copolymer or styrene–butadiene–rubber (SBR) binder.

Latexes are extensively used as nonwoven binders because they are economical, versatile and easily applied, and they are effective adhesives. The versatility of a chemical binder system can be seen by considering a few of the factors involved in formulating such a system. The chemical composition of the monomer determines stiffness and softness properties, strength, water affinity (hydrophilic–hydrophobic balance), elasticity, and durability. The type and nature of functional side-groups determine solvent resistance, adhesive characteristics, and cross-linking nature. The type and quantity of surfactant used influence the polymerization process and application method. The ability to incorporate additives such as colorants, water repellents, bacteriastats, flame retardants (qv), wetting agents, lubricants, and catalysts to enhance curing expands this inherent versatility even further.

Chemical binders are applied to webs in amounts ranging from about 5 to 60 wt %. In some instances when clays (qv) or other weight additives are included, add-on levels can approach or even exceed the weight of the web. Waterborne binders are applied by spray, saturation, print, and foam methods.

A general objective of each method is to apply the binder material in a manner sufficient to interlock the fibers and provide chemical and mechanical properties sufficient for the intended use of the fabric.

Spray bonding is used for fabric applications which require the maintenance of high loft or bulk, such as fiberfill and air-laid pulp wipes. The binder is atomized by air pressure, hydraulic pressure, or centrifugal force, and is applied to the upper surfaces of the web in droplet form using a system of nozzles. Lower web surface binder addition is accomplished by reversing web direction on a second conveyor and passing the web under a second spray station. After each spraying, the web is passed through a heating zone to remove water. The binder is cured, or cross-linked, upon passage through a third heating zone. Drying and curing is frequently done in a three-pass oven. Binder addition levels commonly range from 30 to 60% of the fiber weight.

Saturation bonding is used in conjunction with processes that require rapid binder addition, such as card-bond systems, and for fabric applications that require strength and maximum fiber encapsulation, such as carrier fabrics. Fiber encapsulation is achieved by totally immersing the web in a binder bath or by flooding the web as it enters the nip point of a set of pressure rolls. Excess binder is removed by vacuum or roll pressure. There are three variations of saturation bonding: screen, dip-squeeze, and size-press. Screen saturation is used for medium weight nonwovens, such as interlinings. Dip-squeeze saturation is used for web structures with strength sufficient to withstand immersion without support. Size-press saturation is used in high speed processes, such as wet-laid nonwovens. Through-air ovens or perforated drum dryers are used to remove water and cure the resin. Binder addition levels range from 20 to 60% of fiber weight.

In print bonding, binder is applied in predetermined areas or patterns. This method is used for fabric applications that require some areas of the fabric to be binder-free, such as wipes and coverstocks. Many lightweight nonwovens are print bonded. Printing patterns are designed to enhance strength, fluid transport, softness, hand, and drape. Print bonding is most often carried out with gravure rolls arranged as shown in Figure 8**a**. Binder addition levels are dependent on both engraved area and depth, and binder–solids level. Increased pattern versatility can be achieved by using rotary screen rolls arranged as shown in Figure 8**b**. Drying and curing are carried out on heated drums or steam-heated cans.

Foam bonding is used when low water and high binder–solids concentration levels are desired. The basic concept involves using air as well as water as the binder carrier medium. Foam-bonded nonwovens generally require less drying and curing energy because less water is used. The foam is generated by concurrently aerating and mechanically agitating the binder compound. Air/binder dilutions (blow ratios) range from 5 to 25. The addition of a stabilizing agent to the binder solution causes the foam to resist collapse during application and curing, and yields a fabric with enhanced loft, hand, and resilience.

Nonstabilized foams are referred to as froths; froth-bonded fabrics are similar in properties to some saturation-bonded nonwovens. Typical foams used as nonwoven binder solutions have a consistency similar to shaving cream. Application methods include knife-edge layering onto a horizontal web surface

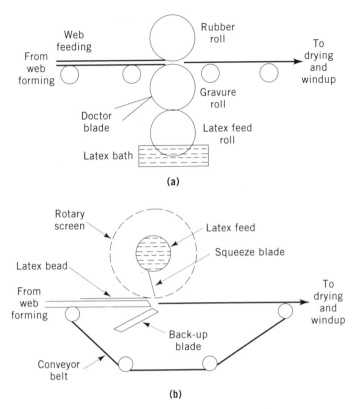

Fig. 8. Print bonding methods where (**a**) is gravure and (**b**) is rotary screen printing.

followed by vacuum penetration, and saturation and penetration of a vertical web surface using a horizontal-nip pad. Drying and curing is carried out in ovens, drum dryers, or steam cans.

Thermal Bonding. In thermal bonding, heat energy is used to activate an adhesive, which in turn flows to fiber intersections and interlocks the fibers upon cooling. The adhesive may be individual fibers, portions of individual fibers, or powders. The use of fusible fibers as bonding agents was commercialized in the 1940s (18). Fusible powders have been used in nonwoven interlinings since the mid-1960s. Flat fabrics, such as coverstock, constitute most of the thermal-bonded nonwovens consumed; however, thermally bonded highloft structures are used for a number of applications which require layer-to-layer consolidation or a heat-sensitive finishing operation. In addition to fiber-to-fabric nonwovens, thermal bonding is used to consolidate spunbonds, meltblowns, dry-laid pulps, textured films, and combination nonwovens. Advantages of thermal bonding include low cost, the general availability of new binder materials and machinery, and process and product enhancement.

Thermal bonding is achieved as the result of a sequence of three events: heating, flowing, and cooling. An adhesive component, distributed in a nonwoven web in the form of a unicomponent fiber, bicomponent fiber, or powder particle, is subjected to heat. For binder fibers and powders, initial heat softens the binder surface and expands its contact area with other fibers; additional heat induces binder flow, resulting in molten binder–fiber wetting and broader binder-to-fiber contact. As the adhesive approaches its melting point, its surface softens, and contact areas with more stable fibers expand further to form potential bonding sites. Upon melting, the adhesive, now in liquid form, becomes attached to a network fiber. It then flows along the network fiber into a crossing of two or more fibers, or forms an adhesive bead. Upon cooling, the adhesive solidifies and forms a bond at each fiber contact.

In addition to the melt-flow properties of the adhesive, individual bond strength is a function of the percentage of fiber surface area joined or shared at fiber intersections, the heating and cooling times, and bonding temperature. Bond effectiveness is also dependent on binder distribution and binder concentration. Fabric strength, resilience, softness, and drape are affected by individual bond strength, bond placement, and total bonded area. A properly produced thermal-bonded nonwoven can approach the idealized nonwoven structure, namely, one in which individual fibers are connected at crossings with each other.

Three basic methods of heating are used for thermal bonding: conduction, radiation, and convection. Conduction technologies include direct contact with a heated surface and ultrasonic welding. Direct contact heating is done with heated calender rolls. For area bonding or surface glazing, smooth rolls are used. For point bonding, patterned or embossed rolls are used. Thermal calendering is most efficient in terms of heat loss, but heavy roll pressures tend to destroy fabric loft.

Ultrasonic bonding concentrates heat energy even more efficiently than calendering and has the additional advantage of ease of pattern change. An illustration of the basic elements of an ultrasonic bonding unit is given in Figure 9. In this bonding method, a web is placed between a high frequency (~20,000 Hz) oscillator or horn, and a patterned roll. As the waves pass through the web and are concentrated on the raised points of the patterned roll, sound energy is converted to thermal energy beams, much the same way mechanical

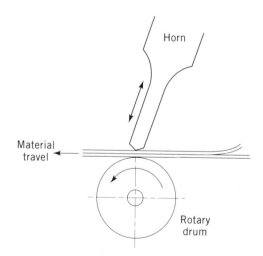

Fig. 9. Basic elements of ultrasonic bonding.

energy is converted to thermal energy at the blacksmith's anvil. The horn acts as a hammer, the raised surfaces of the patterned roll serve as individual anvils, and the web receives thermal energy at the raised areas. Portions of the web that lie between patterned areas and the horn are thus exposed to heat, and bonding or welding occurs in patterns corresponding to those on the surface of the roll.

Radiation heating concentrates fiber bonding on the fabric surface. For lofty or thick structures, this effect yields a bond intensity gradient throughout the fabric thickness. Radiant heating systems are used mostly for applications which require instant heating and concentrated heating zones.

Convection heating methods pass heated air through the nonwoven web and are used to bond many medium and heavy weight nonwovens. Two common commercial configurations are multizone through-air ovens and compact through-air ovens. Multizone ovens transport the nonwoven web through heating and cooling zones on a flat conveyor, with production speed and dwell time requirements being accommodated by increasing oven length. Compact through-air ovens use felt or perforated belts to guide the webs around perforated drums. In these systems, hot air is recirculated through the fabric, drum, and heat exchanger by low speed radial fans. The belt guide conveyor serves to stabilize the nonwoven batt during heating, and also controls fabric loft and shrinkage.

From an energy standpoint, modern thermal bonding, ie, web consolidation with no heat requirement for water removal, is very efficient. For example, in comparing only material and process requirements for thermal bonding, chemical bonding, and hydroentanglement, the energy requirements for the thermal bonding method could be as significant as 16 times lower than the others. Manufacturing lines for thermal-bonded nonwovens also require less floor space and operate at higher production rates. Thermal-bonded nonwovens are generally softer and drier, have greater strength per unit weight, and are absorbent and porous because of smaller bonding points.

Finishing

Commercial nonwoven fabrics are shipped from manufacturing plant to customer in the form of rolls of varying dimensions to accommodate the fabric end use application or subsequent conversion processes. Slitting and winding are finishing processes common to all nonwoven manufacturing methods. Roll width is determined at the slitting operation, and roll length is determined at the winding operation.

The fabric may also be given one or more of a number of other finishing treatments, either in tandem with web formation and bonding or off-line as a separate operation, as a means of enhancing fabric performance or aesthetic properties. Performance properties include functional characteristics such as moisture transport, absorbency, or repellency; flame retardancy; electrical conductivity or static propensity; abrasion resistance; and frictional behavior. Aesthetic properties include appearance, surface texture, and smell.

Generically, nonwoven finishing processes can be categorized as either chemical, mechanical, or thermomechanical. Chemical finishing involves the application of chemical coatings to fabric surfaces or the impregnation of fabrics with chemical additives or fillers. Mechanical finishing involves altering the texture of fabric surfaces by physically reorienting or shaping fibers on or near the fabric surface. Thermomechanical finishing involves altering fabric dimensions or physical properties through the use of heat and pressure.

Finishing may also be viewed as another means for providing nonwovens with additional application-dependent chemical and/or physical properties. Finishing processes bring about value-added fabrics with technically sophisticated properties for specific end use applications.

Chemical Finishing. For many nonwovens, chemical finishing is an extension of the binder application process through the use of technology associated with fabric coating. In most instances, the coating process is applied to enhance the properties of the nonwoven; however, in some applications, the nonwoven is used as a carrier to transmit the properties of the coating material. The coating may be applied as a continuous covering or as a pattern; it is most frequently applied in aqueous solution form. With many nonwoven substrates, special care must be taken because of the delicate nature of the structure itself or the arrangement of fibers on or near the fabric surface.

A number of different methods are used to coat nonwovens depending on the viscosity requirements of the coating material and the amount and location of coating desired. Knife-over-roll (blade coating), reverse-roll, air-knife, wire-wound rod, transfer-roll, rotary screen, and slot-die coating methods are used to apply continuous coatings to single surfaces (see COATING PROCESSES). Double-surface coating of relatively nonporous nonwovens with high viscosity materials can be achieved by using dip saturators or size presses with gapped or low pressure squeeze rolls. Impregnation of substantially porous nonwovens can be achieved by using the same equipment at higher roll pressures. Patterned coatings or decorative printing can be achieved with the use of gravure rolls or rotary screens.

Transfer roll, rotary screen, saturation, size press, and gravure apparatus are similar to those used for resin bonding. Reverse-roll coaters are similar in configuration to gravure print bonding apparatus, but differ in the surface patterning and direction of rotation of the applicator roll. The amount of material applied when using this method is controlled by adjusting the relative speeds of the applicator roll and the rate of fabric passage through the coating system.

In knife-over-roll or blade coating, the coating material is placed on the fabric surface behind a knife, or doctor blade, and metered according to the gap set between the blade and the fabric surface. This method is used to apply thick coatings of highly viscous materials such as pastes, plastisols, or foams.

Air-knife coating is a high speed process used to apply continuous coatings of relatively low viscosity materials onto nonwovens with irregular surfaces. The principal components of this system are illustrated in Figure 10. Following an initial application, the coating material is metered by air impingement.

Wire-bar (Mayer) coating is used to uniformly coat lightweight material applications. As in air-knife coating, the material is applied initially at a first station, but in this system, the coating material is metered and leveled by a wire-wound rod. Coating weight and uniformity are controlled by changing wire thickness and pitch on the metering rod.

Transfer-roll or flexographic coating is used to apply continuous coatings of low or medium viscosity materials at high speeds. This system is particularly

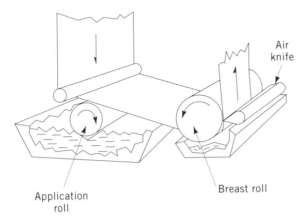

Fig. 10. Components of an air-knife coating apparatus.

suitable for coating stiff or irregularly surfaced nonwovens and for applying abrasives. With high viscosity materials and appropriately designed gravure rolls, flexographic coaters can be used as pattern applicators for decorative prints.

Rotary-screen coating is also used to apply either continuous or discontinuous coatings to nonwovens. The screen itself is a nickel sleeve, perforated according to a mesh size which corresponds to the size and number of holes per unit area of surface. A material supply tube and squeegee blade are fitted inside the screen. The coating material, in the form of a paste or foam, is forced by the blade through the perforations of the rotating screen onto the nonwoven. For a given coating material, coating weight and penetration are controlled by varying mesh size and squeegee pressure. Patterned coatings or printed designs can be achieved by blocking out selected perforations.

Slot-die or extrusion coating involves the application of molten polymer resin through a slot die directly onto the surface of a nonwoven. Upon extrusion, the resin is smoothed and quenched by a cooling roll. Coating weight is controlled by slot size and extrusion rate.

Mechanical and Thermomechanical Methods. These methods provide nonwovens with patterned surface structures, enhance the surface texture of nonwovens, or both. Patterned surfaces may be established by embossing, by compressive shrinkage (creeping), and for needle-felted nonwovens by creating loops or pile. Surface textures, ranging from flat and smooth to raised and leveled, may be created or altered by calendering, sueding, napping, polishing, brushing, or shearing. In general, mechanical finishing processes operate at speeds slower than web-consolidation processes and, consequently, are carried out off-line or as separate batch processes.

Smooth surfaces are normally established by calendering, a process which subjects the fabric at the nip point(s) of two or more rolls to the influence of controlled time, temperature, and pressure. When calendering is used as a thermal-bonding process, the rolls are of the same dimension and composition and are independently driven. However, when calendering is used as a fabric finishing operation, the rolls are frequently of different dimensions and composition and are not always independently driven.

Specific terms have been designated according to the function and composition of various rolls. Steel rolls that impose pressure, transmit heat, and emboss a pattern onto the fabric are known as pattern rolls. Flexible surface rolls that transport the fabric and permit pressure transmission to the fabric are termed bowl rolls or bowls. Bowl rolls are usually larger in diameter than pattern rolls. The material used to make these types of rolls is chosen according to the depth of surface smoothness to be placed on the fabric being calendered, and must be compatible with the pattern roll. Cellulose pulp, cotton, wool, cotton–wool mixtures, corn husk, and various polymer materials are used as fillers for the roll surface compound.

Paper-filled bowls are used for deep embossing patterns, because the short fiber length permits the pattern of the pattern roll to be effectively transferred to the bowl roll, thus forming a matched indentation or female surface to receive the fabric and accomplish precisely defined surface patterns on the fabric. In general, cotton-filled bowls are hard and dense, wool-filled bowls are more resilient but susceptible to picking fibers from the fabric because of the scales on the wool, and corn husk-filled bowls are more resilient than cotton bowls but weaker and less durable. Polymer-filled bowls can be compounded to accommodate a wide range of resilience, strength, and durability requirements. Calendering machinery designations correspond to the mechanical action applied to the fabric.

Calender designations include embossing calenders, friction calenders, and compaction calenders. Most embossing calenders are fitted with a main pattern roll and either one or two bowl rolls which are positively driven by the pattern roll through interconnecting gearing. Pattern roll design textures range from cire' (polished) to schreinering (lined) to deeply cut ornamentation. In friction calendering, the rubbing action is accomplished by operating the pattern rolls at higher rates than their bowl counterparts. Compaction calendering establishes desired fabric thicknesses or caliper through adjustable gapping or roll spacing.

Sueding is a mechanical finishing process in which fibers on the surface of a lubricated fabric are cut by the abrasive action of a sanding roll operating at relatively high speed. The cut fibers are oriented in the direction of sand roll rotation and protrude about a millimeter from the surface. The primary components of a sueding machine are the guiding system, sanding roll, support roll, and roll spacing structure and control.

The napping process mechanically raises fibers to the surface of a lubricated fabric by withdrawing the fibers from the interior of the fabric. A planetary napping machine configuration is shown in Figure 11; basic components include a series of working rolls wound with hooked wire and a fabric guiding system. The working rolls are operated in a direction opposite the fabric, and at surface speeds greater than the fabric passage speed. The napping action takes place as the wires of the working rolls penetrate the fabric, withdraw fibers, and form a nap of raised fibers on the surface of the fabric. Depending on wire design, wire wrapping pattern, roll arrangement, number of rolls, and relative roll rotation and direction, nappers can be used to produce a wide range of either loop or velour surface effects.

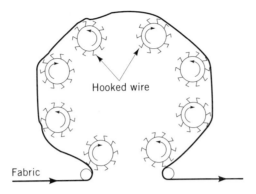

Fig. 11. Planetary napping machine configuration.

Polishing is a thermomechanical process that aligns the pile of a raised fabric surface. Polishing machine components include a guiding system consisting of a tension blanket and a spirally grooved heated cylinder. The mechanical action of the rotating edge of the roll groove against the tensioned fabric surface results in a static electrification of the pile fibers, which in turn aligns the fibers in a parallel orientation. Rotation of the spiral roll in the direction that momentarily entraps fibers in the grooves results in a raised, parallel pile surface. Rotation of the spiral roll in the opposite direction results in a flat, parallel pile surface.

Brushing is a mechanical finishing process that lifts fibers to the fabric surface and aligns the raised fibers along the machine direction of the fabric. Brushing machinery is similar in configuration to both sueding and napping machinery, but the composition of the working roll is different. Straight-wire clothing is used in brushing machine rolls. As the working roll rotates against the fabric surface, the straight wire withdraws and orients fibers along the direction of fabric passage through the machine. The length of fiber withdrawn is determined by the gap adjustment between the working and support rolls.

Shearing cuts raised fibers to uniform heights. Fabric shearing generally follows a brushing operation and consists of subjecting the fabric surface to a series of spirally wound shearing blades rotating over a stationary ledger blade. The working elements of a shearing machine are similar in configuration to a reel-type lawn mower. In operation, the fabric is guided under the shear blades while the pile is held in a raised position by vacuum. As the fabric passes a shearing point, the raised fibers strike the ledger blade and are cut by the rotating shear blades. Cut pile height is controlled by adjustment of the distance between the fabric guide and the rotating blades.

The width and thickness of thermoplastic nonwovens can be altered by heating and stretching the fabric across its width. Two common types of apparatus used are tenter frames and biaxial roll spreaders. Both devices expand fabric width by controlled spreading. Tenter frame spreading is accomplished by placing the fabric edges on a series of pins between two guide chains which are set at different widths at the entrance and exit points of a heating chamber or oven. As the fabric passes through the oven, the heat softens the fibers and allows the fabric to be stretched along its width. Biaxial roll spreaders use heated rolls to stretch the fabric along both the length and width directions. By adjusting the relative input and output speeds, fabric thickness can be either proportionally increased or decreased.

A mechanical process used to provide surface texture to needle-punched nonwovens is termed needle structuring. In this process, forked needles and special needle loom configurations move groups of fibers to the surface of pre-needled fabric. The fork needle has dimensions similar to felting needles used to mechanically bond nonwovens, but employs a recessed prong to transport the fibers to the fabric surface. The fundamental difference between a structuring and a consolidating needle loom is the bedplate configuration. Structuring needle looms use bars (lamella) or grooved cylinders oriented along the machine direction at spacings corresponding to the pitch of the structuring pattern cylinders as the bedplate surface.

Two basic effects, ribs and velours, can be achieved. Rib geometry results when the needle prong openings are placed 90° to the direction of fabric passage through the machine. Velour geometry results when the needle prong openings are placed parallel to the direction of fabric passage through the machine. Loop height or velour depth depends on depth of needle penetration. Loop thickness depends on fork width. Loop density depends on the number of needles per lamella row and the amount of fabric advance per stroke of the needle board. Pattern effects can be achieved by structuring two layers of different colored fabrics, by arranging the needles in a geometric pattern and controlling the fabric advance in a staggered manner, or by selectively raising and lowering the lamella in coordination with the fabric advance. Velour surfaces can also be made by placing a stiff brush in the bedplate position.

Production Trends

The nonwoven fabrics industry is relatively young, but has experienced rapid growth, as illustrated by the data depicted in Figure 12, and is international in scope. In the 1990s, about half of worldwide nonwoven fabric production takes place in North America, with one-third in Europe, and one-eighth in Japan. New nonwoven enterprises are in the works throughout Asia and South America; about two-thirds of all nonwovens are made from fibers, and one-third are made directly from polymers. Some 1.1 million tons of nonwoven roll goods were produced in the United States in 1992.

Production of nonwoven fabrics according to manufacturing technology is as follows: highloft, 25.5%; spunbond, 17.27%; needle-punch, 10.04%; bonded pulp, 8.23%; card (thermal bond), 7.43%; hybrids, 7.23%; card (resin bond), 5.62%; spunlace, 5.22%; wet-laid, 5.02%; meltblown, 4.42%; stitchbond, 2.61%; and porous films, 1.41%. The majority of card–resinbond and card–thermal-bond fabrics were used as coverstock; interlinings, wipes, and, carrier sheets accounted for most of the remainder. More than half of the highloft volume was used in furniture and sleeping applications; filtration, apparel, insulation, health care, and geotextile products accounted for most of the remainder. Stitchbond fabrics were used in bedding, shoes, and a variety of coated products. Automotive trim and geotextile applications consumed 50–60% of all needle-punch fabrics;

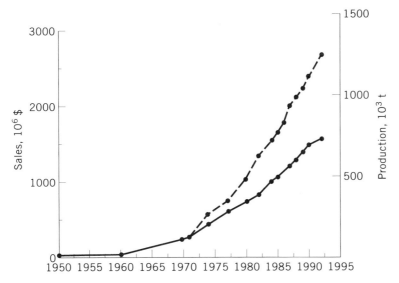

Fig. 12. Growth of the nonwoven fabrics industry where (– • –) represents 10^6 \$ and (–•–), 10^3 t.

other significant applications included filtration, bedding, home furnishings, and coating.

In 1992, as much as two-thirds of all spunlace fabrics were used in medical products; other applications included wipes, industrial apparel, interlinings, absorbent components, filtration, and coating. Medical product applications also accounted for about one-third of all wet-laid nonwovens; other applications included tea bags, meat casings, filter media, battery separators, and wipes. Most bonded-pulp fabrics were used as wipes or absorbent components.

The largest yardage application for spunbonds was coverstock; other significant uses included geotextiles, roofing, carpet backing, medical products, filtration, furniture, and packaging. About one-half of all meltblown nonwoven roll goods were used in filtration and medical applications; other applications included sorbents, wipes, and sanitary products. Porous film applications included coverstock, medical products, and laminating media. Nonwoven hybrids were used in absorbent products, wipes, filtration, and barrier applications.

Product Applications

Consumption of nonwoven roll goods is often reported in two broad areas, according to product application: disposables and durables. In general, disposable products account for ~85% of the volume and 60% of the value of nonwoven roll goods consumption (19). Items within each disposable or durable product category reflect the diversity of applications which utilize nonwovens.

Data reflecting the usage of nonwovens in disposable products are summarized in Table 3. Coverstock is the nonwoven fabric placed on the user's side of sanitary absorbent products such as baby diapers, nappy liners, adult diapers, incontinence devices, and feminine hygiene products. Medical and surgical products include protective wrap for hospital items distributed through the central

Table 3. Usage of Nonwoven Fabrics

Application	Market share, %
Disposable products	
absorbent covers	25
medical/surgical	19
wipes	18
filters	16
packaging	9
apparel (nonmedical)	6
laundry aids	4
other	3
Durable products	
coating substrates	18.5
geotextiles and roofing	17
interlinings	14.5
bedding and home furnishings	11.5
carpet components	7.5
automotive trim	7
electronic components	5.5
durable papers	4
agricultural fabrics	2
other	12.5

supply room; surgical drapes, packs, and gowns; other protective products such as face masks, caps, aprons, bibs, and shoe covers; absorbent products such as surgical dressings and sponges; and other hospital products such as isolation gowns, scrub suits, examination gowns, sheets, drawsheets, shrouds, underpads, tapes, bedding, and bed linens.

Nonwoven wipe categories include products for babies and adults, the food service and electronics industries, medical and clean room applications, industrial cleaning, computer diskettes, and household products such as dusters, tea towels, shoe cleaning cloths, towelettes, and hand towels. Nonwoven fabrics are used to filter air, water, petroleum (qv), food, and beverages. Nonwovens loaded with abrasives, cleansers, or finishes can be found in a variety of products used by many industries and in many homes to scour or polish. Also, a majority of garments designed to protect industrial workers and consumers from hazardous environments are made from nonwoven fabrics.

Durable product application data are also given in Table 3. Geotextiles, for example, include fabrics for earth work and soil stabilization, roads, building foundation and surfacing, dam and dike reinforcements, canal and landfill lining, landscaping, drainage, and sport field covering. In agriculture nonwovens are used as protective or capillary mats, shading, and windbreaks on commercial farms, horticultural establishments, forestry plots, and home gardens.

Nonwovens are also used in aircraft as reinforcement media for composites and lightweight insulation. In electronic components, nonwovens are used as battery separators, floppy disk liners, and wire, cable, and tape insulation. Automotive and other transportation vehicle applications include moldable carpet backings, headliners, and interior trim. In the construction of homes and commercial buildings, nonwovens are used for roofing, insulation, and water-impermeable wrappings. Furniture, bedding, and other home uses for nonwovens include dust covers, spring insulators, arm and back supports, decking and ticking, flanges and quilt backings, carpet backings and underlays, blankets, wall coverings, window shades, and highloft for padding stuffing, cushioning, and insulation. Most all items of apparel are enhanced by fusible or sewn-in nonwoven interlinings. Many shoe constructions, book cloth backings, and other coated or laminated products use nonwovens as substrates.

As the nonwoven fabrics industry has matured and technology has become publicly available, emphasis in the various sectors of the industry has changed. In the 1990s, some portions of the nonwovens industry are technology driven, whereas others are market driven. A number of firms are proprietary technology based, and others are turn-key plant operations. Some corporations are commodity roll-goods producers, and others are more oriented to niche market, high value-added products. Many nonwoven produces continue the quest for new markets and more opportunities to compete with textiles, paper, and plastics.

BIBLIOGRAPHY

"Staple Fibers" under "Nonwoven Textile Fabrics" in *ECT* 3rd ed., Vol. 16, pp. 104–124, by A. Drelich, Chicopee Division, Johnson & Johnson.

1. N. Hyde, *National Geographic* **173**(5), 552–591 (1988).
2. H. G. Lauterbach, *Text. Res. J.* **25**, 143–149 (1955).
3. G. E. Linton, *Applied Basic Textiles*, Duell, Sloan and Pearce, New York, 1966.
4. H. Inagaki, *Nonwovens Asia 89*, Miller Freeman, San Francisco, Calif., 1989.
5. R. Yang, *J. Nonwovens Res.* **1** (1989).
6. R. Johnson, *Robert Johnson Talks It Over*, Johnson & Johnson, New Brunswick, N.J., 1949.
7. J. N. Balboni, *INDA-TEC '93*, INDA, Cary, N.C., 1993.
8. F. H. Osborne, C. H. Dexter, *Textile-Like Patterned Nonwoven Fabrics and Their Production*, Windsor Locks, Conn., 1975.
9. E. A. Vaughn, *Nonwovens World Factbook 1991*, Miller Freman, San Francisco, Calif., 1990.
10. V. A. Wente, *Industrial Eng. Chem.* **48**, 1342 (1956).
11. R. R. Buntin and D. T. Lohkamp, *TAPPI J.* **56**, 74 (1973).
12. *ASTM Designation D1117-80*, American Society for Testing and Materials, Philadelphia, Pa.
13. *ISO 9092:1988*, International Standards Organization, Bethesda, Md.
14. E. A. Vaughn, *Nonwoven Fabric Primer and Reference Sampler*, INDA, Cary, N.C., 1992.
15. E. A. Vaughn, *J. Nonwovens Res.* **4**, 1 (1992).
16. *ASTM Designation D2475-77*, American Society for Testing and Materials, Philadelphia, Pa.
17. U.S. Pat. 3,485,706 (Dec. 23, 1969), F. J. Evans (to Du Pont).
18. A. Drelich, *Nonwovens World*, 45–60 (May–June 1986).
19. J. R. Starr and co-workers, *The Nonwoven Fabrics Handbook*, INDA, Cary, N.C., 1992.

E. A. VAUGHN
Clemson University

FABRICS, SPUNBONDED

In 1990 the global production of spunbonded fabrics reached a record 307,000 metric tons with an annual growth rate of between 6 and 8% (1). Spunbonded fabrics are distinguished from other nonwoven fabrics in their one-step manufacturing process which provides either a complete chemical-to-fabric or polymer-to-fabric process. In either instance the manufacturing process integrates the spinning, laydown, consolidation, and bonding of continuous filaments to form a fabric. Commercialization of this process dates to the early 1960s in the United States and Western Europe (2,3) and in the early 1970s in Japan (4). Most of the first plants constructed are still in operation attesting to the usefulness of the method. New production plants continue to be built (2,3) to supply the growing demand (Table 1).

The large investment required for a spunbonded plant ($15–120 million, 1986 U.S. dollars) (5) is offset by their high productivity. Spunbonded production was originally limited to Western Europe, the United States, and Japan, but considerable activity has begun in other areas. Production lines, mainly nonproprietary, have been installed in China, Taiwan, Brazil, Argentina, Israel, and other countries that previously did not participate in the technology. A number of ownership changes occurred in the United States and Europe, as the strategies of companies committed to the technology evolved (2).

Early marketing efforts for spunbonded fabrics centered on their substitution for existing, ie, woven, textile fabrics. Generally, success was achieved in areas where only functionality was important. Extremely slow progress has occurred in areas where textile-like aesthetics are required. Nevertheless, spunbonded fabrics are recognized as a unique class of materials within the general category of nonwoven fabrics (see NONWOVEN FABRICS, STAPLE FIBERS).

The area of largest growth for spunbonded fabrics has been disposable diaper coverstock which accounts for approximately 50% of the U.S. coverstock market. Forecasts for the future growth of spunbonded fabrics continue to be favorable as consumption in both durable and disposable areas continues to grow. Growth is forecast to generally exceed the growth of all nonwovens, which itself is expected to grow at 7% per annum (1). In addition to diaper coverstock and hygiene, growth is anticipated in geotextiles, roofing, carpet backing, medical wrap, and durable paper applications such as envelopes (6).

New plant construction will bring increased capacity to a level which will depend on real growth to keep sales abreast with production. It is anticipated that consolidation of ownership will continue and that the trend to specialized businesses supporting a plant facility will also continue. Pressures from environmental issues could change the cost of final products as well as mandate the use of post-consumer waste resin as feedstock for production.

Although producers have benefited from the favorable prices for crude oil, a sudden increase in these prices or decrease in availability of resin feedstocks would adversely impact both profitability and growth. There appear to be no new fiber technologies that would radically change the manner in which spunbonded structures are produced. Any serious challenges to existing markets will likely come from film, foam, or advances in alternative technologies within a specific market segment.

Table 1. Spunbond Producers

Company	Polymer base	Technology base
	North America	
American Nonwovens	PP	NWT
Amoco Fabric & Filters	PP/PE	self-developed
Atlas Corp.	PP	Impianti
BASF	PET	Akzo license
Bonlam SA de CV (Mexico)	PP	Impianti
Du Pont	HDPE	self-developed (Tyvek)
Fiberweb	PP	Lurgi and self-developed
Cerex	Nylon	via Monsanto
Freudenberg	PET	self-developed
Hoechst	PET	self-developed[b]
Kimberly-Clark	PP	Lurgi and self-developed
Poly-Bond	PP	Reicofil
Polyfelt	PP	Lurgi and self-developed bonding
Reemay, Inc.	PP, PET	via Du Pont
Veratec (Canada)	PP	Reicofil
	Western Europe	
Akzo	PET/PA bicomponent	self-developed
Amoco	PP	Reicofil
Chemie Linz	PP	Lurgi
Corovin	PP	Lurgi
Don & Low	PP	Reicofil
Du Pont	PP, HDPE	self-developed
Terram Ltd.	PP/PE bicomponent	via ICI
Fabriano Soft SRL	PP/PET	self-developed
Ferma Gizeh Werk GmbH	PP	Reicofil
Fiberweb (Sweden)	PP	Lurgi and self-developed
Freudenberg	PET, Nylon	self-developed
Hoechst	PET	self-developed[a]
Fiberweb (Neuberger)	PP	self-developed
Polyfelt GmbH	PP	Lurgi and self-developed
Polyfelt France	PET/PP	needlepunch
Silver Plastics	PP	Reicofil
Fiberweb (Sodoca)	PP	Lurgi
Terbond	PET	via Toyobo
	Japan	
Asahi	PP, PET, HDPE, nylon, Cupra	self-developed
Idemitsu	PP	Reicofil
Kanebo	PU	self-developed
Mitsui	PP	Lurgi
Nippon Nonwovens	PP	Reicofil
Teijin	PP, PET	self-developed
Toray	PET	self-developed
Toyobo	PET	self-developed
Unitika	PET, nylon	self-developed

337

Table 1. (Continued)

Company	Polymer base	Technology base
South America		
Fitesa SA	PP	Impianti
Industrias Mami	PP	Reicofil
Kami Nonwovens	PP	Reicofil
Kaymac	PET	b
Novatex	PP	Impianti
Providencia	PP	Reicofil
Rhodia SA	PET	self-developed
Telares	PP	Reicofil
Other areas		
Avgol (Israel)	PP	Reicofil
Cheil Synthetics (S. Korea)	PP	Reicofil
Freudenberg Far Eastern (Taiwan)	PET	self-developed
Hanil (S. Korea)	PET/PP	via Kobe
K-C (Australia)	PP	Lurgi and self-developed
Kolon Industries (Korea)	PET/PP	self-developed/Reicofil
Nan Ya (Taiwan)	PP	Lurgi
VEB Chemiefaserkombinat	PP/PE	self-developed
VEB Textile	nylon-6	self-developed
Yuhan-Kimberly	PP	Lurgi and self-developed

[a]Began with Rhône Poulenc license.
[b]Bidim process.

General Characteristics

Fabric Structure. Spunbonded fabrics are filament sheets made through an integrated process of spinning, attenuation, deposition, bonding, and winding into roll goods. The fabrics are made up to 5.2 m wide and usually not less than 3.0 m in order to facilitate productivity. Fiber sizes range from 0.1 to 50 dtex although a range of 2–20 dtex is most common. A combination of thickness, fiber fineness (denier), and number of fibers per unit area determines the fabric basis weight which ranges from 10–800 g/m^2; 17–180 g/m^2 is typical.

Most spunbonded processes yield a sheet having planar–isotropic properties owing to the random laydown of the fibers (Table 2). Unlike woven fabrics, spunbonded sheets are generally nondirectional and can be cut and used without concern for higher stretching in the bias direction or unraveling at the edges. It is possible to produce nonisotropic properties by controlling the orientation of the fibers in the web during laydown. Although it is not readily apparent, most sheets are layered or shingled structures with the number of layers increasing with higher basis weights for a given product. Fabric thickness varies from 0.1 to 4.0 mm; the range 0.2–1.5 mm represents the majority of fabrics in demand. The method of bonding greatly affects the thickness of the sheets, as well as other characteristics. Fiber webs bonded by thermal calendering are thinner than the same web that has been needle-punched, because calendering compresses the structure through pressure whereas needle-punching moves fibers from the x–y plane of the fabric into the z (thickness) direction.

Table 2. Physical Properties of Spunbonded Products

Product	Basis weight, g/m²	Thickness, mm	Tensile strength,[a] N[b]	Tear strength,[a] N[b]	Mullen burst, kPa[c]	Bonding method
Accord	69		144 MD 175 XD	36 MD 40 XD	323	point thermal
Bidim	150		495	280	1545	needle-punch
Cerex	34	0.14	135 MD 90 XD	40 MD 32 XD	240	chemically induced area
Colback	100	0.6	300[d]	120		area thermal (sheath–core)
Corovin	75		130	15		point thermal
Lutradur	84	0.44	225 MD 297 XD	85 MD 90 XD	598	copolymer area thermal
Polyfelt	137		585	225	1445	needle-punch
Reemay	68	0.29	225 MD 180 XD	45 MD 50 XD	330	copolymer area thermal
Terram	137	0.7	850	250	1100	area thermal (sheath–core)
Trevira	155		630 MD 495 XD	270 MD 248 XD	1512	needle-punch
Typar	103	0.305	540 MD 495 XD	207 MD 235 XD	825	undrawn segments area thermal
Tyvek	54	0.15	4.6[e] MD 5.1[e] XD	4.5 MD 4.5 XD		area and point thermal

[a] MD = Machine direction; XD = cross direction.
[b] Unless otherwise noted. To convert N to pound-force, divide by 4.448.
[c] To convert kPa to psi, multiply by 0.145.
[d] 300 N/5 cm = 34.5 ppi.
[e] N/mm; to convert N/mm to ppi, divide by 0.175.

The structure of traditional woven and knit fabrics permits the fibers to readily move within the fabric when in-plane shear forces are applied, resulting in a fabric which readily conforms in three dimensions. Because calender bonding of a spun web causes some of the fibers to fuse together, thus giving the sheet integrity, the structure has a relatively stiff hand or drape compared to traditional textile fabrics. This is a result of the immobilization of fibers in the areas of fiber-to-fiber fusion. The immobilization may be moderated by limiting the bonds to very small areas (points) or by entangling the fibers mechanically or hydraulically. Saturation bonding of spun webs with chemical binders such as acrylic emulsions can bond the structure throughout and result in very stiff sheets. This technique is used to provide dimensional stability to certain structures whereby the emulsion binder functions as a nonthermoplastic component within the thermoplastic matrix.

Other approaches include powder bonding, although this method may be more suitable for bonding nonwoven fabrics made from staple fibers (7,8) (see NONWOVEN FABRICS, STAPLE FIBERS).

Fabric Composition. The method of fabric manufacture dictates many of the characteristics of the sheet, but intrinsic properties are firmly established by the base polymer selected. Properties such as fiber density, temperature resistance, chemical and light stability, ease of coloration, surface energies, and others are a function of the base polymer. Thus, because nylon absorbs more moisture than polypropylene, spunbonded fabrics made from nylon are more water absorbent than fabrics of polypropylene.

The majority of spunbonded fabrics are based on isotactic polypropylene and polyester (Table 1). Small quantities are made from nylon-6,6 and a growing percentage from high density polyethylene. Table 3 illustrates the basic characteristics of fibers made from different base polymers. Although some interest has been seen in the use of linear low density polyethylene (LLDPE) as a base polymer, largely because of potential increases in the softness of the final fabric (9), economic factors continue to favor polypropylene (see OLEFIN POLYMERS, POLYPROPYLENE).

Polypropylene. Isotactic polypropylene is the most widely used polymer in spunbonded production because it is the least expensive fiber-forming polymer that provides the highest yield (fiber per weight) and covering power owing to its low density. Isotactic polypropylene is only ca 70% the density of most types of polyesters and thus equivalent yields of fiber require a greater weight of more expensive polyester. Considerable advances have been made in the manufacture of

Table 3. Fibers for Spunbonded Nonwoven Fabrics

Fiber type	Breaking tenacity, N/tex[a]	Elongation, %	Specific gravity	Moisture regain,[b] %	Approximate melt point, °C
polyester	0.17–0.84	12–150	1.38	0.4	248–260
nylon-6,6	0.26–0.88	12–70	1.14	4.0	248–260
polypropylene	0.22–0.48	20–100	0.91	~0.0	162–171

[a]To convert N/tex to gf/den, multiply by 11.3.
[b]At 21°C and 65% rh.

polypropylene resins and additives since the first spunbonded polypropylene fabrics were commercialized in the 1960s. Unstabilized polypropylene fibers are readily degraded by uv light, but dramatic improvements in additives permit years of outdoor exposure to occur before fiber properties are significantly affected (10).

Polypropylene fibers are neither dyeable by conventional methods nor readily stained because dye receptor sites do not naturally exist along the molecular backbone. However, some spunbonded polypropylene fabrics are colored by the addition of a pigment to the polymer melt wherein the pigment becomes encased within the fiber interior. Advantages to this method include higher resistance to fading and bleeding and ease of reproducibility of color shades from lot to lot. A key disadvantage is the generation of small to large quantities of off-quality production during the transitions into and out of a particular color. A delustering pigment, eg, TiO_2, is often added to polypropylene as it almost always is with the manufacture of nylon fibers.

Most off-quality or scrap polypropylene fibers may be repelletized and blended in small percentages with virgin polymer to produce first-grade spunbonded fabrics. The economics are of great importance in a process where high yields are required in order to be competitive. Some manufacturing equipment directly recycles edge-trim back into the extruder where it is blended back into the polymer melt (see FIBERS, OLEFIN).

Polyester. This fiber has several performance advantages versus polypropylene, although it is less economical. Polyester can produce higher tensile strength and modulus fabrics that are dimensionally stable at higher temperatures than polypropylene. This is of importance in selected applications such as roofing. Polyester fabrics are easily dyed and printed with conventional equipment which is of extreme importance in apparel and face fabrics although of lesser importance in most spunbonded applications (see FIBERS, POLYESTER).

Nylon. Spunbonded fabrics have been made from both nylon-6 and nylon-6,6 polymers. Because nylon is more costly and highly energy intensive, it is less economical than either polyethylene or polypropylene. Although a considerable body of knowledge exists in the preparation of nylon polymers, such as end group control, it has been of little advantage in spunbonded fabric production. Nylon-6,6 spunbonded fabrics have been commercially produced at weights as low as 10 g/m^2 with excellent cover and strength. Unlike the olefins and polyesters, fabrics made from nylon absorb water quite readily through hydrogen bonding between the amide group and water molecules (see POLYAMIDES).

Polyethylene. Traditional melt spun methods have not utilized polyethylene as the base polymer because the physical properties obtained have been lower compared to those obtained with polypropylene. Advances in polyethylene technology may result in the commercialization of new spunbonded structures having characteristics not attainable with polypropylene. Although fiber-grade polyethylene resin was announced in late 1986 (11,12), it has seen limited acceptance because of higher costs and continuing improvements in polypropylene resin technology (see OLEFIN POLYMERS, POLYETHYLENE).

Flashspun high density polyethylene fabrics have been commercial since the 1960s; however, this is a proprietary and radically different process of manufacturing a spunbonded fabric, more technically challenging to produce, and highly capital intensive.

Polymer Combinations.   Some fabrics are composed of combinations of polymers where a lower melting polymer functions as the binder element. The binder element may be a separate fiber interspersed with higher melting fibers (13), or the two polymers may be combined in one fiber type (14). In the latter case, the so-called bicomponent fiber may have the lower melting portion as a sheath covering a core of the higher melting polymer (Fig. 1). Bicomponent fibers can also be spun whereby the two polymers are extruded side by side. The polymer composition of the binder element in such structures may be either polyethylene, nylon-6, and polyester copolymers typically modified by lowering the terephthalic acid content by substitution with isophthalic acid.

Polyurethane.   A type of spunbonded structure has been commercialized in Japan based on thermoplastic polyurethanes (15). This represents the first commercial production of such fabrics, although spunbonded urethane fabrics have been previously discussed (16). The elastomeric properties claimed are unique for spunbonded products and appear to be well suited for use in apparel

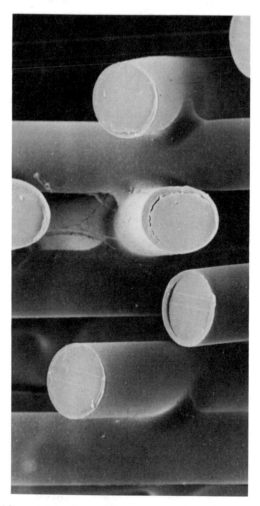

Fig. 1.   Microstereo view of the cross section of skin core filaments.

and other applications requiring stretch and recovery. Polyurethanes are also candidates for processing by the meltblown process.

Fabric Properties.   There is an almost unlimited number of ways to characterize spunbonded fabrics. Many tests in use were originally developed for the characterization of textiles and paper products. When taken together, properties such as tensile, tear and burst strength, toughness, elongation to break, basis weight, thickness, air porosity, dimensional stability, and resistance to heat and chemicals are often sufficient to uniquely describe one product. This is because these properties reflect both the fabric composition and its structure, the latter being defined by a manufacturing process unique to that fabric. Compare, for example, the differing shapes of the generic stress–strain curves of thermally bonded and needle-punch bonded fabrics (Fig. 2). The shape of each curve is largely a function of the freedom of the filaments to move when the fabric is placed under stress and is thus a function of fabric structure.

Diverse applications for the fabric sometimes demand specialized tests such as for moisture vapor, liquid transport barrier to fluids, coefficient of friction, seam strength, resistance to sunlight, oxidation and burning, and/or comparative aesthetic properties. Most properties can be determined using standardized test procedures which have been published as nonwoven standards by INDA (9). A comparison of typical physical properties for selected spunbonded products is shown in Table 2.

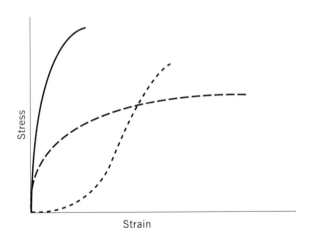

Fig. 2.   Typical stress–strain curves of nonwoven fabrics, where (—) is woven; (---), thermally bonded nonwoven; and (-·-), needle-punched nonwoven.

Spinning and Web Formation

Spunbonded fabric production couples the fiber spinning operation with the formation of the web in order to maximize productivity. It is the coupling of these two processes that distinguishes the spunbonded process from traditional methods of fabric formation where fiber is first spun and collected, then formed into a fabric by a separate process such as weaving. If the bonding device is placed in line with spinning and web formation, the web is converted into bonded fabric in one step (Fig. 3). In some arrangements, the web is bonded off-line in a separate step which appears at first to be less efficient; however, this offers the advantage of being more flexible if more than one type of bonding is to be performed on the web being produced.

Spinnerette Process. The basic spinning process is similar to the production of continuous filament yarns and utilizes similar extruder conditions for a given polymer (17). Fibers are formed as the molten polymer exits the ≥100 tiny holes (ca 0.2 mm) of each spinnerette where it is quenched by chilled air. Because a key objective of the process is to produce a relatively wide (eg, 3 m) web, individual spinnerettes are placed side by side in order that sufficient fibers be generated across the width. This entire grouping of spinnerettes is often called a block or bank, and in commercial production it is common for two or more blocks to be used in tandem in order to increase the coverage and uniformity of laydown of the fibers in the web.

A number of producers have begun to utilize large rectilinear spinnerettes in lieu of multiple small individual ones. In effect, the spinning plate is slightly wider than the desired web and a continuous curtain of filaments is formed

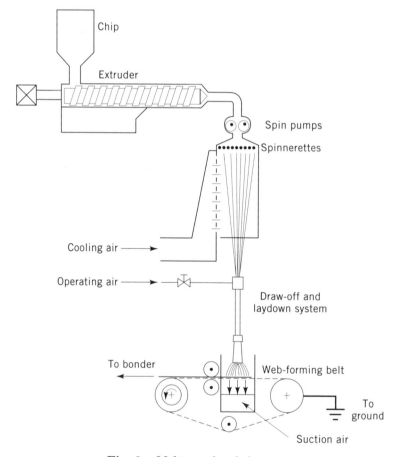

Fig. 3. Melt spunbonded process.

providing uniformity from point to point relative to multiple side by side spin-nerettes in a block.

Prior to deposition on a moving belt or screen, the molten polymer threads from a spinnerette must be attenuated to orient the molecular chains of the fibers in order to increase fiber strength and decrease extendibility. This is accomplished by hauling the plastic fibers off immediately after they have exited the spinnerette. In practice this is done by accelerating the fibers either mechanically (18) or pneumatically (17,19,20). In most processes, the fibers are pneumatically accelerated in multiple filament bundles; however, other arrangements have been described wherein a linearly aligned row(s) of individual filaments is pneumatically accelerated (21,22).

In traditional textile spinning, some orientation of fibers is achieved by winding up the filaments at a rate of approximately 3200 m/min to produce the so-called partially oriented yarns (POYs) (23). The POYs can then be mechanically drawn in a separate step to achieve maximum strength. In spunbonded production filament bundles are partially oriented by being pneumatically accelerated at speeds of 6000 m/min or greater (20,24). Accelerating the filaments at such great speeds not only achieves a partial orientation but results in extremely high rates for web formation, particularly for lightweight structures (eg, 17 g/m^2). The formation of wide webs at high speeds results in a high efficiency of manufacture.

For many applications this partial degree of orientation imparts a sufficient increase in strength and decrease in extendibility to make the final bonded fabric perfectly functional, eg, diaper coverstock. However, some applications, such as geotextiles (qv) and primary carpet backing, demand that the filaments achieve a very high tensile strength and low degree of extension. This requires subsequent additional attenuation, such as the mechanical drawing of filaments, a process usually accomplished over heated rolls with a typical draw ratio of ca 3.5:1 (18). After drawing, the filaments are pneumatically deposited onto a moving belt or screen. Because drawing rolls cannot normally dispatch filaments as fast as pneumatic jets, the web-forming process is usually less rapid, although the resulting web has greater physical strength.

The pneumatic deposition of the filament bundles onto the moving belt results in formation of the web. A pneumatic gun uses high pressure air to move the filaments through a constricted area of lower pressure but higher velocity, as in a venturi tube. Pneumatic jets used in spunbonded production have been described (17,24). Unfortunately, the excellent filament uniformity coming out of the spinnerette is lost when the filaments are consolidated going through a gun.

In order for the web to achieve maximum uniformity and cover, it is desirable that the individual filaments be separate from each other prior to reaching the belt. Failure to sufficiently separate individual filaments results in the appearance of "ropes" in the web. One method used to effect this state of separation is to induce an electrostatic charge onto the bundle while still under tension and prior to pneumatic deposition. The charge may be induced either triboelectrically or by applying a high voltage charge, the former being a result of the rubbing of the filaments against a grounded but conductive surface. The level of electrostatic charge on the filaments must be at least 30,000 esu/m^2 of filament surface area (17) to be effective. After deposition onto the moving belt it is necessary to discharge the filaments; this is usually accomplished by bringing the filaments in contact with a conductive grounded surface. In some cases, the deposition belt is made of conductive wire and connected to ground. The electrostatic repulsion method has the advantage of being relatively simple and reliable. Producing webs by spinning rectilinearly arranged filaments through a so-called slot jet reduces or eliminates the need for such bundle-separating devices (21,22), because the filament bundles are not collapsed en route to the belt as they are in a pneumatic gun.

Other routes to reachieving filament separation have been described and rely on mechanical or aerodynamic forces to affect separation. Figure 4 illustrates one method which utilizes a rotating deflector plane to force the filaments apart while depositing the opened filaments in overlapping loops (25). After the splayed filaments fall to the deposition surface or forming screen, a suction from below the disposition surface holds the fiber mass in place.

For some applications, it is acceptable or desirable to lay down the filaments in a random fashion without orienting the filament bundles with respect to the direction of the laydown belt (24). However, it is sometimes desirable to control the directionality of the splayed filaments on the laydown belt in order to achieve a particular characteristic in the final fabric. Directionality can be controlled by traversing the filament bundles either mechanically (20,26) or aerodynamically (18,27) as they travel downward toward the collecting belt. The aerodynamic method consists of supplying alternating pulses of air on either side of the filaments as they emerge from the pneumatic jet. By properly arranging the spinnerette blocks and the directing jets, laydown can be achieved predominately in the desired direction. Figure 5 illustrates the production of a web with predominately machine and cross-machine direction filament laydown (18). It is possible to generate highly ordered laydown patterns by oscillating

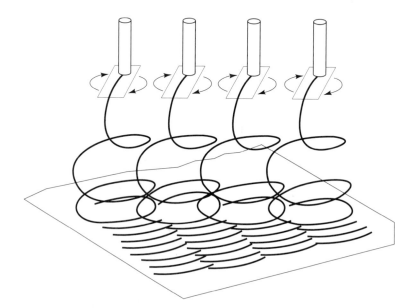

Fig. 4. Deflector plane for separation of filaments.

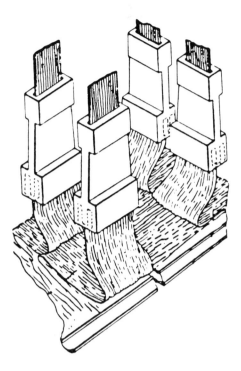

Fig. 5. Web production with predominantly machine and cross-machine direction.

filament bundles between closely spaced plates to achieve a high degree of parallelism (27).

If the laydown belt is moving and filaments are rapidly traversed across this direction of motion, the filaments are deposited in a zigzag or sine wave pattern on the surface of the moving belt. The effect of the traverse motion on the coverage and uniformity of the web have been described mathematically (28,29). The relationships between the collecting belt speed, period of traverse, and the width of filament curtain being traversed determine the appearance of the formed web upon the laydown belt. Figure 6 illustrates the laydown for a process where the collecting belt travels a distance equal to the width of the filament curtain x, during one complete period of traverse across a belt width y.

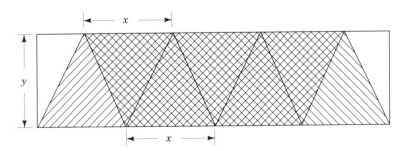

Fig. 6. Laydown pattern diagram.

If the belt speed is v_b and the traverse speed is v_t, the number of layers deposited, z, is calculated by the formula, $z = (x \cdot v_t)(y \cdot v_b)$. It can be seen that if the traverse speed is twice the belt speed and if x and y are equal, then a double coverage will occur over all areas of the belt.

Curtain Spin Process. An alternative to the use of multiple spinnerettes per bank is the so-called curtain spin process which utilizes a single plate the width of the desired web which has been drilled with holes for fiber formation. The advantage to this approach is that it results in a uniform distribution of filaments within the curtain of continuous fibers produced from the spinning plate. The use of the single uniform distribution of filaments within the curtain of continuous fibers is produced from the spinning plate. The use of the single spinning plate automatically places the fibers in a uniformly distributed array and thereby presents a curtain of high uniformity filaments to the fiber attenuation mechanism.

By comparison, the multiple spinnerette per bank process requires additional effort prior to laydown in order to compensate for the gaps between the individual spinnerettes. Failure to present a filament array to the laydown screen, which is not uniformly distributed, can result in spot-to-spot variations in fiber density and a web that has the appearance of blotch.

In general, once the curtain of filaments has been produced, it is necessary to attenuate the filaments in order to provide strength and resistance to deformation. The most commonly practiced approach is to utilize a single slot, which is at least the width of the curtain, at a point below the spinning plate and above the laydown screen. There are two practical approaches taken. The first utilizes the injection of low pressure air at a point above the slot so that the fibers attain sufficient acceleration in the slot to provide adequate draw (22) (Fig. 7). The second utilizes a low pressure vacuum below a venturi to provide the pressure differential required for sufficient acceleration and resulting attenuation (30).

One of the limitations of the curtain/slot draw process is that the amount of fiber attenuation is constrained due to the short distance generally allowed between the spinnerette and the venturi slot and the use of relatively low pressure air for drawing so as not to induce high turbulence in the area of the laydown. In practical terms this has made the process difficult to adapt for the production of polyester fabrics which inherently require much higher fiber acceleration to attain the desired polyester fiber properties.

Bonding

Many methods can be used to bind the fibers in the spun web. Although most procedures were originally developed for use with nonwoven staple fibers, three were adapted for use with continuous filaments: mechanical needling, thermal, and chemical/binder. Thermal and chemical/binder methods may bond the web by fusion or adhesion of fibers using either large or small regions, generally referred to as area bonding and point bonding, respectively. Point bonding results in the fusion of fibers at discrete points with fibers remaining relatively free in between the point bonds. Other methods that are used with staple fiber webs but which are not routinely used with continuous filament webs are stitchbonding (29,31), ultrasonic fusing (8,32), and hydraulic entanglement (33). Hydraulic

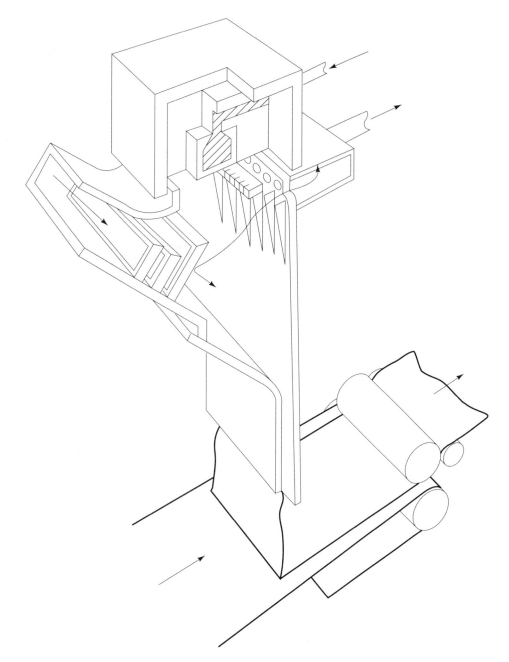

Fig. 7. Curtain spinning process (22).

entanglement has the potential to produce the most radically different continuous filament structures, however, it has the disadvantage of being a more costly and complex bonding process.

Of the three standard bonding methods used in spunbonded manufacturing, mechanical needling, also called needle-punching or needle-bonding, is the simplest and least expensive. Although it is the oldest process, it continues to be widely used. Significant improvements in throughput and flexibility have resulted in the sales growth of needle-bonded fabrics, particularly in geotextiles (qv). An excellent review of mechanical needling technology has been published (29).

In the needle-punching process, a continuous filament web is subjected to barbed needles which are rapidly passed through the plane of the moving spun web (see NONWOVEN FABRICS, STAPLE FIBERS). The needles pass in and out of the web at frequencies up to 2200 strokes per minute which can result in as many as 500 penetrations per cm^2 depending on the needle density and the line speed, typically between 5 and 25 m/min (34). The effect of this operation is to interlace the fibers and thus bond the structure together, relying only on the mechanical entanglement and fiber-to-fiber friction. The fabric produced tends to be more conformable and bulky than fabrics bonded by thermal or chemical/binder methods. Because the fibers have freedom to move over each other, the fabric is easily deformed and exhibits a low initial modulus (Fig. 2).

The principal variables in needle-punching are the needle design, punch density, and depth of punch. Considerable research has been conducted on the shape and design of the needles and how this affects the interlacing of the fibers (29). Needling produces a fabric which is 100% fiber with no points or areas of fusion or melting, thus it is easily adapted to most fiber webs and requires less precise control than thermal bonding. In addition it is the only bonding method suitable for the production of spunbonded fabrics of very high basis weights, eg, 800 g/m^2. It is, however, only suitable for the production of uniform fabrics $\geq$100 g/m^2 because needling tends to concentrate fibers in areas resulting in loss of visual uniformity at lower weights.

Unlike mechanical needling, both thermal and chemical/binder bonding depend on fiber-to-fiber attachment as the means of establishing fabric integrity. It is the degree and extent of attachment which determines many of the fabric qualities, most notably the hand or softness. Because point bonding can be accomplished using as little as 10% bonding area, ie, 90% unbonded area, such fabrics are considerably softer than area-bonded structures. Fiber mobility is retained, in part or in total, outside the areas of the point bonds. Thermal bonding is far more common than chemical/binder bonding and is generally more economical because the latter method can still require a thermal curing treatment as the final step. Both area and point thermal bonding are rapid processes having line speeds in excess of 300 m/min during production of lightweight fabrics.

Area thermal bonding can be accomplished by passing the spun web through a source of heat, usually steam or hot air. Prior to entering the bonding area the spun web may be consolidated by passing it under compressional restraint through a heated prebonding area which adds integrity to the web (13). While in the bonder the consolidated web is exposed to hot air or pressurized steam which causes fusion to occur between some, but not all, of the fiber crossover points. Complete fusion leads to a paper-like structure with low resistance to tearing. The spun web may contain small percentages, typically 5–30%, of a lower melting fiber (13), or the filaments may contain undrawn segments that are lower melting than the drawn or matrix segment (26). Heterofilament structures utilize a lower melting covering (sheath) on the outside of the filaments to effect fusion. Both polyethylene and nylon-6 have been used as the lower melting sheath in commercial spunbonded products.

The use of steam is generally limited to polypropylene and polyethylene fusion because impractical pressures are required to reach the temperature levels, eg, >200°C, required for bonding polyesters. In general, greater temperature control is required for area bonding polypropylene than for other polymers because the temperature difference between the matrix and binder fibers can be only 3°C (26).

Whereas thermal area bonding uses temperature as the variable to a great degree and relies on sophisticated web structures containing binder fibers, thermal point bonding utilizes both temperature and pressure to affect fiber-to-fiber fusion. Thus it is a simpler approach to bonding because it does not require the web to contain lower melting fibers or segments and is less demanding of the technology required to produce the web. Point bonding is usually accomplished by passing a preheated consolidated web through heated nip rolls, one of which contains a raised pattern on its surface (Fig. 8). When bonding, polypropylene roll temperatures generally do not exceed 170°C; however, pressures on the raised points are quite high, preferably 138–310 MPa (20,000–45,000 psi) (35). The degree of bonding between the points can be controlled by varying the ratio of heights of the raised points to the depth of the web (36). Typically only 10–25% of the surface available for bonding is converted to fused, compacted areas of bonding.

Optimum conditions of pressure and temperature are dependent on many variables including but not limited to the nature of the web, line speed, and engraved pattern. Optimized conditions are best obtained through detailed investigations and much experience. Even subtle changes in any of these variables can result in significant changes in the properties of the finished fabric (7,8).

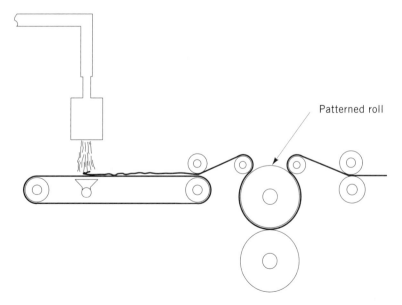

Fig. 8. Pattern bonding roll at the end of a spunbonding line.

Because engraved point bonding rolls can be as wide as 5 m, the problem of maintaining uniform pressure across the width must be addressed. Small differences in pressure across the width can produce an unacceptably variable product. Hydraulic pressure is applied at the ends to the roll causing a slight deflection that results in less pressure being applied in the center compared to the ends. A number of solutions to this problem have been devised (37), the most common being cambering wherein the roll diameter decreases slightly from the center to the ends.

Chemical/binder bonding is used less frequently than thermal bonding in the production of spunbonded fabrics, whereas the opposite has been true for staple fiber nonwovens. Resin binders are occasionally used with spunbonded webs to achieve special characteristics which are unattainable thermally (38). In a typical case, acrylic resin(s) are applied to saturate the web, excess resin is removed by nip rolls, and the wet web is passed through a drying oven to remove excess water and cure the resin which tends to concentrate at fiber–fiber junctions. By curing the resin, a thermoset binder conveys high thermal dimensional stability to the web for applications such as roofing.

Resin binders may alternatively be applied in discreet points in a pattern so as to immobilize fewer fibers and produce a softer fabric; however, it is difficult to accurately control the diffusion of the resin, and the drying step requirements make it less attractive than thermal bonding.

Chemical bonding with hydrogen chloride gas has been used with spun webs of nylon-6,6 to commercially produce spunbonded nylon fabrics (39). In this bonding process the activating hydrogen chloride gas is passed over web fibers held in close contact by tension. The hydrogen chloride disrupts hydrogen bonds between the polymer chains and forms a complex with the amide group. When the gas is desorbed the process reverses, this time with new hydrogen bonds formed between polymer chains in different fibers. This basic method has been further refined to permit only the formation of pattern bonds, whereby fiber mobility is retained between the bonded areas yielding a softer hand to the bonded fabric (40).

Bonding a web by any means allows for certain generalizations. If the web is highly bonded, most of the fibers are bonded to another fiber. The resulting structure is relatively stiff, paper-like, and has higher tensile and modulus but lower resistance to tear propagation. On the other hand, if the web is only slightly bonded fewer fiber-to-fiber bonds are present and the structure is more conformable with lower tensile and modulus but higher resistance to tear propagation. Additionally, webs which are only slightly bonded exhibit low surface abrasion resistance. In comparing area to point bonding, greater varieties of structures are achievable through point bonding because of the various bonding roll patterns available. The expense associated with the manufacture of the pattern roll generally dictates the careful selection of the pattern, however.

Meltblown Fabrics

Meltblown fabrics differ from the traditional spunbonded fabrics by having lower fiber denier (fineness) and by usually being composed of discontinuous filaments. Although meltblown fabrics are not generally referred to as spunbonded, the integration of spinning, attenuation, laydown, and bonding during the production of meltblown webs describes a process traditionally defined as spunbonding. The inherent fiber entanglement often makes additional bonding unnecessary, however. Fibers produced by melt blowing are often very fine, having typical diameters of 3 μm (41,42), smaller by nearly an order of magnitude than traditional spunbonded fibers. The fibers are extremely fine and largely unoriented, causing the webs to be quite weak and easily distorted. Most thermoplastic polymers are meltblown, but the majority of commercial products are high melt flow polypropylene.

In the manufacture of meltblown fabrics, a special die is used in which heated, pressurized air attenuates the molten polymer filament as it exits the orifice of the dye or nozzle (Fig. 9). Air temperatures range from 260–480°C with sonic velocity flow rates (43).

The rapidly moving hot air greatly attenuates the fibers as they exit from the orifices to create their small diameters. The fibers are relatively weak and deposited on the forming screen as a random entangled web that may be thermally point bonded to improve strength and appearance. The web may also be deposited onto a conventional spun web, then thermally bonded. Sandwich structures have also been created with the meltblown web in the middle between two conventional spunbonded webs (44). Other materials, eg, cellulosics, have been blended into the meltblown filament stream to yield a meltblown structure with a unique combination of properties (45). Mixtures of meltblown and crimped bulking fibers have been sold as thin thermal insulation for use in outdoor clothing and gear (46). Meltblown technology has also been adapted to produce nontraditional spunbonded fabrics, such as elastomeric webs (47).

The great quantity of very fine fibers in a meltblown web creates several unique properties such as large surface areas and small (<1 μm) pore sizes. These have been used in creating new structures for hospital gowns, sterile wrap, incontinence devices, oil spill absorbers, battery separators, and special requirement filters. It is expected that much innovation will continue in the design of composite structures containing meltblown webs.

Flash Spun Fabrics

The process of producing spunbonded webs by flash spinning is a radical departure from the conventional melt spinning approach. In melt spinning, a molten polymer is extruded through a spinnerette containing $\sim$100 tiny holes. This produces a fiber bundle containing $\sim$100 fibers, each typically 15–50 μm in diameter. The fibers within a bundle are separate from each other until the bonding operation connects some or all of them.

By contrast, flash spinning begins with a 10–15% polymer solution prepared by dissolving a solid polymer, such as high density polyethylene, with a suitable solvent, such as trichlorofluoromethane or methylene chloride (48). The solution is heated to approximately 200°C, pressurized to $\sim$4.5 MPa (653 psi), and the pressurized vessel is connected to a spinnerette containing a single hole. When the pressurized solution is permitted to expand rapidly through the hole, the low boiling solvent is instantaneously flashed off leaving behind a three-

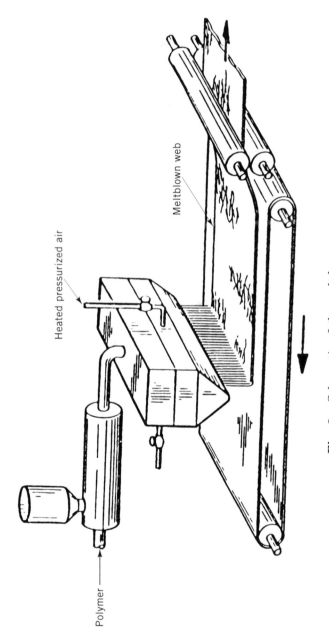

Polymer

Heated pressurized air

Meltblown web

Fig. 9. Schematic of the meltdown process.

dimensional film–fibril network referred to as a plexifilament. The three-dimensionality results from the cross-linking interconnection of the fine fibers which produces a film thickness of 4 μm or less (48). Thus many individual but interconnected fibers are created from a single-hole spinnerette.

It is believed that bubbles form rapidly as the pressurized solution undergoes depressurization during spinning and the bubbles may grow and rupture thus forming the plexifilamental network (48). Gases that are effectively insoluble in the solvent may be added to the pressurized solution in order to facilitate high rates of bubble nucleation.

When a multiplicity of single-hole spinnerettes are assembled across a width, the plexifilaments produced can form a wide web that can be thermally bonded to produce a flat sheet structure (49). The web-forming procedure is ameliorated by use of a baffle which deflects the stream of plexifilaments after exiting the spinnerette.

Unlike fine fibers prepared by melt blowing, the plexifilaments from flash spinning are substantially oriented and possess relatively high tenacities (0.08 N/tex (>1 g·f/den)). The plexifilaments scatter light effectively as a result of high surface areas (ca 2 m^2/g) and thus form opaque webs. In addition, the fineness of the plexifilament fibrils also results in a web structure of exceptional softness. Webs are either area or point bonded to yield paper- or cloth-like aesthetics, respectively. The paper-like sheets are used as durable papers and may be printed using conventional inks (qv) and printing equipment, whereas the point-bonded structures are very soft and find use in disposable protective clothing.

Flash spinning is the most complex and sophisticated method for manufacturing spunbonded fabrics. Although the process has been in use since the 1960s, the need to spin heat and pressurize solutions under precise conditions has resulted in only two companies (Du Pont and Asahi) practicing the technology as a route to spunbonded products. New solvents are being developed for spinning because chlorofluorocarbons, the traditional solvents, are being replaced with more environmentally friendly solvents (50).

Test Methods

Spunbonded fabrics are characterized by standardized test procedures originally developed for textile fabrics and paper products. The Association of the Nonwoven Fabrics Industry (INDA) has published a list of test procedures (Table 4) which are routinely used in determining specific physical characteristics of spunbonded and other nonwoven fabrics. Many tests are established for the evaluation of nonstrength related properties such as washability, stiffness, and softness. Great strides have been made in the test methodology used to evaluate the hand of materials for textile applications such as clothing. A methodology and equipment, permitting quantitative evaluation of fabric hand, have been developed (52).

As applications are developed, the need for new end use specific test methods grows. Geotextile uses are a good example of how a large new application requires the design of new test methods (53). In addition to break, stretch, tear, and burst resistances described in Table 4, geotextile fabrics are tested for puncture, maximum opening, permittivity, and asphalt retention, according

Table 4. INDA Test Methods[a]

Property	Description	IST Number
absorbency	amt of liquid absorbed and speed of absorption	10.1–3
abrasion	resistance of nonwovens to being worn away	20.1–5
bursting strength	force to rupture nonwoven under water pressure	30.1
electrostatic properties	amt of charge that can build up on a sample	40.1–2
optical properties	opacity: resistance to light being passed brightness: whiteness	60.1–2
permeability	ease of air or water vapor passage under pressure	70.1–2
repellency	resistance of nonwovens to wetting and penetration after exposure to water, salt solutions, alcohol, and hydrocarbon solvents and oils	80.1–9
bacterial	resistance of a nonwoven to penetratration by bacteria in a salt solution under water pressure	
stiffness		
cantilever	tendency for a nonlimp nonwoven to droop as it is pushed over the edge of a surface	90.1
curly	ability of a heavy, stiff nonwoven to push a pendulum aside as it is moved past it	90.2
Handle-O-Meter	ability of a soft, lightweight nonwoven to flex and not drag as it is pushed through an opening	90.3
tear	resistance of a nonwoven to continue to tear after being cut and pulled from both sides	100.1–3
breaking load and elongation	force to break a nonwoven when it is pulled from both ends; extent of stretching before breaking	110.1–4
seam breaking	force needed to break a seam holding two pieces of nonwoven together when the sample is pulled from both ends	
bond strength of laminates	force to separate a nonwoven from another material after they have been laminated together	
internal bond strength	force to pull a nonwoven fabric into two plies	
thickness	how thick a nonwoven is when it is held between a weight and a surface	120.1–2
coefficient of friction	drag when a nonwoven is slid over itself or over a polished surface	140.1
dry cleaning and laundering	shrinkage, loss of strength, ability to be peeled apart experienced by a single fabric or laminate	150.1
linting	extent of particles loosened from nonwoven as it is bent and flexed in air stream	160.1
extraction	amt of material leached out of nonwoven after exposure to hot solvents	190.1

to IST 180.1–.9. The puncture test notes the resistance to being punctured by a probe with either a flat or spherical tip. Maximum opening measures the largest size glass beads that can pass through a fabric, thereby reflecting the size of soil particles that can be stopped by a geotextile. Permittivity is how fast water, at a given pressure, passes through a geotextile. Asphalt retention is judged by how much asphalt cement is left in a geotextile after it is dipped in the cement and allowed to drain, and what change in area the geotextile undergoes.

Long-term applications also demand test methodology on the aging characteristics of spunbonded fabrics. Roofing applications, for example, require that the saturated fabrics retain their strength for many years despite a hostile environment. By heating the fabric at several different temperatures higher than the expected nominal conditions, and measuring the time it takes to observe a significant property change, for instance loss of 50% tensile strength, effects can be plotted to permit some extrapolation back to expected nominal conditions (54). The importance of aging tests will increase as more long-term applications are developed for synthetic fabrics. The Swedish Building Institute has developed heat aging tests and standards for films and fabrics used in building construction.

In medical applications, many test procedures have been developed for screening the efficiency of fabrics to block the passage of viruses, blood-borne pathogens, etc.

Overall, the test methods published by INDA (Table 4) continue to be the general tests used to characterize fabrics; however, specific market applications often generate special test procedures to fulfill unique needs.

Applications for Spunbonded Fabrics

Uses for spunbonded fabrics have traditionally been segmented into durable and disposable categories. In the early 1970s consumption of spunbondeds was predominately for durable uses such as carpet backing, furniture, bedding, and geotextiles. By 1980, however, disposable applications accounted for an increasingly large percentage due to the acceptance of lightweight (eg, 17 g/m^2) spunbonded polypropylene fabrics as a coverstock for diapers and incontinence devices (6). In the 1990s, the use of new diaper and training pants designs have increased the demand for lightweight fabrics far beyond earlier prediction.

Both the durable and disposable markets for spunbondeds have experienced dramatic growth ($\sim$6%/y). Significant areas of durable growth have been in the building and construction industries where spunbondeds are used in geotextiles and roofing membranes (see BUILDING MATERIALS). Growth has also been achieved in primary carpet backing in automotive carpets and carpet tiles, where moldability and high dimensional stability, respectively, were achieved through the use of spunbondeds.

With the possible exception of geotextiles and housewrap, however, there have been virtually no new markets established as a result of the special characteristics of spunbonded fabrics. Growth has come about in an evolutionary fashion where spunbonded fabrics were substituted for woven fabrics, other nonwoven fabrics (including knits), paper or film in previously existing applications, or where the cost–property relationship has permitted an extension of an existing application, such as the redesign of diapers. The principal contributions that spunbondeds have made in these markets generally have been attractive economics, or improved processability and performance in the final product. This combination has greatly accelerated the use of the products within an application and consequently contributed to the growth of specific markets. General market opportunities for nonwovens have been reviewed (6,7,54).

Of the four basic polymer types available in spunbonded form, ie, polypropylene, polyethylene, polyester, and nylon, both polyester and nylon are more costly polymer forms than either of the olefins. It is possible for this cost advantage to be offset by other factors, such as production of the fabric in lighter unit weight, but in general olefin-based products have an economic advantage for an equivalent weight fabric. In some applications, however, this cost disadvantage is moot if the olefin-based product cannot perform properly. An example of this is in roofing membranes where a key requirement is dimensional stability to hot bitumen at temperatures approaching 200°C, which is above the melting point of both polypropylene and polyethylene but well within the performance limits of polyester. To a great extent this one property, ie, higher temperature resistance, largely differentiates the opportunities for polyester spunbondeds versus olefinic counterparts. Polyester fibers also exhibit higher modulus and more flexible dyeing, but these properties seem to be of little advantage in the market of the 1990s for spunbonded fabrics.

SPUNBONDED MARKETS: DURABLE APPLICATIONS

A summary of 1990s markets for nonwoven fabrics in the United States and western Europe is shown in Tables 5 and 6 and Figure 10. Approximately 25% of total global production for 1990 was estimated as being spunbonded. In the United States this represented nearly 182,000 t of spunbonded production with volume growth projected at 8–9% per year through the mid-1990s (6). The principal spunbonded durable applications center around housing, construction, and automotive applications although there are other smaller areas.

Housing. One of the first applications for a spunbonded product was the use of spunbonded polypropylene in primary carpet backing. First introduced in the mid-1960s as a replacement for woven jute, it is used in the 1990s in carpets and holds a unique position in applications which require isotropic planar properties for dimensional stability such as printed or patterned carpets. The finer fiber versus woven ribbons or jute also allows tufting needles to penetrate with little deflection where fine-gauge tufting is desired. Finally, because the spunbonded backing is bonded at many fiber junctions, it offers the advantage of maintaining clean edges after cutting or trimming, making it attractive for use in small rugs where the unraveling feature of woven ribbon backings can be a concern. Although the first spunbonded primary carpet backing was made from polypropylene, other spunbonded products based on polyester and polyester–nylon were later commercialized as tuftable carpet backing products.

An extremely successful application for spunbonded fabrics is in the area of furniture, bedding, and home furnishings. In furniture construction the use of lower cost spunbonded fabrics has become routine, whereas in the 1970s woven sheeting dominated the market. Spunbondeds are used in hidden areas requiring high strength and support in chairs, sofas, and other seating. The bottoms of chairs are often covered with dust covers made of spunbonded fabrics because of the nonfraying characteristics, high porosity, excellent cover, and low cost. An

Table 5. Markets for Nonwoven Fabrics[a]

Item	1980	1990	1995	2000
nonwoven disposable shipments, 10^6 $	3,380	8,739	12,730	17,900
consumer and household products				
diapers	1,454	3,620	4,820	6,320
feminine hygiene	776	1,705	2,520	3,490
incontinence products	58	375	745	1,400
nonwoven fabric softeners	93	335	535	770
premoistened towelettes	74	276	415	560
nonwoven towels	67	178	255	350
other consumer/household	34	140	250	405
Total	*2,556*	*6,629*	*9,540*	*13,295*
medical and surgical products				
OR/OB products	286	705	1,045	1,475
diapers	35	62	80	100
patient/exam gowns	15	37	55	80
sheets/pillowcases	7	23	35	50
other medical/surgical	167	325	430	545
Total	*510*	*1,152*	*1,645*	*2,250*
miscellaneous products				
fabric filter media	115	386	665	1,030
protection/career garments	101	245	355	495
industrial/commercial wipes	64	187	275	385
other miscellaneous	34	140	250	445
Total	*314*	*958*	*1,545*	*2,355*
nonwoven fabrics used, 10^6 m^2 [b]	5,796	10,271	12,223	14,297

[a]Ref. 55.
[b]To convert m^2 to yd^2, multiply by 1.196.

inherent resistance to rot and mildew versus natural fabrics also adds to the popularity of spunbonded fabrics in home uses.

In bedding, spunbondeds are used as spring insulators, spring wrapping in mattress construction, dust covers under box springs, and facing cloth for quilting. Home furnishing uses include mattress pad covers where the spunbonded fabric serves as the top and bottom of a sandwich structure with a middle layer of fiberfill and fastened by ultrasonic quilting. Draperies also have used spunbonded fabrics wherein the lightweight fabric serves as a stitching medium for use with stitchbonding equipment. Spunbonded fabrics are also used in blinds, both vertical and horizontal, wherein the fabric, which must be extremely uniform, is saturated with colored resins to form opaque and optionally pleatable blinds.

A fascinating and growing application for spunbonded fabrics is the air infiltration barrier whereby the penultimate vertical surfaces of old or newly constructed houses are covered with a layer of spunbonded fabric followed by the application of the ultimate external sheathing such as siding or masonry. The objective is to construct a barrier to the infiltration of air into the wall cavity and to the insides of homes, thus lowering the cost of heating and cooling. Tests conducted by the National Bureau of Standards and the National Associ-

Table 6. Markets for Nonwovens in Western Europe, 1990[a]

Use	% of Total
coverstock	26.3
civil engineering/building	20.7
furniture and bedding	8.67
wipes	8.4
interlinings	5
medical/surgical	4.4
liquid filtration	4
shoe and leather goods	3.6
electrical, electronic, and abrasive	2.7
automotive	2
garments	1.7
air and gas filtration	1.6
agriculture	1
other	10

[a] Ref. 56.

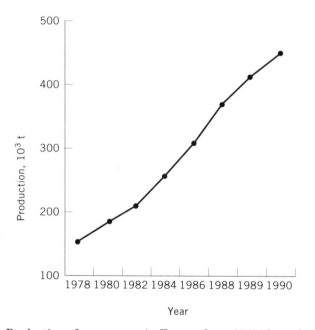

Fig. 10.   Production of nonwovens in Europe from 1978 through 1990 (57).

ation of Homebuilders confirmed the effectiveness of the air infiltration barrier concept as a means of lowering the cost of heating a home (58). Certain spunbonded fabrics are well suited for this application because they possess a unique combination of properties required for functionality. These include resistance to the penetration of liquid water and low porosity to air currents, but with a simultaneously high transport of moisture vapor. In a winter climate warm moist air from inside the house can penetrate through the wall cavity and to the outside. If the air barrier material is not sufficiently permeable to moisture vapor, condensation can occur inside the wall cavity where damage from moisture can occur. In addition, the effective R-value of the insulation (eg, fiber glass) inside the wall cavity is diminished by the presence of liquid and solid water. The combination of water and air current resistance combined with breathability to moisture vapor and high tensile and tear strength is a difficult combination of properties to assemble. Spunbonded technology provides these characteristics in economical form.

Automotive. Uses for nonwovens in automobiles have grown from a rather modest beginning in the 1970s to a position of significance (59). Although needle-punched nonwoven fabrics have been used in large-area applications, such as backing for vinyl seats and landau tops, spunbonded fabrics have historically been utilized in lower volume applications such as labels for seat belts, spring insulators, listings in seats, and as coated fabrics for ducting. Spunbonded polyester has become accepted as a tuftable backing in molded carpets where the use of spunbonded backing allows for greater molding precision, improved dimensional stability, and resistance to puncture. Newer applications include headliners, which are often complex composites that can be molded into sophisticated shapes. Lightweight spunbonds are used as sound insulators in between dashboard components, and as the base fabric in interior door panels and sun visors.

Roofing. Material acceptance in roofing applications has changed significantly since the mid-1970s, particularly for spunbonded products. The market opportunity is extremely large and is thought to exceed 1.86×10^8 m^2 for commercial buildings (flat roofing) in the United States alone. Much of the development for roofing applications was done in Europe and slowly became accepted in the United States. Although fiber glass fabrics have been the largest volume nonwoven consumed in roofing, spunbonded polyester and polypropylene have made considerable penetration (60). A significant difference between glass and polyester is the ability of polyester to flex and stretch without damage to the filaments. Because rooftops are known to expand and contract with seasonal changes, fabrics of polyester are less susceptible to damage from sudden temperature fluctuations which induce rapid dimensional changes.

Spunbonded polyester is basically a carrier for bituminous waterproofing membrane. Here spunbonded fabric is saturated with bitumen and serves to provide integrity and dimensional stability to the bitumen. As bitumen coatings modified with elastomeric polymers, such as atatic polypropylene (APP) or sequenced butadiene–styrene (SBS), became accepted as improvements over unmodified bitumen, changes occurred in the installation and manufacture of membranes. Historically, built up roofs were made *in situ* by mopping hot bitumen into organic felts which had been placed on the roof decking. In the 1990s the roof membrane is manufactured under tightly controlled conditions in a factory distant from the site of application. The spunbonded fabric is typically saturated with modified bitumen by dipping into tanks of hot bitumen which are heated up to 200°C. Excess bitumen is metered off and the cooled surfaces are coated with a release material such as talc to prevent blocking together on the roll. The composite is packaged into rolls approximately 1-m wide and 50-m long. The rolls are then shipped to the job site and applied to the flat roof surface by slowly unrolling while heating the underside to tackiness with a propane torch to enable it to adhere to the roof deck. Spunbonded polyester is also used in the so-called cold roof method, typically used for roof maintenance. In this method, a cold mastic is applied over a fiber glass base sheet, followed by more mastic, another layer of polyester, more mastic, and a final topcoat.

In Europe, bitumen-coated spunbonded polypropylene fabrics are widely accepted for rooflinings under concrete, clay, or ceramic tiles for pitched roof construction. In this use, the spunbonded fabric is a critical element of the membrane because the rooflining is draped between roof rafters and depends on the strength of the spunbonded for self-support during the life of the roof. The bitumen coating renders the spunbonded waterproof and allows it to shed any water that might leak between the tiles during snow and rainstorms. Spunbonded fabrics coated with nonbituminous materials such as acrylics have also been used in Europe. Rooflinings represent a considerable opportunity for spunbonded fabrics in Europe and in the sunbelt areas of the United States.

Geotextiles. Nonwoven fabrics have played an important part in the development of geotextile applications. Both needle-punch fabrics manufactured from staple fibers and spunbonded continuous filament products have found worldwide acceptance based on field performance. In 1992, it is estimated that the United States consumed approximately 192×10^6 m^2 of geotextiles (61). Many fabric manufacturers have dedicated considerable effort to the marketing of their products in order to participate in this growth area. Geotextile fabrics function by being porous to water but not to the fines of the soil, thereby permitting them to effectively separate or partition soil fines from other elements. For example, in the construction of a new road the geotextile can separate the subsoil from the gravel or aggregate. By maintaining this separation, the aggregate is not driven into the subsoil base by the weight of vehicles nor are soil fines pumped up into the aggregate since the geotextile filters out their passage. However, water is freely transported through the fabric enabling proper drainage without buildup of hydrostatic pressures. Thus the road resists rutting and sustains the weight of traffic more effectively while permitting proper drainage of water through the fabric (62). In drainage ditches, perforated drainage pipes are often wrapped with a geotextile prior to installation to prevent them from becoming clogged.

Spunbonded fabrics are effective filters in that they are layered structures of relatively fine fibers, the three-dimensional structure of which creates a torturous path. Even relatively thin spunbonded fabrics (eg, 0.2–0.25 mm) present a significant challenge to the passage of soil fines and are suitable for use in some filtration applications. The porosity of geotextile fabrics is classified by means of several procedures such as flux (volume flow/area per time) and equivalent opening size (EOS), which is a measure of the apparent pore size of the openings in the fabric. The flux measures the porosity to liquid water, and the EOS measures the porosity to solid particles of a known diameter. Literature is available on limitations of particular styles of fabrics within an application (63).

Growing university research ensures that users and specifiers will continue to become more sophisticated in their methodology and more demanding of manufacturers. Excellent textbooks are available for both students and practicing engineers (53,64).

Other Durable Applications. Other durable applications such as interlinings and coating/laminating substrates do not appear to offer much near-term opportunity for growth for spunbonded fabrics. In interlinings, however, spunlaced nonwovens have received wide acceptance because of the outstanding drape and softness previously unavailable from any other fabric.

Spunbonded fabrics have a relatively small percentage of the coated fabric market which is dominated by other nonwovens. Needle-punched nonwovens offer more of the bulk and resiliency required for functionality in automotive and furniture seating.

Many filtration requirements are fulfilled by spunbonded structures and a growing but technically complex market has developed since the 1970s (65).

SPUNBONDED MARKETS: DISPOSABLE APPLICATIONS

The outlook for spunbonded disposable applications indicates a 5–7% compounded annual growth forecast (6). Much of the U.S. spunbonded plant capacity installed or announced since the 1980s has been aimed at satisfying increased demand for disposable applications. Key markets are coverstock for diapers, training pants, and incontinence devices, surgical gowns, medical sterilization wrap, protective clothing, and envelopes (see Table 5).

Diapers and Incontinence Devices. The use of spunbonded fabrics as coverstock for diapers and incontinence devices has grown dramatically since 1980 and by 1992 consumption exceeded 57,000 t in the United States (6). A coverstock functions as a one-way medium through which body fluids are transported into the absorbent core. The laminar structural feature of the coverstock helps keep the skin of the user dry and comfortable. Although 17 g/m^2 spunbonded polyester was once extensively used in diaper coverstock it has been supplanted largely by an equivalent weight spunbonded polypropylene. A summary of disposable diaper sales history is provided in Figure 11.

Changes in diaper design have made disposable diapers among the most highly engineered disposable products in the world, and involve not only coverstock but the nature of the absorbent layer and design of the diaper itself. The use of form-fitting legs with leg cuffs for leak protection, as well as refastenable closures, has accelerated the acceptance of disposable diapers for both infants and adults. Spunbonded coverstock is also widely used in feminine napkins and to a limited extent in tampons. The conversion of eastern European countries to market economies has created demand for disposable diapers and consequently spunbonded polypropylene coverstock.

The uses of spunbonded fabrics as coverstock in diapers and other personal absorbent devices will most likely remain unchallenged for the near term. Virtually any other nonwoven production method appears to be at a cost disadvantage opposite spunbonded polypropylene. There have been composite products developed from meltblown and spunbonded combinations, where areas of either improved hydrophobicity or hydrophilicity are desired. These products can be produced on-line at relatively low additional cost and offer high value to diaper manufacturers. Any competitive threat is likely to come from advances in film technology such as large improvements in perforated film used in segments of absorbent product applications, particularly sanitary napkins.

Medical Markets. In medical applications, progress has been made in the substitution of traditional materials with higher performing spunbondeds (67). Historically, flash spunbonded polyethylene was the first 100% spunbonded to find wider acceptance in medical uses such as disposable operating room gowns,

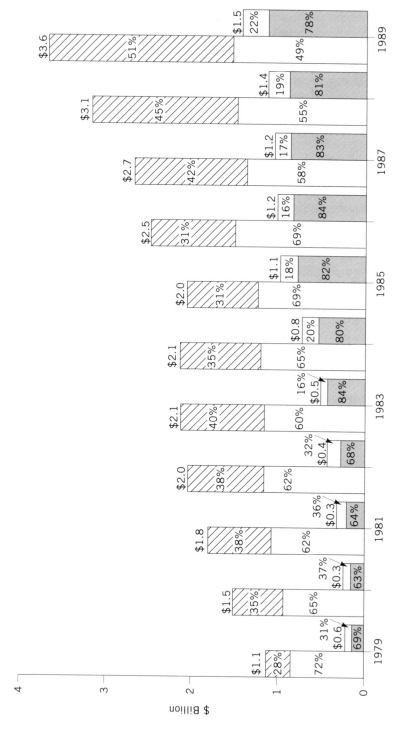

Fig. 11. Diaper sales for the Proctor & Gamble Co., where ☑ is foreign and ☐ is domestic, and the Kimberly-Clark Corp., where ☐ is foreign and ▨ is domestic (66).

shoe covers, and sterilizable packaging. Other spunbonded fabrics of polypropylene or nylon found some acceptance as cellulosic composites with the lightweight spunbonded serving to add physical strength to the composite. More recently composite structures of spunbonded polypropylene and meltblown polypropylene fibers have gained acceptance in operating room gown applications and sterilizable (CSR) wrap, although structures made of spunlaced polyester–cellulose are also widely used (67).

Operating room gowns worn by members of the surgical team place very high demands on fabric properties. Key requirements include breathability for comfort, low noise, resistance to fluid penetration, low particle generation (linting), sterilizability, and impermeability to bacteria. Woven cotton fabric gowns were worn for many years but had to be reused because of high cost. This required the added expense of laundering the garments and the need to decide when each garment was no longer suitable for use in the operating room. Several studies comparing disposable and reusable fabrics have been conducted in an attempt to correlate the effect of fabric linting with post-operative infections. Although no correlation has yet been established, some studies have demonstrated that single-use fabrics generate significantly fewer particles than cotton fabrics (63). Other studies have indicated that the rate of post-operative wound infection is reduced with the use of high barrier spunbonded olefin gowns and drapes. Concern for viral transmission (eg, AIDS) has increased the demand for higher barrier fabrics without loss of comfort.

Medical devices are often sterilized after the nonsterile device is sealed in a package. A part of this package, often the lid, is made from flashspun or spunbonded/meltblown fabric because it possesses the unique property of permitting the sterilizing gas of ethylene oxide to pass through while remaining impenetrable to bacteria. These fabrics are manufactured to tightly controlled standards to ensure the highest resistance to bacterial penetration. The superiority of a spunbonded fabric to the alternative coated papers has been reviewed (68).

Spunbonded fabrics have been utilized as shoe covers in the operating room. The covers are usually sewn with an elastic band at the top to allow them to be held snugly in place. The fabric requirements are toughness, some porosity for comfort, nonlinting, resistance to slippage, and a nonstatic characteristic. In order to achieve the last property, the fabric is usually treated with an antistatic coating. Failure to use a nonstatic fabric may cause sparks to be generated in the operating room environment which could damage sensitive electronic devices or lead to fire. Other medical applications for spunbonded fabrics include head covers, face masks, drapes, and other uses requiring barrier properties.

Protective Clothing. A large-volume area for spunbonded fabrics is the disposable protective clothing market (69). To a great extent the demand for high performance disposable protective clothing has tracked high technology manufacturing and environmental demands. The manufacture of particulate sensitive electronic components such as integrated circuits resulted in the construction of clean rooms where the generation of particulate was in part controlled through the use of nonlinting yet comfortable clothing made from spunbonded fabrics. Because the spunbonded fabric is made of continuous filaments, practically no linting results. At the same time the structure allows the passage of moisture and air, thus helping the wearer to remain comfortable. The spunbonded garment is worn over other clothing, therefore the maximum pore size must be sufficiently small to prevent the passage of lint and other particles through the garment and into the clean room.

The 1980s saw the removal of large quantities of asbestos (qv) from buildings, creating the need for clothing which could not be penetrated by small asbestos fibrils, yet was inexpensive enough to permit daily disposal. Certain spunbonded and spunbond/meltblown laminate fabrics demonstrate excellent resistance to asbestos penetration from particles as small as 0.5 μm (70).

Similarly the handling of hazardous materials has prompted the need for affordable, disposable protective clothing (69). Once exposed to toxic waste, pesticides, or radioactive materials, the clothing itself is transformed into a hazardous material and must be disposed of to prevent spreading of contamination. Garments that demand the utmost in protection at the lowest price are often made by extrusion coating, laminating spunbonded fabrics with polyethylene, or laminating the fabric to poly(vinylidene chloride) film.

Packaging. Packaging applications for spunbonded fabrics are for the most part a specialty area in which paper products or plastic films do not adequately perform. A clear example of this is medical sterile packaging, discussed as a medical application. One of the largest packaging applications for spunbonded fabric is high performance envelopes. Although lightweight tear- and puncture-resistant envelopes are not in demand by individual consumers, both large and small businesses have found spunbonded envelopes to outperform those made from conventional paper products. The lighter weight of the spunbonded envelope allows for postal savings, and future increases in postal rates may make the use of spunbonded envelopes even more attractive. Some corporations specializing in overnight delivery have successfully used spunbonded envelopes with excellent results. Coated spunbonded fabrics are used as the outerwrap of coils of steel and aluminum where they outperform alternative materials such as films and papers.

Other Disposable Applications. There are many other disposable applications for spunbonded fabrics including fiber bale wrap, metal core wrap, wipes, and clothes dryer fabric softener sheets.

A significant percentage of U.S. staple fiber production is packaged in bales of extrusion coated spunbonded fabrics, so treated to render the fabric impervious. Synthetic fibers have been shipped worldwide in this manner with great success.

Many types of spunbonded fabrics are utilized in the very large wipe market. Some are treated with surfactants or coatings which enhance the performance of the fabric. Most uses benefit from the low cost and low linting nature of the fabric.

A reverse-wipe application is seen in the clothes dryer fabric softener sheet wherein the spunbonded fabric is coated with a complex combination of compounds that are released into the environment of a hot clothes dryer to soften and perfume the clothes, as well as provide an antistatic quality. The spunbonded sheet, which must be made of polyester or nylon for temperature resistance, provides a simple and cost-effective medium to store the chemical compounds prior to release in the dryer.

BIBLIOGRAPHY

"Spunbonded" under "Nonwoven Textile Fabrics" in *ECT* 3rd ed., Vol. 16, pp. 72–104, by K. Porter, ICI Fibres.

1. G. Najour, *Nonwovens World* **5**(6), 22 (Aug. 1990).
2. R. L. Smorada, *INDA J. Nonwovens Res.* **3**(4) (Fall 1991).
3. R. G. Mansfield, *Nonwovens Indust.*, 26–34 (Feb. 1985).
4. *Nonwovens World*, 36–42 (May–June 1986).
5. *Eur. Chem. News* **45**, 21 (Nov. 25, 1985).
6. J. R. Starr, *Nonwovens Indust.*, 38–41 (June 1992).
7. *Nonwovens World* **5**(1), 16 (Jan. 1990).
8. *Nonwovens Indust.* **24**(10), 54–63 (Oct. 1993).
9. *Guide to Nonwoven Fabrics*, INDA, Cary, N.C., p. 24.
10. F. Gugumus, *Third International Conference, Polypropylene Fibers and Textiles*, Oct. 4–6, 1983, University of York, U.K.
11. *Nonwovens Indust.*, 166 (Oct. 1986).
12. Z. P. Jezic, "Recent Developments in Polyethylene Fiber Grade Resins," *Insight*, 87 (Sept. 1987).
13. U.S. Pat. 3,989,788 (Nov. 2, 1976), L. L. Estes, A. F. Fridrichsen, and V. S. Koshkin (to Du Pont).
14. Brit. Pat. 1,157,437 (July 9, 1969), B. L. Davies (to ICI Ltd.).
15. Y. Ogawa, *Spunbonded Technology Today 2*, Miller-Freeman, San Francisco, Calif., 1992, p. 123.
16. U.S. Pat. 3,439,085 (Apr. 15, 1969), L. Hartmann (to C. Freudenberg).
17. U.S. Pat. 3,338,992 (Aug. 29, 1967), G. A. Kinney (to E. I. du Pont de Nemours & Co., Inc.).
18. U.S. Pat. 3,991,244 (Nov. 9, 1976), S. C. Debbas (to E. I. du Pont de Nemours & Co., Inc.).
19. Brit. Pat. 1,436,545 (May 19, 1976), J. Brock (to ICI Ltd.).
20. U.S. Pat. 4,017,580 (Apr. 12, 1977), J. Barbey (to Rhône-Poulenc).
21. U.S. Pat. 3,502,763 (Mar. 24, 1970), L. Hartmann (to C. Freudenberg).
22. U.S. Pat. 4,405,297 (Sept. 20, 1983), D. W. Appel and M. T. Morman (to Kimberly-Clark Corp.).
23. U.S. Pat. 3,771,307 (Nov. 13, 1973), D. G. Petrille (to E. I. du Pont de Nemours & Co., Inc.).
24. U.S. Pat. 3,692,618 (Sept. 19, 1972), O. Dorschner, F. Carduck, and C. Storkebaum (to Metallgesellschaft AG).
25. U.S. Pat. 4,163,305 (Aug. 7, 1979), V. Semjonow and J. Foedrowitz (to Hoechst AG).
26. U.S. Pat. 3,322,607 (May 30, 1967), S. L. Jung (to E. I. du Pont de Nemours & Co., Inc.).
27. Brit. Pat. 2,006,844 (Oct. 10, 1978), P. Ellis and R. Gibb (to ICI Ltd.).
28. Brit. Pat. 1,231,066 (Oct. 4, 1968), C. H. Weightman (to ICI Ltd.).
29. H. Külter, in J. Lunenschloss and W. Albrecht, eds., *Non-Woven Bonded Fabrics*, Ellis Horwood Ltd., Chichester, U.K., 1985, p. 178.
30. U.S. Pat. 4,813,864 (Mar. 21, 1989). H. Balk (to Reifenhauser GmbH).
31. D. L. Heydt, *12th Technical Symposium*, INDA, May 22–23, 1984.
32. G. Flood, *Nonwovens Indust.*, 30–35 (Apr. 1986).
33. D. F. Beaumont, *Nonwovens World* **1**(3), 76–80 (1986).
34. E. Fehrer, *Can. Tex. J.*, 67–74 (Dec. 1985).
35. U.S. Pat. 3,855,046 (Dec. 17, 1974), P. B. Hansen and L. B. Pennings (to Kimberly-Clark Corp.).
36. U.S. Pat. 3,855,045 (Dec. 17, 1974), R. J. Brock (to Kimberly-Clark Corp.).

37. D. H. Muller and S. Barnhardt, *Nonwovens Report Int.*, 19–24 (Mar. 1986).
38. U.S. Pat. 4,125,663 (Nov. 14, 1978), P. Eckhart (to Hoechst AG).
39. U.S. Pat. 3,542,615 (Nov. 24, 1970), E. J. Dobo, D. W. Kim, and W. C. Mallonee (to Monsanto Corp.).
40. U.S. Pat. 3,322,607 (May 30, 1967), S. L. Jung (to E. I. du Pont de Nemours & Co., Inc.).
41. L. C. Wadsworth and A. M. Jones, *Nonwovens Indust.*, 44–51 (Nov. 1986).
42. S. Sullivan, *Nonwovens Indust.*, 38–41 (Feb. 1993).
43. U.S. Pat. 3,972,759 (Aug. 3, 1976), R. R. Buntin (to Exxon).
44. U.S. Pat. 4,374,888 (Feb. 22, 1983), S. R. Bornslaeger (to Kimberly-Clark Corp.).
45. U.S. Pat. 4,100,324 (July 11, 1978), R. A. Anderson, R. C. Sokolowski, and K. W. Ostermeier (to Kimberly-Clark Corp.).
46. U.S. Pat. 4,118,531 (Oct. 3, 1978), E. R. Hauser (to 3M Co.).
47. Y. Ogawa, *Nonwovens World*, 79–81 (May–June 1986).
48. U.S. Pat. 3,081,519 (Mar. 19, 1963), H. Blades and J. R. White (to E. I. du Pont de Nemours & Co., Inc.).
49. U.S. Pat. 3,442,740 (May 6, 1969), J. C. David (to E. I. du Pont de Nemours & Co., Inc.).
50. U.S. Pat. 5,081,177 (Jan. 14, 1992), H. Shin (to E. I. du Pont de Nemours & Co., Inc.).
51. *Nonwoven Fabrics Handbook*, INDA, Cary, N.C., pp. 77–79.
52. S. Kawabata, *The Standardization and Analysis of Hand Evaluation*, 2nd ed., Textile Machinery Society of Japan, Osaka, Japan, 1980.
53. R. M. Koerner, *Designing with Geosynthetics*, Prentice-Hall, Englewood Cliffs, N. J., 1990, p. 15.
54. J. D. M. Wisse and S. Birkenfeld, *2nd International Conference on Geotextiles*, Las Vegas, Nev., 1982.
55. *Nonwovens Indust.*, 52–53 (Oct. 1991).
56. *Nonwovens Indust.*, 30 (May 1992).
57. Ref. 56, p. 27.
58. *Du Pont Tyvek Literature*, Bulletin #E-60658.
59. D. Kamat, *Nonwovens Indust.*, 28–29 (Feb. 1992).
60. M. Jacobson, *Nonwovens Indust.*, 32–33 (Aug. 1990).
61. E. Noonan, *Nonwovens Indust.*, 34 (Feb. 1993).
62. Ref. 53, p. 12.
63. Ref. 53, p. 43; S. P. Scheinberg, J. H. O'Toole, and S. K. Rudys, *Partic. Microb. Cont.* (July/Aug. 1983).
64. A. A. Balkema, in R. Veldhuijzen Van Zanten, ed., *Geotextiles and Geomembranes in Civil Engineering*, Rotterdam/Boston, Mass., 1986.
65. H. Sandstedt, *Nonwovens Indust.* 40–44 (Feb. 1990).
66. J. Salzman, *Nonwovens World*, 36–38 (Jan. 1990).
67. D. K. Lickfield, *Int. Nonwoven J.* **6**(2), 37–41 (1994).
68. Th. Mengen and A. Jordy, *8th Symposium of the Austrian Society for Hygiene, Microbiology and Preventative Medicine*, Vienna, Austria, 1984.
69. P. Hanna, *Nonwovens Indust.*, 24–26 (Apr. 1991).
70. *Arthur D. Little Report* #C87352 (May 1982).

RONALD SMORADA
Reemay, Inc.

NOREPINEPHRINE.   See EPINEPHRINE AND NOREPINEPHRINE.

NOVOLOID FIBERS. See PHENOLIC RESINS.

NUCLEAR MAGNETIC RESONANCE. See MAGNETIC SPIN RESONANCE.

NUCLEAR REACTORS

INTRODUCTION

The nuclear reactor is a device in which a controlled chain reaction takes place involving neutrons and a heavy element such as uranium. Neutrons are typically absorbed in uranium-235 [15117-96-1], ^{235}U, or plutonium-239 [15117-48-3], ^{239}Pu, nuclei. These nuclei split, releasing two fission fragment nuclei and several fast neutrons. Some of these neutrons cause fission in other uranium nuclei in a sequence of events called neutron multiplication. The fission fragments are stopped within the nuclear fuel, where their kinetic energy becomes thermal energy. The thermal energy is removed by a cooling agent and converted into electrical energy in a turbine-generator system. Many of the fission fragments are radioactive, releasing radiation and decay heat. Some of the radioactive materials have useful purposes; others form nuclear waste (see NUCLEAR REACTORS, WASTE MANAGEMENT).

Nuclear reactors as a source of heat energy and radiation were the outgrowth of World War II defense applications. Research and development was pursued on several fronts in the Manhattan Project. Success of a graphite and uranium pile built and tested at the University of Chicago in 1942 prompted construction of production reactors at Hanford, Washington, to accumulate plutonium for the atomic bomb. A second approach to obtaining weapons material involved uranium isotope separation methods. Two techniques were successful: the electromagnetic process at the University of California and gaseous diffusion at Columbia University. Oak Ridge, Tennessee, became the enriched-uranium production center, utilizing both methods. At the same time, knowledge was gained at Los Alamos, New Mexico, about conditions for controlled chain reactions in uranium and plutonium assemblies (see DIFFUSION SEPARATION METHODS; NUCLEAR REACTORS, ISOTOPE SEPARATION).

The Manhattan Project culminated in the use of nuclear weapons. After the war, the U.S. Atomic Energy Commission (AEC), the predecessor of the Nuclear Regulatory Commission (NRC) and the Department of Energy (DOE), was formed. The AEC led U.S. research and development programs on nuclear naval vessels and central station power plants, in cooperation with industry. Excellent accounts of the history of the nuclear enterprise have been provided, including the period 1939–1961, during which the designs of newer reactors came into being (1).

A variety of nuclear reactor designs is possible using different combinations of components and process features for different purposes (see NUCLEAR REACTORS, REACTOR TYPES). Two versions of the lightwater reactors were favored: the pressurized water reactor (PWR) and the boiling water reactor (BWR). Each requires enrichment of uranium in ^{235}U. To assure safety, careful control of coolant conditions is required (see NUCLEAR REACTORS, WATER CHEMISTRY OF LIGHTWATER REACTORS; NUCLEAR REACTORS, SAFETY IN NUCLEAR FACILITIES).

Power Generation

The principal application of the nuclear reactor is as a heat source for electrical power generation. Growth in the relative contribution of nuclear energy to the electricity supply of the United States since 1949 is shown in Figure 1. As of 1995 there were 109 nuclear power reactors in operation in the United States, generating almost 100 GW of electrical power. Outside of the United States, there

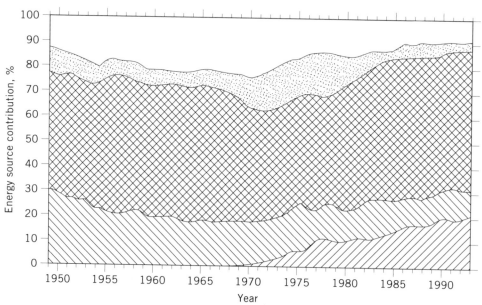

Fig. 1. U.S. utility net electricity generation, where ▨ represents coal, ▨ hydro power and other energy sources, ☐ natural gas, ▨ nuclear, and ▨ petroleum. Data from Reference 2.

were 313 reactors producing 236 GW. Table 1 lists power reactors by country (3). Table 2 gives the worldwide distribution by reactor type (3). The United States and Europe have the greatest number of nuclear facilities. There are few in South America or Africa and none in Australia. The fraction of total electricity that is derived from nuclear reactors varies greatly among countries. Notable approximate figures are France, 75%; Japan, 30%; and the United States, 21%. Some of the characteristics of the PWR and BWR, ie, the pressurized lightwater reactor and the boiling water reactor, which are the most widely used reactor types, are given in Table 3.

Safety. A large inventory of radioactive fission products is present in any reactor fuel where the reactor has been operated for times on the order of months.

Table 1. World Nuclear Power Plants[a]

Nation	Operative units	Power, net MWe	Total number of units	Power, net MWe
Argentina	2	935	3	1,627
Belgium	7	5,527	7	5,527
Brazil	1	626	3	3,084
Bulgaria	6	3,420	6	3,420
Canada	22	15,442	22	15,442
China	2	1,800	5	3,300
Cuba	0	0	2	834
Czech Republic	4	1,632	6	3,412
Finland	4	2,310	4	2,310
France	56	57,623	61	64,033
Germany	21	22,703	21	22,703
Hungary	4	1,729	4	1,729
India	9	1,834	16	3,874
Japan	47	36,946	54	43,692
Kazakhstan	1	135	1	135
Korea	9	7,220	16	13,083
Lithuania	2	2,760	2	2,760
Mexico	1	654	2	1,308
the Netherlands	2	507	2	507
Pakistan	1	125	2	425
Philippines	0	0	1	605
Romania	0	0	5	3,100
Russia	25	19,799	29	23,174
Slovakia	4	1,632	8	3,296
Slovenia	1	620	1	620
South Africa	2	1,840	2	1,840
Spain	9	7,085	15	12,832
Sweden	12	10,002	12	10,002
Switzerland	5	2,985	5	2,985
Taiwan, China	6	4,884	6	4,884
Ukraine	14	12,095	20	17,795
United Kingdom	34	11,540	35	12,728
United States	109	99,510	116	107,994
Total	*422*	*335,920*	*494*	*395,060*

[a]Ref. 3. Courtesy of *Nuclear News*.

Table 2. Worldwide Nuclear Power Units by Reactor Type[a]

Reactor type	Units in operation	Power, net MW	Total number of units	Power, net MW
pressurized lightwater reactors (PWR)	243	214,234	286	253,872
boiling lightwater reactors (BWR)	91	74,941	99	83,243
gas-cooled reactors, all types	36	12,239	36	12,239
heavy-water reactors, all types	33	18,645	49	26,540
graphite-moderated lightwater reactors (LGR)	15	14,785	16	15,710
liquid-metal-cooled fast-breeder reactors (LMFBR)	3	928	7	3,308

[a]Ref. 3. Courtesy of *Nuclear News*.

Table 3. Characteristics of Reactors

Parameter	Reactor type	
	PWR	BWR
heat power, MWt	3425	3579
electrical power, MWe	1150	1220
coolant temperatures, °C	292 (326)[a]	216 (285)[a]
pressure, MPa[b]	15.5	7.0
reload fuel, wt % ^{235}U	4.0–5.0	3.5–3.8

[a]In (out).
[b]To convert MPa to psia, multiply by 145.

In steady state, radioactive decay heat amounts to about 5% of fission heat, and continues after a reactor is shut down. If cooling is not provided, decay heat can melt fuel rods, causing release of the contents. Protection against a loss-of-coolant accident (LOCA), eg, a primary coolant pipe break, is required. Power reactors have an emergency core cooling system (ECCS) that comes into play upon initiation of a LOCA.

Nuclear power has achieved an excellent safety record. Exceptions are the accidents at Three Mile Island in 1979 and at Chernobyl in 1986. In the United States, safety can be attributed in part to the strict regulation provided by the Nuclear Regulatory Commission, which reviews proposed reactor designs, processes applications for licenses to construct and operate plants, and provides surveillance of all safety-related activities of a utility. The utilities seek continued improvement in capability, use procedures extensively, and analyze any plant incidents for their root causes. Similar programs intended to ensure reactor safety are in place in other countries.

A technique called probabilistic safety assessment (PSA) has been developed to analyze complex systems and to aid in assuring safe nuclear power plant operation. PSA, which had its origin in a project sponsored by the U.S. Atomic Energy Commission, is a formalized identification of potential events and consequences leading to an estimate of risk of accident. Discovery of weaknesses in the plant allows for corrective action.

Reactors are designed to be inherently safe based on physical principles, supplemented by redundant equipment and special procedures. Nuclear power benefits from the application of the concept of defense in depth, ie, by using fuel form, reactor vessel, building containment, and emergency backup procedures to ensure safety.

The accident in 1979 at Three Mile Island Unit 2 (TMI-2), although highly publicized and very costly to clean up, resulted in minimum hazard to the public. The design included a thick steel reactor vessel and a tight containment building. The incident resulted from mechanical failure compounded by misinterpretation of events by the operating crew. The TMI-2 accident, which prompted a number of improvements in equipment and procedures, also led the nuclear industry to create the Institute of Nuclear Power Operations (INPO), a self-regulatory organization. The INPO maintains extensive safety-related databases, conducts power plant visits, and oversees operator training programs.

The steam explosion of the Chernobyl reactor in Ukraine in 1986 caused scores of immediate deaths and released large amounts of radioactivity, with resultant contamination and radiation exposure. The accident occurred because of inadequate inherent safety, improper operating practices, and lack of containment. The Chernobyl accident resulted in some design and operation changes in the reactor, making it less vulnerable in future operation. Countries of the former USSR have been encouraged by the International Atomic Energy Agency (IAEA) and the United States to shut down the reactors, but as of early 1995 demands for electrical power have prevented such action.

The public perceives the risk of nuclear power to be much greater than that determined by experts (4). Among explanations for the discrepancy are the belief in the possibility of a disaster and the association of reactors with weapons. Living 50 years within five miles of a nuclear power plant has been shown to be comparable in terms of risk to smoking 1.4 cigarettes during the same period (5).

Environmental Aspects

In contrast to power plants using fossil fuel, nuclear reactor plants emit no compounds of carbon, nitrogen, or sulfur, and thus do not contribute to acid rain, ozone layer depletion, or global warming (see AIR POLLUTION; ATMOSPHERIC MODELING). Emissions of radioactive materials during regular operations are within regulatory requirements based on medical knowledge. These emissions do include radionuclides of the noble gases xenon and krypton, which readily disperse throughout the atmosphere. Small quantities of soluble radionuclides are released into lakes or streams that provide very large dilution factors. Plant and animal life are monitored regularly at such facilities. On the other hand, the potential, however small, of radioactive contamination of the environment in case of a reactor accident in which containment is breached does exist.

As the result of many years of nuclear reactor research and development and weapons production in U.S. defense programs, a large number of sites were contaminated by radioactive materials. A thorough cleanup of this residue of the Cold War is expected to extend well into the twenty-first century and cost many billions of dollars. New technologies are needed to minimize the cost of the cleanup operation.

Wastes.   Nuclear reactors produce unique wastes because these materials undergo radioactive decay and in so doing emit harmful radiation. Spent nuclear fuel has fission products, uranium, and transuranic elements. Plans call for permanent disposal in underground repositories. Geological studies are in progress at the Yucca Mountain site in Nevada. Until a repository is completed, spent fuel must be stored in water pools or in dry storage casks at nuclear plant sites.

Nuclear wastes are classified according to the level of radioactivity. Low level wastes (LLW) from reactors arise primarily from the cooling water, either because of leakage from fuel or activation of impurities by neutron absorption. Most LLW will be disposed of in near-surface facilities at various locations around the United States. Mixed wastes are those having both a hazardous and a radioactive component. Transuranic (TRU) waste containing plutonium comes from chemical processes related to nuclear weapons production. These are to be placed in underground salt deposits in New Mexico (see ACTINIDES AND TRANSACTINIDES).

Mill tailings are another form of nuclear waste. The residue from uranium ore extraction contains radium, the precursor of short-lived radon and its daughters. Piles of tailings must be properly covered.

Other wastes are expected to arise from the decontamination and decommissioning of existing nuclear facilities. These include reactors at the time of life extension or at the end of their operating life. Whereas technologies are available for waste disposal, as of this writing (ca 1995) there is much public resistance to the establishment of disposal facilities.

Economic Aspects

In the early years of reactor development, electricity from nuclear sources was expected to be much cheaper than that from other sources. Whereas nuclear fuel cost is low, the operating and maintenance costs of a nuclear facility are high. Thus on average, electric power from coal and nuclear costs about the same.

Optimism about economic growth in the period 1960–1975 led to a large number of reactor orders. Many of these were canceled even after partial completion in the period after the 1974 oil crisis, as the result of a reduction in energy demand. Inflation, high interest rates, long construction periods, and regulatory delays resulted in severe cost overruns. Moreover, the reactor accidents of TMI and, later, Chernobyl produced an atmosphere of public concern. As a consequence, there is a general reluctance in the financial community to support the construction of new nuclear plants.

Resources

Predictions in the 1960s of the growth in nuclear power indicated the need for recycling (qv) of nuclear fuels. Radionuclides involved are uranium-235, uranium-238 [24678-82-8], and plutonium-239. This last is produced by neutron absorption in the reactions:

$$^{238}_{92}\text{U} + ^{1}_{0}n \longrightarrow ^{239}_{92}\text{U}$$

$$^{239}_{92}\text{U} \longrightarrow ^{239}_{93}\text{Np} + ^{0}_{-1}e$$

$$^{239}_{93}\text{Np} \longrightarrow ^{239}_{94}\text{Pu} + ^{0}_{-1}e$$

Uranium-239 [13982-01-9] has a half-life of 23.5 min; neptunium-239 [13968-59-7] has a half-life of 2.355 d. Recycling or reprocessing of spent fuel involves separation of plutonium from uranium and from bulk fission product isotopes (see NUCLEAR REACTORS, CHEMICAL REPROCESSING).

Uranium resources were originally expected to be rapidly depleted in a growing economy. There were, however, ample supplies of uranium as of 1995.

The breeder reactor, which would produce and burn plutonium and gradually increase the inventory of fissionable material, requires reprocessing of nuclear fuel. As of 1995 only limited research and development was in progress on breeder reactors, mainly in France and Japan.

The importance of nuclear power for meeting growing U.S. energy needs in an environmentally sound manner has been highlighted (6). The role of nuclear power for the world in the twenty-first century has also been discussed (7).

In the hope of stimulating interest in the building of nuclear power plants, the nuclear industry is designing advanced lightwater reactors. These are of two types, known as simplified and enhanced safety. The first takes advantage of knowledge gained in the operation of previous nuclear reactor designs. It has lower (ca 600 MW) power levels than the 1200 MW reactors of the 1970s and 1980s. The second uses passive features such as natural convection and the force of gravity for enhanced safety. The U.S. government is funding limited development of liquid-metal and gas-cooled advanced reactors.

BIBLIOGRAPHY

"Nucleonics" in ECT 1st ed., Vol. 9, pp. 515–547, by E. B. Ashcraft, Westinghouse Electric Corp.; "Nuclear Reactors" in ECT 1st ed., Suppl. 1, pp. 519–614, by H. H. Hausner, Penn-Texas Corp.; J. M. Fanto, Consulting Engineer; G. M. Roy, General Electric Co.; A. Strasser, W. Arbiter, and J. M. McKee, Nuclear Development of America; "Introduction" under "Nuclear Reactors" in ECT 2nd ed., Vol. 14, pp. 74–75, by D. E. Ferguson, Oak Ridge National Laboratory; in ECT 3rd ed., Vol. 16, pp. 138–142, by W. B. Lewis, Queen's University.

1. R. Rhodes, The Making of the Atomic Bomb, Simon and Schuster, New York, 1986; R. G. Hewlett and J. M. Holl, Atoms for Peace and War 1953–1961: Eisenhower and the Atomic Energy Commission, University of California Press, Berkeley, Calif., 1989.
2. Annual Energy Review 1993 DOE/EIA-0384(93), U.S. Department of Energy, Washington, D.C., July 1994, p. 233.
3. Nucl. News, 62 (Mar. 1994).
4. B. Fischhoff, S. R. Watson, and C. Hope, in T. S. Glickman and M. Gough, eds., Readings in Risk, Resources for the Future, Washington, D.C., 1990, pp. 30–41.
5. R. Wilson, in Ref. 3, pp. 55–59.
6. R. Rhodes, Nuclear Renewal: Common Sense About Energy, Penguin Books, New York, 1993.

7. C. Starr, *Electr. Perspect.*, 22 (Jan. 1993).

General References

R. L. Murray, *Nuclear Energy*, 4th ed., Pergamon Press, Oxford, U.K., 1993.
A. V. Nero, Jr., *A Guidebook to Nuclear Reactors*, University of California Press, Berkeley, Calif., 1979.
R. A. Knief, *Nuclear Engineering: Theory and Technology of Commercial Nuclear Power*, Taylor & Francis, Bristol, Pa., 1992.
J. Weisman, ed., *Elements of Nuclear Reactor Design*, 2nd ed., Robert E. Krieger Publishing Co., Malabar, Fla., 1983.
V. N. Shah and P. E. MacDonald, *Aging and Life Extension of Major Light Water Reactor Components*, Elsevier, Amsterdam, the Netherlands, 1993.
S. Villani, *Isotope Separation*, American Nuclear Society, La Grange Park, Ill., 1976.
P. Cohen, *Water Coolant Technology for Power Reactors*, American Nuclear Society, La Grange Park, Ill., 1980.
R. L. Murray, *Understanding Radioactive Waste*, Battelle Press, Columbus, Ohio, 1994.
M. W. Golay and N. E. Todreas, *Sci. Amer.*, 82–89 (Apr. 1990).

RAYMOND L. MURRAY
Consultant

NUCLEAR FUEL RESERVES

Lightwater reactors, the primary type of nuclear power reactor operated throughout the world, are fueled with uranium dioxide [*1344-57-6*], UO_2, enriched from the naturally occurring concentration of 0.71% uranium-235 [*15117-96-1*], ^{235}U, to approximately 3% ^{235}U (1). As of this writing all civilian nuclear fuel has been produced by enriching natural uranium (see DIFFUSION SEPARATION METHODS). An additional source of enriched uranium for civilian nuclear reactor fuel is expected to become available from the dismantlement of nuclear weapons from the stockpiles of the United States and the Commonwealth of Independent States (former Soviet Union). The de-enrichment of highly enriched uranium (HEU) affords the potential for an additional, readily available reserve for civilian nuclear fuel (see also URANIUM AND URANIUM COMPOUNDS).

Uranium Mineral Resources

The Organization for Economic Cooperation and Development's Nuclear Energy Agency (OECD/NEA) and the International Atomic Energy Agency (IAEA) estimate uranium resources in four cost categories: $40/kg U or less; $80/kg U or less; $130/kg U or less; and $260/kg U or less (2). Previous NEA/IAEA evaluations employed only the last three cost categories (3–6). Costs include the direct costs of mining, transporting, and processing the uranium ore; the costs of associated environmental and waste management; the costs where applicable of maintaining nonoperating production units; in the case of ongoing projects, any capital costs which remain unamortized; the capital cost of providing new production units where applicable, including the costs of financing; indirect costs such as office overheads, taxes, and royalties where applicable; and future exploration and development costs wherever required for further ore delineation to the stage where it is ready to be mined. Sunk costs, such as the costs for geologic exploration and land acquisition, are not usually taken into consideration for the determination of uranium costs for these four categories (2).

The U.S. Department of Energy (DOE) and the NEA/IAEA employ similar terms to classify uranium resources, as (7) reasonably assured, estimated additional (EA), or speculative. The NEA/IAEA divides the estimated additional resources into two types, EAR-I and EAR-II, describing known resources and undiscovered ones, respectively (8).

Reasonably assured resources (RAR) include uranium that occurs in known mineral deposits of delineated size, grade, and configuration in which the estimated quantities could be recovered within the given range of production costs using proven mining and processing technologies (8). EAR-I includes uranium deposits in addition to RAR that are inferred from direct geological evidence, extensions of well-explored deposits, or in deposits in which geologic continuity has been established but insufficient data exist to permit classification as RAR (8). EAR-II includes uranium in addition to EAR-I that is expected to occur in deposits for which evidence is primarily indirect and which are believed to exist in well-defined geological trends (8). The speculative resources (SR) category includes uranium that is thought to exist, primarily on the basis of indirect evidence and geologic extrapolations, in deposits that are deemed discoverable with existing exploration techniques (9).

Geochemical Nature and Types of Deposits. The crust of the earth contains approximately 2–3 ppm uranium. Alkalic igneous rock tends to be more uraniferous than basic and ferromagnesian igneous rocks (10). Elemental uranium oxidizes readily. The solubility and distribution of uranium in rocks and ore deposits depend primarily on valence state. The hexavalent uranium ion is highly soluble, the tetravalent ion relatively insoluble. Uraninite, the most common mineral in uranium deposits, contains the tetravalent ion (11).

A classification system for the principal types of uranium ore deposits was revised by the IAEA in 1988–1989 (12). This system assigns uranium resources to various categories on the basis of geological setting. There are 15 main categories of uranium ore deposits arranged according to approximate economic significance.

The first six geologic ore types, together with selected types from the seventh category, are considered conventional resources. These categories represent a majority of the uranium-producing geologic formations worldwide as of 1992. Very low grade resources, which are not economic as of the mid-1990s, or from which uranium is only recoverable as a minor by-product, are considered unconventional resources (13).

Unconformity-Related Deposits. Deposits of the unconformity-related type occur spatially close to significant unconformities. These deposits usually developed during the period about 1800–800 million years ago in intracratonic basins. Deposits also developed during Phanerozoic time. Examples of unconformity-related deposits include the ore bodies at Cluff Lake, Key Lake, and Rabbit Lake in northern Saskatchewan, Canada, and those in the Alligator Rivers area in northern Australia (12).

Sandstone Deposits. A majority of the sandstone ore deposits are contained in rocks that were deposited under marginal marine or fluvial conditions. Lacustrine and eolian sandstones are also mineralized, but uranium deposits are much less common in these rocks. Host rocks containing uranium are almost always medium-to-coarse-grained, poorly sorted sandstones containing pyrite and organic matter of plant origin. The sediments are commonly associated with tuffs. Unoxidized deposits of this type consist of pitchblende and coffinite in arkosic and quartzitic sandstones. Upon weathering, secondary minerals such as carnotite, tuyamunite, and uranophane are formed. The Tertiary, Jurassic, and Triassic sandstones of the Western Cordillera of the United States account for most of the U.S. uranium production. Cretaceous and Permian sandstones are important host rocks in Argentina. Other important uranium deposits are found in carboniferous deltaic sandstones in Niger, in Permian lacustrine silt-stones in France, and in Permian sandstones of the Alpine region. The deposits in Precambrian marginal marine sandstones in Gabon in Africa have also been classified as sandstone deposits (14).

Quartz-Pebble Conglomerate Deposits. Known quartz-pebble conglomerate ores are restricted to a specific period of geologic time. These ore types occur in basal Lower Proterozoic beds unconformably situated above Archaean basement rocks composed of granitic and metamorphic strata. A number of commercial deposits are located in Canada and South Africa. Some subeconomic occurrences have been reported in Brazil and India (14).

Vein Deposits. The vein deposits of uranium are those in which uranium minerals fill cavities such as cracks, fissures, pore spaces, breccias, and stock-works. The dimensions of the openings have a wide range, from the narrow pitchblende-filled cracks, faults, and fissures in some of the ore bodies in Europe, Canada, and Australia to the massive veins of pitchblende at Jachymov, Czech Republic (15).

Breccia Complex Deposits. Deposits of the breccia complex group were developed in Proterozoic continental regimes during anorogenic periods. The host rocks include felsic volcanoclastics and sedimentary rocks. The ores generally contain two phases of mineralization, an earlier strata bound and a later transgressive one. The principal ore deposit of this type is found at the Olympic Dam site in South Australia. Deposits which may also belong to this category occur in Zambia, Zaire, and the Aillik Group in Labrador, Canada (15).

Intrusive Deposits. Deposits included in the intrusive deposit type are those associated with intrusive or anatectic rocks of different chemical composition, eg, alaskite, granite, monzonite, peralkaline syenite, carbonatite, and pegmatite. Examples include the uranium occurrences in the porphyry copper deposits such as Bingham Canyon and Twin Butte in the United States, the Rossing Deposit in Namibia, and Ilimaussaq deposit in Greenland, Palabora in South Africa, and the deposits in the Bancroft area, Canada (15).

Phosphorite Deposits. Sedimentary phosphorites contain low concentrations of uranium in fine-grained apatite. Uranium of this type is considered an unconventional resource. Significant examples of these uranium ore types include the U.S. deposits in Florida, where uranium is recovered as a by-product, and the large deposits in North African and Middle Eastern countries (16).

Collapse Breccia Pipe Deposits. The primary occurrence of collapse breccia pipe deposits is in circular, vertical pipes filled with down-dropped fragments. Uranium is concentrated in the permeable breccia matrix and in the accurate fracture zones enclosing the pipe. An example of this type of deposit is found in the Arizona Strip in Arizona (16).

Volcanic Deposits. Uranium deposits of volcanic deposits type are stratabound and structure-bound concentrations in acid volcanic rocks. Uranium is commonly associated with molybdenum, fluorine, etc. Examples are the uranium deposits in Michelin, Canada; Nopal I in Chihuahua, Mexico; Macusani in Peru; and numerous deposits in China and the CIS (16).

Surficial Deposits. Uraniferous surficial deposits may be broadly defined as uraniferous sediments, usually of Tertiary to recent age which have not been subjected to deep burial and may or may not have been calcified to some degree. The uranium deposits associated with calcrete, which occur in Australia, Namibia, and Somalia in semiarid areas where water movement is chiefly subterranean, are included in this type. Additional environments for uranium deposition include peat and bog, karst caverns, as well as pedogenic and structural fills (15).

Metasomatite Deposits. Included in the metasomatite deposit grouping are uranium deposits in alkali metasomatites, eg, albitites, aegirinites, and alkali–amphibole rocks, commonly intruded by microcline granite. Examples are the deposits in Espinharasin, Brazil, and Ross Adams, Alaska, as well as the Zheltye Vody deposit in Krivoy Rog area in Ukraine (16).

Metamorphic Deposits. Uranium deposits belonging to the metamorphic class occur in metasediments or metavolcanics generally, without direct evidence of post-metamorphic mineralization. Examples include the deposits at Forstau, Austria (16).

Lignite. Deposits generally classified as unconventional uranium resources occur in lignite and in clay or sandstone immediately adjacent to lignite. Examples are uraniferous deposits in the Serres Basin, Greece, North and South Dakota in the United States, and Melovoe in the CIS (17) (see CLAYS; LIGNITE AND BROWN COAL).

Black Shale Deposits. Low concentrations of uranium occur in carbonaceous marine shales. These resources are also considered unconventional. Examples include the uraniferous alumshale in Sweden and the Chattanooga Shale in the United States, but also the Chanziping deposit of the argillaceous–carbonaceous–siliceous–pelitic rocks type in the Guangxi Autonomous Region in China and the deposit of Gera-Ronneburg in the eastern portion of Germany (17).

Other Deposits. Those deposits which cannot be classified as one of the previous 14 deposit types are called other. These include the uranium deposits in the Jurassic Todilto Limestone in the Grants district in New Mexico (17).

Uranium Reserves

Domestic. Estimates of U.S. uranium resources for reasonably assured resources, estimated additional resources, and speculative resources at costs of $80, $130, and $260/kg of uranium are given in Table 1 (18). These estimates include only conventional uranium resources, which principally include sandstone deposits of the Colorado Plateaus, the Wyoming basins, and the Gulf Coastal Plain of Texas. Marine phosphorite deposits in central Florida, the western United States, and other areas contain low grade uranium having 30–150 ppm U that can be recovered as a by-product from wet-process phosphoric acid. Because of relatively low uranium prices, on the order of $20.67/kg U (19), *in situ* leach and by-product plants accounted for 76% of total uranium production in 1992 (20).

Foreign. The OECD/NEA and IAEA have issued annual reports on world uranium resources, production, and demand since the mid-1960s (2–6). NEA/IAEA data for reasonably assured and estimated additional resources at costs of $80 and $130/kg uranium are given in Table 2 (21). These estimates incorporate data from both former world outside centrally planned economies (WOCA) and non-WOCA nations. A summary of other known uranium resources with and without cost range estimates is provided in Table 3 (22). These resources total about 1.4×10^6 t and include estimates that are not strictly consistent with standard NEA/IAEA definitions.

Estimates of speculative resources (SR) at $130/kg uranium and those having an unassigned cost range are provided in Table 4 (23). These resources, which total about 11.28×10^6 t, would be in addition to the reasonably assured and estimated additional resources. Estimates of uranium resources from unconventional and by-product sources are presented in Table 5 (24). These resources total about 7×10^6 t for phosphates, 0.013×10^6 t for nonferrous ores, 0.016×10^6 t for carbonates, and 0.014×10^6 t for lignites. These would be in addition to the reasonably assured resources, estimated additional resources, and the speculative resources (24).

Table 1. U.S. Uranium Resources[a]

Uranium cost category, $/kg	Resource category[b]			
	RAR	EAR	SR	Total
80	114	850	502	1466
130	370	1314	889	2573
260	588	1893	1352	3833

[a]Ref. 18.
[b]In 1000 metric tonnes U as of Dec. 31, 1992.

Resources

As of the beginning of 1993, reasonably assured resources (RAR) recoverable at costs of $80/kg U or less were estimated at 1.53×10^6 t U. Estimated additional resources (EAR) in the same cost category were about 1.769×10^6 t U. Total RAR and EAR, recoverable at costs of $130/kg U or less, were estimated at 2.205×10^6 t U and 3.540×10^6 t U, respectively, as of the beginning of 1993. There

Table 2. Uranium Resources, t × 10³ Uᵃ

Country	RAR				EAR			
	$80/kg U	Percent of total	$130/kg U	Percent of total	$80/kg U	Percent of total	$130/kg U	Percent of total
Algeria	26	1.7	26	1.2	0	0.0	0	0.0
Argentina	4.6	0.3	7.3	0.3	2.6	0.1	3.2	0.1
Australia	462	30.2	517	23.5	272	15.4	394	11.1
Austria			0	0.0	0.7	0.0	1.7	0.0
Brazil	162	10.6	162	7.3	94	5.3	94	2.7
Canada	277	18.1	397	18.0	81	4.6	223	6.3
Central African Republic	8	0.5	16	0.7	0	0.0	0	0.0
Chile			0	0.0	0	0.0	0.76	0.0
Colombia			0	0.0	0	0.0	11	0.3
Czech Republic	15.85	1.0	22.25	1.0	7.15	0.4	53.35	1.5
Denmark	0	0.0	27	1.2	0	0.0	16	0.5
Finland			0	0.0	0	0.0	0	0.0
France	19.85	1.3	33.65	1.5	3.55	0.2	7.93	0.2
Gabon	9.78	0.6	14.43	0.7	1.3	0.1	9.6	0.3
Germany	0	0.0	3	0.1	0	0.0	4	0.1
Greece	0.3	0.0	0.3	0.0	6	0.3	12	0.3
Hungary	0.62	0.0	1.13	0.1	1.32	0.1	28.66	0.8
India			0	0.0	0	0.0	13.03	0.4
Indonesia	0	0.0	5.42	0.2	0	0.0	2.15	0.1
Italy	4.8	0.3	4.8	0.2	0	0.0	1.3	0.0
Japan	0	0.0	6.6	0.3	0	0.0	0	0.0
Kazakhstan			0	0.0	0	0.0	380	10.7
Korea, Republic of	0	0.0	11.8	0.5	0	0.0	3	0.1
Mexico	0	0.0	1.7	0.1	0	0.0	3.4	0.1
Namibia	80.62	5.3	96.62	4.4	30	1.7	53	1.5
Niger	159.17	10.4	165.82	7.5	303.27	17.1	343.27	9.7

Table 2. (Continued)

	RAR				EAR			
Country	$80/kg U	Percent of total	$130/kg U	Percent of total	$80/kg U	Percent of total	$130/kg U	Percent of total
Peru	1.79	0.1	1.79	0.1	1.72	0.1	1.86	0.1
Portugal	7.3	0.5	8.7	0.4	2.95	0.2	2.95	0.1
Russian Federation			0	0.0	48	2.7	145	4.1
Slovenia	0	0.0	1.8	0.1	1.45	0.1	2.61	0.1
Somalia	0	0.0	6.6	0.3	0	0.0	3.4	0.1
South Africa	144.4	9.4	240.84	10.9	54.92	3.1	83.01	2.3
Spain	17.85	1.2	39	1.8	4.2	0.2	4.2	0.1
Sweden	2	0.1	4	0.2	1	0.1	6.3	0.2
Thailand	0	0.0	0.01	0.0	0	0.0	0	0.0
Turkey	9.13	0.6	9.13	0.4	0	0.0	0	0.0
Ukraine			0	0.0	0	0.0	3.9	0.1
United Kingdom	0	0.0	0	0.0	0	0.0	1	0.0
United States	114	7.4	369	16.7	846	47.8	1308	37.0
Uzbekistan			0	0.0	0	0.0	290	8.2
Vietnam			0	0.0	5	0.3	5.4	0.2
Zaire	1.8	0.1	1.8	0.1	1.7	0.1	1.7	0.0
Zambia			0	0.0	0	0.0	22	0.6
Zimbabwe	1.8	0.1	1.8	0.1	0	0.0	0	0.0
Total	*1530.66*	*100.0*	*2204.29*	*100.0*	*1769.83*	*100.0*	*3539.68*	*100.0*

[a]Ref. 21.

Table 3. Other Known Resources, t × 10³ U[a]

Country	Cost range							
	$80/kg U	Percent of total	$130/kg U	Percent of total	Unassigned	Percent of total	Total	Percent of total
Chile					0.30	0.1	0.30	0.02
China					72.10	18.3	72.10	5.02
India					66.36	16.8	66.36	4.62
Kazakhstan	417.50	58.1	512.30	49.2			512.30	35.69
Mongolia	19.00	2.6	80.00	7.7			80.00	5.57
Romania					26.00	6.6	26.00	1.81
Russian Federation	219.60	30.6	299.70	28.8			299.70	20.88
Ukraine	62.20	8.7	148.90	14.3			148.90	10.37
Uzbekistan					230.00	58.2	230.00	16.02
Total	718.30	100.0	1040.90	100.0	394.76	100.0	1435.66	100.0

[a]Ref. 22.

Table 4. Speculative Resources, t × 10³ U[a]

Country	Cost range					
	$130/kg U	Percent of total	Unassigned	Percent of total	Total	Percent of total
Australia	na	na	2,600–3,900	32.61–42.06	2,600–3,900	26.06–34.58
Canada	700.00	34.9		19.09	700.0	6.21
China			1,770.00		1,770.00	15.70
Colombia	217.00	10.8			217.00	1.93
Denmark	50.00	2.5	10.00	0.11	60.00	0.53
Egypt			15.00	0.16	15.00	0.13
Germany	0.00	0.0	61.50	0.66	61.50	0.55
Greece			6.00	0.06	6.00	0.05
Italy			10.00	0.11	10.00	0.09
Mexico			10.00	0.11	10.00	0.09
Mongolia			1,390.00	14.98	1,390.00	12.33
Peru	26.45	1.3			26.45	0.24
Portugal	0.00	0.0	1.50	0.02	1.5	0.01
Romania			15.00	0.16	15.00	0.13
South Africa			1,113.50	12.01	1,113.50	9.87
Ukraine	na	na	235.40	2.54	235.40	2.09
United Kingdom	1.00	0.05	1.00	0.01	2.00	0.02
United States	885.00	42.4	461.00	4.97	1,346.00	11.94
Venezuela			163.00	1.76	163.00	1.45
Vietnam	100.00	5.0	110.00	1.19	210.00	1.86
Zimbabwe	25.00	1.2	0.00	0.0	25.00	0.22
Total	2,004.45	100.0	7,972.90–9,272.90	100.00	9,977.35–11,277.35	100.00

[a]Ref. 23.

Table 5. Unconventional and By-Product Resources, t $\times$ 10^3 U^a

Country	Phosphates	Percent of total	Nonferrous	Percent of total
Brazil[b]	28.0	0.40		
Chile	0.6	0.01	4.5	35.71
Colombia	20.0	0.28		
Egypt	35.0	0.50		
Finland[c]				
Greece[d]	0.5	0.01		
India	1.7	0.02	6.6	52.38
Jordan	100.0	1.42		
Kazakhstan	64.4	0.91		
Mexico	151.0	2.14	1.0	7.94
Morocco	6526.0	92.38		
Peru	20.0	0.28	0.5	3.97
Syria	60.0	0.85		
Thailand	0.5–1.5	0.01		
United States	14.0	0.20		
Venezuela	42.0	0.59		
Vietnam[e]				
Total	7064.2	100.0	12.6	100.0

[a]Ref. 24.
[b]Also has 13,000 t of U from carbonates.
[c]Has 2,500 t of U from carbonates and from 3,000–9,000 t from lignites.
[d]Also has 4,000 t of U from lignites.
[e]Has 500 t of U from lignites.

remains good potential for the discovery of additional uranium resources of the conventional type, as reflected by estimates of speculative resources (SR). Based on reported estimates, this potential is about 13 $\times$ 10^6 t U. Nearly two-thirds of this potential occurs in Australia, Canada, South Africa, and the United States; over one-quarter occurs in China, Mongolia, and the CIS. There are also large tonnages of unconventional resources of uranium, most of which are associated with marine phosphate deposits. Production from such resources has been declining, and is limited, both geographically and in terms of output (25).

Production

Cumulative production in countries outside the former USSR, Eastern Europe, and China since the late 1930s has totaled about 1 $\times$ 10^6 t U. A majority of this production came from the United States, Canada, Germany, Namibia, Niger, and South Africa. In addition, some 218,500 t U, 102,245 t U, 16,700 t U, and 16,850 t U have been produced, respectively, in the former GDR, former Czechoslovakia, Hungary, and Romania. It is estimated that about 72,000 t U have been produced in Kazakhstan. Reliable cumulative production data for the rest of the CIS, other Eastern European countries, and China, however, are not available (26).

Uranium production in 1992 of 36,246 t U was only about 63% of world reactor requirements of 57,182 t U; the remainder, 20,950 t U, was met from inventory drawdown. The worldwide production shortfall has developed since 1990 when production exceeded reactor requirements by about 1000 t U (27).

World reactor-related requirements are expected to increase from 57,182 t U in 1992 to about 75,673 t U by the year 2010. Some utilities are expected to continue to meet their requirements by purchasing or drawing on excess inventory. Annual uranium production should remain below actual requirements until some target level of stocks is reached (27).

Capability. The projected production capability over the period 1993 through 2010 is about 720,500 t U, about 61% of the projected total demand over the same period. The total annual production capability is expected to reach a maximum of about 48,200 t U in the year 2000, when projected total world demand should be about 63,500 t U. The total annual production capability is then projected to fall steadily to about 30,500 t U in 2010 (27).

Supply Projections. Additional supplies are expected to be necessary to meet the projected production shortfall. A significant contribution is likely to come from uranium production centers such as Eastern Europe and Asia, which are not included in the capability projections (27). The remaining shortfall between fresh production and reactor requirements is expected to be filled by several alternative sources, including excess inventory drawdown. These shortfalls could also be met by the utilization of low cost resources that could become available as a result of technical developments or policy changes, production from either low or higher cost resources not identified in production capability projections, recycled material such as spent fuel, and low enriched uranium converted from the high enriched uranium (HEU) found in warheads (28).

Once all technical and political problems are resolved, reactor-grade uranium produced from HEU warhead material could contribute significantly to meeting the anticipated fresh uranium production shortfall. This source, however, is not expected to have a significant impact until the year 2000 or later. The discovery of new low cost resources is not expected to make a significant contribution to production until after the year 2005 because of the very low level of uranium exploration and the relatively long lead times required to develop new production centers (29).

Demand. The demand for uranium in the commercial sector is primarily determined by the requirements of power reactors. At the beginning of 1993, there were 424 nuclear power plants operating worldwide, having a combined capabity of about 330 GWe. Moderate but steady growth is projected for nuclear capacity to the year 2010. The capacity in 2010 is expected to be about 446 GWe (29).

World annual uranium requirements in 1993 were estimated at about 58,382 t natural uranium equivalent. Reactor-related requirements are expected to rise about 1015 t/yr on the average, reaching 75,700 t U total requirements in the year 2010. The cumulative aggregate world uranium requirements for the period 1993–2010 are estimated to be about 1.185×10^6 t U metal (29).

Economic Aspects

The most significant economic aspect of the uranium market in 1991 and 1992 was oversupply. Excess inventory, as well as uranium produced by nontraditional suppliers to the world market led to all uranium market prices decreasing. During 1992 the average annual spot market price, as indicated by the Nuexco Exchange Value (NEV), reached an all-time low of U.S. \$20.67/kg U (U.S. \$7.95/lb U_3O_8) (30). The price of uranium on the spot market in October 1994 had increased to U.S. \$23.53/kg U (U.S. \$9.05/lb U_3O_8). By April 1995 the price was U.S. \$29.64/kg U (U.S. \$11.40/lb U_3O_8) (31).

Alternative Sources

Low Grade Resources. *Seawater.* The world's oceans contain ca 4×10^9 t of uranium (32). Because the uranium concentration is very low, approximately 3.34 ppm, vast amounts of water would be required to recover significant amount of uranium metal, ie, 10^6 m^3 of seawater for each metric ton of U. Significant engineering development and associated environmental concerns have limited the development of an economic means of uranium extraction from seawater (32) (see OCEAN RAW MATERIALS).

Mill Tailings. Recovery efficiency in ore processing is not 100%. Accumulated mill tailings contain tens of thousands of metric tons of uranium. Whereas improved techniques have been employed to recover some of this uranium, the recovery rate is still generally low, approximately 35–50%, owing to metallurgical problems and economic considerations (33).

HEU De-Enrichment. Highly enriched uranium (HEU), initially enriched to $>93\%$ ^{235}U, for use in research, naval reactors, and nuclear weapons, may be de-enriched and fabricated into fuel for civilian nuclear reactors. An estimate of the world inventory of highly enriched uranium in the nuclear weapons states is provided in Table 6 (34).

An agreement between the United States and Russia led to a commitment in 1994 by the United States to buy 500 metric tons of Russian HEU, which has been converted to low enriched uranium (LEU). The HEU must come from dismantled nuclear weapons before it is converted to LEU. The sale of converted HEU to the United States is to be carried out on a timetable in which no less than 10 t are to be converted in each of the first five years of the agreement and no less than 30 t in each year thereafter (35). In all, the agreement would last for 20 years if only these minimums were sold each year.

De-enrichment of HEU from approximately 93% ^{235}U to 3% ^{235}U can be accomplished using the depleted tails from the original enrichment process. These tails contain on the average 0.20% ^{235}U. The de-enrichment of 1 t of HEU

Table 6. Estimates of HEU, t of U[a]

Country	HEU	Percent of total
China	15	1.15
France	15	1.15
United Kingdom	10	0.76
CIS	720	54.96
United States	550	41.98
Total	*1310*	*100.00*

[a]Ref. 34.

uses 32 t of tails, yielding approximately 33 t of fuel having an enrichment of 3% ^{235}U. Producing the same amount of 3% enriched uranium from natural sources would require approximately 180 t of natural uranium metal. Therefore, 1 t of HEU is equivalent to 180 t of natural uranium.

The amount of HEU that becomes available for civilian use through the 1990s and into the twenty-first century depends on the number of warheads removed from nuclear arsenals and the amount of HEU in the weapons complex that is already outside of the warheads, ie, materials stockpiles and spent naval reactor fuels. An illustrative example of the potential amounts of weapons-grade materials released from dismantled nuclear weapons is presented in Table 7 (36). Using the data in Table 7, a reduction in the number of warheads in nuclear arsenals of the United States and Russia to 5000 warheads for each country results in a surplus of 1140 t of HEU. This inventory of HEU is equivalent to 205,200 t of natural uranium metal, or approximately 3.5 times the 1993 annual demand for natural uranium equivalent.

Toxicology of Uranium

The two primary effects associated with the introduction of uranium species into the human body are the development of cancer, primarily from radiation-induced tissue damage in the lung, and renal damage accompanied by possible kidney failure owing to uranium ingestion (37). Soluble uranium compounds affect the respiratory system, liver, blood, lymphatics, kidneys, skin, and bone marrow. The insoluble uranium compounds affect the skin, bone marrow, and lymphatics. Lung cancer incidence has been noted to increase in uranium mine workers, especially those who smoke. Acute chemical toxicity produces damage primarily to the kidneys in the form of necrosis of the renal tubular epithelium, leading ultimately to kidney failure in acute cases (38).

The U.S. Nuclear Regulatory Commission (NRC) regulates the protection of the health and safety of the public by issuing standards (39). In addition, the U.S. Environmental Protection Agency (EPA) issues drinking water regulations (40) that address radioactive contamination of drinking water supplies. The maximum allowable effluent concentrations for release of uranium to air, water, and sewer systems depends on the individual isotope of uranium. Concentration limits for the naturally occurring uranium isotopes, ^{234}U, ^{235}U, and ^{238}U, range from 1.11×10^{-7} to 1.85×10^{-9} Bq/mL (3×10^{-12} to 5×10^{-14} μCi/mL) for releases to air to 0.011 Bq/mL (3×10^{-7} μCi/mL) for releases to water. The EPA regulations for uranium in drinking water limit the concentration to levels causing 40 μSv (4 mrem) total body or organ dose equivalents per year (41). These doses are calculated on the basis of 2 L/d drinking water intake and result in a maximum uranium concentration of about 3 μg/L (38).

Handling of soluble uranium compounds requires appropriate clothing to prevent skin contact and eye protection to prevent any possible eye contact. Protective clothing requirements for insoluble uranium compounds should prevent repeated or prolonged skin contact. Eye protection for use in handling insoluble uranium compounds should prevent any possibility of eye contact. Respirators should always be worn to prevent inhalation of uranium dust, fumes, or gases (38).

Table 7. HEU Available from Dismantled Warheads, t[a,b]

Number of warheads per arsenal	United States			Russia			U.K., France, and China			
	A	B	Percent of total	A	B	Percent of total	A	B	Percent of total	Total, column B
30,000		265	49	40	255	47		20	4	540
20,000		265	38	205	420	59		20	3	705
15,000	60	325	39	280	495	59		20	2	840
10,000	135	400	40	355	570	58		20	2	990
5,000	210	475	42	430	645	56		20	2	1,140
0	285	550	42	505	720	55	20	40	3	1,310

[a]Ref. 36.
[b]A = weapon-grade uranium that would be released from warheads after reducing to warhead number in left-hand column; B = value in column A plus weapon-grade uranium already held outside warheads.

BIBLIOGRAPHY

"Nuclear Fuel Reserves" under "Nuclear Reactors" in *ECT* 3rd ed., Vol. 16, pp. 143–150, by J. A. Patterson, U.S. Department of Energy.

1. *World Nuclear Capacity and Fuel Cycle Requirements 1993*, Energy Information Administration, U.S. Department of Energy, DOE/EIA-0436(93), Washington, D.C., Nov. 1993, p. 81.
2. *Uranium Resources, Production and Demand*, Joint Report of OECD Nuclear Energy Agency and International Atomic Energy Agency, OECD Publications Service, Paris, 1994, p. 17.
3. *Uranium Resources, Production and Demand*, Joint Report of OECD Nuclear Energy Agency and International Atomic Energy Agency, OECD Publications Service, Paris, 1992.
4. *Uranium Resources, Production and Demand*, Joint Report of OECD Nuclear Energy Agency and International Atomic Energy Agency, OECD Publications Service, Paris, 1990.
5. *Uranium Resources, Production and Demand*, Joint Report of OECD Nuclear Energy Agency and International Atomic Energy Agency, OECD Publications Service, Paris, 1988.
6. *Uranium Resources, Production and Demand*, Joint Report of OECD Nuclear Energy Agency and International Atomic Energy Agency, OECD Publications Service, Paris, 1986.
7. *Uranium Industry Annual 1992*, Energy Information Agency, U.S. Department of Energy, DOE EIA-0478(92), Washington, D.C., Oct. 1993, p. 30.
8. Ref. 2, p. 15.
9. Ref. 2, p. 16.
10. K. H. Wedepohl, ed., *Handbook of Geochemistry*, Vol. II-5, Springer-Verlag, New York, 1969, pp. 92-D-1, 92-D-4.
11. *Formation of Uranium Deposits, Proceedings of IAEA Symposium, Athens, Greece, May 6–10, 1974*, International Atomic Energy Agency STI/PUB/394, UNIPUB, Inc., New York, pp. 142–145.
12. Ref. 2, p. 295.
13. Ref. 2, p. 20.
14. Ref. 2, pp. 295–296.
15. Ref. 2, p. 296.
16. Ref. 2, p. 297.
17. Ref. 2, p. 298.
18. Ref. 2, pp. 24–29.
19. Ref. 2, p. 35.
20. Ref. 2, p. 267.
21. Ref. 2, pp. 24, 25, 28.
22. Ref. 2, p. 27.
23. Ref. 2, p. 29.
24. Ref. 2, p. 30.
25. Ref. 2, p. 10.
26. Ref. 2, p. 11.
27. Ref. 2, p. 12.
28. F. Von Hippel, M. Miller, H. Feiverson, A. Diakov, and F. Berkhout, *Sci. Amer.*, 44–49 (Aug. 1993).
29. Ref. 2, p. 13.
30. Ref. 2, p. 9.
31. *Nucl. News*, 17 (May 1995).

32. M. Benedict, T. Pigford, and H. Levi, *Nuclear Chemical Engineering*, McGraw-Hill Book Co., Inc., New York, 1981, pp. 261–264.
33. *Fuel and Heavy Water Availability*, Report of Working Group 1, International Nuclear Fuel Cycle Evaluation, Vienna, Austria, International Atomic Energy Agency STI/PUB/534, UNIPUB, Inc., New York, 1980, pp. 174–175.
34. D. Albright, F. Berkhout, and W. Walker, *World Inventory of Plutonium and Highly, Enriched Uranium 1992*, Oxford University Press, Oxford, U.K., 1993, p. 198.
35. P. Passell, *New York Times*, A1, C5 (June 8, 1994).
36. Ref. 34, p. 208.
37. M. O. Amdur, J. Doull, and C. D. Klaassen, eds., *Casarett and Doull's Toxicology: The Basic Science of Poisons*, Fourth ed., Pergamon Press, New York, 1991, pp. 671–672.
38. E. Hodgson, R. B. Mailman, and J. E. Chambers, eds., *Macmillan Dictionary of Toxicology*, Macmillan Press Ltd., London, 1988, pp. 909–911.
39. *Code of Federal Regulations*, Title 10, Part 20, Standards for Protection Against Radiation, Washington, D.C., 1995, Appendix B.
40. *Code of Federal Regulations*, Title 40, Part 141, National Primary Drinking Water Regulations, Washington, D.C., 1995.
41. *Code of Federal Regulations*, Title 40, Part 141.16(b), Washington, D.C., 1995.

DANIEL B. BULLEN
Iowa State University

WATER CHEMISTRY OF LIGHTWATER REACTORS

As of 1994 there were 105 operating commercial nuclear power stations in the United States (1) (see POWER GENERATION). All of these facilities were light, ie, hydrogen–water reactors. Seventy-one were pressurized water reactors (PWRs); the remainder were boiling water reactors (BWRs).

In a PWR, a closed circuit of high pressure, high temperature water transfers heat from the reactor core to once-through or recirculating U-tube steam generators, as shown in Figure 1 (see HEAT-EXCHANGE TECHNOLOGY). The steam (qv), which is produced on the secondary side of the steam generator, is used to drive a turbine generator. In contrast to fossil-fired steam-generating equipment, steaming in a PWR occurs on the outside of the boiler tubes, ie, the secondary side, where a large number of tube-to-tube support plate and tube-to-tubesheet crevices exist. Extensive corrosion has been observed in such crevices and beneath sludge piles on the tube–tube support plates and tubesheet (2). Control of secondary chemistry is critical if high concentration of aggressive chemicals and accelerated local corrosion are to be avoided (see CORROSION AND CORROSION CONTROL; WATER, INDUSTRIAL WATER TREATMENT).

Boron, in the form of boric acid, is used in the PWR primary system water to compensate for fuel consumption and to control reactor power (3). The concentration is varied over the fuel cycle. Small amounts of the isotope lithium-7 are added in the form of lithium hydroxide to increase pH and to reduce corrosion rates of primary system materials (4). Primary-side corrosion problems are much less than those encountered on the secondary side of the steam generators.

In a BWR (Fig. 2) steam is generated in the reactor core and used directly to drive a turbine generator. Steam generators are not employed in modern BWRs, although several early units were furnished with such equipment as a source of low pressure steam. Radiation levels during operation near the turbine, condenser, and feedwater heaters are higher than in a PWR as a result of steam transport of short-lived activation products. Other than this consideration, the power-generating cycles of PWRs and BWRs are reasonably similar. No reactivity or pH control additives are used in BWRs. As a result, the corrosion behavior of the materials of construction is dependent primarily on coolant oxygen concentrations which are governed by the radiolytic decomposition rates of water and steam–water equilibrium relations. In some cases, these parameters are controlled by the addition of dissolved hydrogen to the feedwater entering the reactor vessel.

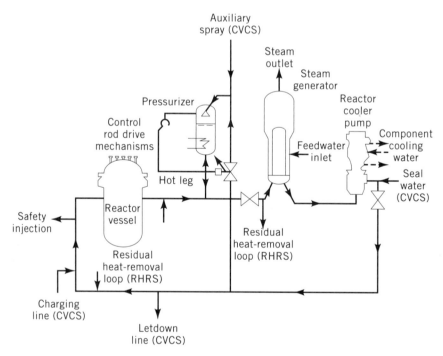

Fig. 1.   Pressurized water reactor (PWR) coolant system having U-tube steam generators typical of the 3–4 loops in nuclear power plants. PWR plants having once-through steam generators contain two reactor coolant pump-steam generator loops. CVCS = chemical and volume-control system.

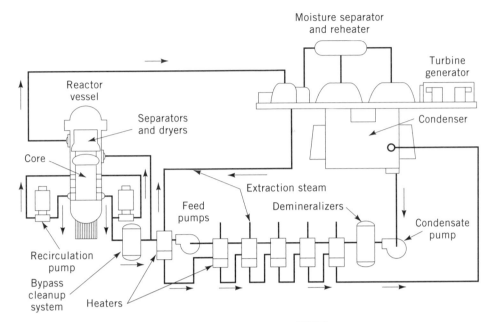

Fig. 2. Boiling water reactor (BWR) system.

Pressurized Water Reactors

Primary System. In a PWR, reactor coolant is circulated continuously at ca 300°C and 15 MPa (150 bar). Austenitic stainless steel is employed for the main system piping, or cladding, reactor vessel cladding, pressurizer cladding, pumps (qv), valves, and auxiliary piping. The steam generator tubing used in modern U.S. nuclear facilities is Alloy 600, ca 75 wt % Ni, 8 wt % Fe, and 16 wt % Cr. However, most of the tubing in replacement units is Alloy 690, ca 55 wt % Ni, 8 wt % Fe, and 30 wt % Cr. The steam generator tubing accounts for ca 75% of the total area exposed to the primary coolant and Zircaloy 4 and stainless steel account for ca 20 and 5%, respectively (5).

The primary water specifications for a PWR are given in Table 1 (4). Rigid controls are applied to the primary water makeup to minimize contaminant ingress into the system. In addition, a bypass stream of reactor coolant is processed continuously through a purification system to maintain primary coolant chemistry specifications. This system provides for removal of impurities plus fission and activated products from the primary coolant by a combination of filtration (qv) and ion exchange (qv). The bypass stream also is used both to reduce the primary coolant boron as fuel consumption progresses, and to control the ^7Li concentrations.

Oxygen is a prime factor in the corrosion of system materials and the release, activation, and redeposition of activated corrosion products. Dissolved hydrogen is maintained to promote rapid recombination of the oxygen whether radiolytically formed or introduced into the coolant from other sources, thereby minimizing corrosion rates.

Table 1. PWR Primary System Specifications[a,b]

Parameter[c]	Normal value
O_2, ppb	<5
H_2,[d] mL/kg H_2O	25–50
Cl, ppb	<50
F, ppb	<50
SO_4, ppb	<50
suspended solids, ppb	<350
SiO_2, ppb	e

[a] Ref. 4.
[b] Values are for normal power operation. Conductivity, pH, and concentrations of lithium and boron are plant specific and vary over the fuel cycle according to the control scheme used. See Fig. 3.
[c] ppb = parts per billion.
[d] At STP.
[e] Monitored in fresh makeup water. Should be <100 ppb.

The quantity of boric acid maintained in the reactor coolant is usually plant specific. In general, it ranges from ca 2000 ppm boron or less at the start of a fuel cycle to ca 0 ppm boron at the end. Most plants initially used 12-month fuel cycles, but have been extended to 18- and 24-month fuel cycles, exposing the materials of construction of the fuel elements to longer operating times. Consequently concern over corrosion problems has increased.

The reactor coolant pH is controlled using lithium-7 hydroxide [72255-97-1], ^7LiOH. Reactor coolant pH at 300°C, as a function of boric acid and lithium hydroxide concentrations, is shown in Figure 3 (4). A pure boric acid solution is only slightly more acidic than pure water, 5.6 at 300°C, because of the relatively low ionization of boric acid at operating primary temperatures (see BORON COMPOUNDS). Thus the presence of lithium hydroxide, which has a much higher ionization, increases the pH ca 1–2 units above that of pure water at operating temperatures. This leads to a reduction in corrosion rates of system materials (see HYDROGEN-ION ACTIVITY).

The pH control scheme is plant specific. It is selected on the basis of impacts on fuel and materials integrity and radiation field control. The pH at 300°C should be maintained above 6.9, and from a radiation control point of view, at 300°C a pH up to 7.4 is desirable. However, the high lithium values, required for pH values between 6.9 and 7.4 at 300°C, during the initial part of the fuel cycle, have the potential of accelerating the corrosion of Zircaloy fuel cladding and of increasing the susceptibility of mill-annealed Alloy 600 tubing used in some recirculating steam generators (RSGs) to primary-side stress corrosion cracking (4,6,7).

In some plants high silica, SiO_2, levels (up to the ppm range) have been found in the primary water. The source is the boroflex material used in the fuel storage racks in the spent fuel pool. The SiO_2 leaches into the pool water, then finds its way into the primary system when the primary water comes in contact with the pool water during refueling and similar maintenance. The SiO_2 can only be effectively removed by reverse osmosis (qv).

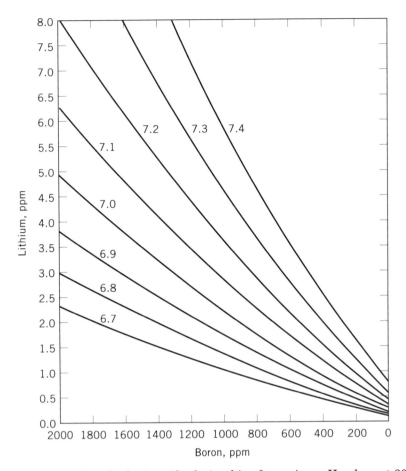

Fig. 3. Lithium hydroxide–boric acid relationships for various pH values at 300°C (4).

A schematic of the radiation contamination for a PWR is shown in Figure 4. Whereas numerous attempts have been made to model this process, there has been only limited success (8–11). Trends in the radiation field regarding time for RSG channel heads are shown in Figure 5 (12). The fields vary over a wide range, but reach a peak between four and six years. The two principal nuclides leading to out-of-core shutdown radiation levels are cobalt-58 [13981-38-9], [58]Co, and cobalt-60 [10198-40-0], ^{60}Co, (4,5) (Table 2). Studies show that electropolishing of the plenum surfaces of RSGs before they go into service can reduce the deposition rate of cobalt, and consequently, the rate of radiation buildup (13). The electropolishing smoothes the microscopic surfaces of the plenums and reduces the sites where the cobalt can deposit.

Secondary System. The water quality specifications for the feedwater and blowdown water in a recirculating steam generator (RSG) and the feedwater for a once-through steam generator (OTSG) are given in Table 3 (14).

Recirculating Steam Generator. The corrosion performance of many RSGs in commercial power stations in the United States has been marginal (2). Many tube bundles have had to be replaced. Many tubes have been plugged or sleeved with inserts as a result of excessive corrosion on the secondary side.

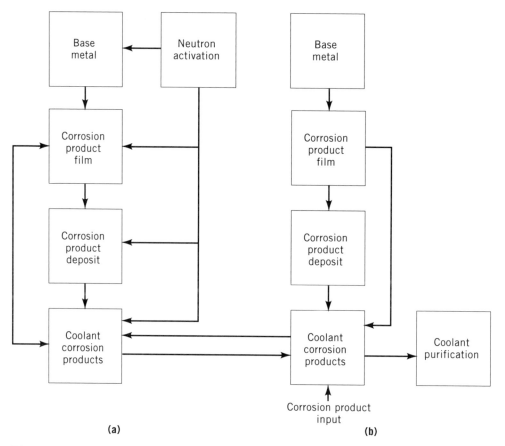

Fig. 4. Contamination process schematic for PWR where (**a**) is inside the core and (**b**), outside the core.

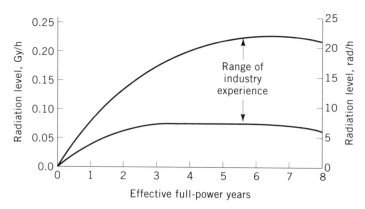

Fig. 5. RSG channel head shutdown radiation fields (12).

Table 2. Corrosion Products Contributing to PWR Shutdown Radiation Levels[a]

			Gamma rays		
Isotope	Half-life	Source	Energy, MeV	Abundance, %	Disintegration, MeV
^{58}Co	71 d	$58_{Ni(n,p)}$	0.511	0.3	0.98
			0.810	0.99	
			0.865	0.01	
			1.67	0.01	
^{60}Co	5.27 yr	$59_{Co(n,\gamma)}$	1.17	1.0	2.5
			1.33	1.0	

[a]Ref. 3.

Table 3. PWR Steam Generator Water Specifications[a,b]

		Recirculating	
Parameter[c]	Once-through feedwater	Feedwater[d]	Blowdown water
components, ppb			
N_2H_4	20[e]	<100	
O_2	≤3	<5	
Na	≤3		ca 2
Cl	≤5		ca 2
SiO_2	≤10		
total Fe	≤5	<5	
total Cu	≤1	<1	
SO_4	≤3		ca 3
other	f	g	g,h
cation conductivity			
at 25°C, μS/cm			
corrected[i]	≤0.2	≤0.2	
ammonia AVT[j]			0.15
nonammonia AVT[j]			0.5

[a]Ref. 14.
[b]Values correspond to normal power operation.
[c]ppb = parts per billion.
[d]The pH is plant specific, depending on additive used and secondary system materials. Feedwater generally should be equivalent to pH = 9.3 at 25°C for ammonia and carbon steel equipment.
[e]Value given is minimum.
[f]Organics related to use of amines for pH control. Value is plant specific.
[g]Boron related to boric acid treatment of ~5–10 ppm B in feedwater.
[h]For caustic crevice environment, a plant-specific chemical impurity molar ratio <0.5 is defined, eg, Na:Cl molar ratio <0.5.
[i]Corrected for the presence of organics such as acetates and formates.
[j]AVT = all-volatile treatment.

Sodium phosphate treatment was originally used for pH control and conditioning of boiler water (15). Initially, a boiler water concentration range, but not a sodium-to-phosphate (Na:PO_4) molar ratio, was specified. Some utilities opted to control the phosphates at various levels within the allowable range (16,17). During condenser in-leakage, the phosphate and calcium ion reactions depleted the steam generator water of soluble phosphate, increasing the Na:PO_4 molar ratio and forming free caustic. This resulted in an increased rate of Alloy 600 steam generator tubing stress corrosion cracking. Although recommended phosphate control parameters were modified, corrosion attack continued. For example, wastage or local wall thinning was observed. An all volatile treatment (AVT) employing ammonia (qv) was then used for pH control of the feedwater. As of this writing (ca 1995) the treatment of choice is AVT (Table 3).

Shortly after the conversion to AVT, inward deformation of the Alloy 600 tubing at the tube-to-tube support plate interface region was detected. This phenomenon, commonly referred to as denting (18), results from excessive corrosion of carbon steel in the crevice region between the tube and the tube support plate; as denting progresses, tube support plate cracking can occur. The tube stress corrosion cracking also continued. Newer support plate materials and modified support plate designs were used in later units and in replacement units. In denting, the accelerated corrosion of the carbon steel tube support appears to result from the ingress of chlorides and other contaminants, such as O_2 and Cu^{2+}

Stress corrosion cracking, prevalent where boiling occurs, concentrates corrosion products and impurity chemicals, namely in the deep tubesheet crevices on the hot side of the steam generator and under deposits above the tubesheet. The cracking growth rates increase rapidly at both high and low pH. Either of these environments can exist depending on the type of chemical species present.

Possible remedial and preventive actions are as follows (14). (1) Use of elevated hydrazine [302-01-2], N_2H_4, treatment where hydrazine in the range of several hundred ppb is added to the feedwater thus producing a reducing atmosphere in the steam generator. (2) Boric acid treatment where boric acid in the range of 5–10 ppm B is added to the feedwater. This reduces the growth rate of corrosion products that cause denting and possibly helps arrest tube stress corrosion cracking. (3) Alternative amine treatment where, for example, morpholine [110-91-8] or ethanolamine [141-43-5] (ETA), is added in place of ammonia to the feedwater to produce a more uniformly basic environment throughout the secondary system. The transport of corrosion products into the steam generators is thus reduced. (4) Molar ratio management where the ratio of cations to anions in the steam generator water is controlled reducing the possibility of forming environments in boiling crevices and underdeposit areas which are conducive to accelerating the stress corrosion cracking (SCC) of Alloy 600. For caustic SCC environments the desired molar ratio, eg, Na:Cl, is <0.5.

Before any remedial or preventive actions are implemented, an evaluation should be conducted as to applicability to the specific plant. The evaluation should continue while the actions are in progress. The main action should be to take measures to reduce the ingress of contaminants into the steam generator by using more reliable materials, such as in the condenser tubes, to reduce leakage. Contaminant control equipment, such as full-flow condensate demineralizers, should also be employed.

Once-Through Steam Generator. The corrosion of OTSGs has not been as extensive or as serious a problem as that of RSGs. Design and operating differences between these systems may be responsible for the corrosion differences. For example, the main fraction of impurities entering the OTSG via the feedwater is transported from the OTSG by the steam. In RSG designs, impurities concentrate in the bulk water. Moreover, full-flow condensate demineralizers, which significantly reduce impurity ingress into the system, especially from condenser tube leaks, are employed at all OTSG PWRs. Condensate demineralizers are not employed at most RSG PWRs.

Feedwater quality control is the main method for controlling the chemistry environment in OTSGs (Table 3). Whereas the all-volatile treatment (AVT) utilizing ammonia was used first, a switch has been made to morpholine or other amines for better distribution in the steam plant (14). The amount of corrosion products (mainly iron) carried into the OTSGs has been greatly reduced.

Most of the Alloy 600 outer diameter tube corrosion has occurred in the region of the upper tubesheet near the open lane, ie, an untubed lane across the middle of the steam generator (16,17). The steam carries entrained droplets of water through the open lane to the upper tubesheet region where the droplets dry out and concentrate the chemicals. Long tube inserts have been used to sleeve tubes in this region where wall defects have been detected.

OTSGs also experience deposition of material on the flow areas in the tube support plates which causes an increase in pressure drop and eventual reductions in plant power production.

The possible remedial and preventive actions are hot soaks and drains during cooldown to help remove soluble deposited material, chemical cleaning to remove corrosion products and reduce the pressure drop (see METAL SURFACE TREATMENTS), and reduced corrosion product transport into OTSG using amines other than ammonia in feedwater (14).

Boiling Water Reactors

BWRs operate at ca 7 MPa (70 bar) and 288°C. Some of the coolant passing through the core is converted into steam which is separated from the water with equipment inside the reactor vessel (Fig. 2). The steam goes to the turbine generator while the water is recirculated back to the bottom of the core. A side stream is continuously purified using demineralizers and filters to control the water quality of the reactor water. Full-flow condensate demineralizers are also used to control the ingress of impurities into the reactor water from the steam plant.

The BWR water chemistry parameters are given in Table 4 (19). Originally, no additives were made to feedwater–condensate or the primary water. The radiolytic decomposition of the fluid produced varying concentrations of O_2 in the reactor vessel, ranging from about 200 ppb O_2 in the reactor recirculation water to about 20 ppm O_2 in the steam. Stoichiometric amounts of hydrogen were also produced, ie, 2 mL H_2 for each mL of O_2. Feedwater O_2 was about 30 ppb, hence the radiolytic decomposition of the water was a primary factor in determining the behavior of materials in the primary system and feedwater systems.

Some of the earlier BWR units had feedwater heaters having copper alloy tubes. The environment of high oxygen and neutral pH water led to high copper concentrations in the feedwater and to undesirable deposits on the fuel and inlet fuel nozzles (20). In some instances, the copper deposits resulted in an increase in core pressure drop and necessitated plant power reduction. The copper alloys

Table 4. BWR Normal Water Chemistry Values[a,b]

Parameter[c]	Median value
Reactor water	
conductivity at 25°C, μS/cm	ca 0.11
Cl, ppb	ca 1
SO_4, ppb	ca 2
Zn, ppb	$5-10^d$
electrochemical potential, V	[e]
O_2 in recirculation water, ppb	[f]
SiO_2, ppb	<100
Feedwater/condensate	
feedwater conductivity at 25°C, μS/cm	0.06
feedwater total Fe, ppb	ca $1-3^g$
feedwater total Cu, ppb	ca $0.1-0.2^g$
O_2, ppb	ca 30

[a]Ref. 19.
[b]Values given correspond to normal power operation.
[c]ppb = parts per billion.
[d]Consistent with plant program for zinc injection.
[e]Corrective action should be initiated when value is > -0.23 V against the standard hydrogen electrode (SHE). Plant-specific values should be established for protection of stainless steels and nickel-based critical components.
[f]Plant specific. In range of 200 ppb without H_2 addition and in range of 10 ppb with H_2 addition.
[g]Values depend on plant-specific design and operating procedures for steam plant.

were eliminated from the feedwater system in subsequent plants and most existing plants.

Many instances of intergranular stress corrosion cracking (IGSCC) of stainless steel and nickel-based alloys have occurred in the reactor water systems of BWRs. IGSCC, first observed in the recirculation piping systems (21) and later in reactor vessel internal components, has been observed primarily in the weld heat-affected zone of Type 304 stainless steel.

Three factors are important in such cracking: total stresses, sensitization of heat-affected weld zones, and chemical species such as oxygen and hydrogen peroxide in the water. In some cases for recirculation piping, induction heating of welds subsequent to fabrication has been used to change the stress of the inner pipe wall from tensile to compressive. This change reduces the likelihood of IGSCC. Type 316NG stainless steel has been used in replacement and repair situations to eliminate the tendency for sensitization.

Laboratory experiments have shown that IGSCC can be mitigated if the electrochemical potential (ECP) could be decreased to -0.230 V on the standard hydrogen electrode (SHE) scale in water with a conductivity of 0.3 μS/cm (22). This has also been demonstrated in operating plants. Equipment has been developed to monitor ECP in the recirculation line and in strategic places such as the core top and core bottom, in the reactor vessel during power operation.

The desired ECP value is not obtainable unless H_2 is injected into the feedwater. For example, using no hydrogen addition, the recirculation line O_2 is in the range of 200 ppb and the corresponding ECP is in the vicinity of zero volts. The amount of H_2 required is plant specific. Experiments at one BWR indicated that the required feedwater H_2 concentration ranged from 2.4 ppm in the recirculation line, to 1.4 and 1.9 in the core bottom and top, respectively (23). Measurements at the same relative location for several plants indicate that required H_2 ranges between 0.4 and 1.6 ppm (22). Some plants use material specimens inside the reactor vessel and recirculation line to monitor the effectiveness of the ECP control by measuring the crack growth rate.

Whereas addition of hydrogen to feedwater helps solve the O_2 or ECP problem, other complications develop. An increase in shutdown radiation levels and up to a fivefold increase in operating steam plant radiation levels result from the increased volatility of the short-lived radioactive product nitrogen-16, ^{16}N, (7.1 s half-life) formed from the coolant passing through the core. Without H_2 addition, the ^{16}N in the fluid leaving the reactor core is in the form of nitric acid, HNO_3; with H_2 addition, the ^{16}N forms ammonia, NH_3, which is more volatile than HNO_3, and thus is carried over with the steam going to the turbine.

Figure 6 (24) is a flow diagram of radioactive contamination in BWR. Although iron is the main corrosion product entering the reactor water, as for PWRs, ^{60}Co is the long-term source of shutdown radiation in BWRs. Some operating BWRs were noted to have shutdown radiation levels substantially lower than those of other plants having similar operating conditions and water chemistry. The common denominator for these low radiation levels was the use of tubes containing zinc in the condensers. Confirmation was made via subsequent studies where zinc oxide was added to the feedwater to produce zinc concentrations in the range of 5–10 ppb in the reactor water (Fig. 7) (12). It appears that whereas the Zn inhibits the deposition of cobalt, there are also some side effects. Zinc addition produces ^{65}Zn hot spots within the plant. Zn-65 has a 244 d half-life.

The cobalt deposition rate on new, replacement, or decontaminated recirculation piping surface has been reduced by pretreating the piping using an atmosphere of oxygenated wet steam to form an oxide film (25). Studies have been conducted for both PWRs and BWRs to reduce the cobalt content of materials used in the nuclear parts of the plants, particularly in hardened and wear surfaces where cobalt-base alloys ($\approx 50\%$ Co) are used (26). Some low cobalt materials have been developed; however, the use of the materials is limited to replacement parts or new plants.

Economic Aspects

Water chemistry is important to the safe and reliable operation of a nuclear power plant. Improper conditions can lead to equipment and material failures which in turn can lead to lengthy unscheduled shutdown periods for maintenance (qv) and repair operations. Water chemistry can also have an impact on the radiation levels during both power operations and shutdown periods. These affect the ability of personnel to perform plant functions.

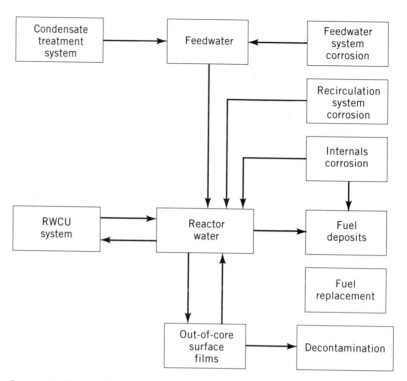

Fig. 6. Impurity flow paths of BWR radioactive contamination (24). RWCU = reactor water cleanup system.

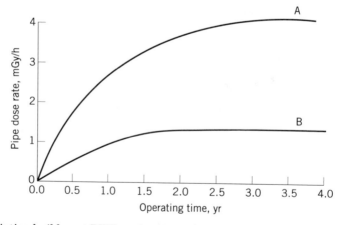

Fig. 7. Radiation buildup at BWRs using normal water chemistry, A, and zinc additions, B (12). To convert mGy to mrad, multiply by 100.

BIBLIOGRAPHY

"Chemical Technology of Nuclear Reactors" under "Nuclear Reactors" in *ECT* 2nd ed., Vol. 14, pp. 75–85, by R. N. Lyon, Oak Ridge National Laboratory; "Water Chemistry of Light-Water Reactors" under "Nuclear Reactors" in *ECT* 3rd ed., Vol. 16, pp. 150–161, by S. G. Sawachka and W. L. Pearl, NWT Corp.

1. "World Nuclear Performance," *McGraw-Hill Nuclear Publications*, Vol. 6, McGraw-Hill Book Publishing Co., New York, pp. III-7–III-9.
2. "Steam Generator Performance Degradation," Report NP-7524, Electric Power Research Institute, Palo Alto, Calif., Sept. 1991.
3. P. Cohen, *Water Coolant Technology of Power Reactors*, American Nuclear Society, LaGrange Park, Ill., 1980.
4. "PWR Primary Water Chemistry Guidelines: Revision 2," Report NP-7077, Electric Power Research Institute, Palo Alto, Calif., Nov. 1990.
5. "PWR Primary Shutdown and Startup Chemistry Guidelines," Report TR101884, Electric Power Research Institute, Palo Alto, Calif., 1993.
6. "Stress Corrosion Cracking of Alloy 600," Report NP-2114-SR, Electric Power Research Institute, Palo Alto, Calif., Nov. 1981.
7. "Proceedings: 1987 EPRI Workshop on Mechanisms of Primary Water Intergranular Stress Corrosion Cracking," Report NP-5987M, Electric Power Research Institute, Palo Alto, Calif., Sept. 1988.
8. "The CORA II Model of PWR Corrosion-Product Transport," Report NP-4246, Electric Power Research Institute, Palo Alto, Calif., 1985.
9. S. M. Ali, "An Updated Version of Computer Code CORA II for Estimation of Corrosion Product Mass and Activity Migration in PWR Primary Circuits and Related Experimental Loops," *Fourth International Conference on Water Chemistry of Nuclear Systems*, Bournemouth, U.K., Oct. 1986, pp. 107–109.
10. M. Metge, P. Beslu, and A. Lalet, "Cobalt Sources in PWR Primary Systems—Pactole Prediction," in Ref. 9, pp. 71–74.
11. "Evaluation of Cobalt Sources in Westinghouse-Designed Three- and Four Loop Plants," Report NP-2681, Electric Power Research Institute, Palo Alto, Calif., Oct. 1983.
12. "Progress in Radiation Control Technology," Report NP-6708, Electric Power Research Institute, Palo Alto, Calif., Feb. 1990.
13. R. A. Shaw, H. Ocken, and C. J. Wood, "Recent Progress in LWR Radiation Field Control," in Ref. 9, pp. 49–55.
14. "PWR Secondary Water Chemistry Guidelines—Revision 3," Report TR-102133, Electric Power Research Institute, Palo Alto, Calif., 1993.
15. W. D. Fletcher and L. F. Picone, *Proceedings of the 33rd International Water Conference*, Pittsburgh, Pa., 1971, p. 151.
16. R. Garney, *Nuclear Energy* **18**, 117 (1979).
17. S. J. Green and J. P. N. Paine, *Nuclear Technol.* **55**, 10 (1981).
18. "Causes of Denting," Report NP-3275, Electric Power Research Institute, Palo Alto, Calif., Dec. 1883.
19. "BWR Water Chemistry Guidelines—1993 Revision, Normal and Hydrogen Water Chemistry," Report TR-103515, Electric Power Research Institute, Palo Alto, Calif., Feb. 1993.
20. R. Gasparini and E. Ioannilli, *Proc. Am. Power Conf.* **33**, 776 (1971).
21. "Proceedings: Seminar of Countermeasures for Pipe Cracking in BWRs," Report WS-79-174, Electric Power Research Institute, Palo Alto, Calif., May 1980.
22. E. Indig and J. L. Nelson, *Corrosion* **47**, 202–209 (Apr. 1991).
23. C. Sacco, "Quad Cities Unit 2 Incore Hydrogen Addition Test Results," presented at *1994 INPO-EPRI Chemistry Managers Workshop*. Atlanta, Ga., Oct. 1994.

24. "BWR Normal Water Chemistry Guidelines—1986 Revision," Report NP4946-SP, Electric Power Research Institute, Palo Alto, Calif., Sept. 1988.

25. R. H. Asay, "Preconditioning of Reactor Piping," in Ref. 9, pp. 131–138.

26. H. Ocken, *Nuclear Technol.* **68**, 18–27 (Jan. 1985).

JACK H. HICKS
Consultant

ISOTOPE SEPARATION

The high cost of isotope separation has limited the use of separated isotopes in nuclear reactors to specific cases where substitutes that do not involve separated isotopes are not available. The most important example is that of uranium-235 [15117-96-1], ^{235}U, the most abundant naturally occurring fissionable material (see URANIUM AND URANIUM COMPOUNDS). Other isotopes that are separated for nuclear use are deuterium, ^{2}H or D, which as deuterium oxide [7789-20-1], D_2O, is used as a neutron moderator in water-cooled reactors; tritium, ^{3}H or T, which is produced in nuclear reactors and is a probable fuel in future thermonuclear or fusion reactors (see DEUTERIUM AND TRITIUM; FUSION ENERGY; RADIOISOTOPES); boron-10 [14798-12-0], ^{10}B, which has a high neutron cross section and is used in control rods and safety devices for nuclear reactors; and lithium-7 [13982-05-3], ^{7}Li, which is used in reactor cooling-water systems because of its low thermal neutron cross section.

Uranium-235

The separation of the isotope ^{235}U, which occurs in natural uranium to the extent of 0.72 atomic %, is discussed in detail elsewhere (see DIFFUSION SEPARATION METHODS). Uranium-235 is concentrated mostly by means of the gaseous diffusion process using uranium hexafluoride [7783-81-5], UF_6, as the process gas. The separation effect in this process arises from relatively more of the lighter gas molecules passing through a porous membrane, called a diffusion barrier, per unit time than of the heavier gas molecules. The effective single-stage separation factor is proportional to the ratio of mean thermal velocities of the molecules or inversely proportional to the ratio of the square root of their molecular weights. That factor is quite small, ie, somewhat less than the theoretical value of 1.0043. A cascade containing more than a thousand separation stages is therefore required to enrich uranium to reactor grade (3–4 atomic %) and several thousand stages are required to produce highly enriched material. The first large gaseous diffusion cascade was built in the early 1940s at Oak Ridge, Tennessee (1). Additional plants were built in the 1950s at Oak Ridge, Tennessee; Paducah, Kentucky; and Portsmouth, Ohio. Gaseous diffusion cascades for uranium enrichment have also been built in the United Kingdom, France, the former USSR, China, and, more recently, in Argentina.

The gaseous diffusion process is energy-intensive. As a consequence of this and other economic considerations, as of this writing (ca 1995) most facilities for uranium enrichment utilize the gas centrifuge process. Gas centrifuge cascades for the enrichment of uranium, using UF_6 as the process gas, have been built and operated in the former USSR, the United Kingdom, the Netherlands, Germany, Japan, Pakistan, India, Brazil, and China. A small aerodynamic uranium enrichment facility, based on vortex tube technology and using a mixture of UF_6 and hydrogen as the process gas, has been built and operated by South Africa, but it too is energy-intensive. A chemical exchange process for uranium enrichment, based on the isotopic distribution between uranium compounds in the +3 and +4 valence states, has been developed and tested in France (2), and an ion-exchange (qv) process for uranium enrichment based on the isotopic distribution between uranium compounds in the +4 and +6 valence states has been developed and tested by Asahi Chemical Co. in Japan (3). Several laser photochemical processes (4) have also been developed (see LASERS; PHOTOCHEMICAL TECHNOLOGY). The atomic vapor laser isotope separation method (AVLIS) developed in the United States is also ready for deployment. However, as of the mid-1990s the supply of enriched uranium exceeds demand. New facilities are not expected until the demand for enriched uranium increases. General discussions of uranium enrichment are available (4–6).

Deuterium

The use of deuterium in nuclear reactors is discussed in detail elsewhere (see DEUTERIUM AND TRITIUM). Fast neutrons, obtained by the fission of ^{235}U, must be slowed to propagate the chain reaction or to react with ^{238}U to produce plutonium (see PLUTONIUM AND PLUTONIUM COMPOUNDS). Heavy water, D_2O, is one of the most efficient of those substances that are able to accomplish this purpose. Methods of isotope separation include catalytic exchange between H_2 and H_2O, electrolysis, water distillation, hydrogen distillation, and various chemical-exchange systems (7,8). Quantities of D_2O have been produced by catalytic exchange electrolysis and by vacuum distillation of water. The hydrogen sulfide–water, dual-temperature exchange process is used to produce multiton quantities of heavy water (9–13). Canada has built a number of large plants to furnish heavy water as a coolant and moderator of their CANDU (Canada–Deuterium–Uranium) power reactors. Other smaller plants in various countries employ water electrolysis, H_2–NH_3 exchange, water distillation, and liquid-hydrogen distillation (14) for deuterium concentration. Many plants use combinations of the above methods. Deuterium separation has been reviewed (15–17).

Tritium

Tritium is radioactive and decays with a half-life of 12.26 yr.

$$^3_1H \longrightarrow {}^3_2He + \beta^-$$

However, it is produced in the upper atmosphere by cosmic rays (18,19) (see RADIOACTIVITY, NATURAL).

Tritium is produced in heavy-water-moderated reactors and sometimes must be separated isotopically from hydrogen and deuterium for disposal. Ultimately, the tritium could be used as fuel in thermonuclear reactors (see FUSION ENERGY). Nuclear fusion reactions that involve tritium occur at the lowest known temperatures for such reactions. One possible reaction using deuterium produces neutrons that can be used to react with a lithium blanket to breed more tritium.

$$\mathrm{^{3}_{1}H} + \mathrm{^{2}_{1}H} \longrightarrow \mathrm{^{4}_{2}He} + \mathrm{^{1}_{0}}n + 17.6\,\mathrm{MeV}$$

Tritium can be produced in a manner requiring very little isotopic separation by the reaction of neutrons with ^{6}Li.

$$\mathrm{^{6}_{3}Li} + \mathrm{^{1}_{0}}n \longrightarrow \mathrm{^{3}_{1}H} + \mathrm{^{4}_{2}He}$$

This method is used in nuclear reactors at the plant at Savannah River, Georgia, to produce quantities of tritium. Final purification from hydrogen and deuterium can be accomplished by any of the methods used for deuterium separation. Separation factors for tritium from hydrogen are larger than for deuterium from hydrogen, and those for separation of tritium from deuterium are smaller than the deuterium–hydrogen factors. Complex mixtures of all three isotopes of hydrogen are problematic because of the numerous mixed species present, eg, HD, HT, and DT. A plant has been built and is in operation for the simultaneous extraction of hydrogen and tritium from heavy water (20). In this plant, low temperature rectification of liquid hydrogen at 150 kPa (1125 torr) is used to accomplish isotopic fractionation.

Boron-10

Boron-10 has a natural abundance of 19.61 atomic % and a thermal neutron cross section of 3.837×10^{-25} m^2 (3837 barns) as compared to the cross section of ^{11}B, 5×10^{-31} m^2 (0.005 barns). Boron-10 is used at 40–95 atomic % in safety devices and control rods of nuclear reactors. Its use is also intended for breeder-reactor control rods.

Examination of possible systems for boron isotope separation resulted in the selection of the multistage exchange-distillation of boron trifluoride–dimethyl ether complex, $BF_3 \cdot O(CH_3)_2$, as a method for ^{10}B production (21,22). Isotope fractionation in this process is achieved by the distillation of the complex at reduced pressure, ie, 20 kPa (150 torr), in a tapered cascade of multiplate columns. Although the process involves reflux by evaporation and condensation, the isotope separation is a result of exchange between the liquid and gaseous phases.

$$^{11}BF_3 \cdot O(CH_3)_2 \text{ (l)} + {}^{10}BF_3 \text{ (g)} \Longleftrightarrow {}^{10}BF_3 \cdot O(CH_3)_2 \text{ (l)} + {}^{11}BF_3 \text{ (g)}$$

The equilibrium constant for this reaction is ca 1.022 at 100°C. The ^{10}B concentrates in the liquid phase (23). However, the vapor phase contains ca 40% undissociated complex, which lowers the effective single-stage separation factor to ca 1.014.

A plant built by Amoco (Standard Oil of Indiana) and operated from 1944–1946 produced ca 50 kg of enriched ^{10}B (24). A larger plant, constructed in 1953 by the Hooker Electrochemical Co. (Model City, New York) produced ca 3400 kg of ^{10}B enriched to >90 atomic % (25,26). In 1977, the Eagle-Picher Co. reconstructed and operated the Model City boron isotope separation plant at Quapaw, Oklahoma. In addition, a separation cascade involving the boron trifluoride–dimethyl ether exchange-distillation process was designed and built for Eagle-Picher by Sulzer Brothers. This plant has a capacity of >2 metric tons ^{10}B as metal per year enriched to 95 atomic %. Tails concentration is ca 5 atomic % ^{10}B. There are two identical separation cascades. Each cascade consists of two distillation columns, which operate in series. The first column is 50.8 cm ID and has a packed height of 42 m; the second column is 30.5 cm ID and has a packed height of 27.4 m. Equilibrium time to achieve product level is ca 14 d. Either ^{10}B or ^{11}B can be supplied at >99 atomic % isotopic enrichment. The operating pressure is 27 kPa (200 torr), and the distillation boilup temperature is 96°C.

The search for a system with less decomposition and a higher separation factor has been summarized (27–29). The most promising system is the BF_3–anisole system, in which BF_3 (g) exchanges with the anisole [100-66-3] (methyl phenyl ether)·BF_3 complex (l) (30):

$$C_6H_5OCH_3 \cdot {}^{11}BF_3 \text{ (l)} + {}^{10}BF_3 \text{ (g)} \rightleftharpoons C_6H_5OCH_3 \cdot {}^{10}BF_3 \text{ (l)} + {}^{11}BF_3 \text{ (g)}$$

The equilibrium constant for this reaction is 1.029 at 25°C and reflux may be accomplished by decomposing the complex with heat and by absorbing BF_3 in anisole. This system, in contrast to the dimethyl ether–BF_3 system, is a true gas–liquid exchange system and contains no associated complex in the gas phase (30).

Lithium-7

The normal abundance of lithium-7 is 92.44 atomic %. Because of its low thermal neutron absorption cross section, ie, 3.7×10^{-30} m^2, highly enriched ^{7}Li as lithium hydroxide, LiOH, is used to control the pH in pressurized water reactors. Many possible systems have been investigated for the separation of lithium isotopes (31). Lithium-7 has been produced in the United States by chemical exchange between lithium hydroxide and lithium amalgam.

BIBLIOGRAPHY

"Isotope Separation" under "Nuclear Reactors" in *ECT* 2nd ed., Vol. 14, pp. 85–87 by J. S. Drury, Oak Ridge National Laboratory; in *ECT* 3rd ed., Vol. 16, pp. 161–165, by G. M. Begun, Oak Ridge National Laboratory.

1. H. Smyth, *Atomic Energy for Military Purposes*, Princeton University Press, Princeton, N.J., 1945, Chapt. X, p. 172.
2. J. M. Lerat, in P. Louvet, P. Noe, and Soubbaramayer, eds., *Proceedings of the Second Workshop on Separation Phenomena in Liquids and Gases, Versailles, July, 1989*, Centre d'Études Nucléaires de Saclay and Cité Scientifique Parcs et Technopoles Ile de France Sud, Massy, France, 1989, pp. 555–594.

3. M. Seko, T. Miyake, K. Inada, and K. Takeda, *Nucl. Technol.* 50, 178 (1980).

4. S. Villani, ed., *Uranium Enrichment, Topics in Applied Physics*, Vol. 35, Springer-Verlag, New York, 1979.

5. H. London, ed., *Separation of Isotopes*, George Newnes, Ltd., London, 1961.

6. M. Benedict, T. Pigford, and H. Levi, *Nuclear Chemical Engineering*, 2nd ed., McGraw-Hill Book Co., Inc., New York, 1981, Chapt. 14.

7. G. M. Murphy, H. C. Urey, and I. Kirshenbaum, eds., *Production of Heavy Water, National Nuclear Energy Series*, Div. III, Vol. 4F, McGraw-Hill Book Co., Inc., New York, 1955.

8. G. M. Murphy, H. C. Urey, and I. Kirshenbaum, eds., *Commercial Production of Heavy Water, National Nuclear Energy Series*, Div. III, Vol. 4E, U.S. Atomic Energy Commission Technical Information Service, Oak Ridge, Tenn., 1951 (declassified 1960).

9. *Ibid.*, p. 143.

10. J. S. Spevack, *The Concentration of Deuterium by the S-Process, Report A-393*, Columbia University, War Research Department, New York, Dec. 3, 1942.

11. U.S. Pats. 2,787,526 (April 2, 1957); 2,895,803 (July 21, 1959), J. S. Spevack (to the U.S. Atomic Energy Commission and J. S. Spevack).

12. W. P. Bebbington and V. R. Thayer, *Chem. Eng. Prog.* **55**(10), 70 (1959).

13. J. F. Proctor and V. R. Thayer, *Chem. Eng. Prog.* **58**(4), 53 (1962).

14. K. D. Timmerhaus, D. H. Weitzel, and T. M. Flynn, *Chem. Eng. Prog.* **54**(6), 35 (1958).

15. T. F. Johns, in Ref. 5, Section II-5.2, pp. 123–142; E. Glueckauf, in Ref. 5, Section II-5.6, pp. 209–248.

16. S. Villani, *Isotope Separation*, American Nuclear Society, Hillsdale, Ill., 1976, Chapts. 9, 10, 11, and 12.

17. H. K. Rae, ed., *ACS Symp. Ser.* **68**, (1978).

18. E. A. Evans, *Tritium and Its Compounds*, 2nd ed., John Wiley & Sons, Inc., New York, 1974.

19. D. G. Jacobs, *Sources of Tritium and Its Behavior Upon Release to the Environment, TID-24635*, U.S. Atomic Energy Commission, Oak Ridge, Tenn., 1968.

20. P. Pautrot and M. Damiani in Ref. 17, pp. 163–171.

21. M. Kilpatrick, C. A. Hutchison, Jr., E. H. Taylor, and C. M. Judson in G. M. Murphy, ed., *Separation of the Boron Isotopes, National Nuclear Energy Series*, Div. III, Vol. 5, USAEC Technical Information Service, Oak Ridge, Tennessee, 1952 (declassified 1957 as TID-5227).

22. U.S. Pat. 2,796,330 (June 18, 1957), R. H. Crist and I. Kirshenbaum (to the United States Atomic Energy Commission).

23. A. A. Palko, G. M. Begun, and L. Landau, *J. Chem. Phys.* **37**, 552 (1957).

24. A. L. Conn and J. E. Wolf, *Ind. Eng. Chem.* **50**, 1231 (1958).

25. C. H. Chilton, *Chem. Eng.* **64**(5), 148 (1957).

26. G. T. Miller, R. J. Kralik, B. A. Belmore, and J. S. Drury, *Paper 1836, Proc. 2nd UN International Conference on the Peaceful Uses of Atomic Energy, Geneva, Sept. 1–13, 1958*, Vol. 4, United Nations, New York, 1958, p. 585.

27. A. A. Palko and J. S. Drury, *Adv. Chem. Ser.* **69**, 40 (1969).

28. A. A. Palko, *The Chemical Separation of Boron Isotopes, ORNL-5418*, Oak Ridge National Laboratory, Oak Ridge, Tenn., June 1978.

29. N. N. Seryugova, O. V. Uvarov, and N. M. Zhavoronkov, *At. Energ.* **9**(8), 614 (1960).

30. A. A. Palko, R. M. Healy, and L. Landau, *J. Chem. Phys.* **28**, 214 (1958).

31. A. A. Palko, *Ind. Eng. Chem.* **51**, 121 (1959).

EDWARD VON HALLE
Consultant

CHEMICAL REPROCESSING

The process of separating the components of irradiated nuclear reactor fuel into several streams, usually uranium, plutonium, and wastes, is called chemical reprocessing. Within the context of the nuclear fuel cycle, synonyms for chemical reprocessing include reprocessing, fuel reprocessing, separations, and chemical separations. The term separations is also used to describe isotopic separation techniques, which fall outside the scope of this article. At times additional products have been recovered from the waste stream. Both flow sheet selection and the resulting economics depend on the context in which the reprocessing is performed. This context is commonly referred to as the fuel cycle. The fuel cycles of interest as of this writing (ca 1995) are illustrated in Figure 1. Even before the end of the Cold War, essentially all nations possessing nuclear weapons had stopped or drastically curtailed production of nuclear weapons materials. Thus weapons-related fuel cycles are not of interest herein, except for historical background and in the case where weapons materials may be recycled as a source of energy.

The raw material for nuclear reactor fuel, uranium, exits the mining–milling sequence as uranium oxide. Because of its color, it is called yellow cake. The yellow cake is converted to uranium hexafluoride and enriched in ^{235}U (see DIFFUSION SEPARATION METHODS; NUCLEAR REACTORS, ISOTOPE SEPARATION). The energy required by the enrichment process is measured in terms of separative work units (SWU). The enriched uranium hexafluoride is then reduced to the oxide and formed into pellets, which are fabricated into reactor-fuel elements (see NUCLEAR REACTORS, NUCLEAR FUEL). The cost of reactor fuel is thus the sum of the costs for yellow cake, the conversion, enrichment, and fabrication. Other fuel-cycle costs are associated with the waste streams originating at each of the fuel-cycle facilities (see NUCLEAR REACTORS, WASTE MANAGEMENT).

The classical fuel cycle is based on the recovery and recycle of energy values contained in spent fuel, where energy values are defined in terms of the four components of reactor fuel cost. The classical fuel cycle begins with an initial reactor charge of low enriched uranium (approximately 2.5–3.5% ^{235}U). During reactor operation, the amount of ^{235}U is reduced by the fission reactions which occur, a phenomenon referred to as burn-up, but at the same time ^{238}U is absorbing neutrons to create plutonium. As a result, at the time of discharge the irradiated fuel rods contain recoverable energy in the form of both residual fissile uranium and plutonium. The uranium and plutonium are chemically separated from the fission products for recycle in the form of mixed uranium and plutonium oxide (MO_x) fuel. During each recycle, the energy level of the recycled uranium is maintained either by blending with more highly enriched uranium (>3.5% ^{235}U), or by routing the recovered uranium through a conversion and enrichment step and adding virgin low enriched uranium ($\sim 3.5\%$ ^{235}U) to make up for the depleted ($\sim 0.2\%$ ^{235}U) uranium tails stream.

As the recycled fuel composition approaches steady state after approximately four cycles (1), the heat and radiation associated with ^{236}U and ^{238}Pu require more elaborate conversion and fuel fabrication facilities than are needed for virgin fuel. The storage, solidification, packaging, shipping, and disposal considerations associated with wastes that result from this approach are primarily

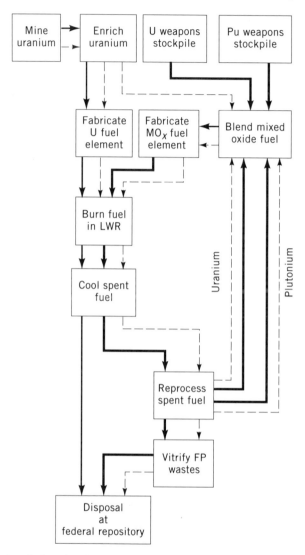

Fig. 1. Alternative fuel cycles for nuclear fuel, where (– – –) corresponds to the classical fuel cycle, (——) the throwaway fuel cycle, and (▬▬) the recycle weapons fuel cycle. FP = fuel processing; LWR = lightwater reactor.

concerned with the relatively short-lived fission products. The transuranic isotopes are recycled to and burned in the reactors. Commercial reactor fuel is being reprocessed and recycled in the U.K., France, Japan, India, and the former Soviet Union. Essentially all other nuclear nations, except the United States, have contracted to have these services performed.

A variation of the classical fuel cycle is the breeder cycle. Special breeder reactors are used to convert fertile isotopes into fissile isotopes, which creates more fuel than is burned (see NUCLEAR REACTORS, REACTOR TYPES). There are two viable breeder cycles: ^{238}U/^{239}Pu, and ^{232}Th/^{233}U. The thorium fuels were, however, not in use as of 1995. A breeder economy implies the existence of both breeder reactors that generate and nonbreeder reactors that consume the fissile material. The breeder reactor fuel cycle has been partially implemented in France and the U.K.

The throwaway fuel cycle does not recover the energy values present in the irradiated fuel. Instead, all of the long-lived actinides are routed to the final waste repository along with the fission products. Whether or not this is a desirable alternative is determined largely by the scope of the evaluation study. For instance, when only the value of the recovered yellow cake and SWU equivalents are considered, the world market values for these commodities do not fully cover the cost of reprocessing (2). However, when costs attributable to the disposal of large quantities of actinides are considered, the classical fuel cycle has been the choice of virtually all countries except the United States.

The recycle weapons fuel cycle relies on the reservoir of SWUs and yellow cake equivalents represented by the fissile materials in decommissioned nuclear weapons. This variation impacts the prereactor portion of the fuel cycle. The post-reactor portion can be either classical or throwaway. Because the availability of weapons-grade fissile material for use as an energy source is a relatively recent phenomenon, it has not been fully implemented. As of early 1995 the United States had purchased highly enriched uranium from Russia, and France had initiated a modification and expansion of the breeder program to use plutonium as the primary fuel (3). All U.S. reactor manufacturers were working on designs to use weapons-grade plutonium as fuel.

History

The nuclear fuel cycle was developed in the United States during World War II. Two paths were pursued: a uranium weapon, which required the development of a means of separating the fissionable and fertile isotopes of uranium (enrichment); and a plutonium weapon, which required the manufacture and recovery of plutonium (reactors and chemical reprocessing). Since that time, many techniques have been developed for recovering uranium and plutonium from spent fuel. The first large-scale process for recovering plutonium from irradiated nuclear fuel was based on a bismuth phosphate precipitation reaction in Hanford, Washington, ca 1944. This was essentially a throwaway fuel cycle, because everything except the plutonium was routed to the waste stream. In an effort to reduce both the demand for uranium and the problems associated with the large waste stream, liquid–liquid solvent extraction processes were developed that recovered both the plutonium and uranium. A portion of the uranium was recycled for enrichment (see also EXTRACTION, LIQUID–LIQUID).

The first solvent extraction plant, REDOX, utilized methyl isobutyl ketone [108-10-1] (hexone) as the solvent (4). Aluminum nitrate nonahydrate [7784-27-2] (ANN) was used as the salting agent to enhance uranium and plutonium extraction. An advantage of the process was that a simple distillation (qv) step removed radiation degradation products from the solvent. Disadvantages of this process include the large quantities of unrecoverable aluminum nitrate in the wastes, the potential fire hazard from the hexone, and poor ruthenium decontamination (5). During this same period, the British developed the BUTEX process, which used dibutoxy diethyl ether as the extractant at Windscale in 1952. No salting agent was required, resulting in a much smaller waste stream. The Windscale facility was also the first to use mechanical decladding. In the American plants fuel cladding was chemically removed from the fuel.

The American facilities also differed fundamentally from the British facilities in regard to maintenance philosophy. The American plants were designed to employ remote maintenance, ie, to remove and replace equipment using shielded cranes operating inside the shielded structure. The British developed a contact approach based on simplified designs for equipment downstream of the fission product removal step. The British approach has been used at all commercial facilities.

An improved solvent extraction process, PUREX, utilizes an organic mixture of tributyl phosphate solvent dissolved in a hydrocarbon diluent, typically dodecane. This was used at Savannah River, Georgia, ca 1955 and Hanford, Washington, ca 1956. Waste volumes were reduced by using recoverable nitric acid as the salting agent. A hybrid REDOX/PUREX process was developed in Idaho Falls, Idaho, ca 1956 to reprocess high burn-up, fully enriched (97% ^{235}U) uranium fuel from naval reactors. Other separations processes have been developed. The desirable features are compared in Table 1.

Commercial Experience.   The Nuclear Fuel Services (NFS) plant in West Valley, New York, is the only separations facility in the United States to have reprocessed commercial power reactor fuel (7). This facility was a significant improvement over the government-owned PUREX process, in that a mechanical head-end eliminated the chemical decladding step and the resulting liquid waste stream. After successfully processing the entire available backlog of U.S. commercial power reactor fuel (ca 250 t) as well as a small amount of government-

Table 1. Separations Processes for Chemical Reprocessing[a,b]

| Process | Feature | | | | | | Equipment | |
	Flexibility	Continuous operation	Fission-product separation	U/Pu separation	Ease of waste handling	Size	Remote handling	Corrosion resistance
precipitation								X
solvent extraction	X	X	X	X	X	X	X	X
ion exchange	X		X	X	X			X
zone melting/ slagging			X		X	X		
molten salt/ pyrometallurgical			X			X		
fluoride volatility			X	X		X		

[a]Ref. 6.
[b]X represents desirable process characteristic.

owned fuel from the Hanford reactor, the facility was shut down for modification to increase its capacity. Changing licensing requirements prevented the plant from restarting. As of this writing, it is being decontaminated and decommissioned.

A facility designed to process one metric ton per day of discharged or spent fuel was built by General Electric in Morris, Illinois, to demonstrate their proprietary Aquafluor process (8). This process had the potential for significant cost savings over the classic PUREX process in that the solvent extraction cycles for uranium purification were eliminated. Final uranium purification was to be accomplished during conversion to the uranium hexafluoride product. Technical problems encountered during facility cold checkout, coupled with changing market conditions and an unstable licensing environment, all contributed to cancellation of the project. As of 1995 the plant's fuel storage basin is being used to store lightwater reactor (LWR) fuel.

Plans to construct reprocessing facilities were also pursued by Exxon, Atlantic Richfield, and Allied-General Nuclear Services (AGNS) using variations of the PUREX process. Only the AGNS project proceeded through construction and checkout. The AGNS flow sheet was unique in that it used an electrolytic reduction technology and a nitric oxide oxidation step for plutonium valance adjustments (9). As a result, the high level liquid waste contained only fission-product salts. This reduced the volume of high level waste and simplified its vitrification, packaging, and transportation. Cold check-out runs in the 1500-t/yr AGNS facility were completed in 1981, but the plant never entered hot operation, owing to an unstable licensing environment combined with changing market conditions.

As of 1995, there were no nuclear fuel reprocessing plants operating in the United States. Other nuclear nations have constructed second- or third-generation reprocessing facilities. These nations have signed the nuclear non-proliferation treaty, and the facilities are under the purview of the International Atomic Energy Agency (IAEA).

The UP2 plant located at La Hague, France, began processing gas-cooled reactor and fast breeder reactor fuels in 1965, and in 1987 began reprocessing LWR fuel for utilities located throughout Europe and Japan. A newer, 800-t/yr plant, UP3, was constructed at the same site and was commissioned in 1990. The UP2 plant was subsequently refurbished and upgraded to the same capacity, with the added ability to handle MO_x and high burn-up fuels. The facility was renamed UP2-800 and was brought on line in 1994. The site capacity should gradually increase to 1600 t/yr of LWR fuel. Between 1950 and 1991, 80% of the power reactor fuel reprocessed worldwide was reprocessed in France (10).

Site preparation for the British thermal oxide reprocessing plant (THORP) at Sellafield, Seascale Cumbria, began in 1983, and the plant began hot operation in 1994. British Nuclear Fuels Limited (BNFL), the owners of the 1200-t/yr plant, have contracted to process spent fuel from utilities throughout Europe and Japan. The THORP facility, the third reprocessing plant to be built at the Sellafield site, was conceived in the mid-1970s when it became apparent that a new facility would be required to process LWR fuel from U.K. power reactors. The design and construction time scale of the reprocessing plant is typical of nuclear projects in the regulatory environment of the latter twentieth century.

The features incorporated were thus developed and tested more than a decade before the plant began hot operation (11).

During the early 1970s, the French assisted a Japanese company, PNC, in the design of a small, 150-t/yr reprocessing facility which was built at Tokai-mura primarily to gain experience in the back-end of the fuel cycle. The plant was commissioned in 1977 and continues to operate. Based on this experience, the Japanese have built an 800-t/yr reprocessing plant at Rokashamura that is very similar to the French UP3 facility. As of 1994 it was undergoing testing and start-up (11). Smaller, 100-t/yr plants are also in operation in India, at Prefre and Kalpakkam (12).

Reprocessing Strategy

The radioactive isotopes associated with spent fuel emit substantial amounts of α-, β-, and γ-radiation as a result of radioactive decay. These radioisotopes (qv) have half-lives, $t_{1/2}$, of up to thousands of years. The storage of spent fuel prior to reprocessing allows the shorter lived isotopes to decay, thus reducing the costs associated with shielding and heat removal. These savings are balanced by the higher costs associated with maintaining a large spent fuel inventory. Under the economic conditions that existed during design of the AGNS facility, the optimum cooling time was 180 d for fuel having a burn-up of 33×10^3 MW·d/t. The average fuel processed thus far, however, has a substantially lower burn-up and has been cooled for a number of years.

A second consideration affecting reprocessing strategy is the preservation of separative work units (SWU), a measure of the relative value of enriched uranium. Fuel is campaigned through a reprocessing facility in batches having similar discharge enrichments. However, a single utility rarely has enough fuel of the same enrichment and burn-up to avoid SWU degradation. Batches are therefore a mix of similar discharge enrichments from a number of utilities, and the utility/reprocessor differences in fissile measurements become important. The amount of fissile material in spent fuel is calculated by the utility based on reactor physics models. The reprocessor determines fissile content based on analysis of the dissolver solutions and the dissolver solids. Reconciliation is an ongoing effort. The process steps involved in reprocessing irradiated fuel are illustrated in Figure 2.

Fuel Characteristics. Historically, chemical reprocessing of irradiated fuel was developed specifically to handle the U.S. government's defense-related fuels. These were of two types. There were the very low enrichments and burn-ups, eg, 0.9% ^{235}U and 2000 MW·d/t, as was used for plutonium production, and very high enrichments and burn-ups, eg, 97% ^{235}U and 10^5 MW·d/t, as is used in naval reactors. In both cases, metallic uranium is used. The low burn-up fuels are clad in aluminum; the higher burn-up fuels are clad in zirconium.

By contrast, uranium fuels for lightwater reactors fall between these extremes. A typical pressurized water reactor (PWR) fuel element begins life at an enrichment of about 3.2% ^{235}U and is discharged at a burn-up of about 30×10^3 MW·d/t, at which time it contains about 0.8 wt % ^{235}U and about 1.0 wt % total plutonium. Boiling water reactor (BWR) fuel is lower in both initial enrichment and burn-up. The uranium in LWR fuel is present as oxide pellets, clad in

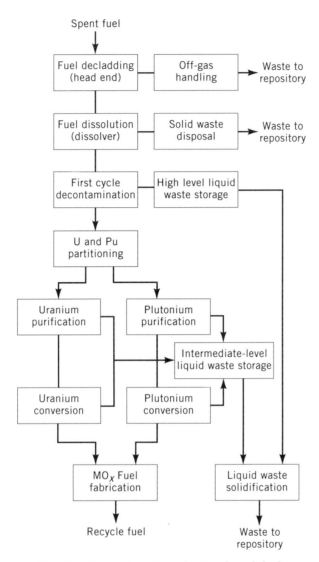

Fig. 2. Processing steps for irradiated fuel.

zirconium alloy tubes about 4.6 m long. The tubes are assembled in arrays that are held in place by spacers and end-fittings.

MO_x fuel designed for use in existing LWRs is typically exposed to burn-ups greater than 40×10^3 MW·d/t. The discharged MO_x fuel has essentially the same uranium enrichment as uranium oxide fuel, but has a greater total amount of plutonium.

Canadian reactors (CANDU) are fueled using natural uranium. The discharged fuel contains small amounts of plutonium, but the fissile uranium content is below that of natural uranium. Therefore, the irradiated fuel is not normally considered a candidate for economic reprocessing.

Head-End. *Fuel Decladding.* A power reactor fuel assembly consists of uranium or mixed uranium/plutonium oxide pellets enclosed in a zirconium metal tube which is nominally 1.27 cm in diameter by about 4.6 m long, referred to as a fuel pin. The fuel pins are then arranged in an array nominally 17 × 17 for a PWR fuel assembly, or 6 × 6 for a BWR fuel assembly. The pins are held in place by massive metal end fittings and the configuration is stabilized by stamped sheet-metal spacer grids located at intervals along the length of the pins. Depending on the fuel design, some of the LWR pin locations are occupied by instrument tubes or burnable poisons. The British Advanced Gas Reactor (AGR) fuel assemblies contain graphite moderators.

All operating facilities shear the spent fuel elements into segments several centimeters long to expose the oxide pellets to nitric acid for dissolution. This operation is often referred to as chop-leach. The design and operation of the shear is of primary importance because (*1*) the shear can be the production bottleneck, and (*2*) the shear is the point at which tritium and fission gases are released.

Fuel Dissolution. In the American and British plants, LWR fuel pieces typically fall directly from the shear into a dissolver basket, which fits inside the dissolver vessel. A soluble poison such as gadolinium is added to the nitric acid to prevent criticality. The massive end fittings are sometimes separated from the fuel pieces before the latter enter the dissolver. The French have installed continuous rotary dissolvers in the UP3 and UP2-800 plants at La Hague. The units each consist of a drum rotating within a geometrically favorable slab tank (13).

Fast reactor fuel assemblies are shrouded with a relatively heavy metal envelope. This envelope is removed before shearing by either laser cutting (14) or stress cracking (15).

The rate (kinetics) and the completeness (fraction dissolved) of oxide fuel dissolution is an inverse function of fuel burn-up (16–18). This phenomenon becomes a significant concern in the dissolution of high burn-up MO_x fuels (19). The insoluble solids are removed from the dissolver solution by either filtration or centrifugation prior to solvent extraction. Both financial considerations and the need for safeguards make accounting for the fissile content of the insoluble solids an important challenge for the commercial reprocessor. If hydrofluoric acid is required to assist in the dissolution, the excess fluoride ion must be complexed with aluminum nitrate to minimize corrosion to the stainless steel used throughout the facility. Also, uranium fluoride complexes are inextractable and formation of them needs to be prevented.

Chemical Separation. A reprocessing facility typically utilizes multiple extraction/reextraction (stripping) cycles for the recovery and purification of uranium and plutonium. For example, a co-decontamination and partitioning cycle is followed by one or more cycles of uranium and plutonium purification. The basic process is illustrated in Figure 3.

Chemistry. Chemical separation is achieved by countercurrent liquid–liquid extraction and involves the mass transfer of solutes between an aqueous phase and an immiscible organic phase. In the PUREX process, the organic phase is typically a mixture of 30% by volume tri-*n*-butyl phosphate (solvent) and a normal paraffin hydrocarbon (diluent). The latter is typically dodecane or a high grade kerosene (20). A number of other solvent or diluent systems have been investigated, but none has proved to be a substantial improvement (21).

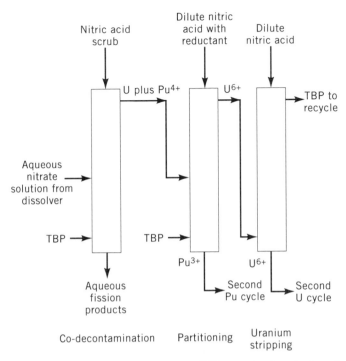

Fig. 3. Basic PUREX process where TBP = tri-*n*-butyl phosphate.

The distribution of highly extractable solutes such as U^{6+} and Pu^{4+} between the aqueous and organic phases is strongly dependent upon the nitrate anion concentration in the aqueous phase. This salting effect permits extraction or reextraction (stripping) of the solute by controlling the nitric acid concentration in the aqueous phase. The distribution coefficient, D, of the solute is expressed as

$$D = \frac{\text{concentration in organic solution}}{\text{concentration in aqueous solution}} \tag{1}$$

which is influenced by many factors. For instance, the salting effect of the anion (nitrate) plays a role as does the approach to the saturation of the solvent (loading), competition of the solutes for free solvent, and temperature.

Two ions a and b can be separated by countercurrent extraction as long as the ratio of the distribution coefficients, that is, the separation factor Ω, is not unity:

$$\Omega_{a-b} = \frac{D_a}{D_b} \neq 1 \tag{2}$$

The solubility of the actinides in the organic phase, the right-hand side of equation 3, is achieved by the weak complexes that tri-*n*-butyl phosphate [126-73-8] (TBP), $(n\text{-}C_4H_9O)_3PO$, forms with the neutral metal nitrates:

$$UO_2^{2+} + 2\,NO_3^- + 2\,TBP \rightleftharpoons UO_2(NO_3)_2 \cdot 2TBP \tag{3}$$

The equilibrium constant for uranium, K_U, is

$$K_U = \frac{[UO_2(NO_3)_2 \cdot 2TBP]_o}{[UO_2^{2+}]_a[NO_3^-]_a^2[TBP]_o^2} \tag{4}$$

where the subscripts a and o correspond to organic and aqueous, respectively. The distribution coefficient, therefore, is

$$D_U = K_U[NO_3^-]_a^2[TBP]_o^2 \tag{5}$$

where the brackets indicate activities. A similar expression can be written for NO_3^- and plutonium:

$$H^+ + NO_3^- + TBP \rightleftharpoons HNO_3 \cdot TBP \tag{6}$$

$$Pu^{4+} + 4\,NO_3^- + 2\,TBP \rightleftharpoons Pu(NO_3)_4 \cdot 2TBP \tag{7}$$

Each of the actinides has more than one valence state and most tend to hydrolyze. Uranium in nitric acid is usually hexavalent and stable as uranyl ion, UO_2^{2+}, although U^{4+} can be prepared under strong reducing conditions. Plutonium in nitric acid can exist in the +3, +4, and +6 valence states. It can be oxidized readily to the extractable +4 valence using NO_2^- prior to extraction; it can also be reduced using Fe^{2+} or hydroxylamine, or electrolytically to the less extractable +3 valence. This characteristic is exploited to transfer Pu^{4+} and U^{6+} from the aqueous phase to the organic phase, or to selectively strip the plutonium back into the aqueous phase, leaving the uranium in the organic. The selective stripping of plutonium from the organic phase back into the aqueous phase is called partitioning (Fig. 3).

Computer simulation programs for process design optimization have been developed for the PUREX process utilizing these relationships (22). A subroutine has also been developed which describes the behavior of fission products (23).

First-Cycle Decontamination. In the first cycle, clarified dissolver solution containing the uranium, plutonium, other transuranics, fission products, and excess nitric acid are contacted with the organic phase in the extraction column. A strong nitric acid stream is introduced at the top of the column to remove (scrub) residual impurities that extract or are entrained in the organic phase. Approximately 99.9% of the uranium and plutonium nitrates transfer to the organic phase, and conversely, 99.9% of the fission products and nitric acid remain in the aqueous phase. This acidic aqueous solution from the first cycle of the separations process is the high level liquid waste stream. It is collected in underground storage tanks pending solidification (see NUCLEAR REACTORS, WASTE MANAGEMENT).

Relative values for the various TBP distribution coefficients of the feed constituents are as follows:

Species	D
U^{6+}	8.1
Pu^{4+}	1.55
Pu^{6+}	0.62
HNO_3	0.07
Zr	0.02
Ce^{3+}	0.01
Ru	0.01
Pu^{3+}	0.008
Nb	0.005
rare earths	0.002
combined β-emitters	0.001
Cs	>0.0001

These variations permit the separation of other components, if desired. Additional data on uranium, plutonium, and nitric acid distribution coefficients as a function of TBP concentration, solvent saturation, and salting strength are available (24,25). Algorithms have also been developed for the prediction of fission product distributions in the PUREX process (23).

Uranium–Plutonium Partitioning. The uranium and plutonium are separated in the partitioning column by reducing the plutonium to a less extractable valence state. The plutonium nitrate transfers back to the aqueous phase and the uranium remains with the organic.

Historically, ferrous sulfamate, $Fe(NH_2SO_3)_2$, was added to the HNO_3 scrubbing solution in sufficient excess to ensure the destruction of nitrite ions and the resulting reduction of the Pu^{4+} to the less extractable Pu^{3+}. However, the sulfate ion is undesirable because sulfate complexes with the plutonium to complicate the subsequent plutonium purification step, adds to corrosion problems, and as SO_2 is an off-gas pollutant during any subsequent high temperature waste solidification operations. The associated ferric ion contributes significantly to the solidified waste volume.

Hydroxylamine is used as a substitute for the ferrous sulfamate (26). These systems are called salt-free flow sheets. The main purpose is to ease the problems associated with the processing and storage of the liquid waste streams (27). Another approach is to use an electropulse column to electrolytically produce U^{4+} to reduce Pu^{4+} to Pu^{3+} on a continuous basis (28,29). The half reactions for the flow sheets are

$$U^{6+} + 2\,e^- \xrightarrow[N_2H_4]{} U^{4+} \tag{8}$$

$$2\,Pu^{4+} + U^{4+} \xrightarrow[N_2H_4]{} 2\,Pu^{3+} + U^{6+} \tag{9}$$

During electrolytic reduction of U^{6+} to U^{4+} and the subsequent reduction of Pu^{4+} to inextractable Pu^{3+} hydrazine is added as a holding agent to destroy excess nitrite ions and prevent reoxidation of U^{4+} and Pu^{3+} to their higher valence states.

In the uranium-stripping step, a dilute HNO_3 solution is used to remove uranium from the organic phase. The aqueous effluent containing the uranium product is evaporated to a concentrated uranyl nitrate $UO_2(NO_3)_2$ solution that is adjusted to about $1.8 M$ uranium and $1.0 M$ HNO_3 and fed to the second uranium purification cycle.

Uranium Purification.   Subsequent uranium cycles provide additional separation from residual plutonium and fission products, particularly zirconium–niobium and ruthenium (30). This is accomplished by repeating the extraction/stripping cycle. Decontamination factors greater than 10^6 at losses of less than 0.1 wt % are routinely attainable. However, ruthenium can exist in several valence states simultaneously and can form several nitrosyl–nitrate complexes, some for which are extracted readily by TBP. Under certain conditions, the nitrates of zirconium and niobium form soluble compounds or hydrous colloids that complicate the liquid–liquid extraction. Silica-gel adsorption or one of the similar liquid–solid techniques may also be used to further purify the product streams.

Improvements in the ability to control operating conditions and in contractor designs have allowed a steady reduction in the number of purification steps required. The THORP facility, commissioned as of 1994 in the U.K., uses only a single purification step.

Plutonium Purification.   The aqueous feed for the second plutonium cycle is typically prepared by adding HNO_3 and an excess of sodium nitrite, $NaNO_2$, to destroy the excess reductant and oxidize the Pu^{+3} to the more extractable Pu^{4+}. An alternative approach which reduces the amount of salt in the liquid waste involves absorbing nitrogen tetroxide, N_2O_4, as a substitute for the $NaNO_2$:

$$N_2O_4 + H_2O \longrightarrow 2\,H^+ + NO_3^- + NO_2^- \tag{10}$$

The excess nitric acid is recovered during waste concentration and is recycled for fuel dissolution (12).

After the second extraction/stripping cycle, the plutonium is concentrated by evaporation or by preferential adsorption (qv) on ion-exchange resins. As in the case for uranium, the newer facilities, such as THORP, use only a single purification step.

Product Conversion.   *Uranium.*   The uranium product from the PUREX process is in the form of uranyl nitrate which must be converted to some other chemical depending on anticipated use. One route to MO_x fuel is to mix uranium and plutonium nitrates and perform a coprecipitation step. The precipitate is heated to form a mixed U–Pu oxide. Another approach is to convert the uranium to an oxide, either directly (31) or via an ammonium diuranate intermediate precipitation step (32), followed by a U–Pu oxide powder blending step. The resulting mixed oxide can then be reenriched by blending with more highly enriched uranium. The option of fluorinating recycle uranium to UF_6 for recycle to a cascade-type enrichment process (that is, diffusion or centrifuge) is not usually attractive because of the presence of ^{236}U. As of this writing, the ^{236}U problem had been addressed for laser enrichment. The value of the uranium product at this point is a function of the residual separative work (SWU) it contains and the cost of disposal. If the uranium has ^{235}U concentrations in excess of natural uranium, it can be recycled in lieu of virgin feedstock. This approach has both economic and has environmental benefits in that less uranium mining is required.

Plutonium. The plutonium nitrate product must be converted to MO_x fuel if it is to be recycled to lightwater reactors. Whether from a plutonium nitrate solution or a mixed U/Pu nitrate solution, the plutonium is typically precipitated as the oxalate and subsequently calcined to the oxide for return to the fuel cycle (33).

By-Products. The PUREX process is efficient at separating uranium and plutonium from everything else in the spent fuel. Within the high level waste stream are a number of components which have, from time to time, been sufficiently interesting to warrant their recovery. The decision to recover a particular isotope is usually based on a combination of market incentives and desired waste reduction (qv).

Neptunium has been recovered during the reprocessing of defense-related fuels. The ^{237}Np is recycled back to a reactor where it is transmuted to ^{238}Pu. The ^{238}Pu has seen application as a long-lived isotopic heat source. Plutonium-238 is most useful in space programs, but is also of interest as part of a proliferation-resistant fuel cycle (2).

The 30-yr half-lives of ^{137}Cs and ^{90}Sr make these attractive isotopic heat or radiation sources. However, ^{144}Ce and ^{134}Cs are shorter lived alternatives. Each has been recovered and used in various quantities. Rhodium, palladium, and technetium have also been recovered for potential catalytic or precious metal applications (34,35).

Waste Handling. *Off-Gas Treatment.* Before the advent of the shear, the gases released from the spent fuel were mixed with the entire dissolver off-gas flow. Newer shear designs contain the fission gases and provide the opportunity for more efficient treatment. The gaseous fission products krypton and xenon are chemically inert and are released into the off-gas system as soon as the fuel cladding is breached. Efficient recovery of these isotopes requires capture at the point of release, before dilution with large quantities of air. Two processes have been developed, a cryogenic distillation and a Freon absorption. Neither method has seen widespread use (36).

Tritium differs from the noble gases. Tritium is very reactive and readily combines to form water. The voloxidation process, which controls the point at which the oxidation occurs and recovers the resulting tritiated water, has been developed. As of this writing it has not yet been used.

A large portion of the ^{14}C can be released from the reprocessing facility as CO_2. Thus a recovery process has been developed using an alkaline absorbent column (37).

In the reprocessing environment there are many ruthenium compounds, some of which are gaseous. Some reprocessing approaches, notably the REDOX process, require a ruthenium removal step in the off-gas system. The PUREX process maintains ruthenium in one of its nonvolatile states.

In contrast, iodine recovery is both highly developed and universally used throughout the industry. Iodine can be absorbed on solid sorbents or in an off-gas scrub of caustic, mercuric nitrate solution, or hyperazeotropic nitric acid (qv) (38). The last process, called Iodox, is effective, particularly for trace amounts of organic iodides. Stable periodic acid forms as a precipitate in concentrated nitric acid, and can be stored as a slurry, pending decay of ^{131}I. The periodic acid is converted into a stable compound, for instance barium iodate, and packaged for terminal disposal. Solid absorbers, eg, silver nitrate deposited on Berl saddles or silver zeolite granules, are reliable and effective (39). The disposal of silver reactors has a negligible environmental impact. These are, however, costly and have the potential for forming highly reactive compounds if operated at excessively high temperatures.

Hulls Handling. After the fuel has been dissolved, the residual pieces of zirconium cladding, referred to as hulls, are rinsed and removed from the dissolver vessel. The decay of activation products provides sufficient heat to ensure drying of the hulls and preclude hydrogen formation caused by the radiolysis of water.

The hulls are the first material removed from the process to merit concern in respect to safeguards. These represent a possible route for special nuclear material (SNM) to be clandestinely removed from the system. Special instrumentation is used to assay the hulls containers for residual SNM. This step not only monitors the dissolver for proper functioning, but also serves an accountancy function as a material access area portal and enables the satisfaction of contractual requirements regarding process losses (40).

The volume of hulls generated is nominally 62 m^3/t of fuel, which is about 10 times the actual volume of metal. Whereas they are not yet in commercial use, both compaction and melting processes are being developed to improve waste handling economics (41).

High Level Waste Management. The technical challenges related to the interim storage, solidification, and disposal of irradiated fuel using the throwaway fuel cycle (Fig. 1) are driven by the nuclear properties of the actinides and the configuration of the irradiated fuel. The technical challenges of the high level waste from the recycle option are driven by the nuclear properties of the fission products, and the solidification of a liquid. The United States is wrestling with the daunting environmental challenges associated with disposing of large quantities of actinides. Most other nations have elected to address the more tractable problems associated with fission-product disposal. The U.S. experience with high level wastes (HLW) is thus limited to the interim storage of spent LWR fuel and liquid wastes from the weapons program (see NUCLEAR REACTORS, WASTE MANAGEMENT).

In the United States, liquid HLW from the reprocessing of defense program fuels was concentrated, neutralized with NaOH, and stored in underground, mild steel tanks pending solidification and geologic disposal (see TANKS AND PRESSURE VESSELS). These wastes are a complex and chemically active slurry. Suspended in the supernatant liquid are dissolver solids which never went into solution, insoluble reaction products which formed in the tank, and salts which have exceeded their solubility limit. The kinetics of many of the reactions taking place are slow (years) so that the results of characterization attempts are sometimes transient. This is a problem made more difficult by the lack of homogeneity, which frustrates representative sampling (42). After the first few months, the decay heat in the HLW is dominated by ^{137}Cs and ^{90}Sr, which are found in the supernatant liquid and sludge, respectively. To ease the problems of heat removal and mitigate the consequences of spills, these isotopes can be removed and stored separately (43). The cesium capsules have been used as radiation sources for food preservation, and the strontium capsules are available for heat sources. Removal of these radioisotopes has simplified waste management and may position some of the waste for a less costly management option than geologic disposal.

By contrast, HLW from LWR fuel reprocessing is stored in cooled, well-agitated, stainless steel tanks as an acidic nitrate solution having relatively few solids. Modern PUREX flow sheets minimize the addition of extraneous salts, and as a result the HLW is essentially a fission-product nitrate solution. Dissolver solids are centrifuged from the feed stream and are stored separately. Thus the HLW has a low risk of compromising tank integrity and has a favorable composition for solidification and disposal (11).

Reprocessing Equipment

Fuel Shear. At the NFS plant in West Valley, New York, fuel assemblies were transferred from the fuel storage pool via an underwater conveyor and fed horizontally into the head-end cell, where the end fittings were removed by an abrasive disk saw and placed in a scrap drum (9). A hydraulic pneumatic feed pusher advanced the fuel into the shear. The shear chopped the fuel into preselected lengths ranging from 1.25–5 cm. The blade was driven by a hydraulic ram which developed a force adjustable between 250 and 320 t. The hydraulic power unit for the shear-blade drive and gag actuators was located outside the process cell. The shear blade was arranged to cut diagonally across the fuel bundle. The throat of the shear housing had a 23-cm opening and could accommodate bundles up to 232 cm^2 and up to 4.9 m long. Gamma detectors automatically stopped the shear when either the basket was full or the chute became plugged. An inert gas sweep and a fire-retardant injection system were also part of the design. The capability to cool the fuel bundle in the shear magazine was provided but never needed.

The NFS shear successfully processed irradiated LWR fuel between 1966 and 1972. Among the lessons learned were that the chute that channeled sheared fuel pieces to the dissolver basket was susceptible to bridging, the gag drives required a disproportionate amount of maintenance, and insufficient throat clearance led to jamming of chopped fuel.

The French chop-leach efforts began at La Hague in 1970. The first implementation was a vertically fed shear installed at the UP2 facility. The design has proven robust and the unit continues to operate as of this writing (1995). A horizontally fed unit of French design which extended this technology to include BWR fuel and the diversion of end fittings was installed in the PNC facility at Tokai-mura, Japan. It went into hot operation in 1977, and is operative as of 1995. The same technology was used in the AGNS shear which was also manufactured by the French (44). The UP3 plant at La Hague and the more recently upgraded UP2 facility continue to use this technology.

The AGNS shear was designed to cut the entire fuel assembly, thus eliminating the need for prior removal of end fittings. The necessity of physically moving sheared fuel to different processing locations was eliminated by installing multiple multicycle dissolvers (45). Specialized equipment facilitated hulls removal. Fission gases, released during the shearing operation, were routed into the vessel vent system by an inert gas sweep at the cutting blade, and at this point, provisions were made to install noble gas recovery equipment when a suitable technology was demonstrated (12).

The General Electric shear was a small unit. The fuel assembly end fittings were mechanically removed in a separate operation prior to shearing. The fuel rods were pulled from the fuel assembly frame and were inserted into the shear feed chamber as a single layer of rods. The shear feed chamber was then sealed to the shear, and the rods were cut into 7.6-cm lengths. The shear and feed chamber were vented to the dissolver, trapping the fission gases without having to handle the extra gas volumes associated with a sweep (10).

The thermal oxide reprocessing plant (THORP) at Sellafield, U.K., uses a hydraulically operated shear to cut fuel rods into pieces between 2.5 and about 10 cm in length. Because of wear on the shear blade, the moving blade and gag are remotely replaceable as a module. The hydraulic equipment for this, the feed envelope charge ram, and the shear incremental feed ram, are mounted outside the biological shielding. Modular redundancy allows off-line maintenance and high equipment availability (46).

The Japan Nuclear Fuel Service Co. reprocesses LWR fuel in facilities which take advantage of French shear and dissolver designs, German iodine removal technology, and British reduced-pressure evaporation.

Reprocessing facilities use equipment capable of shearing entire fuel assemblies. This accounts for a significant portion of the reprocessing cost in terms of size of the equipment needed and the consequent size of the cell required to house it. Shear blade maintenance requirements are also important. There are active programs aimed at reducing these costs. For instance, an operation involving only a single fuel pin would require a significantly smaller shear and cell to house it. Less oxide dust and cladding fines might also be anticipated. However, a fuel disassembly step with its attendant equipment and the space to house it must be added (7).

A number of alternatives to shearing have been investigated wherein the cladding is breached by lasers (qv), plasma torches (see PLASMA TECHNOLOGY), or inductive heating (47). In one proposal the oxide pellets are crushed while still inside the cladding and the fuel is tipped or vibrated out into a dissolver (48). Another approach is to chemically embrittle the cladding, which can then be crushed along with the oxide pellets before being fed to the dissolver (49). The crushing step would substitute for both the shear and the subsequent hulls volume reduction step.

Liquid–Liquid Contactors. A variety of contactors have been developed for liquid–liquid extraction processes. These designs include equilibrium-stage contactors such as mixer–settlers and centrifugal contactors, and differential contactors, such as packed and sieve-tray pulse columns. Each design has advantages and disadvantages; for example, mixer–settlers and centrifugal contactors are compact and require substantially less cell space than do columns. Centrifugal contactors have the additional advantage of high throughput capacities with minimal liquid holdup, ie, short residence times, and can perform over a wide range of organic-to-aqueous flow ratios. They are, however, subject to pluggage by entrained solids, and the high mixing energies can create stable emulsions in the presence of contaminates such as silicates. Other disadvantages include the difficulties of remotely maintaining rotating equipment. Compact annular centrifugal contactors are under development for application in the next generation of reprocessing plants (50).

Pulse columns have proven to be reliable, but have long residence times, ie, high volumetric holdup, which result in radiation damage to the solvent when reprocessing high burn-up fuels. They also must be housed in large shielded cells, are subject to flooding, and have only limited capacity if a geometrically safe configuration is required (51). The THORP facility in the U.K. uses conventional nozzle plate pulse columns having hafnium components which act as neutron absorbers to ensure nuclear safety (52). The French plants at La Hague use annular, baffel plate columns to maintain criticality safety (53).

The overall decontamination factor of fission products and other impurities in the process is often limited, not by mass-transfer relationships, but by hydrodynamic characteristics, such as the entrainment of one liquid phase in the other or poor solvent quality. Mono- and dibutyl phosphate degradation products, which are caused by radiation and chemical damage to TBP, form complexes with plutonium and with some of the fission products (54–57). Good extraction equipment designs provide efficient separation of the phases between contactors. Sequential washing of the organic phase using Na_2CO_3 and HNO_3, or alternatively treatment with a macroreticular resin, enhances the solvent quality (58).

Liquid Waste Storage Tanks.   Reprocessing of high burn-up, short-cooled LWR fuel required improvements in the design of the liquid waste storage tanks. Modern tanks are fabricated from stainless steel, and are placed in underground, stainless steel-lined concrete vaults. Advancements in the reprocessing flow sheet minimize salt concentrations and preclude discharge of dissolver solid to the storage tanks. The tanks are equipped with pneumatically activated, liquid-discharge pulsers to prevent the settling of heat-emitting solids on the bottom of the tank. The tanks are also equipped with air-lift circulators which maintain the solids in suspension and enhance heat transfer to the cooling coils. The coils are suspended from the tank ceiling to minimize interference with performance of the pulsers (11,59).

A mathematical model of the operating characteristics of a modern HLW storage tank has been developed (60). This model correlates experimental data for the rate of radiolytic destruction of nitric acid, the rate of hydrogen generation owing to radiolysis of water, and cooling coil heat transfer. These are all functions of nitric acid concentration and air-lift circulator operation.

The capacity of any specific tank configuration, in terms of metric ton equivalents, is determined by one of three parameters. (1) The solubility of waste salts. Precipitates can settle and cause thermal hot spots, which in turn can result in accelerated corrosion rates. Thus it is important to maintain the tank contents below the solubility limits. (2) The density of the waste solution. The tank footings are designed for a specific loading which translates to a waste solution density. As fuel burn-ups change or as process improvements are made, it is possible for the waste concentration to exceed the allowable density before the solubility limits are reached. (3) The decay heat-removal capacity. The heat-removal capacity of the tank is a function of the cooling-coil area and the overall heat-transfer coefficient. As the fuel burn-up or age changes, the specific heat generation rate of the waste changes. This, coupled with potential fouling of the cooling coils, could result in a heat-removal limitation, even when concentration and density are well below their limits. Waste tank operations consists in managing each of the above parameters so as to maximize the capacity of the available waste storage facilities.

BIBLIOGRAPHY

"Chemical Reprocessing" in *ECT* 2nd ed., under "Nuclear Reactors," Vol. 14, pp. 91–102, A. T. Gresky, Oak Ridge National Laboratory; in *ECT* 3rd ed., Vol. 16, pp. 173–183, by W. E. Unger, Oak Ridge National Laboratory.

1. W. R. Waltz, W. L. Godfrey, and A. K. Williams, *Int. Nucl. Technol.* (Dec. 1980).
2. R. H. Rainey, W. D. Burch, M. J. Haire, and W. E. Unger, *Fuel Cycle for the 80's Conference*, CONF-800943, Gatlinburg, Tenn., 1980, pp. 155–158.
3. *Nucleonics Week*, (July 7, 1994).
4. S. Lawroski and M. Levenson, *The Redox Process—A Solvent Extraction Reprocessing Method for Irradiated Uranium*, TID-7534, USAEC, Oak Ridge, Tenn., 1957.
5. W. L. Godfrey, R. Y. Dean, and J. C. Stouffer, *Recovery of Aluminum Nitrate Non-ahydrate from REDOX Acid Waste*, HW-82771-P3, General Electric Co., Dec. 1965.
6. P. Buchan to James Schlesinger, *Response To Query*, British Nuclear Fuels, Risley, U.K., July 4, 1994.
7. *Safety Analysis Report, NFS Reprocessing Plant*, Docket-50-201, Nuclear Fuel Services, Inc., Rockville, Md., 1973.
8. *Safety Evaluation of the Midwest Fuel Recovery Plant*, General Electric Co., Docket No. 50-268, United States Atomic Energy Commission, Washington, D.C., 1972.
9. *Final Safety Analysis Report—Barnwell Nuclear Fuel Plant, Separations Facility*, Docket-50-332, Allied-General Nuclear Services, Barnwell, S.C., 1973.
10. Cogema, *The Reprocessing Division*, Compagnie Générale Des Matières Nucléaires, Vélizy-Villacoubly Cedex, France, Apr. 1992.
11. *BNFL Welcomes THORP Go-Ahead*, BNFL Press Release, British Nuclear Fuels plc, Risley, U.K., Dec. 15, 1993.
12. J. Lovett, IAEA, personal communication, 1994.
13. C. Bernard, J. P. Moulin, P. Pradel, and M. Viala, *Global '93*, 57–62 (1993).
14. J. D. Frew and co-workers, *International Conference on Current Status and Innovations Leading to Promising Plants*, AESJ, PNC, and JAPC, Kyoto, Japan, Oct. 1991.
15. M. Viala, M. Tarnero, and M. Bourgeois, in Ref. 14, pp.
16. F. Baumgartner, *Kerntechnic*, **18**(6), 245 (1976).
17. G. Koch, *Kerntechnic*, **18**(6), 253 (1976).
18. W. S. Groenier, R. H. Rainey, and S. B. Watson, *Ind. Eng. Chem. Process Des. Devel.* **18**, 385 (1979).
19. A. L. Uriarte and R. H. Rainey, *Dissolutions of High Density UO_2, PuO_2, and UO_2–PuO_2 Pellets in Inorganic Acids*, ORNL-3695, Oak Ridge National Laboratory, Oak Ridge, Tenn., 1965.
20. *Purex Technical Manual*, HW-31000 (unclassified), United States Department of Energy, Washington, D.C., 1955.
21. W. D. Arnold and D. J. Crouse, *Evaluation of Alternative Extractants to TBP-1*, ORNL/TM-7536, Oak Ridge National Laboratory, Oak Ridge, Tenn., 1980.
22. S. B. Watson and R. H. Rainey, *Modifications of the SEPHIS Computer Code for Calculating the Purex Solvent Extraction System*, ORNL/TM-5123, Oak Ridge National Laboratory, Oak Ridge, Tenn., 1975.
23. AGNS Staff, *Engineering Evaluations of Alternatives for Processing Uranium-Based Fuels, Studies and Research Concerning the Barnwell Nuclear Fuels Plant (BNFP)*,

AGNS-1040-3.1-32, National Technical Information Service (NTIS), Springfield, Va., 1978.

24. R. F. Fleming, *A Compilation of Physical and Chemical Properties of Materials and Streams Encountered in the Chemical Processing Department*, HW-57386 (unclassified), United States Department of Energy, Washington, D.C., 1958.

25. A. F. Krigins, *A Compilation of Physical and Chemical Properties of Materials and Streams Encountered in the Chemical Processing Department—Addendum to HW-57386*, ARH-724 ADD (unclassified), U.S. Department of Energy, Washington, D.C., July 24, 1968.

26. I. S. Denniss and C. Phillips, *Solvent Extraction 1990*, Elsevier Science Publishing Co., Inc., New York, 1992, pp. 549–554.

27. I. S. Denniss and A. P. Jeapes, *Fourth International Conference on Nuclear Fuel Reprocessing and Waste Management, RECOD '94*, London, 1994.

28. A. F. Cermac and R. C. Spaunburgh, *Actinide Separations Symposium*, American Chemical Society, Honolulu, Hawaii, 1979.

29. H. Schneider, F. Baumgartner, H. Goldacking, and H. Hansberger, *Electrolytic Techniques in the Purex Process*, KFK-2082, Kernfonchungszentrum Karlsruhe, Germany, 1975; Eng. trans., ORNL-tr-2999, Oak Ridge National Laboratory, Oak Ridge, Tenn.

30. D. L. Pruett, *Radiochim. Acta* **27**, 115 (1980).

31. J. M. Leitnaker, M. L. Smith, and C. M. Fitzpatrick, *Conversion of Uranium Nitrate to Ceramic Grade Oxide for the Light-Water Breeder Reactor Process Development*, ORNL-4755, Oak Ridge National Laboratory, Oak Ridge, Tenn., 1972.

32. J. L. Woolfrey, *Preparation and Calcination of Ammonium Uranates—A Literature Survey*, AEC/TM-476, Australian Atomic Energy Commission, Lucas Heights, Australia, Sept. 1968.

33. *Westinghouse Plutonium Fuels Demonstration Progress Report*, Edison Electric Institute, ECAP-4167, Westinghouse Plutonium Fuels Development Laboratory, New York, 1970.

34. A. U. Blackham and J. Palmer, *Technetium as a Catalyst in Organic Reactions*, AT945-1-2017, Atlantic Richfield Hanford Co., Richland, Wash., July 1967.

35. M. H. Campbell, *Summary Report: Loading Technetium-99 on IRA-401 and Measurement of Product Purity*, Atlantic Richfield Hanford Co., Richland, Wash., June 1967.

36. J. R. Merriman, M. J. Stephenson, B. E. Kanak, and D. K. Little, *International Symposium on Management of Gaseous Wastes from Nuclear Facilities*, IAEA-SM-245/53, International Atomic Energy Agency (IAEA), Vienna, Austria, 1980.

37. D. W. Holladay and G. L. Haag, *15th DOE Nuclear Air Cleaning Conference*, Boston, Mass., 1978.

38. D. W. Holladay, *A Literature Survey: Methods for the Removal of Iodine Specia from Off-Gas and Liquid Waste Streams of Nuclear Power and Nuclear Fuel Reprocessing Plants, with Emphasis on Solid Sorbents*, ORNL/TM-6350, Oak Ridge National Laboratory, Oak Ridge, Tenn., 1979.

39. G. J. Raab and co-workers, *Operating Experience Using Silver Reactors for Radioiodine Removal in the Hanford Purex Plant*, ARH-SA-67, Atlantic Richfield Hanford Co., Richland, Wash., June 1970.

40. W. L. Godfrey, *Safety and Analysis Report—Caisson Storage of Hulls and High-Level General Process Trash*, GEN-007 and NT75-212, Allied-General Nuclear Services, Barnwell, S.C., May 1975.

41. P. Lederman, P. Miquel, and B. Boullis, *Fourth International Conference, Nuclear Fuel Reprocessing and Waste Management, RECOD '94*, London, 1994.

42. W. L. Godfrey and D. G. Bouse, *Tank Sludge Characterization*, Waste Management Research Abstracts, IAEA, Vienna, Austria, 1971.

43. W. L. Godfrey and D. J. Larkin, *Chem. Eng.* (July 13, 1970).

44. C. Byerly, Numatec, personal communication, 1994.
45. W. S. Groenier, *Equipment for the Dissolution of Core Material from Sheared Power Reactor Fuels*, ORNL/TM-3194, Oak Ridge National Laboratory, Oak Ridge, Tenn., 1971.
46. R. W. Asquith, P. I. Hudson, and M. Astill, *International Conference on Nuclear Fuel Reprocessing and Waste Management: RECOD '87*, Paris, 1987.
47. Fr. Pat. FR2 667 533A1 (1992), B. Roger.
48. Y. Takashima and co-workers, *International Conference on Nuclear Fuel Reprocessing and Waste Management, RECOD '87*, Paris, 1987.
49. M. Nakatauka, *Nucl. Technol.* **103**, 426–433 (1993).
50. R. A. Leonard and co-workers, *Development of a 25cm Annular Centrifugal Contactor*, ANL-80-15, Argonne National Laboratory, Chicago, Ill., June 1980.
51. L. Burkhart, *A Survey of Simulated Methods for Modeling Pulsed Sieve-Plate Extraction Columns*, UCRL-15101, Ames Laboratory, Iowa State University, Ames, Iowa, 1979.
52. C. Phillips, *Solvent Extraction Conference*, Toulouse, France, 1985.
53. W. Fournier and co-workers, in Ref. 26, pp. 747–752.
54. Z. Nowak, *Wukleonia* **18**, 447 (1973).
55. E. V. Barelko and I. P. Solyanina, *Sov. Atom. Energy* **35**, 898 (1973).
56. P. G. Clay and M. Witort, *Radiochem. Radioanal. Lett.* **19**(2), 101 (1974).
57. L. P. Sokhina, F. A. Bogdanov, A. S. Solovkin, E. G. Telerin, and W. N. Shesterikov, *Russian J. Inorg. Chem.* **21**, 1358 (1976).
58. W. W. Schulz, *Macroreticular Anion Exchange Resin Cleanup for TBP Solvents*, ARH-SA-129, Atlantic Richfield Hanford Co., Richland, Wash., 1972.
59. D. W. Cleelland, *Proceedings of the Symposium on the Solidification and Long-Term Storage of Highly Radioactive Wastes*, USAEC, Richland, Wash., 1966.
60. J. C. Hall, *Proceedings of the Waste Management and Fuel Cycle Symposium*, University of Arizona, Tucson, 1978, pp. 371–387.

W. L. GODFREY
J. C. HALL
G. A. TOWNES
BE Incorporated

REACTOR TYPES

The minimum ingredients of a nuclear reactor, where the basic reactions are nuclear rather than chemical, are neutrons and a fuel such as uranium, the atoms of which can undergo fission. Products of the fission process, in order of importance, are (*1*) heat energy, originally in the form of kinetic energy of particles; (*2*) neutrons, originally of high energy, which can be slowed to lower energies; (*3*) radionuclides, originally as fission fragments and collectively called fission products; (*4*) beta and gamma radiation, released in the fission process and by decay of fission products, which contributes both to heat and hazard; and

(5) neutrinos, which play no role, because of the ease with which these penetrate matter. The distribution of energy among these products of fission is as follows, for a total of 200 MeV (1).

Particle	Energy, MeV
fission fragments	166
neutrons	5
prompt γ-rays	7
fission product γ-rays	7
beta particles	5
neutrinos	10

Reactor Components

Several components are required in the practical application of nuclear reactors (1–5). The first and most vital component of a nuclear reactor is the fuel, which is usually uranium slightly enriched in uranium-235 [15117-96-1] to approximately 3%, in contrast to natural uranium which has 0.72% ^{235}U. Less commonly, reactors are fueled with plutonium produced by neutron absorption in uranium-238 [24678-82-8]. Even more rare are reactors fueled with uranium-233 [13968-55-3], produced by neutron absorption in thorium-232 (see NUCLEAR REACTORS, NUCLEAR FUEL RESERVES). The chemical form of the reactor fuel typically is uranium dioxide, UO_2, but uranium metal and other compounds have been used, including sulfates, silicides, nitrates, carbides, and molten salts.

The second important component is the cooling agent or reactor coolant which extracts the heat of fission for some useful purpose and prevents melting of the reactor materials. The most common coolant is ordinary water at high temperature and high pressure to limit the extent of boiling. Other coolants that have been used are liquid sodium, sodium–potassium alloy, helium, air, and carbon dioxide (qv). Surface cooling by air is limited to unreflected test reactors or experimental reactors operated at very low power.

The third component is the moderator, a substance containing light elements such as hydrogen, deuterium, or carbon. Because low (ca 0.025-eV) energy neutrons are much more effective in causing fission than high (ca 2-MeV) energy neutrons, such a medium is desirable to slow neutrons by causing multiple collisions. The lighter the nuclear target, the greater is the energy loss per collision. Thermal reactors are those having a moderator that brings neutrons down to energies comparable to the thermal agitation of atoms. Fast reactors do not have a moderator, and because there is limited slowing, neutrons remain at around 1 MeV.

The fourth component is the set of control rods, which serve to adjust the power level and, when needed, to shut down the reactor. These are also viewed as safety rods. Control rods are composed of strong neutron absorbers such as boron, cadmium, silver, indium, or hafnium, or an alloy of two or more metals.

The fifth component is the structure, a material selected for weak absorption for neutrons, and having adequate strength and resistance to corrosion. In thermal reactors, uranium oxide pellets are held and supported by metal tubes, called the cladding. The cladding is composed of zirconium, in the form of an alloy called Zircaloy. Some early reactors used aluminum; fast reactors use stainless steel. Additional hardware is required to hold the bundles of fuel rods within a fuel assembly and to support the assemblies that are inserted and removed from the reactor core. Stainless steel is commonly used for such hardware. If the reactor is operated at high temperature and pressure, a thick-walled steel reactor vessel is needed.

The sixth component of the system is the shield, which protects materials and workers from radiation, especially neutrons and gamma rays. Concrete is commonly used, augmented by iron and lead for gamma rays and water for fast neutrons.

Uses

The primary use of a nuclear reactor is as a compact alternative heat source, wherever that is needed. Because a reactor is a complex system, it is economical as a power source only if it produces fairly large blocks of power, especially for steady electric power in the 500–1500-MWe range (6–9). If, however, expense is not a factor, a reactor could be used to provide electricity in small units to remote locations in which it is difficult to supply ordinary fuels. Electrical energy from reactors can be used to provide mechanical energy for propulsion of ocean vessels, including submarines, aircraft carriers, cargo ships, and icebreakers. A reactor can provide heat only, for a variety of applications, including process steam production, district heating, desalination of water, and direct propulsion of space vehicles.

The neutrons in a research reactor can be used for many types of scientific studies, including basic physics, radiological effects, fundamental biology, analysis of trace elements, material damage, and treatment of disease. Neutrons can also be dedicated to the production of nuclear weapons materials such as plutonium-239 from uranium-238 and tritium, ^{3}H, from lithium-6. Alternatively, neutrons can be used to produce radioisotopes for medical diagnosis and treatment, for gamma irradiation sources, or for heat energy sources in space.

Classification

Nuclear reactors can be classified in a variety of ways: by purpose or use, key components, method of heat extraction, role in application, neutron energy, power level or neutron flux, arrangement of materials, stage of development, and manufacturer (10,11). Reactors are used for heat power, electrical power, propulsion, training, neutron production for basic research or materials testing, and for radioisotope production, including weapons materials. U.S. commercial power plants are called lightwater reactors because ordinary water is used for neutron slowing and heat removal. Canadian reactors use deuterium oxide, ie, heavy water, and some European reactors use graphite.

Most nuclear reactors use a heat exchanger to transfer heat from a primary coolant loop through the reactor core to a secondary loop that supplies steam (qv) to a turbine (see HEAT-EXCHANGE TECHNOLOGY). The pressurized water reactor is the most common example. The boiling water reactor, however, generates steam in the core.

The role of the reactor may be either as a converter, which produces some plutonium by neutron absorption in uranium-238 but depends on uranium-235 for most of the fission, or as a breeder, which contains a large amount of plutonium and produces more fissile material than it consumes. Breeding is also possible using uranium-233 produced by neutron absorption in thorium-232.

The characteristic neutron energy, ie, that at which most of the fission occurs, may be thermal, fast, or intermediate. In thermal reactors, those which have a moderator, the typical energy is on the order of a fraction of an electron volt (eV). Fast reactors, in which little neutron slowing occurs, operate with the neutrons in the MeV range. Intermediate reactors are those that operate on epithermal neutrons, those above thermal but considerably slowed from fission energy. The reactor power level or neutron flux is another classification system. For example, a low power reactor, high flux reactor, or 1200-MW version all aid in identifying a given system.

Another category is that of arrangement. Historically, studies were made of homogeneous reactors, where the fuel was solid metal, or in solution or slurry. All modern reactors are heterogeneous. The fuel and moderator are in distinctly separate zones. The term heterogeneous is also applied to an alternative arrangement of materials in a fast breeder reactor. Also in the fast breeder, there are two arrangements of the coolant: the loop, in which fluid circulates through the core and heat exchanger; and the pot, a large reservoir in which the core sits.

Reactors may be experimental, test, prototype, demonstration, or commercial. In the United States, the four principal companies that have designed and built most of the power reactors are Westinghouse Corp., General Electric Co., Babcock and Wilcox Co., and Combustion Engineering. There are several important foreign manufacturers as well, such as Framatome in France; Atomic Energy of Canada, Ltd. in Canada; Mitsubishi and Toshiba in Japan; Siemens AG in Germany; and UKAEA in the United Kingdom.

Herein reactors are described in their most prominent application, that of electric power. Five distinctly different reactors, ie, pressurized water reactors, boiling water reactors, heavy water reactors, graphite reactors, and fast breeder reactors, are emphasized. A variety of other applications and types of reactors also exist. Whereas space does not permit identification of all of the reactors that have been built over the years, each contributed experience of processes and knowledge about the performance of materials, components, and systems.

Research and Development

The chronology of the development of nuclear reactors can be divided into several principal periods: pre-1939, before fission was discovered (12); 1939–1945, the time of World War II (13–15); 1945–1963, the era of research, development, and demonstration (16–18); 1963–mid-1990s, during which reactors have been deployed in large numbers throughout the world (10,18); and extending into the twenty-first century, a time when advanced power reactors are expected to be built (19–23). Design of nuclear reactors has been based on a combination of theory, measurement of basic and derived parameters, and experiments with complete systems (24–27).

The nuclear chain reaction can be modeled mathematically by considering the probable fates of a typical fast neutron released in the system. This neutron may make one or more collisions, which result in scattering or absorption, either in fuel or nonfuel materials. If the neutron is absorbed in fuel and fission occurs, new neutrons are produced. A neutron may also escape from the core in free flight, a process called leakage. The state of the reactor can be defined by the multiplication factor, k, the net number of neutrons produced in one cycle. If k is exactly 1, the reactor is said to be critical; if $k < 1$, it is subcritical; if $k > 1$, it is supercritical. The neutron population and the reactor power depend on the difference between k and 1, ie, $\delta k = k - 1$. A closely related quantity is the reactivity, $\rho = \delta k/k$. If the reactivity is negative, the number of neutrons declines with time; if $\rho = 0$, the number remains constant; if ρ is positive, there is a growth in population.

The mathematical model most widely used for steady-state behavior of a reactor is diffusion theory, a simplification of transport theory which in turn is an adaptation of Boltzmann's kinetic theory of gases. By solving a differential equation, the flux distribution in space and time is found or the conditions on materials and geometry that give a steady-state system are determined.

A key parameter in determining the possibility of a self-sustained chain reaction is the value of k for an infinite medium, k_∞. In the four-factor formula,

$$k_\infty = \epsilon p f \eta$$

the succession of events in the neutron cycle are shown, where ϵ represents the fast fission factor; p, the resonance escape probability; f, thermal utilization; and η, the reproduction factor (28,29). A finite assembly called k-effective, k_e, is defined as

$$k_e = k_\infty \mathcal{L}$$

where $\mathcal{L}$ is a nonleakage probability, which depends on neutron slowing and diffusion properties in reactor materials and on the size and shape of the reactor. Critical experiments in which fuel is gradually accumulated determine the critical mass of uranium or the number of fuel assemblies.

The analysis of steady-state and transient reactor behavior requires the calculation of reaction rates of neutrons with various materials. If the number density of neutrons at a point is n and their characteristic speed is v, a flux $\phi = nv$ can be defined. Where an effective area of a nucleus as a cross section σ, and a target atom number density N, a macroscopic cross section $\Sigma = N\sigma$ can be defined, and the reaction rate per unit volume is $R = \phi\Sigma$. This relation may be applied to the processes of neutron scattering, absorption, and fission in balance equations leading to predictions of k_e, or to the determination of flux distribution. The consumption of nuclear fuels is governed by time-dependent differential equations analogous to those of Bateman for radioactive decay chains. The rate of change in number of atoms N owing to absorption is as follows:

$$dN/dt = -\phi\Sigma_a$$

Greater detail in the treatment of neutron interaction with matter is required in modern reactor design. The neutron energy distribution is divided into groups governed by coupled space-dependent differential equations.

The simplest model of time-dependent behavior of a neutron population in a reactor consists of the point kinetics differential equations, where the space-dependence of neutrons is disregarded. The safety of reactors is greatly enhanced inherently by the existence of delayed neutrons, which come from radioactive decay rather than fission. The differential equations for the neutron population, n, and delayed neutron emitters, c_i, are

$$\frac{dn}{dt} = (\rho - \beta)\frac{n}{\Lambda} + \sum \lambda_i c_i$$
$$\frac{dc_i}{dt} = \beta_i \frac{n}{\Lambda} - \lambda_i c_i$$

where the reactivity $\rho = \delta k_e / k_e$ and the neutron cycle time is Λ. Decay constants are λ_i, and delayed fractions are β_i, where $i = 1, 2, \ldots, 6$, and β is the sum of β_i over i. A fraction β of only 0.65% of the neutrons having an average delay of approximately 13 s greatly slows down the transient response to a change in k_e. In this simple reactor kinetics model, neglecting temperature effects, a positive reactivity gives, for long times after application, an exponential increase in neutron number with time, where $n(t)$ is proportional to $\exp(t/T)$, and T is the asymptotic period.

More generally, the neutron number density and the reactor power distribution are both time- and space-dependent. Also, there is a complex relation between reactor power, heat removal, and reactivity.

Operation of a reactor in steady state or under transient conditions is governed by the mode of heat transfer, which varies with the coolant type and behavior within fuel assemblies (30). Qualitative understanding of the different regimes using water cooling can be gained by examining heat flux, q'', as a function of the difference in temperature between a heated surface and the saturation temperature of water (Fig. 1).

In region A of Figure 1, transfer is by convection between heated fuel surfaces and contacting water. In region B, evaporation of water into bubbles occurs at points. The bubbles detach and rise owing to buoyancy, carrying vapor and the heat of vaporization with them. As the temperature difference increases into region C, a transition occurs because the vapor that begins to coat the surface is a poorer conductor of heat than was the liquid in the previous region. This region is unstable because of intermittent wetting. Finally, film boiling, or post-dryout heat transfer (region D), takes place, where sufficient wall temperatures have developed to cause the heat flux again to increase.

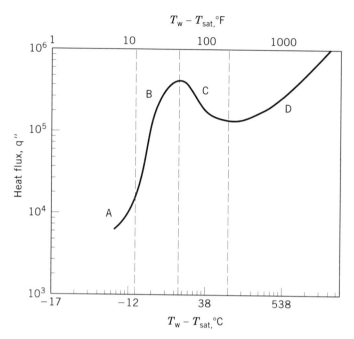

Fig. 1. Variation of heat flux, q″, with temperature difference between heated wall, T_w, and saturation temperature of water, T_{sat}, in regions where A represents convection; B, nucleate boiling; C, transition; and D, film boiling.

The mathematical formulation of forced convection heat transfer from fuel rods is well described in the literature. Notable are the Dittus-Boelter correlation (26,31) for pressurized water reactors (PWRs) and gases, and the Jens-Lottes correlation (32) for boiling water reactors (BWRs) in nucleate boiling.

Designs are sought that maximize inherent safety by achieving a net negative reactivity feedback, where any tendency for the power of the reactor to increase and raise temperatures results in a counteracting effect. Several mechanisms are available. For a reactor in which neutron multiplication depends strongly on moderation, thermal expansion of the fluid or the creation of steam bubbles results in reduced neutron thermalization and greater neutron leakage. Both effects provide negative reactivity feedback. For a reactor having considerable uranium-238 content, a fuel temperature rise changes the rate of interaction between neutrons and the fuel in the resonance region. This is the Doppler effect. Such effects are quantified by measurements of various coefficients of reactivity, eg, temperature, power, Doppler, or void. For safety, the net coefficient must be sufficiently negative at all times.

Account must be taken in design and operation of the requirements for the production and consumption of xenon-135 [14995-12-1], ^{135}Xe, the daughter of iodine-135 [14834-68-5], ^{135}I. Xenon-135 has an enormous thermal neutron cross section, around 2.7×10^{-18} cm^2 (2.7×10^6 barns). Its reactivity effect is constant when a reactor is operating steadily, but if the reactor shuts down and the neutron flux is reduced, xenon-135 builds up and may prevent immediate restart of the reactor.

Several of the reactor physics parameters are both measurable and calculable from more fundamental properties such as the energy-dependent neutron cross sections and atom number densities. An extensive database, Evaluated Nuclear Data Files (ENDF), has been maintained over several decades. There is an interplay between theory and experiment to guide design of a reactor, as in other engineering systems.

The results of design studies, calculations, and experiments for the reactor types selected for investigation in the post-World War II period were collected in the multivolume proceedings of several international conferences at Geneva (24). The U.S. Atomic Energy Commission (AEC) sponsored the publication of individual books between 1958 and 1964 describing the status of several reactor types (9,18,33–36).

Graphite Reactors

The first nuclear reactor made was composed of graphite, the only moderator available at that time for use with natural uranium. Reactors for the production of plutonium during World War II and for power in the United Kingdom also utilized carbon in the form of graphite. A modern helium-cooled graphite reactor has been tested. A distinct advantage of having carbon as the moderator is that it provides the ability to use natural uranium as fuel, avoiding the necessity of expensive and power-absorbing enrichment facilities.

The First Reactor.   When word about the discovery of fission in Germany reached the United States, researchers thereafter found that (1) the principal uranium isotope involved was uranium-235; (2) slow neutrons were very effective in causing fission; (3) several fast neutrons were released; and (4) a large energy release occurred. The possibility of an atom bomb of enormous destructive power was visualized.

At about the same time, the artificial isotope plutonium-239 [15117-48-3] was discovered and was recognized as also being fissionable. This led to the conjecture that a controlled chain reaction might be achieved and that neutrons could be used to produce enough plutonium for a weapon. Experiments were conducted during the Metallurgical Project, centered at the University of Chicago, and led by Enrico Fermi. Subcritical assemblies of uranium and graphite were built to learn about neutron multiplication. In these exponential piles the neutron number density decreased exponentially from a neutron source along the length of a column of materials. There was excellent agreement between theory and experiment.

A larger assembly, ie, one that might be self-sustaining, or critical, was built. Of special importance was the need for graphite of sufficiently high purity, because traces of boron would absorb neutrons and prevent multiplication. The pile was constructed of 37 layers of graphite blocks where chunks of uranium oxide and uranium metal alternated with the layers of graphite only. To control the reaction and provide safety in case of accident, a set of neutron-absorbing rods and an emergency cadmium solution were provided. On December 2, 1942 the reactor was brought to critical and the power allowed to increase to a few hundred watts (37). Safety aspects of that experiment are highlighted in a 1988 article (38).

The Hanford Production Reactors. On the basis of the success of the first reactor experiment, construction of several plutonium production reactors began at Hanford, Washington. These reactors used graphite as moderator, but because high power levels were involved in producing the required amounts of plutonium, water cooling of the fuel was provided. Graphite blocks were pierced by holes lined with aluminum tubes, into which aluminum-canned uranium metal cylinders known as slugs were placed. After a period of operation, the slugs could be pushed along and discharged.

Chemical processing or reprocessing (39) of the fuel to extract the plutonium and uranium left a residue of radioactive waste, which was stored in underground tanks. By 1945, the reactors had produced enough plutonium for two nuclear weapons. One was tested at Alamogordo, New Mexico, in July 1945; the other was dropped at Nagasaki in August 1945.

A second approach to the production of weapons material was the uranium electromagnetic separation process, based on research at the University of California and production facilities at Oak Ridge, Tennessee. In a two-stage process, uranium of over 90% U-235 was obtained for use at Hiroshima.

Magnox and AGR Reactors. The greatest use of graphite has been in the United Kingdom (40,41). In the period 1956–1960 the first eight 50-MWe reactors were built and put into operation at Calder Hall and Chapelcross. These used natural uranium in the form of metal rods, clad with a magnesium alloy, Magnox, and using carbon dioxide as coolant. The reactor cores were very large and contained as many as 10,000 fuel channels. Many of these reactors, in operation as of the mid-1990s, are said to be rather inefficient in converting heat to electricity, but are especially reliable. Later, the United Kingdom and France built advanced gas-cooled reactors (AGR), which were much more compact. These latter used slightly (ca 2% ^{235}U) enriched uranium as fuel.

Table 1 contains technical data for the newer plants of the Magnox and AGR type. These are operated in the United Kingdom by Nuclear Electric plc. The electrical power output of the AGR is almost three times that of the Magnox, whereas its core volume is less than half as large.

Sodium Graphite Reactor. A reactor cooled by liquid sodium and moderated by graphite can take advantage of excellent heat-transfer features and low neutron absorption, permitting use of low enrichment uranium (33). The sodium reactor experiment (SRE) and the Hallam, Nebraska nuclear power facility (HNPF), both designed by Atomics International (AI), were the only U.S. examples of this reactor type. These were part of the Atomic Energy Commission's power reactor demonstration program. The 75-MWe HNPF reactor used liquid sodium as coolant. Its fuel consisted of slugs of uranium–molybdenum alloy in Zircaloy 2 tubes, as 18-rod clusters. Design details are available (18). A few tests of uranium carbide fuel were made. The Hallam reactor operated from 1962 to 1964, when the thin stainless steel cans for the graphite blocks developed leaks that admitted sodium. In that brief period, many technical problems were solved, the reactor concept was tested successfully, and the ability to manage large volumes of sodium was demonstrated (42).

High Temperature Gas-Cooled Reactors. The high temperature gas-cooled reactor (HTGR) uses graphite as moderator, but has an unusual type of fuel (31,43). As produced by General Atomics of San Diego, California, the

Table 1. Operating Graphite Reactors of the U.K.[a]

Parameter	Reactor site	
	Oldbury	Heysham 2
reactor type	Magnox	AGR
startup date	Jan. 1968	July 1988
number of reactors	2	2
electrical output, MWe	435	1230
coolant	CO_2	CO_2
coolant temperatures, °C		
in	220	298
out	365	635
gas pressure, MPa[b]	2.52	4.57
pressure vessel, inside dimensions		
diameter, m	23.5	20.25
height, m	18.3	21.87
core dimensions		
diameter, m	14.2	9.46
height, m	9.8	8.31
fuel type	metal rods	UO_2 pellets
fuel can material	Magnox	stainless steel

[a]Courtesy of Nuclear Electric plc.
[b]To convert MPa to psia, multiply by 145.

highly enriched (93%) fuel consists of coated spherical particles of diameter about 1 mm. As shown in Figure 2, the kernel is a sphere of uranium dioxide, uranium carbide, or mixtures of these with silicon or aluminum. Kernels are prepared by a series of chemical processes and heat treated. Several thin coatings are applied, consisting of pyrolytic carbon or silicon carbide, or a combination of the two. These layers prevent fission products from escaping from the kernel, even when the temperature is as high as 1000°C. Solid fuel rods are fabricated from the coated particles and a carbon binder, and inserted into holes in large hexagonal graphite blocks. The prisms also have holes for passage of coolant. Stacks of blocks form the large core, measuring several meters in each direction. The core is located within a large prestressed concrete reactor vessel.

The coolant for the HTGR is helium. The helium is not corrosive; has good heat properties, having a specific heat that is much greater than that of CO_2; does not condense and can operate at any temperature; has a negligible neutron absorption cross section; and can be used in a direct cycle, driving a gas turbine with high efficiency.

The highest power of a reactor of the HTGR type was 330 MWe in Fort St. Vrain, Colorado. The reactor, started in 1979, had many technical problems, including helium leaks, and did not perform up to expectations. It was shut down in 1989.

Chernobyl. The most well-known graphite-moderated reactor is the infamous Chernobyl-4, in Ukraine. It suffered a devastating accident in 1986 that spread radioactivity over a wide area of Europe.

The 950-MWe RBMK reactor of the type used at Chernobyl-4 has graphite for the moderator and slightly enriched (2%) uranium as oxide canned in a

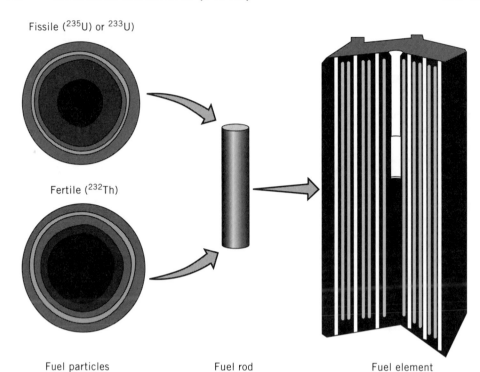

Fissile (^{235}U) or ^{233}U)

Fertile (^{232}Th)

Fuel particles   Fuel rod   Fuel element

Fig. 2. Fuel for high temperature gas-cooled reactor. Fissile material is coated with carbon and silicon carbide, fertile material with carbon. Particles mixed with carbon form fuel rods inserted in graphite blocks. Courtesy of General Atomics.

zirconium–niobium alloy for fuel. The bundles of fuel rods are inside 8.6-cm diameter pressure tubes in which light water is brought to boiling. The core is cylindrical, 7 m high and 12 m wide, pierced by 1661 vertical fuel channels and 222 control and safety channels. An overhead refueling machine allows fuel insertion and removal during operation.

Chernobyl-4 was completely destroyed in a violent explosion in 1986. The roof of the reactor building blew off and the graphite caught on fire and released radioactive fuel into the atmosphere. A number of workers were killed and the public was exposed to radiation. The accident was caused by a combination of reactor design features and operational errors: performance of an experiment that bypassed the safety equipment, without evaluation of hazardous consequences; removal of all safety rods during the course of the experiment in order to raise the reactor power; inadequate speed of control by the neutron-absorbing rods; and an inherently unsafe design, having a positive temperature coefficient of reactivity. It is generally believed that the consequences of the Chernobyl accident would have been far less serious if there had been a containment of the type used in Western reactors, rather than simple confinement by a conventional building (44).

Other RBMK reactors are still in operation in the former USSR. Some changes improving the prospects of safe operation have been made.

The Hanford N Reactor. The Hanford N reactor was built in 1964 for purposes of plutonium production during the Cold War. It used graphite as moderator, pierced by over 1000 Zircaloy 2 tubes. These pressure tubes contained slightly enriched uranium fuel cooled by high temperature light water. The reactor also provided 800 MWe to the Washington Public Power Supply System. This reactor was shut down in 1992 because of age and concern for safety. The similarity to the Chernobyl-type reactors played a role in the decision.

Pressurized Water Reactors

The development of the pressurized lightwater reactor (PWR) involved studies of many types of reactors, including a gas-cooled power reactor, a fast breeder reactor, an aircraft propulsion reactor, and a high flux reactor for radiation testing. This last, called the Materials Testing Reactor (MTR), was authorized by the AEC in 1948 to test structural materials and fuels under high radiation conditions. It was built and put into operation in 1952 at the National Reactor Testing Station (NRTS) (now the Idaho National Engineering Laboratory) in Idaho, through the cooperation of Argonne National Laboratory (Illinois) and Oak Ridge National Laboratory (Tennessee). Fuel plates were sandwiches of aluminum and uranium–aluminum metal alloy. Water served as moderator and coolant. The reactor had a beryllium reflector to enhance thermal neutron flux.

Another reactor that was approved for development was a land-based prototype submarine propulsion reactor. Westinghouse Electric Corp. designed this pressurized water reactor, using data collected by Argonne. Built at NRTS, the reactor used enriched uranium, the metal fuel in the form of plates. A similar reactor was installed in the submarine *Nautilus*.

The experience and capability of the Westinghouse Bettis Laboratory were then applied to designing and constructing the first full-scale commercial power reactor, the 60-MWe Shippingport, Pennsylvania reactor of Duquesne Light Co. The core of the Shippingport reactor (18,34) was composed of two types of fuel. There were 32 "seed" assemblies of highly enriched (90%) uranium alloyed with zirconium and clad with Zircaloy 2 (a 98.3% Zr alloy having 1.45% Sn and 0.05% Ni), in the form of plates 3.175 mm (1/8 in.) thick. There were 113 blanket assemblies, each of 120 fuel rods composed of natural uranium as UO_2 pellets in Zircaloy 2 tubes. The seed region was in the shape of a square ring, with blanket fuel both inside and outside the ring. Neither type of fuel could sustain a chain reaction by itself, the seed because of excessive neutron leakage, the blanket because of the low uranium-235 content. Together these formed a critical system, and about the same amount of power was produced by each type of fuel. The use of Zircaloy tubes filled with uranium dioxide pellets has become standard for the industry.

Modern PWRs have power levels far higher than 60 MWe, are considerably larger, and are much more complex. Figure 3 shows a schematic diagram of the reactor vessel, heat exchanger, turbine-generator, and other equipment. Also shown is a schematic of the containment, a large concrete and steel structure capable of withstanding a significant excess pressure from accidental release of hot steam from the reactor vessel.

The key feature of the pressurized water reactor is that the reactor vessel is maintained above the saturation pressure for water and thus the coolant-moderator does not boil. At a vessel pressure of 15.5 MPa (2250 psia), high

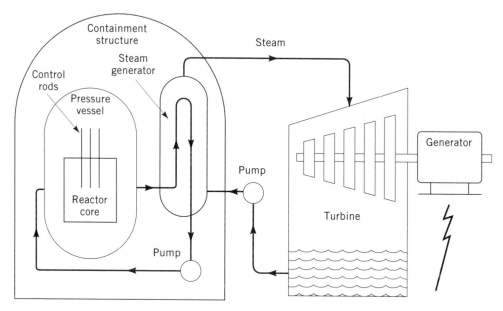

Fig. 3. Schematic of a pressurized water reactor system. Fission heat is extracted by the lightwater coolant. The steam drives the turbine-generator. Courtesy of the Nuclear Energy Institute.

water temperatures averaging above 300°C can be achieved, leading to acceptable thermal efficiencies of approximately 0.33.

About half of the world's nuclear power plants are from Westinghouse Electric Corp. or its licensees. One Westinghouse PWR design is the four-loop Model 412 (45). To maintain the pressure on the coolant-moderator in the reactor vessel at the required 15.5 MPa (2250 psia), an auxiliary device called a pressurizer is used. For the Model 412, the pressurizer is a container with a length of 16.1 m and a diameter of 2.3 m that is connected directly to the reactor vessel. The pressurizer has electrical heaters that can heat the water to raise the pressure, and a spray head that injects cold water to lower the pressure.

The steel reactor vessel must be quite thick, eg, 203 mm (8 in.), to withstand the pressure. Figure 4 shows a cutaway view of the pressure vessel, showing the relationship of fuel, control rods, and coolant passages; Table 2 gives the corresponding data of the Westinghouse Model 412; the fuel assembly, consisting of an array of 264 fuel rods and 25 vacant spaces is shown in Figure 5. Typically uranium of three different enrichments is initially loaded into the reactor. The configuration shown in Figure 6, where the highest enrichment is placed on the outside and a checkerboard arrangement of two other enrichments is put in the center, favors uniform power and burn-up.

The Model 412 PWR uses several control mechanisms. The first is the control cluster, consisting of a set of 25 hafnium metal rods connected by a spider and inserted in the vacant spaces of 53 of the fuel assemblies (Fig. 6). The clusters can be moved up and down, or released to shut down the reactor quickly. The rods are also used to (*1*) provide positive reactivity for the startup of the reactor from cold conditions, (*2*) make adjustments in power that fit

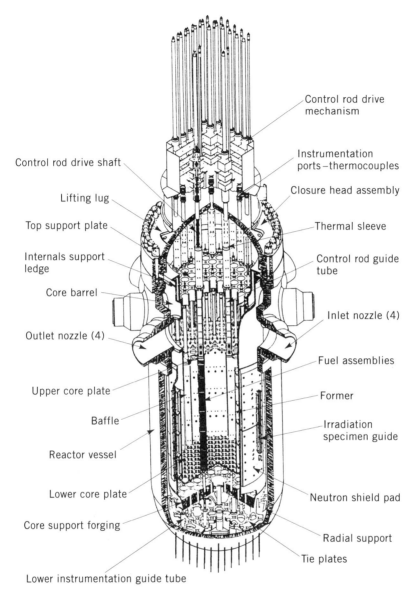

Fig. 4. Cutaway view of the Model 412 four-loop pressurized water reactor vessel (46). Courtesy of Westinghouse Electric Corp.

the load demand on the system, (3) help shape the core power distribution to assure favorable fuel consumption and avoid hot spots on fuel cladding, and (4) compensate for the production and consumption of the strongly neutron-absorbing fission product xenon-135. Other PWRs use an alloy of cadmium, indium, and silver, all strong neutron absorbers, as control material.

The second control mechanism is the soluble reactor poison boric acid [10043-35-3], H_3BO_3. Natural boron contains 20% boron-10 [14798-12-0], ^{10}B, which has a thermal neutron cross section of ca 4.0×10^{-25} m^2 (4000 barns). As

Table 2. Westinghouse Model 412 Pressurized Water Reactor[a]

Parameter	Value
thermal power, MW	3425
electrical power, MWe	1150
reactor vessel ID, m	4.394
primary system pressure, MPa[b]	15.5
coolant flow rate, kg/s	17,438
coolant temperatures, °C	
inlet	291.9
outlet	325.8
rise	33.9
steam pressure, MPa[b]	6.9
fuel dimensions, mm	
fuel rod OD	9.14
Zircaloy-4 cladding thickness	0.572
diametral gap	0.157
UO_2 pellet diameter	7.844
lattice pitch	12.60
fuel assembly array	17×17
rods per assembly	264[c]
number of assemblies in core	193
rods per core	50,952
fuel total weight, kg	81,639
core dimensions, m	
effective diameter	3.38
fuel height	3.658

[a]Ref. 47.
[b]To convert MPa to psia, multiply by 145.
[c]25 spaces are taken by control rods, burnable poison rods, or neutron sources, or are plugged.

fuel is consumed and fission products build up during a year or so of operation, the concentration of boron is adjusted by dilution. Starting from an initial value of around 2000 ppm, the boron concentration goes to zero at the end of the cycle.

The third control is by use of a fixed burnable poison. This consists of rods containing a mixture of aluminum oxide and boron carbide, included in the initial fuel loading using the vacant spaces in some of the fuel assemblies that do not have control clusters. The burnable poison is consumed during operation, causing a reactivity increase that helps counteract the drop owing to fuel consumption. It also reduces the need for excessive initial soluble boron. Other reactors use gadolinium as burnable poison, sometimes mixed with the fuel.

In the startup of a reactor, it is necessary to have a source of neutrons other than those from fission. Otherwise, it might be possible for the critical condition to be reached without any visual or audible signal. Two types of sources are used to supply neutrons. The first, applicable when fuel is fresh, is californium-252 [13981-17-4], ^{252}Cf, which undergoes fission spontaneously, emitting on average three neutrons, and has a half-life of 2.6 yr. The second, which is effective during operation, is a capsule of antimony and beryllium.

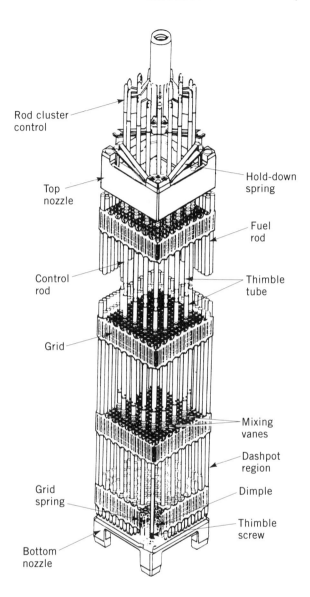

Labels on figure:
Rod cluster control
Top nozzle
Control rod
Grid
Grid spring
Bottom nozzle
Hold-down spring
Fuel rod
Thimble tube
Mixing vanes
Dashpot region
Dimple
Thimble screw

Fig. 5. Fuel assembly of Model 412 PWR having a 17 × 17 array of rods (48). Courtesy of Westinghouse Electric Corp. (48).

Antimony-123 [*14119-16-5*], ^{123}Sb, is continually made radioactive by neutron absorption. The product antimony-124 [*14683-10-4*], ^{124}Sb, is radioactive, has half-life 60 d, and its gamma rays cause beryllium-9, ^{9}Be, to emit neutrons.

An engineered safety system is provided to protect against hazard from a loss-of-coolant accident (LOCA). If a coolant pipe should break, causing a drop in pressure in the vessel, an emergency core cooling system (ECCS) begins supplying auxiliary water from storage tanks to continue cooling the core. A

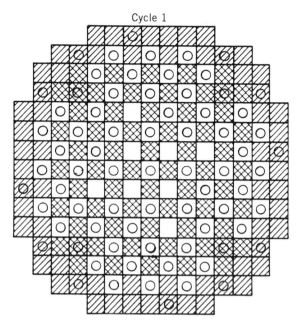

Fig. 6. Initial fuel loading of PWR showing the three enrichments where (▨) represents 3.10 wt % ^{235}U; (▨) 2.60; (□) 2.10; and (◯) the cluster openings of control rods (49). Courtesy of Westinghouse Electric Corp.

water spray system in the containment helps condense steam, and cooling fans go into operation.

A PWR can operate steadily for periods of a year or two without refueling. Uranium-235 is consumed through neutron irradiation; uranium-238 is converted into plutonium-239 and higher mass isotopes. The usual measure of fuel burnup is the specific thermal energy release. A typical figure for PWR fuel is 33,000 MW·d/t. Spent fuel contains a variety of radionuclides (50):

Isotope	Percent
^{235}U	0.81
^{236}U	0.51
^{238}U	94.3
^{239}Pu	0.52
^{240}Pu	0.21
^{241}Pu	0.10
^{242}Pu	0.05
fission products	3.5

The original fuel contained 3.3 wt % uranium-235 and 96.7 wt % uranium-238.

Boiling Water Reactors

Water is also used as moderator-coolant in the boiling water reactor (BWR). The principal distinguishing feature from the PWR is that in the BWR steam is produced in the core and delivered directly to a steam turbine for the generation of electricity, eliminating the need for a heat exchanger. Benefits are that the system is simpler than a PWR and the capital equipment cost is lower. The flow diagram of Figure 7 shows a direct cycle BWR. A dual-cycle BWR is one where some heated water from the core goes through a heat exchanger.

Initial studies for the BWR were made at Argonne National Laboratory in the early 1950s (6,35,51). The first experiments used electrical heating of metal plates and tubes immersed in a water bath. Stable boiling, a high velocity of steam, and a very short time necessary for steam bubbles to form were observed. These results encouraged planning for a boiling water experiment involving fuel.

A series of tests were performed at the AEC's National Reactor Testing Station in Idaho, starting in 1953. The reactor was situated outdoors, and was operated remotely. The core of the first version had fuel assemblies of aluminum and enriched uranium plates of the Materials Testing Reactor (MTR) type, installed in a water tank. One of the five control rods could be ejected downward and out of the core by spring action upon interruption of a magnet current. The other four rods could be dropped into the core to terminate an excursion in power. Among the findings of this, the BORAX program, were that the reactor had a high degree of inherent safety; operation was more stable than had been expected, although it was possible to induce oscillations in power; high heat-transfer rates from fuel surfaces to steam were possible; the reactor operated well

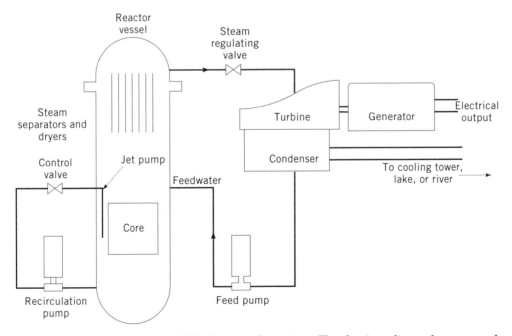

Fig. 7. Flow diagram of a BWR direct-cycle system. The demineralizers, heaters, and one recirculation loop are omitted.

at either atmospheric or elevated pressures; and there was little radioactivity in the steam, although two radionuclides were formed by neutron absorption in oxygen, that is, oxygen-19 and nitrogen-16 having half-lives of 29 and 7 s, respectively. An excellent description of the BORAX equipment, experimental results, and boiling water reactors in general is available (35).

The next facility in the evolution of the BWR was the experimental boiling water reactor (EBWR), built at Argonne National Laboratory (51). This reactor went critical in 1956 and eventually reached a power of 100 MWt. Operating at a power twice that of BORAX, it was more stable. Another BWR having still higher pressure, the Vallecitos boiling water reactor (VBWR), was operated in California by General Electric Co. A succession of central station power reactors appeared: 200 MWe Dresden (Illinois) in 1960, 63 MWe Humboldt Bay (California) in 1963, and 610 MWe Oyster Creek (New Jersey) in 1969.

Figure 8 shows a cutaway of the reactor vessel of the General Electric Co.'s model BWR/6 (52). Table 3 lists numerical data about this reactor.

In the operation of the BWR, water is caused to flow past the fuel rods by a system consisting of two recirculation pumps (only one is shown in Fig. 7), and a number of internal jet pumps. The jet pump has no moving parts, but consists of a high pressure (driving) stream that converges as it is injected along the axis of a low pressure (suction) stream. The two streams merge and are allowed to expand in a diffuser section of the pump. In simplest terms, the high pressure stream drags along the low pressure one. The operating efficiency of a jet pump is the ratio of energy gain of suction flow to energy loss of driving flow.

As the water rises in the reactor core, the steam void volume fraction increases. To remove water droplets from steam, two devices are installed in the upper part of the BWR reactor vessel. The first is an array of steam separators, which consist of tubes having vanes inside giving the steam–water mixture a spinning motion. This allows centrifugal force to carry the water to the walls, where it flows back down. The second is the steam dryer, a bank of vanes through which the stream passes. Any remaining water condenses on the vanes, flows into a trough, and goes down toward the core.

The reactor is equipped with a set of cross-shaped control rods. These are inserted into the core from the bottom. The position can be controlled automatically or manually for both start-up and power adjustments during operation. The reactor is started from cold conditions by moving the control rods to change the reactivity and by varying the flow of water in the recirculation loops. The role of the rods in relation to the negative steam void coefficient of reactivity is as follows. If the flow is increased, steam is swept out more rapidly, reducing the void fraction and thus giving a positive reactivity, which causes a power increase. This creates more steam and a negative reactivity. The reactor stabilizes at a higher power level.

For safety in case of accident, an unusual containment system is employed. Early models of BWRs used large dry containments, such as those used for many PWRs. Modern versions such as General Electric's BWR/6 Mark III instead make use of a pressure-suppression system (Fig. 9). This is designed to accommodate a loss-of-coolant accident (LOCA) such as a pipe break, which releases steam and water from the reactor vessel and tends to build up pressure in the containment drywell. The pressure is relieved by allowing steam, water, and

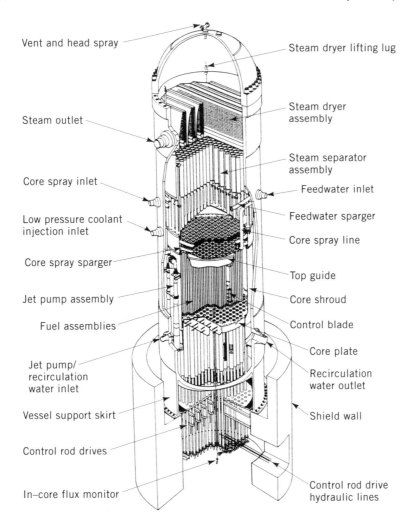

Vent and head spray

Steam dryer lifting lug

Steam outlet

Steam dryer assembly

Steam separator assembly

Core spray inlet

Feedwater inlet

Low pressure coolant injection inlet

Feedwater sparger

Core spray line

Core spray sparger

Top guide

Jet pump assembly

Core shroud

Fuel assemblies

Control blade

Core plate

Jet pump/ recirculation water inlet

Recirculation water outlet

Vessel support skirt

Shield wall

Control rod drives

In–core flux monitor

Control rod drive hydraulic lines

Fig. 8. Cutaway view of the Model BWR/6 pressure vessel (52). Courtesy of GE Nuclear Energy.

air to flow through a vent into a large pool of water, where the steam condenses. Above the pool is the wetwell, a large volume that receives air from the drywell. The water in the pool also serves as a reservoir for makeup cooling of the core after a LOCA. Two features not shown in the simplified diagram are vacuum breakers to allow flow from the wetwell to the drywell if spray cooling is used to reduce drywell pressure, and a quencher that conducts steam from the reactor vessel directly into the suppression pool.

Heavy Water Reactors

A heavy water reactor (HWR) uses deuterium oxide, D_2O, also called heavy water, as moderator. There has been relatively little experience using commercial heavy water moderated power reactors in the United States. Early experimental

Table 3. Design Data for Model BWR/6, General Electric Co.[a]

Parameter	Value
reactor power, MW	
thermal	3579
electric	1220
reactor vessel pressure, MPa[b]	7.17
temperature, °C	
coolant	288
fuel[c]	1871
linear thermal output,[c] kW/m	44
initial fuel enrichment in U-235, wt %	1.7–2.0
fuel rods, OD, mm	12.27
Zircaloy 2 cladding thickness, mm	0.81
number of fuel assemblies	748
number of B₄C control rods	177
reactor vessel dimensions, m	
height	22
diameter	6

[a]Courtesy of GE Nuclear Energy. For a more complete list, see Refs. 11 and 52.
[b]To convert MPa to psia, multiply by 145.
[c]Value given is maximum.

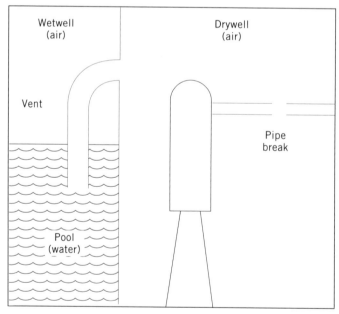

Fig. 9. Principle of pressure suppression containment. Steam from a broken pipe escapes from the drywell through a vent and is condensed in the water pool.

reactors were operated at Argonne National Laboratory in the 1950s. Additionally, a prototype power reactor was built in the 1960s, sponsored by the Carolinas–Virginia Power Associates. It provided valuable experience about reactor design and operation for the organizations involved. Several heavy water production reactors were operated at the Savannah River Plant. The most successful application of heavy water reactors has been in Canada.

Savannah River Production Reactors. For the production of weapons plutonium and tritium, the U.S. Government operated heavy water reactors for approximately 35 years at the Savannah River Plant (SRP) in South Carolina. Design, construction, and operation of the five reactors and related equipment were under the leadership of E. I. du Pont de Nemours & Co., Inc. Heavy water produced on-site was used as both moderator and coolant. Reactor designs were based in part on the early Argonne reactors. The moderator was contained in large (5-m) diameter stainless steel reactor vessels. Primary coolant passed through 600 vertical tubes in a hexagonal array, and then through a heat exchanger. The secondary loop was ordinary river water. The maximum power achieved in any reactor was 2915 MWt.

Many redundant safety features were provided at the SRP. These included a moderator dump tank, gadolinium nitrate solution as emergency absorber, continuously running diesel generators, and a 95×10^6-L (25×10^6-gal) elevated water tank for each reactor, for assurance of cooling.

Over the years, a variety of fuel types were employed. Originally, natural uranium slugs canned in aluminum were the source of plutonium, while lithium–aluminum alloy target rods provided control and a source of tritium. Later, to permit increased production of tritium, reactivity was recovered by the use of enriched uranium fuel, ranging from 5–93%.

Fuel assemblies became much more sophisticated, eventually consisting of concentric tubes made from an outer sheath, three fuel tubes, and an inner lithium-target tube, thus having four coolant channels. Locally developed extrusion techniques were used.

Among other isotopes produced at SRP were uranium-233 for breeder research, cobalt-60 [10198-40-0] for irradiators, plutonium-238 for spacecraft such as *Voyager* and lunar research power supplies, and californium-252 as a fast neutron source. The accomplishments of Du Pont at SRP are well chronicled (53).

When the Du Pont contract ended in 1989, Westinghouse took over. Plans for upgrading the existing reactors and constructing a new production reactor have been abandoned.

The CANDU Reactors. The Canadian deuterium uranium (CANDU) reactors are unique among power reactors in several respects. Heavy water is used as moderator; natural uranium having ^{235}U isotopic content of 0.72 wt % is used as fuel, rather than the typical 2–4 wt % ^{235}U for lightwater reactors; the heavy water coolant flows through pressure tubes passing through the moderator tank; and continuous refueling is performed.

There are several hundred pressure tubes, each containing bundles of 28 fuel rods, 50 cm long. The coolant is at a pressure of around 10 MPa (1450 psia) and the D_2O is at 310°C. Headers on each side of the vessel collect and return coolant from all the tubes. The 4-mm wall-thickness zirconium–4.5% niobium alloy pressure tubes are surrounded by heavy water moderator at much lower temperature and pressure. The reactor vessel, called a calandria, is a large cylinder 8.5 m in diameter and 6 m long, oriented horizontally. Because heavy water is expensive, costing around $50/kg, negligible leakage is mandatory.

Refueling is done without shutting the reactor down, reducing outage times. Refueling machines are located at each end of the reactor vessel. A steam generator having light water in the secondary side supplies steam to the turbine. Figure 10 shows a cutaway view of the reactor building and its contents.

CANDU has a unique negative pressure containment that functions if an accident such as a cooling-water pipe break should occur in the reactor building, resulting in a release of steam, hot water, and radioactive material. The increased pressure is relieved into a vacuum building, maintained at nearly zero pressure. The building is 50 m in diameter and in height, having one meter thick walls and roof. Inside is a large emergency storage tank that provides a water spray to quench the hot vapor and wash out radioactivity.

The Canadian nuclear power program was developed by Atomic Energy of Canada, Ltd. (AECL) of Ottawa. The CANDU reactors provide a large fraction of the electricity of that region. The price of electricity produced is much lower than in any other country because of low fuel cost, high capacity factor, and efficient operation. All of the 22 heavy water reactors are in continuous use, producing 15,442 MW of electrical power. Most of these are operated in the province of Ontario by Ontario Hydro. Pickering, Bruce, and Darlington are multiple reactor stations having 8, 8, and 4 reactors, respectively. The latest reactor was put in operation in 1993 and has a power of 881 MWe. AECL has supplied heavy water reactors to Pakistan, India, Argentina, Romania, and South Korea. The use of spent fuel from lightwater reactors as fuel for the CANDU is being explored.

Early research and development is described in a symposium proceedings (54). The status of the CANDU program as of 1975 is given in Reference 55 and a brief history may be found in a more recent publication of the American Nuclear Society (37).

A variant of the HWR is the Fugen reactor developed by Japan. This reactor is heavy water-moderated but lightwater-cooled. It is fueled by mixed uranium–plutonium oxides.

Fast-Breeder Reactors

Breeding of nuclear fuel was recognized as having a potentially important impact on the availability of energy resources as soon as plutonium was discovered. The most likely nuclear reaction involved the absorption of a neutron in uranium-238 to form plutonium-239, a fissile nuclide with a half-life of approximately 24×10^3 yr. Fission in plutonium-239 by fast neutrons gives rise to about three fast neutrons per absorption. Thus in a reactor using plutonium as fuel the chain reaction can be maintained and enough neutrons are left over to produce more fuel than is burned. At the same time, by consuming uranium-238 instead of merely burning uranium-235 as in converter reactors, the amount of uranium ore needed to produce a given energy is reduced by a factor as large as 50, thus extending the practical life of the uranium resource for thousands of years.

Full advantage of the neutron production by plutonium requires a fast reactor, in which neutrons remain at high energy. Cooling is provided by a liquid metal such as molten sodium or NaK, an alloy of sodium and potassium. The need for pressurization is avoided, but special care is required to prevent

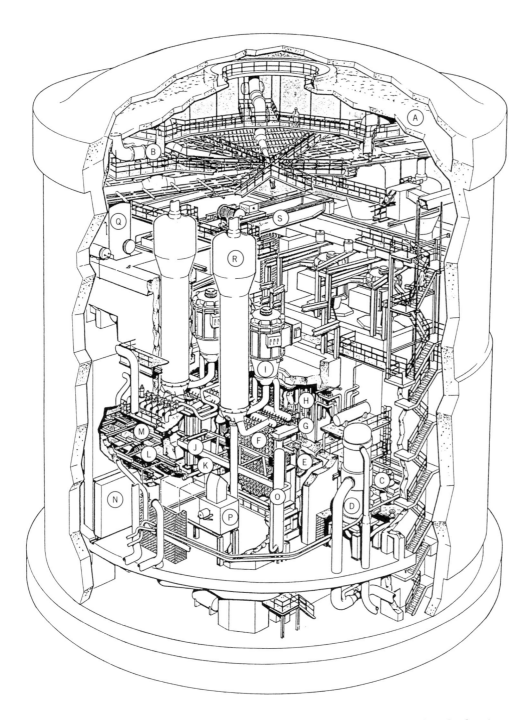

Fig. 10. Cutaway view of containment building of CANDU reactor where A is the dousing water tank; B, dousing water valves; C, moderator pump; D, moderator heat exchanger; E, feeder cabinets; F, reactor face; G, reactor; H, reactivity mechanism; I, heat transport system pump; J, fueling machine bridge; K, fueling machine carriage; L, fueling machine catenary; M, fueling machine maintenance lock; N, fueling machine maintenance lock door; O, end shield cooling water delay tank; P, vault cooler; Q, pressurizer; R, steam generator; and S, steam generator room crane. Courtesy of Atomic Energy of Canada, Ltd.

leaks that might result in a fire. A commonly used terminology is liquid-metal fast-breeder reactor (LMFBR).

An important parameter of any breeder is the breeding ratio (BR) defined as the ratio of the fissile atoms produced to the fissile atoms consumed and given by the simple relation

$$BR = \eta - 1 - \ell$$

where η is the number of neutrons per absorption and ℓ is the number of neutrons lost by leakage and nonfuel absorption. Values of BR around 1.2 are regarded as excellent. A related quantity is the breeding gain (BG) where

$$BG = BR - 1$$

In the evaluation of these parameters, the chain of plutonium isotopes produced and consumed must be taken into account. Successive neutron captures create plutonium-239, -240, -241, and -242. Isotopes having odd mass number are fissile, the others are not.

An extensive theoretical, experimental, and computational knowledge base for fast breeders has been developed. A compact review of key design concepts, analytic methods, and data is available (56). Two types of cooling systems for fast breeders have been employed. The first is the loop, in which the liquid metal is circulated by pump through the reactor vessel and an external heat exchanger. The other is the pot or pool, in which the heat exchanger and pump are in a tank with the reactor core. There are advantages and disadvantages of each arrangement.

There are two ways to locate fissile and fertile materials to achieve breeding. One is the homogeneous arrangement, in which all fissile fuel such as Pu is located in a core and the fertile material such as natural or depleted U is outside in a breeding blanket. The other is the heterogeneous arrangement, having concentric rings of fertile materials within a larger core. The second technique has improved breeding gain and safety.

Most fast reactors that use Na or NaK as coolant utilize an intermediate heat exchanger (IHX) that transfers heat from the radioactive core coolant to a nonradioactive liquid-metal coolant loop, which has the reactor's steam generator. This helps minimize the spread of contamination in the event of a leak or fire.

The first experimental breeder reactor (EBR-I), which was the first reactor to generate electricity on a practical basis, went into operation in 1951 at the National Reactor Testing Station in Idaho. After the first reactor was damaged by a power excursion, EBR-II was put into operation in 1961 (57). As of early 1995 it continued to operate very well.

As a part of the power demonstration program of the AEC in the 1950s, the Enrico Fermi fast breeder reactor (Fermi-1) was built near Detroit by a consortium of companies led by Detroit Edison. Fermi-1 used enriched uranium as fuel and sodium as coolant, and produced 61 MWe. It suffered a partial fuel melting accident in 1966 as the result of a blockage of core coolant flow by a metal plate. The reactor was repaired but shut down permanently in November 1972 because of lack of funding. Valuable experience was gained from its operation, however (58).

The United States continued fast-breeder reactor research and development with the building of the fast flux test facility (FFTF) at Hanford and the SEFOR reactor in Arkansas (59). The next planned step was to build a prototype power reactor, the Clinch River fast-breeder plant (CRFBP), which was to be located near Oak Ridge, Tennessee.

Prospects in the United States for deploying breeders on a large scale were bright when it was believed that rich uranium ore would be quickly exhausted as use of nuclear power expanded. The expected demand for uranium was not realized, however. Moreover, the utilization of breeders requires reprocessing (39). In 1979 a ban was placed on reprocessing in the United States. A dampening effect on development of that part of the fuel cycle for breeder reactors resulted. The CRFBP was canceled and France and Japan became leaders in breeder development.

One of the most advanced versions of a LMFBR is the French SuperPhénix, located at Creys-Malville (60). Partners in development are Electricité de France and firms of Italy, Germany, Belgium, the Netherlands, and the United Kingdom. It is a pool-type system using sodium coolant and a small core surrounded by a breeding blanket. The reactor was shut down for some time for repairs of a sodium leak. A few of its pertinent features are listed in Table 4 (61). SuperPhénix is expected to become a research and demonstration facility having an emphasis on burning plutonium and possibly other actinides.

The MONJU fast-breeder reactor is located on the northern coast of Japan. It is the result of a long-term research and development program led by Hitachi, Ltd. and integrates the work of several companies. Japan is totally dependent on foreign sources of uranium and seeks to make effective use of resources through a plutonium recycle. For earthquake protection and ease of maintenance, the loop-type arrangement was chosen. Specially designed mechanical snubbers were used to support piping. The MONJU core is very compact. It is 93 cm high, 180 cm in diameter, and has only 2340 L volume. It has four concentric regions using a triangular (hexagonal) arrangement (Fig. 11). The inner core has 108 assemblies and 19 control rods, the outer core has 90 assemblies, the blanket has 172 assemblies, and the shield has 324 assemblies. This reactor uses mixed uranium and plutonium oxides for fuel and uranium metal for blanket. The electrical output is 280 MW. Design is underway on a demonstration fast-breeder reactor (DFBR) to go into operation around the year 2000.

The only other fast-breeder reactors in operation in the world are the 233 MWe Phénix in France, the 135 MWe BN-350 in Kazakhstan, and the 560 MWe BN-600 Beloyarskiy in Russia.

Other Reactors

The Natural Reactor. Some two billion years ago, uranium had a much higher (ca 3%) fraction of ^{235}U than that of modern times (0.7%). There is a difference in half-lives of the two principal uranium isotopes, ^{235}U having a half-life of 7.08×10^8 yr and ^{238}U 4.43×10^9 yr. A natural reactor existed, long

Table 4. Data for the SuperPhénix[a]

Parameter	Value
power, MW	
thermal	3000
electric	1200
sodium coolant temperatures, °C	395–545
coolant flow rate, kg/s	16,900
reactor vessel dimensions, m	
height	19.5
diameter	21
core dimensions	
fuel height, m	1.0
diameter, m	3.7
volume, L	10,766
core fuel composition, wt %	
UO_2	83
PuO_2	17
peak flux, $(cm^2 \cdot s)^{-1}$	6.5×10^{15}
fuel pin diameter, mm	8.5
cladding, type[b]	316
pin pitch (triangular), mm	9.8
pins per assembly	271
number of core assemblies	384
control material	B_4C
number of control assemblies	24
blanket fuel	UO_2
average fuel burnup, MW·d/kg	44
refueling interval, d	320
Doppler coefficient	−0.0086
breeding ratio	1.25
doubling time, yr	23

[a]Ref. 61.
[b]Stainless steel.

before the dinosaurs were extinct and before humans appeared on Earth, in the African state of Gabon, near Oklo. Conditions were favorable for a neutron chain reaction involving only uranium and water. Evidence that this process continued intermittently over thousands of years is provided by concentration measurements of fission products and plutonium isotopes. Useful information about retention or migration of radioactive wastes can be gleaned from studies of this natural reactor and its products (12).

Homogeneous Aqueous Reactors.   As a part of the research on neutron multiplication at Los Alamos in the 1940s, a small low power reactor was built using a solution of uranium salt. Uranyl nitrate [36478-76-9], $UO_2(NO_4)_2$, dissolved in ordinary water, resulted in a homogeneous reactor, having uniformly distributed fuel. This water boiler reactor was spherical. The ^{235}U mass was quite low, approximately 1 kg.

The Los Alamos water boiler served as a prototype for the first university training reactor, started in September 1953 at North Carolina State College.

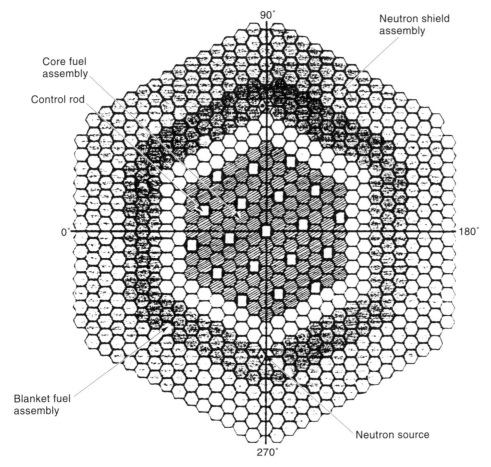

Fig. 11. Reactor core of MONJU, the Japanese fast-breeder reactor. Courtesy of Power Reactor and Nuclear Fuel Development Corp.

The cylindrical reactor core used uranyl sulfate [1314-64-3], UO_2SO_4, and cooling water tubes wound inside the stainless steel container. A thick graphite reflector surrounded the core.

The homogeneous aqueous reactor was studied extensively at Oak Ridge in the 1950s. The objective was to develop a circulating-fuel power reactor that would be easy to refuel, have no temperature limitations on materials, and allow continuous extraction of fission products. The homogeneous reactor experiment-1 (HRE-1) had a core having a 45.72-cm (18 in.) diameter stainless steel spherical vessel containing highly enriched uranium as uranyl sulfate in light water. Because radiation in the water caused dissociation, a flame recombiner was used. The remaining fission gases, xenon and krypton, were held for decay. Operation at powers up to 1.6 MW in the period 1952–1954 was quite satisfactory (18,36). The large negative temperature coefficient of reactivity provided a high degree of stability. Maintenance of the highly radioactive system was performed using long-handled tools and temporary shielding.

The homogeneous reactor experiment-2 (HRE-2) was tested as a power-breeder in the late 1950s. The core contained highly enriched uranyl sulfate in heavy water and the reflector contained a slurry of thorium oxide [1314-20-1], ThO$_2$, in D$_2$O. The reactor thus produced fissile uranium-233 by absorption of neutrons in thorium-232 [7440-29-1], the essentially stable single isotope of thorium. Local deposits of uranium caused reactivity excursions and intense sources of heat that melted holes in the container (18), and the project was terminated.

Aircraft Reactors. As early as World War II, the U.S. Army Air Force considered the use of a nuclear reactor for the propulsion of aircraft (62–64). In 1946 the nuclear energy for propulsion of aircraft (NEPA) program was set up at Oak Ridge, under Fairchild Engine and Airplane Corp. Basic theoretical and experimental studies were carried out. The emphasis was on materials. A high temperature reactor was built and operated successfully. It had beryllium oxide [1304-56-9], BeO, moderator and nickel tubes, through which ran a molten salt fuel consisting of fluorides of Na, Be, and U.

In 1950, a new program, aircraft nuclear propulsion (ANP), was begun by General Electric Co. in Cincinnati, Ohio, and the National Reactor Testing Station in Idaho. A reactor was built to test the heating of air for a turbojet. The first heat-transfer reactor experiment (HTRE-1) consisted of an aluminum water-filled tank through which many tubes passed. Metal fuel elements were placed in the tubes and air pumped through to a ground-based turbojet engine. Operation for 150 h led to energy of 5000 MWh. In a later version, beryllium oxide moderator was used to achieve higher temperature operation.

Another program under Pratt & Whitney in Hartford, Connecticut, was designed to test the indirect cycle. It involved basic studies leading to the design of a liquid-metal-cooled reactor, having a heat exchanger to air for a turbojet. The advantage claimed was high temperature operation without contamination of the air.

The launching of the Russian satellite *Sputnik* in 1957 galvanized the United States, and there were recommendations for expansion of the aircraft program. As late as 1959, the Air Force was optimistic about the use of nuclear reactors for aircraft (64). However, the projects were canceled in 1961 (16). The technology of jet aircraft had advanced, reducing the need for a long-range but slow nuclear bomber. There was the unsolved problem of adequately shielding the crew of the vehicle without excessive weight, and concerns about radioactive contamination in case of an airplane accident (67). Subsequently, the programs were reoriented to materials studies related to advanced high temperature reactors for space (63).

Naval Reactors. The possibility of using nuclear energy to propel ocean vessels was discussed as soon as fission and the chain reaction were discovered. After the end of World War II serious consideration was given to nuclear-propelled ships. These would not depend on air from the surface, would be extremely silent, could travel at high speed, and could remain submerged for long distances because of the large energy yield from nuclear fuel. The translation of such concepts into a working propulsion system was effected quickly. Then-Captain Hyman G. Rickover, selected in 1946 as one of the U.S. naval officers to study nuclear technology at Oak Ridge, took charge of the group, collected all information that might be relevant to the goal of a nuclear submarine, and promoted a program of engineering development. There were two approaches toward the achievement of a nuclear submarine. One was a water-moderated and water-cooled pressurized reactor; the other was a liquid-metal-cooled intermediate neutron energy reactor. A land-based prototype submarine power plant called Mark I was built and tested at the National Reactor Testing Station. Argonne National Laboratory provided scientific data and Bettis Laboratory of Westinghouse Electric Corp. supplied engineering expertise.

A great deal of technical information was needed on the behavior of materials under severe conditions of temperature and radiation. Stainless steel was selected for structures, but for the fuel cladding zirconium, a rare metal, was chosen because of its very low thermal neutron absorption cross section and its resistance to corrosion by hot water (see ZIRCONIUM AND ZIRCONIUM COMPOUNDS). Methods were developed for extraction of zirconium from ore, of removal of the strong accompanying absorber hafnium, and of fabrication into desired shapes. Eventually hafnium was found useful for control rods in place of an alloy of silver, cadmium, and indium. Special seals were needed to prevent leakage of water and radioactivity. Extensive radiation-shielding studies assured safety of the crew in the cramped quarters of a submarine (65,66). Voluminous technology handbooks have been written on these findings (67–69).

Following a successful test of the Mark I, construction of the first nuclear submarine *N.S. Nautilus* was begun in 1953, *Nautilus* made a trip of over 114,824 km (62,000 nautical mi) submerged from the United States to the British Isles in 1955, and served as a model for the fleet of over 100 submarines of the U.S. Navy. Several nuclear-powered aircraft carriers and guided missile cruisers were also built. The carrier *Enterprise*, launched in 1961, is propelled by eight PWRs. The ship, having a length of 342 m, carries 75 aircraft and over 5000 personnel (70).

Maritime Reactors.   Nuclear power has had limited use for propulsion of merchant ships, largely because of economic reasons, although public reaction has also played a role (71). The construction of a nuclear-powered merchant ship was proposed in the 1950s to demonstrate the U.S. interests in peace. The AEC and the Maritime Administration sponsored the design and construction of the *N.S. Savannah*. Babcock & Wilcox supplied the reactor and New York Shipbuilding Corp. constructed the ship. The cargo–passenger vessel was almost 183 m long and was powered by a 80-MWt pressurized water reactor. The core was fueled with 4.4% enriched uranium dioxide in stainless steel tubes, operating at 12.1 MPa (1750 psi). Steam was supplied to a turbine that drove the ship's propeller. Seawater was used to condense steam. Sustained speeds of the ship were 21 knots.

Launched in 1959, *N.S. Savannah* operated very well. Starting in 1962, it made a goodwill voyage around the world. It was able to travel a distance of several times the earth's circumference on one fuel loading. However, the ship was not competitive economically with oil-powered merchant ships. The shielding was quite adequate, so that the reactor was safe. Nonetheless the vessel was opposed by antinuclear groups and the *N.S. Savannah* was eventually retired and put on display in Charleston, South Carolina. In 1994, the ship was transferred to Norfolk, Virginia, to be held in reserve.

The Russian icebreaker *Lenin*, launched in 1959, had three 90 MWt PWRs, one of which was a spare. It operated for many years in the Arctic Ocean.

Package Power Reactors.   Several small, compact power reactor plants were developed during the period 1957–1962 by the U.S. Army for use in remote locations. Designed by Alco Products, Inc., the PWRs produced electrical power of about 1 MW along with space heat for military bases. The first reactor, SM-1, was operated at Fort Belvoir, Virginia. Others were located in Wyoming, Greenland, Alaska, and Antarctica. The fuel consisted of highly enriched uranium as the dioxide, dispersed in stainless steel as plates or rods. Details are available in Reference 18.

Space Reactors.   Two quite different applications of reactors in space have been studied: one for electrical power of a spacecraft mission, and the other for propulsion of spacecraft. Both applications are for long missions where solar power is inadequate or chemical propulsion is impractical (72,73).

The AEC sponsored research in the program known as Systems for Nuclear Auxiliary Power (SNAP) as early as the 1950s. Most of the systems developed involved the radioisotope plutonium-238 as a heat source for a thermoelectric generator. Such electrical supplies permitted radio transmission to earth from spacecraft such as *Pioneer* and *Voyager*.

Several actual reactors have been built to provide the heat that can be converted into electricity. One of them, SNAP-10A, was flown in space in 1965. It was placed in 1300-km radius orbit by the launch vehicle Agena. The power unit remained subcritical until the orbit was reached, at which time automatic control took over to start the reactor. Its core was composed of stainless steel-clad rods containing a homogeneous mixture of zirconium hydride [7704-99-6], ZrH_2, and uranium-235, serving as both fuel and moderator. Liquid-metal coolant NaK circulated by an electromagnetic pump passed between the rods and to a Si–Ge thermoelectric generator producing about 500 W of power. It operated successfully for 43 d until a nonreactor-related failure occurred. A twin reactor on the ground operated at full power for 10,000 h.

Future space missions of long duration and long distance, eg, flights to Mars and back, would need a solid-core nuclear rocket. The key measure of effectiveness of a rocket for propulsion is the specific impulse, defined as the ratio of thrust to mass flow rate of propellant. Whereas a nuclear reactor cannot produce a gas temperature as high as a chemical fuel can, the former can use the light element hydrogen as coolant–propellant, instead of the heavier products of combustion.

Reactors for direct propulsion of spacecraft were built and tested in the ROVER project from 1959–1973. These used uranium carbide as fuel and graphite as moderator. Liquid hydrogen (qv) served as coolant. The hydrogen was volatized and exhausted from a nozzle as a propellant (Fig. 12). The most successful reactor was the nuclear engine for rocket vehicle application (NERVA), which operated at a very high (4000 MW) power for a time of 12 min. The ROVER project was canceled in 1973. More recently, NASA-sponsored studies advocated use of a nuclear rocket for a Mars visit around the year 2018 (75). Such a voyage appears unlikely as of this writing.

Research and Training Reactors.   Research reactors generally fall in one of three categories: an experimental reactor to test a concept, a high flux reactor

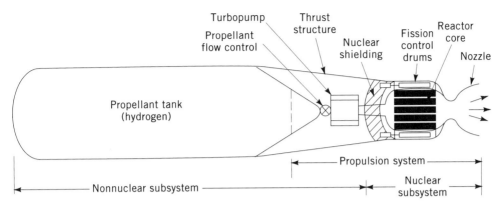

Fig. 12. Schematic of a nuclear rocket (74). Liquid hydrogen is heated by a reactor and expelled as a gas through the nozzle.

dedicated to basic research, or a reactor used primarily for educational purposes. Reactors at universities or laboratories may be used for purposes such as production of radioisotopes (qv), the study of radiation effects, neutron activation analysis, measurements of reactor properties and behavior, and the teaching of nuclear engineering students. Some of these reactors have been in operation for 30 yr or more without incident. Many university reactors have been shut down for economic reasons. The International Atomic Energy Agency (IAEA) has a catalog of research reactors (76) and includes critical assemblies as well.

The determination of critical size or mass of nuclear fuel is important for safety reasons. In the design of the atom bombs at Los Alamos, it was crucial to know the critical mass, ie, that amount of highly enriched uranium or plutonium that would permit a chain reaction. A variety of assemblies were constructed. For example, a bare ^{235}U metal sphere was found to have a critical mass of approximately 50 kg, whereas a natural uranium reflected ^{235}U sphere had a critical mass of only 16 kg.

The reactor Lady Godiva was constructed at Los Alamos for the study of dynamic behavior of a supercritical assembly. This was an unreflected, essentially spherical metal structure of uranium enriched to 93% and having a density of approximately 18 g/cm^3. The reactor had no moderator, thus the neutrons causing fission were fast, ca 1 MeV. Cooling was only by conduction through the metal and convection in air at the surface. The sudden application of a positive reactivity to such a reactor causes it to rise in power rapidly, and the strong negative temperature feedback effect causes the power to peak and drop back, giving rise to a pulse or burst. The neutrons that accompany the power were used to irradiate detecting equipment related to weapons. Although large (ca 10^{17}) bursts of neutrons are created and the instantaneous power levels are very high (thousands of MW), the time span of the pulse is very short (a few μs) and the heat energy (a few J) is modest.

In the early 1950s, the Argonaut research and training reactor was designed and built by Argonne National Laboratories. It was subsequently adopted by several U.S. universities. Its initial purpose was for nuclear studies conducted by scientists and engineers from many countries as a part of the Atoms for Peace program. The reactor consisted of a ring of plate-type fuel assemblies having graphite fillers interspersed among them. Water within the fuel boxes provided self-limiting safety. Graphite served as internal and external reflector. A peak power of 10 kW for short times was possible.

One of the early popular low power research and training reactors was the AGN-201, supplied by Aerojet General Nuclear. This is a homogeneous solid fuel reactor, consisting of a mixture of polyethylene and uranium at 20% enrichment in ^{235}U. The core ^{235}U loading is around 0.7 kg and the core volume is 12 L. The reflector is graphite and the core has no cooling. Power is limited to 5 W, giving a thermal neutron flux of around 10^8 $(cm^2 \cdot s)^{-1}$, sufficient for certain experiments.

A number of pool, also called swimming pool, reactors have been built at educational institutions and research laboratories. The core in these reactors is located at the bottom of a large pool of water, 6 m deep, suspended from a bridge. The water serves as moderator, coolant, and shield. An example is the Ford nuclear reactor at the University of Michigan, started in 1957. The core is composed of fuel elements, each having 18 aluminum-clad plates of 20% enriched uranium. It operates at 2 MW, giving a thermal flux of 3×10^{13} $(cm^2 \cdot s)^{-1}$. The reactor operates almost continuously, using a variety of beam tubes, for research purposes.

A variant on the pool reactor is the tank type, in which a limited volume of water surrounds the core. The TRIGA reactor, marketed throughout the world by General Atomic of San Diego, California, is an example (77). The reactor core contains many fuel rods immersed in water, and has a graphite reflector. The fuel is a mixture of zirconium hydride and 20% enriched uranium, clad with stainless steel. Control rods are boron carbide. TRIGA Mark II is capable of steady operation at 250 kW or pulsing to 250 MW for a brief time. The steady thermal flux is approximately 10^{13} $(cm^2 \cdot s)^{-1}$. The core fuel loading is 2.7 kg in a volume of 63 L. Included in experimental facilities are a rotating specimen rack and various beam tubes.

The Slowpoke reactor, supplied by Atomic Energy of Canada, Ltd., is installed in several Canadian universities. It is a lightwater moderated reactor having 93% enriched uranium. Various features make it inherently safe and give it its name. The core is composed of aluminum-clad fuel rods 0.473 cm in diameter and 22 cm long. Its reflector is beryllium and it has a single cadmium control rod. Cooling is by natural convection. The power level is 20 kW, giving a thermal flux of 1×10^{12} $(cm^2 \cdot s)^{-1}$. A larger version, the Slowpoke demonstration reactor (SDR), is designed for district heating. It provides thermal power of 2 MW.

Advanced Power Reactors. Most of the U.S. nuclear reactors were built in the twenty-year period 1964–1984. Many are approaching their design life. Increased attention is being given to the aging of components, using preventive maintenance, and replacement of parts and assemblies to extend the life of facilities (22). At the same time, the U.S. nuclear industry, in cooperation with the Department of Energy, is developing several advanced reactor designs to supplant and supplement the existing reactors (23). Advantage is being taken of the experience and knowledge gained in more than 30 yr of lightwater reactor operation, and features that provide inherent passive safety are being included wherever feasible (20). By incorporating simplicity, economy, and improved safety, the advanced reactors are being designed to be attractive to the public, to utility management, and to the financial community (21).

The advanced lightwater reactor (ALWR) program involves several reactor manufacturers, such as Westinghouse, General Electric, and ABB Combustion Engineering as well as organizations such as the Electric Power Research Institute (EPRI) and the Institute of Nuclear Power Operations (INPO). This program has two separate thrusts: a large, evolutionary-design ALWR consisting of improved reactors in the 1200 MWe range; and a small, revolutionary design that emphasizes passive or inherent safety, in the 600 MWe class. As they are ready, the designs are being submitted to the Nuclear Regulatory Commission for review and certification.

An example of the large reactor is the advanced boiling-water reactor of General Electric Co., designed and built in collaboration with the Japanese companies Hitachi and Toshiba. Goals for the reactor are reduced damage frequency by an order of magnitude, simplification of design, reduced costs of construction, fuel, and operation, and reduced radiation exposure and waste. New features are internal recirculation pumps, modern electronics, use of inert nitrogen in the containment to prevent hydrogen explosion, steel lining for the reinforced concrete containment, new control rod drives, a reactor vessel having forged rings instead of welded plates, and a backup gas turbine generator in addition to diesels.

The General Electric simplified boiling-water reactor (SBWR) of lower (600 MWe) power features natural circulation of the coolant rather than the usual forced circulation. Use is made of a water reservoir and pools for emergency cooling of the reactor and the containment building air.

An example of the large reactor concept as applied to the PWR is the System 80+ of ABB Combustion Engineering, designed in conjunction with Duke Engineering Services. System 80+ is an extension of System 80 that embodies several features, such as safer design, simpler design, greater reliability, and enhanced operability. It has a large spherical steel containment building, gravity feed for the emergency water, hydrogen control, and decay-heat removal capability. It also has an advanced control console that emphasizes the application of human engineering. Construction times of only four years are expected.

The Westinghouse AP600 is a pressurized-water reactor of 600 MWe capacity, of the passive safety type (78). The system has far fewer pumps, pipes, valves, and ducts than current designs. It depends greatly on passive natural processes such as gravity, natural circulation, convection, evaporation, and condensation. As sketched in Figure 13, it has an emergency core cooling system that does not require pumps or electric power. Two large water tanks are located above the reactor. If a loss-of-coolant accident (LOCA) occurs when the reactor is still under pressure, water is driven by pressurized nitrogen into the core. If the reactor pressure is lost, gravity produces water flow from a tank at atmospheric pressure. In the event steam generators are not operable, natural circulation to the large water tank above the reactor removes decay heat. The metal containment is kept cool to condense vapor released in a LOCA by air drawn through a chimney at the top and by a gravity-fed water spray.

Lower power ratings of reactors provide greater flexibility for a utility to add power generation to a system (19). The AP600 uses prefabricated modules

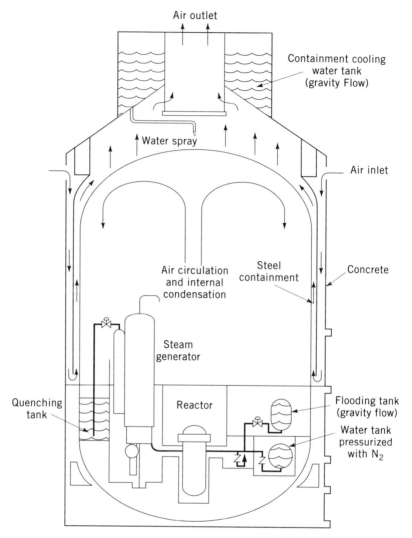

Fig. 13. Proposed advanced PWR design having passive safety features (79). Courtesy of Westinghouse Electric Corp.

to shorten the construction time. Thus construction and operating costs are expected to be competitive with coal-fired plants.

ALWRs are expected to be deployed in the United States and in Asian counties. However, France will use improved versions of standard reactors, considering them to be amply safe and economical. The reactors were modified after the Three-Mile Island-2 (TMI-2) accident. The company Framatome that has built most of the reactors of France is associated with Babcock & Wilcox in the United States. The new Framatome 1500 MWe N4 PWR is an extension of the successful four-loop units of 1300 MWe originally designed by Westinghouse. Full emphasis is given to safety, economy, and reliability. More severe design criteria than those in the former model have been adopted.

BIBLIOGRAPHY

"Fast Breeder Reactors" under "Nuclear Reactors" in *ECT* 3rd ed., Vol. 16, pp. 184–205, by P. Murray, Westinghouse Electric Corp.

1. R. L. Murray, *Nuclear Energy*, 4th ed., Pergamon Press, Oxford, U.K., 1993.
2. R. A. Knief, *Nuclear Engineering: Theory and Technology of Commercial Nuclear Power*, 2nd ed., Hemisphere Publishing Corp., Washington, D.C., 1992.
3. D. J. Bennet and J. R. Thomson, *The Elements of Nuclear Power*, 3rd ed., Longman Scientific & Technical, Essex, U.K., 1989.
4. G. Kessler, *Nuclear Fission Reactors*, Springer-Verlag, New York, 1983.
5. R. L. Murray, *Introduction to Nuclear Engineering*, 1st ed., Prentice-Hall, Inc., Englewood Cliffs, N.J., 1954, 2nd ed., 1961.
6. W. H. Zinn, F. K. Pittman, and J. F. Hogerton, *Nuclear Power, U.S.A.*, McGraw-Hill Book Co., Inc., New York, 1964.
7. W. Hafele, *Scientific American*, 136 (Sept. 1990).
8. R. K. Lester, *Scientific American*, 31 (Mar. 1986).
9. J. R. Dietrich and W. H. Zinn, *Solid Fuel Reactors*, Addison-Wesley Publishing Co., Inc., Reading, Mass., 1958.
10. *World Nuclear Industry Handbook 1993* (annual issuance), Nuclear Engineering International, Sutton, Surrey, U.K.
11. A. V. Nero, Jr., *A Guidebook to Nuclear Reactors*, University of California Press, Berkeley, Calif., 1979.
12. G. A. Cowan, *Scientific American*, 36 (July 1976).
13. H. DeW. Smyth, *Atomic Energy for Military Purposes*, Princeton University Press, Princeton, N.J., 1945. Latest reprint, Stanford University Press, Palo Alto, Calif., 1989.
14. R. Rhodes, *The Making of the Atomic Bomb*, Simon and Schuster, New York, 1986.
15. R. G. Hewlett and F. Duncan, *Atomic Shield: A History of the United States Atomic Energy Commission, Vol. II 1947–1952*, U.S. Atomic Energy Commission, Washington, D.C., 1972.
16. R. G. Hewlett and J. M. Holl, *Atoms for Peace and War 1953–1961: Eisenhower and the Atomic Energy Commission*, University of California Press, Berkeley, Calif., 1989.
17. F. G. Dawson, *Nuclear Power: Development and Management of a Technology*, University of Washington Press, Seattle, Wash., 1976.
18. R. L. Loftness, *Nuclear Power Plants: Design, Operating Experience and Economics*, D. Van Nostrand Co. Inc., Princeton, N.J., 1964.
19. *Small and Medium Reactors, I. Status and Prospects, II. Technical Supplement*, Nuclear Energy Agency of the Organisation for Economic Co-Operation and Development, Paris, 1991.
20. L. S. Tong, *Principles of Design Improvement for Light Water Reactors*, Hemisphere Publishing Corp., New York, 1988.
21. M. W. Golay and N. E. Todreas, *Scientific American*, 82 (Apr. 1990).
22. V. N. Shah and P. E. McDonald, eds., *Aging and Life Extension of Major Light Water Reactor Components*, Elsevier Science Publishers, Amsterdam, the Netherlands, 1993.
23. *Strategic Plan for Building New Nuclear Power Plants*, Nuclear Energy Institute Executive Committee, Washington, D.C., 1994 (annual update).
24. *Proceedings of the United Nations International Conference on the Peaceful Uses of Atomic Energy, Geneva*, 1955, United Nations, New York, 1956.
25. R. L. Hellens and G. A. Price, in L. E. Link, ed., *Reactor Technology: Selected Reviews—1964*, U.S. Atomic Energy Commission, Oak Ridge, Tenn., 1964, pp. 529–609.
26. M. M. El-Wakil, *Nuclear Heat Transport*, American Nuclear Society, La Grange Park, Ill., 1978.

27. J. R. Lamarsh, *Introduction to Nuclear Reactor Theory*, Addison-Wesley, Reading, Mass., 1972.
28. Ref. 13, pp. 60–62.
29. Ref. 13, pp. 132–134.
30. N. E. Todreas and M. S. Kazimi, *Nuclear Systems I: Thermal Hydraulic Fundamentals*, 1989, *II: Elements of Thermal Hydraulic Design*, 1990, Hemisphere Publishing Corp., New York.
31. G. Melese and R. Katz, *Thermal and Flow Design of Helium-Cooled Reactors*, American Nuclear Society, La Grange Park, Ill., 1984.
32. R. T. Lahey, Jr. and F. J. Moody, *The Thermal-Hydraulics of a Boiling Water Nuclear Reactor*, 2nd ed., American Nuclear Society, La Grange Park, Ill., 1993.
33. C. Starr and R. W. Dickinson, *Sodium Graphite Reactors*, Addison-Wesley Publishing Co., Inc., Reading, Mass., 1958.
34. *The Shippingport Pressurized Water Reactor*, Addison-Wesley Publishing Co., Reading, Mass., 1958.
35. A. W. Kramer, *Boiling Water Reactors*, Addison-Wesley Publishing Co., Inc., Reading, Mass., 1958.
36. J. A. Lane, H. G. MacPherson, and F. Maslan, ed., *Fluid Fuel Reactors*, Addison-Wesley Publishing Co., Inc., Reading, Mass., 1958.
37. *Controlled Nuclear Chain Reaction: The First 50 Years*, American Nuclear Society, La Grange Park, Ill., 1992.
38. R. L. Murray, *Nucl. News*, 105–106 (Aug. 1988).
39. W. Bebbington, *Scientific American*, 30 (Dec. 1976).
40. R. M. E. Diamant, *Atomic Energy*, Ann Arbor Science, Ann Arbor, Mich., 1982.
41. D. R. Poulter, ed., *The Design of Gas-Cooled Graphite-Moderated Reactors*, Oxford University Press, London, 1963.
42. S. D. Strauss, *Nucleonics* **24**, 63 (1966).
43. H. Agnew, *Scientific American*, 55 (June 1981).
44. D. R. Marples, *Chernobyl and Nuclear Power in the USSR*, St. Martin's Press, New York, 1986.
45. *The Westinghouse Pressurized Water Reactor Nuclear Power Plant*, Westinghouse Electric Corp., Water Reactor Divisions, Pittsburgh, Pa., 1984.
46. Ref. 45, p. 34.
47. Ref. 45, pp. 3, 16, and 21.
48. Ref. 45, p. 20.
49. Ref. 45, p. 27.
50. Ref. 1, p. 302.
51. *The EBWR Experimental Boiling Water Reactor*, ANL-5607, Argonne National Laboratory, U.S. Atomic Energy Commission, Washington, D.C., 1957.
52. *BWR/6 General Description of a Boiling Water Reactor*, General Electric Co., Nuclear Energy Group, San Jose, California, 1980.
53. W. P. Bebbington, *History of Du Pont at the Savannah River Plant*, E. I. du Pont de Nemours & Co., Inc., Wilmington, Del., 1990.
54. "Heavy-Water Power Reactors," *Proceedings of the International Atomic Energy Agency Symposium*, Sept. 11–15, 1967, IAEA, Vienna, Austria, 1968.
55. H. C. McIntyre, *Scientific American*, 17 (Oct. 1975).
56. K. Wirtz, *Lectures on Fast Reactors*, American Nuclear Society, La Grange Park, Ill., 1978.
57. C. E. Stevenson, *The EBR-II Fuel Cycle Story*, American Nuclear Society, La Grange Park, Ill., 1987.
58. E. P. Alexanderson, ed., and H. A. Wagner, ed. dir., *Fermi-1 New Age for Nuclear Power*, American Nuclear Society, La Grange Park, Ill., 1979.

59. P. V. Evans, ed., *Fast Breeder Reactors*, Proceedings of the London Conference on Fast Breeder Reactors of the British Nuclear Energy Society, May 17–19, 1966, Pergamon Press, Oxford, U.K., 1967.

60. G. Vendreyes, *Scientific American*, 26 (Mar. 1977).

61. A. E. Waltar and A. B. Reynolds, *Fast Breeder Reactors*, Pergamon Press, Oxford, U.K., and New York, 1981, pp. 780–781.

62. *Aircraft Nuclear Propulsion Program*, Hearing before the Subcommittee on Research and Development of the Joint Committee on Atomic Energy, July 23, 1959. U.S. Government Printing Office, Washington, D.C., 1959.

63. *Annual Report to Congress, 1961*, U.S. Atomic Energy Commission, Washington, D.C., 1961.

64. K. F. Gantz, ed., *Nuclear Flight: The United States Air Force Programs for Atomic Jets, Missiles, and Rockets*, Duell, Sloan and Pearce, New York.

65. T. Rockwell, *The Rickover Effect: How One Man Made a Difference*, Naval Institute Press, Annapolis, Md., 1992.

66. R. G. Hewlett and F. Duncan, *Nuclear Navy 1946–1962*, University of Chicago Press, Chicago, Ill., 1974.

67. C. B. Jackson, ed., *Liquid Metals Handbook*, 3rd ed., Superintendent of Documents, Washington, D.C., 1955.

68. *Naval Reactor Physics Handbook*, Vol. I, A. Radkowsky, ed.; Vol. II, S. Krasik, ed.; Vol. III, J. R. Stehn, ed., Superintendent of Documents, Washington, D.C., 1964.

69. T. Rockwell, III, ed., *Reactor Shielding Design Manual*, McGraw Hill Book Co., Inc., New York, 1956.

70. R. Sharpe, ed., *Jane's Fighting Ships 1994–1995*, Jane's Information Group, Alexandria, Va., 1993.

71. *Nuclear Merchant Ships*, Maritime Transportation Research Board, National Academy of Sciences, Washington, D.C., 1974.

72. S. Aftergood, D. W. Hafemeister, O. F. Prilutsky, J. R. Primack, and S. M. Rodionov, *Scientific American*, 42 (June 1991).

73. J. A. Angelo, Jr. and D. Buden, *Space Nuclear Power*, Orbit Book Co., Malabar, Fla., 1985.

74. H. F. Crouch, *Nuclear Space Propulsion*, Astronuclear Press, Granada Hills, Calif., 1965.

75. *America at the Threshold: Report of the Synthesis Group on America's Space Exploration Initiative*, Superintendent of Documents, Washington, D.C., 1991.

76. *Directory of Nuclear Research Reactors*, International Atomic Energy Agency, Vienna, Austria, 1989.

77. I. Mele, M. Ravnik, and A. Trkov, *Nucl. Technol.* **105**, 37, 52 (1994).

78. *Westinghouse AP600 Executive Summary*, Westinghouse Electric Corp., Pittsburgh, Pa.

RAYMOND L. MURRAY
Consultant

WASTE MANAGEMENT

Radioactive wastes are generated in all parts of the fuel cycle supporting nuclear electric power including mining and milling of uranium ore, chemical conversion, isotope separation, fuel fabrication, nuclear reactor operation, spent fuel storage, and waste disposal. Successful management of wastes from nuclear reactors is vital to continued use of nuclear power and assurance of waste safety should aid in improving public acceptance. Many reactor wastes are radioactive. Disintegration (decay) of various materials results in the release of high energy radiation requiring protective measures. Half-lives of reactor products range from fractions of a second to billions of years.

Classification of wastes may be according to purpose, distinguishing between defense waste related to military applications, and commercial waste related to civilian applications. Classification may also be by the type of waste, ie, mill tailings, high level radioactive waste (HLW), spent fuel, low level radioactive waste (LLW), or transuranic waste (TRU). Alternatively, the radionuclides and the degree of radioactivity can define the waste. Surveys of nuclear waste management (1,2) and more technical information (3–5) are available.

Sources

Radioactivity occurs naturally in earth minerals containing uranium and thorium (see RADIOACTIVITY, NATURAL). It also results from two principal processes arising from bombardment of atomic nuclei by particles such as neutrons, ie, activation and fission. Activation involves the absorption of a neutron by a stable nucleus to form an unstable nucleus. An example is the neutron reaction of a neutron and cobalt-59 to yield cobalt-60 [10198-40-0], ^{60}Co, a 5.26-yr half-life gamma-ray emitter. Another is the absorption of a neutron by uranium-238 [24678-82-8], ^{238}U, to produce plutonium-239 [15117-48-3], ^{239}Pu, as occurs in the fuel of a nuclear reactor. Fission occurs when a neutron is absorbed by uranium-235 [15117-96-1]. One typical reaction is as follows:

$$^{235}\text{U} + {}^1_0n \longrightarrow {}^{90}\text{Kr} + {}^{144}\text{Ba} + 2\,{}^1_0n$$

Many of the uranium fission fragments are radioactive. Of special interest are technetium-99 [14133-76-7] and iodine-129 [15046-84-1] having half-lives of 2.13×10^5 yr and 1.7×10^7 yr, respectively. Data on all isotopes are found in Reference 6 (see also RADIOISOTOPES).

Radioactive waste is characterized by volume and activity, defined as the number of disintegrations per second, known as becquerels. Each radionuclide has a unique half-life, $t_{1/2}$, and corresponding decay constant, $\lambda = 0.693/t_{1/2}$. For a component radionuclide consisting of N atoms, the activity, A, is defined as

$$A = N\lambda$$

Activities and existing and projected volumes of all types of radioactive waste are listed in Reference 7.

Most uranium ore has a low, ca 1 part in 500, uranium content. Milling involves physical and chemical processing of the ore to extract the uranium. The mill tailings, which release gaseous radon-222 [13967-62-9], ^{222}Ra, half-life 3.82 d, are placed in large piles and covered to prevent a local health problem.

Uranium oxide [1344-57-6] from mills is converted into uranium hexafluoride [7783-81-5], UF_6, for use in gaseous diffusion isotope separation plants (see DIFFUSION SEPARATION METHODS). The wastes from these operations are only slightly radioactive. Both uranium-235 and uranium-238 have long half-lives, 7.08×10^8 and 4.46×10^9 yr, respectively. Uranium enriched to around 3 wt % is shipped to a reactor fuel fabrication plant (see NUCLEAR REACTORS, NUCLEAR FUEL RESERVES). There conversion to uranium dioxide is followed by pellet formation, sintering, and placement in tubes to form fuel rods. The rods are put in bundles to form fuel assemblies. Despite active recycling (qv), some low activity wastes are produced.

Uranium dioxide fuel is irradiated in a reactor for periods of one to two years to produce fission energy. Upon removal, the used or spent fuel contains a large inventory of fission products. These are largely contained in the oxide matrix and the sealed fuel tubing.

Spent fuel can be stored or disposed of intact, in a once-through mode of operation, practiced by the U.S. commercial nuclear power industry. Alternatively, spent fuel can be reprocessed, ie, treated to separate the uranium, plutonium, and fission products, for re-use of the fuels (see NUCLEAR REACTORS, CHEMICAL REPROCESSING). In the United States reprocessing is carried out only for fuel from naval reactors. In the nuclear programs of some other countries, especially France and Japan, reprocessing is routine.

Water as coolant in a nuclear reactor is rendered radioactive by neutron irradiation of corrosion products of materials used in reactor construction. Key nuclides and the half-lives in addition to cobalt-60 are nickel-63 [13981-37-8] (100 yr), niobium-94 [14681-63-1] (2.4×10^4 yr), and nickel-59 [14336-70-0] (7.6×10^4 yr). Occasionally small leaks in fuel rods allow fission products to enter the cooling water. Cleanup of the water results in LLW. Another source of waste is the residue from applications of radionuclides in medical diagnosis, treatment, research, and industry. Many of these radionuclides are produced in nuclear reactors, especially in Canada.

Weapons materials from production reactors were accumulated during the Cold War period as a part of the U.S. defense program. Prominent were tritium, ie, hydrogen-3, having a $t_{1/2}$ of 12.3 yr, and plutonium-239, $t_{1/2} = 2.4 \times 10^4$ yr. The latter constitutes a waste both as a by-product of weapons fabrication in a waste material called transuranic waste (TRU), and as an excess fissionable material if not used for power production in a reactor.

A number of legislative actions govern the management of radioactive waste. The Atomic Energy Acts of 1946 and 1954 charged the Atomic Energy Commission with maintaining national nuclear defense and developing peaceful uses of the atom. The National Environmental Policy Act of 1969, which created the U.S. Environmental Protection Agency (EPA), initiated the requirement for an environmental impact statement (EIS) to be prepared for federal facilities. In 1980 the Low-Level Radioactive Waste Policy Act assigned responsibility for LLW disposal to the states. The Nuclear Waste Policy Act (NWPA) of 1982 set forth the schedule and procedure by which high level waste would be managed by the U.S. Department of Energy. Each of these acts have been amended (8).

Treatment

Several modes of waste management are available. The simplest is to dilute and disperse. This practice is adequate for the release of small amounts of radioactive material to the atmosphere or to a large body of water. Noble gases and slightly contaminated water from reactor operation are eligible for such treatment. A second technique is to hold the material for decay. This is applicable to radionuclides of short half-life such as the medical isotope technetium-99m ($t_{1/2}$ = 6 h), the concentration of which becomes negligible in a week's holding period. The third and most common approach to waste management is to concentrate and contain. Various processes are applied to minimize volume and to prevent or delay access of water to the contents of waste containers (9,10).

Low Level Waste Treatment.   Methods of treatment for radioactive wastes produced in a nuclear power plant include (1) evaporation (qv) of cooling water to yield radioactive sludges, (2) filtration (qv) using ion-exchange (qv) resins, (3) incineration with the release of combustion gases through filters while retaining the radioactively contaminated ashes (see INCINERATORS), (4) compaction by presses, and (5) solidification in cement (qv) or asphalt (qv) within metal containers.

All processes in a nuclear plant, in a treatment facility, or at a disposal site are governed by rules of the U.S. Nuclear Regulatory Commission (NRC) (11). Radiation protection, ie, the limits on radiation dose to workers and the public are specified. Exposure is maintained as low as reasonably achievable (ALARA).

Limits on concentration of radioactive content in air and water are specified in Part 20 of Reference 11. For example, the concentration limit of gamma-emitting cesium-137 [10045-97-3], $t_{1/2}$ = 30 yr, in water is 0.037 Bq/mL (1 × 10^{-6} μCi/mL). For beta-emitting tritium, $t_{1/2}$ = 12.3 yr, in air the limit is 0.0037 Bq/mL (1 × 10^{-7} μCi/mL). The NRC classifies wastes according to half-life and concentration in terms of activity per cubic centimeter. Classes A, B, and C are generally in increasing level of long-term hazard and requirement for integrity of disposal. Materials classed as LLW but of high concentration are designated as greater-than-class-C (GTCC). Wastes of finite but negligible activity are regarded as being below regulatory concern (BRC). Mixed wastes, those having a hazardous material or feature plus a radioactive component, are subject to regulation by both the NRC and EPA. Standards differ significantly.

Radiation dose limits at a disposal site boundary are specified by the NRC as 25 × 10^{-5} Sv/yr (25 mrem/yr), a small fraction of the average radiation exposure of a person in the United States of 360 × 10^{-5}/Sv/yr (360 mrem/yr). Protection against nuclear radiation is fully described elsewhere (12).

Nuclear utilities have sharply reduced the volume of low level radioactive waste over the years. In addition to treating wastes, utilities avoid contamination of bulk material by limiting the contact with radioactive materials. Decontamination of used equipment and materials is also carried out. For example, lead used for shielding can be successfully decontaminated and recycled using an abrasive mixture of low pressure air, water, and alumina.

Recycling techniques for radioactive material are evolving, especially for scrap metals. Large quantities of scrap are expected from decommissioning operations (13). Advantage is being taken of successful German and Japanese experience. One important use of recycled steel from the nuclear industry is for waste containers and canisters. A slight residual radioactivity is immaterial.

Spent Fuel Treatment. Spent fuel assemblies from nuclear power reactors are highly radioactive because they contain fission products. Relatively few options are available for the treatment of spent fuel. The tubes and the fuel matrix provide considerable containment against attack and release of nuclides. To minimize the volume of spent fuel that must be shipped or disposed of, consolidation of rods in assemblies into compact bundles of fuel rods has been successfully tested. Alternatively, intact assemblies can be encased in metal containers.

Reprocessing as practiced outside the United States, involves chopping fuel rods into small pieces, leaching out the uranium oxide by nitric acid (qv), and applying suitable solvents to separate the uranium, plutonium, and fission products (14). A vitrification process is applied to fission product wastes which are mixed with pulverized glass, heated to melting by an electric current, and the mixture poured into a metal canister for solidification, safe storage, and ultimate disposal.

Special chemical treatment can isolate the nuclides of intermediate half-life, ie, cesium-137 and strontium-90 [*10098-97-2*], ^{90}Sr, $t_{1/2}$ 29 yr (15). These provide most of the radioactivity, radiation, and heat during the early years of a disposal facility. Solid-phase extraction methods using macrocyles and membranes promise to yield waste of low volume and activity (see INCLUSION COMPOUNDS; MEMBRANE TECHNOLOGY). The separated intensely radioactive chemicals can be placed in separate storage or made available for industrial radiation use.

Storage and Transport

Storage. Storage of spent fuel assemblies in deep water pools at reactor sites serves several safety functions. Cooling by water prevents the fuel from melting from decay heat. The shielding effect of the water provides protection for workers in the vicinity from gamma-radiation. Moreover, the separation of assemblies by racks prevents a chain reaction from occurring. The reinforced concrete pools are designed to withstand earthquakes. Water passes through a heat exchanger to maintain a constant temperature, and the purity of the water is assured by use of a demineralizer.

The pools of most reactors were designed for a limited number of fuel assemblies. As a consequence, pools are filling up. The accumulation of U.S. spent fuel over time for commercial sources is shown in Figure 1. The material awaits permanent disposal. Alternatives to pool storage are being sought. One relatively inexpensive alternative is dry storage in large sealed concrete containers. The fuel is inserted into the containers, water drained off, and helium, an inert gas, is added. Arrays of these can be held on a thick concrete pad out in the open.

The Nuclear Waste Policy Act of 1982 specified that DOE would begin accepting spent fuel from nuclear utilities in 1998. To meet that target date, DOE would have to utilize existing storage facilities at federal laboratories and

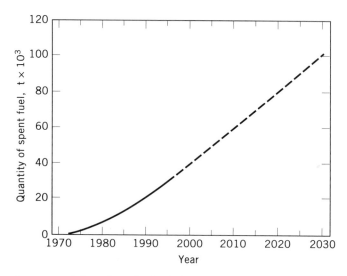

Fig. 1. Spent commercial nuclear fuel in the United States where (— — —) represents projected quantities (1). Courtesy of Battelle Press.

possibly construct additional storage. As a part of the Nuclear Waste Policy Act, consideration was to be given to building a monitored retrievable storage (MRS) facility. Its role would be to receive fuel from utilities, store it for as long as needed, package it for safe burial, and ship it to a disposal site (16). The Department of Energy continues to seek a host for such a facility. Some interest has been expressed by Native American tribes of the western United States.

Waste by-products of the operation of plutonium-producing reactors beginning in World War II have been stored in underground tanks at Hanford (Washington state) and the Savannah River Plant (South Carolina). Some single-walled tanks (Hanford) have leaked. A remediation program for tank storage is in place and mechanical stirring pumps have removed the potentially explosive hydrogen from these Hanford tanks.

Plutonium itself, generated by production reactors during the Cold War for weapons purposes, may become a waste by national policy (17). The plutonium would then be disposed of along with other high level wastes. Alternatively, the plutonium could serve as fuel for reactors that generate electric power, for example the developmental integral fast reactor (see NUCLEAR REACTORS, REACTOR TYPES), or could be bombarded by neutrons produced by high energy charged particles from an accelerator (see also PLUTONIUM AND PLUTONIUM COMPOUNDS).

Low level waste with its generally smaller radioactivity level can be stored in suitable containers in buildings. Protective shielding and handling equipment are required.

Transport.   In the United States waste transportation is regulated by the NRC and the Department of Transportation (DOT). Packaging and shipping must be in conformity with comprehensive rules. Shipping container classes are defined in accord with the amount of radioactivity involved. A letter code for radioactivity, distinct from that for waste class, is used. Type A containers involve a minimum of protection, whereas type B containers must withstand a series of events, including specified impacts with hard surfaces or spikes, exposure to high temperature fire, and long immersion in water. Spent fuel casks are of type B. For the movement of spent fuel, computer tracking systems are used. State radiological safety units are informed of shipments of spent fuel and other high activity radioactive materials so that these units may respond in case of accident.

A multipurpose canister (MPC) is planned for the transportation, storage, and disposal of spent fuel, minimizing the amount of handling required. In the design of the container, factors being considered are fuel assembly size, weight, enrichment, amount of burnup, and age. Production of canisters must fit the schedule of fuel acceptance (18).

The safety record for transport of radioactive materials including spent fuel and wastes is excellent. Information about transportation of radioactive materials including waste is managed by DOE. Codes such as RADTRAN that can calculate public radiation dose owing to the passage of shipments have been developed. The maximum dosage from such shipments is a very small fraction of the typical annual radiation dose from all other sources.

Disposal

The disposal of radioactive waste is governed by rules of the NRC and the EPA (19). NRC regulations differ for low level waste and for high level waste, including spent fuel (20).

Isolation of radioactive wastes for long periods to allow adequate decay is sought by the use of multiple barriers. These include the waste form itself, the primary containers made of resistant materials, overpacks as secondary layers, buffer materials, concrete vaults, and finally the host rock or soil. Barriers limit water access to the waste and minimize contamination of water supplies. The length of time wastes must remain secure is dependent on the regulatory limit of the maximum radiation exposure of individuals in the vicinity of the disposal site.

Performance assessments are predictions of radioactivity releases, the rate of transfer of contaminants through various media, and the potential for hazard to the public. These are based on a combination of experimental data obtained in the process called site characterization and detailed computations about radionuclides and their effects. The progressive attack on the metal or ceramic waste container, the diffusion of water into the waste form, the leaching of the radioactive compounds, diffusion out, and washing away of radionuclides are all considered.

Relevant hydrological fundamentals are utilized (21) to take account of the complex interaction of physical and chemical processes involving soil or rock, water, and contaminant. Attention is paid to uncertainties in calculated results.

Models for transport distinguish between the unsaturated zone and the saturated zone, that below the water table. There the underground water moves slowly through the soil or rock according to porosity and gradient, or the extent of fractures. A retardation effect slows the motion of contaminant by large factors in the case of heavy metals. For low level waste, a variety of dose calculations are made for direct and indirect human body uptake of water. Performance assessment methodology is described in Reference 22.

High Level Waste.   Many studies have been made of possible modes of disposal for high level waste including spent fuel. Some of the techniques considered are a very deep hole, a remote island, mountainside, subseabed, the Antarctic ice sheet, and delivery to space. As of this writing (ca 1995) the preferred method is deep underground burial in the floor of a mined cavity. Several types of rock have been considered as potential hosts to the waste, including granite, salt, basalt, and tuff. A history of high level waste disposal prior to 1987 is available (23). In the 1970s an investigation of the use of salt as disposal medium was undertaken, but terminated when it was found that there had been extensive drilling in the region.

Regulations on high level radioactive waste management have traditionally been provided by the NRC and a number of specific requirements on the character of a disposal site are spelled out (19,20). The generic regulations planned by the EPA have been delayed by court actions and federal law requirements. Wastes are to be placed at least 300 meters below the earth's surface. The waste form must be free of liquid, noncorrosive, and noncombustible. The container is to remain intact for between 300 and 1000 years. The travel time of groundwater prior to waste emplacement is preferred to be greater than 10,000 years, but not less than 1000 years. The repository must be placed where there are no attractive resources and far from population centers. Wastes are to be retrievable for a period of 50 years. Finally, releases of radionuclides from the repository must be less than figures specified by EPA (19). Typical limits are 3.70×10^{12} Bq/10^3 t (100 Ci/10^3 t) of heavy metal.

Regulations include guidelines on geologic conditions. Of special interest is the stability of the geology against faulting, volcanic action, and earthquakes. The repository is to be located in an arid region, where the water table is quite low. The host rock is to have a suitable porosity and a low hydraulic conductivity.

Tuff, a compressed volcanic material, is the primary constituent of Yucca Mountain, near Las Vegas, Nevada, the site selected by Congress in 1987 for assessment for spent fuel disposal. An underground laboratory, to consist of many kilometers of tunnels and test rooms, is to be cut into the mountain with special boring equipment to determine if the site is suitable for a repository.

Site characterization studies include a surface-based testing program, potential environmental impact, and societal aspects of the repository. Performance assessment considers both the engineered barriers and the geologic environment. Among features being studied are the normal water flow, some release of carbon-14, and abnormal events such as volcanic activity and human intrusion. The expected date for operation of the repository is 2013.

The geologic aspects of waste disposal (24–26), proceedings of an annual conference on high level waste management (27), and one from an annual conference on all types of radioactive waste (28) are available. An alternative to burial is to store the spent fuel against a long-term future energy demand. Uranium and plutonium contained in the fuel would be readily extracted as needed.

Low Level Waste.   The NRC 10CFR61 specifies the nature of the protection required for waste containers (20). Class A wastes must meet minimum standards, including no use of cardboard, wastes must be solidified, have less than 1% liquid, and not be combustible, corrosive, or explosive. Class B wastes must meet the minimum standards but also have stability, ie, these must retain size and shape under soil weight, and not be influenced by moisture or radiation. Class C wastes must be isolated from a potential inadvertent intruder, ie, one who uses unrestricted land for a home or farm. Institutional control of a disposal facility for 100 years after closure is required.

The traditional method of disposal of low level radioactive waste has been shallow land burial, consisting of filling a deep trench and covering it with a layer of earth (29). The trend in time of annual volume of LLW per reactor in the United States is shown in Figure 2. Three of the original six commercial sites were closed owing to leaks of radioactivity. Two initiatives resulted: stricter regulations (20) and the 1980 Low Level Radioactive Waste Policy Act which called for each state to be responsible for wastes generated within its borders, but recommended the formation of compacts among states to build regional disposal facilities. Figure 3 shows the groupings that have been formed. Each compact is seeking to implement the law. The historical background of low level waste management is available (30,31).

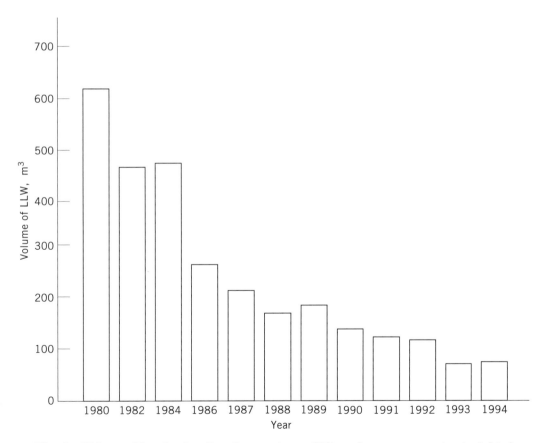

Fig. 2. Volume of low level radioactive waste per U.S. nuclear power reactor (weighted industry median). The decrease over the period 1980–1994 was more than a factor of seven. Courtesy of Institute of Nuclear Power Operations.

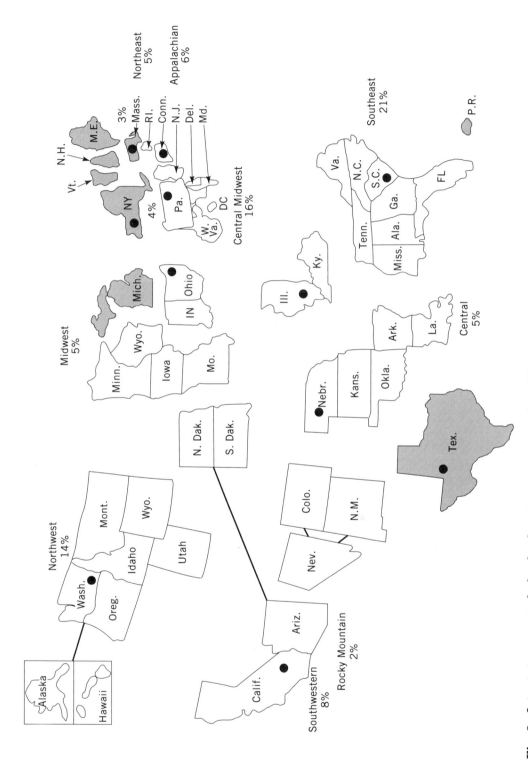

Fig. 3 Interstate compacts for low level waste management where (▦) represents unaffiliated states and (●), host sites. The percentages of total U.S. LLW are also given. Courtesy of Battelle Press.

Shallow land burial is planned for disposal facilities in arid regions, notably in California and Texas. Public demand at locations in humid regions led to the adoption of greater confinement designs, which provide additional protection. Included are concrete vaults, overpacks, multiple-layered caps, and enhanced monitoring systems. The main features of a design by Chem-Nuclear Systems, Inc., for the proposed North Carolina facility intended to serve eight states of the Southeast, are shown in Figure 4.

Methods to control infiltration of water into low level waste disposal facilities are being studied. Three techniques that may be employed separately, in sequence, or in conjunction are use of a resistive layer, eg, clay; use of a conductive layer, involving wick action; and bioengineering, using a special plant cover.

Funding for developing commercial waste disposal facilities is to come from the waste generators. In the case of spent fuel disposal, a Nuclear Waste Fund is accumulating based on an assessment of one mill per kilowatt-hour of electricity. For low level wastes, surcharges on waste disposal and direct assessments of utilities have been imposed.

Transuranic Waste. Transuranic wastes (TRU) contain significant amounts (>3,700 Bq/g (100 nCi/g)) of plutonium. These wastes have accumulated from nuclear weapons production at sites such as Rocky Flats, Colorado. Experimental test of TRU disposal is planned for the Waste Isolation Pilot

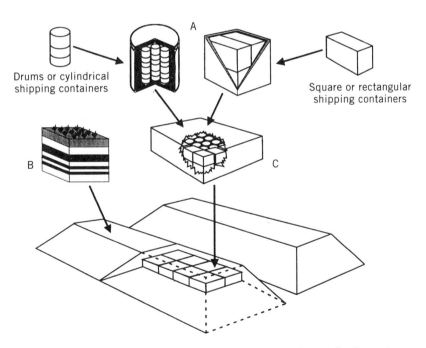

Drums or cylindrical shipping containers

Square or rectangular shipping containers

Fig. 4. Integrated vault technology for low level waste disposal where A represents waste containers that are placed in concrete overpacks and sealed with grout; B, closed modules covered with a multiple-layer earthen cover, to direct water away from modules, and short rooted vegetation for erosion control; and C, overpacks placed in reinforced concrete modules which are closed with a reinforced concrete roof. Courtesy of Chem-Nuclear Systems, Inc.

Plant (WIPP) site near Carlsbad, New Mexico. The geologic medium is rock salt, which has the ability to flow under pressure around waste containers, thus sealing them from water. Studies center on the stability of structures and effects of small amounts of water within the repository.

Other fuel besides that from U.S. commercial reactors may be disposed of in the ultimate repository. Possibilities are spent fuel from defense reactors and fuel from research reactors outside of the United States. To reduce the proliferation of nuclear weapons, the United States has urged that research reactors reduce fuel enrichment in uranium-235 from around 90 to 20%. The latter fuel could not be used in a weapon. The United States has agreed to accept spent fuel from these reactors.

Environmental Issues

Progress toward the disposal of nuclear wastes has been slow for a number of reasons. One is public opposition, epitomized by the not in my back yard (NIMBY), syndrome. Much of the public is fearful of nuclear reactors, radioactivity, and radiation (32). Concern was heightened by the Three Mile Island and Chernobyl accidents. Others in the public oppose siting of disposal facilities on economic grounds, believing that business, tourism, and property values would be jeopardized. Some people are concerned with equity when a local population must host a site that benefits the state, region, or nation. Efforts by the nuclear community to maintain positive dialogue are often successful, especially when the public is involved in discussions at an early stage of a project, and is fully and correctly informed of plans and developments (33,34). Educational material on radioactive wastes is available (35,36).

There is an enormous amount of research literature about nuclear waste management, and nuclear scientists and engineers generally are convinced that wastes can be disposed of safely. Delays in disposal can also be attributed to an accepted national policy and to the administration of programs. Until the late 1980s, treatment and disposal of defense wastes were of lower priority than accumulating the needed weapons material. Efforts in the 1970s to find a suitable site for HLW disposal in salt were unsuccessful. Changes in national policy and governmental plans since the 1970s, including the demand for more information about geology and hydrology, have delayed projects. The costs associated with long-term storage of spent fuel or low level waste are high. Moreover, the nuclear industry recognizes that continued and extended use of nuclear power depends on acceptance of waste treatment by the public and the financial community, which depends on the demonstrated ability of industry and government to meet challenges in a safe and an economical manner (37,38).

BIBLIOGRAPHY

"Waste Management" under "Nuclear Reactors" in *ECT* 2nd ed., Vol. 14, pp. 102–108, by J. O. Blomeke, Oak Ridge National Laboratory; in *ECT* 3rd ed., Vol. 16, by J. O. Blomeke, Oak Ridge National Laboratory.

1. R. L. Murray, *Understanding Radioactive Waste*, 4th ed., Battelle Press, Columbus, Ohio, 1994.
2. *Radioactive Waste Management: An IAEA Source Book*, International Atomic Energy Agency, Vienna, Austria, 1992.
3. R. E. Berlin and C. C. Stanton, *Radioactive Waste Management*, John Wiley & Sons, Inc., New York, 1989.
4. Y. S. Tang and J. H. Saling, *Radioactive Waste Management*, Hemisphere, New York, 1990.
5. F.-S. Lau, *Radioactivity and Nuclear Waste Disposal*, Research Studies Press, Letchsworth, Hertsfordshire, U.K., 1987.
6. D R. Lide, ed., *CRC Handbook of Chemistry and Physics*, 75th ed., 1994–1995, CRC Press, Boca Raton, Fla., 1994; N. Holden, "Table of the Isotopes," pp. 11-35–11-139.
7. *Integrated Data Base for 1993: U.S. Spent Fuel and Radioactive Waste Inventories, Projections, and Characteristics*, DOE/RW-0006, Rev. 9, Oak Ridge National Laboratory, Oak Ridge, Tenn., Feb. 1994.
8. "Atomic Energy Act of 1946," Public Law 585, 79th Congress, 2nd Session, Aug. 1, 1946; "Atomic Energy Act of 1954," Public Law 703, 83rd Congress, 2nd Session, Aug. 20, 1954; "Low-Level Radioactive Waste Policy Act," Public Law 99-240, 96th Congress, 2nd Session, Dec. 22, 1980; and "Nuclear Waste Policy Act of 1982," Public Law 97-425, 97th Congress, 2nd Session, Jan. 7, 1983, in *United States Statutes at Large*, U.S. Government Printing Office, Washington, D.C.
9. M. W. Carter, A. A. Moghissi, and B. Kahn, *Management of Low-Level Radioactive Waste*, Vols. 1 and 2, Pergamon Press, New York and Oxford, 1979.
10. A. A. Moghissi, H. W. Godbee, and S. A. Hobart, *Radioactive Waste Technology*, American Society of Mechanical Engineers, New York, 1986.
11. *Code of Federal Regulations, Energy, Title 10, parts 1-199*, U.S. Government Printing Office, Jan. 1, 1994.
12. J. Shapiro, *Radiation Protection: A Guide for Scientists and Physicians*, 3rd ed., Harvard University Press, Cambridge, Mass., 1990.
13. R. G. Post, ed., *WM'94, Proceedings of the Symposium on Waste Management*, Laser Options, Tuscon, Ariz., 1994, pp. 1615–1639.
14. W. Bebbington, *Sci. Am.*, 30 (Dec. 1976).
15. R. R. Jackson, *Nuclear Technol.* **32**, 10–15 (1977).
16. *The Role of the Monitored Retrievable Storage Facility in an Integrated Waste Management System*, DOE/RW-0238, Office of Civilian Radioactive Waste Management, U.S. Department of Energy, Washington, D.C., 1989.
17. *Management and Disposal of Excess Weapons Plutonium*, National Academy Press, Washington, D.C., 1994.
18. Ref. 13, pp. 2003–2007.
19. *Code of Federal Regulations*, Title 40, part 191, U.S. Government Printing Office, Washington, D.C.
20. Ref. 11, Title 10, part 61; Ref. 11, Title 10, part 60 and part 960.
21. R. A. Freeze and J. A. Cherry, *Groundwater*, Prentice-Hall, Englewood Cliffs, N.J., 1979.
22. J. E. Till and H. R. Meyer, eds., *Radiological Assessment, A Textbook on Environmental Dose Analysis*, NUREG/CR-3332, U.S. Nuclear Regulatory Commission, Washington, D.C., 1983; *Disposal of Radioactive Waste: Review of Safety Assessment Methods*, Nuclear Energy Agency, Paris, 1991.
23. L. Carter, *Nuclear Imperatives and Public Trust: Dealing with Radioactive Waste*, Resources for the Future, Washington, D.C., 1987.
24. I. S. Roxburgh, *Geology of High-Level Nuclear Waste Disposal: An Introduction*, Chapman and Hall, London and New York, 1987.

25. N. A. Chapman and I. G. McKinley, *The Geological Disposal of Nuclear Waste*, John Wiley & Sons, Ltd., Chichester, U.K., 1987.

26. K. B. Krauskopf, *Radioactive Waste Disposal and Geology*, Chapman and Hall, London and New York, 1988.

27. *International High-Level Radioactive Waste Conference, Las Vegas, Nev., 1994*, American Nuclear Society, LaGrange Park, Ill., American Scoeity of Civil Engineers, New York.

28. R. G. Post, ed., *WM'95, Proceedings of the Symposium on Waste Management*, Laser Options, Tucson, Ariz., 1995.

29. *The Shallow Land Burial of Low-Level Radioactively Contaminated Solid Waste*, Committee on Radioactive Waste Management, National Academy of Sciences, Washington, D.C., 1976.

30. *Directions in Low-Level Radioactive Waste Management: A Brief History of Commercial Low-Level Radioactive Waste Disposal*, DOE/LLW-103, Rev. 1, The National Low-Level Waste Management Program, INEL, Idaho Falls, Idaho, Aug. 1994.

31. E. L. Gershey, R. C. Klein, E. Party, and A. Wilkerson, *Low-Level Radioactive Waste: From Cradle to Grave*, Van Nostrand Reinhold, New York, 1990.

32. S. Weart, *Nuclear Fear: A History of Images*, Harvard University Press, Cambridge, Mass., 1988.

33. M. R. English, *Siting Low-Level Radioactive Waste Disposal Facilities: The Public Policy Dilemma*, Quorum Books, New York, 1992.

34. M. E. Burns, *Low-Level Radioactive Waste Regulation: Science, Politics, and Fear*, Lewis Publishers, Chelsea, Mich., 1988.

35. The League of Women Voters Education Fund, *The Nuclear Waste Primer: A Handbook for Citizens*, Lyons & Burford, Publishers, New York, 1993.

36. *Science, Society, and America's Nuclear Waste*, DOE/RW-0361, U.S. Department of Energy, Washington, D.C., 1992, Units 1–4.

37. *Nuclear Power: Technical and Institutional Options for the Future*, National Academy Press, Washington, D.C., 1992.

38. B. L. Cohen, *The Nuclear Energy Option: An Alternative for the 90s*, Plenum Press, New York, 1990.

RAYMOND L. MURRAY
Consultant

SAFETY IN NUCLEAR POWER FACILITIES

Nuclear energy is a principal contributor to the production of the world's electricity. As shown in Table 1, many countries are strongly dependent on nuclear energy. For some countries, more than half of the electricity is generated by nuclear means (1,3). There were 424 nuclear power plants operating worldwide as of 1995. Over 100 of these plants contributed over 20% of the electricity in the United States (see also POWER GENERATION).

Table 1. Nuclear Power Units in Operation Worldwide[a]

Nation	Number of units[b]	Net power, MWe[b]	Percent of total electricity generated[c,d]
Belgium	7	5,527	59
Bulgaria	6	3,420	34
Canada	22	15,439	16
China	3	2,100	
Czech Republic	4	1,632	
Finland	4	2,310	33
France	55	57,373	73
Germany	21	22,715	32
Hungary	4	1,729	46
India	9	1,620	2
Japan	49	38,859	24
South Korea	9	7,220	48
Lithuania	2	2,760	
Russia	25	19,799	
Slovakia	4	1,632	
South Africa	2	1,840	6
Spain	9	7,084	35
Sweden	12	10,075	52
Switzerland	5	3,025	40
Taiwan	6	4,884	38
Ukraine	14	12,095	
United Kingdom	34	11,540	19
United States	109	99,238	21
other[e]	9	3,602	
Totals	*424*	*337,518*	

[a]Ref. 1.
[b]Status as of Dec. 31, 1994.
[c]Status as of Jan. 1, 1992.
[d]Ref. 2.
[e]Nations having < 1000 MWe capacity are combined under "other."

Safety has played a dominant role in the ability to generate electricity by nuclear means. When energy is released from the atom by fission, ie, breaking apart the nucleus of a heavy element such as uranium or plutonium, two lighter nuclei called fission products are also produced. Most of these fission products are radioactive. Thus, as energy is generated, highly radioactive materials, which can be harmful to living organisms if not kept under strict control, are produced.

All safety provisions must meet strict limitations on allowable levels of radiation exposure, assuring that neither the public nor the plant workers are harmed as a result of operation of the nuclear power plant. Control of radioactive materials must be effected by (1) careful design and testing of the integrity and reliability of the components and systems which contain and control the radioactive materials in the nuclear power plant; (2) fabrication, installation, and construction of these components to meet high quality standards; and (3) thorough training of plant operators to assure that the systems and components function as designed and integrity is maintained.

The design of safety systems and components must also provide tolerance for human fallibility in achieving strict controls, ie, protect against design mistakes, equipment failure, and operational error. Redundancy and diversity are provided for key safety functions. If one component fails, another component, either a duplicate or in different form, is available to carry out that function. In this fail-safe design, a failure of any device should lead to a stable condition. Automatic devices are provided to shut down or reduce power or inject coolant in the event of component failure or operator error. Finally, passive safety features are provided. The design utilizes physical laws which intrinsically bring the system to a stable state when an abnormal condition ensues. An important example of such a passive safety feature is a nuclear fuel system which has an intrinsic characteristic that causes the fission reaction rate to reduce if power or temperature increases. Thus, runaway power excursions are prevented by the laws of physics, not by the operation of equipment or the actions of people.

Whereas these design measures provide the primary assurance of protection from the harmful effects of radiation, additional protection is provided in the unlikely event that the integrity of the systems or components break down. The entire portion of the nuclear plant containing radioactive material is enclosed in a strong containment building. If a release of radioactivity from the plant were to occur, the radioactive material would be captured within the containment building. As a further precaution, these processes are subject to continual cross-checks by people separate from those engaged in the design, fabrication, construction, and operational processes. Cross-checks in the form of design reviews, inspections, operations, and safety audits are carried out. The U.S. nuclear industry has set up the Institute of Nuclear Power Operations (INPO) to establish operations and training standards and to audit nuclear plant operations for compliance with those standards. In addition, a totally independent body responsible to the public, the U.S. Nuclear Regulatory Commission (NRC), establishes overriding safety regulations and monitors compliance.

Safety provisions have proven highly effective. The nuclear power industry in the Western world, ie, outside of the former Soviet Union, has made a significant contribution of electricity generation, while surpassing the safety record of any other principal industry. In addition, the environmental record has been outstanding. Nuclear power plants produce no combustion products such as sulfuric and nitrous oxides or carbon dioxide (qv), which are significant causes of acid rain and global warming (see AIR POLLUTION; ATMOSPHERIC MODELING).

The accident at the Three Mile Island (TMI) plant in Pennsylvania in 1979 led to many safety and environmental improvements (4–6). No harm from radiation resulted to TMI workers, to the public, or to the environment (7,8), although the accident caused the loss of a 2×10^9 investment. The accident at the Chernobyl plant in the Ukraine in 1986, on the other hand, caused the deaths of 31 workers from high doses of radiation, increased the chance of cancer later in life for thousands of people, and led to radioactive contamination of large areas. This latter accident was unique to Soviet-sponsored nuclear power. The Soviet-designed Chernobyl-type reactors did not have the intrinsic protection against a runaway power excursion that is required in the rest of the world, nor was there a containment building (9–11).

Basic Safety Principles

The three fundamental safety objectives advocated by the International Atomic Energy Agency (IAEA) (12) for all nuclear power plants worldwide are (1) to protect individuals, society, and the environment by establishing and maintaining in nuclear power plants an effective defense against radiological hazard; (2) to ensure in normal operation that radiation exposure within the plant as well as that resulting from any release of radioactive material from the plant is kept as low as reasonably achievable (ALARA) and below prescribed limits, and to ensure mitigation of the extent of radiation exposures owing to accidents; and (3) to prevent accidents in nuclear plants with high confidence; to ensure that, for all accidents taken into account in the design of the plant, even those of very low probability, radiological consequences, if any, would be minor; and to ensure that the likelihood of severe accidents with serious radiological consequences is extremely small.

IAEA also defines the fundamental responsibilities for nuclear power plant safety as ultimately resting with the operating organization (12). Designers, suppliers, constructors, and regulators are also responsible for their separate activities. Responsibility is reinforced by the establishment of a safety culture, ie, "the personal dedication and accountability of all individuals engaged in any activity which has a bearing on the safety of nuclear power plants" (13). Safety design of nuclear power plants is founded on the defense-in-depth concept, which provides multiple levels of protection to both the public and the workers, in the form of physical barriers and levels of implementation of the associated defenses. Each of the multiple physical barriers prevents the release of radioactive materials, but all envelop a given number of the others so that if an inner barrier fails, the next outer barrier holds back the radioactive material (8,12,14). Figure 1 shows both the physical barriers and the multiple levels of protection in conceptual form. The first barrier is the nuclear fuel rod which heats up as fission occurs. The fuel rod is made up of corrosion-resistant ceramic and uranium oxide pellets, placed in tubes called cladding (see NUCLEAR REACTORS, NUCLEAR FUEL RESERVES), comprising the second barrier which surrounds the nuclear fuel. The cladding is made of a metal alloy, usually Zircaloy, which is highly corrosion-resistant. The third barrier is a steel pressure boundary, consisting of the reactor pressure vessel. All the core is placed within this vessel, as is the main coolant piping which contains the cooling water that takes the heat from the fuel and transfers it to provide the electricity. The fourth barrier is the containment building, a massive reinforced concrete or steel structure within which is placed the nuclear portion of the power plant's generation system, called the nuclear steam supply system.

The reliability of the physical barriers is assured by implementation of multiple levels of defense-in-depth, characterized by a sequence of concentric design features and operational defenses against the release of radiation from the plant. The first level is the design, fabrication, and construction of the plant to high quality standards together with its reliable operation and maintenance within the prescribed operational bands. The second level of defense is comprised of systems and operating procedures which control abnormal conditions, ie, transients beyond the prescribed bands, so that the basic integrity of the system is maintained. The third level of defense-in-depth is the provision of backup

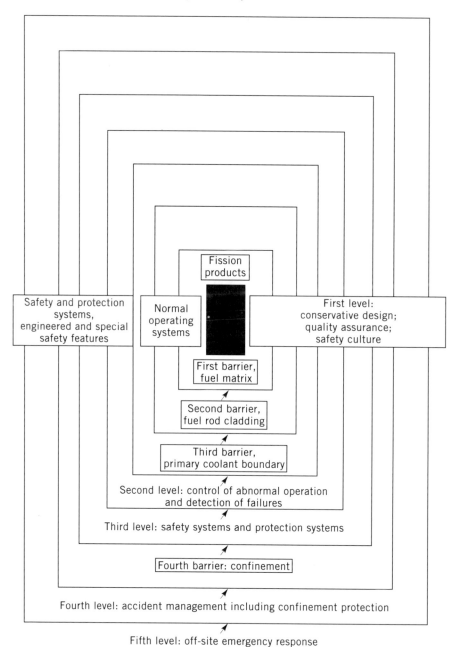

Fig. 1. The relation between physical barriers and levels of protection in defense-in-depth design of a nuclear facility (12). Courtesy of IAEA.

systems and emergency operating procedures that become operative in the event that there is a loss of integrity or a loss of a basic function of the normal nuclear systems, assuring that the radioactive material is not released from the nuclear systems. In each of these first three levels, a separate layer of multiple protection is provided through redundancy and diversity.

The fourth level of defense-in-depth is activated if all of the previous levels fail and radioactivity is released from the power-generating system. This level consists of containment systems and accident management processes that prevent the dissemination of radioactivity to the atmosphere even if it is released from the nuclear systems. The fifth level is the provision for emergency planning outside the plant boundary in the highly unlikely event that all of the first four levels of defense were to fail.

The defense-in-depth process requires that each physical barrier be designed conservatively using substantial margins against failure, on-line monitoring instrumentation, off-line inspections to detect incipient failures, and highly trained operators and maintenance personnel guided by prudent procedures. In particular, the containment building is designed to withstand external assaults from earthquakes, hurricanes, tornadoes, floods, and flying objects such as crashing airplanes. The safety of the nuclear plant and the integrity of its containment must also be maintained in the event of aggression from terrorists or saboteurs. Stringent security measures are provided at each plant to meet such a challenge. These include large around-the-clock guard forces, modern detection and alarm systems, and vehicle intrusion barriers. Strict personnel checks, emphasis on professional discipline, and the redundant, fail-safe design of the safety systems provide protection against internal sabotage. A prioritization process applies in the design of the barriers and the provisions for defense-in-depth that is based on the principle "prevention first." Thus, no compromise is permitted in the design integrity of the first three barriers and the first two levels of defense-in-depth which prevent an accidental release.

Safety Design

Design Features. Design safety features (8,14,15) are utilized at each of the concentric safety barriers. The most important of these safety features apply at the innermost barriers in what is called the reactor core. A cylindrical arrangement of bundles of nuclear fuel rods are arranged to cause fission and spaced to permit the flow of cooling water through them. Sustained fission is possible because a chain reaction can be established. When a neutron is absorbed by the uranium-235 nucleus causing it to split, 2.44 neutrons on the average are also released. Some of these additional neutrons escape or are absorbed in uranium and other materials in the fuel without causing fission, but if just one of them is absorbed in another uranium-235 nucleus, fission is self-sustained. This is called criticality or a critical mass. If an average of more than one neutron is absorbed, the rate of fission increases, and vice versa (see NUCLEAR REACTORS, REACTOR TYPES).

First Barrier. The rate of fission must be kept under strict control so as to prevent a runaway power excursion, ie, an excessive increase in fission rate. Control is carried out in two ways: one, intrinsic to the chain reaction, involves a negative coefficient of reactivity; the other is external, through use of control rods. The fission rate is dependent upon the temperature of the fuel and the temperature and density of the coolant. Fuel composition and absorber materials, ratio of fuel to coolant, and geometrical arrangement of the fuel and the fuel rods can be designed so that the fission rate decreases as temperature, coolant density, or power increases. This intrinsic feature can be designed into the fuel system, ie, the core, to cause the fission rate to slow down when temperature, steam content, or power increases. This is called a negative temperature, void, or power coefficient of reactivity. Thus, when an incipient transient in temperature or power occurs in a core having a negative coefficient of reactivity, the physical processes governing the fission rate slow the excursion down to prevent a runaway condition. This basic safety design is a requirement in the United States and the rest of the Western world.

The external means of controlling the fissioning rate is through use of control rods. Metal rods composed of strong neutron absorbers can reduce or cut off the chain reaction. These can be inserted into the core to reduce or stop fission or alternatively pulled out of the core to start or increase fission. The control rods are moved by remote control by the operator for normal power control. Redundant and diverse radiation detectors and temperature sensors are installed in the core to signal control rod mechanisms to insert the rods automatically so as to keep the power excursions within the allowable band whenever limiting conditions are reached. Insertion into the core to stop the chain reaction completely is rapid under these circumstances to prevent damage to the fuel. There is a chance that deficiencies in meeting the high quality standards of nuclear fuel manufacture would cause a loss of integrity of the fuel rods. To be on the alert for such an event, radiation monitors check the level of radioactivity in the coolant water to detect incipient fuel rod failure, which would show itself by leakage of fission products from the fuel rods into the coolant stream. For further assurance, on-line monitors detect incipient fuel failure.

Cutting off the chain reaction removes concern that a runaway power excursion involving rapid melting of fuel rods can occur, but does not eliminate the possibility of slow fuel melting. The fission products generated in the fuel rods during power operation continue to emit radioactive particles which are converted to thermal energy in the fuel rods. Although the energy generated is orders of magnitude smaller than that at full power, the fuel rods can slowly heat up to melting temperature unless cooling is maintained. Thus, continued reliability of both the reactor cooling system (second barrier) and back-up emergency cooling provisions (third barrier) is essential during plant shutdown.

Second Barrier.   Safety design features at the second barrier involve the primary coolant circuit. These are derived from adherence to rigorous standards in the selection of materials and in conservative design of the coolant system and coolant pressure boundaries. The conservatism is provided by designing for forces, pressures, temperatures, fluid conditions, radiation levels, thermal transients, and fatigue cycles, that are higher than are expected during power operation. This difference between the design levels and the actual levels is called margin.

Margin is provided in the coolant system by designing to keep the peak temperatures in the fuel rods well below the fuel melt temperature, to keep peak coolant temperatures in a range of stable operation, to assure adequate coolant pumping capacity, and to provide reliable component cooling. To assure the integrity of the coolant pressure boundaries, margin is provided by designing these boundaries to withstand pressure surges and steady-state pressures substantially higher than the operating pressure, to prevent fatigue failure of the pressure boundaries after many thermal cycles, to maintain integrity even after substantial loss of ductility occurs for those parts of the system that are exposed to intense radiation, and to withstand the shocks and stresses caused by earthquakes of magnitudes at the upper limit of that which would be expected at the plant site.

Another design feature is the provision for on-line coolant leakage monitors that would signal incipient pressure boundary failure. Radiation monitors are installed in the containment building to detect airborne radioactivity, which would signal incipient loss of integrity of some part of the pressure boundary. Extensive inspection requirements are also stipulated. Sensitive ultrasonic devices are used to check the condition of the piping (see PIPING SYSTEMS). Samples of the reactor pressure vessel material are removed from the reactor zone periodically to be tested for ductility loss. Eddy current detectors are used to inspect the steam (qv) generator tubes for the occurrence of cracking (see NONDESTRUCTIVE EVALUATION).

Third Barrier. At the level of the third barrier, the key design feature is the provision of backup cooling systems which continue to cool the core in the event of a significant loss of integrity which would disable the normal cooling functions of the primary circuit. Separate and redundant coolant injection systems are provided: (*1*) the normal coolant recharging systems which replenish the primary coolant circuit; (*2*) gas-pressurized accumulator tanks which force water under high pressure into the primary coolant circuit; and (*3*) safety coolant injection systems which pump water at high pressure into the primary coolant circuit from separate reservoirs.

Fourth Barrier. The design feature of the fourth barrier is the containment building. It is designed to withstand the high temperatures, pressures, and radiation resulting from a severe accident entailing fuel meltdown. Supplementary features are utilized to reduce the consequences of such severe accident conditions: spray systems are installed to reduce the containment temperature; catalytic devices are provided to absorb airborne fission products; igniters are installed to burn off hydrogen gas emitted during an accident before the deflagration temperature is reached; interlocks and alarms are activated to assure that containment hatches are appropriately closed; and periodic testing of the leak-tightness of the containment is carried out.

Assessment. It is important to verify that safety is actually being achieved by monitoring the operations of the nuclear plants. Both the U.S. NRC (16–19) and the INPO (20,21) perform key roles in this process. Each organization periodically sends teams to every nuclear power plant in the United States to assess the safety of the operations. Each plant is given a performance rating, backed up by detailed critiques identifying operational strengths as well as weaknesses. The plant is then required to follow up on any corrective actions indicated by these safety audits. The NRC ratings are made public and have significant power in motivating corrective action when that is needed.

Audits by INPO and the U.S. NRC are a culmination of a high degree of self-auditing by the plant operators and the utilities themselves, often assisted by special third-party safety review boards set up to help carry out safety assessments (21). Self-auditing and self-criticism are essential to the process. These reflect the fundamental reactor safety principle that the owner-operator of the plant has the ultimate responsibility for plant safety.

Another element of safety monitoring is the requirement stipulated by the NRC that each utility report any operational event which is out of the ordinary or has safety implications. These licensing event reports (LERs) are placed in the public record (19). Both INPO and the NRC evaluate the LERs and inform all the utilities of any event that has broad safety significance to the industry. In particular, if one of these field events is judged to be the precursor of a serious accident, all operating nuclear plants are made fully aware of its implications. Thus, steps can be taken to prevent the event from occurring, and if the event does occur, the plant should be prepared to stop the progress of the event before the stage of a severe accident is reached (19,20). The methodology INPO used to make these evaluations and communicate them to the nuclear plants (22,23) was developed by the Electric Power Research Institute (EPRI), the collaborative R&D arm of the U.S. electric utilities. There was a reduction of a factor of 10 in the number of these events from 1985, 2.4 events per plant, to 1993, 0.24 events per plant.

A further assessment is carried out through the definition and measurement of industry-average performance indexes relating to safety. These indexes have been established by the utilities, working with INPO, EPRI, and the suppliers (24). Each index bears on some aspect of safe operation of the nuclear power plant, ie, industrial safety accident rate, unplanned automatic scrams, collective radiation exposure, plant capability factor, and unplanned capability loss factor. Five-year goals are established for average performance of all U.S. plants for each of these performance indexes. A substantial improvement has been made in all of these indexes since the early 1980s. The goals which were set in 1990 to be achieved by 1995 were either met prior to 1995 or are expected to be met by the end of that year. International performance indexes very similar to those utilized in the United States have been established for nuclear plants elsewhere in the Western world. Measurement of performance against these indexes also shows significant improvement of reactor performance worldwide.

The World Association of Nuclear Operators (WANO) has been formed, consisting of nuclear plant operators over the entire world who have pledged to assist each other in the achievement of safe operations (25). There are four centers from which this international program is administered: one in the United States in Atlanta, Georgia, operated by INPO; one in Paris operated by Electricité de France; one in Moscow operated by the Ministry of Nuclear Power; and one in Tokyo operated by the Central Research Institute for the Electric Power Industry (CRIEPI). Through this mechanism, teams of operators from the U.S., Western Europe, and Asia visit CIS plants to share safety experience and know-how, and similarly, plant personnel from Russian and Eastern European nuclear units visit European, Asian, and U.S. plants.

Comparative Risks. All efforts are directed to reducing to an extremely low level the chance of a severe nuclear accident that would harm the public. The question of how low a level this should be remains. A safety goal stipulated by the U.S. NRC states that (*1*) the risk of prompt fatality to an average individual in the vicinity of a nuclear power plant that might result from reactor accidents should not exceed 0.1% of the sum of prompt fatality risks resulting from other accidents to which members of the U.S. population are generally exposed, and

(2) the risk of cancer fatalities to the population in the area near a nuclear power plant in operation should not exceed 0.1% of the sum of cancer fatality risks resulting from all other causes. On that basis, the U.S. NRC established that the chance of an accident that would initiate melting of the core should be less that 1 in 10,000 per reactor year of operation, and that the containment systems should assure another factor of 10 reduction in the probability of harming the public.

All U.S. plants are assessed for the ability to meet this goal by carrying out what are called probabilistic safety assessments (PSAs). These are sophisticated failure analyses which estimate the probability that a sequence of failures could result in incipient core melting, called the core degradation frequency (CDF). The detailed evaluation by this process of every nuclear power plant in the United States has led the U.S. NRC to judge that existing operating plants are adequately safe. These assessment methods require a significant amount of detailed knowledge as to the causes of failure and the characteristics a severe accident would have, were one to occur. Extensive experiments and analyses of severe accident scenarios have been carried out. For example, an experimental reactor (LOFT) at the Idaho National Engineering Laboratory was driven to core melting to measure the rates of release and release characteristics of the fission products. Not inconsequential among these experiments has been the evaluation of the TMI accident. Although approximately 50% of the TMI core melted and some of the molten fuel slumped into the bottom of the reactor vessel causing damage, the molten fuel was held within the vessel (26).

Analyses and experimental results used to assess the consequences of a severe potential accident have resulted in substantially reduced estimates of severe accident consequences. Comparing estimates made by the U.S. Atomic Energy Agency (27) in 1977 with those reported by the U.S. NRC (18,28) in 1990 shows that improved knowledge and plant modifications have reduced the core damage frequency by a factor of 3–15, depending on reactor type. Additionally, the fractions of radioactive species that would be released are lower by a factor of 10–100,000, depending on the radioactive species.

The NRC safety goal can be evaluated by comparison to the risks from accidents incurred from other human activities (Fig. 2) (29). The safety goal and the safety record of the nuclear power industry indicate much lower societal risks from commercial nuclear power than from a wide range of other common human activities.

If possible comparisons are focused on energy systems, nuclear power safety is also estimated to be superior to all electricity generation methods except for natural gas (30). Figure 3 is a plot of that comparison in terms of estimated total deaths to workers and the public and includes deaths associated with secondary processes in the entire fuel cycle. The poorer safety record of the alternatives to nuclear power can be attributed to fatalities in transportation, where comparatively enormous amounts of fossil fuel transport are involved. Continuous or daily refueling of fossil fuel plants is required as compared to refueling a nuclear plant from a few truckloads only once over a period of one to two years. This disadvantage applies to solar and wind as well because of the necessary assumption that their backup power in periods of no or little wind

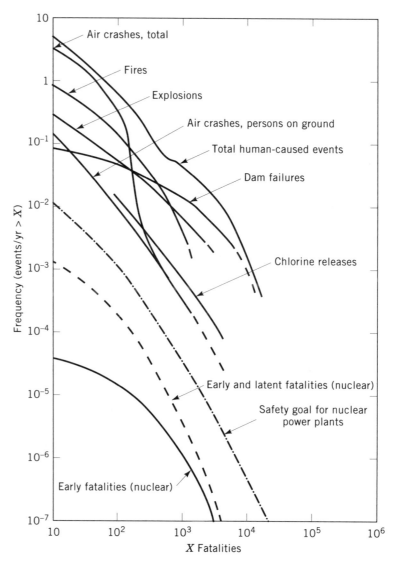

Fig. 2. Frequency of fatalities owing to human-caused events (—) and those caused by nuclear reactor accidents (– – –) together with proposed nuclear power plant safety goals (– · – ·).

or sun is from fossil-fuel generation. Now death or serious injury has resulted from radiation exposure from commercial nuclear power plants in the United States (31).

Safety Characteristics of the Nuclear Power Plant

The Reactor. The nuclear power plant reactor types used to produce electricity worldwide are listed in Table 2 (see NUCLEAR REACTORS, REACTOR TYPES). The lightwater reactor (LWR) is representative of commercial nuclear power

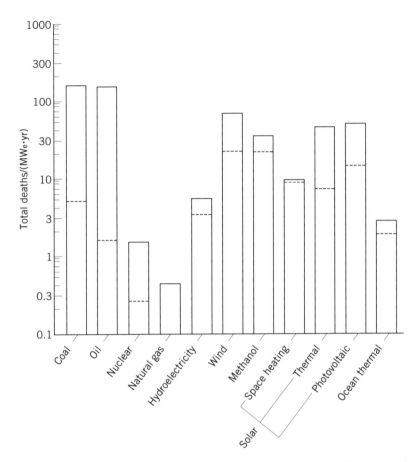

Fig. 3.   Total deaths per 100 MWe·yr as a function of energy system. The space above the dashed line in each bar represents the range of uncertainty in each estimate. Courtesy of the Atomic Energy Control Board, Canada.

Table 2. Nuclear Power Units by Reactor Type Worldwide[a]

Nuclear reactor	Number of units	Net power, MWe $\times 10^3$
lightwater reactor		
pressurized water reactor (PWR)	245	215.7
boiling water (BWR)	92	75.9
gas-cooled reactor	35	11.7
heavy-water reactor	34	18.5
graphite-moderated lightwater reactor	15	14.8
liquid metal-cooled fast breeder reactor	3	0.9

[a]Ref. 1.

plants. About 80% of the plants worldwide, and all of those in the United States, are LWR (1). The LWR uses ordinary or light water as distinguishable from heavy or deuterated water to transfer the heat generated from fission in the nuclear fuel assemblies, called the core, to make steam. The steam turns the turbine generators, which produce the electricity. Two versions of the LWR are utilized: the two-circuit pressurized water reactor (PWR) in which the steam is made by circulating the heated water from the core into heat exchangers (or steam generators), which causes water in a second circuit to be brought to a boil to provide the steam, and the one-circuit boiling water reactor (BWR) in which the steam is generated in the core itself and goes directly to the turbine (8,14,32).

The Chernobyl reactor type, designated RBMK, and built only in the former Soviet Union, is also cooled using ordinary water. The water is circulated through fuel tubes inserted in a large graphite block. This reduces the energy of the neutrons to increase the interactions to cause fission. The remaining systems utilized worldwide are gas-cooled, heavy water-cooled, and liquid metal-cooled. All of the nuclear reactors designed to meet Western world standards have safety features conceptually similar to those of the LWRs.

Pressurized Water Reactor. Figure 4 is a schematic of the PWR power plant. The reactor core made of the nuclear fuel assemblies is installed within the reactor pressure vessel, into which coolant water is pumped. The water passes through the core and is heated. The heated water flows through the tubes of a steam generator and transfers its heat to water on the other side of the tubes. The water turns to steam and is fed to a steam turbine causing the turbine to rotate. The turbine shaft is connected to the shaft of an electrical generator, causing the generator shaft to rotate, and generating electricity. A separate tank, called the pressurizer, provides the ability to accommodate volume changes in the primary circuit as well as to maintain a constant pressure. Associated with the steam

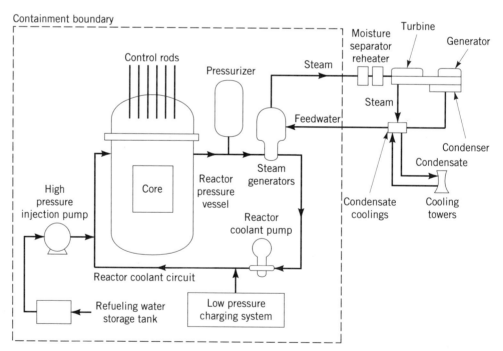

Fig. 4. Pressurized water reactor (PWR) power plant schematic. Courtesy of Westinghouse Electric Corp.

turbine are conventional systems which enhance the efficiency of heat extraction into work and recirculate the condensed water and unused steam back to the steam generator as feedwater. The reactor pressure vessel containing the core along with the reactor coolant circuit, reactor coolant pumps, pressurizer, steam generators, and emergency cooling systems are installed within the containment building. When the power plant is shut down for refueling or repair, a residual heat-removal system continues to cool the core, which would otherwise heat up from fission-product radioactive decay.

To assure continued cooling of the fuel in the event of loss of normal cooling, the emergency core cooling system is brought into play. Water can be immediately supplied from the accumulator tanks which are constantly under gas (nitrogen) pressure. A high pressure injection system can pump water from the refueling water storage tank into the reactor coolant circuit. Water can also be pumped from the refueling water storage tank to spray nozzles at the top of the containment building to deep the temperature of the containment building under control. These systems are activated automatically and are installed in multiple form to provide redundancy. Emergency electric power is provided from independent diesel generators when power from the reactor plant or off-site power is unavailable.

Control of the core is affected by movable control rods which contain neutron absorbers; soluble neutron absorbers in the coolant, called chemical shim; fixed burnable neutron absorbers; and the intrinsic feature of negative reactivity coefficients. Gross changes in fission reaction rates, as well as start-up and shutdown of the fission reactions, are effected by the control rods. In a typical PWR, ca 90 control rods are used. These, inserted from the top of the core, contain strong neutron absorbers such as boron, cadmium, or hafnium, and are made up of a cadmium–indium–silver alloy, clad in stainless steel. The movement of the control rods is governed remotely by an operator in the control room. Safety circuitry automatically inserts the rods in the event of an abnormal power or reactivity transient.

Chemical shim control is effected by adjusting the concentration of boric acid dissolved in the coolant water to compensate for slowly changing reactivity caused by slow temperature changes and fuel depletion. Fixed burnable poison rods are placed in the core to compensate for fuel depletion. These are made of boron carbide in a matrix of aluminum oxide clad with Zircaloy. As the uranium is depleted, ie, burned up, the boron is also burned up to maintain the chain reaction. This is another intrinsic control feature. The chemical shim and burnable poison controls reduce the number of control rods needed and provide more uniform power distributions.

Boiling Water Reactor. The BWR differs from the PWR primarily in that the steam of the former is generated in the reactor core and sent directly to the steam turbine. This permits elimination of the steam generators and secondary coolant circuit, as shown in Figure 5. There are other differences associated with the core and its controls, as well as in the containment system. There are over 700 fuel assemblies in a BWR core, each containing 62 fuel rods. The core diameter is 4.7 m, the length 3.8 m. The movable neutron-absorbing elements are in the form of blades which insert between fuel assemblies, rather than in rods which insert within the fuel assemblies in the PWR. Containment heat removal is

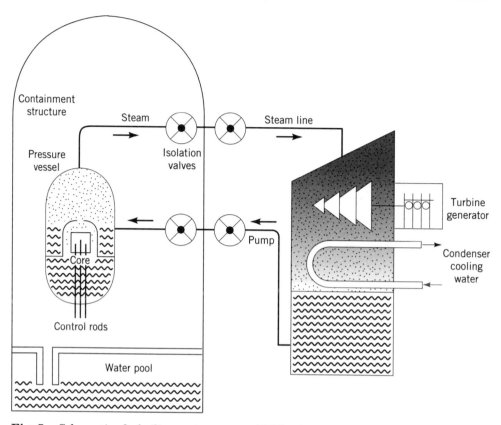

Fig. 5. Schematic of a boiling water reactor (BWR) plant. Courtesy of Atomic Industrial Forum, Inc.

aided in the BWR by venting any steam issuing from a pipe break, if such were to occur, to a pool of water located at the bottom of the containment building. The resultant steam condensation would then reduce the pressure and temperature within containment.

The Nuclear Fuel Cycle.   Fuel for a nuclear power plant is provided and dispositioned through the nuclear fuel cycle (33), shown in Figure 6. The disposition process has two options, recycling, also called chemical reprocessing, or throwaway. In the former the spent fuel is reprocessed in a chemical plant, where residual uranium and plutonium are separated from the fission products (see NUCLEAR REACTORS, CHEMICAL REPROCESSING). The uranium and plutonium are then recycled to fabricate mixed uranium–plutonium oxide pellets to be used in subsequent reactor refuelings. The fission product waste is then vitrified, ie, converted to a gaseous form, encapsulated in metal casks, and sent to a permanent repository (see NUCLEAR REACTORS, WASTE MANAGEMENT).

The recycling option is being utilized outside the United States. Whereas the technology of this option has been completely demonstrated in the United States, the economics have not been favorable. Moreover, concerns have been raised as to the diversion of the plutonium to weapons use. Thus, the throwaway option is the only one in use in the United States as of this writing.

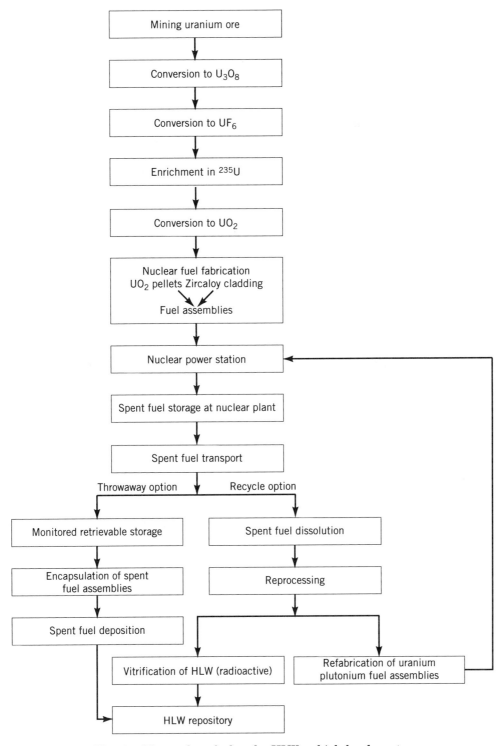

Fig. 6. The nuclear fuel cycle. HLW = high level waste.

The safety principles and criteria used in the design and construction of the facilities which implement the nuclear fuel cycle are analogous to those which govern the nuclear power plant. The principles of multiple barriers and defense-in-depth are applied with rigorous self-checking and regulatory overview (17,34). However, the operational and regulatory experience is more limited.

One feature of reprocessing plants which poses potential risks of a different nature from those in a power plant is the need to handle highly radioactive and fissionable material in liquid form. This is necessary to carry out the chemical separations process. The liquid materials and the equipment with which it comes in contact need to be surrounded by 1.5–1.8-m thick high density concrete shielding and enclosures to protect the workers both from direct radiation exposure and from inhalation of airborne radioisotopes. Rigid controls must also be provided to assure that an inadvertent criticality does not occur.

Shielding protection entails design engineering and installation similar to that provided in a nuclear power plant. Additionally, to protect against exposure to airborne radioactivity, controlled air ventilation and air cleaning is provided. The air flows progressively from clean areas to contaminated ones and is then filtered before being discharged.

The principal methods for preventing criticality are limitations on the mass of the fuel being handled, the equipment size, the concentration of nuclear material in solution, minimization of the presence of water or plastic that would reduce the margin to reaching a critical mass, and the addition of a neutron absorber, eg, cadmium or boron, either in solution or as a solid packing in vessels. Borosilicate glass rings are used as a neutron-absorber packing for tanks that contain many times the critical mass of fuel solutions. At least two and sometimes more of these independent methods usually are employed at fuel-processing facilities to prevent criticality. In addition, control of other parameters individually or in combination permit the safe handling of quantities many times the critical mass (35).

In fuel fabrication facilities handling plutonium, Pu, protection must be afforded against both inadvertent criticality and inhalation of airborne particles. Plutonium, if inhaled or ingested, is very harmful. Thus, air-flow controls are employed. Fabrication operations are carried out in glove boxes, ie, tightly sealed enclosures maintained at lower pressure than the surroundings, so that any leakage is into the glove box. Ventilation air for the boxes is cleaned through high efficiency particulate air filters. Inadvertent criticality is an even more sensitive safety issue in plutonium fuel fabrication because a criticality accident would emit lethal levels of radiation near the unshielded glove boxes. Strict control of Pu quantities, therefore, is enforced, limiting the amount of Pu handled in a single operation to less than that needed to start a chain reaction.

The sum total of risks of the nuclear fuel cycle, most of which are associated with conventional industrial safety, are greater than those associated with nuclear power plant operation (30,35–39). However, only 1% of the radiological risk is associated with the nuclear fuel cycle so that nuclear power plant operations are the dominant risk (40). Public perception, however, is that the disposition of nuclear waste poses the dominant risk.

Spent Fuel and Radioactive Waste.   The basic safety objective governing radioactive waste management is to protect the public and the workers from radiation, at a minimum meeting federal regulatory standards for maximum allowable radiation dosage. This protection is provided for both spent fuel and plant radioactive wastes during the transfer and treatment processes at the plants, temporary storage of wastes at the plants, and during transportation from the plants to storage sites and repositories (41). Plant radioactive wastes arise from the coolant systems of the nuclear power plants; from auxiliary process wastes; from the enrichment, reprocessing, and fuel fabrication facilities; and from the decommissioning of these facilities. These wastes, called low level radwastes (LLW), are of lower radiation intensity than those arising from the nuclear fuel, called high level radwastes (HLW) (see NUCLEAR REACTORS, WASTE MANAGEMENT).

The safety objective for the final storage facilities for the low level wastes and permanent repository for the spent fuel or high level wastes is to isolate the radioactive materials from the biosphere, ie, to package the waste in rugged containers and bury these containers in stable geologic formations far from ground water so that they do not come in contact with humans directly or indirectly. The specific means of meeting this objective can vary in regard to whether the spent fuel throwaway or recycling option is being utilized.

For the throwaway option in the United States, planning as of the mid-1990s is to store the spent fuel at the reactor site initially in water pools where the relatively high decay heat is removed by natural circulation of the water in the pool. After about 10 years, the spent fuel is moved to a dry storage facility where natural circulation of air provides sufficient cooling, all the while awaiting transfer to either an interim centralized storage facility or a permanent repository (42). Once removed from the water pool, the spent fuel assemblies are placed in a stainless steel or titanium container, called a multipurpose canister, which provides an inner shell from which the fuel assembly need never be removed again. This shell is inserted into various other overpacks of concrete or steel, depending on whether the fuel is being stored on site, is being transported, or is being placed in the permanent repository.

In keeping with the overall safety principles, the spent fuel repository is designed using concentric barriers (43). The first barriers are the same as those for the nuclear power plant, ie, the solid, corrosion-resistant ceramic fuel pellets and the Zircaloy cladding which surrounds the pellets. The next barrier is the canister or inner shell, which becomes a permanent element and within which the fuel assemblies are placed. The third barrier is the overpack of concrete or steel. This set of barriers makes up the engineered package. The last barrier is the geologic surroundings within which the engineered package is buried. Even after 1000 years or more at a good site, when the integrity of the engineered package may become reduced, the dry, impermeable ground formation should contain the radioactive material with high probability for indefinite periods of time.

The fundamental safety criterion (44) for the permanent repository is that the engineered package retain complete integrity for at least 1000 yr. Rigorous engineering criteria and testing is needed to provide that assurance. The level of radioactivity is greatly reduced after 1000 yr and the need for complete integrity thus alleviated. Extensive scientific studies of the impact of geologic change as well as probabilistic safety assessments have been performed to develop estimates of the probabilities that the radioactive material would enter the biosphere. The half-lives of the radioactive species (the time it takes for radioactive species to diminish by a factor of 2 through radioactive decay) are important characteristics of the evaluations. Half-lives determine the remaining content of the radioactive material in the repository over time. However, half-life in itself is not the dominant characteristic of concern. Otherwise, stable nonradioactive toxic wastes which have infinite half-lives would be risk-dominant. Rather, it is the toxic and chemical characteristics in combination with the radioactivity which determine the radioisotopes (qv) of dominant risk.

In all of the transportation and storage steps, sensitive radiation monitors are located at and around the spent fuel to detect incipient leakage. Whenever such leakage is detected, steps are taken to repair the defect. Even for the permanent repository, radiation monitoring should be kept up indefinitely and provision made for retrieving the spent fuel for a period of at least 50–100 years to effect repairs. Based on the experience gleaned in that initial period, a decision would then be made as to whether the repository were fully suitable as permanent. The first 50–100 years could be considered as interim storage, either as the first phase of a permanent facility, or alternatively as a separate interim storage facility. This latter, termed a monitored retrievable storage (MRS) facility, has been constructed in Sweden. Storage of up to 5000 metric tons of spent fuel has been initiated within the facility, called CLAB. It is in an underground manmade rock cavern about 40 m deep. Sweden treats CLAB as a separate interim storage facility. A permanent repository is under development (45).

The primary issue is to prevent groundwater from becoming radioactively contaminated. Thus, the property of concern of the long-lived radioactive species is their solubility in water. The long-lived actinides such as plutonium are metallic and insoluble even if water were to penetrate into the repository. Certain fission-product isotopes such as iodine-129 and technicium-99 are soluble, however, and therefore represent the principal although very low level hazard. Studies of Yucca Mountain, Nevada, tentatively chosen as the site for the spent fuel and high level waste repository, are underway (44).

The high level waste of the recycling (qv) option is made up primarily of fission products having only residual amounts of plutonium and other actinides following the reprocessing (46). The fission-product wastes come from the chemical reprocessing plant in liquid form and have to be converted to a solid. Vitrification of the waste is planned, so that the first barrier in radioactive containment design is a highly corrosion-resistant glass. The vitrified form is in pellets or logs stored in stainless steel or titanium canisters, which in turn are installed in an overpack to make up the engineered package. This package could then be buried in much the same manner as planned for the spent fuel. Repository safety advantages exist in this option in that the bulk of the long-lived plutonium has been removed.

In addition, processes are under development to separate the other long-lived actinides from the fission products and recycle these materials into the reactor. This recycling is most effective in a liquid metal-cooled reactor because in its high energy neutron spectrum, neutrons are not absorbed appreciably by the actinides, and thus the efficiency of the chain reaction is maintained. By contrast, efficiency would be poor in a lightwater-cooled reactor, which has a low energy neutron spectrum, and actinides become strong neutron absorbers.

Other improvements could be made to the waste by converting the soluble fission products into insoluble forms. A promising pyrometallurgical reprocessing method is under development for actinide separations (47).

If the economics of recycling were improved, that option would become preferable for spent fuel because the permanent repository issues of the residual fission products would be simpler. The economic value of the energy generated from the recycled plutonium and uranium would substantially allay the costs of the repository as compared to the spent fuel throwaway option.

Another safety issue to be considered which might be exacerbated in the reprocessing option is that the plutonium generated in power reactors, called reactor-grade plutonium because it is made up of a variety of plutonium isotopes, contains plutonium-241, which is subject to spontaneous fission (8). The mixture of isotopes makes it extremely difficult to build an effective nuclear weapon. However, an explosive device could be built using this mixture if control of detonation is sacrificed (48).

When reactor-grade plutonium is left in spent fuel, the large size of the fuel assemblies and the lethal radiation fields make it extremely difficult to divert the material covertly. Once the reactor-grade plutonium is separated in the commercial reprocessing option, however, the radiation barrier is almost eliminated, and in certain steps of the process the plutonium is in powder or liquid form, which is much more easily diverted than large, bulky fuel assemblies. This issue is under study (49,50) and strict standards of control of separated reactor-grade plutonium have been instituted (51). In the United States, the Nuclear Nonproliferation Act of 1978 was passed to strengthen control over export trade of plutonium-bearing components by U.S. industry. In addition, under the Nonproliferation Treaty, which most larger nations have signed, the IAEA monitors plutonium from power reactors so as to detect covert diversion.

Concern about the potential diversion of separated reactor-grade plutonium has led to a reduction in U.S. governmental support of development of both plutonium recycle and the liquid metal reactor. This latter ultimately depends on chemical reprocessing to achieve its long-range purpose of generating more nuclear fuel than it burns in generating electricity.

Radiation Exposure and Health Standards

In the United States, each person receives an average of about 0.0036 Sv/yr of radiation exposure. About 0.003 Sv comes from natural background sources such as radon gas, cosmic rays, and radioactive elements present in the air, soil, and rocks. Another 0.0006 Sv come from other sources, primarily medical treatments and consumer products. Nuclear utility workers may be exposed to an additional occupational radiation exposure. In any given year, about 50% of nuclear industry workers receive no measurable radiation. The remaining 50% of the workers are exposed to an additional average 0.003 Sv/yr, for a total of 0.0066 Sv annually (52). To assure that this additional radiation exposure is not harmful, several measures are taken: (1) standards are set for the maximum allowable radiation dosage by national and international commissions of radiation health experts (53–55) and incorporated into federal regulations (56); (2) dosimeters, ie, radiation monitors are worn by all workers potentially exposed to occupational radiation and accurate records are kept of the accumulated exposure to each worker to assure that maximum allowable levels are not exceeded; and (3) the concept of keeping occupational radiation exposure to as low a level as reasonably achievable (ALARA) is practiced so that a relatively small number of workers come close to the maximum allowable levels.

High levels of radiation exposure received over a short period of time (minutes to hours) can cause both near- and long-term effects. Near-term effects include radiation sickness or death. Long-term effects predominantly involve the incidence of cancer (57). Studies of radiation exposure to researchers in the early twentieth century; Japanese atomic bomb survivors, patients undergoing medical radiation exposures, and radiation accidents have led to standards of maximum allowable radiation exposure (55,58). Short-term exposures of several sieverts are required to cause severe radiation sickness or death; exposures of tenths of sieverts may induce cancer in humans. For lower exposures, the risk of cancer or genetic effects is difficult to assess. The number of these effects seen in exposed individuals is about the same as the number occurring in people who receive only normal background radiation exposure.

Most of the data on radiation health effects have come from medical monitoring of Japanese atomic bomb survivors. For survivors who received radiation exposures up to 0.10 Sv, the incidence of cancer is no greater than in the general population of Japanese citizens. For the approximately 1000 survivors who received the highest radiation doses, ie, >2 Sv, there have been 162 cases of cancer. About 70 cases would have been expected in that population from natural causes. Of the approximately 76,000 survivors, as of 1995 there have been a total of about 6,000 cases of cancer, only about 340 more cases than would be expected in a group of 76,000 Japanese citizens who received only background radiation exposure (59).

Thousands of studies of radiation and its risks have been conducted. Yet there is no conclusive evidence that low levels of radiation exposure cause either cancer or birth defects. The nuclear industry operates on the conservative ALARA approval, assuming that any exposure involves some risk.

For radiation doses <0.5 Sv, there is no clinically observable increase in the number of cancers above those that occur naturally (57). There are two risk hypotheses: the linear and the nonlinear. The former implies that as the radiation dose decreases, the risk of cancer goes down at roughly the same rate. The latter suggests that risk of cancer actually falls much faster as radiation exposure declines. Because risk of cancer and other health effects is quite low at low radiation doses, the incidence of cancer cannot clearly be ascribed to occupational radiation exposure. Thus, the regulations have adopted the more conservative or restrictive approach, ie, the linear hypothesis. Whereas nuclear industry workers are allowed to receive up to 0.05 Sv/yr, the ALARA practices result in much lower actual radiation exposure.

Reduction in occupational radiation exposure is portrayed in Figure 7 (60). In the decade between 1983 and 1993, the annual total radiation dosage received by U.S. nuclear plant workers dropped by 54% whereas the annual MW·yr of electricity generated increased by 51%. Thus, the annual ratio of total occupational radiation exposure to total electricity generated dropped by almost a factor of 5. This achievement can be credited in part to improved management

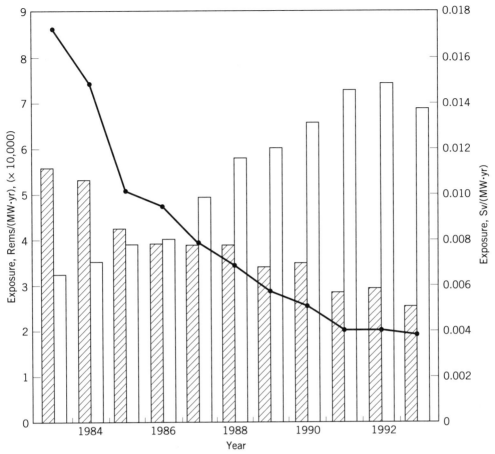

Fig. 7.   U.S. nuclear power plant occupational radiation exposure, where (▨) corresponds to total radiation exposure, (□) to the electricity generated, and (─●─) to the radiation exposure per unit of electricity 5(Sv/(MW·yr)) (60). Courtesy of the Electric Power Research Institute.

practices, but a series of technological innovations have also made a significant contribution (60).

The dominant sources of residual radiation in the primary circuit outside the reactor core in nuclear plants are cobalt isotopes: ^{60}Co and ^{58}Co form by neutron absorption in ^{59}Co and ^{58}Ni. These last two species are naturally occurring isotopes in commonly used plant construction materials. The processes of transport, activation, and deposition of cobalt-containing corrosion products in the PWR primary system is shown in Figure 8. Similar processes apply to the BWR primary circuit. Technological approaches to reduce this residual radioactive cobalt are as follows. (*1*) Minimize the cobalt impurities in the structural materials, replacing the high cobalt hardfacing alloys where practicable. Development of a cobalt-free hardfacing alloy and preparation of cost-effective materials procurement specifications that minimize cobalt content both contribute to significant reductions in ^{60}Co in the primary circuit (61). (*2*) Precondition out-of-

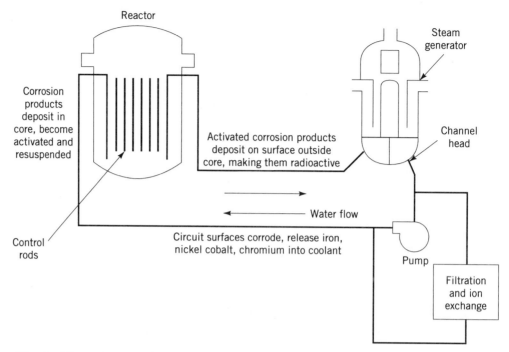

Fig. 8. The activation of cobalt-containing corrosion products in a PWR primary circuit. See text. Courtesy of the Electric Power Research Institute.

core primary circuit surfaces to minimize the release of corrosion products and the resuspension of radioactive species. Protective surface films can be provided by electropolishing and preoxidizing as well as by electroplating (qv) a thin film of chromium. (*3*) Specify and control primary water chemistry to minimize corrosion and the transport of corrosion products into the core, the disposition and subsequent activation of these products, and resuspension in the coolant. Coolant chemistry guidelines have been developed which specify the allowable levels of impurities, the addition of lithium in the PWR coolant to maintain the proper pH in the presence of boric acid, and the injection of hydrogen and addition of zinc in the BWR systems. (*4*) Remove the residual radiation in the out-of-core primary circuit by decontamination. Several decontamination processes, such as CITROX, CANDECON, and LOMI, have been developed. The last, LOMI has been the most widely used.

Safety of Future Reactors

Substantial research and development is ongoing to define the characteristics of improved lightwater-cooled nuclear power plants (62–65). The safety area is no exception.

The development of computer capabilities in hardware and software, related instrumentation and control, and telecommunication technology represent an opportunity for improvement in safety (see COMPUTER TECHNOLOGY). Plant operators can be provided with a variety of user-friendly diagnostic aids to assist in plant operations and incipient failure detection. Communications can be more rapid and dependable. The safety control systems can be made even more reliable and maintenance-free. Moreover, passive safety features to provide emergency cooling for both the reactor system and the containment building are being developed.

The Electric Power Research Institute (EPRI) initiated the advanced light-water reactor (ALWR) program because lightwater reactors are expected to continue to be used (66,67). The ALWR program, supported by electric utilities in the United States, Europe and Asia, the U.S. suppliers, and the U.S. Department of Energy, is developing large-sized (1350-MWe) reactors and smaller sizes (600-MWe) of the passive safety feature type (68). Both PWR and BWR units are being developed.

Three designs are under development: the PWR System 80+ designed by ABB-CE, the advanced BWR (ABWR) designed by GE, and the Advanced PWR (APWR) designed by Westinghouse. Other evolutionary plants are under development by international firms, although not under the sponsorship of the EPRI ALWR Program. Electricité de France is sponsoring a 1400-MWe reactor plant called the N-4; Nuclear Electric in the United Kingdom is sponsoring a 1350-MWe plant developed jointly with Westinghouse; and Siemans of Germany and Framatome of France are jointly developing a 1350-MWe advanced PWR called the EPR. The evolutionary reactors are based on the same design concept as is used in the lightwater reactors of the mid-1990s. Many significant improvements have been made, such as selection of alloys having more corrosion resistance, eg, Inconel 690, for steam generator tubes; a high pressure system for the removal of decay heat; and the reactor vessel materials and weldments chosen to reduce radiation embrittlement and shielded to reduce the fast neutron fluency.

Two smaller-sized passive reactors are under development in the United States: a PWR called AP600 (69), designed by Westinghouse, and a BWR called SBWR (70), designed by GE. These designs combine the improvements of the larger reactors with passive emergency cooling features. A schematic of the nuclear steam supply system and the containment system AP600 (69) is shown in Figure 9 (63). The power train of the AP600 uses proven technology: a UO_2-fueled core and field-proven plant components. The burden on the equipment and systems has been reduced by increasing design margins through reductions in coolant temperature, flow rate, and core power density and by selecting higher quality materials and more robust components. Passive cooling in the AP600 is provided by a passive emergency core cooling system (ECCS) and a passive containment cooling system. The passive ECCS consists of a combination of cooling water sources: gravity drain of water (from two core make-up tanks and a large refueling water storage tank suspended above the level of the core) and water ejected from two accumulator tanks under nitrogen pressure. If a feedwater supply interruption renders the steam generators inoperable, core decay heat is removed through a passive residual heat exchanger located in the refueling water storage tank. This transfers core decay heat to the refueling water by natural circulation.

Containment integrity is ensured by cooling the containment shell through evaporation of water that is gravity-fed from a large tank located above the containment. The heat is ultimately removed to the atmosphere by a natural

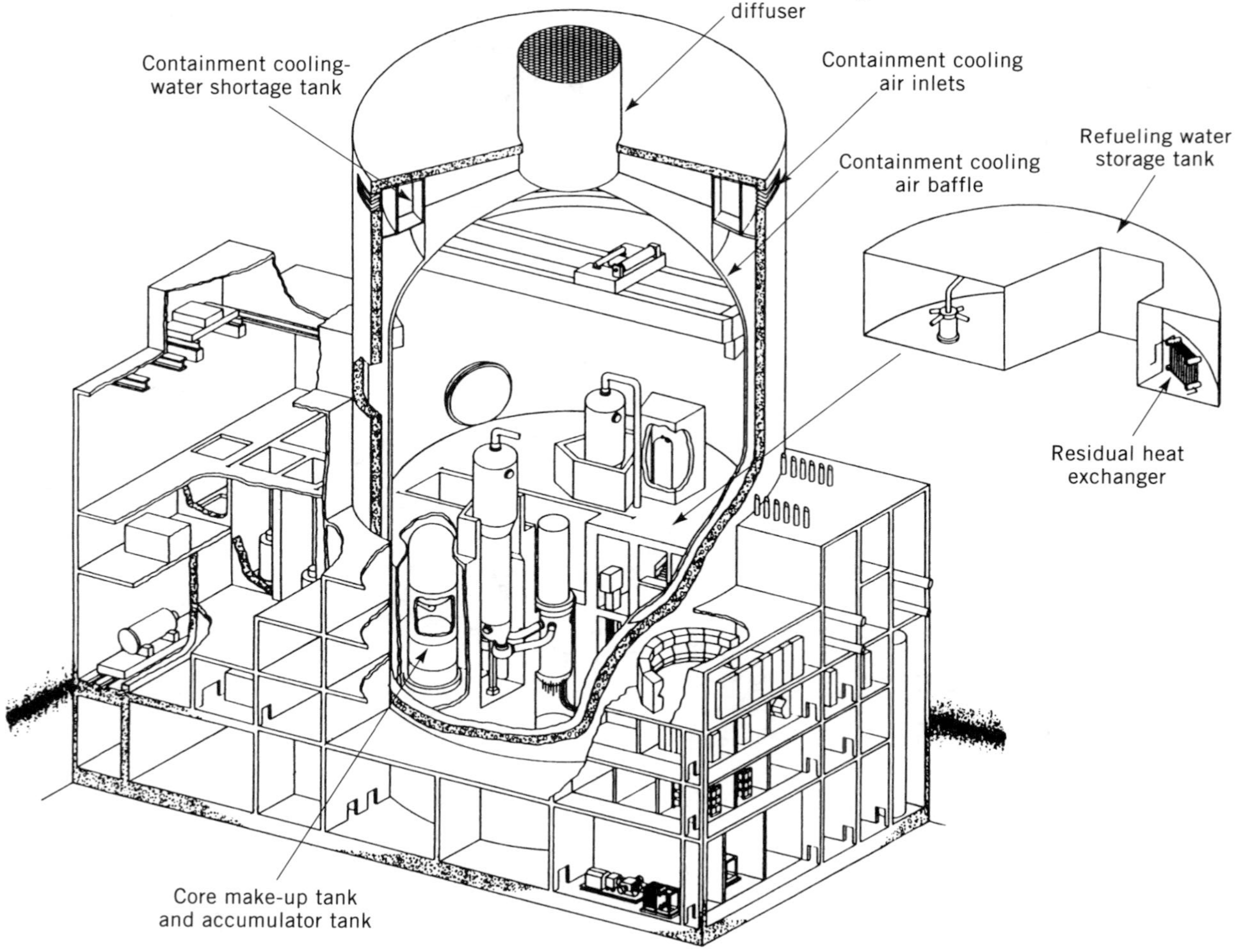

Fig. 9. Sketch showing AP600 passive plant configuration (65). Courtesy of the American Association for the Advancement of Science (63).

circulation air system. Only active automatic valve operations, ie, no operator action and no pump, diesel, or fan operations, are required to provide emergency core cooling and containment cooling after a significant energy release into containment from the maximum loss-of-coolant accident.

Safety objectives have been established to make both ALWRs even safer than the plants of the early 1990s and safer than required by the safety goals established by the U.S. NRC. The ALWR safety objectives are that there would be only one chance in 10×10^4 per reactor-year that a severe accident would be initiated, a factor of 10 better than the U.S. NRC safety goal. Mitigation of the accident through the containment systems would reduce the risk by another factor of 10, so that the chance that the radiation dose at the boundary of the plant would be as high as 0.25 Sv, the level below which there is no clinically observable effect, would be one in 1×10^6. An additional objective has been set to limit the level of occupational radiation exposure. No more than 1.00 Sv/yr occupational exposure should be received by all the workers in each plant, an average of about 0.001 Sv/yr. Improved performance objectives have also been set to provide an additional power margin. This places less burden on both the equipment and operators in running the plant, resulting in increased reliability and lower operating and maintenance costs.

Another overall objective of the ALWR Program is to achieve standardization of families of plants in design, construction, and operation. Two fundamental bases for that standardization are common owner–operator (utility) requirements and common regulatory requirements. The common utility requirements are contained in substantial detail in utility requirements documents (URD) which have been developed by EPRI, working jointly with experienced utility personnel and the reactor designers. These have been approved by the U.S. NRC. The URD applies the operational experience of existing plants to define optimum plant characteristics from the operators' standpoint, which in turn will govern the entire design, operating procedures, and configuration control during plant life. The common regulatory requirements are specified through a standardization process defined by the U.S. NRC (72) which provides for certification of a design from a safety standpoint. This can be used in replicate on many sites and also provides for early site approval so that an approved site can be matched with a certified design and a combined construction and operating license obtained. Successful implementation of standardization is a significant contributor to safety because the regulator, the operator, and the supporting industry are then focusing resources and sharing experience on a small number of plant designs and operational processes.

The design and testing of these developmental plants are well underway. The evolutionary designs have received final design approval from the U.S. NRC. As of 1995 evolutionary ALWR designs very similar to these, by the same U.S. manufacturers, are already being built in Asia: the GE-Advanced BWR in Japan and the ABB-CE System 80+ PWR in South Korea. The N-4 PWR is under construction in France, and the Sizewell PWR has completed construction and has initiated operation in the U.K.

Two other more advanced systems offer some promise for nuclear energy. One such advanced system is the gas-cooled reactor, which can operate at significantly higher temperatures than existing designs and therefore can be used to provide the energy for many process heat applications in industry. This reactor is a principal advance over the gas-cooled reactors of the twentieth century listed in Table 2. The second system is cooled with liquid metal, ie, sodium, and can produce more nuclear fuel than it burns as it generates electricity. Reactors such as the liquid-metal type should be needed to generate abundant nuclear fuel for the production of electricity (52).

Nuclear power plants of the future are to be designed and operated with the objective of better fulfilling the role as a bulk power producer that, because of reduced vulnerability to severe accidents, should be more broadly accepted and implemented. Use of these plants could help stem the tide of environmental damage caused by air pollution from fossil-fuel combustion products (64).

BIBLIOGRAPHY

"Safety in Nuclear Facilities" under "Nuclear Reactors" in *ECT* 2nd ed., Vol. 14, pp. 108–115, by C. E. Guthrie, Oak Ridge National Laboratory; in *ECT* 3rd ed., Vol. 16, pp. 216–238, by W. B. Cottrell, Oak Ridge National Laboratory.

1. *Nucl. News*, (Mar. 1995), p. 42.
2. *World Nuclear Industry Handbook*, Nuclear Engineering International, Sutton, U.K., 1995.
3. *Nuclear Power Reactors in the World*, 1994 ed., International Atomic Energy Agency, Vienna, Austria, 1994.
4. *TMI-2 Lessons Learned Task Force Status Report and Short-Term Recommendations*, NRC Report NUREG-0578, U.S. Nuclear Regulatory Commission, Washington, D.C., July 1979.
5. *NRC Action Plan Developed as a Result of the TMI-2 Accident*, Report No. NUREG-0600, U.S. Nuclear Regulatory Commission, Washington, D.C., May 1980.
6. *Clarification of TMI Action Plan Requirements*, Report No. NUREG-0737, U.S. Nuclear Regulatory Commission, Division of Licensing, Washington, D.C., Nov. 1980.
7. *The Report of the President's Commission on the Accident at Three Mile Island: The Need for Change; The Legacy of TMI*, U.S. Government Printing Office, Washington, D.C., Oct. 1979.
8. R. A. Knief, *Nuclear Engineering—Theory and Technology of Commercial Nuclear Power*, Hemisphere Publishing Corp, Washington, D.C., 1992.
9. *The Chernobyl Accident*, International Atomic Energy Agency (IAEA) Safety Series No. 75, INSAG-1, IAEA, Vienna, Austria, 1987.
10. *The Chernobyl Accident Update*, IAEA Safety Series No. 75, INSAG-7, IAEA, Vienna, Austria, 1992.
11. T. Moore and D. Dietrich, *EPRI J.* **12**(4), 4 (June 1987).
12. *Basic Safety Principles for Nuclear Power Plants*, IAEA Safety Series #75, INSAG-3, IAEA, Vienna, Austria, 1988, pp. 6–8.
13. *Ibid.*, p. 11.
14. R. L. Murray, *Nuclear Energy*, 3rd ed., Pergamon Press, Elmsford, N.Y., 1988.
15. J. J. Taylor and E. E. Kintner, *The Evolution of Self-Stabilization in Nuclear Power Development: 50 Years with Nuclear Fission*, American Nuclear Society, La Grange, Ill., 1989.
16. *Code of Federal Regulations*, Title 10, Part 50, Domestic Licensing of Production and Utilization Facilities, Section 50.36, Technical Specifications, Washington, D.C., Jan. 1, 1980.

17. *Code of Federal Regulations*, Title 40, Protection of Environment, Part 190, Environmental Radiation Protection Standards for Nuclear Power Operations, Washington, D.C., 1976.
18. *Severe Accident Risks: An Assessment of Five Nuclear Power Plants*, NUREG-1150, Vol. 1, U.S. Nuclear Regulatory Commission, Washington, D.C., Dec. 1990.
19. *Analysis and Evaluation of Operational Data—1993 Annual Report: Reactors*, NUREG-1272, Vol. 8, No. 1, U.S. Nuclear Regulatory Commission, Washington, D.C., Nov. 1994.
20. *Performance Objectives and Criteria for Operating and Near-Term Operating License Plants*, INPO 90-015, Institute of Nuclear Power Operations, Atlanta, Ga., Aug. 1990.
21. J. V. Rees, *Hostages of Each Other: The Transformation of Nuclear Safety Since Three Mile Island*, University of Chicago Press, Chicago, Ill., 1994.
22. *Summary Report: Screening and Evaluation of License Event Reports for 1979*, Nuclear Safety Analysis Center Report, NSAC-2, Palo Alto, Calif., 1980.
23. W. S. Grant and co-workers, *Handbook for Nuclear Power Plant Self-Assessment Programs*, NSAC-170, EPRI, Palo Alto, Calif., July 1991.
24. *INPO 1994 Annual Report*, Institute of Nuclear Power Operations, Atlanta, Ga., 1995.
25. *WANO 1991–1993 Biennial Review*, World Association of Nuclear Operators Coordinating Centre, London, 1994.
26. *What If—A Study of Severe Core Damage Events*, NP 3001, EPRI, Palo Alto, Calif., 1989.
27. *Reactor Safety Study: An Assessment of Accident Risks in U.S. Commercial Nuclear Power Plants*, Report WASH-1400 (NUREG-75/014), U.S. Nuclear Regulatory Commission, Washington, D.C., Oct. 1975.
28. H. J. C. Kouts and co-workers, *Special Committee Review of the NRC's Severe Accident Risk Report*—NUREG 1420, U.S. Nuclear Regulatory Commission, Washington, D.C., 1990.
29. S. Levine, *Nucl. Safety* **21**(6), 718 (Nov.–Dec. 1980).
30. H. Inhaber, *Risks of Energy Production*, Report AECD-1119, Rev. 2, Atomic Energy Control Board, Ottawa, Canada, Nov. 1978; H. Inhaber, Science **203**, 718 (Feb. 24, 1979).
31. H. W. Bertini, *Descriptions of Selected Accidents That Have Occurred at Nuclear Reactor Facilities*, Report ORNL/NSIC-176, Oak Ridge National Laboratory, Oak Ridge, Tenn., Apr. 1980.
32. S. Glasstone and A. Sesonske, *Nuclear Power Engineering*, D. Van Nostrand Co., Inc., New York, 1963.
33. R. G. Wymer and B. L. Vondra, *Light-Water Reactor Nuclear Fuel Cycle*, CRC Press, Boca Raton, Fla., 1981.
34. *American National Standard for Nuclear Criticality in Operations with Fissionable Materials Outside Reactors*, ANS Standard N 16.1-1975, American Nuclear Society, LaGrange Park, Ill., 1975.
35. *Energy and the Environment*, Council on Environmental Quality, Executive Office of the President, Washington, D.C., Aug. 1973.
36. *Comparative Risk-Cost-Benefit Study of Alternative Sources of Electrical Energy*, Report WASH-1224, U.S. Atomic Energy Commission, Washington, D.C., Dec. 1974; *Nucl. Sa* **17**(2), 171 (1976).
37. L. B. Lave and L. C. Freeburg, *Nucl. Safety* **14**, 409 (Sept.–Oct. 1973).
38. R. W. Gotchy, *Health Effects Attributed to Coal and Nuclear Fuel Cycle Alternatives*, Report NUREG-0332, U.S. Nuclear Regulatory Commission, Washington, D.C., 1977.
39. AMA Council on Scientific Affairs, *J. Amer. Med. Ass.* **240**, 2193 (Nov. 10, 1978).
40. Science Applications, Inc., *Status Report on the EPRI Fuel Cycle Accident Risk Assessment*, Report EPRI-NP-1128, Electric Research Power Institute, Palo Alto, Calif., July 1979.

41. N. Tsoulfandidis and R. G. Cochran, Nucl. Technol. **93**(3), 263 (Mar. 1991).

42. E. R. Gilbert and co-workers, Nucl. Technol. **89**, 141161 (Feb. 1990).

43. R. G. Cochran and N. Tsoulfanidis, *The Nuclear Fuel Cycle—Analysis and Management*, American Nuclear Society, LaGrange Park, Ill., 1988.

44. *A Proposed Public Health and Safety Standard for Yucca Mountain*, A Report prepared for the National Academy of Sciences Committee on Technical Bases for Yucca Mountain Standards, EPRI TV-104012, Electric Power Research Institute, Palo Alto, Calif., Dec. 1994.

45. *Central Interim Storage Facility for Nuclear Fuel—CLAB*, Swedish Nuclear Fuel and Waste Management Company (SKB), Stockholm, Sweden, 1986.

46. F. J. Rahn and co-workers, *A Guide to Nuclear Power Technology: A Resource for Decision Making*, Wiley-Interscience, New York, 1984.

47. C. E. Till, *The Liquid-Metal Reactor: Overview of the Integrated Fast Reactor Rationale and Basis for Its Development*, Presented to National Academy Sciences Committee on Future Nuclear Power, Argonne National Laboratory, Chicago, Ill., Aug. 1989.

48. *Management and Disposition of Excess Weapons Plutonium*, Committee on International Security of Arms Control, National Academy of Sciences, National Academic Press, Washington, D.C., 1994.

49. *Nuclear Proliferation and Civilian Nuclear Power-Report of the Nonproliferation Alternative Systems Assessment Program*, Executive Summary, DOE Report DOE/NE-0001, U.S. Department of Energy, Washington, D.C., June 1980.

50. *International Nuclear Fuel Cycle Evaluation*, Summary Vol., Report No. STI/PUB/534, International Atomic Energy Agency, Vienna, Austria, Mar. 1980.

51. Physical Protection of Plants and Materials, *Code of Federal Regulations*, Title 10, Part 73, U.S. Government Printing Office, Washington, D.C., rev. Jan. 1, 1980.

52. *Radiation at Nuclear Power Plants—What Do We Know About Health Risks?* EPRI, Palo Alto, Calif., Dec. 1994.

53. *Maximum Permissible Body Burdens and Maximum Permissible Concentration of Radionuclides in Air and in Water for Occupational Exposure*, Report No. NCRP, No. 22, National Council on Radiation Protection and Measurement, Washington, D.C., 1959.

54. *Health Effects Of Exposure to Low Levels of Ionizing Radiation*, Report of Committee on the Biological Effects of Radiation (BEIR Report V), National Academy Press, Washington, D.C., 1990.

55. *Risk Associated With Ionizing Radiation*, Annals of the International Committee on Radiological Protection, Vol. 22, No. 1, Pergamon Press, Oxford, U.K., 1991.

56. *Code of Federal Regulations*, Title 10, Energy, Part 20, Standards for Protection Against Radiation, Washington, D.C., rev. May 21, 1991.

57. H. Behling and co-workers, "Health Risks Associated with Low Doses of Radiation," TR-104070, Electric Power Research Institute, Palo Alto, Calif., Aug. 1994.

58. "Recommendations of the International Commission on Radiological Protection," International Commission on Radiological Protection, Publication 26, Pergamon Press, Oxford, U.K., 1977.

59. Y. Shimizu and co-workers, "Life Span Study Report 11, Part 1, Comparison of Risk Coefficients for Site-Specific Cancer Mortality," Technical Report RERF-TR-12-87, Radiation Effect Research Foundation, Hiroshima, Japan, 1987.

60. C. J. Wood, Prog. Nucl. Energy **23**(1), 35–80 (1990).

61. *Cobalt Reduction Guidelines: Revision 1*, TR-103296, Electric Power Research Institute, Palo Alto, Calif., Dec. 1993.

62. M. W. Golay and N. E. Todreas, *Advanced Light-Water Reactors, Sci. Amer.* **262**(4), 8289 (Mar. 1990).

63. J. J. Taylor, *Science* **244**, 318325 (Apr. 1989).

64. J. J. Taylor, *Electr. J.* **4**(1) (Jan. 1991).

65. *Nuclear Power–Technical and Institutional Options for the Future*, Report of National Academy Committee on Future Nuclear Power Development, National Academy Press, Washington, D.C., 1992.

66. J. Santucci and J. J. Taylor, *Safety, Technical and Economic Objectives of the Electric Power Institute's Advanced Light-Water Reactor Programme*, IAEA-SM-332/II.1, Proceedings of International Symposium on Advanced Nuclear Power Systems, Seoul, Korea, Oct. 1993.

67. J. Douglas, *EPRI J.* **19**(8), (Dec. 1994).

68. *The Next Generation*, Nuclear Industry, U.S. Council for Energy Awareness, Washington, D.C., Jul–Aug. 1988.

69. R. Vijuk and H. Bruschi, *Nucl. Eng. Int.* **33**(412), 22 (Nov. 1988).

70. B. Wolfe and D. R. Wilkins, paper presented at ANS Topical Meetings on the Safety of the Next Generation of Power Reactors, Seattle, Wash., May 1988.

71. *Code of Federal Regulations*, Title 10, Part 52, "Early Site Permits, Standard Design Certifications, and Combined Licenses for Nuclear Power Plants," Washington, D.C.

J. J. TAYLOR
Electric Power Research Institute

NUCLEIC ACIDS

Nucleic acids are polymeric materials formed from nucleotides and essential to all organisms. Deoxyribonucleic acid (DNA), most often a double-helical biopolymer, encodes the genetic information contained in each cell (1,2). Ribonucleic acid (RNA) constitutes a more diverse class of biopolymers that are able to adopt both helical and other more complex tertiary structures. The structural diversity of RNAs enables these molecules to carry out a variety of intracellular functions, including transmitting the genetic message to the site of protein synthesis (see PROTEINS) (3). Both DNA and RNA interact with a host of other molecules, eg, proteins, drugs, as well as other RNAs and DNAs. The specificity of these interactions is thought to be related to local sequence-dependent structural variation.

Development of techniques to synthesize oligonucleotides, ie, short, well-defined sequences of DNA or RNA, has provided the opportunity to study nucleic acid structure in detail. In addition, oligonucleotides have proved invaluable in analytical procedures used in genetic engineering (qv), protein engineering (qv), affinity chromatography, and forensics, as well as in medicine (see CHROMATOGRAPHY; FORENSIC CHEMISTRY). The unique ability of nucleic acids to bind to self-complementary sequences has been exploited in the design of oligonucleotide probes and in antisense drug strategies.

DNA Structure

The structure of DNA is characterized by its primary sequence, secondary helical structure, and higher order structure or topology. The primary sequence of DNA refers to the atomic connectivities required to construct the polynucleotide chain. The helical conformation of these polynucleotide chains constitutes the secondary structure of DNA. Sequence-dependent structural diversity and flexibility are important DNA characteristics and play a crucial role in biological processes. The organization of helical DNA in topologically distinct three-dimensional conformations represents the higher order structure. Higher order structural features, in particular the supercoiling of DNA, are thought to have a profound influence on the dynamic processes and biology of nucleic acids within living cells.

Primary and Secondary Structure. The DNA double helix was first identified by Watson and Crick in 1953 (4). Not only was the Watson-Crick model consistent with the known physical and chemical properties of DNA, but it also suggested how genetic information could be organized and replicated, thus providing a foundation for modern molecular biology.

The primary structure of DNA is based on repeating nucleotide units, where each nucleotide is made up of the sugar, ie, 2′-deoxyribose, a phosphate, and a heterocyclic base, N. The most common DNA bases are the purines, adenine (A) and guanine (G), and the pyrimidines, thymine (T) and cytosine (C) (see Fig. 1). The base, N, is bound at the 1′-position of the ribose unit through a heterocyclic nitrogen.

The nucleotides are linked together via the phosphate groups, which connect the 5′-hydroxyl group of one nucleotide and the 3′-hydroxyl group of the next to form a polynucleotide chain (Fig. 1a). DNA is not a rigid or static molecule; rather, it can adopt a variety of helical motifs.

A-DNA and B-DNA. In A- or B-form DNA, two self-complementary polynucleotide strands associate with one another to form a right-handed double helix. The two polynucleotide chains are antiparallel. One chain extends in the 5′→3′ direction, the other in the 3′→5′ direction. The phosphate groups of each polynucleotide chain lie on the surface of the double helix, constituting a phosphodiester backbone, which defines two concave regions on the surface of the double helix known as the minor and major grooves. The polynucleotide chains are associated

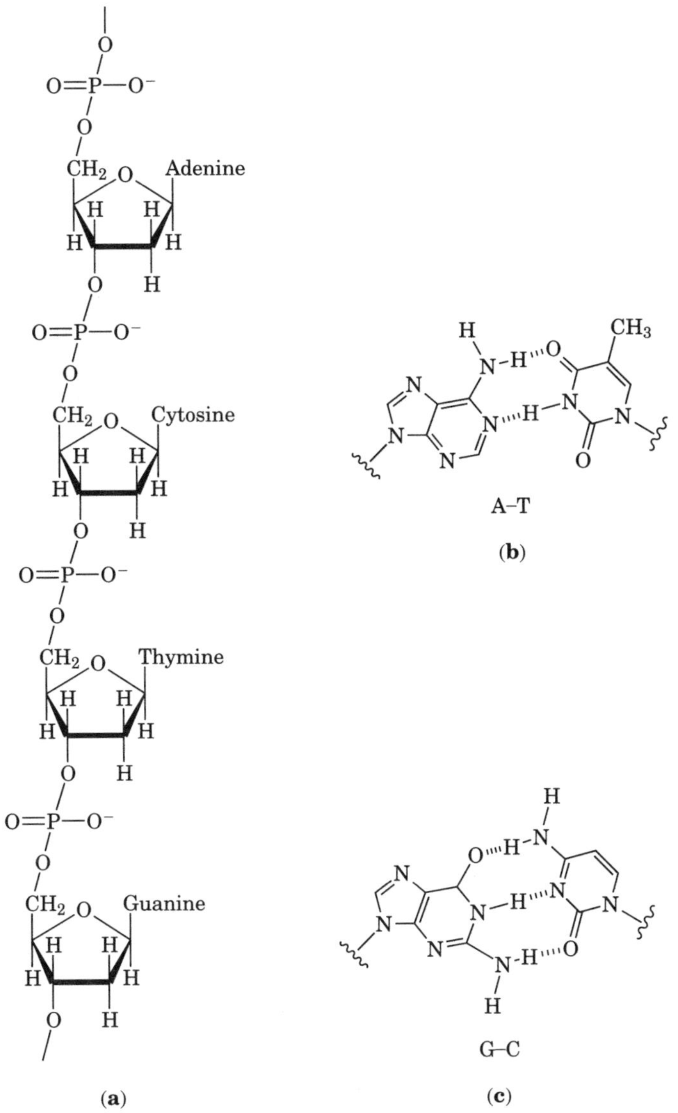

Fig. 1. Elements of DNA structure: (**a**) a deoxypolynucleotide chain, which reads d(ACTG) from $3' \longrightarrow 5'$ or d(GTCA) from $3' \longrightarrow 5'$; and (**b**) and (**c**) the Watson-Crick purine–pyrimidine base pairs, A–T and G–C, respectively, where $\longrightarrow\!\xi$ represents attachment to the deoxyribose.

with one another via hydrogen bonding between complementary A–T or G–C base pairs (Fig. 1) lying perpendicular to the helical axis at the center of the helix. This double-helical structure is known as B-DNA and is the conformation adopted by most of the DNA found in cells.

The original structural model for B-DNA was based on x-ray diffraction of heterogeneous DNA fibers (4). Because heterogeneous DNA was used, the

Watson-Crick model characterizes the average structure of B-form DNA, which has helical parameters of 10.5 base pairs per turn of the helix, a rise of 0.34 nm per base pair, and a helical width of 2 nm. In this double helix, the hydrophobic bases are oriented toward the interior of the molecule where the bases are stabilized by hydrogen bonding and base-stacking interactions with one another. The hydrophilic sugar–phosphate backbone characterizes the surface of the double helix and gives B-DNA the water-solubility properties required in a biological environment.

Other helical DNA structures have been identified from fiber diffraction data. Most important among them is A-form DNA, which is characteristic of DNA fibers in a low (75%) humidity environment and is the common helical conformation adopted by double-stranded RNA. The A-DNA helix (2.6 nm across) is wider than B-form and requires 11 base pairs per turn of the helix. Moreover, the base pairs are tilted and off center with respect to the helix axis. Unlike B-DNA, where the minor and major grooves resemble each other in depth, the A-DNA minor groove is wide and shallow, and its major groove is narrow and deep. The characteristic difference between these two helices, however, is the conformation of the sugar ring. Sugar puckering is $C_{2'}$-endo for B-DNA and $C_{3'}$-endo for A-DNA.

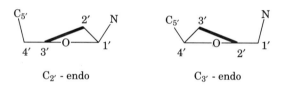

$C_{2'}$ - endo $C_{3'}$ - endo

Many biological mechanisms are suggested by the structure of the double helix. Transcription and translation of DNA are specific events mediated by proteins (qv) that recognize unique sites within a DNA chain which can be millions of base-pairs long. According to the Watson-Crick model, the only feature distinguishing one DNA region from another is the base-pair sequence, which suggests that DNA binding proteins recognize specific DNA sites by direct readout of the sequence. These proteins were thought to take advantage of characteristic hydrogen-bonding sites located in the major groove of the helix and occurring on the purine and pyrimidine bases. Although this mode of recognition is important, sequence-dependent structural variation within the DNA helix may also be a factor. Structural studies of oligonucleotides have revealed that double-stranded helical DNA is structurally diverse. This diversity appears to have a sequence-dependent component (5).

Elucidation of the helical structural variation requires single-crystal x-ray analysis of well-defined DNA sequences (see X-RAY TECHNOLOGY). Such studies became feasible in the late 1970s when methods were developed for synthesizing short oligonucleotides having the purity required to obtain crystals suitable for structural analysis. The first single-crystal structure of B-form DNA was of the self-complementary dodecamer, d(CGCGAATTCGCG), in 1980 (Fig. 2**b**). Since that time, numerous other B-form oligonucleotides have been characterized by this method (5). Although generally consistent with a B-DNA conformation, the structures of these oligonucleotides show considerable variation in local helical parameters. For example, using crystallographic data derived from 12 B-form oligonucleotides, the average helical twist angle between consecutive base pairs is 36.1°, which is virtually identical to the Watson-Crick value. Actual twist angles, however, range from 24–51° for individual base steps. Similarly, the rise per base pair averages 0.34 nm, but actual values range from 0.25–0.44 nm; and roll angles between base pairs average +0.6°, but range from −18° to +16°. Significant variations are also apparent in helical parameters referred to as propeller, buckle, and inclination. Based on comparisons of several crystal structures, twist, rise, and roll appear to vary in a sequence-dependent manner. Too few sequences have been examined, however, to define a set of rules for this variation.

Some oligonucleotides adopt an A-form helical structure (Fig. 2**a**) (5). The average structural parameters have been found consistent with the fiber diffraction model, but, as for B-form DNA, considerable variation is apparent among individual base pairs.

Another technique often used to examine the structure of double-helical oligonucleotides is two-dimensional ^{1}H nmr spectroscopy (see MAGNETIC SPIN RESONANCE). This method relies on measurement of the nuclear Overhauser effects (NOEs) through space to determine the distances between protons (6). The structure of an oligonucleotide may be determined theoretically from a set of interproton distances. As a result of the complexities of the experiment and data analysis, the quality of the structural information obtained is debated. However, nmr spectroscopy does provide information pertaining to the structure of DNA in solution and can serve as a complement to the structural information provided by crystallographic analysis.

DNA Bending and Flexibility. Another important sequence-dependent variation in oligonucleotide structure is localized bending of the helical axis. Sequence-specific DNA bending, the subject of a great deal of research, can be recognized by site-specific DNA binding proteins. One intrinsically bent oligonucleotide is the decamer, d(CATGGCCATG), which in the crystal structure has a helix bent by 23° over the central four base pairs (7). Other examples of bent DNA include oligonucleotides having phased dA·dT tracts inserted at regular intervals within the sequence. These dA·dT tracts impart a regular curvature to the molecule either by being intrinsically curved themselves or by spacing curved junctions between dA·dT tracts and ordinary B-DNA at regular intervals (8). The nature of sequence-dependent bending of the B-DNA helix is controversial: intrinsic bending in GC sequences is thought not to be a factor in the bending of DNA-containing phased dA·dT tracts. However, bending of the helical axis may be important in orienting long stretches of DNA to allow interactions between nonconsecutive sequences, and in site-specific protein binding.

In addition, many proteins, including histones, endonucleases, and specific regulatory proteins, can induce bending of the helical axis (9). For example,

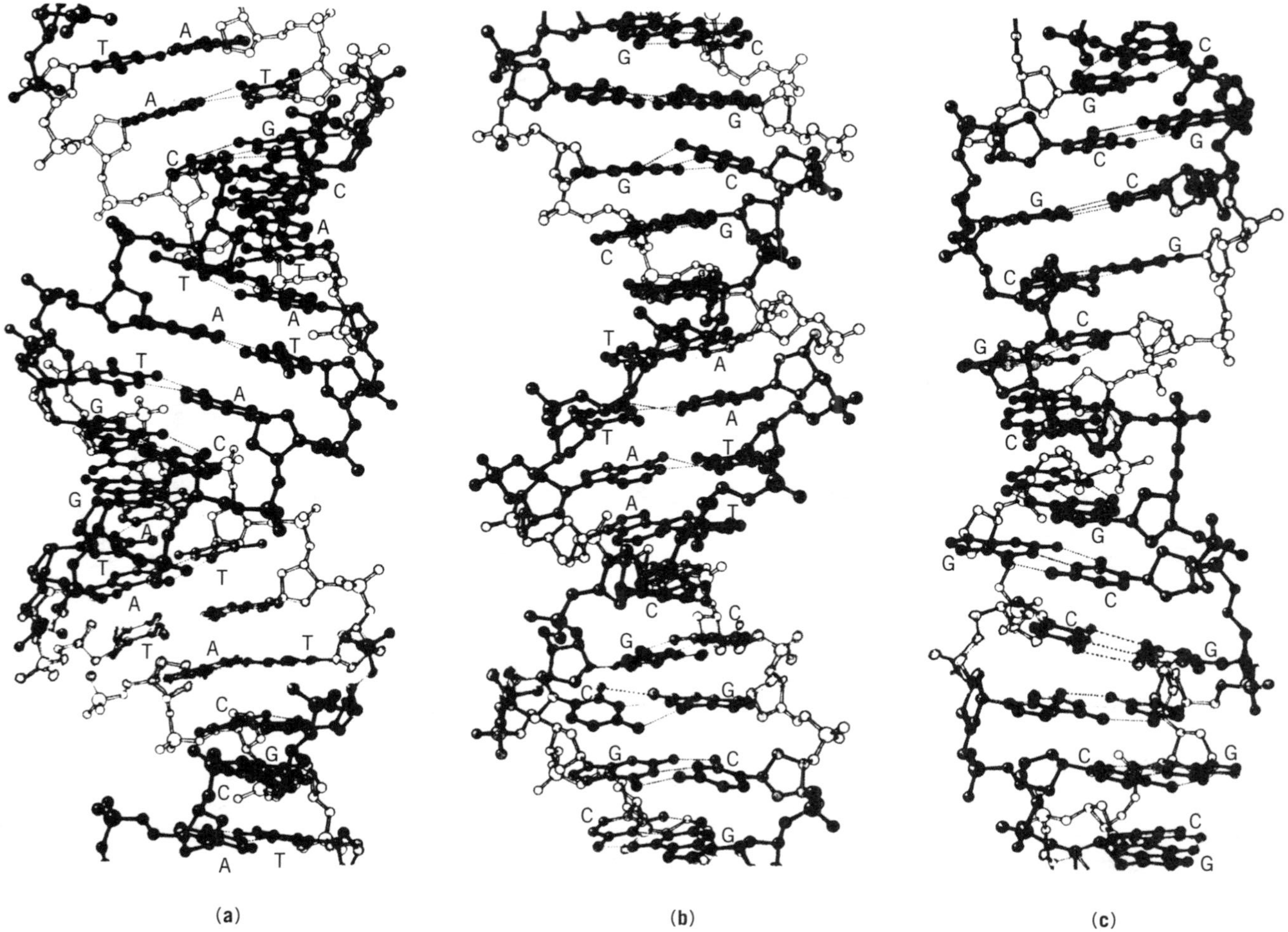

Fig. 2. Comparison of three forms of DNA: (a) A-DNA, (b) B-DNA, and (c) Z-DNA.

(a)

(b)

(c)

when the endonuclease *Eco*R1 binds specifically to d(GAATTC), bends or kinks are induced at two positions along the DNA helix. The first kink is centered at the (AT)·(AT), dislocating the helix by 12°. The second, involving a roll toward the minor groove, occurs at the flanking (CG)·(CG) sequences and bends the helix by 20–40°. A more dramatic example of protein-induced bending is provided by the regulatory protein cap and its operator DNA sequence, where two 40°-kinks are introduced at symmetrically disposed (CA)·(TG) base pairs in the bound protein-DNA complex. Investigations of modified oligonucleotides indicate that protein-induced bending might result from charge neutralization by cationic amino acids of the bond protein (10). Electrophoretic experiments using oligonucleotides, where phosphate linkages on one side of the DNA were replaced by neutral methylphosphonate groups, demonstrate that DNA bends toward the uncharged surface in a process that is driven by repulsions between the anionic phosphate groups on the opposite side of the DNA. These oligonucleotides mimic the charge environment found when a protein binds to DNA and suggests that phosphate neutralization plays a significant role in DNA bending by bond proteins. Additionally, certain sequences may be more flexible than others. A protein-DNA complex requiring helical deformation for optimal binding has a higher affinity for a flexible sequence. Whereas specific protein-DNA hydrogen-bonding interactions are important, the ability of DNA to accommodate a ligand through bending and flexibility of the helix is important in stabilizing site-specific interactions.

Alternative DNA Helices.   Z-DNA.   In addition to A- and B-form DNA, several other helical conformations have been identified. Among these, the most well-studied is Z-DNA, a left-handed helix first characterized by x-ray crystallographic analysis of the oligonucleotides d(CGCGCG) and d(CGCG) (Fig. 2c) (5). Other alternating purine–pyrimidine sequences, in particular alternating CG sequences, have been shown to adopt the Z-conformation at high ionic strength (11).

This helix is named for its zigzagging sugar–phosphate backbone, caused by the syn–anti conformation adopted by the glycosyl linkages to G and C, respectively. The Z-helix requires 12 base pairs per turn and has a diameter of 1.8 nm; the major groove of the helix is essentially flat, and the minor groove is narrow and deep. The biological significance of the Z-conformation is controversial.

Short segments of poly(dG–dC) incorporated within plasmids, or circular DNA, adopt the Z-conformation under negative superhelical stress. This left-handed DNA may be important in genetic control. On the other hand, the structural alteration of the helix required in a B-to-Z transition within a plasmid is radical, and would involve either a multistep mechanism or the complete melting and reformation of helix. The improbability of such transitions has led to questions concerning the feasibility of a biological role for Z-DNA.

Triple Helices.   Triple helices occur when a double helix, containing only purine bases on one strand, binds to a third polynucleotide containing only polypyrimidines. The polypyrimidine strand binds on one side of the major groove of the double helix, forming Hoogsteen hydrogen bonds with the purine bases along one side of the duplex (12).

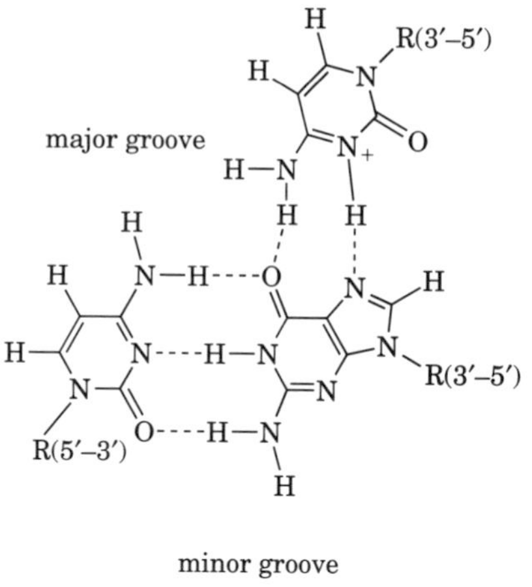

C – G – C+

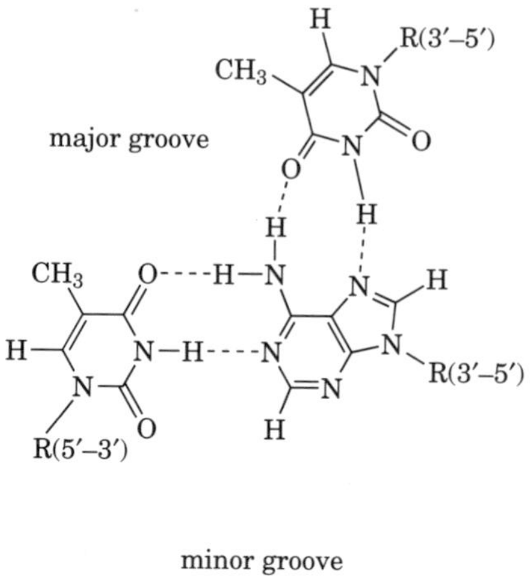

T – A – T

Mirror-repeat homopurine–homopyrimidine sequences within plasmid DNA can adopt triple-helical conformations under superhelical stress. These structures, referred to as H-DNA, occur when a polypurine–polypyrimidine duplex dissociates and the polypyrimidine strand folds back to form a triple helix with a duplex section of DNA. The other polypurine sequence remains single stranded. Sequences able to adopt this conformation occur within eukaryotic chromosomes, which suggests a biological role for H-DNA. In addition, short oligonucleotides can be designed to target specific DNA sites via triple-helix formation. This strategy is being investigated in an effort to synthesize sequence-specific regulators of gene expression (12).

Cruciforms. Cruciforms require a double-stranded DNA sequence having an inverted base-repeat called a palindrome. Cruciform formation involves looping out of the DNA, and replacing the interstrand base pairs with intrastrand ones to generate a pair of hairpin structures (Fig. 3). The hairpin structure, consisting of a stem and a single-stranded loop, is inherently less stable than the fully hydrogen-bonded interstrand duplex. These structures may be induced, however, in negatively supercoiled plasmid DNA; and cruciform extrusion has been used to study supercoiling in closed-circular plasmids (13). Although the kinetics of cruciform extrusion indicate that the process is too slow for these structures to have a biological role, the presence of palindromic sequences within regulatory regions in the gene does suggest that cruciforms may be significant.

Antiparallel Quadruple Helices. Eukaryotic chromosomes are linear DNA molecules that end in a series of short guanine tracts regularly distributed along the strand oriented in the $5' \rightarrow 3'$ direction. These guanine-rich sequences, called telomeres, are highly conserved and necessary for chromosome function. Structural studies of these sequences *in vitro* demonstrate that the guanine tracts form intramolecular four-stranded complexes in which the guanine bases associate with one another via Hoogsteen-type hydrogen bonds (14). It is not known, however, whether these structures are biologically relevant.

DNA Topology. DNA must be highly compacted and organized in the cellular environment. For example, the *E. coli* chromosome, 1.5-mm long, must be

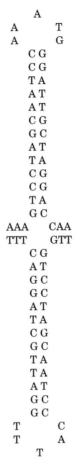

Fig. 3. Example of cruciform formation from an inverted-repeat sequence of DNA which occurs in the plasmid ColE1. The bolded bases are paired with bases of the same strand.

5' AAA**GTCCTAGCAATCC**AAATG**GGATTGCTAGGAC**CAA 3'
3' TTT**CAGGATCGTTAGG**TTTAC**CCTAACGATCCTG**GTT 5'

contained within a cell having a diameter of less than 1 μm. All naturally occurring DNA molecules, including plasmids, chromosomes, as well as mitochondrial and chloroplast DNA, can achieve compaction in ways that depend in part on supercoiling. Whereas compaction is the most obvious objective of supercoiling, it plays a more important role in mediating DNA-associated biological processes. Natural DNA molecules are negatively supercoiled. Supercoiled DNA is a high energy conformation, and this energy contributes to the binding of proteins involved in replication, transcription, and recombination. In addition, supercoiling may induce the formation of some of the higher energy helical structures, including Z-DNA, cruciforms, and H-DNA (15).

DNA Supercoiling. Supercoiling is a topological property of closed-circular DNA molecules. Circular DNA molecules can exist in various conformations differing in the number of times one strand of the helix crosses the other. These different isomeric conformations are called topoisomers and may be characterized in terms of the linking number, *Lk*. A linear DNA molecule having *N* base pairs and h base pairs per turn of the helix, if joined end to end, has the following:

$$Lk° = N/h \tag{1}$$

where $Lk°$ refers to the linking number of the relaxed state of the circular molecule. If torsion is applied either to unwind or to overwind the DNA helix before the ends of the linear DNA are joined, the linking number of the circular DNA changes. The linking difference, ΔL, is as follows:

$$\Delta L = Lk - Lk° = Lk - N/h \tag{2}$$

When the helix is unwound prior to rejoining the ends, ΔL is negative and the circular DNA molecule is negatively supercoiled. Likewise, when the DNA molecule is overwound, ΔL is positive and the circular DNA is positively supercoiled.

The linking difference can be considered in terms of changes in the twist, Tw, and writh, Wr, of the molecule. Twist describes how the DNA strands are coiled around each other; writh describes the coiling of the helical axis. The change in linking number is related to changes in twist and writh:

$$\Delta L = \Delta Tw + \Delta Wr \tag{3}$$

In a DNA molecule, constraints imposed by duplex structure resist unwinding and overwinding of the helix about its axis, ie, they resist changes in the twist of the helix. In other words, energy is required to change the number of base pairs per helical turn. The circular DNA molecule may accommodate a change in linking number by coiling the helical axis, which is equivalent to a change in writh.

Relaxed closed-circular DNA may be represented as shown in Figure 4a. The black line on the surface of the circle indicates that $\Delta Lk = 0$. Laboratory tubing is often used as a model for DNA supercoiling. If the circular tubing is opened at one point and the end is twisted around four times in a left-handed sequence before the tubing is reconnected, the linking number changes to $\Delta Lk = \Delta Tw = -4$ (Fig. 4b). ΔLk can also be accommodated by a change in writh, $\Delta Wr = -4$, where there is no change in winding about the axis. When this occurs, the axis coils around and crosses itself four times (Fig. 4c). In an actual model, however, a change in linking number can generate some change in both twist and writh (Fig. 4d).

Electron micrographs of supercoiled DNA *in vitro* demonstrate that supercoiled molecules have a structure intermediate between Figures 4c and 4d, where changes in the linking number are accompanied by approximately a 75% change in Wr and a 25% change in Tw. This conformation is called plectonemic supercoiling.

Many naturally occurring DNA molecules, including plasmids, bacterial chromosomes, and mitochondrial as well as chloroplast DNA, are examples of closed-circular DNA in which much of the linking difference is manifested in plectonemic supercoiling. In the cellular environment, however, DNA is organized by binding to the surface of proteins. When proteins interact with DNA,

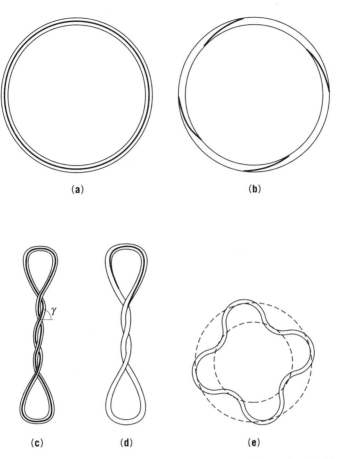

Fig. 4. Laboratory tubing model for supercoiling in closed-circular DNA: (a) relaxed DNA, $\Delta LK = 0$; (b) $\Delta LK = \Delta Tw = -4$; (c) $\Delta LK = \Delta Wr = -4$; (d) plectonemic super-coil, $\Delta LK = -4$; and (e) toroidal winding of DNA. See text.

the supercoiled structure is better described by a toroidal model (Fig. 4e), where the DNA helix is wound around the surface of a torus created by DNA-binding proteins. This model is particularly important in the organization of eukaryotic chromosomes where the DNA is bound to nucleosomes and organized in a solenoid structure.

    Chromatin.   Chromosomal DNA, found in the nucleus of eukaryotic cells, is bound to histone proteins to form the nucleoprotein complex called chromatin (16). Chromatin assumes an organized structure based on sequentially repeating units called nucleosomes, which serve the dual purpose of condensing the chromosomal DNA, and organizing it so that it is accessible to the cellular machinery required for transcription, replication, and mitosis. A nucleosome consists of a core of eight histone proteins. The DNA binds to the histones by wrapping the helical axis twice around the histone core to form a left-handed toroidal super-coil. Nucleosomes along a DNA strand resemble a string of beads, a structure known as the 10-nm fiber.

The native form of chromatin in cells assumes a higher order structure called the 30-nm filament, which adopts a solenoidal structure where the 10-nm filament is arranged in a left-handed coil (Fig. 5). The negative supercoiling of the DNA is manifested by writhing the helical axis around the nucleosomes. Chromatin structure is an example of toroidal winding; whereas eukaryotic chromosomes are linear, the chromatin structures, attached to a nuclear matrix, define separate closed-circular topological domains.

Biological Consequences of Supercoiled DNA.   The free energy associated with negative supercoiling facilitates a number of biological mechanisms (17). Many proteins unwind the helix upon binding to DNA, counteracting the negative twisting and writhing associated with supercoiling. Unwinding proteins include those involved in DNA replication and transcription (18). In both of these processes, the double helix must be opened to facilitate biosynthesis of either DNA or RNA. Although these mechanisms involve complex multistep processes, DNA supercoiling is an important component.

The extrusion of cruciforms, the transition of B- to Z-DNA, and H-DNA formation all occur upon the unwinding of the helical axis, and are promoted by negative supercoiling of the DNA. Supercoiling can also bring together two or more sequences separated by hundreds or thousands of base pairs. Regulation of gene expression often involves several proteins that bind to multiple sites separated by distances precluding direct interaction. Genetic recombination also involves alignment of distant DNA sequences. Remote DNA sequences can interact with one another by bending the DNA helix to form a loop connecting the two sites (19), or by a sliding mechanism where oriented supercoils act to bring two sites together.

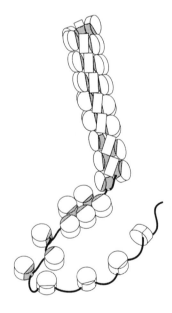

Fig. 5.   Solenoid model of the 30-nm filament of chromatin, where the disks represent nucleosomes and the dark line unbound DNA.

Knots and Catenanes. Closed-circular DNA helices can cross over one another three or more times to form topological knots. These structures are not common, but have been found to occur naturally in some bacteriophage DNAs.

Catenanes are formed when two or more closed-circular DNAs are linked together to form a chain. Catenanes were first isolated in human mitochondrial DNA and have since been identified in a number of biological systems. These structures often occur as intermediates during the replication of circular DNA molecules.

Both catenanes and knots can bring together remote DNA sequences and may be important in transcription regulation and genetic recombination (15).

RNA Structure

RNA has a variety of functions within a cell; for each function, a specific type of RNA is required. Messenger RNA (mRNA) serves as intermediaries for carrying genetic messages from the DNA to the ribosomes where protein synthesis takes place. Ribosomal RNA (rRNA) serves both structural and functional roles in the ribosome; it is diverse, both in terms of its size and structure. Transfer RNAs (tRNAs) are small molecules that have a central role in protein synthesis (20). Other RNA molecules, called ribozymes, function as enzymes to catalyze chemical transformations (21). Although ribozymes most often catalyze cleavage of the RNA phosphodiester backbone, they have also been shown to participate in cleavage of DNA, replication of RNA, and reactions with phosphate monoesters. Other RNAs are associated with enzymes to form riboprotein complexes involved in many biological processes. The multifunctional character of RNA, particularly the involvement of RNA in enzymatic processes, has led to the hypothesis that life on earth evolved from RNA, and that RNA had both the genetic and catalytic functions commonly associated with DNA and proteins, respectively (3).

Primary and Secondary Structures of RNA. The primary structure of RNA is similar to that of DNA, but with a few notable exceptions. First, in RNA, instead of thymine, the pyrimidine base uracil (U) occurs, forming a complementary base pair with adenine in regions of double-stranded RNA (Fig. 6). Also, a wide variety of ribonucleotides having modified or minor bases are found in naturally occurring RNA, one of the most common of which is pseudouridine. In human tRNAs, as many as 25% of the bases are nonstandard. Over 80 modified bases have been characterized in naturally occurring tRNA; although the role of base modification is not clear, it may be important for biological recognition (20).

The other important feature of the primary structure of RNA is the presence of the 2'-hydroxyl group in ribose. Although this hydroxyl group is never involved in phosphodiester linkages, it does impose restrictions on the helical conformations accessible to double-stranded RNA.

RNAs are single-stranded molecules that fold, allowing different regions of the ribonucleotide to form distinct secondary structural elements (22). When self-complementary regions of the RNA strand are aligned, duplex regions, which may have Watson-Crick base pairs, are formed. In contrast to DNA, double-stranded regions in RNA are much more likely to have unusual base-pairing between noncomplementary bases and to incorporate non-Watson-Crick base-pairing. Owing to the steric requirements of the 2'-hydroxyl group on the ribose

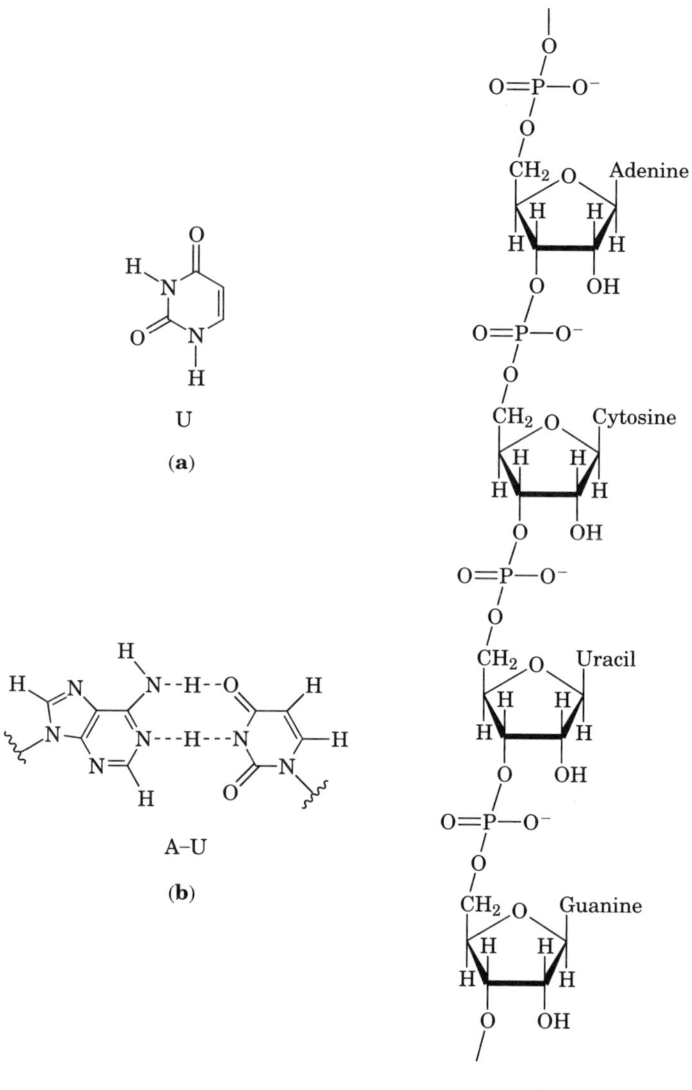

Fig. 6. Elements of RNA structure: (**a**) uracil; (**b**) a Watson-Crick A–U base pair; and (**c**) a polynucleotide chain which reads from $5' \rightarrow 3'$ ACUG, and from $3' \rightarrow 5'$ GUCA.

sugar, these duplex regions are constrained to an A-form helix, ie, a $3'$-endo sugar conformation. Although double-stranded RNA has the general features of an A-form helix, actual duplex characteristics, such as rise per base pair, groove dimensions, and base pair displacement from the helical axis, may vary.

Extensive duplex RNA regions exist, but this folding back of the strand and noncomplementary sequences necessarily give rise to single-stranded RNA regions. These latter regions form structures such as hairpins, internal loops, bulges, and junctions (Fig. 7). A hairpin, similar to one side of a DNA cruciform, forms when a single strand doubles back, creating a duplex stem and a single-

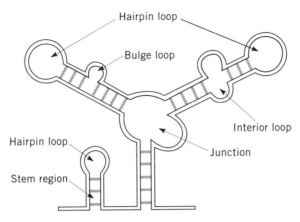

Fig. 7. Structural features of folded RNA where ▓ represents base pairing.

stranded loop. Hairpin loop regions range in length from 2–10 ribonucleotides. Internal loops form when base mismatches occur on both strands on an otherwise self-complementary RNA duplex. Whereas these regions are called loops, the noncomplementary bases actually adopt unusual base-pairing schemes known as wobble, reverse-wobble, Hoogsteen, reverse-Hoogsteen, and reverse-Watson-Crick hydrogen bonds (Fig. 8). These looped regions retain an overall A-form helical geometry, but are more open and flexible than A-form DNA, facilitating binding interactions with proteins and other RNAs. Bulge loops occur when noncomplementary bases are present on one strand of an otherwise complementary duplex. Bulges have been shown to introduce a bending in the structure of RNA and can serve as recognition site for protein binding (23). Branched loops or junctions connecting multiple hairpin structures are the functional domains of ribozymes. All of these structural elements are combined and folded to yield RNAs having diverse tertiary structures.

Tertiary Structure of RNA. The functional diversity of RNA is directly related to its structural diversity. In contrast to DNA, RNA molecules are synthesized as single-stranded polynucleotides that fold to give complex tertiary structures. These structures, which incorporate hairpins, loops, bulges, and junctions between single-stranded and double-stranded regions, exhibit long-range interactions within the folded tertiary structure. Long-range intramolecular interactions serve to stabilize the three well-characterized RNA structures.

Transfer RNA. Over 2000 sequences of tRNAs have been determined and several tRNA crystal structures have been solved (20). In addition, the overall structures, as well as some specific nucleotides within the tRNA sequences, are highly conserved. The secondary structure (Fig. 9**a**) conforms to a clover leaf model. Crystal structures show that tRNA is an L-shaped molecule characterized by stem and loop regions (Fig. 9**b**). These structural regions participate in specific interactions with mRNAs, rRNAs, and proteins.

Transfer RNAs participate in protein synthesis by translating the genetic information encoded in the base sequence of mRNA. Each tRNA is for a specific amino acid, which is covalently attached to the 3′-terminus to give an aminoacyl-tRNA (see AMINO ACIDS). The anticodon, a base triplet contained in the anticodon

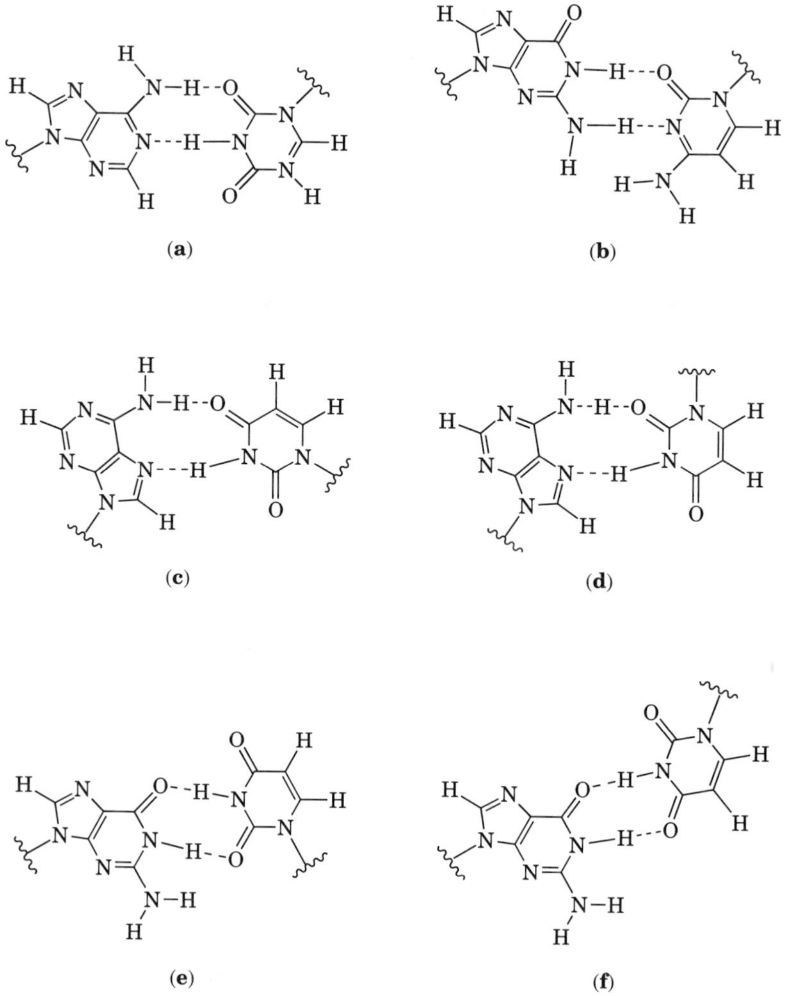

Fig. 8. Non-Watson-Crick base pairs occurring in double-stranded RNA where —⧛
represents the site of attachment to the sugar: (**a**) A–U reverse-Watson-Crick; (**b**) G–C
reverse-Watson-Crick; (**c**) A–U Hoogsteen; (**d**) A–U reverse-Hoogsteen; (**e**) G–U wobble;
and (**f**) G–U reverse-wobble.

loop of the tRNA, recognizes the corresponding mRNA triplet codon and forms
a complex by complementary base-pairing. When this complex forms, the amino
acid is transferred from the 3′-end of the tRNA to the growing polypeptide chain.
Amino-tRNA synthetases, the enzymes responsible for attaching the amino acids
to the tRNA, recognize specific sites along the acceptor stem and the anticodon
loop, ensuring fidelity in protein synthesis. Transfer RNA interacts with as many
as 30–40 proteins and participates in other processes in addition to protein
synthesis. The conserved structure of this molecule is essential for its biological
function.

   Hammerhead Ribozyme.   A small RNA molecule that catalyzes cleavage of
the phosphodiester backbone of RNA is known as the hammerhead ribozyme.

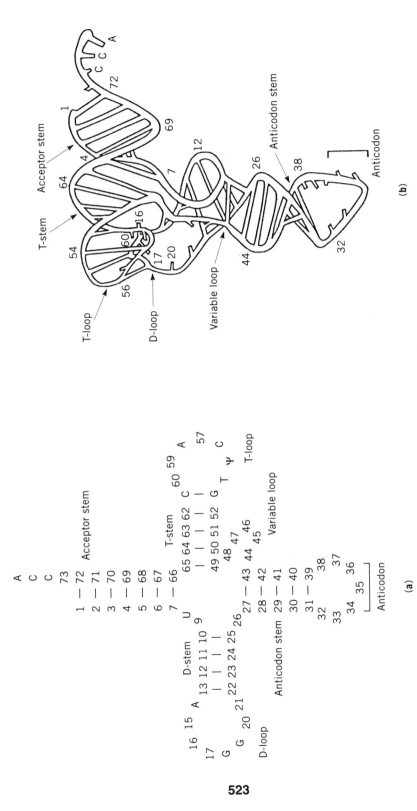

Fig. 9. Structures of tRNA: **(a)** secondary structure of tRNA, where conserved base pairs are indicated by lines; and **(b)** tertiary structure.

523

This ribozyme occurs naturally in certain viruses where it facilitates a site-specific self-cleavage at the phosphate and generates a 2'3'-cyclic phosphate and a 5'-hydroxyl terminus. The reaction requires a divalent metal ion, such as Mg^{2+} or Mn^{2+}, as a cofactor. Whereas the hammerhead ribozyme does not cleave DNA, it does form a complex with a complementary DNA target sequence, the crystal structure of which provides insight into the structural requirements for hammerhead activity (Fig. 10) (24). The ribozyme has a wishbone-shaped tertiary structure and three A-form helical stems. The stems meet at a central core encompassing a conserved CUGA sequence similar to sequences found in the anticodon loops of tRNA. The second domain at the base of stem 2 is a potential site for the coordination of a divalent metal ion. Stems 1 and 3 are bound to the substrate DNA sequence and hold the DNA at a bent position at the cleavage site. The ability of the RNA to orient the substrate cleavage site is probably a critical element in catalysis (24).

ColE1 Regulation by RNA Hairpins.   Replication of the *E. coli* plasmid ColE1 is regulated by two short RNA molecules and a protein in a system that provides an example of the unique structural elements accessible to RNA molecules. Multidimensional heteronuclear nmr spectroscopy has been used to characterize the complex formed between the two RNAs (25). Each of the RNA molecules fold back on the other to form a pair of hairpin structures. There are seven unpaired bases in the loop (25). The two RNA hairpins interact to form a complex stabilized by base-pairing between the two looped domains. The loop–loop helix of the complex is sharply bent and binds specifically to the regulatory protein.

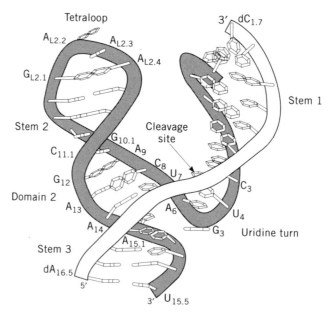

Fig. 10.   Three-dimensional structure of the hammerhead ribozyme (shaded cord) bound to a substrate oligonucleotide (nonshaded cord). The uridine turn occurs at a CUGA sequence that is highly conserved (24).

Oligonucleotide Synthesis

Synthetic oligonucleotides are widely used in scientific investigations. Most synthetic oligonucleotides are produced for use as primers in the polymerase chain reaction (PCR), a widely used analytical technique having commercial applications in diagnostic medicine, genetic engineering (qv), and forensics (see FORENSIC CHEMISTRY) (26). A large volume of oligonucleotides are also synthesized for use as primers in DNA-sequencing (27). Demands for sequencing primers have increased rapidly to support large-scale DNA mapping and sequencing efforts such as the human genome project (28). Although the quantities of oligonucleotides used are relatively small, these materials are essential in many areas of basic research.

In molecular biology, synthetic oligonucleotides are used as linkers in gene-cloning (29) and to introduce site-directed mutations in genes (30). Synthetic oligonucleotides are required for structural (31), biochemical, and biophysical studies of DNA and RNA. Oligonucleotides also are important for examining the association of proteins and small molecules, eg, for intercalating drugs with nucleic acids on a molecular level (32).

The first procedures for oligonucleotide synthesis, typically carried out in solution, made use of *H*-phosphonate and phosphotriester chemistry. These approaches are useful in some large-scale syntheses and in syntheses of various oligonucleotide analogues. Most modern procedures, however, are based on solid-phase phosphoramidite chemistry (33). Automated oligonucleotide synthesizers are commercially available, as are the required reagents and phosphoramidites. Together these permit the rapid production of custom oligonucleotides and oligonucleotide analogues. The chemistry involved in the three most frequently used oligonucleotide syntheses is described herein.

Unmodified Oligonucleotides. *The Phosphoramidite Method.* The phosphoramidite method is the most efficient and widely used protocol for syntheses of both unmodified and modified oligonucleotides (34). It can be carried out in either a manual or automated way by using commercially available oligonucleotide synthesizers; oligonucleotides having chain lengths up to about 200 nucleotides may be obtained (35,36). Phosphoramidite technology has been used to generate both nanomolar quantities of DNA or RNA for use in genetics experiments, and micromolar quantities of oligonucleotide for biophysical and structural investigations.

The synthetic scheme typically involves chain-extending addition of protected mononucleotides to a nucleoside bound covalently at the 3′-hydroxyl to an inert silica-based solid support, such as controlled pore glass (Fig. 11). The initial base-protected 5′-O-dimethoxytrityl (DMT) deoxynucleoside is linked to the solid support via the reaction of a silica-bound amino-silane and the *p*-nitrophenylester of the 3′-succinylated nucleoside, yielding a 3′-terminal nucleoside attached to the solid support (**1**) (Fig. 11). Chain elongation requires the removal of the 5′-DMT protecting group.

The chain is extended in the 3′→5′ direction by tetrazole-catalyzed addition of the 5′-hydroxyl group to a protected 3′-phosphoramidite. The β-cyanoethyl-N,N-diisopropylphosphoramidite (**2**) is most commonly used for this purpose. It is prepared by reaction of a protected deoxynucleoside and bis(diisopropylamino)-

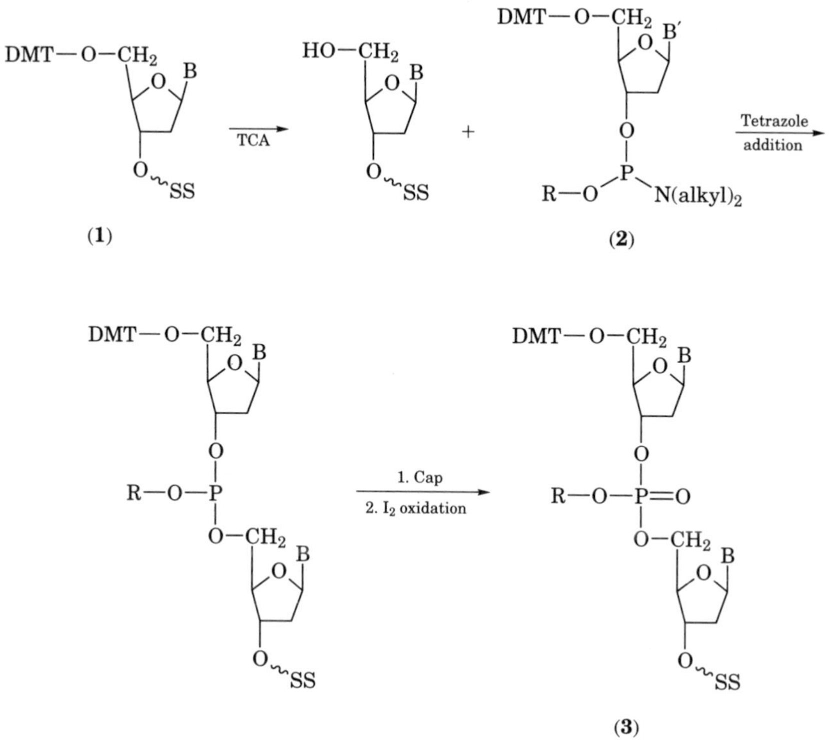

Fig. 11. The amidite method for synthesizing oligodeoxynucleotides, where B and B′ represent one of the protected bases thymine, N^4-benzoylcytosine, N^6-benzoyladenine, or N^2-isobutyrylguanine; $-$N(alkyl)$_2$ is $-$N(CH(CH$_3$)$_2$)$_2$, $-$N⟋◯⟍O, $-$N⟋◯, or $-$N⟋◯ ; DMT is p-(dimethoxytrityl); R is NCCH$_2$CH$_2$ or CH$_3$; ⁓⁓ SS is silica (CH$_2$)$_3$NH$-$C$-$(CH$_2$)$_2$$-C=$O on a solid support; and TCA is trichloroacetic acid. See text.

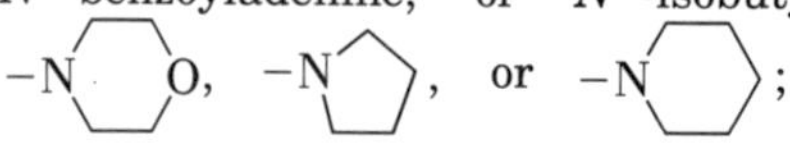

2-cyanoethoxyphosphine, using diisopropylammonium tetrazolide as a catalyst. The coupling reaction generates a phosphite triester and is followed by capping, which involves the acylation of unreacted 5′-hydroxyl groups to prevent these from reacting in the next condensation cycle. Chain elongation at these hydroxyls can generate oligonucleotides having one base deleted from the target sequence, thus complicating the purification of the product oligonucleotide. The phosphite triester is oxidized using I$_2$ to form the stable phosphate triester (**3**). Treatment with acid frees the 5′-hydroxyl of the growing chain so that chain elongation can proceed by repeating the condensation step and using the appropriate protected deoxynucleoside phosphoramidite. The coupling reaction has been optimized to produce yields of >99% in ~8 minutes for a synthetic cycle.

The exocyclic nitrogens on cytosine and the purine bases must be protected during the synthesis. The search is ongoing for protecting groups that are subject to fewer side reactions and that can be removed more easily in the final deprotection step (34).

When the synthesis is complete, the protecting groups are removed and the oligonucleotide liberated from the solid support by treatment with ammonium hydroxide. The oligonucleotide is then purified either by polyacryamide gel electrophoresis (page) (34) or, more commonly, by high performance liquid chromatography (hplc) (37) (see CHROMATOGRAPHY; ELECTROSEPARATIONS). The products may be characterized by Maxam and Gilbert sequencing, which involves partial base-specific degradation of the oligonucleotide followed by page (see GENETIC ENGINEERING, PROCEDURES). The sequence can be verified by determining the order of mobility of the fragments obtained in each base-specific degradation reaction (38). Mass spectrometry (qv) is another method that is particularly attractive for characterizing shorter oligonucleotides (33).

The H-Phosphonate Method. In the late 1980s, an early synthesis of oligonucleotides based on *H*-phosphonates (**4**) (Fig. 12), reemerged (39). This method eliminates one of the steps in the coupling reaction and offers advantages in the preparation of certain modified oligonucleotides. The procedure involves the activation of 5′-DMT-deoxynucleoside 3′-*H*-phosphonates with a sterically hindered carbonyl chloride, followed by condensation with the 5′-hydroxyl group of a protected nucleoside. The phosphite diester (**5**) is stable, and following the

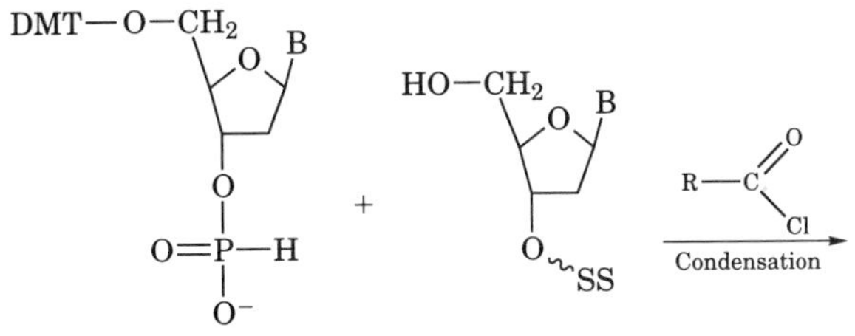

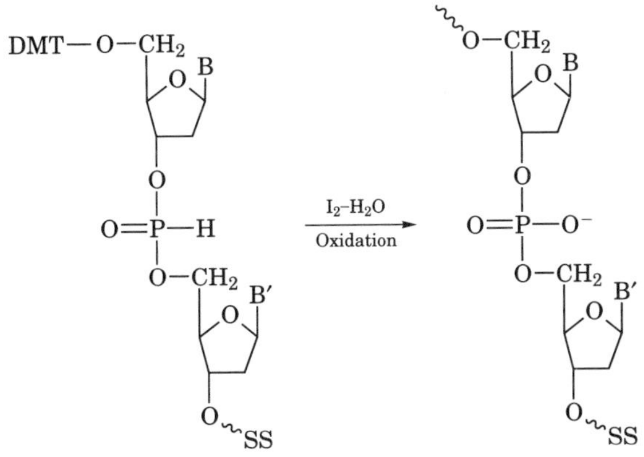

Fig. 12. *H*-Phosphonate method where R—C(=O)Cl is pivaloyl chloride or adamantoyl chloride; B, B′, DMT and 〜〜〜 SS are as defined in Figure 11.

required capping and deprotection steps, chain elongation can proceed until the desired oligomer is obtained. Upon completion of the synthesis, the phosphite diester linkages are converted to phosphodiesters (**6**) by a single oxidation using *tert*-butyl hydroperoxide or iodine. In addition to simplifying the oxidation, this method eliminates the need for a phosphate protecting group. Because coupling reactions proceed in high yield, the method has been used successfully to synthesize oligonucleotides in excess of 100 nucleotides in length.

   The Phosphotriester Method. The phosphotriester method is sometimes used in large-scale (>100 mg) syntheses of oligonucleotides (40). Like the other methods, synthesis proceeds in the $3' \to 5'$ direction when the $3'$-hydroxyl is bound to a solid support. The oligonucleotide is obtained by sequential condensation of protected nucleoside $3'$-phosphodiesters (**7**). The $5'$-hydroxyl group of the growing chain has a 3-nitro-1,2,4-triazolide (NT) of an arenesulfonic acid as a condensation reagent. Commonly used arenesulfonic acids include 2,4,6-triisopropylbenzesulfonic acid (TIPS) and 8-quinolinesulfonic acid (QSNT).

(**7**)

The time required for each coupling cycle in the phosphotriester method compares favorably to that in the phosphoramidite method, but yields for the former method are generally lower (typically around 98%). Although the coupling yields are lower, this synthetic method has the advantage of generating the phosphotriester directly, thereby eliminating the oxidation step in each coupling cycle.

   Oligoribonucleotides. The same general methods used in oligodeoxynucleotide synthesis can be applied to the synthesis of oligoribonucleotides, provided that the $2'$-hydroxy group is suitably protected. When solid-phase phosphoramidite chemistry is used, the protecting group must be resistant to the acidic reagents needed to deprotect the $5'$-hydroxy in the chain elongation step. A strategy that makes use of a pair of protecting groups having different acid stabilities has been developed (41). For example, the more stable

1-(2-chloro-4-methylphenyl)4-methoxy-4-piperidinyl (CTMP) group is used to protect the 2′-position, and a labile 9-phenyl-9-xanthenyl (pixyl) group is used at the 5′-position. Mildly acidic conditions can remove the pixyl group in the chain-elongation step. The other groups may be cleaved at a pH of 2 when the synthesis is complete. This protocol has been used in conjunction with phosphoramidite technology to prepare 19-mer oligoribonucleotides. Coupling yields are ca 95%. Another useful protecting group for the 2′-hydroxy functionality is 1-(2-fluorophenyl)-4-methoxypiperidin-4-yl (Fpmp) (42). This group can be removed under mildly acidic conditions (pH 3, 25°C) and is particularily useful in preparing oligoribonucleotides that are sensitive to either hydrolytic cleavage or 3′- to 2′-migration of the internucleotide phosphate linkages. The 2′-hydroxy can also be protected by using 2-nitrobenzyl, which can be removed at the end of the synthesis by photolysis at a pH of 3.5. This group, however, is difficult to remove completely and caution must be exercised to prevent light exposure during synthesis.

If large quantities of oligoribonucleotides are needed, the phosphotriester method carried out in solution is the method of choice (Fig. 13). A 2′,5′-protected ribonucleoside (8) is treated with 2-chlorophenyl phosphordi-(1,2,4-triazolide) (9), followed by triethylamine, to give (10). The 3′→5′ phosphodiester linkage is obtained upon reaction of (10) and a 2′-protected ribonucleoside in a ~80% yield. Using the phosphotriester method, long (~30 nucleotides in length) oligoribonucleotides have been synthesized by successive coupling of tri- and tetranucleotide

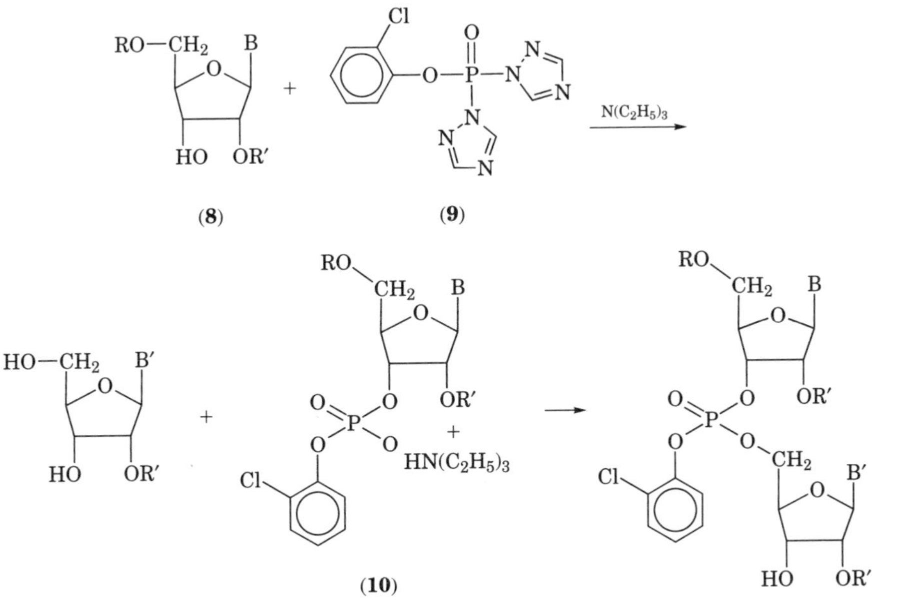

Fig. 13. Phosphotriester method for oligoribonucleotide synthesis, where R is CHB'_2 ; R′ is CH_3O ; and B and B′ are protected bases as defined in Fig. 11.

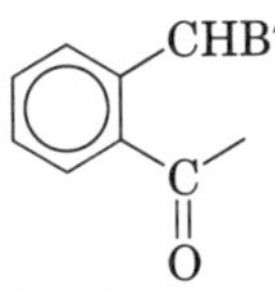

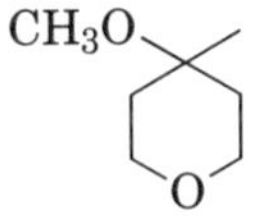

blocks. The low yields in this reaction, however, make solid-phase synthesis preferable for smaller quantities.

Synthesis of small (<100 nucleotides in length), unmodified RNAs can also be carried out by using the enzyme T7 RNA polymerase (43). T7 RNA polymerase transcribes RNA from a double-stranded DNA template only if a 17-base-pair promoter region is positioned before the sequence to be transcribed. The promoter serves as a recognition and binding site for the enzyme. When using this method, the complementary DNA oligonucleotides are synthesized according to standard procedures. RNA synthesis is carried out by incubating the DNA template with ribonucleosidetriphosphates and T7 RNA polymerase in an appropriate buffer system. The method generates large quantities of RNA from a small amount of DNA.

Modified Oligonucleotides. Much of the interest in modified oligonucleotides is related to use as antisense agents (44–47). Antisense agents are typically short (15–30 base pairs in length) oligonucleotides having sequences that are complementary to coding or regulatory regions within mRNA, although some antisense oligonucleotides have also been designed to target DNA. The antisense sequence recognizes and binds to a complementary sequence via the formation of a double-stranded duplex that have normal Watson-Crick base-pairing. Antisense oligonucleotides can inhibit gene expression at the translational level (47). The potential to design oligonucleotides having the ability to recognize and inhibit specific genes makes the antisense approach promising in the development of new therapeutic agents. In addition, antisense oligonucleotides can be used in research to elucidate gene function by providing a mechanism for regulating a gene artificially.

The antisense mechanism is a natural mechanism used in both prokaryotic and eukaryotic cells to regulate gene expression (48). In one example of the naturally occurring processes, the genome encodes antisense sequences complementary to cellular mRNA. When these sequences are transcribed, antisense RNA hybridizes with the complementary mRNA sequences to form a double-stranded RNA–RNA duplex. Formation of the hybrid duplex inhibits expression of the gene corresponding to that particular mRNA by an enzyme-mediated process, in which cellular enzymes recognize and chemically modify the double-stranded RNA sequences. When modified, the mRNA cannot be translated. Thus, naturally occurring antisense mechanisms repress gene expression by interfering with the translation of mRNA.

Although all natural antisense oligonucleotides are short RNA sequences, most of the synthetic antisense oligonucleotides are deoxyoligonucleotides. In the design of an effective antisense oligonucleotide, several factors must be considered. First, the oligonucleotide must be specific, binding with high affinity to a single sequence within the target RNA. Statistically, a 17-base-pair sequence is expected to occur only once in the human genome; therefore, an oligonucleotide of this size range should have the required specificity.

A second consideration is stability within the cellular environment. All cells produce nucleases, ie, enzymes that catalyze the rapid degradation of unmodified oligonucleotides via cleavage of the phosphodiester bonds. 3'-Exonucleases are particularly problematic. These enzymes degrade DNA and RNA nonspecifically in a stepwise manner that begins at the 3'-phosphodiester linkage. Thus all unmodified oligonucleotides are degraded too rapidly to be used effectively as therapeutic agents. A significant research effort has been directed toward discovering chemical modifications that can increase the nuclease resistance of the oligonucleotide backbone.

An effective therapeutic agent must also have the ability to reach its target sequence *in vivo*. Bioavailability requires that the antisense oligonucleotide be able to pass through the cell membrane, and that it have a low affinity for nontarget cellular compartments and, in animal systems, nontarget organs. Cell membranes are lipophilic and designed to be barriers against large anionic molecules, although there is a natural mechanism for intercellular transport of anionic oligonucleotides. In order to enhance membrane transport, antisense oligonucleotides are frequently modified by covalent attachment of carrier molecules or lipophilic groups.

Antisense oligonucleotides are usually designed to inhibit gene expression by interfering with the translation of mRNA. One mechanism for this type of inhibition involves binding the oligonucleotide to the translation-initiation sequences of the mRNA, which prevents ribosome association and protein synthesis. Another potential mechanism involves hybrid formation at some other sequence within the mRNA, thus impeding translocation of the ribosome along the mRNA strand by steric blocking (49). These two mechanisms are based on blocking a sequence of RNA or DNA by double-stranded duplex formation using a specific antisense oligonucleotide.

A less specific mechanism based on the action of Rnase H, an enzyme catalyzing single-strand cleavage of RNA, may be predominate for unmodified, thioate and dithioate oligonucleotides (50). In the Rnase H mechanism, the duplex formed by the antisense oligonucleotide and the target RNA is a substrate for Rnase H. The enzyme cleaves the RNA at the complexes site rendering the RNA vulnerable to further degradation and inactivation by cellular exonucleases. The oligonucleotide, which is probably not a substrate for Rnase H, can target multiple copies of complementary RNA. Where applicable, antisense oligonucleotide action mediated by the Rnase H mechanism has been shown to be a potent inhibitor of gene expression.

Modified oligonucleotides can also be designed for binding to double-stranded DNA by forming a triple helix. Triple-helix formation is a sequence-specific interaction in which the single-stranded oligonucleotide binds in the major groove of homopurine/homopyrimidine DNA targets via the formation of Hoogsteen or reverse-Hoogsteen hydrogen bonds. If the complementary DNA target is within a gene or a regulatory region associated with a specific gene, triple-helix formation can interfere with the transcription of the gene by the same steric mechanism discussed for antisense agents. As of this writing (ca 1995), this approach, termed the antigene strategy, is an active area of research (51). One focus of this research involves extending the range of recognition sequences. Specificity of the Hoogsteen base pairs and steric considerations strongly favor triple-helix formation along homopurine/homopyrimidine DNA tracts. Base modifications and novel oligonucleotides are being developed in an effort to make triplex formation viable for other DNA target sequences. However, despite good specificity, triple helices appear to be less stable and slower to form than double helices. Oligonucleotides have thus been modified by using substituents designed to enhance binding, eg, intercalators or cationic groups.

Artificial endonucleases, ie, molecules able to cleave double-stranded DNA at a specific sequence, have also been developed. These endonucleases can be obtained by attaching a chemically reactive group to a sequence-specific oligonucleotide. When the oligonucleotide is bound to its complementary sequence, the activation of the reactive group results in double-stranded DNA cleavage.

Oligonucleotides can also inhibit gene expression at the transcriptional level by binding to a single-stranded or open sequence of DNA. In this mechanism, the antigene oligonucleotide is designed to be complementary to a regulatory sequence preceding a gene. For example, an oligonucleotide complementary to the *lac* operator sequence (repressor protein binding site) has been found to inhibit specifically β-galactosidase synthesis in *E. coli* (33). Normally, expression of the lactose-metabolizing enzyme, β-galactosidase, is blocked at the transcriptional level by the repressor binding to the operator sequence. In the presence of a lactase metabolite, the repressor is converted to a nonbinding form and dissociates from the operator, which results in the transcription of the gene and β-galactosidase production. However, in the presence of the antisense oligonucleotide, the synthesis of β-galactosidase is inhibited. The antisense oligonucleotide can then act as a repressor by binding to an open or single-stranded region within the operator.

Although development of modified oligonucleotides as antisense and antigene agents is a principal focus of research in the 1990s, there are many other interesting applications that have greater immediate commercial significance. Included among these are applications using nucleic acid probes, which are oligonucleotides that have been modified by the attachment of a detectable chemical group (52). Probes can be designed to recognize RNA or DNA sequences characteristic of specific eukaryotic genes, viruses, or bacteria. Several analytical and diagnostic procedures have been developed based on the hybridization of the probe with its target sequence and the subsequent detection of the hybrid by the group attached to the oligonucleotide. Probes, particularly useful in automated sequencing protocols, may contain fluorescent groups, phosphors, radioactive tracers (qv), etc. In addition, probes can be designed to help elucidate the structure of biological molecules. DNA-binding molecules, including intercalators, alkylating agents, and photosensitive molecules, have also been linked to oligonucleotides as a way of directing a drug to a specific DNA sequence. These modifications often enhance binding as well.

Modification of the Phosphodiester Backbone. Oligonucleotides having modified phosphate backbones have been extensively studied (46). Because altering the backbone makes derivatives generally more resistant to degradation by cellular nucleases, these materials have the potential to be more resilient antisense drugs.

Changing the phosphodiester backbone frequently reduces the anionic character of the oligonucleotide. This can have two consequences. First, the water solubility of the oligonucleotide is reduced, making the modified polymer less capable of forming a duplex structure that is appropriately hydrated. In this case, the modified oligonucleotide binds less well to its complementary sequence than the unmodified oligonucleotide. In contrast, an analogue having a less anionic, neutral, or cationic character is expected to experience less electrostatic repulsion by the phosphate groups on the target sequence. This effect enhances the nonspecific component of binding.

One often encountered but usually undesirable consequence of backbone modification is the introduction of a chiral center at the phosphorus. Although chiral syntheses are being explored, all commonly used synthetic methods yield a mixture of diastereomers having either R_p or S_p configuration. In an oligonucleotide having n backbone linkages, this gives rise to 2^n isomers that vary in their mechanism of action, cellular transport, as well as pharmacokinetics (53). Some of the most frequently encountered backbone modifications are listed in Table 1.

Phosphorothioates. All three synthetic approaches applicable to unmodified oligonucleotides can be adapted for synthesis of phosphorothioates (**11**) (33,46). If all of the phosphodiester linkages in an oligonucleotide are to be replaced with phosphorothioates, the *H*-phosphonate method for coupling, followed by oxidation with S_8 in carbon disulfide and triethylamine in the final step, is the most straightforward method.

The phosphoramidite method may also be used, again by changing the oxidant from I_2 to S_8. Another approach makes use of various sulfur transfer agents to produce phosphorothioate analogues from phosphoramidite building blocks. The phosphoramidite method offers the flexibility of introducing a thiophosphate at various positions within the oligonucleotide. A variation of the

Table 1. Phosphate Analogue Structures[a]

Phosphodiester analogues	Structure numbers	X	Y	Z
phosphorothioate	(**11**)	S^-	O	O
phosphorodithioate	(**12**)	S^-	S	O
methylphosphonate	(**13**)	CH_3	O	O
(difluoromethyl)phosphonate	(**14**)	CHF_2	O	O
(trifluoromethyl)phosphonate	(**15**)	CF_3	O	O
(aminomethyl)phosphonate	(**16**)	$CH_2NH_3^+$	O	O
phosphoramidate	(**17**)	NHR^b	O	O
phosphotriester	(**18**)	OR^b		O
5′-deoxy-5′-methylphosphonate[a]	(**19**)	O^-	O	CH_2

[a]B = B′ = thymidine.
[b]R = H or alkyl groups.

phosphoramidite method, using KSeCN as the oxidizing agent, yields the phosphoroselenide derivative.

Phosphorothioates, among the first modified oligonucleotides to be investigated, are similar to naturally occurring phosphodiesters in that each linkage retains a formal negative charge. The magnitude of the charge is reduced, however, because of the nature of the sulfur atom. As a result, phosphorothioates are less soluble in water and have somewhat lower thermal stability than the phosphodiester homologues. These sulfur-containing oligonucleotides are more, but not completely, resistant to the nucleases that have evolved and which are specific for phosphodiesters. Moreover, these modified nucleotides are transported across the cell membrane, albeit two to five times less well than the unmodified counterparts.

The properties of the thioate derivatives have led to considerable interest in exploring their potential as antisense agents. Indeed, phosphorothioates having C-5 propyne-modified cytodine and uridine are active antisense agents at nanomolar concentrations when microinjected into cells (54). In addition, the incorporation of a single phosphorothioate linkage within an oligonucleotide provides a nucleophilic site that can be alkylated with fluorophores, spin labels, and other reporter groups (55).

Chirality at the phosphorus is an unavoidable problem in all phosphorothioate syntheses. The phosphoramidite method produces a mixture of both the R_p and the S_p diastereomers having a small excess of the R_p isomer (53). Although some progress has been made in the chiral synthesis of dinucleoside phosphorothioates, low yields have limited the utility of these approaches. The chiral center may be eliminated by replacing the other, nonbridging oxygen with sulfur. Avoidance of the chirality problem is one reason for the interest in phosphorodithioates.

Phosphorodithioates.   Phosphorodithioates (**12**) can be obtained by using variations of the phosphoramidite, *H*-phosphonate, and phosphodiester methods, but a modified phosphoramidite method is the most flexible approach (34). Phosphoramidites are replaced by 3′-phosphorothioamidites that yield thiophosphite triesters in the condensation reaction. Using S_8 as the oxidizing agent, phosphorodithioates have been obtained in coupling yields ranging from 96–99%. No other changes in the standard protocol are necessary.

The phosphorodithioates DNA derivatives have been shown to bind specifically to complementary DNA or RNA sequences to form stable adducts. Because they are also highly resistant to degradation by cellular exonucleases, these derivatives can be useful both for applications in research and as therapeutic drugs. Phosphorodithioate DNA has been shown to stimulate Rnase H activity in nuclear cell extracts and is a potent inhibitor of HIV type-1 reverse transcriptase (56).

Another important feature is the reactivity of the phosphorodithioates with alkylating agents. When one phosphorodithioate linkage is incorporated within an oligonucleotide, it can serve as a point of attachment for reporter groups bearing an alkylating function.

Methylphosphonates.   Initially synthesized by using a modified phosphotriester method, methylphosphonates (**13**) are most efficiently made by a modified phosphoramidite approach. Using methylphosphonoamidites as coupling reagents, yields run to about 96–97% (57). Again, the phosphoramidite method affords the flexibility of positioning a methylphosphonate linkage at any point in the oligonucleotide chain. Using the phosphotriester approach, the fluoro-substituted structures (**14**) and (**15**) (58), as well as (aminomethyl)phosphonates (**16**) (59), have been obtained. However, the approach is cumbersome and often involves the incorporation into the oligonucleotide of protected dimers having the desired linkage. Alkylphosphoramidite intermediates are under development (60).

The methylphosphonates differ from the phosphodiesters and phosphorothiolates in that the methyl derivatives are uncharged and are thus less water soluble. Moreover, compared to the naturally occurring phosphodiesters, the methylphosphonates form slightly less stable duplexes with complementary DNA and RNA sequences. This effect has been ascribed to the inevitable chirality problem; that is, if one isomer binds less well, the overall binding is decreased. Methylphosphonates can enter cell membranes by a passive mechanism and are completely resistant to nucleases.

Efforts to devise efficient synthetic approaches to the fluoromethyl derivatives have been undertaken as a way to increase binding by mimicking the electronegativity of the naturally occurring phosphodiester linkage while maintaining the membrane transport enhancement obtained with the methyl derivative (60). The (aminomethyl)phosphonate derivatives, bearing a positive charge, have been shown to form more stable duplexes with their complementary sequences. In addition, these compounds are expected to be transported more efficiently across cell membranes owing to the cationic character (59).

Phosphoramidates and Phosphate Triesters. Even though all three synthetic approaches can be adapted to obtain phosphoramidate intranucleotide linkages (**17**), the *H*-phosphonate and phosphoramidite methods are the most convenient. As in the case of the phosphorothioates, the *H*-phosphonate chemistry provides the most direct route to all-phosphoramidate oligonucleotides. In this approach, the phosphite diester oligonucleotide is synthesized according to the standard procedure and oxidized with primary or secondary amines in the final step to give the desired phosphoramidate. The phosphoramidite method may also be used, but again the oxidation step must be modified (61). When methylphosphite triesters are oxidized with I_2 in the presence of alkylamines, the methyl protecting group is eliminated to give the phosphoramidate.

Phosphate triesters (**18**) are intermediates in both the phosphotriester and phosphoramidite methods, and under appropriate conditions for deprotection of the bases and cleavage of the support, can be obtained directly by using these approaches. The ethyl and isopropyl esters have been obtained directly by using the phosphoramidite method because these are stable during the normal deprotection procedure (62). By changing the oxidizing agent to S_8, both amidate and triester thiolates can be obtained.

Both phosphoramidate and phosphate triester derivatives have been used as linkers to attach reporter groups to oligonucleotides. These derivatives are not entirely resistant to nucleases and they possess a chiral center. They have not been widely investigated as antisense drugs.

5′-Deoxy-5′-Methylphosphonates. Thymidine oligonucleotide analogues, in which all of the 5′-oxygen atoms are replaced by methylene groups, have been prepared by means of a solid support synthesis (63). These derivatives were designed to be similar in geometry to the natural phosphodiester linkage, but are more resistant to nucleases as a result of the chemical modification. In addition, they are achiral at the phosphorus center.

Phosphate-Free Backbones. Several modified oligonucleotides have been reported in which the bridging phosphate group has been replaced with a phosphate-free linker (33,46). These derivatives have been investigated in an effort to develop achiral and nuclease-stable analogues having good hybridization and cellular permeation characteristics. Some criteria for designing these oligomers are structural similarity to the natural phosphodiester backbone, conformational flexibility to allow duplex formation, water solubility, and stability at physiological pH. Because syntheses of these analogues are varied and often challenging, an approach sometimes used is to form a dimer that contains the novel linkage and incorporate the dimer into an oligonucleotide having normal phosphate linkages.

Examples of several novel linkages are shown in Figure 14. Early attempts to synthesize nonphosphate linkages have produced siloxanes (**20**), carbonates (**21**), carboxymethyl esters (**22**), and acetamidates (**23**) (46). All of these derivatives have characteristics, including poor water solubility, stability, and binding to complementary oligonucleotides, that limit biological relevance. The carbamates (**24**), though relatively insoluble, are more stable at physiological pH. Cytosine-containing hexanucleotides incorporating the carbamate linkages have been shown to form stable complexes with complementary DNA (64). In addition, the covalent attachment of a polyethylene glycol group to the 5'-terminus of the oligonucleotide has been effective in increasing water solubility.

Analogues using formacetal (**25**) and thioformacetal (**26**) linkages have also been synthesized, and the pyrimidine oligodeoxynucleosides containing the latter linkage have been shown to have high affinity and specificity for RNA and double-stranded DNA (65). Oligonucleotides incorporating thioethers (**27**) also have been investigated (66). The thioether analogues were synthesized as dimers and incorporated into oligonucleotide dodecamers by using solid-phase synthesis. These oligonucleotides bind specifically to RNA but not to double-stranded DNA. The thermal stability of thioether-containing duplexes are lower than those of the unmodified DNA hybrids.

More recently, analogues having peptide-based linkers have been explored. A thymidine glycine copolymer (**28**) has been synthesized in high yield and forms a stable duplex with its cognate single-stranded partner (46).

Peptide Nucleic Acids. In another approach, the entire deoxyribose phosphate backbone is replaced by an achiral polyamide backbone. These analogues, referred to as peptide nucleic acids (PNAs) (**29**), were based on computer-modeling studies which suggested that an N-(2-aminoethyl)glycine backbone is structurally homomorphous to the deoxyribose phosphate backbone (67). The oligomers are easily constructed by using the Merrifield solid-phase synthesis for peptides. Although PNAs appear to form duplex structures with complementary RNAs or DNAs, their primary binding target is double-stranded DNA, through the formation of a triple helix. Specifically, homopyrimidine PNAs form highly stable (PNA)$_2$–DNA triplexes in which one strand of the DNA duplex is displaced. PNA binding to double-stranded DNA has been shown to inhibit

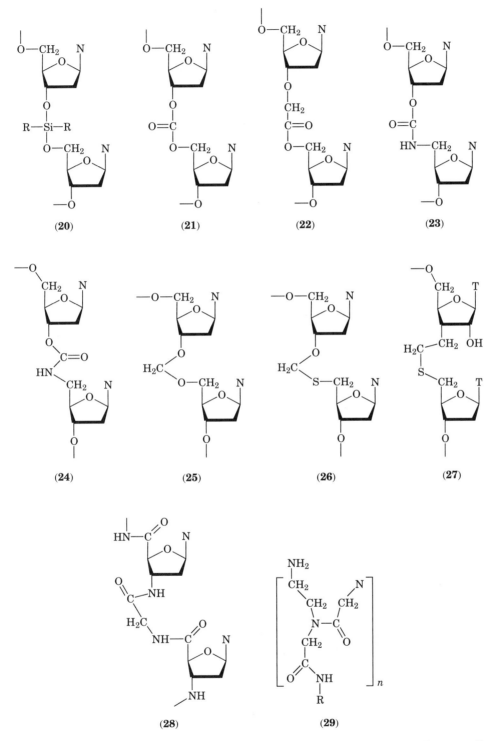

Fig. 14. Nonphosphate backbone-modified oligonucleotides where N is a heterocyclic base and R is an alkyl group. See text.

the association of proteins that recognize the DNA target. In addition, *in vitro* translation experiments indicate that PNA–DNA complexes can block RNA polymerase, as evidenced by the detection of truncated protein products. These results indicate that PNAs are attractive candidates for drugs designed to interfere with gene transcription. Also, these modified analogues have potential for use in the design of triple-helix-forming antigene oligonucleotides that recognize sequences including, but not limited to, homopurine/homopyrimidines (68).

Sugar Modifications. The 2′-*O*-alkylribonucleotides are a group of oligonucleotide analogues that have potential as antisense drugs and probes (69). These derivatives are easily obtained by solid-phase synthesis using the phosphoramidite method. Oligo(2′-*O*-alkylribonucleotides) having short, unbranched alkyl groups are resistant to nucleases, chemically stable, and capable of hybridizing with complementary RNA to form duplexes of greater stability than unmodified DNA or RNA oligonucleotides. In contrast to the 2′-alkyl-nucleosides, the 2′-amido-nucleotides form neither specific nor stable duplexes with either RNA or DNA (70).

The α-anomeric form of a 2′-deoxyribose, which has the base inverted with respect to the natural β-anomeric form, can be synthesized by using the phosphoramidite method; sugar modification renders the derivatives nuclease-resistant. These analogues form parallel duplexes with complementary RNA targets. A 12-mer α-anomeric oligonucleotide has been shown to inhibit de novo HIV infection by a nonspecific rather than an antisense mechanism (71). In addition, α-oligonucleotides form triple helices with double-stranded DNA, which are slightly less stable than those formed by the natural β-oligomers (51). The binding can be improved by attaching intercalating agents to the ends of the oligonucleotide making these compounds potential regulators of gene transcription.

Oligonucleotide Bioconjugates

Although many molecules, including proteins and small intercalating and groove-binding ligands, bind to RNA and DNA, only nucleic acids are able to bind with the high specificity required to recognize a single sequence within the 3×10^9 base-pairs of the human genome. The unique specificity of oligonucleotides can be exploited to direct a multitude of other chemical agents to a sequence of interest by attaching these agents to oligonucleotides through molecular linkers. Numerous examples of these modified oligonucleotides, or bioconjugates, have been reported, and have found applications in many areas (33,44,51,52). For example, intercalating, cross-linking, alkylating, and DNA-cleaving groups can be directed to unique sequences by attaching these moieties to the appropriate oligonucleotides. Oligonucleotides labeled with fluorescent or other detectable groups provide nucleic acid probes that can be used to screen a large pool of DNA for a specific sequence. Cationic or lipophilic groups can be attached to improve the binding and bioavailability of antisense oligonucleotides. Bioconjugates have also been widely used in research because they enable scientists to learn more about the structure and function of nucleic acids and ligands that bind to them.

Several strategies have been devised to attach various chemical groups to oligonucleotides. Groups can be attached to the 3′- or 5′-terminus of the oligonucleotide, along the backbone through the phosphate or the 2′-hydroxy group of ribose, or to modified purines or pyrimidines (45).

Incorporation of Conjugate Groups into Oligonucleotides. A phosphorothioate incorporated at any point within an oligonucleotide reacts with alkylhalides to provide an easy method for attaching a chemical group. The alkylhalides generally used for this purpose require a linker region, L, attached to the molecule of interest. Examples of alkyl halides used to attach o-phenanthroline (**30**), psoralen (**31**), and ellipticine (**32**) are provided in Figure 15 (51). Similarly, nucleophilic reaction of a binuclear platinum complex (**33**) having a phosphorothiolate group has yielded an oligonucleotide conjugate able to react with a guanine base.

Many bioconjugates can be obtained from phosphoramidite derivatives (45). In this approach, the phosphoramidite derivative of the conjugate group reacts with the 5′-end of the oligonucleotide. Among the numerous phosphoramidites that have been used to functionalize oligonucleotides are derivatives incorporating biotin (**34**), fluorescein (**35**), lipids (**36**), and cholesterol (**37**). Nonnucleotide phosphoramidites are also available which may be extended by following the reaction with a growing oligonucleotide chain and allowing the incorporation of a conjugate group at various positions within oligonucleotide.

Applications of Bioconjugates. Reactive chemical groups attached to oligonucleotides can be used to affect sequence-specific modifications of DNA. For example, both copper–phenanthroline (72) and iron-EDTA (73) have been linked to triple-helix-forming oligonucleotides. These groups can be activated by using either thiol or hydrogen peroxide, respectively, to generate a hydroxyl radical in situ. The hydroxyl radical, a diffusible reactive species, is capable of cleaving the phosphodiester DNA backbone. Thus, when bound to a specific DNA or RNA target sequence, these reagents can act as artificial endonucleases. Ellipticine (**32**) can be activated photochemically to yield another reactive group capable of DNA cleavage.

Another synthetic approach to a site-specific DNA- or RNA-cleavage reagent is to chemically link a naturally occurring nonspecific nuclease to an oligonucleotide. The first example of this type of artificial endonuclease is an oligonucleotide conjugate with staphylococcal nuclease. This adduct has been shown to cleave target DNA in a 75% yield over the nonspecific sites (74). Rnase H has also been linked to oligonucleotides in an effort to obtain a site-specific RNA nuclease (75).

Modified oligonucleotides can be used to cross-link DNA sequences via a reactive group tethered to an oligonucleotide. When irradiated with uv light, psoralens (**31**) reacts with thymine bases, and the reaction yields a cross-link if the thymine residues are adjacent to each other on opposite strands. Psoralen linked to oligonucleotides have been shown to induce site-specific cross-links in vitro (51).

Alkylating agents, such as N-2-chloroethyl-N-methylaminobenzylamide, tethered to oligonucleotides can link oligonucleotides and target DNA sequences. Also, the binuclear platinum oligonucleotide conjugate (**33**) can alkylate DNA through a reaction with guanine heterocyclic nitrogens (76). cis-Platinum, a similar platinum complex, is an important drug used in cancer treatment.

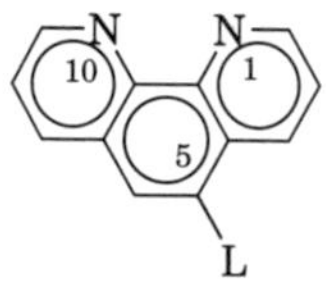

L =—NHCO(CH$_2$)$_n$Br

(30)

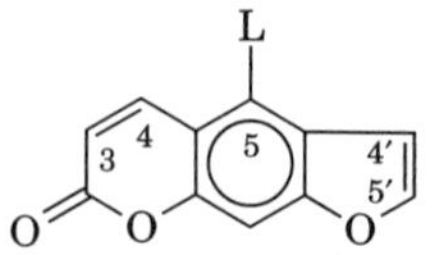

L =—O(CH$_2$)$_2$I

(31)

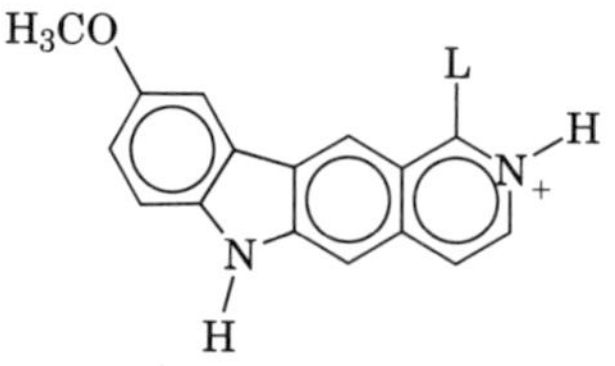

L =—NH(CH$_2$)$_3$NHCO(CH$_2$)$_n$Br

(32)

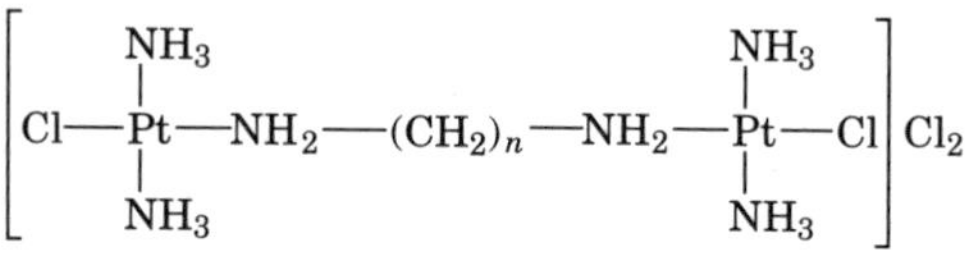

(33)

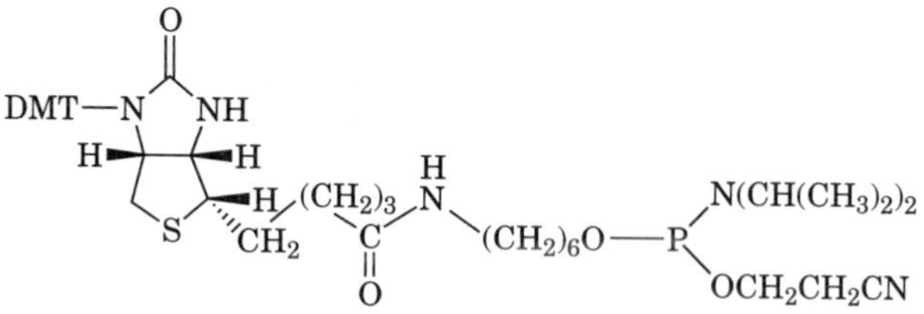

(34)

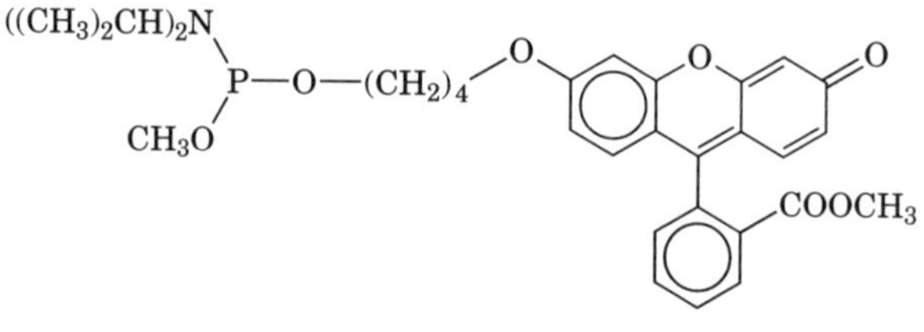

(35)

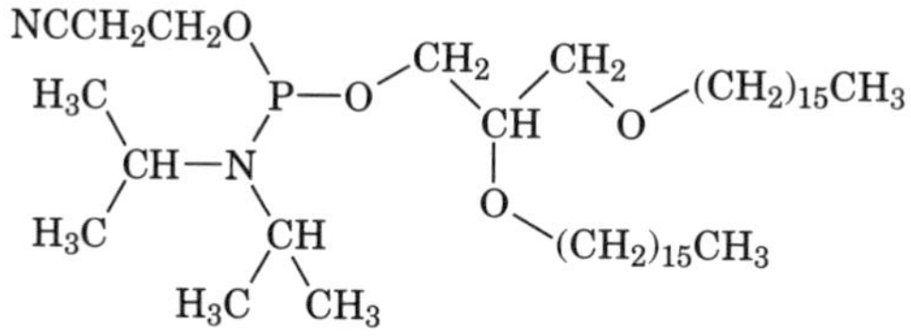

(36)

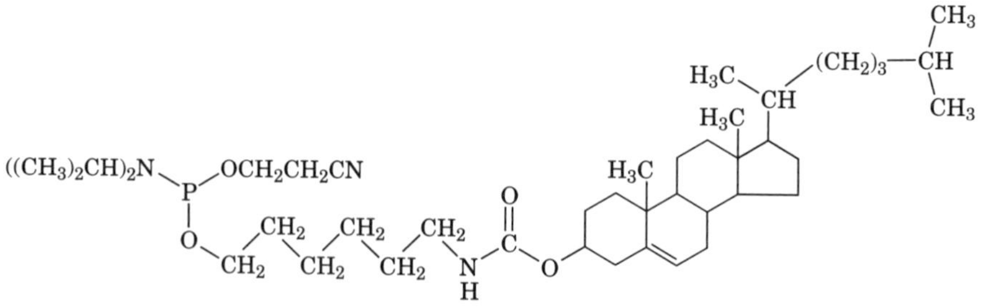

(37)

Fig. 15. Structures of conjugate groups linked to oligonucleotides. DMT = dimethoxytrityl.

Several fluorescent dyes, including fluorescein (**35**), have been conjugated to oligonucleotides to generate hybridization probes. When bound to target nucleic acid sequences, the fluorescent labels permit the visualization of the hybrid. For example, electron microscopy can be used to determine the location of nucleic acid sequences bound to the probe among various compartments in a cell (45,52) (see MICROSCOPY). Fluorescent probes can also be used to detect fragments in DNA-sequencing and finger-printing methods. The biotinylated (**34**) conjugates form highly colored chromophores in the presence of the enzyme avidin (or streptavidin), and can be used in place of the fluorescent groups in the applications described above.

Many dye–oligonucleotide conjugates have therapeutic uses. The dye acts as a photosensitizer able to generate singlet oxygen photochemically. Singlet oxygen causes damage to the DNA target by a mechanism that results in the cleavage of the phosphodiester backbone. Compounds of this type are used in photodynamic treatment of tumors and, as of this writing, are being evaluated in clinical trials (77).

In a related application, fluorescent donor (5-carboxyfluorescein) and acceptor (5-carboxytetramethylrhodamine) groups were attached to the 3'- and 5'-ends of several model oligonucleotides representative of the hammerhead ribozyme, a catalytic oligoribonucleotide, in order to characterize its three-dimensional structure. By studying the efficiency of fluorescence resonance energy transfer (fret) between the donor and acceptor groups, distance information can be extracted which defines the positions of the donor and acceptor groups. These experiments provide information regarding the geometry of the ribozyme, a Y-shaped substance made up of three double-stranded helical regions (78).

Oligonucleotides have also been modified with various groups in an effort to expedite their transport across cell membranes. Conjugation with lipids (**36**) and cholesterol (**37**) has been shown to increase intracellular concentrations of conjugated oligonucleotides as well as inhibition of gene expression (79,80). Novel oligonucleotide–peptide conjugates have been prepared, in which antisense oligonucleotides complementary to an HIV-1 nucleic acid sequence were joined to peptides at the 3'-end. The peptides used in the experiment were derived from HIV transmembrane glycoprotein gp41, which can serve as a carrier for facilitating membrane-transport of the antisense oligonucleotide (81,82).

BIBLIOGRAPHY

1. W. Saenger, *Principles of Nucleic Acid Structure*, Springer-Verlag, New York, 1984.
2. R. L. P. Adams, J. T. Knowler, and D. P. Leader, *The Biochemistry of Nucleic Acids*, 10th ed., Chapman and Hall, New York, 1986.
3. R. F. Gesteland and J. F. Atkins, eds., *The RNA World*, Cold Spring Harbor Laboratory Press, Cold Spring Harbor, New York, 1993.
4. J. D. Watson and F. H. Crick, *Nature* **171**, 737 (1953).
5. R. E. Dickerson, in D. M. J. Lilley and J. E. Dahlberg, eds., *Methods in Enzymology*, Vol. 211, Academic Press, Inc., San Diego, Calif., 1992, pp. 67–111.
6. J. Feigon and co-workers, in Ref. 5, pp. 235–253.
7. D. S. Goodsell and co-workers, *Proc. Natl. Acad. Sci. USA* **90**, 2930 (1993).
8. T. E. Haran, J. D. Kahn, and D. M. Crothers, *J. Mol. Biol.* **244**, 135 (1994).

9. R. E. Harrington and I. Winicov, in W. E. Cohen and K. Moldave, eds., *Progress in Nucleic Acids Research and Molecular Biology*, Vol. 47, Academic Press, Inc., San Diego, Calif., 1994, pp. 195–270.

10. J. K. Strauss and L. Maher III, *Science* **266**, 1829 (1994).

11. B. H. Johnston, in D. M. J. Lilley and J. E. Dahlberg, eds., *Methods in Enzymology*, Vol. 211, Academic Press, Inc., San Diego, Calif., 1992, pp. 127–158.

12. N. T. Thuong and C. Helene, *Angew. Chem. Int. Ed. Engl.* **32**, 666 (1993); M. D. Frank-Kamenetskii and S. M. Mirkin, *Ann. Rev. Biochem.* **64**, 65 (1995).

13. R. Bowater, F. Aboul-Ela, and D. M. J. Lilley, in D. M. J. Lilley and J. E. Dahlberg, eds., *Methods in Enzymology*, Vol. 212, Academic Press, Inc., San Diego, Calif., 1992, pp. 105–120.

14. W. I. Sundquist, in F. Eckstein and D. M. J. Lilley, eds., *Nucleic Acids and Molecular Biology*, Vol. 5, Springer-Verlag, Berlin, 1991, pp. 1–24.

15. A. D. Bates and A. Maxwell, *DNA Topology*, Oxford University Press, Oxford, 1993; N. R. Cozarelli and J. C. Wang, eds., *DNA Topology and its Biological Effects*, Cold Spring Harbor Laboratory Press, Cold Spring Harbor, 1990.

16. K. E. van Holde, *Chromatin*, Springer-Verlag, New York, 1989.

17. S. D. Levine, in F. Eckstein and D. M. J. Lilley, eds., *Nucleic Acids and Molecular Biology*, Vol. 8, Springer-Verlag, Berlin, 1994, pp. 119–132.

18. D. C. Chen, R. P. Bowater, and D. M. J. Lilley, in Ref. 17, pp. 147–166.

19. N. D. F. Grindley, in Ref. 17, pp. 236–267; R. Schleif, *Annu. Rev. Biochem.* **61**, 199 (1992).

20. D. Soll, in Ref. 3, pp. 157–175.

21. A. M. Pyle, *Science* **261**, 709 (1993).

22. J. R. Wyatt and I. Tinoco, in Ref. 3, pp. 465–496.

23. M. Zacharias and P. J. Hagerman, *J. Mol. Biol.* **247**, 486 (1995).

24. H. W. Pley, K. M. Flaherty, and D. B. McKay, *Nature* **372**, 68 (1994).

25. J. P. Marino and co-workers, *Science* **268**, 1448 (1995).

26. M. J. McPerson, P. Quirke, and G. R. Taylor, *PCR, A Practical Approach*, Oxford University Press, Oxford, 1991.

27. J. Sambrook, E. F. Fritsch, and T. Maniatis, *Molecular Cloning: A Laboratory Manual*, Cold Spring Harbor Laboratory Press, Cold Spring Harbor, New York, 1989, 13.7.

28. J. C. Murray and co-workers, *Science*, **265**, 2049 (1994).

29. K. Itakura, J. J. Rossi, and R. B. Wallace, *Annu. Rev. Biochem.* **53**, 323 (1984).

30. D. Shortle, D. DiMaio, and D. Nathans, *Annu. Rev. Genet.* **15**, 265 (1981).

31. O. Kennard and W. N. Hunter, *Q. Rev. Biophys.* **22**, 327 (1989).

32. S. C. Harrison and A. K. Aggarwal, *Annu. Rev. Biochem.* **59**, 933 (1990).

33. E. Uhlmann and A. Peyman, *Chemical Reviews* **90**, 543 (1990).

34. M. H. Caruthers and co-workers, *Methods in Enzymology*, Vol. 154, Academic Press, Inc., New York, 1987, pp. 287–313; M. H. Caruthers and co-workers, *Methods in Enzymology*, Vol. 211, Academic Press, Inc., New York, 1992, pp. 3–20.

35. S. J. Horvath and co-workers, *Methods in Enzymology*, Vol. 154, Academic Press, Inc., New York, 1987, pp. 314–326.

36. L. E. Sindelar and J. M. Jaklevic, *Nucleic Acids Res.* **23**, 982 (1995).

37. T. Brown and D. J. S. Brown, *Methods in Enzymology*, Vol. 211, Academic Press, Inc., New York, 1992, pp. 20–35.

38. A. M. Maxam and W. Gilbert, *Methods in Enzymology*, Vol. 65, Academic Press, Inc., New York, 1980, p. 499.

39. B. C. Froehler and M. D. Matteucci, *Tetrahedron Lett.* **27**, 469 (1986).

40. E. Sonveaux, *Bioorg. Chem.* **14**, 274 (1986).

41. C. B. Reese and co-workers, in F. Eckstein and D. M. J. Lilley, eds., *Nucleic Acids and Molecular Biology*, Springer-Verlag, Berlin, 1989, pp. 164–180.

42. D. C. Cipaldi and C. B. Reese, *Nucleic Acids Res.* **22**, 2209 (1994).
43. J. F. Milligan and O. C. Uhlenbeck, *Methods in Enzymology*, Vol. 180, Academic Press, Inc., New York, 1989, pp. 51–62.
44. J. A. H. Murray, ed., *Antisense RNA and DNA*, Wiley-Liss, Inc., New York, 1992; U. Englisch and D. H. Gauss, *Angew. Chem. Int. Ed. Engl.* **30**, 613 (1991).
45. S. L. Beaucage and R. P. Iyer, *Tetrahedron* **49**, 1925 (1993).
46. R. S. Varma, *Synlett.*, 621 (1993).
47. J. F. Milligan, M. D. Matteucci, and J. C. Martin, *J. Med. Chem.* **36**, 1993 (1993).
48. R. W. Simmons and N. Klecker, *Annu. Rev. Genet.* **22**, 567 (1988).
49. C. Boiziau and co-workers, *Nucleic Acids Res.* **19**, 1113 (1991).
50. M. A. Bonham and co-workers, *Nucleic Acids Res.* **23**, 1197 (1995).
51. N. T. Thuong and C. Helene, *Angew. Chem. Int. Ed. Engl.* **32**, 666 (1993).
52. J. L. McInnes and R. H. Symons, in R. H. Symons, ed., *Nucleic Acid Probes*, CRC Press, Inc., Boca Raton, Fla., 1989, pp. 33–80.
53. A. Wilk and W. J. Stec, *Nucleic Acids Res.* **23**, 530 (1995).
54. C. A. Stein and Y. C. Cheng, *Science* **261**, 1004 (1993).
55. J. A. Fidanza, H. Ozaki, and L. W. MaLaughlin, *J. Amer. Chem. Soc.* **114**, 5509 (1992).
56. W. S. Marshall and M. H. Caruthers, *Science* **259**, 1564 (1993).
57. T. Loschner and W. J. Engels, *Nucleosides Nucleotides* **7**, 729 (1988).
58. G. M. Blackburn and M.-J. Guo, *Tetrahedron Lett.* **34**, 149 (1993).
59. R. Fathi and co-workers, *Bioconjugate Chem.* **5**, 47 (1994).
60. M. Mayer, I. Ugi, and W. Richter, *Tetrahedron Lett.* **36**, 2047 (1995).
61. A. Jager, M. Levy, and S. M. Hecht, *Biochemistry* **27**, 7237 (1988).
62. P. Guga and co-workers, *Nucleosides Nucleotides* **6**, 111 (1987).
63. T. Szabo, A. Kers, and J. Stawinski, *Nucleic Acids Res.* **23**, 893 (1995).
64. E. P. Stirchak, J. E. Summerton, and D. D. Weller, *Nucleic Acids Res.* **17**, 6129 (1989).
65. R. J. Jones and co-workers, *J. Org. Chem.* **58**, 2983 (1993).
66. B. Meng and co-workers, *Angew. Chem. Int. Ed. Engl.* **32**, 729 (1993).
67. P. E. Nielsen, M. Egholm, and O. Buchardt, *Bioconjugate Chem.* **5**, 3 (1994).
68. M. Egholm and co-workers, *Nucleic Acids Res.* **23**, 217 (1995).
69. A. I. Lamond and B. S. Sproat, *FEBS* **325**, 123 (1993).
70. C. Hendrix and co-workers, *Nucleic Acids Res.* **23**, 51 (1995).
71. F. Debart and co-workers, *Nucleic Acids Res.* **20**, 1193 (1992).
72. J. C. Francois and co-workers, *J. Biol. Chem.* **264**, 5891 (1989).
73. P. A. Beal and P. B. Dervan, *Science* **251**, 1360 (1991); S. A. Strobel and P. B. Dervan, *Science* **249**, 73 (1991).
74. D. Pei, D. R. Corey, and P. G. Schultz, *Proc. Natl. Acad. Sci. USA* **87**, 9858 (1990).
75. Y. Uchiyama and co-workers, *Bioconjugate Chem.* **5**, 327 (1994).
76. E. S. Gruff and L. E. Orgel, *Nucleic Acids Res.* **19**, 6849 (1991).
77. U. Moller, F. Schubert, and D. Cech, *Bioconjugate Chem.* **6**, 174 (1995).
78. T. Tuschl and co-workers, *Science* **266**, 785 (1994).
79. A. M. Krieg and co-workers, *Proc. Natl. Acad. Sci. USA* **90**, 9858 (1993).
80. C. MacKellar and co-workers, *Nucleic Acids Res.* **20**, 3411 (1992).
81. S. Soukchareun, G. Tregear, and J. Haralambidis, *Bioconjugate Chem.* **6**, 43 (1995).
82. R. W. Sobol and co-workers, *J. Biol. Chem.* **270**(11), 5963–5978 (Mar. 17, 1995).

JILL REHMANN
Fordham University

NUCLEOSIDE ANTIBIOTICS.   See ANTIBIOTICS, NUCLEOSIDES AND NUCLEOTIDES.

NUTS

Nuts, as generally defined, are hard-shelled seeds enclosing a single edible oily kernel. Most of the common nuts fall within this classification. Many species, however, differ greatly in size, structure, shape, composition, and flavor. A more technical definition of a nut is a hard, one-seeded fruit developed from many carpels united to form a compound ovary, the woody pericarp of the shell remaining closed at maturity. Examples are the beechnut, butternut, chestnut, hickory nut, pecan, and walnut, which are all true nuts. The word nut, moreover, is applied indiscriminately to many seeds, fruits, and tubers having hard coverings, as well as to single oily or starchy kernels such as those of the almond, Brazil nut, cashew, chufa, coconut, lichee, peanut, piñon, and pistachio, which do not fit the more technical definition. Although in common parlance, the term nut implies edibility (1), nuts of many kinds are grown not only for food but also as sources of oil and for medicinal, ornamental, and other uses. Table 1 lists many nuts from various parts of the world.

With the exception of peanuts, most of the important nuts from around the world are borne on trees, many of them from native seedlings. Among the latter group are the beechnut, Brazil nut, butternut, chestnut, filbert, hickory nut, pecan, pine nut, and black walnut. The pecan, English walnut, filbert, and almond are the four principal edible tree nuts produced in the United States, where the term English walnut is used synonymously with the Persian or Carpathian walnut (2).

Physical Characteristics

Fruits and nuts vary considerably in structure. The kernel may be single and firm as in acorn, almond, chestnut, coconut, filbert, macadamia, piñon, pistachio, and tung; or it may be formed in distinct halves as in the hickory, pecan, butternut, peanut, and walnut; it may also be clustered within an outer husk as in the Brazil nut; or, oddly, formed outside of the fleshy part of the fruit as in the cashew. The shuck or husk covering the nut may be an extraneous growth such as the acorn cup of the oak or the spiny burr of the chestnut, a husk or fringe as in the hazel, or a prickly involucre as in the beechnut. The shuck of the pecan or hickory nut has definite sutures and splits into four parts when matured; shucks of the walnut and macadamia, however, remain entire throughout their growth. Brazil and paradise nuts are true seeds, borne in a hollow dry fruit that contains a number of the triangular nuts. The covering of the almond, ginkgo, and lichee is the dried pulp of the fruit. The shells encasing the kernels are usually hard cellular structures of varying thickness. Some are cracked easily, eg, English walnut, improved pecan, almond, chestnut, filbert, peanut, and pistachio; others can be broken only by a sharp blow, eg, eastern black walnut, hickory nut, Brazil nut, coconut, macadamia, and cashew. The cashew nut has a double outer shell containing an oily acrid fluid of a phenolic nature which has important industrial uses. The general shape of most nuts is globular, although some of these, such as the Brazil nut, pili, and beechnut, are triangular; other nuts are elongated or flattened like almond and pistachio nut, or comma-shaped like cashew nut, or curiously formed, like buffalo horns, for instance, in the case of water chestnut.

Table 1. Botanical Classification, Occurrence, and Uses of Nuts Worldwide

Common name	Botanical name	Principal geographic occurrence	Uses
Edible tree nuts			
acorn			
oak	*Quercus spp*	Europe, North America	feed for animals animals
edible	*Pasania cornea*	China	food
almond			
sweet	*Prunus amygdalus*	southern Europe, North America	food and oil
bitter	*Prunus amygdalus*	southern Europe	food and oil
beechnut			
American	*Fagus grandifolia*	North America	food and oil
European	*Fagus sylvatica*	Europe	food and oil
Brazil (pará)	*Bertholletia excelsa*	South America	food and oil
bread	*Brosimum alicastrum*	West Indies	food
buriti	*Mauritia flexuosa*	tropical America	food and oil
butternut	*Juglans cinerea*	North America	food
cashew	*Anacardium occidentale*	India, Africa	food and oil
chestnut			
American	*Castanea dentata*	North America	food
European	*Castanea sativa*	Europe	food
Chinese	*Castanea mollissima*	China, North America	food
Japanese	*Castanea crenata*	Japan	food
chinquapin	*Castanea pumila*	North America	food
coconut	*Cocos nucifera*	tropics	food and oil
filbert, hazel, cob	*Corylus avellana*	Europe, North America	food
hazel	*Corylus americana*	North America	food
hickory	*Carya* spp.	North America	food and oil
jojoba	*Simmondsia californica*	southwest U.S.	food
Java almond (Kanari)	*Canarium commune*	Sunda Isles	food and oil
lunau	*Otophora fruticosa*	Philippines	food
Luzon nut	*Canarium buconicum*	Pacific islands	food
macadamia	*Macadamia ternifolia*	Australia, Hawaii	food
palm	*Elaeis guineensis*	tropical Africa	oil
paradise	*Lecythis ollaria*	South America	food and oil
pecan	*Carya illinoensis*	North America	food and oil
pignolia	*Pinus spp*	southern Europe	food
pine, araucarian	*Araucaria angustifolia*	Chile	food
piñon (pine)	*Pinus edulis and other spp*	southwest U.S.	food and oil
pistachio	*Pistacia vera*	southern Europe, U.S.	food
torreya (kaya)	*Torreya nucifera*	China, Japan	food
tropical almond	*Terminalia catappa*	southern Asia	food and tannin
walnut			
eastern, black	*Juglans nigra*	eastern North America	food and oil
northern, California, black	*Juglans hindsii*	northern California	food and oil
Japanese	*Juglans sieboldiana*	Asia	food
Persian (English)	*Juglans regia*	southern Europe, North America	food and oil
Edible ground and aquatic nuts			
ar or earth	*Bunium flexuosum*	western Europe	food
bambarra groundnut	*Voandzeia subterranea*	tropics, Africa	food

Table 1. (*Continued*)

Common name	Botanical name	Principal geographic occurrence	Uses
chufa (ground almond)	*Cyperus esculentus*	southern Europe, North America	feed for animals
fox	*Euryale ferox*	eastern India	food and medicine
ground	*Apios americana*	North America	food
ground	*Panax trifolium*	North America	food
hawk	*Bumium bulbocastanum*	western Europe	food
horned chestnut			
China	*Trapa bicornis*	Asia	food
Europe	*Trapa natans*	southern Europe	food
peanut	*Arachis hypogaea*	tropics, Asia, U.S.	food and oil
singhara	*Trapa bispinosa*	Asia, Africa	food
water chestnut	*Eleocharis dulcis*	China	food and starch
water chinquapin	*Nelumbo lutea*	North America	food

Nonedible tree nuts

ben	*Moringa oleifera*	tropical Asia	oil for perfume
bladder	*Staphylea trifolia*	U.S.	ornamental
bomah	*Pycnocoma macrophylla*	Africa	tanning
bonduc	*Guilandina bonduc*	India	medicine and beads
clearing	*Strychnos potatorum*	India	clearing water
coquilla	*Attalea funifera*	South America	turnery and buttons
cumara	*Dipteryx odorata*	South America	perfume
horse chestnut	*Aesculus hippocastanum*	Europe, U.S.	starch
ivory (tagua)	*Phytelephas seemanni*	Central America	turnery and buttons
manketti	*Ricinodendron rautanenii*	southwest Africa	oil for soap and paint
marking	*Semecarpus anacardium*	Asia, India	ink and varnish
murumuru	*Astocaryum murumuru*	Brazil	oil
oiticica	*Licania rigida*	Brazil	paint
physic	*Jatropha curcas*	tropical America	medicine
poison	*Strychnos nux-vomica*	India, Australia	medicine
portia	*Thespesia populnea*	tropics	oil and illuminant
snake	*Ophiocaryon paradoxum*	Guyana	curiosity
soap	*Sapindus saponaria*	tropics	soap
soft lumbang	*Aleurites triloba*	South Sea islands	paint and varnish
tallow	*Sapium sebiferum*	China	substitute for tallow
tucum	*Astrocaryum rulgare*	Brazil	oil
tung (China wood)	*Aleurites fordii*	China or southern U.S.	paint and varnish

Nuts with secondary or supplemental food uses

babassu	*Orbignya martiana*	Brazil	oil for soap, margarine shortening
betel	*Areca cathecu*	South Sea islands	chewing
candlenut (kukui)	*Aleurites moluccana*	Hawaii and the Philippines	food and drying oil
hyphaene	*Hyphaene crinita*	Malagasy Republic	spice
kola	*Kola acuminata*	western Africa	stimulant in beverage
nutmeg	*Myrustica fragrans*	East Indies	spice
oyster	*Telfairia pedata*	eastern Africa	food
ravensara	*Agathophyllum aromaticum*	Malagasy Republic	spice
sunflower	*Helianthus annuus*	Europe, U.S.	food
watermelon	*Citrullus vulgatis*	Africa, Asia, U.S.	food

Chemical Composition

Most nuts for commercial use are characterized by high oil and protein contents (see PROTEINS) as well as a low percentage of carbohydrates (qv). However, some varieties, mostly inedible tree nuts such as acorn, horse chestnut, and chufa, contain at least as much sugar and/or starch as protein. The edible water chestnut is also in this category, as is the cashew nut, which contains starch in addition to a rich store of oil. The proximate composition of a number of nuts and of some nut products are given in Table 2 (3).

The edible portion of the nut ranges from as much as 73% by weight in the peanut to less than 15% in the butternut. The double shell of the cashew nut comprises about 70% of the nut. The unhusked coconut is roughly 43% fibrous husk, 15% shell, 30% kernel, and 12% milk; in contrast, the husked coconut is 27% shell, 55% kernel, and 18% milk. Moreover, the kernels of most oily nuts contain about 3% moisture; those of starchy nuts contain 50% or more water. The protein content of nut kernels can range from about 26% in the peanut and pignolia to 2.5% or less in the Chinese chestnut and lunau. A number of the most common edible nuts contain about 60% oil or fat. Pecan, pili, and macadamia, for instance, are all rich in fatty constituents and contain over 70% fat. In general, nuts are rather low in carbohydrates, especially starches. The nitrogen-free extract usually includes all constituents (mostly carbohydrates) not present in the ether extract (fats), protein (N $\times$ 6.25), crude fiber, and ash. Ash constituents are usually less than 3% in the kernels of the nuts and consist largely of the oxides of potassium, calcium, magnesium, sodium, iron, sulfur, silicon, and phosphorus; they also contain smaller amounts of the mineral forms of manganese, copper, and zinc.

In the following discussion of chemical constituents, unless reported as approximate, the percentages given are representative rather than absolute, since they are based on the analysis of a limited number of samples. All results are reported on a moisture-free basis.

Proteins.   Most edible nuts contain a high protein content; they also supply adequate amounts of various essential amino acids (qv) for growth and the maintenance of body tissues. The main proteins present in the following nut kernels are different types of globulins: glutelin (acorn), amandin (almond), excelsin (Brazil nut), castanin (chestnut), edestin (coconut), corylin (hazel nut and filbert), arachin and conarachin (peanut), and juglansin (butternut and walnut). Hydrolysis products, such as amino acids and ammonia, of some of these globulins are given in Table 3 (4,5). Ginkgoin, the glubolin of ginkgo, contains 60% of the seed's nitrogen and a large percentage of tryptophan; other proteins in the ginkgo kernel are albumin, glutelin, and prolamine. Feeding experiments indicate that arachin is deficient in tryptophan, in methionine, and possibly also in isoleucine (7). Conarachin, on the other hand, is an excellent protein for growth when fed as the only source of protein (8,9).

The average composition of several nut globulins is about 51% carbon, 6.9% hydrogen, 19.0% nitrogen, 0.8% sulfur, and 22.0% oxygen; total nitrogen in the globulin ranges from 18.4% in Brazil nut excelsin to 1.0% in almond and filbert globulins. From nitrogen distribution studies of several nut globulins, juglansin from English walnut was found to be the lowest in monoamino nitrogen

Table 2. Composition of Nuts[a]

Name	Refuse, wt %	Water, wt %	Protein, g	Fat, g	Total carbohy-drate, g	Fiber, g	Ash, g	Calcium, mg	Phosphor-us, mg
acorns, raw		27.9		23.9	40.8	2.6	1.4	41.0	79
almond	49								
dried, blanched		5.4	20.4	52.5	18.5	2.3	3.1	2.5	532
dry roasted, unblanched		3.0	16.3	51.6	24.2	4.9	4.9	282.0	548
meal, partially defatted		7.2	39.5	18.3	28.9	2.3	6.1	424.0	914
beechnut, dried	39	6.6	6.2	50.0	33.5	3.7	3.7	1.0	
Brazil nut, dried unbleached	50	3.3	14.3	66.2	12.8	2.3	3.3	176.0	600
butternuts, dried		3.3	2.5	57.0	12.1	1.9	2.7	53.0	446
cashew nut, dry roasted		1.7	15.3	46.4	32.7	0.7	4.0	45.0	490
chestnut, European	19								
fresh, raw, peeled		52.0	1.6	1.3	44.2	1.0	1.0	19.0	38
dried, peeled		9.0	5.0	3.9	78.4	5.0	3.6	64.0	137
coconut									
cream, expressed liquid		53.9	3.6	34.7	6.6		1.1	11.0	122
meat, fresh		47.0	3.3	33.5	15.2	4.3	1.0	14.0	113
meat, dried and unsweetened		3.0	6.9	64.5	23.7	5.3	1.9	26.0	206
meat, dried, sweetened, and shredded		12.6	2.9	35.5	47.7	2.2	1.4	15.0	107
milk, expressed		67.6	229.0	23.8	5.5		0.7	16.0	100
hazelnut (filberts) dried, unblanched	53	5.4	13.0	62.6	15.3	3.8	3.6	188.0	312
hickory nut	80	2.7	12.7	64.4	18.3	3.2	2.0	61.0	336
macadamia nut, dried	69	1.7	7.3	76.5	12.9	1.7	1.7	45.0	200
peanut									
kernels, dried		6.7	25.7	49.2	16.2	4.9	2.3	58.0	383
butter, added salt		13.0	28.5	51.1	15.8	3.3	3.3	33.0	374
flour, defatted		7.8	52.2	0.6	34.7	4.1	4.8	140.0	1290
salted, oil-cooked		2.0	26.8	49.2	18.5	2.4	3.6	18.5	506
pecans, dried	56	4.8	7.8	67.6	18.2	1.6	1.6	36.0	291
pilinut		0.8	10.8	79.6	4.0	2.8	2.9	145.0	575
pinenut									
pignolia		6.7	24.0	50.7	14.2	0.8	4.4	26.0	508
piñon		5.9	11.6	61.0	19.3	4.7	2.3	8.0	35
pistachio nut, dried	70	3.9	20.6	43.4	24.8	1.9	2.4	135.0	503
walnuts, black	78	4.4	24.4	56.6	12.1	6.5	2.6	58.0	464
walnuts, English or Persian	55	3.7	14.3	61.9	18.3	4.6	1.9	94.0	317

[a]100 g portions are used for the calculations; Ref. 3.
[b]To convert J to cal, divide by 4.184.

(amino nitrogen in the alkali filtrate) but highest in lysine and arginine nitrogen. Protein precipitated from the alkali extract of pecan flour (10) contains notably higher arginine nitrogen and lower histidine nitrogen than that from the ground pecan kernel without previous extraction to remove the oil (11). The closest similarity between proteins from seeds of unrelated plants is presented by corylin from filbert and juglansin from Persian walnut, which are distinguished only by a slight difference in the nitrogen content (see SOYBEANS AND OTHER OILSEEDS.)

Table 2. (*Continued*)

Iron, mg	Sodium, mg	Potassium, mg	Magnesium, mg	Vitamin A, IU	Thiamine, mg	Riboflavin, mg	Niacin, mg	Ascorbic acid, mg	Fuel value, MJ[b]
	0	539	62		0.11	0.12	1.80	0.00	1.54
3.7	10	750	286		0.16	0.68	3.20	0.60	2.45
3.8	780	770		0	0.13	0.60	2.80	present	2.46
8.5	7	1400		0	0.32	1.68	6.30	present	1.71
									2.41
	2	600	225		1.00	0.12	1.60	0.70	2.74
4.0	1	446	237						2.56
6.0	16	565		0	0.20	0.20	1.40	0.00	2.41
0.9	2	484	30		0.14	0.02			0.82
2.4	37	991			0.35	0.05			1.55
2.3	4	325		0	0.03	0.00	0.90	2.80	1.38
2.4	20	356		0	0.07	0.02	0.50	3.30	1.48
3.3	37	543	90	0	0.06	0.10	0.60	1.50	2.76
1.9	262	337	50	0	0.03	0.02	0.50	0.70	2.10
1.6	15	263	37	0	0.03	0.00	0.80	2.80	0.96
3.3	3	445	285	67	0.50	0.11	1.10	1.00	2.64
2.1	1	436	173						2.75
1.8	7	329	117	9	0.22	0.11	2.00	0.00	3.01
3.2	16	717	180	0	0.66	0.13	14.20	0.00	2.37
1.8	469	685	175		0.15	0.11	13.40	0.00	2.47
2.1	16	1290	370		0.70	0.48	27.00	0.00	1.37
1.9	432	703	188	0	0.29	0.10	14.80	0.00	2.43
4.8	1	392	128	128	0.85	0.13	0.89	2.00	2.79
3.5	3	507		41	0.91	0.09	0.52		3.01
9.2	4	599			0.81	0.19	3.57		2.15
3.1	72	628	234	29	124.00	0.22	4.40	2.00	2.38
6.8	6	1093	158	233	0.82	0.17	108.00		2.42
3.1	1	524	202	296	0.22	0.11	0.70		2.54
2.4	10	502	169	124	0.39	0.15	1.05	3.20	2.68

The amino acid content of a variety of nuts is given in Table 4. Amino acid data have also been published on the Japanese chestnut, ginkgo, pine nut (13), acorn, breadnut, chirauli nut, gabon nut, hazel sterculia, kola, marking oyster, pistachio, and tropical almond (14). As many as 23 free amino acids have been identified in the English walnuts (15). Brazil nuts, on the other hand, have been shown to be one of the richest sources of sulfur-containing amino acids (16). The 2S albumin is about 30% methionine and cysteine. All three classes of Brazil nut proteins (2S, 7S, and 11S) have been characterized and sequenced (17–19).

Oil. Most nuts are characterized by a high oil or fat content, usually about 60%, but in certain varieties of pecans it can get as high as 76% (see FATS

Table 3. Products from Hydrolysis of Nut Proteins[a,b]

Hydrolysis product	Amandin (almond)	Excelsin (Brazil nut)	Edestin (coconut)	Globulin (spruce)	Arachin (peanut)	Conarachin (peanut)
alanine	1.40	2.33	4.11	1.80	4.11	
ammonia	3.70	1.80	1.57		2.03	1.90
arginine	11.85	16.02	15.92	10.90	13.51	14.60
aspartic acid	5.42	3.85	5.12	1.80	5.25	
cystine	0.85	1.84	1.54		1.08	3.00
glutamic acid	23.14	12.94	19.07	7.80	16.69	
glycine	0.51	0.60		0.60		
histidine	1.58	1.47	2.42	0.62	1.88	1.83
leucine	4.45	8.70	5.96	6.20	3.88	
lysine	0.70	1.64	5.80	0.25	4.98	6.04
methionine					0.67	2.12
phenylalanine	2.53	3.55	2.05	1.20	2.60	
proline	2.44	3.65	5.54	2.80	1.37	
serine		0.00	1.76	0.10	5.20	4.99
tryptophan	1.37	2.59	1.25	present	0.88	2.13
tyrosine	1.12	3.03	3.18	1.70	5.50	
valine	0.16	1.51	4.21	present	1.13	
Total	*61.22*	*65.50*	*79.50*	*27.77*	*70.76*	*36.61*

[a] Refs. 4–6.
[b] Values are % of total.

AND FATTY OILS). For the most part, oil is contained in the kernel or embryo of the seed, though it can also occur in the flesh of the ginkgo fruit and in the endosperm of coconut, palm, and pine nuts. Relative amounts of some fatty acids present in a few types of nuts are given in Table 5. Considerable variations in the percentages of fatty acids have also been reported in both pecan and peanut oils from a variety of sources. (Table 6). (For main physical characteristics and the composition of nut oils, see FATS AND FATTY OILS and VEGETABLE OILS.)

Carbohydrates.   Carbohydrates from pecan kernel (21), coconut meal (22), English walnut (23), European chestnut (24), chufa, and the ivory nut (25) all include the following ingredients: reducing sugar, sucrose, raffinose, mannitol, dextrin, pentosan, amyloid, starch, cellulose (qv), tannin, gum (qv), wax, resin, and materials yet to be identified (see CARBOHYDRATES; RESINS, NATURAL; SUGAR). Mannitol is the main sugar derivative in the milk of young coconuts (see SUGAR ALCOHOLS); however, it is not present in the milk of the mature nut, where the principal sugar is sucrose. Saccharose, on the other hand, is the predominant sugar in cashew nuts (26). The starch in peanuts (27) decreases during roasting.

Minerals.   Nuts are considered to be a good source of minerals essential for nutrition, supplying elements of copper, manganese, iron, and sulfur (see MINERAL NUTRIENTS). The values for the mineral constituents of many nuts shown in Table 2 are averages of available analytical data. Values for the mineral content of the peanut kernel (28) and ash constituents in the macadamia kernel (29) and cashew (26) have also been reported. Chufa nuts have a high silicon content.

Some nut trees accumulate mineral elements. Hickory nut is notable as an accumulator of aluminum compounds (30); the ash of its leaves contains up to

Table 4. Essential Amino Acid Content of Nuts[a,b]

Amino acid	Almond	Brazil nut	Butternut	Cashew	Chestnut, European	Coconut	Filbert	Peanut	Pecan	Pilinut	Pinenut, pignola	Pinenut, piñon	Pistachio	Walnut, black
arginine	2.495	2.390	4.862	1.741	0.173	0.546	2.155	3.456	1.105	1.516	4.668	2.251	2.186	3.661
histidine	0.558	0.402	0.808	0.399	0.067	0.077	0.327	0.748	0.227	0.255	0.575	0.277	0.536	0.680
isoleucine	0.866	0.601	1.179	0.731	0.095	0.131	0.568	0.997	0.322	0.483	0.933	0.450	0.975	0.978
leucine	1.552	1.187	2.199	1.285	0.143	0.247	1.100	1.928	0.520	0.890	1.730	0.834	1.677	1.704
lysine	0.666	0.541	0.770	0.817	0.143	0.147	0.399	0.992	0.292	0.369	0.901	0.434	1.278	0.721
methionine	0.227	1.014	0.611	0.274	0.057	0.062	0.162	0.263	0.186	0.395	0.430	0.207	0.381	0.473
phenylalanine	1.113	0.746	1.442	0.791	0.102	0.169	0.686	1.467	0.409	0.497	0.919	0.443	1.184	1.107
threonine	0.739	0.460	0.940	0.592	0.086	0.121	0.448	0.743	0.253	0.407	0.761	0.367	0.722	0.730
trytophan	0.358	0.260	0.366	0.237	0.027	0.039	0.216	0.310	0.199		0.303	0.146	0.283	0.322
valine	1.028	0.911	1.541	1.040	0.135	0.202	0.662	1.161	0.386	0.701	1.241	0.598	1.410	1.286

[a]Refs. 3 and 12.
[b]Values given are in g/100 g of edible food.

Table 5. Oil and Fatty Acid Content of Commercial Nuts[a,b]

Nut	Oil, %[c]	16:0	16:1	18:0	18:1	18:2	20:0	20:1	22:0	24:0
almond	46.6	6.50		1.17	63.82	28.42	0.04	0.05	0	0
Brazil nut	62.9	12.84		8.72	27.94	50.21	0.21	0.06	0.02	0
cashew	47.1	10.07		9.26	60.67	19.05	0.55	0.18	0.10	0.12
filbert	58.6	4.57		1.85	79.17	14.18	0.07	0.15	0	0
macadamia	65.8	8.43	17.90	3.02	61.21	2.50	2.62	2.81	0.8	0.32
pecan	70.2	6.41		2.44	55.87	34.93	0.09	0.25	0.01	0
pistachio	43.4	11.82		1.05	51.68	35.00	0.08	0.33	0.06	0
sunflower	43.6	5.38		3.83	15.59	74.00	0.21	0.12	0.70	0.20

[a]Values for fatty acids are % of total fatty acid content.
[b]Fatty acids are represented as x:y, where x is the carbon chain length and y is the number of double bonds.
[c]On a basis of total dry weight.

37.5% of Al_2O_3, compared with only 0.032% of aluminum oxide in the ash of the English walnut's autumn leaves. As an accumulator of rare-earth elements, hickory greatly exceeds all other plants; their leaves show up to 2296 ppm of rare earths (scandium, yttrium, lanthanum, dysprosium, holmium, erbium, thulium, ytterbium, and lutetium). The amounts of rare-earth elements found in parts of the hickory nut are kernels, at 5 ppm; shells, at 7 ppm; and shucks, at 17 ppm. The kernel of the Brazil nut contains large amounts of barium in an insoluble form; when the nut is eaten, barium dissolves in the hydrochloric acid of the stomach.

Deficiencies in minor mineral elements have been observed in the growing of nut crops; the main deficiencies are zinc in pecans, English walnuts, and almonds, and boron in peanuts and English walnuts. The disorder of pecan known as rosette used to threaten the growing crop until the cause was found to be a deficiency in zinc, which could be corrected by applying zinc salts to the tree by various means. For that matter, small almond leaf can also be corrected by zinc treatment (31). In studies on the deficiencies of several elements in pecan trees grown in sand cultures, no single abnormality is found to be sufficiently characteristic for identifying the deficiency of an element in the tree (32). Boron deficiency in peanut usually causes hollow heart; in English walnuts it causes shoots to die back (snakeheads) and nuts to fail to set on the tree. Such conditions can be corrected by applying 1.8–2.7 kg of borax for each English walnut tree; indeed, so great is the response to boron application that a walnut orchard previously noted for poor production has resulted in the setting of so many nuts that the trees were finally broken down by their weight (an average of 5.6 metric tons of green nuts per hectare) (33). Filberts, on the other hand, have never shown any deficiency symptom or response to boron treatment. In an intensive study of the boron content of almond and English walnut tree parts, the kernels of the almond were found to contain 15.6–57.7 ppm of boron, the shells 21.8–102.5 ppm, the husks 36.4–440.5 ppm, and the leaves 30.4–53.6 ppm (34). The walnut was found to contain 7.5–16.3 ppm of boron in the kernels, 13.6–15.4 ppm in the shells, 21.7–79.0 ppm in the husks, and 35.9–516.0 ppm in the leaves. Much of the boron in the almond husk is water soluble; that in the walnut husk is water insoluble.

Table 6. Fatty Acid Analysis by Gas–Liquid Chromatography of Pecan Oils and Peanut Oils[a,b]

Type of oil	Palmitic acid	Stearic acid	Oleic acid	Linoleic acid	Linolenic acid	Arachidic acid	Iodine value (IV) glc	Wijs
Pecan oil								
Big Z	5.9	2.9	59.7	30.3	1.6	1.6	108.0	110.0
Curtis	6.1	3.1	71.8	18.1	0.8	0.8	95.9	98.3
Frotcher	6.5	2.5	63.9	25.5	1.3	1.3	102.5	105.9
Mobile	7.1	2.7	56.8	31.4	1.8	1.8	107.9	111.2
Moneymaker	6.3	2.6	51.0	37.8	1.7	1.7	114.0	118.0
Randall	6.0	2.9	61.3	28.2	1.5	1.5	105.2	107.2
Stuart	6.6	2.2	68.8	21.0	1.1	1.1	98.8	101.3
Success	4.5	2.9	76.5	13.5	1.3	1.1	92.6	94.5
Tesche	6.1	3.0	63.4	25.7	1.3	1.3	102.4	105.3
Van Deman	5.1	2.9	74.6	16.0	0.9	0.9	94.2	96.2
Average	*6.0*	*2.8*	*64.8*	*24.8*	*1.3*	*1.3*	*102.2*	*104.8*
Peanut oil								
Dixie Runner	9.3	3.1	58.3	22.4	1.1	0.7	92.0	91.1
Early Runner	9.7	3.1	55.6	24.7	1.1	1.0	93.5	92.9
Florida 393-47	7.3	4.3	63.5	17.7	1.4	1.1	88.9	87.4
Florigiant	10.3	4.1	55.7	22.9	1.6	0.7	91.8	90.3
Georgia Bunch 119-20	11.1	4.1	54.4	23.2	1.4	1.1	90.7	90.7
North Carolina 2	10.5	4.3	51.9	26.2	1.3	0.6	93.4	92.3
Spanette	109.0	3.7	52.5	25.7	1.3	0.7	93.4	92.0
Spanish 18-38-47	12.9	4.5	43.1	32.5	1.5	0.7	97.5	101.8
Virginia 56R	9.6	4.4	54.5	23.8	1.4	0.9	92.3	90.8
Virginia Bunch G2	10.1	3.7	55.7	23.7	1.3	0.7	92.4	91.3
Virginia Runner G-26	10.3	4.7	52.3	25.1	1.5	0.9	92.9	92.0
Average	*19.1*	*4.0*	*54.3*	*24.4*	*1.4*	*0.8*	*92.6*	*92.1*

[a]Ref. 20.
[b]Values are % of total fatty acid.

Vitamins.   Most nuts contain a good supply of vitamins A and B_1 but are usually lacking in other vitamins (qv). Table 2 lists the vitamins in various nuts and their parts. Red palm oil is an important source of beta-carotene; it contains 30–40% of this vitamin. Peanut oil contains 0.03–0.05% of tocopherol (vitamin E). Immature English walnuts, mainly the pulp, contain a large amount of ascorbic acid (vitamin C), around 1300–3000 mg/kg, a much greater amount than is found in any other plant part except rose hips, which contain about 5000 mg/kg of ascorbic acid. Methods for extracting vitamin C from green walnuts have also been developed (35). Considerable vitamin C (>2000 mg/kg) has been reported in the buds and male catkins of English walnuts (36). Unfortunately, as the immature fruit ripens and the walnut forms, the vitamin also disappears. However, the large amounts of vitamin C apparently present in walnuts may in part be due to faulty analyses, as other reducing substances may have been present that interfered with the analyses. Thus, by isolating ascorbic acid in a new analysis, the large amounts of vitamin C previously claimed to be in green husks of walnuts could not be confirmed. In addition, vitamin C can be found in mature hazel nuts and almonds (37).

A considerable amount of vitamin B_1 (thiamine) has been found in the red skins of Virginia-type peanuts (38); vitamin B_1 has also been found in almonds and filberts, at levels of 114 and 570 mg/kg, respectively (39).

The nicotinic acid content of several nuts has been reported (in mg/kg) as follows: chestnut, 200; hazel nut, 600; almond, 1600; and sunflower seed, 5000 (40). The results of analyses for pantothenic acid are (in mg/kg): hazel nut, 380; almond, 75; sunflower seed, 620; and walnut, 600. Nuts also contain more biotin than most fruits and vegetables.

Phytin and Phospholipids.   Phytin, an important organic phosphorus compound, as well as lecithin (qv) and cephalin, both phospholipids, are present in nuts. Phytin in food occurs mainly as a mixed calcium and magnesium salt of inositolphosphoric acid, $Ca_5Mg(C_6H_{12}O_{24}P_6 \cdot 3H_2O)$, and supplies phosphorus and calcium in a readily available form. The phytic acid phosphorus content in various nut kernels (in % of total P) is almond, 82%; Brazil nut, 86%; American chestnut, 18%; hazel nut, 74%; filbert, 83%; peanut, 57%; and English walnut, 42%. The phosphorus present in the protonated form of phytin, $C_6H_{18}O_{24}P_6$, is almond, 2.57%; Brazil nut, 2.96%; filbert, 1.66%; hickory, 1.57%; peanut, 1.67%; pecans, 1.46%; English walnut, 1.43%; and eastern black walnut, 2.03% (41). Peanuts are found to contain 0.5% of lecithin and cephalin; piñons 1.0% of lecithin. Choline is present in pecan, peanut, and peanut meal at levels of 0.53, 1.70, and 2.52 mg/g, respectively. Phytosterol is present in the oil of beechnut to the extent of 0.18% of total weight, in hazelnut 0.50%, and in piñon 0.40%. Cholesterol is present in detectable quantities in cashew nut oil (26).

Glycosides and Alkaloids.   Saponins are sapogenin glycosides that usually contain triterpene or sterol nuclei (see STEROIDS). These are mainly amorphous, water-soluble compounds, and can form colloidal solutions that foam readily and reduce surface tension (Fig. 1). A saponin occurs in the pulp of soapnut to the extent of about 30% of the dry weight; horse chestnut contains 10–14% of bitter principle in the form of a saponin, argyrin, and a glycoside, esculin [531-75-9], $C_{15}H_{16}O_9$ (**1**). Esculin is hydrolyzed by enzyme to glucose and esculetin [305-01-1], $C_9H_6O_4$ (**2**). Amygdalin [29883-15-6], $C_{20}H_{27}NO_{11}$, (**3**)

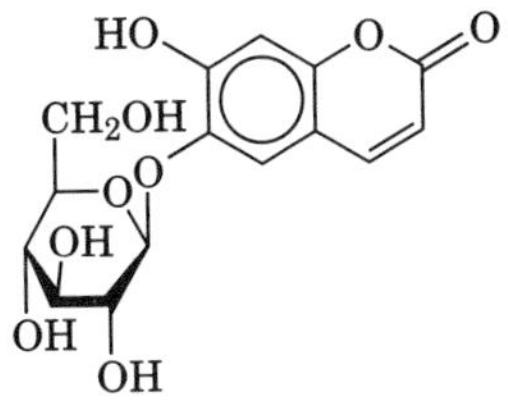

Fig. 1. Glycosides (**1**, **3**, and **7**), their hydrolysis products (**2** and **4**), and alkaloids (**5** and **6**) derived from nuts and seeds. See text.

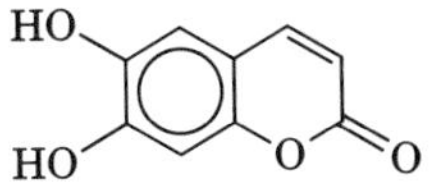

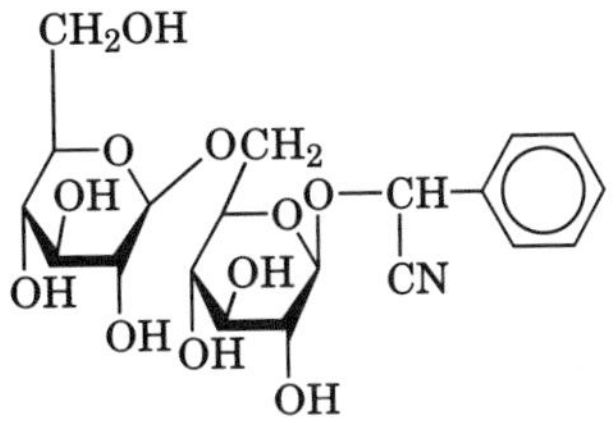

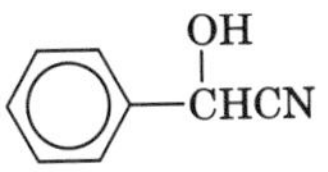

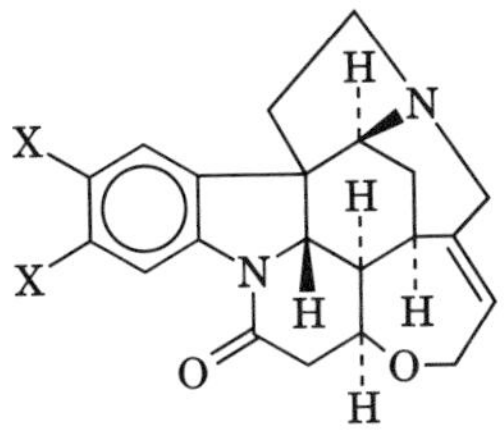

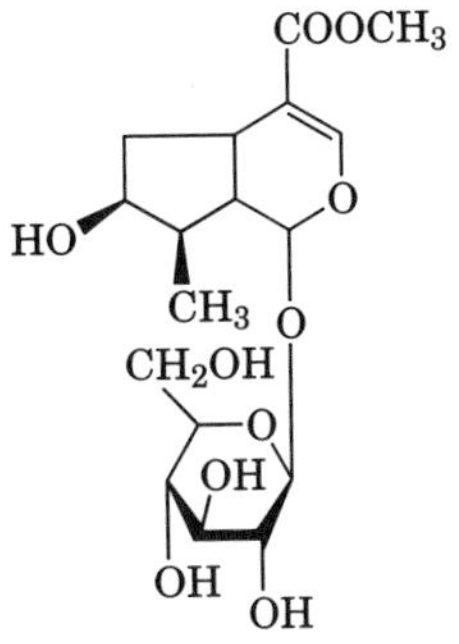

is the most important of the cyanogenetic glycosides that yield hydrogen cyanide on hydrolysis. It is found in all seeds of the rose family but more abundantly in the seed of bitter almond (2.5–3.5%); in addition, it is found in the seeds of apple, apricot, cherry, peach, pear, plum, and quince. Hydrolysis of amygdalin catalyzed by enzymes amydalase and prunase yields glucose and benzaldehyde as well as hydrocyanic acid. Intermediary products are mandelonitrile glucoside (prunasin) and mandelonitrile [532-28-5], C_8H_7NO (**4**). *Strychnus nuxvomica*, or poison nut, has as its chief constituent the alkaloids strychnine [57-24-9], $C_{21}H_{22}N_2O_2$ (**5**) (0.4%) and brucine [357-57-3], $C_{23}H_{26}N_2O_4$ (**6**) (90.2%) (see ALKALOIDS); it also contains strychnic acid, sugar, fat, and the glycoside loganin [18524-92-2], $C_{17}H_{26}O_{10}$ (**7**). The Betel nut contains at least four alkaloids, of which arecoline is the primary constituent (42,43). Arecoline is an effective anthelminthic and vermifuge (42).

Tannins. Tannins are amorphous astringent substances that produce colors, eg, inks (qv) with ferric salts, and precipitate gelatin from solution (tanning). They are hydrolyzed by acids into various products such as sugar (*d*-glucose) and hydroxy acids. The process of tanning, involving both physical and chemical reactions, is the conversion of a hydrophilic gel into a relatively nonhydrophilic one, eg, leather (qv). Tannins occur abundantly in certain nuts and parts of nut trees. The wood and bark of the chestnut are especially rich sources of commercial tannin. The integument of the pecan contains tannin at up to 25% of total weight, the outer shell, 6–8%; the almond hull, in contrast, contains only 4.5% of tannin. Both pecan shell waste and almond hulls from processing plants are a source of commercial vegetable tannin (44,45). The tannins of the English walnut are composed chiefly of four polyphenolic esters of glucosides (46).

Toxic Constituents. The seed coat of European beechnut contains an unidentified toxic substance that makes the feeding of beechnut cake to certain farm animals hazardous (47). A toxic concentration (up to 4000 ppm) of barium found in some Brazil nut kernels (30,48) has been reported to cause illness in children who ingest the nuts. Bethel nuts have also been shown to include constituents that are carcinogenic, embryotoxic, and immunotoxic (49–51). The liquid from cashew nutshells contains anacardic acid [18654-17-6], $C_{22}H_{32}O_3$ (**8**), a brown crystalline substance (47), and cardol, $C_{21}H_{32}O_2$ (**9**), a dark-brown phenolic oil; both of these substances are toxic, irritating, and can produce blisters on the skin similar to those caused by poison ivy (Fig. 2). Juglone [481-39-0], $C_{10}H_8O_3$ (**10**), has yellow-brown crystals (52) that stain the skin and occurs chiefly as its colorless reduction product, α-hydrojuglone [481-40-3], $C_{10}H_6O_3$ (**11**), in all green and growing parts of walnut trees and the unripe hulls of the nut. Juglone released from black walnuts can stunt the growth of nearby plants (Fig. 2). Some nuts contain glucosides that are disagreeable or toxic to some people; for example, bitter almonds contain cyanogenetic glucoside that releases hydrogen cyanide. The presence of allergens has been confirmed in Brazil nut, coconut, filbert, pecan, peanut, almond, and walnut. Tannins may also be disagreeable to some people for their unpleasant taste (53). A detailed study of the amino acid content of allergens isolated from almond, Brazil nut, filbert, peanut, and black walnut has been reported (54). Peanut seeds through a phenolic glycoside in the skin (55) can inhibit iodine uptake by humans (56). The glycoside is preferentially iodinated and deprives the thyroid of available iodine (57) (see FOOD TOXICANTS NATURALLY OCCURRING; IODINE AND IODINE COMPOUNDS).

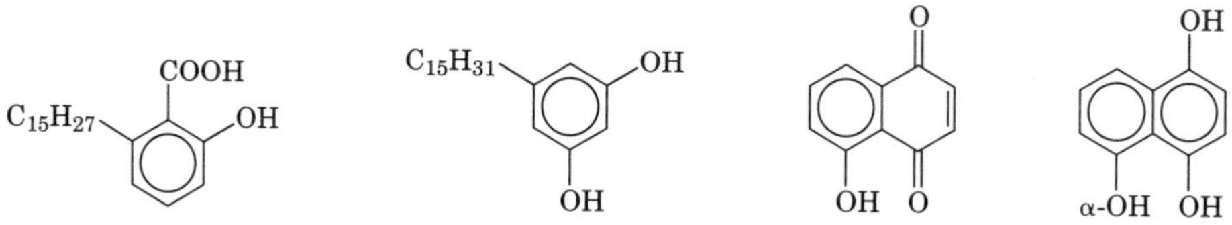

Fig. 2. Toxic compounds found in cashew nuts (**8** and **9**) and in walnuts and walnut trees (**10** and **11**). See text.

Nutritive Value of Nuts

Nuts are rich in protein and fat; most commercially important nuts supply about 28 MJ/kg (6600 kcal/kg) of kernel, more than most other foods (Table 2); cereals supply about 15 MJ/kg (3640 kcal/kg); meats about 7.5 MJ/kg (1790 kcal/kg); and fresh fruits less than 2.8 MJ/kg (660 cal/kg). The energy content in kilojoules (1 food calorie or cal = 4.184 kJ) per kilogram of food is calculated by the following formula:

$$171.6[\% \text{ protein} + \% \text{ total carbohydrates } (N\text{-free extract} + \text{ fiber})]$$
$$+ 389.3 \ (\% \text{ fat}) \qquad (1)$$

One kilogram of nut kernels is equal in energy value to 5.1 kg of bread, 8.21 kg of steak, 27.1 kg of white potatoes, or 33.1 kg of oranges. A human male doing ordinary work requires 12.5 MJ/d (3000 kcal/d) for energy production: 75 g of protein, 85 g of fat, 0.67 g of calcium, 1.44 g of phosphorus, and 1.5 mg of iron. Thus, 45 g of oily nuts can supply all the food energy needed each day, ie, about 40% of the protein, 60% of the phosphorus, 30% of the calcium and iron, and four times the requirement for fat. Nuts, as a rule, contain considerable vitamin A and vitamin B complexes and are sources of iron, copper, manganese and sulfur, though they usually lack vitamin C and calcium. The peanut is an especially rich source of niacin. Nuts also contain the essential amino acids tryptophan, phenylalanine, lysine, valine, and leucine. Processing peanuts and other nuts by roasting or cooking in oil significantly reduces their vitamin content. For instance, cooking macadamia nuts in hot oil at 135°C for 12–15 minutes can cause a loss of 16% of the original thiamine content (58); similarly, modern commercial methods of roasting or processing peanuts in hot oils can destroy about 70–80% of the thiamine content (38).

The reaction of edible nuts in the diet is mainly basic, although some nuts, such as peanut, pecan, and English walnut, can also provoke an acid reaction. In a research experiment, young white rats were fed several nuts to obtain digestion coefficients of the protein and fat. Digestion coefficients for protein, in percentages, were as follows: pine nut, 95.12; almond, 90.17; peanut, 89.43; hazel nut, 82.91; and walnut, 79.49. For fat, the results, also in percentages, were the following: almond, 96.13; peanut, 95.50; hazel nut, 87.22; pine nut, 82.93; and walnut, 67.77 (59). For cashews, the chemical score of 59–67 is not high, but the protein score is high (60).

Nuts are a very concentrated food and may cause discomfort if they are not well masticated; digestibility, of course, can be increased if they are finely ground. Rather than as a basic diet, nuts are used chiefly as appetizers and seasoning ingredients to salads, cakes, candies, and cookies, or as emergency rations for persons expending great amounts of energy. However, a peanut-butter sandwich with 30 g (2 tablespoon) of peanut butter, a 237-mL (8 ounce) glass of milk, and an orange can supply about one-third of all the recommended daily allowances except those for iron and vitamin A (61). Incorporating moderate quantities of walnuts into a cholesterol-lowering diet can also decrease serum levels of total cholesterol and improve the lipoprotein profile in an average person (62). The advantages of nuts over meats are that the former do not have to be prepared for eating and that they are free from waste products such as uric acid formed from purines (as in meats). Also, nuts are relatively sterile; that is, they are free from putrifactive bacteria and parasites. A slight drawback is that mold is sometimes present. Another drawback is that the handling of nuts after shelling may introduce some microbiological contamination to nut kernels. But the chief disadvantage of nuts is that they do not supply the bulk for propulsion in the digestive system as meat and other foods are able to provide.

Chemical Changes during Development and Storage

Development. Several investigations have been made on the chemical changes taking place during the development of various nuts; for instance, in macadamia (63) (see Table 7), pecan (65), almond (66), English walnut (67), eastern black walnut (see Table 8), and tung (69). The levels of sugars and other carbohydrates decrease rapidly during oil synthesis, so that by the time the kernel is mature, most of the carbohydrates have disappeared. Since the decrease in carbohydrate levels in the kernel accounts for only a small proportion of the oil synthesized, it follows that the materials formed into oil come from other parts of

Table 7. Carbohydrate, Nitrogen, and Oil Levels in Macadamia Embryos as a Function of Age[a,b]

Component	Days after flowering				
	90	111	136	185	215
total sugar	7.54	27.28	22.98	9.60	5.80
reducing sugar	1.47	3.21	1.07	0.41	0.30
sucrose	6.07	24.07	21.91	9.19	5.50
total nitrogen	4.88	3.04	2.19	1.72	1.70
soluble nitrogen	2.92	1.13	0.61	0.33	0.27
insoluble nitrogen	1.96	1.91	1.58	1.39	1.43
acid-hydrolyzable matter	4.54	4.88	3.85	2.56	2.16
soluble solids in 80% alcohol	60.10	39.92	28.36	14.82	9.88
ether- and alcohol-insoluble matter	36.43	28.88	23.69	17.89	16.68
petroleum-ether extract	3.46	31.19	47.94	67.28	73.44

[a]Ref. 64.
[b]Values are percentages of total dry weight.

Table 8. Seasonal Change in Composition of Black Walnut Kernels[a,b]

Component	June 15	July 15	July 29	August 12	August 26
total solids	3.510	4.128	20.487	38.325	68.110
ether extract	0.123	0.556	8.043	19.592	42.176
total nitrogen	0.182	0.290	0.818	0.892	1.044
amino nitrogen	0.111	0.145	0.290	0.312	0.466
protein nitrogen	0.071	0.145	0.528	0.580	0.578
pentosans		present	0.775	0.758	0.940
ash		0.502	1.336	1.321	1.570
potassium	0.312	0.253	0.485	0.675	
calcium	0.266	0.048	0.072	0.071	0.044
phosphorus	0.127	0.086	0.055	0.091	0.051
magnesium		0.011	0.079	0.098	
acidity	acid	acid	neutral	neutral	neutral

[a]Ref. 68.
[b]Values are % of total dry weight.

the tree. For example, studies of the tung fruit at different stages of development show that even though all the carbohydrates are utilized for synthesis of oil in the kernel, a large part of the substances necessary for oil formation are brought in from outside the nut (69). The fat content in the eastern black walnut kernel increases out of proportion to any decrease in other constituents (68).

A systematic sampling of pecan nuts, at nine weekly intervals during kernel development, to determine changes in the oil, protein, and mineral constituents reveals that the most critical period in the filling of the kernel in the Northern Hemisphere is the three weeks prior to September 15, during which time 64% of the total oil, 71% of the protein, 43% of the ash constituents, and 63% of the dry weight are formed (70). Potassium, comprising 73% of all mineral elements in the shuck, accumulates at a rapid rate in the shuck and kernel during the early filling stage; likewise, phosphorus increases rapidly in the kernel during this stage. The vitamin C content of English walnut during development of the nut was found to increase to a maximum of 15% forty days after blooming and then to decrease to 1–2% at maturity (71); vitamin C made up 52% of the weight of an immature nut's hull, 32% of the tissue under the shell, 15% of the shell, and 1% of the kernel.

Respiration of the developing peanut is very rapid during fat synthesis but declines before maturity (72); studies of respiration indicate another decline about 8–10 hours after the harvest of peanuts (73).

Storage. Investigation of the effects of moisture, heat, and darkness on raw kernels of macadamia nuts found the stability of the kernels to decrease with increasing moisture content and increasing storage temperature (74). At 1.4% moisture, according to the investigation, only very small changes were found in flavor and chemical composition after 16 months, regardless of storage conditions. At high moisture levels and storage temperatures, total sugar levels decreased but reducing sugar and free fatty acid levels increased; light, on the other hand, had no effect on stability. The investigation found that stability is greatest at 1.4% moisture; but at 2.3% and 4.3% moisture, $-18°C$ storage is recommended for kernel stability. A different study (75) on the roasted macadamia kernel gave results similar to those in the raw nut research. Moreover, an earlier study (76) indicated that supplemental protection by the application of antioxidants, elimination of oxygen, and shielding from ultraviolet radiation can further enhance kernel stability of the macadamia nut.

For the English walnut, the optimum moisture level for storage is 3.2–3.8%. Stability of the shelled nut is significantly reduced at higher or lower moisture levels (77,78). In addition, the skin colors of nuts are also important. A procedure was developed (77) to estimate the darkening rate of English walnuts during storage, which consists of the extraction of walnut kernels for 30 minutes with 65% aqueous methyl alcohol at reflux temperature, filtration, and photometric estimation at 400 μm of the absorbance of the clarified solution. An arbitrary numerical color scale has been devised to express the color values of shelled walnuts.

Because most edible nuts are high in oil, rancidity is one of the first signs of deterioration. There are two types of rancidity: oxidative, in which aldehydes and ketones are formed by the addition of molecular oxygen, and hydrolytic, in which hydrolysis of the glycerides forms free fatty acids. In nuts of the hydrolytic type, the glycerides hydrolyze in the presence of moisture and a fat-splitting enzyme. High temperature and humidity, as well as sunlight, favor the development of rancidity in nut kernels; consequently, the best storage conditions consist of low temperature, low humidity, and little or no light.

Considerable research has been done on the storage of nuts, especially of pecans, which have a high free fatty acid content (79). Most pecans are refrigerated at some time during storage and marketing. They may be held at 3°C for one year, at −4.5°C for three years, or at −17.5°C for more than five years (80). Most nuts cannot be held for more than three or four months at ordinary temperatures, especially during the summer, without developing rancidity; however, storage experiments have shown that nuts and nut products can be kept two to five years longer under refrigeration at 0–5°C (80). For instance, peanuts stored in the shell at 4.4°C have remained perfect for as many as six years with no detectable chemical changes (81). Nut kernels sealed in tins by a vacuum process also keep well (see VACUUM TECHNOLOGY). On the other hand, various tests with nuts stored in glass, plastic, or containers of other kinds, and sealed with nitrogen, carbon dioxide, or other gases have proved satisfactory only when the nuts are held under refrigeration. Slight ventilation must be provided in containers storing chestnuts because of the respiration of nuts.

An edible and nutritive coating of zein protects tree nuts and peanuts from developing rancidity, staleness, and sogginess during storage. Zein coatings, applied as alcohol solutions, contain an acetylated monoglyceride and antioxidants. The zein prevents the transfer of oil from the nut, whereas the monoglyceride prevents the transfer of moisture into the nut and at the same time plasticizes the zein. Antioxidants, on the other hand, prevent rancidity of oil on the nut surface at the time of coating. In an experiment, both coated and uncoated pecans were stored at 37.9°C. After one month the uncoated pecans were rancid, but the coated pecans stayed fresh for three months at 37.9°C and for six more months at room temperature. Similar results have been obtained with walnuts, almonds, peanuts, as well as nuts incorporated in chocolate bars (82).

Nuts are susceptible to infestation by weevils in storage, especially in warm weather. Infestation is usually prevented or controlled by placing the nuts in cold storage or by fumigating them either in the open with gas, such as methyl bromide or hydrogen cyanide, or under vacuum with a mixture of carbon disulfide and carbon dioxide. For example, phosphine and methyl bromide can be used to control insects in farmer's stock (in-shell) and shelled peanuts in transit (83), whereas pyrethrins are used to control insects in shelling plants. Malathion is also used to control insects in stored products but is in the process of being deregistered by the USDA (84). Methyl bromide used at the rate of 35 g/m³ (1.0 g/ft³) or liquid hydrogen cyanide at 18 g/m³ (0.5 g/ft³) for eight hours at 16°C or above is sufficient to kill all insects and rodents. In contrast, a 0.48-kg/m³ (13.6 g/ft³) dosage of an ethylene oxide–carbon dioxide mixture for 1–2 hours provides excellent control in vacuum fumigation of nut kernels. However, the carbon disulfide–carbon dioxide mixture has proved unsatisfactory for pecan, Brazil nut, and cashew (85) (see INSECT CONTROL TECHNOLOGY).

Although the storage of peanut butter is influenced by temperature (86), the extent of roasting, hydrogenation, and addition of salt have little effect on product stability if packages are sealed properly. This is because oxygen in the headspace is the main factor in reducing stability; consequently, peanut butter for the retail trade is packed in air-tight containers or under vacuum. Metalloproteins as well as iron and copper salts are the chief catalysts of fatty acid peroxidation in peanut butter; however, certain chelating agents (qv) can retard or reduce this catalytic effect (87). Autoxidation in peanut butter is relatively rapid at first but soon decreases in speed so that staleness and rancidity develop slowly over an extended period of storage. The product can remain in good condition for two years at room temperature, provided that the oil has not separated and oxygen is not present for free-radical reactions.

Aflatoxins. Mycotoxins are toxic compounds elaborated by fungi. The Greek words *mykes* means fungus and *toxikon* poison. Although mycotoxins have been know to exist for many years, they have been studied less intensively than bacterial toxins in foods. Interest in the effects of mycotoxins was renewed in recent years after the discovery that aflatoxins were the causative agents in turkey X disease (88), which killed thousands of turkeys in England in 1960. Poultry (eg, ducks, turkeys, and chickens) are especially susceptible to aflatoxins, which induce liver damage (89). In humans, aflatoxins are carcinogenic, hepatotoxic, and teratogenic. Dietary intake of aflatoxin-contaminated products (maize and peanuts) is associated with hepatitis, Indian childhood cirrhosis, and liver cancer (90).

Aflatoxins are fluorescent toxic factors elaborated by the common mold *Aspergillus flavus* during its growth on nuts or grains. Peanuts may be invaded by toxigenic strains of *A. flavus* present in the soil, which allows aflatoxins to be produced during the pre- and post-maturity stages in the field, during post-harvest drying, or in storage (91). Over 18 aflatoxins belong to the mycotoxin group. The two principal toxins are aflatoxins B_1 and G_1 (92), so named because they showed blue and green fluorescence, respectively. Aflatoxin B_1 is the most common of the aflatoxins (93) as well as the most potent hepatocarcinogen known (94).

Toxins have similar structures and form a unique group of highly oxygenated, naturally occurring heterocyclic compounds that fluoresce when exposed to fluorescent light (95). This fluorescence is used for detection in thin-layer chromatographic separations (96). Increasingly, aflatoxins are detected, using high performance liquid chromatography (hplc) (97,98) or enzyme-linked immunosorbent assay (elisa) (99,100) (see CHROMATOGRAPHY, ANALYTICAL METHODS; IMMUNOASSAY). Tolerances for total aflatoxin or aflatoxin B_1 range from 1 to 50 $\mu g/kg$ (97). In the United States, the allowed level of total aflatoxin for products intended for human consumption, dairy animals, or immature animals is 15 $\mu g/kg$ (97,101). Levels greater than 100 $\mu g/kg$ may be acceptable for breeding cattle, breeding swine, finishing swine, and feedlot beef cattle (101).

Although other mycotoxins have been and are being discovered, aflatoxins retain a position of importance because of their high toxicity and common natural occurrence in such foods as cereal grains, oilseeds, and oilseed meals stored under adverse conditions. Almost all agricultural commodities and foods support the development of aflatoxins if conditions are favorable for the growth of *A. flavus* (102). Severe and prolonged drought, for instance, promotes aflatoxin formation in peanuts (103). Mathematical models have been developed to describe mold growth and aflatoxin formation under field conditions (104).

Four main approaches are used to control the problem of aflatoxin in peanuts: (1) the development of cultivars that resist the invasion of aflatoxin-producing fungi; (2) cultural practices minimizing insect damage that facilitates fungal invasion; (3) detoxification of contaminated nuts and their products; and (4) separation of contaminated nuts (105). Generally, most aflatoxin production occurs during the harvest period after the nuts have been dug and begun to dry but before the moisture level best suited for storage is obtained. As a result, nuts must pass through this critical moisture zone quickly. Storage of the nuts under proper temperature and humidity conditions can prevent further contamination (70).

One effective step to ensure a wholesome product is to divert from edible use any contaminated lots of nuts as early in the food processing chain as possible. In the United States, the peanut industry has been a leader in detecting and diverting mycotoxin-containing peanuts, which are kept out of the food and feed channels. In addition, sampling instructions (106) for in-shell and shelled peanuts require that three 21.8-kg samples of shelled peanuts be ground and tested for certification. Such a large sample size is required because of the difficulties in obtaining a representative sample of peanuts for aflatoxin analysis (107). However, the necessity for a large sample size also makes the sampling of expensive nuts such as the cashew almost prohibitive.

Although aflatoxins are stable during heating and cooking (108), they are sensitive to strong acids and bases at elevated temperatures. For example, alkali-refining and washing destroys aflatoxins in peanut oil (109,110) as does ammonia fumigation for corn (111). For powdered materials, eg, corn flour or peanut cake, treatment with $NaClO_2$ or $(NH_4)_2S_2O_8$ at 60°C followed by washing is effective in degrading and removing aflatoxin (112). However, rheological properties are affected unfavorably by this method (113). Biological detoxification using *Flavinobacterium aurantiacum* is also effective for milk, oil, peanut butter, peanuts, corn, or peanut milk but unfortunately dyes the product a bright orange-pink (108,112–115). Fungicides and food preservatives may also be used to control microbial growth (115–117). Separation of damaged aflatoxin-containing nuts by flotation or electronic sorting has also been found effective for peanuts (112,118). Most of these methods, however, have not been approved by the FDA or USDA for production of human or animal food.

Processing

Shelling, Cracking, and Bleaching. Very elaborate machinery is used for cleaning, grading, bleaching, cracking, and packing edible tree nuts such as pecan, English walnut, filbert, and almond. For other nuts, the processing remains at least in part a hand operation.

Shelled in India, cashews are heated in cashew nutshell liquid before being shelled to make the shells brittle and to extract the toxic liquid from the shell; this cashew nutshell liquid or oil, which should not be confused with the fatty oil from the kernel, is recovered for many industrial uses (119), including its use in sweet-tasting salad oil (47). After heating, the nuts are cracked and the inner seed coat is removed by hand. They are then packed and shipped in such a way that cashew kernels are protected against spoilage and insect infestation. Tin cans, for instance, are vacuumed through a small hole in the lid and charged with an inert gas such as carbon dioxide or nitrogen.

Babassu presents a problem in cracking. The shell of this nut is extremely hard. Because a cracking machine has yet to be perfected, most of the extracting is done by hand. Macadamia also has a hard shell and requires the use of special cracking machinery (68). In contrast, although piñon nuts are hard-shelled, they are very small (2600–4400 per kg). The invention of a shelling and separating machine has greatly increased the commercial value of this nut in the United States. Brazil nuts are soaked in water for 24 hours and then placed in boiling water for 3–5 minutes prior to cracking, which is done in hand-operated machines. Special care must be taken not to damage the kernels. In some cases, the hulls of English walnuts are loosened after harvest by exposure to ethylene (1:1000 parts of air), a process said to preserve the natural color of the kernel. This treatment prevents hulls from sticking to and staining the shells. Dehydration of English walnuts by artificial heat has to a large extent superseded sun-drying. After hulling and washing, the walnuts are placed in bins and dried to less than 10% water content by warm air not over 43°C.

Bleaching of almonds is usually accomplished by exposure to steam for 20 minutes and then to sulfurous fumes in an airtight chamber for 20–30 minutes (burning sulfur at the rate of 1.5 kg sulfur per metric ton of almonds). English walnuts are bleached for market by dipping for 5–10 seconds in 100 liters of water, combined with a small amount of sulfuric acid. Pecans (in-shell) are sometimes bleached in sodium hypochlorite solution and occasionally dyed; however, for the most part they are marketed in their natural color. Large in-shell Virginia-type peanuts are sometimes bleached with sodium hypochlorite solution, dried, and whitened by sprinkling with marble dust (see BLEACHING AGENTS).

Blanching, Salting, and Roasting. Some of the nut kernels processed for market and for various nut products are blanched, ie, the skin or membrane covering the white meats is removed. Pecan meats, in contrast, are not blanched. Almonds are soaked in hot water until the skin slips off readily; they are dehydrated subsequently. Salted almonds are prepared by immersion in hot peanut or coconut oil at 149°C followed by salting (120). Since hot-water treatment is insufficient to loosen and remove the skins from the wrinkled surface of English walnuts, the kernels of these nuts are blanched by immersion in lye solution followed by a dilute acid rinse; this process yields a white nonastringent product (10). In an improved glycerol–alkali process for blanching almonds, Brazil nuts, filberts, and other nuts, which better preserves the flavor and texture of the nuts, the kernels are passed through a heated solution of 7.4 g of glycerol and 45 g of sodium carbonate per liter of water; the skins are removed with a stream of water; the kernels are dipped in a weak citric acid solution to neutralize any alkali retained by the kernels (121). Pistachio nuts are salted and roasted with the shell partially open; however, the method of treatment to make the shell milk-white is proprietary. In addition, several plants in the United States and Europe are processing peanuts that are salted in the shell before roasting. Usually, the peanuts are impregnated with salt by using a saturated brine solution and vacuum (122). A great increase in peanut consumption in the 1990s has been due to the popularity of dry roasted peanuts. The shelled peanuts are heated for blanching; then blanched and coated with a mixture of proprietary ingredients consisting of gum, zein, acetylated monoglycerides, spices, an antioxidant, and a coloring agent; finally the coated peanuts are roasted for maximum flavor. These vacuum-packed products have excellent flavor and long shelf life.

Peanut seed cotyledons consist of epidermal, vascular, and parenchymal tissues. Processors are chiefly concerned with the parenchymal cells that form the majority of the cotyledon. The main subcellular organelles of the parenchymal cells are lipid bodies, protein bodies, and starch grains (Fig. 3). Seed maturity, environmental conditions, and processing all significantly affect peanut seed microstructure (123). For instance, drought stress can result in cracking and fissuring of parenchymal tissue; both oil-cooking and oven-roasting can result in disruption of the cytoplasmic network, bursting of lipid bodies, and distension of the protein bodies, but cause little damage to starch grains. Microscopic examinations of raw and roasted peanuts, cashews, and almonds found cell walls to be less plastic and more brittle (124).

Peanut Butter. By federal regulation, at least 90% of commercial peanut butter consists of shelled roasted peanuts that are ground and blended with salt, sweeteners, and emulsifiers. No artificial flavors and sweeteners, chemical preservatives, natural or artificial color, purified vitamins, or minerals are allowed (125). To meet consumer demand for low fat products, several U.S. manu-

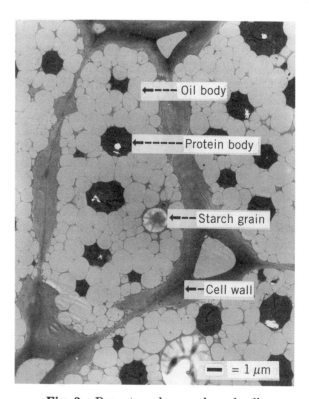

Fig. 3. Peanut seed parenchymal cell.

facturers have created products in which the peanut content has been partially replaced by maltodextrins, corn syrup solids or similar starches, and soy protein (126–128). As of the mid-1990s the FDA is still considering modifying the standard of identity, in accordance with the Nutrition Labeling and Education Act of 1990 (129), to permit the adding of vitamin as well as the display of nutrient content claim (eg, "reduced fat") to an existing standardized term (peanut butter). This has met with opposition from growers' associations. In a 1994 ruling, the FDA upheld the existing standard.

Commercial manufacture of peanut butter varies considerably. The influence of roasting time on sensory attributes and on chemical measurements of flavor components can be found in Reference 39 (see FLAVOR CHARACTERIZATION). Shelled peanuts are heated to an internal temperature close to 145°C (81) to obtain the proper roasted flavor. At this temperature, free amino acids and sugar in the peanut react to form pyrazines, the main roasted-flavor components (130). The peanut is then quickly cooled to stop the cooking at a definite point to produce a uniform product. Next, blanching is used to remove the peanut skins and hearts. Peanut hearts (embryonic axes) should be removed from the kernels prior to peanut butter manufacture because they impart a gray color and bitter flavor to the product (109); likewise, defective nuts should also be removed. Finally, a coarse or medium grind is made and the ingredients added and blended, in which both salt (about 2%) and sugar (0–6%) impart flavor. Oil separation in peanut butter products can be prevented by adding partially hydrogenated vegetable oils at levels of 3–5%; it may also be controlled by keeping the product at about 10°C (109). The final grind, at a temperature sufficient to obtain a melt of the stabilizer, produces the desired texture: smooth, creamy, or chunky; old-fashioned peanut butter, however, does not contain a stabilizer. Air is removed by vacuum, and the mixture is cooled to about 32°C before packaging to ensure longer shelf life and proper crystallization of the fat. The vacuum or gas-flushed final package should remain undisturbed for about 12–24 hours to allow complete crystallization which prevents cracking, shrinking, or pulling away from the container.

Nut Products. Peanut products include peanut flour, lipoprotein, protein, milk, and partially defatted peanuts (109). Pecan butter is made from dry roasted meats, ground to a very fine state, and mixed with salt (2% of final weight), hydrogenated fat (1.5%), and the antioxidant butylated hydroxyanisole (BHA) (131).

The press cake and ground kernels from bitter almond and other fruit pits contain glycoside amygdalin. Hydrolysis of amygdalin is accomplished by heating the cake or powder at 50°C with an excess of water; steam is passed through, and oil of bitter almonds is driven off together with hydrocyanic acid. Benzaldehyde thus collects in the condensate as a heavy lower layer, which is then drawn off. If the oil is to be used for flavoring, it is refined by heating with slaked lime and iron salt or with sodium bisulfite.

Research has shown that ascorbic acid can be produced from hulls of immature walnuts by extracting the hull with 0.2% sulfur dioxide solutions, and purifying the extract by adsorption on and elution from anion-exchange resins (see ION EXCHANGE). Eluates from the anion-exchange step are concentrated, purified by organic solvent fractionations, decolorized, and crystallized (35).

Uses

Nuts and Nut Products.   Nuts are used mainly as edible products and marketed either with or without the shell, as the demand requires. The most popular nuts in the shell are English walnut, filbert, almond, Brazil nut, peanut, pistachio, and the improved, or paper-shell, pecan; the most popular salted and roasted nut kernels include these as well as the cashew, macadamia, and pignolia. Each year more nuts are shelled in centrally located plants and marketed as meats. The annual per capita consumption of shelled tree nuts in the United States rose from 0.78 kg in 1987 to 1.00 kg in 1992 (132). Similarly, the annual per capita consumption of shelled peanuts has also continued to increase from about 1.4 kg at the turn of the century to about 2.7 kg in 1950, culminating with 3.18 kg in 1989 and 2.9 kg in 1992 (132).

For convenience, meats are rated and sold by sizes and grades. Nut kernels are used extensively in confections and in the baking industry as ingredients in pies, cakes, cookies, and other products. Various nut products include salted and roasted nuts, nut butters such as peanut and almond butter, macaroon paste and powder from almonds, specialty oils, and various confections, notably pralines, peanut brittle, and nut bars (124). Filberts are also widely used in the baking industry (133,134). Hazelnut butters may include added margarine to improve flavor and acceptability (135). Nuts such as pistachio, walnut, pecan, and almond are used extensively in the manufacture of ice cream. Nut flours, manufactured to a limited extent, offer a satisfactory supplement to wheat flour. Peanut flour manufactured from the finer grades of peanut meal is wholesome and nutritious, containing over four times as much protein, eight times as much fat, and nine times as many mineral ingredients as wheat flour. Acorn flour has been used extensively in Europe and is still used to a limited extent in making bread in the southwestern United States. Yeasted peanut butter containing 20% pasteurized brewer's yeast is said to have superior dietetic value because it contains a significant amount of the vitamin B complex group.

Nuts have many uses, both industrial and domestic. For instance, the ivory nut, or tagua, is a source material for the manufacture of buttons and turnery articles. The kola nut supplies ingredients for popular cola beverages in the United States (see CARBONATED BEVERAGES). *Strychnos nux-vomica* provides the important medicine and poison, strychnine. The areca or betel nut is chewed by the Indian and Malayan people as a narcotic: a slice of the nut is placed in a leaf of the pepper plant (*Piper betle*) together with a pinch of lime; the mixture is an acrid, astringent narcotic that dyes the mouth red, blackens and destroys the teeth. The areca nut contains, among other alkaloids, arecoline, an active anthelminthic widely used in veterinary practice for the treatment of tapeworm infections.

Peanuts.   Its popularity, together with the obvious potential for increased worldwide production, seems to make peanuts a prime candidate for meeting world food needs (136). However, peanuts are one of the three most important oil-bearing seed crops in the world, and the majority of their world production is crushed for the manufacture of peanut oil. Most of the countries with high levels of peanut production have very limited process and storage facilities for the manufacture and handling of other peanut products. The United States is an exception. More than half of the peanuts grown in the United States are used as peanut-food products. Only the surplus and mold-contaminated peanuts are crushed for oil. In 1993, for example, peanut oil amounted to only 2.9% of the total farmers' stock. Future patterns of distribution, utilization, and export are expected to depend to a large extent on changes in federal support and acreage policies.

Nearly half of the U.S. domestic food consumption of peanuts in 1993 was as peanut butter; salted peanuts, at 27.3%, and peanut candy, at 23.9% made up the other half (137). Although the per capita domestic peanut consumption in the United States has increased steadily, the consumption in recent years has not kept pace with production. Domestic food use of peanuts has been confined almost entirely to roasted peanuts. A number of investigations and developmental efforts are being made to extend the use of nonroasted peanut products such as flour and meal flakes. As of the mid-1990s, market outlets for these latter products are neither sizable nor firmly established. The food-use patterns emphasize the uniqueness and demand for products having a distinct roasted-peanut flavor. The development of the desired flavor as well as the storage stability of such flavor in peanut-food products are therefore important.

Partially Defatted Nuts. There is considerable demand for nuts and nut products of reduced fat content. Almond meal and peanut meal are examples of products having low fat content achieved by pressing oil from the nuts and by grinding the cake. Much of the flavor is in the oil; defatted nuts are thus less tasty.

The process for the defatting of whole peanuts can be found in References 109 and 138. Whole kernels are roasted at 305°C for eight minutes, or partly roasted at 215°C, and then extracted with hexane at room temperature for various amounts of time, after which the rate of oil removal is determined. Fully roasted nuts lose 81% of their oil content and have the best appearance after extraction for 120 hours. Solvent removal requires forced draught or vacuum conditions for 9–10 hours of drying. Salting is achieved by dipping into saturated brine or, preferably, by dipping into water followed by sprinkling with salt.

A process has been developed (139) whereby up to 80% of the oil can be removed from whole, raw peanuts without the use of solvent. In this process, the blanched peanuts are brought to a proper moisture content, pressed mechanically, and then reshaped or reconstituted by dipping in hot water; subsequently they can be roasted and salted, or used in confections or other formulations. Defatted peanuts may also be ground into meal and added to cookies, cakes, and many other products, where they impart a distinctly nutty flavor and crunchy texture. On the other hand, the resulting high grade oil is refined and employed in cooking and industrial products. This process can also be used for pecans, walnuts, almonds, Brazil nuts, cashews, and other nuts (140–142).

Defatted peanuts are high in protein, low in moisture, contain only 20% of the naturally occurring fat, and have better stability than whole peanuts. Monosodium glutamate (MSG) has been used as a flavor enhancer for defatted nuts, but the result has not been entirely satisfactory as the addition of MSG produces a meaty rather than nutty flavor. This meaty flavor is more compatible with salted butter and nuts than with candy.

Oil. Tung and oiticia are sources of quick-drying oils for the paint and varnish industry (see DRYING OILS). Coconut, babassu, and palm oils are used chiefly for the manufacture of margarine, soap, shaving cream, cosmetics (qv), and other domestic products. Walnut oil, a fine specialty oil having drying qualities, is used in the preparation of artists' colors (see PIGMENTS). Peanut oil is used as a lubricant and in shaving creams, shampoos, and cosmetics. It is also a good source of edible oil in the manufacture of shortening and margarine. Sweet almond oil is transparent, consisting chiefly of triolein. It has important uses as a laxative, in treating bronchitis and colds, and in fine soaps and cosmetics. Because most edible tree nuts are considered luxury food items, production of oil from these is usually not an economical practice except for specialty oils.

A considerable quantity of oil can be extracted from waste material from shelling and processing plants, eg, the inedible kernels rejected during shelling and fragments of kernels recovered from shells. About 300 t of pecan oil and 300–600 t of English walnut oil are produced annually from such sources. The oil is refined and used for edible purposes or for the production of soap; the cake is used in animal feeds (see FEEDS AND FEED ADDITIVES). Fruit-pit oils, which closely resemble and are often substituted for almond oil, are produced on a large scale for cosmetic and pharmaceutical purposes (143). For instance, leaves, bark, and pericarp of walnut may be used to manufacture vitamin C, medicines, dyes and tannin materials (144).

Cashew nutshell liquid or oil, obtained by heat treatment, contains about 10% cardol and 90% cardanol. The latter is a vesicant oily liquid from the decarboxylation of anacardic acid and its polymers. Formerly a waste product of the cashew kernel industry in southern India, cashew nutshell oil has become a valuable raw material in the manufacture of many industrial products. For instance, its condensation with formaldehyde and other chemicals yields resins that can be used in many unique combinations for the manufacture of brake linings, clutch and other friction facings; insulating and waterproof varnishes and coatings, lamination resins, molding compositions, oil and acid-proof cement, industrial floor tile, typewriter rolls, and numerous other products. Varnishes made with these resins are resistant to both acids and alkalies and possess unusual resistance to the softening action of mineral oils. These varnishes are used to coat paper for bottle-cap liners and for many other waterproofing and insulation purposes. Because of its high heat resistance, polymerized liquid is used as a potting compound for magnetoelectric machine armatures in airplanes. Subsequent treatment of the polymerized product with formaldehyde, trioxymethylene, paraformaldehyde, or furfuraldehyde at room temperature for 24–72 hours can produce a substantially infusible solid resin that retains high binding power at elevated temperatures without softening (145). In addition, hydrogenation of the phenols in oil produces many useful products, such as a cosolvent for rotenone in the preparation of insecticides. The hydrogenated oil, however, does not have the vesicant action of the original liquid. Cashew nutshell oil is also used in India for protecting wood and paper against termites as well as for waterproofing and preserving fishnets, textiles, and lumber (see COATINGS; WATERPROOFING AND WATER/OIL REPELLANCY).

Meal. The meal or press cake from oil extraction of pecan, walnut, almond, and other nuts is usually bitter because it contains skins and pieces of shell; when refined, however, it can be used in the baking industry and, more commonly, in animal feed. Almond, peanut, babassu, and other nut cakes and meals are extensively used in such feeds; in addition, ivory nut meal, a by-product of the button industry, also provides a valuable animal feed since most of its cell wall carbohydrate (92.5% mannose) can be converted into hexose sugar or its equivalent. Indeed, sheep and cows can digest and utilize as much as 84% of the dry matter and 92% of the nitrogen-free extract. Various methods have been developed for preparing mannose from the ground-meal waste from button manufacture. Attempts have also been made to utilize the horse chestnut, which contains a high percentage of starch. The conversion of starch by enzymes and fermentation, however, is made difficult by the presence of saponins which make up 10–14% of the raw meal.

Shells and Hulls. Nutshell waste from shelling and processing plants is used extensively for a variety of purposes. Pecan and English walnut shells reduced to flour of various mesh sizes can be used as soft grit in blasting metals; as an ingredient for plastic fillers, battery cases, molding resin forms, and industrial tile; as an insecticide diluent; and for cleaning fur. Radio horns made from walnut-shell flour seem to filter vibrations more effectively than loudspeakers made from other materials. Walnut-shell flour, which has a high lignin content, has been used on a commercial scale for some time as a diluent or spreader for agricultural insecticides. The presence of cutin (5%) is responsible for its capacity to absorb and retain toxic agents. When used in molding resins, the cutin in the shell flour reduces resin absorption while the lignin (qv) helps bind the molded product. These two in conjunction tend to reduce materially the proportion of resin necessary for making the molding compound (146). A practical floor covering has been developed from a mixture of water-insoluble aluminum, nutshell flour, pigment, and resinous materials. It is applied to a flexible base and covers the whole area with a coloring and binding material to fill any pores.

About half of the annual production of walnut-shell and pecan-shell flour, >1000 t/yr, is used in the plastics industry alone. Special machinery is required for grinding and transmitting the ground material because the dust from grinding is subject to spontaneous explosion. However, this hazard has largely been overcome by applying a fine spray of water to the shell flour as it leaves the grinder. Waste material from the pecan-shelling industry has shown commercial possibilities for the recovery and production of oil and tannin shell flour as well as activated charcoal (45). Tannic acid produced from pecan shells is used by the tanning industry and by the oil industry to control the viscosity of drilling muds (see PETROLEUM). Because the United States has to import two-thirds of the tannin used in the leather industry, the extraction of this material from nutshells is of vital importance. After the extraction of tannin, the spent shells are used for making activated charcoal. Pecan shell, for instance, is an excellent material for the production of activated charcoal (147). Charcoal produced from pecan shells not only proves more effective for odor removal than commercial carbons used in water treatment, but also has comparable decolorizing power as charcoals made from oat hulls and corn cobs (148). In addition, activated charcoal made by treating the pecan shell charcoal with concentrated hydrochloric acid, washing it free of acid, and heating it in an electric furnace for four hours at 800–1000°C in an atmosphere of carbon dioxide, has the same decolorizing effect on aqueous solutions of azo dyes as activated charcoal sold for water treatment. Much of the walnut shell charcoal is sold to poultry producers; some is used as a filtrant for high grade vinegar.

Table 9. United States Production, Imports, and Exports of Shelled Nuts, 1969–1993[a,b]

Year	Almonds		Filbert			Macadamia nuts		Pecans		
	Production	Imports	Production	Imports	Imports in-shell[c]	Production	Exports	Production	Imports	Exports
1969	55,339	136	3,024	2,016				41,024	118	884
1972	56,700	91	4,148	3,744				33,222	15	1,039
1975	72,576	220	4,953	3,854				44,779	1,193	1,660
1977	112,947	92	9,601	3,447				35,324	232	1,637
1979	140,339	63	10,623	2,118				31,442	300	1,539
1981	152,283	54	12,012	2,494				50,627	220	1,352
1983	89,578	67	6,700	4,013				40,310	2,244	1,548
1985	90,324	185	20,101	1,780				36,488	5,181	968
1987	236,844	174	6,426	1,442	3,863	3,554	236	45,213	4,839	1,469
1989	170,635	92	4,738	2,671	7,157	4,448	1,120	50,399	3,729	3,549
1990	229,805	49	5,101	3,776	10,116	4,367	1,493	76,701	11,382	6,641
1991	172,300	76	7,063	2,304	6,172	4,434	1,866	44,391	6,989	6,426
1992	192,593	96	7,888	3,288	8,808	3,852	1,866	27,675	11,314	5,928
1993	170,161	112	10,534	1,749	4,685	3,919	1,866	61,184	3,732	6,718

Table 9. (Continued)

Pistachio			Walnuts			Tree nuts			Peanuts		
Production	Imports	Exports	Production	Imports	Exports	Production	Imports	Exports	Production	Imports	Exports
			40,214	105	2,609	137,962	61,454	37,431	1,150,055	152	45,383
			44,521	586	6,236	140,734	81,090	49,329	1,485,432	201	174,903
			75,968	54	19,839	192,943	75,343	89,214	1,749,591	132	144,498
			69,658	66	7,574	204,827	39,702	87,028	1,690,120	24	322,862
			75,462	98	18,541	229,246	45,507	109,862	139,039	52	278,374
			81,630	37	23,614	275,856	34,561	104,407	100,132	1,729	139,896
			72,197	31	17,713	191,422	54,656	81,860	200,750	640	197,799
			79,453	43	24,109	285,422	56,436	146,670	321,178	725	269,365
5,441	808	1,295	76,250	175	22,112	373,730	49,434	159,104	255,156	727	152,117
6,729	793	1,927	73,004	53	24,968	296,539	63,387	167,774	191,727	551	215,662
15,694	318	3,574	67,482	35	23,851	358,852	74,051	181,163	170,044		
9,509	93	6,124	78,543	31	27,017	316,240	63,849	196,990			
24,396	147	10,094	60,957	3,003	21,769	317,360	86,084	184,448			
23,144	149	10,824	76,354	746	22,395	345,296	75,725	176,170			

[a]Refs. 132 and 153.
[b]Values are in metric tons.
[c]Numbers represent raw data.

Various methods of home-dyeing cotton and wool materials using natural dyes made from hulls of butternut, hickory nut, pecan, eastern black walnut, and English walnut have been described (149). As far back as during the Civil War, butternut hulls have been used to furnish the yellow dye for uniforms of the Confederate troops. More recent attempts have been made to manufacture yellow and brown dyes from filbert shells on a commercial scale. The hulls are treated with copper sulfate and concentrated nitric acid to produce a yellow color, with ferrous sulfate to produce olive-green, or with ammonia to produce ruby-red (150) (see DYES AND DYE INTERMEDIATES; DYES, NATURAL).

Peanuts have been studied extensively for the development of products ranging from dyes and ink to artificial wool; peanut shells or hulls have been used in the manufacture of lacquers, linoleum, dynamite, guncotton, celluloid, artificial leather, photographic film, cellophane, and rayon. Peanut shells are also used as a stock feed by mixing ground hulls and molasses in the proportion of 80% hulls and 20% molasses; the resulting product has a protein content of 6–7%. Moreover, peanut shells, combined with an asphalt binder, can also produce insulation block. Finally, an artificial vegetable fiber that resembles wool is produced from peanut protein; it can be used for making clothing and other fabrics by weaving with it an equivalent amount of wool (151).

World Production

Although nuts have been a staple food in many countries for generations, their status in the United States as a chief food crop is relatively recent. The main supplier of English walnuts, filberts, and almonds had been Europe. However, pecans and black walnuts are indigenous to North America, and the United States is the principal producer of pecans. Other U.S. nuts, such as beech, butternut, white walnut, American chestnut, chinquapins, hickory, piñon, and northern California black walnut, are utilized mainly for local consumption. Chestnuts and chinquapins are susceptible to the chestnut blight fungus, *Endothia parasitica*, which has virtually destroyed the American chestnut (152).

Foreign production of almonds, filberts, and walnuts is concentrated in southern Europe, especially in countries with mild, dry climates bordering the Mediterranean Sea. In some of these countries, only part of the total production is on a commercial basis. In addition to orchards and groves, nut-bearing trees are found growing in scattered plantings amid field and other tree crops, along roads, and in yards. For this reason, and for the additional reason that crop-reporting methods have heretofore been inadequate, data on areas used for the growing of nuts are of limited usefulness. More meaningful are the estimates of production. Unless otherwise indicated, data in Table 9 and in the following discussion are expressed in metric tons (t) on a shelled basis. The ton and dollar value figures (153) relating to production are an average of 1990 and 1991 values.

Commercially important nuts in world trade include almond, Brazil nut, cashew, chestnut, coconut (copra), filbert, macadamia, palm nut, peanut, pecan, pignolia, pistachio, and English walnut. Coconut, palm nut, peanut, as well as babassu, oiticia, and tung, are important sources of oil for soap, paint, varnish, as well as many other domestic and industrial uses.

Almonds. Almonds are closely related to peach, plum, apricot, and other stone fruits (154). The United States is the biggest supplier of almonds in the world. The annual production of 299,400 t in 1990–1991, grown primarily in California, was worth $591,560,000. This volume represents a doubling of the almond production in the United States from 1976 to 1990. As the next two biggest suppliers, Spain produces a yearly average of 55,000 t of almonds, Italy 21,000 t. Turkey, Morocco, and Portugal produce most of the remainder of the world almond crop, at 11,000 t/yr. Production in these last countries has changed little from the 1935–1939 average, except for Portugal, which underwent dramatic production fluctuations.

Since 1951, the majority of imported almonds have come from Spain and Turkey. Production of almonds in southern Italy has declined by nearly 50% owing to competition from Spain, the United States, and Turkey. Efforts have been made, however, to expand the Italian almond industry (155). The United States is a primary exporter of almonds. Shelled almonds are used primarily by candy manufacturers and nut salters. In-shell almonds are sold as mixtures of the in-shell nut through retail outlets.

Cashew and Brazil Nuts. Although cashews are native to northeastern Brazil, they have been introduced to India, Tanzania, Kenya, and Mozambique, which became their principal producers (156). The United States imports over half of the world exports of cashews, mostly from India, because cashews produced in Africa are exported to India for shelling. In fact, almost all of the processing, a complicated procedure requiring much hand labor, is done in India. Most of the cashews are used by the nut-salting trade, but these nuts are becoming increasingly popular in the confectionery trade as well.

The principal producer and exporter of the Brazil nut is Brazil, where the trees grow wild. However, small quantities of the shelled nuts are also imported from Bolivia, Peru, and other South American countries. In-shell Brazil nuts are sold in in-shell nut mixtures, primarily during the Christmas holiday season. Shelled Brazil nuts are used by nut salters and candy manufacturers. Total imports of both cashews and Brazil nuts depends largely on the price these nuts bring in New York markets.

Coconut. In 1988, total coconut production was 36,802,000 t, of which 81% was produced in Asia, mainly in Indonesia and the Philippines (157). The coconut is essentially a crop of the lowland tropics (157). On the average, five nuts are required to produce 1 kg of copra, the dried endosperm of the nut. Copra is further processed to obtain coconut oil and copra meal. To produce coconut milk, which is an emulsion of coconut oil and water, grated fresh coconut meat is mixed with hot water and pressed (157). Either poles having an attached sickle-shaped knife or monkeys (158) may be used for harvesting.

English or Persian Walnuts. Walnuts are produced in more countries than any of the other tree nuts but production figures are unreliable. The United States is the leading producer of walnuts, at 234,913 t in 1990–1991 (valued at $279,720,000). Production, mostly in California and Oregon, is nearly one-half of the world total. Many countries produce substantial quantities of walnuts. The USDA considers estimates of the commercial crop more reliable than the total yearly production figures, which in 1990–1991 were 152,500 t in China, 66,000 t in Turkey, 17,000 t in India, 13,000 t in France, and 12,000 t in Italy.

Walnuts, both in-shell and shelled, may be imported from many countries, eg, France, Italy, China, Turkey, and India. Most imported walnuts are smaller than domestic walnuts and are used by the confectionery and baking industries.

Filberts or Hazelnuts. Filberts are native to the forests of Europe and are widely grown in the Balkans (159). Turkey and Italy control a significant part of the hazelnut market (160). In 1990–1991, Turkey produced 350,000 t, about two-thirds of the world's filbert crop; Italy and Spain produced 140,000 and 12,000 t, respectively. Filberts are imported to the United States from Turkey, Spain, and Italy. Turkey is by far the biggest U.S. supplier. The U.S. production of 23,000 t/yr, valued at $18,519,000, is produced primarily in Oregon and Washington. Filberts are susceptible to the Eastern filbert blight fungus, *Anisogramma anomala*, which is not present in the Pacific Northwest (161). The nut-salting trade provides the largest demand for imported shelled filberts. In Europe, filberts (known as hazelnuts) are used in preference to peanuts. Turkish nuts have a strong flavor but also more flavor defects compared to the milder tasting Italian and Spanish filberts (162).

Macadamia. Macadamia nuts are native to Queensland, Australia, but are also grown commercially in South Africa and Guatemala. In the United States, production of macadamia nuts in Hawaii has increased threefold since 1971. In 1991–1992, yearly in-the-shell production was 49,500 t, valued at $34,650,000. Macadamia nuts are also grown in small plantings in California.

Peanuts. Since 1963, domestic peanut production (132,153) has doubled, according to the 1989–1991 average, to 685,292 t/yr of shelled peanuts, valued at $1,186,818. This is 7.6% of the world production and is grown on only 3.9% of the total world area planted with peanuts. Most of the U.S. production goes into the domestic edible trade and one-fifth is crushed for oil (153). In the United States, Georgia leads the peanut production (41%), followed by Alabama (15%), North Carolina (12%), and Texas (11%). Asia produces 15,730,000 t/yr, or 73.6% of the world's total, and Africa 15.4%. South America, the native land of the peanut, only produces less than 6% of the world total. The principal producers of peanuts are India (34.3%), China (29.8%), the United States (7.6%), Indonesia (4.2%), Sengal (3.2%), Myanmar (formerly Burma) (2.4%), and Argentina (2.2%). The highest yields, 2.4 and 2.2 t/ha, are obtained in the United States and Argentina, respectively. Peanut exports have doubled in the 1980s primarily because of excellent quality and low aflatoxin content. During 1989–1990, 215,662 t of shelled peanuts, a yearly average, along with 6,918 t of peanut oil (133) and 44,216 t of roasting stock (153), were exported, mostly to Europe, Canada, and Japan.

Pecans. Pecan is the most important horticultural crop native to North America (163). The United States is the only substantial producer of pecan, despite the fact that pecans have been introduced into Australia, Israel, South Africa, and Argentina. The principal producing states, from high to low, are Georgia, Texas, New Mexico, Louisiana, Alabama, Oklahoma, Mississippi, North Carolina, South Carolina, Florida, Arkansas, and California. Domestic production of 135,597 t/yr in 1990–1991 was valued at $308,954,500 (153).

Pistachios. Pistachios have been commercially produced in the United States since the late 1960s (164). The United States has displaced Turkey as the number two producer; Iran is the number one producer.

BIBLIOGRAPHY

"Nuts" in *ECT* 1st ed., Vol 9, pp. 547–567, by H. E. Hammar, U.S. Department of Agriculture; in *ECT* 2nd ed., Vol. 14, pp. 122–150, by J. G. Woodroof and C. T. Young, University of Georgia College of Agriculture; in *ECT* 3rd ed., Vol. 16, pp. 248–276, by C. T. Young, North Carolina State University.

1. J. A. Duke, *CRC Handbook of Nuts*, CRC Press, Inc., Boca Raton, Fla., 1989.
2. H. I. Forde, in R. A. Jaynes, ed., *Nut Tree Culture in North America*, Northern Nut Growers Assoc., Hamden, Conn., 1979, p. 84–97.
3. M. A. McCarthy and R. H. Matthews, *Agriculture Handbook*, Vols. 8–12, 1984.
4. N. Raica, J. Heimann, and A. R. Kennern, *Agric. Food. Chem.* **4**, 704 (1956).
5. E. J. Conkerton, E. J. St. Angelo, and A. J. St. Angelo, in *CRC Handbook of Processing and Utilization in Agriculture*, Vol. II: Part 2, *Plant products*, CRC Press, Inc., Boca Raton, Fla., 1983, p. 157–185.
6. C. O. Johns and D. B. Jones, *J. Biol. Chem.* **28**, 77 (1916); **30**, 33 (1917); **36**, 491 (1918).
7. D. B. Jones, *Proc. 1st Ann. Meet. Nat. Peanut Council* **1**, 31 (1941)
8. H. D. Baerenstein, *J. Biol. Chem.* **122**, 781 (1937).
9. L. Randoin and J. Boisselot, *J. Bull. Soc. Chem. Biol.* **25**, 250 (1943).
10. F. A. Cajori, *J. Biol. Chem.* **49**, 389 (1921).
11. E. H. Nollau, *J. Biol. Chem.* **21**, 611 (1915).
12. M. L. Orr and B. K. Watt, *Home Econ. Res. Rep.* **4**, (1968).
13. M. N. Rao and W. Polacchi, *Food Composition Table for Use in East Asia*, FAO. 1972.
14. *FAO Nutr. Stud.* **24**, (1970).
15. L. B. Rockland and B. Nobe, *Agric. Food Chem.* **12**, 528 (1964).
16. S. M. Sun, F. W. Leung, and J. C. Tomic, *J. Agric. Food Chem.* **35**, 232 (1987).
17. C. Ampe and co-workers, *Eur. J. Biochem.* **159**, 597 (1986).
18. S. M. Sun, S. B. Altenbach, and F. W. Leung, *Eur. J. Biochem.* **162**, 477 (1987).
19. D. L. Weller, *J. Food Biochemistry* **13**, 353 (1989).
20. R. B. French, *J. Am. Oil Chem. Soc.* **39**, 176 (1962).
21. W. G. Friedemann, *J. Am. Oil Chem. Soc.* **39**, 176 (1920).
22. A. L. Winton and K. B. Winton, *The Structure and Composition of Food*, Vol. 1, John Wiley & Sons, Inc., New York, 1932, p. 385.
23. L. Jurd, *J. Org. Chem.* **21**, 759 (1956).
24. Ref. 17, p. 209.
25. Ref. 17, p. 388.
26. G. Lercker and U. Pallotta, in *Cashew Research and Development: Proceedings of the International Cashew Symposium*, Indian Society for Plantation Crops, 1984, p. 184–195.
27. A. M. Henry, *J. Assoc. Off. Agric. Chem.* **37**, 845 (1954).
28. J. D. Guthrie and co-workers, *Agric. Ind. Chem.* **61**, 41 (1944) (mimeographed by U.S. Dept. of Agriculture Southern Regional Research Laboratory, New Orleans, 86 pp.).
29. P. Guest, *Proc. Am. Soc. Hortic. Sci.* **41**, 61 (1942).
30. W. O. Robinson and G. Edington, *Soil Sci.* **60**, 15 (1945).
31. M. Wood, *Calif. Agric. Extension Circ.* **103**, 65 (1947).
32. A. O. Alben, H. E. Hammar, and B. G. Sitton, *Proc. Am. Soc. Hortic. Sci.* **41**, 53 (1942).
33. R. E. Stephenson, *Better Fruit* **41**(11), 20,24 (1947).
34. A. R. C. Haas, *Proc. Am. Soc. Hortic. Sci.* **46**, 69 (1945).
35. A. A. Klose and co-workers, *Ind. Eng. Chem.* **42**, 387 (1950).
36. R. Melville, J. G. Organ, and E. M. James, *Biochem. J.* **39**(2), XXV (1945).

37. L. Rossi and co-workers, *Rev. Assoc. Bioquim. Arg.* **14**, 30 (1947).
38. L. E. Booher, *Natl. Peanut Council Proc.* **1**, 27 (1941).
39. G. Fabriani and M. A. Spadoni, *Quad. Nutria.* **10**, 88 (1947).
40. D. P. James, *Brit. J. Nutr.* **6**, 341 (1952).
41. H. P. Averill and C. C. King, *J. Am. Chem. Soc.* **48**, 724 (1926).
42. A. J. Mujumdar, A. H. Kapadi, and G. S. Pendse, *J. Plant. Crops* **7**, 69 (1979).
43. J. L. Huang and M. J. McLeish, *J. Chromatography* **475**, 447 (1989).
44. W. V. Cruess, J. H. Kilbuck, and E. Hahl, *Chemurgic Dig.* **6**(13), 197 (1947); J. R. Fleming, *Peanut J. Nut World* **26**(11), 24 (1947).
45. L. Jurd, *J. Am. Chem. Soc.* **78**, 3445 (1956).
46. M. van Eekelen and P. J. van der Laan, *Voeding* **4**, 1 (1945).
47. A. S. Haagen-Smit, *Proc. Acad. Sci. Amsterdam* **34**, 165 (1931).
48. W. Seaber, *Analyst* **58**, 575 (1933).
49. S. Shahabuddin and co-workers, *Indian J. Exp. Biol.* **18**, 1493 (1980).
50. N. M. Shivapurkyar and co-workers, *Indian J. Exp. Biol.* **18**, 1159 (1980).
51. A. Sinha and A. Rama Rao, *Toxicology* **37**, 315 (1985).
52. L. H. MacDaniels and D. L. Pinnow, *Proc. 67th Annual Report of the Northern Nut Growers' Assoc.* **67**, 114 (1976).
53. G. Borgstrom, *Proc. 55th Ann. Rep. Northern Nut Growers' Assoc.* **55**, 60 (1964).
54. J. R. Spies and co-workers, *J. Am. Chem. Soc.* **73**, 3995 (1951).
55. V. Sreenivasa and co-workers, *J. Nutr.* **61**, 87 (1957).
56. M. A. Greer and E. B. Asherwood, *Endrocrinology* **43**, 105 (1948).
57. I. E. Liener, in *Chemistry and Biochemistry of Legumes*, E. Arnold, London, 1982, p. 217–258.
58. C. D. Miller and L. Louis, *Food Res.* **6**, 547 (1941).
59. O. Moreiras and A. Pujol, *An. Biomatol.* **13**, 9 (1961).
60. G. Piva and E. Santi, in *Cashew Research and Development, Proceedings of the International Cashew Symposium*, Indian Society for Plantation Crops, Kasagarod, Kerala, Ind., 1984, p. 177–183.
61. B. Owens, *Virginia-Carolina Peanut News* **25**(4), 11 (1979).
62. J. Sabaté and co-workers, *N. Engl. J. Med.* **328**, 603 (1993).
63. W. W. Jones, *Plant Physiol.* **14**, 755 (1939).
64. W. Jones and L. Shaw, *Plant Physiol.* **18**, 1 (1943).
65. C. J. B. Thor and C L. Smith, *J. Agric. Res.* **50**, 97, (1935).
66. C. Vallee, *Compt. Rend.* **136**, 114 (1903).
67. M. LeClere du Salbon, *Compt. Rend.* **123**, 1084 (1896).
68. F. M. M'Clenahan, *J. Am. Chem. Soc.* **31**, 1093 (1909).
69. R. M. Sell and co-workers, *J. Agric. Res.* **73**, 319 (1946).
70. C. Golumdio and M. M. Kulik, in *Aflatoxin*, Academic Press, Inc., New York, 1969, p. 309–332.
71. A. A. Klose and co-workers, *Plant Physiol.* **23**, 139 (1948).
72. R. U. Schenk, *Crop Sci.* **1**, 103 (1961).
73. Ibid., p. 162.
74. C. Cavaletoo and co-workers, *Food Technol.* **20**(8), 108, (1966).
75. A. de la Cruz and co-workers, *Food Technol.* **20**(9), 123 (1966).
76. L. B. Rockland, *California Macadamia Society Yearbook* **8**, 30 (1962).
77. L. B. Rockland, P. C. Slodwski, and E. B. Luchsinger, *Food Technol.* **10**(2), 113 (1956).
78. L. B. Rockland, D. M. Swarthout, and R. A. Johnson, *Food Technol.* **15**(3), 112 (1961).
79. R. C. Wright, *U.S. Dept. Agric. Techn. Bull.* **80** (1961).
80. J. G. Woodroof and E. K. Heaton, *Ga. Agric. Exp. Stn. Tech. Bull.* **80**, (1961).
81. C. T. Young and K. T. Holley, *Ga. Agric. Exp. Stn. Bull.* **41**, (1965).
82. H. B. Cosler, *Manuf. Confect.* **38**, 15 (1958).

83. L. M. Redlinger and R. Davis, in H. E. Pattee and C. I. Young, eds., *Peanut Science and Technology*, American Peanut Research and Education Society, Yoakum, Tex, 1982, p. 520–570.

84. F. H. Arthur, personal communication with author, 1994.

85. E. A. Black and R. I. Cotton, *U.S. Dept. Agric. Circ.* **369**, 38 (1935).

86. J. G. Woodroof, in *Commodity Storage Manual*, The Refrigeration Research Foundation, Washington, D.C., 1974.

87. A. J. St. Angelo and R. L. Ory, *J. Am. Oil Chem. Soc.* **52**, 38 (1975).

88. W. P. Blount, *J. Br. Turkey Fed.* **9**(2), 55 (1961).

89. J. D. Reed and O. B. Kasali, in *Aflatoxin Contamination of Groundnut: Proceedings of the International Workshop*, ICRISAT, Patancheru, Ind., 1989, p. 31–36.

90. R. V. Bhat, Ibid., p. 19–29.

91. V. K. Mehan, in P. S. Reddy, ed., *Groundnut*, Indian Council of Agricultural Research, New Delhi, India, 1988, p. 526–541.

92. B. J. Wilson and W. Hayes, in *Toxicants Occurring Naturally in Foods*, National Academy of Sciences, Washington, D.C., 1973, p. 372–423.

93. U. L. Diener, R. E. Pettit, and R. J. Cole, in Ref. 83, p. 486–519.

94. G. N. Wogan, *Cancer Res.* **27**, 2370 (1967).

95. J. A. Miller, in Ref. 92, p. 507–549.

96. *Official Methods of Analysis, 15th ed.*, Assoc. of Official Analytical Chemists, Washington, D.C., 1990, p. 1184–1190.

97. J. Gilbert and M. J. Shepherd, *Food Add. Contaminants* **2**, 191 (1985).

98. J. Gilbert and co-workers, *Food Add. Contaminants* **8**, 305 (1991).

99. *Official Methods of Analysis, First Supplement*, Association of Official Analytical Chemists, Arlington, Va., 1990, p. 46–50.

100. H. Chu, in Ref. 89, p. 161–172.

101. *Food Chem. News* **36**(8), 62 (1994).

102. U. L. Diener, in *Peanuts-Culture and Uses*, American Peanut Research and Education Association, Stillwater, Okla., 1973, p. 523–557.

103. R. J. Cole and co-workers, in Ref. 89, p. 279–287.

104. R. E. Pitt, *J. Food Protect.* **56**, 139 (1993).

105. T. O. M. Nakayama, in Ref. 89, p. 203–207.

106. *Milled Peanuts Inspections*, USDA Fresh Products Branch, 1979.

107. T. B. Whitaker, T. W. Dickens, and R. J. Monroe, *J. Am. Oil Chem. Soc.* **51**, 214 (1974).

108. M. P. Doyle and co-workers, *J. Food Protect.* **45**, 964 (1982).

109. J. G. Woodroof, *Peanuts: Production, Processing, and Products*, Avi Publishing Co., Westport, Conn., 1973.

110. W. A. Parker and D. Melnick, *J. Am. Oil Chem. Soc.* **43**, 635 (1966).

111. W. P. Norred, *J. Food Protect.* **45**, 972 (1982).

112. R. W. Beaver, *Trends Food Sci. Tech.* **2**, 170 (1991).

113. H. J. Ciegler and co-workers, *Appl. Microbiol.* **14**, 934 (1966).

114. Y. Y. Hao and R. E. Brackett, *J. Food Science* **53**, 1384 (1988).

115. D. Y. Y. Hao and co-workers, in Ref. 89, p. 141–151.

116. S. Raisuddin and J. K. Misra, *Food Add. Contaminants* **8**, 707 (1991).

117. A. A. Arino and L. B. Bullerman, *J. Food Protection* **56**, 718 (1993).

118. D. L. Park and B. Liang, *Trends Food Sci. Tech.* **4**, 334 (1993).

119. M. T. Harvey and S. Caplan, *Ind. Eng. Chem.* **32**, 1306 (1940).

120. *Manual of Instruction*, California Almond Grower's Exchange, San Francisco, Calif., 1928.

121. G. Leffingwell and M. A. Lesser, *Peanut J. Nut World* **25**(10), 25 (1946).

122. S. R. Cecil and J. G. Woodroof, *Ga. Agric. Exp. Stn. NS.* **68**, 1959.
123. C. T. Young and W. E. Schadel, *Food Structure* **9**, 317 (1990).
124. Y. S. Rybakova and K. Konditer, *Prom.* **2**, 12 (1958).
125. *Fed. Reg.*, **29**, (1964).
126. C. Nail, *Peanut Farmer* **19**(3), 14 (1994).
127. *Food Chemical News* **35**(44), 51; **35**(50), 43 (1994).
128. *Food Labeling News* **2**(12), 26 (1993); **2**(15), 21 (1994).
129. N. H. Mermelstein, *Food Tech.* **47**(2) 81 (1993).
130. F. R. Eirich and E. K. Rideal, *Nature* **146**, 541,551 (1940).
131. E. K. Heaton and J. G. Woodroof, *Ga. Agric. Exp. Stn. Circ.* **19**, (1960).
132. USDA Economic Research Service, *Fruit and Tree Nuts Situation and Outlook Report*, USDA Economic Research Service, 1993.
133. R. Karney, *Proc. 78th Ann. Meet. Nut Growers Soc. Oreg. Wash. B.C.* **78**, 32 (1993).
134. P. Shannon, *Proc. 78th Ann. Meet. Nut Growers Soc. Oreg. Wash. B.C.* **78**, 38 (1993).
135. M. Villarroel and co-workers, *Plant Foods for Human Nutrition* **44**, 131 (1993).
136. A. L. Shewfelt and C. T. Young, *J. Food Sci.* **42**, 1148 (1977).
137. National Agricultural Statistics Service, *Peanut Stocks and Processing*, USDA, 1993.
138. J. Pominski, E. L. Patton, and J. J. Spadaro, *J. Am. Oil Chem. Soc.* **41**(1), 66 (1964).
139. H. L. E. Vix and co-workers, *J. Food Process. Mark.* **26**, 80 (1965).
140. H. L. E. Vix, J. Pearce, Jr., and J. J. Spadaro, *Peanut J. Nut World* **46**(4), 10 (1967).
141. H. L. E. Vix, J. Pominski, and H. M. Pearce, Jr., *Peanut J. Nut World* **46**(5), 10,18 (1967).
142. H. L. E. Vix, J. Pominski, and J. J. Spadero, *Peanut J. Nut World* **46**(6), 10 (1967).
143. M. L. Dewan, M. C. Nautiyal, and V. K. Sah, *Nut Fruits for the Himalayas*, Concept Publishing Co., New Delhi, India, 1991, p. 118.
144. A. F. Zarubin, *Reclamation and Development of Walnut and Fruit Forests in Southern Kirghizia*, Israel Program for Scientific Translations, Jerusalem, 1968, p. 92–93.
145. U.S. Pat. 2,165,140 (July 4, 1939), M. T. Harvey (to the Harvel Corp.).
146. L. C. Baron, *Peanut J. Nut World* **28**(1), 77 (1948).
147. T. R. McElhenney, B. M. Becker, and P. B. Jacobs, *Iowa State Coll. J. Sci.* **16**, 227 (1942).
148. T. H. Whitehead and H. Warshaw, *Ga. Inst. Technol. Eng. Exp. Stn. Bull.* **4**, 6 (1938).
149. M. S. Furry and B. M. Viemont, *U.S. Dept. Agric. Misc. Publ.* **230**, 28 (1935).
150. E. H. Wiegand, *Chemurgic Dig.* **2**(17), 143 (1943).
151. J. V. Sherman and S. L. Sherman, *The New Fibers*, Van Nostrand Reinhold Co., Inc., New York, 1946, p. 537.
152. R. A. Jaynes, in Ref. 2, p. 111–127.
153. *Agricultural Statistics*, USDA, 1993.
154. D. E. Kester, in Ref. 152, p. 148–162.
155. E. Senesi, A. Rizzolo, and S. Sarlo, *Ital. J. Food Sci.* **3**, 209 (1991).
156. D. Johnson, *J. Plant. Crops* **1**, 1 (1973).
157. G. J. Persley, *Replanting the Tree of Life*, CAB International, Wallingford, Oxon., U.K., 1992.
158. A. H. Green, *Coconut Production: Present Status and Priorities for Research*, World Bank Technical Paper, Washington, D.C., 1991.
159. I. Kovacevic, *The Basic Principles of the Cultivation and Selection of Hazelnuts*, NOLIT, Belgrade, 1962.
160. T. Warner, *Proc. 71st Ann. Meet. Nut Growers Soc. Oreg. Wash. & B.C.* **71**, 37 (1986).
161. H. B. Lagerstedt, in Ref. 152, p. 128–147. P. H. Thompson, in Ref. 152, p. 188–202.
162. C. Lauer, *Proc. 71st Ann. Meet. Nut Growers Soc. Oreg. Wash. B.C.* **71**, 43 (1986).
163. F. R. Brison, *Pecan Culture*, Texas Pecan Growers Association, College Station, Tex., 1986.

164. N. D. Vietmeyer and W. Reid, *Proc. 77th Ann. Report Northern Nut Growers Assoc.* **77**, 39 (1986).

General References

F. R. Brison, *Pecan Culture*, Texas Pecan Growers Assoc., College Station, Tex., 1986.

M. L. Dewan, M. C. Nautiyal, and V. K. Sah, *Nut Fruits for the Himalayas*, Concept Publishing, New Delhi, India, 1992, J. A. Duke, *CRC Handbook of Nuts*. CRC Press, Inc., Boca Raton, Fla., 1989.

J. F. Hutchinson, in *Horticultural Reviews* Vol. 9, Van Nostrand Reinhold Co., Inc., New York, 1987, p. 273–349.

ICRISAT, *Aflatoxin Contamination of Groundnut: Proceedings of the International Workshop*, ICRISAT Center, Patancheru. India, 1989.

R. A. Jaynes, ed., *Nut Tree Culture in North America*, Northern Nut Growers Assoc., Hamden, Conn., 1979.

H. E. Pattee and C. T. Young, eds., *Peanut Science and Technology*, American Peanut Research and Education Society, Yoakum, Tex., 1982.

P. S. Reddy, ed., *Groundnut*, Indian Council of Agricultural Research, New Delhi, India, 1988.

D. K. Salunke and co-workers, *World Oilseeds: Chemistry, Technology and Utilization*, Van Nostrand Reinhold Co., Inc., New York, 1992.

J. G. Woodroof, *Peanuts: Production, Processing, Products*, Avi Publishing, Westport, Conn., 1983.

J. G. Woodroof, *Tree Nuts: Production, Processing, Products*, Avi Publishing, Westport, Conn., 1979.

CLYDE T. YOUNG
North Carolina State University

NYLON. See POLYAMIDES, FIBERS.

OCEAN RAW MATERIALS

The ocean is host to a variety and quantity of inorganic raw materials equal to or surpassing the resources of these materials available on land. Inorganic raw materials are defined here as any mineral deposit found in the marine environment. The mineral resources are classified generally as industrial minerals, mineral sands, phosphorites, metalliferous oxides, metalliferous sulfides, and dissolved minerals and include geothermal resources, precious corals, and some algae. The resources are mostly unconsolidated, consolidated, or fluid materials which are chemically enriched in certain elements and are found in or upon the seabeds of the continental shelves and ocean basins. These may be classified according to the environment and form in which they occur (Table 1) and with few exceptions are similar to traditional mineral deposits on land.

Global Ocean Resources

The nature and distribution of the apparent global marine mineral resources are shown in Figure 1. Although quantitative determinations are for the most part speculative, several publications have addressed the issue in some detail (3–5). Moreover, geological research and exploration support many of the forecasts that the oceans may contain mineral deposits of even greater potential than those found onshore (6). Comparative values for apparent marine mineral resources for 86 mineral commodities (7) have been tracked by the U.S. Bureau of Mines. Alternative marine resources exceeding existing land resources exist for each mineral except platinum, asbestos, graphite, and quartz crystal (8).

The terms ore, mineral resources, and ore reserves appear commonly in the literature and are often misunderstood. As defined by the American Geological Institute (AGI) in 1980, ore is a natural aggregate of mineral materials which can be mined at a profit (9). Prices of metals and other mineral-derived commodities fluctuate owing to a variety of technical, economic, political, and social factors such as engineering capabilities, supply and demand, wars and embargoes, or needs and fashions. Elasticity of price can be seen for gold, cobalt, copper, silver,

Table 1. Classification of Global Marine Mineral Resources[a]

	Resource classification		
Location	Unconsolidated	Consolidated	Fluid
	Conshelf		

seabed	industrial materials
	  sand and gravel
	  shell sands
	  aragonite
	  coral sands
	mineral sands
	  magnetite
	  ilmenite
	  rutile
	  chromite
	  monazite
		outcrops
		  exposures of veins,
		    etc
			seawater
			  magnesium
			  sodium
			  uranium
			  bromide and salts of
			    26 other elements

subseabed	mineral sands
	  gold
	  platinum
	  cassiterite
	  gem stones
	bedded deposits
	  phosphorites
		vein, stratified, dis-
		  seminated or mas-
		  sive deposits
		  coal
		  phosphates
		  carbonates
		  potash
		  ironstone
		  limestone
		  metal sulfides
		  metal salts
			freshwater springs

	Ocean basins		

seabed	muds or oozes
	  metalliferous
	  carbonaceous
	  siliceous
	  calcareous
	  baritic
	nodules
	  manganese
	  cobalt
	  nickel
	  copper
		crusts
		  phosphorite
		  cobalt
		  manganese
		mounds and stacks
		  metal sulfides
			seawater
			  magnesium
			  sodium
			  uranium
			  bromine and salts of
			    26 other elements

subseabed		vein, stockwork, strat-
		  bound or massive
		  deposits
		  metal sulfides
			hydrothermal fluids

[a]Ref. 1.

and other metals. Ore reserves are currently economic mineral resources that have been measured, indicated, or inferred from acquired geological data. Only those reserves which are measured or indicated are used in the evaluation of a mineral property. There are few ore reserves in the marine environment other than those of operating mines because by definition the deposit must be

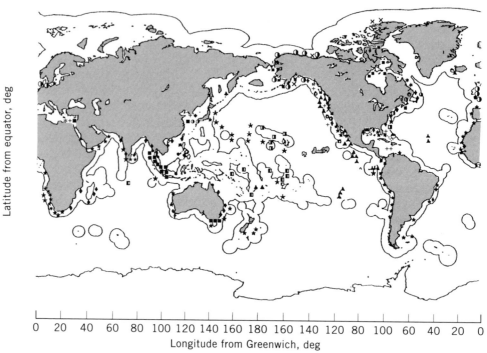

Fig. 1. Global distribution of seabed mineral deposits, where × represents chromite; + barite; ◆ titanium, zirconium, hafnium, and thorium; • tin; ■ gold, platinum, and silver; ◑ sand and gravel; ◓ shell, calcium carbonate; ∗ gems; ▲ marine polymetallic sulfides; ★ phosphorites; ◨ cobalt crusts; S sulfur; and ◻ ferromanganese nodules. Adapted from Ref. 2.

currently mineable. As of this writing (ca 1995) most marine mining technology is in the development stage. Mineral resources include all potentially recoverable minerals in a specified area, including measured ore reserves and undiscovered speculative deposits. Most marine mineral deposits are speculative and are classed as mineral resources. A more definitive discussion of the principles of the resource/reserve classification for minerals and forecasting methods used in the United States is given in Reference 10 (see also MINERALS RECOVERY AND PROCESSING).

Unconsolidated Deposits

Deposits that can be recovered without having to use explosives or other primary energy sources to break up the material in place are called unconsolidated deposits. These may be found stratified or disseminated as surficial or subsurface deposits on the continental shelf or in deep ocean basins.

Continental Shelf.   A variety of terms may be applied to deposits in the continental shelf (conshelf) including beach, bank, alluvial, strand-line, heavy mineral, or placer. These contain mostly industrial materials, mineral sands, or precious metals (Table 1).

Industrial Minerals.   Industrial minerals are bulk materials recovered for use directly as an industrial commodity rather than for their metal content.

These include sand and gravel derived from glacial and alluvial sources which are predominantly siliceous; biogenic carbonate materials such as shell, coral, and carbonate sands; aragonite [14791-73-2], a denser and harder form of calcium carbonate, commonly of marine origin; salt; sulfur; and precious corals. A key distinguishing feature of the industrial mineral class is that a large percentage (60–100%) of the deposits is composed of saleable material and only simple beneficiation may be required to prepare the commodity for market. Within the United States the most important commodity appears to be unconsolidated deposits of sand and gravel for use in coastal protection, beach replenishment, and industrial construction. Phosphorites are treated as a special case because of the maturity and complexity of the U.S. phosphate industry.

Sand and Gravel. Sand and gravel are terms used for different size classifications of unconsolidated sedimentary material composed of numerous rock types. The principal constituent of sand is normally quartz, although other minerals such as rock fragments or carbonates may be present, and may even be dominant, as is the case in tropical reef-derived sands. Gravel, because of its larger size, usually consists of multigrained rock fragments. Sand is generally defined as material that passes through a No. 4 mesh (4.75-mm) U.S. Standard seive and is retained on a No. 200 mesh (74-μm) U.S. Standard seive. Gravel is sized in the range of 4.75–76.2 mm in diameter. Resources of sand and gravel in the United States Exclusive Economic Zone (EEZ) appear to be not only adequate, but almost ubiquitous. More than two trillion metric tons of resources may be available (11). However, in the outer continental shelf (OCS), ie, U.S. seabeds seaward of the territorial limits of the States, it is doubtful if there are any known deposits that could be classed as commercial reserves as of this writing (1995). To evaluate a deposit for commercial use, it is necessary to know the specifications of the material needed for sale in a particular market. Similarly, it is necessary to ascertain that sufficient tonnage of saleable material exists within the deposit to sustain a profitable operation, ie, not only the areal extent, but the volume of the deposit, the grain size, the nature and quality of the grains and the percentage and location of unsaleable material within the deposit must all be ascertained. Of the vast quantities that are indicated as resources, a considerably lesser amount is expected to be mineable. Nevertheless, substantial mineable resources of sand and gravel aggregates exist off the U.S. coasts (12). Deposits of pure silica [7631-86-9] (qv) sand, valuable for optical and other special purposes, are found in certain environments. Methods for the exploration of sand and gravel deposits are both conventional and satisfactory. Many references to such exploration technology are available (13–15). Some 20% of total sand and gravel production in Japan (16) and 15% in the United Kingdom (5) is obtained from offshore as of the mid-1990s.

Biogenic Materials. Living creatures form shells, reef corals, foraminiferal (f-sands), and associated carbonate sands. Deposits of oyster, clam, and mollusc shells are commonly found in temperate waters in embayments or near the coast (17). Reef corals form coastal deposits in tropical or subtropical waters (18).

Aragonite. Calcium carbonate is a common deposit in shallow tropical waters as a constituent of muds, or in the upper part of coral reefs where it precipitates from carbon dioxide-rich waters supersaturated with carbonate from intense biological photosynthesis and solar heating. Deposits of oolitic aragonite,

$CaCO_3$, extending over 250,000 km^2 in water less than 5 m deep are mined for industrial purposes in the Bahamas for export to the United States (19).

Precious Corals. One important deep seabed resource having worldwide distribution is precious coral. The industry extends worldwide, but the richest beds are found on seamounts in the western North Pacific Ocean and the western Mediterranean Sea. The U.S. Bureau of Mines lists corals in its mineral statistics reports although most countries, including the United States, regulate the industry as fisheries. Precious coral fisheries have existed in the Mediterranean Sea since ancient times. Beds of commercial density were not developed elsewhere until the early nineteenth century off Japan. In 1966 about 95% of the world's coral was dredged off Japan, Okinawa, and Taiwan. Overfishing led to stock depletion. Intensive exploration led to discoveries off Hawaii at the junction of the Emperor Seamounts and the Hawaii Ridge system, about 800 km west of Midway. In 1980 these corals represented about 90% of the world's precious coral production. Throughout history, precious coral production has followed the pattern of exploration, discovery, exploitation, and depletion. Although the production of pink coral from Midway was about 140×10^3 kg in 1983, only about 10% of the beds lie within the United States EEZ and the Pacific fishery is virtually unregulated. There has been no production of precious corals from within U.S. waters as of 1995.

Mineral Sands. Mineral sands is the generic industrial term for sands containing gold, precious gemstones, or minerals of tin, ie, cassiterite [*1317-45-9*], SnO_2; titanium, ie, rutile [*1317-80-2*] TiO_2; leucoxene [*1358-95-8*] (altered rutile); ilmenite [*12168-52-4*], $FeTiO_3$; and other metals derived from the breakdown of rocks by weathering. Development activities for these minerals are well summarized (20). Other terms commonly used in the literature are placer, alluvial, and heavy mineral deposits. The principal distinction between industrial minerals and mineral sands is that the commercially valuable portion of the latter is rarely more than a few percent of the amount dredged. The waste product

Table 2. Examples of Metal Grades and Values for Crusts and Nodules[a]

Component	Price, $/kg	MPM[b] crusts				Line Island Ridge crusts			
		% Dry	% Wet	$/t Wet	$/m²	% Dry	% Wet	$/t Wet	$/m²
Co	27.56	0.83	0.54	148.82	2.98	1.00	0.65	179.14	3.58
Ni	4.98	0.49	0.32	15.94	0.32	0.55	0.36	17.93	0.36
Cu	1.77	0.07	0.05	0.89	0.02	0.07	0.05	0.89	0.02
Mo	10.58	0.06	0.04	4.23	0.08	0.06	0.04	4.23	0.08
Mn[e]	1.52	25.0	16.25	247.00	4.94	25.0	16.25	247.00	4.94
[f]	0.58	25.0	16.25	94.25	1.89	25.0	16.25	94.25	1.89
Total									
[e]				416.88	8.34			449.19	8.98
[f]				264.13	5.29			296.44	5.93

[a]Ref. 24.
[b]MPM = mid-Pacific mountains.
[c]*MIDPAC 81* cruise.
[d]Abyssal manganese nodules from the nodule belt area in the northeast Pacific.
[e]99.95% Mn.
[f]Ferromanganese (78% Mn) recalculated to 100% Mn basis.

from these operations is therefore generally over 95% of the total mined and processed. For the special cases of gold and platinum, the amount of saleable metal is measured in parts per million (ppm) and for diamonds in parts per billion (ppb). For all mineral sands or placers the beneficiation processes are usually based on the physical differences between the ore mineral and the waste materials and rarely employ chemical additives. Mineral sands are widely distributed throughout the world generally in coastal areas, eg, tin has been mined offshore in southeast Asia since 1910 (1), but the occurrence of many mineral sands deposits also has been indicated in the U.S. OCS (21). In the early 1990s significant discoveries were made of high quality diamonds in water depths over 100 m off the coast of Namibia and South West Africa. This discovery sparked a minor diamond rush there and in Canada, where geologic conditions were similar. The occurrence of continentally derived gold placers in submarine sand channels at depths greater than 2500 m has been postulated, based on new side scan imagery of the U.S. EEZ off southeastern Alaska (22). This offers an interesting potential for future deep seabed activities.

Phosphorites and Glauconite. Phosphorites, or marine apatites, $Ca_5(F,Cl, OH,1/2CO_3)(PO_4)_3$, are commonly, though not predominantly, found as oolitic sands or as nodules in areas of upwelling of phosphorus-rich deep water (23). Glauconite [1317-57-3] is an authigenic form of mica (qv) having a high potassium content, ie, 2.5–8.5% K_2O, and some rubidium, lithium, and boron. It is widely distributed and commonly found as a beach sand. It is often associated with phosphorites in both ancient and recent marine sediments. Its occurrence as rounded dark-green grains less than 1 mm in diameter have earned it the sobriquet green sands.

Ocean Basins. Ocean basins are primarily formed from oceanic basalts and may be interspersed with continental remnants, ridges, seamounts, or volcanic islands rising from the depths. Average water depth is around 4000 m but the most significant mineralization is generally found at 5000 m for manganese nodules, 4000 m for biogenic oozes, and 3000 m for hydrothermal metalliferous sulfides. The area is poorly explored, however.

Metalliferous Oxides. Marine oxide deposits primarily of manganese and iron are termed metalliferous oxides. These ubiquitous seabed deposits contain potentially commercial quantities of manganese (20–30%), copper, nickel,

Table 2. (Continued)

MIDPAC[c]				Nodule[d]			
% Dry	% Wet	$/t Wet	$/m^2	% Dry	% Wet	$/t Wet	$/m^2
0.79	0.51	140.56	2.81	0.24	0.16	44.10	0.44
0.49	0.32	15.94	0.32	1.21	0.79	39.34	0.39
0.07	0.04	0.71	0.01	1.00	0.65	11.51	0.12
0.06	0.04	4.23	0.08	0.04	0.03	3.17	0.03
24.6	15.99	243.05	4.86	25.2	16.38	248.98	2.49
24.6	15.99	92.74	1.85	25.2	16.38	95.00	0.95
		404.49	*8.08*			*347.10*	*3.47*
		254.18	*5.07*			*193.12*	*1.93*

and cobalt (<3% combined). Deposits occur in all oceans, primarily as discrete potato-sized nodules overlying soft sediments in water depths between 4000 and 6000 m. The nodules, commonly referred to as manganese nodules or polymetallic nodules, are primarily valued for the nickel content (Table 2). Nodules apparently form directly on the seabed through chemical precipitation from seawater and sediment pore-water fluids. Iron and manganese are the two most common transition metals in the earth's crust. The precipitation process, which seems to occur everywhere on the seabed, where the bottom water is oxygenated, represents the result of the transport of these metals from terrigenous rocks. Such deposits seem to form commonly where natural sedimentation rates are not high enough to overwhelm the chemical accretion processes. Extensive research has been conducted to elucidate and quantify the deposit formation processes because of the prevalence of these deposits in the oceans of the world and the apparent links to the global processes of weathering, primary production, and transport of metals through the world's marine ecosystems.

Large deposits of manganese nodules having relatively large concentrations of nickel, copper, and cobalt are found on the seabed surface in the principal oceanic basins. Deposits in several areas have been investigated for commercial recovery potential, including the Clarion-Clipperton region of the Northeastern Tropical Pacific approximately 5–20° North latitude, 110–155° West longitude (25,26); in the Central Indian Ocean south of India approximately 5–10° South latitude, 75–88° East longitude (27); in the EEZ of the Cook Islands in the South Pacific approximately 15–25° South latitude, 155–165° West longitude (28); and the Blake Plateau off the Atlantic coast of the United States approximately 30–35° North latitude, 65–80° West longitude (29).

Other excellent summaries describing what is known of the formation processes and geological setting of these deposits have also been assembled (30–32). Listings of nodule compositions for these and other areas have been compiled (33,34). For many sites these data are available as computer databases (qv) maintained by the NOAA National Geophysical Data Center in Boulder, Colorado.

Biogenic Materials.   Deep ocean calcareous or siliceous oozes are sediments containing ≥30% of biogenic material. Foraminifera, the skeletal remains of calcareous plankton, are found extensively in deep equatorial waters above the calcium carbonate compensation depth of 4000 to 5000 m. Similar deposits of radiolaria or diatoms composed of siliceous skeletal remains are widespread in more temperate areas in deep water below 5000 m. The deposits may be very pure. The diatoms recovered from deposits on land are used as fillers or filter materials or as a source of high quality carbonate or silica (see DIATOMITE).

Metalliferous Sulfides.   Hydrothermal discharges are found extensively in the vicinity of active oceanic plate boundaries. Precipitates from these discharges may form sediments or muds highly enriched in sulfides of lead, zinc, copper, silver, and gold. A prime example of these metalliferous muds is a series of deposits in the Red Sea overlain by hot pools of metalliferous brines. The Atlantis Deep deposit was discovered in 1976 in water depth of around 2200 m. The sediments, overlying oceanic basalt, are 2–25 m thick and cover approximately 60 km^2. They are estimated to contain 30×10^6 t Fe, 2.5×10^6 t Zn, 50×10^3 t Cu, and 9×10^3 t Ag. The sediment is compacted to the consistency of shoe polish at depth. On the surface it is quite soft and the water content can be as high as 200%. The wet density is about 1.25 g/cm^3, the shear strength from 60–250 N/m^2. To mobilize the required daily metals throughput in a full-scale operation it would be necessary to extract 10×10^3 t/d of *in situ* mud, creating 100×10^3 t/d of slurry in the process. This slurry would then need to be diluted with an equal amount of seawater before it could be pumped to the ship. Recovered as a relatively dilute slurry containing not more than 100 g/L solids of fine particle size (80% at ≤2 μm), the metalliferous muds are unlike any of the feeds usually supplied to mineral-processing facilities. The muds contain only 3–6% zinc, up to 1% copper, and about 50 g/t silver; thus concentration prior to extractive metallurgy is essential (see METALLURGY, EXTRACTIVE METALLURGY). Further diluting the slurry to limit the solids content to less than 30 g/L immediately prior to the slurry entering the flotation (qv) circuit poses no particular problems on the mining vessel. Direct leaching and thermal beneficiation are too costly.

Authigenic barium sulfate or barite [13462-86-7] is found in relatively high concentrations in sediments covering active diverging oceanic plate boundaries. It occurs as rounded masses containing up to 75% $BaSO_4$ or as a dispersed constituent of the sediment. Its origins are uncertain, but it is likely that it is associated with hydrothermal actions.

Consolidated Deposits

Consolidated deposits are those which occur as solid masses upon or within the structure of the seabeds. These may be removed only by fracturing, fluidizing, or dissolving the materials to be recovered.

Continental Shelf. Most consolidated mineral deposits found on the continental shelf are identical to those found on land and are only fortuitously submerged. Exceptions include those laid down in shallow marine seas or basins in earlier geochemical environments such as bedded ironstones, limestones, potash, and phosphorites.

Phosphorites. Phosphorites are of special significance. These constitute an already established and somewhat complex industry and a variety of technologies proposed for offshore mining. The deposits are most commonly bedded marine rocks of carbonate fluorapatite in the form of laminae, nodules, oolites, pellets, and skeletal or shell fragments. The commercial term phosphate rock includes phosphatized limestones, sandstones, shales, and igneous rocks containing apatite. Marine phosphorites, widely distributed in continental margins, vary in character depending on their genesis. They are found consolidated as crusts and indurated sands, sands in shallow basins, submerged plateaux, on the slopes of islands, and in tropical lagoons. Extensive bedded deposits are indicated offshore of the eastern United States. Marine phosphorites commonly contain minor quantities of uranium–thorium, platinum, cadmium, and rare earths (35,36).

Ocean Basins. Known consolidated mineral deposits in the deep ocean basins are limited to high cobalt metalliferous oxide crusts precipitated from seawater and hydrothermal deposits of sulfide minerals which are being formed in the vicinity of ocean plate boundaries. Technology for drilling at depth in the seabeds is not advanced, and most deposits identified have been sampled only within a few centimeters of the surface.

Metalliferous Oxides. Cobalt crusts, commonly referred to as ferromanganese crust, manganese crust, or high cobalt manganese crust, are relatively rich in cobalt, nickel, and platinum. These are found on hard-substrate seabed elevations, such as seamounts, and the submerged flanks of islands. They occur as encrustations up to 40 cm, but are more commonly 3–5 cm thick, on the exposed rocks of island slopes, seamounts, or submerged plateaux in water depths between 800–2400 m. Research into the genesis, composition, and distribution of these deposits has been significant since the early 1980s (37). The crusts, primarily evaluated on the basis of cobalt content, are commonly associated with platinum and phosphorites. Two general sources for the metals seem to dominate: pore-water solutions and colloidal suspensions within the host sediments, and seawater itself. Manganese, nickel, and copper seem to be relatively more important components in the pore-water source; cobalt and iron are relatively more important in the seawater source. The crust deposits seem to be composed almost entirely of material derived directly from seawater. Growth rates vary between $1-20 \times 10^6$ mm/yr and are generally at the lower rate.

Because crust deposits are derived almost completely from direct precipitation from seawater, these offer a potentially important record of the ocean's chemical environment during the past 80 to 100 million years. Work on Pacific deposits (38–40) indicate that the layers of deep ocean seamount deposits show significant differences related to regional, and perhaps global, changes in the marine chemical environment. Future research on the crusts can be expected to produce at least a partial history of seabed chemistry. As the links between the deposit formation and seawater chemistry become more fully understood, it is quite possible that the stratigraphic record of crust deposition could permit a better understanding of the constraints on variability of oceanographic parameters. Of particular interest are the long-term oceanic rates of absorption of CO_2, heavy metals, and other pollutants. The association of crusts with the rare-earth metals has been described (39).

Metalliferous Sulfides. Hydrothermal mineral deposits are formed by the action of circulating waters in seabed rocks which dissolve, transport, and redeposit elements and compounds in the earth's crust. Most of the familiar hard rock mineral lodes of tin, copper, barite, lead, silver, and gold on land or in the submerged continental shelves are of this type. More recently discovered, and potentially a significant resource, are deep seabed hydrothermal sulfide deposits (41), commonly referred to as metalliferous or polymetallic sulfides (PMS). These deposits may carry sulfides of almost any of the metallic elements, including copper, lead, zinc, silver, and gold. They are formed in the vicinity of active diverging plate boundaries throughout the world's oceans. Figure 2 indicates their known distribution in the Pacific. Knowledge of the grades, sizes, occurrences, and settings of these marine deposits is rudimentary, but consistent with the notion that this type of deposit should ultimately be as important as, or more important than, any other type of marine deposit (43–45).

Deposits which are forming are frequently characterized by venting streams of hot (300°C) mineralized fluid known as smokers. These result in the local formation of metalliferous mud, rock chimneys, or mounds rich in sulfides. In the

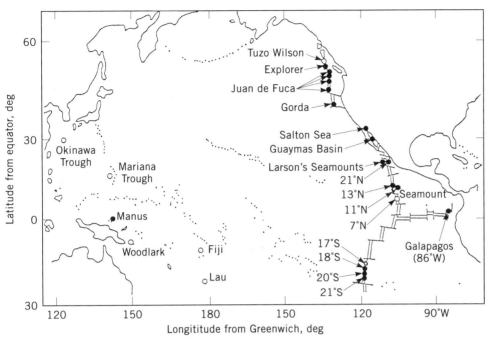

Fig. 2. Distribution of (•) known and (○) suspected metalliferous sulfide deposits and active hydrothermal vents in the Pacific Ocean (42).

upper fractured zone or deep in the rock mass beneath the vents, vein or massive sulfide deposits may be formed by the circulating fluids and preserved as the crustal plates move across the oceans. These off-axis deposits are potentially the most significant resources of hydrothermal deposits, even though none has yet been located.

Fluid Deposits

Fluid deposits are defined as those which can be recovered in fluid form by pumping, in solution, or as particles in a slurry. Petroleum products and Frasch process sulfur are special cases. At this time no valid distinction is made between resources on the continental shelf and in the deep oceans. However, deep seabed deposits of minerals which can be separated by differential solution are expected to be amenable to fluid mining methods in either environment.

Dissolved Minerals. The most significant source of minerals for sustainable recovery may be ocean waters which contain nearly all the known elements in some degree of solution. Production of dissolved minerals from seawater is limited to fresh water, magnesium, magnesium compounds (qv), salt, bromine, and heavy water, ie, deuterium oxide. Considerable development of techniques for recovery of copper, gold, and uranium by solution or bacterial methods has been carried out in several countries for application onshore. These methods are expected to be fully transferable to the marine environment (5). The potential for extraction of dissolved materials from naturally enriched sources, such as hydrothermal vents, may be high.

Minerals Recovery

Technology.   There are four basic methods of mining solid minerals: scraping the surface, excavating a pit or trench, removal through a borehole in the form of a slurry or fluid, and tunneling into the deposit (Fig. 3). All deposits on land are mined by one or more adaptations of these methods. Marine mining is amenable to the same basic approaches whether on the continental shelf or in the deep seabeds. Each mining method has variations that may be tailored to a specific situation. Most of the deposit types can be mined by more than one method. Similarly, any one method can be applied to more than one deposit type. These systems are described in a number of publications (1,5).

Environmental Considerations.   A significant advantage that appears to accrue from the recovery of mineral raw materials from the marine environment is the apparently benign effects of these activities when compared to the recovery of the same materials from land. Although a great deal of work remains to confirm this assumption (8), limited testing and monitoring on existing producing operations indicate an environmental advantage to ocean mining (47,48). Regulatory constraints regarding limits for potential toxic contaminants are given in Table 3.

Economic Aspects.   To be useful the raw materials must be recoverable at a cost not greater than the cost of similar terrestrial materials. These costs must include transportation to the point of sale. Comparative costs of recovery are strongly influenced by secondary environmental or imputed costs, such as legal costs or compensatory levies.

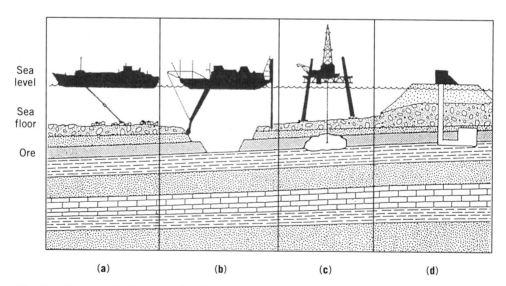

Fig. 3.   The four basic methods of mining: (**a**) scraping, (**b**) excavating, (**c**) fluidizing, and (**d**) tunneling (46).

Table 3. Coastal Water Quality Criteria for Toxic Substances Other Than Biocides[a]

Substance	Maximum acceptable concentrations		Minimum risk threshold, μg/L[d]
	96 h LC_{50}[b,c]	μg/L[c]	
aluminum	1/100	1500	200
antimony	1/50	200	
arsenic	1/100	50	10
barium	1/20	1000	500
beryllium	1/100	1500	100
boron	1/10		5000
bromine[e]			
cadmium[f]	1/100	10	0.2
chromium[g]	1/100	100	50
copper	1/100	50	10
fluorides	1/10	1500	500
iron		300	50
lead[g]	1/50	50	10
manganese	1/50	100	20
mercury[h]	1/100	1.0	
molybdenum	1/20		
nickel	1/50	100	2
phosphorus	1/100	0.1	
selenium	1/100	10	5
silver	1/20	0.5	
thallium[i]		100	50
uranium	1/100	500	100
vanadium	1/20		
zinc	1/100	100	20
cyanides[j]	1/10	10	5
detergents	1/20	200	
phenolics	1/20	100	
phthalate esters		0.3	
PCBs[k,l]		0.002	
sulfides[l,m]	1/10	10	5

[a]Ref. 49. [b]Maximum acceptable concentration figures expressed as fractions of 96 h LC_{50} for most sensitive species in given area. The 96 h LC_{50} is that concentration of a substance which kills 50% of the test species within 96 h under standard bioassay conditions. [c]Data are U.S. EPA official criteria where available; National Academy of Sciences (NAS) data used where EPA data not available. [d]NAS data, for concentrations "below which there is a minimal risk of deleterious effects." [e]The maximum acceptable concentration for free (molecular) bromine is 0.1 mg/L; for ionic bromate, 100 mg/L. [f]In the presence of copper or zinc in concentrations of 1 mg/L or more, the minimum risk threshold should be lower by a factor of 10. [g]In oyster growing areas, the minimum risk threshold should be lower. [h]According to NAS, "Fish-eating birds should be protected if mercury levels in fish do not exceed 0.5 mg/g." [i]1/20 of the 20-d LC_{50}. [j]Marine and estuarine aquatic and wildlife criteria not available; freshwater criteria are used (by EPA). [k]According to EPA, "The maximum acceptable concentrations of PCB in any sample consisting of a homogenate of 25 or more whole fish of any species that is consumed by fish-eating birds and mammals, within the size range consumed is 0.5 mg/kg on a net weight basis." [l]Data supplied by NAS. [m]These concentrations are valid only if salt water pH is between 6.5–8.5.

BIBLIOGRAPHY

"Ocean Raw Materials" in *ECT* 2nd ed., Vol. 14, pp. 150–170, by W. F. McIlhenny, The Dow Chemical Co.; in *ECT* 3rd ed., Vol. 16, pp. 277–296, by W. F. McIlhenny, Dow Chemical U.S.A.

1. M. J. Cruickshank, in *Mining Engineering Handbook*, Society of Mining Engineers, Littleton, Colo., 1992, Chapt. 28, pp. 1985–2028.
2. P. Hoagland and J. M. Broadus, *Seabed Commodity and Resource Summaries*, Technical Report WHOI 87–43, Woods Hole Oceanographic Institute, Woods Hole, Mass., 1987.
3. M. J. Cruickshank, in *Proceedings of the International Symposium on Minerals and the Environment*, Vol. 18, paper No. 20, Institution of Mining and Metallurgy, London, 1974.
4. V. E. McKelvey and co-workers, *Subsea Mineral Resources*, Bulletin 1689-A, U.S. Geological Survey, Woods Hole, Mass., 1986.
5. F. C. F. Earney, *Marine Mineral Resources*, Routledge, London, 1990.
6. M. J. Cruickshank and R. Kincaid, in *Minerals, Materials and Industry: Proceedings of the Fourteenth Congress of the Council of Mining and Metallurgical Institutions, Edinburgh, Scotland, July 2–6, 1990*, Institution of Mining and Metallurgy, London, 1990, pp. 197–206.
7. M. J. Cruickshank, *Technological and Environmental Considerations in the Exploration and Exploitation of Marine Minerals*, Ph.D. dissertation, University of Wisconsin, Madison, 1978.
8. M. J. Cruickshank and C. L. Morgan, *Synthesis and Analysis of Existing Information Regarding Environmental Effects of Marine Mining*, consulting report to Continental Shelf Associates for U.S. MMS Contract No. 14-35-0001-30588, U.S. Dept. of the Interior, Washington, D.C., 1992.
9. *Dictionary of Geological Terms*, American Geological Institute (AGI), Washington, D.C., 1980.
10. *Principals of a Resource/Reserve Classification for Minerals*, Circular 831, U.S. Department of the Interior, Washington, D.C., 1980.
11. Ref. 4, 106 pp.
12. M. J. Cruickshank, *Mar. Mining* **7**, 149–163 (1988).
13. L. E. Parkinson, *Underwater Sci. Technol. J.*, 63–68 (June 1970).
14. M. J. Cruickshank and R. W. Marsden, eds., *SME Engineering Handbook*, Society of Mining Engineers, New York, 1973, pp. 1–200.
15. E. H. Macdonald, *Alluvial Mining: The Geology, Technology and Economics of Placers*, Chapman and Hall, New York, 1983.
16. H. Okamura, *J. Mining Min. Inst. Jpn.* (1992); Engl. summary in unpub. proceedings of 14th Joint Meeting of the U.S.–Japan Cooperative Program in Natural Resources, Marine Mining Panel, Honolulu, Hawaii, June 6, 1992.
17. A. H. Bouma, *Shell Dredging and Its Influence on Gulf Coast Environments*, Gulf Publishing Co., Houston, Tex., 1976.
18. S. J. Dollar, *Sand Mining in Hawaii: Research, Restrictions, and Choices for the Future*, UNIHI-SEAGRANT-79-01, Sea Grant Program, University of Hawaii, Honolulu, 1979.
19. *Carib* 1:np (1978).
20. Ref. 5, 387 pp.
21. A. E. Grosz and E. C. Escowitz, in W. F. Tanner, ed., *Proceedings of the 6th Symposium on Coastal Sedimentology*, Florida State University, Talahassee, Fla., 1983, pp. 231–242.
22. M. R. Dobson, *Mar. Mining* **9**, 495–506 (1990).

23. M. J. Cruickshank and T. J. Rowland, *Mineral Deposits at the Shelf Break*, Publication No. 33, Society of Economic Paleontologists and Mineralogists, 1983, pp. 429–436.

24. P. G. Telecki, M. R. Dobson, J. R. Moore, and U. von Stackelberg, eds., *Marine Minerals*, Vol. 194, C. D. Reidel, Dordrecht, the Netherlands, 1987.

25. J. L. Bischoff and D. Z. Piper, eds., *Marine Geology and Oceanography of the Pacific Manganese Nodule Province*, Plenum Press, New York, 1979.

26. P. Halbach, G. Friedrich, and U. von Stackelberg, eds., *The Manganese Nodule Belt of the Pacific Ocean*, Stuttgart, Ferdinand Enke Verlag, 1988.

27. A. B. Valansangkar and N. V. Ambre, *Ocean Technology Perspectives*, Publications and Information Directorate, New Delhi, India, pp. 827–841.

28. A. L. Clark and co-workers, *Cook Islands Manganese Nodule Resource Assessment*, Pacific Islands Development Program, Honolulu, Hawaii, Feb. 28, 1993.

29. F. T. Manheim, *Science* **232**, 601–611 (1986).

30. N. A. Wogman, K. Chave, R. K. Sarem, eds., *Inter-University Program of Research on Ferromanganese Deposits of the Ocean Floor*, Seabed Assessment Program, Washington, D.C., 1973.

31. D. S. Cronan, *Underwater Minerals*, Academic Press, London, 1980.

32. P. G. Teleki, M. R. Dobson, J. R. Moore, and U. von Stackelberg, eds., *Marine Minerals*, NATO ASI Series, Vol. 194, C. D. Riedel Publishing Co., the Netherlands, 1987, p. 588.

33. J. M. Botbol and G. I. Evenden, *U.S. Geolog. Surv. Bull.*, 1863 (1989).

34. Japan International Cooperation Agency, in *Metal Mining Agency of Japan*, South Pacific Applied Geoscience Commission (SOPAC), Suva, Fiji, 1991.

35. F. T. Manheim, R. M. Prat, and P. V. McFarlin, *Soc. Econ. Paleontol. Mineralog.* **SP 29**, 117–137 (1980).

36. S. Riggs and co-workers, in Ref. 32, pp. 9–27.

37. F. T. Manheim and D. M. Lane-Bostwick, eds., in *U.S. Geological Survey, Open File Report 89-020*, 1989; *Final Environmental Impact Statement*, EIS/EA MMS 90-0029, 2 vols., U.S. Dept. of the Interior, Washington, D.C., Aug. 1990.

38. P. Halbach and D. Puteanus, *Earth Planet. Sci. Lett.* **68**, 73–87 (1984).

39. E. H. DeCarlo, *Marine Geol.* **98**, 449–467 (1991).

40. J. R. Hein and co-workers, *Paleoceanography* **7**(1), 63–77 (1992).

41. G. R. McMurray, *Proceedings of the Gorda Ridge Symposium, May 1987*, Springer-Verlag, New York, 1990.

42. S. D. Scott, in Ref. 32, pp. 189–204.

43. R. A. Koski, W. R. Normark, and J. L. Morton, *Mar. Mining* **4**(2), 147–164 (1985).

44. A. Malahoff, *Marine Mining: A New Beginning*, State of Hawaii, Honolulu, 1985, pp. 31–61.

45. P. A. Rona, *Marine Mining* **5**(2), 117–145 (1985).

46. M. J. Cruickshank and co-workers, *Marine Mining on the Outer Continental Shelf*, Report 87-0035, U.S. Dept. of the Interior, Washington, D.C. 1987.

47. P. C. Rusanaowski, *Alluvial Mining*, Institution of Mining and Metallurgy, London, 1991, pp. 586–601.

48. H. E. Thiel, E. J. Foell, and G. Schriever, *Berichte aus dem Zentrum für Meeres- und Klimaforshung der Universität Hamburg*, No. 26, Institut für Hydrobiologie und Fisserelwiessenschaft Hamburg, Hamburg, Germany, 1991.

49. J. Clark, *Coastal Ecosystems*, The Conservation Foundation, Washington, D.C., 1974.

MICHAEL J. CRUICKSHANK
University of Hawaii at Manoa

OCHER. See Pigments, organic.

OCTANE NUMBER. See Gasoline and other motor fuels.

OCTOATES. See Carboxylic acids; Desiccants; Driers and metallic soaps.

ODOR CONTROL. See Air pollution control methods; Odor modification.

ODOR MODIFICATION

Olfaction

Although the nose houses and protects the cells that perceive odor, it does not directly participate in odor perception. The primary function of the nose is to direct a stream of air into the respiratory passages. While this function is occurring, a small fraction of the inhaled air passes over the olfactory epithelium, located 5–8 cm inside the nasal passages. This olfactory area occupies about 6.45 cm^2 (one square inch) of surface in each side of the nose.

Olfaction begins when an odorant stimulates the olfactory receptor cells, triggering the opening or closing of the ion channels, which in turn convert this stimulus to an electrical response to the olfactory bulb and ultimately to other parts of the brain. The olfactory neurons send messages to the olfactory bulbs, structures about the size and shape of peach pits, located on the underside of the large overhanging frontal lobes of the cerebrum. Humans have tiny olfactive lobes in comparison with those of most other animals. Olfactory messages do not stop at the olfactive lobes; they also travel to brain regions involved in cognition, emotion, and other activities. These messages are subject to very little rational control. They bypass the thalamus, a group of brain cells that processes and edits most sensory information. The direct route to behavioral areas of the brain may be responsible for olfactory memory. These brain cells are proteinaceous material, and may consist of as many as 1000 different receptor sites. Olfactory neurons are the only nerve cells in the body that regenerate, replacing themselves once every month or two. Thus, people whose olfactory neurons are accidentally damaged, eg, by a blow to the head, may regain their sense of smell when new neurons grow and reestablish links with the proper regions of the brain. The fact that there is a gene responsible for odor receptor proteins was discovered in 1991 (1).

Perception of odor is therefore a physical mechanism by which information is processed in the brain. The nasal fossae are able to detect an unlimited number of dissimilar odor stimuli, sometimes from remote distances and sometimes in dilutions less than one part per trillion of air. In the case of the eye and the ear, the perceiving apparatus is confronted only with a limited and precise range of vibrations in receiving light and sound waves. For taste, there are only four primary stimuli. In contrast, the varieties of odor stimuli, which require evaluation and identification by the brain, are thought to number in tens or possibly hundreds of thousands. Day by day new odors are appearing, all of which are immediately accepted and sorted within the seemingly unlimited categories of the olfactory brain. The brain not only recognizes this information, but evaluates it, sorts it, and associates it with experiences, events, likes, and dislikes (Fig. 1).

The vomeronasal organ (VNO), located in the nose, is a small chemical sensing structure associated with odors and behavioral effects. The vomeronasal system, which is made up of the VNO and a portion of the brain's limbic system, is structurally independent of the olfactory and *nervous terminalis* systems in the nose. It may, however, interact with these systems in a manner dependent on prior experience or learning, and therefore be directly related to the association of smells and experiences. This independent chemosensory system in the nose

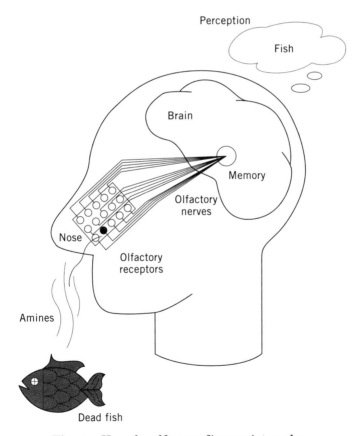

Fig. 1. How the olfactory fingerprint works.

may prove to open doors to new learning associated with the sense of smell and human behavior.

According to the physical theory of olfaction, odorous substances emit radiations of high frequency that directly transmit their energy or vibration to pigment granules in the olfactory receptors. The energy involved in olfactory stimulation is the characteristic molecular vibration of each specific odorous substance; differences in odor are therefore dependent on differences in the wavelength of the radiation emitted. To substantiate this theory, it has been demonstrated that certain odorous substances in the gaseous state have absorption bands in the ultraviolet region of the spectrum (0.20–0.36 μm), whereas the absorption bands of nonodorous compounds fall outside this range (2).

According to the chemical theory of olfaction, the mechanism by which olfaction occurs is the emittance of particles by the odorous substances. These particles are conveyed to the olfactory epithelium by convection, diffusion, or both, and directly or indirectly induce chemical changes in the olfactory receptors.

Researchers at the Monell Center (Philadelphia, Pa.) are using a variety of electrophysical and biochemical techniques to characterize the ionic currents produced in taste and olfactory receptor cells by chemical stimuli. These studies are concerned with the identification and pharmacology of the active ion channels and mode of production. One of the techniques employed by the Monell researchers is that of "patch clamp." This method allows for the study of the electrical properties of small patches of the cell membrane. The program at Monell has determined that odors stimulate intracellular enzymes to produce cyclic adenosine 3′,5′-monophosphate (cAMP). This production of cAMP promotes opening of the ion channel, allowing cations to enter and excite the cell. Monell's future studies will focus on the connection of cAMP, and the production of the electrical response to the brain. The patch clamp technique also may be a method to study the specificity of receptor cells to different odors, as well as the adaptation to prolonged stimulation (3).

There is still some debate regarding the degree of ability in humans to experience the sense of smell. Most people can distinguish between smells that are common odors and perceive the same type of smell. Although a certain smell can be perceived as agreeable or disagreeable to different people, the majority of humans have a fairly uniform sense of smell. The general population can detect the aroma of an apple pie baking in the oven; however, only those who are trained in the study of olfaction are able to detect the components of that aroma, ie, cinnamic aldehyde.

In 1986, the National Geographic Society, in cooperation with the Monell Center, conducted a worldwide survey of the sense of smell. Over 10 million survey forms were sent to readers of the Society's journal, of which close to 1.5 million forms were completed and returned. With responses to 40 demographic and 42 odor-related questions, the results constitute the largest set of data on human olfaction (4).

Both genetics and experience can influence one's olfactive capabilities. The inability of an individual to smell a certain odor is known as anosmia. According to the National Geographic Survey, 40–50% of the population cannot smell androsterone. For those who can smell androsterone, the odor perception ranges from sweet and perfumey to sweaty and urinous. Studies at the Monell Center have determined that genes play a significant role in determining whether a person can smell androsterone, and possibly how it is perceived (5). However, the ability to smell androsterone can be induced in some people, who initially cannot smell it, by repeated daily exposure to the compound.

There are certain well-established facts about olfaction (6). All normal people can smell. People suffering from brain lesions, injured olfactory nerve, or obstructed nasal passages may be anosmic. Cases of preferential anosmia, ie, ability to sense certain smells and not others, are not well established. Such cases occur, but little is known of them.

Some substances are odorous, others are not. Humans can smell at a distance; if one smells the roses in a garden, it is not ordinarily considered that part of the rose is in contact with the nose. Substances of different chemical constitution may have similar odors. Substances of similar constitution usually have similar odors, eg, in a homologous series; nevertheless, even stereoisomers may have different odors. Substances of high molecular weight are usually inodorous and often nonvolatile and insoluble. The quality as well as the strength of odor may change on dilution.

The sense of smell is rapidly fatigued. Fatigue for one odor does not affect the perception of other dissimilar odors, but will interfere with the perception of similar odors. Two or more odorous substances may cancel each other out; this compensation means that two odorous substances smelled together may be inodorous.

Odor travels downwind. Many animals have a keener sense of olfaction than humans. Insects have such extraordinary keenness of smell that it may be a different modality of the chemical sense from that known to humans.

Odors

Odors have been classified according to Carolus Linnaeus, the eighteenth century Swedish botanist who proposed seven odoriferous qualities: aromatic, fragrant, musky, garlicky, goaty, repulsive, and nauseous. Later in the twentieth century, ethereal (fruity) and empyreumatic (burnt organic matter), together with subdivisions of Linnaeus' classification, were added. In the 1990s, researchers concentrate less on categorizing odors, and more on how people detect and interpret them. Although the average person can name only a handful of common odors, this limitation results from memory retrieval failure, rather than a failure to detect the differences.

It was thought previously that there were no inborn odor preferences; that these are learned from experience. However, studies at the Monell Center have indicated that flavors consumed by a mother and transmitted into the milk influence the feeding behavior of her infant. When mothers consume garlic, their infants feed longer than when no garlic is consumed (7).

Odors are measured by their intensity. The threshold value of one odor to another, however, can vary greatly. Detection threshold is the minimum physical intensity necessary for detection by a subject where the person is not required to identify the stimulus, but just detect the existence of the stimulus. Accordingly, threshold determinations are used to evaluate the effectiveness of different treatments and to establish the level of odor control necessary to make a product acceptable (8). Concentration can also produce different odors for the same material. For example, indole (qv) in low concentrations has the smell of jasmine and a low threshold of perception. In high concentrations, it has a strong odor of feces and α-naphthylamine as well as a considerably higher threshold of perception.

Evaluation Methodologies

Industry has standardized procedures for the quantitative sensory assessment of the perceived olfactory intensity of indoor malodors and their relationship to the deodorant efficacy of air freshener products. Synthetic malodors are used for these evaluation purposes. These malodors should be hedonically associated to the "real" malodor, and must be readily available and of consistent odor quality. These malodors should be tested in various concentrations and be representative of intensities experienced under normal domestic conditions.

Panelists are trained to evaluate malodor intensity and the degree of modification. It is important that the panelist be able to smell through any extraneous odor(s), such as the fragrance of the product, to evaluate the efficacy of products making elimination or neutralization claims, as opposed to the phenomenon of masking.

Specific scales may be used to rate the perception of intensity of (1) the malodor, and (2) the malodor along with an odorous material designed to modify the malodor. Rating scales may consist of numerical assignments to words, eg, from 0 = "no odor" to 10 = "very strong odor." These same type of scales may be used to describe both the hedonic acceptability of the net result, ie, from 0 = "very unpleasant" to 10 = extremely pleasant," and the degree of modification, ie, from 0 = "does not modify" to 10 = "complete elimination or cover-up."

Protocols allow for at least two vessels to test both the untreated control malodor and the treated malodor(s). The untreated control malodor is used as a reference point for the maximum malodor score. When panelists evaluate each of the two or more vessels, they must wait a period of time (usually 30–60 s) for recovery from adaptation before smelling the next vessel. This step is always repeated between each evaluation.

Modification

Masking. Masking can be defined as the reduction of olfactory perception of a defined odor stimulus by means of presentation of another odorous substance without the physical removal or chemical alteration of the defined stimulus from the environment. Masking is therefore hyperadditive; it raises the total odor level, possibly creating an overpowering sensation, and may be defined as a reodorant, rather than a deodorant. Its end result can be explained by the simple equation of $1 + 1 = >2$ (Fig. 2a).

An olfactive evaluation of this phenomenon produces the following outcome:

$$\text{intensity rating malodor} = 6 \text{ (moderately strong)}$$

$$\text{intensity rating malodor} + \text{odorous material} = 8 \text{ (strong)}$$

$$\text{degree of modification} = 8 \text{ (good masking agent)}$$

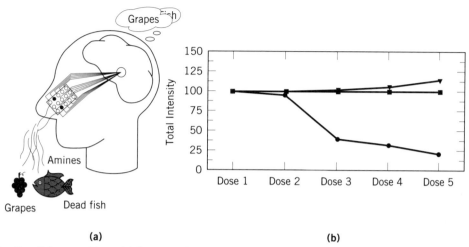

(a)   (b)

Fig. 2. Odor masking: (**a**) how masking agents work, and (**b**) masking curve for tobacco odor, where (▼) is perceived odor strength (treated); (■) is perceived odor strength (untreated); and (●) is tobacco odor, malodor reduction.

Odor masking does little or nothing to control malodors; it merely covers them up (Fig. 2**b**). Many materials used in masking odors are aldehydes, which are very chemically reactive and usually comprise the top note of a fragrance. Odor masking has had a long and colorful history. It gave birth to eau de cologne, devised to mask the malodor that was presumed to carry the plague. Odor masking is used in many areas of household, industrial, and institutional use via products that mask such malodors as pet smells, smoke, cooking, and numerous other odors. The forms by which masking is executed vary, and can be solid, liquid, and aerosol.

Counteraction. Counteraction, sometimes referred to as neutralization, occurs when two odorous substances are mixed in a given ratio and the resulting odor of the mixture is less intense than that of the separate components. The acceptable term to describe this occurrence is compensation. Materials that can accomplish this are basically organic odors which are highly polarized, have a strong affinity for each other, and may also have a low vapor pressure. Some of these molecules have the ability to compensate physiologically for certain malodor materials; others to react chemically with them. Counteraction occurs when the compensating substrate is able to form a coordinate bond with osmophoric sites unique to malodor molecules, such as amino- and thio- moieties. The result is overall reduction in odor; the malodor is transformed into an acceptable state, often with some residual freshening odor. This result lowers the total odor perception and can be exemplified by $1 + 1 = <2$ (Fig. 3).

An olfactive evaluation of this phenomenon produces the following outcome:

intensity rating malodor = 6 (moderately strong)

intensity rating malodor + odorous material = 5 (slightly strong)

degree of modification = 8 (good counteractancy agent)

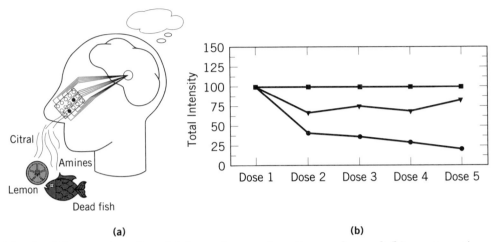

(a)   (b)

Fig. 3. Odor counteraction: (**a**) how odor counteraction works, and (**b**) counteraction curve, where (▼) is perceived odor strength (treated); (■) is perceived odor strength (untreated); and (●) is kitchen odor, malodor reduction.

It is unlikely that two odors when combined will cancel each other and result in no odor, ie, $1 + 1 = 0$; there is always some residual odor. However, reduction of an odor by an oxidation process can destroy the odor molecule permanently and leave no residual odor.

Commercial Aspects

Translating odor modifiers into consumer products results in forms, such as solids, liquids, and aerosols, for a market defined as products "for the nose." This includes products that cover up or eliminate odors, perfume the home, or cleanse the air. Such products thus defined were reported to have sales in 1992 of just under $2 billion. The categories of this market can be broken out as traditional air fresheners, cat litter products, aroma care, air purification, and disinfectant in both consumer and industrial applications.

This article deals primarily with the traditional air freshener category; some product trade names, manufacturers, and forms are listed in Table 1. The active ingredients of these pleasant odors are fragrances composed of both synthetic and natural essential oils. The products themselves can be described as either active (instant action, such as an aerosol), or passive (solids or liquids). The most recent advancement in this product category is that of electrically powered air fresheners. Consumer expectation of the efficacy of these products is increasing, putting greater demand on marketers to develop products that meet their needs. Passive product longevity claims by marketers range anywhere from 30 to 45 days. Active (aerosol) products' lifespan is minutes.

Behavior Modification by Odor

Although odorous materials no doubt impact each other, much discussion centers around the ability of odorous materials to influence human behavior. In articles

Table 1. Air Freshener Odor Modifiers

Manufacturer	Trade name	Form
	Industrial	
Walex Products Co.	Metazine, Exodor	liquid
Union Carbide	Abscents	powder
Belle-Aire Fragrances Inc.	Ordenone	liquid
	Consumer	
Armour-Dial	Renuzit	aerosol, solid, electric
Reckitt & Colman	Wizard, Airwick, Love My Carpet, Lysol	aerosol, electric, solid, carpet powder, liquid
Church & Dwight	Arm & Hammer	carpet powder
S. C. Johnson	Glade	carpet powder, aerosol, solid, electric, liquid
Scotts Liquid Gold	Touch of Scent	aerosol

ranging from scientific journals to trade magazines, there is discussion on the potential of fragrances, ie, essential oils, to affect people's moods, their ability to focus and maintain attention, to relax and sleep, and even their sexual capability.

Consider these findings from studies that assess whether fragrance is a powerful mood-altering substance (9):

In scented stores, shoppers look at merchandise longer.

Whiffs of peppermint and lily of the valley improve the alertness of people engaged in monotonous tasks.

A lemon scent pumped through the air-conditioning system of a Japanese office building increased workers' productivity.

Pleasant odors of flowers, fruit, or food can relieve the depression of hospital patients, whether they are aware of the scents or not.

People say they have an enhanced self-image and greater confidence when they think others like their fragrance.

The words aromatherapy, aromachology, and aromakinetics are coinages of the 1990s. Aromatherapy, once based on a tradition of folklore and herbal medicine, is being investigated scientifically.

A technique known as contingent negative variation (CNV) measures brain-wave reaction to olfaction. CNV is being conducted by universities as well as fragrance manufacturers (10). These types of studies have shown the effect of materials such as lavender and nutmeg in reducing stress or anxiety, and the ability of oils such as peppermint to stimulate brainwave activity. CNV research was incorporated into the development of the fragrance for a consumer personal care product launched in the late 1980s.

Airlines, such as Air New Zealand and Virgin Air, give their long-distance passengers after-flight regulator travel kits, which contain vials of scents labeled "awake" and "asleep." Sloan Kettering Hospital (New York) has conducted studies using the scent of heliotropin to reduce the anxiety of patients undergoing magnetic resonance imaging therapy.

The interrelationship between fragrance and psychology has been the subject of systematic investigation only recently. Consumer moods can be calculated in both positive and negative directions, and changes can be measured subjectively after exposure to fragrance. Personality traits, such as extrovert or introvert, may in some ways also be determined by fragrance preference.

Odors play a much greater role in human behavior than previously thought. The sense of smell provides a direct link with the function of the brain; therefore, the further study of olfaction can only advance the learning of causes and effects of stimuli to the brain.

The future in research will certainly lead to a better understanding of how odors are recognized, sorted, and classified. Studies promise, among other things, to determine whether perceptually similar, but structurally different, odors share the same class of receptor proteins, whether responses to odors can be modified, and possibly why olfactory neurons regenerate but other neurons do not.

Further to this, it should be possible to better understand how odors affect human moods and performance. Studies involving autistic children, severely mentally handicapped persons, and traumatic and stress-related situations where odors are being used to affect behavior may open many more doors to the understanding (4).

BIBLIOGRAPHY

"Odor Control" in *ECT* 2nd ed., Vol. 14, pp. 170–178, by A. Turk, The City College of the City University of New York; "Odor Modification" in *ECT* 3rd ed., Vol. 16, pp. 297–306, by R. H. Buckenmayer, Airwick Industries, Inc.

1. G. J. Dorland, *Spray Technol. Market.*, 16 (Mar. 1992); originally published in *Cell* (Apr. 1991).
2. C. P. McCord and W. N. Witheridge, *Odors Physiology and Control*, McGraw-Hill Book Co., Inc., New York, 1949, p. 20.
3. G. H. Gold, T. Nakamura, and G. Lowe, *Neurosci. Res.* (12), 127–134 (1990).
4. Technical data, Monell Chemical Center, Philadelphia, Pa.
5. G. K. Beauchamp, *Perfumer Flavor.*, 3 (Mar./Apr. 1993).
6. R. W. Moncrieff, *The Chemical Senses*, Leonard Hill, London, 1967.
7. J. A. Menella and G. K. Beauchamp, *Pediatrics* (88), 737–744 (1991).
8. F. A. Fuzzalari, ed., *Compilation of Odor and Taste Threshold Values Data*, American Society for Testing and Materials, Philadelphia, Pa., 1978, p. 1.
9. M. Menter, *Redbook*, 30 (Jan. 1993).
10. C. Warren and S. Warrenburg, *Perfumer Flavor.*, 12 (Mar./Apr. 1993).

General References

C. P. McCord and W. N. Witheridge, *Odors Physiology and Control*, McGraw-Hill Book Co., Inc., New York, 1949.
S. Lord, *Vogue*, 171 (Dec. 1991).
H. T. Lawless, *Chem. Senses Flavor*, **14**(3), 349–360 (1989).
H. C. Zwaardemaker, *Arch. Anat. Physical*, 423–432 (1900).
R. W. Moncrieff, *Export Rev. Br. Drug Chem. Ind.* (180), 27–29 (1955); (182), 29–35 (1955); (184), 31–37 (1955).
B. J. Willis, *Perfumer Flavor.*, 1 (July/Aug. 1993).
J. E. Amoore, *Molecular Basis of Odor*, Charles C. Thomas, Springfield, Ill., 1970.

M. G. J. Beets, in E. C. Carterette and M. P. Friedman, eds., *Handbook of Perception*, Vol. 6A, Academic Press, Inc., New York, 1978, pp. 245–255.

W. S. Cain, in E. C. Carterette and M. P. Friedman, eds., *Handbook of Perception*, Vol. 6A, Academic Press, Inc., New York, 1978, pp. 197–229.

Encyclopaedia Brittanica, 15th ed., s.v. "chemoreception," Chicago, Ill., 1988.

Encyclopaedia Brittanica, 15th ed., s.v. "sensory perception, human," Chicago, Ill., 1988.

K. Kukla, *Ann. N. Y. Acad. Sci.* **116**, 676–681 (1964).

E. R. Weaver, *Nat. Bur. Stand U.S. Circ.* (885.84) 491 (1950).

YVETTE BERRY
Reckitt & Colman Inc.

OIL SANDS.   See TAR SANDS.

OILS, ESSENTIAL

The volatile etherial fraction obtained from a plant or plant part by a physical separation method is called an essential oil. The physical method involves either distillation (including water, steam, water and steam, or dry) or expression (pressing). For the most part, essential oils represent the odorous part of the plant material, and therefore these oils have traditionally been associated with the fragrance and flavor industry (see PERFUMES). Since essential oils frequently occur as a very small percentage by weight of the original plant material, the processing of large quantities is often required to obtain usable amounts of oil. As a result, expression of an essential oil is only employed in those cases where both the form of the natural plant material, such as a citrus peel, and the quantity of oil present make the process feasible.

It has frequently been observed that the aroma of an essential oil is substantially different from that of the plant before processing. Because this phenomenon is largely the result of the treatment of the plant material with heat or hot water, various other methods have evolved over the years in an attempt to obtain a concentrate of the volatiles which more truly represents the aroma of the original. With the exception of the method of expression, almost all of these involve treatment of the plant material with one or more organic solvents (or mixtures thereof) followed by concentration of the extracted solute. Solvent extraction frequently yields, in addition to the volatile oil, various quantities of semi- or nonvolatile organic material such as waxes (qv), fats, fixed oils, high molecular weight acids, pigments (qv), and even alkaloidal material. However, because solvent extraction often results in a product with superior and more representative odor properties to that of a distilled oil, many natural products critically important to the flavor and fragrance industry are available as various extracts in addition to an essential oil.

Some of the commonly used botanical extracts include the following.

Absolute. This is concentrated extract obtained by treatment of a concrete or other hydrocarbon-type extract of a plant or plant part with ethanol. It is usually liquid and should be totally soluble in alcohol. By this method, waxes, hydrocarbons (including terpenoid), as well as most of the odorless material of the concrete are removed from the extract.

Absolute Oil. This is the steam distillable portion of an absolute. Frequently, the absolute oil possesses superior odor properties to that of the corresponding essential oil.

Aroma Distillate. Used by the flavor industry, aroma distillates are the product of continuous extraction of the plant material with alcohol at temperatures between ambient and 50°C followed by steam distillation, and, lastly, concentration of the combined hydro–alcoholic mixture. On cooling, terpenes often separate from the aroma distillate and are removed.

Concrete. Hydrocarbon extracts of plant tissue, concretes are usually solid to semisolid waxy masses often containing higher fatty acids such as lauric, myristic, palmitic, and stearic as well as many of the nonvolatiles present in absolutes.

Infusion. Infusion botanical extracts are tinctures that have been concentrated by either total or partial removal of the alcohol by distillation.

Oleoresin. Natural oleoresins are exudates from plants, whereas prepared oleoresins are solvent extracts of botanicals, which contain oil (both volatile and, sometimes, fixed), and the resinous matter of the plant. Natural oleoresins are usually clear, viscous, and light-colored liquids, whereas prepared oleoresins are heterogeneous masses of dark color.

Pommade. These are botanical extracts prepared by the enfleurage method wherein flower petals are placed on a layer of fat which extracts the essential oil. This method is applied to low odored flowers, which do not yield appreciable oil on steam or water distillation, or flowers of valuable but delicate odor (such as jasmin), which are destroyed on such treatment. Pommades, as such, are seldom used by the industry at present (ca 1995), but are further processed to provide more concentrated extracts such as absolutes. Absolutes, being alcohol-soluble, are much more convenient forms for the perfumer.

Resin and Resinoid. Natural resins are plant exudates formed by the oxidation of terpenes. Many are acids or acid anhydrides. Prepared resins are made from oleoresins from which the essential oil has been removed. A resinoid is prepared by hydrocarbon extraction of a natural resin.

Tincture. This is prepared by aqueous alcoholic extraction of the raw plant material. Since the extract is not further concentrated, the plant extract is not exposed to heat.

Essential oils are isolated from various plant parts, such as leaves (patchouli), fruit (mandarin), bark (cinnamon), root (ginger), grass (citronella), wood (amyris), heartwood (cedar), gum (myrrh oil), balsam (tolu balsam oil), berries (pimento), seeds (dill), flowers (rose), twigs and leaves (thuja oil), and buds (cloves).

Exceptions to the simple definition of an essential oil are, for example, garlic oil, onion oil, mustard oil, or sweet birch oils, each of which requires enzymatic release of the volatile components before steam distillation. In addition, the physical process of expression, applied mostly to citrus fruits such as orange, lemon, and lime, yields oils that contain from 2–15% nonvolatile material. Some flowers or resinoids obtained by solvent extraction often contain only a small portion of volatile oil, but nevertheless are called essential oils. Several oils are dry-distilled and also contain a limited amount of volatiles; nonetheless they also are labeled essential oils, eg, labdanum oil and balsam oil Peru. The yield of essential oils from plants varies widely. For example, nutmegs yield 10–12 wt % of oil, whereas onions yield less than 0.1% after enzymatic development.

The function of the essential oil in the plant is not fully understood. Microscopic examination of plant parts that contain the oil sacs readily shows their presence. The odors of flowers are said to act as attractants for insects involved in pollination and thus may aid in preservation and natural selection. Essential oils are almost always bacteriostats and often bacteriocides. Many components of essential oils are chemically active and thus could participate readily in metabolic reactions. They are sources of plant metabolic energy, although some chemists have referred to them as waste products of plant metabolism. Exudates, which contain essential oils, eg, balsams and resins, act as protective seals against disease or parasites, prevent loss of sap, and are formed readily when the tree trunks are damaged.

The cultivation of essential oil-bearing plants has kept pace with modern agricultural methods. Hybrids are grown to yield oils of specific odor, flavor, or other properties. New plants have been developed, eg, lavandin, and new essential oils are isolated and evaluated every year. Very few are commercialized. New growing areas for specific oils are opened that offer quality or economic advantages, such as soil conditions, irrigation, and the availability of labor. Most oils are prepared close to their source so as to provide access to the freshly harvested plant material and to keep processing costs low. Exceptions include the absolutes and other extracts of natural oleoresins, such as myrrh, olibanum, and labdanum, which tend to maintain their odor quality over a longer period of time. A few oils are still produced under very primitive conditions.

Rectified oils have been redistilled to improve a particular property or characteristic, such as flavor or aroma. For example, natural oil of peppermint is frequently rectified to remove dimethyl sulfide, which has a powerful and objectionable cooked vegetable note deleterious to the use of the oil in crème de menthe liqueurs. Distillation is also used to remove psoralens, which are harmful photosensitizing agents present in natural bergamot oil. Color may be removed, eg, from cassia oil, by vacuum steam distillation. A desirable component, such as 1,8-cineole (eucalyptol) 85% in eucalyptus oil, may be concentrated further by distillation to remove a forerun (topping).

Concentrated or folded oils are processed by various physical means to remove wholly or partly undesirable or nonflavor components, such as terpenes or sesquiterpenes, which have poor alcohol and water solubility, very low flavor value, and poor stability. Although this group, for the most part, comprises citrus oils with high terpene contents which cause clouding in drink applications, other oils such as spearmint are included. The processing methods include fractional distillation, topping, solvent extraction, countercurrent extraction, supercritical extraction, thin-film evaporation, and molecular distillation. In some cases, both distillation and solvent extraction are needed for complete removal of terpenes.

Thus, such oils as tangerine-terpeneless, lemon-sesquiterpeneless, or orange-80% terpeneless are processed oils. Some oils, particularly citrus, are folded or concentrated to reduce the terpene content to a designated level, ie, when half of the volatile constituents of the oil are removed, their removal is said to double the concentration and the oil is then called twofold. Although termed concentration, this process is, nevertheless, not merely a concentration in the ordinary sense, since the flavor body of the concentrate is always weaker than that of the complete essential oil, demonstrating that valuable products are lost in the course of removing the terpenes. In the past, a distinction was made between terpeneless and sesquiterpeneless oils, but this distinction has been abandoned since it is only fractional distillation which can practically remove monoterpenes without removing sesquiterpenes at the same time.

Aroma chemicals are isolates, or chemically treated oils or components of oils. Some components are removed physically, others chemically. In most cases, they are further purified by distillation. For example, Bois de Rose (rosewood) oil may be distilled to isolate linalool, which may be then further treated chemically to yield derivatives such as linalyl acetate, an important fragrance ingredient and a primary component in its own right of lavender and lavandin oils. Vetiver oil Haiti, although containing only 70% alcohols, is treated with acetic anhydride, then carefully distilled to include valuable odor components in the distillate, even though they may not be esters.

A number of other valuable aroma chemicals can be isolated from essential oils, eg, eugenol from clove leaf oil, which can also, on treatment with strong caustic, be isomerized to isoeugenol, which on further chemical treatment can be converted to vanillin (qv). Sometimes the naturally occurring component does not require prior isolation or concentration, as in the case of cinnamaldehyde in cassia oil which, on direct treatment of the oil by a retro-aldol reaction, yields natural benzaldehyde (qv). This product is purified by physical means.

Economic Aspects

Essential oils are used as flavoring and fragrance agents in every possible application. Combinations have raised greatly the total sales volume; eg, mint and cinnamon are used in toothpaste, mouthwash, or lozenges. Combinations can be found in every fragranced product, such as fine fragrances, soaps, detergents, room fresheners, paper, printing ink, paint, candles, condiments, floor polishes, etc. Convenience foods and frozen foods are flavored best by essential oils or oleoresins. Although citronella oil was used as such as an insect repellant, synthetic repellants have, for the most part, taken their place. Flavor essential oils are encountered in baked goods, snack foods, soft drinks, liqueurs, tobacco, sauces, gravies, salad dressings, and other food products.

In 1993, the United States imported nearly 22×10^6 kg of essential oils at a total value of almost $\$190 \times 10^6$, an increase over 1992 of ca 2.3×10^6 kg and \$935,000. Table 1 lists the quantities and values of 35 imported essential oils. The United States exports seven principal essential oils: orange, lemon, peppermint, spearmint, cedarwood, clove, and nutmeg. The latter two are not grown in the United States but are imported as dried spice, processed for oil, and then exported.

Table 1. U.S. Imports of Specified Essential Oils, 1993[a]

Essential oils	CAS Registry Number	Amount, kg	Value,[b] $\times 10^3$ $
anise oil	[8007-70-3]	49,455	400.8
bergamot oil	[8007-75-8]	37,821	2,362.9
caraway oil	[8000-42-8]	10,878	411.8
cassia oil	[8007-80-5]	385,158	16,477.1
cedarwood oil	[8000-27-9]	338,179	1,693.6
citronella oil	[8000-29-1]	885,843	3,955.2
citrus oils, other		358,230	2,866.6
clove oils		462,770	1,486.7
cornmint		248,841	1,157.7
eucalyptus oil	[8000-48-4]	454,152	1,859.2
geranium oil	[8000-46-2]	64,251	2,924.9
grapefruit oil	[8016-40-4]	178,501	1,331.4
jasmin oil	[8022-96-6]	10,716	1,967.8
lavender oil (including spike[c])	[8000-28-0]	417,518	6,253.5
lemon oil	[8008-56-8]	1,406,479	23,028.6
lemongrass oil		67,796	479.4
Bois de Rose oil		30,088	700.6
lime oil	[8008-26-2]	756,724	13,267.9
mint oil, other varieties		79,498	859.6
nutmeg oil	[8008-45-5]	109,520	935.6
onion and garlic oils		73,563	5,219.6
orange oil	[8008-57-9]	11,908,627	16,205.6
orris oil	[8002-73-1]	3,387	2,752.7
patchouli oil	[8014-09-3]	390,100	7,398.6
peppermint oil	[8006-90-4]	146,739	2,558.1
petitgrain oils		76,507	2,010.1
pine oil	[8002-09-3]	2,117	12.3
rose oil (Attar of Roses)	[8007-01-0]	2,504	6,666.0
rosemary oil	[8000-25-7]	63,894	717.5
sandalwood oil	[8006-87-9]	31,052	3,280.9
sassafras oil		250,880	1,027.0
spearmint oil	[8008-79-5]	318,487	3,019.5
vetiver oil	[8016-96-4]	48,434	2,795.3
ylang ylang oil (cananga[d])	[8006-87-3]	45,105	2,865.4
other essential oils		2,214,041	46,671.7
Total		*21,927,855*	*187,621.2*

[a]Ref. 1.
[b]All values are fob country of origin.
[c]CAS Registry Number [8016-78-2].
[d]CAS Registry Number [68606-83-7].

Composition

The volatile components of essential oils, for the most part, are made up of relatively low molecular weight ($\leq \sim$300–350) organic molecules of carbon, hydrogen, and oxygen, and occasionally nitrogen and sulfur. Less frequently, chlorine and bromine are found, eg, in seaweed volatiles. By far the largest class of natural volatiles of plants is the terpenes, which consist of head-to-tail condensation products of unsaturated five-carbon isoprene units. The simplest terpenes are monoterpenes with 10 carbon atoms. They may be aliphatic, alicyclic, or bi- or tricyclic, with varying degrees of unsaturation up to three double bonds. Sesquiterpenes contain three isoprene units, diterpenes four, triterpenes six, etc. Diterpenes (C-20) or larger units are rarely found in volatile oils, but may be extracted from the botanical. Oxygenated terpenes (terpenoids) usually accompany terpenes in essential oils, but are often present in a lower total percentage (see TERPENOIDS).

Other commonly occurring chemical groups in essential oils include aromatics such as β-phenethyl alcohol, eugenol, vanillin, benzaldehyde, cinnamaldehyde, etc; heterocyclics such as indole (qv), pyrazines, thiazoles, etc; hydrocarbons (linear, branched, saturated, or unsaturated); oxygenated compounds such as alcohols, acids, aldehydes, ketones, ethers; and macrocyclic compounds such as the macrocyclic musks, which can be both saturated and unsaturated.

An essential oil may contain >200 components, and often the trace substances ($\leq$ ppm) are essential to the odor and flavor of the oil. The absence or decreased presence of even one component may be cause for odor or flavor rejection of the oil. The same species of plant grown in different parts of the world usually contains the same chemical components, but the relative percentages may be different. Climatic and topographical conditions affect plant chemistry and can alter the essential oil content both qualitatively and quantitatively. For example, Bulgaria and Turkey both grow *Rosa damascena* Mill. for the production of rose oil. Each contains the same constituents, but the relative percentages are somewhat different although not sufficiently so to cause rejection of one over the other. Synthetic versions of several of the key aroma ingredients of rose oil, such as *cis*- and *trans*-rose oxides (**1**), β-phenethyl alcohol [60-12-8], (**2**), and *trans*-β-damascenone [23726-93-4] (**3**), are available commercially.

(1)   (2)   (3)

Analytical Methods

Early methods for the determination of the quality of an essential oil involved such subjective techniques as odor and color comparison with respect to a standard retention sample. Gradually, analytical techniques evolved to include specific gravity, refractive index, distillation range, and iodine number determination, and gas–liquid chromatography (glc) analysis of essential oils provided quick and routine determination of qualitative and quantitative profiles of the volatile constituents of an oil. It became obvious that many essential oils were much more complex than originally believed. Many structures, both simple and complicated, were worked out by natural products chemists before the availability of chromatography. Some of the early structures needed revision as purer isolates became available and as more sophisticated tools for structure elucidation were developed, including ir and uv spectroscopy, mass spectrometry (ms), and ^{1}H- and ^{13}C-nmr. With the advent of capillary glc, the more sensitive flame ionization detection (FID), and the marriage of this separation technique with the total ion current detector of the mass spectrometer (ms/gc), baseline separation of most volatile components of an oil was achieved. Not only can trace constituents present at levels below 1% in the oil be readily identified, but new, previously undetected components have become evident. To aid in their structure elucidation, new, more powerful, and selective characterization tools such as Fourier-transform gc/ir, high resolution ms/gc (by means of which an empirical formula may be obtained), chemical ionization ms/gc (which can provide the molecular weight of a molecule), and ftnmr and cryogenic nmr with much improved sensitivity have been developed and are being routinely used.

A component with critical odor properties may be present in an oil at ultratrace levels for which a discreet glc peak cannot be readily assigned. This is often the case with molecules containing heteroatoms such as nitrogen and, particularly, sulfur. In this case, the use of olfactory detection involving glc sniff runs is required to pinpoint the elution time of the trace constituent. Thus equipped, the natural product "detective" can zero in on the compound and, through preconcentration methods or the use of ultrasensitive spectroscopy tools, sufficient information can frequently be obtained for a preliminary postulation of structure, which must then be confirmed through synthesis. Through these methods, such powerfully odored and important new aroma and flavor molecules as *p*-menthene-8-thiol [*71159-90-5*] (**4**) in grapefruit (**2**) and 4-methoxy-2-methyl-2-butanethiol [*94087-83-9*] (**5**) in black currant bud oil (**3**) have been made available to the flavorist and perfumer.

(**4**)   (**5**)

Along with the refinement in separation and characterization methodology, an increasing awareness has evolved on the part of fragrance chemists of the role played by chirality in the odor properties of certain organic molecules. With the ability to separate, synthesize in high optical purity, and assign absolute configurations to enantiomers, the relative odor contributions of individual optical isomers can frequently be determined. The results are often surprising, as in the case of "methyl jasmonate" from jasmin oil, lemon peel, boronia, etc. Following the isolation and characterization of methyl *epi*-jasmonate from the pheromone glands of the male oriental fruit moth (4) and the disclosure that this is the only epimer occurring in ripe lemon peels (5), the odor thresholds of the stereoisomers of methyl jasmonate [91905-97-4] were determined (6). It was found that (±)-methyl epi-jasmonate [53369-26-9] (6) is more than 400 times stronger than (±)-methyl jasmonate [20073-13-6] (7), the principal ingredient in both commercial synthetic methyl jasmonate and in jasmin absolute. Also, (+)-methyl epi-jasmonate [42536-97-0] is more than four times stronger than (±)-methyl epi-jasmonate (6). It was, therefore, concluded that the (−)-methyl jasmonate [1211-29-6] and (+)-methyl jasmonate [78609-06-0] contribute almost nothing to the odor of the commercial mixture (7). It is apparent that it is the small amount of methyl epi-jasmonate which occurs both in commercial methyl jasmonate and in various natural oils and extracts which is the main contributor to the typical lemony-jasmin odor of these materials. Moreover, since it has been demonstrated (5) that methyl epi-jasmonate is the thermodynamically less stable form, its predominance in the living plant and its conversion to the more stable methyl jasmonate during harvesting and processing cannot be excluded. Indeed, it has been found that methyl epi-jasmonate occurs in high concentration in the volatiles of living *Cymbidium kanran* orchid but is transformed to methyl jasmonate on picking (8).

(6)   (7)

Commercial Essential Oils

The information given for the commercial oils represents data critically selected from published reliable sources or, in some cases, analyses performed in the authors' laboratories on authentic oil samples. These data are provided as a general guide to composition only and are not meant to be exclusive of analytical results obtained by other researchers on similar oils. Where the literature provides ranges of composition, these have, in several cases, been included. Many variables affect the composition of essential oils.

Rose.   Of all the natural oils, rose is probably the most desired material used in the fine fragrance industry. For years chemists have tried to unravel the mystery of the odor-donating components of this high priced natural material. Simple glc analysis shows that nine components constitute nearly 89% of the total volatiles of rose otto (9) (see Table 2).

Table 2. Constituents of Otto of Rose Bulgarian

Component	CAS Registry Number	Structure number	Percent
citronellol	[106-22-9]	(8)	36
paraffin hydrocarbons[a]			25
geraniol	[624-15-7]	(9)	17
β-phenethyl alcohol	[60-12-8]	(2)	3
linalool	[78-70-6]	(10)	2
ethanol	[64-17-5]		2
eugenol methyl ether	[93-15-2]	(11, R = CH₃)	1.5
eugenol	[97-53-0]	(11, R = H)	1.3
geranial[b]	[141-27-5]		1
Total			*88.8*

[a]Nonadecane [629-92-5], 1-nonadecene [18435-45-5], heneicosane [629-94-7], heneicosene [27400-79-9], eicosane [112-95-8], and eicosene [27400-78-8].
[b]3,7-Dimethyl-*trans*-2,6-octadienal; see also geraniol (9).

(8)   (9)   (10)   (11)

Of all these, probably β-phenethyl alcohol (2) comes closest to the odor of fresh rose petals; however, mixing all these components does not reproduce the total fine character of the natural oil. It has been determined that a number of trace constituents representing less than 1% of the volatiles are critical to the development of the complete rose fragrance (10). These include *cis*- and *trans*-rose oxide (1), nerol oxide (12), rose furan (13), *para*-menth-1-en-9-al (14), β-ionone (15), β-damascone (16), and β-damascenone (3).

(12)   (13)   (14)

(15)   (16)

Sometimes a skilled perfumer detects a sandalwood-musky note in authentic Bulgarian otto of rose. This note has been identified (11) as the trace ingredient, 7-methoxy-3,7-dimethyl-2-octanol [41890-92-0] (17), which has been commercially available for some time as Ossyrol (trademark of Bush, Boake, Allen Inc). This compound had never before been identified in nature, but demonstrates how, sometimes, synthetic fragrance chemists can anticipate nature.

$$\underset{\substack{|\\ CH_3}}{\overset{\substack{OH\\|}}{CH_3CH}}CHCH_2CH_2CH_2\underset{\substack{|\\ CH_3}}{\overset{\substack{OCH_3\\|}}{CCH_3}}$$

(17)

Jasmin. "If the rose is the queen of flowers, the jasmin is the fairest and prettiest princess. The two together reign supreme in the world of flowers as well as in the world of perfume" (12). As in the case of rose, jasmin has been the subject of many investigations, and more than 95 compounds have been identified to date in extracts of *Jasminum officinale* L. var. *grandiflorum*, the source of commercial jasmin oil, concrete, and absolute. The principal components are shown in Table 3 (13).

(18)   (19)   (20)

Of the 11 compounds which constitute approximately 86% of jasmin volatiles, only benzyl acetate, *cis*-jasmone (18), and methyl jasmonate possess the characteristic odor of jasmin. Trace components including *cis*-jasmin lactone [34686-71-0] (20) (0.9%) and methyl *epi*-jasmonate (6) (0.1%) are the key contributors to the jasmin odor.

As early as 1967, IFF chemists (11), in an in-depth study of jasmin absolute, identified an ultratrace amount of one of the key compounds in the entire

Table 3. Constituents in *Jasminum officinale*[a]

Component	CAS Registry Number	Structure number	Percent
phytol[b]	[150-86-7]		
isophytol[c]	[505-32-8]		27
benzyl benzoate	[120-51-4]		16
benzyl acetate	[140-11-4]		15
benzyl alcohol	[100-51-6]		6
linalool		(10)	6
geranyl linalool[d]	[1113-21-9]		4
cis-jasmone	[488-10-8]	(18)	4
methyl jasmonate			3
eugenol			2
indole	[120-72-9]	(19)	2
cis-3-hexenyl benzoate	[25152-85-6]		2
methyl linolenate[e]	[301-00-8]		2
Total			*89*

[a]Ref. 13.
[b]3,7,11,15-Tetramethyl-*trans*-2 hexadecen-1-ol.
[c]3,7,11,15-Tetramethyl-1-hexadecen-3-ol.
[d]3,7,11,15-Tetramethyl-1,6,10,14-hexadecatetraen-3-ol.
[e]Methyl (Z,Z,Z)-9,12,15-octadecatrienoate.

fragrance repoitoire, hydroxycitronellal [107-75-5] (21). This chemical has been used for many years in almost every "white flower" fragrance to give a very diffusive and lasting lily-of-the-valley and jasmin note, but this represents the only known identification of the compound in nature. This illustrates that the human nose can often predict the presence of a molecule well before the instrumentation becomes sufficiently sensitive to detect it.

$$\underset{\text{(21)}}{(CH_3)_2 \overset{\overset{\text{OH}}{|}}{C}CH_2CH_2CH_2 \overset{\overset{\text{CH}_3}{|}}{C}HCH_2CHO}$$

Orange Flower (Neroli) Oil. "The rose we call the queen of flowers, the jasmin the fairest and prettiest princess, but the orange flower is the most fragile and dainty of our royal family of flowers. If the rose stirs our memories, the jasmin our hopes, the orange flower stirs sentiments—sentiments most romantic!" (14). Commercial neroli oil [8016-38-4] is obtained by steam distillation of the freshly picked blossoms of the bitter orange *Citrus aurantium* L. subspecies *amara*, which is cultivated in Mediterranean countries as well as in Haiti and several other tropical countries. More than 125 components have been identified in the oil; the principal ones are shown in Table 4 and Figure 1.

Of the 10 constituents which represent nearly half the oil of neroli, only linalool (10) can be said to contribute directly to the characteristic aroma of

Table 4. Components of Neroli Oil[a]

Component	CAS Registry Number	Structure number	Percent
linalool		(**10**)	36.0
β-pinene	[127-91-3]	(**22**)	16.0
limonene	[138-86-3]	(**23**)	11.6
linalyl acetate	[115-95-7]	(**24**)	5.8
trans-β-ocimene	[3779-61-1]	(**25**)	5.1
α-terpineol	[98-55-5]	(**26**)	4.0
trans-nerolidol	[7212-44-4]	(**27**)	3.9
geranyl acetate	[16409-44-2]	(**28**)	2.8
geraniol		(**9**)	2.4
myrcene	[123-35-3]	(**29**)	1.8
sabinene	[3387-41-5]	(**30**)	1.2
Total			*90.6*

[a]Ref. 15.

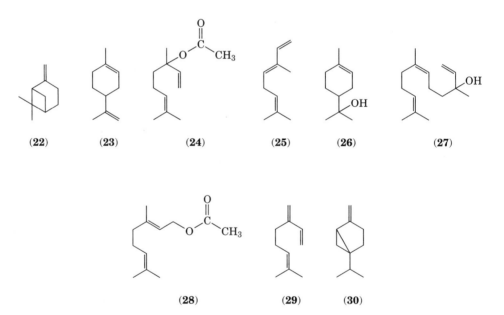

Fig. 1.   Structures of some components of neroli oil.

orange flower oil. In 1977, IFF chemists performed an in-depth analysis of this oil and identified three simple terpenic compounds, each present at less than 0.01%, α-terpenyl methyl ether [1457-68-0] (**31**), geranyl methyl ether [2565-82-4] (**32**), and linalyl methyl ether [60763-44-2] (**33**) (11). The latter two compounds possess green floral-citrus aromas and have been known to perfumery for some time; α-terpenyl methyl ether (**31**) has been called the orange flower ether by IFF chemists owing to its characteristic odor.

CH$_3$ ... CH$_2$OCH$_3$ ... CH ... CH$_2$ ... CH$_2$ ... CH ... C ... H$_3$C ... CH$_3$

CH$_3$ ... OCH$_3$ ... H$_3$C ... C ... CH$_3$

CH$_3$ ... C—OCH$_3$ ... CH ... CH$_2$ ... CH$_2$ ... CH$_2$ ... CH ... C ... H$_3$C ... CH$_3$

(31)   (32)   (33)

Lavender and Lavandin. Lavender and lavandin [8022-15-9] are considered together both because of their somewhat similar odors and because they are botanically related. Lavender oil is obtained by steam distillation of the flowering plant *L. angustifolia* Mill., the so-called English Lavender, whereas lavandin comes from the corresponding parts of *L. hybrida* Reverchon, a cross between true lavender and spike lavender *L. latifolia* Med. Both plants are cultivated for their oil around the Mediterranean region. Each species exists in a large number of varieties and cultivars. Ranges for the principal constituents of a number of varieties of lavender and lavandin are provided in Table 5 (16).

Table 5. Comparison of Constituents of Oils of Lavender and Lavandin[a]

Component	CAS Registry Number	Structure number	Percent in lavender[b]	Percent in lavandin
3-octanone	[106-68-3]		nd–2	nd–<1
eucalyptol	[470-82-3]	(34)	trace–5	6–22
limonene		(23)	trace–1	<1–2
cis- and *trans-β*-ocimenes[c] and *γ*-terpinene[d]			<1–11	1–6
linalool		(10)	6–51	34–41
camphor	[76-22-2]	(35)	nd–trace	<1–8
lavandulol	[498-16-8]	(36)	trace–8	<1–1
terpinen-4-ol and lavandulyl acetate	[562-74-3] [25905-14-0]	(37)	<1–30	2–7
linalyl acetate		(24)	7–57	2–31
borneol[e]	[507-70-0]		<1	1–2
α-terpineol		(26)		

[a]Ref. 16.
[b]nd = not detected.
[c]*trans-β*-Ocimene [3779-61-1] (25) = 3,7-dimethyl-(*E*)-1,3,6-octatriene; *cis-β*-ocimene [3338-55-4] = 3,7-dimethyl-(*Z*)-1,3,6-octatriene.
[d]*γ*-Terpinene [99-85-4] = 1-methyl-4-isopropyl-1,4-cyclohexadiene.
[e]The reduction product of camphor (35), ie, *endo*-1,7,7-trimethyl bicycloheptane[2.2.1]-2-ol.

(34)   (35)   (36)   (37)

Generally speaking, lavandin oil is characterized by a lower quantity of esters, the most critical of which for odor quality is linalyl acetate (**24**). The French specification organization (AFNOR) has fixed provisional standards for both oils (17) (Table 6).

Because it has been long recognized that trace components play an important role in determining the quality of these oils, a large number of detailed investigations of the trace constituents of both oils have been undertaken. Some of the compounds that have been determined to contribute important notes include coumarin [91-64-5] (**38**); unbelliferone methyl ether [531-39-9] (hernerin) (**39**); 4-vinyl-4-methyl-γ-butyrolactone [1073-11-6] (**40**); methyl *cis*- and *trans*-3-methyl-2-(3'-methyl-2-butenyl)-3-cyclohexenyl ketone [73019-21-3] (Diels-Alder adducts of ocimene and methyl vinyl ketone); as well as a series of compounds related to the last named (11).

Table 6. Provisional AFNOR Specifications for Lavender and Lavandin Oils

Component	CAS Registry Number	Lavender, %	cv. *abrialis*	cv. *grosso*
3-octanone	[106-68-3]	2 max	[a]	[a]
eucalyptol[b]	[470-82-3]	1.5 max	6–11	4–7
limonene	[138-86-3]	0.5 max	[a]	[a]
cis-β-ocimene	[3338-55-4]	4–10	1.5–4	[a]
trans-β-ocimene	[3779-61-1]	2–6	3–7	[a]
linalool	[78-70-6]	25–38	30–38	25–35
camphor[c]	[76-22-2]	<0.6	7–11	6–8
lavandulol	[498-16-8]	0.3 min	0.5–1.5	0.3–0.5
terpinen-4-ol	[562-74-3]	2–6	<1	2–4
α-terpineol	[98-55-5]	1 max	[a]	[a]
linalyl acetate	[115-95-7]	25–45	20–30	28–38
lavandulyl acetate	[25905-14-0]	2 min	1–2	1.5–3

[a]No specifications.
[b]1,8-Cineole [470-82-3].
[c]The reduction product of camphor (**35**), ie, *endo*-1,7,7-trimethyl bicycloheptane[2.2.1]-2-ol.

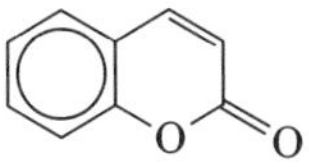

(38) 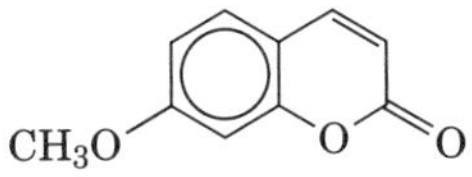 (39) 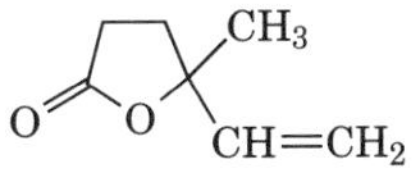 (40)

Geranium Oil. This oil is produced in China, Egypt, India, Réunion Island, Morocco, and Algeria by steam distillation of the leaves and branches of *Pelargonium* sp. Geranium oil Réunion is also called Geranium Bourbon from the old name for the island (Ile de Bourbon). Because the importance of geranium oil in perfumery cannot be overemphasized, it has been the subject of numerous investigations. The relative percentages of the principal constituents of geranium oil vary appreciably depending on species, source, and growth conditions, but a representative profile of a typical Geranium Bourbon is shown in Table 7 (18).

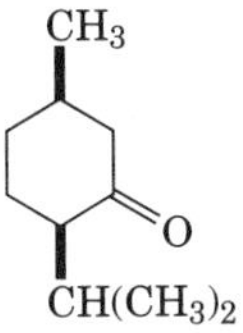

(41) 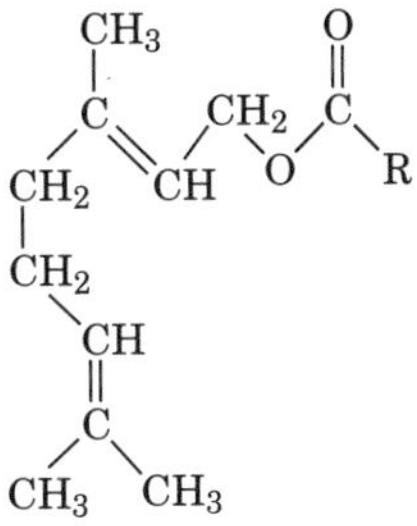 (42) 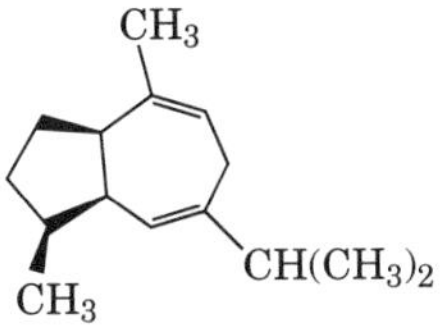 (43) 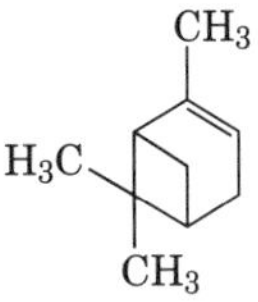 (44)

In 1961, chemists (19) isolated a minor component of geranium oil, characterized it as rose oxide (**1**), and reported that this compound contributed to the characteristic geranium odor. Rose oxide (**1**) is manufactured by the photooxidation of citronellol (**8**).

Table 7. Components of Geranium Bourbon Oil[a]

Component	CAS Registry Number	Structure number	Percent
citronellol		(**8**)	21
geraniol		(**9**)	18
linalool		(**10**)	11
isomenthone	[491-07-6]	(**41**)	8
citronellyl formate	[105-85-1]		8
geranyl formate	[61759-63-5]	(**42**), R = H	7
6,9-guaiadiene	[37839-64-8]	(**43**)	5
geranyl butyrate	[106-29-6]	(**42**), R = n-propyl	2
geranyl tiglate	[7785-33-3]	(**42**), R = 1-methylpropenyl	1
α-pinene	[80-56-8]	(**44**)	1
Total			82

[a]Ref. 18.

Part of the characteristic odor of geranium oil is described as peppery. In 1964, another trace constituent of the oil, furopelargone [*1143-45-9, 1143-46-0*] (**45**), was characterized and found to possess the typical peppery note (20). It has been demonstrated that furopelargone (**45**) can be made through the photooxidation of 6,9-guaiadiene (**43**), a component of geranium oil (11).

(**45**)

Citronella Oil. This is commercially produced by steam distillation of either of two related species of *Cymbopogon* grasses, *C. nardus* for the Ceylon type and *C. winterianus* for the Java type, which is also cultivated in Taiwan, Guatemala, Honduras, and Haiti. The oil finds significant usage in industrial fragrancing and as a source of citronellal (**46**), geraniol (**9**), and citronellol (**8**) for use in their own right or as raw materials for chemical conversions, including the manufacture of the critical fragrance compound, hydroxycitronellal (**21**). The comparative composition of the principal ingredients of typical samples of the two oils is given in Table 8 (21).

Table 8. Comparison of the Ingredients of Citronella Oils[a]

Component	CAS Registry Number	Structure number	*C. nardus*, %	*C. winterianus*, %
geraniol		(**9**)	21	23
citronellol		(**8**)	6	11
citronellal	[*106-23-0*]	(**46**)	6	35
cis-methyl isoeugenol[b]	[*6380-24-1*]		10	
and *trans*-methyl isoeugenol[c]	[*6379-72-2*]			
limonene		(**23**)	9	3
camphene	[*79-92-5*]	(**47**)	7	d
borneol			5	d
geranyl acetate		(**28**)	2	5
β-cubebene	[*13744-15-5*]	(**48**)	4	2
elemol	[*639-99-6*]	(**49**)	1	3
eugenol		(**11**)		3
citronellyl acetate	[*150-84-5*]		1	2
Total			72	87

[a]Ref. 21.
[b]1,2-Dimethoxy-4(*Z*-1-propenyl)benzene [*6380-24-1*].
[c]1,2-Dimethoxy-4(*E*-1-propenyl)benzene [*6379-72-2*].
[d]Trace amounts.

(46)   (47)   (48)   (49)

From a perfumery point of view, trace ingredients again play an important role in the typical aroma of citronella oil. In 1980, IFF chemists isolated and characterized two nonterpenic compounds, melonol (2,6-dimethyl-5-hepten-1-ol [4234-93-9]) and melonal (2,6-dimethyl-5-heptenal [106-72-9]), with interesting green-melon odors (10). Although neither of these had previously been reported in nature, melonal has been a product of commerce for a number of years.

Bergamot Oil. Obtained by expression from the rind of the inedible fruit *Citrus bergamia* Risso, bergamot oil is a high volume fragrance material. The trees are primarily cultivated for their oil in the Calabria region of Italy, Increasingly, however, quality oil from northwestern Africa is becoming available. The composition of the principal components of a typical Calabrian oil from the 1988 season is shown in Table 9 (22).

Bergamot oil has undergone a large number of investigations with regard to its trace components. In 1964, Firmenich chemists (23) claimed that *cis*-jasmone (**18**) and *cis*-5-oceten-2-one [19093-20-0] were important to the bergamot odor. In 1969, a series of novel trace bifunctional monoterpenoid molecules were characterized (24). Of these, two esters had very interesting citrus odors; they are the *cis*- and *trans*-1-acetates [64777-00-0, 64777-01-1] of 2,6-dimethyl-2,7-octadiene-1,6-diol.

Lime Oil. This oil is obtained from the fruit *Citrus aurantifolia* Swingle; the Key, Mexican, or West Indian lime; or *C. latifolia* Tanaka, the Persian lime, either by steam distillation or expression. Either the entire crushed fruit or only

Table 9. Components of a Typical 1988 Calabrian Bergamot Oil[a]

Component	Structure number	Percent
limonene	(**23**)	42
linalyl acetate	(**24**)	27
γ-terpinene		8
linalool	(**10**)	7
sabinene	(**30**)	1
myrcene	(**29**)	1
α-pinene	(**44**)	1
Total		*87*

[a] Ref. 22.

the peel may be used, depending on the specific properties desired. A typical commercial distilled lime oil contains the constituents shown in Table 10 (25).

(**50**)   (**51**)   (**52**)   (**53**)

Because of its more interesting odor and flavor properties, many perfumers and flavorists prefer to use a terpeneless lime oil. A typical analysis is shown in Table 10 (26).

Table 10. Components of Typical Commercial Lime Oils[a]

Component	CAS Registry Number	Structure number	Distilled oil, %	Terpeneless oil, %
limonene		(**23**)	45	
γ-terpinene			11	
terpinolene	[586-62-9]	(**50**)	9	
α-terpineol		(**26**)	9	46
eucalyptol		(**34**)	5	
1,4-cineole	[470-67-7]	(**51**)	3	
α-farnesene[b]	[21499-64-9]		2	
para-cymene[c]	[99-87-6]		2	
α-pinene		(**44**)	1	
β-pinene		(**22**)	1	
myrcene		(**29**)	1	
borneol			1	4
terpinen-4-ol		(**37**)	1	5
decanal	[112-31-2]		1	
β-caryophyllene	[13877-93-5]	(**52**)	1	
β-bisabolene	[495-61-4]	(**53**)	1	
γ-terpineol	[586-81-2]	(**54**)		7
terpinen-1-ol	[586-82-3]	(**55**)		5
trans-β-terpineol	[7299-40-3]	(**56**)		4
α-bergamotene	[17699-05-7]	(**57**)		4
α-fenchyl alcohol	[512-13-0]	(**58**)		3
cis-β-terpineol	[7299-41-4]	(**59**)		1
linalool		(**10**)		1
geranial				0.4
Totals			94	80.4

[a]Refs. 25 and 26.
[b]3,7,11-Trimethyl-(E,E)-1,3,6,10-dodecatetraene.
[c]p-Isopropyltoluene.

(54)   (55)   (56)   (57)   (58)   (59)

From West Indian lime oil, a trace low boiling constituent, 1-methyl-1,3-(or 1,5 [1489-57-2])-cyclohexadiene has been characterized (27). This compound, which possesses an intense and characteristic lime aroma, was later confirmed to be the 1,3-isomer [1489-56-1] (11). This compound can easily be made in a biomimetic way through the reaction of citral [5392-40-5] (3,7-dimethyl-2,6-octadienal) with citric acid (28,29).

Orange Oil. Orange oil, sweet, in terms of the quantity produced, ranks number one among the citrus oils. It is primarily obtained by expression from the peel of *Citrus sinensis* (L.) Osbeck. The United States, Cyprus, Guinea, Israel, and Brazil are the principal producers; specialty orange oils are produced in many other countries. A typical cold-pressed Valencia orange oil has the gross composition shown by the following (30). Neral [106-26-3] has the systematic name (Z)-5,7-dimethyl-3,6-heptadienal.

Component	Structure number	Percent
limonene	(23)	95
myrcene + octanal	(29)	2
α-pinene	(44)	0.4
linalool	(10)	0.3
decanal		0.3
sabinene	(30)	0.2
geranial		0.1
neral		0.1
dodecanal		0.1
Total		*98.5*

Decanal, a well-known fragrance and flavor item, gives, in dilution, the characteristic odor impression of sweet orange. In 1965, the trace constituents, α- and β-sinensal (**60,61**), were isolated from orange oil (31). According to one authority, "(E,E,E)α-sinensal [17909-77-2] (**60**) has an orangelike note, whereas (E,E)β-sinensal [3779-62-2] (**61**) is dominated by a strong metallic-fishy undertone and at high concentrations, causes an unpleasant odor sensation. (E,E,Z)α-Sinensal [61432-64-2] (**62**) (E,E,Z-2,6,10-trimethyl-2,6,9,11-dodecatetraenal) has the same odor character as (E,E,E) but is the weakest of the three aroma compounds (32). Many elegant syntheses of the sinensals have been reported in the chemical literature; however, none is practical.

(60)   (61)   (62)

Grapefruit Oil.   The grapefruit is a relatively recent arrival on the citrus scene, having presumably evolved from an earlier species, possibly the West Indian shaddock. The fruit is still undergoing refinement, and commercial grapefruit oil, obtained by expression of the peel of *Citrus paradisi* Macfayden, is much milder and sweeter than that available in the marketplace only a few years ago. Primary producers of grapefruit oil include the United States, Israel, Brazil, the West Indies, and Nigeria. A typical Israeli oil has the following composition (33):

Component	Structure number	Percent
limonene	(23)	93
myrcene	(29)	2
α-pinene	(44)	1
sabinene	(30)	0.4
octanal		0.3
β-pinene	(22)	0.2
decanal		0.2
trans-β-ocimene	(25)	0.1
γ-terpinene		0.1
Total		*97.3*

In 1964, the presence of nootkatone [*4674-50-4*] (**63**), a known molecule, was confirmed in grapefruit oil (34). Some have considered nootkatone responsible for the characteristic aroma of grapefruit oil, but highly purified (+)-nootkatone (**63**) actually displays very weak odor properties. This was also observed in 1983 in a report that grapefruit oil contains a series of nootkatone analogues, one of which, (+)-8,9-didehydronootkatone [*5090-63-1*] (**64**), displays a particularly valuable grapefruit aroma similar to, but definitely stronger than, that of (+)-nootkatone (**63**) itself (35). Just prior to this work, investigators characterized the true character-donating component of grapefruit oil and juice as *p*-menthene-8-thiol (**4**). This compound is present in fresh grapefruit juice (in which it naturally occurs at or below the ppb level) (36).

(63)   (64)

Sandalwood Oil, East Indian.   The use of sandalwood oil for its perfumery value is ancient, probably extending back some 4000 years. Oil from the powdered wood and roots of the tree *Santalum album* L. is produced primarily in India, under government control. Good quality oil is a pale yellow to yellow viscous liquid characterized by an extremely soft, sweet–woody, almost animal–balsamic odor. The extreme tenacity of the aroma makes it an ideal blender–fixative for woody-Oriental–floral fragrance bases. It also finds extensive use for the codistillation of other essential oils, such as rose, especially in India. There the so-called attars are made with sandalwood oil distilled over the flowers or by distillation of these flowers into sandalwood oil. The principal constituents of sandalwood oil are shown in Table 11 (37) and Figure 2.

The principal component, *cis-α*-santalol (**65**), has controversial odor properties; *cis-β*-santalol (**66**) contributes most to the odor of sandalwood. A number of trace constituents have been characterized, two of which (**76**) [*59300-43-5*] and (**77**) [*63569-02-8*] are worthy of mention for their very clean sandalwood notes (38).

(76)   (77)

Table 11. Constituents of Sandalwood Oil, East Indian[a,b]

Component	CAS Registry Number	Structure number	Percent
cis-α-santalol	[*115-71-9*]	(**65**)	50
cis-β-santalol	[*77-42-9*]	(**66**)	21
epi-β-santalol	[*42495-69-2*]	(**67**)	4
α-bergamotol	[*88034-74-6*]	(**68**)	4
α-santalal	[*13827-97-9*]	(**69**)	3
cis-lanceol	[*10067-28-4*]	(**70**)	2
trans-β-santalol	[*37172-32-0*]	(**71**)	2
β-santalene	[*511-59-1*]	(**72**)	1
epi-β-santalene	[*25532-78-9*]	(**73**)	1
cis-nuciferol	[*78339-53-4*]	(**74**)	1
spirosantalol	[*117020-19-6*]	(**75**)	1
Total			*90*

[a] Ref. 37.
[b] See Fig. 2.

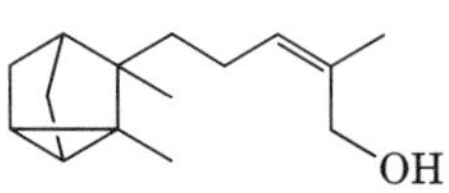

(65) 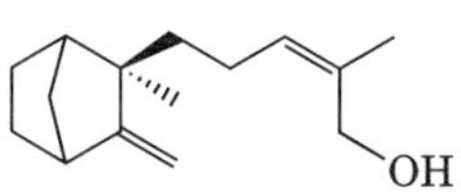   (66) 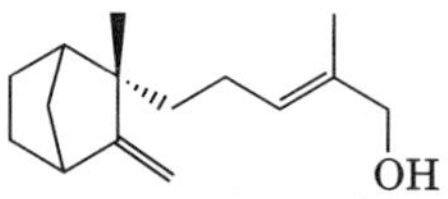   (67)

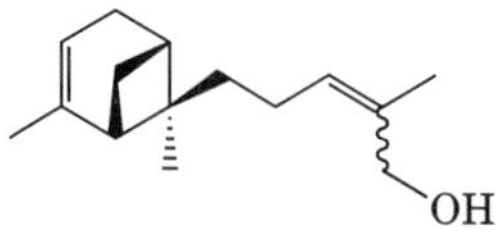

(68) 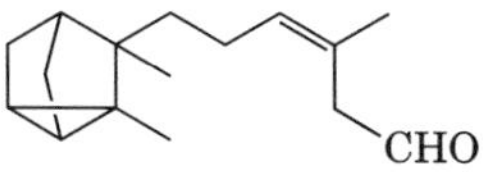   (69) 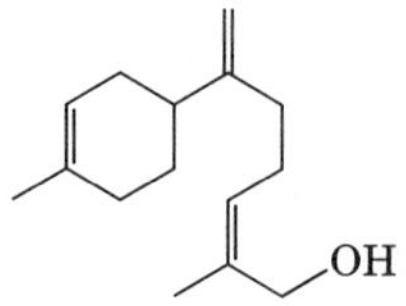   (70)

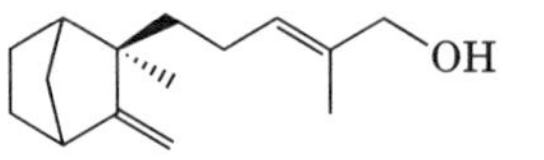

(71) 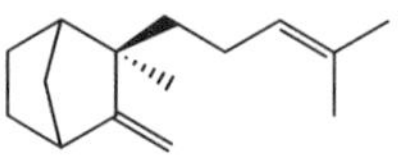   (72)    (73)

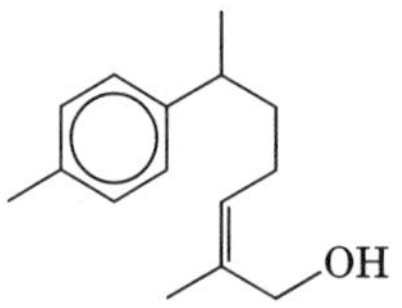

(74) 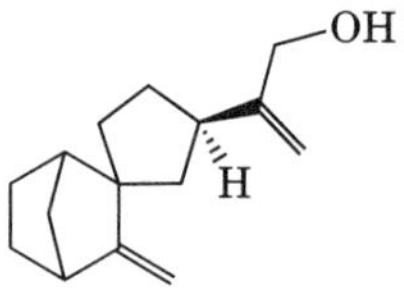   (75)

Fig. 2.   Constituents of East India sandalwood oil. See Table 11.

Patchouli Oil.   This oil is obtained by the steam distillation of the dried leaves of a small plant *Pogostemum cablin* Benth. which is cultivated primarily in Southeast Asia, Indonesia, and the Philippines. It is of particular interest to the fragrance industry because of its unique woody, herbaceous, earthy aroma. Oil which is produced locally generally has substantially different odor properties from that made from exported leaves. However, whether the oil is the product of native or European-American distillation, the odor of the oil improves significantly upon aging. A full yield of oil is only obtained if the walls of the oil glands are first ruptured, preferably by means of a controlled fermentation achieved by stacking or baling of the dried leaves. The principal constituents of a commercial Javan oil are shown in Table 12 (39) and Figure 3.

Patchouli oil is one of a class of unique essential oils consisting mostly of sesquiterpenoid material, of which patchouli alcohol constitutes 30%. It has been reported (40) that patchouli alcohol (**78**) is totally odorless. However, patchouli alcohol has been synthesized, proving that this compound indeed possesses the basic odor of patchouli oil (41). Moreover, as early as 1966, an in-depth analysis of patchouli oil derived from both dried and green leaves was performed

Table 12. Constituents of a Commercial Javan Patchouli Oil[a,b]

Component	CAS Registry Number	Structure number	Percent
patchouli alcohol	[5986-55-0]	(78)	30
α-bulnesene	[3691-11-0]	(79)	17
α-guaiene	[3691-12-1]	(80)	14
seychellene	[20085-93-2]	(81)	9
α-patchoulene	[560-32-7]	(82)	5
β-caryophyllene		(52)	4
β-patchoulene	[514-51-2]	(83)	2
δ-cadinene	[483-76-1]	(84)	2
pogostol	[21698-41-9]	(85)	2
nor-patchoulenol	[41429-52-1]	(86)	1
caryophyllene oxide	[13877-94-6]	(87)	1
nortetrapatchoulol		(88)[c]	
Total			*87*

[a]Ref. 39.
[b]See Fig. 3.
[c]Only 0.00% by weight.

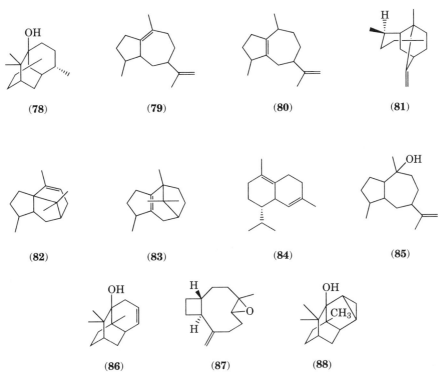

Fig. 3. Constituents of a commercial Javan oil.

and identified two novel character-donating components, norpatchoulenol (**86**) and nortetrapatchoulol (**88**) (41). The latter compound (**88**), present in the oil to the extent of only 0.001% by weight, is the strongest-odored of all. It has been reported elsewhere that norpatchoulenol (**86**) is the true carrier of the odor of patchouli oil (40).

Vetiver Oil. This oil is also a high volume fragrance material, prized for its precious woody note. Like patchouli, vetiver, obtained by steam distillation of the dried root of the grass *Vetiveria zizanoides* Staph., consists almost entirely of sesquiterpenic (C-13–C-15) compounds. The grass, a tall perennial which is native to India, Indonesia, and Sri Lanka and grows wild throughout southeast Asia, is cultivated for its oil in southern India, Indonesia (Java), Malaysia, the Philippines, Japan, Réunion, Africa, Haiti, the Dominican Republic, South America, the West Indies, and Mauritius. For the most part, distillation takes place in the field. Réunion and Haiti are the largest producers of vetiver oil. It is an amber to dark-brown viscous liquid with a sweet, very heavy, woody–earthy odor, reminiscent of roots and wet soil, with a rich, precious-woody undertone. Vetiver oil is used extensively in perfumery for fougère, chypre, oriental, moss, and woody notes. The oil is also used for the manufacture of a commercial product, vetiver acetate [52771-09-2], an acetylated vetiver oil possessing a faint, fresh-sweet, slightly woody odor. Some of the principal constituents of a Haitian vetiver oil have been reported (42) and are shown in Table 13 and Figure 4.

Galbanum Oil. This material is a natural oleoresin collected from several species of *Ferula* which grow wild in Iran and countries of Asia Minor such as Lebanon. Although both a hard and a soft galbanum are offered commercially, only the soft variety is of interest to perfumery. Soft galbanum is a dark amber-colored to yellowish grayish olive brown viscous liquid with a powerful green-woody, almost balsamic–resinous odor. The oil is obtained by steam or steam–water distillation of the oleoresin in either Europe or the United States and is a colorless to pale olive-yellow mobile liquid with an intensely green fresh-leafy odor which dries to a woody, balsamic, bark-like aroma with musk-like undertones. The pine-like topnote is frequently removed by topping to enrich the soft, tenacious green-woody note so important to perfumery. The principal components of untopped galbanum oil [8023-91-4] shown in Table 14 are, not surprisingly, terpene hydrocarbons (43). Galbanum oil contains a wide va-

Table 13. Some Components of a Haitian Vetiver Oil[a]

Component	CAS Registry Number	Structure number	Percent
zizanol	[28102-79-6]	(**89**)	13.4
vetiselinenol	[28102-68-3]	(**90**)	11.2
cyclocopacamphenol	[30810-34-5]	(**91**)	6.6
β-eudesmol	[95529-67-2]	(**92**)	5.5
β-vetivone	[18444-79-6]	(**93**)	5.2
α-vetivone	[15764-04-2]	(**94**)	3.2
elemol		(**49**)	2.3
10-*epi*-γ eudesmol	[1209-71-8]	(**95**)	2.2
Total			57.6

[a]See Fig. 4.

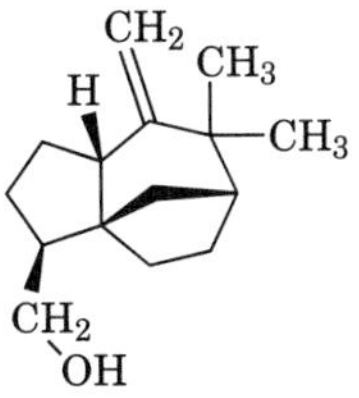

(89) 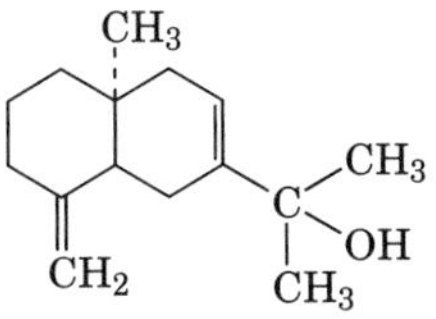 (90) 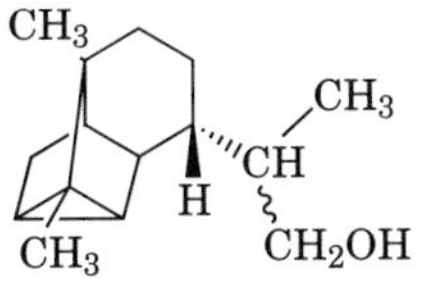 (91) 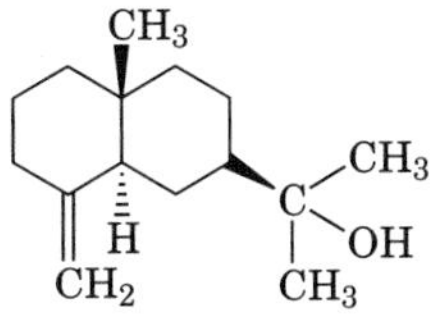 (92)

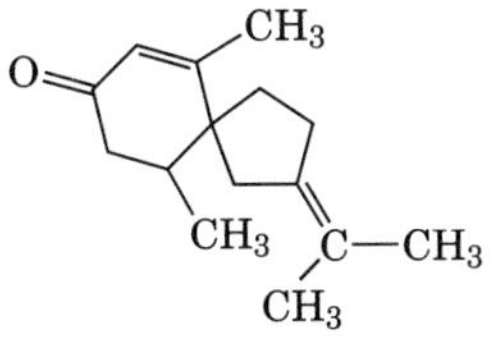

(93) 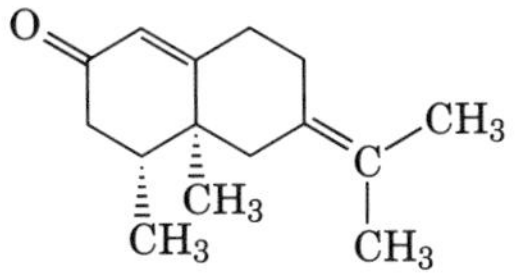 (94) 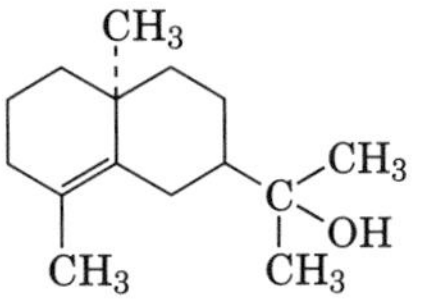 (95)

Fig. 4. Some of the principal constituents of a Haitian vetiver oil. See Table 13.

riety of compounds, including terpenoids (qv), nonterpenoids, olefins, and even sulfur-containing molecules. Of all of these, only a few contribute to the typical galbanum character.

The presence in galbanum oil of the isomeric 1,3,5-undecatrienes was almost simultaneously announced by three groups of investigators (44–46), although one group (45) claims to have isolated and characterized the compounds as early as 1963. Of the four possible geometrical isomers, 1,3-*trans*,5-*cis*-undecatriene [*51447-08-6*] and 1,3-*trans*,5-*trans*-undecatriene [*19883-29-5*], have been identified in galbanum oil. Of these, the 1,3-*trans*,5-*trans*-isomer possesses the intense green galbanum odor, whereas the 1,3-*trans*,5-*cis*-isomer has a fatty galbanum note (11).

A second compound which was identified as important in the aroma of galbanum oil was 2-isopropyl-4-methylanisole [*31574-44-4*]. This compound also possesses a powerful green galbanum odor (47).

Table 14. Components of Untopped Galbanum Oil

Component	Structure number	Percent
β-pinene	(**22**)	60
α-pinene	(**44**)	14
Δ³-carene		4
limonene	(**23**)	4
myrcene	(**29**)	3
α-thujene		1
sabinene	(**30**)	1
para-cymene		1
cis-β-ocimene		1
Total		*89*

Another commercially available compound which is important to the galbanum aroma is 2-methoxy-3-isobutylpyrazine [24683-00-9] (48), which possesses an intense green pepper odor.

In 1969, α-methylcyclopentadecanolide [4459-57-8] was isolated from galbanum (49). In 1978, three additional α-methyl macrolides, α-methylcyclododecanolide [71736-24-8], α-methylcyclotridecanolide [27198-63-6], and α-methylcyclotetradecanoline [32539-85-8], were characterized as galbanum constituents (50). These four undoubtedly contribute to the musky undertone of galbanum.

Myrrh. This is a typical natural oleo-gum-resin consisting of two-thirds water-soluble gum and one-third alcohol-soluble resin. It occurs naturally in the trunks of small trees of various *Comiphora* species which grow in eastern Africa north of the equator and in southwestern Arabia. It is processed either to an absolute or to an essential oil by various manufacturers in Europe or the United States. Myrrh [8016-37-3] has been prized since ancient times for its rich, sweet, warm, balsamic, incense-like fragrance. Until recently, the chemical composition of myrrh has been little studied, and early investigators alluded to the complex sesquiterpenoid nature of the volatile portion without providing further detail. The important aroma components are furanosesquiterpenoids which can be related to elemol (49) and β-eudesmol (92). These include the novel furanocadinane [115526-32-4] (96) (12% in volatiles), which possesses a deep rich leathery, sweet, warm, balsamic incensey aroma characteristic of myrrh, and the dihydropyrocurzerenone [52557-06-9] (97) (>1% in volatiles), which was described as having a rich, heavy, resinous, sweet incense note characteristic of myrrh (51).

(96)   (97)

Oakmoss. Extracts of oakmoss are extensively used in perfumery to furnish parts of the notes of the fougère or chypre type. The first step in the preparation of an oakmoss extract is treatment of the lichen *Evernia prunastri* (L.) Ach., collected from oak trees mainly in southern and central Europe, with a hydrocarbon solvent to obtain a concrete. The concrete is then further processed by solvent extraction or distillation to more usable products, of which absolutes are the most versatile for perfumery use. A definitive analysis of oakmoss volatiles was performed in 1975 (52). The principal constituents of a Yugoslav oakmoss are shown in Table 15 (53). A number of phenolic compounds are responsible for the total odor impression. Of these, methyl β-orcinol carboxylate is the most characteristic of oakmoss.

Table 15. Components of a Yugoslav Oakmoss[a]

Component	CAS Registry Number	Structure number	Percent
ethyl everninate	[6110-36-7]	(**98**, R = CH$_3$; R' = H)	18
methyl β-orcinol carboxylate	[4707-47-5]	(**98**, R =H; R' = CH$_3$)	16
orcinol monomethyl ether	[3209-13-0]	(**99**, R = CH$_3$)	16
orcinol	[504-15-4]	(**99**, R = H)	13
ethyl hemmatomate	[39503-14-5]	(**98**, R = H; R' = CHO)	5
ethyl orsellinate	[2524-37-0]	(**98**, R = R' = H)	4
atranol	[526-37-4]	(**100**; R = H)	3
chloroatrinol	[57074-21-2]	(**100**; R = Cl)	1
Total			76

[a]Ref. 53.

(**98**)　　　　　(**99**)　　　　　(**100**)

Tonquin Musk. This particular natural musk is obtained from the scent glands of the male musk deer, *Moschus moschiferus*, which is native to the Himalayas region and southern China. Because of its characteristic animal, civety, musky odor, it is a highly prized perfumery ingredient which has been used since ancient times. As it is still common practice to kill the animal in order to remove the gland, more than 70,000 deer are slaughtered annually to produce 0.5–1.5 t of musk. As a result of this, the possession or use of natural Tonquin musk or its extracts is forbidden in many parts of the world, including the United States. Therefore, there has been intense interest in the identification of the chemicals responsible for the characteristic odor of Tonquin musk. Investigation of musk began in 1906 with the isolation (54) followed by the characterization (55) of the principal odor constituent of Tonquin musk as muscone (**101**) (β-methylcyclopentadecanone [541-91-3]).

(**101**)　　　　　(**102**)

Since muscone (**101**), by itself, does not reproduce the total odor impression of this musk, IFF chemists (56) as early as 1971 in an analysis of tincture of

Tonquin musk, reported a series of macrocyclic ketones (Table 16) which play a key role in creating the characteristic odor of this musk (11). The introduction of a double bond into a macrocyclic ketone (eg, **102**) changes the odor from flowery musk to animal musk.

Ambergris. Known from antiquity, ambergris [8038-65-1], a substance of animal origin formed in the stomach or intestine of the sperm whale (*Physeter macrocephalus* L.) has been prized both as a perfume and as a drug. In modern usage, it has been employed only in perfumery, but this is no longer possible owing to laws forbidding the importation of gray ambergris or any other whale product. Few reports exist on the constituents of ambergris; however, it has been reported (57) that the compounds shown in Table 17 play a key role in the total odor impression of ambergris. Of these, Ambrox (**106**), which is commercially available, has the finest ambergris odor.

(103)   (104)

Table 16. Macrocyclic Ketones of Tonquin Musk

Component	CAS Registry Number	Odor description
muscone[a]	[541-91-3]	powerful soft, sweet animal musk
β-methylcyclotridecanone	[61415-11-0]	strong floral musk with an animal character
cyclotetradecanone	[3603-99-4]	angelica-like musk
5-cyclopentadecen-1-one	[35720-58-2]	very animal, sweet, ambrette musk
14-methyl-5-cyclopentadecen-1-one[b]		strong musky, civet notes

[a]Structure (**101**).
[b]Structure (**102**).

Table 17. Aroma Constituents of Ambergris

Component	CAS Registry Number	Structure number	Odor description
γ-homocyclo geranyl chloride[a]	[72892-58-1]		ozony-seawater
α-ambrinol	[41199-19-3]	(103)	moldy-animal-fecal
γ-dihydroionone	[13720-12-2]	(104)	weak tobacco
γ-ambronal	[72892-63-8]	(105)	seawater
Ambrox[b]	[3738-00-9]	(106)	moist, creamy, soft, persistant amber odor with velvety effect

[a]1,1-Dimethyl-2-(2-chloroethyl)-3-methylene cyclohexane.
[b]Ambrox is the registered trademark of Firmenich SA, Switzerland.

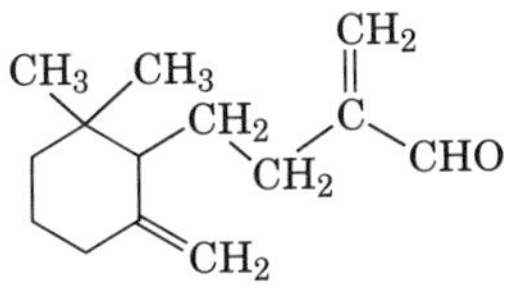

(**105**) 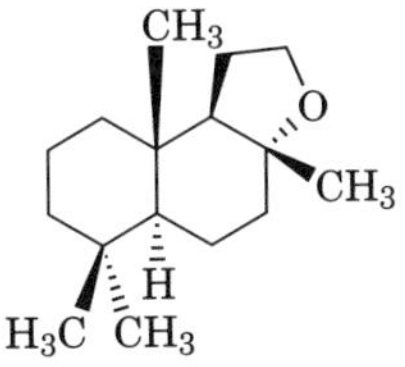   (**106**)

Tobacco. Among all the natural products of interest to the fragrance and flavor industry, tobacco oil [*8037-19-2*] probably enjoys the largest volume of usage throughout the world by the general population. Whereas the greatest part is directly consumed as smoking products, tobacco extracts find limited use in fragrance and flavor formulations owing to their ability to supply interesting and unique aromatic, sweet, floral, pungent, or hay-like notes. As a result of tobacco's position in the flavor industry, interest has been generated in elucidating the chemical nature of the components present in the various tobacco types, but not until recently has the flavor contribution of the key tobacco constituents been definitively reported. IFF chemists (58) performed a comparative analysis of the four tobacco types, ie, Virginia, burley, Turkish, and black, and evaluated the contribution to tobacco smoke flavor of a large number of the volatile constituents. They reported that, although many of the components were compatible with tobacco flavor and aroma to varying degrees, probably the key chemicals are those shown in Table 18:

Table 18. Tobacco Constituents

Constituent	CAS Registry Number	Structure number	Odor description
methylthiodamascone[a]	[*68697-67-6*]		rich sweet tobacco aromatics, smooth harshness
methylthiodamascenone	[*68697-66-5*]	(**107**)	
solanone[b]	[*40286-47-3*]		hay–tea aromatics, body and mouth-coating effects
norsolanadione[c]	[*35953-21-0*]		burley, cocoa aromatics
megastigmatrienone (isomer mix)	[*13215-88-8*]	(**108**)	rich, pungent tobacco aromatics, fullness, and body

[a]See structure (**107**). This 4,5-position is saturated.

[b]5-Isopropyl-8-methyl-6,8-nonadien-2-one.

[c]5-Isopropyl-3-nonene-2,8-dione.

(107)   (108)

Of these, the megastigmatrienone isomer mix (108) possesses the truest tobacco aroma and taste; solanone is equally important for its hay–tea aromatics and its body and mouth-coating effect.

Osmanthus. The tiny, powerfully fragrant, creamy yellow to orange flowers of *Osmanthus fragrans* (Thunb.) Lour, a tree native to the Himalayas, China, and southern Japan, also known as sweet or fragrant olive, are employed for the production of a concrete, absolute, pommade, or infusion, the use of which is, however, traditionally limited to the local area. An interesting comparison of the volatile concentrates of two different colors of osmanthus, light-cream and golden-orange, was reported in 1986 (59). The principal constituents of each are shown in Table 19.

Table 19. Comparative Composition of Components of the Volatile Concentrates of Two Different Osmanthus Flowers

Component	CAS Registry Number	Structure number	In creamy, %	In golden-orange, %
cis-linalool oxide (furanosyl)	[5989-33-3]	(109)	2.4	21.1
trans-linalool oxide (furanosyl)	[34995-77-2]	(110)	3.5	14.6
γ-decalactone	[706-14-9]		5.3	11.7
1-decanol	[112-30-1]		0.6	9.2
farnesol[a]	[4602-84-0]		6.0	0.2
β-ionone	[79-77-6]	(15)	1.2	5.9
eugenol			3.4	1.9
undecanoic acid	[112-37-8]		0.4	3.1
linalool		(10)	2.1	2.4
benzyl alcohol			1.2	0.2
α-ionone	[6901-97-9]	(111)	0.2	0.9
nonanal	[124-19-6]		0.7	0.1
Total			27.0	71.3

[a]3,7,11-Trimethyl-2,6,10-dodecatrien-1-ol.

(109)   (110)   (111)

When a comparative analysis of the headspace volatiles of living and picked osmanthus flowers was performed by the dynamic headspace trapping method using Tenax GC, even more dramatic differences were observed, shown in Table 20 (60).

It is interesting to compare data for headspace volatiles of the living vs picked and both of these with respect to the direct analysis of the extracts shown in Table 19.

Table 20. Differences Between the Volatiles of Living vs Picked Osmanthus[a]

Compound	CAS Registry Number	Structure number	In living flower volatiles, %	In harvested flower volatiles, %	
				Air-purged	Nitrogen-purged
4-hydroxy-4-methyl-2-pentanone	[123-42-2]		1.3	1.3	4.7
trans-2-hexenal	[6728-26-3]			1.1	
cis-3-hexenol	[928-96-1]			1.1	
2-(2-methoxyethoxy)ethanol	[111-77-3]			1.0	
γ-butyrolactone	[96-48-0]				4.4
cis-3-hexenyl acetate	[3681-71-8]		0.3	1.5	6.1
para-cymene[b]					8.5
cis-/trans-linalool oxide (furanosyl)		(109,110)	2.0	2.5	10.0
cis-3-hexenyl butyrate	[16491-36-4]		0.2	1.2	4.2
geraniol		(9)	15.0	11.4	18.0
geranial			2.1	1.0	
β-damascenone		(3)	0.1		
dihydro β-ionone	[17283-81-7]	(112)	4.7	13.8	8.0
γ-decalactone			7.0		
dihydro β-ionol	[3293-47-8]	(113)	0.5		
jasmin lactone		(20)	0.2		
β-ionone		(15)	13.2	5.7	4.0
4-keto β-ionone	[27185-77-9]	(114)	1.9		
Total			48.5	41.5	67.9

[a]Ref. 60.
[b]p-Isopropyltoluene.

(112)   (113)   (114)

Olibanum. As in the case of myrrh, olibanum (frankincense) is a natural oleo gum resin. It is obtained from the bark of various species of *Boswellia*, of the family Burseraceae, native to southern Arabia, northeastern Africa, and the mountainous regions of western India. Two commercial brands of olibanum oil [8016-36-2], called Aden and Eritrean, exist. These gums usually result from two species *Boswellia carterii* Bird. (Arabia and Somalia) and *B. frereana* Bird. (Somalia). The volatiles profiles of oils from the two gums are remarkably different, as shown in Table 21 (61). Several saturated and unsaturated γ-butyrolactones with strong coumarinic odors which are key contributors to the characteristic aroma of olibanum have been reported (62): 3-isopropyl-2-butenyl lactone [10547-89-4], γ-isopropyl-γ-butyrolactone [38624-29-2], γ-isopropenyl-γ-butyrolactone [22616-89-3], and α,γ,γ-trimethyl-γ-butyrolactone [2610-96-0].

Table 21. Volatile Components of Two Brands of Olibanum[a]

Component	CAS Registry Numbers	Structure numbers	In Aden, %	In Eritrean, %
α-pinene		(44)	43.0	4.6
camphene		(47)	2.0	1.1
β-pinene		(22)	1.5	
sabinene		(30)	1.0	
hexyl acetate	[142-92-7]			1.5
ortho-cymene[b]	[527-84-4]		0.5	
limonene/eucalyptol		(23)/(34)	7.0	5.8
para-cymene[b]			7.5	
linalool		(10)		2.5
campholenic aldehyde	[4501-58-0]	(115)	1.5	
verbenone	[80-57-9]	(116)	6.5	
octyl acetate	[112-14-1]		1.5	52.0
Total			72.0	67.5

[a]Ref. 61.
[b]Cymene = isopropyltoluene.

(**115**)   (**116**)

Amyris Oil. Obtained by steam distillation of the wood of *Amyris balsam-ifera* L., the so-called West Indian sandalwood which is indigenous to northern South America, Central America, and the West Indies, amyris oil [*8015-65-4*] is a pale yellow to brownish yellow viscous oil with a slightly oily-sweet and occasionally peppery balsamic woody note. It finds use as a blender and fixative for soap fragrances. The volatile constituents, which are primarily hydrocarbon and oxygenated sesquiterpenes, are shown in Table 22 and Figure 5 (63).

Anise Oil. Steam-distilled from the dried crushed fruits of *Pimpinella anisum* L., which is cultivated around the world, anise oil is a water-white to very pale yellow mobile liquid with an intensely sweet and clean odor reminiscent of the ripe crushed fruit. The principal constituent is *trans*-anethole [*4180-23-8*] (4-(*trans*-1-propenyl)anisole). It once found quite extensive use as an important ingredient in licorice candy, cough-drops, baked goods, pharmaceutical preparations, pipe tobacco, etc. However, for many applications, the oil has been largely replaced by synthetic anethole, frequently made from estragole [*7677-68-1*] (4-(2-propenyl)anisole) isolated from American pine oil. In regard to cost, the natural anise oil cannot compete with the synthetic anethole, but its flavor is still preferred by many connoisseurs for its fine and delicate sweetness, its rich body, and bouquet. The toxic cis isomer of anethole, 4-(*cis*-2-propenyl)anisole [*140-67-0*], always exists as a trace component where *trans*-anethole occurs in nature.

Anise Oil, Star. Steam-distilled from the fresh or partly dried whole or crushed fruits of *Illicium verum* Hook f., anise star oil [*84650-59-9*], was once

Table 22. Components of Amyris Oil[a]

Component	CAS Registry Number	Structure number	Percent
valerianol	[*20489-45-6*]	(**117**)	22
β-eudesmol		(**92**)	17
10-*epi*-γ-eudesmol		(**95**)	11
elemol		(**49**)	10
selina-3,7(11)-diene	[*6813-21-4*]	(**118**)	4
epi-α-selinene	[*70560-75-7*]	(**119**)	3
ar-curcumene[b]	[*4176-06-1*]		3
β-bisabolene		(**53**)	2
α-agarofuran	[*5956-12-7*]	(**120**)	1
dihydroagarofuran[c]	[*5956-09-2*]		1
α-acoradiene	[*24048-44-0*]	(**121**)	1
Total			76

[a]Ref. 63. See Fig. 5.
[b]2-Methyl-6-(*p*-methylphenyl)-2-heptene.
[c]Saturated rings (see structure **120**).

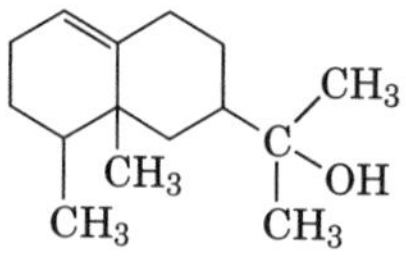

Fig. 5. Components of amyris oil.

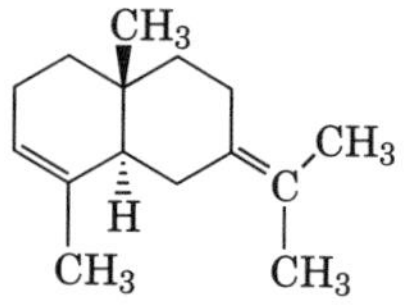

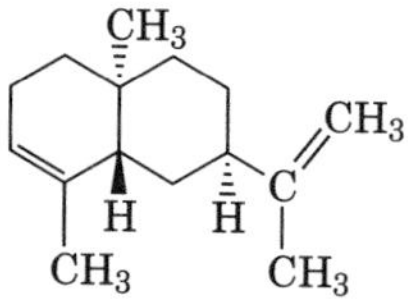

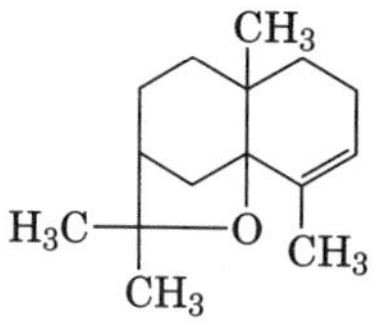

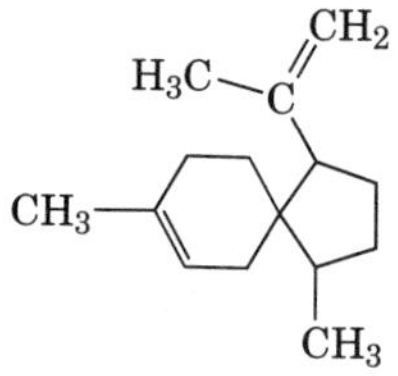

extensively produced in regions of southeastern Asia, where it is native. Its post-World War II production has not regained its earlier levels. The oil is a pale yellow or almost water-white liquid with an intensely sweet odor. Owing to the high anethole content, it is also strongly reminiscent of true anise oil. Crude anise star oil is frequently rectified prior to distribution or use, bringing the anethole content up to 90–95%. The principal constituents of Vietnamese and Chinese crude star anise oils have been compared and are shown in Table 23 (64).

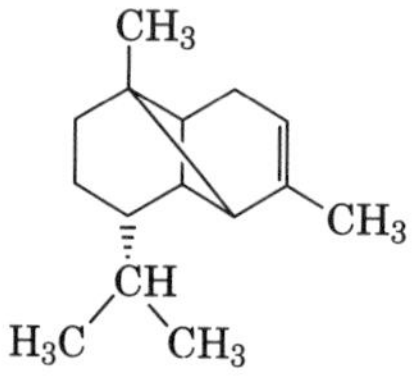

Sweet Basil Oil. A large number of species and varieties of the herb basil are cultivated throughout the world both for their culinary use and as sources of essential oils. The true sweet basil oil, however, which is steam-distilled from the flowering tops of the species *Ocimum basilicum* L. in France and in the United States, is readily distinguishable from the many other types by its chemical content. The absence of camphoraceous notes and the presence of a perfect odor balance between linalool (**10**) and estragole is characteristic of the sweet basil oil [*8015-73-4*]. It is usually a pale yellow to almost colorless mobile liquid, with a

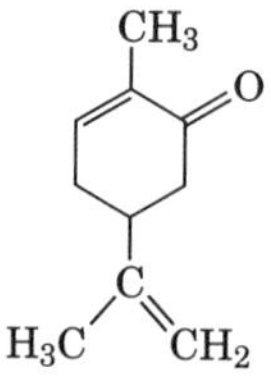

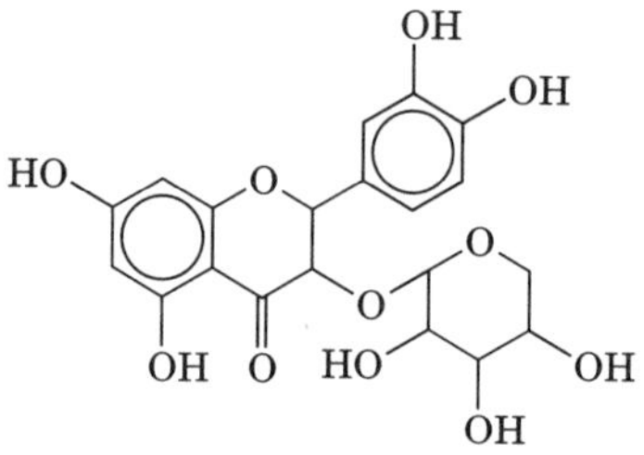

Table 23. Comparison of Ingredients of Vietnamese and Chinese Crude Anise Oils[a]

Component	CAS Registry Number	Structure number	In Vietnamese, %	In Chinese, %
terpene hydrocarbons			15.8	3.0
α-copaene	[3856-25-5]	(122)	0.1	0.1
linalool		(10)	1.1	0.6
α-bergamotene/ β-caryophyllene		(57)/(52)	0.6	0.6
cis-β-farnesene[b]	[28973-97-9]		0.1	0.1
estragole[c]			0.8	5.5
α-terpineol		(26)	0.4	0.2
carvone	[99-49-0]	(123)	d	d
β-bisabolene		(53)	0.1	0.2
cis-anethole[e]			0.3	0.2
trans-anethole[e]			78.6	82.7
para-anisaldehyde[f]	[123-11-5]		0.5	1.7
methyl anisate[g]	[121-98-2]		d	d
para-methoxyphenyl- 2-propanone	[122-84-9]		0.2	0.2
trans-methylisoeugenol				0.1
feniculine	[22255-13-6]	(124)	0.6	5.1
cinnamic alcohol	[104-54-1]		d	0.1
Total			*83.4*	*96.3*

[a] Ref. 64.
[b] 7,11-Dimethyl-3-methylene-(Z)-1,6,10-dodecatriene.
[c] 4-(2-Propenyl)anisole; anisole = methoxybenzene.
[d] Trace.
[e] cis- and trans-4(1-Propenyl)anisole.
[f] p-Methoxybenzaldehyde.
[g] Methyl p-methoxybenzoate.

sweet-spicy, slightly green odor, fresh with a faint balsamic woody undertone and a lasting sweetness. Although the low production volume makes this a relatively expensive oil, its great strength makes it very useful and generally applicable in fine perfumery and flavor work. The principal components of sweet basil oil, as well as, for comparison, the volatiles profile in the headspace over living sweet basil plant, are shown in Table 24 (65).

(125)

Bay Oil. Steam distillation of the leaves of the tree *Pimenta racemosa* (Mill) which is indigenous to certain islands of the West Indies, particularly

Table 24. Components of Sweet Basil Oil and Headspace of Living Sweet Basil[a]

Component	CAS Registry Number	Structure number	In distilled oil, %	In headspace, %
estragole	[7677-68-1]		82	45
limonene/1,8-cineole		(23)/(34)	4	1
linalool	[78-70-6]	(10)	3	27
trans-β-ocimene	[3779-61-1]	(25)	1	2
α-bergamotene	[17699-05-7]	(57)	1	2
β-pinene	[127-91-3]	(22)	1	
eugenol	[97-53-0]		1	
eugenol methyl ether	[93-15-2]	(11)	1	
myrcene	[123-35-3]	(29)		4
β-elemene	[515-13-9]	(125)		2
1-octen-3-ol	[3391-86-4]			1
Totals			94	84

[a]Ref. 65.

Dominica and Puerto Rico, is called bay or bay leaf oil. The same source was used in the past to produce Bay Rum in which rum was distilled over the leaves. Bay oil [8006-78-8] is a yellowish to dark brown mobile liquid with a fresh-spicy, sometimes medicinal odor with a lasting sweet-balsamic undertone. The oil finds extensive use in hair tonics, after-shave lotions as well as other men's-type fragrances. There is little or no use by the flavor industry. The range of components for a number of bay leaf oils is shown in Table 25 (66).

   Bitter Orange Oil. The cold-expressed oil from the peel of the nearly ripe fruit of *Citrus aurantium* L. subsp. amara is called bitter orange oil [68916-04-1]. Although it is related botanically to bergamot, there are substantial odor and flavor differences between the two. The tree is cultivated primarily in Spain, Guinea, and the West Indies, and the odor and flavor vary considerably from region to region. Cold-pressed bitter orange oil is a pale yellow to dark yellow or pale brown mobile liquid with a peculiar rich, fresh, dry aroma with a lasting sweet-floral undertone. The oil is the main ingredient of the orange

Table 25. Components of Bay Leaf Oil[a]

Component	Structure number	Percent
eugenol		38–75
chavicol[b]		11–21
myrcene	(29)	14–32
1,8-cineole	(34)	0.2–2
trans-β-ocimene	(25)	0.4–2
3-octanone		1
limonene	(23)	1
1-octen-3-ol		1
β-caryophyllene	(52)	1
α-terpineol	(26)	1
geranyl acetate	(28)	1

[a]Ref. 66.
[b]4-(2-Propenyl)phenol [501-92-8].

or triple-sec liqueur flavors and as an intensifier for sweet orange-flavored drinks. In perfumery it finds extensive use in colognes and fine fragrances of the chypres, fougère, and fresh aldehydic citrus type. A comparison of the volatiles of Spanish and Italian oils of the same season has been performed and the principal ingredients are shown in Table 26 (67).

Black Pepper. Black pepper oil [8006-82-4] is obtained by steam distillation of the dried ripe fruit of a vine-like plant *Piper nigrum* L. native to southern and southeastern India and possibly the Sunda islands. Primary centers for cultivation are Indonesia, India, Malaya, and Indochina. Little if any pepper oil is distilled in the local areas of production. The bulk of the black pepper grown is utilized as a condiment. The oil is an almost water-white to greenish gray mobile liquid which becomes more viscous on aging. Its odor, which is described as fresh, dry-woody, and warm-spicy, is quite reminiscent of the dried fruit. However, as the pungent principles of black pepper are not steam-distillable, the flavor is totally lacking in this quality. Two grades of pepper oil, light and heavy, are distinguished by the pepper trade. The former consisting of distillation foreruns has the typical flavor of freshly ground peppercorns but without tenacity and deteriorates rapidly. The latter, containing the higher boiling components, although inferior with respect to the naturalness of the aroma, possesses superior stability and tenacity. Components of black pepper oil are shown in Table 27 (68). An oleoresin and concrete, which contain not only the oil but the pungent principles as well, are also produced.

Bois de Rose Oil. Although early Essential Oil Association standards (1959, 1963) gave several possible botanical sources for Bois de Rose (or rosewood) oil, it was later concluded that this oil is obtained from wood of the evergreen tree *Aniba rosaeodora* var. *amazonica* Ducke. which grows wild in the Amazon basin of Brazil and Peru (69). It is steam-distilled locally to yield a colorless-to-pale yellow liquid with a refreshing sweet-woody, floral-spicy odor. The topnote varies widely with source and quality, but is usually described as camphoraceous-peppery, reminiscent of nutmeg and cineole-eucalyptus terpenes. The oil is used either in its original form in flavors and perfumery or rectified to enrich the

Table 26. Comparison of Components of Spanish and Italian Bitter Orange Oils[a]

Component	Structure number	In Spanish, %	In Italian, %
limonene	(**23**)	94.34	92.43
myrcene	(**29**)	1.81	2.07
α-terpineol	(**26**)	0.57	0.06
α-pinene	(**44**)	0.45	0.59
β-pinene	(**22**)	0.30	0.90
linalyl acetate	(**24**)	0.28	0.78
octanol		0.18	0.17
linalool	(**10**)	0.15	0.37
decanal		0.14	0.19
octanal		0.13	0.24
geranyl acetate	(**28**)	0.13	0.09
nootkatone	(**63**)	0.08	0.03
Total		*98.56*	*97.92*

[a]Ref. 67.

Table 27. Principal Components of Black Pepper Oil[a]

Component	Structure number	Percent
sabinene	(**30**)	19
limonene	(**23**)	18
β-caryophyllene	(**52**)	15
β-pinene	(**22**)	10
α-pinene	(**44**)	9
Δ^3-carene		5
β-phellandrene[b]/1,8-cineole		4
myrcene	(**29**)	2
α-phellandrene[c]		2
para-cymene		1
terpinen-4-ol	(**37**)	1
trans-β-farnesene[d]		1
piperitone[e]		1
Total		88

[a]Ref. 68.
[b]3-Methylene-6-isopropylcyclohexene [555-10-2].
[c]2-Methyl-5-isopropyl-1,3-cyclohexadiene [99-83-2].
[d]7,11-Dimethyl-3-methylene-(*E*)-1,6,10-dodecatriene [18794-81-6].
[e]3-Methyl-6-isopropyl-2-cyclohexen-1-one [89-81-6].

principal component, linalool (**10**), which finds extensive use in perfumery. A typical commercial Bois de Rose oil [8015-77-8] made in 1988 contains the components shown in Table 28 (70).

Cananga Oil. The flowers of *Cananga odorata* Hook. f. et Thomson are used for the production of both cananga oil [68606-83-7] and ylang-ylang oil. In the case of cananga oil, a small quantity (10–30%) of leaves and twigs are also usually added to the still to facilitate even transfer of steam throughout the otherwise glutinous mass. It has been determined that a botanical difference

Table 28. Components of Bois de Rose Oil[a]

Component	Structure number	Percent
linalool	(**10**)	78
α-terpineol	(**26**)	6
geraniol	(**9**)	2
linonene/1,8-cineole	(**23**)/(**34**)	2
cis-linalool oxide	(**109**)	1
trans-linalool oxide	(**110**)	1
geranial		1
β-pinene	(**22**)	1
α-pinene	(**44**)	1
neral		1
Total		94

[a]Ref. 70.

does exist between trees used for each oil: forma *macrophylla* for cananga oil and forma *genuina* in the case of ylang. The oils may be differentiated in other ways: cananga oil is produced only in the northern and western parts of the island of Java in Indonesia from wild-growing trees using primitive stills, whereas ylang is obtained primarily from cultivated trees grown in the Comoro islands and Madagascar, with smaller amounts coming from Haiti, Réunion, and some of the French South Pacific islands. Cananga is a so-called total or complete essential oil in that, during the water distillation of the flowers, no fractions or cuts are taken; a total oil is removed. The yield and general odor quality of cananga oil is inferior to that of ylang. This is primarily a function of the distillation process, in which the flowers are packed tightly in the still. The flowers of *C. odorata* are delicate and easily lose their fragrance when damaged. Also, generally speaking, less care is taken in the choice of blossoms for the cananga oil still than for ylang. Java cananga oil is a yellow to orange-yellow or greenish yellow, somewhat viscous liquid possessing a sweet-floral, balsamic, and tenacious odor. The initial impression is woody-leathery, followed by a fresh-floral undertone. Cananga gives a much heavier odor impression than does ylang. It is generally assumed that most of the compounds found in ylang are also present in cananga, but in different proportions. Cananga generally contains more sesquiterpenes and sesquiterpene alcohols and a smaller quantity of esters than does ylang. Very few in-depth studies of cananga oil have been performed. Cananga finds use in soap and fine perfumery.

Caraway Oil. Produced by steam distillation of the crushed ripe seeds of *Carum carvi* L., caraway oil finds extensive use in food flavors, alcoholic liqueurs, and cheeses. The small herb, which grows wild in many regions of the world, is extensively cultivated in the Netherlands, Poland, Denmark, and parts of the former USSR. The Netherlands is the largest producer of the oil, which is offered commercially in two grades: crude or natural, and double-rectified or redistilled. The former is the direct distillate obtained from the fruits and is a pale yellow-to-brownish mobile liquid with a strong odor reminiscent of the fruit but with fatty-harsh undertones. It has a burning, warm, biting taste. The redistilled oil is colorless to pale yellow with a stronger and less fatty odor. The flavor is less sweet and more biting than the crude. Although (+)-carvone (**126**) is recognized as the odor-impact molecule of caraway, many lesser constituents play an important role in the total aroma and taste effect. The principal components of a Dutch oil are shown in Table 29 (71).

      (**126**)   (**127**)   (**128**)

Table 29. Components of a Dutch Caraway Oil[a]

Component	CAS Registry Number	Structure number	Percent
(+)-carvone	[2244-16-8]	(126)	50
limonene		(23)	47
dihydrocarveol	[619-01-2]	(127)	0.6
cis-dihydrocarvone	[3792-53-8]	(128)	0.5
myrcene		(29)	0.4
trans[b]-carveol	[1197-07-5]	(129)	0.3
trans-dihydrocarvone	[69424-02-8]	(128)	0.2
cis[b]-carveol	[2102-59-2]	(129)	0.1
perillaldehyde	[2111-75-3]	(130)	0.1
linalool		(10)	0.1
sabinene		(30)	0.1
cis-/trans-1,2-epoxylimonene	[1195-92-2]	(131)	0.2
Total			99.6

[a]Ref. 71.
[b]The OH group is cis or trans to the propenyl group.

(129)   (130)   (131)

The oil possesses moderate antibacterial and strong antifungal properties. Thus the application of the oil to the crust of cheese could serve to prevent the formation of mycotoxin in the cheese. The optical purity of the carvone in caraway has been determined using a chiral gc column (72). It was found to be $(R)(+) = 97.64\%$ and $(S)(-) = 2.36\%$.

Cardamom Oil. One of the oldest essential oils known, cardamom oil [8000-66-6] is steam-distilled from the seeds of *Elettaria cardamomum* Maton, a plant of the ginger family which grows wild and also is cultivated in India, Sri Lanka, and in Central America, primarily in Guatemala. India consumes more than 50% of the cardamom spice produced each year. The oil is an almost colorless to pale yellow to light brown mobile liquid which darkens when exposed to sunlight. The topnote of cardamom oil is warm-spicy, aromatic, at first penetrating camphoraceous-cineole-like or medicinal as in eucalyptus. The drydown becomes balsamic-woody, sweet and almost floral with extreme tenacity. A comparison of the constituents of a commercial oil and of the headspace volatiles of living ripe cardamom seed is shown in Table 30 (73).

Cassia Oil. Also known as Chinese cinnamon oil, cassia oil is the steam distillation of the dried leaves, twigs, and bark of *Cinnamomum cassia* Blume, tall slender trees which grow in the southeastern areas of China and some parts of Vietnam and India. The bark of the trees is known throughout the world as

Table 30. Comparison of Components of the Headspace of Living Ripe Cardamom Seed and Commercial Oil[a]

Component	Structure number	In headspace, %	In commercial oil, %
1,8-cineole	(34)	19.7	44.7
α-terpinyl acetate	(132)	17.4	23.6
myrcene	(29)	12.5	0.1
geraniol	(9)	7.1	0.4
linalool	(10)	6.2	4.9
geranial		5.7	0.2
α-terpineol	(26)	3.2	1.2
sabinene	(30)	2.9	1.4
neral		1.7	0.4
terpinolene	(50)	1.0	
Total		*77.4*	*76.9*

[a]Ref. 73.

cinnamon bark or cassia-cinnamon, not to be confused with true cinnamon bark Ceylon (Sri Lanka) which is a completely different species. The oil, which is produced in local stills by a water distillation, is a crude dark-brown liquid with a strong spicy, warm, woody-resinous aroma with an intensely sweet-balsamic undertone. It finds extensive use in the flavoring of soft drinks and as a raw material for the manufacture of natural benzaldehyde. Although some oil is purified by filtration and rectification in China, most material is shipped crude either for use as is or for cleanup or concentration by the user. Adulteration with synthetic cinnamaldehyde can still be a problem, but the use of modern analytical techniques such as isotope analysis has definitely lessened the risk of it to the user. Some companies go so far as to have representatives at the site of the distillation to oversee the process and personally seal the drums of crude oil. The headspace volatiles of living cassia leaf have been analyzed for comparison with those of picked leaf and of an authentic commercial oil (Table 31) (74).

(133)　　　　　(134)

Cedarleaf Oil. Also known commercially as thuja oil [8007-20-3], cedarleaf oil is steam-distilled almost exclusively from the leaves and twigs of the Eastern or Northern White Cedar *Thuja occidentalis* L., which grows abundantly in the northeastern United States and eastern Canada. The oil is distilled locally mostly in New York State, Vermont, and Quebec. It is a colorless to pale greenish yellow mobile liquid with an intensely sharp and quite fresh camphoraceous aroma. Although the principal constituent, α-thujone, is considered to be a skin irritant and somewhat poisonous, the low level required in perfumes permits its use.

Table 31. Comparative Analysis of Cassia Leaf Volatiles[a]

Component	CAS Registry Number	Structure number	In living, %	In picked, %	In commercial oil, %
trans-2-hexenal			4.0	0.8	
cis-3-hexenol			1.5	3.0	
trans-2-hexenol	[2305-21-7]		4.3	3.5	
n-hexanol	[25917-35-5]		1.2	1.1	
benzaldehyde	[100-52-7]		0.2	0.4	0.9
benzyl alcohol			0.4	0.1	
salicaldehyde	[90-02-8]		0.4	0.2	0.2
β-phenethyl alcohol		(2)	2.1	0.1	0.4
hydrocinnamaldehyde	[104-53-0]	(133)			0.2
2-methyl benzofuran	[4265-25-2]	(134)			0.4
cis-cinnamaldehyde	[57194-69-1]		0.8	0.8	0.2
2-methoxybenzaldehyde	[135-02-4]			0.6	0.6
phenethyl acetate	[103-45-7]				0.4
trans-cinnamaldehyde	[14371-10-9]		47.4	68.8	72.8
cinnamic alcohol			20.6	0.3	0.3
coumarin		(38)	7.9	4.4	1.7
2-methoxy cinnamaldehyde	[1504-74-1]		1.0	1.4	11.5
4-methoxy cinnamaldehyde	[1963-36-6]		4.1	12.3	
2-methoxy cinnamyl acetate	[1504-61-6]				1.6
Total			95.7	98.0	91.1

[a]Ref. 74.

It finds occasional use in fine fragrances of the chypre or fougère type and in some consumer products. The analysis of a number of cedarleaf oils has been performed, and the range for the principal constituents is shown in Table 32 (75).

(135)   (136)   (137)   (138)

Cedarwood Oil. A large number of different types of cedarwood oil are produced. The largest volume oils are "Texas-type" from *Juniperus ashei* Buchh.

Table 32. Range of Composition for Primary Components of Thuja Oil[a]

Component	CAS Registry Number	Structure number	Percent
α-thujone	[546-80-5]	(135)	31.2–47.1
β-thujone	[471-15-8]	(136)	8.1–11.7
fenchone	[1195-79-5]	(137)	6.5–14.7
bornyl acetate	[76-49-3]	(138)	3.2–5.5
terpinen-4-ol	[562-74-3]	(37)	1.9–5.7
α-pinene	[80-56-8]	(44)	1.7–3.6
camphor	[76-22-2]	(35)	1.5–3.0
α-terpinyl acetate	[80-26-2]	(132)	1.3–2.5
sabinene	[3387-41-5]	(30)	1.1–2.3
limonene	[138-86-3]	(23)	1.0–3.3
γ-terpinene	[99-85-4]		1.0–1.9

[a]Ref. 75.

(Texan cedarwood oils [68990-83-0]), "Virginia" from *Juniperus virginiana*, and "Chinese" from several varieties of trees of Japanese and Chinese origin. Cedarwood oils are used to provide woody fragrance notes and as fixatives for perfumes. The principal constituents of oils from *J. ashei* and *J. virginiana* are shown in Table 33 (76).

(139)   (140)   (141)

(142)   (143)

Table 33. Constituents of Oils from *J. ashei* and *J. virginiana*[a]

Component	CAS Registry Number	Structure number	Percent in *J. ashei*	Percent in *J. virginiana*
thujopsene	[470-40-6]	(139)	60.4	27.6
cedrol	[77-53-2]	(140)	19.0	15.8
α-copaene		(122)	2.8	6.3
α-cedrene	[469-61-4]	(141)	1.8	27.2
β-cedrene	[546-28-1]	(142)	1.6	7.7
widdrol	[6892-80-4]	(143)	1.1	1.0
Total			86.7	85.6

[a]Ref. 76.

Roman Chamomile Oil. Steam distilled from the ligulate florets of *Chamaemelum nobile* (L.) All. (*Anthemis nobilis* L.), a member of the Dog Fennel or Chamomile family which is native to the Azores, northern Africa, and western Europe where it is extensively cultivated, Roman chamomile oil is a pale blue mobile liquid (when fresh but fades with aging) with a sweet herbaceous, fruity-warm, tealeaf-like aroma. The odor is extremely diffusive but with little tenacity. The pale blue color is due to the presence of a high boiling hydrocarbon, chamazulene, which only distills in trace amounts in the true essential oil. The oil has a bitter and medicinal flavor which finds some application in certain types of alcoholic beverages. However, it finds frequent but sparing use in perfumes where it contributes a warm, fresh, natural note difficult to achieve by other means. The oil, the principal constituents of which are shown in Table 34, possesses an extremely high ester value (77).

(**144**)   (**145**)   (**146**)

Cinnamon Bark Oil. True cinnamon bark oil, ie, "Ceylon," is steam-distilled (occasionally water-distilled) from the dried inner bark of the shoots of coppiced bushes of *Cinnamonum zeylanicumi* Nees, primarily in Sri Lanka but also in India, Burma, Indochina, and several Indonesian islands. In each of these

Table 34. Components of Roman Chamomile Oil[a]

Component	CAS Registry Number	Structure number	Percent
3-methylpentyl isovalerate	[35852-41-6]		21.0
methallyl angelate	[61692-78-2]	(**144**)	15.8
3-methylpentyl isobutyrate	[53082-58-9]	(**144**)	11.9
2-methylbutyl angelate	[61692-77-1]	(**144**)	7.8
pinocarveol	[5947-36-4]	(**145**)	4.8
pinocarvone	[16812-40-1]	(**146**)	4.3
3-methylbutyl angelate	[10482-55-0]	(**144**)	4.1
isobutyl angelate	[7779-81-9]	(**144**)	3.4
α-pinene		(**44**)	3.0
3-methylpentyl methacrylate	[113615-00-2]		2.6
3-methylbutyl isobutyrate	[2050-01-3]		2.2
2-methylbutyl isobutyrate	[2445-69-4]		1.7
methallyl methacrylate	[816-74-0]		1.4
3-methylpentyl 2-methylbutyrate	[83783-89-5]		1.3
2-methylbutyl methacrylate	[60608-94-8]		1.2
3-methylpentyl acetate	[35897-13-3]		1.0
Total			87.6

[a]Ref. 77.

areas it grows wild but also is cultivated for its oil and as a spice. (Coppicing involved the felling of a mature tree and repeated harvesting of the shoots which grow from the stump.) The quality of the bark and its oil is strongly affected by climatic and soil conditions as well as the age of the coppiced trees and the handling before and during curing and processing. Although some distillation takes place locally, the best grades of oil are produced in Europe and the United States. Because an appreciable quantity of the important volatiles are soluble in the distillation waters, extraction of the dissolved volatiles after separation of the oil layer and use of skillful distillation techniques are necessary to obtain the best oils.

Ceylon cinnamon bark oil is a pale yellow-to-dark or brownish yellow liquid of extremely powerful, diffusive, warm-spicy, sweet, and tenacious odor. The best oils have a dry-powdery-dusty, but warm, uniform and lasting dry-out note. The flavor is distinctly sweet and spicy, and the best oils are five to ten times stronger in flavor than ordinary cinnamon bark oils. The oil finds extensive use in flavors for food and candy, baked goods and beverages, dental preparations, mouth rinses and gargles, and as a masking agent for pharmaceuticals with strongly unpleasant medicinal flavors. This essential oil ranks among the most powerful natural antiseptic germicides. Cinnamon bark oil is much appreciated by perfumers for its warm dry spiciness, immediate sweetness, and tremendous diffusive power, as well as its ability to blend well with woody-oriental and olibanum notes. A typical sample of Sri Lankan cinnamon bark oil has the components shown in Table 35 (78).

(**147**)   (**148**)

Citronella Oil. Citronella oil is steam-distilled from citronella grass, of which some 30 varieties exist growing both wild and cultivated throughout southern Asia, including China and Taiwan and the islands of Indonesia, as well as in northern Australia. The grasses are also grown in East Africa, South and Central America, the West Indies, Madagascar, the Comoro islands, and the Seychelles. Two types of the oil exist commercially: Java-type and Ceylon (Sri Lankan). By far, the largest-volume oil is citronella-Java, obtained from *Cymbopogon winterianus* Jowitt, the so-called Maha Pengiri grass cultivated in Taiwan, Hainan, Java, Malaysia, Guatemala, and Honduras. One advantage of the Maha Pengiri grass is that it yields up to twice as much essential oil as does the Ceylon-type which comes from *C. nardus* L. It also contains more of the components important for the production of synthetic or semisynthetic perfumery materials, eg, citronellal (**46**), which may be transformed into menthol [*89-78-1*] (*trans*-2-isopropyl-*cis*-5-methylcyclohexanol), or hydroxycitronellal (**21**), geraniol

Table 35. Components of Sri Lankan Cinnamon Bark Oil[a]

Component	CAS Registry Number	Structure number	Percent
trans-cinnamaldehyde			72.0–82.2
cinnamyl acetate	[103-54-8]		3.2–3.7
eugenol			1.1–13.3
β-caryophyllene		(52)	1.0–2.0
linalool		(10)	0.7–1.1
cinnamyl alcohol			0.5–0.6
benzyl benzoate			0.4–1.0
α-terpineol/borneol			0.4–0.6
ortho-methoxy cinnamaldehyde			0.3–0.8
para-cymene			0.3–0.4
α-humulene	[6753-98-6]	(147)	0.2–0.6
cis-cinnamaldehyde			0.1–1.0

[a] Ref. 78.

(**9**), and citronellol (**8**), which may be used as such, or after conversion to esters. Components of a *C. winterianus* oil are shown in Table 36 (79).

Clove Bud Oil. Clove bud oil is water- (and sometimes steam-) distilled from the dried flower buds of a slender, medium-sized, cultivated tropical tree, *Syzygium aromaticum* L., also known as *Eugenia caryophyllata* Thunb. and *Jambosa caryophyllus* K. Spreng, which originated and still grows wild in the Moluccas island group, particularly the island of Amboyna, in the eastern Indonesian archipelago. The cultivation of clove trees is at least 2000 years old. The bud oil is the largest of the essential oils obtained from parts, including the stems as well as twigs and leaves, of the clove tree. Although most of the clove leaf and clove stem oil is distilled locally, clove bud oil is, for the most part, produced in European and U.S. distilleries. Madagascar, Zanzibar (Tanzanian), or Comoro clove buds are used for the distillation; Amboyna cloves are usually sold as the highest grade of the whole spice. During steam distillation of clove buds, eugenol acetate is hydrolyzed to eugenol. Thus, water distillation is preferred.

Table 36. Components of *C. winterianus* Citronella Oil[a]

Component	Structure number	Percent
citronellal	(49)	32.0
geraniol	(9)	20.0
citronellol	(8)	14.5
geranyl acetate	(28)	8.0
citronellyl acetate	(121)	5.0
limonene	(23)	4.0
elemol	(49)	2.5
2,6-dimethyl-5-heptenal		1.5
linalool	(10)	1.5
3,7-dimethyl-3,6-octadien-1-ol		1.5
neral		1.0
Total		*81.5*

[a] Ref. 79.

The total eugenol content of water-distilled oil is usually greater than 90%, but more than 10% of this is in the form of the acetate, which is partly responsible for the characteristic odor of quality clove bud oil. For some time it was believed in the industry that sesquiterpenes such as β-caryophyllene (**52**) were formed during both water- and steam-distillation of clove buds. By means of the dynamic headspace analysis of authentic clovebuds it has now been unequivocally shown that β-caryophyllene is present as such in the untreated clove bud (79). A comparison of the headspace constituents of both whole and crushed clove buds with a commercial oil is shown in Table 37 (80).

Clove bud oil is frequently used in perfumery for its natural sweet-spicy note but the greatest application is in the flavor area in a large variety of food products, including spice blends, seasonings, pickles, canned meats, baked goods, ready-made mixes, etc. As in the case of cinnamon bark oil, its well-known antiseptic properties make it ideal for application in mouth washes, gargles, dentifrices, and pharmaceutical and dental preparations. Candy, particularly chewing gum, is also flavored with clove bud oil in combination with other essential oils.

Coriander Oil. Crushed fully ripe fruits (seeds) of *Coriandrum sativum* L., a small herb native to southeastern Europe, yield a colorless to pale yellow steam-distilled oil with a delightfully sweet, peppery-woody aromatic odor with a floral-balsamic undertone. The herb grows not only wild but is cultivated throughout

Table 37. Comparison of the Headspace Components of Whole and Crushed Clove Bud with a Commercial Oil[a]

Component	CAS Registry Number	Structure number	In headspace of whole buds, %	In headspace of crushed buds, %	In oil, %
eugenol[b]		(**11**, R = H)	85.1	51.3	86.8
β-caryophyllene		(**52**)	6.4	29.0	2.6
trans-β-farnesene[c]			1.0		
vanillin[d]	[121-33-5]		0.9	0.8	
eugenyl acetate	[93-28-7]		0.6	0.4	9.7
benzyl acetate			0.5	0.2	
isoeugenol[e]	[97-54-1]		0.2		
furfural	[98-01-1]		0.2	0.3	f
5-methyl-2-hexyl acetate	[145119-71-7]		0.2	1.0	f
2-nonanone	[821-55-6]		0.2	0.3	
carvone		(**123**)	0.2		
γ-cadinene	[1460-97-5]	(**148**)	0.1	0.3	
benzyl alcohol			0.1		
benzaldehyde			0.1	0.1	
Total			95.8	83.7	99.1

[a]Ref. 80.
[b]2-Methoxy-4-(2-propenyl)phenol.
[c]7,11-Dimethyl-3-methylene-(*E*)-1,6,10-dodecatriene.
[d]4-Hydroxy-3-methoxybenzaldehyde.
[e]2-Methoxy-4-(1-propenyl)phenol.
[f]Trace amounts.

the world, and distillation takes place mainly in Europe. A tremendous quantity of fruit is grown in areas such as India for use as is in curry flavors. The flavor of coriander oil [8008-52-4] is mild, sweet, and spicy-aromatic but also warm and slightly burning. It is used extensively in flavors for alcoholic beverages, candy, tobacco, pickles, meat sauces, seasonings, etc. In perfumery, its warm and sweet notes are useful in oriental and "white-flower" fragrances. A comparison of the components of an authentic coriander seed oil with the headspace volatiles over freshly crushed ripe coriander seeds has been made and is shown in Table 38 (81).

Cornmint Oil.   The steam-distilled oil of different varieties of *Mentha arvensis* L., a native of China and Japan, is known as cornmint or Japanese mint in the United States. In the rest of the world, it is erroneously called peppermint oil and, indeed, it has a history of use in the adulteration of true peppermint oil. It is extensively cultivated in regions of China, Japan, Brazil, Taiwan, Australia, Africa, and India. The principal ingredient of the oil is menthol (70–75%), a large proportion of which can be easily recovered simply by cooling the oil and filtering. So-called dementholized cornmint oil is an item of commerce, and a typical analysis of a Japanese version is shown in Table 39 (82).

(149)   **(150)**   **(151)**

Table 38. Comparison of Components of Authentic Coriander Seed Oil and Headspace Volatiles over Freshly Crushed Ripe Coriander Seeds[a]

Component	Structure number	In oil, %	In freshly crushed ripe coriander seeds, %
linalool	(10)	73.3	63.2
γ-terpinene		5.4	0.8
α-pinene	(44)	5.0	0.3
camphor	(35)	4.8	
limonene	(23)	2.5	0.9
geranyl acetate	(28)	2.2	15.6
citronellal	(46)		1.3
para-cymene		1.6	1.3
citronellyl acetate			1.2
trans-β-ocimene	(25)	b	1.0
camphene	(47)	1.0	
myrcene	(29)	0.6	3.2
terpinolene	(50)	0.5	0.2
Total		*96.9*	*86.0*

[a]Ref. 81.
[b]Trace amounts.

Table 39. Components of a Dementholized Japanese Cornmint Oil[a]

Component	CAS Registry Number	Structure number	Percent
menthol (isomers)			38.0
menthone[b]	[89-8-5]		26.3
isomenthone[c]	[491-07-6]		7.3
limonene		(23)	7.1
piperitone[d]	[89-81-6]		3.8
menthyl acetate	[89-48-5]	(149)	3.4
α-pinene		(44)	2.0
β-pinene		(22)	2.0
β-caryophyllene		(52)	1.3
pulegone	[89-82-7]	(150)	1.3
Total			*91.5*

[a]Ref. 82.
[b]*trans*-2-Isopropyl-5-methylcyclohexanone.
[c]*cis*-2-Isopropyl-5-methylcyclohexanone.
[d]3-Methyl-6-isopropyl-2-cyclohexen-1-one.

Eucalyptus Oil (Cineole-Rich). More than 700 species of the genus *Eucalyptus* are known, with the large number of varieties and cultivars making this an enormous class of botanicals. Australia is considered the home of the eucalyptus, although little cultivation of trees for their commercial oils is undertaken there. Vast plantations of various species of eucalyptus, frequently grown for their timber as well as the oil, exist in Brazil, Africa, Central America, and other regions such as Spain. From a pharmacological perspective, the so-called 1,8-cineole- or eucalyptol-rich oils, obtained by steam distillation of the fresh or partly dried leaves, are of greatest commercial importance. Varieties of this type include *E. globulus* Labill., the "blue gum" or Tasmanian eucalyptus; *E. smithii* R. T. Baker, the "gully gum" tree; *E. dives* Schauer. var. C, the "broadleaf peppermint;" *E. leucoxylon* F. J. Muell.; and the "white" eucalyptus. The oils are colorless and mobile, and they are frequently rectified immediately following distillation to improve their keeping qualities and to remove unwanted low boiling aldehydes. The main use of these type oils is in pharmaceutical preparations, cough drops, vaporizers, gargles, toothpastes, and as germicides. A pharmaceutical-grade eucalyptus oil must meet the following specifications: as eucalyptol minimum 70%, it must be free of phellandrene and soluble in five volumes of 70% ethanol. A typical commercial Spanish *E. globulus* oil has the components shown in Table 40 (83).

Some of the other eucalyptus oils of commercial importance include the Chinese eucalyptus, a camphor/cineole-type oil; *E. citriodora* Hook., a citronellal-type oil; *E. staigeriana* F.v. Muel., a citral-type oil; and *E. macarthuri* H. Deane & Maiden, a geranyl acetate-type oil.

Ginger Oil. The product of the steam distillation of the dried, unpeeled, freshly ground rhizomes of *Zingiber officinale* Roscoe, a native of the tropical coastal regions of India, ginger oil [8007-08-7] is a pale yellow to light amber-colored mobile liquid which noticeably resinifies on aging. Ginger is cultivated in most tropical and subtropical regions; for the production of essential oil, the Nigerian and Jamaican gingers are preferred. The rhizomes are shipped to Europe and the United States for distillation. Although the odor of the oil

Table 40. Components of Spanish *E. globulus* Oil[a]

Component	Structure number	Percent
1,8-cineole	(**34**)	71.3
α-pinene	(**44**)	14.4
limonene	(**23**)	2.7
trans-pinocarveol	(**145**)	2.6
para-cymene		1.8
aromadendrene	(**151**)	1.1
α-fenchyl alcohol/pinocarvone	(**58**)/(**146**)	1.0
Total		*96.9*

[a]Ref. 83.

is considerably dependent on geographical source, generally it is described as having a warm, fresh-woody-spicy and somewhat lemony topnote with a heavy and tenacious, sweet, rich, balsamic-floral undertone. The exceptional pungent quality of the fresh or dried ginger spice is not a characteristic of the essential oil, but is retained in the oleoresin, which is also a product of commerce. The oil finds use both in perfumery and flavor work. In the former, it adds a spicy-sweetness in oriental and some floral fragrances and is finding increasing use for the spicy-trend in men's fragrances. The most important flavor area is for baked goods such as cookies and spice cakes. A comparison of the headspace volatiles of fresh ginger rhizome and a commercial oil has been performed, and differences are shown in Table 41 (84). The relatively high percentage of citral (geranial/neral) in the headspace accounts for the pronounced lemony odor of the freshly cut rhizome.

(**152**)   (**153**)

Juniper Oil. The best juniper oil [*8012-91-7*] is obtained from the steam distillation of the ripe crushed, dried berries of *Juniperus communis* L., a shrub which grows wild in many regions of Europe, Asia, Africa, and North America. However, most commercial juniperberry oil comes from the fermented fruits as a by-product of flavors for alcoholic beverages such as gin, brandy, liquors, cordials, and sloe-gin. This represents the actual commercial juniperberry oil, since very little true juniperberry oil is produced. A comparison of the headspace volatiles of ripe juniperberries (85) with an authentic, freshly prepared juniperberry oil (86) is shown in Table 42.

Table 41. Comparison of the Differences Between the Headspace Volatiles of Fresh Ginger Rhizome and a Commercial Oil[a]

Component	CAS Registry Number	Structure number	In headspace, %	In oil, %
α-zingiberene	[7785-34-3]	(152)	15.2	34.4
α-farnesene[b]			13.7	6.0
geranial			11.7	0.7
ar-curcumene[c]			11.3	4.8
β-sesquiphellandrene	[20307-83-9]	(153)	8.0	11.8
β-phellandrene[d]			4.1	8.0
β-bisabolene		(53)	3.8	6.2
neral			3.6	0.5
2-undecanone	[112-12-9]		3.4	0.8
myrcene		(29)	2.3	0.7
2-nonanone			1.8	0.3
2-nonanol	[628-99-9]		1.2	0.5
camphene		(47)	[e]	5.8
α-pinene		(44)		1.9
Total			*91.1*	*82.4*

[a]Ref. 84.
[b]3,7,11-Trimethyl-(*E,E*)-1,3,6,10-dodecatriene.
[c]2-Methyl-6-(*p*-methylphenyl)-2-heptene.
[d]3-Methylene-6-isopropylcyclohexene.
[e]Trace amounts.

Table 42. Comparison of Headspace Volatiles of Ripe Juniperberries with an Authentic Juniperberry Oil

Component	CAS Registry Number	Structure number	In headspace, %	In oil, %
limonene		(23)	65.0	8.7
myrcene		(29)	13.7	8.5
sabinene		(30)	4.3	1.7
cis-3-hexenol			2.9	
β-caryophyllene		(52)	2.1	7.2
α-elemene	[5951-67-7]	(154)	1.6	
terpinolene		(50)	1.4	0.4
8-*para*-cymenol	[68279-51-6]	(155)	1.1	
α-pinene		(44)	1.0	20.0
β-elemene		(125)	0.8	
methyl salicylate[a]	[119-36-8]		0.8	
borneol				8.0
germacrene D	[23986-74-5]	(156)		7.0
α-humulene		(147)	0.5	3.9
Total			*95.1*	*65.4*

[a]Methyl 2-hydroxybenzoate.

(154)   (155)   (156)

Labdanum Oil.   Labdanum is the natural oleoresin obtained from various species of *Cistus* including particularly *C. ladanifer* L., commonly called laudanam, native to the Mediterranean region from Portugal to France and northwestern Africa. Various extracts and oils derived from labdanum are commercially available. These include absolutes, concretes, resinoids, and oils. The oils, called cistus oils, are of two types: the so-called cistus oil which is either steam-distilled directly from the crude gum or obtained by mixed solvent extraction, and true cistus oil obtained by steam distillation of the entire herb, including leaves, stems, and flowering tops. So-called cistus oil is an amber-colored, viscous liquid with a powerful and tenacious warm, sweet, animalic odor. True cistus oil is a pale orange liquid with a peculiar, warm, herbaceous, ambery odor and much lower tenacity than the so-called cistus oil. However, it has immense power in its topnote. Only limited quantities of the true cistus oil are produced. Both oils find specialty uses in perfumery. The principal components of a so-called cistus oil are shown in Table 43 (87).

(157)   (158)   (159)   (160)

(161)   (162)   (163)   (164)

Lavender Oil Spike.   Also known as spike oil, lavender oil spike is obtained by steam distillation of the flowering tops of *Lavandula latifolia* Vill., which

Table 43. Components of Cistus Oil[a]

Component	CAS Registry Number	Structure number	Percent
α-pinene		(44)	10.8
viridiflorene	[21747-46-6]	(157)	8.4
viridiflorol	[552-02-3]	(158)	6.3
trans-pinocarveol		(145)	3.4
ledol	[577-27-5]	(159)	3.3
bornyl acetate		(138)	2.1
α-phellandren-8-ol	[1686-20-0]	(160)	1.9
α-gurjunene	[489-40-7]	(161)	1.7
α-para-dimethylstyrene[b]	[1195-32-0]		1.6
terpinen-4-ol		(37)	1.5
β-phellandren-8-ol	[65293-09-6]	(162)	1.5
para-cymene			1.5
alloaromadendrene	[25246-27-9]	(163)	1.4
pinol	[2437-97-0]	(164)	1.4
eugenol			1.3
acetophenone	[98-86-2]		1.4
limonene/eucalyptol		(23)/(34)	1.3
camphene		(47)	1.0
Total			*51.8*

[a]So-called custus oil. Ref. 87.
[b]p-Isopropenyltoluene.

grows wild and is also cultivated throughout the Mediterranean region, with most production in Spain and France. There was a time when spike oil, particularly Spanish lavender oil [8016-78-2], found extensive use in soap perfumery; however, with the introduction of the less expensive lavandin oil, a hybrid of true lavender (*L. angustifolia*) and *L. latifolia*, utilization of spike oil in perfumery has been reduced appreciably. French spike oil, a more delicate version, still finds considerable use in functional and industrial perfumery. A comparison of the volatiles of a commercial Spanish oil and a lab-distilled version of the same has been made and components are shown in Table 44 (88).

Lemon Oil. The cold-pressed oil of the lemon (*Citrus limon* L.) is, after orange oil, the most important citrus oil. Although it is most likely native to the East India–Burma region, lemon cultivation rapidly spread throughout the tropical and subtropical world to the point that, at present, the largest lemon (and lemon oil) industry is in California; main production centers are Cyprus, Italy, Guinea, Brazil, Tunisia, Mexico, Israel, India, Pakistan, Spain, Jamaica, China, Australia, and South Africa. Expressed lemon oil is a pale yellow to greenish yellow mobile liquid with a light, fresh, sweet odor of varying tenacity, depending on the method of espression. The oil usually requires stabilization against oxidation, although Cyprus oil possesses remarkable keeping qualities. Lemon oil finds wide application in both fine and functional perfumery, for its refreshing sweet-fruity note, and flavors, where a concentrated or sesquiterpeneless lemon oil is preferred. Lemon oils have been extensively analyzed. A comparison of some components of a Sicilian and a California oil is shown in Table 45 (89).

Table 44. Components of a Lab-Distilled and Commercial Lavender Spike Oil[a]

Component	Structure number	In lab-distilled oil, %	In commercial oil, %
1,8-cineole	(**34**)	36.3	34.9
linalool	(**10**)	30.3	18.9
camphor	(**35**)	8.0	15.0
borneol		2.8	1.4
α-terpineol	(**26**)	2.6	1.1
coumarin	(**38**)	2.4	0.6
caryophyllene oxide	(**87**)	2.4	1.9
limonene	(**23**)	1.0	1.2
8-*para*-cymenol	(**155**)	1.0	[b]
cis-/trans-linalool oxide (furanosyl)	(**109**)/(**110**)	0.5	12.7
Total		*87.3*	*87.7*

[a]Ref. 88.
[b]Trace amounts.

Table 45. Comparison of Components of Sicilian and California Expressed Lemon Oils[a]

Component	Structure number	In Sicilian oil, %	In California oil, %
limonene/eucalyptol/ *cis*-β-ocimene	(**23**)/(**34**)/	65.2	65.7
β-pinene	(**22**)	10.5	11.1
γ-terpinene		8.9	8.3
geranial		2.1	1.2
α-pinene	(**44**)	1.8	1.8
sabinene	(**30**)	1.8	1.9
myrcene	(**29**)	1.6	1.6
neral		1.3	0.7
Total		*93.2*	*92.3*

[a]Ref. 89.

Nutmeg Oil. The nutmeg, the fruit of the tree *Myristica fragrans* Houtt., which originated in the East Indies, represents another plant product with significant use in the dried natural state as a spice and a much lesser use as an oil or extract. Considerable acreage in Indonesia and, more recently, in the West Indies has been devoted to the cultivation of nutmeg. In both areas, nutmeg oil is distilled locally, although large quantities of the oil are also produced in Europe and the United States. The oil finds extensive use in both perfumery, in the modern "spicy" perfumes and in men's fragrances, and in flavor work, although a terpeneless oil is generally preferred in the latter case. As such, nutmeg oil is a significant spice additive in tomato ketchup. It is produced by steam or steam–water distillation of the freshly chopped fruits. Mace, the dried outer husk of the fruits, may be included. Some distillers prefer to remove a substantial portion of the fixed oil by hydraulic pressing prior to distillation. Nutmeg essential oil is a water-white to pale yellow mobile oil with a light, fresh, warm-spicy, and aromatic odor with a terpeny topnote and a woody, warm-sweet dryout. Nutmeg oil has been extensively analyzed, and a comparison of the components of several oils is shown in Table 46 (90).

(**165**)   (**166**)

Oregano Oil Spanish.   Steam-distilled from the dried flowering herb *Thymus capitatus* (L.) Hoffmanns & Link (also known as *Coridothymus capitatus* (L.) Reichb. f.), which grows wild in the Mediterranean region, Spanish oregano (or origanum oil [*8007-11-2*]) oil is one of the largest-volume oregano oils of commerce. This oil is distilled locally and is a dark brownish red to purple to orange-colored mobile liquid with a strong tar-like, herbaceous, and refreshing odor. The topnote is slightly green-camphoraceous, herbaceous, and the body is rich, dry-woody, and phenolic. On dryout, the aroma remains dry-woody and phenolic. The flavor is somewhat burning, warm-phenolic, and herbaceous; it is pleasant only in high dilution. Fine perfumery uses oregano oil for its powerfully refreshing notes and its spicy-herbaceous effect. Soap perfumery utilizes the medicinal note of its principal phenolic component, carvacrol [*499-75-2*] (2-methyl-5-isopropylphenol). For flavor work, the milder oregano oils such as marjoram or thyme are usually preferred. The composition of an oregano oil of Spanish origin has been analyzed, and components are shown in Table 47 (91).

Table 46. Comparative Analysis of the Components of Various Nutmeg Oils[a]

Component	CAS Registry Number	Structure number	In Sri Lankan (water-distilled), %	In Indonesian (steam-distilled), %
sabinene		(**30**)	28.0	16.5
α-pinene		(**44**)	13.9	21.7
terpinen-4-ol		(**37**)	8.9	5.9
β-pinene		(**22**)	8.7	14.5
myristicin	[*607-91-0*]	(**165**)	3.3	5.5
α-thujene			2.5	2.7
myrcene		(**29**)	2.1	2.0
linalool		(**10**)	2.1	0.6
limonene		(**23**)		5.8
γ-terpinene				5.2
safrole	[*94-59-7*]	(**166**)		1.6
α-terpineol		(**26**)		1.1
eugenol				0.7
citronellol		(**8**)		0.2
Total			*69.7*	*86.0*

[a]Ref. 90.

Table 47. Components of Spanish Oregano Oil[a]

Component	Structure number	Percent
carvacrol		80.2
γ-terpinene		4.0
para-cymene		3.9
β-caryophyllene	(52)	3.5
α-thujene		2.2
myrcene	(29)	1.3
α-phellandrene		1.0
Total		96.1

[a]Ref. 91.

Orris. Steam-distillation of the aged (3-yr) peeled, dried, pulverized rhizomes of the decorative garden perennial *Iris pallida* Lam. yields a waxy, cream-colored mass known as orris butter or orris concrète. Fresh rhizomes are practically odorless. This material melts at body temperature and possesses a woody, fatty-oily, violet-like odor with a sweet, floral, warm, and fruity undertone. The wax, which accounts for 85–90% of the concrete, is myristic acid which, because it can cause problems in perfumery and handling, is usually removed by alkali washing in an alcoholic solution. This process yields the highly desirable, from a perfumery standpoint, but very expensive, orris absolute. Although most of the cultivation and curing of the plant material takes place in Italy, the bulk of the processing occurs in France. Most of the absolute is used in fine perfumery, although traces are effective in fruit and rum flavors. The volatiles composition of a commercial orris absolute is shown in Table 48 (92).

The irones (**167,168,169**), which constitute slightly more than 75% of the volatiles, are primarily responsible for the fine odor of the natural material. For this reason and because of the high cost of orris absolute, synthetic versions of the irones have been commercialized. Of the possible irone structures, the γ-isomer (**168**) possesses the best fragrance properties.

Table 48. Volatiles Composition of Orris Absolute[a]

Component	CAS Registry Number	Structure number	Percent
α-irone (two isomers)	[79-69-6]	(167)	60.7
γ-irone	[79-68-5]	(168)	14.9
methyl myristate[b]	[124-10-7]		10.0
ethyl myristate[b]	[124-06-1]		4.0
β-irone	[472-46-8]	(169)	2.1
ethyl laurate[c]	[106-33-2]		1.6
Total			93.3

[a]Ref. 92.
[b]Myristic acid = tetradecanoic acid.
[c]Ethyl dodecanoate.

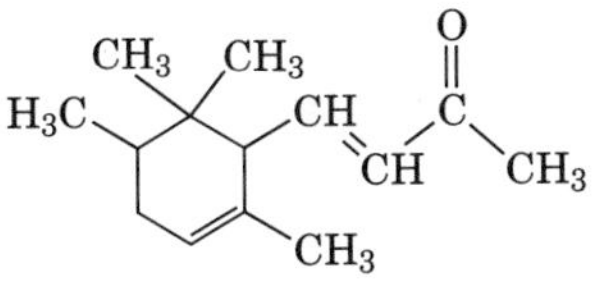

(167)

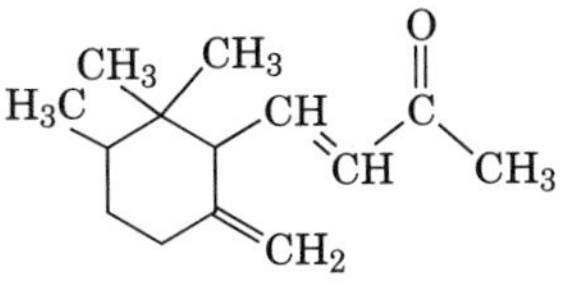

(168)

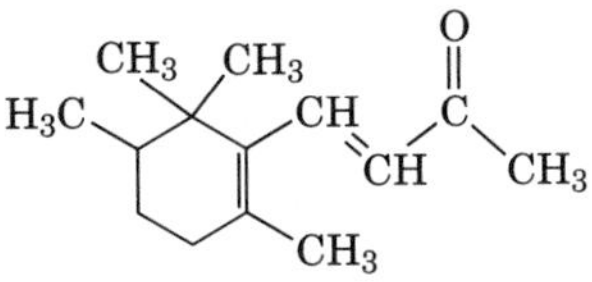

(169)

Palmarosa Oil. The grass *Cymbopogon martini* Staph., which grows wild in India from Bombay to the Himalayas, yields, on steam or water distillation, an essential oil with a pale yellow-to-olive color possessing a sweet floral-rosy odor. This oil is the best natural source of geraniol (**9**) which can be isolated either for use as is or for the preparation of other fragrance and flavor molecules. Apart from its use as a geraniol source, palmarosa oil (East Indian geranium oil [*8014-19-5*]) is used in many perfumes, particularly soap fragrances where its greater tenacity is a highly desirable quality. Components of palmarosa oils of various regions is shown in Table 49 (90).

Peppermint Oil. True peppermint oil is steam-distilled from the partially dried herb of various cultivars of *Mentha* X *piperita*, a nonfertile hybrid of *M. aquatica* L., known as watermint, and spearmint, *M. spicata* L., which are native to southern Europe. During the nineteenth century, cuttings of *M. piperita* were brought to the United States, which is the world's largest producer of peppermint oil. Production areas include Indiana, Michigan, Ohio, Wisconsin, Oregon, and

Table 49. Components of Palmarosa Oil of Different Origins[a]

		Percent in			
Component	Structure number	Brazilian oil	Guatemalan oil	Indian oil	Madagascan oil
geraniol	(**9**)	80.9	60.0	76.2	84.0
nerol[b]			14.8	1.5	0.2
geranyl acetate	(**28**)	12.4	17.3	9.1	8.0
linalool	(**10**)	2.2	0.8	3.9	3.0
γ-terpinene		1.4		0.9	1.7
elemol	(**49**)	0.4		1.5	0.3
limonene	(**23**)	0.1		1.7	0.1
myrcene	(**29**)	44		0.1	0.2
Total		97.9	92.9	94.9	97.5

[a]Ref. 90.
[b]3,7-Dimethyl-2,6-octadien-1-ol [*106-25-2*].

Washington state. It is important to harvest the peppermint plant at the early blooming stage so as to minimize the content of menthofuran, a characteristic but undesirable ingredient which lends a kerosene-like aroma to the oil. The essential oil must be rectified to remove water and all unpleasant-smelling or -tasting fractions, including bitter, sulfury-weedy foreruns and resinous-oily residues. Natural peppermint oil (unrectified) is a pale yellow to olive liquid possessing a strong, fresh, grassy-minty odor and a deep sweet-balsamic undertone with a clean dryout. Even the odor smells "cool." Rectified oils are water-white and free from weedy topnotes and any harsh-resinous aftertaste. Peppermint oil finds occasional use in perfumery to provide lift and freshness, but the principal use is in the flavor area for candies, chewing gums, liqueurs, oral hygiene products, etc. In 1988, the comparative analysis of the volatiles of the headspace of living peppermint, partially dried picked peppermint, and commercial peppermnt oil were reported, and differences are shown in Table 50 (93).

The isomers of 1,3,5-undecatriene in living plant volatiles are partly responsible for the unique, fresh natural-green character of the living herb.

Petitgrain Bigarade Oil. Orange flower water [8030-28-2], obtained by steam distillation of the leaves, stems, and twigs of the tree *Citrus aurantium* subsp. *amara*, is called petitgrain bigarade oil. The tree is cultivated in almost all of the mild-temperate, semitropical, and tropical regions of the world; the best-quality oil comes from trees grown and processed in the south of France. Petitgrain bigarade oil is a pale yellow to amber-colored mobile liquid with a pleasant, fresh-floral, sweet odor reminiscent of orange flowers with a slightly woody-herbaceous undertone and faint sweet-floral dryout. It is used in per-

Table 50. Comparative Analysis of Peppermint Volatiles[a]

Component	CAS Registry Number	Structure number	In living plant, %	In picked plant, %	In commercial oil, %
hexanal	[66-25-1]			0.1	
cis-3-hexenal	[6789-80-6]			0.5	
trans-2-hexenal				0.8	
trans-2-hexenol	[2305-21-7]			1.4	
cis-3-hexenol				1.4	
hexanol	[25917-35-5]			0.5	
2,4-hexadienal	[142-83-6]			0.1	
1-octen-3-ol				2.0	
eucalyptol		(34)			5.7
menthone[b]			0.2	12.7	18.1
isomenthone[c]			9.6	7.7	2.3
menthofuran	[494-90-6]		49.7	26.3	5.2
menthol (isomers)			[d]	4.7	48.0
pulegone		(150)	1.6	24.5	1.7
1,3,5-undecatriene (isomers)			0.6		
Total			*61.6*	*81.6*	*81*

[a]Ref. 93.
[b]trans-2-Isopropyl-5-methylcyclohexanone.
[c]cis-3-Isopropyl-5-methylcyclohexanone.
[d]Trace amounts.

fumery mainly for its refreshing sweet-floral notes, and in fruit and honey flavors at low levels. The constituents of petitgrain bigarade oil are shown in Table 51 (94).

Pimento Berry Oil. The pimento or allspice tree, *Pimenta dioca* L. (syn. *P. officinalis*, Lindl.), a native of the West Indies and Central America, yields two essential oils of commercial importance: pimento berry oil and pimenta leaf oil. The leaf oil finds some use in perfumery for its resemblance to clove leaf and cinnamon leaf oils as a result of its high content of eugenol. Pimento berry oil is an item of commerce with extensive application by the flavor industry in food products such as meat sauces, sausages, and pickles, and moderate use in perfumery, where it is used primarily as a modifier in the modern spicy types of men's fragrances. The oil is steam-distilled from dried, crushed, fully grown but unripe fruits. It is a pale yellow liquid with a warm-spicy, sweet odor with a fresh, clean topnote, a tenacious, sweet-balsamic-spicy body, and a tea-like undertone. A comparative analysis of the headspace volatiles of ripe pimento berries and a commercial oil has been performed and differences are shown in Table 52 (95).

Pine Oil. This oil is obtained by extraction and fractionation or by steam distillation of the wood of *Pinus palustris* Mill. and other species. Most of the oil is produced in the southeastern United States. The composition of the oil depends on the fractions chosen, but the chief constituents are terpene alcohols, mainly terpineol. Pine oil finds use as a germicide in disinfectants and soaps; as an ingredient in insecticides, deodorants, polishes, sweeping compounds, and cattle sprays; and as raw material for the manufacture of perfumery-grade terpineol [8000-41-7], anethole [104-46-1], fenchone (**137**), and camphor (**35**).

Rosemary Oil. This oil is steam-distilled from the flowers and leaves (twigs are added in Spanish distillations) of the shrub *Rosmarinus officinalis* L., which grows wild throughout the Mediterranean region. For the most part, distillation is performed locally, with the highest quality oils coming from Spain, France, and Tunisia. The oil is a pale yellow to almost water-white mobile liquid with a strong, fresh, woody, herbaceous and minty forest-like odor. The

Table 51. Principal Components of Petitgrain Bigarade Oil[a]

Component	Structure number	Percent
linalyl acetate	(**24**)	44.1
linalool	(**10**)	11.7
trans-linalool oxide (furanosyl)	(**110**)	7.1
cis-linalool oxide (furanosyl)	(**109**)	5.4
methyl anthranilate[b]		3.5
geranyl acetate	(**28**)	1.7
α-humulene	(**147**)	1.4
α-terpineol	(**26**)	1.3
α-terpinyl acetate	(**132**)	1.1
myrcene	(**29**)	1.0
geraniol	(**9**)	1.0
Total		79.3

[a]Ref. 94.
[b]Methyl 2-aminobenzoate [134-20-3].

Table 52. Comparative Analysis of the Headspace Volatiles of Ripe Pimento Berries and a Commercial Oil[a]

Component	Structure number	In headspace, %	In commercial oil, %
eugenol		31.3	71.4
myrcene	(29)	27.0	0.3
trans-β-ocimene	(25)	8.4	0.3
eugenyl methyl ether	(11)	7.5	8.6
β-caryophyllene	(52)	7.4	7.9
limonene	(23)	1.8	0.2
linalool	(10)	1.5	0.1
terpinolene	(50)	1.2	0.2
Total		*86.1*	*89.0*

[a]Ref. 95.

body note is clean, woody-balsamic drying out to a pleasant dry-herbaceous and tenacious bittersweet note. It finds extensive use both in fine perfumery and functional products such as room deodorants, household sprays, insecticides, and disinfectants. A comparative analysis has been performed of the headspace volatiles of both living and picked rosemary and a high quality commercial oil, and components are shown in Table 53 (96).

Sage Oil, Dalmatian. Several sage oils are produced commercially, each from a different species of plant. Sage oil [8022-56-8], Dalmatian is steam-distilled from the dried leaves of wild *Salvia officinalis* L. growing primarily in the former Yugoslavia; sage oil, Spanish comes from steam distillation of

Table 53. Comparative Analysis of Headspace Volatiles of Living and Picked Rosemary and a Commercial Oil[a]

Component	Structure number	In living, %	In picked, %	In commercial oil, %
para-cymene		19.8	13.7	1.7
limonene	(23)	14.1	14.3	1.0
myrcene	(29)	9.5	11.1	1.7
verbenone	(116)	7.2	6.0	0.1
linalool	(10)	7.1	7.6	0.8
α-terpineol	(26)	5.0	4.6	1.4
β-caryophyllene	(52)	5.4	4.6	2.9
terpinen-4-ol	(37)	3.3	3.2	0.5
geraniol	(9)	3.2	1.7	
γ-terpinene		2.7	2.6	0.5
1,8-cineole	(34)	2.0	0.7	44.5
bornyl acetate	(138)	2.0	2.3	1.3
eugenyl methyl ether	(11)	2.0	1.9	
borneol		1.5	1.4	2.6
α-pinene	(44)	1.0	1.3	13.3
camphor	(35)	0.2		10.1
β-pinene	(22)	0.2	0.1	7.6
camphene	(47)	0.1	0.4	5.8
Total		*86.3*	*77.5*	*86.8*

[a]Ref. 96.

wild growing *S. lavandulifolia* Duhl. in Spain; and sage (clary) oil is steam-distilled from the flowering tops and folliage of *S. sclarea* L. originally from the Mediterranean, but now cultivated in central and eastern Europe, England, Morocco, and the United States. Sage oil, Dalmatian is a pale-yellow, mobile liquid with a strong fresh warm-spicy, herbaceous, and camphoraceous odor with a pleasant, sweet-herbaceous and tenacious dryout. The oil finds extensive use as a flavor material for liqueurs, canned meats, sauces, pickles, sausages, etc. In fine perfumery, it is used for its power and tenacity in both men's and women's fragrances. A comparison of the headspace volatiles of living *Salvia officinalis* and a commercial sage oil, Dalmation has been performed, and the components are shown in Table 54 (97).

   Sage (Clary) Oil. Also known commercially as essence sauge sclarée clary sage oil [*8016-63-5*] is steam-distilled from the flowering tops and folliage of *Salvia sclarea* L., which is native to the Mediterranean region but is cultivated extensively in central and eastern Europe, the Crimea and Caucasus regions, as well as in England, Morocco, and the United States. The finest oils traditionally come from France, England, and Morocco. Clary sage oil is a colorless to pale yellow to olive mobile liquid with a tenacious sweet-herbaceous odor becoming soft and somewhat reminiscent of ambra in the undertone. The dryout note has also been described as tobacco-like, balsamic, or tea-like. The oil is used in fine perfumery as a modifier for other naturals, such as bergamot, and to provide soft, ambra notes. It is particularly effective in oriental-type fragrances. It is also used in flavorings for liqueurs and wine, and in grape flavors. A comparison of the volatiles of U.S., French, and Russian oils has been performed, and components are shown in Table 55 (98).

Table 54. Comparison of Headspace Volatiles of Living *Salvia officinalis* and Commercial Sage Oil, Dalmatian[a]

Component	Structure number	In headspace, %	In commercial oil, %
thujone/isothujone	(**136**)	27.7	30.0
cis-3-hexenyl isovalerate		12.1	
limonene	(**23**)	7.1	5.9
cis-3-hexenyl acetate		5.4	
para-cymene		2.4	0.8
myrcene	(**29**)	2.2	0.8
β-caryophyllene	(**52**)	1.7	10.4
hexyl valerate		1.5	
camphor	(**35**)	0.8	21.9
camphene	(**47**)	0.1	6.9
1,8-cineole	(**34**)	0.2	5.3
α-pinene	(**44**)	0.5	3.3
borneol			3.0
bornyl acetate	(**138**)	0.5	2.1
β-pinene	(**22**)	0.4	1.8
Total		*62.6*	*92.1*

[a]Ref. 97.

Table 55. Comparative Chemical Composition of Commercial Clary Sage Oils[a]

Component	Structure number	In U.S. oil, %	In French oil, %	In Russian oil, %
linalyl acetate	(24)	44.9–53.4	49.0–73.6	45.3–61.8
linalool	(10)	20.3–28.6	9.0–16.0	10.4–19.3
α-terpineol	(26)	3.1	0.2–0.6	1.2–2.5
germacrene D	(156)	2.6–3.6	1.6–2.0	0.7–2.0
geranyl acetate[b]	(28)	1.9–3.2	0.3–0.5	0.8–1.2
geraniol	(9)	1.7–3.3	0.1–0.3	0.6–1.2
myrcene	(29)	1.3–1.7	0.1–0.2	0.3–0.5
neryl acetate		1.0–1.7	0.2–0.3	0.4–0.6

[a]Ref. 98.
[b]3,7-Dimethyl-3,6-octadienyl acetate.

Spearmint Oil (Native).   Produced by steam distillation of the flowering tops of *Mentha spicata* Huds., native spearmint oil is one of the largest-volume essential oils of the United States. The plant, a native of Europe, was introduced into the United States in the nineteenth century, and the flavor immediately became popular there and has remained so, whereas it has never been as popular in the countries of its origin. The main producing areas are the same as for true peppermint. Distillation is carried out in the fields and either sold as such or rectified to the buyer's specifications. The crude oil obtained directly from the stills is called natural spearmint oil. It is a pale olive to yellow mobile liquid with a warm, biting, spicy-herbaceous, bitter quality. Rectified oils are less bitter with a more burning-biting, sweeter-balsamic taste. The oil finds its primary use in flavors for toothpaste, chewing gum, candy, and mouthwashes, where it blends well with peppermint. There is some use of it in perfumery for its herbaceous-green effect. The principal and characteristic constituent is *l*-carvone with a completely different odor and taste from that of the *d*-isomer. Spearmint oil has been the subject of thorough analysis, and in 1990 a comparison of the headspace volatiles of living and picked spearmint with those of a commercial oil was reported (99); differences are shown in Table 56.

Tagetes Oil.   Steam distillation immediately after flowering of the above-ground parts of *Tagetes minuta* L. (syn. *T. glandulifera* Schrank.), a native of South America but abundantly growing wild in sub-Saharan Africa and also found in Australia, Europe, Asia, and the United States, yields a dark yellow to orange-yellow mobile liquid which solidifies on exposure to air, daylight, and moisture. An absolute is also made from the concrete. The oil is distilled in Africa and France. The main components of tagetes oil are 4-octenones (or dienes or trienes) (Table 57) (100).

The branched-chain polyunsaturated ketones tagetone and ocimenone, the odor impact compounds in tagettes, are unstable and presumably responsible for the resinification of the oil.

Thyme Oil.   The water- and steam-distilled essential oil obtained from the partially dried herb of the wild-growing *Thymus vulgaris* L., *T. zygis* L., or related species is called thyme oil [8007-46-3]. Various species of thyme grow abundantly in the Mediterranean region as well as central and eastern Europe.

Table 56. Comparison of Headspace Volatiles of Living vs Picked Spearmint and Commercial Oil[a]

Component	In headspace of live plant, %	In headspace of picked plant, %	In commercial oil, %
carvone (**123**)	24.0	70.0	63.0
limonene (**23**)	17.7	1.8	21.4
dihydrocarvone[b]	0.7	2.6	0.1
hexanal	0.5	c	
hexanol		2.3	0.1
menthone/isomenthone			1.2
menthol			1.7
1,3,5-undecatriene (isomers)	0.5		

[a]Ref. 99.
[b]2-Methyl-5-isopropenylcyclohexanone [7764-50-3].
[c]Trace amounts.

Table 57. Components of Tagetes Oil[a]

Common name	Systematic name	CAS Registry Number	Percent
cis-β-ocimene	2,6-dimethyl-cis-2,5,7-octatriene		41.6
trans-ocimenone	2,6-dimethyl-trans-2,5,7-octatrien-4-one	[33746-45-1]	12.5
dihydrotagetone	2,6-dimethyl-7-octen-4-one	[1879-00-1]	10.9
tagetone	2,6-dimethyl-2,7-octadien-4-one	[6752-80-3]	7.9
cis-ocimenone	2,6-dimethyl-2,5,7-octatrien-4-one	[33746-71-3]	7.0
limonene	4-isopropenyl-1-methylcyclohexene		3.9
Total			*83.8*

[a]Ref. 100.
[b]Structure (**23**).

In addition, the plant is cultivated in many parts of the world as a dried culinary herb. Spain is a primary center for distillation. Thyme oil is a brownish orange-red liquid with a rich, powerful, sweet, warm-herbaceous odor, very aromatic and spicy. The flavor is rich, warm, biting, spicy-herbaceous. In perfumery, the oil finds primary use in functional products such as soap fragrances and only trace use in fine perfumery. The principal use of thyme oil is in flavors for sauces, dressings, pickles, canned meats, etc, and in pharmaceutical preparations including mouth washes, gargles, dentifrices (qv), cough syrups, and lozenges because of its excellent germicidal properties. A comparative analysis of the headspace volatiles over living and picked *T. vulgaris* and a commercial oil was performed and differences are shown in Table 58 (101).

Turpentine Oil. The world's largest-volume essential oil, turpentine [8006-64-2] is produced in many parts of the world. Various species of pines and balsamiferous woods are used, and several different methods are applied to obtain the oils. Types of turpentines include dry-distilled wood turpentine from dry distillation of the chopped woods and roots of pines; steam-distilled wood turpentine which is steam-distilled from pine wood or from solvent extracts of

Table 58. Differences Between Headspace Constituents of Living and Picked *Thymus vulgaris* and a Commercial Oil[a]

Component	CAS Registry Number	Structure number	In headspace of live plant, %	In headspace of picked plant, %	In commercial oil, %
para-cymene			29.7	48.6	30.0
thymol[b]	[89-83-8]		15.2	9.0	39.7
cis-3-hexenyl acetate			11.2	0.1	
1-octen-3-ol			8.2	8.0	
linalool		(**10**)	3.3	5.2	6.8
myrcene		(**29**)	3.2	2.5	1.4
β-caryophyllene		(**52**)	1.8	3.1	1.2
carvacrol[c]			1.5	0.9	1.0
hexyl acetate			1.5		
limonene		(**23**)	1.3	1.1	1.8
geranyl isovalerate[d]	[109-20-6]		1.0		
α-pinene			0.1		3.1
α-terpineol				0.6	1.3
Total			*78.0*	*79.1*	*86.3*

[a]Ref. 101.
[b]2-Isopropyl-5-methylphenol.
[c]2-Methyl-5-isopropylphenol.
[d]3,7-Dimethyl-2,6-octadienyl-3-methylbutanoate.

the wood; and sulfate turpentine, which is a by-product of the production of sulfate cellulose. From a perfumery standpoint, steam-distilled wood turpentine is the only important turpentine oil. It is rectified to yield pine oil, yellow or white as well as wood spirits of turpentine. Steam-distilled turpentine oil is a water-white mobile liquid with a refreshing warm-balsamic odor. American turpentine oil contains 25–35% β-pinene (**22**) and about 50% α-pinene (**44**). European and East Indian turpentines are rich in α-pinene (**44**) with little β-pinene (**22**), and thus are excellent raw materials for the production of terpineol. β-Pinene (**22**) is used for the production of nerol, geraniol (**9**), and linalool (**10**), as well as further derivatives of these important chemicals.

Wintergreen Oil. Water distillation of the leaves of *Gaultheria procumbens* L. yields an oil which consists of essentially one chemical constituent, methyl salicylate. Because of this, the oil has been almost totally replaced by the synthetic chemical. Natural oil of wintergreen [68917-75-9] is a pale yellow to pinkish colored mobile liquid of intensely sweet-aromatic odor and flavor. The oil or its synthetic replacement find extensive use in pharmaceutical preparations, candy, toothpaste, industrial products, and in rootbeer flavor. In perfumery, it is used in fougère or forest-type fragrances.

Ylang Ylang. The oil produced by water- or water-and-steam-distillation of freshly early morning-picked blossoms from cultivated plantings of *Cananga odorata* Hook f. et Thomson, a tree native to Indonesia and the Philippines, is called ylang-ylang. Extensive plantings of this tree exist in the Comoro islands, Madagascar, the French South Pacific, the West Indies, and Réunion. Most ylang-ylang oil comes from the Comoro islands and Madagascar (the Nossi-Bé region). Distillation is carefully performed in small stills to minimize crushing of the delicate blossoms. Several fractions (called ylang extra, first, second, third, etc) are taken over different periods of time. The time for making the cuts is frequently determined by measuring the specific gravity of the oil which has come over. Ylang-ylang extra usually represents the first 30–45% of the total distillate. It is a pale yellow oil with a powerful, floral, and intensely sweet odor and a cresylic and benzoate topnote of limited tenacity. On dryout, it becomes more pleasant, soft, sweet, slightly spicy, and balsamic-floral. This fraction is used mainly in sophisticated perfumes of the floral and heavy-oriental type. "First" and "second" cuts are in-between quality, and are rarely used as such. Ylang-ylang "third" is a yellowish oily liquid of sweet-floral and balsamic-woody odor with a tenacious and sweet-balsamic undertone. This oil finds use in soap perfumery and less expensive consumer-type fragrances. The chemical composition of first, second, third, and extra grades of ylang-ylang oil from the Comoro islands has been reported (102), and the components are shown in Table 59.

Safety and Regulatory Aspects

Essential oils possess a variety of biological properties which may result in varying responses by humans on exposure. An important factor in these effects is the dose to which one is exposed. Thus, essential oils may have both beneficial and toxic effects, depending on their dose. The potential for biological effects from essential oils is not surprising; many botanical species are known to contain substances that possess biological properties, and their identification has contributed significantly to knowledge of biochemistry and physiology as well as the development of therapeutic agents, eg, quinine and digitalis.

The toxicities of many essential oils have been reported in monographs published by the Research Institute for Fragrance Materials (RIFM) (Table 60). Most essential oils used by the flavor and fragrance industries are relatively nontoxic or slightly toxic on acute oral or dermal exposure, and are considered safe when used at levels present in consumer products. In general, the levels of fragrances and flavors in consumer products, and thus the levels of any essential oil ingredients, are relatively low. For example, a fragrance oil may typically be used in a soap at 0.5%. The oil may contain 5% of orange oil distilled. The final concentration of orange oil distilled in the soap therefore is 0.025%.

Because essential oils are used predominantly by the flavor and fragrance industries, these commercial oils must undergo the same scientific scrutiny as all other flavor and fragrance substances and must be in compliance with all applicable health, safety, and environmental regulations. Guidelines and regulations on the use of essential oils in fragrances differ from those applying to essential oils used in flavors.

The industry supports several key organizations which strengthen scientific criteria and develop guidelines for the safe use of essential oils as flavor or fragrance ingredients (see FLAVORS AND SPICES). The Research Institute for Fragrance Materials (RIFM) is an internationally recognized scientific organization

Table 59. Major Components of Ylang-Ylang Oils[a]

Component	CAS Registry Number	Structure number	In first, %	In second, %	In third, %	In extra, %
meta-cresyl acetate[b]/*para*-cresyl acetate	[122-46-3]/[140-39-6]		29.0	21.4	28.7	15.9
benzyl acetate/*ortho*-cresyl acetate[c]	[533-18-6]		16.1	32.3	36.6	29.9
linalool		(10)	12.2	6.1	2.1	11.0
para-cresyl methyl ether[d]	[104-93-8]		6.5	2.7	0.7	8.4
benzyl benzoate			5.4	6.5	5.5	5.2
eugenol/cinnamyl alcohol			3.0	2.4	1.8	5.6
nerol			2.0	2.9	3.0	1.3
farnesyl acetate	[29548-30-9]		2.0	2.0	2.9	1.6
α-humulene		(147)	1.7	2.6	1.8	0.9
farnesol			1.5	1.6	1.5	2.0
benzyl salicylate	[118-58-1]		1.2	1.1	1.9	1.9
geraniol		(9)	1.1	0.9	0.8	0.7
γ-terpinene			0.9	0.2		
cinnamaldehyde/ *meta*-cresol[e]/*para*-cresol[e]	[108-39-4]/[106-44-5]		0.9	0.1	0.7	0.9
methyl anthranilate			0.8	0.9	1.0	0.4
benzyl alcohol			0.4	0.2	0.1	0.1
Total			*84.4*	*89.9*	*89.1*	*85.8*

[a]Ref. 102.
[b]*m*- and *p*-Methylphenyl acetate.
[c]*o*-Methylphenyl acetate.
[d]*p*-Methylphenyl methyl ether (*p*-methylanisole).
[e]*m*- and *p*-Methylphenol.

Table 60. Toxicities of Some Essential Oils[a]

Essential oil	LD$_{50}$, g/kg[b] Oral	Dermal	Irritation[c]	Sensitization[d] dose, wt %	Reference
anise	2.25	>5	e	2.0	104
basil, sweet	1.4	>5[f]	e	4.0	105
bergamot, expressed	>10	>20[f]	e	30.0	106
bergamot, rectified	>10	>20	e	30.0	107
cedarwood, Texas	>5	>5	negative	8.0	108
cinnamon leaf, Ceylon	2.65	>5	strong	10.0	109
citronella	>5	4.7	moderate	8.0	110
galbanum	>5	>5	slight	4.0	111
geranium	>5	2.5	moderate	20.0	112
lavender	>5	>5	slight	16.0	113
lemon, distilled	>5	>5	slight	15.0	114
lemon, expressed	>5	>5	moderate	10.0	115
orange, expressed	>5	>5	moderate	8.0	116
rose, Bulgarian	>5	2.5	moderate	2.0	117
sage, Spanish	>5	>5	negative	8.0	118
spearmint	≈5	>5	moderate	4.0	119
thyme, red	4.7	>5	severe	8.0	120

[a]Phototoxicity is negative for all oils except bergamot, expressed, for which it is severe.
[b]The LD$_{50}$ is the statistically derived dose of a substance which results in death in 50% of a population of animals. The LD$_{50}$ is expressed in milligrams of substance per kilogram of body weight. The species used to determine the oral and dermal LD$_{50}$ are the rat and rabbit, respectively.
[c]Irritation results are based on the substance applied full strength to a test population for a 24-h period.
[d]Sensitization results are based on a human maximization test (103) using a petrolatum vehicle. The effect is expressed as the number of panelists responding over the total number of panelists tested and was 0/25 except for spearmint (0/32). That is, at the dose indicated, the oils were not irritating when tested in a 48-h closed patch test in humans.
[e]Not reported in the monograph (103).
[f]LD$_{50}$ values expressed as milliliter of substance per kilogram of body weight (mL/kg).

that collects, generates, and disseminates information on the safety of fragrance ingredients, including essential oils. This information may originate from published or unpublished sources, or be generated through RIFM's ongoing research program. The findings are peer reviewed by an expert panel of academicians and published in the scientific literature.

The activities of RIFM are harmonized with those of the International Fragrance Association (IFRA) and the International Organization of the Flavour Industry (IOFI). These organizations are concerned with all aspects of safety evaluation and regulation in the industry, into which it has introduced self-regulatory discipline. Both IFRA and IOFI have formulated and continue to update their respective "Codes of Practice" for both the fragrance and flavor industries. These Codes set forth manufacturing, safety evaluation, and usage guidelines to assure the quality, safety, and legality of flavor and fragrance ingredients, including essential oils. For example, the IFRA Code of Practice restricts essential oils containing psoralens, thereby reducing concerns about their potential dermal phototoxicity. In addition, the United States Flavor Extracts

Manufacturers Association (FEMA), the U.S. Food and Drug Administration (FDA), the Council of Europe (CoE), the Food and Agricultural Organization of the United Nations (FAO), and the World Health Organization (WHO) monitor usage and evaluate the safety of essential oils when they are used in foods and flavors.

Many essential oils have been designated by the FDA or by the expert panel of FEMA as Generally Recognized As Safe (GRAS) for their intended use in foods and flavors. The use and safety of these GRAS substances are continuously being reviewed and the list of GRAS substances updated. New essential oils intended to be used as a flavor ingredient must undergo extensive safety evaluations and scrutiny by one or more of these groups of experts before they may be used in flavors.

Many countries have adopted chemical substance inventories in order to monitor use and evaluate exposure potential and consequences. In the case of essential oils used in many fragrance applications, these oils must be on many of these lists. New essential oils used in fragrances are subject to premanufacturing or premarketing notification (PMN). PMN requirements vary by country and predicted volume of production. They require assessment of environmental and human health-related properties, and reporting results to designated governmental authorities.

Essential oils are also influenced by legislation that regulates specific products that may contain these oils, eg, the U.S. Food, Drug, and Cosmetic Act and the European Community Cosmetic Directive. An example of an environmental issue that affects essential oils used in fragrances is the regulatory trend to reduce atmospheric release of volatile organic chemicals (VOCs) from consumer products and other sources. Fragrances and their essential oil ingredients are recognized as unique and essential components of consumer products, and are given specific exemptions from most VOC regulations. Essential oils would not be anticipated to be of environmental concern, considering that they originate from botanical sources. Thus, natural processes exist to degrade essential oils and recycle their components effectively in the environment.

BIBLIOGRAPHY

"Oils, Essential" in *ECT* 1st ed., Vol. 9, pp. 569–591 by E. E. Langenau, Fritzsche Brothers, Inc.; in *ECT* 2nd ed., Vol. 14, pp. 178–216, by Max Stoll, Firmenich & Cie.; in *ECT* 3rd ed., Vol. 16, pp. 307–332 by J. A. Rogers, Jr., Fritzsche, Dodge & Olcott, Inc.

1. U.S. Department of Commerce, Horticultural and Tropical Products Division, FAS/USDA, Washington, D.C., Apr. 1994.
2. E. Demole, P. Enggist, and G. Ohloff, *Helv. Chim. Acta* **65**(6), 1785 (1982).
3. J. Rigaud, P. Etievant, R. Henry, and A. Latrasse, *Sci. Aliments* **6**(2), 213 (1986).
4. T. C. Baker, R. Nishida, and W. L. Roelofs, *Science (Washington, D.C.)* **214**, 1359 (1981).
5. R. Nishida and T. E. Acree, *J. Agric. Food Chem.* **32**(5), 1001 (1984).
6. T. E. Acree, R. Nishida, and H. Fukami, *J. Agric. Food Chem.* **33**(3), 425 (1985).
7. *Ibid.*, p. 425.
8. B. D. Mookherjee, R. Trenkle, and R. Wilson, *Aerosol Age*, 20 (May, 1989).

9. Technical data, B. D. Mookherjee and R. Trenkle, International Flavors & Fragrances Inc., Union Beach, N.J., 1967.
10. G. Ohloff, *Perfumer Flavor.* **3**, 11 (1978).
11. B. D. Mookherjee and R. A. Wilson in *On Essential Oils*, Synthite Industrial Chemicals Private Ltd., Synthite Valley, Kolenchery, India, 1986, pp. 281–329.
12. J. Jessee, *Perfume Album*, R. Krieger, Huntington, N.Y., 1974, p. 10.
13. Technical data, B. D. Mookherjee and R. W. Trenkle, International Flavors & Fragrances Inc., Union Beach, N.J., 1967.
14. Ref. 12, p. 18.
15. M. N. Boelens and A. Oporto, *Perfumer Flavor.* **16**(6), 1 (1991).
16. Technical data, B. D. Mookherjee and R. W. Trenkle, International Flavors & Fragrances Inc., Union Beach, N.J., 1985.
17. B. Lalande, *Perfumer Flavor.* **9**(2), 117 (1984).
18. M. Guerere and F. DeMarne, *Ann. Fals. Exp. Chim.* **78**(837), 183 (1985).
19. Y. Naves, D. Lamparsky, and P. Ochsner, *Bull. Soc. Chim. France*, 645 (1961).
20. G. Lucas, J. Ma, J. McCloskey, and R. Wolff, *Tetrahedron* **20**, 1789 (1964).
21. Technical data, B. D. Mookherjee and R. A. Wilson, International Flavors & Fragrances Inc., Union Beach, N.J., 1985.
22. G. Lamonica, I. Stagno D'Alcontres, M. G. Donato, and I. Merenda, *Chimicaoggi* **8**(5), 59 (1990).
23. E. Sundt, B. Willhalm, and M. Stoll, *Helv. Chim. Acta* **47**(2), 408 (1964).
24. B. D. Mookherjee, "The Identification of Bifunctional Compounds in Bergamot Oil," paper presented at *the 158th National Meeting of the American Chemical Society, Sept. 7–12, 1969, New York.*
25. S. Inoma, Y. Miyagi, and T. Akieda, *Kanzei Chuo Bunsekishoho* **29**, 87 (1989).
26. A. Fleisher, G. Biza, N. Secord, and J. Dono, *Perfumer & Flavorist* **12**(2), 57 (1987).
27. E. Kovats, *Helv. Chim. Acta* **46**(7), 2705 (1963).
28. U.S. Pat. 3,988,432 (Oct. 26, 1976), R. Steltenkamp (to Colgate-Palmolive Co.).
29. Technical data, B. D. Mookherjee and R. Santangelo, International Flavors & Fragrances Inc., Union Beach, N.J., 1975.
30. J. D. Vora, R. F. Matthews, P. G. Crandall, and R. Cook, *J. Food Sci.* **48**, 1197 (1983).
31. K. Stevens, R. Lundin, and R. Teranishi, *J. Org. Chem.* **30**, 1690 (1965).
32. G. Ohloff, *Fortsch. Chem. Org. Naturst.* **35**, 473 (1983).
33. M. H. Boelens, *Perfumer Flavor.* **16**(2), 17 (1991).
34. W. D. MacLeod, Jr. and N. Buigues, *J. Food Sci.* **29**, 565 (1964).
35. E. Demole and P. Enggist, *Helv. Chim. Acta* **66**(5), 1381 (1983).
36. E. Demole, P. Enggist, and G. Ohloff, *Helv. Chim. Acta* **65**(6), 1785 (1982).
37. E. J. Brunke and F. J. Hammerschmidt, *Dragoco Rpt.* (4), 107 (1988).
38. U.S. Pat. 4,014,823 (Mar. 29, 1977), B. D. Mookherjee, V. Kamath, and E. Shuster, (to International Flavors & Fragrances Inc.).
39. B. M. Lawrence, *Perfumer Flavor.* **15**(2), 77 (1990).
40. P. Teisseire, P. Maupetit, and B. Corbier, *Recherches (R.B.D.)* **19**, 8 (1974).
41. B. D. Mookherjee, K. Light, and I. Hill, in B. D. Mookherjee and C. J. Mussinan, eds., *Essential Oils*, Allured Publishing Corp., Wheaton, Ill., 1981, p. 247.
42. S. Lemberg and R. B. Hale, "Vetiver Oils of Different Geographical Origins," Paper No. 117 presented in *the VII International Congress of Essential Oils, Oct. 7–11, 1977, Kyoto, Japan.*
43. Technical data, B. D. Mookherjee and R. A. Wilson, International Flavors & Fragrances Inc., Union Beach, N.J., 1985.
44. Y. Chretien-Bessiere, J. Garnero, L. Benezet, and L. Peyron, *Bull. Soc. Chim. France*, (1), 97 (1967).
45. P. Teisseire, B. Corbier, and M. Plattier, *Recherches* **16**, 5 (1967).

46. Y. R. Naves, *Bull. Soc. Chim. France*, (9), 3152 (1967).
47. Technical data, B. D. Mookherjee and R. Santangelo, International Flavors & Fragrances Inc., Union Beach, N.J., 1975.
48. A. Bramwell, J. Burrell, and G. Riezebos, *Tetrahedron Lett.* (37), 3215 (1969).
49. Y. Naves, *Parfum Cosmet. Savons*, **12**(2), 586 (1969).
50. R. Kaiser and D. Lamparsky, *Helv. Chim. Acta* **61**(7), 2671 (1978).
51. R. A. Wilson and B. D. Mookherjee, "Characterization of Aroma Donating Components of Myrrh," paper presented at *the IXth International Congree of Essential Oils, Mar. 13–17, 1983, Singapore.*
52. R. ter Heide and co-workers, *J. Agric. Food Chem.* **23**(5), 950 (1975).
53. Technical data, B. D. Mookherjee and R. A. Wilson, International Flavors & Fragrances Inc., Union Beach, N.J., 1984.
54. H. Walbaum, *J. Prakt. Chem.* **73**, 488 (1906).
55. L. Ruzicka, M. Stoll, and H. Schinz, *Helv. Chim. Acta* **9**, 249 (1926).
56. B. D. Mookherjee and R. W. Trenkle, "Volatile Constituents of Tincture of Tonquin Musk," paper presented at the *VIIIth International Congress of Essential Oils, Oct. 12–17, 1980, Cannes, Grasse, France.*
57. B. D. Mookherjee and R. Patel, "Isolation and Identification of Volatile Constituents of Tincture of Ambergris," paper presented at *the VIIth International Congress of Essential Oils, Oct. 7–11, 1977, Kyoto, Japan.*
58. R. A. Wilson, B. D. Mookherjee, and J. Vinals, *Tob. Rep.*, 42 (Oct. 1983).
59. V. T. Gogiya, L. G. Kharebava, R. V. Gogiya, and E. B. Gvatua, *Rastit. Resur.* **22**, 243 (1986).
60. B. D. Mookherjee, R. W. Trenkle, and R. A. Wilson, *J. Ess. Oil Res.* **2**, 85 (1989).
61. E. Klein and H. Obermann, paper presented at the *VIIth International Congress of Essential Oils, Oct. 7–11, 1977, Kyoto, Japan.*
62. B. D. Mookherjee and R. A. Wilson, *Perfumer & Flavorist* **15**(1), 27 (1990).
63. B. Corbier, C. Ehret, and P. Maupetit, paper presented at *La Chemie des Terpenes*, Apr. 24–25, 1986, Grasse, France.
64. F. de Maack, D. Brunet. J.-C. Malnati, and J. Estienne, *Ann. Fals. Expert Chim.* **75**, 357 (1982).
65. Technical data, B. D. Mookherjee and R. A. Wilson, International Flavors & Fragrances Inc., Union Beach, N.J., 1991.
66. L. Peyron, J. Acchiardi, D. Bignotti, and P. Pellerin, "The Berries of *Pimenta Dioca*," Paper No. 128 presented at *the VIIIth International Congress of Essential Oils, Oct. 12–17, 1980, Cannes, France.*
67. M. H. Boelens, *Perfumer Flavor.* **16**(2), 17 (1991).
68. B. M. Lawrence, *Perfumer Flavor.* **10**(2), 52 (1985).
69. B. M. Lawrence, *Perfumer Flavor.* **9**(5), 87 (1984).
70. F. Buccellato, *Perfumer Flavor.* **13**(4), 35 (1988).
71. A. M. Janssen, Ph.D. dissertation, Rijksuniversiteit, Leiden, the Netherlands, 1989.
72. W. A. König, R. Krebber, P. Evans, and G. Bruhn, *J. High Res. Chromatog.* **13**, 328 (1990).
73. Technical data, B. D. Mookherjee and R. A. Wilson, International Flavors & Fragrances Inc., Union Beach, N.J., 1988.
74. Technical data, B. D. Mookherjee and R. A. Wilson, International Flavors & Fragrances Inc., Union Beach, N.J., 1987.
75. B. M. Lawrence, *Perfumer Flavor.* **15**(6), 56 (1990).
76. R. P. Adams, *Econ. Bot.* **41**, 48 (1987).
77. N. A. Shaath, S. Dedeian-Johnson, and P. M. Griffin, *Proceedings of the 11th International Congress of Essential Oils, Fragrances, and Flavors*, Vol. 4, Oxford and IBH Publishing Co. Pvt. Ltd., New Delhi, India, 1989, p. 207.

78. C. Vernin, G. Vernin, J. Metxger, and L. Pujol, *Parfum. Cosmet., Arôm.* (93), 85 (1990).
79. M. Manzoor-i-Khuda, M. Rahman, M. Yusaf, and J. Chowdhury, *Bangladesh J. Sci. Ind. Res.* **21**(1/4), 70 (1986).
80. Technical data, B. D. Mookherjee and R. A. Wilson, International Flavors & Fragrances Inc., Union Beach, N.J., 1988.
81. Technical data, B. D. Mookherjee and R. A. Wilson, International Flavors & Fragrances Inc., Union Beach, N.J., 1987.
82. J. A. Retamar and E. C. DeRiscala, *Rivista Ital.* **52**, 127 (1980).
83. D. Garcia-Martin and M. C. Garcia-Vallejo, in Ref. 61, p. 362.
84. Technical data, B. D. Mookherjee and R. A. Wilson, International Flavors & Fragrances Inc., Union Beach, N.J., 1988.
85. Technical data, B. D. Mookherjee and R. A. Wilson, International Flavors & Fragrances Inc., Union Beach, N.J., 1987.
86. A. Proença and O. L. R. Roque, *J. Ess. Oil Res.* **1**, 15 (Jan.–Feb., 1989).
87. Technical data, B. D. Mookherjee and R. A. Wilson, International Flavors & Fragrances Inc., Union Beach, N.J., 1990.
88. J. de Pascual Teresa and co-workers, *Planta Med.* **55**, 398 (1989).
89. T. S. Chamblee and co-workers, *J. Agric. Food Chem.* **39**, 162 (1991).
90. S. R. Srinivas, *Atlas of Essential Oils*, the Bronx, N.Y., 1986.
91. B. M. Lawrence, *Perfumer Flavor.* **9**(5), 41 (1984).
92. Technical data, B. D. Mookherjee and R. A. Wilson, International Flavors & Fragrances Inc., Union Beach, N.J., 1989.
93. B. D. Mookherjee, R. A. Wilson, R. W. Trenkle, M. J. Zampino, and K. P. Sands, in R. Teranishi, R. G. Buttery, and F. Shahidi, eds., *Flavor Chemistry: Trends and Developments*, ACS Symposium Series 388, American Chemical Society, Washington, D.C., 1989, p. 176.
94. Z.-K. Lin, Y.-F. Hua, and Y.-H. Gu, *Acta Botan. Sinica* **28**(6), 635 (1986).
95. Technical data, B. D. Mookherjee and R. A. Wilson, International Flavors & Fragrances Inc., Union Beach, N.J., 1991.
96. Technical data, B. D. Mookherjee and R. A. Wilson, International Flavors & Fragrances Inc., Union Beach, N.J., 1992.
97. Technical data, B. D. Mookherjee and R. A. Wilson, International Flavors & Fragrances Inc., Union Beach, N.J., 1992.
98. B. M. Lawrence, *Perfumer & Flavorist* **15**(4), 71 (1990).
99. B. D. Mookherjee, R. W. Trenkle, and R. A. Wilson, *Pure Appl. Chem.* **62**(7), 1357 (1990).
100. Technical data, B. D. Mookherjee and R. A. Wilson, International Flavors & Fragrances Inc., Union Beach, N.J., 1985.
101. Technical data, B. D. Mookherjee and R. A. Wilson, International Flavors & Fragrances Inc., Union Beach, N.J., 1987.
102. J. M. Derfer, *Perfumer Flavor.* **3**(1), 45 (1978).
103. A. M. Kligman, *J. Invest. Derm.* **47**, 393 (1966).
104. *J. Food Chem. Toxicol.* **11**, 865 (1973).
105. Ref. 104, p. 867.
106. Ref. 104, p. 1031.
107. Ref. 104, p. 1035.
108. *J. Food Chem. Toxicol.* **14**, 711 (1976).
109. *J. Food Chem. Toxicol.* **13**, 749 (1975).
110. Ref. 104, p. 1067.
111. *J. Food Chem. Toxicol.* **16**, 765 (1978).
112. Ref. 108, p. 451.

113. *J. Food Chem. Toxicol.* **12**, 727 (1974).
114. Ref. 113, p. 725.
115. Ref. 113, p. 733.
116. Ref. 109, p. 913.
117. Ref. 108, p. 857.
118. Ref. 111, p. 871.
119. Ref. 113, p. 1003.

General Reference

S. Arctander, *Perfume and Flavor Materials of Natural Origin*, 1960.

BRAJA D. MOOKHERJEE
RICHARD A. WILSON
International Flavors & Fragrances, Inc.

OIL SHALE

Oil shale is a sedimentary mineral that contains kerogen, a mixture of complex, high molecular weight organic polymers. The solid kerogen is a three-dimensional polymer that is insoluble in conventional organic solvents. Upon heating, kerogen decomposes to form gas composed of hydrogen (qv), low molecular weight hydrocarbons (qv), and carbon monoxide (qv); liquids, composed of water and shale oil; and a solid char residue.

Oil shale deposits were formed in ancient lakes and seas by the slow deposition of organic and inorganic remains. The geology and composition of the inorganic minerals and organic kerogen components of oil shale vary with deposit locations throughout the world (1) (see also FUEL RESOURCES; PETROLEUM).

Reserves

Estimates of oil shale deposits by continent are given in Table 1 (2). Characteristics of many of the world's best known oil shales are summarized in Table 2 (3,4). Oil shale deposits in the United States occur over a wide area (Table 3). The most extensive deposits, covering ca 647,000 km^2 (250,000 mi^2), are the Devonian-Mississippian shales of the eastern United States (5). The richest U.S. oil shales are in the Green River formation of Colorado, Utah, and Wyoming. Typical mineral and organic analyses for Green River oil shale are given in Table 4.

The Green River formation includes an area of ca 42,720 km^2 (16,500 mi^2), and in-place reserves are ca $(0.5 - 1.1) \times 10^{12}$ m^3 $((3 - 7) \times 10^{12}$ bbl) of which ca 80% are federally owned. The richest portion (85% of the reserve) of the Green

Table 1. Shale Oil Resources[a], 10^9 m^3 [b]

Geographic area	Total resource[c,d]			Marginal or submarginal resources[d]		
	21–42	42–104	104–417	21–42	42–104	104–417
Africa	71,500	12,700	636	small	small	14
Asia	93,800	17,500	874		2	11
Australia and New Zealand	15,900	3,200	159		small	small
Europe	22,260	4,100	223		1	6
North America	41,400	8,000	477	350	254	99
South America	33,400	6,400	318		119	small
Total	*278,260*	*51,900*	*2,687*	*350*	*376*	*130*

[a]Ref. 2.

[b]To convert m^3 to bbl, divide by 0.159.

[c]Includes oil shale in known resources, in extensions of known resources, and in undiscovered but anticipated resources.

[d]Numbers represent shale oil yield range in L/t. To convert L/t to gal/short ton, multiply by 0.2397.

River formation is in the Piceance Basin of Colorado. The deposits in Utah and Wyoming contain 10% and 5% of the reserves, respectively (6).

The Parachute Creek member contains the majority of the oil shale in the Piceance Creek Basin and is ca 580 m thick at the depositional center of the basin. The members of the Green River formation and the thickness of the various zones are indicated in Figure 1. Organic and saline mineral contents increase toward the depositional center of the basin. The rich Mahogany zone extends across the Piceance Basin and into the Uinta Basin in eastern Utah. In addition to its high contents of organic matter, the Parachute Creek member contains large reserves of nahcolite [15752-47-3], NaHCO$_3$, and dawsonite [12011-76-6], NaAl(OH)$_2$CO$_3$, which are present in the deepest parts of the basin (see ALKALI AND CHLORINE PRODUCTS, SODIUM CARBONATE).

Analytical and Test Methods

Sample preparation for the modified Fischer assay technique, a standard method to determine the liquid yields from pyrolysis of oil shale, is necessary to achieve reproducible results. A 100-g sample of >230 μm (65 mesh) of oil shale is heated in a Fischer assay retort through a prescribed temperature range, eg, ca 25.5–500°C, for 50 min and then soaked for 20 min. The organic liquid which is collected is the Fischer assay yield (7). The Fischer assay is not an absolute method, but a qualitative assessment of the oil that may be produced from a given sample of oil shale (8). Retorting yields of greater than 100% of Fischer assay are possible.

A total material balance assay is a Fischer assay in which the retort gases are collected. A complete material balance closure and yields in excess of those expected from Fischer assay results are achieved. More complete descriptions of both the Fischer assay and the Tosco material balance assay methods have been reported (9).

Table 2. Properties of World Wide Oil Shales[a]

Property	Timahdit	Irati	Nagoorin	Kentucky	Maoming	Colorado	Condor	Alpha	New Brunswick	Israeli	Kunkersite
Fischer assay, %											
oil weight	6–9	6–12	14.1	5.3	9.7	16.5	6.3	52.0	6–12	6.2	28.6
water, bound	2.1–2.7	0.2–2.1	6.9	1.9	3.8	1.0	1.9	4.0	0.9–1.4	2.8	2.5
spent shale	85–88	83–90	72.4	90.0	82.0	78.6	87.3	33.0	91.1–84.5	87.4	62.7
gas + loss	2.8–3.7	2–4	6.6	2.8	4.5	3.8	4.5	11.0	2.0–2.1	3.6	6.2
Other properties											
moisture, wt %	6.7–9.8	0.2–6	23.2	2.8	11.3	0.7	7.7	2.8	5.4–6.7	8.1	5.8
specific gravity	1.88–1.99	1.9–2.1	1.47	2.22	1.73	1.94	2.05	1.16	2.32–1.97	1.57	1.60
gross heating value, J/g[b]	5.230–6.904	5.439–6.987	11.950	5.791	8.577	9.113	4.728	30.669	3.766–6.908	4.209	16.07
total carbon, wt %	14.78–19.46	12–17	25.67	12.82	18.74	23.45	10.50	70.54	10.75–16.58	15.5	36.8
total hydrogen, %	1.9–2.0	0.9–2.4	3.7	1.5	2.9	2.9	1.7	8.39	1.4–2.2	1.6	4.3
total sulfur, %	2.1–2.7	3.9–5.6	1.0	4.4	1.6	1.1	0.9	1.4	0.9–1.0	2.9	2.0
nitrogen, %	0.46–0.63	0.3–1.9		0.3	1.3	0.6	0.3	1.0			0.1
loss on ignition, at 950°C, %	31.4–38.9	20–24	41.9	21.3	32.1	38.0	18.6	91.7	21.1–28.5		56.2
Ash composition, wt %											
SiO_2	31.6–37.5	5.0–5.6		64.8	57.2	45.2	73.2	53.4	54.8–55.7		33.2
Fe_2O_3	3.5–5.8	7.6–9.8		10.7	12.2	5.5	8.1	9.9	6.8–5.7		6.6
Al_2O_3	8.6–13.0	9.8–12.6		12.5	19.5	2.3	12.1	24.3	17.9–15.0		8.9
CaO	15.7–26.7	1.3–3.9		1.9	1.1	18.9	2.0	3.4	8.9–13.8		33.7
MgO	5.6–7.4	2.0–3.7		0.6	0.8	17.4	1.0	4.1	6.1–3.7		9.5
Fischer assay oil											
specific gravity, 20°C	0.962	0.906	0.918	0.926	0.890	0.902	0.895	0.905	0.880	0.980	0.958
total carbon, wt %	78.73	84.60	83.40	84.95	84.81	84.21	84.72	84.32	85.6	80.8	83.4
hydrogen, %	9.69	12.50	11.37	11.85	11.65	11.29	12.54	11.89	12.3	10.4	10.7
sulfur, %	6.33	1.10	1.16	1.40	0.52	0.92	0.46	1.72	0.6	5.0	0.7
nitrogen, %	1.52	0.90	1.18	1.12	2.60	1.78	1.30	0.69	1.1	1.2	0.1
gross heating value, J/g[b]	40.074	42.547	43.070	41.773	42.447	42.723	42.677	42.539	43.932	39.748	39.790

[a]Ref. 3.
[b]To convert J to cal, divide by 4.184.

676

Table 3. Shale Oil Resources of the United States[a], 10^9 m[3][b]

	Total resource[c,d]			Marginal or submarginal resources[d]		
Geographical area	21–42	42–104	104–417	21–42	42–104	104–417
Green River Formation, ie, Colorado, Utah, and Wyoming	636	445	191	318	223	83
central and eastern United States	318	159		32	32	0
Alaskan deposits	large	32	40	small	small	small
other	21,300	3,537	80		small	small
Total	*22,254*	*4,173*	*311*	*350*	*254*	*83*

[a]Ref. 2.
[b]To convert m³ to bbl, divide by 0.159.
[c]Includes oil shale in known resources, in extensions of known resources, and in undiscovered but anticipated resources.
[d]Numbers represent shale oil yield range in L/t. To convert L/t to gal/short ton, multiply by 0.2397.

Table 4. Composition of Green River Oil Shale[a,b]

Material	Composition, wt %
mineral (inorganic, 85 wt % of total)	
carbonates	40.8
feldspars	17.8
quartz	12.8
clays	11.0
pyrite and analcite	2.6
kerogen (organic, 15 wt % of total)	
carbon	11.7
hydrogen	1.5
nitrogen	0.3
sulfur	0.15
oxygen	1.35

[a]Ref. 1.
[b]Shale oil yield of 104 L/t (25 gal/short ton).

General Properties

Kerogen Decomposition. The thermal decomposition of oil shale, ie, pyrolysis or retorting, yields liquid, gaseous, and solid products. The amounts of oil, gas, and coke which ultimately are formed depend on the heating rate of the oil shale and the temperature–time history of the liberated oil. There is little effect of shale richness on these relative product yields under fixed pyrolysis conditions, as is shown in Table 5 (10).

Numerous kinetic mechanisms have been proposed for oil shale pyrolysis reactions (11–14). It has been generally accepted that the kinetics of the oil shale pyrolysis could be represented by a simple first-order reaction (kerogen → bitumen → oil), or

$$\text{sequential A} \longrightarrow \text{B} \longrightarrow \text{C} \tag{1}$$

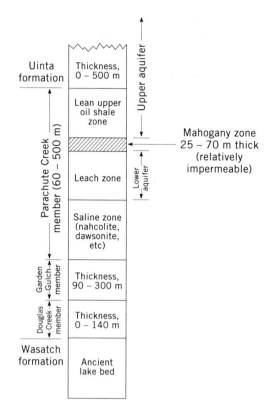

Fig. 1. Green River formation, Colorado.

Table 5. Conversion of Kerogen by Fischer Assay[a]

Component	Grade of shale, L/t					
	43.8	111.4	151.5	238.3	257.9	312.9
oil, wt %	51	65	69	66	69	71
gas, wt %	14	12	11	12	12	11
organic residue, wt %	35	23	20	22	19	18

[a]Ref. 10.

This sequential first-order reaction adequately describes the kinetics of pyrolysis of the Green River oil shale in western United States. Additional kinetic studies (15,16) indicate that sequential reactions are inadequate to describe the kinetic reactions for the thermal decomposition of oil shales worldwide. First, there is no well-defined chemical induction time as predicted by first-order reactions. Secondly, kerogen decomposition is a complex array of thermal reactions involving a variety of organic materials, water, and gases, such as CO and CO_2, as well as hetero-atom reactions involving nitrogen, sulfur, and oxygen. It is impossible to define the process using simple individual reactions. The kinetic reactions can best be described using a global approach that encompasses the sequential first-order reaction (eq. 1) as well as

$$\text{parallel} \qquad A \xrightarrow{\quad} C \qquad\qquad (2)$$

and

$$\text{alternate} \qquad \begin{array}{c} A \longrightarrow B \\ \diagdown \ \diagup \\ C \end{array} \qquad\qquad (3)$$

Temperature and Product Yields.   Most oil shale retorting processes are carried out at ca 480°C to maximize liquid product yield. The effect of increasing retort temperature on product type from 480 to 870°C has been studied using an entrained bed retort (17). The oil yield decreased and the retort gas increased with increased retorting temperature; the oil became more aromatic as temperature increased, and maximum yields of olefinic gases occurred at about 760°C. Effects of retorting temperatures on a distillate fraction (to 300°C) are given in Table 6.

Carbonate Decomposition.   The carbonate content of Green River oil shale is high (see Table 4). In addition, the northern portion of the Piceance Creek basin contains significant quantities of the carbonate minerals nahcolite and dawsonite. The decomposition of these minerals is endothermic and occurs at ca 600–750°C for dolomite, 600–900°C for calcite, 350–400°C for dawsonite, and 100–120°C for nahcolite. Kinetics of these reactions have been studied (19). Carbon dioxide, a product of decomposition, dilutes the off-gases produced from retorting processes at the above decomposition temperatures.

Table 6. Effect of Retorting Temperature on Product Type[a,b]

Retorting temperature of distillate (reduced to 300°C)	Saturates, vol %	Olefins, vol %	Aromatics, vol %
537°C	18	57	25
649°C	7.5	39.5	53
760°C	0	2.5	97.5
871°C	0	0	100
gas combustion	30	50	20
simulated *in situ*	41	37	22
in situ	59	16	25
median U.S. crude	60–100	<5	0–40

[a]Ref. 18.
[b]Colorado oil shale.

Retorting

Oil shales are solid minerals, impervious to the flow of fluids, and are generally situated in deposits below the earth's surface. Therefore, several process steps must be undertaken to produce crude shale oil. In the case of the commonly used above-ground retorting (AGR), these steps involve mining, crushing, and heating (see MINERALS RECOVERY AND PROCESSING). The grade (volume of oil per weight of rock) of most oil shales is low (see Table 2), and large amounts of the oil shale rock must be processed to produce crude shale oil. Depending on the grade, 2 to 25 metric tons of oil shale must be processed to produce one cubic meter of crude shale oil (0.4–4.6 short tons per barrel of crude shale oil). In order to eliminate the costs of mining and material handling, direct underground retorting (*in situ* retorting) has been considered as an alternative to the conventional AGR.

Historically, direct combustion has been employed in which some of the organic matter of the kerogen is combusted to provide the heat necessary for retorting. Although these direct heat (DH) processes do not require a supplemental source of fuel, some of the kerogen is consumed and the gaseous products of the kerogen decomposition are diluted with the products of combustion. In order to obviate these shortcomings, indirect heat (IH) processes were developed in which the heat required for retorting was supplied by hot gases or solids that were heated externally. However, the IH processes do not utilize any of the solid residual carbon or char resulting from kerogen decomposition and they do require an external source of fuel.

There are numerous means of classifying the many processes that have been employed to retort oil shale. In addition to the types of retorting, the retorting process can be classified by the type of feed used and by the flows within the retort. Types of oil shale feed can be classified as coarse, >5 mm (>0.25 in.) or fine, <65 mm (<0.25 in.). The flows within the retort can be classified as concurrent, ie, all materials flowing in the same direction, or countercurrent, ie, the solid oil shale flowing in one direction, the air and gases in the opposite direction. The coarse feed systems usually result in the disposal of the raw shale fines as waste. The fine feed systems obviate the latter problem but result in increased crushing costs and greater environmental impacts from particulate emissions during the material handling operations. A list of most of the oil shale retorting processes in use worldwide since the 1940s is provided in Table 7.

Retorting processes consist of several well-defined steps, or zones, within the retort, as shown in Figure 2 for the batch-process Nevada-Texas-Utah (NTU) retort, the forerunner of most of the technologies listed in Table 7. For DH systems the zones are the oil shale preheating or off-gas oil-mist cooling zone; the pyrolysis zone, where the solid organic kerogen is converted into gases, oil mists and vapors, and residual carbon; the combustion zone, where carbon is burned to provide heat; and the shale cooling zone, where the retorted shale is cooled and the incoming air is preheated.

Above-Ground Retorting. AGR processes can be grouped into DH or IH processes. Numerous design configurations as well as a variety of heat-transport mediums have been used in the indirect heated processes (Table 7).

Gas Combustion Retort. The continuous gas combustion retort (GCR) has been modeled after the earlier batch-operation NTU retort. Although the term

Table 7. Retorting Technologies

Technology	Country	Heating process[a,b]	Feed	Flow[c]
		Above-ground retorting		
Chevron	United States	DH	fine	
FBC	Israel	DH	fine	CC
Fuschun	China	DH	coarse	CC
Galoter	Russia	IH	coarse	CO
Gas combustion	United States	DH	coarse	
Kiviter	Russia	IH and DH	coarse	CO and CC
LLNL/HRS	United States	IH (ash)	fine	CO
Lurgi	United States	IH (ash)	fine	CO
Paraho DH	United States	DH	coarse	CC
Paraho IH	United States	IH (gas)	coarse	CC
Petrosix	Brazil	IH (gas)	coarse	CC
Superior	United States	IH (gas)	coarse	
Taciuk	Australia	IH (gas)	fine	CC
TOSCO II	United States	IH (solids)	fine	CC
Unishale A	United States	DH	coarse	CC
Unishale B	United States	IH (gas)	coarse	CC
		In situ retorting		
Equity BX	United States	IH (steam)		
IGT	United States	IH (H$_2$/steam)		CO
LOFRECO	United States	DH		
MultiMineral	United States	DH		
RISE	United States	DH		
VMIS	United States	DH		

[a]DH = direct heat; IH = indirect heat.
[b]Heat-transfer medium is given in parentheses.
[c]CC = countercurrent; CO is concurrent.

"gas combustion" has been applied to this process, it is a misnomer in that, in a well-designed and properly operated system, the residual char on the retorted shale supplies much of the fuel for this process. The GCR is the forerunner of most continuous AGR processes (Table 7).

PETROSIX. The PETROSIX technology is operated in the IH mode using hot recycle gas as the heat-transport medium. The PETROSIX retort has only one level of heat input, uses countercurrent flows, and uses a circular grate to control the flow of solids (Fig. 3). The PETROSIX has been operated by Petrobras (Brazil) since the 1950s and is one of the few retorting processes producing shale oil in 1995.

Paraho. The Paraho retorting technology is similar to the PETROSIX technology except that it can be operated in the direct heat (DH) mode. The unique feature of the Paraho technology is the two levels of heat input (Fig. 4). In the IH mode, the air blower shown in Figure 4 is replaced by a recycle gas heater. The Paraho DH operation has been carried out near Rifle, Colorado since the 1970s; operations to produce asphalt (qv) from shale oil are continuing.

Tosco II. The Tosco II retorting technology, developed by The Oil Shale Corp., represents IH technology employing concurrent flow using fine-sized

Combustion air

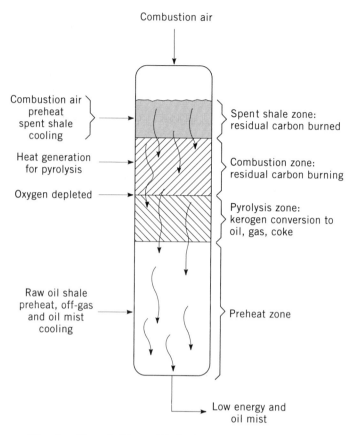

Combustion air
preheat
spent shale
cooling

Spent shale zone:
residual carbon burned

Heat generation
for pyrolysis

Combustion zone:
residual carbon burning

Oxygen depleted

Pyrolysis zone:
kerogen conversion to
oil, gas, coke

Raw oil shale
preheat, off-gas
and oil mist
cooling

Preheat zone

Low energy and
oil mist

Fig. 2. Nevada-Texas-Utah (NTU) retort schematic.

feedstock. It was tested in Colorado from the late 1950s until the early 1980s. The unique feature of the Tosco II process is the use of ceramic balls as the heat-transport medium (Fig. 5). These 125-mm (0.5-in.) balls, larger than the finely crushed shale feed, are separated from the retorted shale, recycled, reheated, and reused in the process.

UNISHALE B. The UNISHALE process, like the Paraho process, uses lump feed and countercurrent flows, and can be operated in either the DH or IH mode. The UNISHALE B process is an IH process that uses hot recycled gas as the heat-transport medium (Fig. 6). The unique feature of the UNISHALE processes is the rock pump. The solids move upward through the retort as the vapors are moving downward. The rock pump was used in the UNISHALE technology at Parachute, Colorado to produce more than 0.64×10^6 m^3 (four million barrels) of crude shale oil. Operations were shut down in 1991.

Lurgi. The Lurgi process, developed by Lurgi-Ruhrgas GmbH for mild gasification of coal, has been tested as an oil shale retort. The Lurgi process is similar to the Tosco II process in that it uses finely divided oil shale as a feedstock and employs indirect heating and concurrent flows. The unique feature of the Lurgi process is the use of hot combusted retorted shale as the heat-transport medium (Fig. 7) and retorting is carried out by mixing the hot

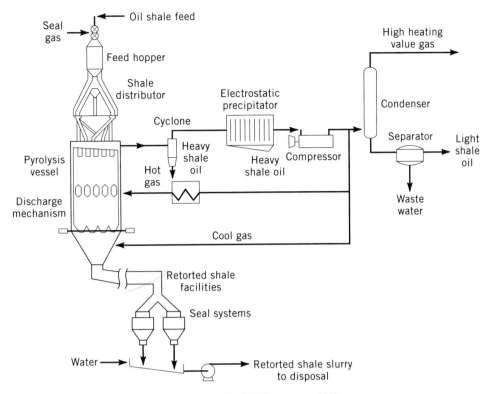

Fig. 3. PETROSIX process (20).

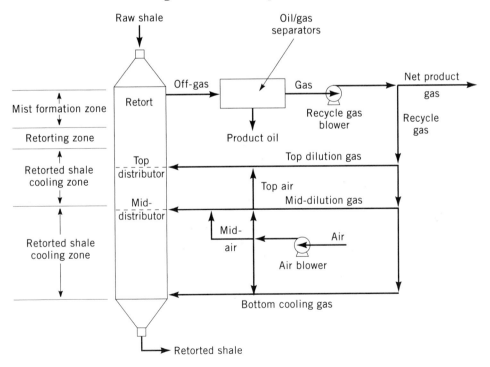

Fig. 4. Paraho DH process (21).

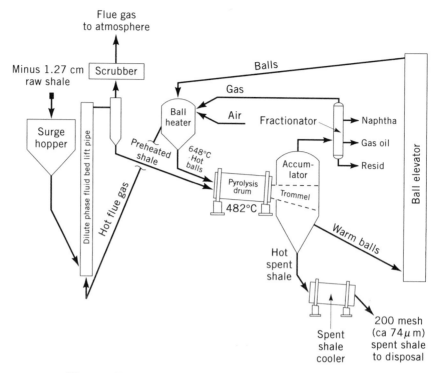

Fig. 5. Tosco II process (20). The resid is residual oil.

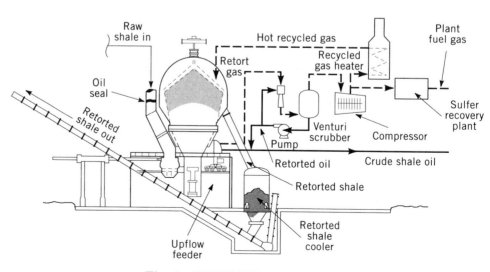

Fig. 6. UNISHALE B process (20).

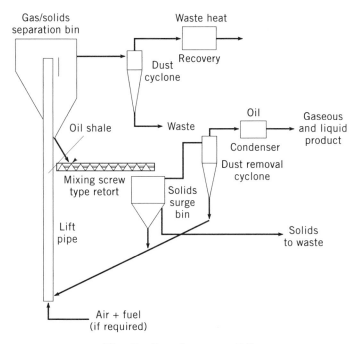

Fig. 7. Lurgi process (20).

combusted shale and raw oil shale in a screw conveyor. Because a portion of the
retorted shale is combusted to supply the process heat, the Lurgi process utilizes
more of the organic matter in the oil shale than other IH processes (see Table 7).

   Superior.   The Superior retort is different from all the other AGR processes
in that it consists of a slowly rotating circular grate instead of the vertical
shafts, rotating drums, or screw conveyors used by other technologies (Fig. 8).
The Superior technology is an adaptation of the circular grate system used to
calcine limestone (see LIME AND LIMESTONE). Raw oil shale is loaded onto the
rotating circular grate which transports the solids through the same zones as
shown in Figure 2: solids preheat, retorting, char combustion, shale cooling, and
air preheating. Because the gas composition varies within each of these zones,
each zone is separated by a baffle screen as the solids are transported around
the circle. The Superior retort was tested in Colorado in the late 1970s.

   In Situ **Retorting.**   True *in situ* retorting has been considered as a means of
avoiding the costs of mining, crushing, and surface disposal of spent shale, and
the associated environmental impacts of AGR. However, the impervious nature of
the oil shale formation and the overburden pressures have prevented true *in situ*
operations. Shale oil yields, the amount of oil produced divided by the theoretical
amount estimated to be in the oil shale rock, for *in situ* retorting are usually
half that experienced with AGR retorting. A true *in situ* experiment, using
drilling and resource fracturing procedures typical of conventional petroleum
development, was tried by the Energy Research Development Administration (a
forerunner of the U.S. Department of Energy) in 1975 in Rock Springs, Wyoming.
No significant yields of shale oil were produced (22). Other true *in situ* tests were
conducted using the Equity BX superheated steam process in Colorado, and Dow

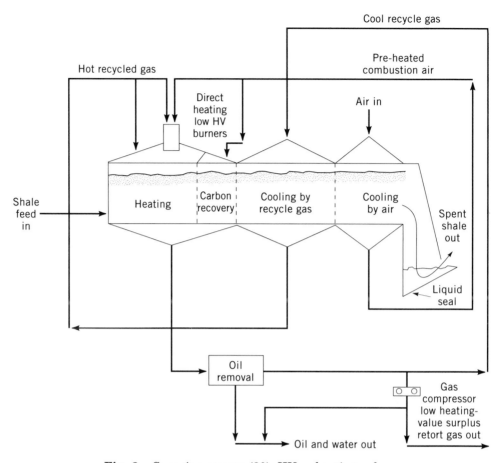

Fig. 8. Superior process (20). HV = heating value.

hot air process in Michigan; neither produced significant yields of oil shale. It appears that true *in situ* retorting is not a practical approach for the thick strata of oil shale normally situated deep below the surface.

LOFRECO. The LOFRECO process, developed by Geokinetics, Inc. (Utah) is a true *in situ* process. It is limited to relatively thin deposits of oil shale situated beneath a relatively thin overburden (Fig. 9). The LOFRECO process has been successful because the oil shale is rubblized in place by raising the overburden. Retorting consists of direct combustion horizontally through the rubblized formation, similar to that shown for the NTU retort (see Fig. 2). Although the costs of mining and materials handling are obviated, the LOFRECO process causes significant surface disturbance and results in oil yields significantly lower than those obtained in the better controlled AGR processes.

VMIS. The Vertical Modified *in situ* (VMIS) process consists of constructing an underground retort of rubblized oil shale within the deep, thick deposits situated in the Piceance Basin in western Colorado. In order to provide space for the rubblization without upheaval of the overburden, a portion of the oil shale is mined out and taken to the surface (Fig. 10). Retorting is carried out in the

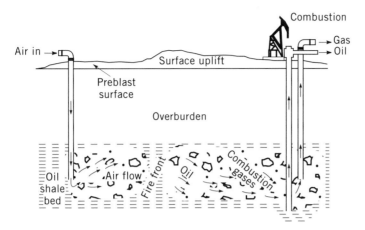

Fig. 9. LOFRECO process (20).

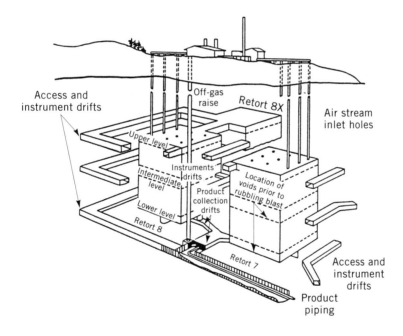

Fig. 10. VMIS process (23).

DH mode exactly as shown in Figure 2. Steam and air are pumped into the top of the VMIS retort, combustion proceeds down through the rubblized bed, and oil and gas are pumped out from the bottom. Although yields are significantly lower than those attained by AGR processes, the VMIS has demonstrated that modified *in situ* can produce shale oil from thick deposits situated deep below the surface. These operations have resulted in approximately one-fifth of the oil shale being mined out to provide space for the VMIS rubblizing and retorting. A project involving both VMIS and AGR processing, to utilize the mined-out shale, had been planned, but was canceled in 1991 (24).

Crude Shale Oil

Properties.   The composition of shale oil has depended on the shale from which it was obtained as well as on the retorting method by which it was produced. Properties of shale oils from various locations are given in Table 8. A comparison of a Green River shale oil and a Michigan Antrim shale oil, retorted under similar conditions, is given in Table 9.

Shale oil contains large quantities of olefinic hydrocarbons (see Table 8), which cause gumming and constitute an increased hydrogen requirement for upgrading. Properties for crude shale oil are compared with petroleum crude in Table 10. High pour points prevent pipeline transportation of the crude shale oil (see PIPELINES). Arsenic and iron can cause catalyst poisoning.

The primary difference in shale oils produced by different processing methods is in boiling point distribution. Rate of heating, as well as temperature level and duration of product exposure to high temperature, affect product type and yield (28). Gas combustion processes tend to yield slightly heavier liquid products because of combustion of the lighter, ie, naphtha, fractions.

Carbon-to-hydrogen weight ratios for typical hydrocarbon fuels are natural gas (methane), 3; gasoline, 6; crude oil, 6–7; shale oil, 7–8; Green River kerogen, 7; diesel and fuel oil, 8; residual oil, 10; and coal and coke, 12. A typical Green River shale oil contains 40 wt % hydrocarbons and 60 wt % organic compounds which contain nitrogen, sulfur, and oxygen. The nitrogen occurs in ring compounds such as pyridines and pyrroles, as well as in nitriles, and these materials comprise 60 wt % of the nonhydrocarbon organic components. Another 10 wt % of these components is comprised of sulfur compounds which exist as thiophenes and some sulfides and disulfides. The remaining 30 wt % is oxygen compounds occurring as phenols and carboxylic acids (1).

Upgrading Shale Oil.   Crude shale oil has a high (~2 wt %) content of organic nitrogen which acts as a catalyst poison, contains a large (20–50 wt %) atmospheric residuum fraction, and has a high (>5°C) pour point (29,30). Prerefining crude shale oil to produce a synthetic crude that is compatible with typical refineries generally is necessary (31–33). Prerefining to reduce organic nitrogen content to low levels consists usually of either a delayed coking step of the crude shale oil or residuum fraction, followed by one or more hydrogenation steps, or a more severe direct hydrogenation of the crude shale oil. Conditions for the hydrogenations are ca 400°C, 13.8 MPa (2000 psi) hydrogen partial pressure, and up to 356 standard cubic meters of hydrogen uptake per cubic meter of shale oil (2000 ft^3 at STP/bbl) (34). The nitrogen and sulfur are converted to ammonia and elemental sulfur, and the hydrogen content of the oil is increased. Upgraded shale oil is a desirable refinery feedstock. It is paraffinic and characterized by low residuum, nitrogen, and sulfur (see FEEDSTOCKS).

Shale oil has been refined to produce gasoline, kerosene, jet fuel, and diesel fuel (33). Different procedures have been tested to produce different product states, eg, hydrotreating followed by hydrocracking for jet fuel production,

Table 8. Properties of Oils Produced from Shales

Country or company	Retort	Sp gr	°API	N, wt %	S, wt %	Analysis of distillate, wt %[a]		
						Saturates	Olefins	Aromatics
Australia, Glen Davis[b]	Pumpherston	0.828	27.9	0.52	0.56	42	39	19
Brazil, Tremembe[b]	gas combustion	0.919	22.5	1.06	0.68	23	41	36
France[b]								
Autun	Pumpherston	0.931	20.5	0.90	0.51	33	36	31
Severac	Marcecaux	0.925	21.5	0.53	3.0	30	32	38
Severac	Petit	0.959	16.0	0.65	3.40	25	20	55
St. Hilaine	Lantz	0.908	24.3	0.54	0.61	31	44	25
Scotland[b]	Pumpherston	0.874	30.4	0.77	0.35	42	39	19
South Africa, Ermelo	Salermo	0.906	24.7	0.85	0.64	35	44	21
Spain, Puertollano[b]	Pumpherston	0.901	25.6	0.68	0.40	51	27	22
Sweden, Kvarntorp[b]	Rockesholm	0.977	13.3	0.68	1.65	12	24	64
United States								
Colorado	gas combustion	0.943	18.6	2.13	0.69	27	44	29
Colorado	Pumpherston	0.900	25.7	1.57	0.77	30	38	32
Superior Shale Oil[c,d]		0.630	0.93	2.0	0.8	25	25	50
Rundle Shale Oil[c,e]		0.636	0.91	0.99	0.41	48	2	50
Israeli Shale Oil[f]		0.623	0.955	1.2	7.1			

[a]Boiling at 315°C.
[b]Ref. 4.
[c]Ref. 25.
[d]Initial boiling point to 204°C.
[e]Whole oil.
[f]Ref. 26. Also contains 79.8 C, 9.7 H, and 2.2 wt % O.

Table 9. Comparison of Colorado and Michigan Antrim Shale Oils[a]

Property	Colorado	Michigan
naphtha, vol %	6.8	3.5
light distillate, vol %	24.9	41.1
heavy distillate, vol %	43.6	38.6
residuum, vol %	23.9	16.3
specific gravity (°API)	0.911 (23.8)	0.934 (20.0)
pour point, °C	10	−15
hydrogen, wt %	12.5	11.1
carbon, wt %	84.7	83.6
nitrogen, wt %	1.6	0.7
sulfur, wt %	0.8	3.5

[a]Ref. 27.

Table 10. Comparison of Green River Crude Shale Oil and Median U.S. Crude[a]

Property	Tosco II	U.S. Bureau of Mines Gas combustion	*In situ*	Simulated *in situ*	Union Oil Co. A	Median U.S. crude
distillation boiling point, °C[b]						
ibp[c] 200	18	6	11–15	7	5	30
200–315	24	19	41–48	31	20	22
315–480	34	38	27–35	46	40	28
>480	24	37	9–14	17	35	20
pour point, °C	−1[d], 15[e]	21–28	−1 to 5	10–15	32	< −15
specific gravity	0.927	0.934	0.892	0.910	0.940	0.850
°API	21	20	27	24	19	35
nitrogen, wt %	1.9	1.5–2.1	1.4–1.8	1.6	2.0	0.09
sulfur, wt %	0.7	0.8	0.7	0.6–0.9	0.9	0.6
oxygen, wt %	0.8	1.7			0.9	
viscosity, mm^2/s (=cSt)						
at 37°C	22	59	8–15	21	46	6
100°C	04	07			06	
saturates, vol %		30	59	41		60–100
olefins, vol %		50	16	37		<5
aromatics, vol %		20	25	22		0–40
carbon-to-hydrogen ratio		7–8				5–7
arsenic, ppm		40				<0.03

[a]Refs. 1, 27.
[b]Values represent % of product.
[c]ibp = initial boiling point.
[d]After a patented heat treatment which temporarily reduces pour point.
[e]No heat treatment.

hydrotreating followed by fluid catalytic cracking for gasoline production, and coking followed by hydrotreating for diesel fuel production. Production of military fuels from the refining of 1590 m^3 (10,000 bbl) of Paraho crude shale oil at the Gary Western refinery in Colorado has been reported (35) and 15,900 m^3

(100,000 bbl) of Paraho shale oil has been processed under a U.S. Navy contract, at Sohio's Toledo refinery (33).

At the Parachute Creek Project, Unocal designed and operated an oil shale upgrading unit to prerefine crude shale oil into syncrude, ie, upgraded shale oil (36). The unit was designed to handle 1600 m³ (10,000 bbl) of crude shale oil per stream day. More than 650,000 m³ (four million barrels) of syncrude were produced. Results of the Unocal shale oil upgrading process are given in Table 11. The syncrude is compared with Arabian light crude oil in Table 12. Although Arabian light is considered a premium crude oil among petroleum refiners, the Unocal shale oil syncrude shows improvements in each of the characteristics listed. Production of the conventional fuels and lubricants using Unocal shale oil syncrude as the refinery feedstock is less difficult and less costly than using Arabian light crude oil (see LUBRICATION AND LUBRICANTS).

Table 11. Properties of Shale Oil during the Unocal Upgrading Process[a]

Raw shale oil from retort	After particulates removal	After arsenic removal	After unicracking	Syncrude
gravity, °API	22	22	25	38
pour point, °C	23.9	23.9	26.6	<3.9
particulates, ppm	300	0	0	0
arsenic, ppm	25	25	0	0
distillation, °C				
initial	65.5			433
maximum	590			538

[a]Ref. 36.

Table 12. Properties of Shale Oil Syncrude and Arabian Light Crude[a]

Property	Syncrude	Arabian Light
gravity, °API	40	34
sulfur, ppmwt	5	17,000
nitrogen, ppmwt	60	800
carbon residue, wt %	0.05	3.6
heavy metals, ppmwt		20
distillation, vol %		
X–538°C	100	85
538°C +		15

[a]Ref. 36.

Alternative Uses

Oil shale is an energy resource that produces a liquid fuel that can be used to replace conventional crude oil or petroleum. However, the costs associated with processing oil shale into conventional refined products are significantly greater than that of processing conventional crude oil. In order to develop the oil shale resource, other uses have been considered. These functions include direct combustion to produce process heat for power generation (qv), direct gasification of the oil shale geological deposit, and special petrochemical production.

Direct Combustion. Direct combustion of oil shale has been used to produce heat for power generation at specific sites and is being carried out in Estonia and Israel.

In Estonia, most of the rich oil shale, 209 L/t kukersite, is burned as a solid fuel to produce electric power (37). The kukersite, although technically an oil shale, is actually similar to a high ash, low grade coal (qv) ideally suited for this use.

In the remote Negev desert region of Israel, oil shale is being burned in a fluidized-bed combustor to supply process heat and produce electric power. Unlike the Estonian kukersite, this oil shale is lean, about 63 L/t, but is being mined to access an underlying phosphate deposit (26).

Gasification. For significant conversion of shale oil or oil shale to gaseous products, considerable hydrogen must be used. Hydrogasification is the main process under consideration for gasification of oil shale. Hydroretorting of oil shale has been studied extensively (38,39). Gasification of Colorado oil shale in hydrogen and synthesis gas has been carried out at 524–760°C and at up to 38.3 MPa (5540 psig). Another study involves conventional oil shale retorting followed by gasification of the resulting shale oil in a fluidized bed (40). The hydrogen is supplied by gasification of the coke on the spent shale and from the gasifier. Other hydrogasification work includes the processing of Green River and eastern Devonian oil shales (41,42). Because of their character, the eastern Devonian shales produce less product oil, ca 35 wt % Fischer assay, than do Green River shales. The hydrogasification process is claimed to recover about 90 wt % of the kerogen content. Additional oil shale gasification research at the Laramie Energy Technology Center (U.S. DOE) has been reported (43–45).

Petrochemicals Production. *Early Consideration.* In the 1950s, the U.S. Bureau of Mines studied pyrolysis of both oil shale and shale oil for the production of light olefins. High temperature retorts were used to determine the effect of continuing the cracking, which is begun when the kerogen is converted to shale oil (46). Low temperature shale oils are low in aromatic content, thus one goal has been the production of an aromatic-rich naphtha. High temperature retorting of this type on Green River oil shales has the disadvantage of the additional energy requirements of the endothermic carbonate decomposition, with over 50 wt % decomposition occurring at 815°C. This effect has little importance for low carbonate oil shale, eg, the eastern U.S. Antrim oil shales (41). Comparison of thermal cracking of conventional shale oil to high temperature retorting under the same conditions illustrates that naphtha production is enhanced considerably by high temperature retorting (46).

Utilization of shale oil products for petrochemical production has been studied (47–51). The effects of prerefining on product yields for steam pyrolysis of shale oil feed and the suitability of Green River shale oil as a petrochemical feedstock were investigated. Pyrolysis was carried out on the whole oil, vacuum distillate, and mildly, moderately, and severely hydrogenated vacuum distillates.

Specialty Chemicals. Specialty chemicals have been considered an economically attractive means of using the oil shale resource. These specialty chemicals consist of high value, niche market items (52) that utilize the high concentration of heteroatoms (nitrogen, sulfur, and oxygen) found in most crude shale oils (see Table 8). The use of shale oil with its complex, high molecular weight, low pour point resid materials, and high concentration of various functional groups could be used to produce waxes, aromatic lubricating oils, sulfonate feeds, substitutes for coal-tar acids and bases, resins, and special organic intermediates (53). Revenue that may be achieved from shale oil would be greatly enhanced if used in their production.

Shale Oil Asphalt. The New Paraho Corp. has been producing asphalt (qv) made from crude shale oil (54,55). This shale oil asphalt, SOMAT, represents a specialty product that utilizes many of the properties of crude shale oil that would reduce its value as a refinery feedstock, ie, low pour point, high boiling point, and large quantities of heteroatoms and organic functional groups, especially basic nitrogen groups. These properties reduce the value of crude shale oil as a refinery feedstock, but tend to produce an improved asphalt. SOMAT is superior to conventional petroleum asphalt (AC-10) in terms of tensile strength retained after freeze–thaw cycles (Table 13).

Shale oil asphalt also meets or exceeds the performance of the improved, but more costly, polymer-based asphalt. Since 1989, more than 8 km (5 mi) of test strips of SOMAT have been placed on various roadways in seven U.S. states. The SOMAT has demonstrated marked improvement over conventional petroleum-based asphalt. These assessments are continuing. As of this writing (ca 1995), Paraho is preparing preliminary designs for an oil shale facility to produce about 325 m^3/d (2000 bbl/d) of the shale oil modifier used to produce SOMAT (56).

Table 13. Tensile Strength Retained, %[a]

	Number of freeze–thaw cycles			
Asphalt	1	3	5	9
AC-10	85	79	72	48
SOMAT	85	96	87	84

[a] Ref. 56.

Environmental Issues

The plans to develop a commercial oil shale industry in the three-state region of Colorado, Utah, and Wyoming in the 1970s raised the possibility of significant adverse environmental, health, safety, and socioeconomic (EHSS) impacts. Processing oil shale to produce oil on a large-scale commercial basis requires a large amount of mining, crushing, material transport, and disposal operations.

Adverse EHHS impact could result from uncontrolled, or inadequately controlled, large-scale oil shale operations. Without controls, significant amounts of dust, ie, particulates, would be produced. Because the gas produced from kerogen breakdown contains significant amounts of hydrogen sulfide and ammonia, uncontrolled release, or direct combustion with no control technology, could pose adverse health impacts and air pollution. The liquids produced from retorting operations, ie, process water and crude shale oil, contain significant levels of toxic metals, suspected or known carcinogens, and other hazardous materials. Discharge of this water would thus require treatment. Combusting and/or refining the crude shale oil would also require adequate treatment and environmental controls. The large quantities of materials involved in oil shale development means that disposal of the retorted shale poses special problems. Proper controls are needed to avoid significant air pollution from dust emissions, and surface and groundwater contamination from leaching and runoff. The amount of water required for commercial oil shale operations poses water quality impact on the semiarid region of Colorado, Wyoming, and Utah. Engineering technology was thus developed for oil shale operations. Most predictions of significantly adverse EHHS impacts (Table 14) were based on assumptions from earlier foreign operations and impacts from similar industries (57–61), and were not realized during the 1975–1990 oil shale boom.

Air Pollution. Particulates and sulfur dioxide emissions from commercial oil shale operations would require proper control technology. Compliance moni-

Table 14. Estimate of Colony Oil Shale Project Emissions Production, kg/h[a,b]

Source	SO$_2$	NO$_x$	Solid particulates	Hydro- carbons	Carbon monoxide
crushing and conveying					
primary crusher dust collection system	0	0	27	0	0
final crusher dust collection system	0	0	136	0	0
fine-ore storage dust collection system	0	0	27	0	0
pyrolysis and oil recovery unit					
preheat systems	529	2172	51	128	17.4
steam superheater ball circulation systems	9	56	103	0.2	1.0
processed shale moisturizing systems	0	0	116	0	0
hydrogen unit					
reforming furnaces	146	244	4.9	0.8	4.6
gas–oil hydrogenation unit					
reactor heaters	10	54	1.1	0.4	0.45
reboiler heater	2.7	17.7	0.4	0.04	0.3
naphtha hydrogenation unit					
reactor heater	2.2	4	0.09	0.01	0.09
sulfur recovery unit					
sulfur plants with common tail- gas plant	29				
delayed coker unit					
heater	21	35	0.7	0.14	0.64
utilities					
boilers	51	246	5.4	1.8	1.4
Total	*799.9*	*2828.7*	*472.59*	*131.39*	*25.88*

[a]Ref. 57.
[b]Using no control technology.

toring carried out at the Unocal Parachute Creek Project for respirable particulates, oxides of nitrogen, and sulfur dioxide from 1986 to 1990 indicate a +99% reduction in sulfur emissions at the retort and shale oil upgrading facilities. No violations for unauthorized air emissions were issued by the U.S. Environmental Protection Agency during this time (62).

Water Quality. All commercial oil shale operations require substantial quantities of water. All product water is treated for use and operations are permitted as zero-discharge facilities. In the Unocal operation, no accidental releases of surface water have occurred during the last four years of sustained operations from 1986 to 1990. The Unocal Parachute Creek Project compliance monitoring program of ground water, surface water, and process water streams have indicated no adverse water quality impacts and no violations of the Colorado Department of Health standards (62).

Solids. Proper handling and disposal techniques can obviate potential problems associated with the solid waste-retorted shale. Retorted shale disposal and revegetation have posed no adverse environmental impacts at the Unocal Parachute Project (62). Earlier studies carried out using Paraho and Lurgi retorted shales indicated that these materials behave as low grade cements (63,64) and can be engineered and compacted into high density materials (Fig. 11) and water impervious structures (Table 15).

Health and Safety. Much of the adverse health issue publicity involving risks of exposures to carcinogens, such as benzo-α-pyrene, have been based on recorded exposures of Scottish oil shale workers that took place nearly 100 years ago. It is believed that the increase in cancer was due more to poor personal hygiene than exposure to shale oil. Industrial hygiene monitoring and health surveys indicate no significantly increased health risks among oil shale workers (62).

Socioeconomics. Impact from recurring boom-and-bust cycles typified many of the earlier mining developments in the western United States (65). However, this and the Oil Shale Trust Fund, established by the legislation that set up the Federal prototype leases (C-a, C-b, U-a, and U-b), has provided

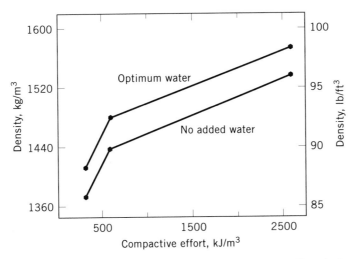

Fig. 11. Density of retorted shale vs compactive effort (64).

Table 15. Permeability of Retorted Shale[a]

Compaction	Loading, kPa[b]	Permeability cm/s × 10⁻⁶	
		No water	Optimum water
standard, 593 kJ/m³[c]	345	43.0	6.8
	690	29.3	1.4
	1380	19.1	0.8
heavy, 2693 kJ/m³[c]	345	38.1	1.1
	690	32.4	0.6
	1380	25.2	0.1

[a]Ref. 64.
[b]To convert kPa to psi, multiply by 0.145.
[c]To convert kJ to kcal, divide by 4.184.

funds to local towns and counties to assist in the construction and upgrading of infrastructures. These infrastructures are needed to accommodate the work force needed to construct and operate the large oil shale processing facilities. Further, companies such as Unocal have provided socioeconomic grants to these counties (66).

Economic Aspects

As of 1995, there were only a few commercial oil shale facilities operating in the world. These facilities are located in countries where the economic, political, and environmental requirements for commercial oil shale development are met. There are commercial oil shale facilities in Brazil, China, Estonia, and Israel. No commercial oil shale facilities have existed in the United States because the costs of shale oil processing exceed those associated with conventional petroleum crude processing.

In the United States, estimates of oil shale retorting have ranged from $113/m³ ($18/bbl) to $567/m³ ($90/bbl). The lower estimate is based on using the actual costs of constructing a commercial oil shale retorting facility and using a high grade (>125 L/t (>30 gal/short ton)) of oil shale (3). The higher estimate is based on conservative estimates utilizing unproven or noncommercial oil shale retorting technology.

The estimated costs for upgrading crude shale oil range from $38/m³ ($6/bbl) to $63/m³ ($10/bbl). However, the resulting upgraded shale oil is superior to most conventional crude petroleum and is more valuable as a refinery feedstock (67). The costs for upgrading crude shale oil depend on the upgrading techniques, ie, hydrotreating and coking or hydrotreating and fluid catalytic cracking. However, the greatest economic factor in oil shale upgrading is the amount of hydrogen required as reflected in the concentration of heteroatoms such as nitrogen, sulfur, and oxygen.

The commercial production of shale oil as an alternative energy source has not been economically feasible. As of 1995, all commercial oil shale operations in the world (Petrobras, Brazil; PAMA, Israel; The Chinese Petroleum Corp., Fushun and Maoming, China; Kivioli Oil Shale Processing Plant, Kohtla-Jarve,

Estonia) receive some sort of economic incentives or assistance from the countries in which they are operating.

The first stage of the Stuart oil shale project near Gladstone, Australia, 6000 t/d (6600 short tons/d), is scheduled to be constructed by Southern Pacific Petroleum. Financial assistance from the Australian government, consisting of special depreciation incentives and exemption of gasoline taxes equivalent to about U.S. $1.91/m³ of crude shale oil ($12.00/bbl) has been assured (68).

As a result of the 1980 Energy Security Act, the United States Synthetic Fuels Corp. (USSFC) was established to provide financial assistance in the development of alternative energy sources to reduce the dependence on foreign petroleum to meet the needs for liquid fuels. More than $15 billion was authorized for financial assistance to those projects having the potential for producing about 318,000 m³/d of crude shale oil (two million bbl/d) by 1992. Whereas letters of intent to negotiate for assistance were authorized by USSFC for three oil shale projects, ie, Cathedral Bluffs, Occidental Oil Shale, Inc. (Rio Blanco, Colorado); Seep Ridge, Geokinetics, Inc. (Vernal, Utah); and Parachute Creek, Union Oil Co. (Parachute, Colorado) (69), the only oil shale project to receive financial assistance from the USSFC was the $654 million Parachute Creek project which received about $114.5 million assistance in the form of a price guarantee of $56–73/bbl from July 1983 until Union Oil Co. ceased operations in Parachute in June 1991. Less than one-third of the available funds had been utilized (70).

The Deficit Reduction Act of 1984 significantly curtailed the available funding for the USSFC, which ceased operations in 1985.

Commercial Operations

The number of commercial oil shale operations worldwide has decreased significantly since the decade 1975–1985 and are producing only a fraction of the world's liquid fuels' needs. Most commercial oil shale operations have been scaled back.

PETROSIX Operations in Brazil. Petroleo Brasilerio (Petrobras) has a dedicated facility to produce crude shale oil from the Irati formation in southern Brazil. The facility is called the Oil Shale Industrialization Superintendency (SIX) and uses the PETROSIX retorting technology (see Table 7 and Fig. 3).

During its 40-year development, three different sizes of PETROSIX retorts have been operated on a continuous basis: a 1.83-m (6-ft) diameter demonstration plant; a 5.49-m (18-ft) diameter Prototype Unit (UPI); and a 10.97-m (36-ft) diameter Industrial Module (MI). Within the SIX facility are numerous pilot plants available for retorting coarse-sized oil shale, fines utilization, and oil shale upgrading (3,67).

The UPI and MI retorts are processing 7000 t/d (7700 short tons/d) of Irati shale to produce 24,381 m³/d (3870 bbl) of shale as well as 80 t (55 short tons) of LPG, 132 t (145 short tons) of clean fuel gas, and 98 t (108 short tons) of sulfur. The SIX plant has reached its design rate (Table 16) in an energy efficient manner with a high on-stream (operating) factor.

Pilot-plant studies have been conducted by Petrobras on many of the different oil shales from around the world. Tests indicate that many oil shales could be processed using the PETROSIX retorting technology (67) (Fig. 12).

Table 16. MI-Main Operation Data[a]

energy consumption/energy produced	0.38
oil yields, %	87–90
gas yields, %	140–150
operation factor, %	88–90
retorting rate, kg/(h·m²)	2300–2900

[a]Ref. 3.

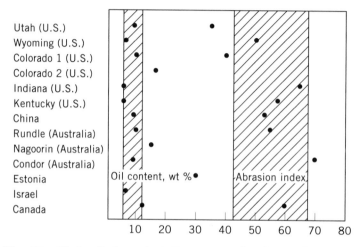

Fig. 12. Shale oil characteristics compared to Irati oil shale (3).

Oil Shale Operations in Israel. Oil shale, the only fossil fuel resource in Israel, is being used to generate electric power. The oil shale feed stock, typical of the low grade Israeli oil shale (see Table 2), is situated in a deposit overlying phosphate ore. The oil shale operations are being carried out because the oil shale has to be mined to obtain the phosphate ore.

A circulating fluidized-bed boiler, using raw shale oil as a feedstock, is being used to supply process heat for the phosphate operations and to operate a 100-MW power plant. Scale-up in the 1990s should increase the electric power generation to 1000 MW (71).

United States. In 1980, Unocal began constructing the Parachute Creek Project, designed to produce 1600 m³ (10,000 bbl) of upgraded shale oil per day. The project included a conventional underground room-and-pillar mine, the Unishale B (see Table 7) retort, and a special Unocal upgrading facility. Plant startup occurred in 1986, and daily shale oil production reached 1100 m³/d (7000 bbl/d). By 1991, total production exceeded 0.6 × 10⁶ m³ (four million barrels). However, the Parachute Creek Project was shut down in mid-1991 for economic reasons.

The New Paraho Corp. has been conducting research on asphalt derived from shale oil, SOMAT, at its pilot plant (Rifle, Colorado) (54,55). It is the only active oil shale operation in the United States as of 1995. New Paraho is continuing its pilot operations while designing a commercial facility to produce SOMAT. The economics appear promising (56).

Other Oil Shale Operations. As of this writing, commercial production of shale oil is still being conducted in the People's Republic of China and Estonia. However, production rates continue to dwindle owing to the availability of conventional petroleum and other sources of energy as well as continued worldwide energy conservation.

Commercial shale oil production in the People's Republic of China is still being carried out in Fushin and Maoming. However, production in both areas is declining because of continued emphasis on conventional petroleum and coal. Annual shale oil production in China is about 10×10^6 m^3 (1.6×10^6 barrels) (72).

Shale oil production in the former Soviet Union is also declining. The only significant shale oil operations are in Estonia. Most of the rich (208 L/t (50 gal/short ton)) Estonian oil shale is combusted directly as fuel.

Plans are underway to develop commercial shale oil operations in Australia. Southern Pacific Petroleum, N.L. is planning a commercial oil shale project utilizing the Stuart deposit (Brisbane, Australia). Favorable economics are attained by tax incentives to the Stuart project in the form of increased depreciation write-offs and exempting excise tax for gasoline produced from shale oil. In Stage 1 of the Stuart project, expected to be operational in 1996, surface mining is to be used with AGR Taciuk retorting (see Table 7) to produce about 675 m^3 (4250 barrels) of hydrotreated naphtha and fuel oil per stream day. In the full-scale, Stage 3 of the Stuart project, daily production is estimated to be nearly 10,000 m^3 (about 60,000 barrels) of upgraded shale oil syncrude (73).

BIBLIOGRAPHY

"Shale Oil" in *ECT* 1st ed., Vol. 12, pp. 207–221, by B. Guthrie and H. M. Thorne, Bureau of Mines, U.S. Dept. of the Interior; in *ECT* 2nd ed., Vol. 18, pp. 1–20, by R. E. Gustafson, Cameron and Jones, Inc.; "Oil Shale" in *ECT* 3rd ed., Vol. 16, pp. 333–357, by P. F. Dickson, Colorado School of Mines.

1. S. Siggia and P. Uden, eds., *Analytical Chemistry Pertaining to Oil Shale and Shale Oil*, National Science Foundation Grant Number GP 43807, June 24–25, 1974, pp. 11–13.
2. D. C. Duncan and V. E. Swanson, *U.S. Geol. Surv. Circ.* **523**(7), 9 (1965).
3. D. L. Bachman and co-workers, *1993 Eastern Oil Shale Symposium*, Lexington, Ky., Nov. 16–19, 1993.
4. H. M. Thorne and co-workers, *U.S. Bur. Mines Inf. Circ.* **8216**, 6 (1964).
5. G. L. Baughman, ed., *Synthetic Fuels Data Book*, 2nd ed., Vol. 4, Cameron Engineers (Division of The Pace Co.), Denver, Colo., 1978, pp. 67–104.
6. *Oil Shale and the Environment*, EPA-600/9-77-033, U.S. Environmental Protection Agency, Washington, D.C., Oct. 2, 1977.
7. ASTM D3904, American Society for Testing and Materials, Philadelphia, Pa., 1980.
8. R. N. Heistand, *Am. Chem. Soc. Div. Fuel Chemis. Preprint* **21**(6), (1976).
9. L. Goodfellow and M. T. Atwood, *Colo. Sch. Mines Q.* **69**, 205 (Apr. 1974).
10. K. E. Stanfield, *U.S. Bur. Mines Rep. Invest.* **4825**, (1951).
11. V. D. Allred, *Colo. Sch. Mines Q.* **62**, 657 (1967).
12. J. J. Cummins and W. E. Robinson, *U.S. Bur. Mines Rep. Invest.* **7620**, 1 (1972).
13. R. L. Braun and A. J. Rothman, *Fuel* **54**, 129 (Apr. 1975).
14. J. H. Campbell and co-workers, paper presented at *The 10th Oil Shale Symposium Proceedings*, Colorado School of Mines Press, July 1977, pp. 148–165.
15. A. K. Burnham and co-workers, *Comp. Stud. Various Oil Shales* **34**(1), 36–42 (Mar. 1989).

16. A. K. Burnham, *Relationship Between Hydrous and Ordinary Pyrolysis*, UCRL-JC-114130, Lawrence Livermore National Laboratory, Livermore, Calif., June 1993.
17. S. S. Tihen, *Ind. Eng. Chem.* **47**, 464 (1955).
18. G. U. Dineen, "Effect of Retorting Temperature on the Composition of Shale Oil," *AlChE Meeting*, Denver, Colo., Aug. 1962.
19. J. H. Campbell and A. K. Burnham, paper presented at *The 11th Oil Shale Symposium Proceedings*, Colorado School of Mines Press, Golden, Colo., Nov. 1978, pp. 242–259.
20. E. M. Piper, *6th IIASA Resource Conference*, Luxenborg, Austria, Colorado School of Mines Press, Golden, Colo., 1981.
21. J. B. Jones, Jr., and R. N. Heistand, *Proceedings, 12th Oil Shale Symposium*, Colorado School of Mines Press, Golden, Colo., 1979, pp. 184–195.
22. Energy Development Consultants, "Oil Shale in the United States, 1981," *EDC*, Golden, Colo., 1980.
23. A. Stevens and R. L. Zahradnik, *Proceedings, 16th Oil Shale Symposium*, Colorado School of Mines Press, Golden, Colo., 1983, pp. 267–268.
24. L. Dockter, *Colo. Sch. Mines Q.* **83**(4), 120–122 (1988).
25. R. F. Crane, in Ref. 21, pp. 1–16.
26. A. H. Pelofsky and co-workers, in Ref. 21, pp. 32–42.
27. A. Long, Jr., N. W. Merriam, and C. G. Mones, in Ref. 15, pp. 120–135.
28. N. D. South and co-workers, *Colo. Sch. Mines Q.* **71**, 153 (Oct. 1976).
29. R. E. Poulson, C. M. Frost, and H. B. Jensen, *ACS Div. Pet. Chem. Prepr.* **17**(2), 175 (1972).
30. M. T. Atwood, paper presented at *The ACS Symposium: Fuels of the Future*, Dallas, Tex., Apr. 4–13, 1973.
31. C. M. Frost, R. E. Poulson, and H. B. Jensen, *ACS Div. Pet. Chem. Prepr.* **17**(2), 156 (1972).
32. D. P. Montgomery, *Ind. Eng. Chem. Prod. Res. Dev.* **7**, 274 (1968).
33. R. F. Sullivan, B. E. Stangeland, and H. A. Frumkin, *ACS Div. Pet. Chem. Prepr.* **22**, 998 (1977).
34. V. F. Yesavage, C. F. Griswold, and P. F. Dickson, paper presented at *The 180th National ACS Meeting*, San Francisco, Calif., Aug. 25–29, 1980.
35. H. Batrick and co-workers, *Final Report—The Production and Refining of Crude Shale Oil into Military Fuels*, Applied Systems Corp. for Office of Naval Research, Washington, D.C., Aug. 1975.
36. C. P. Reeg, A. C. Randle, and J. H. Duir, *Proceedings, 23rd Oil Shale Symposium*, Colorado School of Mines Press, Golden, Colo., 1990, pp. 68–95.
37. J. D. Baker and C. O. Hook, in Ref. 21, pp. 26–31.
38. H. F. Feldman and co-workers, *Inst. Gas Technol. Res. Bull.* **36** (Aug. 1966).
39. S. A. Weil and co-workers, paper presented at *The 167th National ACS Meeting*, Los Angeles, Calif., Apr. 1974.
40. U.S. Pat. 3,703,052 (1972), H. R. Linden (to Institute of Gas Technology).
41. F. C. Schora and co-workers, *Hydrocarbon Process.* **56**(4), 107 (1977).
42. F. C. Schora, "The Application of the IGT Hydroretorting Process to Eastern Shale," *Energy Topics—a Supplement to IGT Highlights*, Institute of Gas Technology, Chicago, Ill., May 9, 1977.
43. E. L. Burwell and I. A. Jacobson, Jr., "Concurrent Gasification and Retorting of Oil Shale—a Dual Energy Source," *Rocky Mountain Regional Meeting*, SPE 5535, Society of Petroleum Engineers (SPE), AIME, Denver, Colo., Apr. 7–9, 1975.
44. E. L. Burwell and I. A. Jacobson, Jr., *U.S. Bur. Mines Tech. Prog. Rep.* **85**, 1 (Nov. 1974).
45. E. L. Burwell and I. A. Jacobson, *Colo. Sch. Mines Q.* **71**, 139 (Oct. 1976).
46. H. W. Sohns and co-workers, *Ind. Eng. Chem.* **47**, 461 (1955).
47. C. F. Griswold, V. F. Yesavage, and P. F. Dickson, *ACS Div. Fuel Chem. Prep.* **21**, 207 (1976).

48. E. A. Fritzler, V. F. Yesavage, and P. F. Dickson, *Proceedings, The Second Pacific Chemical Engineering Conference*, American Institute of Chemical Engineers, New York, 1977, p. 542.

49. P. D. Smith, P. F. Dickson, and V. F. Yesavage, *ACS Div. Pet. Chem. Prepr.* **23**, 756 (1978).

50. C. F. Griswold, A. Ballut, H. R. Kavianian, P. F. Dickson, and V. F. Yesavage, *Chem. Eng. Prog.* **75**(9), 78 (1979).

51. C. F. Griswold, A. Ballut, H. R. Kavianian, P. F. Dickson, and V. F. Yesavage, *Energy Commun.* **6**(2), 153 (1980).

52. J. E. Sinor, "Niche Market Assessment for a Small Western Oil Shale Project," Final Report, DE-FC-86MC11076, U.S. Dept. of Energy, Washington, D.C., 1979.

53. J. E. Bunger and A. V. Deveni, *Proceedings, 25th Oil Shale Symposium*, Colorado School of Mines Press, Golden, Colo., 1992, pp. 281–294.

54. L. A. Lukens, *Proceedings, 22nd Oil Shale Symposium*, Colorado School of Mines Press, Golden, Colo., 1989, pp. 196–206.

55. L. A. Lukens and M. A. Plummer, *Colo. Sch. Mines Q.*, 108–114 (1988).

56. L. A. Lukens, *Col. Sch. Mines Q.*, 115–119 (1988).

57. *An Environmental Impact Analysis for a Shale Oil Complex at Parachute Creek, Colorado*, Vol. I, II, III, Colony Development, Denver, Colo., 1974.

58. *Detailed Development Plan*, Vol. 1 and 2, Oil Shale Lease Tract C-b, C-b Shale Oil Project, 1976.

59. *Modifications to Detailed Development Plan*, Oil Shale Tract C-b, C-b Shale Oil Venture, 1977.

60. *Project Independence Oil Shale Task Force Report*, FEA, Washington, D.C., 1974, p. 154.

61. T. L. Thoem and E. F. Harris, in Ref. 19, pp. 1–9.

62. J. B. Benton, *FUEL* **71**, 238–242 (Feb. 1992).

63. J. P. Fox, *Proceedings of the 13th Oil Shale Symposium*, Colorado School of Mines Press, Golden, Colo., 1980, pp. 131–139.

64. R. N. Heistand and W. G. Holtz, in Ref. 63, pp. 140–150.

65. A. Gulliford, *Boomtown Blues: Colorado Oil Shale, 1885–1985*, University Press of Colorado, Boulder, 1989.

66. J. Evans, *Colo. Sch. Mines Q.*, 103–105 (1988).

67. E. M. Piper and co-workers, *Proceedings of the 25th Oil Shale Symposium*, Colorado School of Mines Press, Golden, Colo., 1992, pp. 221–242.

68. *1992 Annual Report*, Southern Pacific Petroleum N.L., 1993.

69. *1984 Annual Report*, United States Synthetic Fuels Corp., 1985.

70. *Oil Gas J.*, 38 (Apr. 8, 1991).

71. J. Yerushalmi, *Proceedings of the 1992 Eastern Oil Shale Symposium*, IMMR Press, Lexington, Ky., 1993, p. 367.

72. C. Du, *Proceedings of the 18th Oil Shale Symposium*, Colorado School of Mines Press, Golden, Colo., 1985, pp. 210–215.

73. B. C. Wright, *Alternate Energy '89 Proceedings*, Council on Alternate Fuels, Washington, D.C., 1989, pp. 175–194.

EDWIN M. PIPER
Piper Designs LLC

ROBERT N. HEISTAND
Consultant

OLEFIN FIBERS.   See FIBERS, OLEFIN.

OLEFIN POLYMERS

POLYETHYLENE

INTRODUCTION

Polyethylene (PE) is a generic name for a large family of semicrystalline polymers used mostly as commodity plastics. PE resins are linear polymers with ethylene molecules as the main building block; they are produced either in radical polymerization reactions at high pressures or in catalytic polymerization reactions. Most PE molecules contain branches in their chains. In very general terms, PE structure can be represented by the following formula:

$$(CH_2 — CH_2)_x — branch_1 —(CH_2— CH_2)_y — branch_2—(CH_2 — CH_2)_z — branch_3 \ldots$$

where the $-CH_2-CH_2-$ units come from ethylene, and x, y, and z values can vary from 4 or 5 to over 100. This allows the industry to produce a large variety of PE resins with different molecular weights and branching characteristics.

Chain Length.   The total number of monomer units (which approximately equals $x + y + z + \ldots$ in the above formula) in PE chains is called the degree of polymerization. It can vary from small (about 10–20 in PE waxes) to very large (over 100,000 for PE of ultrahigh molecular weight (UHMW)). Consequently, the molecular weights of PE resins can range from several hundred to several million.

Branching.   In high pressure ethylene polymerization processes, the branches are formed spontaneously according to peculiarities of radical polymerization reactions (see OLEFIN POLYMERS, LOW DENSITY POLYETHYLENE). These branches are either linear or branched alkyl groups. Their lengths vary widely, sometimes even within a single polymer molecule, and can be both short (from methyl to isooctyl group, collectively known as short-chain branching) and long, up to several thousands of carbon atoms (long-chain branching). In catalytic polymerization processes, the branches are introduced deliberately by copolymerizing ethylene with α-olefins (see OLEFIN POLYMERS, LINEAR LOW DENSITY POLYETHYLENE). The structure of these branches is determined by the type of olefin used in the copolymerization reaction. All branches are the same if only one α-olefin is used, but there can be several different types if two or more α-olefins are copolymerized with ethylene. All these branches are usually called short-chain branches. However, some polymerization processes utilizing metallocene catalysts can also introduce long-chain branches in PE molecules.

Some PE molecules, on the other hand, contain no branches at all. From a chemical standpoint, such resins can be regarded as polymethylene, $H-(CH_2)_{\overline{n}}-H$. In other PE resins, the number of branches is significant; the weight content of ethylene units in PE resins of low density can be as low as 70–75%, and even lower in some radical ethylene–vinyl acetate copolymers.

Uniformity of Branching Distribution. The distributions of branches among different polymer molecules in PE resins can be quite different, depending on the method of PE production. Some PE resins have uniform branching distributions, which means that any given polymer molecule in these resins contains the same relative fraction of branches as all others. PE prepared in high pressure processes and most ethylene copolymers produced with metallocene catalysts belong to this group. In contrast, PE resins produced with heterogeneous titanium- and chromium-based catalysts are mixtures of polymer molecules with very different branching degrees. Some of the macromolecules in these mixtures contain very few branches, whereas other polymer molecules may contain a relatively large number of them.

History

The first polymer with a polymethylene structure was synthesized by von Pechmann in 1898 from diazomethane (1). Since then, there have been four milestones in the history of PE polymers as commercial plastics.

First, in 1935, Perrin at the ICI laboratories in the United Kingdom, discovered that ethylene could be polymerized, at a very high pressure, into a solid semicrystalline material with a melting point of approximately 120°C (2). This discovery was the starting point in the manufacture of low density PE (LDPE), which began in the United Kingdom in 1938 and in the United States in 1943. The polymerization reaction proceeds by a free-radical mechanism in supercritical ethylene at high pressures of around 60–350 MPa (10,000–50,000 psi) and temperatures of 200–350°C. This reaction also makes possible the synthesis of ethylene copolymers with α-olefins as well as with other polar monomers, the most important one being vinyl acetate.

Second, in the early 1950s, Hogan and Bank at Phillips Petroleum Co., discovered (3,4) that ethylene could be catalytically polymerized into a solid plastic under more moderate conditions: at a pressure of 3–4 MPa (435–580 psi) and temperature of 70–100°C, with a catalyst containing chromium oxide supported on silica (Phillips catalysts). PE resins prepared with these catalysts are linear, highly crystalline polymers of a much higher density of 0.960–0.970 g/cm³ (as opposed to 0.920–0.930 g/cm³ for LDPE). These resins, or HDPE, are currently produced on a large scale. (see OLEFIN POLYMERS, HIGH DENSITY POLYETHYLENE).

Third, in 1953, Ziegler in Germany discovered that PE resins of a similarly high density, ca 0.945–0.960 g/cm³, could be prepared under even milder conditions, ie, at atmospheric pressure and temperatures of 50–100°C, in the presence of catalyst systems containing titanium halides and alkylaluminum compounds (Ziegler catalysts) (5–7). Versatile in nature, Ziegler catalysts can easily copolymerize ethylene and α-olefins with vinyl double bonds. Depending on the amount of α-olefin in the copolymerization reaction, these polymers can range from highly crystalline to completely amorphous. As a consequence, their densities may vary widely, from 0.960 to 0.880 g/cm³. The discovery of Ziegler brought to life a large industry that manufactures PE resins with a broad range of properties, each tailored to a particular application.

Finally, in 1976, Kaminsky and Sinn in Germany discovered a new family of catalysts for ethylene polymerization. These catalysts (ie, Kaminsky catalysts) contain two components: a metallocene complex, usually a zirconocene, and an organoaluminum compound, methylaluminoxane (8,9). These catalysts and their various later modifications enable the synthesis of ethylene copolymers with a high degree of branching uniformity. Formally classified as MDPE, LLDPE, or VLDPE, the resins thus produced have a number of properties that set them apart from common PE resins in terms of performance and application.

Classification of PE resins

Historically, the classification of PE resins has developed in conjunction with the discovery of new catalysts for ethylene polymerization as well as new polymerization processes and applications. The classification (given in Table 1) is based on two parameters that could be easily measured in the 1950s in a commercial environment with minimum instrumentation: the resin density and its melt index. In its present state, this classification provides a simple means for a basic differentiation of PE resins, even though it cannot easily describe some important distinctions between the structures and properties of various resin brands.

Over a period of about 50 years, the science of polymer chemistry has developed a comprehensive means of polymer characterization techniques. In

Table 1. Commercial Classification of Polyethylenes

Designation	Acronym	Density, d, g/cm³
high density polyethylene	HDPE	≥ 0.941
ultrahigh molecular weight polyethylene[a]	UHMWPE	0.935–0.930
medium density polyethylene	MDPE	0.926–0.940
linear low density polyethylene	LLDPE	0.915–0.925
low density polyethylene[b]	LDPE	0.910–0.940
very low density polyethylene	VLDPE	0.915–0.880

[a]Linear polymer with molecular weight of over 3×10^6.
[b]Produced in high pressure processes.

the case of PE, these parameters include the composition, molecular weight, and compositional distribution. The composition of ethylene copolymers is usually measured by ^{13}C-nmr, ^{1}H-nmr, or ir techniques.

Molecular weights of polymers are determined by the weight–average molecular weight, $\overline{M}_w$, and the number–average molecular weight, $\overline{M}_n$. The molecular weight distribution of polymers (MWD) is usually represented as $\overline{M}_w/\overline{M}_n$. The $\overline{M}_w$ and $\overline{M}_n$ values of PE resins are measured by high temperature gel permeation chromatography (gpc).

The compositional distribution of ethylene copolymers represents relative contributions of macromolecules with different comonomer contents to a given resin. Compositional distributions of PE resins, however, are measured either by temperature-rising elution fractionation (tref) or, semiquantitatively, by differential scanning calorimetry (dsc). Table 2 shows some correlations between the commercially used PE characterization parameters and the structural properties of ethylene polymers used in polymer chemistry.

Table 2. Correlations between Industrial and Scientific Parameters Describing PE Characteristics

PE parameter used in industry	Correlation with structural properties
melt index (MI or I_2): weight of molten resin flowing at 190°C for 10 min through a 2.095-mm diameter die at 2.16-kg load	MI is an approximate measure of the weight–average molecular weight[a]: $\log_{10}(\mathrm{MI}) = K_1 - K_2 \times \log_{10}(\overline{M}_w)$
melt flow ratio (MFR or I_{21}/I_2): the ratio of two MI values measured at 21.6 and 2.16-kg loads	MFR value is an approximate measure of the width of MWD, for LLDPE[b]: $\overline{M}_w/\overline{M}_n \sim 0.24(I_{21}/I_2) - 2.4$
density, d, g/cm^3	PE density is a function of its crystallinity (CR)[c]: $1/d = \mathrm{CR}/d_{\mathrm{cr}} + (1 - \mathrm{CR})/d_{\mathrm{am}}$

[a]Both K_1 and K_2 depend on molecular weight distribution (MWD).
[b]Ref. 10.
[c]d_{cr}(1.00 g/cm^3) and d_{am}(0.852–0.862 g/cm^3) are densities of crystalline and amorphous phases.

Synthesis Technologies

A variety of technological processes is used for polyethylene manufacture.

Polymerization in Supercritical Ethylene. If a polymerization process is carried out at a high ethylene pressure and a temperature above the PE melting point (110–140°C, depending on the PE grade), the mixture of supercritical ethylene and the PE melt serves as a polymerization medium. This process can accommodate radical initiators, Ziegler catalysts, and metallocene catalysts. In the case of radical polymerization, an ethylene pressure of 120–300 MPa (17,400–43,500 psi) is required to produce PE with a sufficiently high molecular weight at 150–300°C. These reactions are initiated by oxygen or organic peroxides.

Polymerization in Solution or Slurry.   Many hydrocarbon solvents dissolve PE at elevated temperatures of 120–150°C. Polymerization reactions in solution require, as their last step, the stripping of solvent. A variety of catalysts can be used in these processes.

Polymerization in a hydrocarbon slurry (usually a light-saturated hydrocarbon) was the first commercial polymerization process to utilize Phillips and Ziegler catalysts. These processes enjoy high popularity because of their versatility.

Polymerization in the Gas Phase.   Many polymerization catalysts can be adapted for use in the gas phase. A gas-phase reactor contains a bed of small PE particles that is agitated either by a mechanical stirrer or by employing the fluidized-bed technique. These processes are economical because they do not require solvent recirculation streams.

Control of PE Properties

The tailoring of PE properties in commercial processes is achieved mostly by controlling the density, molecular weight, MWD, or by cross-linking. Successful control of all reaction parameters enables the manufacture of a large family of PE products with considerable differences in physical properties, such as the softening temperatures, stiffness, hardness, clarity, impact, and tear strength.

Density.   The density (crystallinity) of catalytically produced PE is primarily determined by the amount of comonomer (α-olefin) in ethylene copolymer. This amount is easily controlled by varying the relative amounts of ethylene and the comonomer in a polymerization reactor. In contrast, the density of PE produced in free-radical processes is usually controlled by temperature.

Molecular Weight.   PE mol wt (melt index) is usually controlled by reaction temperature or chain-transfer agents. Reaction temperature is the principal control method in polymerization processes with Phillips catalysts. On the other hand, special chemical agents for chain transfer are required for both radical reactions and catalytic reactions with Ziegler catalysts and metallocene catalysts.

Some applications require PE with a very high molecular weight: nearly 10 times that of common PE materials. These resins are essentially nonbranched and require special catalysts, synthesis, and fabrication techniques.

Molecular Weight Distribution.   In industry, the MWD of PE resins is often represented by the value of the melt flow ratio (MFR) as defined in Table 2. The MFR value of PE is primarilly a function of catalyst type. Phillips catalysts produce PE resins with a broad MWD and their MFR usually exceeds 100; Ziegler catalysts provide resins with a MWD of a medium width (MFR = 25–50); and metallocene catalysts produce PE resins with a narrow MWD (MFR = 15–25). If PE resins with especially broad molecular weight distributions are needed, they can be produced either by using special mixed catalysts or in a series of connected polymerization reactors operating under different reaction conditions.

Cross-Linking.   A variety of PE resins, after their synthesis, can be modified by cross-linking with peroxides, hydrolysis of silane-grafted polymers, ionic bonding of chain carboxyl groups (ionomers), chlorination, graft copolymerization, hydrolysis of vinyl acetate copolymers, and other reactions.

The Market

PE resins command a wide range of applications, both as commodity resins and, in many cases, as specialty polymers. Their uses include numerous film grades of LDPE, HDPE, and LLDPE for bags and packaging; coatings for paper, metal, wire, and glass; household and industrial containers such as bottles for various fluids, ie, water, food products, detergents, liquid fuels, etc; toys; and different types of pipe and tubing. Because of its versatility, PE has become the largest commercially manufactured polymer in the world. The 1994 world capacities were 1.7×10^7 metric tons of LDPE, 1.3×10^7 t of LLDPE and MDPE, and 1.4×10^7 t of HDPE. The same numbers for the United States were, respectively, 3.7, 3.8, and 4.5 million tons.

Properties of the three most important types of PE (LDPE, HDPE, and LLDPE) are described in the following articles.

BIBLIOGRAPHY

1. R. Raff and E. Lyle, in R. Raff and K. W. Doak, eds., *Crystalline Olefin Polymers*, John Wiley & Sons, Inc., New York, 1965, Part I, Chapt. 1.
2. Brit. Pat. 471,590 (Sept. 6, 1937), E. W. Fawcett, R. O. Gibson, M. H. Perrin, J. G. Paton, and E. G. Williams (to Imperial Chemical Industries, Ltd.).
3. Belg. Pat. 530,617 (Jan. 24, 1955), J. P. Hogan and R. L. Bank (to Phillips Petroleum Co.).
4. U.S. Pat. 2,825,721 (Mar. 4, 1958), J. P. Hogan and R. L. Bank (to Phillips Petroleum Co.).
5. Ger. Pat. 973,626 (Apr. 14, 1960; prior. Nov. 18, 1953), K. Ziegler, H. Breil, H. Martin, and E. Holzkamp.
6. K. Ziegler, *Kunststoffe* **45**, 506 (1955).
7. Belg. Pat. 533,362 (May 5, 1955), K. Ziegler.
8. W. Kaminsky and co-workers, *Angew. Chem. Int. Ed. Engl.* **15**, 630 (1976).
9. H. Sinn and W. Kaminsky, *Adv. Organomet. Chem.* **18**, 99 (1980).
10. T. E. Nowlin, Y. V. Kissin, and K. P. Wagner, *J. Polym. Sci. Part A:* **26**, 755 (1988).

YURY V. KISSIN
Mobil Chemical Company

LOW DENSITY POLYETHYLENE

The first high molecular weight crystalline polyolefin was produced in 1933 by Imperial Chemical Industries, Ltd. through the high pressure process. Initial production was targeted for use in specialized applications such as insulation for high voltage cable. Since that time, the number of applications of various grades of homopolymers and copolymers have expanded to cover many areas once dominated by paper (qv), glass (qv), steel (qv), and other polymers.

The molecular weight of low density polyethylene (LDPE) ranges from waxy products at about 500 mol wt to very tough products at about 60,000 mol wt. One unique feature of LDPE, as opposed to high density polyethylene (HDPE) or linear low density polyethylene (LLDPE), is the presence of both long- and short-chain branching along the polymer chain. Another important feature of LDPE is its ability to incorporate a wide range of comonomers that can be polar in nature along the polymer chain. Disadvantages of LDPE include the high capital investment for commercial plant construction, engineering problems related to high pressure operation, and high energy costs in production (see HIGH PRESSURE TECHNOLOGY).

LDPE, also known as high pressure polyethylene, is produced at pressures ranging from 82–276 MPa (800–2725 atm). Operating at 132–332°C, it may be produced by either a tubular or a stirred autoclave reactor. Reaction is sustained by continuously injecting free-radical initiators, such as peroxides, oxygen, or a combination of both, to the reactor feed.

Traditionally, LDPE has been defined as homopolymer products having a density between 0.915–0.940 g/cm^3 (products having a density above 0.940 g/cm^3 are considered HDPE). However, with the commercialization of LLDPE via the fluidized-bed or solution processes, this distinction is no longer valid; these newer processes, utilizing Ziegler-type catalysts as well as olefinic comonomers, are capable of producing polyethylene in the density range of 0.900–0.964 g/cm^3. Furthermore, there has been development of high pressure polyethylene utilizing Ziegler catalysts resulting in a new class of polymers, with densities ranging from 0.870 to 0.960 g/cm^3. These products are in commercial production by several principal LDPE producers (1). To blur even further the distinctions based on density, this new class of polymers, generally referred to as very low density polyethylene (VLDPE), is under development or has already been introduced in the early 1990s by producers utilizing the solution and fluidized-bed processes.

Properties

The mechanical properties of LDPE fall somewhere between rigid polymers such as polystyrene and limp or soft polymers such as polyvinyls. LDPE exhibits good toughness and pliability over a moderately wide temperature range. It is a viscoelastic material that displays non-Newtonian flow behavior, and the polymer is ductile at temperatures well below 0°C. Table 1 lists typical properties.

Structure. The physical properties of LDPE depend on the molecular weight, the molecular weight distribution, as well as the frequency and distribution of long- and short-chain branching (2).

The short side chain branching frequency is inversely proportional to polymer crystallinity. Short branches occur at frequencies of 2–50 per 1000 carbons in chain length; their corresponding crystallinity varies from 35 to 75%. Directly proportional to the polymer density, crystallinity can be calculated by the following formula,

$$C = \frac{\rho_C}{\rho} \cdot \frac{\rho - \rho_A}{\rho_C - \rho_A}$$

where C = wt % crystallinity, ρ = measured polymer density, ρ_A = amorphous density, and ρ_C = crystalline density (2). The crystalline density is usually

Table 1. Properties of Low Density High Pressure Polyethylene

Property	ASTM Method	LDPE[a]
tensile yield stress, kPa[b]	D638	80–180
yield elongation, %	D638	10–40
tensile ultimate stress, kPa[b]	D638	100–170
ultimate elongation, %	D638	100–700
secant modulus of elasticity at 1% strain, kPa[b]	D638	900–5000
hardness, Rockwell	D785	D41–D60
dart drop 38-μm thickness, at g/25 μm	D1709	50–300
low temperature brittleness at F_{50}[c], °C	D746	<-76°C
dielectric constant at 60 Hz	D150	2.25–2.35
density, g/cm^3 [d]	D1505	0.912–0.940
refractive index, n_D^{25} [d]	D542	1.51
thermal expansion, 10^{-5} cm/cm per °C[d]	D696	10–22
narrow angle scatter	D1746	4–80
haze, %	1003	40–50
gloss, %	2457	0–80
water absorption 24 h, %[d]	D570	<0.02

[a]LDPE homopolymers in the 0.2–150 melt index or 100,000–20,000 mPa·s(=cP) viscosity range. Specialty polymers such as greases and waxes or highly cross-linked polymers are not included.
[b]To convert kPa to psi, divide by 6.895.
[c]F_{50} = number of hours at which 50% fail.
[d]Taken at 25°C.

taken as 1.00 g/cm^3, whereas amorphous density is equal to 0.85 g/cm^3. As polymer density increases, certain properties increase also; such properties include heat softening point, yield strength, stiffness, impermeability to gases and liquids, and film drawdown. In contrast, some properties decrease as polymer density increases; for instance, transparency, freedom from haze, low temperature brittleness resistance, resistance to environmental stress crack, and film impact strength (3).

Molecular Weight Distribution. MWD offers a general picture of the range of long, medium, and short molecular chains in the polymer; the broad molecular weight distribution in LDPE, however, is attributed only to the presence of long branches on the polymer molecule (4). LDPE may have molecules that range in length from a few thousand carbons to a million or more carbons. Melt viscosity is directly related to the average molecular weight of the polymer, and is commonly determined by using an extrusion rheometer that yields a viscosity value called the melt index. Widely accepted in the industry, this is a convenient method of estimating viscosity behavior. Nonetheless, because LDPE exhibits non-Newtonian flow behavior, the test cannot distinguish between resistance to flow contributed by viscous and elastic effects, shear initiation, or die entry effects (2). Consequently, values generated by this method must be used cautiously.

With increasing molecular weight, certain properties increase: melt viscosity, abrasion resistance, tensile strength, resistance to creep, flexural stiffness, resistance to brittleness at low temperature, shrinkage, warpage, and film impact strength. On the other hand, increasing mol wt results in reduced film transparence, freedom from haze, and gloss; drawdown rate; neck-in and beading; and adhesion (3).

Melt Index or Melt Viscosity.   Melt index describes the flow behavior of a polymer at a specific temperature under specific pressure. If the melt index is low, its melt viscosity or melt flow resistance is high; the latter is a term that denotes the resistance of molten polymer to flow when making film, pipe, or containers. ASTM D1238 is the designated method for this test.

Product Density.   Density is measured by using a density gradient column. This method is described in ASTM D1505.

Film Clarity.   Slight haziness is a characteristic of all polyethylene resins. It may be caused either by a surface roughness which diffuses light passing through the film (surface roughness is a function of extrusion conditions and the fundamental structure of the polymer), or it may be caused by the partly crystalline structure of the polymer which has a larger index of refraction than the surrounding amorphous material. Film haze is determined by using a standard pivotal sphere hazemeter in accordance with ASTM D1003, a method for determining the light-transmitting properties or the wide angle light-scattering properties of the film. A second method, ASTM D1746, may also be used to calculate narrow angle light scatter, ie, a measure of the film's freedom from distortion when viewing distant objects through it. Still a third method, ASTM D2457, may be used to measure the specular gloss (ie, sheen, sparkle, or luster) at 60°; that is, the luminous reflectance of the plane surface of the film.

Impact Strength.   The impact resistance of film (ie, its resistance to shock impact) must be high in many film applications, eg, in chemical packaging bags, in tarpaulins used in the construction field, and, more generally, in heavy gauge film. Whenever impact strength is the paramount requirement, a lower density homopolymer or a copolymer must be used. Its impact strength can be calculated by ASTM D1709.

Stress Crack Resistance.   Failure caused by environmental stress cracking may be attributed to stored stresses acquired in the molding or extruding operation. These dormant stresses may release themselves by cracking under the combined influence of an adverse environment and polyaxial stretching during use. Polyethylene of narrow molecular weight distribution tends to crack less under environmental stress. This is because resistance to environmental stress cracking increases with increasing molecular weight. For this reason, products with high molecular weight are particularly suitable for electrical application such as cable coating, in which contact under stress with potentially aggressive chemicals sometimes occurs. The bent strip test, ASTM 1693, is commonly used for testing resistance to stress cracking.

Yield Strength.   Yield strength is the highest tensile force or stress to which a plastic molded shape or film may be submitted and still return to its original shape; elongation is the extension of the sample at the moment of rupture. For obvious reasons, yield and tensile strength at rupture are crucial properties for processors and converters. Yield strength, tensile strength, and elongation are all functions of the basic molecular properties. In other words, a higher density LDPE homopolymer has higher yield strength, slightly lower tensile strength, and lower elongation; in contrast, a higher average molecular weight polymer has higher tensile strength and slightly higher yield strength. The testing method is described in ASTM D882.

Low Temperature Brittleness. Brittleness temperature is the temperature at which polyethylene becomes sufficiently brittle to break when subjected to a sudden blow. Because some polyethylene end products are used under particularly cold climates, they must be made of a polymer that has good impact resistance at low temperatures; namely, polymers with high viscosity, lower density, and narrow molecular weight distribution. ASTM D746 is used for this test.

Film Appearance. It is important to minimize film imperfections or defects when LDPE is blown into film. The common defects are arrowheads, pinpoint gels, gels or "fish eyes", and oxidized gels or colored specks. Arrowheads are lines that meet at a rounded angle and point in the direction of extrusion. Pinpoint gels are small, round imperfections that look as if the film had been pricked with a pin; they are usually uniformly dispersed and numerous enough to cause a grainy appearance in the film. Clear in appearance, gels or fish eyes are round or elliptical imperfections that are usually randomly dispersed and that can vary greatly in size from barely visible to as large as 1.5 mm or more in diameter. Yellow-brown in color, oxidized or colored specks are imperfections that appear as flakes of material or as gels with discolored centers (3).

Film imperfections are one of the more serious quality problems for both the producer and the film converter. Although gels of sufficiently low population and size present an aesthetic problem that does not affect the end use of the film, in large quantities and size they can impair the ability to produce blown film at reduced thickness or gauge. Moreover, in the case of extrusion coating or laminating operations, they can also cause microholes in the surface and destroy the barrier properties of the coated material.

Film imperfections can arise from many sources. Most commonly, they are the result of contamination in the reaction system, in post-reactor handling, during shipping and unloading, or in the end users equipment. Gels originating within the reaction system usually have considerably higher molecular weight than the base polymer. This type of gel can be formed either in microregions of the reactor at low temperature or in stagnant areas of the system, where at high temperature and long residence time the polymer thermally cross-links to a high molecular weight.

Even though there is no single test to quantify gel types and populations, equipment utilizing laser scanning techniques with computer processing has come into wide use since the 1980s. Microscopes equipped with heating stages and special lighting are also used to identify gel characteristics as well as possible sources of origin.

Chemical Resistance. LDPE is highly resistant to penetration by most chemically neutral or reactive substances. This is a property of prime importance for all kinds of packaging applications. Because of its high impermeability, polyethylene containers can store and transport many kinds of chemicals without leak hazards. Likewise, easily spoiling foods such as vegetables or meats can be shelved and sold in polyethylene bags without the danger of water infiltration from the outside or irreplaceable moisture being lost from the inside; exchange of gases through the film can also be kept to a minimum. In addition, LDPE is resistant to penetration from most polar liquids, water, aqueous acids, and alkalies, as well as most metal plating solutions. However, it can be easily penetrated by nonpolar liquids such as hydrocarbons, and animal and vegetable oils (3).

Electrical Properties. LDPE's electrical properties make it extremely well suited for wire and cable insulation for electrical power supplies at high transmission, lower domestic voltage, and high frequency, very high frequency, or ultrahigh frequency applications in electronics (see INSULATION, ELECTRIC). As a dielectric, ie, electric insulator, LDPE for all practical purposes does not transmit electrical current. Because polyethylene allows so little energy to be absorbed in wire insulation carrying very high frequency radio impulses, it has become the universal insulating material for television lead-in wire. Its dielectric strength ranges from 18 to 39 V/μm. Because polyethylene is also moisture resistant, its insulating qualities are not measurably affected by humidity or direct exposure to water. For example, water absorption on a 12.7-mm thick specimen in 24 hours is less than 0.01 wt % as determined by ASTM D570.

Manufacture

LDPE is produced in either a stirred autoclave or a tubular reactor; total domestic production, divided between the two systems at 45% for tubular and 55% for autoclave, is estimated to be 3.4 million metric tons per year (5). Neither process has gained a clear advantage over the other, although all new or added capacity production in the 1990s has been through the autoclave.

Stirred Autoclave. The stirred autoclave consists of a cylindrical vessel with a length:diameter ratio of 4:1–18:1. The vessel is usually stirred by multiple paddle arrangements of proprietary design. The reaction chamber is partitioned into one to five zones and uses baffles to create a group of reactors in series. Although initiator and/or feed gas is usually introduced at several locations along the length of the reactor, it is common for each zone to have independent initiator injection and temperature control. Pressure is held constant by using a modulating valve at the reactor exit. The reactor residence time can vary from as little as 10 to over 60 seconds. The reactor wall thickness, which is required to contain pressure in the range of 207 MPa (30,000 psi), severely limits the removal of heat from the vessel. Accordingly, the unreacted monomer must act as a heat sink for the heat of polymer formation, hence limiting conversion per pass through the reactor to about 22%. Nevertheless, due to end product requirements, conversion in commercial vessels may be as low as 9%.

Tubular Reactor. The tubular reactor is essentially a long double-pipe heat exchanger. The process side tube's internal diameter may be as wide as 6.4 cm, its length as long as 1800 m (4). A cooling jacket is utilized to remove a portion of the reaction heat, which in turn results in operation at a conversion per pass of 35% or lower. Most commercial reactors control reaction pressure with a cycle valve that opens periodically. This type of pressure control not only creates pressure pulses and surges in process side velocity that can reduce reactor wall polymer contamination, but also optimizes heat transfer to the jacket. In tubular reactors, the tubes are operated either with a single front feed entry or in combination with one or more side entries. Peroxide-type initiators and/or oxygen are used to start reaction at one or more points along the reactor. Thus, the tubular reactor resembles a series of alternating polymerization and cooling zones.

Recycle and Polymer Collection. Due to the incomplete conversion of monomer to polymer, it is necessary to incorporate a system for the recovery and recycling of the unreacted monomer. Both tubular and autoclave reactors have similar recycle systems (Fig. 1). The high pressure separator partitions most of the polymers from the unreacted monomer. The separator overhead stream, composed of monomer and a trace of low molecular weight polymer, enters a series of coolers and separators where both the reaction heat and waxy polymers are removed. Subsequently, this stream is combined with fresh as well as recycled monomers from the low pressure separator; together they supply feed to the secondary compressor.

The polymer stream exiting the high pressure separator is fed to a second low pressure separator, where most of the remaining monomer is removed; only then is the overhead stream from this vessel fed to a multiple stage compressor where the pressure is increased from around 2 kPa (15 mm Hg) to incoming feed pressure. Here a small purge stream of less than 2% of the reactor feed is taken to prevent accumulation of feed stream impurities such as methane and ethane. Finally, the stream is combined with fresh monomer and compressed to high pressure recycle pressure in the range of 17–31 MPa (2500–4500 psi). Comonomer and/or telogens are added at this point.

Molten polymer from the low pressure separator is gravity-fed to an extruder, melt homogenizer, or gear pump capable of forcing the product through a die with multiple holes. Polymer strands exiting the die are cut under water into

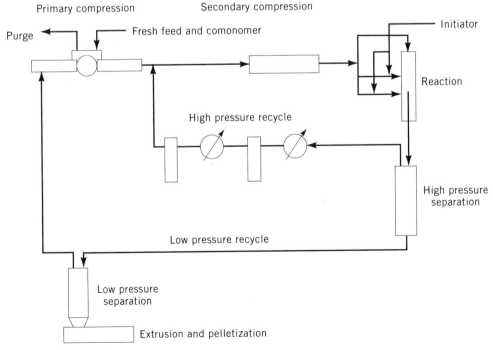

Fig. 1. Autoclave reaction and recycle system.

pellets of approximately 3 mm in diameter. Afterward, these pellets are flushed by a circulating water stream from the pellet chamber into the centrifugal drier, where the pellets are separated from the water and the polymer is crystallized to solid form.

Additives. Compounds are often added to the polymer at the extruder or melt homogenizer. Common additives are antioxidants (qv), thermal stabilizers, slip agents, antiblock agents, and uv stabilizers (qv). Popular antioxidants are 2,6-di-*t*-butyl-*p*-cresol (BHT), octadecyl 3,5-di-*tert*-butyl-4-hydroxyhydrocinnamate, and tetrakis[methylene(3,5-di-*tert*-butyl-4-hydroxycinnamate)]methane. Ultraviolet stabilizers are usually in the form of benzophenones such as 2-hydroxy-4-*n*-octylbenzophenone. A common antiblock agent is diatomaceous silica; the optimum particle size is in the range of 5–15 μm. Agents such as this permit layers of film to be easily separated. Slip agents, in contrast, lower the coefficient of friction between layers of film and therefore permit the two surfaces to slide easily against one another. Common slip agents are erucamide, oleamide, and steramide. About 2–3% of carbon black (qv) of very fine particle size (10 μm) is utilized to protect against the hazard of photo oxidation, the main component of natural outdoor weathering. Finally, cling control agents such as polybutenes are used in stretch wrap applications.

Blending and Purging. Polymer pellets are air-conveyed to holding silos where they are purged with heated, filtered air and then blended to ensure uniformity. Purging is necessary for removing the entrapped monomer and comonomer. Residual monomer may be estimated by applying Henry's law at the existing low pressure separator conditions. However, in older handling systems, it is often necessary to introduce a second melt blending step for the introduction of additives and homogenization of product. In order to avoid the more energy-intensive second step, most modern systems utilize melt homogenization prior to pelletization. To that end, improved reactor control and polymer-type scheduling have also reduced the need for a second melt homogenization step.

Packaging. Most LDPE is packaged in railway hopper cars of 63.5 t capacity, although trucks of 20 t capacity are also used. Conversely, smaller quantities are packaged either in 500-kg boxes or in 25-kg paper or polyethylene bags.

Environmental Considerations. Good progress has been made in reducing emissions during production. Ethylene is by far the largest fugitive gas; purge from the recycle system is commonly recycled to the ethylene cracking unit for recovery. Leak gas from the compression system also is recovered and reinjected to the low pressure recycle system; naphtha-like initiator solvent likewise can be recovered and recycled, just as compressor lubricants are commonly collected for refining and comonomers recovered and recycled to the reaction system. Due to improved operation techniques and control equipment, runaway reactions or "decomps" that are vented to the atmosphere have also decreased substantially. In addition, many plants utilize particulate and noise reduction devices on the emergency venting systems.

Many units have waste heat recovery systems that generate low pressure steam from reaction heat. Such steam is often employed to drive adsorption refrigeration units to cool the reactor feed stream and to increase polymer conversion per pass, an energy-saving process that reduces the demand for electrical power.

Pellet degassing, however, still presents a problem. The volume of purge air required to degas the pellets makes recovery of ethylene from the purge stream using conventional methods uneconomical. A partial solution is to install melt-processing equipment capable of removing more unreacted ethylene and comonomer from the polymer prior to pelletization. Because this is the single largest source of emissions in the high pressure reaction system, demand is great for the development of a separator capable of operating in vacuum and a pump able to transfer oxygen-free volatiles back to the low pressure recycle system.

Reaction Kinetics

Chain initiation is similar to other vinyl polymerizations; the main reaction steps are outlined in Figure 2. Propagation or chain growth occurs through the successive addition of polymer radical to ethylene units (6). In commercial polymerization processes, the maximum polymerization rate is controlled by the rate constants for propagation, chain transfer to monomer, and transfer to polymer. As temperature increases, these rate constants increase according to the Arrhenius equation, where A_p = collision frequency factor, E_p = energy of activation for propagation, k_p = propagation rate constant, R = gas constant, and t = temperature (2).

$$k_p = A_p e^{-E_p/Rt}$$

Chain growth of polymer radicals is terminated by coupling or disproportionation; polymer radicals can also be terminated by chain transfer with solvents. Although intermolecular transfer leads to long-chain branching, both long and short branches are formed by abstracting a hydrogen from the polymer chain; close correlation therefore exists between the two types of branching in commercial LDPE. The free-radical abstracts a hydrogen atom bound to the fifth atom away from the radical site (7); this "backbiting" or 1,5-hydrogen transfer thus accounts for the short-chain branching in LDPE. Furthermore, it has been shown that the extent of short-chain branching varies inversely with reaction pressure but directly with reaction temperature (8), and is independent of chain-transfer agents, conversion, initiator type or amount, and flow rate in a continuous reactor.

Chain length can be controlled by varying reaction pressure, temperature, or by adding a comonomer or telogen. More specifically, chain length can be increased by increasing pressure and decreasing temperature; both pressure and temperature can also affect the molecular weight distribution and the degree of branching. However, it is usually preferred to control chain length with either a comonomer or a chain-transfer agent. Common chain-transfer agents are listed in Table 2 (6).

Both propylene and isobutylene are comonomers that are incorporated along the chain, resulting in additional short-chain branching. One important factor in controlling polymer crystallinity is the choice of chain-transfer agent. Ethane and methane, for example, are inefficient agents whose presence in the monomer feed stream must be considered in reaction control.

Propagation
$$C_2H_4 + R_n^{\cdot} \xrightarrow{k_p} R_{n+1}^{\cdot}$$

Termination by coupling
$$R—CH_2^{\cdot} + R'—CH_2^{\cdot} \xrightarrow{k_t} R—CH_2—CH_2—R'$$

Termination by disproportionation
$$R—CH_2^{\cdot} + R'—CH_2—CH_2^{\cdot} \xrightarrow{k_t} RCH_3 + R'—CH{=}CH_2$$

Chain transfer with transfer agent
$$R^{\cdot} + SA \xrightarrow{k_S} RA + S^{\cdot}$$

Chain transfer to monomer

$$R—CH_2—CH_2^{\cdot} + C_2H_4 \underset{k'_m}{\overset{k_m}{\rightleftharpoons}} \begin{array}{l} R—CH_2—CH_3 + C_2H_3^{\cdot} \\ R—CH{=}CH_2 + C_2H_5^{\cdot} \end{array}$$

Intramolecular H transfer
$$R—CH_2—CH_2—CH_2—CH_2—CH_2^{\cdot} \xrightarrow{k_b} R—\overset{\cdot}{C}H—C_4H_9$$

Intermolecular H transfer
$$R + R'—CH_2—R'' \xrightarrow{k_1} RH + R'—\overset{\cdot}{C}H—R''$$

Beta scission

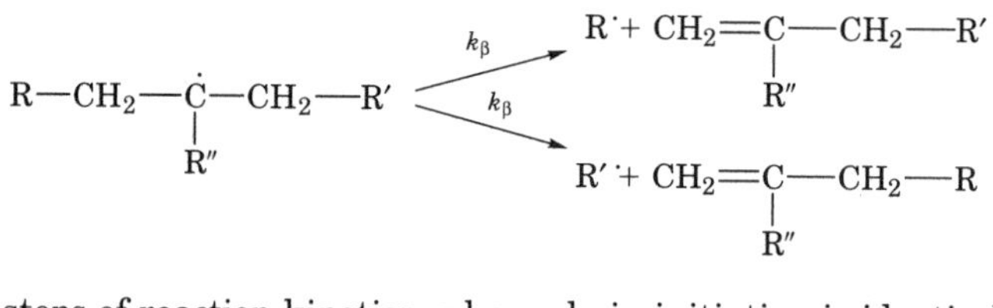

$$R—CH_2—\underset{\underset{R''}{|}}{\overset{\cdot}{C}}—CH_2—R' \underset{k_\beta}{\overset{k_\beta}{\rightleftarrows}} \begin{array}{l} R^{\cdot} + CH_2{=}\underset{\underset{R''}{|}}{C}—CH_2—R' \\ R'^{\cdot} + CH_2{=}\underset{\underset{R''}{|}}{C}—CH_2—R \end{array}$$

Fig. 2. Main steps of reaction kinetics, where chain initiation is identical to other vinyl polymerizations.

Initiators. The degree of polymerization is controlled by the addition rate of initiator(s). Initiators (qv) are chosen primarily on the basis of half-life, the time required for one-half of the initiator to decay at a specified temperature. In general, initiators of longer half-lives are chosen as the desired reaction temperature increases; they must be well dispersed in the reactor prior to the time any substantial reaction takes place. When choosing an initiator, several factors must be considered. For the autoclave reactor, these factors include the time permitted for completion of reaction in each zone, how well the reactor is stirred, the desired reaction temperature, initiator solubility in the carrier, and

Table 2. Common Chain-Transfer Agents

Type	Chain-transfer constant
propane	0.007
n-butane	0.010
ethane	0.0006
methane	0.0002
propylene	0.015
isobutylene	0.016
hydrogen	0.0159
aldehydes	0.197

the cost of initiator in terms of active oxygen content. For the tubular reactors, an additional factor to take into account is the position of the peak temperature along the length of the tube (9).

The degree of polymerization is controlled by the rate of addition of the initiator. Reaction in the presence of an initiator proceeds in two steps. First, the rate-determining decomposition of initiator to free radicals. Secondly, the addition of a monomer unit to form a chain radical, the propagation step (Fig. 2) (9). Such regeneration of the radical is characteristic of chain reactions. Some of the more common initiators and their half-life values are listed in Table 3 (10).

The thermal decomposition of organic peroxides results in the formation of free radicals. Addition of free radicals to the double bond of a vinyl monomer such as ethylene results in the generation of another radical. Propagation continues in this manner until the monomer is exhausted except for the strong tendency of radicals to act in pairs to form a paired-electron covalent bond with subsequent loss in radical activity. Most initiators have efficiencies in the range of 60–80%. Loss of efficiency is attributed in large part to the recombination of radical pairs before they move apart in the reaction medium (8). Some peroxides, for instance, can generate radicals in LDPE production in the 100–300°C temperature range; most of them have bond-dissociation energies between 125–167 kJ/mol (30–40 kcal/mol).

Polymerization using oxygen is not well understood; it is known that oxygen copolymerizes with ethylene to form peroxidic copolymers (10). Other free-radical generators such as azo compounds and carbon–carbon compounds have found only limited use in the synthesis of LDPE.

Table 3. Common Initiators and Their Half-Life Values

Type	Temperature for a half-life of 1 s, °C	Activation energy, kJ/mol[a]	Active oxygen, %
di-*sec*-butyl peroxydicarbonate	133	126.27	6.83
tert-butyl peroxyneodecanoate	155	104.72	6.55
tert-butyl peroxypivalate	172	105.89	9.18
tert-butyl peroxy-2-ethylhexanoate	182	122.42	6.95
tert-butyl peroxyisobutyrate	188	123.68	9.99
tert-butyl peroxybenzoate	224	133.93	8.24
di-*tert*-butylperoxide	247	130.74	10.94

[a]To convert kJ/mol to kcal/mol, divide by 4.184.

Copolymers. Although many copolymers of ethylene can be made, only a few have been commercially produced. These commercially important copolymers are listed in Table 4, along with their respective reactivity coefficient (see CO-POLYMERS. The basic equation governing the composition of the copolymer is as follows, where M_1 and M_2 are the monomer feed compositions, r_1 and r_2 are the reactivity ratios (6).

$$\frac{d[M_1]}{d[M_2]} = \frac{[M_1]r_1[M_1] + [M_2]}{[M_2][M_1] + r_2[M_2]}$$

Alpha-olefins are used as comonomers for controlling polymer density and molecular weight. Vinyl acetate is by far the most common comonomer; it is commercially marketed with incorporation levels ranging from less than 1 wt % to as high as 50 wt %. Since at lower incorporation levels the copolymers exhibit high impact strength and clarity, they are commonly used in laminating applications. In contrast, copolymers of the acrylates exhibit more temperature stability, better impact strength, and are softer to the touch; they are thus used in making disposable gloves. As a rule, both types of copolymers are used in packaging operations from ice bags to produce bags where high impact strength is required. In addition, copolymers of acrylic acid also show good adhesion to aluminum; the laminated products made from these copolymers thus find applications in retort bags for microwave ovens. Silanes, on the other hand, are used to produce intermediate copolymers that can react with compounds of sodium to produce ionomers (qv); which are cross-linked by exposure to moisture during electrical cable coating operations to provide tough abrasive resistant coatings (11). Not surprisingly, wire coated in this manner is often used in electrical wiring harnesses for automotive and aeronautical applications. Still another copolymer, that of carbon monoxide, is used to manufacture products where uv degradability is desired or required, such as in can carrier rings.

Although the reaction rate of ethylene and various copolymers differs substantially, the reaction constants can be established by using an arbitrary

Table 4. Copolymers of Commercial Interest

Comonomer	Reactivity coefficient	
	r_1	r_2
vinyl acetate	1.0	0.9
methyl acrylate	0.05	8
ethyl acrylate	0.04	15
n-butyl acrylate	0.03	4
acrylic acid	0.08	4
methacrylic acid	0.10	4
vinyltriethoxysilane	1.0	0.9
carbon monoxide	[a]	[a]
propylene	3.1	0.8
1-butene	3.4	0.86
isobutylene	3.6	0.7

[a] Carbon monoxide does not obey general copolymerization models.

value of 1 for ethylene (5). Thus, a value of 0.1 would indicate that the comonomer reacts at 10 times the rate of ethylene. However, the wide range of reaction rates can present problems not only in determining the comonomer content of the final product but also in producing a homogeneous product (4,6).

Applications

LDPE is used in a wide range of applications, the largest segment of which is taken up by end uses requiring processing into thin film (see FILMS AND SHEETING). Table 5 lists usage and estimated volume for the United States in metric tons per year (12). Table 6 lists the United States LDPE capacity by company (5).

Blown Film.   Blown film has a number of advantages over flat film. For instance, in blown film extrusion, molecular orientation is achieved in both the machine and transverse direction, the relative degree depending primarily on the drawdown ratio (the die opening to film gauge) and the blow-up ratio. The result is a film with more uniform strength in both directions. Moreover, with proper extrusion conditions, the physical properties of the film are equal in both the machine and transverse direction; such even distribution provides maximum toughness. Another advantage of blown film or tubing is the absence of a lengthwise seam. Figure 3 illustrates the blown film process.

In blown film extrusion, the molten polymer enters a ring-shaped die either through the bottom or from the side. It is then forced around a mandrel inside

Table 5. Applications of LDPE in 1993

Application	Volume, t/yr
Film	
food packaging	544,199
nonfood packaging	472,215
stretch and shrink film	91,125
carry-out bags	55,242
trash and can liners	125,476
other nonfood packaging	292,017
Total	*1,580,274*
Other	
extrusion coating	374,977
injection molding	166,958
wire and cable	97,412
compounding	83,840
adhesives and sealants	68,860
sheet[a]	39,466
blow molding	35,901
extruded products	55,505
exports	564,926
resale	207,887
miscellaneous	91,026
Total	*1,786,759*

[a]Thickness >0.3 mm (12 mils).

Table 6. Manufacturers and Capacities of LDPE in 1993

Company	Capacity, t/yr
Quantum	705,215
Dow	487,528
Chevron	417,234
Westlake	362,812
Du Pont	346,939
Exxon	279,819
Eastman	294,785
UCC	37,172
Mobil	226,757
Rexene	185,941
Lyondel	63,494

the die, shaped into a sleeve, and extruded through the die opening in the form of a comparatively thick-walled tube. While still in the molten state, the tube is expanded to a "bubble" or hollow cylinder of desired diameter and corresponding lower film thickness. The expansion of the tube is produced by the pressure of internal air admitted through the center of the mandrel. The bubble, after a few meters of free suspension, is flattened between two nip rolls, whose take-off speeds range from 10 to 90 m/min. Gauges as thin as 9 μm can therefore be achieved through this process for use in such applications as garment bag film. Another important factor is the height of the frost line above the die. The frost line is the ring-shaped zone where the bubble frequently becomes frosty due to the drop in temperature to the melting range of the polymer. Frost line control is thus essential for determining the molecular orientation of the melt in the machine and transverse directions, as well as certain physical properties of the film, such as tear, tensile, and impact strengths. The frost line is usually controlled by the volume of cooling air blown against the bubble; raising the frost line thus gives the film more time to solidify, resulting in a gloss with a smoother surface and higher clarity. Conversely, too high a frost line results in film blocking or sticking when rolled up (3).

Flat Film Extrusion.   In flat film extrusion, the melt is extruded through a long slot in a "T" or coat hanger-type die, past the die lands. In this setup, the polymer melt is forced into the slot die at its center; it reaches the slot opening by way of a manifold and over the lands. For this reason, the inside surface of the die and the lands must be precision-machined and well polished. In chill roll of cast film extrusion the hot polymer melt extruded through the die slot is cooled by the surface of two or more water-cooled chill or casting rolls. The principal advantages of film casting are substantial improvements in the film's transparency, freedom from haze, improved gloss, and other optical properties.

Extrusion Coating.   In extrusion coating, a thin film of molten polymer is pressed onto or into the substrate. Coating thickness may range from 6.5 μm or less to more than 100 μm. In polymer lamination, a related operation, two or more substrates, such as paper or aluminum foil, are combined by using the polymer film as adhesive and moisture barrier. In order to coat a substrate, the polymer must be extruded through a narrow slit in the extrusion coating die by

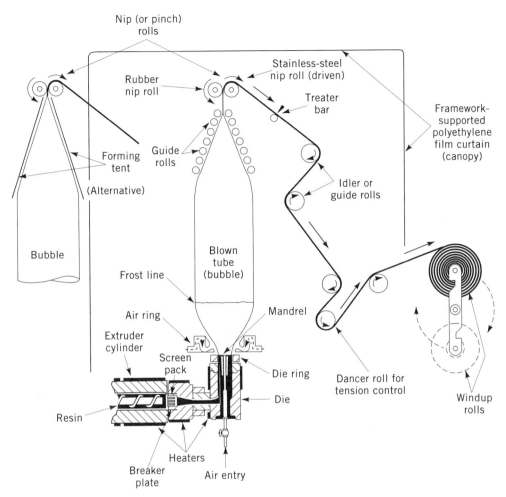

Fig. 3. Blown film extrusion procedure. Courtesy of Quantum Chemical Co.

an extruder screw. The molten film is then drawn down into the nip below the die and between two rolls: the driven, water-cooled chill roll and the rubber-covered pressure roll. While coming in contact with the faster moving substrate, the hot film is drawn out to the desired thickness and forced into the substrate, where both layers are pressed together by the two rolls (Fig. 4). The pressure exerted by two rolls is usually in the range of 9–18 kg per linear cm. The hot film shrinks (neck-in) at the edges. Neck-in is the difference between the hot melt width at the die face and the coating width on the substrate; it is undesirable since it forces more than the usual trimming of the substrate edges and increases the loss of material. However, neck-in can be reduced by lowering the polymer melt index, density, or by increasing the elasticity.

Molding Applications. Molding is accomplished by three different methods: blow molding, injection molding, and rotational molding, although the use of LDPE in these applications has been declining since the introduction of LLDPE on the market.

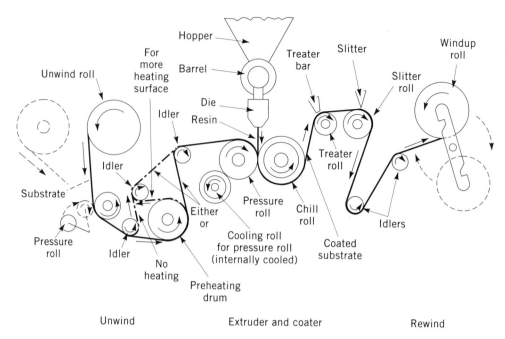

Fig. 4. Extrusion coating procedure.

Adhesives and Sealants. Dominated by copolymers, adhesives and sealants remain somewhat of a specialty market. These polymers usually contain high copolymer content and low viscosity, and often require blending with other compounds prior to final application. Their uses are numerous, eg, as seals for bottled drinks, as tie layers between incompatible polymers, or as automotive adhesives.

Health and Safety Factors

LDPE is nontoxic, and is commonly used in food packaging (qv) where Food and Drug Administration requirements, such as 21CFR in the case of thin films, must be met. It is also used in packaging pharmaceuticals and other medical applications such as iv bags and tubes.

Waste management of LDPE can be approached in several different ways, ie, via recycling, energy recovery, biodegradability, increased packaging efficiency, or uv degradability. Ultraviolet degradable grades for can carriers are available to processors; these products are mandated in several states.

Environmental Degradability. Degradation and stabilization of polymeric materials have been studied extensively for many years, mainly with the aim to extend the service life of polymers. However, it has become increasingly clear that, for certain end uses at least, plastics with a limited lifetime are preferable because they help to solve problems of litter and waste disposal. These types of plastics are subject to biodegradation, which occurs when microorganisms such as fungi or bacteria secrete enzymes and chemically break down the polymer structure into small fractions that can be digested by other microorganisms.

Biodegradability has been approached in several areas. One attempt is to add fillers (qv) such as starches into the product. The U.S. Agricultural Department has been studying biodegradable polyolefins prepared by adding starch as a filler since the mid-1980s (13). Although the incorporation of starch into polyethylene has a severely negative impact on product properties, possible application areas may yet be found in the agricultural field.

Another approach is to replace petrochemical-based polymers with polymers made from carbohydrates (14). Unfortunately, approaches of this type have yet to produce economically competitive polymers.

BIBLIOGRAPHY

"High Pressure (Low and Intermediate Density) Polyethylene" in *ECT* 2nd ed., under "Olefin Polymers," Vol. 14, pp. 217–241, by P. L. Clegg, Imperial Chemical Industries Ltd.; in *ECT* 3rd ed., under "Olefin Polymers," Vol. 16, pp. 402–420, by R. L. Boysen, Union Carbide Corp.

1. L. Pebsworth, "Retrofit of Autoclave Reactors for the Production of HDPE to VLDPE," *Annual Meeting*, AIChE, New York, Nov. 15, 1987.
2. R. A. V. Raff and K. W. Doak, *Crystalline Olefin Polymers*, John Wiley & Sons, Inc., New York, 1965, pp. 307,495,682.
3. *Petrothene Polyolefins: A Processing Guide*, 5th ed., Quantum Chemical Co., Cincinnati, Ohio, 1986.
4. C. Kiparissides, G. Verros, and J. M. S. Rex, *Macromol. Chem. Phys.* **C33**(4), 437–527 (1993).
5. Chemical Data Inc., *Monthly Petrochemical and Plastics Analysis* (Jan. 1994).
6. G. A. Moritmer and P. Ehrlich, *Adv. Polymer Sci.* **7**, 386–448 (1970).
7. M. J. Roedel, *J. Am. Chem. Soc.* **75**, 6110 (1953).
8. Y. Tatsukami, *Macromol. Chem.* **181**, 1107–1114 (1980).
9. G. Luft and H. Bitsch, *J. Macromol. Sci.* **11**, 1089–1112 (1977).
10. F. Billmeyer, *Text Book of Polymer Science*, 3rd ed., Wiley-Interscience, New York, 1984.
11. U.S. Pat. 4,689,369 (Aug. 25, 1987), I. Ishivo and co-workers (to Mitsubishi Petrochemical Co., Ltd.).
12. Unpublished data, Society of the Plastics Industry, Inc., Washington, D.C., Mar. 1994.
13. F. Otey and W. Done, *Proceedings of the Symposium on Degradition of Plastics*, Society of the Plastics Industry, Inc., Washington, D.C., June 10, 1987.
14. D. Lloyd, *Symposium on Degradition of Plastics*, Society of the Plastics Industry, Inc., Washington, D.C., June 19–21, 1987.

General References

H. Oosterwijk and H. Van Der Bend, *Akzo Chemie America Bulletin*, Initiations Seminar, 1980, New York, pp. 87–35.

LLOYD W. PEBSWORTH
Polyethylene Technology

HIGH DENSITY POLYETHYLENE

High density polyethylene (HDPE) is defined by ASTM D1248-84 as a product of ethylene polymerization with a density of 0.940 g/cm³ or higher. This range includes both homopolymers of ethylene and its copolymers with small amounts of α-olefins. The first commercial processes for HDPE manufacture were developed in the early 1950s and utilized a variety of transition-metal polymerization catalysts based on molybdenum (1), chromium (2,3), and titanium (4). Commercial production of HDPE was started in 1956 in the United States by Phillips Petroleum Co. and in Europe by Hoechst (5). HDPE is one of the largest volume commodity plastics produced in the world, with a worldwide capacity in 1994 of over 14×10^6 t/yr and a 32% share of the total polyethylene production.

The term HDPE embraces a large variety of products differing predominantly in molecular weight, molecular weight distribution (MWD), and crystallinity.

Molecular Weight. The range of molecular weights of commercially produced HDPE is wide, from several hundreds for polyethylene (PE) waxes to several millions for ultrahigh molecular weight PE resins (UHMWPE). A parameter that is widely accepted, easily measured, and which provides information on molecular weight, is the rheological parameter called the melt index. The melt index value, which is reciprocally related to mol wt, gives the weight of a molten resin flowing at 190°C for 10 min through a capillary die under a particular melt pressure (see ANALYTICAL METHODS). Different HDPE resins have melt indexes ranging from over 500 (low molecular weight polymers) to less than 0.001.

Molecular Weight Distribution. The width of the molecular weight distribution (MWD) of PE resins is usually represented by the ratio of the weight-average and the number-average molecular weights, $\overline{M}_w/\overline{M}_n$. In industry, the width of MWD is represented by the rheological parameter called the melt flow ratio (MFR) or shear response. MFR is the ratio of two melt indexes measured at two melt pressures that differ by a factor of 10. The range of MFR values for commercial HDPE resins is wide, from a low value of around 25 for injection molding resins with a narrow MWD, to over 150 for some HDPE film resins with a broad MWD.

Crystallinity and Density. Crystallinity and density of HDPE resins are derivative parameters; both depend primarily on the extent of short-chain branching in polymer chains and, to a lesser degree, on molecular weight. The density range for HDPE resins is between 0.960 and 0.941 g/cm³. In spite of the fact that UHMWPE is a completely nonbranched ethylene homopolymer, due to its very high molecular weight, it crystallizes poorly and has a density of 0.93 g/cm³.

Molecular Structure and Chemical Properties

HDPE is a linear polymer with the chemical composition of polymethylene, $(CH_2)_n$. Its name, polyethylene, reflects the principal method of production: ethylene polymerization by various processes. Depending on application, HDPE molecules either have no branches at all, as in certain injection molding and blow molding grades, or contain a small number of branches which are introduced by copolymerizing ethylene with α-olefins, eg, ethyl branches in the case of 1-butene and n-butyl branches in the case of 1-hexene. One of the chain ends in an HDPE molecule is a methyl group; the other chain end can be either a methyl group or a double bond (usually the vinyl group).

The number of branches in HDPE resins is low, at most 5 to 10 branches per 1000 carbon atoms in the chain. Even ethylene homopolymers produced with some transition-metal based catalysts are slightly branched; they contain 0.5–3 branches per 1000 carbon atoms. Most of these branches are short, methyl, ethyl, and n-butyl (6–8), and their presence is often related to traces of α-olefins in ethylene. The branching degree is one of the important structural features of HDPE. Along with molecular weight, it influences most physical and mechanical properties of HDPE resins.

As with nearly all other polymers, HDPE resin is a collection of polymer chains of different lengths, varying from short, with molecular weights of 500–1000, to very long, with molecular weights of over 10 million. Relative contents of chains with different lengths (ie, the shape and width of MWD) depend mostly on production technology and on the type of catalyst used for polymerization. The MWD width of HDPE resins can be tailored to specific applications.

Reactivity. HDPE is a saturated linear hydrocarbon and, for this reason, exhibits very low chemical reactivity. The most reactive parts of HDPE molecules are the double bonds at chain ends and tertiary CH bonds at branching points in polymer chains (9–11). Because its reactivity to most chemicals is reduced by high crystallinity and low permeability, HDPE does not react with organic acids or with most inorganic acids, including HCl and HF. Furthermore, its high stability toward HF makes it the most suitable container material for HF solutions. Concentrated solutions of H_2SO_4 (>70%) at elevated temperatures slowly react with HDPE with the formation of sulfo-derivatives. HDPE can be nitrated at room temperature with concentrated HNO_3 (~ 50%) and its mixtures with H_2SO_4. Under more severe conditions, at 100–150°C, these acids decompose the polymer and produce mixtures of organic acids. HDPE is also stable in alkaline solutions of any concentration as well as in solutions of all salts, including oxidizing agents such as $KMnO_4$ and $K_2Cr_2O_7$.

At room temperature, HDPE is not soluble in any known solvent, but at a temperature above 80–100°C, most HDPE resins dissolve in some aromatic, aliphatic, and halogenated hydrocarbons. The solvents most frequently used for HDPE dissolution are xylenes, tetralin, decalin 1,2,4-trimethylbenzene, o-dichlorobenzene, and 1,2,4-trichlorobenzene.

Thermal Degradation. HDPE is relatively stable under heat. Chemical reactions at high temperature in the absence of oxygen become noticeable only above 290–300°C. Thermocracking of HDPE is a free-radical C–C bond scission reaction. The reaction reduces the resin molecular weight, introduces vinyl double bonds in polymer chains, and produces low molecular weight hydrocarbons. Pyrolysis in an inert atmosphere becomes significant at 500°C; it produces mostly waxes, low molecular weight alkanes, alkenes, and dienes.

Thermooxidative Degradation. At elevated temperatures, oxygen attacks HDPE molecules in a series of radical reactions. These reactions reduce the molecular weight of HDPE and introduce oxygen-containing groups, such as hydroxyl and carboxyl groups, into polymer chains. Other oxidation products are low molecular weight compounds such as water, aldehydes, ketones, and alcohols. Oxidative degradation in HDPE is initiated by impurities, which are mainly catalyst residues containing transition metals, eg, titanium and chromium. Since thermooxidative degradation can occur during pelletization and processing of HDPE resins, molten resins should be protected from oxygen attack during these operations. The protection is provided by antioxidants (radical inhibitors) such as naphthylamines or phenylenediamines, hindered phenols, quinones, and alkyl phosphites, which are used in 0.1–1.0 wt % concentration (12).

Many commercial processes involving surface dyeing and printing (eg, on film and containers) employ thermooxidation as a pretreatment step. Although dyes adhere poorly to HDPE surfaces, their adhesion can be improved by thermooxidation of the surface layer, that is, by treatment with an open flame or in a strong electric field. Both procedures create polar groups on the HDPE surface, thus making it accessible for dyeing.

Photooxidative Degradation. Even though degradation of HDPE initiated by oxygen and light resembles thermooxidative degradation, it proceeds at a much lower temperature. Light with a wavelength below 400 nm initiates radical reactions that result in the reduction of molecular weight, formation of double bonds and organic peroxides in polymer chains, and evolution of low molecular weight compounds such as alcohols, aldehydes, and ketones. Although HDPE itself absorbs uv radiation poorly, the polar products of its degradation are able to react with oxygen much faster and accelerate additional radical reactions. Photooxidative degradation of HDPE causes aging, development of surface cracks, brittleness, change in color, and drastic deterioration of mechanical and dielectric properties. The reaction can be slowed down or prevented by utilizing light stabilizers that protect the resin and absorb uv radiation. The preferred light stabilizers are carbon black (qv) (2–4%) for HDPE and esters of salicylic acid or derivatives of benzotriazole or benzophenone (0.1–0.5%) for colorless articles. Photooxidative degradation is the principal process responsible for gradual disintegration of discarded PE litter.

The chemical industry manufactures a large number of antioxidants (qv) as well as uv stabilizers (qv) and their mixtures with other additives used to facilitate resin processing. These companies include American Cyanamid, BASF, Ciba–Geigy, Eastman Chemical, Elf Atochem, Enichem, General Electric, Hoechst–Celanese, Sandoz, and Uniroyal, among others. The combined market for these products in the United States exceeded $900 million in 1994 and will reach $1 billion in the year 2000.

Crystalline Structure and Physical Properties

HDPE is a semicrystalline plastic, whose crystallinity varies from 40 to 80%, depending on the degree of branching and molecular weight. Polymer chains in crystalline HDPE have a flat zigzag configuration (13,14). The principal crystalline form of HDPE is orthorhombic, with a density of 1.00 g/cm^3 and the cell parameters $a = 0.740, b = 0.493$, and $c = 0.2534$ nm. The polymer chains are aligned in the c-axis direction. The cross section of the HDPE unit cell is shown in Figure 1. HDPE can also exist in a second, unstable pseudomonoclinic modification with the cell parameters $a = 0.405, b = 0.485, c = 0.254$ nm, $\alpha = \beta = 90°$,

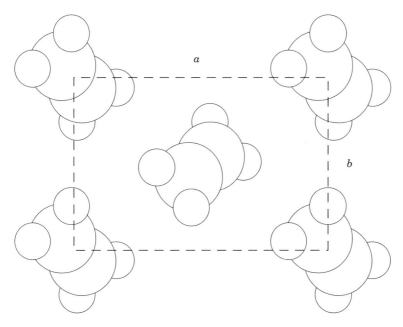

Fig. 1. Orthorhombic unit cell of HDPE crystal.

$\gamma = 105°$, and a density of 0.965 g/cm^3. This modification is formed during many types of low temperature processing, including manufacture, stretching, and calendering of film and sheet. Therefore, the pseudomonoclinic form is usually present in articles made of HDPE. This modification is stable only at a temperature of 50° or below; annealing at 80–100°C restores the orthorhombic cell.

Morphology. HDPE crystallizes from the melt under typical conditions as densely packed morphological structures known as spherulites. Spherulites are small spherical objects (usually from 1 to 10 μm) visible only under high magnification. They are composed of even smaller structural subunits: rod-like fibrils that spread in all directions from the spherulite centers, filling the spherulite volume. These fibrils, in turn, are made up of the smallest morphological structures distinguishable, small planar crystallites called lamellae (Fig. 2). These crystallites contain folded polymer chains that are perpendicular to the lamella plane and tightly bend every 5 to 15 nm. Lamellae are interconnected by a few polymer chains which pass from one lamella, through a small amorphous region, to another (Fig. 2). These connecting chains, or tie molecules, are ultimately responsible for mechanical integrity and strength of all semicrystalline polymer materials. Crystalline lamellae offer the spherulites rigidity and account for their high softening temperature, whereas the amorphous regions between lamellae provide flexibility and high impact strength to HDPE articles.

Phase Transitions. The extrapolated equilibrium melting point of orthorhombic HDPE crystals is 146–147°C, the equilibrium heat of fusion is 4.01 kJ/mol (0.96 kcal/mol) (15). Actual measurements of slowly crystallized samples give the highest melting point, T_m, at 133–138°C. The melting point is partially affected by molecular weight: a decrease in molecular weight from 1×10^6 to

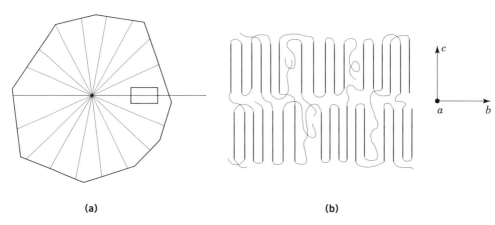

(a)   (b)

Fig. 2.   Structure of (**a**) HDPE spherulite and (**b**) the lamella.

4×10^4 results in a slight decrease in the melting point from 137 to 128°C. Melting points of commercial HDPE resins decrease nearly linearly with the branching degree, ie, one branch per 1000 carbon atoms in the chain reduces the melting point by approximately 1°C.

Determination of the glass-transition temperature, T_g, for HDPE is not straightforward due to its high crystallinity (16–18). The glass point is usually associated with one of the relaxation processes in HDPE, the γ-relaxation, which occurs at a temperature between -100 and -140°C. The brittle point of HDPE is also close to its γ-transition.

Crystallization Kinetics.   Crystallization of HDPE proceeds in two separate stages. During the first stage, HDPE crystallizes rapidly, accounting for 50–70% of the final crystallinity, CR_∞. The kinetics of this stage is described by the Avrami equation (19,20). In its simplified version, the equation gives the dependence of the crystallinity degree, CR, on time as $CR/CR_\infty = kt^n$, where all three parameters, CR_∞, k, and n, depend on molecular weight. For example, the n value decreases from around 4 to 2 with increasing molecular weight (19,20). The k value, however, goes through a maximum: an increase in molecular weight up to around 150,000 is accompanied by an increase in k, but when the increase in molecular weight goes beyond this point, k begins to decrease. Finally, the CR_∞ value decreases with molecular weight from 80% for HDPE with a very low molecular weight (waxes) to around 40% for UHMWPE. The dependence of the rapid crystallization rate on temperature is complex. When the temperature decreases from T_m to 70°C, k progressively increases because the formation rate of the smallest seed crystallites is also increasing. However, at lower temperatures, the rapid crystallization process is impeded by slow diffusion of polymer molecules from the overcooled melt to crystallization sites.

The stage of rapid crystallization ends when crystallinity reaches approximately 0.5–0.7, ie, when HDPE is 50–70% crystallized. After that, the crystallization rate falls drastically. The slower process that follows is known as secondary crystallization (20), which can be described by the empirical equation

$CR = A + B \cdot \log(t)$. Secondary crystallization can be accelerated if HDPE is heated.

Orientation. Most articles made of HDPE, including film, fiber, pipes, and injection-molded articles, exhibit some degree of molecular and crystal orientation (21). In some cases, orientation develops spontaneously; for example, during melt flow into a mold and its subsequent crystallization. When blown HDPE film and fiber are manufactured, orientation can be introduced deliberately by stretching.

Two patterns of orientation can develop in articles made of HDPE. When film or fiber is stretched uniaxially below the melting point, the c-axis of the crystals is always oriented in the stretching direction. The degree of orientation increases with the stretching ratio and can approach 100% (21,22). A similar type of orientation is developed during the crystallization of a strongly oriented HDPE melt, eg, in the injection molding process. A second orientation pattern is produced when HDPE melt is slightly stretched at the outset of crystallization, a condition typical for the production of blown film. The film formed under such conditions exhibits a significant degree of orientation of the crystal a-axis in the machine direction of the film (23,24).

Optical Properties. Owing to the high crystallinity of HDPE, most thick-walled articles made from HDPE resins are opaque. Significant surface roughness can also add to the opacity. Thin HDPE film, in contrast, is translucent, but its transparency is significantly lower than that of LDPE or LLDPE film. The ultraviolet transmission limit of HDPE is around 230 nm.

Electrical Properties. From a chemical standpoint, HDPE is a saturated aliphatic hydrocarbon and hence a good insulator. Its electrical characteristics are given in Table 1. Because polymer density and molecular weight affect electrical properties only slightly, HDPE is widely used for wire and cable insulation.

Permeability. HDPE is poorly permeable to water and inorganic gases and only slightly more so to organic compounds, whether liquids or gaseous. At 25°C and 101.3 kPa (1 atm), its permeability values, in $mol/(m \cdot s \cdot PPa)$, are for water, 6; for nitrogen, ~ 0.1; for oxygen, ~ 0.33; and for carbon dioxide, ~ 1.3. To convert these units to $(cc \cdot cm)/cm^2 \cdot s \cdot cm$ Hg, multiply by 3×10^{-10} (see BARRIER POLYMERS).

Mechanical Properties. The principal mechanical properties are listed in Table 1. The features of HDPE that have the strongest influence on its mechanical behavior are molecular weight, MWD, orientation, morphology, and the degree of branching, which determines resin crystallinity and density.

The stress–strain curve of HDPE recorded under unidirectional stretching is shown in Figure 3. At low deformations, around 0.5% of its initial length, a sample elongates elastically; if the stress is lifted, it rapidly returns to its original length. Further strain precipitates irreversible changes and develop, at some point called the yield point, a "neck," ie, a region with a clearly visible mechanical deformation. Polymer molecules in the neck region are highly oriented in the stretching direction. After the yield point is reached, the whole sample is gradually transformed into a highly oriented neck at an almost constant stress until all material becomes highly stretched. Finally, the oriented structure breaks if

Table 1. Physical, Thermal, Electrical, and Mechanical Properties of HDPE

Property	Highly linear	Low degree of branching[a]
	Physical	
density, g/cm^3	0.962–0.968	0.950–0.960
refractive index, n_D at 25°C	1.54	1.53
	Thermal	
melting point, °C	128–135	125–132
brittleness temperature, °C	−140 to −70	−140 to −70
heat resistance temperature, °C	~122	~120
specific heat capacity, kJ/(kg · K)[b]	1.67–1.88	1.88–2.09
thermal conductivity, W/(m · K)	0.46–0.52	0.42–0.44
temperature coefficient		
of linear expansion	$(1–1.5) \times 10^{-4}$	$(1–1.5) \times 10^{-4}$
of volume expansion	$(2–3) \times 10^{-4}$	$(2–3) \times 10^{-4}$
heat of combustion, kJ/g[b]	46	46
	Electrical	
dielectric constant at 1 MHz	2.3–2.4	2.2–2.4
dielectric loss angle, 1 kHz–1 MHz	$(2–4) \cdot 10^{-4}$	$(2–4) \cdot 10^{-4}$
volume resistivity, Ω·m	$10^{17}–10^{18}$	$10^{17}–10^{18}$
surface resistivity, Ω	10^{15}	10^{15}
dielectric strength, kV/mm	45–55	45–55
	Mechanical	
yield point, MPa[c]	28–40	25–35
tensile modulus, MPa[c]	900–1200	800–900
tensile strength, MPa[c]	25–45	20–40
notch impact strength, kJ/m^2 [d]	~120	~150
flexural strength, MPa[c]	25–40	20–40
shear strength, MPa[c]	20–38	20–36
elongation, %		
at yield point	5–8	10–12
at break point	50–900	50–1200
hardness		
Brinell, MPa[c]	60–70	50–60
Rockwell	R55, D60–D70	

[a] 2–3 CH$_3$ per 1000 carbons.
[b] To convert J to cal, divide by 4.184.
[c] To convert MPa to psi, multiply by 145.
[d] To convert kJ/m^2 to ft · lbf/in.^{2}, divide by 2.10.

additional stress is applied. This ultimate stress is defined as the tensile strength (Table 1).

Effect of Molecular Weight and MWD. Both mol wt and MWD significantly influence the mechanical properties of HDPE. All morphological elements of HDPE (lamellae, radial fibrils, and spherulites) are held together as a single entity by two types of forces: van der Waals forces between the neighboring stretches of polymer chains in crystals, and tie molecules that pass from one structural element to another. Only polymer molecules of a sufficient length can serve as tie molecules; without them, spherulites can easily disintegrate. For this

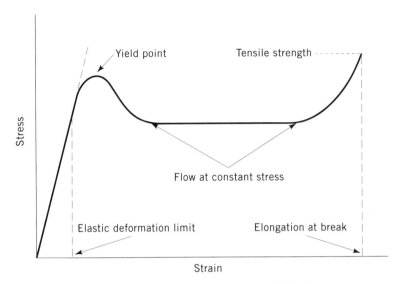

Fig. 3. Stress–strain curve of HDPE.

reason, low molecular weight HDPE is brittle and may break even at a low strain of ~10% without neck development. When molecular weight of HDPE reaches the range of 80,000–1,200,000, typical for commercial use, a sufficient amount of tie molecules between lamellae are formed. Stretching of articles made from such resins usually causes neck development. The yield point of such polymers is almost independent of molecular weight. Further increase in molecular weight is accompanied by a steady decrease of elongation at the break point, from 1200–1500% to 200–300%, and by a significant increase in the tensile strength, from 35–40 to ~60 MPa (8700 psi). UHMWPE (with mol wt of 1.5×10^6 to 4×10^6) does not develop a neck; instead, it elongates uniformly by 200–400%. The tensile strength of such polymers is very high, around 60–70 MPa.

Molecular weight also affects the impact strength of articles made of HDPE. Low molecular weight samples are brittle, but when molecular weight increases, the HDPE impact–stress resistance increases and becomes very high for HDPE resins with a molecular weight of $0.5–1 \times 10^6$. In addition, molecular weight can influence HDPE properties indirectly through its effect on crystallization kinetics and final crystallinity, CR_∞.

The effect of molecular weight on HDPE's physical and mechanical properties is especially considerable for UHMWPE resins with molecular weights of 3×10^6 or more. Crystallinity of such resins is typically around 40–45%. These resins have a unique combination of properties: the highest impact toughness, even at cryogenic temperatures; high resistance to cyclical fatigue; the highest abrasion resistance among all thermoplastic materials; excellent corrosion and environmental stress–crack resistance; a low coefficient of surface friction (non-stick surface); and resistance to radiation.

Effect of Branching. When the branching degree in HDPE increases, its crystallinity and the thickness of its crystalline lamellae decrease. This change brings about significant alterations in the mechanical properties of HDPE, two of the most strongly affected are tensile strength and tensile elongation. HDPE resins with increased branching degree are softer and more elastic. An increase in the branching degree from 2 to 10 per 1000 carbon atoms results in a decrease of the resin tensile strength from approximately 60 to 25 MPa (3625 psi), but an increase of its tensile elongation from 850 to over 1100%. Table 2 shows the combined effect of molecular weight and branching degree on mechanical properties of four HDPE and two LDPE resins. The differences in mechanical properties stem from the differences both in the nature of branches and the degree of branching, ie, linear HDPE contains almost no short-chain branches, slightly branched HDPE features variable, controlled short-chain branching, and LDPE has a combination of short- and long-chain branching.

Effect of Orientation. Orientation has a strong effect on the mechanical properties of HDPE. Articles with the same cross-section made of highly oriented HDPE are approximately 10 times stronger than those made from a nonoriented polymer. This effect is explained by the theory that the mechanical strength of a polymer is determined by the number of intercrystalline links: the tie chains anchored in adjacent crystallites and binding them together (Fig. 2). Because these links are few, intercrystalline boundaries are the weakest elements of the polymer structure. However, since the process of polymer stretching and the dismantling of its original morphological elements is accompanied by a significant increase in the number of intercrystalline chains, polymer strength thus increases greatly. Similarly, orientation significantly increases polymer rigidity; thus the elastic modulus of highly oriented HDPE filaments is increased about six times (25). For this reason, if technology allows for the utilization of highly oriented HDPE materials, as in fiber, ropes, and film bands, the introduction of oriented structures can offer significant advantages. In the case of HDPE film, for instance, the best balance of strength is achieved if most molecules are oriented in the film plane.

Table 2. Effects of Molecular Weight and Branching on Mechanical Properties of Polyethylenes

Property	Linear HDPE		Branched HDPE		LDPE	
	Parameters					
branching degree per 1000 carbons	~1	~1	~3	~3	20	20
melt index, g/10 min	5	1.1	6	0.9	7	1.0
density, g/cm^3	0.968	0.966	0.970	0.955	0.918	0.91
	Properties					
yield point, MPa[a]	33	31	29	29		
strain at yield point, %	9	9	20	20		
tensile strength, MPa[a]	20	30	22	30	8.5	10.5
maximum elongation, %	900	990	1000	1000	500	500
elastic modulus, MPa[a]	1550	1400	1000	900	500	400
impact strength, kJ/m^2[b]	9	50	2	30		

[a]To convert MPa to psi, multiply by 145.
[b]To convert kJ/m^2 to ft·lbf/in.^{2}, divide by 2.10.

Special processing techniques allow for a substantial increase in the number of intercrystalline links in HDPE resins with high molecular weights. Such technologies include low temperature extrusion of solid HDPE at around 100°C and 200–300 MPa (29,000–43,500 psi), and continuous casting of film or fiber from solution, a method known as gel spinning (21,26,27). The cast film or fiber are subsequently cold-stretched up to 40 times, which produces a high concentration of tie molecules. Fibers manufactured from UHMWPE by gel spinning are transparent and their polymer chains are almost perfectly oriented. These fibers exhibit an ultrahigh modulus, up to 100 GPa (14.5×10^6 psi), and a high tensile strength, 500–600 MPa (72,500–87,000 psi) (21,28–32).

Catalysts for HDPE Production

HDPE resins are produced in industry with several classes of catalysts, ie, catalysts based on chromium oxides (Phillips), catalysts utilizing organochromium compounds, catalysts based on titanium or vanadium compounds (Ziegler), and metallocene catalysts (33–35). A large number of additional catalysts have been developed by utilizing transition metals such as scandium, cobalt, nickel, niobium, molybdenum, tungsten, palladium, rhodium, ruthenium, lanthanides, and actinides (33–35); none of these, however, are commercially significant.

Chromium Oxide Catalysts. Phillips catalysts polymerize ethylene to high molecular weight HDPE with broad MWDs and nearly linear chains (34,36). These catalysts are based on Cr(6+) and supported on inert porous substrates. The best kind of support is silica; others include alumosilicates with a low alumina content, $AlPO_4$, and oxides of Ti, Zr, Ge, and Th. The principal requirements for supports are chemical stability at high temperatures and high porosity. All preferred supports have a specific surface area greater than 300 m^2/g and a pore volume greater than 1 cm^3/g. Another important feature is their low mechanical strength, which permits the breakup of catalyst particles during ethylene polymerization.

Catalyst synthesis consists of two main steps. In the first, a particulate support is impregnated with an aqueous solution of chromic acid or an organic solution of either chromium acetate or nitrate. In the second step, the solvent is removed from the catalyst by evaporation at elevated temperatures, after which the dry catalyst is activated by calcination at 500–850°C in a dry oxidizing environment. At these temperatures, remaining water and oxygen leave the catalyst surface while the precursors of active sites, silyl chromate groups with Cr(6+), are formed (37). The catalyst is purged with inert gas and degassed in vacuum. A typical chromium content in the finished catalysts ranges from 0.1 to 5 wt %. In some cases, activated catalysts are subsequently reduced with carbon monoxide or hydrogen at 300–500°C.

Numerous modifications of chromium-based catalysts have been made through the introduction of various additives, the most effective of which are titanium alkoxides (38,39). These additives apparently reduce surface silyl chromate moieties to chromium titanates, which are then oxidized to titanyl chromates. These catalysts offer a better control of the resin molecular weight (39).

Most chromium-based catalysts are activated in the beginning of a polymerization reaction through exposure to ethylene at high temperature. The activation step can be accelerated with carbon monoxide. Phillips catalysts operate at 85–110°C (38,40), and exhibit very high activity, from 3 to 10 kg HDPE per g of catalyst (300–1000 kg HDPE/g Cr). Molecular weights and MWDs of the resins are controlled primarily by two factors, the reaction temperature and the composition and preparation procedure of the catalyst (38,39). Phillips catalysts produce HDPE with a $\overline{M}_w/\overline{M}_n$ ratio of about 6–12 and MFR values of 90–120.

Organochromium Catalysts.   Several commercially important catalysts utilize organochromium compounds. Some of them are prepared by supporting bis(triphenylsilyl)chromate on silica or silica–alumina in a hydrocarbon slurry followed by a treatment with alkylaluminum compounds (41). Other catalysts are based on bis(cyclopentadienyl)chromium deposited on silica (42). The reactions between the hydroxyl groups in silica and the chromium compounds leave various chromium species chemically linked to the silica surface. The productivity of supported organochromium catalysts is also high, around 8–10 kg PE/g catalyst (800–1000 kg PE/g Cr).

Ziegler Catalysts.   These ethylene-polymerizing catalyst systems consist of two components. The first, called catalyst, contains a derivative of titanium or vanadium; the second component, called cocatalyst, is an organoaluminum compound (33,35,43), most often triethylaluminum. Although every Ziegler catalyst system is capable of ethylene polymerization, most commercially important catalysts are specially designed for this purpose, and nearly all of them are heterogeneous catalysts comprised of solid transition-metal compounds. One early example includes a catalyst system containing finely particulated mixed crystals, $TiCl_3 \cdot 0.33AlCl_3$, and triethylaluminum as cocatalyst (44). At present, supported catalysts are preferred for ethylene polymerization. Numerous inorganic and organic supports are employed, both inert (eg, $MgCl_2$, graphite, carbon black, PE, and polystyrene) and supports containing reactive chemical groups. These groups either are present on the surface of the support (eg, hydroxyl groups in silica, alumina, magnesium oxide, and titanium oxide) or are inherent to the support chemical structure (such as Mg(OH)Cl, polymers with reactive OH, NH, and COOH groups) (45–48). Many commercial processes require polymerization catalysts with a special morphology, preferably round particles of a uniform size. This kind of special morphology can be achieved by preparing support particles with a desired shape and size, usually through spray drying, or with special precipitation techniques.

The active constituents of all Ziegler catalysts are titanium or vanadium compounds; ie, $TiCl_4$, $TiCl_3$, $Ti(OR)_4$, VCl_4, $VOCl_3$, or $VO(OR)_3$. These compounds are either attached to the support surface or deposited into their pores. Examples of such catalysts include dispersions of $TiCl_3$ or VCl_3 on the surface of $MgCl_2$ or coprecipitates of Ti and V chlorides with $MgCl_2$ within pores of carrier particles. If the supports possess reactive groups such as hydroxyl groups, catalyst components can be attached to the support surface through oxygen bridges, similarly to chromium oxide catalysts. However, subsequent reaction with a cocatalyst, an organoaluminum compound, breaks the complexes and forms highly dispersed transition-metal halogenates (33,49).

Heterogeneous Ziegler catalysts operate at temperatures in the 70–110°C range and at pressures of 0.1–2 MPa (15–300 psi). The polymerization reactions are carried out in an inert liquid medium (eg, hexane, isobutane) or in the gas phase. Molecular weights of HDPE obtained with these catalysts without any additional control are usually quite high and make such resins unsuitable for processing by conventional techniques. To lower the molecular weight, hydrogen is widely employed as a chain-transfer agent.

Metallocene Catalysts. Metallocene catalyst systems (Kaminsky catalysts) contain two components (35). The first is a metallocene complex of zirconium, titanium, or hafnium, which usually contains two cyclopentadienyl rings. The second component is either an organoaluminum compound, methylaluminoxane, or a perfluorinated boron-aromatic compound. The most attractive feature of Kaminsky catalysts is that they are able to produce ethylene copolymers with high compositional uniformity. Although this feature is important for LLDPE resins, it is relatively insignificant in the case of copolymers with a low comonomer content, such as slightly branched HDPE. Another distinguishing feature of Kaminsky catalysts is high sensitivity of the polymer molecular weight to the presence of hydrogen in the reaction medium; this is a useful feature in the manufacture of HDPE resins for injection molding.

Bimetallic Catalysts. Several HDPE applications require resins with especially broad, sometimes bimodal, molecular weight distributions. Such material cannot be produced simply by mixing pellets of two resins with vastly different molecular weights. Instead, these resins must be obtained either through special technological processes employing two or more polymerization reactors in a series, or in a single reactor, by making use of bimetallic catalysts. The idea behind these catalysts is to combine, within each catalyst particle, two catalyst components that operate under identical conditions but produce polymers with radically different molecular weights. Several binary catalyst combinations of this type have been developed (50–55).

Polymerization Processes

All technologies employed for catalytic polymerization processes in general are widely used for the manufacture of HDPE. The two most often used technologies are slurry polymerization and gas-phase polymerization. Catalysts are usually fine-tuned for a particular process.

Slurry (Suspension) Polymerization. This polymerization technology is the oldest used for HDPE production and is widely employed because of process engineering refinement and flexibility. In a slurry process, catalyst and polymer particles are suspended in an inert solvent, ie, a light or a heavy hydrocarbon. Modern slurry processes utilize highly active catalysts and eliminate the need for de-ashing the polymer product. This process is capable of producing the full range of HDPE resins, from low molecular weight waxes to resins with high molecular weights, including UHMWPE. Slurry technology accounts for nearly 60% of all PE produced worldwide.

Slurry processes are classified into four types (56). Two of them employ loop reactors with a diluent circulating at high speed through a long circular pipe (they differ by the solvent type: low boiling vs high boiling). A third type is the continuous stirred tank reactor with a high boiling diluent; a fourth is the liquid pool process in which polymerization takes place in a boiling light diluent (propane or butane).

Some slurry processes use continuous stirred tank reactors and relatively heavy solvents (57); these are employed by such companies as Hoechst, Montedison, Mitsubishi, Dow, and Nissan. In the Hoechst process (Fig. 4), hexane is used as the diluent. Reactors usually operate at 80–90°C and a total pressure of 1–3 MPa (10–30 psi). The solvent, ethylene, catalyst components, and hydrogen are all continuously fed into the reactor. The residence time of catalyst particles in the reactor is two to three hours. The polymer slurry may be transferred into a smaller reactor for post-polymerization. In most cases, molecular weight of polymer is controlled by the addition of hydrogen to both reactors. After the slurry exits the second reactor, the total charge is separated by a centrifuge into a liquid stream and solid polymer. The solvent is then steam-stripped from wet polymer, purified, and returned to the main reactor; the wet polymer is dried and pelletized. Variations of this process are widely used throughout the world.

One of the most efficient implementations of the slurry process was developed by Phillips Petroleum Co. in 1961 (Fig. 5). Nearly one-third of all HDPE produced in the 1990s is by this process. The reactor consists of a folded loop with four long (~50 m) vertical runs of a pipe (0.5–1.0 m dia) connected by short horizontal lengths (around 5 m) (58–60). The entire length of the loop is jacketed for cooling. A slurry of HDPE and catalyst particles in a light solvent (isobutane or isopentane) circulates by a pump at a velocity of 5–12 m/s. This rapid circulation ensures a turbulent flow, removes the heat of polymerization, and prevents polymer deposition on the reactor walls.

The reactor operates at total pressures of 3–4.5 MPa (450–700 psig) and temperatures of 70–110°C. The concentration of polymer particles in the slurry is maintained at around 20–25 wt %. Ethylene and a comonomer (if used), solvent, and catalyst components are continuously charged to the reactor. The average residence time of polymer particles in the reactor ranges from 0.5 to 2.5 h. The polymer slurry is concentrated in settling legs to about 60 wt % and continuously removed from the reactor. Ethylene conversion is very high (95–98%) so no recovery is needed. The solvent is separated from hot polymer by flashing in a tank and then recycled. Nearly dry particulate polymer is dried to completion and pelletized.

A wide variety of chromium oxide and Ziegler catalysts was developed for this process (61,62). Chromium-based catalysts produce HDPE with a relatively broad MWD; other catalysts provide HDPE resins with low molecular weights (high melt indexes) and resins with a narrower MWD (63,64).

Gas-Phase Polymerization. Gas-phase polymerization, the most recent technology for HDPE manufacture, was introduced by Union Carbide in 1968 and has been widely licensed since then throughout the world (Unipol process). Several other firms have independently developed gas-phase technology; among them, British Petroleum, Naphthachimie, Amoco, and BASF. A gas-phase reactor contains a bed of dry PE particles. This bed is agitated by a high velocity gas stream, that also fluidizes it, or by mechanical devices such as baffles. Gas-phase processes account for over 20% of the world PE capacity. The technology is economical and flexible; it can accommodate a large variety of solid and supported catalysts capable of ethylene polymerization at relatively low pressure.

The Union Carbide gas-phase process is suitable for the production of both HDPE and LLDPE (57,65–68). A flow diagram is shown in Figure 6 (69).

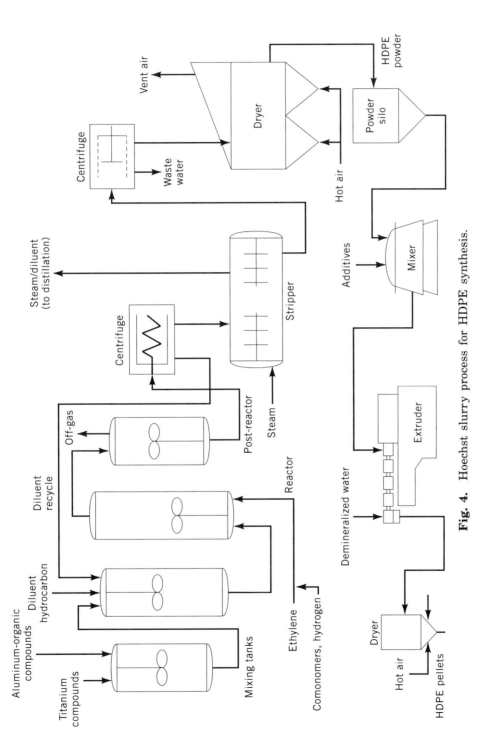

Fig. 4. Hoechst slurry process for HDPE synthesis.

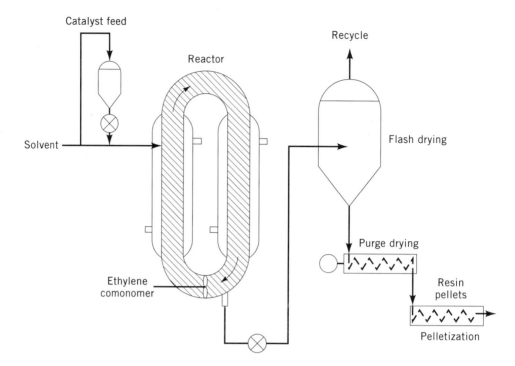

Fig. 5. Phillips slurry process for HDPE synthesis.

The reactor is a tall (up to 30 m) cylindrical tower with a length-to-diameter ratio of ~7. Uniform fluidization of the polymer bed is achieved by ethylene flowing through a perforated distribution plate at the reactor bottom. Rapid gas circulation ensures removal of polymerization heat. Both fluidization and heat removal require rapid recycling of ethylene through the reactor; as a result, the conversion of ethylene in a single pass is quite low, only about 2%. Nitrogen is often charged in the reactor for precise control of polymerization parameters. Unreacted ethylene enters the disengagement zone (a large-diameter extension at the top of the reactor), separates from the entrained polymer particles, and is then cooled, compressed, and recycled. The reactor is usually maintained below 110°C to prevent formation of sticky polymers detrimental to fluidization.

A solid catalyst is continuously fed to the reactor above the distribution plate, either as dry powder or as concentrated slurry in high boiling mineral oil. If a cocatalyst is used to activate the catalyst, its solution is fed into the ethylene stream below the plate. A comonomer (α-olefin) and, in some cases, a small amount of light solvent, are charged into the reactor at the same point. The two most frequently used olefins for controlling resin density are 1-butene and 1-hexene. Although catalyst particles increase in size as polymerization proceeds, they retain their spherical shape. Polymer particles with an average size of ~500 μm are removed from the bed through a system of valves; subsequently, they are transferred into a series of bins for removing unreacted ethylene and comonomer before they are pelletized.

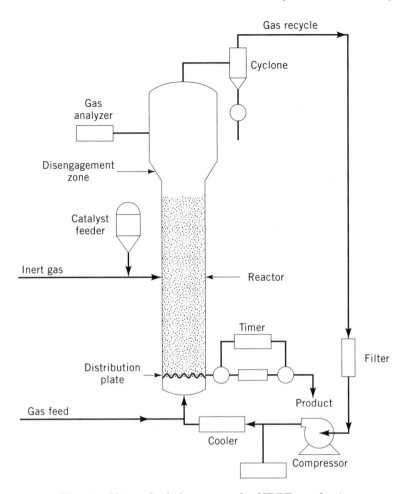

Fig. 6. Union Carbide process for HDPE synthesis.

The fluidized-bed reactors are highly versatile and can accommodate many types of polymerization catalysts. The list of catalysts includes chromium oxide-based catalysts, organochromium catalysts, most types of heterogeneous Ziegler catalysts (both supported and unsupported), and metallocene catalysts (39,50,65,69,70). Naphthachimie has developed a similar gas-phase process, which employs a proprietary, highly active Ziegler catalyst based on magnesium, titanium, and a trioctylaluminum cocatalyst (71–73).

Two modifications of the fluidized-bed reactor technology have been developed. In the first, two gas-phase fluidized-bed reactors connected to one another have been used by Mobil Chemical Co. and Union Carbide to manufacture HDPE resins with broad MWD (74,75). In the second development, a combination of two different reactor types, a small slurry loop reactor followed by one or two gas-phase fluidized-bed reactors (Spherilene process), was used by Montedision to accommodate a Ziegler catalyst with a special particle morphology (76,77). This catalyst is able to produce PE resins in the form of dense spheres with a diameter of up to 4–5 mm; such resins are ready for shipping without pelletization.

Another type of gas-phase technology for ethylene polymerization has been developed independently by Amoco and BASF. The Amoco process utilizes a horizontally positioned cylindrical compartmentalized stirred-bed reactor in Figure 7 (78,79). Within each compartment, the bed of polymer particles is agitated by mixing blades turning at low speed. The polymer bed is maintained in a subfluidized state by introducing gases (recycle stream, make-up ethylene, hydrogen, and α-olefin comonomer) through the inlets at the bottom. Catalyst components are sprayed into the polymer bed from other inlets positioned at the top of the reactor. Heat of polymerization is removed from the polymer bed by vaporizing light hydrocarbons, butane, or isopentane, which are also sprayed into the reactor from the top inlet ports. As the polymer bed grows in volume, it overflows from one compartment to another before it is finally discharged from the reactor through a series of locks. Gaseous effluents from the reactor are separated into a recycle-gas stream (ethylene, hydrogen, and comonomer) and a recycle-solvent stream before they are returned to the reactor. This process provides a much higher ethylene conversion per pass, although it also has a more complicated hydrocarbon recovery and recycling procedure.

The Amoco reactor operates at 70–80°C and ~2 MPa (300 psi) reactor pressure. The existence of several partially isolated compartments allows a

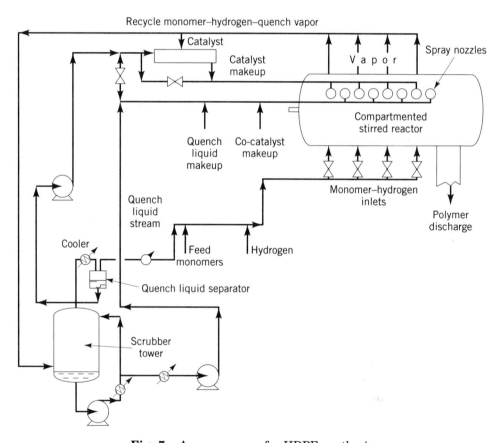

Fig. 7. Amoco process for HDPE synthesis.

semi-independent control of temperature as well as comonomer and hydrogen concentrations within each section, which in turn offers a substantial control of the molecular weight and MWD of resins. Amoco technology also accommodates a large variety of polymerization catalysts, including Phillips and Ziegler catalysts.

BASF developed a vertical stirred bed gas-phase reactor with a large baffle stirrer, a process that utilizes supported Ziegler catalysts. In this set-up, the monomer, the catalyst components, and hydrogen are continuously fed into the reactor containing a bed of vigorously agitated PE particles. A stream of polymer particles is then removed from the reactor, separated from the gas phase in a cyclone, while the gas stream is circulated through a cooling loop. Reaction parameters in this process are similar to those employed in the Amoco horizontal gas-phase reactor.

Solution Polymerization. Two solution polymerization technologies are practiced. Processes of the first type utilize heavy solvents; those of the second use molten PE as the polymerization medium (57). Polyethylene becomes soluble in saturated C_6–C_9 hydrocarbons above 120–130°C. Because the viscosity of HDPE solutions rapidly increase with molecular weight, solution polymerization is employed primarily for the production of low mol wt resins. Solution process plants were first constructed for the low pressure manufacture of PE resins in the late 1950s; they were later extensively modified to make their operation economically competitive.

Several commercial processes utilize the concept of solution polymerization. The Du Pont process uses cyclohexane as a solvent and is shown in Figure 8 (80,81). The Dow solution process follows a similar scheme, using a single continuous stirred-tank reactor (80,82). Ethylene, a solvent (a mixture of saturated hydrocarbons), catalyst components, and hydrogen are continuously fed into the reactor. The polymerization reaction proceeds at a total pressure of around 5–10 MPa (725–1450 psi) and a temperature of 150–200°C for 5–10 minutes. The maximum polymer concentration in solution is based on the resin molecular weight. When a low molecular weight polymer is produced, the polymer concentration can reach 30–35 wt %. Hot polymer solution is discharged from the reactor into a flash tank where most of the solvent is vaporized for recycling. Finally, the remaining solvent (about 3–5%) is removed during pelletization in a devolatilizing extruder, combined with the makeup solvent, and returned to the reactor.

An independent development of a high pressure polymerization technology has led to the use of molten polymer as a medium for catalytic ethylene polymerization. Some reactors previously used for free-radical ethylene polymerization at a high pressure (see OLEFIN POLYMERS, LOW DENSITY POLYETHYLENE) have been converted to accommodate catalytic polymerization, both stirred-tank and tubular autoclaves operating at 30–200 MPa (4,500–30,000 psig) and 170–350°C (57,83,84). CdF Chimie uses a three-zone high pressure autoclave at zone temperatures of 215, 250, and 260°C (85). Residence times in all these reactors are short, typically less than one minute.

Most catalysts for solution processes are either completely soluble or pseudo-homogeneous; all their catalyst components are introduced into the reactor as liquids but produce solid catalysts when combined. The early Du Pont process employed a three-component catalyst consisting of titanium tetrachlo-

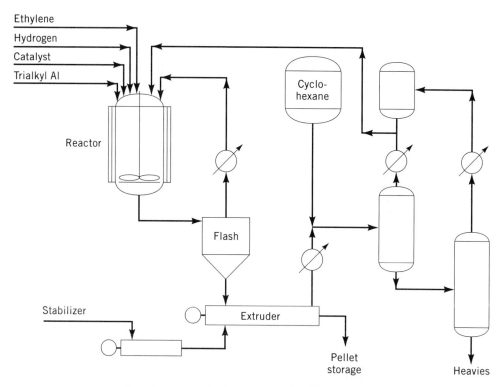

Fig. 8. Du Pont solution process for HDPE synthesis.

ride, vanadium oxytrichloride, and triisobutylaluminum (80,81), whereas Dow used a mixture of titanium tetrachloride and triisobutylaluminum modified with ammonia (86,87). Because processes are intrinsically suitable for the use of soluble catalysts, they were the first to accommodate highly active metallocene catalysts. Other suitable catalyst systems include heterogeneous catalysts (such as chromium-based catalysts) as well as supported and unsupported Ziegler catalysts (88–90).

Processes for HDPE with Broad MWD. Synthesis of HDPE with a relatively high molecular weight and a very broad MWD (broader than that of HDPE prepared with chromium oxide catalysts) can be achieved by two separate approaches. The first is to use mixed catalysts containing two types of active centers with widely different properties (50–55); the second is to employ two or more polymerization reactors in a series. In the second approach, polymerization conditions in each reactor are set drastically differently in order to produce, within each polymer particle, an essential mixture of macromolecules with vastly different molecular weights. Special plants, both slurry and gas-phase, can produce such resins (74,91–94).

PROCESSING

Most high density polyethylene processing technologies require the melting of HDPE. Typical HDPE melt viscosities are between 1,000 and 100,000 Pa·s (10,000 – 10^6 P); the melt viscosity of HDPE strongly depends on temperature and on the resin molecular weight. Some resins can have a visosity 250 times greater than that of others (95). Most HDPE in the 150–250°C range (a typical processing range) are non-Newtonian liquids; their effective viscosity is significantly reduced (up to 6–7 times) when the melt flow speed is increased (95). The effect of temperature on the HDPE melt viscosity is described by an exponential dependence similar to the Arrhenius equation with an activation energy of 25–29 kJ/mol (6–7 kcal/mol).

Because of its low melting point and high chemical stability, HDPE is easily processed by most conventional techniques (94–98); specialized techniques have also been described (99–107) (see PLASTICS, PROCESSING).

Injection Molding.   Molten HDPE is injected at 200–260°C and 70–140 MPa (10,000–20,000 psi) into a steel mold (99–101). An injection-molding machine consists of two units: an injection unit (an extruder) and a clamp unit. The clamp unit has two principal parts, a fixed and a movable platen, held together and forming a cavity into which the HDPE melt is injected. HDPE pellets are fed by a hopper into a single-screw extruder and, when they are melted, pushed through a nozzle into the mold. The mold is rapidly filled with the melt, then held under pressure produced by the extruder screw for a short period of time so it can be completely filled with resin. Upon completion of the injection cycle and cooling, the mold is opened and the article is removed. Duration of the molding cycle depends on the resin viscosity and the rate of polymer crystallization. Because HDPE crystallizes rapidly, molding cycles are short, typically from 10 to 30 seconds. Molds can accommodate up to 50 or more articles that are formed in a single shot.

Blow Molding.   This is the largest volume processing technique for HDPE and is used for the manufacturing of large quantities of articles of a relatively simple configuration, such as bottles and containers (102,103). A thick-walled tube of HDPE melt, called a parison, is extruded downward through a circular die and embraced by a doubly split metal mold of a generally cylindrical shape. As the bottom part of the mold pinches the parison, the bag of a molten resin is blown by air pressure to assume the internal surface of the mold; an article is thus produced. In the first step of the process, the molten resin tube remains for a short time in a free-hanging state. To prevent parison from sagging under its own weight, resins with high molecular weight and melt viscosity are used.

Rotational Molding.   One drawback of the blow-molding process for the manufacture of large containers is that it has a limited application due to the instability of long and heavy parisons. To circumvent this limitation, most large containers are produced with a specialized technique called rotational molding. A machine for rotational molding contains from one to several dozens large split metal molds attached to a complex mechanism which rotates the molds in two perpendicular planes. A load of HDPE powder is introduced into each mold, and the whole mold assembly is moved into an oven. The molds rotate inside the oven at speeds ranging from 10 to 40 rpm. The polymer powder melts and is uniformly distributed on the internal surface of each mold by adhesion to the metal wall. After the containers are formed, the mold assembly is removed from the oven and cooled; molds are opened at this point to remove the articles.

Extrusion.   In general, extrusion is the process of forcing a polymer melt through a die (104,105). Typical extrusion applications include initial resin pelletization after manufacture and production of film, sheet, pipe, tubing, and insulated wire. The HDPE extrusion temperature is around 150°C, the pressure 40–50 MPa (5800–7250 psi). An extrusion production line usually consists of an extruder (mono- or twin-screw) with a die at the end, a cooling and shaping device, a pulling device (a roller), and a cutter.

Pipes and Tubing.   Pipes (>1 cm dia) and tubing (<1 cm dia) are produced by extruding the HDPE melt through a circular die (104,105). Immediately after leaving the die, the molten tube enters a vacuum calibrator, where it is forced against sizing rings and cooled. The extrusion rate and the drawdown ratio (the ratio of cross-sectional areas of the die opening and the tube wall) are adjusted with the help of mechanical pulling devices. In pipe manufacture, the die openings are large, the resistance to polymers flowing through the die is therefore low, and drawdown ratios do not exceed 1.1. Pipes with a large diameter can be produced by this method.

Blown Film.   A continuous roll of HDPE film is produced by extrusion of the polymer melt through a large diameter circular die with a narrow gap of about 1.0 mm; the diameter of the die ring can range from 40 to 100 cm (106,107). The extruded thin-walled tube of molten polymer rises vertically and is pressurized with air. Nearly all HDPE film is manufactured on high stalk film lines. At first, when the cylindrical tube of molten polymer exits the circular die, it has a sufficient melt strength to resist expansion and so retains its original diameter, which is close to the die diameter, up to a height of 0.6–1.0 m. Subsequently, the cooled melt yields to the internal air pressure; it rapidly expands to a blow-up ratio of from 3:1 to 4:1 and immediately solidifies into a film. Blow-up ratio (BUR) is the ratio of the final diameter of the film tube to the die diameter. This technique depends on the presence of a sufficient amount of high molecular weight fraction in the resin, which provides the needed melt strength during the first stage of film formation. The film, 0.007–0.125-mm thick, is air-cooled and rolled.

Processing of UHMWPE.   Because of high molecular weight and extremely high melt viscosity, UHMWPE cannot be readily processed by any technique involving melt extrusion or thermoforming. Instead, these resins are processed either by compression molding into sheet, block, and precision parts; by ram extrusion into board, rods, pipe, and profiles; or by forging into parts of complex configuration. After the sheet is manufactured, a modified metal-stamping process is employed to cut different parts. The sheet is heated to ~20°C above its melting point and placed in the stamping die. Die pressure is maintained for a short time to allow the article to cool before removal.

UHMWPE is also used for the manufacture of ultrahigh strength fiber by gel spinning (108,109). UHMWPE resin, usually with a mol wt of $2–3 \times 10^6$, is dissolved at 160–200°C in high boiling solvent such as decalin. To produce fiber, the 5–10 wt % solution is pushed through a spinnerette, a plate with round holes, 0.5–1.0 mm in diameter. The gel fiber is cooled in an air flow, then stretched from 10 to 40 times, dried in a drawing oven, and wound. The process produces strong fibers with tensile strength of 3–4 GPa or 3–4 N/tex (35–45 g/den) and tensile modulus of nearly 150 GPa.

Recycling of HDPE.   Polyolefins, including HDPE, are the second most widely recycled thermoplastic materials after PET (110). A significant fraction of articles made from HDPE (mostly bottles, containers, and film) are collected from consumers, sorted, cleaned, and reprocessed (110–113). Processing of post-consumer HDPE includes the same operations as those used for virgin resins: blow molding, injection molding, and extrusion.

Economic Aspects

Large quantities of pelletized HDPE are shipped by rail in hopper cars with a capacity of 80–100 t. Smaller orders are distributed in corrugated cardboard boxes (1.0 × 1.2 × 0.9 m) with a capacity of 450–500 kg of resin.

HDPE is one of the largest commodity plastics manufactured worldwide. Dynamics of HDPE production is represented by the following data indicating both the existing and projected demand (t/yr) (114):

Year	1983	1990	1995	2005
Demand, 10^3 t	6,400	11,440	14,000	20,000–21,000

The current and projected HDPE capacities are shown in Table 3, and producers of resins in Table 4. In most cases, an accurate estimation of the total HDPE volume is complicated by the fact that a large number of plants also use the same reactors for manufacture of HDPE or LLDPE. UHMWPE is produced in the United States (Himont and American Hoechst), in Japan (Asahi), and in Germany (Hoechst); worldwide capacity is approximately 45,000 tons. The use of post-consumer (recycled) HDPE is gradually increasing in volume. The growth of recycling programs is driven principally by economics (110,114); it has increased from a mere 60,000 tons in 1989 to 350,000 tons in 1994 and is expected to increase to 1.4 million t in the year 2000 (115).

In the case of most commodity-grade HDPE resins, the cost of raw materials, primarily ethylene, but also comonomer, hydrogen, solvents, catalyst components, and additives, amounts to about 75–80% of total production costs, and can reach nearly 90% for commodity-grade resins produced in the gas-phase process. Utility costs, including power, cooling water, steam, and fuel, amount to approximately 5%; operating costs (primarily capital-related maintenance) and overhead expenses each account for an additional 5%. In general, capital investment costs are the single largest expense. The most capital-intensive processes are the solution processes, which entail a higher operating pressure and incur expenses associated with solvent recovery, purification, and recycling. The lowest capital

Table 3. World Production of HDPE, 10^3 t/yr

Region	1986	1989	1992
North America	3,250	4,090	4,540
Latin America	390	530	650
western Europe	2,080	2,240	2,575
eastern Europe	660	710	900
Far East and Australia	1,775	2,230	2,540
Middle East and Africa	420	470	530
Total	*8,575*	*10,270*	*11,735*

Table 4. Worldwide Production of HDPE in 1994 by Country and Company

Country	Company	Capacity, 10^3 t/yr
The Americas		
United States		4800
	Chevron	390
	Dow Chemical	135
	Exxon Chemical	120
	Fina Oil	160
	Formosa Plastics	200
	Himont	21
	Hoechst	24
	Mobil Chemical	170
	Occidental Petroleum	680
	Paxon Polymer	545
	Phillips Petroleum	820
	Quantum Chemical	765
	Solvay Polymers	650
	Union Carbide	120
Canada		270
Brazil		370
Mexico		200
Venezuela		100
Europe and Africa		
Belgium		650
France		200
Germany		820
Italy		410
Netherlands		225
Norway		130
Portugal		130
Russia		620
Spain		315
Sweden		225
United Kingdom		250
Bulgaria, Czech Republic, Poland, Romania, Yugoslavia		500
South Africa		160
Asia and the Pacifics		
Australia		125
China		480
India		80
Iran		60
Japan		1250
Kazakhstan		70
Saudi Arabia		1050
Singapore		160
South Korea		1115
Taiwan		310
Thailand		260
Turkey		60

investment is required by gas-phase polymerization. On account of a high relative contribution of raw material and energy costs to the overall HDPE production, the net relative production cost for HDPE in different processes increases in the following general range as gas phase, 1; loop slurry, 1.025; solution, 1.047; high pressure solution, 1.12; stirred-tank slurry, 1.13.

Different grades of HDPE, intended for different applications, command different prices. The 1994 list prices for large-volume grades of HDPE in the United States were \$0.70–0.80/kg and in Europe \$1.00–1.30/kg; in mid-1995 they increased by ca 40%. The price of UHMWPE is about \$2.0–2.6/kg.

Specifications, Standards, and Quality Control

According to ASTM D1248, HDPE materials are divided into various classifications based on properties. Two of the most easily measured characteristics are density and melt index; the former determines the type of HDPE, the latter its category. Other characteristics, predominantly color, are specified by class. HDPE with a nominal density of 0.941–0.959 g/cm^3 belongs to Type III, those with a nominal density higher than 0.960 are classified as Type IV. Although formally belonging to Type II PE, some high molecular weight resins with density lower than 0.941 g/cm^3 and UHMWPE with a density of ~0.930 g/cm^3 are also represented as HDPE due to their low branching degree.

Five categories of PE are specified with respect to their melt index measured according to ASTM D1238: for category 1, the melt index is >25 g/10 min; for 2, 10–25; for 3, 1–10; for 4, 0.4–1.0; and for 5, <0.4 g/10 min.

The three classes of PE, designated as A, B, and C, specify the color, amount, and type of antioxidants and other additives. Class A refers to naturally colored PE, Class B includes white or black polymer, and Class C covers weather-resistant black polymer containing no less than 2% carbon black. Typical characteristics of resins used for film manufacture, injection molding, and blow molding are given in Table 5.

Table 5. Typical Characteristics of HDPE Commercial Resins

Property	Injection molded	Blow molded	Film	UHMWPE
melt index, g/10 min	5–50	0.30–0.40	0.06–0.08	<0.01
melt flow ratio	22–25	100–110	100–180	
density, g/cm^3	0.950–0.960	0.950–0.960	0.948–0.955	0.930
color	natural	natural	natural	natural
Vicat softening point, °C	126	133		135
brittleness temperature, °C		< −118		< −200
tensile strength, MPa[a]	27–28	29–30	23	40–42
elongation, %	10–20	250	400	300
flexural stiffness, MPa[a]	1015	1005–1010		
hardness, Shore D	67	68		67
environmental stress-cracking resistance	poor	good		excellent

[a]To convert MPa to psi, multiply by 145.

Analytical and Test Methods

Structure.   Commonly evaluated structural parameters of HDPE include the type of branching, degree of branching, type and degree of unsaturation, degree of oxidation, and crystallinity. The type and degree of short-chain branching are measured by nmr and ir methods. ASTM D2238-92 employs ir techniques for estimating the total methyl group content. Numerous modifications of these methods are described in the literature (116–118). The ir spectroscopic method also allows identification and quantitative measurement of minor structural chain features, primarily unsaturation (ASTM D3124-72) and carbonyl groups. Crystallinity of HDPE is measured by ir, x-ray diffraction, dsc, and density measurements (119). The x-ray method, which is most popular, involves the recording of an x-ray diffractogram of a thick polymer film and the measurement of three areas (S): those under two crystalline reflections ($\langle 110 \rangle$ and $\langle 200 \rangle$) and under the amorphous halo. Crystallinity is calculated as $[S\langle 110 \rangle + 1.43S\langle 200 \rangle]/[S\langle 110 \rangle + 1.43S\langle 200 \rangle + 0.69S\langle halo \rangle] \times 100(\%)$ according to ASTM D1324.

Calorimetry (dsc) measurements give the melting point and the heat of fusion, ΔH_f. Crystallinity of HDPE is calculated from ΔH_f as $100 \times (\Delta H_f)/(\Delta H_f{}^0)$ where $\Delta H_f{}^0$ for a 100% crystalline material is 280–300 J/g (66–72 cal/g).

The direction of orientation in HDPE film and the orientation degree can be determined by either the x-ray method, ir polarization spectroscopy, acoustical methods, or birefringence (21,22,120,121).

Density.   HDPE density is measured by the flotation method in density gradient columns according to ASTM D1505-85. Suitable liquid systems include 2-propanol–water (0.79–1.00 g/cm^3) and 2-propanol–ethylene glycol (0.79–1.11 g/cm^3). The method is simple but has a significant drawback: long crystal annealing time, over 50 h, is required for a precise measurement. The correlation between density, d, and crystallinity, CR, is given by $1/d = CR/d_{cr} + (1 - CR/d_{am}$, where the density of the crystalline phase, d_{cr}, is 1.00 g/cm^3 and the density of the amorphous phase d_{am}, is 0.852–0.862 g/cm^3. Ultrasonic and solid-state nmr methods have also been developed, offering a rapid estimation of HDPE crystallinity and density both in pelletized and granular forms.

Molecular Weight and Rheological Properties.   Molecular weight and MWD are determined by gel permeation chromatography (gpc) or by a viscosimetric method (ASTM D1601-86 and D2857-87). In gpc, polymer molecules in solution in o-dichlorobenzene or 1,2,4-trichlorobenzene at 125–145°C are separated according to their molecular weight in a series of columns containing cross-linked polystyrene gel. This technique provides the simultaneous determination of the number-average and weight-average molecular weights as well as detailed information on the MWD of HDPE. Standard ASTM procedures for the measurement of rheological, dielectric, and mechanical properties of HDPE are given in Table 6. The parameter that is generally accepted in industry as a measure of PE viscosity and an indicator of its molecular weight is the melt index, which is measured by using a melt index tester or an extrusion plastometer. The melt index is defined as the amount of polymer melt discharged at 190°C for 10 minutes through a capillary 2.095 mm in diameter and 8 mm in length under a load of 2.16 kg. The melt index is reciprocally related to the melt viscos-

Table 6. ASTM Methods for HDPE Testing

Measurement	ASTM Method
conditioning of plastics samples	D618
density	D1505, D792
sample preparation	D2839
by ultrasonic method	D4883
melt index (flow rate of melt)	D1238
preparation of compression-molded samples	D1928
brittleness temperature by impact	D746
impact resistance	D1709, D4272A
flexural properties	D790
tensile properties	D638
Vicat softening point	D1525B
hardness	D785, D2240
environmental stress-cracking	D1693
thermal stress-cracking	D2951
dielectric properties	D150, D257, D1531
spectroscopic measurements	
branching	D2238
unsaturation	D3124
copper-catalyzed oxidation	D3895

ity. The molecular weight distribution of HDPE resins is usually characterized in industry by the ratios of melt indexes measured in the same apparatus using different loads, ie, 2.16, 10.16, and 21.6 kg. The commonly used ratios are I_{21}/I_2 (the melt flow ratio, MFR) and I_{10}/I_2. Both of these ratios closely correlate with the M_w/M_n value.

Mechanical Properties of HDPE Film. Two specialized tests exist for the performance evaluation of HDPE film. The first parameter is the dart impact strength (ASTM D4272-90, Method A), which determines the film overall strength. To measure the dart impact strength, a dart with a round tip (38.1 mm dia) and thin stem is dropped from the height of 66 cm on a piece of stretched film 12.7 cm in diameter. The weight of the dart can be varied by putting metal loads on its stem; the average weight of the dart sufficient to penetrate the film in half of the tests is its dart impact strength. This value varies from around 100 g for film of poor quality to over 500 g for high grade film.

The second parameter, tear strength, describes the film resistance to tear propagation. It is measured with a special apparatus, the Elmendorf Tear Tester (ASTM D1922), and defined as the weight of a loaded pendulum capable of tearing a notched piece of film. Two values are usually measured for each film sample. One determines tear propagation in the machine direction of the film, the other in the transverse direction.

Chemical Properties. HDPE is chemically stable, therefore very few analyses and tests related to its chemical properties are carried out routinely. An important parameter of HDPE employed for manufacture of household containers is its environmental stress-cracking resistance (ESCR). When a strained HDPE specimen is exposed to an aqueous solution of a detergent or an organic liquid, the polymer eventually develops numerous cracks which penetrate through its thickness and, in time, fails. Early crack failure is unacceptable in HDPE bottles with soaps and commercial detergents, or in pipes and containers carrying organic liquids. According to ASTM D1693-70, environmental stress-cracking is examined by exposing bent notched strips of molded HDPE to various environments (eg, soap solutions in water, organic solvents) at 50 or 100°C for different periods of time, and followed by visual examination of the samples. Another ESCR test, specifically designed for bottles, is based on ASTM D2561-70. Resistance to thermal stress-cracking is examined by exposing film strips wrapped on a metal mandrel to hot air (100°C) for 48, 96, and 168 hours (ASTM D2951-71). The induction period in thermooxidative decomposition of HDPE is carried out according to ASTM D3895-92. The test is conducted in pure oxygen at 200°C. The length of the induction period depends on the presence of antioxidants or copper deactivators in HDPE.

Health and Safety Factors

HDPE by itself is a safe plastic material on account of its chemical inertness and lack of toxicity. Consequently, film and containers made from HDPE are used on a large scale in food and drug packaging. Moreover, HDPE has been used in prosthetic devices including hip and knee joint replacements (122). All these applications underscore polymer safety. If articles made of HDPE contain fillers, processing aids, and colorants, their toxic effects must be estimated separately.

Nevertheless, HDPE can present health hazards when it burns. Heavy smoke, fumes, or potentially toxic decomposition products can result from incomplete combustion. Irritation of the skin, eyes, and mucous membranes of the nose and throat during thermocutting of HDPE has been reported and is attributed to the presence of acrolein and formaldehyde (123). Large-scale fire testing has shown that the products formed from HDPE present no greater hazard than those from cellulosic materials, wood, felt, or rubber (124–127). The highest volume of volatile fatty acids, formaldehyde, and acrolein was measured from smoldering combustion at 300–400°C (128); consequently, toxicity of HDPE pyrolysis products depends on temperature and the heating rate (124–131).

A significant part of HDPE is collected from consumers for recycling; uncollected HDPE can be disposed of by landfill or incineration. In landfill, HDPE is completely inert, degrades very slowly, does not produce gas, and does not leach any pollutants into groundwater. When incinerated in commercial or municipal facilities, HDPE produces a large amount of heat (the same as heating fuel) and therefore should constitute less than 10% of the total trash.

Uses

Table 7 gives a general distribution of HDPE applications in the three largest markets: the United States, Europe, and Japan.

Blow-molded products represent the biggest use of HDPE resins, at around 40%. Packaging applications account for by far the greatest share of this market. These include such products as bottles (especially for milk, juice, and soap), housewares, toys, pails, drums, and tanks. Expected annual average growth

Table 7. General Distribution of HDPE Applications, %

Application	United States	Europe	Japan
blow molding	41.3	41.1	16.4
injection molding	28.9	26.2	16.4
film	11.8	14.0	36.0
pipe	8.3	10.5	5.2
fiber and tape	1.7	2.9	12.7
other	8.0	5.3	13.3

rates for these applications are about 5–6%. Injection-molding products are the second largest application, with approximately 20% of the HDPE market. These products include housewares, toys, food containers, pails, crates, and cases.

Film is the third most important application for HDPE resins, accounting for nearly 15% of the total HDPE market. Its share of the market is increasing rapidly as it gradually replaces paper and glass and competes with LLDPE film for many uses. Bags made from HDPE are commonly used in food stores, supermarkets, fast-food outlets, department stores, and for garbage. Two different types of HDPE are used for film manufacture. The resins of the first type have a broad unimodal MWD and are produced with organochromium catalysts. Those of the second type have a bimodal MWD and are manufactured in special processes in two reactors which are combined in series and operate under different conditions. The properties of these two types of film resins are compared in Table 8 (see also FILM AND SHEETING MATERIALS).

HDPE pipes of various diameters represent another important use, amounting to about 10% of total HDPE products and exhibiting rapid growth. These pipes are used for transporting water, sewer wastes, and gas, on account of their chemical inertness, high corrosion stability, and good environmental stress-cracking resistance; they are also widely used in the chemical industry. Other significant applications include wire and cable coatings, foam, insulation for coaxial and communication cables, as well as those areas where high resistance to oil and chemicals is desirable. For the most part, applications of recycled HDPE duplicate those for virgin resins and include a broad spectrum of blow-molded and injection-molded products, such as detergent and motor oil bottles, garbage bags, refuse containers, strapping, drainage pipes, and plastic lumber (112,113).

Table 8. Film Properties of Two Types of HDPE with Different MWDs

MWD shape[a]	Unimodal	Bimodal
melt flow ratio	~100	130–180
dart impact strength[b], g (25-μm film)	150–180	300–500
tear strength, g[c]		
MD	10	15–25
TD	150	100

[a]MD = machine direction; TD = transverse direction.
[b]Av wt of dart sufficient to break film in 50% of tests (ASTM D4272-90, Method A).
[c]Weight of a loaded pendulum capable of tearing a notched piece of film.

Low molecular weight HDPE with molecular weight of several thousand (waxes) is widely used for paper coatings, spray coatings, emulsions, printing inks, crayons, and wax polishes. Waxes are also used as additives to butyl rubber and various higher molecular weight PE grades for improving melt flow characteristics, hardness, and resistance to abrasion and grease.

Ultrahigh Molecular Weight PE. UHMWPE possesses a unique combination of mechanical and technological properties and enjoys a variety of special applications (132). These applications benefit from such properties as low friction (solid lubricant), wear resistance (protection of metal surfaces), excellent chemical stability, as well as radiation and neutron resistance. UHMWPE has the highest neutron resistance among thermoplastic and thermoset materials. Uses of UHMWPE extend from chemical processing to the food and beverage industries, foundries, and the lumber industry; from the electrical industry, eg, parts of batteries, to medical implants; from mining and mineral processing to sewage treatment, paper, recreational equipment, textiles, and transportation. Use of UHMWPE in food processing and handling is covered by FDA Regulation 21CFR177–1520 and USDA acceptance for meat, poultry, and food processing. Graphite fibers, powered metals, glass fibers, and beads can all improve the stiffness of UHMWPE and increase its heat-deflection temperatures. As a result, the filled resins of this type compete with some rigid engineering plastics. Gel-spun fibers produced with UHMWPE are so strong, ~2.6 N/tex (30 g/den), that they are used in bullet-proof lightweight jackets for the military and police, and in nonwoven compositions for the manufacture of hard and soft armor (133,134).

BIBLIOGRAPHY

"High-Density (Linear) Polyethylene" in *ECT* 2nd ed., under "Olefin Polymers," Vol. 14, pp. 242–259, by J. P. Hogan, Phillips Petroleum Co., and R. W. Myerholtz, Amoco Chemical Corp.; "Ziegler Process Polyethylene" in *ECT* 2nd ed., under "Olefin Polymers," Vol. 14, pp. 259–282, by W. E. Gloor, Hercules Inc.; "Linear (High Density) Polyethylene" in *ECT* 3rd ed., under "Olefin Polymers," Vol. 16, pp. 421–433, by J. P. Hogan, Phillips Petroleum Co.; "Ziegler Process Polyethylene" in *ECT* 3rd ed., under "Olefin Polymers," Vol. 16, pp. 433–452, by E. Paschke, Hoechst AG.

1. U.S. Pat. 2,692,257 (May 17, 1952), A. Zletz (to Standard Oil Co. of Indiana).
2. Belg. Pat. 530,617 (Jan. 24, 1955), J. P. Hogan and R. L. Banks (to Phillips Petroleum Co.).
3. U.S. Pat. 2,825,721 (Mar. 4, 1958), J. P. Hogan and R. L. Banks (to Phillips Petroleum Co.).
4. Ger. Pat. 973,626 (Apr. 14, 1960), K. Ziegler and co-workers (to K. Ziegler).
5. H. R. Sailors and J. P. Hogan, *J. Macromol. Sci. Chem.* **A15**, 1377 (1981).
6. J. Spavacek, *Polymer*, **19** 1149 (1978).
7. D. H. Ahlstrom, S. A. Liebman, and K. B. Abbas, *J. Polym. Sci. Polym. Chem. Ed.* **14**, 2479 (1976).
8. S. A. Liebman and co-workers, *J. Macromol. Sci. Chem.* **A17**, 935 (1982).
9. H. Zweifel and S. Moss, *Polym. Degr. Stabil.* **25**, 217 (1989).
10. H. Zweifel and co-workers, *Angew. Makromol. Chem.* **176/177**, 215 (1990).
11. W. O. Drake and K. D. Cooper, *SPE Polyolefins VIII International Conference*, Houston, Tex., Feb. 1993, p. 414.

12. F. H. Winslow and W. L. Hawkins, in R. A. V. Raff and K. W. Doak, eds., *Crystalline Olefin Polymers*, Pt. II, Wiley-Interscience, New York, 1965, p. 361.
13. H. V. Boenig, *Polyolefins: Structure and Properties*, Elsevier, Amsterdam, the Netherlands, 1966.
14. R. L. Miller, in R. A. V. Raff and K. W. Doak, eds., *Crystalline Olefin Polymers*, Pt. 1, Wiley-Interscience, New York, 1965, p. 577.
15. J. N. Hay, *Polymer* **22**, 718 (1981).
16. U. Gaur and B. Wunderlich, *Macromolecules* **13**, 445 (1980).
17. R. Lam and P. H. Geil, *J. Macromol Sci. Phys.* **B20**, 37 (1981).
18. J. Dechter and co-workers, *J. Polym. Sci. Polym. Phys. Ed.* **20**, 641 (1982).
19. I. Ergoz, J. G. Fatou, and L. Mandelkern, *Macromolecules* **5**, 147 (1972).
20. J. N. Hay and P. J. Mills, *Polymer* **23**, 1380 (1982).
21. *Int. J. Polym. Mater.* **22** (1–4) (1993).
22. R. S. Stein, in B. Ke, ed., *Newer Methods of Polymer Characterization*, Wiley-Interscience, New York, 1964, p. 155.
23. A. Keller and I. Sanderman, *J. Polym. Sci.* **15**, 31, 133, 137 (1955).
24. T. Nagasawa, T. Matsumura, and T. Hochino, *Appl. Polym. Symp.* **20**, 295 (1973).
25. J. R. Dees and J. S. Sprueill, *J. Appl. Polym. Sci.* **18**, 1053 (1974).
26. U.S. Pat. 4,413,110 (Nov. 1, 1983), S. Kavesh and D. C. Prevorcek (to AlliedSignal).
27. U.S. Pat. 4,422,993 (Dec. 23, 1983), P. Smith and P. J. Lemstra (to AlliedSignal).
28. R. S. Porter, *Polymer News* **4**, 184 (1978).
29. A. E. Zachariades, W. T. Mead, and R. S. Porter, *Chem. Rev.* **80**, 351 (1980).
30. A. E. Zachariades and R. S. Porter, *J. Macromol. Sci. Phys.* **B19**, 377 (1981).
31. I. M. Ward, *Polym. Eng. Sci.* **24**, 724 (1984).
32. Y. Ohta, H. Sugiyama, and H. Yasuda, *J. Polym. Sci. Polym. Phys. Ed.* **32**, 261 (1994).
33. Y. V. Kissin, *Isospecific Polymerization of Olefins with Heterogeneous Ziegler-Natta Catalysts*, Springer-Verlag, New York, 1985.
34. T. E. Nowlin, *Prog. Polym. Sci.* **11**, 29 (1985).
35. H. Sinn and W. Kaminsky, *Adv. Organomet. Chem.* **18**, 99 (1980).
36. M. P. McDaniel and co-workers, *J. Catal.* **82**, 98,110,118 (1983).
37. J. P. Hogan, *J. Polym. Sci. Polym. Chem. Ed.* **8**, 2637 (1970).
38. T. J. Pullukat, M. Shida, and R. E. Hoff, in R. P. Quirk, ed., *Transition Metal Catalyzed Polymerizations: Alkenes and Dienes*, Harwood Academic Publishers, New York, 1983, p. 697.
39. U.S. Pat. 5,096,868 (March 17, 1992), J. T. T. Hsieh and J. C. Simondsen (to Mobil Oil Co.).
40. M. P. McDaniel, in Ref. 38, p. 713.
41. W. L. Carrick and co-workers, *J. Polym. Sci, A-1* **10**, 2609 (1972).
42. D. L. Beach and Y. V. Kissin, in J. I. Kroschwitz, ed., *Encyclopedia of Polymer Science and Engineering*, Vol. VI, 2nd ed., John Wiley & Sons, Inc., New York, p. 454.
43. Y. V. Kissin, in N. P. Cheremisinoff, ed., *Encyclopedia of Engineering Materials*, Part A, Vol. I, Marcel Dekker, Inc., New York, 1988, p. 103.
44. *Chem. Week*, 39 (June 18, 1975).
45. J. C. W. Chien and J. T. T. Hsieh, *J. Polym. Sci. Polym. Chem. Ed.* **14**, 1915 (1976).
46. V. A. Zakharov and Y. I. Yermakov, *Cat. Rev. Sci. Eng.* **19**, 63 (1979).
47. F. S. Dyachkovsky and A. D. Pomogailo, *J. Polym. Sci. Polym. Symp.* **68**, 97 (1980).
48. A. Munoz-Escalona, in Ref. 38, p. 323.
49. T. Ris, M. Dahl, and O. H. Ellestadt, *J. Mol. Catal.* **18**, 203 (1983).
50. U.S. Pat. 5,032,562 (July 16, 1991), F. Y. Lo, T. E. Nowlin, and P. P. Shirodkar (to Mobil Oil Co.).
51. U.S. Pat. 4,562,170 (Dec. 31, 1985), V. Graves (to Exxon).
52. U.S. Pat. 4,663,404 (May 5, 1987), R. Invernizzi and F. Marcato (to Enichimica).

53. F. Masi and co-workers, *Makromol. Chem. Suppl.* **15**, 147 (1989).
54. C. Chabrand and co-workers, *Proceeding of SPO Third Int. Business Forum Specialty Polyolefins*, 1993, p. 55.
55. J. H. Schut, *Plast. Technol.*, 27 (Nov. 1993).
56. *Polyolefins through the 80s: A Time of Change*, SRI International, Menlo Park, Calif. 1983.
57. K.-Y. Choi and W. H. Ray, *J. Macromol. Sci. Rev. Macromol. Chem., Phys.* **C25**, 57 (1985).
58. U.S. Pat. 3,152,872 (Oct. 13, 1964), J. S. Scoggin and H. S. Kimble (to Phillips Petroleum Co.).
59. U.S. Pat. 3,374,211 (Mar. 19, 1968), S. J. Marwil and R. G. Wallace (to Phillips Petroleum Co.).
60. U.S. Pat. 4,121,029 (Oct. 17, 1978), H. B. Irwin and F. T. Sherk (to Phillips Petroleum Co.).
61. U.S. Pat. 2,825,721 (Mar. 4, 1958), J. P. Hogan and R. L. Banks (to Phillips Petroleum Co.).
62. U.S. Pat. 3,622,521 (Nov. 23, 1971), J. P. Hogan and D. R. Witt (to Phillips Petroleum Co.).
63. U.S. Pat. 3,622,522 (Nov. 23, 1971), B. Horvath (to Phillips Petroleum Co.).
64. U.S. Pat. 3,947,433 (Mar. 30, 1976), D. R. Witt (to Phillips Petroleum Co.).
65. U.S. Pat. 4,003,712 (Jan. 18, 1977), A. R. Miller (to Union Carbide).
66. U.S. Pat. 4,011,382 (Mar. 8, 1977), I. J. Levine and F. J. Karol (to Union Carbide).
67. U.S. Pat. 4,255,542 (Mar. 10, 1981), G. L. Brown, D. F. Warner, and J. H. Byon (to Union Carbide).
68. I. D. Burdett, *Chemtech*, 616 (Oct. 1992).
69. U.S. Pat. 4,302,565 (Nov. 24, 1981), G. L. Goeke, B. E. Wagner, and F. J. Karol (to Union Carbide).
70. A. B. Furtek, *Proceedings of Worldwide Metallocene Conference, MetCon '93*, Houston, Tex., 1993, p. 125.
71. U.S. Pat. 4,260,709 (Apr. 7, 1981), D. C. Durand and P. M. Mangin (to Naphthachimie).
72. U.S. Pat. 3,878,124 (Apr. 15, 1975), D. C. Durand, M. Avaro, and P. M. Mangin (to Naphthachimie).
73. U.S. Pat. 3,954,909 (May 4, 1976), L. J. Havas and P. M. Mangin (to Naphthachimie).
74. Can. Pat. 2,078,655 (Mar. 21, 1993), A. H. Ali, R. O. Hagerty and S. C. Ong (to Mobil Oil Co.).
75. K. C. H. Yi and N. J. Maraschin, *Maack'90 International Business Conference*, Zurich, Nov. 1990.
76. B. Miller, *Plast. World* **89**, 46 (May 1992).
77. *Chem. Eng.*, 89,66 (1982).
78. U.S. Pats. 3,957,448 (May 18, 1976) and 3,965,083 (June 22, 1976), J. L. Jezl, E. F. Peters, and J. W. Shepard (to Standard Oil Co.).
79. U.S. Pat. 3,971,768 (July 27, 1976), E. F. Peters and co-workers (to Standard Oil Co.).
80. J. N. Short, in Ref. 38, p. 651.
81. J. P. Forsman, *Hydrocarbon Process.* **51**, 132 (Nov., 1972).
82. *Plast. Technol.* **26**(2), 39 (1980).
83. J. P. Machon, R. Hermant, and J. P. Houzeaux. *J. Polym. Sci. Polym. Symp.* **52**, 107 (1975).
84. J. P. Machon in Ref. 38, p. 639.
85. U.S. Pat. 4,105,842 (Aug. 8, 1978), A. Nicco and J. P. Machon (to CdF Chimie).
86. U.S. Pat. 3,644,325 (Feb. 22, 1972), R. F. Roberts, Jr. (to Dow Chemical Co.).
87. A. Montagna and J. C. Floyd, in Ref. 70, p. 171.

88. Ger. Offen. 2,831,581 (Feb. 1, 1979), L. J. M. A. Van de Leemput and H. W. Van de Loot (to Stamicarbon B.V.).

89. U.S. Pat. 4,209,603 (June 24, 1980), L. J. M. A. Van de Leemput (to Stamicarbon B.V.).

90. U.S. Pat. 4,067,822 (Jan. 10, 1978), D. E. Gessell, G. L. Dighton, and L. D. Valenzuela-Bernal (to Dow Chemical Co.).

91. U.S. Pat. 4,307,209 (Dec. 22, 1981), Y. Morita and N. Kashiwa (to Mitsui Petrochemical).

92. U.S. Pat. 4,420,592 (Dec. 13, 1983), A. Kato, J. Yoshida, and R. Yamamoto (to Mitsui Petrochemical.)

93. U.S. Pat. 4,414,369 (Nov. 8, 1983), N. Kuroda and co-workers (to Nippon Oil).

94. U.S. Pat. 4,352,915 (Oct. 5, 1982), K. Mashita and K. Hanji (to Sumitomo Chemical).

95. R. I. Tanner, *Engineering Rheology*, Clarendon Press, Oxford, U.K., 1986.

96. J. R. A. Pearson, *Mechanics of Polymer Processing*, Elsevier Science, Ltd., Oxford, U.K., 1985.

97. Z. Tadmor and C. G. Gogos, *Principles of Polymer Processing*, John Wiley & Sons, Inc., New York, 1979.

98. S. Middleman, *Fundamentals of Polymer Processing*, McGraw-Hill Book Co., Inc., New York, 1977.

99. D. V. Rosato and D. V. Rosato, eds., *Injection Molding Handbook*, Van Nostrand Reinhold Co., Inc., New York, 1986.

100. I. I. Rubin, *Injection Molding*, in Ref. 42, Vol. VIII, p. 102.

101. I. I. Rubin, *Injection Molding: Theory and Practice*, Wiley-Interscience, New York, 1972.

102. D. V. Rosato and D. V. Rosato, eds., *Blow Molding Handbook*, Hander Publishers, Munich, 1989.

103. C. Irwin, in Ref. 42, Vol. II, p. 447.

104. S. Levy, *Plastics Extrusion Technology Handbook*, Industrial Press, New York, 1981.

105. G. A. Kruder, in Ref. 42, Vol. VI, p. 571.

106. J. Bristol, *Plastics Films*, Longman, New York, 1983.

107. H. C. Park and E. M. Mount, in Ref. 42, Vol. VII, p. 88.

108. P. Smith and P. J. Lemstra, *J. Mater. Sci.* **15**, 505 (1980).

109. P. J. Lemstra and co-workers, *Developments in Oriented Polymers*, Vol. 2, Elsevier Science Publishing Co., Inc., New York, 1987, p. 99.

110. K. E. Jacobson and R. Eller, *Proceedings from SPE Polyolefins VIII International Conference*, 1993, p. 494.

111. M. E. Dahl, *Proceedings from SPE Recycling Conference*, 1993, p. 17.

112. M. H. Shaver and J. H. Paul, in Ref. 110, p. 183.

113. K. Atkins, in Ref. 110, p. 666.

114. K. W. Swogger and C.-I. Kao, in Ref. 110, 1993, p. 13.

115. P. E. Manders, *Mod. Plast. Encyclopedia*, B-3 (1995).

116. A. Dankovic and L. Fuzes, *J. Appl. Polym. Sci.* **28**, 3707 (1983).

117. M. Ondas, M. Drobny, and E. Spirk, *Int. Polym. Sci. Tech.* **20**, 93 (1993).

118. J. P. Blitz and D. C. McFaddin, *J. Appl. Polym. Sci.* **51**, 13 (1994).

119. D. L. Beach and Y. V. Kissin, *J. Polym. Sci. Polym. Chem. Ed.* **22**, 3027 (1984).

120. Y. V. Kissin and M. L. Friedman, *Polym. Mech.*, (1), 143 (1977).

121. Y. V. Kissin, *J. Polym. Sci. Polym. Phys. Ed.* **30**, 1165 (1992).

122. C. G. Gebelein, *Polymeric Materials and Artificial Organs: ACS Series 256*, American Chemical Society, Washington, D.C., 1984, p. 3.

123. G. Hovding, *Acta. Derm. Venerol.* **49**, 147 (1969).

124. R. L. Kuhn, W. J. Potts, and T. E. Waterman, *J. Combust. Toxicol.* **5**, 434 (1978).

125. R. E. Reinke and C. F. Reinhardt, *76th Annual Meeting, National Fire Protection Association*, Philadelphia, Pa., 1972.

126. W. J. Potts, T. S. Iederer, and J. F. Quast, *J. Combust. Toxicol.* **5**, 408 (1978).
127. T. Morikawa, *J. Combust. Toxicol.* **3**, 135 (1976).
128. J. Michael, J. Mitera, and S. Tardon, *Fire Mater.* **1**, 160 (1976).
129. C. J. Hilado and co-workers, *J. Combust. Toxicol.* **3**, 259,305,381 (1976).
130. C. J. Hilado, J. A. Soriano, and K. J. Kosola, *J. Combust. Toxicol.* **4**, 533 (1977).
131. C. J. Hilado and N. V. Huttlinger, *J. Combust. Toxicol.* **5**, 81,361 (1978).
132. *Directory of Uses for 1900 UHMW Polymer*, Bulliten No. 500–667, HIMONT, U.S.A., Wilmington, Del., Aug. 1985.
133. L. Wilson and co-workers, in Ref. 110, p. 174.
134. U.S. Pat. 4,650,710 (Mar. 17, 1987), G. A. Harpell and co-workers (to AlliedSignal).

YURY V. KISSIN
Mobil Chemical Company

LINEAR LOW DENSITY POLYETHYLENE

The chemical industry manufactures a large variety of semicrystalline ethylene copolymers containing small amounts of α-olefins. These copolymers are produced in catalytic polymerization reactions and have densities lower than those of ethylene homopolymers known as high density polyethylene (HDPE). Ethylene copolymers produced in catalytic polymerization reactions are usually described as linear ethylene polymers, to distinguish them from ethylene polymers containing long branches which are produced in radical polymerization reactions at high pressures (see OLEFIN POLYMERS, LOW DENSITY POLYETHYLENE).

Densities and crystallinities of ethylene–α-olefin copolymers mostly depend on their composition. The classification in Table 1 is commonly used (ASTM D1248-48). VLDPE resins are usually further subdivided into PE plastomers of low crystallinity, 10–20%, with densities in the range of 0.915–0.900 g/cm^3, and completely amorphous PE elastomers with densities as low as 0.860 g/cm^3.

Commercial production of PE resins with densities of 0.925 and 0.935 g/cm^3 was started in 1968 in the United States by Phillips Petroleum Co. Over time, these resins, particularly LLDPE, became large volume commodity products. Their combined worldwide production in 1994 reached 13×10^6 metric t/yr, accounting for some 30% market share of all PE resins; in the year 2000, LLDPE production is expected to increase by 50%. A new type of LLDPE, compositionally uniform ethylene–α-olefin copolymers produced with metallocene catalysts, was first introduced by Exxon Chemical Co. in 1990. The initial production volume was 13,500 t/yr but its growth has been rapid; indeed, in 1995 its combined production by several companies exceeded 800,000 tons.

Table 1. Basic Classification of Copolymers

Resin	Designation	α-Olefin, mol %	Crystallinity, %	Density, g/cm^3
PE of medium density	MDPE	1–2	55–45	0.940–0.926
linear PE of low density	LLDPE	2.5–3.5	45–30	0.925–0.915
PE of very low density	VLDPE	>4	<25	<0.915

The large number of commodity and specialty resins collectively known as LLPDE are in fact made up of various resins, each different from the other in the type and content of α-olefin in the copolymer, compositional and branching uniformity, crystallinity and density, and molecular weight and molecular weight distribution (MWD).

Type of α-Olefin. Four olefins are used in industry to manufacture ethylene copolymers: 1-butene, 1-hexene, 4-methyl-1-pentene, and 1-octene. Copolymers containing 1-butene account for approximately 40% of all LLDPE resins manufactured worldwide, 1-hexene copolymers for 35%, 1-octene copolymers for about 20%, and 4-methyl-1-pentene copolymers for the rest. The type of α-olefin exerts a significant influence on the copolymer properties. Ethylene–1-butene copolymers, for instance, exhibit inferior mechanical properties compared to ethylene copolymers with other α-olefins; consequently, the share of 1-butene-based resins is predicted to decrease to around 30% in 1997, whereas the shares of ethylene copolymers with higher α-olefins will increase. Ethylene copolymers with cycloolefins such as cyclopentene and norbornene have been added to the LLDPE resin group. The content of α-olefin in copolymer varies greatly, from 1 to 2 mol % in MDPE resins to around 20 mol % in PE elastomers.

Compositional (Branching) Uniformity. Two classes of LLDPE resins are on the market. One, introduced in 1990, has a predominantly uniform compositional distribution (uniform branching distribution); that is, all copolymer molecules in these resins have approximately the same composition. Most commercially produced LLDPE resins, in contrast, have pronounced nonuniform branching distributions; there are significant differences in copolymer compositions among different macromolecules in a given resin. For example, a resin containing 3.0 mol % of α-olefin is in fact a mixture of macromolecules with olefin contents ranging from below 0.3 mol % in high mol wt fraction of the resin, to over 20 mol % in fractions with the lowest mol wt. Uniformly and nonuniformly branched LLDPE resins differ significantly in physical and mechanical properties.

Crystallinity and Density. These two parameters, which are closely related, depend mostly on the amount of α-olefin in the copolymer. Both density and crystallinity of ethylene copolymers are also influenced by their compositional uniformity. For example, for LLDPE resins with different α-olefin (1-hexene) content, the density (g/cm^3) is as follows:

LLDPE	2.0 mol %	3.0 mol %	4.0 mol %
uniformly branched	0.920–0.918	0.912–0.908	<0.900
nonuniformly branched	0.930–0.927	0.922–0.920	0.915–0.912

A lower α-olefin content is required in a uniformly branched ethylene copolymer to decrease its crystallinity and density to a given level.

Molecular Weight. The range of molecular weights of commercial LLDPE resins is relatively narrow, usually from 50,000 to 200,000. One accepted parameter that relates to the resin molecular weight is the melt index, a rheological parameter which, broadly defined, is inversely proportional to molecular weight.

A typical melt index range for LLDPE resins is from 0.1 to 5.0, but can reach over 30 for some applications.

The width of molecular weight distribution (MWD) is usually represented by the ratio of the weight–average and the number–average molecular weights, $\overline{M}_w/\overline{M}_n$. In industry, MWD is often represented by the value of the melt flow ratio (MFR), which is calculated as a ratio of two melt indexes measured at two melt pressures that differ by a factor of 10. Most commodity-grade LLDPE resins have a narrow MWD, with the $\overline{M}_w/\overline{M}_n$ ratios of 2.5–4.5 and MFR values in the 20–35 range. However, LLDPE resins produced with chromium oxide-based catalysts have a broad MWD, with $\overline{M}_w/\overline{M}_n$ of 10–35 and MFR of 80–200.

Molecular Structure and Properties

Chain Structure. LLDPE resins are copolymers of ethylene and α-olefins with low α-olefin contents. Molecular chains of LLDPE contain units derived both from ethylene, $-CH_2-CH_2-$, and from the α-olefin, $-CH_2-CHR-$, where R is C_2H_5 for ethylene–1-butene copolymers, n-C_4H_9 for ethylene–1-hexene copolymers, $-CH_2-CH(CH_3)_2$ for ethylene–4-methyl-1-pentene copolymers, and n-C_6H_{13} for ethylene–1-octene copolymers. In a typical copolymer molecule containing 2–4 mol % of α-olefin, the majority of olefin units stand alone in the chain but most ethylene units form long sequences. Usually one of the chain ends in an LLDPE molecule is a methyl group; the other chain end can either be a methyl group (if hydrogen is used as a chain-transfer agent in the polymerization reaction), or a double bond, a vinyl group ($CH_2{=}CH-$) or a vinylidene group ($CH_2{=}C{<}$).

As a rule, LLDPE resins do not contain long-chain branches. However, some copolymers produced with metallocene catalysts in solution processes can contain about 0.002 long-chain branches per 100 ethylene units (1). These branches are formed in auto-copolymerization reactions of ethylene with polymer molecules containing vinyl double bonds on their ends (2).

Compositional Uniformity. LLDPE resins produced with different catalysts vary greatly in their compositional uniformity. The fastest way to evaluate the branching distribution of an LLDPE resin is to measure its melting point. Melting points of copolymers with uniform compositional distributions show a noticeable dependence on their composition and may vary widely; for instance, from ~120°C for copolymers containing 1.5–2 mol % of α-olefin to ~110°C for copolymers containing 3.5 mol % of α-olefin. A copolymer with a nonuniform compositional distribution, in contrast, is a mixture that contains copolymer molecules with a broad range of compositions, from almost linear macromolecules (usually of higher molecular weights) to short macromolecules with quite high α-olefin contents. Melting of such mixtures is dominated by their low branched fractions which are highly crystalline. As a result, the melting points of LLDPE resins with nonuniform branching distributions are not too sensitive to copolymer composition and usually fall in the temperature range of 125–128°C (3). Figure 1 compares the melting curves of two copolymers with the same average compositions but different degrees of compositional uniformity.

Highly branched fractions of nonuniformly branched resins have low molecular weights and are easily soluble, even at room temperature, in saturated

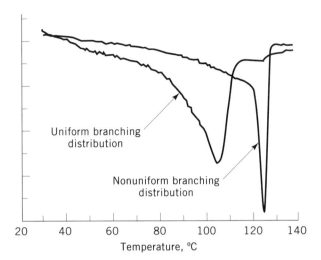

Fig. 1. Melting curves (dsc) of two ethylene–1-hexene copolymers produced in a gas-phase process; one with a uniform branching distribution (1-hexene content 2.5 mol %) and another with a nonuniform branching distribution (1-hexene content 2.8 mol %).

hydrocarbons. These highly branched fractions are called extractables, an excessive amount of which in an LLDPE resin can be detrimental to certain end use properties, especially in food packaging applications.

Chemical Properties and Reactivity. LLDPE is a saturated branched hydrocarbon. The most reactive parts of LLDPE molecules are the tertiary CH bonds in branches and the double bonds at chain ends. Although LLDPE is nonreactive with both inorganic and organic acids, it can form sulfo-compounds in concentrated solutions of H_2SO_4 (>70%) at elevated temperatures and can also be nitrated with concentrated HNO_3. LLDPE is also stable in alkaline and salt solutions. At room temperature, LLDPE resins are not soluble in any known solvent (except for those fractions with the highest branching contents); at temperatures above 80–100°C, however, the resins can be dissolved in various aromatic, aliphatic, and halogenated hydrocarbons such as xylenes, tetralin, decalin, and chlorobenzenes.

Thermal, Thermooxidative, and Photooxidative Degradation. LLDPE is relatively stable to heat. Thermal degradation starts at temperatures above 250°C and results in a gradual decrease of molecular weight and the formation of double bonds in polymer chains. At temperatures above 450°C, LLDPE is pyrolyzed with the formation of isoalkanes and olefins.

Oxidation of LLDPE starts at temperatures above 150°C. This reaction produces hydroxyl and carboxyl groups in polymer molecules as well as low molecular weight compounds such as water, aldehydes, ketones, and alcohols. Oxidation reactions can occur during LLDPE pelletization and processing; to protect molten resins from oxygen attack during these operations, antioxidants (radical inhibitors) must be used. These antioxidants (qv) are added to LLDPE resins in concentrations of 0.1–0.5 wt %, and may be naphthylamines or phenylenediamines, substituted phenols, quinones, and alkyl phosphites (4), although inhibitors based on hindered phenols are preferred.

Photooxidative degradation of LLDPE at ambient temperature under sunlight is also a radical oxidation reaction. It causes change in color and drastic deterioration of mechanical and dielectric properties of LLDPE articles. Photooxidation can be prevented by using light stabilizers that protect resins, absorb uv radiation, and terminate radical chain reactions. A large number of antioxidants and uv stabilizers is manufactured by the chemical industry, including American Cyanamid, BASF, Ciba-Geigy, Eastman Chemical, Elf Atochem, Enichem, General Electric, Hoechst-Celanese, Sandoz, and Uniroyal.

Physical Properties.   LLDPE is a semicrystalline plastic whose chains contain long blocks of ethylene units that crystallize in the same fashion as paraffin waxes or HDPE. The degree of LLDPE crystallinity depends primarily on the α-olefin content in the copolymer (the branching degree of a resin) and is usually below 40–45%. The principal crystalline form of LLDPE is orthorhombic (the same as in HDPE); the cell parameters of nonbranched PE are $a = 0.740$ nm, $b = 0.493$ nm, and c (the direction of polymer chains) = 0.2534 nm. Introduction of branching into PE molecules expands the cell slightly; thus a increases to 0.77 nm and b to around 0.50 nm.

LLDPE rapidly crystallizes from the melt with the formation of spherulites, small spherical objects 1–5 μm in diameter visible only in a microscope. The elementary structural blocks in spherulites are lamellae, small flat crystallites formed by folded linear segments in LLDPE chains, which are interconnected by polymer chains that pass from one lamella to another. These chains, or tie molecules, hold together neighboring lamellae and provide mechanical integrity and strength to spherulites. Crystalline lamellae within spherulites give LLDPE articles necessary rigidity, whereas the large amorphous regions between lamellae, constituting over 60% of the spherulite volume, provide flexibility. The presence of short-chain branches (α-olefin units) in LLDPE molecules is able to inhibit the uniform folding of polymer chains during crystallization. Such inhibition can both decrease the lamella thickness and increase the number of interlamellar tie molecules, thus resulting in a stronger material.

The size of the lamellae for a copolymer of a given composition depends on the degree of branching uniformity. If an LLDPE resin is compositionally uniform, all its macromolecules crystallize poorly due to branching, forming very thin lamellae. Such materials have low rigidity (low modulus) and high flexibility. On the other hand, if an LLDPE resin is compositionally nonuniform, its least-branched components are able to form thicker lamellae; consequently, more branched fractions of the resin remain amorphous and fill the voids between the lamellae. Articles made from such resins are more rigid.

The projected equilibrium melting point of completely linear PE is 146–147°C (5); its highest actual melting point is 133–138°C. In the case of ethylene copolymers with a uniform compositional distribution, the melting point decreases almost linearly with copolymer composition; for instance, introduction of one branch per 1000 carbon atoms in the chain reduces the resin melting point by about 1°C. As a result, melting points of most uniformly branched LLDPE resins are usually in the 100–115°C range. In contrast, the melting points of compositionally nonuniform LLDPE products are not sensitive to the copolymer composition and usually range from 128 to 125°C. The brittle point of LLDPE is low, from −100 to −140°C.

Optical properties of LLDPE resins also depend on the degree of branching uniformity. Resins with a uniform branching distribution make highly transparent film with haze as low as 3–4%. In contrast, film manufactured from compositionally nonuniform copolymers is much more opaque, with haze of over 10–15%; this is due to the presence of large crystalline lamellae consisting of nearly nonbranched PE chains.

Because it is a saturated aliphatic hydrocarbon, LLDPE does not conduct electricity, and so is widely used for wire and cable insulation. LLDPE is poorly permeable to water and inorganic gases and only slightly more so to organic compounds, whether liquid or gas.

Mechanical Properties.   Mechanical characteristics of ethylene copolymers are functions of their structural characteristics.

Content of α-Olefin.   An increase in the α-olefin content of a copolymer results in a decrease of both crystallinity and density, accompanied by a significant reduction of the polymer mechanical modulus (stiffness). For example, the modulus values of ethylene–1-butene copolymers with a nonuniform compositional distribution decrease as shown in Table 2 (6). A similar dependence exists for ethylene–1-octene copolymers with uniform branching distribution (7), even though all such materials are, in general, much more elastic (see Table 2). An increase in the α-olefin content in the copolymers also results in a decrease of their tensile strength but a small increase in the elongation at break (8). These two dependencies, however, are not as pronounced as that for the resin modulus.

Type of α-Olefin.   When resins of the same density and crystallinity are compared, ethylene–1-butene copolymers show noticeably inferior mechanical properties compared to ethylene copolymers with higher α-olefins. These differences are clearly manifested when comparing properties of thin film. Table 3 gives the principal mechanical properties of film manufactured from three grades of LLDPE (ethylene copolymers with 1-butene, 1-hexene, and 1-octene, all with a nonuniform compositional distribution); Table 4 lists various mechanical properties of VLDPE resins. One commonly used measure of film strength is the dart impact strength. An average dart impact strength of a 37-μm film made from ethylene–1-butene copolymer is about 100–150 g, whereas the same parameter for ethylene–1-hexene and ethylene–1-octene copolymers of the same density is about 300–350 g (6).

Table 2. Effect of α-Olefin Content in LLDPE on Modulus

Nonuniform branching distribution			
1-butene content, mol %	0	~3.0	~8.0
crystallinity, %	~60	~40	~25
density, g/cm³	0.96	0.92	0.89
modulus, MPa[a]	~1500	~240	~60
Uniform branching distribution			
1-octene content, mol %	~0.4	~3.3	~18
density, g/cm³	0.940	0.910	0.856
modulus, MPa[a]	~500	~100	1.3

[a]To convert MPa to psi, multiply by 145.

Table 3. Properties of Commercial LLDPE Film of Resins with Nonuniform Compositional Distribution[a]

Property	Copolymer of ethylene and:				
	1-butene		1-hexene		1-octene
density g/cm^3	0.918	0.918	0.918	0.918	0.919
melt index, g/10 min	2.0	1.0	1.0	0.5	1.0
dart impact strength, g[b]	110	150	250	300	350
puncture energy J/mm[c]	60	70	85	94	61
tensile strength,[d] MPa[e]					
MD	33	38	38	43	43
TD	25	31	32	43	34
elongation at break,[d] %					
MD	690	620	570		550
TD	740	760	790		660
modulus,[d] MPa[e]					
MD	210	230			
TD	250	260			

[a]Ref. 6; film thickness 37 μm.
[b]Average weight of the dart sufficient to break the film in 50% of tests (ASTM D4272-90, Method A).
[c]To convert J/mm to ft·lbf/in., multiply by 18.73.
[d]MD = machine direction; TD = transverse direction.
[e]To convert MPa to psi, multiply by 145.

Table 4. Properties of VLDPE Resins[a] from Various Manufacturers

Property	Copolymer of ethylene and 1-butene			Copolymer of ethylene and 1-octene
manufacturer	Mitsui	Exxon	Union Carbide	Dow
compositional distribution	uniform	uniform	nonuniform	nonuniform
density, g/cm^3	0.88	0.88	0.89	0.912
crystallinity, %	20.4	21.2	25.6	42.9
melt index, g/10 min	3.6	3.8	1.0	3.3
melting point, °C	69.5	71.8	118.4	123.8
tensile modulus, MPa[b]	30	25	57	150
dynamic modulus (1 Hz), MPa[b]	38	35	80	310
strain recovery, %	95	98	74	63
haze, %	4	4	46	51

[a]Ref. 6.
[b]To convert MPa to psi, multiply by 145.

Branching Uniformity. Comparison of uniformly and nonuniformly branched ethylene–1-butene copolymers of the same density (Table 4) shows that uniformly branched resins are much more elastic, their tensile modulus is lower, and their strain recovery is nearly complete.

Molecular Weight and MWD. An increase in molecular weight (a decrease of melt index) for both ethylene–1-butene and ethylene–1-hexene resins results in a significant improvement of their mechanical properties, particularly their dart impact strength. Most LLDPE resins for film applications have a relatively narrow MWD: $\overline{M}_w/\overline{M}_n$ values are around 4.0 and MFR values around 28–30. However, MWD width can be reduced even further through a selection of special catalysts and process fine-tuning. When $\overline{M}_w/\overline{M}_n$ is decreased to 3.5 and MFR to 25–27, significant improvement in film properties follows: the dart impact strength improves by 250% and the tear strength by over 25% (9).

LLDPE resins with broad MWD can also have improved mechanical strength. This occurs when the copolymers have a large fraction of molecules with high molecular weights. Such molecules serve as links between polymer crystallites (tie molecules) and strengthen LLDPE spherulites. LLDPE resins with densities of 0.925–0.900 g/cm^3 and melt index of 0.05–0.35 are produced with chromium-based catalysts (10); their MFR values are in the 80–200 range. On account of their higher molecular weights, film manufactured from these resins also has a high dart impact strength.

Effect of Orientation.   Orientation has a significant beneficial effect on the mechanical properties of LLDPE, especially of LLDPE film. During film manufacture, a bubble of polymer melt exits a circular die with a narrow gap and, yielding to the pressure of air charged inside the bubble, immediately expands. The degree of expansion is usually given as the blow-up ratio (BUR), the ratio of the film bubble diameter to the die diameter. Film expansion induces significant orientation of polymer molecules in the film plane and improves its mechanical properties, as the following data for ethylene–1-hexene copolymer with a density of 0.917 g/cm^3 and a melt index of 0.9 demonstrate (9):

blow-up ratio	1.8	2.0	2.2	2.5
dart impact strength, g	170	178	220	280

Catalysts for LLDPE Production

LLDPE resins are produced in industry with three classes of catalysts (11–14): titanium-based catalysts (Ziegler), metallocene-based catalysts (Kaminsky and Dow), and chromium oxide-based catalysts (Phillips).

Ziegler Catalysts.   These catalyst systems account for by far the greatest share of LLDPE resins manufactured. They consist of two components: the first contains as its active ingredient a derivative of a transition metal (usually titanium); the second is an organoaluminum compound (such as triethylaluminum) (11–15). The molar [Al]:[Ti] ratio in these catalyst systems is usually in the 50–500 range. Most commercially important catalysts are heterogeneous; a large number of them are supported. Both inorganic and organic supports are used; the most important are MgCl$_2$, Mg(OH)Cl, silica, alumina, MgO, and polymers with reactive groups such as OH, NH, or COOH (16–20). Titanium compounds used for manufacturing Ziegler catalysts are usually TiCl$_4$ or Ti(OR)$_4$. Catalysts with the highest activity contain combinations of titanium chlorides and MgCl$_2$. Titanium compounds are deposited into the support pores or anchored to the active groups (such as OH) on the support surface. Examples of supported catalysts include TiCl$_3$ on the surface of MgCl$_2$, as well as coprecipitated complexes of Ti chlorides and MgCl$_2$ within pores of inert carrier particles (21).

Typical heterogeneous Ziegler catalysts operate at temperatures of 70–100°C and pressures of 0.1–2 MPa (15–300 psi). The polymerization reactions are carried out in an inert liquid medium (eg, hexane, isobutane) or in the gas phase. Molecular weights of LLDPE resins are controlled by using hydrogen as a chain-transfer agent.

A prerequisite for any catalyst used for LLDPE synthesis is its efficiency in copolymerizing ethylene with α-olefins. Because ethylene has the highest reactivity among all olefins in catalytic polymerization reactions, a relatively high α-olefin concentration in a polymerization reactor is usually required to produce copolymer containing merely 3–4 mol % of α-olefin units in the chains. Reactivities of α-olefins in copolymerization with ethylene depend on two factors: the size of the alkyl groups attached to their double bonds and the type of catalyst (22). Figure 2 shows the dependence of the relative reactivities of linear α-olefins in copolymerization with ethylene on the length of their alkyl groups. Only olefins with the smallest alkyl groups (ie, 1-butene, 1-hexene, and 1-octene) can be copolymerized relatively easily with ethylene under commercial conditions. All solid Ziegler catalysts contain mixtures of several different types of active centers and produce compositionally nonuniform ethylene–α-olefin copolymers. Some active centers in these catalysts exhibit poor copolymerization ability and produce nearly linear PE; others more efficiently incorporate α-olefins into copolymer chains and produce copolymer molecules with a high α-olefin content. The presence of several types of active centers in the catalysts accounts also for a relatively broad copolymer MWD, a fact that can be used to advantage. For instance, specially designed catalysts can be made that produce LLDPE resins with a particularly high molecular weight component and noticeably improved mechanical properties (23).

Dual Ziegler Catalysts. The choice of a particular α-olefin in commercial LLDPE manufacture is dictated by two factors: the desired resin properties,

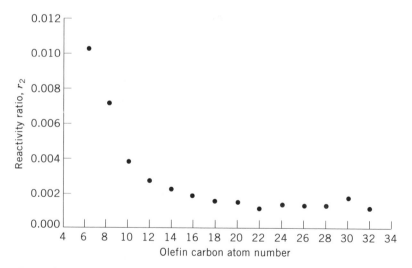

Fig. 2. Dependence of olefin reactivity on its carbon atom number when linear α-olefins are copolymerized with ethylene.

and the cost and availability of an α-olefin. A special type of Ziegler catalysts, dual catalysts, allows the manufacture of ethylene–1-butene LLDPE resins from a single monomer source, ethylene (24). These catalyst systems contain two separate catalyst species, one for ethylene dimerization into 1-butene, usually a combination of $Ti(OR)_4$ and $Al(C_2H_5)_3$, and another for copolymerization of ethylene and the generated 1-butene.

Metallocene Catalysts. Three types of metallocene catalysts are presently used in industry: Kaminsky, combination, and Dow catalysts (see METALLOCENE CATALYSTS (SUPPLEMENT)).

Kaminsky catalysts, the original metallocene catalyst systems, contain two components: methylaluminoxane (MAO) and a metallocene complex of zirconium, titanium, or hafnium with two cyclopentadienyl rings (Fig. 3**a**) (13,25). MAO is an oligomeric organoaluminum compound, represented by the formula $\text{+O—Al(CH}_3)\text{+}_n$, where $n = 4\text{–}20$, and produced by hydrolysis of trimethylaluminum (13,26). A large number of metallocene complexes can be used in Kaminsky catalysts. Their rings contain various alkyl substituents, both linear and cyclic; both rings can be linked together by bridging groups (27). Among the metallocene compounds, zirconium complexes are preferred on the basis of their exceptionally high activity in ethylene polymerization reactions. Kaminsky catalysts can operate at [Al]:[transition metal] ratios of at least 100, but ideally over 1000. Metallocene catalysts are highly sensitive to the presence of hydrogen, which greatly reduces molecular weights of polymers.

MAO is a relatively expensive chemical; its price in 1994 was about \$450/kg of 30 wt % MAO solution, but projected to decrease to about \$200/kg (28). Continuous efforts to replace MAO have resulted in the development of co-catalysts containing mixtures of MAO and trimethylaluminum (29) as well as new cocatalyst types (30,31). Another approach is to prepare MAO directly in a polymerization reactor by co-feeding into it trimethylaluminum and water (32).

Most Kaminsky catalysts contain only one type of active center. They produce ethylene–α-olefin copolymers with uniform compositional distributions and quite narrow MWDs which, at their limit, can be characterized by $\overline{M}_w/\overline{M}_n$ ratios of about 2.0 and MFR of about 15. These features of the catalysts determine their first applications in the specialty resin area, to be used in the synthesis of either uniformly branched VLDPE resins or completely amorphous PE plastomers. Kaminsky catalysts have been gradually replacing Ziegler catalysts in the manufacture of certain commodity LLDPE products. They also facilitate the copolymerization of ethylene with cyclic dienes such as cyclopentene and norbornene (33,34). These copolymers are compositionally uniform and can be used as LLDPE resins with special properties. Ethylene–norbornene copolymers are resistant to chemicals and heat, have high glass transitions, and very high transparency which makes them suitable for polymer optical fibers (34).

Ionic catalyst combinations of metallocene complexes of zirconium or titanium and perfluorinated aromatic boron compounds are ionic (35); the simplest representative is the ion pair shown in Figure 3**b** (36,37). These catalysts do not need any organoaluminum compounds; however, monomers of high purity are required to prevent catalyst poisoning. These catalysts operate over a wide range of temperatures, from -70 to $100°C$. The ethylene copolymers they produce have high compositional uniformity.

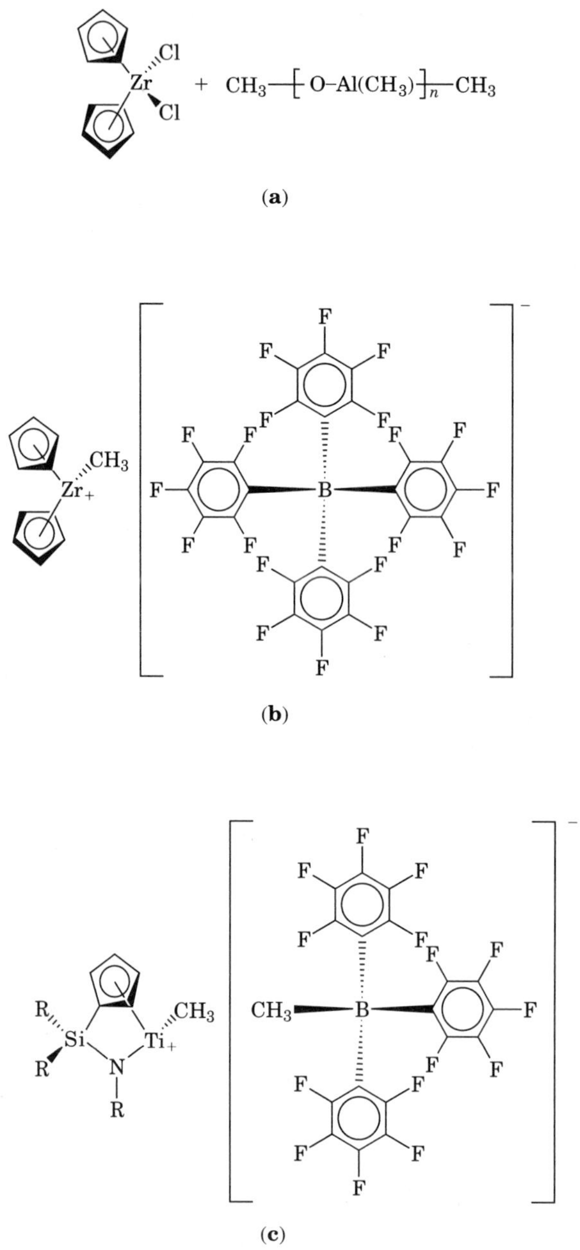

(a)

(b)

(c)

Fig. 3. Metallocene catalyst systems for LLDPE synthesis: (**a**) Kaminsky catalyst; (**b**) cationic catalyst; and (**c**) Dow catalyst.

Dow catalysts, also known as constrained-geometry catalysts (38–40), contain monocyclopentadienyl derivatives of titanium or zirconium. One of the carbon atoms in the cyclopentadienyl ring is additionally linked to the metal atom by a bridge (Fig. 3c). The complexes are converted to polymerization catalysts by reacting with MAO or by forming ionic complexes with noncoordinative anions such as $[B(C_6F_5)_4]^-$ and $[B(C_6F_5)_3CH_3]^-$ (38–40). Dow catalyst systems contain only one type of active center and produce compositionally uniform ethylene–α-olefin copolymers (1); they operate at temperatures up to 160°C.

Dow catalysts have a high capability to copolymerize linear α-olefins with ethylene. As a result, when these catalysts are used in solution-type polymerization reactions, they also copolymerize ethylene with polymer molecules containing vinyl double bonds at their ends. This autocopolymerization reaction is able to produce LLDPE molecules with long-chain branches that exhibit some beneficial processing properties (1,2,38,39). Distinct from other catalyst systems, Dow catalysts can also copolymerize ethylene with styrene and hindered olefins (40).

Chromium Oxide-Based Catalysts. Chromium oxide-based catalysts were originally developed by Phillips Petroleum Co. for the manufacture of HDPE resins; subsequently, they have been modified for ethylene–α-olefin copolymerization reactions (10). These catalysts use a mixed silica–titania support containing from 2 to 20 wt % of Ti. After the deposition of chromium species onto the support, the catalyst is first oxidized by an oxygen–air mixture and then reduced at increased temperatures with carbon monoxide. The catalyst systems used for ethylene copolymerization consist of solid catalysts and co-catalysts, ie, trialkylboron or trialkylaluminum compounds. Ethylene–α-olefin copolymers produced with these catalysts have very broad molecular weight distributions, characterized by $\overline{M}_w/\overline{M}_n$ in the 12–35 and MFR in the 80–200 range.

Polymerization Processes

The technologies suitable for LLDPE manufacture include gas-phase fluidized-bed polymerization, polymerization in solution, polymerization in a polymer melt under high ethylene pressure, and slurry polymerization. Most catalysts are fine-tuned for each particular process.

Gas-Phase Polymerization. The first gas-phase fluidized-bed process for the production of LLDPE was built by Union Carbide in 1977. This technology was widely licensed worldwide (Unipol process), and the resins produced with this technology have gradually replaced high pressure LDPE in many applications. Similar gas-phase processes were also developed by British Petroleum Co. and Naphthachimie. The gas-phase processes are economical, flexible, and wide-ranging in the use of solid and supported catalysts. The first LLDPE products manufactured with gas-phase technology were ethylene–1-butene copolymers. 1-Butene was selected for two reasons: (1) because it was the least expensive comonomer that still allowed the synthesis of LLDPE resins with mechanical properties superior to those of LDPE, and (2) because it had a low dew point and did not condense in a gas-phase reactor. These resins have become the staple of the large volume commodity LLDPE market. Most gas-phase LLDPE producers later modified the technology and introduced a series of more advanced resins (high strength LLDPE) required by the high performance markets. These products utilize higher α-olefins, 1-hexene, and 4-methyl-1-pentene. All these resins are manufactured with Ziegler catalysts and have pronounced nonuniform branching distributions. Properties of two grades of LLDPE film resins produced in gas-phase processes, 1-butene- and 1-hexene-containing copolymers, are compared in Table 3. Impact strength, tear strength, and puncture energy for the film produced from 1-hexene-derived LLDPE resins are substantially higher than those of 1-butene copolymers.

The Union Carbide gas-phase process is suitable for the production of both HDPE and LLDPE (41–45). A flow diagram is shown in Figure 4 (46). The reactor is a tall cylindrical tower (height up to 25 m) with a length-to-diameter ratio of ~7. It usually operates at pressures of 1.5–2.5 MPa (15–25 atm) and temperatures of 70–95°C. The reactor is filled with a bed of dry polymer particles vigorously agitated by a high velocity gas stream, a mixture of ethylene, α-olefin, nitrogen, and hydrogen, which is used for molecular weight control. The gas stream enters the reactor through a perforated distribution plate at the reactor bottom. Rapid circulation serves two purposes, fluidization of the particle bed and the removal of polymerization heat. The unreacted gas stream enters an expanded disengagement zone at the top of the reactor, separates from the entrained polymer particles, and is then compressed, cooled, and recycled. In some cases, the gas mixture can be cooled below its dew point. Fine droplets of liquid α-olefin are carried by the gas stream into the reactor where they rapidly evaporate (44). Operation in the condensing mode increases the heat removing capacity of the circulating stream. Advances in condensing mode technology has greatly increased productivity of fluid-bed reactors (47).

A solid catalyst in the form of small spherical particles is continuously fed into the reactor above the distribution plate, either as powder or a slurry in a high boiling mineral oil. A solution of a co-catalyst, trialkylaluminum compound (usually triethylaluminum), is charged to the bottom of the reactor. Catalyst particles remain in the reactor, on average, for 2.5–4 hours. The particles gradually increase in size due to polymerization. The polymer particles can grow to an av-

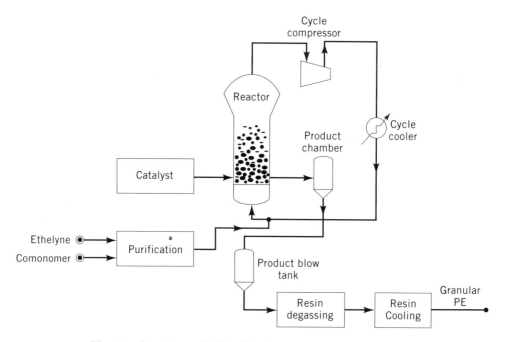

Fig. 4.   Gas-phase fluidized-bed process (Union Carbide).

erage size of 400–700 μm and still retain their spherical shape. Granular resin is continuously removed from the reactor and transferred into a series of vessels, where it is stripped of ethylene and comonomer and then pelletized. Montedison developed a similar gas-phase polymerization technology that combines two different reactor types, a small slurry prereactor followed by one or two gas-phase fluidized-bed reactors (48). This process (Spherilene process) utilizes a special Ziegler catalyst that produces LLDPE resins in the form of large (up to 4–5 mm in diameter) dense spherical particles which do not require post-production pelletization. Commercial production of these resins began in the United States in 1993.

Fluidized-bed reactors are highly versatile and can accommodate many types of polymerization catalysts. Most of the catalysts used for LLDPE production are heterogeneous Ziegler catalysts, in both supported and unsupported forms. The gas-phase process can also accommodate supported metallocene catalysts that produce compositionally uniform LLDPE resins (49–51).

Solution Polymerization.   Two types of solution polymerization technologies are used for LLDPE synthesis. One process utilizes heavy solvents; the other is carried out in mixtures of supercritical ethylene and molten PE as a polymerization medium. Original solution processes were introduced for low pressure manufacture of PE resins in the late 1950s; subsequent improvements of these processes gradually made them economically competitive with later, more advanced technologies.

In the early 1990s, solution processes acquired new importance because of their shorter residence times and ability to accommodate metallocene catalysts. Many heterogeneous multicenter Ziegler catalysts produce superior LLDPE resins with a better branching uniformity if the catalyst residence time in a reactor is short. Solution processes usually operate at residence times of around 5–10 min or less and are ideal for this catalyst behavior. Solution processes, both in heavy solvents and in the polymer melt, are inherently suitable to accommodate soluble metallocene catalysts (52). For this reason, these processes were the first to employ metallocene catalysts for LLDPE and VLDPE manufacture.

Commercial solution processes were developed by Dow Chemical, Du Pont, and Stamicarbon/DSM (45). PE becomes soluble in saturated C_6–C_{10} hydrocarbons above 120–130°C. The Du Pont process (Fig. 5) uses cyclohexane as a solvent (45,53), whereas the Dow solution process uses a mixture of saturated C_8–C_{10} hydrocarbons (1,54). In the Dow process, ethylene and hydrogen are combined in a single stream and introduced into the liquid mixture containing the solvent and a comonomer, 1-octene. The solvent–monomer stream is then continuously fed into the reactor. Catalyst components are premixed and also continuously injected into the reactor. Copolymerization reactions are carried out in the temperature range of 130–250°C, at reactor pressures of 3–20 MPa (30–200 atm), and with an ethylene content of 8–10%. After 5–10 minutes, the product stream leaves the reactor and is discharged into a vessel where the molten polymer is separated from unreacted monomers, hydrogen, and solvent, followed by pelletization (1). Three properties of LLDPE resins are controlled independently: the molecular weight, the copolymer composition, and in the case of metallocene catalysts the degree of long-chain branching. The parameters used to control the Dow process are temperature, pressure, monomer and polymer

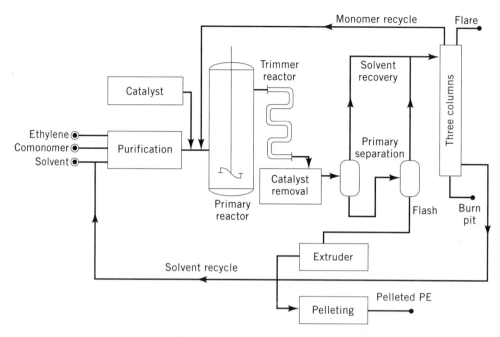

Fig. 5. Solution process (Du Pont).

concentration, residence time, and catalyst concentration. Dow Chemical manufactures two groups of LLDPE products with this process. One group (Dowlex resins) includes ethylene–1-octene resins with a nonuniform branching distribution produced with Ziegler catalysts; the other group (Engage and Affinity resins) are uniformly branched VLDPE resins produced with metallocene catalysts.

The second type of solution polymerization concept uses mixtures of supercritical ethylene and molten PE as the medium for ethylene polymerization. Some reactors previously used for free-radical ethylene polymerization in supercritical ethylene at high pressure (see OLEFIN POLYMERS, LOW DENSITY POLYETHYLENE) were converted for the catalytic synthesis of LLDPE. Both stirred and tubular autoclaves operating at 30–200 MPa (4,500–30,000 psig) and 170–350°C can also be used for this purpose. Residence times in these reactors are short, from 1 to 5 minutes. Three types of catalysts are used in these processes. The first type includes pseudo-homogeneous Ziegler catalysts. In this case, all catalyst components are introduced into a reactor as liquids or solutions but form solid catalysts when combined in the reactor. Examples of such catalysts include titanium tetrachloride as well as its mixtures with vanadium oxytrichloride and a trialkylaluminum compound (53,54). The second type of catalysts are solid Ziegler catalysts (55). Both of these catalysts produce compositionally nonuniform LLDPE resins. Exxon Chemical Co. uses a third type of catalysts, metallocene catalysts, in a similar solution process to produce uniformly branched ethylene copolymers with 1-butene and 1-hexene called Exact resins (56).

Slurry (Suspension) Polymerization. This polymerization technology is the oldest for PE production. Despite the fact that Ziegler catalysts, supported chromium-oxide catalysts, and metallocene catalysts have all been described in the patent literature for LLDPE slurry processes, it is not widely practiced. The main obstacle is the significant swelling of copolymer particles in hydrocarbon solvents that occurs at high temperatures due to the high content of amorphous phase in LLDPE resins. Such swelling can severely limit the accessible range of resin densities and production rates. The degree of swelling depends on the solvent. If relatively heavy solvents (eg, hexane or cyclohexane) are used, the swelling limits the resins to the MDPE range with density of around 0.930 g/cm^3, whereas if very light solvents are used, this range is expanded to around 0.925–0.923 g/cm^3 (1,54,57). In addition, when compositionally nonuniform LLDPE resins are produced in heavy solvents at increased temperatures, the components with the highest α-olefin content dissolve in the solvents.

Phillips Petroleum Co. developed an efficient slurry process used for the production of both HDPE and LLDPE (Fig. 6). The reactor is built as a large folder loop containing long runs of pipe from 0.5 to 1 m in diameter connected by short horizontal stretches of pipe. The reactor is filled with a light solvent (usually isobutane) which circulates through the loop at high speed. A mixed stream containing ethylene and comonomers (1-butene, 1-hexene, 1-octene, or 4-methyl-1-pentene can be used), combined with a recycled diluent and a catalyst slurry, is fed into the reactor. Within it, ethylene copolymer forms discrete particles growing around catalyst particles. Specially formulated chromium oxide-based catalysts are used to produce LLDPE resins with broad MWDs characterized by MFR values in the 80–200 range (10). The reactors operate at total pressures of up to 3 MPa (450 psig) and at temperatures of 60–75°C. The temperature is a crucial operating variable and must be closely controlled to avoid copolymer swelling. The concentration of polymer particles in the slurry is maintained at around 20–25 wt % and should also be carefully controlled to prevent particle agglomeration. After a residence time from 1.5 to 3 hours, the resin precipitates

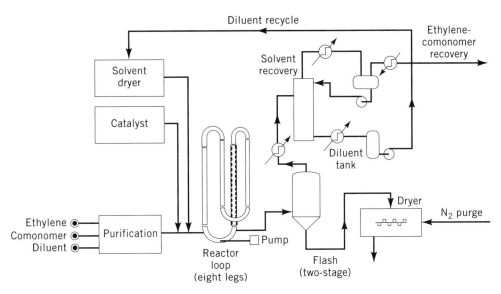

Fig. 6. Slurry process (Phillips Petroleum).

briefly in settling legs at the bottom of the loop and is discharged into a flash tank. Finally, the separated solvent and monomers enter the solvent-recovery system for purification and recycling.

Processing

Rheology of LLDPE. All LLDPE processing technologies involve resin melting; viscosities of typical LLDPE melts are between 5000 and 70,000 Pa·s (50,000–700,000 P). The main factor that affects melt viscosity is the resin molecular weight; the other factor is temperature. Its effect is described by the Arrhenius equation with an activation energy of 29–32 kJ/mol (7–7.5 kcal/mol) (58).

LLDPE melts in the 140–250°C range (a typical processing range) are non-Newtonian liquids; their effective viscosity is significantly reduced when the melt flow speed is increased (59,60). This phenomenon is called shear-thinning; it plays an important role in resin processing. The resins with an expressed shear-thinning capability have decreased viscosities and hence a greatly reduced energy demand at high speed processing. The shear-thinning capability is strongly influenced by the MWD width of a resin: compositionally nonuniform copolymers with MFR values in the 28–30 range can reduce their viscosity by 6–8 times at high processing speeds, whereas most compositionally uniform resins with narrow MWD prepared with metallocene catalysts have a nearly constant melt viscosity (2). This deficiency of metallocene-derived LLDPE and VLDPE resins has been overcome through the use of specially designed metallocene catalysts in a solution process (1,38). The resins produced with this technology have long-chain branching that significantly increases their melt flow ratios and enhances their shear-thinning capability. As a result, such resins have the processibility of LDPE but the strength and toughness of LLDPE (1,2). LLDPE resins produced with chromium oxide-based catalysts have very broad MWDs, with MFR values of 80–200. Their shear-thinning capability is high, and their melt viscosity can be reduced up to 100 times at high processing rates, which allows for the processing of copolymers with quite high molecular weights under standard conditions.

Processing Methods. LLDPE is easily processed by most conventional techniques due to its low melting point and high chemical stability; the principal techniques have been described (60–65).

Film Manufacture. Most LLDPE produced worldwide is made into thin film, either melt blown or cast from the melt. Blown film is produced by extrusion of LLDPE melt through a circular die with a large diameter, up to 100–120 cm, and a narrow gap, usually less than 1 mm (60,66,67). The extruded thin-walled tube of molten polymer is pulled in a vertical direction, where it immediately expands from internal air pressure and forms a bubble of a larger diameter. Typically, the blow-up ratio (BUR), the ratio of the final film tube diameter to the circular-die diameter, is between 1.5:1 and 4:1, thus allowing the formation of a continuous film trunk up to 4 m in diameter. The film, usually 0.007–0.125 mm thick, is air-cooled and rolled. LLDPE cast film is manufactured by depositing polymer melt on a rotating heated drum with a highly polished surface. Compositionally uniform LLDPE resins produced with metallocene catalysts can be easily processed into film by using standard equipment with minor modifications (68,69).

Injection Molding. Injection molding is used for the manufacture of LLDPE articles of complex shapes. An injection molding machine consists of two units, an injection unit (an extruder) and a clamp unit (a mold). LLDPE pellets are fed into a single-screw extruder, and the melt is injected at 200–260°C and 70–140 MPa (10,150–20,300 psi) through a nozzle into the mold (60,62,63). The mold consists of two steel platens, one fixed and one movable. Together they form a cavity into which the resin melt is injected. After the mold is filled with LLDPE melt, it is held under pressure for a short time and then rapidly cooled and opened. The duration of the molding cycle depends on the melt viscosity and the rate of polymer crystallization. Because LLDPE crystallizes rapidly, molding cycles are short, typically from 10 to 30 seconds. Molds usually accommodate up to 50 or more articles formed in a single shot.

Blow Molding. Bottles and simple containers are manufactured in large quantities by the blow molding technique (59,63,64). A thick-walled tube of LLDPE melt is extruded downward through a circular die and embraced by a split metal mold. The bottom part of the mold pinches and closes the tube of molten resin; the closed bubble is then blown by air pressure to conform to the internal surface of the mold into an article. To prevent sagging of the free-hanging tube of molten polymer from under its own weight, LLDPE resins with high molecular weights and high melt viscosities are used in this method.

Rotational Molding. Large containers and some toys are manufactured with a specialized technique called rotational molding. A rotational molding machine contains several large metal molds that can be rotated in two perpendicular planes. A load of fine LLDPE particles is introduced into each mold, and the mold assembly is transferred into an oven. Inside the oven, heated molds rotate at speeds ranging from 10 to 40 rpm. The polymer powder melts and is uniformly distributed on the internal surface of each mold. After the container is formed, the mold assembly is removed from the oven and cooled; at which point the plastic containers are removed.

Extrusion. Extrusion applications include resin pelletization after LLDPE synthesis, and manufacture of thick film, sheet, pipe, tubing, and insulated wire (60,70,71). Extrusion is carried out at around 150°C and 40–50 MPa (5800–7250 psi). Its production line consists of a mono- or twin-screw extruder, a cooling and shaping device, a pulling device (usually a roller), and a cutter. Pipes (>1 cm dia) and tubing (<1 cm dia) are produced by extruding LLDPE melt through a circular die. The molten tube enters a vacuum calibrator where it is pressed by air against sizing rings and cooled. The extrusion rate and the drawdown ratio (the ratio of the cross-sectional area of the die opening to the tube wall) can be adjusted by using mechanical pulling devices. Modified extruders are also used to coat rolls of wire and cable with LLDPE skin.

Shipment, Specifications, and Standards

Large quantities of pelletized LLDPE are shipped by rail in hopper cars with a capacity of 80–100 tons. Smaller amounts are shipped in corrugated cardboard boxes (1.0 × 1.15 × 0.90 m) with a capacity of 450–500 kg of resin.

ASTM D1248-84 classification is based on the two most easily measured characteristics of LLDPE: density and melt index. The former determines the type of a resin and the latter its category. PE resins with a nominal density of 0.926–0.940 g/cm^3 belong to Type II (medium density resins), those with a density of 0.910–0.925 to Type I (low density resins), and resins with a density below 0.910 to Type 0 (very low density resins). The five categories of PE are based on their melt index measured according to ASTM D1238-84. For category 1, the melt index is >25 g/10 min; for 2, it is 10–25; for 3, 1–10; for 4, 0.4–1.0; and for 5, it is <0.4 g/10 min.

Other characteristics, predominantly color, are categorized by three classes of PE, designated as A, B, and C. These classes specify the color, amount, and type of antioxidants and other additives. Class A covers naturally colored PE, Class B white and black polymer, and Class C weather-resistant black polymer containing no less than 2% carbon black.

Economic Aspects

Most LLDPE grades are commodity plastics manufactured in large quantities worldwide. LLDPE production is projected to increase tenfold over the years 1983 to 2005 (72):

year	1983	1990	1995	2005
production, 10^3 t/yr	1,460	4,740	7,760	15,250

Countries producing commodity LLDPE and their capacities, as well as production volumes of some U.S. companies, are listed in Table 5. In most cases, an accurate estimate of the total LLDPE production capacity is complicated by the fact that a large number of plants are used, in turn, for the manufacture of either HDPE or LLDPE in the same reactors. VLDPE and LLDPE resins with a uniform branching distribution were initially produced in the United States by Exxon Chemical Co. and Dow Chemical Co. However, since several other companies around the world have also announced their entry into this market, the worldwide capacity of uniformly branched LLDPE resins in 1995 is expected to reach a million tons. Special grades of LLDPE resins with broad MWD are produced by Phillips Petroleum Co. under the trade name Low Density Linear Polyethylenes or LDLPE.

In the case of commodity-grade LLDPE resins, the total production expenses are dominated by the cost of raw materials, primarily ethylene and the comonomer (especially the more expensive 1-hexene and 1-octene), but also hydrogen, diluent, catalyst, and additives. The combined cost of two monomers usually amounts to about 75–80% of the total production expenses. Other expenses include utilities costs (power, cooling water, steam, and fuel), ~5%; operating costs (maintenance); and overhead expenses. Comparison of investment and operating costs for manufacturing LLDPE is difficult because it depends on the operating policy and marketing strategy of a particular company. The manufacturing cost is influenced by such factors as product mix, marketing strategy, plant layout and design, quality control, environmental requirements and

Table 5. Worldwide Production Capacities[a] for LLDPE in 1994

Country	Company	Capacity, 10^3 t/yr
North America		
United States		5150
	Chevron	200
	Dow Chemical	760
	Eastman Chemical	120
	Exxon Chemical	410
	Formosa Plastics	260
	Himont	200
	Mobil Chemical	550
	Phillips Petroleum	1000
	Quantum Chemical	450
	Solvay Polymers	45
	Union Carbide	1150
Canada		1300
Central and South America		
Argentina		140
Brazil		390
Venezuela		210
Europe		
Austria		80
Belgium		120
Czech Republic		120
France		665
Germany		485
Hungary		160
Italy		360
Netherlands		360
Norway		20
Spain		190
Sweden		75
United Kingdom		125
Asia and the Pacifics		
Australia		90
China		620
India		320
Indonesia		200
Iran		60
Japan		725
Libya		80
Malaysia		200
Pakistan		30
South Korea		660
Taiwan		120
Thailand		140
Africa		
South Africa		80

[a]Includes swing-plant facilities.

safety, and inventory. As a rule, the most capital-intensive process is the solution process, which entails high capital investment (machinery operating at high pressures) as well as expenses for solvent and α-olefin recovery, purification, and recycling. The least capital-intensive process, on the other hand, is the gas-phase process.

As can be expected, different grades of commodity LLDPE containing different α-olefins and intended for different applications command different prices. The mid-1994 list price for large-volume grades of ethylene–1-butene LLDPE resins in the United States ranged from \$0.75 to \$0.80/kg for film resins and from \$0.90 to \$1.00/kg for blow molding and rotomolding resins; the prices were 7–9 cents higher for ethylene–1-octene resins. The European LLDPE prices were 1.00–1.05 DM/kg. The price of uniformly branched VLDPE resins produced with metallocene catalysts is around \$1–3/kg, depending on grade. By mid-1995, the LLDPE prices had increased by 50–55%.

Analytical and Test Methods

Structure and Composition.   Commonly evaluated structural parameters of LLDPE include the type of α-olefin used in copolymerization, content of olefin in the resin, type and degree of unsaturation, degree of oxidation, and crystallinity. The type and content of an olefin (the degree of short-chain branching) is measured by nmr and ir methods (73–75). ASTM method D2238 employs ir techniques for estimating the total methyl group content in resins. Furthermore, ir spectroscopic methods also identify such features as unsaturation (ASTM D3124-72) and the carbonyl groups. A number of methods are used to measure LLDPE crystallinity: x-ray diffraction, ir, dsc, and density measurement (76). Calorimetry measurements give the melting point and the heat of fusion.

Density.   Density of LLDPE is measured by flotation in density gradient columns according to ASTM D1505-85. The most often used liquid system is 2-propanol–water, which provides a density range of 0.79–1.00 g/cm^3. This technique is simple but requires over 50 hours for a precise measurement. The correlation between density (d) and crystallinity (CR) is given by $1/d = CR/d_{cr} + (1 - CR)/d_{am}$, where the density of the crystalline phase, d_{cr}, is 1.00 g/cm^3 and the density of the amorphous phase, d_{am}, is 0.852–0.862 g/cm^3. Ultrasonic methods (Tecrad Co.) and solid-state nmr methods (Auburn International, Rheometrics) have been developed for crystallinity and density measurements of LLDPE resins both in pelletized and granular forms.

Compositional Uniformity.   The degree of compositional uniformity is measured experimentally by temperature-rising elution fractionation (77–79), a very precise but slow and expensive method. A small amount of a resin is dissolved in an aromatic solvent at a temperature of 120–130°C; the polymer is slowly precipitated on an inert support by decreasing temperature. The precipitated polymer is then slowly redissolved in a fresh solvent flowing through the chamber with the residue; subsequently, the concentration of the dissolved polymer is measured as a function of temperature. This method is able to provide the data on the relative amounts of molecules of different compositions in a given LLDPE resin (79). On the other hand, the same property can be measured semiquantitatively but more rapidly, in 20–40 min, by differential scanning calorimetry (dsc) (Fig. 1). Copolymers with a uniform compositional distribution have much lower melting points (usually 95–120°C) than those with a nonuniform distribution (125–128°C) (3). The resins of the latter type contain a significant fraction of a highly branched, low molecular weight component that is soluble in light hydrocarbons. The content of this fraction, called the extractable fraction, is measured according to the FDA analysis procedure CFR177.1520. A piece of LLDPE film is immersed in hexane at 50°C for two hours, then the solvent is evaporated and the extractable fraction is calculated as the weight ratio of the dry residue and the original film sample.

Molecular Weight and Rheological Properties. Molecular weight and MWD of LLDPE resins are measured by high temperature gel-permeation chromatography (gpc), using o-dichlorobenzene or 1,2,4-trichlorobenzene as solvents, or by the viscosimetric method (ASTM D1601-86 and D2857-87). The gpc technique simultaneously determines the number–average and weight–average molecular weights, $\overline{M}_n$ and $\overline{M}_w$, and provides detailed information on MWD of LLDPE.

Standard ASTM procedures for measuring rheological, dielectric, and mechanical properties of LLDPE are the same as for HDPE (see OLEFIN POLYMERS, HIGH DENSITY POLYETHYLENE; Table 6). The measurement of PE viscosity widely used in industry is the melt index, designated as MI or I_2, and carried out in apparatuses such as the melt index tester or extrusion plastometer. I_2 is defined as the amount of polymer melt flowing at 190°C for 10 minutes through an 8 mm-long capillary die with a 0.58-mm diameter under a load of 2.16 kg. I_2 values for most LLDPE resins can range from around 5 to 0.1. The MI value is inversely related to melt viscosity.

Table 6. Film Properties of Standard Ethylene–1-Hexene LLDPE and a Resin with Improved Mechanical Properties[a]

Property	Standard resin	Resin with improved strength
density, g/cm^3	0.917	0.917
melt index, g/10 min	0.9	0.9
melt flow ratio	28	26
FDA extractables, %	3.5	2.5
dart impact strength, g[b]	180	480
tensile strength,[c] MPa[d]		
MD	38	56
TD	33	40
elongation at break,[c] %		
MD	780	565
TD	840	775
Elmendorf tear, g[c,e]		
MD	350	460
TD	750	750

[a]Ref. 9.
[b]Av wt of the dart sufficient to break the film in 50% of tests (ASTM D4272-90, Method A).
[c]MD = machine direction; TD = transverse direction.
[d]To convert MPa to psi, multiply by 145.
[e]25.4-μm film.

The molecular weight distribution of LLDPE resins is usually characterized in industry by the ratios of melt indexes measured in the same apparatus using different loads (2.16, 10.16, and 21.6 kg). The commonly used ratios are I_{21}/I_2 (the melt flow ratio, MFR) and I_{10}/I_2. Both of these ratios approximately correlate with the $\overline{M}_w/\overline{M}_n$ ratio. For example, the dependence between $\overline{M}_w/\overline{M}_n$ and I_{21}/I_2 values for LLDPE resins with a nonuniform branching distribution is $\overline{M}_w/\overline{M}_n \approx 0.24\,(I_{21}/I_2) - 2.4$ (59). The melt index ratios are also sensitive to the presence of long-chain branching in LLDPE resins produced in certain solution processes; I_{10}/I_2 for such resins can increase from 5–6 to 8–12 (1,2).

Mechanical Properties of LLDPE Film.   Film applications account for the largest volume of LLDPE consumption. Several tests have been developed specifically to evaluate film performance; among these, dart impact strength and tear strength are the most frequently used. Based on ASTM D4272-90, Method A, dart impact strength determines the film's overall strength and is evaluated by dropping a dart with a round tip (38.1 mm dia) from the height of 66 cm on a piece of stretched LLDPE film 12.7 cm in diameter. The average weight of the dart sufficient to break the film in half of the tests is its dart impact strength. This value varies from around 100 g for film of poor quality to over 800 g for high grade film.

Tear strength describes the film's resistance to tear propagation and is measured with an Elmendorf tear tester (ASTM D1922-89). Tear strength is determined as the weight of a loaded pendulum that tears a notched piece of film. Two values are usually measured for each film sample. One determines tear propagation in the machine direction of the film (usually the lower value), the other in the transverse direction.

Chemical Properties.   LLDPE is chemically stable. Very few analyses and tests related to its chemical properties are carried out routinely. Resistance to thermal stress-cracking is determined by exposing film wrapped on a metal mandrel to hot (100°C) air for 48, 96, and 168 hours (ASTM D2951-71).

Health and Safety Factors

LLDPE by itself does not present any health-related hazard on account of its chemical inertness and low toxicity. Consequently, film, containers, and container lids made from LLDPE are used on a large scale in food and drug packaging. Some LLDPE grades produced with unsupported metallocene catalysts have an especially high purity due to high catalyst productivity and a low contamination level of resins with catalyst residue. FDA approved the use of film manufactured from these resins for food contact and for various medical applications (80). However, if LLDPE articles contain fillers, processing aids, or colorants, their health factors must then be judged separately.

LLDPE can present a certain health hazard when it burns, since smoke, fumes, and toxic decomposition products are sometimes formed in the process. Exposure to burning LLDPE can cause irritation of the skin, eyes, and mucous membranes of the nose and throat due to the presence of acrolein and formaldehyde (81). Toxicity of LLDPE pyrolysis products depends on temperature, heating rate, and the sample size (82–84).

LLDPE can be disposed of by landfill or incineration. In landfill, the material is completely inert, degrades very slowly, does not produce gas, and does not leach any pollutants into ground water. When incinerated in commercial or municipal facilities, LLDPE produces a large amount of heat (the same as heating fuel) and should constitute less than 10% of the total trash.

Uses

Film. By far the largest application for LLDPE resins (over 60% in the United States) is film. Because LLDPE film has high tensile strength and puncture resistance, it is able to compete with HDPE film for many uses. The toughness and low temperature properties of LLDPE film also exceed those of conventional LDPE. Furthermore, because LLDPE resins exhibit relatively low strain hardening in the molten state and lower extensional viscosity, it can be produced at high rates with little risk of bubble breaks.

The largest single market for LLDPE film is the trash bag market, which exceeds 50,000 t/yr. Typical film properties are shown in Table 3. Ethylene–1-butene copolymers, the cheapest copolymers, exhibit film properties that are inferior compared to those of the high strength ethylene–1-hexene and ethylene–1-octene copolymers. Ethylene–1-hexene resins with especially high film strength (both the dart impact strength and machine-direction tear strength) are produced by Mobil Chemical Co. in a gas-phase process with a modified Ziegler catalyst (9) (Table 6). These resins compete with 1-octene-based LLDPE resins produced in solution processes. Because bags manufactured from thin LLDPE film have excellent tensile strength, puncture resistance, and seal strength at thin gauges, they can be used either for packaging or as garment bags, laundry and dry-cleaning bags, and ice bags. In the last case, LLDPE film has replaced low melt index vinyl acetate copolymers.

Several LLDPE films are currently competing for a special film market: elastic stretch film for packaging. Since the principal requirements for stretch film are high strength, high elongation, and good puncture resistance (85), both blown and die-cast film can be used for this application. Stretch film can be produced from blends, by the coextrusion method, or as a single-layer LLDPE film. The resins based on ethylene–1-hexene and ethylene–1-octene copolymers are particularly suited for these purposes.

A significant volume of LLDPE film is used to manufacture large-size packaging material for food (eg, grocery sacks) and textiles, in addition to such applications as industrial sheeting and agricultural mulch film. LLDPE bags are thinner than LDPE bags; their thickness can be further reduced to around 25 μm, which makes them price-competitive with paper. The bags are easily reused, moisture-resistant, and take up less space.

Several VLDPE grades, especially compositionally uniform resins produced with metallocene catalysts, are used for the manufacture of clear LLDPE film (55,70,86,87) (see Table 4). These resins are also used as cast and blown layers for lamination and heavy-duty film (71). Compositionally uniform LLDPE resins have high toughness, clarity, and purity, as well as low levels of extractable material. These features qualify the resins as replacement for ionomers and ethylene–vinyl acetate copolymers with a high vinyl acetate content. High clarity is necessary in numerous packaging applications. High oxygen-barrier properties of metallocene-derived resins make their film especially attractive for packaging poultry, frozen foods, and vegetables (88). Such properties, moreover, make these resins and their blends with HDPE suitable for producing film for blood bags, surgical disposable bags, and medical gowns (89).

Blending with LLDPE is used to upgrade the properties and improve the processing of conventional LDPE. For example, by adding 25% of ethylene–1-butene LLDPE resin with I_2 of 0.5 to conventional LDPE resin, the dart impact strength of 75 μm film is increased from 490 to 560 g, the puncture strength from 41 to 49 J/mm (770–920 ft·lbf/in.), and the tear strength from 43 to 63 N/mm (246–360 ppi). Compositionally uniform VLDPE resins are used in blends with HDPE, commodity LLDPE, and polypropylene (PP) (70,71,89).

An important property recommending the use of LLDPE in many packaging applications is their sealability. Compositionally uniform resins are especially attractive for such use because their melting and softening points are 15–20°C lower than those of commodity LLDPE resins (Fig. 1).

Injection Molded Articles.   Injection molding is the second largest market for LLDPE, accounting for over 10% of its total consumption (90). Over half of the LLDPE consumed in injection-molding applications is used for housewares (91). LLDPE houseware products are stiffer and more resistant to impact and distortion at elevated temperatures than LDPE; also, their low temperature impact strength is superior to that of PP. Consequently, lids for glass and HDPE jars and containers made from LLDPE have excellent gloss and low warpage. Likewise, garbage cans and industrial containers made of LLDPE exhibit exceptional toughness and can withstand rough treatment. Furthermore, a low brittleness point allows LLDPE to perform well at low temperatures. Table 7 compares the mechanical properties of two resins, ethylene–1-butene LLDPE and high pressure LDPE (the latter was used for houseware applications in the past before being replaced by LLDPE). Injection molding applications for compositionally uniform LLDPE (plastomers) include clear lids for household containers, flexible parts for small appliances, and disposable oxygen face masks (92).

Blow Molded and Rotationally Molded Articles.   What distinguishes LLDPE resins formulated for the blow molding application from similar HDPE resins is their superior environmental stress-cracking resistance and lower gas permeability. These features opened new bottle markets where such properties are important. Blow-molded drum liners made from LLDPE, for instance,

Table 7. Comparison of Injection Molding Grades of LLDPE and LDPE[a]

Property	LLDPE	LDPE
melt index, g/10 min	50	40
density, g/cm^3	0.926	0.924
modulus, MPa[b]	350	210
tensile strength, MPa[b]	12.7	9.3
impact strength at 20°C, J[c]	32.5	4.1
stress-crack resistance at 23°C, min	70	<0.2

[a]Ref. 6.
[b]To convert MPa to psi, multiply by 145.
[c]To convert J to ft·lb, divide by 1.356.

can meet industry standards for low temperature toughness and environmental stress-cracking resistance. Rotational molding also capitalizes on the higher impact strength and increased stress-cracking resistance of LLDPE over both LDPE and HDPE resins. On account of these superior properties, LLDPE is able to compete with more expensive products such as cross-linked and rubber-modified PE. Thus, a large variety of molded articles with a complex configuration is manufactured from LLDPE resins, including toys, large square-edged containers, as well as tanks for agriculture and water treatment.

Pipe and Tubing. The same qualities that make LLDPE attractive for blow molding applications also play a crucial role in its being adapted for pipe manufacture, an area that accounts for about 1% of the LLDPE market. LLDPE pipes provide not only necessary flexibility, high burst strength, and high environmental stress-cracking resistance, but also higher heat-distortion temperature than LDPE and some other HDPE grades. LLDPE tubing is used for drip piping, swimming pool tubing, household hoses, etc. On account of their high purity, transparency, and flexibility, compositionally uniform LLDPE and VLDPE resins have started to replace plasticized poly(vinyl chloride) (PVC) in such specialty markets as medical tubing applications (88).

Wire and Cable Insulation. LLDPE is also widely used for wire and cable coating in electrical and telephone industry (which amounts to 2.5% of LLDPE production). Coating requires such properties as flexibility, strength, low brittleness temperature, high resistance to abrasion, and excellent dielectric properties which are typical for all PE resins. Compositionally uniform VLDPE resins with a narrow MWD produced with metallocene catalysts are especially suitable for this purpose (93,94). LLDPE-coated wire is widely used in overhead medium and low voltage power distribution, in underground power cable insulation, in data and communication networks, and in automotive and appliance wire. A significant volume of LLDPE is also used in such electrical applications as jacketing, molded accessory panels, as well as shields for semiconductors (88).

BIBLIOGRAPHY

"Low Pressure Linear (Low Density) Polyethylene" in *ECT* 3rd ed., under "Olefin Polymers," Vol. 16, pp. 385–401, by J. N. Short, Phillips Petroleum Co.

1. U.S. Pat. 5,272,236 (Dec. 21, 1993), S.-Y. Lai and co-workers (to Dow Chemical Co.).
2. G. W. Knight and S. Lai, *Proceeding of SPE Polyolefins VIII International Conference*, 1993, p. 226.
3. N. Kashiwa, *Proceedings of Worldwide Metallocene Conference, MetCon '93*, Houston, Tex., 1993, p. 237.
4. F. H. Winslow and W. L. Hawkins, in R. A. V. Raff and K. W. Doak, eds., *Crystalline Olefin Polymers*, Pt. II, Wiley-Interscience, New York, 1965, p. 361.
5. J. N. Hay, *Polymer* **22**, 718 (1981).
6. D. E. James, in J. I. Kroschwitz, ed., *Encyclopedia of Polymer Science and Engineering*, Vol. VI, 2nd ed., John Wiley & Sons, Inc., New York, 1987, p. 429.
7. K. Sehanobish and co-workers, *J. Appl. Polym. Sci.* **51**, 887 (1994).
8. G. P. Below and co-workers, *Polym. J.* **6**, 681 (1972).
9. V. J. Crotty, V. Firdaus, and R. O. Hagerty, *Proceedings of SPE Polyolefins VIII International Conference*, 1993, p. 192.

10. U.S. Pat. 5,208,309 (May 4, 1993) and 5,274,056 (Dec. 28, 1993), M. P. McDaniel and E. A. Benham (to Phillips Petroleum Co.).
11. Y. V. Kissin, *Isospecific Polymerization of Olefins with Heterogeneous Ziegler-Natta Catalysts*, Springer-Verlag, New York, 1985.
12. T. E. Nowlin, *Prog. Polym. Sci.* **11**, 29 (1985).
13. H. Sinn and W. Kaminsky, *Adv. Organomet. Chem.* **18**, 99 (1980).
14. A. V. Kryzhanovskii and S. S. Ivanchev, *Polymer Science USSR*, **32**, 1312 (1990).
15. Y. V. Kissin, in N. P. Cheremisinoff, ed., *Encyclopedia of Engineering Materials*, Part A, Vol. I, Marcel Dekker, Inc., New York, 1988, p. 103.
16. J. C. W. Chien and J. T. T. Hsieh, *J. Polym. Sci. Polym. Chem. Ed.* **14**, 1915 (1976).
17. V. A. Zakharov and Y. I. Yermakov, *Cat. Rev. Sci. Eng.* **19**, 63 (1979).
18. F. S. Dyachkovsky and A. D. Pomogailo, *J. Polym. Sci. Polym. Symp.* **68**, 97 (1980).
19. A. Munoz-Escalona, in R. P. Quirk, ed., *Transition Metal Catalyzed Polymerizations: Alkenes and Dienes*, Harwood Academic Publishers, New York, 1983, p. 323.
20. T. Dall-Occo, U. Zucchini, and I. Cuffiani, in W. Kaminsky and H. Sinn, eds., *Transition Metals and Organometallics as Catalysts for Olefin Polymerization*, Springer-Verlag, Berlin, 1988, p. 209.
21. F. J. Karol, K. J. Cann, and B. E. Wagner, in Ref. 20, p. 149.
22. Y. V. Kissin, in Ref. 19, p. 597.
23. U.S. Pat. 5,258,345 (Nov. 2, 1993), Y. V. Kissin, R. M. Mink, and T. E. Nowlin (to Mobil Oil).
24. Y. V. Kissin and D. L. Beach, *J. Polym. Sci., Part A: Polym. Chem.* **22**, 3027 (1984); **24**, 1069 (1986).
25. H. Sinn and co-workers, *Angew. Chem., Int. Ed. Engl.* **19**, 390 (1980).
26. M. S. Howie, *Proceedings of Worldwide Metallocene Conference, MetCon '93*, Houston, Tex., 1993, p. 246.
27. U.S. Pat. 5,324,800 (June 28, 1994), H. C. Wellborn and J. A. Ewen (to Exxon Chemical); W. Kaminsky, *Proceedings of the Worldwide Metallocene Conference, Metcon '95*, Houston, Tex., 1995.
28. N. F. Brockmeier, *Proceedings of Worldwide Metallocene Conference, MetCon '94*, Houston, Tex., 1994, Session 3.
29. J. C. W. Chien and B.-P. Wang, *J. Polym. Sci. Part A: Polym. Chem.* **26**, 3089 (1988).
30. U.S. Pat. 5,086,135 (Feb. 4, 1992), Y. V. Kissin (to Mobil Oil); U.S. Pat. 5,128,295 (July 7, 1992), L. Porri and co-workers (to Enichem Amic SpA).
31. U.S. Pat. 4,258,475 (Nov. 2, 1993), Y. V. Kissin (to Mobil Oil).
32. U.S. Pat. 4,937,217 (June 26, 1990), M. Chang (to Exxon Chemical).
33. W. Kaminsky and R. Spiehl, *Makromol. Chem.* **190**, 515 (1989).
34. W. Kaminsky, in Ref. 26, p. 325.
35. R. F. Jordan, *Adv. Organomet. Chem.* **32**, 325 (1991).
36. X. Yang, C. L. Stern, and T. J. Marks, *J. Am. Chem. Soc.* **113**, 3623 (1991).
37. J. C. W. Chien, W.-M. Tsai, and M. D. Rausch, *J. Am. Chem. Soc.* **113**, 8570 (1991).
38. U.S. Pat. 5,064,802 (Nov. 12, 1991); Eur. Pat. Appl. 416 815 A2 (Aug. 30, 1990), J. C. Stevens and co-workers (to Dow Chemical Co.).
39. B. Story, in Ref. 26, p. 111.
40. J. C. Stevens, in Ref. 26, p. 158.
41. U.S. Pat. 4,003,712 (Jan. 18, 1977), A. R. Miller (to Union Carbide).
42. U.S. Pat. 4,011,382 (Mar. 8, 1977), I. J. Levine and F. J. Karol (to Union Carbide).
43. U.S. Pat. 4,255,542 (Mar. 10, 1981), G. L. Brown, D. F. Warner, and J. H. Byon (to Union Carbide).
44. I. D. Burdett, *Chemtech* 616 (Oct. 1992).
45. K.-Y. Choi and W. H. Ray, *J. Macromol. Sci.-Rev. Macromol. Chem., Phys.* **C25**, 57 (1985).

46. U.S. Pat. 4,302,565 (Nov. 24, 1981), G. L. Goeke, B. E. Wagner, and F. J. Karol (to Union Carbide).
47. U.S. Pat. 5,352,749 (Oct. 4, 1994), M. L. DeChellis and J. R. Griffin (to Exxon Chemical); U.S. Pat. 5,405,922 (Apr. 11, 1995), M. L. DeChellis, J. R. Griffin, and M. E. Muhle (to Exxon Chemical).
48. B. Miller, *Plast. World* 46 (May 1992); *Chem. Eng.*, 89,66 (1982).
49. A. B. Furtek, in Ref. 26, p. 125.
50. C. Chabrand and co-workers, *Proceedings of SPO Third International Business Forum Specialty Polyolefins*, Houston, Tex., Sept. 1993, p. 55.
51. J. H. Schut, *Plast. Tech.* 27 (Nov. 1993).
52. A. Akimoto and A. Yano, in Ref. 27, Session 3.
53. J. P. Forsman, *Hydrocarbon Process.* **51**, 132 (Nov. 1972).
54. K. W. Swogger and C.-I. Kao, in Ref. 9, p. 13.
55. U.S. Pat. 4,105,842 (Aug. 8, 1978), A. Nicco and J. P. Machon, (to CdF Chimie).
56. A. Montagna and J. C. Floyd, in Ref. 26, p. 171.
57. *Plast. Tech.* 63 (Nov. 1993).
58. L. Woo, T. K. Ling, and S. P. Westphal, in Ref. 9, p. 242.
59. T. E. Nowlin, Y. V. Kissin, and K. P. Wagner, *J. Polym. Sci., Part A: Polym. Chem.* **26**, 755 (1988).
60. R. I. Tanner, *Engineering Rheology*, Clarendon Press, Oxford, U.K., 1986.
61. J. R. A. Pearson, *Mechanics of Polymer Processing*, Elsevier, London, 1985.
62. D. V. Rosato and D. V. Rosato, eds., *Injection Molding Handbook*, Van Nostrand Reinhold Co., Inc., New York, 1986.
63. I. I. Rubin, in Ref. 6, Vol. VIII, p. 102.
64. D. V. Rosato and D. V. Rosato, eds., *Blow Molding Handbook*, Hanser Publishers, Munich, Germany, 1989.
65. C. Irwin, in Ref. 6, Vol. 2, p. 447.
66. J. Bristol, *Plastics Films*, Longman, New York, 1983.
67. H. C. Park and E. M. Mount, in Ref. 6, Vol. 7, p. 88.
68. F. Goffreda, in Ref. 27, Session 2.
69. P. C. Wu, *Ibid.*
70. S. Levy, *Plastics Extrusion Technology Handbook*, Industrial Press, New York, 1981.
71. G. A. Kruder, in Ref. 6, Vol. 6, p. 571.
72. R. B. Liberman and D. Del Luca, in Ref. 9, p. 21.
73. A. Dankovic and L. Fuzes, *J. Appl. Polym. Sci.* **28**, 3707 (1983).
74. M. Ondas, M. Drobny, and E. Spirk, *Int. Polym. Sci. Technol.* **20**, 93 (1993).
75. J. P. Blitz and D. C. McFaddin, *J. Appl. Polym. Sci.* **51**, 13 (1994).
76. D. L. Beach and Y. V. Kissin, *J. Polym. Sci., Polym. Chem. Ed.* **22**, 3027 (1984).
77. L. Wild and co-workers, *J. Polym. Sci., Polym. Phys. Ed.* **20**, 441 (1982).
78. J. G. Bonner, C. J. Frye, and G. Capaccio, *Polymer* **34**, 3522 (1993).
79. M. G. Pigeon and A. Rudin, *J. Appl. Polym. Sci.* **51**, 303 (1994).
80. *Plast. Tech.* 11 (July 1993).
81. G. Hovding, *Acta. Derm. Venerol.* **49**, 147 (1969).
82. C. J. Hilado and co-workers, *J. Combust. Toxicol.* **3**, 259,305,381 (1976).
83. C. J. Hilado, J. A. Soriano, and K. J. Kosola, *J. Combust. Toxicol.* **4**, 533 (1977).
84. C. J. Hilado and N. V. Huttlinger, *J. Combust. Toxicol.* **5**, 81, 361 (1978).
85. *Pap. Film Foil Converter* **58**, 49 (1984).
86. *Plast. World* **43**, 54 (Jan. 1985).
87. D. J. Michieles, in Ref. 27, Session 2.
88. D. Smock, *Plast. World*, 16 (Jan. 1994).
89. T. C. Yu and G. J. Wagner, in Ref. 9, p. 539.
90. *Plast. World* **42**, 108 (Oct. 1984).

91. *Plast. World* **41**, 37 (Mar. 1983).
92. J. H. Schut, *Plast. Tech.* 19 (Feb. 1994).
93. U.S. Pat. 5,246,783 (Sept. 21, 1993), L. Spenadel, M. L. Hendewerk, and A. K. Metha (to Exxon).
94. G. Lancaster and co-workers, in Ref. 27, Session 2.

YURY V. KISSIN
Mobil Chemical Company

POLYPROPYLENE

Propylene polymerization processes have undergone a number of revolutionary changes since the first processes for the production of crystalline polypropylene (PP) were commercialized in 1957 by Montecatini in Italy and Hercules in the United States. These first processes were based on Natta's discovery in 1954 that a Ziegler catalyst could be used to produce highly isotactic polypropylene (1). The stereoregular, crystalline polymers produced by this technology had sufficiently attractive economic and property performance that they became significant commercial thermoplastics in a remarkably short period. Consequently, Ziegler and Natta were awarded the Nobel Prize in Chemistry in 1963. Other technologies invented during the same period (2,3), were incapable of achieving commercially viable performance. The tremendous amount of activity in olefin polymerization at this time led to a massive patent interference in the United States, and the award of a U.S. Patent to Phillips for the composition of matter of crystalline polypropylene (4,5). Improvements to the basic Ziegler-Natta catalysts made at Solvay increased the activity and stereoregularity of these catalyst systems, extending their commercial application (6). In most cases, however, supported high yield catalyst systems, invented by Montedison and Mitsui Petrochemical (7), are used in plants that have been constructed since the early 1980s (see CATALYSTS, SUPPORTED). The superior performance of these systems allows the use of simplified processes with dramatic reductions in capital and operating costs. This revolution in the economics of polypropylene production spurred an increase in worldwide capacity in the 1980s. Worldwide production of propylene polymers was 14,381,000 t/yr in 1993 (8), an annual increase of 4.7% from 7,147,000 t/yr in 1978 (9).

Interest has been generated by the advances made in the homogeneously catalyzed polymerization of propylene using metallocenes (10). Polymers produced with these catalyst systems have an unusual balance of properties that may become commercially attractive (see METALLOCENE CATALYSTS (SUPPLEMENT)).

Properties

Structure and Crystallinity. The stereochemistry of propylene polymers was first studied by Natta, who defined three possible structures of polypropylene by the location of the pendent methyl groups relative to the polymer backbone (1,11,12). Isotactic polypropylene consists of molecules in which all methyl groups have the same stereochemistry as a result of all insertions of propylene monomer being identical. Syndiotactic polypropylene is produced by regular alternating stereochemistry of monomer insertion, resulting in alternating locations of the pendent methyl groups. Atactic polypropylene, which is noncrystalline, is the result of nonstereospecific monomer insertion and random location of the pendent methyl groups (Fig. 1). A polymer structure referred to as stereoblock is an intermediate structure consisting of isotactic and atactic segments (13). When Ziegler-Natta catalysts are used, monomer is inserted in a head-to-tail manner with few head-to-head or tail-to-tail inversions (14). The degree of stereoregularity of polypropylene molecules is determined by the use of ^{13}C nmr of the polymer in solution (15). The percentage of (mmmm) pentads is used to describe the percentage of isotacticity in the polymer molecule. Meso (m) insertions produce a polymer with the same methyl configuration, while racemic (r) insertions produce polymers with alternating methyl configuration. Consequently, (rrrr) pentads are syndiotactic. Random insertions sequences, such as (mrrm) produce hetero- or atactic polymers. Information on the macromolecular structure can also be obtained from infrared spectra, which for isotactic polypropylene are affected by vibrations of the regular spiral chain and not by the polymer crystallinity. Bands are detectable at 1167, 997, and 841 cm^{-1} (16). Stereoregularity can be inferred from solubility measurements. Historically, these methods

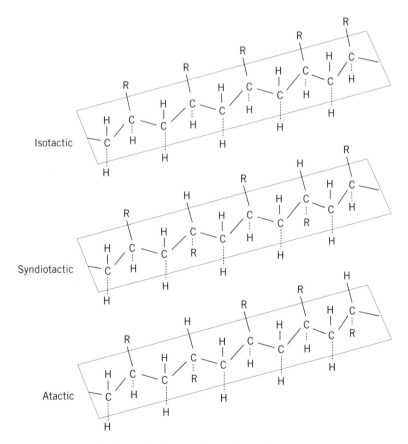

Fig. 1. Polypropylene stereoisomers.

have been used to determine stereoregularity. The isotactic index of the polymer is the fraction insoluble in boiling n-heptane (17). Solubility in xylene, or in decalin (decahydronaphthalene), is also commonly used to describe stereoregularity on a commercial basis (18); it is also possible to separate the polymer into fractions of different crystallinity (19). The use of pulsed proton solid-state nmr has been demonstrated as an alternative to solubility measurements of stereoregularity (20).

Crystallinity of polypropylene is usually determined by x-ray diffraction (21). Isotactic polymer consists of helical molecules, with three monomer units per chain unit, resulting in a spacing between units of identical conformation of 0.65 nm (Fig. 2a). These molecules interact with others, or different segments of the same molecule, to form a monoclinic unit cell containing 12 monomer units having a crystallographic density of 0.936 g/cm^3 (22). The predominant crystalline form of polypropylene is the α-form (23); however, β-, γ-, and smectic

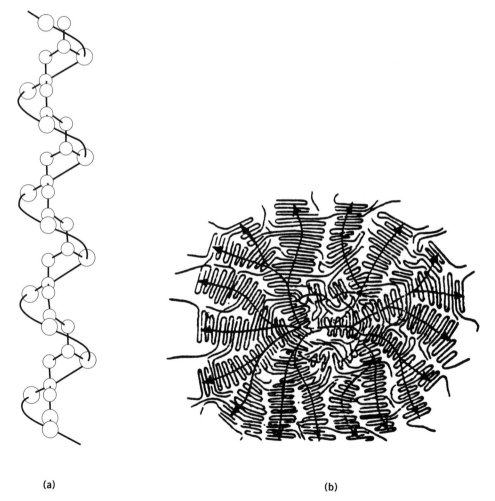

(a)   (b)

Fig. 2. (a) Chain conformation of isotactic polypropylene, and (b) model of a polypropylene spherulite.

forms can also be obtained. Smectic polypropylene, obtained by rapid cooling of the polymer melt at low temperatures, has no wide-range crystalline order (24). The β-form is obtained by rapid cooling at temperatures between 100 and 130°C (25). The γ-form is obtained by crystallization under high pressure (26). Both the β- and γ-forms can be converted to the α-form by heating.

Syndiotactic polypropylene also forms helical molecules; however, each chain unit consists of four monomer units having a spacing of 0.74 nm. The unit cell is orthorhombic and contains 48 monomer units having a crystallographic density of 0.91 g/cm^3 (27).

Polypropylene molecules repeatedly fold upon themselves to form lamellae, the sizes of which are a function of the crystallization conditions. Higher degrees of order are obtained upon formation of crystalline aggregates, or spherulites. The presence of a central crystallization nucleus from which the lamellae radiate is clearly evident in these structures. Observations using cross-polarized light illustrates the characteristic Maltese cross model (Fig. 2**b**). The optical and mechanical properties are a function of the size and number of spherulites and can be modified by nucleating agents. Crystallinity can also be inferred from thermal analysis (28) and density measurements (29).

Molecular Weight. The molecular weight of polypropylene is typically determined by viscosity measurements. The intrinsic viscosity [η] of the polymer in solution is related to the molecular weight, M_v, by the Mark-Houwink equation:

$$[\eta] = KM_v^\alpha$$

where $K = 2.38 \times 10^{-4}$ and $\alpha = 0.725$ for determinations in decahydronaphthalene at 135°C (30). The melt viscosity, or melt flow rate, measured under standard conditions (31), can also be correlated to molecular weight. Because of the distribution of molecular weights in polypropylene, and the relationship of this distribution to important polymer properties, considerable effort has been exerted to measure the molecular weight distribution of these polymers. Typically, size exclusion or gel permeation chromatography is used to obtain a curve of relative weight fraction against molecular weight (32). The dynamic shear behavior of the polymer melt can be used to determine the ratio of weight average, M_w, to number average, M_n, molecular weight (33). High molecular weight components, which can be important in extrusion processes, can be accurately measured in steady shear experiments (34).

Thermodynamic Properties. The thermodynamic melting point for pure crystalline isotactic polypropylene obtained by the extrapolation of melting data for isothermally crystallized polymer is 185°C (35). Under normal thermal analysis conditions, commercial homopolymers have melting points in the range of 160–165°C. The heat of fusion of isotactic polypropylene has been reported as 88 J/g (21 cal/g) (36). The value of 165 ± 18 J/g has been reported for a 100% crystalline sample (37). Heats of crystallization have been determined to be in the range of 87–92 J/g (38).

The value of the glass-transition temperature, T_g, is dependent on the stereoregularity of the polymer, its molecular weight, and the measurement techniques used. Transition temperatures from −13 to 0°C are reported for isotactic polypropylene, and −18 to 5°C for atactic (39,40).

Syndiotactic polypropylene has an ultimate melting point of 174°C, and extrapolated heat of fusion of 105 J/g (25.1 cal/g); both lower than those of isotactic polymer. The heat of fusion of the polymer produced using a metallocene catalyst is reported as 79 J/g (19 cal/g) (41).

Physical Properties and Commercial Specifications. A plethora of polypropylene grades are sold in the United States, including those having variations in melt flow rate, stabilization package, use of filler, and use of comonomers. The basic categories of propylene polymers are homopolymers, impact or heterophasic copolymers, and random copolymers. A number of grades are approved for food contact (42). Many polymers carry a UL94 flammability class HB rating, and some specialized formulations are available with V-0 or V-2 ratings. Standardized test methods, ASTM in the United States, ISO in other countries, are used in commercial specifications. The properties that distinguish polypropylene from high density polyethylene are high heat distortion temperature, stiffness, hardness, and lower density (see OLEFIN POLYMERS, HIGH DENSITY POLYETHYLENE). Polypropylene is resistant to most common solvents.

Properties of various homopolymer grades are given in Table 1. Polymer molecular weights are manufactured to be consistent with the best conditions for each fabrication process, high molecular weights (low melt flow rates) for extrusion, and lower molecular weights (higher melt flow rates) for injection molding. The polymers shown in Table 1 are produced by direct polymerization; however,

Table 1. Properties of Homopolymers[a]

Properties	ASTM test	Extrusion, sheet	General-purpose injection molding		Injection molding thin complex parts	
melt flow, g/10 min	D1238L	0.8	4	12	20	35
density, g/cm^3	D792A-2	0.903	0.903	0.903	0.903	0.902
tensile strength,[b] MPa[c]	D638	35	35	35	32	33
elongation,[b] %	D638	13	12	11	11	12
flexural modulus 1% secant, MPa[c]	D790B	1700	1700	1600	1500	1450
Rockwell hardness, R scale	D785A	95	99	100	100	98
deflection temperature at 455 kPa,[d] °C	D648	95	97	92	91	90
notched Izod impact at 23°C, J/m[e]	D256A	130	40	35	21	32

[a]Ref. 43.
[b]At yield.
[c]To convert MPa to psi, multiply by 145.
[d]To convert kPa to psi, multiply by 0.145.
[e]To convert J/m to ft·lb/in., divide by 53.38.

higher melt flow grades can also be produced by peroxide-initiated degradation of low melt flow polymers. High crystallinity grades provide additional stiffness and higher heat distortion temperature than conventional homopolymer grades (44,45).

Polypropylene polymers are typically modified with ethylene to obtain desirable properties for specific applications. Specifically, ethylene–propylene rubbers are introduced as a discrete phase in heterophasic copolymers to improve toughness and low temperature impact resistance (see ELASTOMERS, ETHYLENE–PROPYLENE RUBBER). This is done by sequential polymerization of homopolymer polypropylene and ethylene–propylene rubber in a multistage reactor process or by the extrusion compounding of ethylene–propylene rubber with a homopolymer. Addition of high density polyethylene, by polymerization or compounding, is sometimes used to reduce stress whitening. In all cases, a superior balance of properties is obtained when the size of the discrete rubber phase is approximately one micrometer. Examples of these polymers and their properties are shown in Table 2. Mineral fillers, such as talc or calcium carbonate, can be added to polypropylene to increase stiffness and high temperature properties, as shown in Table 3.

Table 2. Properties of Impact Resistant Copolymers[a]

Properties	Medium impact		Injection molding, thin parts	High impact		Blush resistance
	Extrusion, sheet	Injection molding		Extrusion, sheet	Injection molding	
melt flow, g/10 min	0.5	4	0.30	2	4	1
density, g/cm^3	0.9	0.899	0.897	0.900	0.901	0.902
tensile strength,[b] MPa[c]	28	28	24	21	21	22
elongation,[b] %	15	10	10	12	10	10
flexural modulus 1% secant, MPa[c]	1270	1300	1050	970	1000	1000
Rockwell hardness, R scale	80	82	80	61	65	75
deflection temperature at 455 kPa,[d] °C	85	85	80	72	75	77
notched Izod impact at 23°C, J/m[e]	530	130	70	640	530	210

[a]Ref. 43.
[b]At yield.
[c]To convert MPa to psi, multiply by 145.
[d]To convert kPa to psi, multiply by 0.145.
[e]To convert J/m to ft·lb/in., divide by 53.38.

Table 3. Properties of Filled Polypropylenes[a]

Properties	ASTM test	Talc filled				Calcium carbonate filled			
		Homopolymer		Copolymer		Homopolymer		Copolymer	
filler, wt %		20	40	20	40	20	40	20	40
melt flow, g/10 min	D1238L	4	4	4	4	4	4	4	4
density, g/cm³	D792A-2	1.05	1.22	1.04	1.22	1.05	1.22	1.03	1.20
tensile strength,[b] MPa[c]	D638	34.5	32	27	25	32	25.5	25.5	21
flexural modulus 1% secant, MPa[c]	D790B	2550	3400	2100	2700	2300	2750	1800	2200
Rockwell hardness, R scale	D785A	98	97	90	85	98	97	88	87
deflection temperature at 455 kPa,[d] °C	D648	125	134	117	127	113	113	98	104
notched Izod impact at 23°C, J/m[e]	D256A	37	37	80	37	48	43	91	69

[a] Ref. 46.
[b] At yield.
[c] To convert MPa to psi, multiply by 145.
[d] To convert kPa to psi, multiply by 0.145.
[e] To convert J/m to ft·lb/in., divide by 53.38.

790

Random insertion of ethylene as comonomer and, in some cases, butene as termonomer, enhances clarity and depresses the polymer melting point and stiffness. Propylene–butene copolymers are also available (47). Consequently, these polymers are used in applications where clarity is essential and as a sealant layer in polypropylene films. The impact resistance of these polymers is slightly superior to propylene homopolymers, especially at refrigeration temperatures, but still vastly inferior to that of heterophasic copolymers. Properties of these polymers are shown in Table 4.

Table 4. Properties of Random Copolymer[a]

Properties	ASTM test	Blow molding	Injection molding[b]
melt flow, g/10 min	D1238L	2	10
density, g/cm^3	D792A2	0.901	0.900
tensile strength,[c] MPa[d]	D638	33	30
elongation,[c] %	D638	14.5	14
flexural modulus 1% secant, MPa[d]	D790B	1400	1200
Rockwell hardness, R scale	D785A	89	88
deflection temperature at 455 kPa,[e] °C	D648	94	80
notched Izod impact at 23°C, J/m[f]	D256A	315	70

[a]Ref. 43.
[b]High clarity resin.
[c]At yield.
[d]To convert MPa to psi, multiply by 145.
[e]To convert kPa to psi, multiply by 0.145.
[f]To convert J/m to ft·lb/in., divide by 53.38.

Manufacture

CATALYSTS

TiCl$_3$-Based Catalysts. Isotactic polypropylene was first synthesized by Natta in 1954, using a catalyst system consisting of TiCl$_4$ and Al(C$_2$H$_5$)$_3$ (1,12,48). This system, based on Ziegler's catalyst for polyethylene, produced a large fraction of polymer with poor structural uniformity and properties. Natta realized that polymer isotacticity was related to uniformity of the catalyst site, and developed solid crystalline TiCl$_3$ catalysts obtained by the reduction of TiCl$_4$. These catalysts, activated with Al(C$_2$H$_5$)$_2$Cl or Al(C$_2$H$_5$)$_3$, dramatically increased the percentage of isotactic polymer. Although all four crystal forms of TiCl$_3$, α, β, γ, and δ, are active as catalysts, the best results were obtained using δ-TiCl$_3$ activated by Al(C$_2$H$_5$)$_2$Cl (49). These catalysts enabled rapid commercialization of the production of isotactic polypropylene by Montecatini in Italy in 1957 and by Hercules in the United States later that year. Improvements in $TiCl_3$ catalyst systems were made by using $AlCl_3$ in solid solution in place of pure $TiCl_3$ (50) and by the use of electron donors, such as carboxylic acid esters, to increase stereoregularity (51) (see CARBOXYLIC ACIDS). Processes based on these catalyst systems required equipment for the removal of the undesirable atactic polymer and the deactivation and separation of catalyst residues.

$TiCl_3$ catalysts produced by the reduction of $TiCl_4$ with $Al(C_2H_5)_2Cl$, and subsequently treated first with an electron donor (diisoamyl ether), then with $TiCl_4$, are highly stereospecific and four to five times more active than δ-$TiCl_3$ (6). These catalysts were a significant advance over the earlier $TiCl_3$ systems, because removal of atactic polymer was no longer required. They are often referred to as second-generation catalysts. The life of many older slurry process facilities has been extended by using these catalysts to produce "clean" polymers with very low catalyst residues.

$MgCl_2$-Supported Catalysts.   Examination of polymerizations with $TiCl_3$ catalysts has established that only a small percentage of titanium located on lateral faces, edges, and along crystal defects is active (52) (see TITANIUM AND TITANIUM ALLOYS). This led to the recognition that much of the catalyst mass acted only as a support, promoting considerable activity aimed at finding a support for active titanium that would not be detrimental to polymer properties.

Magnesium chloride, in active form as a support for $TiCl_4$, has been found to significantly increase catalyst activity, enabling the design of processes in which removal of catalyst residues from the polymer is not required (53). The use of electron donors as stereoregularity control agents reduces formation of the undesirable atactic polymer, giving stereoregularity sufficient for commercial production of polypropylene. Early commercial $MgCl_2$-supported polypropylene catalysts were prepared by the co-milling of $MgCl_2$ with an electron donor, usually ethyl benzoate, either in the presence of $TiCl_4$ or followed by treatment with $TiCl_4$ (7). The presence of the electron donor in the co-milling process promotes the activation of $MgCl_2$; that is, conversion to the disordered δ-form with reduced crystallite size. The δ-form of $MgCl_2$ resembles the active δ-form

Table 5. Evolution of Propylene Polymerization Catalysts

	First generation	Second generation	Third generation
catalyst	$TiCl_3 \cdot \frac{1}{3} AlCl_3$	$TiCl_3$ (treated)	$TiCl_4/MgCl_2/ED^a$
cocatalyst	$Al(C_2H_5)_2Cl$	$Al(C_2H_5)_2Cl$	$Al(C_2H_5)_3$
stereo control modifier			aromatic acid esters
yield, kg PP/g cat.	0.8–1.2	2–5	5
isotactic index, %	88–91	95	92
particle type	irregular powder	regular powder	irregular powder
process requirements	atactic removal, catalyst residue removal	catalyst residue removal, no atactic removal	atactic removal, no catalyst removal

^{a}ED = electron donor.

of $TiCl_3$ (54) and binds the titanium compound to the extent that it cannot be removed by repeated washings with hydrocarbons (qv) or by treatment under vacuum. This strong binding is attributed to the near identical Mg–Cl and Ti–Cl interatomic distances in these crystals (55).

The first commercial $MgCl_2$-supported catalyst systems usually contained ethyl benzoate as a stereoregulating compound (inside donor) in the solid catalyst component and triethyl aluminum, along with methyl *p*-toluate or other aromatic carboxylic acid ester (outside donor) as co-catalyst. Catalysts of this type, referred to as third-generation catalysts, are primarily used in processes that remove the atactic fraction of the polymer. Dramatic increases in catalyst activity and stereoregularity have been obtained by the use of phthalate esters as inside donors and alkoxy silanes as outside donors (56). These newer catalysts, referred to as superactive third-generation catalysts, facilitate the use of processes, using either liquid or gaseous monomer in which neither atactic polymer or catalyst residues are removed. Catalyst manufacturing methods have also improved, producing uniform regularly shaped catalyst particles and improving polymer handling in the polymerization process. The production of highly porous, regularly shaped catalysts, uniquely suited for the production of heterophasic copolymers, represents the fourth generation of these Ziegler-Natta catalysts (57). The evolution of these catalysts is summarized in Table 5. $MgCl_2$-supported catalyst systems that do not require external donors have been disclosed (58,59).

Active catalyst systems have also been produced using silica as a support for $MgCl_2$ and $TiCl_4$ (60). A number of magnesium-containing compounds such as alkylmagnesium (61), magnesium hydrocarbyl carbonates (62), magnesium alkanoates (63), magnesium alkoxides and aryloxides (64), and alkyl magnesium chloride (65), can be used as the source of activated $MgCl_2$ in addition to the alcoholates of $MgCl_2$ or anhydrous $MgCl_2$. In all cases, these systems contain active centers consisting of $TiCl_4$ supported on activated $MgCl_2$.

Metallocene Catalysts. The use of the dicyclopentadienyl titanium dichloride–diethyl aluminum chloride system to catalyze the polymerization of ethylene homogeneously was first reported in 1957 (66). Dramatic improvements in the performance of these metallocene catalysts were obtained by using

Superactive	Fourth generation	Fifth generation
$TiCl_4/MgCl_2/ED^a$	$TiCl_4/MgCl_2/ED^a$	metallocene
$Al(C_2H_5)_3$	$Al(C_2H_5)_3$	alumoxane
alkoxy silanes	alkoxy silanes	
20	>30	>30
98	98	>98
regularly shaped particles	spherical controlled porosity catalysts	
no purification required	no purification required	under development

methylalumoxane as cocatalyst and dicyclopentadienyl zirconium dichloride as catalyst (67). Modification of the zirconocene by the addition of substituents to the cyclopentadiene rings provided the capacity of achieving high molecular weight polyethylene at economical process temperatures (68). The discovery that the catalyst system, consisting of racemic–ethylenebis(indenyl) titanium dichloride and methylalumoxane cyclic oligomers, is stereospecific for the isotactic polymerization of propylene (69), and that isopropyl(cyclopentadienyl-1-fluorenyl) hafnium dichloride can be used in stereospecific syndiotactic polymerization (70) has led to a dramatic increase in research in the homogeneously catalyzed polymerization of propylene (71). Modifications of the organic substituents have resulted in the development of metallocene catalysts capable of producing isotactic polypropylene with similar melting points and molecular weights to commercial Ziegler-Natta propylene (72–74). These catalysts contain only one type of relatively well-characterized active site, producing polymers with narrow molecular weight distributions ($M_w/M_n = 2$). The distribution of stereoirregularities is also more uniform, leading to reduced production of a fraction of polymer that is truly atactic. A number of patents have been issued describing methods of supporting these catalysts on silica, insoluble alumoxanes, or organic polymers (75–77). Successful commercialization of these supported catalyst systems would allow the conversion of facilities used with high yield/high stereoregularity Ziegler-Natta catalysts. Polypropylenes produced with metallocene catalysts are not yet available commercially; however, Fina has sampled some potential customers with syndiotactic polypropylene produced at their Texas plant. Hoechst, which announced the development of a process for isotactic polypropylene using metallocenes (78), has entered into a joint research and patent agreement with Exxon (79). It is expected that this agreement will facilitate the rapid commercialization of this technology.

MECHANISM AND KINETICS

Shortly after the discovery of Ziegler-Natta catalysts, it was suggested that chain propagation occurred by monomer insertion into a metal–carbon bond of the catalyst polarized with a weak negative charge on the carbon (80,81). This basic hypothesis is still widely accepted. A model in which the active site is an atom of an octahedrally coordinated transition metal, with one vacant coordination site, has been proposed (82–86). In the case of TiCl$_3$, this site is formed by the reaction with aluminum alkyl, as shown in Figure 3. Monomer insertion occurs through a first step of monomer coordination to the transition metal, forming a Ti-complex, subsequent weakening of the Ti–C bond, and finally insertion. Subsequently, the vacancy and the growing chain exchange positions because the two positions are not equivalent on the crystal lattice of the catalyst. These phases are repeated at the insertion of each monomer molecule. The stereochemistry of the isotactic insertion of each monomer unit is controlled by the configuration of the catalyst center, as demonstrated by the observations that steric order is maintained when ethylene is inserted and that errors in monomer insertions are not perpetuated (87,88). The stereochemistry of syndiotactic monomer insertion is controlled by the last inserted monomer unit. In this chain end stereochemical control, the

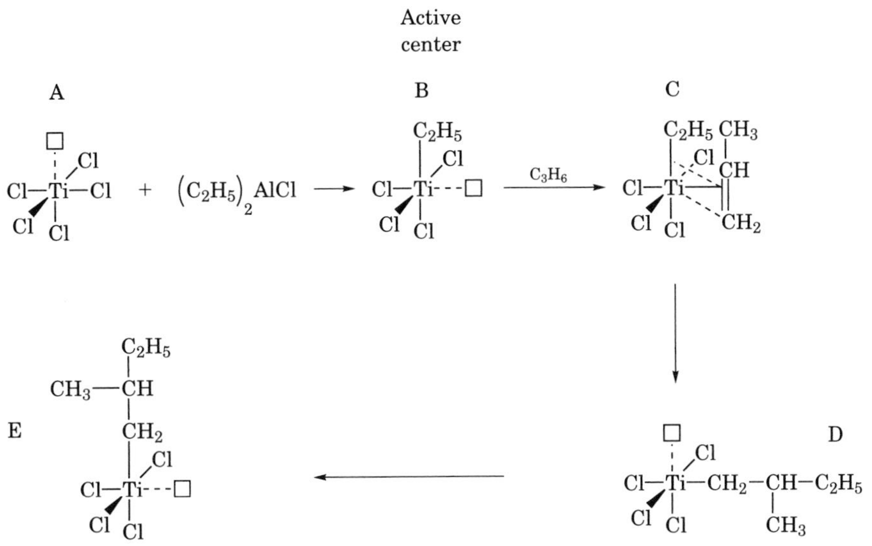

Active
center

A

B

C

$$Cl-\overset{\overset{\square}{|}}{\underset{\underset{Cl}{|}}{Ti}}-Cl \quad + \quad (C_2H_5)_2 AlCl \quad \longrightarrow \quad Cl-\overset{\overset{C_2H_5}{|}}{\underset{\underset{Cl}{|}}{Ti}}---\square \quad \xrightarrow{C_3H_6} \quad Cl-\overset{\overset{C_2H_5 \ CH_3}{|}}{\underset{\underset{Cl}{|}}{Ti}}$$

E

$$CH_3-\overset{\overset{C_2H_5}{|}}{\underset{\underset{Cl}{|}}{CH}}$$

$$\overset{CH_2}{\underset{}{}}$$

$$Cl-\overset{}{\underset{\underset{Cl}{|}}{Ti}}---\square$$

$\longleftarrow$

D

$$Cl-\overset{\overset{\square}{|}}{\underset{\underset{Cl}{|}}{Ti}}-CH_2-\overset{}{\underset{CH_3}{|}}{CH}-C_2H_5$$

Fig. 3. Monometallic mechanism of formation for polymer chain growth on transition-metal catalyst, where (□) represents a coordination vacancy.

steric hindrance of the chain end prevents the insertion of the monomer unit in the same configuration as the previous one (89,90).

The development of soluble metallocene catalysts for the polymerization of olefins (67) has facilitated the study of reaction mechanisms because of the well-defined nature of the active center. These systems are usually co-catalyzed by methylalumoxane, an oligomer of trimethylaluminum; however, aluminum-free systems where the cationic metallocene is stabilized by a borate counterion are also active (91). This lends further credence to the monometallic mechanisms for olefin polymerization. A number of authors have modified the early monometallic models, using a variety of molecular modeling techniques to describe the reaction mechanism (92–95). The activity of superactive third-generation catalyst systems, as a function of process variables, is shown in the following figures (96). In these systems, where polymerization occurs in hexane at 70°C, polymer production increases with time; however, the rate of polymerization decreases (Fig. 4). Increasing temperature (Fig. 5), in the range of 50–80°C, increases polymer production and isotacticity. Isotacticity is also controlled by the addition of electron (outside) donor. As shown in Figure 6, polymerization rate is not adversely affected by electron-donor addition (expressed as Al/donor) required to produce polymers having isotactic indexes between 94 and 98%, the range of greatest commercial interest. The choice of donor compound affects both activity and stereoregularity (97,98). Internal donors interact with the external donor to enhance their effectiveness (99,100). The aluminum alkyl acts as a scavenger for impurities in the polymerization medium, such as water, CO_2, and alcohols, preventing catalyst poisoning. Consequently, quantities used are in excess of that required for alkylation of the catalyst. The aluminum alkyl also acts as a chain-transfer agent, and reduces the transition metal to lower valences, adversely affecting activity. Molecular weight is controlled by the addition of hydrogen.

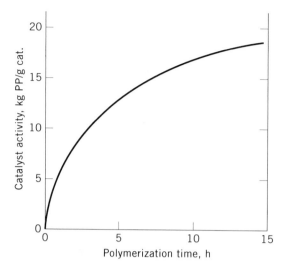

Fig. 4. Activity vs polymerization time. Polymerization occurs in hexane at 343 K (70°C) and 0.7 MPa (7 bar) with a superactive third-generation catalyst (96). Courtesy of Gulf Publishing Co.

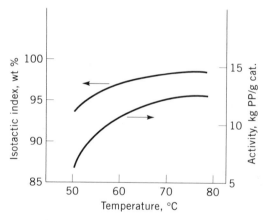

Fig. 5. Activity and isotacticity vs polymerization temperature. Polymerization occurs in hexane at 0.7 MPa (7 bar) for 4 h with a superactive third-generation catalyst (96). Courtesy of Gulf Publishing Co.

A well-known feature of olefin polymerization with Ziegler-Natta catalysts is the replication phenomenon in which the growing polymer particle mimics the shape of the catalyst (101). This phenomenon allows morphological control of the polymer particle, particularly size, shape, size distribution, and compactness, which greatly influences the polymerization processes (102). In one example, the polymer particle has the same spherical shape as the catalyst particle, but with a diameter approximately 40 times larger (96).

Kinetic models describing the overall polymerization rate, R_p, have generally used equations of the following form:

$$R_p = k_p[C^*][M]$$

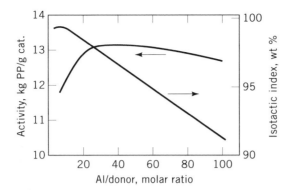

Fig. 6. Activity and isotacticity vs Al/donor ratio. Batch polymerization occurs in diluent hexane at 70°C and 0.7 MPa (7 bar) for 4 h with a superactive third-generation catalyst (96). Courtesy of Gulf Publishing Co.

where k_p is the propagation constant, $[C^*]$ the concentration of active sites, and $[M]$ the concentration of monomer. The increase in polymerization rate during the initiation of polymerization may be the result of progressive activation of new active centers. Decreases in the polymerization rate with time have been attributed to a loss of active centers or a change in their chemical nature (103,104). Determination of the propagation rate constant and the concentration of active sites has been notoriously difficult, and literature results are rarely in agreement. Values reported in Table 6 (105) demonstrate that $MgCl_2$-supported catalysts contain active centers with higher propagation rates than $TiCl_3$ catalysts. Propagation rate constants for metallocene systems are significantly higher than for the titanium chloride systems (74).

Table 6. C^* and k_p for Catalytic Systems Based on Titanium Halides

Catalytic system	Temperature, K	C^*, mmol/ mol Ti	k_p, dm³/ mol·s	Reference
δ-$TiCl_3$·$Al(C_2H_5)_2Cl$	343	28	124	106
δ-$TiCl_3$·0.3$AlCl_3$· $Al(C_2H_5)_3$	343	5.8	100	107
δ-$TiCl_3$·0.3$AlCl_3$· $Al(C_2H_5)_2Cl$	333	12	13	108
δ-$TiCl_3$·0.3$AlCl_3$· $Al(C_2H_5)_2Cl$		13.7–20	5.9–8.2	109
δ-$TiCl_3$·0.3$AlCl_3$· $Al(i$-$C_4H_9)_3$	343	8	90	106
$TiCl_3$ Solvay·$Al(C_2H_5)_2Cl$	333	7.9–27.3	7.6–17.2	109
$TiCl_4$/$MgCl_2(P)$·$Al(C_2H_5)_3$[a,b]	243	5.2	740	106
	343	42	800	106
$TiCl_4EB$/$MgCl_2(P)$· $Al(C_2H_5)_3$[a–c]	343	2.7	870	106
		6.8	1250	106
$MgCl_2$/$TiCl_4$/EB/$Al(C_2H_5)_3$[b,c]	338	80	500	110
$MgCl_2$/$TiCl_4$/EB/$Al(C_2H_5)_3$[b,c]	318	25	970	111

[a]$MgCl_2$ obtained from magnesium alkyl halides; (P) = organic polymer residue present in $MgCl_2$.
[b]Data relating to polymer fractions insoluble in n-heptane.
[c]EB = ethyl benzoate.

The overall performance of Ziegler-Natta polymerization systems has been modeled by a number of investigators. Particle growth and fragmentation was first modeled for TiCl$_3$ catalyst systems; investigators attempted to ascertain if the molecular weight distribution of the polymer was a consequence of diffusional limitations (112,113). Further work has indicated that the source of the broad molecular weight distribution in polypropylene produced from titanium chloride catalysts is most likely the result of a multiplicity of catalyst sites (114). Investigations of the initiation of polymerization, in high yield catalyst systems, indicate that thermal excursions sufficient to fragment the catalyst are possible under industrial polymerization conditions (115–117). The effect of process conditions on the overall performance of the polymerization process has been studied for slurry (118), liquid monomer (119), and gas-phase polymerization processes (120).

EARLY MANUFACTURING PROCESSES

The first commercial processes for the production of polypropylene were batch polymerization processes using TiCl$_3$ catalysts activated by Al(C$_2$H$_5$)$_2$Cl in a hydrocarbon medium. The hydrocarbon, usually hexane or kerosene, maintained the isotactic polypropylene in suspension and dissolved the undesirable atactic fraction. After polymerization, the suspension is treated with alcohol to deactivate and solubilize the catalyst residues, and filtered to separate the residues and atactic fraction from the desirable polymer which is then dried. The alcohol and diluent are recovered by multiple distillations, and the atactic fraction is sold as a by-product. As the demand for polypropylene increased, these batch polymerization processes were rapidly replaced by continuous ones, such as the Hercules process shown in Figure 7. In this process, typical of those used during the 1960s and 1970s, a suspension of TiCl$_3$ catalyst in Al(C$_2$H$_5$)$_2$Cl and kerosene diluent is continuously fed to the first of a series of continuous stirred overflow reactors. Monomer is fed to the first reactors and allowed to react out in the later ones, obviating the requirement for monomer recycle. Typical polymerization temperatures were in the range of 55–70°C, and maximum pressures as high as 0.5 MPa (75 psig). Other similar processes, such as Montedison's, operated at pressures as high as 1.3 MPa (200 psig) with monomer recycle (121). Hydrogen is added to the reactors as required to achieve the desired polymer molecular weight (122). Following polymerization, the slurry is contacted with isopropyl alcohol, then aqueous caustic to decompose and neutralize catalyst residues. The aqueous phase containing the alcohol and catalyst residues is separated from the hydrocarbon, polymer slurry phase. The suspended isotactic polymer is separated from the diluent containing the atactic polymer by continuous filtration or centrifugation, then dried. The alcohol and kerosene are each purified by a series of distillations, then recycled. Atactic polymer is dried using a thin-film evaporator and sold as by-product. The aqueous stream containing catalyst residues is treated prior to disposal of waste water and inorganic solids. The products available from this technology were limited to homopolymers with relatively high molecular weights (MFR < 15 dg/min), random copolymers containing low amounts of ethylene, and impact-resistant copolymers of high molecular weight and low rubber content. Excessive production of soluble polymer causing fouling

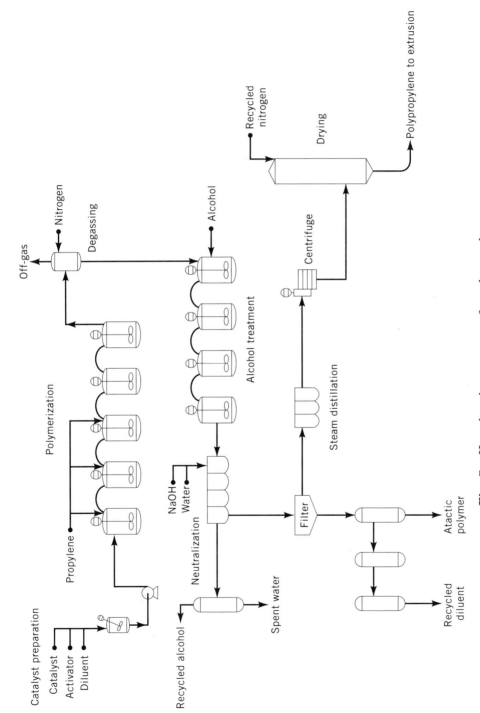

Fig. 7. Hercules slurry process for polypropylene.

of heat-transfer surfaces, was the primary cause of this limitation, more so than the loss of monomer to the production of less valuable by-products. This limitation, and the high energy cost of recycling diluent and alcohol, led to the development of processes that eliminated the need for diluent.

Polymerization in liquid monomer was pioneered by Rexall Drug and Chemical and Phillips Petroleum (United States). In the Rexall process, liquid propylene is polymerized in a stirred reactor to form a polymer slurry. This suspension is transferred to a cyclone to separate the polymer from gaseous monomer under atmospheric pressure. The gaseous monomer is then compressed, condensed, and recycled to the polymerizer (123). In the Phillips process, polymerization occurs in loop reactors, increasing the ratio of available heat-transfer surface to reactor volume (124). In both of these processes, high catalyst residues necessitate post-reactor treatment of the polymer.

Gas-phase polymerization of propylene was pioneered by BASF, who developed the Novolen process which uses stirred-bed reactors (Fig. 8) (125). Unreacted monomer is condensed and recycled to the polymerizer, providing additional removal of the heat of reaction. As in the early liquid-phase systems, post-reactor treatment of the polymer is required to remove catalyst residues (126). The high content of atactic polymer in the final product limits its usefulness in many markets.

Eastman Chemical has utilized a unique, high temperature solution process for propylene polymerization. Polymerization temperatures are maintained above 150°C to prevent precipitation of the isotactic polypropylene product in the hydrocarbon solvent. At these temperatures, the high rate of polymerization decreases rapidly, requiring low residence times (127). Stereoregularity is also adversely affected by high temperatures. Consequently, the product is treated to remove catalyst residues and contains high amounts of atactic polymer.

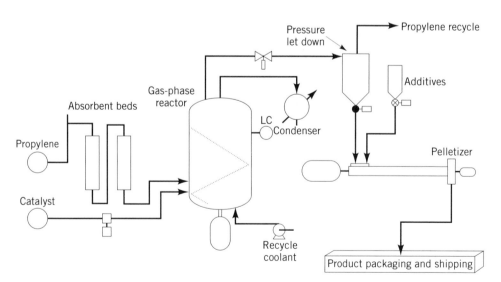

Fig. 8. Novolen process. LC is the level controller (125). Courtesy of the American Chemical Society.

In the 1970s, Solvay introduced an advanced $TiCl_3$ catalyst with high activity and stereoregularity (6). When this catalyst was utilized in liquid monomer processes, the level of atactic polymer was sufficiently low so that its removal from the product was not required. Catalyst residues were also reduced so that simplified systems for post-reactor treatment were acceptable. Sumitomo has developed a liquid monomer process, used by Exxon (United States), in which polymer slurry is washed in a countercurrent column with fresh monomer and alcohol to provide highly purified polymer (128).

Montedison and Mitsui Petrochemical introduced $MgCl_2$-supported high yield catalysts in 1975 (7). These third-generation catalyst systems reduced the level of corrosive catalyst residues to the extent that neutralization or removal from the polymer was not required. Stereospecificity, however, was insufficient to eliminate the requirement for removal of the atactic polymer fraction. These catalysts are used in the Montedison high yield slurry process (Fig. 9), which demonstrates the process simplification achieved when the sections for polymer de-ashing and separation and purification of the hydrocarbon diluent and alcohol are eliminated (121). These catalysts have also been used in retrofitted Rexall (El Paso) liquid monomer processes, eliminating the de-ashing sections of the plant (Fig. 10) (129).

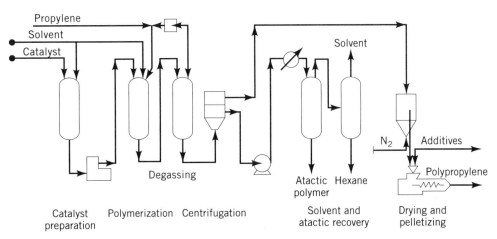

Fig. 9. High yield Montedison polypropylene process (121). Courtesy of Gulf Publishing Co.

CURRENT PROCESSES

The development of superactive third-generation supported catalysts enabled the introduction of simplified processes, without sections for catalyst deactivation or removal of atactic polymer. By eliminating the waste streams associated with the neutralization of catalyst residues and purification of the recycled diluent and alcohol, these processes minimize any potential environmental impact. Investment costs are reduced by approximately one-third over slurry process plants. Energy consumption is minimized by elimination of the distillation of recycled diluent and alcohol. The total plant cost for the production of polymer is less than 130% of the monomer price, when a modern process is used, compared to

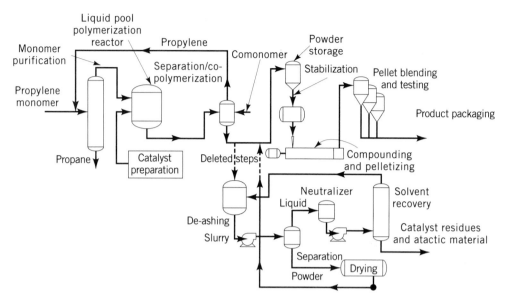

Fig. 10. El Paso liquid propylene process (129).

175% for a slurry process. Consequently, many companies invested in new plants in the 1980s, either increasing capacity or idling plants based on the older, now obsolete processes. Almost all of the plants built since 1985 use one of the advanced, simplified processes (130). The leading process licensors are Technipol (Himont), Mitsui, Union Carbide, and BASF.

The Spheripol process, developed by Himont, Inc., consists of one or more loop-tubular reactors, for the production of homopolymer or random copolymer and one or more fluid-bed gas-phase reactors for the production of the rubber phase for impact-resistant copolymers (Fig. 11). This process takes full advantage of the high yield spherical-form supported catalysts, utilizing the excellent heat-transfer capability of loop reactors to obtain very high specific outputs of over 400 kg PP/h·m^3. After polymerization in the loop reactors, the polymer is separated from the liquid monomer by flashing at a pressure sufficient to allow condensation and recycle of the liquid monomer without recompression. The polymer is then transferred to the gas-phase reactors for the production of the rubber phase of impact-resistant copolymers. Ethylene and propylene are fed to the fluid-bed reactor to produce ethylene–propylene rubber of the desired composition. Unreacted monomer is recycled and cooled using an external heat exchanger. The polymer is then separated from the unreacted monomer at a pressure slightly above one atmosphere, then contacted with steam for complete removal of residual monomer and termination of polymerization (132).

Third-generation high yield supported catalysts are also used in processes in which liquid monomer is polymerized in continuous stirred tank reactors. The Hypol process (Mitsui Petrochemical), utilizes the same supported catalyst technology as the Spheripol process (133). Rexene has converted the liquid monomer process to the newer high yield catalysts. Shell uses its high yield (SHAC) catalysts to produce homopolymers and random copolymers in the Lippshac process (130).

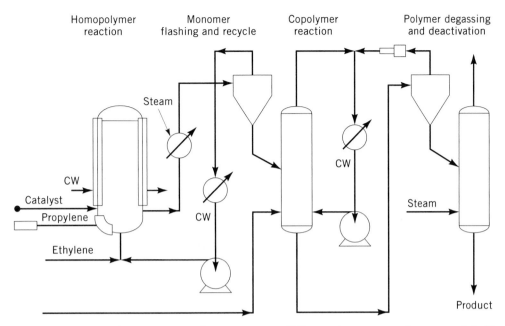

Fig. 11. Spheripol process flow diagram (131). CW = cooling water. Courtesy of Gulf Publishing Co.

The Unipol PP process (Union Carbide), extends technology originally developed for ethylene polymerization to propylene polymerization utilizing Shell (SHAC) catalysts. One large gas-phase fluidized-bed reactor is used for the production of homopolymer and random copolymer; a second, smaller reactor is used to produce the rubber required for impact copolymers. The heat of reaction is removed by cooling the monomer recycle through an external heat exchanger (Fig. 12). The Novolen process (BASF), has been rejuvenated through the use of high yield catalyst technology. Problems associated with low polymer productivity, high levels of atactic polymer, and catalyst residues have been alleviated by the use of these catalysts. Amoco's horizontal stirred-bed gas-phase reactor is the heart of the Amoco/Chisso process. This reactor (Fig. 13) acts as a series of polymerization stages in a single reactor vessel, facilitating the production of broad molecular weight distribution homopolymers and random copolymers. Impact copolymers are produced in a second similar reactor in series as in other processes (135). Montell produces specialty propylene copolymers in the multistage gas-phase Catalloy process (136).

The producers of polypropylene in the United States, western Europe, and Japan are listed in Table 7. Montell is the largest producer in both western Europe (followed by Borealis) and the United States (followed by Amoco).

Processing

Polypropylene is produced in a variety of molecular weights, molecular weight distributions, and crystallinities; consequently, it can be used in most polymer processing technologies. The physical and mechanical properties of polypropylene in the end use product are a function of both the molecular structure and the

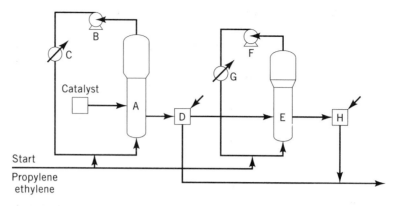

Fig. 12. Unipol PP process where A is the polymerization reactor; B, recycle gas compressor; C, recycle gas cooler; D, product discharge tank; E, impact copolymer reactor; F, recycle gas compressor; G, recycle gas cooler; and H, product discharge tank (134).

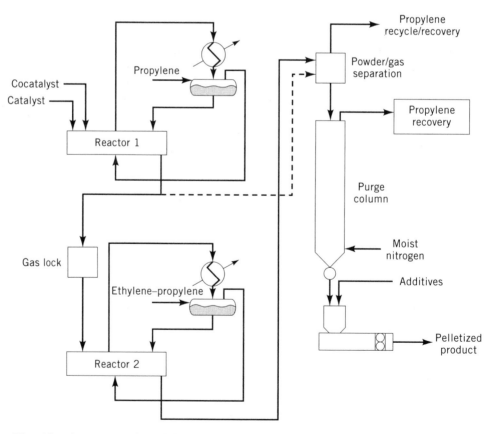

Fig. 13. Amoco gas-phase PP process where (- - -) represents the homopolymer path.

Table 7. Polypropylene Producers and Capacities

Producer	Capacity, 10^3 t/yr	Process
United States,[a] 1994		
Himont[b]	895	Spheripol, Catalloy
Amoco	755	Amoco/Chisso, Amoco gas, slurry
Exxon	485	Sumitomo/Exxon bulk, Spheripol
Fina	435	loop bulk
Aristech	325	Montedison high yield slurry, Spheripol
Shell	245	Unipol, stirred bulk
Formosa[c]	220	BASF
Phillips	215	loop reactor
Solvay	205	loop reactor
Eastman[d]	200	Unipol
Epsilon	165	Unipol
Huntsman	150	Shell slurry
Lyondell	135	stirred bulk (Rexene)
Quantum	135	BASF
Rexene	80	stirred bulk (Rexene)
Novacor	65	Mitsui Hypol
Western Europe,[e] 1992		
Himont[b]	755	Spheripol, Catalloy
Shell[b]	570	Unipol, shell bulk, slurry
Hoechst	535	Spheripol, slurry
Neste[f]	460	Spheripol, slurry
OMV	390	Spheripol
Appryl	350	slurry, bulk
ICI[g]	300	BASF, slurry
BASF[g]	300	BASF
DSM	280	Mitsubishi slurry
Solvay	250	bulk, Unipol
Amoco	200	slurry
Hüls	195	slurry, Unipol
Statoil[f]	180	Rexene bulk, Spheripol
Petrofina	180	Spheripol, Montedison slurry
Repsol	150	Spheripol, Montedison slurry
Beaulieu	150	Unipol
Exxon	140	Sumitomo/Exxon bulk
Japan 1992[e]		
Mitsubishi Petrochemical	320	Montedison
Chisso	215	Amoco/Chisso
Showa Denko	215	Eastman solution, Spheripol
Idemitsu	195	BASF
Mitsui Petrochemical	175	Mitsui slurry, gas phase
Sumitomo	175	Sumitomo gas, Montedison slurry
Tonen	170	Mitsui
Mitsui Toatsu	135	Montedison slurry, Mitsui Toatsu
Tokuyama Soda	115	Slurry, Mitsui gas phase

[a]Ref. 137. [b]Himont and Shell (non-U.S.) merged to form Montell Polyolefins in 1995. [c]Start-up in 1994. [d]Purchased by Huntsman in 1994. [e]Refs. 138 and 139. [f]Neste and Statoil merged Jan. 1994 to form Borealis. [g]BASF purchased the ICI polypropylene business in 1994.

processing conditions. The final crystalline morphology is a function of the melt temperatures, polymer orientation, and cooling temperatures and rates. The most commonly used processes are as discussed in the following (see PLASTICS PROCESSING).

Injection Molding. Injection molded articles are made from homopolymers, random and impact copolymers, and a variety of filled polymers. Conventional machines can be used, without modification. Polymer melt flow rates lower than 4 dg/min and as high as 70 dg/min can be used, depending on the mold geometry, part thickness, and cycle time desired. The higher melt flow materials provide more uniform flow and low cycle times in parts with thin sections. Melt temperatures as low as 200°C can be used with these high flow polymers. Lower melt flow materials generally offer improved toughness when compared to the higher melt flow polymers, and can be used in parts that have thicker cross sections. These polymers must be processed at higher temperatures to obtain good flow properties; melt temperatures as high as 275°C have been used (140).

Blow Molding. Low melt flow polymers are used in blow molding to provide the melt strength required to maintain stability of the parison, ie, a molten, thick-walled tube of melt. High density polyethylene has been more commonly used to form large parts because of its greater stability; however, a number of improved polypropylene grades are suitable for these applications (141). In extrusion blow molding, the extruded parison hangs freely before entering the mold, and low melt temperatures (between 205 and 215°C) are preferred. In injection blow molding, a preform is injection-molded on a steel rod, transferred to a blow molding mold, and blown. Consequently, higher melt temperatures can be used. Stretch blow molding produces a part with biaxial orientation providing increased stiffness, low temperature impact strength, and clarity. The parison is cooled after it is formed by extrusion or injection molding, reheated to the desired orientation temperature, mechanically stretched, then formed into the desired shape by blowing.

Extrusion. Polypropylene is extruded into sheet, usually for subsequent stamping or thermoforming, or into pipes and profiles. Low melt flow rate, high molecular weight resins are used to provide the melt strength required to maintain uniformity. High melt strength polymers, produced by post-polymerization modification of conventional polypropylenes, improve uniformity. Ionomers, produced by grafting acrylates onto polypropylene, have also been used. Extruders with high length-to-diameter ratios, or twin screw extruders, are used to provide good mixing of the melt. Melt temperatures of 230–260°C are used to produce extruded products. Higher temperatures promote degradation, leading to loss of properties, discoloration, and "plate out," the migration of additives and/or low molecular weight polymer to the surface of the part. Lower temperatures reduce throughput and uniformity.

Thermoforming. Historically, polypropylene has not been used in conventional melt-phase thermoforming equipment because of its narrow forming temperature range and the tendency of the melt to sag. Processes such as Shell's solid-phase pressure forming (142) were developed to overcome these problems. The polypropylene sheet is stretched into the mold cavity, at a temperature below the melting point, with a shaped plug. The part is forced against the mold surface by cold air at a pressure of 0.55–0.69 MPa (80–100 psi) to obtain the desired shape. The introduction of newer polypropylene grades with higher melt strength and improved sag resistance has increased the use of polypropylene in conventional melt-phase thermoforming equipment (143). Manufacturers of thermoforming equipment have also modified their processes to effectively utilize polypropylene (144).

FIBERS

Melt Spinning. This process is used to produce a broad range of polypropylene fibers ranging from fine, dtex (one denier) staple to coarse continuous filaments. Homopolymers are almost exclusively used to produce fibers, although copolymer blends are used in some special applications. Processing conditions and polymer melt flow vary with the desired fiber type. In general, higher melt temperatures and lower polymer molecular weights are required to produce finer fibers. Polymer rheology is sometimes controlled by chemical visbreaking, used to reduce viscoelasticity and improve spinning performance (see FIBERS, OLEFIN).

Polypropylene is extruded through a spinnerette containing as few as 10 holes in some continuous filament processes and as many as 2000 for the production of fine-denier staple fiber. The melt expands upon exiting the hole, as a function of its viscoelasticity, because of the well-known die swell phenomenon. The molten filament is then cooled by forced air under an extensional force provided by the take-up reel in vertical chambers as high as 15 m. The tension provided by the take-up reel (melt drawing) provides partial orientation which greatly influences the final properties of the fiber. The fiber is then cold drawn by mechanical stretching of 2 to 10 times its length, usually at a temperature above 70°C, to obtain the desired combination of tenacity and elongation (145). The production of fine-staple fiber by the short spin process has been commercialized (146). Short spin processes generally use low spinning speeds to minimize the space required for quenching. Consequently, the large vertical cooling chambers are eliminated. This has provided a dramatic reduction in the investment required for the production of melt spun fibers.

Slit and Split Films. Thick industrial-grade yarns are often produced by slitting films, providing a less expensive alternative to melt spun fiber. Cast film is slit in the machine direction by parallel rotary knives. The resulting tape can then be cold drawn in an oven in a manner similar to melt spun fibers to produce the final fiber. Knife spacings as low as 1 mm are sometimes used to produce textile fibers, although spacings of 10–35 mm are more common. Fibers from split or fibrillated films are formed by the drawing of polypropylene film to the degree that it splits into numerous fiber-like interconnected tapes. In some processes the draw-induced splitting is mechanically augmented by gears, rollers, or gas jets (145).

Melt Blowing. The melt blowing process uses very high melt flow polymers, sometimes in excess of 400 dg/min, and extrusion and die temperatures above 300°C to produce very fine fibers (<5 μm dia). The flow of hot, low viscosity polymer extruded through a small die is disrupted by high velocity hot air. A large volume of cooling air fed near the die exit quenches the fibers and deposits them on a collecting screen as a mat of entangled fibers. This entanglement is sufficient to maintain the integrity of the web, and thermal bonding is not necessary. Fabrics produced by this process are very soft because of the small fiber diameter. Because the fibers are not highly oriented or bonded, melt blown fabrics usually have low tensile strength (147).

Spun Bonded Fabrics. Spun bonded fabrics are produced by depositing extruded, spun filaments onto a collecting belt in a uniform randomized manner. The fibers are separated during the web laying process by air jets and the collecting belt is usually perforated to prevent the air stream from deflecting and carrying the fibers in an uncontrolled manner. Polymers with melt flow rates above 20 dg/min and narrow molecular weight distributions are used to reduce viscoelastic instability. Thermal bonding, using heated embossing rolls or hot needles, imparts strength to the web by fusing some of the fibers. This process can be combined with the melt blowing process to produce soft, multilayer fabrics with good tensile properties. Spun bonded fabrics are used in a variety of applications requiring nonwoven fabrics (qv), competing with thermally bonded staple fiber.

FILMS

Polypropylene (PP) films were first produced by extrusion casting. Polymer is extruded through a slit or tubular die and quenched by cooling on chill rolls or in a water bath. The rapid quenching below the 70°C transition temperature minimizes orientation by freezing a disordered crystal structure. Cast films can be sealed over a wide range of temperatures and do not shrink in a steam autoclave. Polymers with melt flow rates below 5 dg/min are usually used to maintain the stability of the extrudate. Higher clarity films are produced using random copolymers.

Biaxially oriented polypropylene (BOPP) films have higher stiffness than cast films and consequently can be used in much thinner gauges. Homopolymers are used almost exclusively to provide maximum stiffness and water-vapor barrier. Oriented films are produced by the tenter frame and bubble processes. Most of the newly installed capacity has utilized the tenter frame process, taking advantage of the economics of the large, high capacity units available. In this process, an extruded cast film is oriented by extension through a series of rollers in the machine (longitudinal direction), and stretching with a series of clips or tenter frames in the transverse direction. In the bubble process, a tube is extruded, quenched, and radially expanded by inflation to provide transverse orientation. Axial orientation is provided by simultaneous extension in the machine direction through a series of nip rolls. The tube is then flattened and slit into flat film. In both processes orientation is provided by stretching between 100 and 145°C; heat aging under slight tension at a temperature above the orientation temperature and below the melting point is required to stabilize the structure and minimize shrinkage. Opaque films are produced by introducing microvoids into the film during the orientation. Small particles are blended into the polymer before extrusion; during the extrusion process the polymer expands from the inelastic solid leaving microvoids (148). Opacity is created by the reflection and diffraction of light by these voids. Untreated oriented films are not easily heat sealed; consequently, lower melting random copolymers and terpolymers are often coextruded with the homopolymer to form a heat sealing layer.

STABILIZATION

Polypropylene is subject to attack by oxygen, radiation, and excessive heat causing a loss of molecular weight and physical properties. Stabilizers are added to the polymer to minimize these effects. Small quantities of hindered phenolic antioxidants (qv) are added in the polymerization plant, usually in the drying section, to protect the polymer against degradation during short-term storage. Typically 2,6-di-*t*-butyl-*p*-cresol (BHT) and octadecyl 3,5-di(*t*-butyl-4-hydroxy) hydrocinnamate (Irganox 1076) are used. The bulk of the stabilizer is added during pelletization or fabrication to protect the polymer during processing or in the final application. Stabilizers can also be added to spherical-form polymer, without extrusion, using the Addipol process (Montell) (149). Typical stabilization formulations include a hindered phenolic antioxidant, possibly with a thiodipropionate synergist, a phosphite to provide high temperature melt stabilization, and an acid scavenger such as calcium stearate or dihydrotalcite (150). Hindered phenols limit the propagation of alkyl radicals and the resulting chain scission. Thiodipropionic acid esters act to decompose peroxides formed by polypropylene oxidation. More importantly, the sulfonic acid intermediate of thiodipropionate oxidation acts as a scavenger for the free radicals formed by the decomposition of phenols, increasing the effectiveness of the phenolic antioxidant (151). Phosphites also act to decompose peroxides and are most effective at the temperatures usually encountered in processing. The acid scavenger acts to prevent the reactions between hindered phenols and metal chlorides that can form colored titanium phenolates, as well as minimize equipment corrosion. Protection against ultraviolet radiation is usually provided by a hindered amine light stabilizer (HALS), such as Tinuvin 770. Thermal stabilization of polypropylene has been reviewed (152).

Economic Aspects

Polypropylene is widely used throughout the economy; consequently, the growth in its consumption is related to that of the total economy. Although the growth in consumption is greater than that of the economy, lower growth has occurred in years of economic recession (Table 8). Polypropylene use has increased at a more rapid rate than that of the other thermoplastics, as shown by its increasing share of the overall market (Table 8). The comparative increase in consumption was greatest in the 1970s and 1980s when the relative share of polypropylene use increased from approximately 10% of total thermoplastics in 1974 to 17% in 1990. Consumption in the United States has increased each year since 1982, averaging 7.8%. In western Europe, however, production has declined from 1991 to 1993 because of the economic recession (Table 9), and Japan has also seen a recession-related decline in polypropylene consumption (Table 10). The profitability of polypropylene manufacture is directly related to the difference between monomer and polymer price. As shown in Table 11, the difference and profitability tend to increase with capacity utilization

Table 8. Distribution of Polypropylene Sales and Internal Use, 10^3 t[a]

Application	1985	1987	1989	1990	1991	1992	1993
appliances	52	64	73	80	77	85	101
consumer products	141	195	256	289	294	349	397
rigid packaging	175	240	273	294	318	351	389
transportation	154	147	121	131	120	135	144
other	56	102	86	79	83	85	90
Total injection molding	*578*	*749*	*809*	*873*	*892*	*1,004*	*1,120*
blow molding	47	52	59	65	63	64	69
film	176	254	265	273	303	325	362
sheet, >254 μm	20	37	58	64	58	62	65
fiber and filaments	650	795	888	943	951	1,048	1,111
extruded products[b]	40	57	43	65	46	49	52
resellers and compounders	388	523	620	694	742	858	925
all other uses	96	21	33	26	34	8	11
Total domestic consumption	*1,996*	*2,489*	*2,775*	*3,004*	*3,090*	*3,418*	*3,715*
exports	397	566	545	741	703	431	347
Total consumption	*2,393*	*3,055*	*3,320*	*3,745*	*3,793*	*3,849*	*4,063*
imports					31	41	84
Total thermoplastics	*16,364*	*19,472*	*20,407*	*21,873*	*22,436*	*23,975*	*25,015*
polypropylene, % of total	14.62	15.69	16.27	17.12	16.91	16.06	16.24

[a]Ref. 153.
[b]Includes wire and cable, pipe and conduit, and extrusion coating.

because monomer prices are related to oil prices, whereas polymer prices are related to demand. The low polymer prices seen in the early 1990s were the result of an increase in production capacity beyond demand, and not a decline in polypropylene consumption in the United States. This is, in part, a reaction to the unusually high price differentials in the late 1980s. Another factor in the decline of polypropylene prices in the United States has been the decrease in exports, particularly to the Pacific region. This is a result of the start-up of new polypropylene production plants in countries that formerly imported polypropylene, such as Korea, Thailand, and Malaysia. By the year 2000, polypropylene production is expected to increase much more rapidly in the Pacific region and in Latin America than in North America, Europe, and Japan (Table 9). The changes in the marketplace have led to attempts at consolidation in the industry in Europe, where polymer prices have dropped more precipitously than in the United States (157). The merger of the polyolefin interests of Neste Oy and Statoil was consummated in early 1994 as a new venture, Borealis. BASF purchased the ICI polypropylene business in 1994. Shell and Montedison merged their polyolefins businesses, including Himont, the world's largest polypropylene producer, in April 1995. In the United States, Huntsman acquired the Eastman

Table 9. World Polypropylene Production, 10^3 t[a]

Production	1991	1993	Projected, 2000	Growth, 1985–1992	Rate, 1993–2000
North America	4,096	4,505	5,829	9.8	4.2
United States	3,783	4,079	5,041	9.2	3.4
Canada	283	298	391	9.6	4.5
Mexico	30	128	397		30.0
Latin America	407	513	1,426	11.0	25.4
Western Europe	4,219	4,191	5,966	11.1	6.1
Japan	1,958	2,237	2,763	9.1	3.4
Pacific Rim	1,480	1,944	4,094	21.3	15.8
Middle East/Africa	245	301	730	29.9	20.4
Eastern Europe/CIS	584	690	2,115	2.9	29.5
World Total	*12,988*	*14,381*	*22,923*	*11.0*	*8.5*

[a]Ref. 8.

Table 10. Polypropylene Consumption, 10^3 t

Product	1989	1990	1991	1992	1993
		Western Europe[a]			
film	559	647	685	737	796
molding and extrusion	1,759	1,997	2,101	2,276	2,401
spun fibers	579	661	667	715	754
film fibers (raffia, etc)	377	388	387	402	417
Total	*3,274*	*3,693*	*3,840*	*4,130*	*4,368*
		Japan[b]			
injection molding	930	1,074	1,073	1,106	1,082
film	393	416	426	427	424
flat yarn	87	88	80	77	76
fiber	51	56	63	69	72
blow molding	34	37	39	40	39
extrusion	168	190	190	189	190
miscellaneous	56	1,958	80	89	100
export	69	150	189	218	183
Total	*1,787*	*2,083*	*2,140*	*2,215*	*2,160*

[a]Ref. 154.
[b]Ref. 155.

polypropylene business. Mitsubishi consolidated its chemical businesses to form Mitsubishi Kagaku and has controlling interest in Aristech.

In the United States, fibers and injection molding are the main applications for polypropylene (Table 8), followed by film. In Europe and Japan, injection molding applications predominate (Table 10). This market area is more likely to decline in economic recession, as consumers postpone purchases of appliances and automobiles. Film applications are important in both regions, but fibers are a much less important use for polypropylene in Japan than in other developed regions. The heavy use of polypropylene nonwovens in the manufacture of

Table 11. U.S. Polypropylene Production, Capacity Utilization[a], and Prices[b,c]

Product	1985	1987	1989	1991	1993
polypropylene production, 10^3 t	2,336	3,021	3,290	3,811	4,006
capacity, 10^3 t	2,636	3,120	4,030	4,323	4,455
utilization, %	88.6	96.9	81.6	87.6	89.9
polypropylene,[d] ¢/kg	77	101.2	80.3	80.3	63.8
propylene, ¢/kg	34.1	44	35.2	35.2	29.2
oil $/bbl (Gulf Coast)	29.32	19.24	20.80	19.95	15.25

[a]Ref. 153.
[b]Of polypropylene and monomer vs crude oil.
[c]Ref. 156.
[d]General-purpose homopolymer.

disposable diapers and similar products, and the wide use of polypropylene carpets in the United States, account for the greater consumption of fibers.

Uses

Injection Molding. Polypropylene is extensively used in injection molding because of the wide range of physical properties and melt flow rates available. The principal markets served include transportation (primarily automotive), appliances, consumer products, rigid packaging, and medical products. Polypropylene use has increased in the automotive industry because of the wide availability of high melt flow rate impact copolymers for use in large thin parts, such as interior trim. The availability of materials with higher flexural modulus and heat distortion temperatures has also increased polypropylene use in this market. A number of automobile companies have initiated programs to maximize the use of polypropylene. This facilitates recycling by simplifying the process of disassembly and sorting of the materials from discarded vehicles (158). Consumption of polypropylene in automobile batteries has declined, even though almost all battery cases are produced from polypropylene (see BATTERIES). Extensive use of polymer from recycled batteries has reduced the consumption of virgin materials. Filled or glass-reinforced materials are used in many automobile parts that are covered or in the engine compartment to provide increased stiffness and heat resistance. The softer, more flexible high impact copolymer grades and thermoplastic polyolefins (polypropylene plus ethylene–propylene rubber) are extensively used in automobile bumpers.

Polypropylene can be molded into an integral hinge that can be flexed repeatedly without failure. This property is frequently utilized in the design of container closures, lunchboxes, and similar articles. Polypropylene closures are used on a wide variety of containers, including child-proof caps and screw caps on plastic beverage bottles. Medical devices injection molded from polypropylene include syringes, pans, trays, and a variety of utensils. A large variety of toys, cups, dishes, and other household articles are molded from polypropylene. The development of high melt flow rate copolymer grades for thin-wall injection molding (159) has increased the use of polypropylene in food containers. Random copolymers are used in containers when high clarity is desired; the impact properties of these polymers are acceptable at refrigerator temperatures, but impact modification is required for some freezer applications. Filled compositions and the newer high stiffness grades provide the stiffness and high temperature properties desired in dishwashers and other appliances (160). Molded outdoor furniture is also frequently produced with mineral-filled polypropylene.

Blow Molding. Polypropylene blow molded bottles are used to package a wide variety of products including foods, cleansers, shampoo, pharmaceuticals, and mouthwash. Oriented bottles are used when higher clarity and a moisture barrier is desired. Random copolymers are used to further improve clarity and provide impact resistance for bottles that are refrigerated. Clarifiers are added if additional clarity is desired. Polypropylene provides increased stiffness and heat resistance when compared to high density polyethylene. High melt strength polypropylene blow molding grades have been introduced into a number of large automotive parts, typically blow molded from filled polyethylene (141).

Thermoforming and Extrusion. Improved equipment and polymers have increased the capability to extrude and thermoform polypropylene; however, consumption of polypropylene in these areas has not grown dramatically. Drinking straws are commonly extruded from polypropylene, however most larger diameter tubes, such as pipes and conduits, are predominantly extruded from other thermoplastics. Extruded sheet is thermoformed into food containers and trays; polypropylene is used when microwavability is desired.

Fibers. Polypropylene fibers are extensively used in carpeting. Bulked continuous filament yarns are commonly used in carpet facing in the level loop carpets used in commercial buildings; however, nylon remains the dominant face yarn in tufted household carpets. The advantages of polypropylene carpeting are superior stain resistance and low moisture absorbance. Polypropylene slit tape fibers are heavily used in carpet backing.

Disposable polypropylene nonwoven fabrics are widely used as the coverstock for disposable baby diapers. The expansion of the disposable diaper market throughout the world has been the primary source of growth in the consumption of polypropylene in the fiber market. In addition, nonwoven polypropylene fabrics are used in a variety of other disposable sanitary products, such as baby wipes, adult incontinence, and feminine hygiene products. Use of polypropylene nonwovens in disposable medical apparel, such as surgical gowns, has increased as a means of reducing the spread of infection.

Film. Oriented polypropylene films are widely used in the packaging of snack foods, candy, and other products. The most common method of packaging using these films is the form and fill process in which the package is formed, filled with product, and sealed in a continuous process. Oriented polypropylene films are used because of their stiffness, tensile strength, and resistance to moisture vapor transmission. A barrier to the diffusion of oxygen and other gases is provided by the use of a Saran layer or a metallized coating. Low melting co- or terpolymers are used to provide a layer for efficient heat sealing. The film is often surface treated to obtain good printability. The increased use of oriented films has provided almost all of the growth of consumption in oriented polypropylene films. In 1993, over 290,000 t of polypropylene were used in oriented film (153).

Cast films provide a high clarity, heat sealable film and are primarily used as an overwrap for boxes and other packaging. These films have a lower density than cellophane and provide a longer product shelf life. Properties of polypropylene films are given in Table 12.

Table 12. Properties of Polypropylene Films[a]

Property	Nonoriented	Oriented	ASTM method
tensile strength, MPa[b]	20.7–62.0	172.3–206.8	D882
elongation, %	400–800	60–100	D882
tear strength, N/mm[c]	15.7–129	1.57–2.35	D1922
elastic modulus, MPa[b]	758–965	2206–2620	
impact strength, N·cm[d]	9.8–29.4	49–147	
water vapor transmission rate, at 37°C and 90% rh, nmol[e]/cm^2·s	0.5–0.65	0.3–0.4	E96
gas permeability, at 22°C and 0% rh, mol/m·s·PPa[f]			
O_2	0.168–0.83	0.22	
CO_2	1–2.7	0.48–0.57	D1424

[a]Ref. 161.
[b]To convert MPa to psi, multiply by 145.
[c]To convert N/mm to gf/25 μm, multiply by 2.549; to convert to ppi, divide by 0.175.
[d]To convert N·cm to kgf·cm, divide by 9.8.
[e]nmol/cm^2·s $\approx$ g(H_2O)/100 in.^{2}·d .
[f]To convert mol/m·s·PPa to cc·mil/m^2·d·atm, multiply by 500.

BIBLIOGRAPHY

"Polypropylene" under "Olefin Polymers" in *ECT* 2nd ed., Vol. 14, pp. 282–309, by B. C. Repka, Jr., Hercules, Inc.; in *ECT* 3rd ed., Vol. 16, pp. 453–469, by G. Crespi and L. Luciani, Montedison SpA.

1. G. Natta and co-workers, *J. Am. Chem. Soc.* **77**, 1708 (1955).
2. A. Clark, J. P. Hogan, R. L. Banks, and W. C. Lanning, *Ind. Eng. Chem.* **48**, 1152 (1956).
3. E. F. Peters, A. Zletz, and B. L. Evering, *Ind. Eng. Chem.* **49**, 1879 (1957).
4. U.S. Pat. 4,376,851 (Mar. 15, 1983), J. P. Hogan and R. L. Banks (to Philips Petroleum).
5. "United States District Court for the District of Delaware Civil Action Suit 4319," *U.S. Pat. Q.* **206**, 676 (Jan. 11, 1980).
6. Ger. Offen. 2,213,086 (Oct. 5, 1972), J. P. Hermans (to Solvay & Cie SA).
7. Ger. Offen. 2,643,143 (Nov. 21, 1975), (to Montedison and Mitsui Petrochemical); Ger. Offen. 2,735,672 (Aug. 9, 1976), U. Giannini, E. Albizzati, and S. Parodi (to Montedison and Mitsui Petrochemical). Ger. Offen. 2,822,783 (May 25, 1977), U. Scata, L. Luciani, and P. C. Barbe (to Montedison); U.S. Pat. 4,069,169 (Jan. 17, 1978), N. Kashiwa and A. Toyota (to Mitsui Petrochemical).
8. *Modern Plast.* **70**(12), 18 (1993).
9. *Chem. Week*, **44** (Apr. 15, 1981).
10. W. Kaminsky and H. Sinn, eds., *Transition Metals and Organometallics as Catalysts for Olefin Polymerization*, Springer-Verlag, Berlin, 1988.

11. G. Natta, *Agnew. Chem.* **68**, 393 (1956).
12. G. Natta, *J. Polym. Sci.* **16**, 143 (1955).
13. G. Natta, *J. Polym. Sci.* **34**, 531 (1959).
14. A. Zambelli, P. Locatelli, M. G. Sacchi, and E. Rigamonti, *Macromolecules* **13**, 798 (1980).
15. J. C. Randall, *J. Polym. Sci. Phys. Ed.* **12**, 703 (1974); J. C. Randall, *Polymer Sequence Determination, Carbon-13 NMR Method*, Academic Press, Inc., New York, 1977.
16. M. Peraldo, *Gazz. Chim. Ital.* **89**, 798 (1959).
17. G. Natta, G. Mazzanti, G. Crespi, and G. Moraglio, *Chim. Ind.* **39**, 275 (1957).
18. ASTM Method D5492, American Society for Testing and Materials, Philadelphia, Pa.
19. R. Paukkeri and A. Lehtinen, *Polymer* **34**, 4083 (1993).
20. U.S. Pat. 5,302,897 (Apr. 12, 1994), R. L. Dechene, T. B. Smith, S. A. Marino, J. Tache, and A. Roy (to Auburn International).
21. G. Natta, P. Corradini, and M. Cesari, *Rend. Accad. Naz. Lincei Ser. 8*, **22**, 11 (1957).
22. G. Natta, P. Corradini, and M. Cesari, *Atti. Accad. Naz. Lincei Mem. Classe Sci. Fis. Mat. Nat. Sez 11*, **21**, 365 (1965).
23. G. Natta, M. Peraldo, and P. Corradini, *Atti Accad. Naz. Lincei Mem. Classe Sci. Fis. Mat. Nat.* **26**, 14 (1959).
24. A. Turner-Jones, J. M. Anzlewood, and D. R. Beckett, *Makromol. Chem.* **74**, 134 (1964).
25. J. L. Kardos, A. W. Christiansen, and E. Baer, *J. Polym. Sci. Part A2* **4**, 777 (1966).
26. G. Natta and P. Corradini, *Nuovo Cimento Suppl.* **5**, 40 (1960).
27. G. Natta, L. Porri, P. Corradini, and D. Morero, *Atti Accad. Naz. Lincei Rend. Classe Sci. Fis. Mat. Nat.* **20**, 560 (1956).
28. D. R. Burfield and Y. Doi, *Polymer Comm.* **24**, 48 (1983).
29. H. G. Norholt and H. A. Stutt, *J. Polym. Sci. Phys. Ed.* **16**, 939 (1978).
30. T. G. Schlote, N. L. Meijerink, and H. M. Schloffelers, *J. Appl. Polym. Sci.* **29**, 3763 (1984).
31. ASTM Method D1238/L, American Society for Testing and Materials, Philadelphia, Pa.
32. P. E. Slade, *Techniques and Methods of Polymer Evaluation*, Vol. 4, Marcel Dekker, Inc., New York, 1975.
33. G. Zeichner and P. Patel, paper presented at *The 2nd World Congress of Chemical Engineering*, Montreal, P. Q., Canada, 1981.
34. A. J. P. Franck, *J. Rheol.* **29**, 833 (1985).
35. E. J. Clark and J. D. Hoffman, *Macromolecules* **17**, 878 (1984).
36. R. F. Saraf and R. S. Porter, *Polymer Eng. Sci.* **28**, 842 (1988).
37. U. Gaur and B. Wunderlich, *J. Phys. Chem. Ref. Data* **10**, 1051 (1981).
38. R. Paukkeri and A. Lehtinen, *Polymer* **34**, 4075 (1993).
39. S. Newman and W. P. Cox, *J. Polym. Sci.* **46**, 29 (1960); T. Mijamoto and H. Inagaki, *J. Polym. Sci Part A2* **7**, 963 (1969).
40. J. M. G. Cowie, *Eur. Polym. J.* **9**, 1041 (1973).
41. A. Galambos, M. Wolkowicz, R. Ziegler, and M. Galimberti, *Proc. Am. Chem. Soc. Div. Polym. Mat. Sci. Eng.* **64**, 45 (1991).
42. *Code of Federal Regulations*, Title 21, Section 177.1520, U.S. Printing Office, Washington, D.C.
43. Technical Bulletin 500-725H, Himont, Inc., Wilmington, Del., May 1989.
44. R. B. Lieberman and D. DelDuca, *Proc. SPE Retec: Polyolefins VIII* **21** (1993); *Plast. World* **4**(9), 4 (1989).
45. U.S. Pat. 5,173,540 (Dec. 22, 1992), J. Saito and A. Sampei (to Chisso Corp.).
46. Technical Bulletin 500-742B, Himont, Inc., Wilmington, Del., June 1988.

47. *Mod. Plast.*, **70**(11), 87 (1993).
48. U.S. Pats. 3,112,200 and 3,112,301 (Nov. 26, 1963), (June 8, 1954), (to Montecatini); G. Natta, *Chim. Ind. Milan* **37**, 88 (1955).
49. G. Natta, P. Corradini, and G. Allegra, *J. Polym. Sci.* **51**, 399 (1961).
50. U.S. Pat. 3,128,252 (Aug. 15, 1956), (to Esso Research and Engineering Co.).
51. Brit. Pat. 1,092,390 (Apr. 27, 1966), (to Mitsubishi Petrochemical Co.).
52. L. A. M. Rodriguez and H. M. Van Looy, *J. Polym. Sci. Part A1* **4**, 1971 (1966).
53. Ger. Offen. 1,958,488 (1970), A. Mayr, P. Galli, E. Susa, G. Di Drusco, and E. Giachetti (to Montedison).
54. P. C. Barbe, G. Cecchin, and L. Noristi, *Adv. Polym. Sci.* **81**, 1 (1987).
55. K. Soga, S. Katano, Y. Akimoto, and T. Kogye, *Polym. J.* **5**(2), 128 (1973).
56. Eur. Pat. 0045,977 B1 (Aug. 13, 1980), S. Parodi, R. Nocci, U. Giannini, P. C. Barbe, and U. Scata (to Montedison).
57. P. Galli and J. C. Haylock, *Proc. SPE Retec: Polyolefins VII*, 27 (1991).
58. U.S. Pat. 4,971,937 (1990), E. Albizzati and co-workers (to Himont).
59. U.S. Pat. 5,177,162 (Jan. 5, 1993), M. Matsuura and T. Fujita (to Mitsubishi Petrochemical).
60. U.S. Pat. 4,857,613 (Aug. 15, 1989), R. Zolk, J. Kerth, and R. Hemmerich (to BASF).
61. U.S. Pat. 4,400,302 (Aug. 23, 1983), A. A. Van Der Nat, B. L. Goodall, and S. Willen (to Shell Oil).
62. U.S. Pat. 5,013,702 (May 7, 1991), G. G. Arzoumanidis, N. M. Karayannis, H. M. Khelghatian, S. S. Lee, and B. V. Johnson (to Amoco); U. S. Pat. 4,612,229 (Sept. 16, 1986), G. G. Arzoumanidis, H. M. Khelghatian, and S. S. Lee (to Amoco).
63. U.S. Pat. 4,532,313 (July 30, 1985), A. S. Matlack (to Himont).
64. U.S. Pat. 4,414,132 (Nov. 8, 1983), A. A. Van Der Nat, B. L. Goodall, and S. Willen (to Shell Oil).
65. U.S. Pat. 4,315,835 (Feb. 16, 1982), U. Scata, L. Luciani, and P. C. Barbe (to Montedison).
66. D. S. Breslow and N. R. Newburg, *J. Am. Chem. Soc.* **79**, 5072 (1957).
67. H. Sinn and W. Kaminsky, *Adv. Organomet Chem.* **18**, 99 (1980).
68. U.S. Pat. 5,324,800 (June 28, 1994), H. C. Welborn and J. A. Ewen (to Exxon Chemical Patents).
69. J. A. Ewen, *J. Am. Chem. Soc.* **106**, 6355 (1984).
70. J. A. Ewen, R. L. Jones, A. Razavi, and J. D. Ferrara, *J. Am. Chem. Soc.* **110**, 6255 (1988).
71. K. Soga and M. Terrano, eds., *Catalyst Design for Tailor Made Polyolefins*, Elsevier Science, New York, 1994.
72. U.S. Pat. 5,145,819 (Sept. 8, 1992), A. Winter, M. Antberg, W. Spaleck, J. Rohrman, and V. Dolle (to Hoechst Aktiengessellschaft).
73. B. Rieger, A. Reinmath, W. Roll, and H. H. Brintzinger, *J. Molec. Catal.* **82**, 67 (1993).
74. W. Kaminsky, R. Engehauser, K. Zoumis, W. Spaleck, and J. Rohrmann, *Makromol. Chem.* **193**, 1643 (1992).
75. U.S. Pats. 4,935,397 (June 19, 1990), 5,006,500 (Apr. 9, 1991), M. Chang (to Exxon Chemical Patents).
76. U.S. Pats. 4,923,833 (May 8, 1990), 4,952,540 (Aug. 28, 1990), M. Kioka and N. Kashiwa (to Mitsui Petrochemical Industries).
77. U.S. Pat. 5,169,818 (Dec. 8, 1992), M. Antberg, H.-F. Herrmann, and J. Rohrmann (to Hoechst); U.S. Pat. 5,202,398 (Apr. 13, 1993), M. Antberg, H. Luker, and L. Boehm (to Hoechst).
78. J. H. Schut, *Plastics Technol.* **38**(3), 31 (1992).
79. *Chem. Week*, 16 (June 15, 1994).
80. C. D. Nentizescu, C. Huch, and A. Huch, *Agnew. Chem.* **68**, 438 (1956).

81. D. B. Ludlum, A. W. Anderson, and C. E. Ashby, *J. Am. Chem. Soc.* **80**, 1380 (1958).
82. P. Cossee, *Tetrahedron Lett.* **17**, 12 (1960).
83. P. Cossee, *Proceedings of the International Congress on Coordination Chemistry*, 1961, p. 241.
84. P. Cossee, *J. Catal.* **3**, 80 (1964).
85. E. J. Arlman and P. Cossee, *J. Catal.* **3**, 99 (1964).
86. E. J. Arlman, *J. Catal.* **3**, 89 (1964).
87. A. Zambelli, "Homopolymers and Copolymers of Propylene," in *NMR Basic Principles and Progress*, Vol. 4, Springer, Berlin, 1971, p. 101.
88. W. O. Crains, A. Zambelli, and J. D. Roberts, *Macromolecules* **12**, 330 (1971).
89. P. Locatelli, A. Immirzi, and A. Zambelli, *Makromol. Chem.* **176**, 1121 (1975).
90. A. Zambelli, in J. C. W. Chien, ed., *Coordination Polymerization*, Academic Press, Inc., New York, 1975, p. 25.
91. R. F. Jordan, C. S. Bajgur, R. Willet, and B. Scott, *J. Am. Chem. Soc.* **108**, 7410 (1986).
92. G. Guerra, L. Cavallo, G. Moscardi, M. Vacatello, and P. Corradini, *J. Am. Chem. Soc.* **116**, 2988 (1994).
93. G. Guerra, L. Cavallo, V. Venditto, M. Vacatello, and P. Corradini, *Makromol. Chem. Macromol. Symp.* **69**, 237 (1993).
94. M. Ystenes, *J. Catal.* **129**, 383 (1991).
95. E. P. Bierwagen, J. E. Bercaw, and W. A. Goddard, *J. Am. Chem. Soc.* **116**, 1481 (1994).
96. G. DiDrusco and R. Rinaldi, *Hydrocarbon Proc.* **63**(11), 113 (Nov. 1984).
97. M. C. Sacchi, F. Forlini, I. Tritto, R. Mendici, and G. Zannoni, *Macromolecules* **25**, 5914 (1992).
98. J. V. Sepala and M. Harkonen, *Makromol. Chem.* **192**, 2857 (1991).
99. M. Kioka and N. Kashiwa, *J. Molec. Catal.* **82**, 11 (1993).
100. L. Noristi, P. C. Barbe, and G. Baruzzi, *Makromol. Chem.* **192**, 1115 (1991); P. C. Barbe, L. Noristi, and G. Baruzzi, *Makromol. Chem.* **193**, 229 (1992).
101. C. W. Hock, *J. Polym. Sci. Part A1* **4**, 3055 (1966).
102. P. Galli, L. Luciani, and G. Cecchin, *Ang. Makromol. Chem.* **94**, 63 (1981).
103. T. Keii, *Makromol. Chem.* **183**, 2285 (1982).
104. Y. Doi, M. Murata, K. Yano, and T. Keii, *Ind. Eng. Chem. Prod. Res. Dev.* **21**, 580 (1982).
105. I. Pasquon and U. Giannini, in J. R. Anderson and M. Boudart, eds., *Catalysts, Science and Technology*, Vol. 6, Springer-Verlag, Berlin, 1984.
106. E. Kohn, H. J. L. Shurmans, J. V. Lavendar, and R. A. Mendelson, *J. Polm. Sci.* **58**, 681 (1962).
107. V. A. Zakharov, G. D. Bukatov, and Yu. I. Yermakov, *Polym. Sci. Technol. Coord. Polym.* **19**, 267 (1983).
108. V. Varzelhan, T. F. Burger, and D. J. Stein, *J. Macromol Chem.* **183**, 489 (1982).
109. P. J. T. Tait, *Macromolecular Chemistry*, Vol. 1, London Royal Society of Chemistry, London, 1980, p. 3.
110. U. Giannini, *Makromol. Chem. Suppl.* **5**, 216 (1981).
111. E. Suzuki, H. Tamura, Y. Doi, and T. Keii, *Makromol. Chem.* **180**, 2275 (1979).
112. E. J. Nagel, V. A. Kirilov, and W. H. Ray, *Ind. Eng. Chem Prod. Res. Dev.* **19**, 372 (1980).
113. R. Galvan and M. Tirrell, *Comp. Chem. Eng.* **10**(1), 77 (1986).
114. R. Galvan and M. Tirrel, *Chem Eng. Sci.* **41**(9), 2385 (1986).
115. M. A. Ferrero and M. G. Chiovetta, *Polym. Eng. Sci.* **27**(19), 1436 (1987).
116. *Ibid.*, p. 1446.
117. M. A. Ferrero and M. G. Chiovetta, *Polym. Eng. Sci.* **31**(12), 886 (1991).

118. S. Floyd, R. A. Hutchinson, and W. H. Ray, *J. Appl. Poly. Sci.* **32**, 5451 (1986).
119. J. J. Zacca and W. H. Ray, *Chem. Eng. Sci.* **48**(22), 374 (1993).
120. Ref. 117, p. 904.
121. G. DiDrusco and R. Rinaldi, *Hydrocarbon Process.* **60**(5), 153 (May 1981).
122. U.S. Pat. 3,051,690 (July 29, 1955), E. J. Vandenberg (to Hercules Powder Co.).
123. Brit. Pat. 1,044,811 (Apr. 9, 1962), (to Rexall Drug and Chemical Co.).
124. U.S. Pat. 3,476,729 (Nov. 4, 1969), D. E. Smith, R. M. Keeler, and E. Guenther (to Phillips Petroleum).
125. J. F. Ross and W. A. Bowles, *Ind. Eng. Chem. Prod. Res. Dev.* **24**, 149 (1985).
126. *Oil Gas J.* **68**, 64 (1970).
127. U.S. Pat. 3,423,384 (Jan. 21, 1969), H. G. Hagemeyer and V. K. Park (to Eastman Kodak).
128. A. M. Jones, *Proc. SPE Retec: Polyolefins V*, 33 (1987).
129. C. Cipriani and C. A. Trishman, Jr., *Chem. Eng.* **80**, (Apr. 20, 1981).
130. *Polyolefins—A Global Perspective in a Decade of Change*, Vol. 3, *Technology and Economics*, Chemical Systems, Tarrytown, N.Y., June 1991.
131. *Hydrocarbon Process., Petrochem. Handbook '91*, 173 (Mar. 1991).
132. G. DiDrusco and R. Rinaldi, *Hydrocarbon Process.* **63**(11), 116 (Nov. 1984).
133. *Hydrocarbon Process.* **72**(3), 204 (1993).
134. *Hydrocarbon Process.* **72**(3), 207 (Mar. 1993).
135. N. F. Brockmeier, *Proc. SPE Retec: Polyolefins VII*, 67 (1991).
136. *Plast. World* **51**(6), 12 (1993).
137. *Modern Plast.* **71**(1), 81 (1994).
138. H. D. Schuddemagge, "Challenges of Polypropylene in Europe" presented at *ECMRA Annual Conference*, Brussels, Belgium, 1992.
139. Ref. 130, *Update 1992*, Vol. 3, 1993.
140. *Guide for Injection Molding*, Technical Bulletin AP-017, Himont, Inc., Wilmington, Del., May 1992.
141. J. R. Beren and C. Capellman, "Advances in Polypropylene Blow Molding," paper presented at *SAE International Congress*, Feb. 1994.
142. J. Foster, *Modern Plastics Encyclopedia 1983–1984*, Vol. 60, No. 10A, McGraw-Hill, New York, 1983, p. 328.
143. M. R. Drickman and K. E. McHugh, *Proc. SPE Antec 1992* **1**, 496 (1992).
144. K. T. Johansson, *Proc. SPE Retec: Polyolefins VIII*, 334 (1993).
145. A. Ahmed, *Polypropylene Fibers—Science and Technology*, Elsevier, Amsterdam, the Netherlands, 1982.
146. *Chem. Mark. Rep.*, 7 (Apr. 12, 1993).
147. Ref. 132, p. 436.
148. J. Spitz, *Packaging* **31**(4), 69 (1986).
149. *Chem. Week* **150**(19), 52 (1992).
150. S. Al-Malaika and G. Scott "Thermal Stabilization of Polyolefins," in N. S. Allen, ed., *Degradation and Stabilization of Polyolefins*, Applied Science Publishers Ltd., London, 1983, pp. 247–281.
151. C. Armstrong, M. J. Husbands, and G. Scott, *Eur. Poly J.* **15**, 241 (1979).
152. K. Schwarzenbach, in R. Gachter and H. Muller, eds., *Plastics Additives Handbook*, Hanser, Vienna, 1987, pp. 21–32.
153. *Production and Sales and Captive Use of Thermosetting and Thermoplastic Resins*, Society of the Plastics Industry, Committee on Resin Statistics, Ernst and Young, New York, 1986–1993.
154. *Mod. Plast. Int.* **21**(1), 51 (1991); **22**(1), 45 (1992); **23**(1), 52 (1993); **24**(1), 48 (1994).
155. *Mod. Plast. Int.* **21**(1), 54 (1991); **22**(1), 47 (1992); **23**(1), 53 (1993); **24**(1), 49 (1994).
156. *Plast. World* **44**(1), 86 (1986); **45**(1), 89 (1987); **46**(1), 73 (1988); **47**(1), 86 (1989); **48**(1), 73 (1990); **49**(1), 92 (1991); **50**(2), 72 (1992); **51**(1), 58 (1993); **52**(1), 72 (1994).

157. *Mod. Plast.* **71**(2), 18 (1994).
158. *Plast. News*, 28 (June 17, 1991).
159. G. Hebert, *Proc. SPE Retec: Polyolefins VIII*, 285 (1993).
160. M. S. Bailey and D. Brauer, *Mod. Plast.* **70**(12), 48 (1993).
161. *Packaging* **31**(4), 79 (1986).

General Reference

S. Van der Ven, *Polypropylene and Other Polyolefins: Polymerization and Characterization*, Elsevier Science Publishers, B. V., Amsterdam, the Netherlands, 1990.

RICHARD B. LIEBERMAN
Montell Polyolefins

POLYMERS OF HIGHER OLEFINS

This article describes polymeric materials derived from higher α-olefins, ie, those with carbon atom numbers of four or higher, and polymers of cycloolefins. The polymers may be linear or branched. Crystalline polymers of α-olefins are stereoregular, ie, isotactic or syndiotactic polymers, such as polypropylene (PP). These polymers are produced with a number of different catalysts. For example, heterogeneous Ziegler-Natta catalysts and some soluble bridged metallocene catalysts produce isotactic polymers. Other types of bridged metallocene catalysts produce crystalline syndiotactic polymers whereas nonbridged metallocene catalysts yield amorphous atactic polymers.

The synthesis of isotactic polymers of higher α-olefins was discovered in 1955, simultaneously with the synthesis of isotactic PP (1,2); syndiotactic polymers of higher α-olefins were first prepared in 1990 (3,4). The first commercial production of isotactic poly(1-butene) [9003-29-6] (PB) and poly(4-methyl-1-pentene) [9016-80-2] (PMP) started in 1965 (5).

Higher α-olefins can also be polymerized with cationic initiators to liquid oligomeric materials with isomerized structures. These liquids are manufactured commercially and used as lubricating oils.

Cycloolefins are polymerized by means of two different mechanisms. In the first, catalysts based on tungsten and molybdenum compounds induce ring-opening polymerization (metathesis) of monocycloolefins with the formation of linear elastomers containing regularly spaced double bonds in polymer chains. If cyclodienes are used in this reaction, these catalysts then produce cross-linked resins. In the second mechanism, metallocene-based catalysts polymerize cycloolefins without ring opening into linear, stereoregular, highly crystalline polymers. Polymers of several cycloolefins, polydicyclopentadiene, polyoctenamers, and norbornene elastomers are all produced commercially.

Monomers

Manufacture of Monomers. The monomers of the greatest interest are those produced by oligomerization of ethylene (qv) and propylene (qv). Some olefins are also available as by-products from refining of petroleum products or as the products of hydrocarbon (qv) thermal cracking.

1-Butene. Commercial production of 1-butene, as well as the manufacture of other linear α-olefins with even carbon atom numbers, is based on the ethylene oligomerization reaction. The reaction can be catalyzed by triethylaluminum at 180–280°C and 15–30 MPa (~150–300 atm) pressure (6) or by nickel-based catalysts at 80–120°C and 7–15 MPa pressure (7–9). Another commercially developed method includes ethylene dimerization with the Ziegler dimerization catalysts, $Ti(OR)_4$–AlR_3, where R represents small alkyl groups (10). In addition, several processes are used to manufacture 1-butene from mixed butylene streams in refineries (11) (see BUTYLENES).

4-Methyl-1-Pentene. This olefin is produced commercially by dimerization of propylene in the presence of potassium-based catalysts at 150–160°C and ~10 MPa. Commercial processes utilize several catalysts, such as sodium-promoted potassium carbonate and sodium- and aluminum-promoted potassium hydroxide (12–14) in a fixed-bed reactor. The reaction produces a mixture of C_6 olefins containing 80–85% of 4-methyl-1-pentene.

Other Higher Olefins. Linear α-olefins, such as 1-hexene and 1-octene, are produced by catalytic oligomerization of ethylene with triethylaluminum (6) or with nickel-based catalysts (7–9) (see OLEFINS, HIGHER). Olefins with branched alkyl groups are usually produced by catalytic dehydration of corresponding alcohols. For example, 3-methyl-1-butene is produced from isoamyl alcohol using base-treated alumina (15).

Physical Properties of Monomers. 1-Butene [*106-98-9*] is a colorless, flammable, noncorrosive gas; its physical properties are listed in Table 1, and its thermodynamic properties are available (16). Because 1-butene has a very low flash point, it poses a strong fire and explosion hazard.

4-Methyl-1-pentene [*691-37-2*] is a light, colorless, flammable liquid; its physical constants are also given in Table 1. It is an irritant and, in high concentrations, a narcotic. Like 1-butene, this chemical compound has a low flash point and represents a significant fire hazard when exposed to heat, flame, or oxidizing agents.

Properties of other higher α-olefins and those of some commercially significant cycloolefins are given in Table 2. These monomers are liquids at ambient temperature and pressure. They are highly combustible and can form explosive mixtures with air. The primary health hazards presented by these monomers are associated with inhalation or prolonged skin contact that can cause irritation.

Chemical Properties. Higher α-olefins are exceedingly reactive because their double bond provides the reactive site for catalytic activation as well as numerous radical and ionic reactions. These olefins also participate in additional reactions, such as oxidations, hydrogenation, double-bond isomerization, complex formation with transition-metal derivatives, polymerization, and copolymerization with other olefins in the presence of Ziegler-Natta, metallocene, and cationic catalysts. All olefins readily form peroxides by exposure to air.

Polymer Properties

Chemical properties of most polyolefins resemble those of polypropylene. The resins resist most inorganic or organic acids and bases below 90°C as well as

Table 1. Physical Properties of 1-Butene[a] and 4-Methyl-1-Pentene[b]

Property	1-Butene[c]	4-Methyl-1-pentene[c]
flash point, °C	−150	−7.0
boiling point, °C	−6.25	53.86
melting point, °C	−185.35	−153.64
density, g/cm³	0.6014 (15°C)	0.6637 (20°C)
refractive index, n_D	1.3962 (25°C)	1.3827 (20°C)
critical temperature, °C	147.4	
critical pressure, MPa[d]	4.056	
vapor pressure, kPa[e]	415.7 (38°C)	29.47 (20°C)
		89.34 (50°C)
heat capacity, liquid, at 25°C, J/(mol·K)[f]	89.33	184.4
heat of vaporization at boiling point, J/mol[f]	20.38	27.11
viscosity, MPa·s (=cP)		0.288 (20°C)

[a]Refs. 16 and 17.
[b]Refs. 17 and 18.
[c]Temperature of the measurement is given in parentheses if stated in the reference.
[d]To convert MPa to psi, multiply by 145.
[e]To convert kPa to mm Hg, multiply by 7.5.
[f]To convert J to cal, divide by 4.184.

most salt solutions, solvents, soaps, and detergents. Properties of polyolefins rapidly deteriorate in contact with strong oxidizing agents such as fuming HNO_3, concentrated solutions of chromic acid, and chlorine or bromine solutions. All polyolefins undergo peroxidation, halogenation, and halosulfonation reactions (16).

Commercially produced crystalline polyolefins, PB, and PMP all exhibit high stability to inorganic substances, and excellent resistance in nonoxidative inorganic environments such as aqueous solutions of inorganic salts, acids, and alkalies. PMP easily withstands prolonged boiling and autoclave treatment required for medical and pharmaceutical applications. However, prolonged exposure of polyolefin specimens under stress to some hydrocarbon solvents and aqueous detergent solutions can cause cracks and eventual failure, a phenomenon usually referred to as environmental stress cracking.

Thermal, Thermooxidative, and Photooxidative Degradation. Polymers of α-olefins have at least one tertiary C−H bond in each monomer unit of polymer chains. As a result, these polymers are susceptible to both thermal and thermooxidative degradation. Reactivity in degradation reactions is especially significant in the case of polyolefins with branched alkyl side groups. For example, thermal decomposition of poly(3-methyl-1-butene) starts at ca 300°C and thermooxidative degradation at ca 100°C. The effective activation energy of this process is 60 kJ/mol (14.3 kcal/mol). Thermal degradation of PMP occurs above 280°C, eg, during polymer processing (19), and produces various low molecular weight hydrocarbons, such as propane (~35%), isobutylene (~55%), isobutane (4%), isopentane (2%), and 4-methyl-1-pentene (2.5%). Pyrolysis of polyolefins at 500°C generates mostly light unsaturated hydrocarbons, including monomers.

Table 2. Properties of Higher Olefins[a]

Monomer	Molecular weight	Boiling point, °C	Melting point, °C	Refractive index, n_D^{20}	Density,[b] g/cm³
Linear olefins					
1-pentene	70.1	29.96	−165.22	0.6405	1.3715
3-methyl-1-butene	70.1	20.05	−168.49	0.6272	1.3643
1-hexene	84.2	63.48	−139.83	0.6732	1.3879
3-methyl-1-pentene	84.2	54.17	−153.0	0.6674	1.3842
4-methyl-1-pentene	84.2	53.86	−153.64	0.6637	1.3827
4-methyl-1-hexene	98.2	86.73	−141.46	0.6985	1.4000
5-methyl-1-hexene	98.2	85.31		0.6920	1.3967
1-octene	112.2	121.29	−101.72	0.7149	1.4087
5-methyl-1-heptene	112.2	113.3		0.7164	1.4094
vinylcyclo-hexane	110.2	128		0.8060	1.4458
1-decene	140.3	170.60	−66.28	0.7408	1.4215
Cycloolefins					
cyclopentene	68.1	44.23	−135.08	0.7720	1.4225
cyclopenta-diene	66.1	40.0	−97.2	0.8021	1.4440
cyclooctene	110.2	146	−16	0.846	1.4690
norbornene[c]	94.6	96.0	44−46		
dicyclopenta-diene	132.2	170	−1	1.511	0.986

[a]Ref. 18.
[b]At 20°C.
[c]Bicyclo[2.2.1]-hept-2-ene.

Under thermooxidative conditions, all polyolefins undergo chain scission reactions (20), which result in a decrease of molecular weight and the formation of gaseous products such as formaldehyde (qv), acetone (qv), and various alkanes (21). Amorphous polymers of higher α-olefins are more reactive in thermooxidation reactions than their crystalline analogues, mostly because of a higher rate of oxygen diffusion. The main obstacle for a widespread use of polyolefins is their high reactivity in thermooxidative degradation reactions, especially at the high temperatures required for processing (usually above 260°C) and potential applications.

Thermooxidative degradation of PMP is noticeable even at 140−150°C (22). As with other polyolefins, it reduces molecular weight, causes the formation of hydroxyl and carbonyl groups in polymer chains, and produces various low molecular weight compounds. Because of both a high permeability to oxygen and an abundance of tertiary C−H bonds, PMP is three times more susceptible to this type of degradation than PP. As a result, durability of PMP decreases sharply with increasing temperature: 30,000 h at 120°C, 20,000 h at 130°C, 1000 h at 150°C, and 20 h at 200°C (22,23).

Photooxidative degradation of PMP, especially at wavelengths below 400 nm, also proceeds at a relatively high rate, thus limiting some of its outdoor applications (24). Photodegradation reduces the molecular weight of PMP; as a result, PMP becomes brittle, develops polar groups in polymer chains (carboxylic, ketonic, etc), and produces low molecular weight compounds, methane, ethane, acetone, carbon dioxide, and isobutane.

Both thermooxidation and photooxidation of polyolefins can be prevented by using the same antioxidants as those employed for the stabilization of polypropylene, ie, alkylated phenols, polyphenols, thioesters, and organic phosphites in the amount of 0.2–0.5% (22,25).

Polybutene can be cross-linked by irradiation at ambient temperature with γ-rays or high energy electrons in the absence of air. The performance of articles manufactured from polybutene is only slightly affected by ionizing radiation at doses below 30 kGy (3 Mrad) (26). PMP is also relatively stable to β- and γ-radiation employed in the sterilization of medical supplies (27).

Solubility. Highly crystalline isotactic polyolefins are not soluble in organic solvents at room temperature. However, most amorphous polyolefins and oligomers of α-olefins are easily soluble in saturated and aromatic hydrocarbons at ambient temperature. This difference in solubility can be used to separate amorphous atactic components of polyolefins from crystalline isotactic material in crude polyolefins mixtures. Although crystalline isotactic PMP is insoluble in any organic solvent at 20°C, it can absorb some aliphatic and aromatic hydrocarbons, esters, and halogenated hydrocarbons. The swelling causes loss of mechanical strength and clarity.

Above 100°C, most polyolefins dissolve in various aliphatic and aromatic hydrocarbons and their halogenated derivatives. For example, polybutene dissolves in benzene, toluene, decalin, tetralin, chloroform, and chlorobenzenes. As with other polyolefins, solubility of PB depends on temperature, molecular weight, and crystallinity.

Gas Permeability. Crystalline PMP is relatively highly permeable to various organic and inorganic gases. Permeabilities to oxygen, nitrogen, and light hydrocarbons (C_1–C_4) are 20–30 times higher than those of HDPE (28,29); such high permeabilities are related to PMP's low packing density. Water vapor transmission (2.5-μm film) at 38°C and relative humidity of 90% is 61–69 μmol/(m^2·s) (95–107 g/(m^2·24 h)).

Physical Properties. Table 3 lists physical properties of stereoregular polymers of several higher α-olefins. Crystal cell parameters of these polymers are available (34–36). All stereoregular polyolefins have helix conformations in the crystalline state. Their densities usually range from 0.90 to 0.95 g/cm^3. Crystalline PMP, however, represents an exception; its density is only 0.812–0.815 g/cm^3, lower even than that of amorphous PMP (0.835–0.840 g/cm^3), thus making it one of the lowest densities among plastics.

Polymorphism. Many crystalline polyolefins, particularly polymers of α-olefins with linear alkyl groups, can exist in several polymorphic modifications. The type of polymorph depends on crystallization conditions. Isotactic PB can exist in five crystal forms: form I (twinned hexagonal), form II (tetragonal),

Table 3. Properties of Polyolefins[a]

Polymer[b]	Melting point, °C	Crystal type	Helix type	Crystalline density, g/cm^3
Polymers of α-olefins				
iso-polybutene				
form I	138–142	hexagonal	3_1	0.951
form II	120–130	tetragonal	11_3	0.902
form III	101–109	orthorhombic		0.905
syndio-polybutene				
form I	~50		4_1	
form II	~50		10_3	
iso-poly(1-pentene)	105–115	monoclinic	3_1	0.92
	75–80	pseudoortho-rhombic	4_1	0.90
iso-poly(3-methyl-1-butene)	350	monoclinic	4_1	0.93
iso-poly(1-hexene)	<20	monoclinic	7_2	0.83
iso-poly(3-methyl-1-pentene)	200		7_2	
iso-poly(4-methyl-1-pentene)	235–240	tetragonal	7_2	0.813
syndio-poly(4-methyl-1-pentene)	197		24_7	
iso-poly(1-heptene)	18		3_1	
iso-poly(4-methyl-1-hexene)	188–200	tetragonal	7_2	0.845
syndio-poly(4-methyl-1-hexene)	147			
iso-poly(5-methyl-1-hexene)	110–130	monoclinic	3_1	0.84
iso-poly(1-octene)	~20			
iso-poly(5-methyl-1-heptene)	130	tetragonal	3_1	
iso-poly(vinylcyclo-hexane)	376–385	tetragonal	4_1	0.95
syndio-poly(vinylcy-clohexane)		amorphous		
iso-poly(1-decene)	22–27	side chain		
Polymers of cycloolefins				
diiso-polycyclobutene	485			
diiso-polycyclopentene	395			
polynorbornene	>600			

[a]Refs. 4, 28–33.
[b]*iso* designates isotactic; *syndio* designates syndiotactic.

form III (orthorhombic), form I′ (untwinned hexagonal), and form II′ (37–39). The crystal structures and thermal parameters of the first three forms are given in Table 3. Form II is formed when a PB resin crystallizes from the melt. Over time, it is spontaneously transformed into the thermodynamically stable form I;

at room temperature, the transition takes about one week to complete. Forms I', II', and III of PB are rare; they can be formed when the polymer crystallizes from solution at low temperature or under pressure (38). Syndiotactic PB exists in two crystalline forms, I and II (35). Form I comes into shape during crystallization from the melt (very slow process) and form II is produced by stretching form-I crystalline specimens (35).

PMP can also crystallize in several crystalline forms (28). Form I is produced during crystallization from the melt. This is the only crystalline form present in commercially manufactured articles. Other crystalline modifications are formed when the polymer is crystallized from solution (28).

Thermal Properties. Melting points of stereoregular PO resins depend on the size and shape of side groups in the polymer chains (Table 3). In the case of isotactic polymers of linear α-olefins, the melting points of the crystalline phase rapidly decrease with increasing side-chain length: 175°C for polypropylene, 130–140°C for PB, and 80°C for poly(1-pentene). Isotactic poly(1-hexene) is virtually amorphous at room temperature and isotactic polyolefins with longer linear side groups derive their crystallinity from the side chains rather than from the polymer backbone (40). Their melting points steadily increase with the side-chain length: ~45°C for poly(1-decene), 50°C for poly(1-tetradecene), 68°C for poly(1-hexadecene), and 75°C for poly(1-octadecene). Polymers of α-olefins with branched alkyl groups generally exhibit much higher melting points (Table 3); the melting points of isotactic poly(3-methyl-1-butene) and poly(vinylcyclohexane) are the highest, over 350°C.

Commercial isotactic PB resins melt at ~125°C. The polymer has two low temperature relaxation transitions. One, in the −17 to −25°C range, is attributed to the glass transition in the amorphous phase; the other, at −158°C, to the onset of the local motion of side groups. Crystalline isotactic PMP has a melting point of around 245°C and a high glass-transition temperature of around 50°C. Other thermal transitions in PMP are at 130–180°C (beginning of crystalline phase disordering), at −120 to −150°C (relaxation movements of short-chain segments), and at −250°C. The heat of PMP fusion is 61.65 J/g (14.8 cal/g) (41). Commercial PMP resins are copolymers of 4-methyl-1-pentene and linear α-olefins (C_{10}–C_{16}) of around 3 mol %. These resins have lower melting points (220–225°C); their glass-transition temperatures (20–30°C) and their brittle points are also lower than for pure PMP (42), thus making them more suitable for many applications.

Melt Crystallization. Isotactic PB crystallizes from the melt with spherulitic morphology (43). Directly proportional to molecular weight (44), the PB crystallization rate can be significantly increased by using nucleating agents (45) as well as by stress and pressure (43). Some nucleating agents also increase hardness (qv), modulus, yield strength, and heat distortion temperature of the resin. Syndiotactic PB crystallizes to form I very slowly (eg, two weeks at 20°C (35)), and its crystallization rate can be increased by stretching the resin.

Transformations in the Solid State. From a practical standpoint, the most important solid-state transformation of PB involves the irreversible conversion of its metastable form II developed during melt crystallization into the stable form I. This transformation is affected by the polymer molecular weight and tacticity as well as by temperature, pressure, mechanical stress, and the presence of impurities and additives (38,39). At room temperature, half-times of the transformation range between 4 and 45 h with an average half-time of 22–25 h (39). The process can be significantly accelerated by annealing articles made of PB at temperatures below 90°C, by ultrasonic or γ-ray irradiation, and by utilizing various additives. Conversion of form II to form I is slower in 1-butene-based copolymers containing 7–8 mol % of other α-olefins, both linear (1-hexene, 1-octene, 1-nonene, 1-decene, or 1-dodecene) and branched (3-methyl-1-butene or 4-methyl-1-pentene) (39).

Mechanical Properties.   The side-group type in the polymer chains determines mechanical properties of stereoregular polyolefins. For example, resins with long linear side groups have low crystallinities and exhibit mechanical behavior typical for elastomers. On the other hand, PB and most polymers with branched side groups are highly crystalline and exhibit mechanical properties similar or superior to those of isotactic PP. Excellent mechanical and optical properties contribute to the industrial importance of PB and PMP. Consequently,

Table 4. Properties of Pipe-Grade Polybutene Resin[a]

Property	ASTM test method	Value
mol wt, $\overline{M}_w$		70–75,000
mol wt distribution, $\overline{M}_w/\overline{M}_n$		10–11
melt index, g/10 min	D1238[b]	0.4
crystallinity (form I), %		48–55
density, g/cm³	D792	0.93–0.94
tensile properties		
yield strength, MPa[c]	D638	16–18
break strength, MPa[c]	D638	32–35
elongation to break, %	D638	275–320
modulus, MPa[c]	D638	290–295
flexural modulus, MPa[c]	D790	375–380
notched Izod impact strength, J/m[d]	D256	640–800
hardness, Shore D	D2240	55–65
brittleness temperature, °C	D746	−18
melting temperature, °C	D3418	124–130
Vicat softening point, °C	D1525	112–114
heat deflection temperature, °C		
at 1.82 MPa[e]	D648	54–60
at 0.46 MPa[e]	D648	102–113
linear coefficient of thermal expansion, m/m/°C	D696	0.00013
thermal conductivity, W/(m·K)	C177	0.22
UL flammability	UL94	HB
dielectric constant, 10^3–10^6, Hz	D150	2.53
dissipation factor, 10^3–10^6 Hz	D150	0.0005
water absorption, % in 24 h	D570	<0.03

[a]Form I; Ref. 39.
[b]Condition E.
[c]To convert MPa to psi, multiply by 145.
[d]To convert J/m to ft·lbf/in., multiply by 53.38.
[e]To convert MPa to atm, divide by 0.101.

although poly(3-methyl-1-butene) and poly(vinylcyclohexane) also exhibit good mechanical characteristics, their high melting points, poor oxidative stability, and brittleness still preclude them from industrial application.

Poly(1-butene). Strain–stress testing of crystalline isotactic PB shows that the resin does not have a distinct yield point and that, instead, it undergoes uniform stretching followed by strain-hardening (39). Mechanical properties of two commercial grades of PB are listed in Tables 4 and 5. The resin's ultimate strength and modulus both increase with crystallinity. The most important mechanical properties of PB are excellent creep resistance (39,44,46) and resistance to environmental stress-cracking (11,39,47). Detailed data on morphological, mechanical, and thermomechanical properties and environmental stress cracking resistance of different commercial grades of PB are available (43,44,48).

Poly(4-methyl-1-pentene). Physical, thermal, mechanical, and electrical properties of commercial crystalline PMP are listed in Table 6. PMP resins are tough, rigid plastics with a high tensile modulus and low elongation at break. PMP homopolymer is brittle at low temperatures due to a relatively high glass-transition temperature (Table 3); its brittleness can be reduced by the introduction of 2–3 mol % of a linear α-olefin unit into PMP chains (27). These copolymers have a lower mechanical modulus and a higher elongation at break, up to 80–120%. High temperature mechanical and creep properties of PMP are superior to those of HDPE and PP and the impact strength of PMP is

Table 5. Properties of Polybutene Film[a]

Property	ASTM test method	Value
resin (form I) properties		
melt index, g/10 min	D1238	1.0
density, g/cm^3	D1505	0.91
crystallinity (form I), %		46–52
melting point, °C	D3418	118–120
film properties (50.8-μm film)		
break strength, MPa[b,c]		
MD	D882	44.8
TD	D882	37.9
elongation at break, %	D882	300–320
dart impact strength, g[d]	D1709	350
Elmendorf tear strength, kN/m[e,c]		
MD	D1922	425
TD	D1922	386
optical properties, %		
haze	D1003	16
clarity	D1003	1.5
gloss	D2457	~70
sealing temperature range, °C		155–210

[a]Grade DP 1710; Ref. 39.

[b]To convert MPa to psi, multiply by 145.

[c]MD = machine direction; TD = transverse direction.

[d]Average weight of the dart sufficient to break the film in 50% of tests (ASTM D4272-90, Method A).

[e]To convert kN/m to ppi, divide by 0.175.

Table 6. Properties of Crystalline Poly(4-Methyl-1-Pentene)[a]

Property	ASTM test method	Value
Physical		
density at 20°C, g/cm^3	D792	0.830–0.835
water absorption at saturation, %	D570	0.01
refractive index, n_D^{20}		1.463
Thermal		
melting point, °C		235–240
heat of fusion, J/g[b]		61.65
Vicat softening point, °C	D1525	173–180
glass transition, °C		
homopolymer		~50
copolymer		~30
deflection temperature under		
flexural load, °C		
0.46 MPa[c]	D686	80–90
1.82 MPa[c]	D686	48–50
coefficient of linear expansion	D696	1.17 10^{-4}
per °C		
thermal conductivity, W/(m·K)	C177	0.167–0.172
specific heat capacity, J/(g·K)[b]	C351	2.18
Electrical		
dielectric constant at 25°C,	D150	2.12
10^2–10^6 Hz		
volume resistivity, Ω·cm	D257	>10^{16}
dielectric strength, kV/mm	D149	63–65
dissipation factor at 10 MHz	D150	1.5 10^{-4}
Mechanical		
tensile yield strength, MPa[c]	D638	23–28
tensile break strength, MPa[c]	D638	17–20
elongation at break, %		10–25
bending strength, MPa[c]	D790	25–35
tensile modulus, MPa[c]	D638	1500–2000
flexural modulus, MPa[c]	D790	1300–1800
compression modulus, MPa[c]	D695	800–1200
notch impact strength, kJ/m[d]	D256	100–200
Rockwell hardness		L80–93

[a]Refs. 23, 24, 28, 41, and 49.
[b]To convert J to cal, divide by 4.184.
[c]To convert MPa to psi, multiply by 145.
[d]To convert kJ/m to ft·lbf/in., multiply by 18.73.

two to three times higher than the impact strength of other thermoplastics with comparable light-transmission characteristics, such as polystyrene and acrylics.

Electrical Properties.   All polyolefins have low dielectric constants and can be used as insulators; in particular, PMP has the lowest dielectric constant among all synthetic resins. As a result, PMP has excellent dielectric properties and a low dielectric loss factor, surpassing those of other polyolefin resins and polytetrafluoroethylene (Teflon). These properties remain nearly constant over a wide temperature range. The dielectric characteristics of poly(vinylcyclohexane) are especially attractive: its dielectric loss remains constant between -180 and $160°C$, which makes it a prospective high frequency dielectric material of high thermal stability.

Optical Properties of PMP.   Although all polyolefins are highly opaque, isotactic PMP alone possesses this outstanding feature: it has low haze (1.2–1.5%) and high optical transparency (~90–92%) comparable to that of polystyrene (88–92%) and acrylics (90–92%). High transparency of PMP is a consequence of low optical anisotropy of PMP molecules in the helical conformation; morever, the addition of a small amount of linear α-olefin units of PMP chains can further reduce haze (27). Light transmittance of PMP in the near uv region is also excellent, higher than that of glass and inferior only to quartz. Optical clarity accounts for many applications of PMP.

Properties of Polycycloolefins.   Polymers of monocyclic olefins (cyclopentene, cyclooctene) produced by ring-opening metathesis are linear elastomers. Their properties are somewhat similar to those of poly(cis-1,4-butadiene). Polymers of dicyclopentadiene produced with the same catalysts are heavily crosslinked resins displaying high toughness and tensile strength as well as excellent impact strength at low temperatures (see CYCLOPENTADIENE AND DICYCLOPENTADIENE). Table 7 lists typical properties of polydicyclopentadiene [25038-78-2]. Stereoregular polymers of cycloolefins produced with metallocene catalysts have linear structures containing cycloalkyl rings as parts of their main chains. These polymers are highly crystalline; they have densities greater than 1.0 and extremely high melting points of 395°C for polycyclopentene, 485°C for polycyclobutene, and over 600°C for polynorbornene (32).

Table 7. Properties of Polydicyclopentadiene[a]

Property	Unfilled resin	20% Glass powder
flexural modulus, MPa[b]	1800–2100	2600–2900
tensile strength, MPa[b]	33–35	30–32
tensile elongation, %	80	25
impact strength, J[c]		
23°C	13–16	11–13
−29°C	11–13	11–13
glass transition, °C	127	127

[a]Ref. 50.
[b]To convert MPa to psi, multiply by 145.
[c]To convert J to ft·lbf, multiply by 0.737.

Polymerization of α-Olefins

Ziegler-Natta Catalysts.   All isotactic polymers of higher α-olefins are produced with the same type of heterogeneous, titanium-based Ziegler-Natta catalyst systems as that used for the manufacture of isotactic PP. The catalyst systems have two components, a solid catalyst containing a titanium compound and a co-catalyst containing an organoaluminum compound (51). Both 1-butene and 4-methyl-1-pentene are nearly three to four times less reactive in the polymerization reactions with Ziegler-Natta catalysts than propylene (51,52). Early isospecific Ziegler-Natta catalysts consisted of the highly dispersed δ-crystalline form of $TiCl_3$ in combination with $Al(C_2H_5)_3$ and $Al(C_2H_5)_2Cl$ (51–54); modern, highly active catalysts are supported on $MgCl_2$ and usually prepared by grinding anhydrous $MgCl_2$ and $TiCl_4$ together in the presence of an aromatic ester such as ethyl benzoate (51). These catalysts are employed with cocatalyst mixtures containing $Al(C_2H_5)_3$ or $Al(i\text{-}C_4H_9)_3$ and aromatic esters, ethyl benzoate or ethyl anisate. Another highly active type of the supported catalysts use aromatic diesters (phthalates) during grinding of $MgCl_2$ and $TiCl_4$, with mixtures of $Al(C_2H_5)_3$ and phenylalkoxysilanes as co-catalysts (55–57). Crude polyolefin resins produced with heterogeneous Ziegler-Natta catalysts can be fractionated into crystalline, highly isotactic polymers usually insoluble in boiling n-heptane, fractions of polymer molecules with moderate stereoregularity, and amorphous stereoirregular fractions. The yields of the isotactic fractions depend on the catalytic system (51). For example, crude PMP obtained with the α-$TiCl_3$–$Al(i\text{-}C_4H_9)_3$ catalyst at 70°C contains 61% of isotactic polymer insoluble in boiling n-heptane, 26% of a polymer of lower isotacticity soluble in boiling n-heptane, but insoluble in cold n-heptane, and 13% of amorphous atactic polymer soluble in cold n-heptane (51). A more isospecific δ-$TiCl_3$–$Al(C_2H_5)_2Cl$ system, however, can produce PMP resins that are ~90% insoluble in boiling n-heptane (53).

Polymerization reactions with Ziegler-Natta catalysts are carried out at 40–80°C in pure monomers or in monomer mixtures with aliphatic solvents (hexane, cyclohexane, and heptane). Molecular weights of polymers are controlled by the addition of hydrogen, an effective chain-transfer agent. Some Ziegler-Natta catalysts polymerize linear α-olefins, such as 1-hexane or 1-decene, into linear polymers with ultrahigh molecular weights (58) which are used as drag-reducing agents for hydrocarbon flow.

All higher α-olefins, in the presence of Ziegler-Natta catalysts, can easily copolymerize both with other α-olefins and with ethylene (51,59). In these reactions, higher α-olefins are all less reactive than ethylene and propylene (41). Their reactivities in the copolymerization reactions depend on the size and the branching degree of their alkyl groups (51) (see OLEFIN POLYMERS, LINEAR LOW DENSITY POLYETHYLENE).

Polymerization Processes. Isotactic PB and PMP are produced commercially in slurry processes in liquid monomers or monomer mixtures (optionally diluted with light inert hydrocarbons) at 50–70°C. The first commercial process for PB production used a highly isospecific δ-$TiCl_3$–$Al(C_2H_5)_2Cl$–I_2 catalyst in a stirred-tank reactor (60). Solution of the polymer in unreacted 1-butene was removed from the reactor, de-ashed with water, heated to 130°C, and flush-separated into the monomer and molten polymer. PB can also be manufactured in the gas-phase process (61). The catalysts most suitable for these processes are ester-modified supported $TiCl_4$–$MgCl_2$–$Al(C_2H_5)_3$ systems. After completion of the reaction, the resin is dried and compounded with antioxidants and other additives.

Metallocene Catalysts. Higher α-olefins can be polymerized with catalyst systems containing metallocene complexes. The first catalysts of this type

(Kaminsky catalysts) include common metallocene complexes of zirconium (zirconocenes), such as biscyclopentadienylzirconium dichloride, activated by a special organoaluminum compound, methylaluminoxane. These catalysts polymerize α-olefins with the formation of amorphous atactic polymers. Polymers with high molecular weights are produced at decreased temperatures and have rubber-like properties (62). When the polymerization reactions with the same catalysts are carried out at increased temperatures, eg, 70–100°C, they produce low molecular weight oligomers that can be used as components for synthetic lubricating oils (32).

Zirconocene complexes containing two indenyl or tetrahydroindenyl groups bridged with short links such as $-CH_2-CH_2-$ or $-Si(CH_3)-$ can produce isotactic polymers of higher α-olefins (32). To synthesize syndiotactic PO, bridged zirconocene complexes with rings of two different types are required, one example of which is isopropyl(cyclopentadienyl)(1-fluorenyl)zirconocene. These complexes are used for the synthesis of syndiotactic PB (3,4,35), PMP (4,63), and poly(4-methyl-1-hexene) (4).

Cationic Polymerization Reactions. α-Olefins with linear and branched alkyl groups can be readily polymerized with cationic initiators (see INITIATORS, CATIONIC). For example, olefins containing linear alkyl groups (1-pentene, 1-hexene, 1-octene, 1-decene, etc) and their mixtures are oligomerized by using BF_3, mixtures of BF_3 and alcohols, as well as $AlCl_3$ or $AlBr_3-HBr$ systems at low temperatures with the formation of low molecular weight oils of an irregular structure (64–66). These oligomers are used as base stocks for synthetic lubricating oils (66).

Polymerization of Cycloolefins

Depending on the type of catalyst used, polymerization of cycloolefins proceeds through either ring opening or by opening of the double bond with the preservation of the ring.

Ring-Opening Polymerization. Ring-opening polymerization of cycloolefins in the presence of tungsten- or molybdenum-based catalysts proceeds by a metathesis mechanism (67,68).

$$\left[-(CH_2)_n - R - (CH_2)_m - CH=CH-\right] \longrightarrow \left[-(CH_2)_n - R - (CH_2)_m - CH=CH-\right]_x$$

R stands for a hydrocarbon group such as $-CH=CH-$ in cyclodienes or another cyclic group in dicyclopentadiene and norbornene. For example, *trans*-polyoctenamer is manufactured in two stages. In the first stage, the monomer (cyclooctene) is produced by dimerization of butadiene into cyclooctadiene followed by selective hydrogenation of one double bond. In the next stage, cyclooctene is polymerized with tungsten- or molybdenum-based metathesis catalysts, such as WCl_6, activated by organoaluminum or organotin compounds (69).

Polyoctenamers produced under different conditions have varying [trans]:[cis] ratios, usually from over 5 to 1. For example, two commercial resins TOR 8012 (mol wt 100,000) and TOR 6213 (mol wt 120,000) have ratios of 4.0 and 1.5, respectively. The polymers with the highest trans-content crystallize at 50–60°C (see POLYALKENAMERS). Other properties of TOR resins are as follows.

Resin	Crystallinity, %	Density, g/cm^3	Melting point, °C	T_g, °C
TOR 8012	27	0.910	51	−65
TOR 6213	6	0.890	<30	−55

Dicyclopentadiene is also polymerized with tungsten-based catalysts. Because the polymerization reaction produces heavily cross-linked resins, the polymers are manufactured in a reaction injection molding (RIM) process, in which all catalyst components and resin modifiers are slurried in two batches of the monomer. The first batch contains the catalyst (a mixture of WCl_6 and $WOCl_4$), nonylphenol, acetylacetone, additives, and fillers; the second batch contains the co-catalyst (a combination of an alkylaluminum compound and a Lewis base such as ether), antioxidants, and elastomeric fillers (qv) for better moldability (50). Mixing two liquids in a mold results in a rapid polymerization reaction. Its rate is controlled by the ratio between the co-catalyst and the Lewis base. Depending on the catalyst composition, solidification time of the reaction mixture can vary from two seconds to an hour. Similar catalyst systems are used for polymerization of norbornene and for norbornene copolymerization with ethylidenenorbornene.

Metallocene Catalysts.   Polymerization of cycloolefins with Kaminsky catalysts (combinations of metallocenes and methylaluminoxane) produces polymers with a completely different structure. The reactions proceeds via the double-bond opening in cycloolefins and the formation of C–C bonds between adjacent rings (31,32). If the metallocene complexes contain bridged and substituted cyclopentadienyl rings, such as ethylene(bisindenyl)zirconium dichloride, the polymers are stereoregular and have the *cis*-diisotactic structure.

Processing

Rheology.   Both PB and PMP melts exhibit strong non-Newtonian behavior: their apparent melt viscosity decreases with an increase in shear stress (27,28). Melt viscosities of both resins depend on temperature (24,27). The activation energy for PB viscous flow is 46 kJ/mol (11 kcal/mol) (39), and for PMP, 77 kJ/mol (18.4 kcal/mol) (28). Equipment used for PP processing is usually suitable for PB and PMP processing as well; however, adjustments in the processing conditions must be made to account for the differences in melt temperatures and rheology.

Extrusion.   In extrusion, an article is shaped by forcing a polymer melt through a die. The main applications of this method include the production of film, sheet, pipe, and tubing. Extrusion temperatures and pressures are selected on the basis of the resin melting point and melt viscosity. PB is usually extruded by using the same equipment (single- or twin-screw extruders) as that used for

PP and HDPE, at melt and die temperatures of 170–190°C. PMP is processed on extruders with a high (>25) length-to-diameter ratio at temperatures of 240–300°C and at compression ratios of 2.5–3.5.

Pipe Manufacture.   PB pipe and tubing are fabricated by using single-screw extruders equipped with vacuum- or pressure-calibration devices (70) at melt temperatures of 145–195°C. Properties of pipe-grade PB are given in Table 4. Immediately after exiting the die, the tube of molten resin is cooled first by a water spray and then by passing through a water well. Subsequently, the pipe is vacuum-calibrated; after sizing and cooling, it passes through a puller and is then coiled or cut into straight lengths. Soon after its manufacture, PB pipe shrinks by about 2% due to the onset of the crystal transformation from form II to form I. Newly produced pipes should be stored seven to ten days for the transformation to complete. PB pipes are joined by thermal fusion (socket or butt fusion) or by mechanical fittings (crimp, compression, flare, etc).

Injection Molding.   In this technique, resin pellets are melted in a single-screw extruder and the molten resin is injected through a nozzle into a steel mold, consisting of a fixed and a movable platen held together. After the mold is filled with the resin melt, it is cooled and then opened to remove the article. The duration of the molding cycle depends on the polymer melt viscosity and the rate of polymer crystallization. Molds can usually accommodate several articles formed in a single shot. Injection molding of PB is carried out under conditions similar to those for PP at 145–190°C. Injection molding of PMP is carried out at melt temperatures of 260–330°C, mold temperatures of 30–80°C, injection pressures of around 30 MPa (300 atm), and at relatively low injection rates; molded parts produced by this technique shrink by 1.5 to 3.0%.

Film.   The blown film process is most commonly used in the production of PB film (71). The film is fabricated in this process from resins with melt indexes from 0.3 to 10 g/10 min at a melt temperature of 200–215°C using conventional equipment for the manufacture of LDPE film and chilled air to cool the film. The thin-walled tube of molten PB exits upward from a die with a thin circular gap and is stretched about five times in the machine direction, then expanded in the transverse direction by air pressure. The blow-up ratio of the film, ie, the ratio of the final film bubble diameter to the die diameter, ranges from 2.5 to 3.0 (72). To achieve a necessary level of melt plastification, extruders with single- or two-stage mixing screws with a compression ratio of 2.5:7 and a length-to-diameter ratio of at least 24:1 are preferred (71). Mechanical properties of blown PB film (Table 5) depend on the degree of orientation and other processing parameters. PB film can be sealed at 160–220°C, both on-line (when the polymer is in form II) and after the form II-to-form I transformation. Another technique for the PB film production consists of film casting from the melt on polished chilled rolls and co-extrusion or lamination with other films. Blends of 1-butene–ethylene copolymers with PP can enhance the processibility and properties of PB film (39).

Analytical Methods and Quality Control

Analytical and test methods for the characterization of polyethylene and PP are also used for PB, PMP, and polymers of other higher α-olefins. The ^{13}C-nmr method as well as ir and Raman spectroscopic methods are all used to study the chemical structure and stereoregularity of polyolefin resins. In industry, polyolefin stereoregularity is usually estimated by the solvent–extraction method similar to that used for isotactic PP. Intrinsic viscosity measurements of dilute solutions in decalin and tetralin at elevated temperatures can provide the basis for the molecular weight estimation of PB and PMP with the Mark-Houwink equation, $[\eta] = KM_w^\alpha$. The constants K and α for several polyolefins are given in Table 8.

Tables 4–6 list ASTM methods used for the characterization of PB and PMP. A number of specialized methods were developed for testing particular articles manufactured from polyolefins; several of these determine the performance of PB and PMP film, including the measurement of the film's dart impact strength and tear strength. Dart impact strength is measured by dropping a heavy dart with a round tip on a stretched film. Tear resistance, which reflects the film's resistance to tear propagation, is measured with the Elmendorf tear tester. Two values for the tear strength are usually reported, one in the machine direction of the film and the other in the transverse direction. Pipes manufactured from PB are tested by pressurizing them internally with water; the time-to-burst failure is determined at various temperatures (46). The standard test method for haze and luminous transmittance (ASTM D1003) is used for the measurement of PMP optical characteristics.

Table 8. Mark-Houwink Parameters for Polyolefin Solutions[a]

Polymer	Solvent	Temperature, °C	K	α
polybutene	decalin	115	9.49×10^{-5}	0.74
polybutene	tetralin	100	1.06×10^{-4}	0.76
poly(4-methyl-1-pentene)	o-dichlorobenzene	130	3.95×10^{-5}	0.72
	decalin	135	1.94×10^{-4}	0.81
	decalin	135	7.49×10^{-5}	0.77
poly(1-decene)	tetrahydrofuran	25	5.19×10^{-3}	0.77

[a]Refs. 28, 30, and 51.

Economic Aspects

Linear higher α-olefins with even carbon numbers (1-butene, 1-hexene, 1-octene, 1-decene, etc, up to C_{30+}) are produced commercially by Ethyl, Shell, Chevron, Spolana, Idemitsu, Mitsubishi, and Sasol, which also produces 1-pentene. The global production of these products is approximately 2 million tons. The annual growth of olefin production for the 1990–2000 period is expected to be around 6%. Over 30% of the olefins is used to produce LLDPE resins, the highest market growth of which is 1-hexene and 1-octene (9) (see OLEFIN POLYMERS, LINEAR LOW DENSITY POLYETHYLENE). Most of the olefins with branched alkyl groups are specially synthesized or produced as by-products of petroleum refining; their prices are usually higher than those of linear α-olefins. 4-Methyl-1-pentene is manufactured by Mitsui Petrochemical Co. (22,700 t/yr) and British Petroleum Co. (25,000 t/yr), using propylene dimerization. Dicyclopentadiene is recovered as a by-product from ethylene cracker off-stream by Dow Chemical, Exxon, Shell, and other companies. The 1994 U.S. production was 66,000 tons.

Crystalline isotactic PB is manufactured in the United States by Shell Chemical; volume is 27,000 t/yr, consuming about 9% of the total U.S. production of 1-butene. Several grades are marketed, including extrusion resins for hot- and cold-water pipe and pipes for other purposes, film resins, and hot-melt adhesives. PB is also manufactured in Japan by Mitsui Petrochemical and in Europe by a joint venture of Neste Oy and Idemitsu. Crystalline PMP is manufactured in Japan by Mitsui Petrochemical Industries at 2000 t/yr under the trade name TPX. This resin, a copolymer of 4-methyl-1-pentene and linear $C_{10}-C_{16}$ α-olefin, is produced in several grades that differ in copolymer composition, crystallinity, melting point, and melt index. Such resins are used for injection molding, blow molding, and extrusion. Filling PMP with fiber glass and mica (qv) (filler concentration up to 35–40 wt %) can improve rigidity and increase thermal deflection temperature to 130–140°C.

Polyoctenamer (*trans*-octene rubber) is manufactured in Europe by Hüls AG at 12,000 t/yr under the trade name Vestenamer. Components for the manufacture of polydicyclopentadiene (PDCPD), a liquid molding resin, are produced by Hercules in the United States (13,600 t/yr), by Teijin in Japan (3000 t/yr), and by Shell in Europe, under the trade name Metton; they are also produced by Goodrich in the United States and Japan under the trade name Telene.

U.S. list prices for polyolefins in 1994 were PB film, $3.10; PB pipe for both hot and cold water, $4.50; general-purpose PB and PB adhesives, $2.00–2.50; and PMP, $3.50–4.50/kg.

Health and Safety Factors

Polymers and higher α-olefins are not toxic; their main potential health hazards are associated with residual monomer, antioxidants, and catalyst residues. In particular, PB and PMP are inert materials and usually present no health hazard. PMP is employed extensively for a number of medical and food packaging applications. Several grades conform to FDA regulations and to the health standards of other countries. Flammability of polyolefin resins is equal to that of PP, around 2.5 cm/min (ASTM D635). However, during combustion or pyrolysis, smoke, fumes, and toxic decomposition products are formed and can pose a health hazard.

Uses

Polybutene. The largest share of commercially produced crystalline PB is used for manufacturing pipe and tubing.

Pipe. The advantages of PB in pipe applications include high flexibility, toughness, and high resistance to creep, environmental stress-cracking, wet abrasion, and various chemicals. Pipes manufactured from PB retain their properties (Table 4) at temperatures up to 85°C (47). PB pipe is used in residential and commercial hot- and cold-water plumbing (including chlorine-containing hot water), water wells, water manifolds, and fire sprinklers. Black-pigmented pipe grades are suitable for outdoor use.

Film. Blown film manufactured from PB has a high tensile strength and exhibits good resistance to tear, impact, and puncture (47). Such film also exhibits hard-elastic behavior; that is, it can recover its original length even after extensive stretching. Some properties of PB film are given in Table 5.

Poly(4-methyl-1-pentene). Most PMP applications capitalize on the resin's high optical transparency, excellent dielectric characteristics, high thermal stability, and good chemical resistance. The manufacture of medical equipment comprises about 40% of PMP production, including such articles as hypodermic syringes, needle hubs, blood collection and transfusion equipment, pacemaker parts, blood analysis cells, and respiration equipment. Articles made of PMP can be subjected to all sterilization procedures, such as autoclave sterilization, uv irradiation, dry heat treatment, β- and γ-irradiation, and ethylene oxide treatment. Another important area of PMP use is in chemical and biomedical laboratory equipment, eg, cells for spectroscopic and optical analysis, laboratory ware, and animal cages. Other applications encompass a variety of injection-molded articles, such as caps for enclosures, ink cartridges for printers, light covers, tableware, and sight-glasses. PMP is suitable also for microwave oven cookware and service, and is used in food packaging (qv). In many applications, PMP replaces stainless steel trays and is used for thermal insulating floats, packing rings in gas-scrubbing towers, and barriers for hot sulfuric acid, hydrochloric acid pools, and acid washing tanks. In addition, PMP is also utilized for wire and cable coating (especially in oil well and communication cables) as well as for film and paper coatings with good release properties. Finally, PMP film is used as a separator (release film) in the manufacture of laminates, urethane curing, and printed circuits.

Melt-spun fiber is produced from PMP at 280°C and is drawn around three times in air at 95°C; its fiber strength is 0.18–0.26 N/tex (2–3 g/den), its elongation is around 30%. Melt-spun hollow fibers are also manufactured. PMP has one of the highest permeabilities for gases, and many of its applications capitalize on this property.

Synthetic Lubricating Oils. Liquid oligomers of higher linear α-olefins such as 1-decene are produced with cationic initiators. These viscous liquids are usually called poly(alpha-olefin)s (PAO). They are the most versatile of all synthetic lubricants. They exhibit not only good lubricating properties over a wide temperature range (including excellent low temperature properties), but also high frictional and oxidative stability as well as low volatility, and are miscible with all mineral oils and most synthetic lubricants. For these reasons, PAOs have found wide application as synthetic base oils in the formulation of various lubricants, including lubricating oils for cars, transformer oils, transmission and crankcase fluids, hydraulic fluids, and compressor oils (63,64) (see LUBRICATION AND LUBRICANTS). Mobil Chemical Co. uses a large volume of PAO as a base stock for industrial oils and lubricating oils for cars and trucks, as well as diesel, marine, and gas engines (trade name Mobil 1). Chevron, Ethyl Corp., Neste Chemical, and Uniroyal have started to manufacture similar synthetic lubricants.

Polymers of Cycloolefins. Polyoctenamer elastomers are processed by extrusion, injection molding, and calendering into hoses, rubber coatings, and tire components. They are mostly used as components in rubber-, PVC-, and PS-based compositions. Compounding polyoctenamer with natural rubber and polyisoprene can reduce energy requirements during rubber processing and significantly increase green strength of tire components (73,74).

Cured liquid-molding resins based on polydicyclopentadiene (Metton, Telene) have high strength and rigidity, excellent low temperature (up to −40°C) impact resistance, low moisture absorption, light weight, high chemical resistance, and glossy surface finish. Main applications of these resins are in the manufacture of automotive parts for trucks, snowmobiles, wheel loaders, recreational vehicles, and also in other areas that require toughness and good all-weather impact resistance, such as hopper-car lids, electrical housing, and large size housing for electrochemical cells.

BIBLIOGRAPHY

"Polymers of Higher Olefins" under "Olefin Polymers," in *ECT* 2nd ed., Vol. 14, pp. 309–313, by D. J. Buckley, Esso Research and Engineering Co.; in *ECT* 3rd ed., Vol. 16, pp. 470–479, by R. K. Kochhar, Gulf Oil Chemicals Co., and Y. V. Kissin, Gulf Research & Development Co.

1. G. Natta and co-workers, *J. Am. Chem. Soc.* **77**, 1708 (1955).
2. G. Natta, *J. Polym. Sci.* **16**, 143 (1955).
3. Eur. Pat. Appl. 403,866 (Dec. 27, 1990), T. Asanuma and co-workers (to Mitsui Toatsu Chemicals).
4. Eur. Pat. Appl. 387,609 (Sept. 19, 1990), A. Albizzati, L. Resconi, and A. Zambelli (to Himont).
5. H. R. Sailors and J. H. Hogan, *J. Mackromol. Sci. Chem.* **A15**, 1377 (1981).
6. K. Ziegler and co-workers, *Angew. Chem.* **72**, 829 (1960).
7. W. Keim, *Makromol. Chem., Macromol. Symp.* **66**, 225 (1993).
8. U.S. Pat. 4,293,727 (Oct. 6, 1981), D. L. Beach and J. J. Harrison (to Gulf Oil).
9. D. D. Morris and M. Roberts, *Chem. Week*, 43 (Nov. 18, 1992).
10. U.S. Pat. 2,943,125 (filed July 27, 1955), K. Ziegler and H. Martin.
11. P. W. DeLeeuw, C. R. Lindegren, and R. F. Schimbor, *Chem. Eng. Prog.* 57 (Jan. 1980).
12. Belg. Pat. 611,543 (June 14, 1962), G. W. Alderson, J. K. Hambling, and A. A. Yeo (to British Petroleum Co.).
13. Eur. Pat. Appl. EP57,911 (Aug. 18, 1982), H. Imai, M. Matsuno, and M. Kudoh (to Nippon Oil Co.).
14. U.S. Pat. 4,727,213 (Feb. 23, 1988), C. A. Drake and D. H. Kubicek (to Phillips Petroleum Co.).
15. *Chem. Eng. News*, **63**(16), 27 (1985).
16. I. D. Rubin, *Poly(1-Butene): Its Preparation and Properties*, Gordon & Breach, New York, 1968.
17. N. I. Sax, *Dangerous Properties of Industrial Materials*, 6th ed., Van Nostrand Reinhold Co., Inc., New York, 1984, p. 1902.
18. *Physical Constants of Hydrocarbons C_1 to C_{10}*, ASTM Data Series Publication DS 4A, ASTM, Philadelphia, Pa., 1971.
19. C. H. Bamford and C. H. F. Tipper, eds., *Comprehensive Chemical Kinetics*, Vol. 14, Elsevier, Amsterdam, the Netherland, 1975, p. 42.
20. R. Chandra, *Progress in Polymer Science* **11**, 23 (1985).
21. S. M. Gabbay, S. S. Stivala, and L. Reich, *Polymer* **16**, 741 (1975).

22. S. S. Stivala, J. Kimura, and S. M. Gabbay, in N. S. Allen, ed., *Degradation and Stabilization of Polyolefins*, Elsevier Applied Science Publishers, Ltd., Barking, U.K., 1983, p. 63.
23. *Polymethylpentene (TPX)*, Mitsui Petrochemical Industries Bulletins, 1981, 1984.
24. J. A. Brydson, *Plastic Materials*, 4th ed., Butterworth & Co. (Publishers) Ltd., Kent, U.K., 1982, p. 245.
25. R. P. Singh and A. Syamal, *J. Mater. Sci.* **16**, 3324 (1981).
26. H. Wilski, S. Roesinger, and G. Diedrich, *Kunststoffe* **70**, 221 (1980).
27. H. C. Raine, *J. Appl. Polym. Sci.* **11**, 39 (1969).
28. L. C. Lopez and co-workers, *J. Macromol. Sci., Chem. Phys.* **C32**, 301 (1992).
29. V. I. Kleiner and co-workers, *Polym. Science* **35**, 1403 (1993).
30. J. C. W. Chien and T. Ang, *J. Polym. Sci. Polym. Chem. Ed.* **24**, 2217 (1986).
31. W. Kaminsky and R. Spiehl, *Makromol. Chem.* **190**, 515 (1989).
32. W. Kaminsky and co-workers, in W. Kaminsky and H. Sinn, eds., *International Symposium of Transition Metals and Organometallics as Catalysts for Olefin Polymerization*, Springer Press, Berlin, 1988, p. 291.
33. B. L. Lebedev and co-workers, *Polym. Science* **35**, 871,1641,1650 (1993).
34. H. Tadokoro, *Structure of Crystalline Polymers*, John Wiley & Sons, Inc., New York, 1979, pp. 355–358.
35. C. De Rosa and co-workers, *Macromolecules* **24**, 5645 (1991).
36. B. Wunderlich, *Macromolecular Physics*, Vol. 3, Academic Press, Inc., New York, 1980.
37. P. H. Geil and co-workers, *Soc. Plast. Eng. Tech. Pap.*, 29, 404 (1983).
38. C. Nakafuku and T. Miyaki, *Polymer* **24**, 141 (1983).
39. A. M. Chatterjee, in J. I. Kroschwitz, ed., *Encyclopedia of Polymer Science and Engineering*, Vol. 2, 2nd ed., John Wiley & Sons, Inc., New York, 1985, p. 590.
40. H. V. Boenig, *Polyolefins: Structure and Properties*, Elsevier, Amsterdam, the Netherlands, 1966, p. 17.
41. P. Zoller, H. W. Starkweather, and G. A. Jones, *J. Polym. Sci. Polym. Phys. Ed.* **24**, 1451 (1986).
42. T. Tanigami and co-workers, *J. Appl. Polym. Sci.* **32**, 4491 (1986).
43. J. P. Shaw and M. Gilbert, *J. Mackromol. Sci. Phys.* **B30**, 271 (1991).
44. J. P. Shaw and M. Gilbert, *J. Mackromol. Sci. Phys.* **B30**, 301 (1991).
45. U.S. Pat. 4,321,334 (Mar. 23, 1982) and 4,322,503 (Mar. 30, 1982), A. M. Chatterjee (to Shell Oil Co.).
46. S. G. Kemp, *Plast. Rubber Proc. Appl.* **3**(2), 169 (1983).
47. P. W. Macgregor, *SPE Regional Technical Conference*, TX, Feb. 1981, p. 71.
48. M. Lee and S. Chen, *Angew. Makromol. Chem.* **192**, 57 (1991).
49. H. Saechtling, *International Plastics Handbook*, Hanser Verlag, Munich, Germany, 1983, pp. 156,162.
50. D. S. Breslow, *Polymer Preprints* **ACS31**(2) (1990).
51. Y. V. Kissin, *Isospecific Polymerization of Olefins with Heterogeneous Ziegler-Natta Catalysts*, Springer-Verlag, New York, 1985.
52. V. S. Steinbak and co-workers, *Eur. Polym. J.* **11**, 457 (1975).
53. P. J. T. Tait, in J. C. W. Chien, ed., *Coordination Polymerization*, Academic Press, Inc., New York, 1975, p. 155.
54. J. Boor, Jr., *Ziegler-Natta Catalysts and Polymerizations*, Academic Press, Inc., New York, 1979.
55. U.S. Pat. 4,988,655 (Jan. 29, 1991), K. E. Mitchel, G. R. Hawley, and D. W. Godbehere (to Phillips Petroleum Co.).
56. I. Frolov and co-workers, *Makromol. Chem.* **194**, 2309 (1993).
57. J. J. A. Dusseault and C. C. Hsu, *J. Macromol. Sci. Rev. Macromol. Chem. Phys.* **C32**, 103 (1993).

58. U.S. Pat. 4,945,142 (July 31, 1990), D. E. Gessell and D. P. Hosman (to Conoco).

59. Y. V. Kissin, *Adv. Polym. Sci.* **15**, 91 (1974).

60. A. F. Stancell, A. J. Cavanna, W. D. Eccli, and R. W. Edwards, *Hydrocarb. Proc.* **52**, 129 (1973).

61. Eur. Pat. Appl. 368,631 (May 16, 1990), A. Ahvenainen and co-workers (to Neste Oy).

62. A. R. Siedle and co-workers, *Makromol. Chem. Macromol. Symp.* **66**, 215 (1993).

63. Eur. Pat. Appl. 405,236 (Jan. 2, 1991) (to Mitsui Toatsu Chemicals).

64. J. P. Kennedy, *Cationic Polymerization of Olefins: A Critical Review*, John Wiley & Sons, Inc., New York, 1975; Wo Appls. 89-006,247 (July 13, 1989) and 90-015,080 (Dec. 13, 1990), C. R. Scharf and co-workers (to Lubrizol).

65. Eur. Pat. Appl. 349,276 (Jan. 3, 1990), T. J. Tileo and co-workers (to Ethyl Corp.).

66. R. L. Shubkin, *Synthetic Lubricants and High Performance Functional Fluids*, Marcel Dekker, Inc., New York, 1993.

67. V. Dragutan, A. T. Balaban, and M. Dimonie, *Olefin Metathesis and Ring-Opening Polymerization of Cycloolefins*, John Wiley & Sons, Ltd., Chichester, U.K., 1985.

68. B. M. Novak, W. Risse, and R. H. Grubbs, *Adv. Polym. Sci.* **102**, 47 (1992).

69. C. Quay and J. Leonard, *Polymer Preprints* **ACS31**(1) (1990).

70. *Processing Polybutylene Pipe*, Technical Bulletin SCC:544-81, Shell Chemical Co., Houston, Tex., 1981.

71. *Processing Shell Polybutylene Film Grade Resins*, Technical Bulletin SC. 391–79, Shell Chemical Co., Houston, Tex., 1979.

72. U.S. Pat. 4,906,429 (Mar. 6, 1990), T. Yamawaki and T. Yamada (to Idemitsu Petrochemical and Neste Oy).

73. *Rubber and Plastics News*, 14 (July 9, 1990).

74. P. P. Chattaraj, R. Mukhopadhyay, and D. K. Tripathy, *J. Elast. Plast.* **26**, 74 (1994).

YURY V. KISSIN
Mobil Chemical Company

OLEFINS, HIGHER

Higher olefins are versatile chemical intermediates for a number of important industrial and consumer products, providing a better standard of living with low environmental impact (qv) in many commercial uses. These uses can be characterized by carbon number and by chemical structure.

The even-numbered carbon alpha olefins (α-olefins) from C_4 through C_{30} are especially useful. For example, the C_4, C_6, and C_8 olefins impart tear resistance and other desirable properties to linear low and high density polyethylene; the C_6, C_8, and C_{10} compounds offer special properties to plasticizers used in flexible poly(vinyl chloride). Linear C_{10} olefins and others provide premium value synthetic lubricants; linear C_{12}, C_{14}, and C_{16} olefins are used in household detergents and sanitizers. In addition, many carbon numbers from C_4 to C_{30+} are also utilized in specialty applications such as sizing agents to produce longer-lasting paper. Various uses and production methods of these olefins have been discussed extensively in References 1 and 2. Producers of the even-carbon-number higher olefins are shown in Table 1.

The C_6-C_{11} branched, odd and even, linear and internal olefins are used to produce improved flexible poly(vinyl chloride) plastics. Demand for these branched olefins, which are produced from propylene and butylene, is estimated to be increasing at a rate of 2% per year. However, the growth of the linear α-olefins is expected to slow down to a rate of 5% per year from 1992 to 1997 (3), as opposed to growth rates of 7–10% in the 1980s.

Table 1. World Producers of Linear α-Olefins[a]

Company and location	Capacity, 10^3 t	Process
Albemarle Corp.[b]		
Pasadena, Tex., U.S.	472	Albemarle
Feluy, Belgium	204	Albemarle
Chevron Chemical Co., Cedar Bayou, Tex., U.S.	249	Chevron-Gulf
Gujarat-Godrej Innovative Chemicals, Ankleshwar, India	30	Godrej-Lurgi
Idemitsu Petrochemical Co., Ltd., Ichihara, Chiba Prefecture, Japan	50	Idemitsu
Mitsubishi Kasai Corp., Kurashiki, Ikayama Prefecture, Japan	50	Chevron-Gulf
Nizhnekamskneftekhim, Nizhnekamsk, Russia	181	Albemarle
Shell Chemical Co.,		
Geismar, Louisiana, U.S.	590	Shell (SHOP)
Ellesmere Port, Stanlow, United Kingdom	270	Shell (SHOP)
Spolana, Neratovice, Czech Republic	130	Chevron-Gulf
Total	2226	

[a]Ref. 3.
[b]Formerly Ethyl Corp.

Physical Properties

For a listing of selected physical properties of linear alpha olefins, see Tables 2 and 3. Boiling points for this homologous series increase as the carbon number increases; approximately 30°C is added for each $-CH_2-$ group. Thus, 1-butene boils at −6°C and 1-pentene at about 30°C. Melting points, or freezing points, for that matter, do not behave in as regular a fashion, although an increase of about 20°C per $-CH_2-$ group roughly describes the pattern observed for the lighter olefins, and an increase of 10°C per each added carbon for the heavier liquids. For example, the change from 1-butene to 1-pentene is ∼20°C, whereas the increase from 1-dodecene to 1-tridecene is about 8°C. Typically, densities for the liquid olefins are in the range of 0.7 g/mL at room temperature.

Chemical Properties

The general reactivity of higher α-olefins is similar to that observed for the lower olefins. However, heavier α-olefins have low solubility in polar solvents

Table 2. Properties of C₄ to C₂₀ Linear 1-Olefins

Compound	Mol wt	Density, g/mL 20°C	Viscosity, mm²/s(=cSt) 20°C	Viscosity, mm²/s(=cSt) 100°C	Free energy of formation, kJ/mol[a]	CAS Registry Number
1-butene	56.11	0.6012			72.09	[106-98-9]
1-pentene	70.13	0.6402	0.202		78.67	[109-67-1]
1-hexene	84.16	0.67317	0.39		87.61	[592-41-6]
1-heptene	98.19	0.69698	0.50		96.02	[592-76-6]
1-octene	112.2	0.71492	0.656	0.363	104.4	[111-66-0]
1-nonene	126.2	0.72922	0.851	0.427	112.8	[124-11-8]
1-decene	140.3	0.74081	1.09	0.502	121.3	[872-05-9]
1-undecene	154.3	0.75032	1.38	0.587	129.6	[821-95-4]
1-dodecene	168.3	0.75836	1.72	0.678	138.0	[112-41-4]
1-tridecene	182.3	0.7653	2.14	0.782	146.4	[2437-56-1]
1-tetradecene	196.4	0.7713	2.61	0.894	154.8	[1120-36-1]
1-pentadecene	210.4	0.7765	3.19	1.019	163.3	[13360-61-7]
1-hexadecene	224.4	0.78112	3.83	1.152	171.7	[629-73-2]
1-heptadecene	238.4	0.7852	4.60	1.30	180.1	[6765-39-5]
1-octadecene	252.5	0.7888	5.47	1.46	188.5	[112-88-9]
1-nonadecene	266.5	0.7920 f[b]		1.63	196.9	[18435-45-5]
1-eicosene	280.5	0.7950 f[b]		1.82	205.3	[3452-07-1]

[a]To convert kJ/mol to kcal/mol, divide by 4.184.
[b]f = frozen.

Table 3. Safety Factors of C₄ to C₂₀ Linear 1-Olefins

Compound	Flash point, °C	Flammability limit, vol % Lower	Flammability limit, vol % Upper	Autoignition temperature, °C
1-butene	na	1.6	9.3	383.8
1-pentene	−18.2	1.5	8.7	272.8
1-hexene	−31.2	1.2	9.2	253.0
1-heptene	−0.1	1.1	8.0	263.0
1-octene	20.8	0.9	7.1	229.9
1-nonene	26.8	0.8	6.4	236.8
1-decene	53.3	0.7	5.9	234.9
1-undecene	71.1	0.7	5.6	236.8
1-dodecene	48.8	0.6	5.4	255.0
1-tridecene	79.4	0.6	5.4	236.8
1-tetradecene	109.9	0.5	5.4	234.9
1-pentadecene	112.8	0.4	5.4	na
1-hexadecene	123.8	0.5	5.4	240.0
1-heptadecene	135.0	0.5	5.4	na
1-octadecene	145.8	0.4	5.4	250.0
1-nonadecene	156.8	0.4	5.4	na
1-eicosene	165.8	0.4	5.4	236.8

such as water; consequently, in reaction systems requiring the addition of polar reagents, apparent reactivity and degree of conversion may be adversely affected. Reactions of α-olefins typically involve the carbon–carbon double bond and can be grouped into two classes: (1) electrophilic or free-radical additions; and (2) substitution reactions.

Electrophilic Addition.   In the following example, an α-olefin reacts with a Lewis acid to form the most stable intermediate carbocation. This species, in turn, reacts with the conjugate base to produce the final product. Thus electrophilic addition follows Markovnikov's rule.

$$RCH = CH_2 + HZ \longrightarrow R\overset{+}{C}H - CH_3 + Z^- \longrightarrow RCH - CH_3$$
$$\underset{Z}{|}$$

Free-Radical Addition.   A different outcome is expected in free-radical addition. The reaction of an α-olefin with a typical free radical affords the most stable intermediate free radical. This species, in turn, reacts further to form the final product, resulting in the anti-Markownikov mode of addition.

$$RCH = CH_2 + Br\cdot \longrightarrow R\overset{.}{C}HCH_2Br$$
$$R\overset{.}{C}HCH_2Br + H : Br \longrightarrow RCH_2CH_2Br + Br\cdot$$

Substitution.   In free-radical substitution, the olefin reacts with a free-radical source to form the allyl free radical, which in turn reacts with available reagent to produce both the final product and a new free radical.

$$RCH_2CH = CH_2 + X\cdot \longrightarrow R - \overset{.}{C}H - CH = CH_2 + HX$$

$$R - \overset{.}{C}H - CH = CH_2 + X_2 \longrightarrow R - CH - CH = CH_2 + X\cdot$$
$$\underset{X}{|}$$

Commercial Olefin Reactions.   Some of the more common transformations involving α-olefins in industrial processes include the oxo reaction (hydroformylation), oligomerization and polymerization, alkylation reactions, hydrobromination, sulfation and sulfonation, and oxidation.

Oxo Reaction.   Olefins are well known for their ability to form complexes with various transition metals. The resulting coordination compounds have many uses as catalytic species, including hydrogenation, dimerization, oxidation, isomerization, and hydroformylation. The complexes of olefin–Group VIII metals have wide usage in these hydroformylations, or oxo processes. These processes add carbon monoxide and hydrogen to the starting olefin to form, depending on the conditions employed, aldehydes, carboxylic acids, alcohols, or esters. This procedure can be accomplished in single steps or with sequential modifications carried out after the initial hydroformylation step has occurred (see OXO PROCESS).

Oligomerization and Polymerization. Since an allyl radical is stable, linear α-olefins are not readily polymerized by free-radical processes such as those employed in the polymerization of styrene. However, in the presence of Ziegler-Natta catalysts, these α-olefins can be smoothly converted to copolymers of various descriptions. Addition of higher olefins during polymerization of ethylene is commonly practiced to yield finished polymers with improved physical characteristics.

Alkylation. Benzene and phenol feedstocks are readily alkylated under Friedel-Crafts conditions to prepare extensive families of alkylated aromatics. These materials generally are intermediates in the production of surfactants or detergents such as linear alkylbenzenesulfonate (LABS) and alkylphenolethoxylate (APE). Other uses include the production of antioxidants, plasticizers, and lube additives.

Bromination. 1-Bromoalkanes are produced commercially by the anti-Markovnikow free-radical addition of HBr to α-olefins. These are further reacted with dimethyl amine to produce alkyldimethyl amines, which ultimately are converted to amine products for household cleaning and personal care.

Sulfation and Sulfonation. α-Olefin reactions involving the introduction of sulfur-containing functional groups have commercial importance. As with many derivatives of olefins, several of these products have applications in the area of surfactants (qv) and detergents. Typical sulfur reagents utilized in these processes include sulfuric acid, oleum, chlorosulfonic acid, sulfur trioxide, and sodium bisulfite.

Oxidation. Olefins in general can be oxidized by a variety of reagents ranging from oxygen itself to ozone (qv), hydroperoxides, nitric acid (qv), etc. In some sequences, oxidation is carried out to create a stable product such as 1,2-diols or glycols, aldehydes, ketones, or carboxylic acids. In other instances, oxidation results in the formation of intermediates, eg, ozonides, hydroperoxides, or epoxides that can further react without being isolated. Such quality allows the products of this class of reactions to be diverse and versatile with respect to final application.

Commercial α-Olefin Manufacture

Most linear α-olefins are produced from ethylene. Ethylene-based capacity in 1993 was 2,196,000 t, compared to only 30,000 t for fatty alcohol-based manufacture.

Linear α-olefins were produced by wax cracking from about 1962 to about 1985, and were first commercially produced from ethylene in 1965. More recent developments have been the recovery of pentene and hexene from gasoline fractions (1994) and a revival of an older technology, the production of higher carbon-number olefins from fatty alcohols.

Ethylene oligomerization can be accomplished in the following commercial processes: (*1*) stoichiometric chain growth on aluminum alkyls followed by displacement (Albemarle); (*2*) catalytic chain growth on aluminum alkyls (Chevron-Gulf); (*3*) catalytic chain growth using a nickel ligand catalyst (Shell); and (*4*) catalytic chain growth using a modified zirconium catalyst (Idemitsu). In the Albemarle (formerly Ethyl) process, stoichiometric quantities of aluminum alkyls are used with subsequent displacement of α-olefins from the aluminum, followed by separation of the α-olefins from the aluminum alkyls. In the Chevron-Gulf process, catalytic amounts of aluminum alkyl are used. The operating temperatures are higher than those in the stoichiometric process, thus favoring displacement reactions after a finite amount of chain growth. In the Shell process, a three-phase system is employed, which gives a high linearity at higher carbon numbers. A nickel ligand catalyst dissolved in a solvent forms one liquid phase, the produced olefins form a second liquid phase, and the ethylene forms a third. Once formed, the olefins usually do not engage in further reactions because most of them are not in contact with the catalyst. Shell practices isomerization and disproportionation to produce a narrow range of internal linear olefins for feed to their oxo-alcohol unit. In the Idemitsu process, a zirconium oligomerization catalyst is modified by adding an aluminum alkyl and a Lewis base or an alcohol in a solvent. Variations in the catalyst mix thus offer a variety of carbon-number distributions, some of which resemble those in the catalytic processes, others approaching those in the stoichiometric process. Although operating at lower pressures than the other ethylene-based oligomerizations, the Idemitsu process still produces high quality linear α-olefins.

Vista has offered for license a stoichiometric process, which has not yet been commercialized, although the related primary alcohol process has been described (see ALCOHOLS, HIGHER ALIPHATIC–SYNTHETIC PROCESSES). Other processes, including developments by Dow and Exxon, have been reported in the literature.

Formation of by-products varies with each process; typical by-products are paraffins, linear internal olefins, and vinylidene (branched) olefins.

Olefin distribution in the Albemarle stoichiometric process tends to follow the Poisson equation, where X_p is the mole fraction of alkyl groups in which p ethylene units have been added, and n is the average number of ethylene units added for an equal amount of aluminum.

$$X_p = \frac{n^p e^{-n}}{p!} \tag{1}$$

The olefin distribution in the catalytic processes, on the other hand, tends to follow the Schultz-Flory equation, where X_N equals the number of moles of olefins having carbon number N, X_{N-2} equals the moles of olefins having two carbon numbers lower, and Q_N is a constant depending on the reaction conditions; Q_N can range from 0.4–0.9 but usually equals 0.6–0.8.

$$Q_N = \frac{X_N}{X_{N-2}} \tag{2}$$

Typical distributions of four commercial processes are compared in Figure 1.

Two-Step Ziegler Stoichiometric Process. This commercial two-step process was developed by Albemarle Corp., which has plants in Pasadena, Texas, and Feluy, Belgium (Fig. 2). A plant in the former USSR is also based on the Albemarle technology. In the first step, incoming ethylene is oligomerized

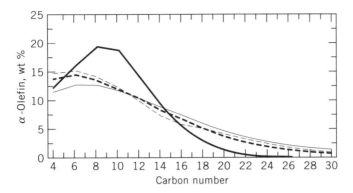

Fig. 1. Typical α-olefin distributions, where (—) represents Albemarle, (----) Chevron, (----) Idemitsu, and (—) Shell.

on triethylaluminum in a continuous stoichiometric reactor at 120–150°C and 14–21 MPa (140–210 atm) for about one hour. These longitudinal coil reactors are designed to remove the large heat of reaction; six reactor shells about eight feet in diameter and containing long tubes immersed in boiling water are used at the Pasadena location. Unreacted ethylene is separated in flash drums and recycled after having been compressed and distilled. The aluminum alkyls are then sent to the second step, in which ethylene is used to displace the olefins and regenerate the triethylaluminum for recycle to step one. Displacement conditions are 280–320°C and 1.0 MPa (10 atm) with minimum contact time. The displacement reactors are heat exchangers designed to heat and cool the streams quickly. The resulting olefins are separated from the aluminum alkyls at various stages in the process, as shown in Figure 2. Some olefins are removed before chain growth, and others after. Trace amounts of aluminum alkyls in the olefins are converted to paraffins by caustic hydrolysis. The mixed olefins are separated into commercial carbon numbers by fractional distillation.

Albemarle has the capability to run additional stoichiometric displacement and growth reactions or various aluminum alkyl separation techniques to create the carbon numbers most in demand while suppressing the rest. For example, butene can be used as a displacing agent to create tri(n-butyl)aluminum, which, when fed to chain growth, releases higher olefins from C_6 through C_{20+} (Feluy design). Similarly, higher olefins can also be fed to stoichiometric reactors to place them back on the aluminum and to create those needed in the marketplace, such as 1-decene. Albemarle continues to revise its processes to meet market demands; the current facility in Pasadena, for example, is vastly more complicated than its initial process design.

One-Step Ziegler Process.   Gulf Research and Development Corp. developed the one-step Ziegler process shown in Figure 3. This process is now owned by Chevron, which has two plants at Cedar Bayou, Texas. Plants based on licensing the technology are operated by Mitsubishi in Japan and at Neratovice in the Czech Republic (see Table 1). Chevron further improved the process around 1990 by reducing paraffin impurities.

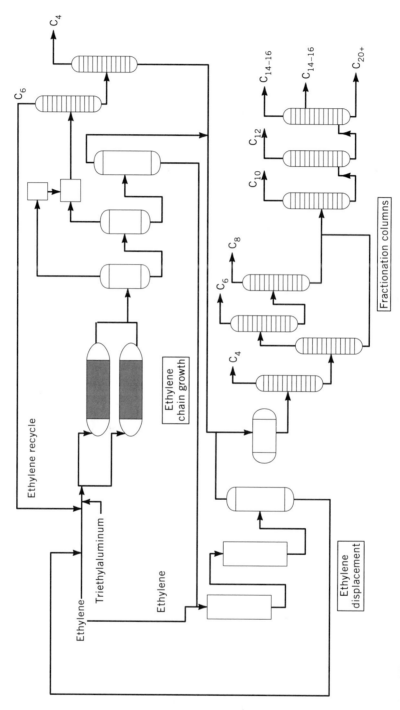

Fig. 2. Albemarle α-olefin schematic (Pasadena, Tex., and Feluy, Belgium; Belgium location includes butene recycle to exhaustion).

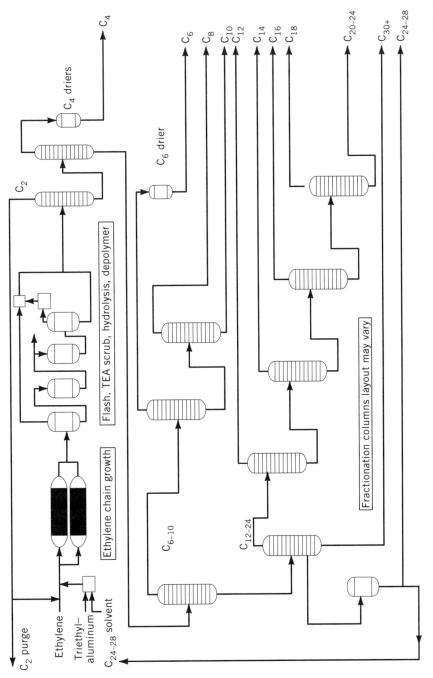

Fig. 3. Chevron α-olefin schematic where TEA = triethylaluminum (Cedar Bayou, Tex.; also Mitsubishi, Japan, and Spolana, Czech Republic). The layout of the fractionation columns may vary.

847

In the one-step Ziegler process, high purity ethylene and a catalytic quantity of triethylaluminum in a solvent are fed to several high pressure continuous reactors. Chevron's original plant, which still operates, had eight reactors, but the new plant has only two (4). The original plant reactors have been described as large horizontal vessels containing a single pipe coil that is immersed in water (5); the new plant reactors are probably similar but much larger. According to the same source (5), α-olefins in the original plant were prepared by polymerizing ethylene at 180–220°C and 14 MPa (140 atm) in a long, narrow reactor that allowed for minimum back mixing and used catalytic amounts of metal alkyls.

At elevated temperatures, the ethylene reacts with the metal alkyl growth products as they are formed, releasing the olefin and the catalyst for further growth reaction. The reaction is exothermic and the temperature can be controlled by regulating the pressure of the steam evolving from the reactor shell. Depending on olefin product quality requirements, ethylene conversion per pass can vary from 30–90 wt %, taking at least 5 minutes of reactor residence time; olefin product distribution can also be varied by controlling the reaction temperature. Since the reactor coils must be cleaned periodically to remove waxy polymer for good temperature control and stable operations, the provision of backup reactors is necessary.

The solvent is C_{24-28} α-olefins recycled from the fractionation section. Effluent from the reactors includes product α-olefins, unreacted ethylene, aluminum alkyls of the same carbon number distribution as the product olefins, and polymer. The effluent is flashed to remove ethylene, filtered to remove polyethylene, and treated to reduce the aluminum alkyls in the stream. In the original plant operation, these aluminum alkyls were not removed, resulting in the formation of paraffins ($\sim$1.4%) when the reactor effluent was treated with caustic to kill the catalyst. In the new plant, however, it is likely that these aluminum alkyls are transalkylated with ethylene by adding a catalyst such as 60 ppm of a nickel compound, eg, nickel octanoate (6). The new plant contains a caustic wash section and the product olefins still contain some paraffins ($\sim$0.5%). After treatment with caustic, crude olefins are sent to a water wash to remove sodium and aluminum salts.

Ethylene is recycled, some of which is purged, to eliminate the accumulation of ethane. The olefins are distilled into even-carbon-number fractions from 4 through 18 and blends of C_{20-22}, C_{24-28}, and C_{30+}. Various blends of the C_6-C_{18} carbon numbers are also available. 1-Butene and 1-hexene are dried to remove moisture; 1-butene is stored in a sphere and the other products are placed in dome-roof rundown and product storage tanks. The 1-octene and higher carbon-number storage tanks have an interconnected nitrogen system that protects quality and reduces emissions. Auxiliary systems include a cooling tower, hot oil heaters, and a reactor-wash solvent system utilizing a xylene mixture. With time, the reactors build up a polymer coat that must be removed with a hot oil wash system.

Shell Higher Olefins Process (SHOP).   In the Shell ethylene oligomerization process (7), a nickel ligand catalyst is dissolved in a solvent such as 1,4-butanediol (Fig. 4). Ethylene is oligomerized on the catalyst to form α-olefins. Because α-olefins have low solubility in the solvent, they form a second liquid phase. Once formed, olefins can have little further reaction because most of them

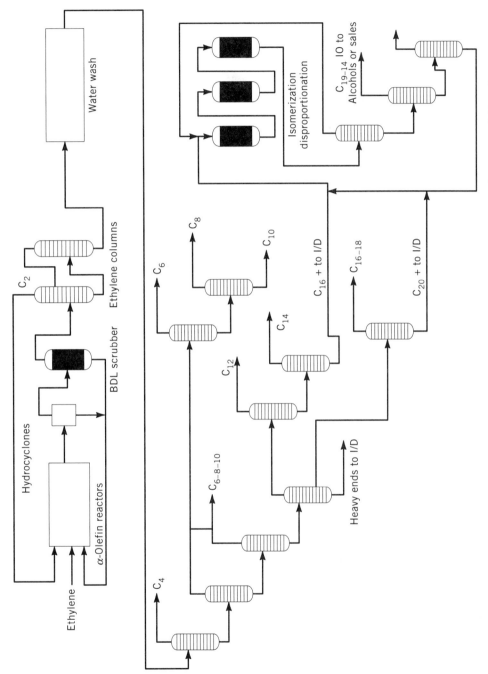

Fig. 4. Shell α-olefin and internal olefins schematic (Geismar, La. and Stanlow, United Kingdom).

are no longer in contact with the catalyst. Three continuously stirred reactors operate at ca 120°C and ca 14 MPa (140 atm). Reactor conditions and catalyst addition rates allow Shell to vary the carbon distribution.

The chain-growth catalyst is prepared by dissolving two moles of nickel chloride per mole of bidentate ligand (BDL) (diphenylphosphinobenzoic acid in 1,4-butanediol). The mixture is pressurized with ethylene to 8.8 MPa (87 atm) at 40°C. Boron hydride, probably in the form of sodium borohydride, is added at a molar ratio of two borohydrides per one atom of nickel. The nickel concentration is 0.001–0.005%. The 1,4-butanediol is used to solvent-extract the nickel catalyst after the reaction.

The reactor outlet is flashed to remove ethylene which is then compressed and recycled; α-olefins are separated from the solvent that contains the catalyst, treated to remove catalyst, and then distilled into commercial fractions. Most of the catalyst in the solvent is recycled but a portion is purged. The catalyst in the purge stream is recovered by reducing the oxidized nickel with boron hydride.

A key portion of the SHOP process is the isomerization–disproportionation (I/D) process in which excess light (C_4–C_{10}) and heavy olefins ($C_{16}+$) are converted to detergent range (C_{11}–C_{15}) odd and even linear internal olefins. For each pass through this system, only 10–15% of the olefins fed are converted to the desired range; consequently, this equipment must be large and feature two distillation columns that are probably >6 m in diameter. In the I/D unit the adsorber bed is probably composed of two molecular sieves (qv), one operating and the other regenerating. The isomerization catalyst is probably a heterogeneous magnesium oxide at 0.4–2.1 MPa (4–21 atm) and 80–140°C, with two beds operating and one regenerating. The disproportionation catalyst is probably rhenium oxide on alumina operating at about 120°C and 1.5 MPa (15 atm), with two operating and one regenerating.

These detergent range (C_{11}–C_{15}) odd and even linear internal olefins are fed to oxo-alcohol plants to produce C_{12}–C_{16} semilinear alcohols. Most of the alcohols are ethoxylated and sold into detergent markets (8). Shell balances carbon numbers by a combination of the ethylene oligomerization extent, I/D unit operation, and alcohol carbon-number composition.

Idemitsu Process. Idemitsu built a 50 t $\times$ 10^3 per year plant at Chiba, Japan, which was commissioned in February of 1989. In the Idemitsu process, ethylene is oligomerized at 120°C and 3.3 MPa (33 atm) for about one hour in the presence of a large amount of cyclohexane and a three-component catalyst. The cyclohexane comprises about 120% of the product olefin. The catalyst includes zirconium tetrachloride, an aluminum alkyl such as a mixture of ethylaluminum-sesquichloride and triethylaluminum, and a Lewis base such as thiophene or an alcohol such as methanol (qv). This catalyst combination appears to produce more polymer ($\sim$2%) than catalysts used in other α-olefin processes. The catalyst content of the crude product is about 0.1 wt %. The catalyst is killed by using weak ammonium hydroxide followed by a water wash. Ethylene and cyclohexane are recycled. Idemitsu's basic α-olefin process patent (9) indicates that linear α-olefin levels are as high as 96% at C_{18} and close to 100% at C_6 and C_8. This is somewhat higher than those produced by other processes.

The Idemitsu process flow diagram is shown in Figure 5. Ethylene, cyclohexane, and a three-component catalyst are fed to a reactor. The ethylene is

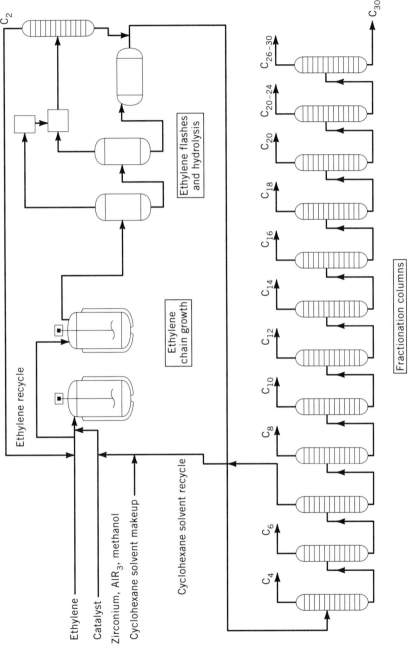

Fig. 5. Idemitsu α-olefin schematic (Japan).

flashed and recycled, the catalyst is deactivated and removed, and the cyclohexane is distilled between the hexene and octene and then recycled. Catalyst usage is small, amounting to less than 0.1% of production capacity. Key parts of the process and components include catalyst preparation, oligomerization reactor, compressors to recycle ethylene, polymer removal, catalyst deactivation and washing, cyclohexane column and recycle storage, distillation columns for single cuts, cooling water system, refrigeration for ethylene column, steam system, and hot oil supply for distillation reboilers. Because of the large amount of solvent recycle and the production of single cuts generated by the Idemitsu process, its utilities are significant.

IFP Process for 1-Butene from Ethylene. 1-Butene is widely used as a comonomer in the production of polyethylene, accounting for over 107,000 t in 1992 and 40% of the total comonomer used. About 60% of the 1-butene produced comes from steam cracking and fluid catalytic cracker effluents (10). This 1-butene is typically produced from by-product raffinate from methyl tert-butyl ether production. The recovery of 1-butene from these streams is typically expensive and requires the use of large plants to be economical. Institut Francais du Petrole (IFP) has developed and patented the Alphabutol process which produces 1-butene by selectively dimerizing ethylene.

Similar to IFP's Dimersol process, the Alphabutol process uses a Ziegler-Natta type soluble catalyst based on a titanium complex, with triethylaluminum as a co-catalyst. This soluble catalyst system avoids the isomerization of 1-butene to 2-butene and thus eliminates the need for removing the isomers from the 1-butene. The process is composed of four sections: reaction, co-catalyst injection, catalyst removal, and distillation. Reaction takes place at 50–55°C and 2.4–2.8 MPa (350–400 psig) for 5–6 h. The catalyst is continuously fed to the reactor; ethylene conversion is about 80–85% per pass with a selectivity to 1-butene of 93%. The catalyst is removed by vaporizing liquid withdrawn from the reactor in two steps: classical exchanger and thin-film evaporator. The purity of the butene produced with this technology is 99.90%. IFP has licensed this technology in areas where there is no local supply of 1-butene from other sources, such as Saudi Arabia and the Far East.

SASOL. SASOL, South Africa, has constructed a plant to recover 50,000 tons each of 1-pentene and 1-hexene by extractive distillation from Fischer-Tropsch hydrocarbons produced from coal-based synthesis gas. The company is marketing both products primarily as comonomers for LLDPE and HDPE (see OLEFIN POLYMERS). Although there is still no developed market for 1-pentene in the mid-1990s, the 1-hexene market is well established. The Fischer-Tropsch technology produces a geometric carbon-number distribution of various odd and even, linear, branched, and alpha and internal olefins; however, with additional investment, other odd and even carbon numbers can also be recovered. The Fischer-Tropsch plants were originally constructed to produce gasoline and other hydrocarbon fuels to fill the lack of petroleum resources in South Africa.

Godrej-Lurgi. Gujarat-Godrej Innovative Chemicals Co. began the production of α-olefins in 1992 in a plant of 30,000 t/yr capacity at Ankleshwar, India, utilizing the Godrej-Lurgi process (11). This is a revival of an old process for producing α-olefins from fatty alcohols, last practiced commercially by Archer-

Daniels-Midland in 1966 in the United States. Since fatty alcohols are usually more expensive than α-olefins, this process is usually not economical; α-olefins from this plant are used to produce α-olefin sulfonate for the Indian surfactant market.

In the Godrej-Lurgi process, olefins are produced by dehydration of fatty alcohols on alumina in a continuous vapor-phase process. The reaction is carried out in a specially designed isothermal multitube reactor at a temperature of approximately 300°C and a pressure of 5–10 kPa (0.05–0.10 atm). As the reaction is endothermic, temperature is maintained by circulating externally heated molten salt solution around the reactor tubes. The reaction is sensitive to temperature fluctuations and gradients, hence the need to maintain an isothermal reaction regime.

The acidic alumina catalyst is very active when fresh; this leads to extensive α-olefin isomerization to internal olefins in the first 100 hours of operation with new catalyst. Thereafter, the desired alpha selectivity of the catalyst bed (ca 93%) is achieved. During prolonged operation, the catalyst tends to lose efficacy due to poisoning. The catalyst bed can be regenerated by increasing the temperature and circulating a nitrogen–oxygen mixture through it to oxidize impurities. The endothermic nature of the reaction can lead to radial temperature gradients as well as capillary condensation of feedstock and products on catalysts at the center of the tubes; consequently, the selection of tube diameter is particularly important in the design of this reactor.

The conversion of fatty alcohols is approximately 99%. The reaction product is then condensed and sent to a distillation column to remove water and high boilers. Typically, α-olefin carbon-number distribution is controlled by the alcohol composition of the reactor feed. The process is currently used to produce C_{16-18} α-olefins from C_{16-18} fatty alcohols. A typical product composition is C_{14} at <5%, C_{16} at 50–70%, C_{18} at 30–50%, C_{20} at <2%, n-α-olefins at >93.0%, total internals at <7.0%, branched olefins (mainly vinylidenes) at <0.5%, fatty alcohols at <1.5%, dienes at <0.1%, and paraffins at <1.0%; aromatics, however, have not been detected.

Wax Cracking. One or more wax-cracked α-olefin plants were operated from 1962 to 1985; Chevron had two such plants at Richmond, California, and Shell had three in Europe. The wax-cracked olefins were of limited commercial value because they contained internal olefins, branched olefins, diolefins, aromatics, and paraffins. These were satisfactory for feed to alkyl benzene plants and for certain markets, but unsatisfactory for polyethylene comonomers and several other markets. Typical distributions were C_{6-9}, 33%; C_{10}, 7%; C_{11-14}, 25%; and C_{15-20}, 35%. Since both odd and even carbon-number olefins were produced, the cost of separations was also increased.

Chlorination and Chlorination–Dehydrochlorination of Paraffins. Linear internal olefins were produced by Shell at Geismar from 1968 to 1988, using the dehydrochlorination of chlorinated linear paraffins, a process that also yields hydrogen chloride as a by-product. To avoid the production of dichloroparaffins, which are converted to diolefins by dehydrochlorination, chlorination of paraffins is typically limited to 10% conversion.

Vista, Huntsman, and other linear alkylbenzene (LAB) producers feed chlorinated paraffins to an alkylation reactor to produce detergent alkylate without prior separation of the unreacted paraffins. Large amounts of paraffins must be recycled in these processes.

Catalytic Dehydrogenation of Normal Paraffins.   Huntsman, Vista, and other linear alkylbenzene producers also use partial dehydrogenation of normal paraffins to produce linear internal olefins for alkylation with benzene. The resulting olefins are highly linear, containing 96 wt % of linear random olefins. Typically, the olefin carbon numbers are limited to the C_{11-14} range. No more than four consecutive carbon numbers may be used or separations become economically impossible. With about 10% conversion per pass, large amounts of paraffins are recycled. Overall yields are in the 85 to 90% range.

A few companies, eg, Enichem in Italy, Mitsubishi in Japan, and a plant under construction at Fushun in China, separate the olefins from the paraffins to recover high purity (95–96%) linear internal olefins (LIO) for use in the production of oxo-alcohols and, in one case, in the production of polylinear internal olefins (PIO) for use in synthetic lubricants (syn lubes). In contrast, the UOP Olex process is used for the separation of olefins from paraffins in the liquid phase over a wide carbon range.

UOP Inc. is the key source of technology in this area, having numerous patents and over 70 units operating worldwide (12). The dehydrogenation catalyst is usually a noble metal such as platinum. For a typical conversion, the operating temperature is 300–500°C at 100 kPa (1 atm) (13); hydrogen-to-paraffin feed mole ratio is 5:1.

Branched Olefins.   Solid phosphoric acid polymerization is the commonly used process for the polymerization of propylene to the corresponding trimer and tetramer, and of propylene–butylene mixtures to heptenes (14). In the UOP process, light olefins are fed to a multibed reactor containing solid phosphoric acid, which is made from a pelletized and calcined mixture of phosphoric acid on kieselguhr. Operating conditions are 175–225°C and >2.8 MPa (>27 atm) (15). In the production of heptenes, some 30 to 40 wt % of the C_3–C_4 olefin feed is converted to useful heptenes; the remainder, which consists of hexenes, octenes, and heavier cuts, are blended into motor gasoline. In either operation, heat removal is achieved by injecting cold reactor effluent between the catalyst beds. Because of the large yields of undesirable olefins, these operations usually are associated with petroleum refining to ensure economical disposal of the by-products.

Dimersol is a commercial process for the dimerization of propylene, butylenes, or a mixture of both, to C_6 and C_8 olefins; this process produces a more linear olefin than the phosphoric acid process. The reaction is conducted at ambient temperature, using a water-soluble catalyst complex (16).

Uses

The principal outlets for higher olefins are in the polymer, surfactant, and detergent industries (1,3) (see also ALCOHOLS, HIGHER ALIPHATIC–SYNTHETIC PROCESSES). Generally, higher olefins are seldom incorporated directly into a product as an ingredient; rather, they are processed through at least one chemical reaction step before appearing in a finished product.

Polymers.   The manufacture of alcohols from higher olefins via the oxo process for use in plasticizers is a significant outlet for both linear α-olefins and branched olefins such as heptenes, nonenes, and dodecenes. These olefins are converted into alcohols containing one more carbon number than the original olefin. The alcohols then react with dibasic anhydrides or acids to form PVC plasticizers. The plasticizers produced from the linear olefins have superior volatility and cold-weather flexibility characteristics, making them an ideal product to use in flexible PVC for automobile interiors.

α-Olefins in comonomers in linear low density polyethylene (LLDPE) and high density polyethylene (HDPE) have grown at a high rate since the 1980s. Butene, hexene, and octene are the principal comonomers used, although BP also uses 4-methyl-1-pentene as a LLDPE comonomer. Because 1-butene and 1-hexene are interchangeable in several LLDPE processes, the choice of comonomer by a particular producer is a function of both economics and desired final properties. In the mid-1990s, SRI estimated the 1992–1997 comonomer growth rates at 11.0% for 1-hexene and 8.3% for 1-octene.

The introduction of metallocene and single-site polyethylene catalyst systems may eventually change the demand for higher olefins used as comonomers. Some sources indicate that their use will increase the demand for comonomers, but others feel that they will reduce comonomer use. At any rate, it is not expected that their introduction will have a significant effect on comonomer growth rates for the period 1992–1996 (see OLEFIN POLYMERS).

Detergents. The detergent industry consumes a large quantity of α-olefins through a variety of processes. Higher olefins used to produce detergent actives typically contain 10–16 carbon atoms because they have the desired hydrophobic and hydrophilic properties.

The oxo process is employed to produce higher alcohols from linear and branched higher olefins. Using a catalyst that is highly selective for hydroformylation of linear olefins at the terminal carbon atom, Shell converts olefins from the Shell higher olefin process (SHOP) to alcohols. This results in a product that is up to 75–85% linear when a linear feedstock is employed. Other oxo processes, such as those employed by ICI, Exxon, and BASF (all in Europe), produce C_{13-15} oxo-alcohols from α-olefin feedstocks; such alcohols have a linearity of about 60%. Enichem, on the other hand, produces oxo-alcohols in Italy from internal olefins and their products have a somewhat lower linearity. These oxo-alcohols are then ethoxylated and/or sulfated to produce nonionic and anionic surfactants for use in consumer and industrial products.

Nonene, or propylene tetramer, is used to alkylate phenol, which is subsequently ethoxylated to produce nonylphenol ethoxylate, an efficient, reliable industrial surfactant.

Linear alkylbenzene (LAB) is produced from C_{10-14} α-olefins and C_{11-15} internal linear olefins. Branched olefins such as propylene tetramer are used to produce branched alkylbenzene (BAB). Higher olefin streams are used to enrich the circulating olefin–paraffin stream in conventional paraffin-based LAB plants in order to boost capacity or to tailor the carbon-number distribution. Linear α- and internal olefins are also used as a direct substitute for propylene tetramer to convert hard alkylate (BAB) plants to linear alkylbenzene with no additional capital investment. As such, they offer the BAB producer an opportunity to convert quickly to LAB for biodegradable surfactants. LAB is converted to an anionic surfactant by sulfonation. The sulfonated product of branched alkylbenzene is not biodegradable and is declining in use worldwide;

it is being replaced by anionic surfactants either based on LAB or on other hydrophobes, such as fatty alcohols.

In addition, C_{14-16} and $_{16-18}$ α-olefins can be directly sulfonated to produce α-olefin sulfonate (AOS). AOS was formerly used as an efficient, reliable surfactant in consumer products in the United States and Japan. However, it has been phased out in the United States and is being replaced in Japan by products based on the methyl esters of palm oil which are deemed more natural than the petrochemically derived α-olefin hydrophobe. An excellent emulsifier, AOS has been tested in a number of enhanced oil-recovery applications. Most of this work, however, is dormant as the 1994 price of oil did not make most enhanced oil-recovery projects profitable. Should the price of oil increase, some of this work would likely be reactivated.

Lighter C_6-C_8 α-olefins and C_8 branched olefins are converted by the oxo process into fatty acids containing one carbon number greater than the starting α-olefin. These fatty acids are then used to produce alkenylbenzenesulfonic acid products which are used in the United States and in Europe as perborate bleach activators in heavy-duty laundry detergents.

Finally, α-olefins find their way into the surfactant and disinfectant market through conversion, first to alkyl dimethylamine, then to benzyl chloride quats (BADMAC) and amine oxides. The former are used broadly as disinfectants, often in combination with cleaning products. The latter is a direct active in consumer and industrial cleaning products.

Lubricants. Lubricants represent a significant and growing outlet for higher olefins. Both basestocks and lube additives are produced from higher olefins by a variety of processes (see LUBRICATION AND LUBRICANTS).

Decene can be oligomerized to produce a variety of high quality synthetic lube basestocks, often called polyalphaolefin oligomer (PAO). The principal component of these basestocks is decene trimer. However, other oligomers such as dimer, tetramer, and pentamer are also present in the various poly-α-olefin oligomer blends. Decene oligomer is typically offered in 2, 4, 6, 8, and 10 mm^2/s(=cSt) blends. Higher viscosity grades (40 and 100 mm^2/s(=cSt)) are produced by a different process, also from decene.

It appears that decene oligomer is growing at approximately 15% per year. At this growth rate, some forecasters fear that the decene demand could ultimately exceed the available supply. However, α-olefin producers, especially Albemarle Corp., can modify their processes to produce additional decene. Since decene oligomer has shown itself to be a versatile basestock, further growth is expected in automatic transmission fluids, gear oils, both hydraulic and industrial lubricants, as well as in the very large crankcase market. In addition, Enichem (Italy) produces an oligomerized internal olefin that has been qualified by AGIP for use in some of their synthetic products.

Hindered esters are also produced from C_6 and C_8 α-olefins. These olefins are converted into C_7-C_9 fatty acids by the oxo process; the acids are then treated with polyols such as pentaerythritol and trimethylol propane to produce hindered esters, which find use in lubricants for jet engines and other high performance applications.

Benzene is alkylated with C_{16-18} and C_{20+} olefins and subsequently sulfonated and neutralized with a dibasic salt such as calcium, magnesium, or barium. These so-called overbased sulfonates are used in crankcase additive packages.

Other Uses. A small but growing outlet for C_{16} and higher linear olefins is the production of alkenylsuccinic anhydride (ASA) for the paper industry. ASA is an effective alkaline sizing agent and competes with alkylketene dimer (AKD) in this application.

Additional uses for higher olefins include the production of epoxides for subsequent conversion into surface-active agents, alkylation of benzene to produce drag-flow reducers, alkylation of phenol to produce antioxidants, oligomerization to produce synthetic waxes (qv), and the production of linear mercaptans for use in agricultural chemicals and polymer stabilizers. Aluminum alkyls can be produced from α-olefins either by direct hydroalumination or by transalkylation. In addition, a number of heavy olefin streams and olefin or paraffin streams have been sulfated or sulfonated and used in the leather (qv) industry.

Health and Safety Factors

Toxicological Information. The toxicity of the higher olefins is considered to be virtually the same as that of the homologous paraffin compounds. Based on this analogy, the suggested maximum allowable concentration in air is 500 ppm. Animal toxicity studies for hexene, octene, decene, and dodecene have shown little or no toxic effect except under severe inhalation conditions. The inhalation LD_{50} for 1-hexene is 33,400 ppm; for these olefins both LD_{50} (oral) and LD_{50} (dermal) are >10 g/kg.

Handling. The main hazard associated with these olefins, especially the lighter homologues, is their low flash point. Table 3 shows flammability limits, autoignition temperatures, and flash points for the series. Although no special precautions are necessary with regard to fire extinguishing, these olefin products should be stored and shipped under an inert atmosphere to maintain product purity. Many applications for α-olefins are adversely impacted by low concentrations of moisture, peroxides, and oxygenates in general. Inhibition with antioxidants (qv) such as hindered phenols and amines should be considered if end uses are compatible with trace amounts of these materials. Finally, special attention should be given to the light olefins ($< C_8$), because the peroxides formed by inadequate handling or storage methods may be concentrated in the heavy residues during distillation. If required, α-olefin manufacturers can provide methods of peroxide removal in olefins.

BIBLIOGRAPHY

"Olefins, Higher" in *ECT* 2nd ed., Vol. 14, pp. 313–335, by R. G. Hay, Gulf Research & Development, Co.; "Olefins, Higher" in *ECT* 3rd ed., Vol. 16, pp. 480–499, by D. G. Demianiw, Gulf Oil Chemicals.

1. G. R. Lappin and J. D. Sauer, eds., *Alpha Olefins Applications Handbook*, Marcel Dekker, Inc., New York, 1989.
2. A. M. Al-Jarallah and co-workers, *Catal. Today* **14**, 1 (1992).
3. C. S. Read, R. Wilhalm, and Y. Yoshida, *SRI Chemical Economics Handbook: Linear Alpha Olefins*, SRI International, Menlo Park, Calif., Oct. 1993.

4. Chevron Permit Applications and Reports to the Texas Air Control Board for Alpha Olefins Plants NAOU-1791 and NAOU-1797, Chevron Research and Technology Co.
5. U.S. Pat. 3,482,000 (Dec. 2, 1969), H. B. Fernald, B. Gwynn, and A. N. Kresge (to Gulf Research and Development).
6. WO Pat. 93/21136 (Oct. 28, 1993), L. W. Hedrich, A. N. Kresge, and R. C. Williamson (to Chevron Research and Technology Co.).
7. Shell Chemical Permit Applications and Reports to the Louisiana Department of Environmental Quality.
8. E. R. Freitas and C. R. Gum, *Chem. Eng. Prog.* **75**(1), 73 (1979).
9. U.S. Pat. 4,783,573 (Nov. 8, 1988) Y. Shiraki, S. Kawanno, and K. Takeuchi (to Idemitsu).
10. J. L. Hennico and co-workers, *Hydrocarbon Process.*, 73–75 (Mar. 1990).
11. N. B. Godrej and co-workers, *Proceedings from* "Alpha Olefins from Oleochemical Raw Materials," *The Third World Detergent Conference*, Montreux, Switzerland, Sept. 1993.
12. Technical data, UOP Inc. Division of AlliedSignal, Des Plaines, Ill., 1966 to 1993.
13. U.S. Pat. 4,000,210 (June 23, 1975), E. E. Sensel and A. W. King (to Texaco Co.).
14. E. K. Jones, in *Advances in Catalysts*, Vol. VIII, Academic Press, Inc., New York, 1956.
15. *Petroleum Ref.* **37**, 270 (1958).
16. J. W. Andrews and co-workers, *Hydrocarbon Process.* **55**(4), 105 (1976).

G. R. LAPPIN
L. H. NEMEC
J. D. SAUER
J. D. WAGNER
Albemarle Corp.

OLIGOSACCHARIDES.   See ANTIBIOTICS, OLIGOSACCHARIDES.

OPERATIONS PLANNING.   See SUPPLEMENT.

OPIATES.   See ALKALOIDS; ANALGESICS, ANTIPYRETICS, AND ANTIINFLAMMATORY AGENTS; HYPNOTICS, SEDATIVES, ANTICONVULSANTS, AND ANXIOLYTICS.

OPIOIDS, ENDOGENOUS

Decades of research in opioid analgesia culminated in the discovery of the endogenous opioid peptides (see ANALGESICS, ANTIPYRETICS, AND ANTIINFLAMMATORY AGENTS; NEUROREGULATORS). Early studies of the structure–activity relationships of opiate alkaloids (qv) had provided evidence of the stereospecificity and antagonist reversibility of opiate action, suggesting that these drugs acted through specific receptors. However, pioneering attempts to demonstrate specific opiate receptors in the brain met with only marginal success, largely because researchers were limited to high ligand concentrations resulting from the low specific activity of opioid ligands available at that time (1). In 1973, stereospecific opioid binding in rat brain was independently demonstrated in three separate laboratories (2–4). These demonstrations relied on comparison of binding by stereoisomers, eg, levorphanol [77-07-6] and its inactive enantiomer, dextrorphan [125-73-5], which differed by four orders of magnitude in their ability to bind to opiate receptors. Radioligands having high specific activity (3.7–14.8 × 10^{11} Bq/mmol (10–40 Ci/mmol)) were essential to these studies (see RADIOACTIVE TRACERS).

The presence of specific opioid receptors in the vertebrate central nervous system suggested the existence of endogenous ligands for these receptors, a hypothesis which received considerable support from the finding that electrical stimulation of specific sites in the rat brain elicited profound analgesia (5). This stimulation-produced analgesia was naloxone [465-65-6] reversible (6), and was subject to tolerance development and to cross-tolerance to morphine [57-27-2] (7). Moreover, a close correlation existed between those brain areas most sensitive to stimulation-produced analgesia and regions containing a high density of opioid receptors (8). These results were most readily explained by the electrically induced release of endogenous substances having morphine-like properties.

Evidence soon emerged that the endogenous opioids were peptides rather than simple morphine-like molecules (9). The first direct evidence for endogenous opioids in brain extracts was provided in 1975 when two pentapeptides were purified that differed only in the carboxyl terminal amino acids (10) (Table 1). These peptides were called methionine- (Met-) and leucine- (Leu-) enkephalin, from the Greek term meaning "in the head."

At the time of the discovery of Met-enkephalin, its sequence was observed to be identical to that of residues 61–65 contained in the C-fragment of the pituitary hormone β-lipotropin [12584-99-5] (β-LPH) (see HORMONES), first isolated

Table 1. Structures of Endogenous Opioid Peptides

Compound	CAS Registry Number	Structure
Pro-opiomelanocortin-derived		
β-endorphin	[60617-12-1]	H-Tyr-Gly-Gly-Phe-Met-Thr-Ser-Glu-Lys-Ser-Gln-Thr-Pro-Leu-Val-Thr-Leu-Phe-Lys-Asn-Ala-Ile-Val-Lys-Asn-Ala-His-Lys-Lys-Gly-Gln-OH
Pro-enkephalin-derived		
Leu-enkephalin	[58822-25-6]	H-Tyr-Gly-Gly-Phe-Leu-OH
Met-enkephalin	[58589-55-4]	H-Tyr-Gly-Gly-Phe-Met-OH
octapeptide		H-Tyr-Gly-Gly-Phe-Met-Arg-Gly-Leu-OH
heptapeptide	[73024-95-0]	H-Tyr-Gly-Gly-Phe-Met-Arg-Phe-OH
Pro-dynorphin-derived		
dynorphin A	[80448-90-4]	H-Tyr-Gly-Gly-Phe-Leu-Arg-Arg-Ile-Arg-Pro-Lys-Leu-Lys-Trp-Asp-Asn-Gln-OH
dynorphin B	[85006-82-2]	H-Tyr-Gly-Gly-Phe-Leu-Arg-Arg-Gln-Phe-Lys-Val-Val-Thr-OH
α-neoendorphin	[777-21-0]	H-Tyr-Gly-Gly-Phe-Leu-Arg-Lys-Tyr-Pro-Lys-OH

in 1964 (11). In 1976, the isolation of a larger peptide fragment, β-endorphin [*60617-12-1*], that also displayed opiate-like activity was reported (12). This peptide's 31-amino-acid sequence comprised residues 61–91 of β-LPH. Subsequently, another potent opioid peptide, dynorphin [*72957-38-1*], was isolated from pituitary (13). The first five amino acids (qv) of this 17-amino-acid peptide are identical to the Leu-enkephalin sequence (see Table 1).

The three principal classes of endogenous opioid peptides share one common characteristic: the pentapeptide structure of enkephalin, either the Met- or Leu-derivatives. Loss of any portion of that structure significantly reduces the affinity of β-endorphin, dynorphin, or enkephalin in binding to opioid receptors. Although at first glance the structures of these pentapeptides do not resemble that of stereotypical opiate alkaloids like morphine, a large number of structure–activity studies have clearly established the structural similarities between enkephalin and morphine (14,15). These similarities are illustrated in Figure 1. For example, the phenolic aromatic ring A of morphine corresponds to the tyrosine residue on enkephalin (regions 1 and 2), whereas the N-terminus of enkephalin corresponds to the *N*-methyl group on morphine (region 3). The C-terminus of enkephalin most closely corresponds to the hydroxyl group on ring C and the ether bridge on morphine (region 4). There is no aromatic moiety on morphine that corresponds to the phenylalanine residue on enkephalin (region 5); however, addition of hydrophobic groups to the corresponding region on morphine (region 5) greatly increases its binding at opioid receptors. These findings suggest that opiate alkaloids and opioid peptides share common structural features which are crucial for high affinity binding at their receptors.

A number of peptides have been discovered that are related to the classical opioid peptides. FMRFamide [*64190-70-1*], which contains the first four amino acids of enkephalin, is biologically active in various invertebrates (16), and FMRFamide-like peptides have also been isolated from mammalian brain. Although these peptides are structurally similar to the enkephalins, they do not bind with appreciable affinities to the opioid receptors. In contrast, the casomorphins, a group of peptides originally isolated from milk, do not contain the enkephalin sequence, yet bind with relatively high affinity to opioid recep-

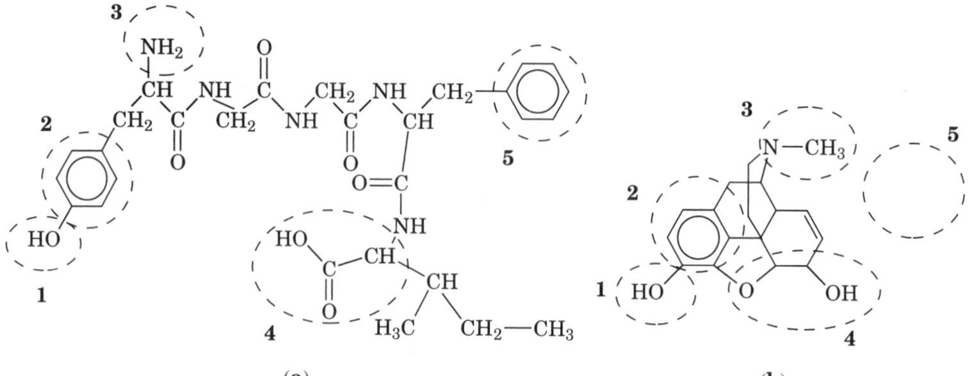

Fig. 1. Structures of two types of opioid agonists where dotted circles surround structural elements common to both compounds: (**a**) Leu-enkephalin and (**b**) morphine.

tors (see MILK AND MILK PRODUCTS) (17). Similarly, a group of unusual D-amino acid-containing peptides isolated from frog skin and termed dermorphins and deltorphins (18) have appreciable affinity for μ- and δ-type opioid receptors, respectively. Another group of peptides that do not contain the enkephalin sequence, but do share some sequence homology with the enkephalins, such as Tyr-MIF-1(H-Tyr-Pro-Leu-Gly-NH$_2$), have been isolated from mammalian brain and show both opioid-like and antiopioid activity (19).

Biosynthesis of the Opioid Peptides

Opioid Precursors. The sequence homology between β-LPH, β-endorphin, and Met-enkephalin suggested that Met-enkephalin might be formed by the proteolytic cleavage of β-endorphin or β-LPH. Likewise, some researchers assumed that dynorphin was a precursor for Leu-enkephalin. Such assumptions were incorrect, however, as demonstrated by experiments showing that the anatomical distributions of β-endorphin and dynorphin were different from that of Met- and Leu-enkephalin (20). Subsequently, the precursor relationships of the various opioid peptides were clarified by the use of cloned complimentary deoxyribonucleic acid (cDNA) techniques (see BIOTECHNOLOGY; GENETIC ENGINEERING). Each opioid peptide is formed by the cleavage of one of three precursor proteins, each of which is encoded by a separate gene (Fig. 2).

Group I contains β-endorphin and its fragments, which arise from the 31,000 mol wt (265 amino acid) pro-opiomelanocortin [66796-54-1] (POMC) (21). POMC is cleaved to yield several peptide products, including an N-terminal fragment, γ-melanocyte-stimulating hormone (γ-MSH) (adrenocorticotrophic hormone [9002-60-2] (ACTH), and β-LPH at the carboxyl terminus (Fig. 2). Cleavage of POMC occurs between pairs of basic amino acids by a trypsin-like proteolytic enzyme. Further processing of POMC is species-dependent. In some systems ACTH$_{1-39}$ is the final product, whereas in others it is further cleaved to yield α-MSH [9002-79-3] and corticotrophin-like intermediate peptide [53917-42-3] (CLIP). β-LPH is cleaved to produce γ-LPH [78065-47-1] and β-endorphin. There is little evidence that further cleavage of β-endorphin to Met-enkephalin occurs. However, tryptic digestion of β-endorphin produces the biologically active intermediates α-endorphin [59004-96-5] (β-endorphin$_{1-16}$) and γ-endorphin [61512-77-4] (β-endorphin$_{1-17}$). The predominant POMC peptide products in the hypothalamus are β-endorphin and α-MSH, neither of which appears to be acetylated to any significant degree (22). Although α-MSH and β-endorphin are also the primary POMC products in the medulla, more than 50% of the peptides are N-acetylated in this region (23). The acetylated form of β-endorphin does not bind to opioid receptors and has no significant analgesic activity (24).

Group II consists of the enkephalins which come from the 267-amino acid precursor pro-enkephalin A [88402-54-4] (Fig. 2). This protein contains four copies of Met-enkephalin, one copy of Leu-enkephalin, and the extended peptides Met-enkephalin-Arg^6-Phe^7 (the last Met-enkephalin sequence in Fig. 2) and Met-enkephalin-Arg^6-Gly^7-Leu^8 (the fourth Met-enkephalin sequence in Fig. 2) (25,26). All of these products are formed by trypsin-like cleavage between pairs of basic residues. The extended enkephalin peptides are further cleaved by carboxypeptidase E (27) to form authentic Met-enkephalin.

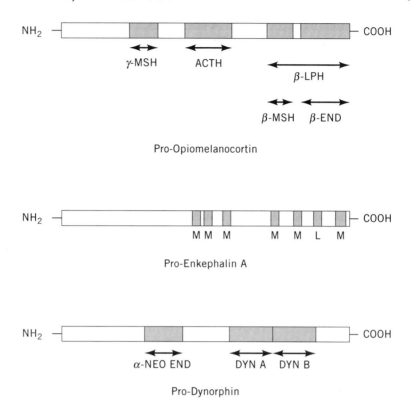

Fig. 2. Schematic drawing of the precursors for opioid peptides. Shaded areas represent the location of sequences of active peptide products which are normally released by trypsin-like enzymes acting on pairs of basic amino acid residues. Precursors are not necessarily drawn to scale. β-END = β-endorphin, L = Leu-enkephalin, M = Met-enkephalin, α-NEO END = α-neoendorphin, and DYN = dynorphin. See Table 1 and text.

The Group III peptides come from the 256-amino acid precursor, pro-dynorphin [88402-55-5] (pro-enkephalin B). This group contains dynorphin A [80448-90-4] and B [85006-82-2] as well as α-neoendorphin [77739-20-9] (Fig. 2), all of which can be further cleaved to form biologically active intermediates, eg, dynorphin A_{1-8} and β-neoendorphin [77739-21-0] (α-neoendorphin$_{1-9}$) (28). The longer of these peptides are relatively basic because of the number of Lys and Arg residues.

Regulation of Biosynthesis and Post-Translational Processing. The biosynthesis of opioid peptides, like that of other neuropeptides, is regulated by factors that influence messenger ribonucleic acid (mRNA) synthesis. A number of studies have examined regulation of POMC synthesis. Both cyclic adenosine monophosphate (cAMP) and calcium have been shown to increase POMC expression in pituitary cultures (29), and corticotrophin-releasing factor [9015-71-8] (CRF) stimulates POMC synthesis in the pituitary (30). This effect of CRF on POMC may be mediated by the immediate early gene *c-fos* (31), a gene product which regulates the expression of other genes. The fact that the CRF response is mimicked by forskolin, 8-bromo-cAMP, and phorbol ester suggests that this effect is elicited via second messenger regulation of the POMC gene

(30). Glucocorticoids inhibit POMC transcription in the pituitary anterior lobe and hypothalamus, and removal of endogenous glucocorticoids by adrenalectomy increases POMC mRNA levels (32,33). These effects are mediated by a negative glucocorticoid response element (nGRE) located on the promoter of the POMC gene (34). A region of the POMC promoter has been identified which binds several regulatory elements that act synergistically to regulate POMC transcription. This region includes the nGRE binding site and an AP-1 site, which binds to immediate-early gene products (35).

Pro-enkephalin mRNA is also under the positive influence of cAMP through the cAMP-responsive promoter element (CRE) (36). In contrast to the POMC system, glucocorticoids increase pro-enkephalin mRNA levels in the adrenal medulla (37). Nicotine also increases adrenal pro-enkephalin, presumably through increased calcium influx (38). Protein kinase C may also regulate proenkephalin expression, probably via multiple mechanisms that include calcium and phosphatidyl inositol pathways (39). The promoter region of the pro-enkephalin gene has a CRE-2 site necessary for cAMP and phorbol ester induction, as well as CRE-1 and AP-2 sites which are necessary for maximal CRE-2 effects (Fig. 3) (40). In addition, immediate-early gene components of the AP-1 complex may regulate CRE. Jun D binds to the CRE sequence either as a homodimer or heterodimer with Fos whereas Jun B inhibits transcription by Jun D (41). *In vivo*, seizures have been correlated with increased levels of Fos and Jun, in the hippocampus, that apparently regulate pro-enkephalin transcription (42). Several AP-1-like sites are also found on the pro-dynorphin gene, although only one functional AP-1 site has been identified (43). There is also evidence that noxious stimuli that increase *c-fos* in spinal neurons also increase pro-dynorphin levels, indicating a potential link between immediate early genes and the dynorphin system (43).

As for many neuropeptides, post-translational modifications are important for opioid peptide function. Such modifications are particularly important for POMC. POMC is glycosylated at the N-terminal portion of the protein prior to

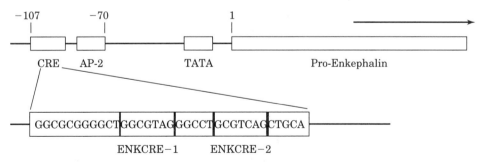

Fig. 3. Representation of promoter sites on the pro-enkephalin gene. The numbers represent the distance in nucleotides from the pro-enkephalin initiation codon; the arrow indicates the direction of transcription. The TATA promoter box occurs immediately before the pro-enkephalin initiation site; the AP-2 site, which binds immediate-early gene products, is 70 nucleotides upstream, and the CRE site, which binds a regulatory protein involved in cAMP induction of mRNA synthesis, is 107 nucleotides upstream from the initiation codon. The expanded section shows that the CRE site actually consists of two elements, ENKCRE-1 and ENKCRE-2, which separately confer cAMP sensitivity to pro-enkephalin mRNA synthesis.

proteolytic cleavage, although the number of glycosylation sites differ among species (44). These cleavages produce ACTH (which may or may not be glycosylated), a glycosylated N-terminal fragment, and β-LPH. Tunicamycin, a glycosylation inhibitor, has been used to determine the importance of glycosylation to normal POMC processing. Pituitary tumor cells treated with tunicamycin are able to process POMC into unglycosylated ACTH and β-LPH, and secrete unglycosylated ACTH and β-endorphin (45). In contrast, tunicamycin treatment of cells from the toad intermediate lobe disrupts POMC processing by the formation of unstable intermediates (46).

Anatomical Distribution and Colocalization of Opioid Peptides

The anatomical distribution of the opioid peptides and their precursors has been mapped in the brain using immunocytochemistry and *in situ* hybridization of corresponding mRNA. The various opioid peptides exhibit different anatomical distributions in brain, and the widespread distribution of opioid immunoreactive fibers suggests opioid involvement in many functional systems. Neurons containing β-endorphin and related POMC-derived peptides, as well as POMC mRNA, are primarily located in the hypothalamic arcuate nucleus (47). A second smaller group is in the nucleus tractus solitarius (NTS) in the medulla (48). Fibers from the hypothalamic neuronal group project extensively throughout the telencephalon, diencephalon, and medial brainstem, whereas NTS-derived fibers are confined primarily to the lateral brainstem (49).

In contrast to the confined localization of β-endorphin neurons, enkephalin and dynorphin cell bodies have a ubiquitous distribution throughout the brain, with both local projections and long fiber pathways (50). Enkephalin and pro-enkephalin mRNA-containing neurons are found in the hippocampus, amygdala, striatum, septum, hypothalamus, thalamus, interpeduncular nucleus, parabrachial nucleus, nucleus locus coeruleus, periaqueductal gray, brainstem raphe and reticular nuclei, NTS, and spinal cord (51). Enkephalin is often colocalized in the same neurons with other peptides and neurotransmitters, including catecholamines, acetylcholine, epinephrine, γ-aminobutyric acid (GABA), serotonin, and substance P (52,53). Although enkephalin has been primarily localized in neurons, it has also been identified in cultured astrocytes, where it may be important in development (54).

Dynorphin is widely distributed throughout the brain, and overlaps in many regions with enkephalin. Dynorphin-containing neurons are found in the hippocampus, central amygdala, striatum, cortex, hypothalamus, periaqueductal gray, brainstem reticular nuclei, NTS, and spinal cord (55). Colocalization of dynorphin with other neurotransmitters and hormones, such as vasopressin in the posterior pituitary and hypothalamic magnocellular neurons (56) and leutinizing hormone and follicle-stimulating hormone in the anterior pituitary (57), has also been reported. Few studies have colocalized the opioid peptides to the same neurons, although enkephalin and dynorphin have been colocalized in neurons in the spinal cord (58).

In addition to the well-defined opioid systems in the central nervous system, the three opioid peptides and their precursor mRNA have also been identified in peripheral tissues. β-Endorphin is most abundant in the pituitary, where it exists in corticotroph cells with ACTH in the anterior lobe and in melanotroph cells with MSH in the intermediate lobe (59). Enkephalin and pre-pro-enkephalin mRNA have been identified in the adrenal medulla (60) and this has been the source of material for many studies of pro-enkephalin synthesis and regulation. Pre-pro-enkephalin mRNA has also been identified in the anterior and posterior lobes of the pituitary (61). mRNA for all three opioid precursors has been identified in the reproductive system (62–64). POMC mRNA and peptide products have been found in the digestive system, kidney, liver, lung, and spleen (62) and pro-enkephalin mRNA has been identified in the heart (65). Pro-enkephalin has also been found in lymphocytes (66), generating interest in possible opioid effects on the immune system.

Receptors for Opioid Peptides

Multiple Opioid Receptors. The concept of multiple opioid receptors was first postulated in 1976 (67). Three distinct opioid receptors were postulated: mu (μ), kappa (κ), and sigma (σ). A fourth type of opioid receptor, the delta (δ) receptor, was postulated in 1977 (68) after discovery of the endogenous opioid peptides. Originally, the prototype agonists for these receptors were morphine [16206-77-2] (μ), ketazocine [36292-69-0] (κ), N-allylnormetazocine (SKF-10,047) [14198-28-8] (σ), and Met- and Leu-enkephalin (δ), although more selective compounds for each receptor type are available. The σ-receptor is no longer thought to be a receptor for the endogenous opioids and is therefore not discussed further herein. The original confusion was in reference to the cross-reactivity of σ-ligands with μ- and κ-opioid receptors. The classification of opioid receptor types is primarily based on the specific affinities displayed by various opioid drugs and peptides in radioligand-binding assays and on the potency of these compounds to inhibit smooth muscle contractions, or to block opioid inhibition in the case of antagonists, in isolated organ preparations such as the guinea pig ileum (μ- and κ-receptors) or the mouse vas deferens (δ-receptors). Confirmation of the original discoveries of multiple opioid receptor types is being obtained by molecular cloning studies. A more accurate reclassification scheme is expected to arise from these studies.

The opioid peptides vary in their binding affinities for the multiple opioid receptor types. Leu- and Met-enkephalin have a higher affinity for δ-receptors than for the other opioid receptor types (68), whereas the dynorphin peptides have a higher affinity for κ-sites (69). β-Endorphin binds with equal affinity to both μ- and δ-receptors, but binds with lower affinity to κ-sites (70). The existence of a β-endorphin-selective receptor, the ϵ-receptor, has been postulated; whether this site is actually a separate β-endorphin-selective receptor or is a subtype of a classical opioid receptor is a matter of controversy (71,72). The existence of opioid receptor subtypes in general is quite controversial although there is some evidence for subtypes of μ- (73), δ- (74), and κ-receptors (72,75), confirmation of which may be obtained by future molecular cloning studies.

Opioid Peptide Analogues and Their Receptor Affinities. In an effort to develop nonaddictive and nontolerance-producing opioid analgesics numerous metabolically stable enkephalin analogues have been synthesized (see PSYCHOPHARMACOLOGICAL AGENTS). The most successful stability-enhancing techniques have included the replacement of naturally occurring L-amino acids with the D-isomer and amidation of the carboxyl terminal residue, to form compounds such as D-Ala^2,Met^5-enkephalinamide [61090-95-7] (76). These derivatives show little promise as nonaddictive analgesics, because they share the tolerance and dependence liabilities of the endogenous opioids (77). However, many enkephalin analogues show remarkable receptor selectivity compared to the naturally occurring peptides. This observation has led to the design of hundreds of analogues having increased selectivity for the multiple opioid receptor types. The principal design strategies include (1) substitution, addition, or deletion of amino acid residues; (2) introduction of conformational restrictions; and (3) modification of peptide bonds. As of this writing, there are a number of commonly used peptide and synthetic opioid ligands that are among the most highly selective agonists and antagonists available for the multiple opioid receptor types. Compared to the native enkephalins, the modified peptide analogues can display increased receptor selectivity for one of three reasons: (1) decreased affinity for other sites along with unchanged affinity for the target site; (2) increased affinity for the target site with no change in affinity for other sites; or (3) a combination of the above.

Among the peptide derivatives that are agonists, D-Ala^2, N-Me-Phe^4, Gly-(CH$_2$OH)^5-enkephalin [78123-71-4] (DAMGO), where Me represents methyl, is highly μ-selective (70). Similar compounds having increased μ-selectivity have been achieved by replacement of residues 4 and 5 of DAMGO with an aliphatic chain (78). Some of the atypical (nonenkephalin-containing), naturally occurring opioid-like peptides also show some degree of μ-selectivity. β-Casomorphin (H-Tyr-Pro-Phe-Pro-Gly-Pro-Ile-OH) [102029-74-3], for example, is moderately μ-selective, and deletions of residues 5–7 produces morphiceptin [74135-04-9], a compound with improved μ-selectivity (17). Further modification of this tetrapeptide to H-Tyr-Pro-N-Me-Phe-D-Pro-NH$_2$ (PLO17) (79) leads to μ-selectivity which is somewhat greater than that of DAMGO. Dermorphin (H-Tyr-D-Ala-Phe-Gly-Tyr-Pro-Ser-NH$_2$) [77614-16-5] is unusual in that it naturally contains D-Ala in the 2-position and a C-terminal carboxamide (80). This compound, as well as many of its tetra- and tripeptide derivatives, including DALDA(H-Tyr-D-Arg-Phe-Lys-NH$_2$), is relatively μ-selective (81). A series of dermorphin analogues with increasing positive charge was synthesized to test the hypothesis that δ-receptors are in a cationic membrane environment from which positively charged ligands are electrostatically excluded (82). Results showed that the μ-selectivity of these analogues increased with increasing number of positive charges so that the peptide with the highest positive charge, [D-Arg^2,Lys^4]dermorphinamide, was 10 times more selective than DAMGO. Other experiments examined the relationship between the Phe^3 residue in dermorphin and the Phe^4 in enkephalin by synthesizing hybrid analogues with Phe in both the 3- and 4-position (83). The prototype Phe^{3,4} analogue H-Tyr-D-Ala-Phe-Phe-NH$_2$ (TAPP) displayed high affinity and selectivity for μ-receptors (82). Nitration in the para position of the aromatic moiety of Phe^3 decreased the affinity of TAPP for μ-receptors, whereas similar nitration of the Phe^4 residue produced an increase in affinity, supporting the contention that the Phe^4 residue of the enkephalins interacts with the μ-receptor in a different fashion than the Phe^3 residue of dermorphin.

The enkephalins are structurally flexible and capable of assuming a number of energetically favorable conformations in aqueous solution. A successful approach to increase receptor selectivity has been to introduce conformational constraints by cyclization of peptides. An example of this approach is the peptide H-Tyr-cyclo[-D-A$_2$bu-Gly-Phe-Leu-], where A$_2$bu is D-α,γ-diaminobutyric acid (84). Cyclization of the γ-amino group of A$_2$bu to the carboxyl terminus conferred significant μ-selectivity compared to the corresponding nonselective linear analogue containing α-aminobutyric acid (85). This comparison provided the first direct demonstration of conformational selectivity among peptide receptor subtypes. Studies have demonstrated a lack of receptor-selectivity when more than one low energy conformation of a constrained, cyclized peptide can be assumed, as has been shown with [D-AlaL^2,L-AlaL^5] enkephalinamide, where AlaL represents lanthionine-containing residues linked by a monosulfide bridge, which displays equal potency in bioassays for μ- and δ-activity (86). Another series of μ-selective cyclized peptides are the cyclic dermorphin tetrapeptides, such as H-Tyr-cyclo[-D-Orn-Phe-Asp]-NH$_2$ and its derivatives (87), which have been used to characterize the importance of the arrangement of the Tyr and Phe aromatic rings to μ-receptor affinity.

The cyclization approach has been extremely successful in the synthesis of highly selective δ-agonists. Substitution of D- or L-penicillamine moieties in the 2 and 5 position of enkephalin has led to compounds having a high degree of δ-selectivity (88). Two of the most selective analogues that have been used extensively as δ-receptor agonists are H-Tyr-cyclo[-D-Pen-Gly-Phe-D-Pen]-OH (DPDPE) and DPLPE (with L-Pen in the 5 position). The δ-selectivity of these compounds was shown to result from steric interference at the μ-site, caused by the presence of the *gem* dimethyl groups in the 2-position side chain (89). Although DPDPE and DPLPE are extremely δ-selective, their absolute affinity for the δ-receptor is low compared to some the δ-selective linear enkephalin analogues. However, this problem has been solved by replacement of Phe^4 with *p*-chlorophenylalanine, which improves the δ-selectivity of the compound by fivefold over DPDPE because of an increase in affinity at the δ-site (90).

Although the natural enkephalins are somewhat δ-selective, a number of linear enkephalin analogues have been synthesized with improved selectivity. One of the earliest peptides analogues employed as a δ-selective ligand was H-Tyr-D-Ala-Gly-Phe-D-Leu-OH (DADLE) (91). Although DADLE is only slightly more selective than the enkephalins, replacement of D-Ala in the 2 position with D-Ser or D-Thr combined with the addition of a Thr residue at the C-terminus to form the analogues H-Tyr-D-Ser (or D-Thr)-Gly-Phe-Leu-Thr-OH (DSLET and DTLET) led to marked improvement in δ-selectivity by reducing the affinity for μ receptors (92). Because improved δ-selectivity was assumed to be due to steric interference at the μ-site by the side chains of residues 2 and 6, compounds were designed with increased bulk in these positions by the addition of *tert*-butyl, t-C$_4$H$_9$, groups to the hydroxyl moieties of Ser or Thr. This strategy produced compounds with greatly increased δ-selectivity, such as H-Tyr-D-Ser(Ot-C$_4$H$_9$)-Gly-Phe-Leu-Thr (or Ot-C$_4$H$_9$Thr)-OH (DSTBUTLET and BUBU) (93). Replacement of D-Ser(Ot-C$_4$H$_9$) with D-Cys(St-C$_4$H$_9$) to form the compound BUBUC has been shown to increase δ-selectivity to a level comparable to that of the cyclic Pen-containing analogues DPD(L)PE (94).

Another class of δ-selective peptides, isolated from extracts of frog skin, is the deltorphins. These compounds are based on the structure H-Tyr-D-Met-Phe-His-Leu-Met-Asp-NH$_2$ [*119975-64-3*] and are approximately equal in δ-selectivity to DPDPE (95). Two analogues, H-Tyr-D-Ala-Phe-Asp-Val-Val-Gly-NH$_2$ (D-Ala^2-deltorphin I) and H-Tyr-D-Ala-Phe-Glu-Val-Val-Gly-NH$_2$ (D-Ala^2-deltorphin II) display greater δ-selectivity than DPDPE owing to their higher δ-receptor affinity (96). These compounds both contain the same N-terminal tripeptide sequence as the μ-selective dermorphins, which underscores the importance of the C-terminal tetrapeptide sequence in conferring δ-selectivity.

The endogenous peptide dynorphin A$_{1-17}$ and its C-terminally degraded fragments dynorphin A$_{1-13}$ and dynorphin A$_{1-9}$ are somewhat κ-selective (70). Several substituted analogues of dynorphin have shown moderate improvement in κ-selectivity, including Ala^8-dynorphin A$_{1-13}$, Trp^8-dynorphin A$_{1-13}$, and D-Pro^{10}-dynorphin A$_{1-13}$ (97). Analogues with C-terminal deletions, such as D-Pro^{10}-dynorphin A$_{1-11}$, have been found to display further improvement in κ-selectivity (98). Replacement of residues 7–15 of dynorphin A with an alternating Lys and Val sequence, along with the substitution of Ser in positions 16 and 17, has produced moderate increases in κ-selectivity (99). Although peptide cyclization has been a successful technique in the development of ligands with improved μ- and δ-selectivity, cyclized dynorphin analogues have proven to be relatively nonselective (100), or more μ- or δ- than κ-selective (101). Thus, the effect of cyclization tends to produce peptide conformations that are not compatible with the κ-receptor binding site.

Among the peptide analogues that are opioid antagonists, the most highly μ-selective are derived from somatostatin [*38916-34-6*]. These cyclic compounds, based on H-D-Phe-cyclo[-Cys-Phe-D-Trp-Lys-Thr-Cys]Thr-ol (SMS-201995) [*83150-76-9*] (102), bear no obvious structural resemblance to the opioid peptides. Most interesting is the lack of an N-terminal Tyr residue, which is common among all the opioid peptides and the atypical opioid-like peptides including β-casomorphin, the dermorphins, and the deltorphins. The presence of another aryl-containing residue, Phe, in place of Tyr may account for the antagonist properties of the somatostatin-based analogues, though this has not been proven. The most μ-selective of these analogues H-D-Tic-cyclo[-Cys-Tyr-D-Trp-Orn-Thr-Pen]-Thr-NH$_2$ (TCTOP) contains another aromatic group, Tic (tetrahydroisoquinoline-3-carboxylic acid), in the 1 position and retains its full antagonist properties (103). The two most commonly used analogues H-D-Phe-cyclo[-Cys-Tyr-D-Trp-Lys-Thr-Pen]-Thr-NH$_2$ (CTP) [*103335-28-0*] and H-D-Phe-cyclo[-Cys-Tyr-D-Trp-Orn-Thr-Pen]-Thr-NH$_2$ (CTOP) [*103429-31-8*] (104) are slightly less μ-selective than TCTOP. However, all of these compounds display reduced affinity for somatostatin receptors compared to the parent compound SMS-201995.

Enkephalin-based antagonists having high δ-selectivity but low δ-affinity have been synthesized by diallylation of the N-terminal α-amino group along with modification of the peptide bond at the 3 to 4 position (105). Replacement of the Gly^2-Gly^3 sequence with α-aminoisobutyric acid (Aib) to form the compound *N,N*-diallyl-Tyr-Aib-Aib-Phe-Leu-OH (ICI 174864) [*89352-67-0*] led to improvement in both δ-receptor affinity and selectivity (106). Conformationally restricted analogues of this compound have shown similar results. Similar modifications of truncated dynorphin peptides, such as N,N-diallyl-Tyr^1,Alb^{2,3},D-Pro^{10}-dynorphin A$_{1-11}$ have produced analogues that are antagonists but are not significantly κ-selective (107). To date there have been no highly κ-selective peptide antagonists developed.

Another recently developed class of δ-antagonists are short-chain (3–4 residues) peptides consisting entirely of aromatic amino acids, and having Tic in the 2 position (108). The most potent and selective of these analogues is H-Tyr-Tic-Phe-Phe-OH (TIPP) which displayed extreme δ-selectivity and improved potency compared to reported values for ICI 174864. Interestingly, replacement of L-Tic with the D-isomer changed the compound to a μ-selective agonist, whereas the amino derivative TIPP-NH$_2$ was a moderately potent μ-agonist with δ-antagonist properties. However, the corresponding tripeptides H-Tyr-Tic-Phe-OH or -NH$_2$ were both δ-selective antagonists. These results provide compelling evidence that intrinsic activity as well as receptor-binding affinity and selectivity can be affected by opioid peptide conformation. Pseudopeptide analogues of the Tic-containing antagonists have been developed which contain a reduced peptide bond between the Tic^2 and Phe^3 residues (109). The compound with the highest potency, H-Tyr-TicΨ[CH$_2$NH]Phe-Phe-OH (TIPP[Ψ]) displayed subnanomolar affinity and the greatest degree of selectivity for the δ-receptor of any ligand yet known.

Although the emphasis of this article is on the opioid peptides, a brief discussion of nonpeptide (alkaloid and synthetic) ligands is appropriate. Among the nonpeptide agonists, the opiate alkaloids, such as morphine, and their synthetic derivatives, such as fentanyl, are relatively μ-preferring. Though not as selective as DAMGO, these have generally equal or greater potency. This is also true of the opiate antagonists, such as naloxone (Fig. 4), which are not as selective as the cyclic somatostatin analogues but tend to be more potent. For δ-receptors, there are few selective nonpeptide ligands available. Naltrindole [*111555-53-4*] (NTI) (Fig. 4) is a δ-selective nonpeptide antagonist (110) that is more potent but less selective than TIPP. Naltrindole has approximately the same affinity as ICI 174864 for non-δ sites, but its affinity for the δ-receptor is orders of magnitude greater (111). The benzofuran analogue of NTI (NTB) and 7-benzylidenenaltrexone (BNTX) discriminate between the putative δ-receptor subtypes, δ_2 and δ_1, respectively (112). The recently developed BW373U86 (Fig. 4) is a δ-selective agonist (113) that has less selectivity but higher affinity for the δ-receptor than DPDPE has. Compared to the linear δ-selective analogue DSLET, BW373U86 is approximately equal in binding affinity but is more potent in functional assays owing to the antagonist-like binding properties of this full agonist (114).

The most highly κ-selective ligands are not of peptide origin. Selective nonpeptide κ-agonists include U50,488δ [*67198-13-4*] (Fig. 4) and related compounds (70,115). One of the most highly selective κ-antagonists is norbinaltorphimine (nor-BNI) (Fig. 4), a dimeric derivative of the opiate antagonist naltrexone (116). In general, the original ketazocine-based benzomorphan ligands are no longer considered to be sufficiently selective for examining κ-sites.

Receptor Structure and Function.   All of the known opioid receptor types belong to the superfamily of G protein-coupled receptors. These receptors reside on the plasma membrane and affect cell physiology by interacting with the

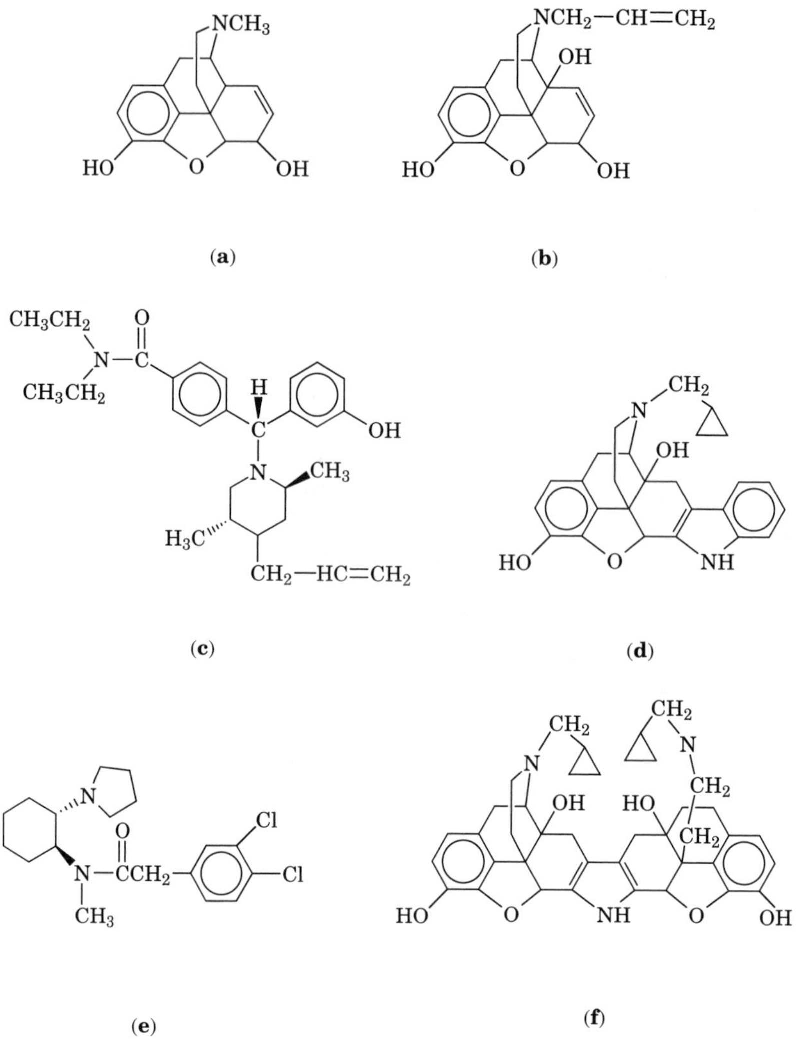

Fig. 4. Structures of several nonpeptide opioid agonists, (**a**) morphine, (**c**) BW373U86, (**e**) U50488 and antagonists, (**b**) naloxone, (**d**) naltrindole, and (**f**) norbinaltorphimine, with specificities at μ-, δ-, κ-opioid receptors.

signal-transducing guanosine triphosphate (GTP)-binding regulatory proteins (G proteins) (117). In most cells, opioid receptors are coupled to G_i and G_o, a class of G proteins that are adenosine diphosphate (ADP)-ribosylated by pertussis toxin. It is the G protein, rather than the receptor itself, that determines which effector(s), an enzyme or ion channel, are affected by receptor activation. The effector activity can be stimulated or inhibited by the receptor, depending on the G protein involved. For example, all opioid receptor types are known to inhibit the activity of adenylyl cyclase (118), the enzyme which converts adenosine triphosphate (ATP) into cyclic-AMP. These receptors also decrease calcium conductance

(119) and increase potassium conductance by direct actions of G proteins on the corresponding channels (120). All of these opioid-induced responses tend to decrease neuronal activity by hyperpolarization or to inhibit neurotransmitter release by blocking depolarization-induced calcium influx. Thus, the opioid peptides are generally considered to be inhibitory neurotransmitters, although excitatory actions have been reported (121,122). Moreover, opioids inhibit the activity of cAMP-dependent protein kinase through their effect on adenylyl cyclase (123). Reviews of opioid receptor-mediated effects on cell biochemistry and physiology are available (124,125).

Despite the knowledge of sequence homology obtained by the cloning of many other G protein-coupled receptors, attempts at cloning the opioid receptors remained unsuccessful until 1992. In that year, two independent reports emerged on the expression cloning of a δ-opioid receptor from NG108-15 cells (126,127), a cell line known to express a high density of δ-opioid receptors (see CELL CULTURE TECHNOLOGY). The cloned receptors, when expressed in COS cells, showed a binding profile expected of a δ-receptor and mediated opioid inhibition of adenylyl cyclase. The cloned δ-receptor contained 371–372 amino acids and showed significant homology to other G protein-coupled receptors, with the characteristic seven transmembrane domains, three intracellular and three extracellular loops, and multiple glycosylation sites on the amino terminal domain. Soon after the reported cloning of the δ-opioid receptor, cloning of the μ- (129–131) and κ- (132–136) opioid receptors, as well as multiple opioid receptor types, were reported (137–139). The multiple opioid receptors share extensive sequence homology with each other, as well as with the somatostatin receptor (Fig. 5). Whether there are multiple subtypes of opioid receptors remains unclear as of this writing.

Biological Activities

Soon after the identification of endogenous opioid peptides, studies were conducted to determine their contribution to physiological function. Morphine was a well-established analgesic drug with central actions mediated by an endogenous anatomical substrate (140). Intracerebral (icv) injection of β-endorphin in mice also elicited naloxone-reversible analgesia (141). β-Endorphin elicited other opiate effects including shivering, pinnae vasodilation, mydriasis, tachypnea (rapid breathing), vocalization, hyperexcitability, and catelepsy (142). Further studies on analgesia demonstrated that β-endorphin was more potent than morphine in eliciting analgesia in a variety of species, including the human (143). Moreover, evidence accumulated to implicate endogenous opioid peptides in mediating stimulation-produced analgesia, especially when stimulation was applied to the periaqueductal gray (PAG), a region rich in opioid peptides and receptors (5). This evidence included (1) partial reversal by naloxone of stimulation-produced analgesia (144), (2) tolerance to repeated stimulation, and (3) cross-tolerance with morphine (145). Another potential analgesic effect of opioid peptides may be stress-induced analgesia, in which noxious or stressful stimuli elicit an analgesic response. This is a complex phenomenon with several neural, including nonopioid, components. However, the findings of partial naloxone reversibility, devel-

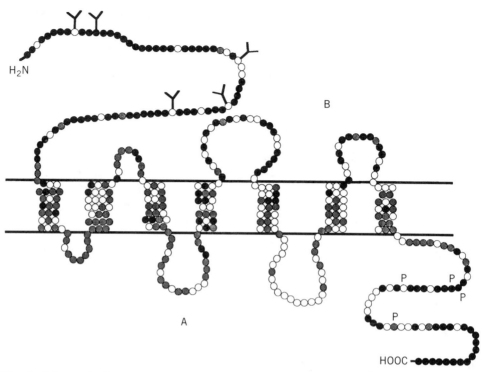

Fig. 5. Schematic diagram of the presumed arrangement of the amino acid sequence for the δ-opioid receptor, showing seven putative transmembrane segments; three intracellular loops, A; three extracellular loops, B; the extracellular N-terminus; and the intracellular C-terminus, where (●) represents amino acid residues common to μ-, δ-, and κ-receptors; (◐), amino acid residues common to all three opioid receptors and other neuropeptide receptors; and (○), other amino acids. Branches on the N-terminal region indicate possible glycosylation sites, whereas P symbols in the C-terminal region indicate possible phosphorylation sites. Adapted from Ref. 128.

opment of tolerance, and an increase in endogenous opioid levels during stress-induced analgesia suggest at least some involvement of opioid peptides (146).

In contrast to the potent, long-lasting analgesic effects of β-endorphin (147), the enkephalins are extremely weak analgesics in laboratory tests. This difference is likely a result of the relatively short (2–3 min) biological half-life of the enkephalins vs the long (2–3 h) half-life of β-endorphin (148). Thus, only transient analgesia has been found in rats, mice, and cats (149) with Met-enkephalin even when administered by icv injection. Not surprisingly, massive doses (320 mg/kg) of Leu-enkephalin administered intravenously produced only weak analgesic activity in mice (150). Nonetheless, the enkephalins may be involved in the physiological response to sensory input from noxious stimuli (nociception) by acting at the spinal level. Enkephalin (151) and opioid receptors (152) have been localized in regions of the spinal cord associated with processing of noxious stimuli. Lesions affecting afferent nociceptive input decrease opioid receptor binding (152,153), indicating that opioid receptors are involved in this system. Enkephalin levels in the spinal cord increase in response to noxious stimuli (154), and increasing enkephalin levels by administration of enkephalinase inhibitors leads to dose-dependent, naloxone-reversible analgesia (155). Application of enkephalin to spinal neurons has also been reported to decrease cell firing in response to noxious stimuli (156). These actions of enkephalin at the spinal level may actually result from modulation of substance P release from primary afferent fibers (157).

Dynorphin may also influence nociception at the spinal level. The levels of prodynorphin mRNA and immunoreactive dynorphin increase in the chronic inflammatory arthritic model (158). Dynorphin$_{1-13}$ also inhibits morphine or β-endorphin-induced analgesia in naive animals and enhances analgesia in tolerant animals, indicating that this peptide may have a regulatory role in opioid analgesia (159). This effect does not appear to be mediated by a classical opioid receptor, since des-tyrosine dynorphin, which does not bind to opioid receptors, also antagonizes morphine analgesia (160).

The finding of analgesic activity for the endogenous opioids created a renewed but short-lived hope that these or related peptides might lead to an analgesic devoid of dependence liability. However, Met-enkephalin and β-endorphin produce symptoms of physical dependence (161) and evidence of tolerance and morphine cross-tolerance in animals and *in vitro* (162). Furthermore, β-endorphin (163,164) and the enkephalins (165) are reinforcing stimuli in behavioral experiments. The effects of these peptides may be mediated in part by disinhibition of mesolimbic dopaminergic neurons (166), which have been implicated in mediating the reinforcing effects of morphine (167). Moreover, drug discrimination experiments in rats indicate that Met-enkephalin is generalized to the narcotic fentanyl (168). Thus, the evidence indicates that the opioid peptides, including at least β-endorphin and the enkephalins, are similar to the opiate alkaloids in their reinforcing properties as well as in their ability to produce tolerance and dependence.

Although many studies have focused on the analgesic effects of opioids, the endogenous opioid peptides have been found to influence a wide range of physiological functions. Opioid peptides and receptors are found in brain areas that influence respiratory and cardiovascular function. Injection of β-endorphin into the NTS results in dose-dependent and naloxone-reversible decreases in mean arterial pressure and heart rate (169). Intracisternal β-endorphin also depresses respiration in a naloxone-reversible manner (170). One aspect of opioid function that has received a great deal of interest is the effect of endogenous opioid systems on immune function (171). Both β-endorphin and Met-enkephalin enhance the cytotoxicity of natural killer cells in a manner that is inhibited by naloxone (172). In contrast, the C-terminal fragment of β-endorphin reduces the activity of natural killer cells; however, this activity is not affected by naloxone (173). Endogenous opioid peptides may also influence reproductive behavior. Studies in rodents with β-endorphin (174) and an enkephalin analogue (175) have demonstrated inhibition of copulatory behavior. POMC mRNA levels are also decreased by both estrogen and testosterone (176). In contrast, estradiol has been shown to increase proenkephalin mRNA levels in the hypothalamus in a manner that coincided with the display of lordosis (177). Another hypothalamic action of opioid peptides is thermoregulation. Hyperthermia occurs after the injection of a μ-agonist, whereas dynorphin decreases temperature by decreasing metabolic rate (178).

Metabolic Inactivation of Opioid Peptides

Several enzymes, none of which are completely specific for the enkephalins, are known to cleave Leu- and Met-enkephalin at various peptide bonds. The main enzymes that degrade enkephalin are zinc metallopeptidases. The first enkephalin-degrading enzyme to be identified, an aminopeptidase which cleaves the amino terminal Tyr–Gly bond (179), has been shown to be aminopeptidase-N (APN) (180). It is a cytoplasmic enzyme which is uniformly distributed throughout the brain. The increased analgesic activity of synthetic enkephalins substituted by D-amino acids at position 2, eg, [D-Ala^2]-Met-enkephalin, is probably the result of increased stability toward this aminopeptidase (181). A second enkephalin-degrading enzyme, enkephalinase B, is a dipeptidylaminopeptidase (DAP) which cleaves the Gly–Gly bond of enkephalin (182). This membrane-bound enzyme has the least overall enkephalin-degrading activity in crude brain homogenates and is uniformly distributed throughout the brain. Enkephalinase A, a dipeptidylcarboxypeptidase which cleaves the Gly–Phe bond of enkephalin (182,183), has been identified as neutral endopeptidase 24.11 (NEP) (184). This membrane-bound enzyme has a K_m for the enkephalins of approximately 20 μM (182,185), and its distribution parallels that of the opioid receptors (186) as well as the enkephalins (187). The finding that administration of thiorphan, a synthetic inhibitor of NEP, produces naloxone-reversible analgesia has provided support for the suggestion that NEP is largely responsible for *in vivo* inactivation of the enkephalins (188) (see ENZYME INHIBITORS). Enkephalin analogues with increased stability toward NEP have been synthesized; the modifications used in making these analogues include N-methylation of the Gly–Phe peptide bond, amidation of the carboxyl terminus, or replacement of the L-Phe with the D-isomer. Such analogues show even more enhanced analgesic activity than the D-Ala^2 analogues (189). Finally, angiotensin-converting enzyme (ACE) also shows enkephalinase activity and, like NEP, cleaves the Gly–Phe bond (190). However, ACE has a low ($\sim$1 mM) affinity for the enkephalins, and several specific ACE inhibitors do not significantly alter the overall enkephalin-degrading activity in brain tissue (191).

Intensive research efforts have focused on the discovery of potent and specific inhibitors of the enkephalin-degrading enzymes for novel analgesic agents. Because the enkephalinases are metallo (Zn) enzymes (192), inhibitor design was based on the synthesis of compounds with a strong metal coordinating group and which display energetically favorable interactions with one or more of the subsites surrounding the catalytic core (193). A potent and specific NEP inhibitor (K_i of 4.7 nM), thiorphan [76721-89-6] (194), produced analgesia on its own and potentiated analgesia elicited by enkephalin analogues (188). Subsequently, a number of modifications were made in order to increase the selectivity and bioavailability of thiorphan (188). Other classes of thiol-based inhibitors, such as the N-mercaptoacetyldipeptides, also show high potencies as NEP inhibitors. Another important class of enkephalinase inhibitors is the N-protected amino acid hydroxamates. These transition-metal chelators display nanomolar K_i values in inhibiting NEP (195), and modifications have led to extremely potent and specific NEP inhibitors (196). Finally, phosphorus-containing dipeptides, such as phosphoramidon [36357-77-4], are also potent inhibitors of NEP (197).

APN inhibitors include substituted aminoethanols (198) and phenyalanine-based compounds (199). Phe-containing dipeptides such as Tyr-Phe-NHOH are highly selective and potent inhibitors of DAP. Various hydroximate- and thiol-containing compounds have also been synthesized as mixed enkephalinase inhibitors. Compounds based on kelatorphan [92175-57-0], for example, potently inhibit NEP, APN, and DAP (195). Development of compounds in which several inhibitors are linked by disulfide bonds has led to systemically active mixed enkephalinase inhibitors that are very potent in antinociceptive tests (200). In addition to their promise as analgesic agents, mixed peptidase inhibitors with specificity for NEP and ACE have been found to possess antihypertensive activity (see CARDIOVASCULAR AGENTS). A review of the design and potential clinical applications of mixed peptidase inhibitors is available (201).

Little is known about metabolic inactivation of β-endorphin and the dynorphins. NEP, and to a lesser extent APN, are only weakly active against β-endorphin (183). Enzymes are known which degrade β-endorphin *in vitro* under nonphysiological conditions (202) or which inactivate β-endorphin by N-acetylation (203). A lack of specific degradative enzymes for these peptides may account for their relatively long half-life *in vivo*, though this has not been definitively established.

Endogenous Opiate Alkaloids

Although the opioid peptides have long been identified as the primary endogenous opioid ligands in brain, several groups have identified the opiate alkaloid morphine and related compounds in the tissues of several species. A nonpeptide opioid has been isolated from toad skin in sufficient quantity for purification and has the same profile as morphine in high performance liquid chromatography (hplc), gas chromatography/mass spectrometry, radioimmunoassay, opiate receptor binding assay and bioassay (204) (see ANALYTICAL METHODS; CHROMATOGRAPHY; IMMUNOASSAY; MASS SPECTROMETRY). A nonpeptide opioid was also identified in bovine brain and adrenal gland, as well as rabbit and rat skin, that corresponded to morphine in hplc analysis. However, the concentration of the compound in these tissues was too low for further purification. Morphine and codeine have been identified in bovine hypothalamus and adrenal gland, as well as rat brain, and the presence of 6-acetylmorphine has been demonstrated in the bovine brain (205). This latter compound is a metabolite of heroin that had not previously been identified in plants or animals. The potential biological importance of 6-acetylmorphine is that it readily enters the central nervous system, where it is then converted to morphine. Thus, it has been speculated that this compound may be a peripherally synthesized hormone that targets the central nervous system. The biological activity of endogenous opiate alkaloids has not been determined, and it is not known how they may interact with endogenous opioid peptides. Although these compounds have been shown to be synthesized *in vivo*, their biosynthetic mechanism(s) and potential physiological significance have yet to be elucidated.

BIBLIOGRAPHY

"Analgesics and Antipyretics" in *ECT* 1st ed., Vol. 1, pp. 851–861, by A. W. Ruddy, Sterling-Winthrop Research Institute, Division, Sterling Drug, Inc.; in *ECT* 2nd ed., Vol. 2, pp. 379–393, by G. deStevens, Ciba Pharmaceutical Co.; "Analgesics, Antipyretics, and Anti-Inflammatory Agents" in *ECT* 3rd ed., Vol. 2, pp. 574–586, by W. F. Michne, Sterling-Winthrop Research Institute; "Opioids, Endogenous" in *ECT* 3rd ed., Suppl. Vol., pp. 574–591, by M. R. Johnson and D. A. Clark, Pfizer, Inc.

1. A. Goldstein, L. I. Lowney, and B. K. Pal, *Proc. Nat. Acad. Sci. USA* **68**, 1742 (1971).
2. L. Terenius, *Acta Pharmacol. Toxicol.* **32**, 317 (1973).
3. C. B. Pert and S. H. Snyder, *Science* **179**, 1011 (1973).
4. E. J. Simon, J. M. Hiller, and I. Edelman, *Proc. Nat. Acad. Sci. USA* **70**, 1947 (1973).
5. D. J. Mayer and J. C. Liebeskind, *Brain Res.* **68**, 73 (1974).
6. H. Akil, D. J. Mayer, and J. C. Liebeskind, *Science* **191**, 961 (1976).
7. D. J. Mayer and R. Hayes, *Science* **188**, 941 (1975).
8. M. J. Kuhar, C. B. Pert, and S. H. Snyder, *Nature* **245**, 447 (1973).
9. J. Hughes, *Brain Res.* **88**, 295 (1975).
10. J. Hughes, T. W. Smith, H. W. Kosterlitz, L. A. Fothergill, B. A. Morgan, and H. R. Morris, *Nature* **258**, 577 (1975).
11. C. H. Li, *Nature* **201**, 924 (1964).
12. C. H. Li and D. Chung, *Proc. Nat. Acad. Sci. USA* **73**, 1145 (1976).
13. A. Goldstein, S. Tachibana, L. I. Lowney, M. Hunkapillar, and L. Hood, *Proc. Nat. Acad. Sci. USA* **76**, 6666 (1979).
14. G. D. Smith and J. F. Griffin, *Science* **199**, 1214 (1978).
15. G. H. Loew and S. K. Burt, *Proc. Nat. Acad. Sci. USA* **75**, 7 (1978).
16. D. A. Price and M. J. Greenberg, *Science* **197**, 670 (1977).
17. K. J. Chang, A. Killian, E. Hazum, P. Cuatrecasas, and J. K. Chang, *Science* **212**, 75 (1981).
18. P. C. Montecucchi, R. de Castiglione, and V. Erspamer, *Int. J. Pept. Prot. Res.* **17**, 275 (1981).
19. J. E. Zadina, A. J. Kastin, L. J. Ge, and V. Brantl, *Life Sci.* **47**, PL25 (1990).
20. S. J. Watson, H. Akil, C. W. Richard, and J. D. Barchas, *Nature* **275**, 226 (1978).
21. S. Nakanishi and co-workers, *Nature* **278**, 423 (1979).
22. C. Gramsch, G. Kleber, V. Hollt, A. Pasi, P. Mehraein, and A. Herz, *Brain Res.* **192**, 109 (1980).
23. R. M. Dores, M. Jain, and H. Akil, *Brain Res.* **377**, 251 (1986).
24. J. F. W. Deakin, J. O. Dostrovsky, and D. G. Smyth, *Biochem. J.*, **189**, 501 (1980).
25. M. Noda and co-workers, *Nature* **295**, 202 (1982).
26. M. Comb, P. H. Seeburg, J. Adelman, L. Eiden, and E. Herbert, *Nature* **295**, 663 (1982).
27. L. D. Fricker, C. J. Evans, F. S. Esch, and E. Herbert, *Nature* **232**, 461 (1986).
28. H. Kakidani and co-workers, *Nature* **298**, 245 (1982).
29. J. P. Loeffler, N. Kley, C. W. Pittius, and V. Hollt, *Endocrinology* **119**, 2840 (1986).
30. H. U. Affolter and T. Reisine, *J. Biol. Chem.* **260**, 15477 (1985).
31. A. L. Boutillier, P. Sassone-Corsi, and J. P. Loeffler, *Mol. Endocrinol.* **5**, 1301 (1991).
32. J. H. Eberwine and J. L. Roberts, *J. Biol. Chem.* **259**, 2166 (1984).
33. S. Beaulieu, B. Gagne, and N. Barden, *Mol. Endocrinol.* **2**, 727 (1988).
34. J. Drouin, M. A. Trifiro, R. K. Plante, M. Nemer, P. Eriksson, and O. Wrange, *Mol. Cell. Biol.* **9**, 5305 (1989).
35. M. Therrien and J. Drouin, *Mol. Cell. Biol.* **11**, 3492 (1991).
36. M. Comb and co-workers, *EMBO J.* **7**, 3793 (1988).
37. C. E. Inturrisi and co-workers, *Mol. Endocrinol.* **2**, 633 (1988).

38. N. Kley, J. P. Loeffler, C. W. Pittius, and V. Hollt, *J. Biol. Chem.* **262**, 4083 (1987).
39. N. Kley, *J. Biol. Chem.* **263**, 2003 (1988).
40. S. E. Hyman, M. Comb, J. Pearlberg, and H. M. Goodman, *Mol. Cell. Biol.* **9**, 321 (1989).
41. L. A. Kobierski, H. M. Chu, Y. Tan, and M. J. Comb, *Proc. Nat. Acad. Sci. USA* **88**, 10222 (1991).
42. J. L. Sonnenberg, F. J. Raucher, J. I. Morgan, and T. Curran, *Science* **246**, 1622 (1989).
43. J. R. Naranjo, B. Mellstrom, M. Achaval, and P. Sassone-Corsi, *Neuron* **6**, 607 (1991).
44. M. A. Phillips, M. L. Budarf, and E. Herbert, *Biochemistry* **20**, 1666 (1981).
45. E. Herbert, M. Budarf, M. Phillips, P. Rose, P. Policastro, and E. Oates, *Ann. New York Acad. Sci.* **343**, 79 (1980).
46. Y. P. Loh and H. Gainer, *Endocrinology* **105**, 474 (1979).
47. F. Bloom, E. Battenberg, J. Rossier, N. Ling, and R. Guillemin, *Proc. Nat. Acad. Sci. USA* **75**, 1591 (1978).
48. D. M. Bronstein, M. K. H. Schafer, S. J. Watson, and H. Akil, *Brain Res.* **587**, 269 (1992).
49. W. H. Pilcher and S. A. Joseph, *Peptides* **7**, 783 (1986).
50. S. J. Watson, H. Khachaturian, H. Akil, D. H. Coy, and A. Goldstein, *Science* **218**, 1134 (1982).
51. J. H. Fallon and F. M. Leslie, *J. Comp. Neurol.* **249**, 293 (1986).
52. E. J. Glazer, H. Steinbusch, A. Verhofstad, and A. I. Basbaum, *J. Physiol. Paris* **77**, 241 (1981).
53. S. Murakami, H. Oamura, G. Pelletier, and Y. Ibata, *J. Comp. Neurol.* **281**, 532 (1989).
54. M. H. Vilijn, P. J. J. Vaysee, R. S. Zukin, and J. A. Kessler, *Proc. Nat. Acad. Sci. USA* **85**, 6551 (1988).
55. S. R. Vincent, T. Hokfelt, I. Christensson, and L. Terenius, *Neurosci. Lett.* **33**, 185 (1982).
56. S. J. Watson, H. Akil, W. Fischli, A. Goldstein, E. Zimmerman, G. Nilaver, and T. B. Van Wisersma Greidanus, *Science* **216**, 85 (1982).
57. H. Khachaturian, T. G. Sherman, R. V. Lloyd, O. Civelli, J. Douglass, E. Herbert, H. Akil, and S. J. Watson, *Endocrinology* **119**, 1409 (1986).
58. C. A. Sasek and R. P. Elde, *Brain Res.* **381**, 8 (1986).
59. F. Bloom, E. Battenberg, J. Rossier, N. Ling, J. Leppaluoto, T. M. Vargo, and R. Guillemin, *Life Sci.* **20**, 43 (1977).
60. O. H. Viveros, E. J. Diliberto, E. Hazum, and K. J. Chang, *Mol. Pharmacol.* **16**, 1101 (1979).
61. M. K. H. Schafer, R. Day, M. R. Ortega, H. Akil, and S. J. Watson, *Neuroendocrinology* **51**, 444 (1990).
62. C. R. DeBold, W. E. Nicholson, and D. N. Orth, *Endocrinology* **122**, 2648 (1988).
63. D. L. Kilpatrick and J. L. Rosenthal, *Endocrinology* **119**, 370 (1986).
64. J. Douglass, B. Cox, B. Quinn, O. Civelli, and E. Herbert, *Endocrinology* **120**, 707 (1987).
65. R. D. Howells, D. L. Kilpatrick, L. C. Bailey, M. Noe, and S. Undenfriend, *Proc. Nat. Acad. Sci. USA* **83**, 1960 (1986).
66. H. Rosen, O. Behar, O. Abramsky, and H. Ovadia, *J. Immunol.* **143**, 3703 (1989).
67. W. R. Martin, C. G. Eades, J. A. Thompson, R. E. Huppler, and P. E. Gilbert, *J. Pharmacol. Exp. Ther.* **197**, 517 (1976).
68. J. A. H. Lord, A. A. Waterfield, J. Hughes, and H. W. Kosterlitz, *Nature* **267**, 495 (1977).
69. A. D. Corbett, S. J. Paterson, A. T. McKnight, J. Magnan, and H. W. Kosterlitz, *Nature* **299**, 79 (1982).

70. H. W. Kosterlitz and S. J. Paterson, *Phil. Trans. R. Soc. Lond.* **B308**, 291 (1985).

71. R. Schulz, M. Wuster, and A. Herz, *J. Pharmacol. Exp. Ther.* **216**, 604 (1981).

72. B. Nock, A. L. Giordano, T. J. Cicero, and L. H. O'Connor, *J. Pharmacol. Exp. Ther.* **254**, 412 (1990).

73. G. W. Pasternak, *Biochem. Pharmacol.* **35**, 361 (1985).

74. A. Mattia, T. Vanderah, H. I. Mosberg, and F. Porreca, *J. Pharmacol. Exp. Ther.* **258**, 583 (1991).

75. D. Paul, C. G. Pick, L. A. Tive, and G. W. Pasternak, *J. Pharmacol. Exp. Ther.* **257**, 1 (1991).

76. C. B. Pert, J. K. Chang, and B. T. W. Fong, *Science* **194**, 330 (1976).

77. E. Wei and H. Loh, *Science* **193**, 1262 (1976).

78. G. Gacel, J. M. Zajac, P. Delay-Goyet, V. Dauge, and B. P. Roques, *J. Med. Chem.* **31**, 374 (1988).

79. K. J. Chang, E. T. Wei, A. Killian, and J. K. Chang, *J. Pharmacol. Exp. Ther.* **227**, 403 (1983).

80. P. C. Montecucchi, R. de Castiglione, and V. Erspamer, *Int. J. Pept. Protein Res.* **17**, 275 (1981).

81. S. Sagan, M. Amiche, A. Delfour, A. Mor, A. Camus, and P. Nicholas, *J. Biol. Chem.* **264**, 17100 (1989).

82. P. W. Schiller, T. M. D. Nguyen, N. N. Chung, and C. Lemieux, *J. Med. Chem.* **32**, 698 (1989).

83. P. W. Schiller, T. M. D. Nguyen, J. DiMaio, and C. Lemieux, *Life Sci.* **33**, 319 (1983).

84. J. DiMaio and P. W. Schiller, *Proc. Nat. Acad. Sci. USA* **77**, 7162 (1980).

85. P. W. Schiller and J. DiMaio, *Nature* **297**, 74 (1982).

86. A. Polinsky, M. G. Cooney, A. Toy-Palmer, G. Osapay, and M. Goodman, *J. Med. Chem.*, 4185 (1992).

87. P. W. Schiller, T. M. D. Nguyen, L. A. Maziak, B. C. Wilkes, and C. Lemieux, *J. Med. Chem.* **30**, 2094 (1987).

88. H. I. Mosberg, R. Hurst, V. J. Hruby, K. Gee, H. I. Yamamura, J. I. Galligan, and T. F. Burks, *Proc. Nat. Acad. Sci. USA* **80**, 5871 (1983).

89. H. I. Mosberg, J. R. Omnaas, and A. Goldstein, *Mol. Pharmacol.* **31**, 599 (1987).

90. G. Thót, T. H. Kramer, R. Knapp, G. Lui, P. Davis, T. F. Burks, H. I. Yamamura, and V. J. Hruby, *J. Med. Chem.* **33**, 249 (1990).

91. C. R. Beddell and co-workers, *Proc. R. Soc. Lond. (Biol.)* **198**, 249 (1977).

92. J. M. Zajac, G. Gacel, F. Petit, P. Dodey, P. Rossignol, and B. P. Roques, *Biochem. Biophys. Res. Commun.* **111**, 390 (1983).

93. G. Gacel, V. Dauge, P. Breuze, P. Delay-Goyet, and B. P. Roques, *J. Med. Chem.* **31**, 1891 (1988).

94. G. Gacel, E. Fellion, A. Baamonde, V. Dauge, and B. P. Roques, *Peptides* **11**, 983 (1990).

95. M. Amiche, S. Sagan, A. Mor, A. Delfour, and P. Nicholas, *Mol. Pharmacol.* **35**, 774 (1989).

96. V. Erspamer and co-workers, *Proc. Nat. Acad. Sci. USA* **86**, 5188 (1989).

97. S. Lemaire, L. Lafrance, and M. Dumont, *Int. J. Pept. Protein Res.* **27**, 300 (1986).

98. J. E. Gairin, C. Gouarderes, H. Mazarguil, P. Alvinerie, and J. Cros, *Eur. J. Pharmacol.* **106**, 457 (1984).

99. C. C. Yang and J. W. Taylor, in J. E. Rivier and G. R. Marshall, eds., *Peptides: Chemistry, Structure, Biology. Proceedings of the 11th American Peptide Symposium*, ESCOM, Leiden, Germany, 1990, p. 346.

100. A. M. Kawasaki and co-workers, in Ref. 99, p. 337.

101. P. W. Schiller, T. M. D. Nguyen, and C. Lemieux, *Tetrahedron* **44**, 733 (1988).

102. R. Maurer, B. H. Gaehwiler, H. H. Buescher, R. C. Hill, and D. Roemer, *Proc. Nat. Acad. Sci. USA* **79**, 4815 (1982).

103. W. Kazmierski and co-workers, *J. Med. Chem.* **31**, 2170 (1988).
104. J. T. Pelton, K. Gulya, V. J. Hruby, S. P. Duckles, and H. I. Yamamura, *Proc. Nat. Acad. Sci. USA* **82**, 236 (1985).
105. A. D. Corbett and co-workers, *Brit. J. Pharmacol.* **83**, 271 (1984).
106. C. W. Thornber, J. S. Shaw, L. Miller, and C. F. Hayward, *Natl. Inst. Drug Abuse Res. Monogr. Ser.* **75**, 181 (1986).
107. J. E. Gairin and co-workers, *Brit. J. Pharmacol.* **95**, 1023 (1988).
108. P. W. Schiller and co-workers, *Proc. Nat. Acad. Sci. USA* **89**, 11871 (1992).
109. P. W. Schiller and co-workers, *J. Med. Chem.* **36**, 3182 (1993).
110. P. S. Portoghese, M. Sultana, H. Nagase, and A. E. Takemori, *J. Med. Chem.* **31**, 281 (1988).
111. P. S. Portoghese, M. Sultana, and A. E. Takemori, *J. Med. Chem.* **33**, 1714 (1990).
112. M. Sofuoglu, P. S. Portoghese, and A. E. Takemori, *Life Sci.* **52**, 769 (1993).
113. K. J. Chang, C. Rigdon, J. L. Howard, and R. W. McNutt, *J. Pharmacol. Exp. Ther.* **267**, 852 (1993).
114. S. R. Childers and co-workers, *Mol. Pharmacol.* **44**, 827 (1993).
115. M. F. Piercey, R. A. Lahti, L. A. Schroeder, F. J. Einspar, and C. Brashun, *Life Sci.* **31**, 1197 (1982).
116. P. S. Portoghese, A. W. Lipkowski, and A. E. Takemori, *Life Sci.* **40**, 1287 (1987).
117. L. Birnbaumer, J. Abramowitz, and A. M. Brown, *Biochim. Biophys. Acta* **1031**, 163 (1990).
118. S. R. Childers, *Life Sci.* **48**, 1991 (1991).
119. J. Hescheler, W. Rosenthal, W. Trautwein, and G. Schultz, *Nature* **325**, 445 (1987).
120. R. A. North, J. T. Williams, A. Suprenant, and J. C. McDonald, *Proc. Nat. Acad. Sci. USA* **84**, 5487 (1987).
121. K. F. Shen and S. M. Crain, *Brain Res.* **491**, 227 (1989).
122. A. R. Gintzler and H. Xu, *Proc. Nat. Acad. Sci. USA* **88**, 4741 (1991).
123. L. M. Fleming, G. Ponjee, and S. R. Childers, *J. Pharmacol. Exp. Ther.* **260**, 1416 (1992).
124. S. R. Childers, in A. Herz, ed., *Opioids I*, Springer-Verlag (Berlin), 1993, p. 189.
125. R. A. North, in Ref. 124, p. 773.
126. C. J. Evans, D. E. Keith, H. Morrison, K. Magendzo, and R. H. Edwards, *Science* **258**, 1952 (1992).
127. B. L. Kieffer, K. Befort, C. Gaveriaux-Ruff, and C. G. Hirth, *Proc. Nat. Acad. Sci. USA* **89**, 12048 (1992).
128. G. R. Uhl, S. R. Childers, and G. Pasternak, *Trends Neurosci.* **17**, 89 (1994).
129. Y. Chen, A. Mestak, J. Liu, J. A. Hurley, and L. Yu, *Mol. Pharmacol.* **44**, 8 (1993).
130. R. C. Thompson, A. Mansour, H. Akil, and S. J. Watson, *Neuron* **11**, 903 (1993).
131. G. X. Xie, A. Miyajima, and A. Goldstein, *Proc. Nat. Acad. Sci. USA* **89**, 4124 (1992).
132. S. Li and co-workers, *Biochem. J.* **295**, 629 (1993).
133. F. Meng and co-workers, *Proc. Nat. Acad. Sci. USA* **90**, 9954 (1993).
134. M. Minami and co-workers, *FEBS Lett.* **329**, 291 (1993).
135. M. Nishi, H. Takeshima, K. Fukuda, S. Kato, and K. Mori, *FEBS Lett.* **330**, 77 (1993).
136. Y. Chen, A. Mestak, J. Liu, and L. Yu, *Biochem. J.* **295**, 625 (1993).
137. K. Yasuda and co-workers, *Proc. Nat. Acad. Sci. USA* **90**, 6736 (1993).
138. K. Fukuda, S. Kato, K. Mori, M. Nishi, and H. Takeshima, *FEBS Lett.* **327**, 311 (1993).
139. K. Raynor, H. Kong, Y. Chen, K. Yasuda, L. Yu, G. I. Bell, and T. Reisine, *Mol. Pharmacol.* **45**, 330 (1993).
140. A. Herz, K. Albus, J. Metys, P. Schubert, and H. Teschemacher, *Neuropharmacology* **9**, 539 (1970).
141. H. H. Loh, L. F. Tseng, E. Wei, and C. H. Li, *Proc. Nat. Acad. Sci. USA* **73**, 2895 (1976).

142. W. Feldberg and D. G. Smyth, *Brit. J. Pharmacol.* **60**, 445 (1977).

143. M. Meglio, Y. Hosobuchi, H. H. Loh, J. E. Adams, and C. H. Li, *Proc. Nat. Acad. Sci. USA* **74**, 774 (1977).

144. H. Akil, D. J. Mayer, and J. C. Liebeskind, *Science* **191**, 961 (1975).

145. D. J. Mayer and R. L. Hayes, *Science* **188**, 941 (1975).

146. J. W. Lewis, J. T. Cannon, and J. C. Liebeskind, *Science* **208**, 623 (1980).

147. J. M. Van Ree, D. DeWied, A. F. Bradbury, E. Hulme, D. G. Smyth, and C. R. Snell, *Nature* **264**, 792 (1976).

148. M. Knight and W. A. Klee, *J. Biol. Chem.* **253**, 3842 (1978).

149. H. H. Loh, L. F. Tseng, E. Wei, and C. H. Li, *Proc. Nat. Acad. Sci. USA* **73**, 2895 (1976).

150. H. H. Buscher, R. C. Hill, D. Romber, F. Cardinaux, A. Closse, D. Hauser, and J. Pless, *Nature* **261**, 423 (1976).

151. E. J. Glazer and A. I. Basbaum, *J. Comp. Neurol.* **196**, 377 (1981).

152. H. L. Fields, P. C. Emson, B. K. Leigh, R. F. T. Gilbert, and L. L. Iversen, *Nature* **284**, 351 (1980).

153. C. Lamotte, C. B. Pert, and S. H. Snyder, *Brain Res.* **112**, 407 (1976).

154. T. L. Yaksh and R. P. Elde, *J. Neurophysiol.* **46**, 1056 (1981).

155. S. Oshita, T. L. Yaksh, and R. Chipkin, *Brain Res.* **515**, 143 (1990).

156. A. W. Duggan, J. G. Hall, and P. M. Headley, *Brit. J. Pharmacol.* **61**, 399 (1977).

157. T. L. Yaksh, T. M. Jessell, R. Gamse, A. W. Mudge, and S. E. Leeman, *Nature* **286**, 155 (1980).

158. E. Weihe, M. J. Millan, V. Hollt, D. Nohr, and A. Herz, *Neuroscience* **31**, 77 (1989).

159. F. C. Tulunay, M. F. Jen, J. K. Chang, H. H. Loh, and N. M. Lee, *J. Pharmacol. Exp. Ther.* **219**, 296 (1981).

160. J. M. Walker, D. E. Tucker, D. H. Coy, B. B. Walker, and H. Akil, *Eur. J. Pharmacol.* **85**, 121 (1982).

161. E. Wei and H. Loh, *Science* **193**, 1262 (1976).

162. A. A. Waterfield, J. Hughes, and H. W. Kosterlitz, *Nature* **260**, 624 (1976).

163. J. M. Van Ree, D. G. Smyth, and F. C. Copaert, *Life Sci.* **24**, 495 (1979).

164. R. Bals-Kubik, T. Shippenberg, and A. Herz, *Eur. J. Pharmacol.* **175**, 63 (1990).

165. J. D. Belluzzi and L. Stein, *Nature* **266**, 556 (1977).

166. S. Iyengar, H. S. Kim, M. R. Marien, D. McHugh, and P. L. Wood, *Neuropharmacology* **28**, 123 (1989).

167. A. G. Phillips and F. G. LePaine, *Pharmacol. Biochem. Behav.* **12**, 965 (1980).

168. F. C. Colpaert, C. J. E. Niemegeers, P. A. J. Janssen, and J. M. Van Ree, *Eur. J. Pharmacol.* **47**, 115 (1978).

169. M. A. Petty, W. DeJong, and D. DeWeid, *Life Sci.* **30**, 1835 (1982).

170. I. R. Moss and E. Friedman, *Life Sci.* **23**, 1271 (1978).

171. M. W. Adler, E. B. Geller, T. J. Rogers, E. E. Henderson, and T. K. Eisenstein, in H. Friedman and co-workers, eds., *Drugs of Abuse, Immunity, and AIDS*, Plenum Press, New York, 1993, p. 13.

172. P. M. Matthews, C. J. Froelich, W. L. Sibbitt, and A. D. Bankhurst, *J. Immunol.* **130**, 1658 (1983).

173. S. A. Williamson, R. A. Knight, S. L. Lightman, and J. R. Hobbs, *Brain Behav. Immun.* **1**, 329 (1987).

174. B. J. Meyerson and L. Terenius, *Eur. J. Pharmacol.* **42**, 191 (1977).

175. B. P. Quarantotti, M. G. Corda, E. Paglietti, G. Biggio, and G. L. Gessa, *Life Sci.* **23**, 673 (1978).

176. M. Blum, J. L. Roberts, and S. Wardlaw, *Endocrinology* **124**, 2283 (1989).

177. A. H. Lauber, G. J. Romano, C. V. Mobbs, R. D. Howells, and D. W. Pfaff, *Mol. Brain Res.* **8**, 47 (1990).

178. C. M. Handler, E. B. Geller, and M. W. Adler, *Pharmacol. Biochem. Behav.* **43**, 1209 (1992).

179. J. M. Hambrook, B. A. Morgan, M. J. Rance, and C. F. C. Smith, *Nature* **262**, 782 (1976).

180. C. Gros, B. Giros, and J. C. Schwartz, *Biochemistry* **24**, 2179 (1985).

181. C. R. Beddell and co-workers, *Proc. Roy. Soc. London Ser. B* **198**, 249 (1977).

182. C. Gorenstein and S. H. Snyder, *Life Sci.* **25**, 2065 (1979).

183. B. Malfroy, J. P. Swerts, A. Guyon, B. P. Roques, and J. C. Schwartz, *Nature* **276**, 523 (1978).

184. R. Matsas, I. S. Fulcher, A. J. Kenny and A. J. Turner, *Proc. Nat. Acad. Sci. USA* **80**, 3111 (1983).

185. M. C. Fournié-Zaluski and co-workers, *Biochem. Biophys. Res. Commun.* **91**, 130 (1979).

186. G. Waksman, E. Hamel, M. C. Fournié-Zaluski, and B. P. Roques, *Proc. Nat. Acad. Sci. USA* **83**, 1523 (1986).

187. M. Pollard and co-workers, *Neuroscience* **30**, 339 (1989).

188. B. P. Roques and co-workers, *Nature* **288**, 286 (1980).

189. D. Roemer and co-workers, *Nature* **268**, 547 (1977).

190. E. G. Erdos, A. R. Johnson, and N. T. Boyden, *Biochem. Pharmacol.* **27**, 843 (1978).

191. A. Arregui, C.-M. Lee, P. Emson, and L. L. Iversen, *Eur. J. Pharmacology* **59**, 141 (1979).

192. C. Gorenstein and S. H. Snyder, *Proc. Roy. Soc. London Ser.* **B210**, 123 (1980).

193. B. P. Roques and M. C. Fournié-Zaluski, in R. S. Rapaka and R. L. Hawks, eds., *Opioid Peptides: Molecular Pharmacology, Biosynthesis and Analysis*, NIDA Research Monograph 70, 1986, p. 128.

194. C. Llorens and co-workers, *Biochem. Biophys. Res. Commun.* **96**, 1710 (1980).

195. R. L. Hudgin, S. E. Charlson, M. Zimmerman, R. Mumford, and P. L. Wood, *Life Sci.* **29**, 2593 (1981).

196. J. F. Hernandez, J. M. Soleihac, B. P. Roques, and M. C. Fournié-Zaluski, *J. Med. Chem.* **31**, 1825 (1988).

197. S. Algeri, M. Altstein, S. Blumberg, G. M. de Simoni and V. Guardabasso, *Eur. J. Pharmacol.* **74**, 261 (1981).

198. W. W. C. Chan, *Biochem. Biophys. Res. Commun.* **116**, 297 (1983).

199. C. Gros, B. Giros, J. C. Schwartz, A. Vlaiculescu, J. Costentin, and J. M. Lecomte, *Neuropeptides* **12**, 111 (1988).

200. F. Noble, J. M. Soleihac, E. Lucas-Soroca, S. Turcaud, M. C. Fournié-Zaluski, B. P. Roques, *J. Pharmacol. Exp. Ther.* **261**, 181 (1992).

201. B. P. Roques, *Biochem. Soc. Trans.* **21**, 678 (1993).

202. B. M. Austen, D. G. Smyth, and C. R. Snell, *Nature* **269**, 619 (1977).

203. S. Zakarian and D. Smyth, *Proc. Nat. Acad. Sci. USA* **76**, 5972 (1979).

204. K. Oka, J. D. Kantrowitz, and S. Spector, *Proc. Nat. Acad. Sci. USA* **82**, 1852 (1985).

205. C. J. Weitz, L. I. Lowney, K. F. Faull, G. Feistner, and A. Goldstein, *Proc. Nat. Acad. Sci. USA* **85**, 5335 (1988).

General References

A. Herz, ed., *Opioids I*, Springer-Verlag, New York, 1993.

J. M. Van Ree, A. H. Mulder, V. M. Weigant, and T. B. Van Wimersma Greidanus, eds., *New Leads in Opioid Research*, Excerpta Medica, Amsterdam, the Netherlands, 1990.

G. W. Pasternak, ed., *The Opiate Receptors*, Humana Press, Newark, N.J., 1988.

DANA E. SELLEY
LAURA J. SIM
STEVEN R. CHILDERS
Wake Forest University

OPTICAL BLEACHES.   See FLUORESCENT WHITENING AGENTS.

OPTICAL DISPLAYS.   See SUPPLEMENT.

OPTICAL FILTERS.   See SUPPLEMENT.

ORGANOLEPTIC TESTING.   See FLAVOR CHARACTERIZATION; ODOR MODIFICATION; PERFUMES.

ORGANOMETALLICS.   See SUPPLEMENT.

OSMIUM.   See PLATINUM-GROUP METALS.

OSMOSIS and OSMOTIC PRESSURE.   See HOLLOW-FIBER MEMBRANES; MEMBRANE TECHNOLOGY; REVERSE OSMOSIS.

OURICUI OIL.   See FATS AND FATTY OILS.

OXALIC ACID

Oxalic acid [144-62-7], HOOC–COOH, or ethanedioic acid, mol wt 90.04, is the simplest dicarboxylic acid. It is soluble in water, and acts as a strong acid (1). This acid does not exist in anhydrous form in nature and is available commercially as a solid dihydrate [6153-56-6], $C_2H_2O_4 \cdot 2H_2O$, mol wt 126.07. The commercial product is packed in polyethylene-lined paper bags or flexible containers. Anhydrous oxalic acid can be efficiently prepared from the dihydrate by azeotropic distillation in a low boiling solvent that can form a water azeotrope, such as benzene and toluene (2).

Oxalic acid was synthesized for the first time in 1776 by Scheele through the oxidation of sugar with nitric acid. Then, Wöhler synthesized it by the hydrolysis of cyanogen [460-19-5] in 1824.

The potassium or calcium salt form of oxalic acid is distributed widely in the plant kingdom. Its name is derived from the Greek *oxys*, meaning sharp or acidic, referring to the acidity common in the foliage of certain plants (notably *Oxalis* and *Rumex*) from which it was first isolated. Other plants in which oxalic acid is found are spinach, rhubarb, etc. Oxalic acid is a product of metabolism of fungi or bacteria and also occurs in human and animal urine; the calcium salt is a principal constituent of kidney stones.

Oxalic acid is used in various industrial areas, such as textile manufacture and processing, metal surface treatments (qv), leather tanning, cobalt production, and separation and recovery of rare-earth elements. Substantial quantities of oxalic acid are also consumed in the production of agrochemicals, pharmaceuticals, and other chemical derivatives.

Physical Properties

The physical and thermochemical constants of anhydrous oxalic acid and oxalic acid dihydrate are summarized in Table 1.

Anhydrous Oxalic Acid. The anhydrous form of oxalic acid is odorless and colorless. It exists in two crystal forms, ie, the rhombic or α-form and the

Table 1. Physical and Thermochemical Properties of Oxalic Acid and its Dihydrate

Property	Value
Oxalic acid, anhydrous, $C_2H_2O_4$	
melting point, °C	
α	189.5
β	182
density d_4^{17}, g/mL	
α	1.900
β	1.895
refractive index, β, n_4^{20}	1.540
vapor pressure (solid, 57–107°C), kPa[a]	$\log_{10} P = -(4726.95/T)$
	$+11.3478$
specific heat (solid, −200 to 50°C), J/g	$C_p^b = 1.084 + 0.0318\,t$
heat of combustion, ΔE_c (at 25°C), kJ/mol[c]	−245.61
standard heat of formation, ΔH_f (at 25°C), kJ/mol[c]	−826.78
standard free energy of formation, ΔG_f (at 25°C), kJ/mol[c]	−697.91
heat of solution (in water), kJ/mol[c]	−9.58
heat of sublimation, kJ/mol[c]	90.58
heat of decomposition, kJ/mol[c]	826.78
specific entropy, S (at 25°C), J/(mol·K)[c]	120.08
logarithm of equilibrium constant, $\log_{10} K_f$	122.28
thermal conductivity (at 0°C), W/(m·K)[d]	0.9
ionization constant	
K_1	6.5×10^{-2}
K_2	6.0×10^{-5}
coefficient of expansion (at 25°C), nL/(g·K)	178.4
Oxalic acid dihydrate, $C_2H_2O_4\cdot2H_2O$	
mp, °C	101.5
density d_4^{20}, g/mL	1.653
refractive index, n_4^{20}	1.475
standard heat of formation, ΔH_f (at 18°C), kJ/mol[c]	−1422
heat of solution (in water), kJ/mol[c]	−35.5
pH (0.1 M soln)	1.3

[a]To convert $\log_{10} P_{kPa}$ to $\log_{10} P_{mm\,Hg}$, add 0.875097 to the constant, T = K.
[b]To convert C_p, J/g, to C_p, cal/g, divide both terms of the equation by 4.184.
[c]To convert J to cal, divide by 4.184.
[d]To convert W/(m·K) to (Btu·in.)/(h·ft^2·°F), divide by 0.1441.

monoclinic or β-form (3). The rhombic crystal is thermodynamically stable at room temperature, but the monoclinic form is metastable or slightly stable. The main difference between the rhombic and monoclinic forms exists in the melting points which are 189.5 and 182°C, respectively (Table 1).

Anhydrous oxalic acid normally melts and simultaneously decomposes at 187°C. Sublimation starts at slightly below 100°C and proceeds rapidly at 125°C; partial decomposition takes place during sublimation at 157°C. Anhydrous oxalic acid is hygroscopic and thus absorbs moisture in the air to form the dihydrate.

Anhydrous oxalic acid is very soluble in polar solvents. The ionization constant K_1 is comparable with those of many mineral acids and is exceeded only by those of a few organic acids; K_2 is approximately the same as the ionization constant of benzoic acid (see Table 1).

Oxalic Acid Dihydrate. Oxalic acid dihydrate is made up of odorless, colorless, monoclinic prisms or granules which contain 71.42 wt % anhydrous oxalic acid and 28.58 wt % water.

When the dihydrate is carefully heated to 100°C it loses its water to give anhydrous oxalic acid. On the other hand, when the dihydrate is heated rapidly or in a sealed tube, it melts at 101.5°C.

The dihydrate is soluble in water. The specific gravities of aqueous oxalic acid solutions are summarized in Table 2. The solubility of the dihydrate in water increases with temperature. Approximate solubility values (S) are given by the following formulas, where $S =$ g $(COOH)_2/100$ g soln and $t =$ °C.

$$0-60°C, \quad S = 3.42 + 0.168\,t + 0.0048\,t^2$$

$$50-90°C, \quad S = 0.33\,t + 0.003\,t^2$$

The dihydrate is very soluble in polar solvents, such as methanol, ethanol, acetone, dioxane, and tetrahydrofuran, but insoluble in benzene, chloroform,

Table 2. Specific Gravities of Various Aqueous Solutions of Oxalic Acid Dihydrate

Weight of dihydrate in soln, g/L[a]	Dihydrate, wt %	Sp gr	Baumé,[b] degrees	Twaddell,[c] degrees
10.04	1	1.0035	0.5	0.70
20.14	2	1.0070	1.0	1.40
30.32	3	1.0105	1.5	2.10
40.56	4	1.0140	2.0	2.80
50.88	5	1.0175	2.5	3.50
61.26	6	1.0210	3.0	4.20
71.72	7	1.0245	3.5	4.90
82.24	8	1.0280	4.0	5.60
92.84	9	1.0315	4.4	6.30
103.5	10	1.0350	4.9	7.00
114.2	11	1.0385	5.4	7.70
125.0	12	1.0420	5.8	8.40
135.9	13	1.0455	6.3	9.10

[a]To convert g/L to lb/gal, divide by 119.8. To convert g/L to lb/ft^3, divide by 16.02.
[b]Sp gr = 145 ÷ (145 − n), where n is Baumé, degrees.
[c]Sp gr = 1.000 + 0.005 n, where n is Twaddell, degrees.

and petroleum ether. Solubility of the dihydrate in diethyl ether (1.47 g/100 g solvent) is different from that of the anhydrous form (23.6 g/100 g solvent).

Reactions

The reactions of oxalic acid, including the formation of normal and acid salts and esters, are typical of the dicarboxylic acids class. Oxalic acid, however, does not form an anhydride.

On rapid heating, oxalic acid decomposes to formic acid, carbon monoxide, carbon dioxide, and water (qv). When it is heated in 96 wt % glycerol solution at 88–121°C, the presence of formic acid in the decomposed product tends to accelerate the decomposition reaction. Formic acid is thus decomposed further to carbon dioxide and water (4). In aqueous solution, it is decomposed by uv, x-ray, or γ-radiation with the liberation of carbon dioxide. Photodecomposition also occurs in the presence of uranyl salts.

Oxalic acid is a mild reducing agent, and is oxidized by potassium permanganate in acid solution to give carbon dioxide and water. This autocatalytic reaction is of great importance in volumetric analysis. Oxalic acid is catalytically reduced by hydrogen in the presence of ruthenium catalyst to ethylene glycol (5), and electronically reduced to glyoxylic acid.

Oxalic acid reacts with various metals to form metal salts, which are quite important as the derivatives of oxalic acid. It also reacts easily with alcohols to give esters. Crystalline dimethyl oxalate is, for example, produced by the reaction of oxalic acid dihydrate and methanol under reflux for a few hours. When oxalic acid is treated with phosphorus pentachloride, oxalyl chloride, ClCOCOCl, is formed (6).

Manufacture

Many industrial processes have been employed for the manufacture of oxalic acid since it was first synthesized. The following processes are in use worldwide: oxidation of carbohydrates, the ethylene glycol process, the propylene process, the dialkyl oxalate process, and the sodium formate process.

Nitric acid oxidation is used where carbohydrates, ethylene glycol, and propylene are the starting materials. The dialkyl oxalate process is the newest, where dialkyl oxalate is synthesized from carbon monoxide and alcohol, then hydrolyzed to oxalic acid. This process has been developed by UBE Industries in Japan as a CO coupling technology in the course of exploring C-1 chemistry.

The sodium formate process is comprised of six steps: (1) the manufacture of sodium formate from carbon monoxide and sodium hydroxide, (2) manufacture of sodium oxalate by thermal dehydrogenation of sodium formate at 360°C, (3) manufacture of calcium oxalate (slurry), (4) recovery of sodium hydroxide, (5) decomposition of calcium oxalate where gypsum is produced as a by-product, and (6) purification of crude oxalic acid. This process is no longer economical in the leading industrial countries. UBE Industries (Japan), for instance, once employed this process, but has been operating the newest dialkyl oxalate process since 1978. The sodium formate process is, however, still used in China.

Oxidation of Carbohydrates.   Oxalic acid is prepared by the oxidation of carbohydrates (7–9), such as glucose, sucrose, starch, dextrin, molasses, etc, with nitric acid (qv). The choice of the carbohydrate raw material depends on availability, economics, and process operating characteristics. Among the various raw materials considered, corn starch (or starch in general) and sugar are the most commonly available. For example, tapioka starch is the Brazilian raw material, and sugar is used in India.

The oxidation of carbohydrates is the oldest method for oxalic acid manufacture. The reaction was discovered by Scheele in 1776, but was not successfully developed as a commercial process until the second quarter of the twentieth century. Technical advances in the manufacture of nitric acid, particularly in the recovery of nitrogen oxides in a form suitable for recycle, enabled its successful development. Thus 150 t of oxalic acid per month was produced from sugar by I. G. Farben (Germany) by the end of World War II.

Monosaccharides such as glucose and fructose are the most suitable as starting materials. When starch is used, it is first hydrolyzed with oxalic acid or sulfuric acid into a monosaccharide, mainly glucose. It is then oxidized with nitric acid in an approximately 50% sulfuric acid solution at 63–85°C in the presence of a mixed catalyst of vanadium pentoxide and iron(III) sulfate.

$$C_6H_{12}O_6 + 12\ HNO_3 \xrightarrow{V_2O_5/Fe^{3+}} 3(COOH)_2 \cdot 2H_2O + 3\ H_2O + 3\ NO + 9\ NO_2$$

$$4\ C_6H_{12}O_6 + 18\ HNO_3 + 3\ H_2O \xrightarrow{V_2O_5/Fe^{3+}} 12(COOH)_2 \cdot 2H_2O + 9\ N_2O$$

The Allied process (Fig. 1) is a typical example of the oxidation of carbohydrates. However, Allied Corp. itself has stopped the production of oxalic acid.

Oxalic acid manufacture via the oxidation of carbohydrates is still actively pursued, especially in China (10–12). In India, processes which produce silica and oxalic acid have been developed (13,14). The raw materials include agricultural wastes, such as rice husks, nut shell flour, corn cobs, baggase, straw, etc.

Ethylene Glycol Process.   Oxalic acid is also prepared by the nitric acid oxidation of ethylene glycol (15–21), and the process is basically the same as in the case of carbohydrates except for the absence of the hydrolyzer (see Fig. 1). In this process, ethylene glycol is oxidized in a mixture of 30–40% sulfuric acid and 20–25% nitric acid in the presence of 0.001–0.1% vanadium pentoxide at 50–70°C to give oxalic acid at more than 93% yield.

An improved process has been developed in Japan by Mitsubishi Gas Chemical, who produces 12,000 t/yr of oxalic acid by this process. Ethylene glycol is oxidized in ca 60% nitric acid at 0.3 MPa (43.5 psi), 80°C with oxygen (16,20). An initiator, such as $NaNO_2$, may be used to help the generation of nitrogen oxides, and a promoter, such as vanadium compounds or sulfuric acid, also may be employed to accelerate the oxidation reaction. The yield of oxalic acid is 90%. The reaction proceeds according to the following equations. Neither nitrogen nor N_2O, that cannot be recovered as nitric acid, is produced in this method.

$$(CH_2OH)_2 + 4\ NO_2 \longrightarrow (COOH)_2 + 4\ NO + 2\ H_2O$$

$$4\ NO + 2\ O_2 \longrightarrow 4\ NO_2$$

Overall : $(CH_2OH)_2 + 2\ O_2 \longrightarrow (COOH)_2 + 2\ H_2O$

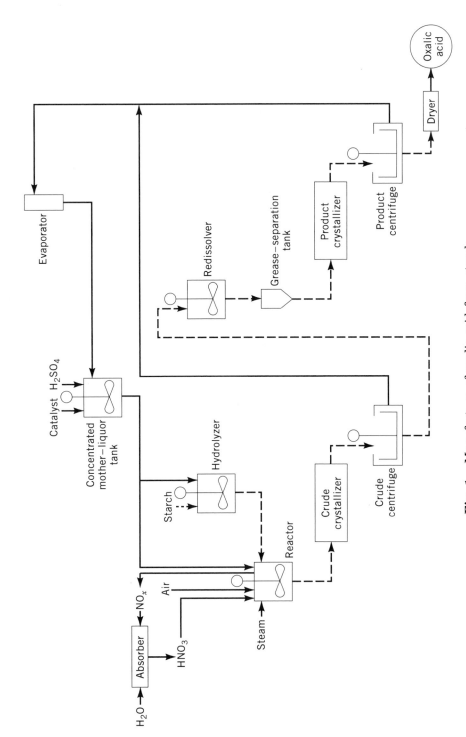

Fig. 1. Manufacture of oxalic acid from starch.

887

Propylene Process.   The oxidation of propylene with nitric acid is a two-step process (22–29). Propylene reacts with liquid NO_2 to produce an intermediate, α-nitratolactic acid, in the first step; the intermediate is then oxidized at higher temperature to form oxalic acid by mixed acid, ie, nitric acid and sulfuric acid. This process does, however, have shortcomings. Because of the use of sulfuric acid, severe equipment corrosion occurs at the oxidation mother liquor recovery step and the concentration step. Nitrogen dioxide is not readily available, and the second oxidation reaction is slow, so that the reactor must be large; during the formation of α-nitratolactic acid, unstable by-products may form which can lead to runaway decomposition or explosion.

Rhône-Poulenc (France) developed a modified version of the process for making either oxalic acid or lactic acid, or both, from propylene. In 1978, 65,000 t/yr of oxalic acid was produced worldwide by this process, although in the 1990s this process is operated only by Rhône-Poulenc. Oxidation reactions of the Rhône-Poulenc process are as follows.

$$CH_3CH{=}CH_2 + 3\,HNO_3 \longrightarrow \underset{\underset{ONO_2}{|}}{CH_3CHCOOH} + 2\,NO + 2\,H_2O$$

$$\underset{\underset{ONO_2}{|}}{CH_3CHCOOH} + \tfrac{5}{2}\,O_2 \longrightarrow (COOH)_2 + CO_2 + HNO_3 + H_2O$$

In the first step, propylene is introduced at 10–40°C into nitric acid, the concentration of which is kept at 50–75 wt % and molar ratio to propylene at 0.01–0.5, and converted into α-nitratolactic acid and lactic acid. α-Nitratolactic acid is oxidized by oxygen in the second step in the presence of a catalyst at 45–100°C to produce oxalic acid dihydrate. The overall yield based on propylene is greater than 90% and the conversion of propylene, 77.5%. The outline of the process is shown in Figure 2. The Rhône-Poulenc process can be characterized by the coproduction of lactic acid.

Dialkyl Oxalate Process.   Oxalic acid is prepared by the hydrolysis of diesters of oxalic acid which are prepared by an oxidative CO coupling reaction. UBE Industries (Japan) commercialized this two-step process in 1978. This is the newest manufacturing process of oxalic acid.

Dialkyl oxalates can be prepared by oxidative CO coupling in the presence of alcohols. The first reported example of the synthesis was in a $PdCl_2$–$CuCl_2$ redox system (30,31).

$$2\,CO + 2\,ROH + PdCl_2 \longrightarrow (COOR)_2 + 2\,HCl + Pd^0$$

$$Pd^0 + 2\,CuCl_2 \longrightarrow PdCl_2 + Cu_2Cl_2$$

$$Cu_2Cl_2 + 2\,HCl + \tfrac{1}{2}\,O_2 \longrightarrow 2\,CuCl_2 + H_2O$$

Overall : $2\,CO + 2\,ROH + \tfrac{1}{2}\,O_2 \longrightarrow (COOR)_2 + H_2O$

This method, however, is not industrially practical because a large amount of dehydrating agent, such as ethyl orthoformate, is required to remove water

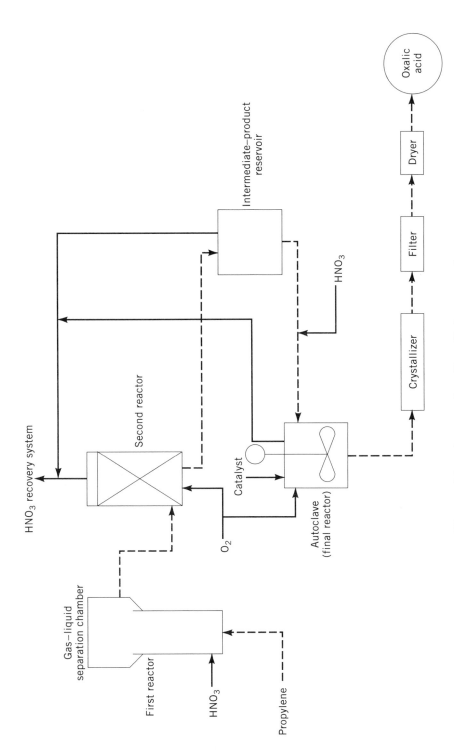

Fig. 2. Manufacture of oxalic acid from propylene.

889

formed in the reaction. Because water is an inhibitor of the reaction, the reaction system has to be kept under substantially anhydrous conditions.

UBE Industries, Ltd. has improved the basic method (32–48). In the UBE process, dialkyl oxalate is prepared by oxidative CO coupling in the presence of alkyl nitrite and a palladium catalyst.

$$2 \text{ CO} + 2 \text{ RONO} \xrightarrow{\text{Pd/C}} (\text{COOR})_2 + 2 \text{ NO}$$

$$2 \text{ NO} + \tfrac{1}{2} \text{ O}_2 + 2 \text{ ROH} \longrightarrow 2 \text{ RONO} + \text{H}_2\text{O}$$

Overall : $2 \text{ CO} + 2 \text{ ROH} + \tfrac{1}{2} \text{ O}_2 \xrightarrow{\text{Pd/C}} (\text{COOR})_2 + \text{H}_2\text{O}$

The dialkyl oxalate thus prepared is hydrolyzed to oxalic acid and the corresponding alcohol.

$$(\text{COOR})_2 + 2 \text{ H}_2\text{O} \longrightarrow (\text{COOH})_2 + 2 \text{ ROH}$$

The alkyl nitrites employed industrially are n-butyl nitrite [544-16-1] (BN) and methyl nitrite [624-91-9] (MN). In 1978, UBE Industries constructed the first plant to use this process (6000 t/yr as oxalic acid), using BN, and the plant is still being operated satisfactorily. The advantage of BN as alkyl nitrite is that it works not only as a reaction component but also as a dehydrating agent of water formed. BN forms an azeotropic mixture with water, so that the water formed is removed by distillation from the reaction system. Other features of this process are that the catalyst system is simple and the catalyst can be easily recovered and recycled, the reaction rate is fast and selectivity of dialkyl oxalate is high, even in the presence of some water (CO_2, CH_4, etc) the catalyst activity is high and the catalyst life is long, and inexpensive construction material can be used for the equipment because the catalyst system does not contain chloride.

The liquid-phase process using BN is outlined in Figure 3 (49). The circulating solution containing BN and n-butanol, and the circulating gas containing CO and O_2 are pressurized and fed to a reactor. In the reactor, di-n-butyl oxalate (DBO) is formed while BN is consumed and equimolar NO is generated. BN is regenerated simultaneously from NO. The overall reaction is exothermic.

As for the selectivity of DBO, the higher the reaction pressure and the lower the reaction temperature, the higher the selectivity. As for the reaction rate, the higher the reaction temperature, the larger the rate. Therefore, the industrial operation of the process is conducted at 10–11 MPa (1450–1595 psi) and 90–100°C. In addition, gas circulation is carried out in order to keep the oxygen concentration below the explosion limit during the reaction, and to improve the CO utilization rate and the gas–liquid contact rate.

The reaction solution is flushed under reduced pressure after it is sent out from the column, to remove CO_2 gas formed as a by-product. The water formed is then removed from the reaction solution by azeotropic distillation with BN, and most of the resultant reaction solution is recycled to the reaction column as the circulating solution. Part of the circulating solution is taken out from the reaction system and processed further to obtain DBO. The catalyst is first filtered, then

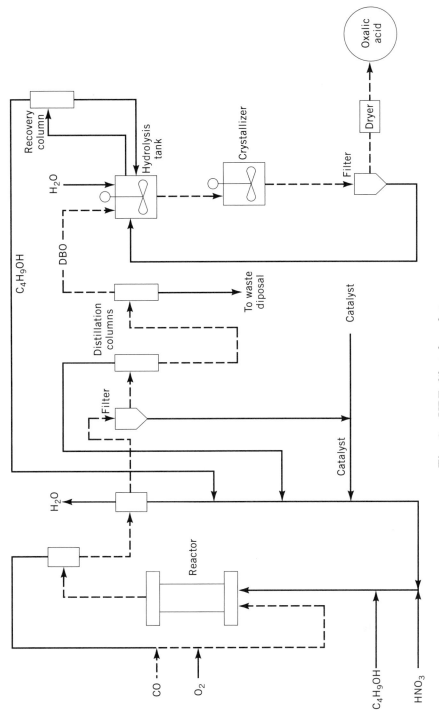

Fig. 3. UBE dibutyl oxalate process.

BN, C_4H_9OH, and by-products are removed from the resultant solution. Purified DBO is thus obtained. The catalyst, BN, and C_4H_9OH are recovered and recycled to the circulating solution. After the make-up C_4H_9OH and nitric acid are added, the circulating solution is pressurized and fed back to the reaction column.

Purified DBO is mixed with the mother liquor from the oxalic acid crystallization, and hydrolyzed at about 80°C into oxalic acid and *n*-butanol. The resultant mixture undergoes phase separation; oxalic acid dihydrate is thus crystallized from the oxalic acid solution, and C_4H_9OH is purified and recycled to the circulating solution.

In addition to the liquid-phase *n*-butyl nitrite (BN) process, UBE Industries has established an industrial gas-phase process using methyl nitrite (50–52). The outline of the process is described in Figure 4 (52). This gas-phase process is operated under lower reaction pressure (at atmospheric pressure up to 490 kPa = 71 psi) and is more economical than the liquid-phase process because of the following reasons: owing to the low pressure operation, the consumption of electricity is largely reduced (~60%); dimethyl oxalate (DMO) formation and the methyl nitrite (MN) regeneration reaction are run separately in different reaction vessels so that side-reactions are suppressed and the yield of DMO increases (98%); and methanol is less expensive than *n*-butanol.

New Synthesis.   Many attempts have been made to synthesize oxalic acid by electrochemical reduction of carbon dioxide in either aqueous or nonaqueous electrolytes (53–57). For instance, oxalic acid is prepared from CO_2 as its Zn salt in an undivided cell with Zn anodes and stainless steel cathodes in acetonitrile containing $(C_4H_9)_4NClO_4$ and current efficiency of >90% (53). Micropilot experiments and a process design were also made.

Economic Aspects

There are five processes employed for the manufacture of oxalic acid as shown in Table 3 (58). Processes are selected depending on geographical conditions.

Supply and demand of oxalic acid in the world market in 1992 are summarized in Table 4 (58). Asia has the largest production capacity, and western Europe the second largest. There is no production of oxalic acid in North America (the United States, Canada, and Mexico) and thus all needs are met by imports. In China, new plants have been constructed and existing plants have been enlarged continuously since 1986. As a consequence, the number of oxalic acid producers exceeds 20, and the total production capacity in China is estimated to be about 100,000 t/yr. China also has the world's largest domestic consumption and export of oxalic acid. India was an importing country until Punjab Chemicals & Pharmaceuticals, a subsidiary of Excel Industries, enlarged its capacity in 1990; now India exports oxalic acid. In Japan, UBE Industries and Mitsubishi Gas Chemical are the only producers. In western Europe, Rhône-Poulenc (France) and DAV (Spain) are significant producers. Although western Europe exports and imports oxalic acid mainly within the region, a fair amount of oxalic acid is also imported from China, India, Brazil, and Taiwan.

Uses of oxalic acid in each region are summarized in Table 5 (58). The demand for agrochemical/pharmaceutical production and for separation/recovery

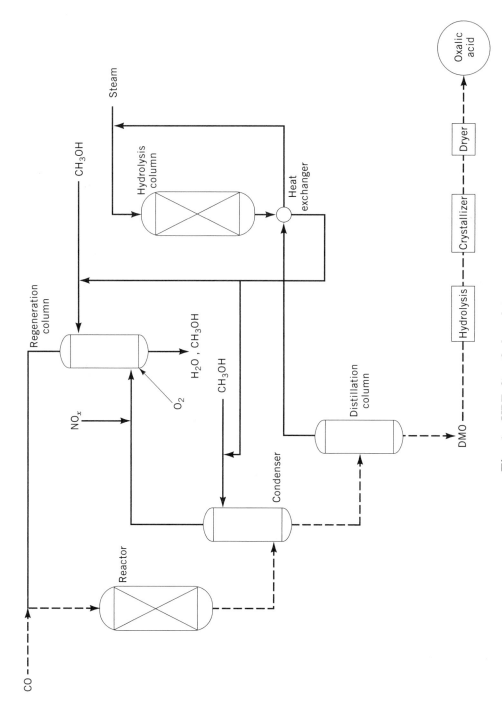

Fig. 4. UBE dimethyl oxalate process.

893

Table 3. Manufacturing Processes of Oxalic Acid

Process	Company	Location
sodium formate		China
dialkyl oxalate	UBE Industries	Japan
propylene	Rhône-Poulenc	France
ethylene glycol	Mitsubishi Gas Chemical	Japan
oxidation of carbohydrates		Brazil, China, Taiwan, India, Korea, and Spain

Table 4. Supply and Demand of Oxalic Acid in the World Market in 1992, tons

Country	Production Capacity	Production Produced	Consumption	Exports	Imports
	North America				
United States	0	0	8,000	0	8,000
Canada, Mexico	0	0	900	0	900
Total	*0*	*0*	*8,900*	*0*	*8,900*
	South America				
Brazil	7,000	2,600	1,000	1,600	0
Venezuela, etc	0	0	100	0	100
Total	*7,000*	*2,600*	*1,100*	*1,600*	*100*
	Europe				
France	8,000	5,000	4,000	1,000	0
Spain, etc	14,000	12,200	14,200	4,700	6,700
Total, Western Europe	*22,000*	*17,200*	*18,200*	*5,700*	*6,700*
Eastern Europe	10,000	9,000	9,500	3,500	4,000
	Asia				
China	100,000	60,000	50,000	10,000	0
Japan	18,000	8,600	10,300	800	2,500
India	20,000	13,000	11,500	3,000	1,500
Korea, etc	12,000	3,700	4,600	2,600	3,500
Total	*150,000*	*85,300*	*76,400*	*16,400*	*7,500*
World Total	*189,000*	*114,100*	*114,100*	*27,200*	*27,200*

of rare-earth elements in each region has been increasing. The use for marble polishing in western Europe is unique to the region.

Analytical and Test Methods

Qualitative Analysis.   Several qualitative analyses can be employed. For example, in the oxamide method (59), oxalic acid is first heated at approximately 200°C with concentrated aqueous ammonia in a sealed tube. When thiobarbituric acid is added and heated to 140°C, a condensed compound of red color forms. The analysis limit is 1.6 μg. In the diphenylamine blue method (59,60), oxalic acid is heated with diphenylamine to form a blue color, aniline blue. The analysis limit is 5 μg.

Table 5. Uses of Oxalic Acid by Region

Use	Amount, %
United States	
separation/recovery of rare-earth elements	20
textile treatment	20
metal treatment	10
bleaching agent	20
leather tanning	10
agrochemical/pharmaceutical production	20
Western Europe	
cobalt production	20
metal treatment	15
agrochemical/pharmaceutical production	15
bleaching agent	15
leather tanning	9
marble polishing	7
separation/recovery of rare-earth elements	5
salts, esters, textile treatment, etc	14
China	
separation/recovery of rare-earth elements	50
agrochemical/pharmaceutical production and metal treatment	50
Japan	
separation/recovery of rare-earth elements	35
agrochemical/pharmaceutical production	20
metal treatment	10
anodic aluminum coatings	4
millet jelly production	6
catalyst	5
salts, esters, and others	20

Quantitative Analysis. Oxalic acid is precipitated as calcium oxalate from a solution containing oxalic acid, and the calcium oxalate obtained is then weighed. If there are no organic substances other than oxalic acid present, oxalic acid can be titrated quantitatively with potassium permanganate.

Shipment, Storage, and Handling

Dry oxalic acid is packed and sold in polyethylene-lined, multilayered 25-kg paper bags or in polyethylene-lined 300–600-kg PVC flexible containers. It should be stored in a cool, dry, ventilated place. For storage of its solutions at ordinary temperature, 316 stainless steel is often used as a construction material.

Oxalic acid is not flammable but its decomposition products, both formic acid and carbon monoxide, are toxic and flammable. Its dust and mist are irritating, especially under prolonged contact. Personnel who handle oxalic acid should wear rubber gloves, aprons, protection masks or goggles, etc, to avoid skin contact and inhalation. Adequate ventilation also should be provided in areas in which oxalic acid dust fumes are present.

Because oxalic acid is toxic and corrosive, neither its crystals nor its solutions should be discarded to the environment without proper treatment. The common treatment methods are acidification, neutralization, and incineration. When oxalic acid is heated slightly in sulfuric acid, it is converted to carbon monoxide, carbon dioxide, and water. Reaction with acid potassium permanganate converts it to carbon dioxide. Neutralization with alkalies, such as caustic soda, yields soluble oxalates. Neutralization with lime gives practically insoluble calcium oxalate, which can be safely disposed of, for instance, by incineration.

Health and Safety Factors

Oxalic acid is caustic and corrosive to humans. The severity of symptoms associated with oxalic acid poisoning is related to the concentration and quantity ingested. With dilute solutions, burning gastrointestinal pain which is experienced with ingestion of oxalic acid crystals may not occur. Ingestion of a strong solution or oxalic acid crystals is accompanied by burning pain in the mouth, esophagus, and stomach, and may cause severe gastroenteritis with symptoms of vomiting, diarrhea, or melena. Oxalic acid removes calcium in the blood, forming calcium oxalate, and severe damage to the kidney may occur because of the insoluble calcium oxalate (61). The minimal fatal dose, LD_{LO} for humans is 71 mg/kg (62), and the mean lethal dose for an adult is 15–30 g (63).

If ingestion occurs, dilute solutions of calcium compounds (calcium lactate, calcium gluconate, etc) or a large amount of milk followed by magnesium sulfate should be administered (63,64).

Uses

Separation and Recovery of Rare-Earth Elements. Because rare-earth oxalates have low solubility in acidic solutions, oxalic acid is used for the separation and recovery of rare-earth elements (65). For the decomposition of rare-earth phosphate ores, such as monazite and xenotime, a wet process using sulfuric acid has been widely employed. There is also a calcination process using alkaline-earth compounds as a decomposition aid (66). In either process, rare-earth elements are recovered by the precipitation of oxalates, which are then converted to the corresponding oxides.

Metal Treatment. The oxalic acid process for anodizing aluminum (67,68) was developed in Japan. Oxalic acid is used as an electrolyte, and the thin aluminum oxide layer forms on the surface of aluminum. The coatings are hard, abrasion- and corrosion-resistant. In addition to oxalic acid, inorganic oxalate salts are also used in coloring anodic coatings (qv). Oxalic acid is a constituent of cleaners that are used for automotive radiators, boilers, and steel plates before phosphating. Many of its industrial cleaning applications are based on its acidity and reducing power which promote dissolution of rust and formation of oxalate coatings on steel (qv). As a chelating agent, oxalic acid forms water-soluble complexes on metal surfaces during cleaning and rinsing.

Bleaching Agent. In pulp bleaching (69), oxalic acid serves as a bleaching agent, but is often used together with other bleaching agents (qv) because of its relatively high cost. Oxalic acid is also used for the bleaching of cork, wood (particularly veneered wood), straw, cane, and natural waxes.

Textile Treatment.   Oxalic acid has various uses in fabric cleaning, application of dyestuff, and modifying properties of cellulose fabrics. Rust stains, which form on fabrics during weaving and finishing, are removed by the chelating action of oxalic acid by forming iron oxalate which is readily washed from the fabric. In laundries, oxalic acid neutralizes excess alkalinity. It also dissolves iron and metallic salts, which could discolor fabrics, and kills bacteria (70). In mordant wool dyeing (71), oxalic acid is used as a reducing or fixing agent. Oxalic acid can be used as a catalyst for cross-linking of textile finishing agents to cellulosic fabrics in the manufacture of permanent press fabrics (72). It is also used in flame proofing for cellulosic fabrics (73–75).

Leather Tanning.   Oxalic acid is used as a pH modifier in leather tanning by tannin and basic chromium sulfate. It also functions as a bleaching agent for leather (qv).

Marble Polishing.   Oxalic acid is used for marble polishing especially in Italy. It not only removes iron veins by forming water-soluble iron oxalate, but also serves as a polishing auxiliary.

Millet Jelly Production.   Starch powder is heated together with oxalic acid and hydrolyzed to produce millet jelly. Oxalic acid functions as a hydrolysis catalyst, and is removed from the product as calcium oxalate. This application is carried out in Japan.

Others.   Oxalic acid is used for the production of cobalt, as a raw material of various agrochemicals and pharmaceuticals, for the manufacture of electronic materials (76–83), for the extraction of tungsten from ore (84), for the production of metal catalysts (85,86), as a polymerization initiator (87–89), and for the manufacture of zirconium (90) and beryllium oxide (91).

Derivatives

SALTS AND COMPLEXES

Oxalic acid forms neutral and acid salts, as well as complex salts.

Ammonium Oxalate.   This salt [1113-38-8], $(NH_4)_2C_2O_4$, mol wt 124.10, exists as a monohydrate [6009-70-7] or in anhydrous form. Anhydrous ammonium oxalate is obtained when the monohydrate is dehydrated at 65°C. The monohydrate is a colorless crystal or white powder, and dissolves in water at 0°C up to 2.17 wt %, and 50°C up to 9.63 wt %. It is slightly soluble in alcohol and insoluble in ether. It is used for textiles, leather tanning, and precipitation of rare-earth elements.

Ammonium Iron(III) Oxalate.   This mixed salt [29696-35-3], $(NH_4)_3$-$[Fe(C_2O_4)_3]$, mol wt 374.04, is produced as an emerald-green crystalline trihydrate [13268-42-3]. It is soluble in water to the extent of 100 g in 100 mL at 25°C and is insoluble in both alcohol and ether. It is used in formulations for anodizing aluminum. The compound is not stable to light. On exposure to light, it is reduced from the ferric to the ferrous ion. It was once used extensively in the manufacture of blueprinting papers, where the development of the blue color results from the photochemical reduction of the oxalate salts containing the ferric iron. The ferrous ion thus formed subsequently reacts with the ferricyanide- or ferrocyanide-coated paper to form the blue pigment, Precision Blue. Blueprinting is now obsolete.

Potassium Hydrogen Oxalate. Potassium acid oxalate [127-95-7], KHC_2O_4, mol wt 146.15, exists as a monohydrate [6100-03-4]. It is of historical interest because it is the salt of sorrel found in vegetation and the first oxalate isolated.

Potassium Oxalate. The monohydrate [6487-48-5], $K_2C_2O_4 \cdot H_2O$, mol wt 184.24, is produced as a colorless crystalline material or a white powder. The anhydrous salt [583-52-8], mol wt 166.22, is obtained when the monohydrate is dehydrated at 160°C. The monohydrate is preferred as a reagent in analytical chemistry and in miscellaneous uses principally because of its high solubility as compared with other simple neutral oxalates; the saturated solution, at 0°C, contains about 20 wt %, and at 20°C, about 25 wt % $K_2C_2O_4$.

Sodium Oxalate. This salt [62-76-0], $Na_2C_2O_4$, mol wt 134.01, is obtained in such high purity and is so stable that it is used as a titrimetric standard. The salt is not very soluble in water; the saturated solution contains 2.6 wt % and 6.1 wt % $Na_2C_2O_4$ at 0°C and 100°C, respectively.

Calcium Oxalate. The monohydrate [5794-28-5], $CaC_2O_4 \cdot H_2O$, mol wt 128.10, is of importance principally as an intermediate in oxalic acid manufacture and in analytical chemistry; it is the form in which calcium is frequently quantitatively isolated. Its solubility in water is very low, lower than that of the other alkaline-earth oxalates. The approximate solubilities of this and several related salts are indicated in Table 6.

Table 6. Solubilities of Alkaline-Earth Oxalates

	Anhydrous salt in sat'd soln, %	
Formula, solid phase	0°C	100°C
$CaC_2O_4 \cdot H_2O$	0.0005	0.0015
$SrC_2O_4 \cdot H_2O$	0.006	0.015
$BaC_2O_4 \cdot \frac{1}{2} H_2O$	0.009	0.021
$MgC_2O_4 \cdot 2H_2O$	0.026	0.041

Nickel Oxalate. This salt, NiC_2O_4, mol wt 146.7, is produced as a greenish white crystalline dihydrate [6018-94-6]. It decomposes by heating at 320°C under vacuum into Ni metal and carbon dioxide. Nickel oxalate is used for the production of nickel catalysts and magnetic materials.

Yttrium Oxalate. This compound [126476-37-7], $Y_2(C_2O_4)_3$, mol wt 441.91, exists as a trihydrate, nonahydrate [7100-75-6], or heptadecahydrate. The compound is used for the production of a red fluorescent material for color television.

ORGANIC DERIVATIVES

Dialkyl Oxalates. Oxalic acid gives various esters. Dialkyl esters, ROOC–COOR, are industrially useful, but monoalkyl esters, ROOC–COOH, are not. The dialkyl esters are characterized by good solvent properties and

Table 7. Physical Properties of Dialkyl Oxalates

Oxalate	CAS Registry Number	Mol wt	Mp, °C	Bp, °C	d_4^{20}
dimethyl	[553-90-2]	118.09	54	163.5	1.148^a
diethyl	[95-92-1]	146.14	−40.6	185.4	1.079
di-n-butyl	[2050-60-4]	202.25	−29.6	245.5	0.9873

^{a}At 54°C.

serve as starting materials in the synthesis of many organic compounds, such as pharmaceuticals, agrochemicals, and fine chemicals (qv). Among the diesters, dimethyl, diethyl, and di-n-butyl oxalates are industrially important. Their physical properties are given in Table 7.

Oxamide. This diamide [471-46-5], $H_2NCOCONH_2$, mol wt 80.07, is sparingly soluble in water and insoluble in various organic solvents. It melts at about 350°C, with accompanying decomposition. Because of the low solubility in water, the compound is granulated and used as a slow-release nitrogen fertilizer. Conventional nitrogen fertilizers (qv), such as ammonium sulfate, urea, ammonium nitrate, and ammonium phosphate, are soluble in water, and thus are easily lost as run-off when it rains. On the contrary, oxamide stays in the soil longer. Therefore, it is gradually decomposed by microorganisms in the soil and utilized by plants for longer periods.

There are three reactions used for the production of oxamide:

$$(COONH_4)_2 \xrightarrow{\Delta} (CONH_2)_2 + 2\,H_2O$$

$$(CN)_2 + 2\,H_2O \longrightarrow (CONH_2)_2$$

$$(COOR)_2 + 2\,NH_3 \longrightarrow (CONH_2)_2 + 2\,ROH$$

The second and third reactions are economical, but the first is not. The second reaction is used in a process where HCN is oxidized to $(CN)_2$ and hydrolyzed in the presence of a strong acid catalyst to give oxamide. The third reaction is employed in a newly developed process where dialkyl oxalates are converted to oxamide by the ammonolysis reaction. This reaction easily proceeds without catalysts and quantitatively gives oxamide as a powder.

Oxalyl Chloride. This diacid chloride [79-37-8], ClCOCOCl, mol wt 126.9, is produced by the reaction of anhydrous oxalic acid and phosphorus pentachloride. The compound vigorously reacts with water, alcohols, and amines, and is employed for the synthesis of agrochemicals, pharmaceuticals, and fine chemicals.

Reduction Products. Glyoxylic acid [298-12-4], HOOCCHO, mol wt 74.04, is produced as aqueous solution by the electrolytic reduction of oxalic acid. It is used for the manufacture of vanillin.

Glycolic acid [79-14-1], $HOOCCH_2OH$, mol wt 76.05, can be obtained by the electrolytic reduction of oxalic acid or the catalytic reduction of oxalic acid with hydrogen in the presence of a ruthenium catalyst. Because of its acidity it is used as a cleaning agent for metal surface treatments and for boiler cleaning. It also serves as an ingredient in cosmetics (qv).

BIBLIOGRAPHY

"Oxalic Acid" in *ECT* 1st ed., Vol. 9, pp. 661–674, by J. C. Pernert, Oldbury Electro-Chemical Co.; in *ECT* 2nd ed., Vol. 14, pp. 356–373, by P. A. Florio and G. R. Patel, Pfister Chemical, Inc.; in *ECT* 3rd ed., Vol. 16, pp. 618–636, by C. A. Bernales, S. E. Bushman, and J. Kraljic, Allied Corp.

1. A. E. Martell and R. M. Smith, *Critical Stability Constants*, Plenum Press, New York, 1974.
2. H. T. Clarke and A. W. Davis, *Organic Syntheses*, 2nd ed., coll. vol. 1, John Wiley and Sons, Inc., New York, 1946, pp. 421–425.
3. R. C. Weast, ed., *CRC Handbook of Chemistry and Physics*, 61st ed., CRC Press Inc., Boca Raton, Fla., 1980.
4. M. A. Haleem and P. E. Yankwich, *Phys. Chem.* **69**, 1729 (May 1965).
5. J. E. Carnahan and co-workers, *J. Am. Chem. Soc.* **77**, 3755 (1955).
6. J. B. Conant and A. H. Blatt, *The Chemistry of Organic Compounds*, 5th ed., The Macmillan Co., New York, 1959, pp. 201–204.
7. U.S. Pat. 2,057,119 (1936), G. S. Simpson (to General Chemical (Allied Chemical)).
8. U.S. Pat. 2,322,915 (1943), M. J. Brooks (to General Chemical (Allied Chemical)).
9. U.S. Pat. 3,536,754 (1970) (to Allied Chemical).
10. Chn. Pat 1,053,228 (1991), Y. Bu and co-workers (to Hebei College of Light Chemical Industry).
11. Chn. Pat. 1,046,323 (1990), J. Sha and co-workers.
12. Chn. Pat. 1,047,854 (1990), X. Jiang and co-workers.
13. Ind. Pat. 164,973 (1989), G. Sen and co-workers (to Council of Scientific and Industrial Research).
14. Ind. Pat. 149,789 (1982) (to Hindustan Lever).
15. U.S. Pat. 3,531,520 (1970), E. V. Obmornov and co-workers (to Novomoskovsk Aniline Dye Plant (USSR)).
16. U.S. Pat. 3,691,232 (1972), E. Yonemitsu and co-workers (to Mitsubishi Gas Chemical).
17. U.S. Pat. 3,678,107 (1972) (to Mitsubishi Gas Chemical).
18. Jpn. Pat. 47-28764-B (1972), E. Yonemitsu and co-workers (to Mitsubishi Gas Chemical).
19. Jpn. Pat. 53-18012-B (1978), E. Yonemitsu and co-workers (to Mitsubishi Gas Chemical).
20. Jpn. Pat. 52-39812-B (1977), E. Yonemitsu and co-workers (to Mitsubishi Gas Chemical).
21. Jpn. Pat. 52-39813-B (1977), E. Yonemitsu and co-workers (to Mitsubishi Gas Chemical).
22. U.S. Pat. 3,081,345 (1963), E. J. Carlson and E. E. Gilbert (to Allied Chemical).
23. Jpn. Pat. 45-17417-B (1970), J. Boichard and co-workers (to Rhône-Poulenc).
24. Jpn. Pat. 45-17657-B (1970) (to Rhône-Poulenc).
25. Jpn. Pat. 45-18842-B (1970), A. Charamel and co-workers (to Rhône-Poulenc).
26. Jpn. Pat. 47-29883-B (1972), A. Charamel and co-workers (to Rhône-Poulenc).
27. U.S. Pat. 3,549,696 (1970) and Brit. Pat. 1,159,066 (1969), J. N. Duroux and L. M. E. Tichon (to Rhône-Poulenc); U.S. Pat. 3,692,830 (1972), Brit. Pat. 1,154,061 (1969), and Fr. Pat. 1,501,725 (1967), A. Charamel and co-workers (to Rhône-Poulenc).
28. Brit. Pat. 1,251,430 (1971), J. C. Jacquemet (to Rhône-Poulenc).
29. Brit. Pat. 1,123,147 (1968) and Fr. Pat. 1,465,640 (1967), J. Boichard and co-workers (to Rhône-Poulenc).
30. U.S. Pat 3,393,136 (1968), D. M. Fenton and co-workers (to Union Oil).
31. D. M. Fenton and P. J. Steinwand, *J. Org. Chem.* **39**, 701 (1974).
32. U.S. Pat. 3,994,960 (1976), T. Yamazaki and co-workers (to UBE Industries).

33. U.S. Pat. 4,229,589 (1980), K. Nishimura and co-workers (to UBE Industries).
34. S. Uchiumi and M. Yamashita, *J. Japan Petrol. Inst.* **25**(4), 197 (1982).
35. Jpn. Pat. 55-42058-B (1980), T. Yamazaki and co-workers (to UBE Industries).
36. Jpn. Pat. 56-12624-B (1981), K. Nishimura and co-workers (to UBE Industries).
37. Jpn. Pat. 56-28903-B (1981), M. Suitsu and co-workers (to UBE Industries).
38. Jpn. Pat. 56-28904-B (1981), M. Suitsu and co-workers (to UBE Industries).
39. Jpn. Pat. 56-28905-B (1981), M. Suitsu and co-workers (to UBE Industries).
40. Jpn. Pat. 56-28906-B (1981), M. Suitsu and co-workers (to UBE Industries).
41. Jpn. Pat. 56-28907-B (1981), M. Suitsu and co-workers (to UBE Industries).
42. Jpn. Pat. 60-13014-B (1985), K. Nishihira and co-workers (to UBE Industries).
43. U.S. Pat. 4,384,133 (1983), H. Miyazaki and co-workers (to UBE Industries).
44. U.S. Pat. 4,410,722 (1983), H. Miyazaki and co-workers (to UBE Industries).
45. U.S. Pat. 4,461,909 (1984), S. Tahara and co-workers (to UBE Industries).
46. U.S. Pat. 4,467,109 (1984), S. Tahara and co-workers (to UBE Industries).
47. U.S. Pat. 4,507,494 (1985), H. Miyazaki and co-workers (to UBE Industries).
48. Eur. Pat. 108,359 (1984), Y. Shiomi and co-workers (to UBE Industries).
49. S. Umemura and H. Miyazaki, *Kagaku Kogyo*, 34 (Jan. 1984).
50. Jpn. Pat. 61-6056-B (1986), S. Tahara and co-workers (to UBE Industries).
51. Jpn. Pat. 61-6057-B (1986), S. Tahara and co-workers (to UBE Industries).
52. Jpn. Pat. 61-26977-B (1986), S. Tahara and co-workers (to UBE Industries).
53. J. Fischer and co-workers, *J. Appl. Electrochem.* **11**(6), 743 (1981).
54. K. Ito and co-workers, *Bull. Chem. Soc. Japan* **58**(10), 3027 (1985).
55. S. Ikeda and co-workers, *Bull. Chem. Soc. Japan* **60**(7), 2517 (1987).
56. B. R. Eggins and co-workers, *Tetrahedron Lett.* **29**(8), 945 (1988).
57. M. Chandrasekaran and co-workers, *Bull. Electrochem.* **8**(3), 124 (1992).
58. Market data, UBE Industries, Ltd., Tokyo, Japan, 1993.
59. F. Feigel, *Spot Tests in Organic Analysis*, 7th ed., Elsevier, New York, 1966, p. 457.
60. F. Feigel and co-workers, *Microchemie* **18**, 272 (1935).
61. M. F. Laker, *Adv. Clin. Chem.* **23**, 259 (1983).
62. N. I. Sax, *Dangerous Properties of Industrial Materials*, 6th ed., Van Nostrand Reinhold Co., New York, 1984.
63. R. E. Gosselin and co-workers, *Clinical Toxicology of Commercial Products*, 4th ed., Section III, 1976, p. 262.
64. *Data Sheet 406, Oxalic Acid*, National Safety Council, Chicago, Ill., 1965.
65. L. A. Sarver and P. H. M. P. Briton, *J. Am. Chem. Soc.* **49**, 943 (1927).
66. Jpn. Pat. 59-11531-B (1984), K. Fukuo and co-workers.
67. S. Werneck and R. Pinner, *The Surface Treatment and Finishing of Aluminum and Its Alloys*, Robert Draper, Ltd., Teddington, U.K., 1972.
68. S. John and B. A. Shenoi, *Met. Finish.* **74**(9), 48, 57 (1976).
69. A. Mita and S. Kashiwabara, *TAPPI Proceedings of the Pulping Conference*, no. 2, Technical Association of the Pulp and Paper Industry, Atlanta, Ga., 1987, p. 401.
70. H. Cohen and G. E. Linton, *Chemistry and Textiles for the Laundry Industry*, Textile Book Publishers, Inc., New York, 1961.
71. L. Diserens, *The Chemical Technology of Dyeing and Printing*, Reinhold Publishing Corp., New York, 1948.
72. U.S. Pat. 3,811,210 (1974), N. A. Cashen and co-workers.
73. U.S. Pat. 3,888,779 (1975), H. C. Tsai (to American Cyanamid).
74. G. Hooper and co-workers, *J. Coat Fabr.* **6**(2), 105 (1976).
75. L. Benisek, *J. Text. Inst.* **67**(6), 226 (1976).
76. Jpn. Pat. 47-38417-B (1972), T. Ishikawa and co-workers (to TDK).
77. Jpn. Pat. 39-3807-B (1964), G. Akashi (to Fuji Photo Film).
78. Jpn. Pat. 55-35371-B (1980), Y. Takahashi and co-workers (to Japan Victor).

79. Jpn. Pat. 55-35372-B (1980), Y. Takahashi and co-workers (to Japan Victor).
80. Jpn. Pat. 57-35853-B (1982), Y. Minagawa and co-workers (to Mitsubishi Kasei).
81. Jpn. Pat. 60-59222-B (1985), Y. Minagawa and co-workers (to Mitsubishi Kasei).
82. Jpn. Pat. 55-28905-A (1980), Y. Minagawa and co-workers (to Mitsubishi Kasei).
83. Jpn. Pat. 63-10690-B (1988), Y. Horibe (to Matsushita Electric Ind.).
84. Jpn. Pat. 52-4247-B (1977), T. Onozaki and co-workers (to Nittetsu Kogyo).
85. Jpn. Pat. 63-10939-B (1988), M. Murata and co-workers (to Fuji Photo Film).
86. Jpn. Pat. 63-22047-A (1988), N. Nojiri and co-workers (to Mitsubishi Petrochem).
87. F. Sandescu and C. I. Simonescu, *Acta Polym.* **37**(1), 7 (1986).
88. U. D. N. Bajpai and S. Rai, *J. Appl. Polym. Sci.* **35**(5), 1169 (1988).
89. A. M. A. Nada and M. A. Yousef, *Acta Polym.* **40**(1), 68 (1989).
90. J. L. Shi and Z. X. Lin, *Solid State Ionics* **32–33**(1), 544 (1989).
91. Ger. Pat. 269,616 (1989), F. Kerbe and co-workers (to Veb Keramische Hermsdorf).

HIROYUKI SAWADA
TORU MURAKAMI
UBE Industries, Ltd.

OXAZINE DYES.   See AZINE DYES.

OXETHANE POLYMERS.   See POLYETHERS.

OXIRANE.   See ETHYLENE OXIDE.

OXO PROCESS

The oxo process, also known as hydroformylation, is the reaction of carbon monoxide (qv) and hydrogen (qv) with an olefinic substrate to form isomeric aldehydes (qv) as shown in equation 1. The ratio of isomeric aldehydes depends on the olefin, the catalyst, and the reaction conditions.

$$RCH{=}CH_2 + CO + H_2 \xrightarrow{\text{catalyst}} RCH_2CH_2CHO + R(CH_3)CHCHO \qquad (1)$$

If a double-bond shift occurs, the number of aldehyde isomers is increased.

Synthesis gas, a mixture of CO and H_2, also known as syngas, is produced for the oxo process by partial oxidation (eq. 2) or steam reforming (eq. 3) of a carbonaceous feedstock, typically methane or naphtha. The ratio of CO to

H_2 may be adjusted by cofeeding carbon dioxide (qv), CO_2, as illustrated in equation 4, the water gas shift reaction.

$$2\,CH_4 + O_2 \longrightarrow 2\,CO + 4\,H_2 \tag{2}$$

$$CH_4 + H_2O \longrightarrow CO + 3\,H_2 \tag{3}$$

$$CO_2 + H_2 \rightleftharpoons CO + H_2O \tag{4}$$

$$2\,CH_4 + CO_2 + O_2 \longrightarrow 3\,CO + 3\,H_2 + H_2O \tag{5}$$

The overall process for producing a 1:1 CO to H_2 ratio by partial methane oxidation and the water gas shift reaction is represented by equation 5.

History

The oxo reaction proceeds most frequently in the presence of a Group 8–10 (VIII) metal catalyst in the liquid phase, most particularly with members of Group 9, the Co–Rh–Ir triad. The earliest catalyst, hydrocobalt tetracarbonyl [*16842-03-8*], $HCo(CO)_4$, was an outgrowth of Fischer-Tropsch investigations carried out prior to World War II on the effect of olefins on hydrocarbon synthesis (1). The hydroformylation reaction, as practiced in the early days using cobalt catalysis, presented formidable requirements of high pressure, containment of the hydrogen, containment of carbon monoxide, and handling of the toxic and unstable metal carbonyls. These conditions were challenging for the experimentalist as well as for the plant operator. However, because the oxo reaction provided a direct route for converting inexpensive olefins into valuable oxygenated building blocks, widespread industrial research and usage occurred throughout Europe, Japan, and the United States.

The search for catalyst systems which could effect the oxo reaction under milder conditions and produce higher yields of the desired aldehyde resulted in processes utilizing rhodium. Oxo capacity built since the mid-1970s, both in the United States and elsewhere, has largely employed tertiary phosphine-modified rhodium catalysts. For example, over 50% of the world's butyraldehyde (qv) is produced by the LP Oxo process, technology licensed by Union Carbide Corp. and Davy Process Technology.

Propylene (qv) [*115-07-1*] is the predominant oxo process olefin feedstock. Ethylene (qv) [*74-85-1*], as well as a wide variety of terminal, internal, and mixed olefin streams, are also hydroformylated commercially. Branched-chain olefins include octenes, nonenes, and dodecenes from fractionation of oligomers of C_3–C_4 olefins as well as octenes from dimerization and codimerization of isobutylene and 1- and 2-butenes (see BUTYLENES).

Linear terminal olefins are the most reactive in conventional cobalt hydroformylation. Linear internal olefins react at less than one-third that rate. A single methyl branch at the olefinic carbon of a terminal olefin reduces its reaction rate by a factor of 10 (2). For rhodium hydroformylation, linear α-olefins are again the most reactive. For example, 1-butene is about 20–40 times as reactive as the 2-butenes (3) and about 100 times as reactive as isobutylene.

Oxo aldehyde products range from C_3 to C_{15}, ie, detergent range, and are employed principally as intermediates to alcohols, acids, polyols, and esters formed by the appropriate reduction, oxidation, or condensation chemistry. The oxo reaction has been the subject of various reviews (4).

The classic challenges in oxo technology are simultaneously to achieve high reaction rate, high selectivity to the desired aldehyde, and to utilize a highly stable catalyst. Since the early 1970s, considerable progress has been made using ligand-modified rhodium catalysts that address these problems. In addition, progress has been made in the development of high reactivity rhodium catalysts for the conversion of internal and mixed-olefin feed streams. These latter are considerably less reactive than simple unsubstituted α-olefins. Development of catalysts which give improved process selectivities to the straight-chain isomer, generally more valuable, and of more efficient ways to recover product from rhodium catalyst solutions, have occurred. Additionally, progress has been made in asymmetric hydroformylation by using chiral ligands as a potential route to chiral pharmaceuticals.

Catalysts

Unmodified Cobalt Process. Typical sources of the soluble cobalt catalyst include cobalt alkanoates, cobalt soaps, and cobalt hydroxide [1307-86-4] (see COBALT COMPOUNDS). These are converted *in situ* into the active catalyst, $HCo(CO)_4$, which is in equilibrium with dicobalt octacarbonyl [10210-68-1]:

$$Co_2(CO)_8 + H_2 \rightleftharpoons 2\,HCo(CO)_4 \tag{6}$$

The mechanism of the cobalt-catalyzed oxo reaction has been studied extensively. The formation of a new C–C bond by the hydroformylation reaction proceeds through an organometallic intermediate formed from cobalt hydrocarbonyl which is regenerated in the aldehyde-forming stage. The mechanism (5,6) for the formation of propionaldehyde [123-38-6] from ethylene is illustrated in Figure 1.

The last step in this mechanism, the product forming hydrogenolysis, has been somewhat controversial. An alternative pathway (7) which involves cleavage by cobalt hydrocarbonyl to form aldehyde has been suggested. Each step in the mechanism is thought to be reversible except the final product forming step. The reverse of this reaction is so slow, it is generally neglected (8).

The rate of hydroformylation increases with increasing hydrogen and decreases with increasing carbon monoxide partial pressures (9), suggesting that rates of hydroformylation would be satisfactory at high H_2 and low CO partial pressures. In industrial practice, however, high pressures of both H_2 and CO are required in order to stabilize the $HCo(CO)_4$ catalyst at the temperatures necessary for practical rates (10). Commercial processes, for example, operate at >24 MPa (3480 psi) and >140°C.

The commercially important normal to branched aldehyde isomer ratio is critically dependent on CO partial pressure which, in propylene hydroformylation, determines the rate of interconversion of the *n*-butyryl and isobutyryl

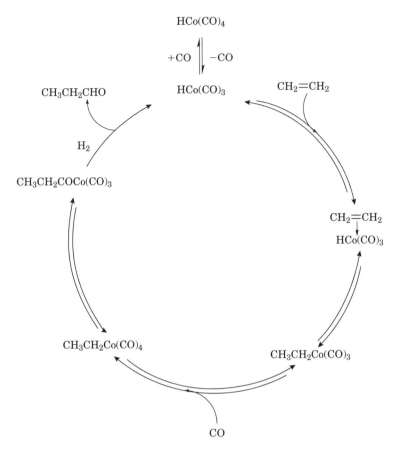

HCo(CO)$_4$

$+CO$ $\Big\updownarrow$ $-CO$

HCo(CO)$_3$

CH$_3$CH$_2$CHO

CH$_2$=CH$_2$

H$_2$

CH$_3$CH$_2$COCo(CO)$_3$

CH$_2$=CH$_2$
HCo(CO)$_3$

CH$_3$CH$_2$Co(CO)$_4$

CH$_3$CH$_2$Co(CO)$_3$

CO

Fig. 1. Mechanism for the unmodified cobalt oxo reaction which produces propionaldehyde from ethylene.

cobalt tetracarbonyl intermediates (11).

$$CH_3CH_2CH_2COCo(CO)_4 \rightleftharpoons (CH_3)_2CHCOCo(CO)_4 \qquad (7)$$

In contrast to triphenylphosphine-modified rhodium catalysis, a high aldehyde product isomer ratio via cobalt-catalyzed hydroformylation requires high CO partial pressures, eg, 9 MPa (1305 psi) and 110°C. Under such conditions alkyl isomerization is almost completely suppressed, and the 4.4:1 isomer ratio reflects the precursor mixture which contains principally the kinetically favored n-butyryl to isobutyryl cobalt tetracarbonyl. At lower CO partial pressures, eg, 0.25 MPa (36.25 psi) and 110°C, the rate of isomerization of the n-butyryl cobalt intermediate is competitive with butyryl reductive elimination to aldehyde. The product n/iso ratio of 1.6:1 obtained under these conditions reflects the equilibrium isomer ratio of the precursor butyryl cobalt tetracarbonyls (11).

Because of its volatility, the cobalt catalyst codistills with the product aldehyde necessitating a separate catalyst separation step known as decobalting.

This is typically done by contacting the product stream with an aqueous carboxylic acid, eg, acetic acid, subsequently separating the aqueous cobalt carboxylate, and returning the cobalt to the process as active catalyst precursor (2). Alternatively, the aldehyde product stream may be decobalted by contacting it with aqueous caustic soda which converts the catalyst into the water-soluble $Co(CO)_4^- Na^+$. This stream is decanted from the product, acidified, and recycled as active $HCo(CO)_4$.

The stringency of the conditions employed in the unmodified cobalt oxo process leads to formation of heavy trimer esters and acetals (2). Although largely supplanted by low pressure ligand-modified rhodium-catalyzed processes, the unmodified cobalt oxo process is still employed in some instances for propylene to give a low, eg, ~3.3–3.5:1 isomer ratio product mix, and for low reactivity mixed and/or branched-olefin feedstocks, eg, propylene trimers from the polygas reaction, to produce isodecanol plasticizer alcohol.

Ligand-Modified Cobalt Process. The ligand-modified cobalt process, commercialized in the early 1960s by Shell, may employ a trialkylphosphine-substituted cobalt carbonyl catalyst, $HCo(CO)_3P(n-C_4H_9)_3$ [20161-43-7], to give a significantly improved selectivity to straight-chain product. The Shell catalyst has vastly improved thermal stability over unsubstituted cobalt hydrocarbonyl and operates at 5–10 MPa (725–1450 psi) of H_2:CO and 160–200°C. The improved stability of the trialkylphosphine-substituted cobalt hydrocarbonyl, however, is offset by a lower hydroformylation activity requiring commensurately higher reaction temperatures. Thus there is a higher tendency of the olefin to undergo hydrogenation to alkane and of the aldehyde products to be hydrogenated to alcohols. Both linear and internal olefins react to yield principally linear alcohols and aldehydes. These products are a consequence of a high rate of isomerization occurring concurrently with hydroformylation and a strong preference for the α-olefinic component to undergo terminal addition with the $HCo(CO)_3P(n-C_4H_9)_3$ catalyst. C_{11}–C_{14} linear olefins, obtained from paraffin cracking or from the Shell Higher Olefins Process (SHOP), are hydroformylated to an 8:1 linear-to-branched isomer ratio, detergent range alcohol product mix in a single step. There has been large industrial usage of the Shell process since the 1960s, particularly for the preparation of detergent range alcohols (see ALCOHOLS, HIGHER ALIPHATIC) (2). 2-Ethyl-1-hexanol can be produced in a single step from propylene by conducting the hydroformylation in the presence of caustic (12).

Ligand-Modified Rhodium Process. The triphenylphosphine-modified rhodium oxo process, termed the LP Oxo process, is the industry standard for the hydroformylation of ethylene and propylene as of this writing (ca 1995). It employs a triphenylphosphine [603-35-0] (TPP) (**1**) modified rhodium catalyst. The process operates at low (0.7–3 MPa (100–450 psi)) pressures and low (80–120°C) temperatures. Suitable sources of rhodium are the alkanoate, 2,4-pentanedionate, or nitrate. A low (60–80 kPa (8.7–11.6 psi)) CO partial pressure and high (10–12%) TPP concentration are critical to obtaining a high (eg, 10:1) normal-to-branched aldehyde ratio. The process, first commercialized in 1976 by Union Carbide Corp. in Ponce, Puerto Rico, has been licensed worldwide by Union Carbide Corp. and Davy Process Technology.

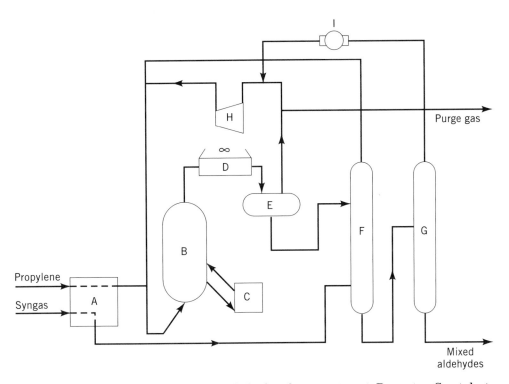

(**1**) X=H

(**2**) X=SO₃Na

The first commercial LP Oxo process flow scheme (Fig. 2) used syngas and propylene feed. Gases are fed to a pretreatment stage to remove poisons such as sulfur compounds, halides, and other harmful impurities. After purification, the gases are fed to the bottom of the reactor containing the catalyst solution consisting of the TTP-modified rhodium complex in butyraldehyde and butyraldehyde condensation products. Product and condensation by-products are removed from the reactor as vapor, a process facilitated by the superficial gas velocities of the gaseous feeds. The unreacted gases are separated from the product and recycled to the reactor. The gas recycle mode of operation may employ multireactors in parallel.

Fig. 2. LP Oxo gas recycle flow scheme: A, feedstock pretreatment; B, reactor; C, catalyst preparation and treatment systems; D, condenser; E, separator; F, stripper; G, stabilizer; H, cycle compressor; and I, stabilizer overhead gas compressor.

In the liquid recycle product recovery mode (Fig. 3), a modification of the initially commercialized technology, aldehyde product is separated in an external vaporizer, effectively decoupling the hydroformylation reaction from product recovery. Decoupling the oxo reaction from the separation step permits operation at milder and more favorable reaction temperatures. Lower temperatures reduce competing side reactions and extend catalyst life. A high degree of conversion is achieved by operating multireactors in series. This obviates the need for recycling unreacted gases.

Mechanism of LP Oxo Reaction. The LP Oxo reaction proceeds through a number of rhodium complex equilibria analogous to those in the Heck-Breslow mechanism described for the ligand-free cobalt process (Fig. 1).

$$\text{HRhCOL}_3 \underset{\text{L}}{\overset{\text{CO}}{\rightleftarrows}} \text{HRh(CO)}_2\text{L}_2 \underset{\text{L}}{\overset{\text{CO}}{\rightleftarrows}} \text{HRh(CO)}_3\text{L} \qquad (8)$$

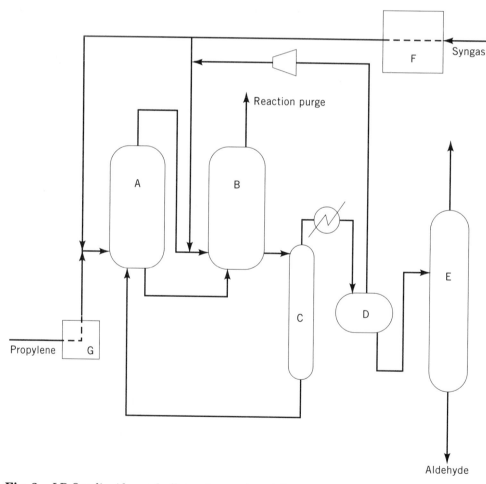

Fig. 3. LP Oxo liquid recycle flow scheme: A and B, reactors; C, vaporizer; D, catchpot; E, stabilizer; F, syngas cleanup; and G, propylene cleanup.

For example, $HRh(CO)_2L_2$, after dissociation of CO, goes on to generate *n*-butyraldehyde as shown in Figure 4 (13,14). A similar cycle could be written for isobutyraldehyde.

The basis of the high normal to isoaldehyde selectivity obtained in the LP Oxo reaction is thought to be the anti-Markovnikov addition of olefin to $HRhCOL_2$ to give the linear alkyl, $Rh(CO)L_2CH_2CH_2CH_2CH_3$, the precursor of straight-chain aldehyde. Anti-Markovnikov addition is preferred in this instance because of fewer unfavorable steric interactions between the straight-chain, ie, less bulky, alkyl substituent and the two bulky phosphine ligands. Conversely, steric constraints are minimal in the singly substituted intermediate, $HRh(CO)_2L$, resulting in a higher proportion of branched alkyl intermediate from Markovnikov addition of olefin to the rhodium hydride, and a commensurately lower selectivity to normal versus branched aldehyde product. From the equilibria depicted in equation 8, it can be seen that a high proportion of the linear aldehyde precursor, $HRhCOL_2$, is favored by a combination of high ligand concentration and low partial pressure.

Rhodium Modified with Ionic Phosphine Ligands. In 1984, Rhône-Poulenc and Ruhrchemie (now Hoechst AG) commercialized a rhodium catalyst

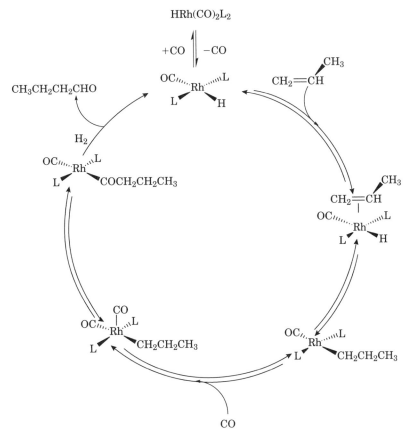

Fig. 4. Mechanism for the TPP-modified rhodium-catalyzed oxo reaction of propylene to *n*-butyraldehyde.

process employing a water-soluble ligand, triphenylphosphine-*m*-trisulfonic acid trisodium salt [*63995-70-0*] (TPPTS) (**2**). Product recovery is achieved by decantation from the aqueous phase containing rhodium and ligand (Fig. 5) (15). An isomer ratio of 20:1 is obtained with the TPPTS-modified rhodium catalyst, but the catalyst activity is significantly lower, so higher temperatures, higher rhodium concentrations, and higher propylene pressures are employed. Table 1 compares the operating conditions of the Hoechst/Rhône-Poulenc and LP Oxo processes. Hydroformylation reactions with a variety of other water-soluble ligands have also been reported (15).

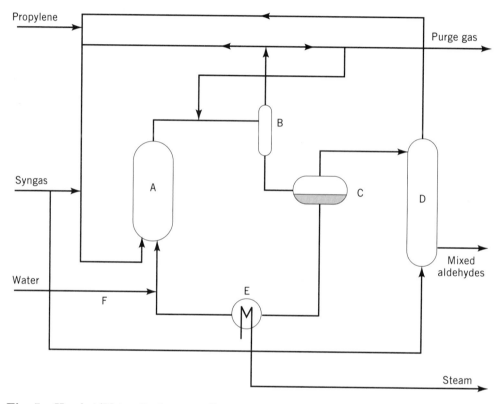

Fig. 5. Hoechst/Rhône-Poulenc oxo flow scheme: A, stirred tank reactor; B, separator; C, phase separator; D, stripping column; E, heat exchanger; and F, water inlet.

Table 1. Comparison of Hoechst/Rhône-Poulenc and LP Oxo Processes

| | HRh(CO)$_2$L$_2$ | |
Parameter	L = TTP	L = TPPTS
temperature, °C	≥90	≥125
pressure, MPa[a]	≥1.4	≥6
Rh, ppm	200–400	300–700
C$_3$H$_6$, MPa[a]	up to 1.1	up to 4.8

[a]To convert MPa to psi, multiply by 145.

Supported aqueous phase (SAP) catalysts (16) employ an aqueous film of TPPTS or similar ligand, deposited on a solid support, eg, controlled pore glass. Whereas these supported catalysts overcome some of the principal limitations experienced using heterogeneous catalysts, including rhodium leaching and rapid catalyst deactivation, SAP catalysts have not found commercial application as of this writing.

Other Rhodium Processes. Unmodified rhodium catalysts, eg, $Rh_4(CO)_{12}$ [19584-30-6], have high hydroformylation activity but low selectivity to normal aldehydes.

A hydroformylation process employing monosulfonated triphenylphosphine rhodium catalysts, soluble in polar organic solvents, which allows the hydroformylation of higher olefins in a single phase, has been reported. These catalysts, which have typical homogeneous catalyst reactivity, can be induced to separate after hydroformylation into nonpolar (product) and polar (catalyst) phases, thereby providing an effective means of catalyst recovery. The practical significance of this technology is that it permits the homogeneous hydroformylation of higher molecular weight and less volatile olefins such as octene, dodecene, styrene, and dienes (16). These materials tend to lose activity by transforming, particularly under acidic conditions, into the poorly soluble cluster, $Rh_6(CO)_{16}$, or under more drastic conditions, into rhodium metal. Mitsubishi Kasei has been able to take advantage of the inherent high reactivity of unmodified rhodium catalysts to convert butene dimers into branched C_9 products, ie, isononyl alcohol (17). Under hydroformylation conditions both CO and a weakly coordinating ligand, triphenylphosphine oxide (TPPO) (**3**), are available for coordination/stabilization of the metal. To provide catalyst stability during the product isolation step, small amounts of TPP are added. The TPP in the stripped catalyst solution is subsequently oxidized to TPPO prior to recycling back to the hydroformylation zone.

(**3**)   (**4**)

Functional Olefin Hydroformylation. There has been widespread academic (18,19) and industrial (20) interest in functional olefin hydroformylation as a route to polyfunctional molecules, eg, diols. There are two commercially practiced oxo processes employing functionalized olefin feedstocks. Allyl alcohol hydroformylation is carried out by Arco under license from Kuraray (20,21). 1,4-Butanediol [110-63-4] is produced by successive hydroformylation of allyl alcohol

[*107-18-6*], aqueous extraction of the intermediate 2-hydroxytetrahydrofuran, and subsequent hydrogenation.

$$
\underset{\text{CH}}{\overset{\text{H}_2\text{C}}{\diagdown}}\text{—}\underset{\text{OH}}{\overset{\text{CH}_2}{\diagup}} + \text{CO} + \text{H}_2 \longrightarrow \underset{\text{O}\quad\text{OH}}{\boxed{}} \xrightarrow{\text{H}_2} \underset{\text{CH}_2\quad\text{CH}_2}{\overset{\text{HO}\quad\text{CH}_2\quad\text{CH}_2}{\diagdown\diagup\diagdown\text{OH}}} \quad (9)
$$

2-Methyl-1,3-propanediol is produced as a by-product. The hydroformylation reaction employs a rhodium catalyst having a large excess of TPP (**1**) and an equimolar (to rhodium) amount of 1,4-diphenylphosphinobutane (DPPB) (**4**). Aqueous extraction/decantation is also used in this reaction as an alternative means of product/catalyst separation.

Kuraray has commercialized a process for producing 3-methyl-1,5-pentanediol [*4457-71-0*] from 3-methyl-3-butenol (**20**).

$$
\text{CH}_2{=}\text{C(CH}_3\text{)CH}_2\text{CH}_2\text{OH} \xrightarrow{\text{two steps}} \text{HOCH}_2\text{CH}_2\text{CH(CH}_3\text{)CH}_2\text{CH}_2\text{OH} \quad (10)
$$

The initial hydroformylation is conducted using tris(2,4-di-*t*-butylphenyl)phosphite (**5**) as ligand.

(**5**)

Hydroformylation Using Other Metals. *Ruthenium.* Ruthenium, as a hydroformylation catalyst (14), has an activity significantly lower than that of rhodium and even cobalt (22). Monomeric ruthenium carbonyl triphenylphosphine species (23) yield only modest normal to branched regioselectivities under relatively forcing conditions. For example, after 22 hours at 120°C, 10 MPa (1450 psi) of carbon monoxide and hydrogen, biscarbonyltristriphenylphosphine ruthenium [*61647-76-5*], Ru(TPP)$_3$(CO)$_2$, at 2000 ppm ruthenium and 1-hexene as substrate, gives only an 86% conversion and a 2.4:1 linear-to-branched aldehyde isomer ratio. At higher temperatures reduced conversions occur. High hydrogen partial pressures increase the reaction rate, but at the expense of increased hydrogenation to hexane. Excess triphenylphosphine improves the selectivity to linear aldehyde, but at the expense of a drastic decrease in rate.

In what may be an example of true cluster catalysis, $[HRu_3(CO)_{11}]^-$ shows good catalytic activity and high regioselectivity using propylene as substrate (24,25). Solvent, CO partial pressure, and temperature are important variables. In monoglyme, at 80°C and starting partial pressures for C_3H_6, CO, and H_2 of 0.034, 0.022, and 0.011 MPa (4.93, 3.19, and 1.60 psi), respectively, the catalyst turnover number, product-to-catalyst ratio, is 34.3 and the n- to isobutyraldehyde ratio is 49.4:1. In acetonitrile solvent, all other things being equal, the turnover number drops to 25.7 and the isomer ratio to 12.1:1.

Platinum. Platinum catalysts that utilize both phosphine and tin(II) halide ligands give good rates and selectivities, in contrast to platinum alone, which has extremely low or nonexistent hydroformylation activity. High specificity to the linear aldehyde from 1-pentene or 1-heptene is obtained using $HPtSnCl_3CO(TTP)$ (26), active at 100°C and 20 MPa (290 psi) producing 95% n-hexanal from 1-pentene.

A further improvement in platinum catalysis is claimed from use of tin(II) halide and phosphine ligands which are rigid bidentates, eg, 1,2-bis(diphenyl-phosphinomethyl)cyclobutane (27). High rates for a product containing 99% linear aldehyde have been obtained. However, a pressure of 10 MPa (1450 psi) H_2:CO is required.

Future Trends. In addition to the commercialization of newer extraction/decantation product/catalyst separations technology, there have been advances in the development of high reactivity oxo catalysts for the conversion of low reactivity feedstocks such as internal and α-alkyl substituted α-olefins. These catalysts contain (as ligands) ortho-t-butyl or similarly substituted arylphosphites, which combine high reactivity, vastly improved hydrolytic stability, and resistance to degradation by product aldehyde, which were deficiencies of earlier, unsubstituted phosphites. Diorganophosphites (28), such as structure (6), have enhanced stability over similarly substituted triorganophosphites.

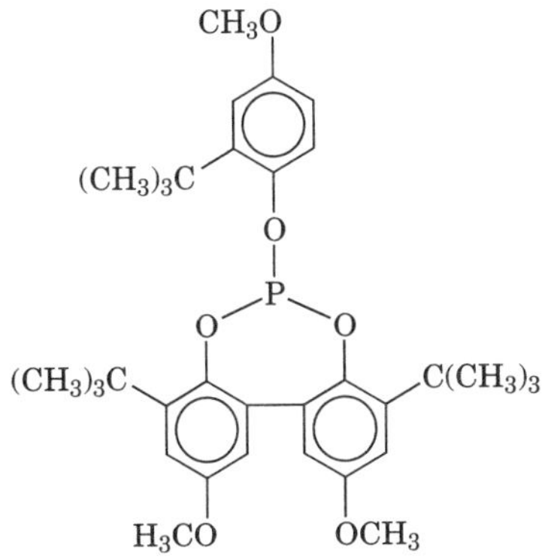

(6)

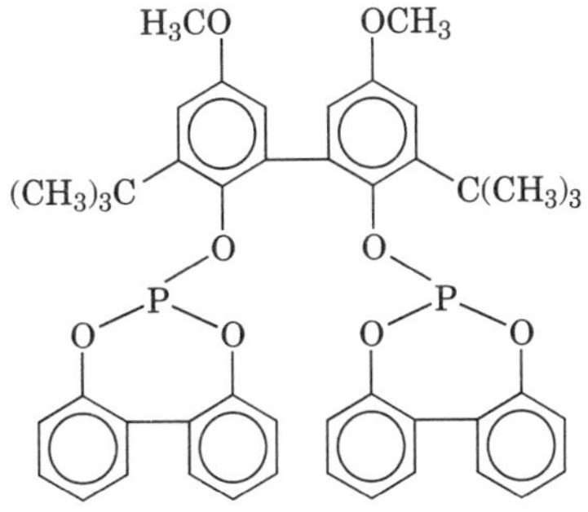

(7)

Bisphosphites such as (**7**) combine excellent reactivity, straight-chain selectivity, and high resistance to the typical phosphite degradation reactions (29). Further, the corresponding oxo catalysts are excellent olefin isomerization catalysts so that high normal-to-branched isomer ratios are obtained even from internal olefins, enabling, in certain instances, the use of inexpensive mixed isomer olefin feedstocks.

Considerable advances in asymmetric hydroformylation, a process which, among other things, provides a potential route to enantiomerically pure biologically active compounds, have occurred. Of particular interest are preparations of nonsteroidal antiinflammatory (NSAI) pharmaceuticals such as Naproxen (**8**) and Ibuprofen (**9**), where the * represents a chiral center.

$$ArCH{=}CH_2 + CO + H_2 \xrightarrow{\text{catalyst}} \underset{(\mathbf{10})}{\overset{\overset{\displaystyle CH_3}{\overset{\displaystyle \vdots}{CH}}}{Ar\underset{*}{\diagup}CHO}} + Ar\diagdown CH_2\diagdown CH_2\diagdown CHO$$

$$\downarrow [O]$$

$$\overset{\overset{\displaystyle CH_3}{\overset{\displaystyle \vdots}{CH}}}{Ar\underset{*}{\diagup}COOH}$$

(**8**) Ar = 6-methoxy-2-naphthyl

(**9**) Ar = 4-iso-butylphenyl

Conceptually at least, these compounds can be obtained via initial enantioselective hydroformylation of the appropriate vinyl aromatic to branched chiral aldehyde and subsequent oxidation.

For such a process to NSAI pharmaceuticals to be commercially practical requires minimally high regioselectivity and stereoselectivity to the chiral–aldehyde intermediate (**10**), combined with retention of optical purity in the subsequent oxidation step. High optical yields (~70–82% ee) of enantiomeric α-methylarylacetaldehydes have been obtained from styrene, p-isobutylstyrene, 2-vinylnaphthalene, and 2-vinyl-6-methoxynaphalene (30) employing a PtCl(SnCl$_3$)L* catalyst, where L* is a chiral ligand, such as (**11**). Even higher (>96%) optical yields were obtained by trapping the aldehyde as the diethyl acetal using triethylorthoformate (31). The regioselectivity of these catalysts to the branched product, eg, 0.5–3.3:1, as well as the overall rates, however, were low. Certain chiral bisphosphite-modified rhodium catalysts (32), however, have been reported to give the desired NSAI pharmaceutical precursors at combined high regioselectivity, rate, and enantioselectivity.

(11)   (12)

High enantioselectivities and regioselectivities have been obtained using both mono- and 1,2-disubstituted prochiral olefins employing chiral phosphine phosphite (33,34) modified rhodium catalysts. For example, *cis*-2-butene in the presence of rhodium and (**12**) (33) gave (*S*)-2-methylbutanal in an optical yield of 82% at a turnover number of 9.84 h^{-1}.

Economic Aspects

Worldwide capacity for oxo process chemicals reached 7.0×10^6 metric tons at the start of 1990 (35). Market share for oxo chemicals is divided between Western Europe (36%), the United States (30%), Eastern Europe (12%), Japan (10%), other Asian countries (8%), and South America and Mexico (4%). U.S. oxo manufacturers, products, and capacities are given in Table 2.

The propylene-based chemicals, *n*- and isobutanol and 2-ethyl-1-hexanol [104-76-7] (2-EH) dominate the product spectrum. These chemicals represent 71% of the world's total oxo chemical capacity. In much of the developed world, plasticizers (qv), long based on 2-EH, are more often and more frequently higher molecular weight, less volatile C$_9$ and C$_{10}$ alcohols such as isononyl alcohol, from dimerized normal butenes; isodecanol, from propylene trimer; and 2-propyl-1-heptanol, from *n*-butenes and aldol addition. Because of the competition from the higher molecular weight plasticizer alcohols, 2-EH and dioctylphthalate [117-81-7] (DOP), its principal derviative, are expected to grow more slowly than the higher molecular weight plasticizers in the 1990s (35). Oxo products other than butyraldehydes are significant only in the United States and Western Europe, representing 41 and 38%, respectively, of these regions' total capacities.

The largest oxo producers in Western Europe are BASF, Hüls, and Hoechst (formerly Ruhrchemie), representing 50–51% of the total regional capacity of 2.527×10^6 metric tons. These companies have the broadest spectrum of products ranging from C$_3$ and C$_4$ adehydes to C$_{13}$ alcohols and acids. However the primary products are *n*- and isobutyraldehyde, at combined capacities of 1.08×10^6 t. The *n*-butyraldehyde goes principally into the manufacture of 2-EH.

The spectrum of oxo products in Japan is far less diverse. Nearly 75% of Japan's total oxo capacity of 733,000 t is dedicated to the hydroformylation of propylene. 2-EH derived from *n*-butyraldehyde is by far the dominant product. Other products include linear alcohols and higher branched alcohols. Additionally, Japan is the world's principal source of branched heptyl alcohol. The three

Table 2. U.S. Oxo Manufacturers[a]

Company (plant location)	Products	Capacity, t × 10^3/yr	Catalyst
Exxon (Baton Rouge, La.)	branched C_6–C_{13} alcohols; linear C_7–C_{11} and C_{13}–C_{15} alcohols	306	Co
Hoechst Celanese (Bay City, Tex.)	propionaldehyde	30	Rh
	n-butyraldehyde	150	
	isobutyraldehyde	16	
Aristech Chemical Corp. (Pasadena, Tex.)	n-butyraldehyde	113	Rh
	isobutyraldehyde	14	
BASF Corp. (Freeport, Tex.)	n-butyraldehyde	107	Rh
	isobutyraldehyde	25	
Eastman (Longview, Tex.)	propionaldehyde	59	Co
	n-butyraldehyde	315	Rh
	isobutyraldehyde	125	
Shell Oil Co. (Deer Park, Tex.)	n-butanol	102	Co–PR$_3$
	2-ethyl-1-hexanol		
	isobutanol		
(Geismar, La.)	C_9–C_{15} linear alcohols	295	Co–PR$_3$
	isononyl alcohol		
Sterling Chemicals, Inc. (Texas City, Tex.)	C_7, C_9, C_{11} linear alcohols	102	Co
Union Carbide Corp. (Texas City, Tex.)	propionaldehyde and valeraldehyde	91	Rh
	n-butyraldehyde	329	
	isobutyraldehyde	33	

[a]Ref. 35.

principal Japanese oxo producers having slightly more than 70% of Japan's total oxo capacity are Mitsubishi Kasei, Kyowa Yuka, and Japan Oxocol.

Uses

n-Propanol and n-propyl acetate account for about 70% of the U.S. propionaldehyde derivative market (see PROPYL ALCOHOLS). These compounds are used principally in flexographic and gravure inks (qv) which require volatile solvents to prevent smearing and ink accumulation on the printing presses (see PRINTING PROCESSES). Some propanol is also converted into n-propylamines which are important pesticide intermediates (see PESTICIDES). n-Propanol is also employed as a precursor for glycol ethers, eg, Union Carbide's Propasol (propoxypropanol), having primary usage in surface coatings applications and flexographic printing inks (see GLYCOLS). The other principal propionaldehyde derivative, propionic acid, is used principally in grain and feed preservative applications (see FEEDS AND FEED ADDITIVES). Sodium and calcium propionates are used in both food and animal feed applications (see FOOD ADDITIVES). Some propionic acid is converted into herbicides (qv) such as Stam (Rohm and Haas) (3′,4′-dichloropropionanilide) and into cellulose acetate propionate, a plastic sheeting and molding precursor.

The highest volume oxo chemical in the United States, n-butyraldehyde, is converted mainly into n-butanol, employed chiefly to produce butyl acrylate and methacrylate (see ACRYLIC ACID AND DERIVATIVES). In contrast, the principal n-butyraldehyde derivative in Europe and Japan is 2-ethylhexanol, the precursor to the poly(vinyl chloride) (PVC) plasticizer, DOP.

1,4-Butanediol [*110-63-4*] (BDO) goes primarily into tetrahydrofuran [*109-99-9*] (THF) for production of polytetramethylene ether glycol (PTMEG), used in the manufacture of polyurethane fibers, eg, Du Pont's Spandex. THF is also used as a solvent for PVC and in the production of pharmaceuticals (qv). Lesser amounts of BDO are employed in the production of polybutylene terephthalate resins and γ-butyrolactone.

The principal C_5 valeraldehyde derivatives, n-amyl and 2-methylbutyl alcohols, are used predominantly to make zinc diamyldithiophosphate lube oil additives (see AMYL ALCOHOLS; LUBRICATION AND LUBRICANTS), which are employed primarily in automotive antiwear applications. Similarly, the n-valerate and 2-methylbutyrate esters of pentaerythritol and trimethylolpropane are used in aeromotive synlube formulations and as refrigerant lubricants.

C_7–C_9 oxo-derived acids are the principal derivatives of the C_7–C_9 oxo aldehydes, and in analogy to C_5 oxo aldehyde market applications, are used chiefly to make neopolyol esters, ie, those based on neopentyl glycol, trimethylolpropane, or pentaerythritol. These synlubes are employed almost entirely in aeromotive applications. Heptanoic acid is also employed to make tetraethylene glycol diheptanoate, a plasticizer used with polyvinyl butyral.

Several alcohols in the C_6–C_{13} range are produced by oxo reactions and are used in both plasticizer and detergent applications. Linear C_{12}–C_{15} alcohols are employed primarily in detergent applications. Slightly more than 50% of the 540,000–590,000 t of domestic U.S. detergent alcohol capacity is produced by hydroformylation of linear olefins from n-paraffins or ethylene chain-growth products. The remainder is produced from natural sources. Detergent alcohols are converted principally into alcohol sulfates, ethoxylates, alcohol ether sulfates, and fatty amines. Only a small (~1%) fraction of the detergent alcohols are destined for direct consumption.

Safety, Health, and Environmental Concerns

Oxo plants employ mixtures of highly toxic, flammable gases under pressure at high temperatures and require strict adherence to established operating safety codes and emergency reporting procedures to local, state, and federal authorities. In the United States, carbon monoxide is classified as both an acute, fire, and sudden release hazard under the Superfund Amendments and Reauthorization Act (SARA) 311/312, requiring strictly maintained documentation of hazards and emergency procedures, eg, maintenance of appropriate Material Safety Data Sheets and reporting procedures in case of accidental release.

The carbon monoxide component of the oxo reactant gases presents the most immediate human health hazard. The OSHA exposure limit for carbon monoxide is 35 ppm (40 mg/m^3) and a ceiling of 200 ppm (229 mg/m^3). Carbon monoxide interrupts the body's normal oxygen metabolic cycle by reacting preferentially with the hemoglobin in the blood, effectively starving the body of required oxygen. The cherry red iron carbonyl heme complex formed is much brighter than the corresponding oxygen complex so that someone overcome with carbon monoxide presents a characteristic flushed appearance and bright red lips. Conventional first-aid procedures require immediate removal from the source. Appropriate fresh air equipment should be worn in the presence of high concentrations of the gas. Artificial respiration is applied if the person overcome has stopped breathing and fresh oxygen is applied to facilitate the release of CO. After initial first aid, medical attention must be sought.

Acute toxicity testing of triphenylphosphine in animals indicates a very low level of physiological activity. The oral LD_{50} for triphenylphosphine in both rats and mice is 0.8–1.6 g/kg; the intraperitoneal LD_{50} is 1.6–3.2 and 0.8–1.6 g/kg, in rats and mice, respectively (36).

BIBLIOGRAPHY

"The Oxo Process" under "Oxo and Oxyl Processes" in *ECT* 1st ed., Vol. 9, pp. 712, by M. Orchin and I. Wender, Bureau of Mines, U.S. Dept. of the Interior; "Oxo Process" in *ECT* 1st ed., Suppl. Vol. 2, pp. 548–564, by J. E. Knapp, Union Carbide Chemicals Co.; in *ECT* 2nd ed., Vol. 14, pp. 373–390, by H. E. Kyle, Union Carbide Corp.; in *ECT* 3rd ed., Vol. 16, pp. 373–390, by I. Kirshenbaum and E. J. Inchalik, Exxon Research and Engineering Co.

1. Ger. Pat. 849,548 (1938), O. Roelen (to Ruhrchemie A. G.); U.S. Pat. 2,327,066 (Aug. 17, 1944), O. Roelen (vested in the Alien Property Custodian).
2. J. Falbe, *New Syntheses With Carbon Monoxide*, Springer Verlag, Berlin, 1980.
3. A. A. Oswald, D. E. Hendriksen, R. V. Kastrup, J. S. Merola, and J. C. Reisch, presented at the *Lubrizol Award Symposium of the 1982 Spring ACS Meeting, Las Vegas, Nev.*, American Chemical Society, Washington, D.C.
4. R. L. Pruett, *Adv. Organometal. Chem.* **17**, 1 (1979).
5. D. S. Breslow and R. F. Heck, *Chem. Ind. (London)*, 467 (1960).
6. R. F. Heck and D. S. Breslow, *J. Am. Chem. Soc.* **83**, 4023 (1961).
7. N. H. Alemdaroglu, J. M. L. Penninger, and E. Oltay, *Monats. Chem.* **197**, 1043 (1976).
8. U.S. Pat. 4,198,352 (July 23, 1980), L. Kim and S. C. Tang (to Shell Oil).
9. G. Natta, R. Ercoli, and S. Castellano, *Chim. Ind. (Milan)* **37**, 6 (1955).
10. J. Berty and E. Oltay, *Chem. Tech. (Leipzig)* **9**, 283 (1957).
11. M. S. Borovikov, I. Kovacs, F. Ungvary, A. Sisak, and L. Marko, *Organometallics* **11**, 1576 (1992).
12. U.S. Pat. 3,278,612 (July 23, 1966), C. R. Greene (to Shell Oil Co.).
13. C. K. Brown and G. Wilkinson, *J. Chem. Soc. (A)*, 1392 (1970).
14. D. Evans, J. A. Osborn, F. H. Jardine, and G. Wilkinson, *Nature* **203**, 1203 (1965).
15. W. A. Herrman and C. W. Kohlpaintner, *Angew. Chem. Int. Ed. Eng.* **32**, 1524 (1993).
16. I. Guo, B. E. Hanson, I. Toth, and M. E. Davis, *J. Organomet. Chem.* **403**, 221 (1991); A. G. Abatjoglou, *209th National ACS Meeting*, Apr. 17, 1995, pp. 25–26.
17. T. Onoda, *ChemTech.*, 34 (1993).
18. C. U. Pittman, Jr. and W. D. Honnick, *J. Org. Chem.* **45**, 2132 (1980).
19. R. V. Korneyeva and co-workers, *Petrol. Chem.* **33**, 236 (1993).
20. N. Y. Yoshimura, Y. Tokitoh, M. Matsumoto, and M. Tamura, *Nippon Kagaku Kaishi*, 119 (1993).
21. *Chem. Mark. Rep.* **241**(5), 5, 20 (1992).
22. K. A. Alekseeva, M. D. Vysotskii, N. S. Imyanitov, and V. A. Rybakov, *Zh. Vses. Khim. Oa.* **22**, 45 (1977).

23. R. A. Sanchez-Delgado, J. S. Bradley, and G. Wilkinson, *J. Chem. Soc., Dalton Trans.*, 399 (1976).
24. G. Suess-Fink and G. F. Schmidt, *J. Mol. Catal.* **42**, 361–366 (1987); G. Suess-Fink and G. Herrmann, *J. Chem. Soc., Chem. Commun.*, 735 (1985).
25. G. Suess-Fink, *J. Organometal. Chem.* **193**, C20 (1980).
26. C. Hsu and M. Orchin, *J. Am. Chem. Soc.* **97**, 3553 (1975).
27. U.S. Pat. 4,229,381 (Oct. 21, 1980), T. Hayashi, Y. Kawabata, I. Ogata, and M. Tanaka (to Agency of Industrial Science and Technology, Japan).
28. U.S. Pat. 4,717,775 (July 8, 1988), E. Billig, A. G. Abatjoglou, D. R. Bryant, J. M. Maher, and R. E. Murray (to Union Carbide Corp.).
29. U.S. Pat. 4,885,401 (Dec. 5, 1989), E. Billig, A. G. Abatjoglou, and D. R. Bryant (to Union Carbide Corp.).
30. G. Parrinello and J. K. Stille, *J. Am. Chem. Soc.* **109**(7), 122 (1987).
31. J. K. Stille, H. Su, G. Parrinello, and L. S. Hegedus, *Organometallics* **10**, 1183 (1991).
32. WO Pat. 9303839 (Apr. 3, 1993), J. E. Babin and G. T. Whiteker (to Union Carbide Corp.).
33. N. Sakai, K. Nozaki, and H. Takaya, *Chem. Commun.*, 395 (1994).
34. N. Sakai, S. Mano, K. Nozaki, and H. Takaya, *J. Am. Chem. Soc.* **115**, 7033 (1993).
35. *Chemical Economics Handbook, Oxo Chemicals Report*, SRI International, Menlo Park, Calif., Jan. 1991 and preliminary 1994 draft.
36. D. W. Fassett, in F. A. Patty, ed., *Industrial Hygiene and Toxicology,* Second ed., Vol. 2, Interscience Publishers, New York, 1972, p. 1918.

ERNST BILLIG
DAVID R. BRYANT
Union Carbide Corporation

OXYGEN

Molecular oxygen [7782-44-7], O_2, is a gaseous element constituting 20.946% (1) of the earth's atmosphere. Oxygen is essential to respiration and life in animals and is formed as a waste product by most forms of vegetation. Oxygen supports the combustion of fuels that supply heat, light, and power and enters into oxidative combination with many materials. The speed of reaction and effectiveness of combination increase with oxygen concentrations greater than that of air. Industry has established a 99.5% purity for the majority of commercial product, although a significant fraction of oxygen is also produced in the 90–93% purity range.

The discovery of oxygen, the development of a clear understanding of the nature of air, and a knowledge of the role of oxygen in combustion and in life processes were historically important scientific achievements. In the 1770s, air and ethers were studied by such scientists as Cavendish, Priestley, and Scheele. Both Scheele and Priestley independently prepared oxygen. Whereas Scheele's work remained unpublished until after his death, Priestley was able to relate his discovery directly to Lavoisier in 1775.

Lavoisier checked Priestley's work and recognizing that air contains mainly two gases, named one vital air and the other azote (nitrogen), the latter not supporting life. Later, vital air became oxygen, from the ability to form acids: *ox*, ie, sharp (taste) and *gen*, to form. In 1777, Lavoisier developed the theory of combustion. His ideas became widely established and were firmly fixed by his textbook, *La Traité Elémentaire de Chemie* (2).

After the discovery of oxygen, Cavendish, who had been the first to make a thorough study of hydrogen in 1766, sparked hydrogen with air and oxygen, thereby producing and proving the nature of water (qv). Cavendish also sparked the mixture of oxygen and air to form nitric oxide and nitrogen dioxide, which was absorbed in aqueous alkali. He also burned the excess oxygen with sulfur, and the resultant sulfur dioxide was absorbed in alkali.

For well over 100 years after its discovery and initial preparation, oxygen was made either chemically or by the electrolysis of water. Early in the twentieth century, Linde and Claude introduced processes for the liquefaction and distillation of air that have since grown into a mature and highly competitive industry. In 1991, over 13.4×10^9 m^3 (4.7×10^{11} ft^3) of oxygen was produced in the United States. About 70×10^9 m^3 (24.7×10^{11} ft^3) was produced worldwide. Usage of oxygen, the second-largest-volume industrial gas in both the United States and the world (nitrogen being the first), is projected to grow by about 5% per year throughout the 1990s. Most of this growth is likely to be in lower purity oxygen, which may be produced by noncryogenic means.

Oxygen in combination with hydrogen forms the waters of the earth's surface (89 wt % O_2). In combination with metals and nonmetals oxygen is contained in well over 98% of rocks, entering into a very large number of known minerals as well as a vast array of organic compounds. Together, free and combined oxygen constitute 46.6% (3) of the mass of the earth's crust, making it the most abundant element. In Earth's geological beginning the atmosphere probably contained little or no oxygen. The subsequent emergence of free molecular oxygen into the atmosphere may have arisen because of dissociation of water by high energy rays from the sun and cosmic rays in the upper atmosphere, followed by differential escape of hydrogen into space (4). The detection by the Apollo 16 moon flight, in April 1972, of a massive cloud of atomic hydrogen enveloping the earth and extending to ca 64,360 km, supports this view. Another thesis explains the emergence of oxygen through evolution of the process of photosynthesis (5,6). Thus, chlorophyll, using the radiant energy of sunlight, converted water and carbon dioxide taken from the atmosphere into a very large number of compounds that are the substance of vegetation. During photosynthesis, oxygen is produced and released into the atmosphere.

In nature, oxygen occurs in three stable isotopic species: oxygen-16 [*14797-70-7*], ^{16}O, 99.76%; oxygen-17 [*13968-48-4*], 0.038%; and oxygen-18 [*14797-71-8*], 0.20% (7). Commercial fractional distillation of water produces concentrations of ^{18}O as high as 99.98%; ^{17}O concentrations up to 55% are also produced. The ^{18}O isotope has been used to trace mechanisms of organic reactions.

Physical Properties

Gaseous oxygen is colorless, odorless, and tasteless. When cooled to 90.188 K it becomes a pale blue liquid, slightly more dense than water. On further cooling, it becomes a blue solid. The blue color probably results from a small equilibrium concentration of associated oxygen molecules. The liquid becomes colorless when passed through a bed of carefully activated silica gel, presumably because the associated molecules are removed. The color returns after some hours of standing. Oxygen is moderately soluble in water.

Temperature, °C	Aqueous solubility of O_2 at STP, L/L
0	0.0489
5	0.0429
10	0.0380
15	0.0342
20	0.0310
25	0.0283
30	0.0261
40	0.0231
50	0.0209
60	0.0195
70	0.0183
80	0.0176
90	0.0172
100	0.0170

Selected physical properties are listed in Table 1.

The oxygen molecule is paramagnetic, having a magnetic moment in accord with two unpaired electrons. The gases with which oxygen is normally associated are diamagnetic. Thus the oxygen content of a mixture of gases can be determined using simple instruments that measure the magnetic properties of the gas.

Chemical Properties

Oxygen reacts with all other elements except the light, rare gases helium, neon, and argon (13). The reactants usually must be activated by heating before the reaction proceeds at appreciable rates, and if the final union releases more than enough energy to activate subsequent portions of both reactants, the overall process may be self-sustaining. The process is known as combustion when light and heat are evolved. For certain elements, such as the alkali metals rubidium and cesium, activation energy provided at room temperatures is sufficient, and chemical reactions become spontaneous upon contact. Other metals such as finely powdered iron and nickel may be made pyrophoric by careful preparation and

Table 1. Physical Properties of Oxygen[a]

Property	Value
triple point	
temperature, K	54.359 ± 0.002
pressure, Pa[b]	146.4
density, g/L	
gas	0.0108
liquid	1306.5
solid	1300
boiling point, at 101.3 kPa, K	90.188
density, g/L	
gas	4.470
liquid	1141.1
melting point, K	54.22
critical point	
temperature, K	154.581
pressure, MPa[c]	5.043
density, g/L	436.1
gas, at 101.3 kPa	
density, g/L	
at 0°C	1.42908
at 21°C	1.327
heat capacity, J/(mol·K)[d]	
C_p, at 25°C	29.40
C_p/C_v, at 26°C	1.396
dielectric constant at 20°C	1.0004947
n_D°	1.0002639
viscosity at 25°C, μPa·s (=cP $\times 10^{-3}$)	20.639
thermal conductivity, at 0°C, mW/(m·K)	2.448
sound velocity, at 0°C, m/s	317.3
liquid	
heat capacity, sat liq, J/(mol·K)[d]	54.317
heat of vaporization, J/mol[d]	6820
viscosity, μPa·s (=cP $\times 10^{-3}$)	189.4
thermal conductivity, mW/(m·K)	149.87
sound velocity, at 87 K, m/s	904.6
surface tension, at 87 K, N/m	13.85×10^{-7}
volume ratio, gas at 21°C to liquid at bp	859.9
solid	
heat of sublimation, J/mol[d]	8204.1
heat capacity, J/(mol·K)[d]	46.40
heat of fusion, J/(mol)[d]	444.5

[a]Refs. 8–12.
[b]To convert Pa to mm Hg, multiply by 0.0075.
[c]To convert MPa to psi, multiply by 145.
[d]To convert joule to cal, divide by 4.184.

reduction of surface oxides. These then ignite spontaneously upon contact with the oxygen in air and continue to react with the evolution of light and heat. Such reactions may be explosive.

The design of all oxygen piping and containment systems must follow specific criteria to ensure that the containment or piping itself does not ignite and subsequently burn in the oxygen. One design criterion (5) limits the velocity of the oxygen in steel pipes so that any foreign material present cannot impact the pipe with sufficient energy to ignite, and in turn ignite the pipe. Oils and other foreign materials may react vigorously with oxygen, which in turn may ignite the oxygen piping or containment systems. All piping components must be thoroughly cleaned (14) to prevent this problem from occurring (see PIPING SYSTEMS).

Most materials must be heated to some particular temperature, called the ignition temperature, to initiate combustion. However, in the presence of moisture, combination reactions with oxygen frequently slowly occur, even at ordinary temperatures. Some examples are found in the respiration of organisms, the rusting and corrosion of metals, the decay of wood, and the hardening of linseed oil. On the other hand, extreme temperatures can be reached, for example, by burning acetylene in the presence of oxygen. The process of cutting iron involves burning away the iron with pure oxygen after the ignition point has been reached by heating with an oxygen–acetylene torch. Whereas noble metals can be oxidized only at very high temperatures, oxides are often formed through an alternative route.

Oxygen usually exhibits a valence of -2 in combination with other chemical elements to form compounds such as oxides. Most elements combine with oxygen, which is highly electronegative, in more than one ratio because of the variety of valences exhibited by the other element, or because of the existence of complicated molecular structures. An extended discussion of oxides is available in the literature (13).

When fuels such as petroleum (qv), natural gas (see GAS, NATURAL), or coal (qv) burn with an excess of oxygen, either as air or pure O_2, the products are heat, carbon dioxide (qv), and water, plus (qv) nitrogen oxides, etc, if air is used. When the amount of oxygen is limited, however, the reaction produces a mixture of carbon monoxide and hydrogen, also known as synthesis gas or syngas. This important gaseous mixture reduces iron (qv) ore to pig iron in a blast furnace. It also serves as the precursor from which methanol (qv) may be synthesized or as a raw material to provide hydrogen for ammonia (qv) synthesis. At lower temperatures, frequently in the presence of catalysts, oxygen reacts with organic chemicals to give oxygenated hydrocarbons (see HYDROCARBON OXIDATION).

Animal metabolism is based on the reactions of oxygen and organic compounds containing carbon, hydrogen, oxygen, and nitrogen and other heteroatoms. Enzymes catalyze these biochemical oxidations, which are accomplished at about 30–40°C and frequently proceed stepwise to produce specific changes in carbohydrates (qv), fats, and proteins (qv).

Oxidation rate is controlled in part by the area available for oxygen contact, eg, iron filings rust far faster than does a solid piece of iron. Violent or explosive reactions take place when combustible, finely divided, and suspended dusts or powders are ignited, as in grain elevators, coal mines, or in a manufacturing plant. These dangerous reactions may be mitigated by controlling the amount of dust or excluding air (oxygen) by dilution with sufficient inert gas (see POWDERS HANDLING). On some metals, such as aluminum, the oxide surface film is adherent and continuous and thus prevents further access of oxygen to the unoxidized metal. Finely divided or thin sheets of aluminum, however, may be explosive in liquid oxygen (15,16).

Manufacture

Commercial oxygen, both gaseous and liquid, at about 99.5% purity is produced by cryogenic distillation in air separation plants. In these plants the air is cleaned, dried, compressed, and refrigerated until it partially liquefies at about 80 K (see CRYOGENICS). The air is then distilled into its components (see Table 2). Commercial gaseous oxygen at about 90–93% purity is produced from air by vacuum swing adsorption (VSA) processes (see ADSORPTION, GAS SEPARATION). The VSA method is the fastest growth portion of oxygen production.

Air was first liquefied and oxygen subsequently separated about 1900. Technologies introduced after World War II substantially improved the air separation process for high purity oxygen production. In the 1960s the production of high volume, low cost oxygen for the manufacture of chemicals and petrochemicals, and especially for the basic oxygen process (BOP) for the manufacture of steel (qv), provided a distinct advantage. Since about 1980 a number of further improvements have occurred in the cryogenic air separation process resulting in significant efficiency, productivity, and reliability gains. The most modern cryogenic plants operate at thermodynamic efficiencies in the vicinity of 35%, a number significantly higher than for other industrial processes, such as petroleum refining.

Table 2. Gaseous Composition of Air [a]

Constituent[b]	Vol %	Boiling point, K
	Fixed components	
nitrogen	78.084 ± 0.004	77.36
oxygen	20.946 ± 0.002	90.18
argon	0.934 ± 0.001	87.28
carbon dioxide	0.033 ± 0.003	194.68[c]
neon	$(1.821 \pm 0.004) \times 10^{-3}$	27.09
helium	$(5.239 \pm 0.05) \times 10^{-4}$	4.215
krypton	$(1.14 \pm 0.01) \times 10^{-4}$	119.81
xenon	$(8.7 \pm 0.1) \times 10^{-6}$	165.04
hydrogen	ca 5×10^{-5}	20.27
	Impurities[d]	
water	0.1–2.8	
methane	1.5×10^{-4}	
carbon monoxide	$(6-100) \times 10^{-6}$	
sulfur dioxide	0.1 to 1.0	
nitrous oxide	5×10^{-5}	
ozone	$(1-10) \times 10^{-6}$	
nitrogen dioxide	$(5-200) \times 10^{-8}$	
radon	6×10^{-18}	
nitric oxide	[e]	

[a]Refs. 1, 11, and 12.
[b]Composition of dry air is constant to an altitude of 20 km.
[c]Sublimation temperature. Liquid CO_2 does not exist at 101 kPa (1 atm).
[d]In ambient air, including dusts, pollen, and local pollutants.
[e]Trace amounts.

Process improvements include (1) improved energy transfer in heat exchangers via improved thermal design of the main heat exchangers; (2) more complex heat integration yielding more optimal distillation column performance; (3) improved expander–compressor efficiencies; (4) reduced pressure drops within distillation columns via improved distillation tray design and the use of packing in place of trays; (5) use of down-flow reboilers; (6) use of high (up to ca 3100 kPa (450 psi)) pressure operating cycles; and (7) improved front-end purification equipment requiring less energy to reactivate (17,18).

Process improvements have been combined into a number of operating cycles. The choice of which cycle to use depends on a number of factors, including the ratio of oxygen:nitrogen produced; the ratio of liquid oxygen:nitrogen vs gaseous oxygen:nitrogen produced; and the final pressure required of the gaseous oxygen and nitrogen.

Improvements in instrumentation and automatic controls, especially the introduction of digital, ie, computerized, control systems, have provided significant gains in personnel productivity as well as in the ability to optimize process parameters to maintain peak efficiencies (see also PROCESS CONTROL). The standardization of (especially) smaller cryogenic air separation plants has further reduced design and construction costs. These types of plants often operate unattended. Even moderately large plants may be unattended during certain periods of operation.

The principal impurity in 99.5% oxygen is argon because of the closeness of the boiling points. There has been a high demand for argon in the 1990s so that the oxygen is further refined (via distillation) to recover the argon. As a result, oxygen purity often rises to 99.8%. This additional refining has also influenced the selection of operating cycles (see also HELIUM GROUP, GASES).

Cryogenic Separation. In the cryogenic air separation process, the ambient intake air is compressed and the moisture and carbon dioxide removed, either by cooling the air or by adsorption systems. In the former, the moisture and carbon dioxide are deposited as both liquid and ice, including solid carbon dioxide. These deposits are sublimed or removed in a second step after switching (reversing) the flows of air and waste nitrogen, thus warming the deposits. In the adsorption method, water, carbon dioxide, and trace impurities, eg, hydrocarbons (qv), are removed by zeolite and silica gel-type adsorbents. The adsorbent bed is regenerated on a periodic basis, typically by flushing with a hot waste oxygen–nitrogen stream while the main air flow is diverted to a second bed. This adsorption process for removal of moisture and carbon dioxide has similarities to the vacuum swing adsorption process for the production of oxygen. The choice of processing technique, ie, reversing exchangers or adsorption, depends on the total output of pure oxygen and nitrogen expected from the unit as well as the size of the unit, but adsorption has largely become the method of choice in newer plants.

Many of the trace impurities within the incoming air are either frozen out and trapped in the reversing exchangers or are removed together with the moisture and carbon dioxide by adsorption. All modern plants also include silica-gel adsorbers in the oxygen-rich liquid circuit as well as in the guard circuit used to control any accumulation of undesirable materials in the oxygen pool of the oxygen distillation column (see ADSORPTION, LIQUID SEPARATION). Silica gel is particularly effective in removing acetylene, which in early plants was responsible for initiating explosions of itself or other hydrocarbons concentrated in the liquid oxygen.

After compression and removal of impurities, the air is cooled in heat exchangers and expanded to low pressure through a turbine, to recover energy, or through a valve. Liquid air, which forms at about 80 K, is separated via a distillation column. The column as well as the heat exchangers and the associated piping are placed within a cold box, which is packed with insulation to minimize heat transfer (qv) between streams and to protect the system from the ambient air external to the cold box.

Adsorptive Separation. A noncryogenic air separation process, which is increasingly employed for small- to moderate-scale oxygen production units, is based on the adsorption of nitrogen (but not oxygen) onto zeolites (see ADSORPTION, GAS SEPARATION) (19). This batch process, known as vacuum swing adsorption (VSA), typically uses two identical switching beds, each containing two strata. The first stratum removes water and carbon dioxide; the second adsorbs nitrogen from the flowing air. In the two-bed system, while unit one is on-stream adsorbing first water and carbon dioxide and then nitrogen from the air, unit two is being evacuated to remove the previously adsorbed nitrogen. The product oxygen is substantially unaffected. After a certain period, the second bed is brought into sequential use, while the first is evacuated, etc. Depending on the operating cycle chosen, the product may be up to about 93% oxygen. The balance is nitrogen and argon. Moisture and carbon dioxide residuals are in the low ppm range. The oxygen is produced at about 24 kPa (3.5 psig) and must be compressed if the oxygen is required at higher pressures. The flow of the oxygen is unsteady but the use of a surge tank or a compressor can even out the oxygen flow.

VSA plants range from small hospital units to very large units producing as much as 229,000 m^3/d (8.0×10^6 ft^3/d). Fully assembled units in moderate sizes can be transported readily. Energy costs of larger units approach the equivalent cryogenic unit. The VSA plant can be started quickly and readily shut down. The largest use for the 90–93% purity, low pressure product is for oxygen enrichment in combustion furnaces. Oxygen may also be produced by pressure swing adsorption (PSA) units. Sales of PSA units, however, are normally for very small applications.

Another noncryogenic technique for oxygen production involves the electrolysis of water. This technique is used only to an insignificant extent. However, if the use of hydrogen energy (qv) grows significantly, hydrogen (qv) production via electrolysis could also grow; concomitantly, relatively large quantities of oxygen would be coproduced. The preferred method for hydrogen production as of 1995 was chiefly from natural gas.

Production, Pipelines, and Shipping

Production. Oxygen production facilities for relatively large users generally fall into one of three categories: (*1*) a captive plant on the oxygen user's property owned and operated by the user; (*2*) a plant owned and operated by an industrial gas company that is on or adjacent to the oxygen user's property (on-site facility), where a long-term contract for the supply of oxygen usually exists between the industrial gas company and the oxygen user; and (*3*) a plant owned and operated by an industrial gas company that supplies oxygen to several users. In the first two cases, the gaseous oxygen is generally supplied to the user site via a pipe. In the last instance gaseous oxygen is carried via a pipeline having branches to the individual industrial users. The pipeline and the central production facility are typically owned by the same industrial gas company.

If an industrial gas company owns the plant, nitrogen may also be supplied to users via additional pipelines. Liquid oxygen, liquid nitrogen, or liquid argon that can be delivered by truck to smaller users or to other industrial gas plants may also be produced. Alternatively, the liquid product could be stored on-site as a backup supply in case the air separation plant is not operating. In this last instance, the liquid oxygen is vaporized and placed in the gaseous oxygen pipeline.

Some oxygen production facilities are built solely to supply liquid oxygen, and generally liquid nitrogen and liquid argon, via truck to generally smaller customers. These plants typically use a process cycle optimized for total liquid production.

In the U.S. air separation industry, the size of a plant is specified by the design capacity for oxygen production which is usually measured in metric tons per day. Air separation plants built in the 1950s favored sizes ranging from 50 t/d ($37,500 \ m^3/d$) to 100 t/d ($75,000 \ m^3/d$). In the 1960s, many facilities built were more than double that size, and a few air separation units (ASUs) were on an order of 1000 t/d ($7.5 \times 105 \ m^3/d$) and even larger. By the end of the 1970s, ASUs rated at 2000 t/d were being completed worldwide. Since that time, however, significant increases in unit capacities have not occurred. The most spectacular assemblage is the six 2000 t/d plants built by L'Air Liquide of France for the SASOL-II coal liquefaction complex in South Africa.

Oxygen is a commodity sold either as a gas under pressure or a very low temperature liquid close to its boiling point. Because of the expense of the containers required, it cannot be stored in more than modest amounts nor can it be shipped profitably more than a few hundred kilometers. Liquid oxygen storage facilities are limited to volumes on the order of $4000 \ m^3$.

Pipelines.   Gas by pipeline is the least expensive way to manufacture and supply oxygen. The energy of refrigeration is recovered by the heat exchangers at the point where ambient-temperature gas exits and ambient-temperature air enters the plant. There is no loss by evaporation, and the costs of truck delivery are eliminated. Between 80 and 90% of all oxygen is transported in gas pipelines. Typical pipeline pressures are up to about 3450 kPa (500 psi) and diameters are usually within the range of 10–30 cm. In certain heavily industrialized areas, multiair separation plants–multicustomer oxygen–pipeline networks or complexes exist (see PIPELINES).

Oxygen pipelines exist extensively in the Gulf Coast and in the southern Lake Michigan areas of the United States. Pipelines covering hundreds of kilometers and linking 10 or more oxygen-producing plants with 24 or more users are among the largest oxygen networks. Oxygen pipelines also exist in Europe as well as in the Pacific Rim (Asian) countries.

Shipping.   Merchant oxygen gas and liquid is transported to the smaller oxygen users by tube (gas) or cryogenic tank (liquid) trailers and railroad car (liquid). Cryogenic liquid tanks or customer stations of appropriate sizes are permanently installed on the premises of large- and small-volume merchant users, including most hospitals. A liquid oxygen tank or customer station may be found outside nearly every hospital.

Liquid storage tanks at customer sites, trailers, and railroad cars are constructed to minimize the inevitable boiloff of the liquid. Generally the liquid oxygen is contained within a tank inside of another tank. The annular space between the tanks is evacuated and filled with a semisolid insulator. Using proper design of the vacuum-jacketed systems, the boiloff rates can be very low. For logistic and economic reasons, liquid oxygen is usually not hauled by truck more than about 800 km, although it can be hauled for longer distances based on acceptable boiloff rates. Liquid oxygen may also be transported for relatively short (usually up to about 400 m) distances via a vacuum-jacketed pipeline system.

If relatively large amounts of liquid oxygen are needed in one location, such as for a rocket launch where the liquid oxygen is one of the propellants, trucks and rail cars are the only practical means of transport. In some cases specially designed barges can also be used.

For users of small merchant quantities, liquid oxygen may be transferred at a distribution station to portable cryogenic containers holding as little as 148 L of oxygen, equivalent to 127 m^3 of oxygen gas. These containers can deliver either gas or liquid. The distributor also has facilities for vaporizing liquid and filling the familiar high pressure oxygen gas cylinders that are widely used, eg, by welders and medical personnel. Liquid distribution stations are equipped with a vaporizer that is heated by ambient air, electrically, or by steam to provide gas under pressure. Unlike liquid nitrogen, very little oxygen is used directly as the liquid for low temperature purposes.

About 80–90% of oxygen that is produced in the United States is distributed by pipeline. Merchant oxygen, distributed in high pressure cylinders and high pressure tube trailers, accounts for only about 1%. The method of shipping the balance of the oxygen, produced as a liquid, is one of at least four possible means. The liquid O_2 may be stored at an air separation plant site then revaporized and placed in a pipeline, hauled by tank truck (or rail car) to customers as merchant liquid oxygen, hauled by tank truck to other air separation plants for eventual distribution, or hauled by tank truck to industrial gas distribution centers where it is vaporized and placed in high pressure cylinders and high pressure tube trailers.

Economic Aspects

Table 3 summarizes annual oxygen production and market value (15). The production figures may be somewhat lower than the actual amounts produced because some of the oxygen produced and directly consumed by some small users may not be included. Very little oxygen enters into export–import trade. Overall, oxygen production increased by a factor of 3.6 between 1962 (ca $4 \times 10^9 \, m^3$) and 1992 (ca $14.5 \times 10^9 \, m^3$). VSA-produced oxygen only accounts for a few percent of total oxygen production. However, it is the fastest growth area of new applications and could reach 5–10% of the total oxygen production by the end of the 1990s.

Table 3. U.S. Production of Oxygen[a]

Year	Production, $m^3 \times 10^6$	Value, $\$ \times 10^3$
1977	11,117	354,127
1978	12,182	432,449
1979	12,925	466,146
1980	12,209	496,312
1981	12,198	594,244
1982	9,881	600,629
1983	9,798	591,028
1994	10,952	653,048
1985	10,099	601,964
1986	10,857	613,101
1987	11,472	620,484
1988	12,805	639,196
1989	13,050	644,607
1990	13,096	662,911
1991	13,377	707,110
1992[b]	14,502	

[a]Refs. 20 and 21.
[b]Figure for 1992 subject to minor revision.

Worldwide annual production of oxygen may be estimated to be about $70 \times 10^9 \, m^3$. The five largest producing areas during the 1980s were Western Europe (including all of Germany), which produced $16 \times 10^9 \, m^3$; Russia (formerly the USSR), $14.8 \times 10^9 \, m^3$; the United States, $14.5 \times 10^9 \, m^3$; Eastern Europe, $9.5 \times 10^9 \, m^3$; and Japan, $9 \times 10^9 \, m^3$. Mexico, Canada, Central and South America, China, and Korea also produced oxygen (3,21–23). The largest users of oxygen worldwide are steel and chemical producers.

In 1991 the average value of the oxygen produced was $707,110,000 representing $0.053/m^3 (15¢/100 ft^3)$. This figure is dominated by 80–90% of the oxygen that was sold via pipeline under long-term contract. Long-term contracts may be on a take-or-pay basis and may also contain energy cost clauses to reflect the energy-intensive nature of producing oxygen. For the smallest customers using liquid oxygen customer stations, liquid cylinders, or high pressure gas cylinders, the cost could be about 10 times this figure, including shipping and equipment rental charges.

Specifications and Analysis

Twelve oxygen grades are defined by the Gas Specification Committee of the Compressed Gas Association (CGA) (24), 10 of which are given in Table 4. The contaminants identified relate to possible residues from the atmosphere as well as particulates or fibers that may have been contributed by the manufacturing process or the distribution system. In addition, government agencies and certain commercial users have developed specifications for individual needs (25). In most cases, these specifications closely parallel the CGA grades.

In general, the products from a given cryogenic air separation unit (ASU) controlled by automatic instruments, sometimes using computer programming, are consistent in quality and contaminants (Table 4). The principal inert constituent in oxygen is argon, which is present because the small difference in boiling points ($\Delta t = 2.90$ K), puts undue demand on the bulk distillation process, (see Table 2). Argon, together with small traces of inert krypton and xenon, slightly dilutes the oxygen and is objectionable in uses where residual gases could interfere, eg, in some electronic processes. Nitrogen may be present, but in smaller quantities than argon. Methane (bp = 111.60 K) and higher hydrocarbons, as well as krypton and xenon, have boiling points higher than oxygen and accumulate in the oxygen liquid.

To improve oxygen purity much above 99.6%, the 99.5% product is redistilled, sometimes after catalytic oxidation of hydrocarbons (methane) followed by removal of the resulting produced carbon dioxide and moisture. Redistillation reduces not only argon and nitrogen but also the traces of krypton and xenon, providing a first step in the recovery of these two gases. The demand for argon has caused many newer (cryogenic) ASUs to be designed to remove most of the argon directly within the ASU, resulting in oxygen purity typically exceeding 99.8%.

The highest purity ($\geq 99.99\%$) oxygen is obtained through further refinement. At 99.99% the impurities total only 100 ppm. This grade of oxygen is used in the manufacture of electronic components, fiber optics (qv), etc, or for gas chromatograph calibration or research applications.

VSA-produced oxygen typically has a purity of 90–93%. The balance is nitrogen, argon, moisture, carbon dioxide, and small amounts of other gases.

Grade A.   Types I and II both represent the requirements of the USP XX (26). The USP tests arose from original formal oxygen specifications made necessary by the low purity and certain contaminants, particularly CO and CO_2, contributed by early chemical and cryogenic manufacturing methods. Containers marked Oxygen-USP must also indicate whether or not the gas has been produced by the air liquefaction process (see also FINE CHEMICALS).

Grades B and C.   Type I B and C and Type II Grade C are typical of pipeline and merchant gas and liquid. These are used for steel refining and synthetic chemical manufacturing, which together account for most oxygen consumption. Water, the only contaminant specified, could freeze out in Type II and cause gas transfer problems as well as chemical manufacturing difficulties.

Grade C, Type II is typical of liquid oxygen used as a rocket propellant oxidizer. Particulate content is limited because of the critical clearances found in mechanical parts of the rocket engine. In addition to water, acetylene and methane are limited because, on long standing, oxygen evaporation could cause concentration of these combustible contaminants to reach hazardous levels.

Other Grades.   Type I and II, Grade D are typical of oxygen purities required for aviators' oxygen masks. The limitation on odor is obvious. Limits are placed on particulates in the liquid as well as on light hydrocarbons, oxides of nitrogen, nitrous oxide, and halogenated solvents. Grade E, Type I typifies oxygen used for purging and pressurizing rocket engines. It may also be specified for more critical merchant uses. Grade F is typical of the oxygen purity required for fuel cells (qv) on space vehicles. In addition, increasing amounts of this grade of oxygen are used in research and fabrication of semiconductors, fiber optics, and similar applications.

Table 4. Compressed Gas Association, Oxygen Commodity Specifications[a]

Limiting characteristics	Type I, gas grades[b]						Type II, liquid grades[b]			
	A	B	C	D	E	F	A	B	C	D
O_2, min %	99.0[c]	99.5	99.5	99.5	99.6	99.995	99.0[c]	99.5	99.5	99.5
inert	d			100[d]			d			d
water			50	6.6	8	1.0		6.6	26.3	6.6
dewpoint, °C			−48.1	−63.1	−62.2	−76.1		−63.3	−53.1	−63.3
THC, as methane[e]					50	1.0			67.7	
methane				50						25
ethane and other hydrocarbons[f]				6						3
ethylene				0.4						0.2
acetylene				0.1		0.05			0.62	0.05
carbon dioxide	300			10		1.0	300			5
carbon monoxide	10					1.0	10			
nitrous oxide				4		0.1				2
halogenated refrigerants				2						1
solvents				0.2						0.1
other by infrared				0.2						0.1
permanent									<1.0	
particulates									mg/L	

[a]Ref. 24.
[b]Minimum contained oxygen and maximum contaminant levels where specified for the several grades of oxygen. Contaminants given in volumes per million (vpm). Low purity oxygen (93%) has not been included.
[c]USP grade.
[d]Material has no odor.
[e]THC = total hydrocarbon content.
[f]As ethane.

Analysis. *Oxygen.* Assay as minimum percentage contained is determined readily by absorption in a suitable oxygen reagent contained in a volumetric apparatus of the Orsat type, particularly the simplified version of the Hempel pipet (27), which is traditionally used in the oxygen industry. When 100 cm³ gas is drawn into the pipet, the unabsorbed inert constituents may be measured directly on a calibrated limb. Modern oxygen plants continuously monitor purity on an instrument of the thermal conductivity or paramagnetic type. These instruments must be calibrated against known standards.

Inert Gases. For oxygen of the highest purity or where grade specifications require limitations, the individual impurities are determined and the sum subtracted from 100%. The main inert constituent is argon, with lesser amounts of nitrogen, methane plus other hydrocarbons, carbon oxides, krypton, and xenon. These gases may be separated and their concentration can be estimated using a gas chromatograph or a quadrupole mass spectrometer.

Table 5. Moisture: Dew Point Conversion[a], Water Vapor per Total Mixture Volume[b]

Dew point, °C	Moisture content	
	Volume, ppm	mg/mL
−78.9	0.65	0.49
−76.1	1.02	0.76
−73.3	1.57	1.17
−70.5	2.4	1.79
−67.8	3.6	2.7
−65.0	5.4	4.0
−62.2	7.9	5.9
−59.4	11.6	8.6
−56.7	16.7	12.5
−53.9	24.0	17.9
−51.1	34.0	25
−48.3	48.0	36
−45.6	67.0	50

[a]Ref. 29.
[b]Table applies equally to air, argon, nitrogen, helium, hydrogen, and neon.

Odor.   Odor is excluded in military applications and in the four grades specified for human breathing. The gas is tested by smelling a flowing stream or by smelling a beaker from which a liquid sample has just evaporated. Oxygen that is produced in modern ASUs employing turbine compressors or nonlubricated piston machines is odorless.

Water.   Oxygen, gas or liquid, exiting the cryogenic ASU contains little or no water. To verify the dryness of gases, electrical hygrometers having scales graduated in parts per million are commercially available. These instruments include a number of models based on the direct amperometric method (28) on the piezoelectric sorption detector. Dew point, the temperature at which frost or condensed water first forms, continues to be a specification. The dew point of dry gases is usually obtained by conversion from the scale reading of the instruments, using the data given in Table 5 (29).

Total Hydrocarbon Content.   The THC includes the methane combined in air, plus traces of other light hydrocarbons that are present in the atmosphere and escape removal during the production process. In the typical oxygen sample, methane usually constitutes more than 90% of total hydrocarbons. The rest may be ethane, ethylene, acetylene, propane, propylene, and butanes. Any oil aerosol produced in lubricated piston compressor plants is also included here.

A flame-ionization, total hydrocarbon analyzer determines the THC, and the total carbon content is calculated as methane. Other methods include catalytic combustion to carbon dioxide, which may be determined by a sensitive infrared detector of the nondispersive type. Hydrocarbons other than methane and acetylene are present only in minute quantities and generally are inert in most applications.

Acetylene has a low solubility in liquid oxygen. Excessive concentrations can lead to separation of solid acetylene and produce accumulations that, once initiated, can decompose violently, detonating other oxidizable materials. Acetylene is monitored routinely when individual hydrocarbons are determined by gas chromatography, but one of the wet classical methods may be more convenient. These use the unique reaction of acetylene with Ilosvay's reagent (monovalent copper solution). The resulting brick-red copper acetylide may be estimated colorimetrically or volumetrically with good sensitivity (30).

Health and Safety Factors

Hazards associated with the use of oxygen derive from the facts that the cryogenic liquid is very cold and increases in volume enormously when it vaporizes, that combustion rates accelerate with oxygen concentrations above 21%, and that the gas is transported and dispensed in high pressure cylinders. Liquid oxygen and other cryogenic liquids can inflict severe damage on human tissues. Protective gloves, garments, and face shields are imperative, and detailed advance instruction in procedures for use are essential before any activity involving liquid oxygen is undertaken (14,31–34). At ambient pressure, the volume of a given quantity of liquid oxygen increases 860 times during the change from the boiling liquid to the gas at ambient temperatures. Hence, if a sample of liquid oxygen is confined to its original volume, a pressure of 87.2 MPa (12,640 psi) develops, bursting all but the strongest containers. Safety considerations demand that any volume where liquid could be trapped must be provided with pressure-relief valves or a bursting disk.

When oxygen from either gas or liquid sources is brought into the presence of oxidizable materials, mixtures ranging from combustible to explosive may result. Oxidizable materials include oils and greases, fingerprints, paint, plastics, rubber materials, carbon, asphalt, wood, clothing, and hair. Some solid materials are porous, eg, wood, clothing, some plastics, asphalt, and charcoal, and may absorb considerable quantities of oxygen. If these are subsequently ignited by an ignition source, very fast combustion or an explosive reaction may ensue. Up to about 1940, sawdust and liquid oxygen were ingredients of some commercial explosives. An ignition source can be as simple as a mechanical impact. Tragic accidents have occurred in which the clothing of workers in confined areas became saturated with oxygen, either accidentally in the oxygen-rich atmosphere or deliberately when the worker dusted himself with an oxygen hose. Ignition of the oxygen-saturated clothing produces an intense flash fire. In oxygen-rich atmospheres, the human body burns vigorously.

Steel cylinders containing any gas under pressure in excess of 173 kPa (25 psig) are covered by U.S. Department of Transportation (DOT) regulations. When these cylinders are moved, the valve cap must be properly secured. In storage and use, the cylinders must be chained or otherwise fastened. If it is excessively shocked, a high pressure cylinder may fracture or shatter. Equipment used in conjunction with oxygen must be kept scrupulously clean. When gas pressures are increased rapidly in the presence of contaminants, the heat of compression in an oxygen atmosphere may be high enough to ignite combustible materials.

Hazards are associated with atmospheres containing either excessive concentrations of oxygen or deficient concentrations of oxygen (34). Accidents and fatalities frequently occur in confined spaces where free access of air into the space is restricted or there is significant in-leakage of inert gases, eg, nitrogen. As a result, an oxygen-deficient atmosphere develops. The atmosphere in any confined space should be monitored.

Uses

There are relatively few substitutes for oxygen in most of its uses. Oxygen cannot be reclaimed commercially or recycled except via the atmosphere. The spectacular increase of oxygen production since World War II has in a very large measure occurred because of availability and its usefulness in steelmaking (see STEEL) and in chemicals and petrochemicals production, high temperature production furnaces, water purification, and hazardous waste destruction. In 1991 the largest uses of oxygen in the United States, accounting for 95% of production, were (21,22) steel, 40%; chemicals, 24%; metal welding and cutting, 7%; coal gasification, 7%; nonferrous metals, 7%; petroleum refining, 6%; and health, 4%.

The practical and economic advantages of the use of oxygen and oxygen-enriched air over air alone, across a number of industries, typically fall into one or more of the following categories. (*1*) Higher combustion or flame or reaction temperatures and shorter flames, yielding faster reactions at nominal reaction temperatures. Typical maximum flame temperatures using oxygen can be as high as 3033 K (5000°F), whereas for air maximum flame temperatures are usually lower by more than 555 K (1000°F). In many applications, higher temperatures permit faster melting or processing. In some processes the desired product or result may not even be achieved at the lower temperatures obtained using air. (*2*) Less waste energy in the gas or vent flue gas. To recover energy, flue gas is often heat-exchanged down to as low a temperature as is practical, but at the end of the heat-exchange process, the residual heat content retained in the flue gas represents a loss of energy. When the nitrogen is removed, the flue gas flow is reduced by about 75%, providing a distinct energy advantage. (*3*) Reduced volume of flue gas or vent gas to treat. When gases must be treated to remove pollutants, reduced volume (no nitrogen) means that the treating devices, eg, scrubber, baghouse, or electrostatic precipitator, may be much smaller in volume and the pollutant to be removed much more concentrated. In some cases lower quality, lower cost fuels or feedstocks may be used because the pollutants may be more easily removed from the flue gas or vent gas. (*4*) Reduced emissions of nitrogen oxides. The elimination of nitrogen from the combustion or process air often results in a significant reduction in the formation of nitrogen oxides such as NO, NO_2, and N_2O. (*5*) The desired product is easier to remove at the end of the process. The reduction or removal of nitrogen from the feed translates to a higher concentration of product in the plant streams and an easier separation to recover the product. (*6*) Economics of air compression. When oxygen or air is required at high, eg, >13,800 kPa (2,000 psi), pressure for a process, but only the oxygen reacts, it may be more economical to separate the air initially and only supply the oxygen at high pressure, saving the cost of compressing the nitrogen. (*7*) Nitrogen in the product is undesirable, yet cannot be removed economically.

Steelmaking. Large amounts of oxygen are used in almost all aspects of the steelmaking process, largely to oxidize the principal contaminants, eg, carbon, phosphorus, and silicon.

The largest use of oxygen for steelmaking is in the basic oxygen process (BOP), also called the basic oxygen furnace (BOF). This process consumes about 75% of all of the oxygen used in the steel industry. Oxygen is added to the molten pig iron via a lance, and the impurities are largely removed via gasification or by the formation of slag. About 0.08 metric tons of oxygen is required for each metric ton of steel produced by the BOF, utilizing advantages (1–3), and (7) outlined. Oxygen is being increasingly used to enrich the hot blast in blast furnaces. Consumption of O_2 of up to 0.10 t/t of pig iron are common.

Electric furnaces, often operated by mini-mills, account for over 35% of steel production. These furnaces operate mostly on scrap steel and produce a wide range of steels, up to high grade stainless steels. Oxygen use, via injection into the molten bath, increases the efficiency of the furnace via advantages (3) and (7). The argon–oxygen-decarburizing (AOD) process for refining stainless steel involves blowing argon–oxygen mixtures into a special AOD vessel. This process follows the electric funance via advantages (3) and (7).

Oxygen is also utilized in a variety of other applications within the steelmaking industry. For example, it is used to reheat furnaces, using oxygen or oxygen enrichment; and to reheat steel more quickly, requiring less fuel gas (advantages 1–4). Also, oxygen scarfing or skimming removes surface defects from steel billets (advantage 1); ladle preheating using fuel and oxygen or oxygen enrichment utilizes advantages (1), (3), and (4).

Nonferrous Metallurgy. In nonferrous metallurgy, oxygen enrichment of the fuel–air flame in reverberatory smelting furnaces and oxygen use in autogeneous smelting save money and time in the smelting of copper (qv), lead (qv), antimony, and zinc. A rapidly growing use for oxygen involves flame enrichment in flash smelters for copper and nickel production from sulfide ores. The addition of oxygen increases productivity, improves sulfur recovery in the form of sulfuric acid, and provides for easier control of emissions via advantages (1–4).

Uranium *in situ* leaching is a more recent development in which an oxidative–complexing solution recovers uranium from low grade, deep-lying deposits that may be water-saturated. Oxygen is finding increased use in gold recovery. Oxygen is used to prepare the less desirable sulfide-based gold ores for refining because of advantage (1).

Oxygen, often the VSA-produced 90–93% purity, is beginning to be used for the recycling of aluminum because of advantages (1–4).

Other Furnace Temperature Applications. Oxygen-enriched combustion within glass (qv) furnaces for fiber glass, bottle glass, and float glass is developing a newer use for oxygen. Oxygen-enriched combustion within cement (qv) kilns and brick kilns is also developing (advantages 1–4). VSA-produced oxygen in the 90–93% purity is often adequate for many of these applications.

Chemical Processing. The use of oxygen in large-volume chemical and petrochemical manufacture is well-established as a result of advantages (3) and (4). Most oxidation reactions are catalytic; many begin with a feedstock initially made catalytically from methane or natural gas.

Syngas is a basic feedstock in the chemical and petrochemical industry and consists of carbon monoxide and hydrogen. Syngas is generally made via partial oxidation by mixing oxygen, natural gas, and steam and passing the mixture over a catalyst (advantages 1 and 5). The hydrogen of syngas may be separated from the carbon monoxide and the carbon monoxide combined with steam and a catalyst to make additional hydrogen as well as carbon dioxide. When the hydrogen is again separated, the resulting two streams of hydrogen represent the main commercial source of hydrogen. This hydrogen is widely used to make ammonia, to hydrotreat petroleum, and for metallurgical applications.

Partial oxidation of natural gas or a fuel oil using oxygen may be used to form acetylene, ethylene (qv) and propylene (qv). The ethylene in turn may be partially oxidized to form ethylene oxide (qv) via advantages (1) and (5). A few of the other chemicals produced using oxygen because of advantages (1) and (5) are vinyl acetate, vinyl chloride, perchloroethylene, acetaldehyde (qv), formaldehyde (qv), phthalic anhydride, phenol (qv), alcohols, nitric acid (qv), and acrylic acid.

Partial oxidation of coal to form either synthetic fuel, syngas, or synthetic natural gas represents a potential use of oxygen (see FUELS, SYNTHETIC). Programs designed to demonstrate these technologies from coal and other solid fuels are being driven primarily by environmental benefits. Advantages (1), (2), and (5) come into play. Oxygen is also used in a growing number of applications within petroleum (qv) refineries including oxygen enrichment of air used to regenerate fluidized catalytic cracking (FCC) catalysts and improving sulfur recovery via the Claus process (35).

Other Industrial Uses. In the pulp (qv) and paper (qv) industries, oxygen is needed to make lime used to prepare or reconstitute white liquor. Oxygen is added to the air–fuel primary burner to enhance the intensity of the flame. The increased heat raises the lime production capacity (advantage 1). Additional benefits are obtained from fuel savings, improved lime reactivity, and control of sulfur odor (36) (advantages 1 and 3). Oxygen is also used for black liquor and white liquor oxidation. Many pulp and paper applications of oxygen are driven by environmental requirements (advantage 3). Oxygen may also be used to delignify and bleach wood chips, increasing the quality and yield of the pulp. The result is lower material costs and the use of an inherently nonpolluting process (advantage 3).

Oxygen, when introduced into fish-farming ponds and pools, particularly trout pools, enables these fish to increase their food intake and hence grow more rapidly (see AQUACULTURE CHEMICALS). Fermentation operations are accelerated by sparging with oxygen (see AERATION, BIOTECHNOLOGY; FERMENTATION).

Oxygen is used to treat municipal wastewater and wastewater from the pulp and paper industry (see AERATION, WATER TREATMENT; WASTES, INDUSTRIAL; WATER). Many of these water applications can use VSA-produced oxygen (advantage 1). Demonstration and development programs are in place that use oxygen to oxidize sludge from municipal waste and burn hazardous wastes and used tires (advantages 1–4).

Medical Applications. The maintenance of arterial blood oxygen at a partial pressure close to the normal level of 13.3 kPa (100 mm Hg) is fundamental to all uses of oxygen in medical and life support. At saturation, each 100 mL of arterial blood holds 19 vol % oxygen. Of this, 0.3 mL are in solution in the plasma; the remainder is chemically bound as oxyhemoglobin. If the hemoglobin is below normal, ie, <14.89 g/100 mL, less oxygen is carried. As blood flows around the body under normal metabolic conditions, 100 mL of blood gives 5 mL of oxygen to the tissue, then returns to the lungs as venous blood carrying a basic 14 mL O_2. Oxygen in solution is 0.13 mL/100 mL blood and the rest is present as reduced hemoglobin. If 100% oxygen is inhaled instead of air, the 0.3 mL oxygen in solution increases to 2 mL and the amount of oxygen taken up by the hemoglobin may be increased (37). Higher concentrations of oxygen are valuable in case of severe anemia and other, similar disorders.

Modern hospitals include a built-in system from which oxygen is conducted to convenient bedside locations and into operating rooms. The supply point is usually an outdoor cryogenic container from which liquid oxygen is withdrawn and vaporized automatically as needed. Modern anesthesia routinely uses oxygen as a component of the gaseous mixture, thereby ensuring an adequate supply for life support and maintaining the diluting effects of gaseous anesthetics.

Oxygen is supplied quite routinely to patients suffering impaired respiratory function as well as in other situations where oxygen is deemed to be useful. The pure oxygen, with humidification, is delivered via a simple double tube (cannula) to a point just inside the nostrils where the 99.5% gas blends with the room air (21% O_2), and is inhaled. The concentration of oxygen that reaches the lungs thus depends on the rate and volume of air inhaled and on the exit flow of oxygen from the cannula, usually one to six L/min.

The use of oxygen in pediatric incubators is an important factor in increasing the survival rate of premature infants who develop cyanosis. However, the use of oxygen is associated with risk of developing the visual defect known as retrolental fibroplasia (38). A careful monitoring of arterial blood oxygen partial pressure is important.

In the past, hyperbaric oxygenation as a medical procedure has received considerable attention. In this treatment the patient is given pure oxygen and may be placed in a pressurized chamber. In effect, the patient may thus receive ≥ 400 kPa (≥ 4 atm) of pure oxygen. Beneficial results in cases of carbon monoxide poisoning, gangrene, severe burns, and other difficulties are often achieved as a result of this treatment.

Oxygen inhalators are used as a first-aid measure for a long list of emergencies, including heart attacks and suffocation, and as a result are carried routinely by rescue squads. Oxygen–helium mixtures have proved beneficial in asthmatic attacks, because these permit more rapid flow of gas into congested areas of the respiratory system.

Life-Support Applications. Exploration of outer space by humans has focused considerable attention on maximum as well as minimum limits in the oxygen content of life-support atmospheres. Above the earth, both the atmospheric pressure and the partial pressure of oxygen decrease rapidly. The oxygen content of air remains constant at 20.946% to an altitude of ca 20 km, after which it decreases rapidly (1).

Human evolution has taken place close to sea level, and humans are physiologically adjusted to the absolute partial pressure of the oxygen at that point, namely 21.2 kPa (159.2 mm Hg), ie, 20.946% of 101.325 kPa (760 mm Hg). However, humans may become acclimatized to life and work at altitudes as high as 2500–4000 m. At the 3000-m level, the atmospheric pressure drops to 70 kPa (523 mm Hg) and the oxygen partial pressure to 14.61 kPa (110 mm Hg), only slightly above the 13.73 kPa (102.9 mm Hg) for the normal oxygen pressure in alveolar air. To compensate, the individual is forced to breathe much more rapidly to increase the ratio of new air to old in the lung mixture.

In passenger aircraft flying at higher altitudes, the cabin is customarily pressurized to an elevation equivalent to ca 2300 m, where the oxygen partial pressure is 16.13 kPa (121 mm Hg), which is equivalent to a sea-level concentration of approximately 16% oxygen. With passengers at rest no discomfort is observed. To provide for an emergency in which the oxygen partial pressure drops dangerously, for example at 11.5 km oxygen partial pressure is 4.46 kPa (4.4% O_2) vs 21.22 kPa (20.95% O_2) sea level equivalents, oxygen masks are provided for each passenger. Hence, using pure oxygen as a supplement, normal physiological exchange continues despite loss of cabin pressure. However, above 11.5 km even pure oxygen is insufficient and the plane is forced to descend.

Where cabin pressure cannot be maintained, as in military aircraft, supplemental oxygen must be provided by mask, and somewhat higher oxygen partial pressures are the rule. The U.S. Air Force has established 56.7 kPa (425 mm Hg) as the maximum oxygen partial pressure supplied to aviators. Exposure to higher partial pressures for more than a few hours leads to pneumonia, difficulties in the central nervous system, and other complications. Similarly, the lower practical limit in military practice has been established at about 16.7 kPa (125 mm Hg). The astronaut's space suit is pressurized with pure oxygen to 34.5 kPa (259 mm Hg), equivalent to 34% oxygen at earth's surface (see OXYGEN-GENERATION SYSTEMS).

Life-support atmospheres supplied in underwater operations require even more careful control. The dangers of excessive pressures of nitrogen and other inert gases are well known. When oxygen is present in excessive amounts and the period of exposure is long, toxic effects develop. The problem thus becomes one of choosing the gas mixture that simultaneously provides mechanical support and becomes the carrier and diluent for the oxygen (39). The borderline incidence of pulmonary toxicity is considered to be 60.8 kPa (456 mm Hg) of oxygen, a <5% reduction in vital capacity for resting humans inhaling pure oxygen continuously in excess of 750 h (38). At 50.66 kPa (380 mm Hg), 50% oxygen at sea level, tolerance times are infinite. Concentrations of oxygen above 284 kPa (2.8 atm) cause difficulties in the central nervous system within a few hours (40).

BIBLIOGRAPHY

"Oxygen" in *ECT* 1st ed., Vol. 9, pp. 713–734, by R. F. Benenati, Polytechnic Institute of Brooklyn; in *ECT* 2nd ed., Vol. 14, pp. 390–409, by A. H. Taylor, Airco Industrial Gases Division, Air Reduction Co., Inc.; in *ECT* 3rd ed., Vol. 16, pp. 653–673, by A. H. Taylor, Airco, Inc.

1. B. A. Mirtov, *Gaseous Composition of the Atmoshere and Its Analysis*, trans., NASA TTF-145, OTS 64-11023, U.S. Dept. of Commerce, Washington, D.C., 1964, p. 22.
2. A. L. Lavoisier, *La Traité Elémentaire de Chemie*, 1789, R. Kerr, trans., *Elements of Chemistry*, 1790; facsimile reprint, Dover Publications, Inc., New York, 1965.
3. W. F. Keyes, in *Mineral Facts and Problems*, U.S. Bureau of Mines, Bulletin 667, Stock No. 024-004-01893-3, U.S. Government Printing Office, Washington, D.C., 1975.
4. R. L. Parker, in M. Fleischer, ed., *Data in Geochemistry*, 6th ed., Geological Survey Professional Paper 440D, U.S. Government Printing Office, Washington, D.C., 1967, Chapt. D, Table 18.

5. J. Schopf, *Endeavour, 55* (May 1975).

6. W. Day, *Genesis of Planet Earth*, House of Talos Publishers, East Lansing, Mich., 1979, pp. 72–76.

7. *Chart of the Nuclides*, 12th ed., General Electric Co., Nuclear Power Systems Division, San Jose, Calif., 1977.

8. L. A. Webber, *Thermodynamic and Related Properties of Oxygen fromTriple Point to 300 K at Pressures to 1000 Bar*, NASA Reference Publication 1011, NBSIR 77–865, National Bureau of Standards, Boulder, Colo., 1977.

9. *Industrial Gas Fact Book*, Air Products and Chemicals, Inc., Allentown, Pa., 1990.

10. H. Stephen and T. Stephen, *Solubilities of Inorganic and Organic Compounds*, Vol. 1, The Macmillan Co., New York, 1963, p. 87.

11. Handbook of Compressed Gases, 3rd ed., Compressed Gas Association, Inc., Van Nostrand Reinhold Co., Inc., 1990.

12. W. Braker and A. Mossman, eds., *Matheson Gas Data Book*, 6th ed., Matheson, Inc., Lyndhurst, N.J., 1980.

13. F. A. Cotton and G. Wilkinson, *Advanced Inorganic Chemistry, A Comprehensive Text*, Wiley-Interscience, New York, 1980.

14. *Cleaning Equipment for Oxygen Service*, Document CGA-4.1, Compressed Gas Association, Arlington, Va., 1984.

15. B. R. Dunbobbin and co-workers, "Oxygen Compatibility of High-Surface-Area Materials," in *Flammability and Sensitivity of Materials in Oxygen-Enriched Atmospheres*, Vol. 5, ASTM STP 1111, American Society for Testing and Materials, Philadelphia, Pa., 1991.

16. C. M. Austin and co-workers, *J. Chem. Educ.* **36**(2), 38 (1959).

17. D. R. Bennett and co-workers, *Cryogenic Air Separation Equipment Design*, AIChE Tutorial on Cryogenic Technology, Houston, Tex., 1993.

18. R. Agrawal and co-workers, "Impact of Low Pressure Drop Structure Packing on Air Distillation," Institute of Conference on Distillation, Birmingham, UK, 1992.

19. S. Sircar, "Air Fractionation by Adsorption," *Separ. Sci. Technol. Chem. Eng.* **23**, 2379 (1988).

20. Current Industrial Reports, Series M28C, U.S. Department of Commerce, Bureau of the Census, Washington, D.C.

21. U.S., *Chemical Industry Statistical Handbook*, Chemical Manufacturers Association, Washington, D.C., 1992.

22. D. Hunter, *Chem. Week,* 30 (Feb. 23, 1994).

23. W. Stowasser, "Oxygen" in *Mineral Facts and Problems*, U.S. Bureau of Mines Bulletin 667, U.S. Government Printing Office, 1980.

24. *Commodity Specification for Oxygen*, Pamphlet G. 4.3, Compressed Gas Association, Inc., Arlington, Va., 1989.

25. *Liquid & Gas, MIL-0-27210E (1977); Propellant MIL-P-25508E (1975)*, Oxygen Specifications: Agencies of U.S. Government, current revisions of Aviator's Breathing Oxygen, NASA, MSFC Spec. 399.

26. *The United States Pharmacopeia XX (USP XX-NFXV)*, The United States Pharmacopeial Convention, Inc., Rockville, Md., 1980, p. 572.

27. W. W. Scott, in N. H. Furman, ed., *Standard Methods of Chemical Analysis*, 6th ed., Vol. 1, D. Van Nostrand Co., Inc., New York, 1962, p. 980.

28. F. A. Keidel, *Anal. Chem.* **31**, 2043 (1959).

29. A. Wexler and R. W. Hyland, N-BS Report to ASHRAE on Project 216, unpublished, Washington, D. C.

30. E. Yasui and H. Suzuki, *J. Chem. Soc. Jpn. Ind. Chem. Sec.* 61(2), 176 (1958).

31. *Industrial Practices for Gaseous Oxygen Transmission and Distribution Piping Systems*, Document CGA-4.4, Compressed Gas Association, Arlington, Va., 1993.

32. *Cryogenics Safety Manual—A Guide to Good Practice*, Safety Panel, The British Cryogenics Council, London, 1970.

33. E. Szymansk and J. Senesksy, *Plant Eng.*, 289 (April 19, 1979); 143 (May 17, 1979); 159 (June 14, 1979).

34. A. Lapin, *Liquid and Gaseous Oxygen Safety Review*, NASA CR-120922 ASRDI, NASA Lewis Research Center, Cleveland, Ohio, 1972 (available from National Technical Service, U.S. Department of Commerce, Springfield, Va.).

35. J. C. Bronfenbrenner and co-workers, *Energy Prog.* **7**(2), 105 (1987).

36. H. B. H. Cooper, Jr., *Black Liquor Oxidation with Molecular Oxidation in a Plug Flow Reactor*, Ph.D. dissertation, University of Washington, Seattle, Wash., 1972.

37. A. B. Vaughan, in *Anaesthetics*, Oxford University Press, London, 1969.

38. L. S. James and J. T. Lanman, *Pediatrics* **57**, 591 (Apr. 1976).

39. J. M. Clark and C. J. Lambertsen, in C. J. Lambertsen, ed., *Underwater Physiology*, The Williams and Wilkins Co., Baltimore, Md., 1967.

40. K. W. Donald, in I. M. Ledingham, ed., The Williams and Wilkins Co., Baltimore, Md., 1965.

JAMES G. HANSEL
Air Products and Chemicals, Inc.

OXYGEN-GENERATION SYSTEMS

Oxygen (qv) generation from oxygen-containing compounds is used for systems for respiratory support in submarines, aircraft, spacecraft, and bomb shelters, as well as in breathing apparatus (1). Convenience and reliability, rather than low cost, are stressed. Discussed herein are systems primarily based on chlorates, perchlorates, peroxides, and superoxides (see CHLORINE OXYGEN ACIDS AND SALTS; PERCHLORIC ACID AND PERCHLORATES; PEROXIDES AND PEROXIDE COMPOUNDS). It does not include oxygen-separation systems or photosynthesis.

Chlorates and Perchlorates

The chlorates and perchlorates of lithium, sodium, and potassium evolve oxygen when heated. These salts may be compounded with a fuel to form a chlorate-based candle that produces oxygen by a continuous reaction. Components include the oxygen-producing material, a fuel, a material which fixes traces of chlorine, and usually an inert binder. Once the reaction begins, oxygen is released from the hot salt by thermal decomposition. A portion of the oxygen reacts with the fuel to produce more heat resulting in production of more oxygen, and so on.

Relevant properties of the chlorates are given in Table 1. Sodium chlorate is generally used. The lithium and potassium salts are more expensive and present

Table 1. Chlorates and Perchlorates as Sources for Oxygen

Substance	Molecular formula	CAS Registry Number	Mp, °C	Decomp., °C[a]	Oxygen density	
					g/g cmpd	g/cm^{3b}
lithium chlorate	LiClO$_3$	[13453-71-9]	129	270	0.53	1.39
sodium chlorate	NaClO$_3$	[7775-09-9]	261	478	0.45	1.12
potassium chlorate	KClO$_3$	[3811-04-9]	357	400	0.39	0.93
lithium perchlorate	LiClO$_4$	[7791-03-9]	247	410	0.60	1.46
sodium perchlorate	NaClO$_4$	[7601-89-0]	471	482	0.52	1.31
potassium perchlorate	KClO$_4$	[7778-74-7]	585	400	0.46	1.17
oxygen	O$_2$					
liquid					1.0	1.14
at 52 MPa^c					1.0	0.58
hydrogen peroxide	H$_2$O$_2$	[7722-84-1]	−89		0.94	1.37
water	H$_2$O		0		0.89	0.80

^{a}Without catalyst.
^{b}Based on crystal densities.
^{c}To convert MPa to psi, multiply by 145.

some problems in maintaining candle integrity during use. Lithium chlorate is also difficult to dry. However, a lithium perchlorate candle of very high oxygen storage density has been described (2).

Chlorate candles are quite stable. While normally hermetically sealed, these candles have been stored uncontained for as long as 20 years, and then operated successfully with no loss of oxygen output. Thus, they are well suited as emergency oxygen-generation systems (3). Chlorate candles also produce oxygen under pressure and, therefore, can be stored in or operated from pressurized cylinders.

Materials and Reactions. Candle systems vary in mechanical design and shape but contain the same generic components (Fig. 1). The candle mass contains a cone of material high in iron which initiates reaction of the solid chlorate composite. Reaction of the cone material is started by a flash powder train fired by a spring-actuated hammer against a primer. An electrically heated wire has also been used. The candle is wrapped in insulation and held in an outer housing that is equipped with a gas exit port and relief valve. Other elements of the assembly include gas-conditioning filters and chemicals and supports for vibration and shock resistance (4).

A fuel provides heat upon reaction with some of the generated oxygen. Whereas a variety of powdered elements, eg, Fe, B, Al, Co, etc, have been used with varying degrees of success, iron is the universal choice for commercial applications. Some of the oxygen from the chlorate decomposition combines with the iron to generate heat to effect decomposition of the chlorate, leaving a mixture

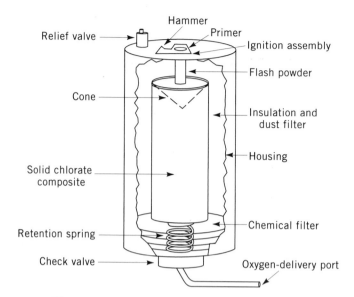

Fig. 1. Cutaway view of generator housing.

of iron oxides (3). Generalized principle reactions are as follows:

$$2\ NaClO_3 \longrightarrow 2\ NaCl + 3\ O_2$$
$$2x\ Fe + y\ O_2 \longrightarrow 2\ Fe_xO_y$$

where Fe_xO_y represents an intermediate between FeO, Fe_3O_4, and Fe_2O_3. An additional endothermic minor reaction is

$$4\ NaClO_3 \longrightarrow 2\ Na_2O + 2\ Cl_2 + 5\ O_2$$

A bleach odor, usually produced during oxygen evolution, is attributed to chlorine-containing compounds, eg, chlorine dioxide or elemental chlorine. These materials are undesirable for breathing oxygen. Thus barium peroxide [1304-29-6] (2–5 wt %) is usually added to the candle mix to scavenge these materials. Some oxygen evolves from this reaction, producing barium chloride (5).

Glass fiber or steel wool are used as binders to improve structural integrity, and account for 4–5 wt % of the candle. The burning zone is reinforced so that the ash and the unused portions cohere. Binders may also enhance oxygen generation. Glass fiber is generally used. In the past, asbestos (qv) was employed, but asbestos catalyzes evolution of chlorine and is unsuitable for breathing apparatus. Steel wool acts as both binder and fuel, but results in a hotter burning candle.

Basic oxides of metals such as Co, Mn, Fe, and Cu catalyze the decomposition of chlorate by lowering the decomposition temperature. Consequently, less fuel is needed and the reaction continues at a lower temperature. Cobalt metal, which forms the basic oxide *in situ*, lowers the decomposition of pure sodium chlorate from 478 to 280°C while serving as fuel (6,7). Composition of a cobalt-fueled system, compared with an iron-fueled system, is 90 wt % NaClO, 4 wt % Co, and 6 wt % glass fiber vs 86% NaClO, 4% Fe, 6% glass fiber, and 4% BaO. Initiation of the former is at 270°C, compared to 370°C for the iron-fueled candle. Cobalt hydroxide produces a more pronounced lowering of the decomposition temperature than the metal alone, although the water produced by decomposition of the hydroxide to form the oxide is thought to increase chlorine contaminate levels. Alkaline earths and transition-metal ferrates also have catalytic activity and improve chlorine retention (8).

Oxygen Purity. Impurities in oxygen-generating systems that have been identified are the chlorine-containing compounds, CO, CO_2, H_2O, and simple organics. All can be minimized by using high purity ingredients and control of moisture, or by gas conditioning. Aside from the chlorate and barium peroxide, materials are degreased by heating at 472°C before manufacture. Iron powders usually are reduced with hydrogen.

Carbon, present in iron or remaining after inadequate degreasing, can form CO or CO_2. Carbon particles may occur in the chlorate if graphite anodes were used in the production process. Additionally, barium peroxide contains carbonate as a contaminant.

Water in the candles influences the evolution of chlorine and chlorine dioxide (9,10). Hydrogen chloride generated from trace impurities in the chlorate can react to produce the same compounds.

$$NaClO_3 + 6\,HCl \longrightarrow NaCl + 3\,H_2O + 3\,Cl_2$$

$$5\,NaClO_3 + 6\,HCl \longrightarrow 6\,ClO_2 + 3\,H_2O + 5\,NaCl$$

Hydrocarbon impurities can be introduced by contamination of the outer surface of the candle if mold release compounds are used. Candles are sometimes shaved to reduce such contamination.

The purity of oxygen from chlorate candles before and after gas filtration is indicated in Table 2. A particulate filter is always used. Filter chemicals are Hopcalite, which oxidizes CO to CO_2; molecular sieves (qv), which remove chlorine compounds; and basic materials, eg, soda lime, which removes CO_2 and chlorine compounds. Other than H_2O and N_2, impurity levels of <1 ppm can be attained. Moisture can be reduced by using a desiccant (see DESICCANTS). Gas purity is a function of candle packaging as well as composition. A hotter burning unit, eg, one in which steel wool is the binder, generates more impurities.

Candle Fabrication. All ingredients must be contaminant-free (especially grease) and the chemical materials must be dry. The oxygen-generating mass is made by mixing and then pressing or casting the ingredients. Care must be exercised to assure thorough mixing or reaction rates can vary throughout

Table 2. Impurity Levels in O$_2$ from Chlorate Candles

Impurity	Quantity, ppm	
	Unfiltered	Filtered
H$_2$O	1000	1000
CO	5–25	1
CO$_2$	50–400	5
Cl$_2$	2–10	0.2
hydrocarbon	2	1

the candle. Shape can be varied as desired, especially if casting is used. With pressing, the shape is limited to some extent, although hydrostatic pressing provides freedom in candle form. All other factors being equal, the rate of oxygen evolution is directly proportional to the cross-sectional area of the unit. Some oxygen evolution rate control can be achieved using graded compositions along the length of the candle (7,11).

Casting is done by heating the ingredients to just above the melting point of the chlorates, eg, 261°C for sodium chlorate. Materials that do not become molten must remain in suspension to maintain a constant composition throughout the cast candle. When the mixed slurry is poured into the mold some settling may occur, depending on the length-to-diameter ratio of the mold. Some items may be cast on their sides, rather than upright to reduce settling in deep molds. Cast items are characterized by high density and maximum resistance to breaking.

Pressing can be done at room temperature using a damp mix or at elevated temperature using finely divided materials. Hydrostatic pressing can be done. For a length-to-diameter ratio of ca 2:1, wet pressing is satisfactory. Units used by the U.S. Navy are ca 17.8 cm in diameter and ca 30.5 cm long, and are produced by pressing a composition containing ca 1.5 wt % water to ca 28 MPa (4060 psi). They are dried to ca 0.5 wt % moisture before packaging for shipment.

Hot pressing has been used, especially for smaller shapes. No moisture is added to the mix, and generally the materials are more finely divided than in cold pressing. The mixture and molds are heated to ca 200°C and pressed as required. Room temperature pressing can sometimes be done using very finely divided materials, eg, ca 0.04-mm (300-mesh) particle size, but the mechanical integrity of the candle is poor. This process is considered for low cost production for special applications. In any pressing technique, density gradations along the length of the candle produce differences in burning rates. Excessive pressure and densification can result in nonfunctioning of the candle.

Reaction can be initiated by several means, all of which depend on delivery of heat at a relatively high temperature to a starting cone. Cartridge-actuated and electric match units are usually used. The former is in the majority. A water-activated unit has been described (12). The heat generated by the starting device initiates reaction in a cone, which is a small amount of candle that is higher in fuel content, eg, 30 wt % iron. Compared to the candle composition, more heat is generated per unit mass and smaller amounts of oxygen are evolved.

Cones usually are pressed. These are easier to ignite than cast cones. Pressure used for cone formation is considerably lower than that used for candle formation to prevent excessive densification. The preformed cone can be added to the mold before pressing or casting the body of the candle, being held in place by the resulting bond. Alternatively, a cavity is pressed into the formed candle, and cone material is added and pressed into place. The bond formed by this method is weak but generally sufficient for most purposes.

Operational Characteristics. Oxygen generation from chlorate candles is exothermic and management of the heat released is a function of design of the total unit into which the candle is incorporated. Because of the low heat content of the evolved gas, the gas exit temperature usually is less than ca 93°C. Some of the heat is taken up within the candle mass by specific heat or heat of fusion of the sodium chloride. The reacted candle mass continues to evolve heat after reaction ends. The heat release during reaction is primarily a function of the fuel type and content, but averages 3.7 MJ/m^3 (100 Btu/ft^3) of evolved oxygen at STP for 4–8 wt % iron compositions.

The oxygen release rate is directly proportional to the cross-sectional area of the candle for a specific composition and also depends on the linear burn rate. Lower fuel contents decrease the burn rate slightly, eg, ca 2 wt % iron is the lower limit for reliable room temperature operation. Low temperature starts require at least 3.5 wt % iron. Another factor is direction of flow of the evolved gas. If the hot oxygen flows over the unburned portion of the candle, as much as 15% rate increases can be produced. The burn time is halved for each 3.4 MPa (500 psi) pressure rise. The highest pressure that can be produced is ca 138 MPa (20,000 psi).

Uses. A primary early use was the incorporation of a small cast candle in a quick-start canister, which was filled with potassium superoxide [12030-88-5] and used in a portable breathing apparatus. The candle rapidly produced an initial supply of oxygen until the superoxide became fully activated, particularly at lower temperatures. Large candles, delivering 3–4 m^3 (120 ft^3) oxygen in 45 minutes, are used in long-duration submergence submarine operation. A furnace holds a stack of two candles; the upper one is ignited, which subsequently ignites the lower one. Together these furnish enough oxygen for 120 people for 1.5 hours.

Special uses include oxygen for fuels intended for long-term unattended service (13). Candles are fired on demand by a pressure sensor in the oxygen accumulator. Chlorate candles were incorporated into backpacks designed for use in the Apollo moon missions (14).

A large-scale use of chlorate candles is oxygen supply in the event of decompression in passenger aircraft (15). Candles meet the requirements of no maintenance (qv), no oxygen leakage, high reliability, and long (15-yr min) storage life. Both percussion cap and electrical ignition are used. Candle sizes vary, supplying up to five passengers from one unit. A system for the U.S. Army is a 12-candle array that maintains pressure in an oxygen accumulator bottle for a helicopter crew. Operation is completely automatic.

In projecting uses for candles, the characteristics to be considered are volumetric oxygen density that is nearly equal to that of liquid oxygen, development of extremely high pressure, high oxygen purity, long storage life with no leak-

Table 3. Chlorate Oxygen-Generator Parameters

Parameter	Value
average composition[a], wt %	
$NaClO_3$	80–85
Fe	3–10
BaO_2	4
O_2 available[b], wt %	40
sp gr of mixture	2.3–2.5
average reaction rate through shape, cm/min	0.64
heat evolution, J/g[c]	837
reaction zone temperature, °C	538
shape of chlorate unit	unlimited
starting method	hot wire, Bouchon cup
cone material[d], wt %	
Fe	30
$NaClO_3$	60
time before O_2 evolved[b], s	1
gas purity, %	>99.8

[a]Remainder is binder.
[b]Value is maximum.
[c]To convert J to cal, divide by 4.184.
[d]Remainder is binder and BaO_2.

age, preprogrammed oxygen delivery, and heat release. General parameters are given in Table 3.

Peroxides and Superoxides

Whereas chlorate candles produce oxygen, devices made with peroxides and superoxides produce oxygen and absorb carbon dioxide. For every volume of oxygen that a person inhales, 0.82 volumes of carbon dioxide are exhaled giving a respiratory coefficient (RQ) of 0.82. An oxygen-generation device used in a closed system should mirror this RQ, ie, absorb 0.82 volumes of carbon dioxide for each 1.0 volume of oxygen liberated. However, even complete recovery of oxygen from the carbon dioxide which is exhaled is not sufficient to sustain human life. Thus, long-term space flight requires additional oxygen, usually supplied by electrolysis of water.

The relevant properties of peroxide and superoxide salts are given in Table 4 (see PEROXIDES AND PEROXY COMPOUNDS, INORGANIC). Potassium peroxide is difficult to prepare and lithium superoxide is very unstable. The ozonides, MO_3, of the alkali metals contain a very high percentage of oxygen, but are only stable below room temperature (see OZONE).

Peroxides. In the presence of lithium peroxide, both water and carbon dioxide react, resulting in evolution of oxygen. The following steps have been postulated (17):

$$Li_2O_2 + 2 H_2O \longrightarrow 2 LiOH + H_2O_2$$

Table 4. Peroxides and Superoxides as Oxygen Sources[a]

Substance	Molecular formula	CAS Registry Number	Quantity of gas, kg/kg cmpd		RQ[c]	Unstable above, °C	Mp, °C
			O_2	CO_2^b			
lithium peroxide	Li_2O_2	[12031-80-0]	0.35	0.96	1.94	315	315
sodium peroxide	Na_2O_2	[1313-60-6]	0.21	0.56	1.94	311–400	596
sodium superoxide	NaO_2	[12034-12-7]	0.43	0.40	0.67	100	d
potassium superoxide	KO_2	[12030-88-5]	0.34	0.31	0.67	145–425	440
calcium superoxide	$Ca(O_2)_2$	[12133-35-6]	0.46	0.42	0.67	200[e]	

[a]Ref. 16.
[b]To form carbonate.
[c]Respiratory quotient.
[d]Forms series of solid solutions; forms Na_2O_2 at 250°C.
[e]Value is estimated.

$$2\ LiOH + CO_2 \longrightarrow Li_2CO_3 + H_2O$$

$$H_2O_2 \longrightarrow H_2O + 0.5\ O_2$$

$$Li_2O_2 + CO_2 \longrightarrow Li_2CO_3 + 0.5\ O_2$$

Because of the delay in decomposition of the peroxide, oxygen evolution follows carbon dioxide sorption. A catalyst is required to obtain total decomposition of the peroxides; 2 wt % nickel sulfate often is used. The temperature of the bed is the controlling variable; 204°C is required to produce the best decomposition rates (18). The reaction mechanism for sodium peroxide is the same as for lithium peroxide, ie, both carbon dioxide and moisture are required to generate oxygen. Sodium peroxide has been used extensively in breathing apparatus.

All the peroxides are colorless and diamagnetic when pure. Traces of the superoxide in technical-grade sodium peroxide impart a yellow color. Storage containers must be sealed to prevent reaction with atmospheric carbon dioxide and water vapor.

Superoxides. The superoxides are colored and paramagnetic: KO_2 is yellow, NaO_2 is orange-yellow, and $Ca(O_2)_2$ is red. In uses as oxygen suppliers and carbon dioxide scrubbers, these materials are demand chemicals, ie, react to the load imposed by generating more oxygen as more water is introduced. Reactions during use, where M is monovalent, are reaction with water to liberate oxygen and then reactions involving carbon dioxide:

$$2\ MO_2 + H_2O \longrightarrow 2\ MOH + 1.5\ O_2$$

$$2\ MOH + CO_2 \longrightarrow M_2CO_3 + H_2O$$

$$2\ MOH + 2\ CO_2 \longrightarrow 2\ MHCO_3$$

Some moisture must be present to promote these reactions. Dry carbon dioxide does not react with the superoxides. One mole of MO_2 yields 1.5 mol oxygen and then absorbs one mole of carbon dioxide to form carbonate, or two moles to form bicarbonate. Thus, if only carbonate is formed, an RQ of 0.67 is reached; if only bicarbonate is formed, an RQ of 1.33 results. The required stoichiometry for an RQ of 0.82 is

$$2\ MO_2 + 1.23\ CO_2 + 0.23\ H_2O \longrightarrow 0.77\ M_2CO_3 + 0.46\ MHCO_3 + 1.5\ O_2$$

By control of the bed temperature and of the moisture content of the inlet gas, some control over the ratio of carbonate to bicarbonate can be obtained.

Potassium superoxide, the most commonly used superoxide of the alkali metals, is produced by spraying the molten metal into dry air. The superoxide is formed as a fine yellow powder which is processed to the desired size. Copper oxychloride is added to activate the superoxide at low temperatures. Sodium superoxide is produced from an open-pore sodium peroxide which is produced by spraying liquid sodium into dry air. Oxygen at 13.8 MPa (2000 psi) and 400°C is used to oxidize the peroxide to the superoxide. The material then can be pressed and regranulated as required. Calcium superoxide has been produced to 65% purity by careful dehydration of $CaO_2 \cdot 2H_2O_2$ (19). Calcium superoxide is of interest for breathing apparatus because calcium hydroxide has a higher melting point than the hydroxides of potassium or sodium, and calcium superoxide is less sensitive to water (20).

Analytical Methods. Analysis of fresh and spent peroxides and superoxides is done by adding the material to water. Approximately 0.1 wt % permanganate is used in the water to decompose the peroxide ion which otherwise forms. The evolved oxygen is measured volumetrically. If the material is spent, the base strength is titrated to a phenolphthalein end point, acidified further, and the carbon dioxide is determined volumetrically.

Uses. The peroxides and the superoxides must be hermetically sealed for storage. The superoxides especially are strong oxidizing agents and should be kept away from grease, oil, and organic materials. In general, sodium peroxide is used more widely than lithium peroxide. A breathing apparatus based on peroxides often is supplied with bottled oxygen, because the peroxides are not very oxygen-weight efficient.

Superoxides are used in breathing applications requiring no auxiliary source of oxygen. Sodium superoxide is ca 10 times as expensive as potassium superoxide because high temperature, high pressure autoclaves are required for its production. Use of calcium superoxide is experimental. Potassium superoxide is the most commonly used superoxide oxygen generator.

Portable breathing apparatus are used by fire departments, damage-control teams, and workers in unbreathable atmospheres (21). The wearer uses a canister containing the chemical, a breathing bag, and a mask. The chemical is packaged as 4.8–9.5-mm (2–4-mesh) granules with glass fiber filters which trap any dust. Approximately three breaths are required to start the chemical reaction and deliver oxygen. Alternatively, a small chlorate candle can be used to deliver oxygen immediately.

In portable breathing apparatus, efficiency is ca 70% in terms of oxygen used by the wearer vs total available; 30% is associated with excessive water vapor which is generated at high work rates and vented to ambient. At low work rates, efficiency is as high as 95%. The same canister of KO_2 sustains the oxygen needs of a person at hard work for an hour or for seven hours if the person is resting. Superoxides are valuable because these evolve only as much oxygen as is required. When a chlorate candle is used, the candle must be sized for the maximum work rate, resulting in waste of oxygen at lower work rates.

Numerous physical forms of the oxygen-generating chemical can be used, eg, granules of various sizes, densities, and porosities or plates pressed from powder. Granules are produced by pressing, grinding, and screening KO_2. Plates are made by pressing the powdery KO_2 to a coherent mass but not so as to impede penetration of the gases (22).

In air conditioning (qv) of closed spaces, a wider latitude in design features can be exercised (23,24). Blowers are used to pass room or cabin air through arrays of granules or plates. Efficiencies usually are 95% or better. The primary limiting factor is the decreased rate of absorption of carbon dioxide. However, an auxiliary small CO_2 sorption canister can be used. Control of moisture entering the KO_2 canister extends the life of the chemical and helps maintain the RQ at 0.82.

Generally the systems operate in the gas phase, but a system in which the superoxides react with liquid-phase water and absorb CO_2 in the resultant KOH solution has been built (25). The RQ is fixed at 0.67.

An unusual application is the use of KO_2 in a closed-cycle diesel system. Oxygen is supplied and CO_2 is removed in a KO_2 bed, through which the exhaust is recycled to the engine. Such a system would supply power in undersea and nonatmospheric operations.

The superoxides compare well with a combination of liquid oxygen or gaseous oxygen systems and a CO_2 scrubber, eg, LiOH. Heat release is somewhat less than the combination system, and the volume of superoxides which is required is roughly equivalent to that of lithium hydroxide.

On-board oxygen-generation (OBOG) systems are used in military aviation. The OBOG systems obviate the use of stored liquid oxygen or high pressure gaseous oxygen. In OBOG systems, oxygen is separated from air using chelating agents (qv), (26,27), ion exchange (qv) (27), barium oxide–dioxide shift (27), or molecular sieves (qv). Commercial units supplying therapeutic oxygen are also available (28,29).

Other Chemical Systems

Regenerative systems that dissociate carbon dioxide to recover the oxygen are of interest to the U.S. space program and in long-duration habitat support. The Bosch process utilizes an iron catalyst for the single-step reaction of hydrogen, carbon dioxide, water, and carbon at 700°C. Oxygen is produced by water electrolysis, and the hydrogen is recycled to the Bosch reactor. Carbon is removed as a solid.

$$CO_2 + 2 H_2 \longrightarrow 2 H_2O + C$$

$$2 H_2O \longrightarrow 2 H_2 + O_2$$

In the Sabatier reaction, methane and water are formed over a nickel–nickel oxide catalyst at 250°C. The methane is recovered and cracked to carbon and hydrogen, which is then recycled:

$$CO_2 + 4 H_2 \longrightarrow CH_4 + 2 H_2O$$

$$CH_4 \longrightarrow C + 2 H_2$$

$$2 H_2O \longrightarrow 2 H_2 + O_2$$

A fused-salt electrolysis process has been demonstrated (30). Carbon dioxide is introduced to the cathode area of a melt of 60 wt % $LiCl$–40 wt % Li_2CO_3 at 550°C. The carbon dioxide reacts with lithium oxide which is produced by electrolysis. Oxygen is released at the anode and carbon plates onto the cathode. The reaction requires a potential of 4.5 V. The reactions are as follows:

Cathode	$4 Li^+ + 4 e^- \longrightarrow 4 Li$
	$4 Li + Li_2CO_3 \longrightarrow C + 3 Li_2O$
Anode	$2 CO_3^{2-} \longrightarrow 2 CO_2 + O_2 + 4 e^-$
	$CO_2 + Li_2O \longrightarrow Li_2CO_3$
Total reaction	$4 Li^+ + 2 CO_3^{2-} \longrightarrow 2 Li_2O + C + CO_2 + O_2$

Although these other chemical systems are feasible chemically, there has been little impetus to solve associated equipment problems. The Bosch reaction appears to be the most promising.

Hydrogen peroxide can be dissociated over a catalyst to produce oxygen, water, and heat. It is an energetic reaction, and contaminants can spontaneously decompose the hydrogen peroxide. Oxygen from water electrolysis is used for life support on submarines.

Health and Safety

Peroxides, superoxides, and chlorates are oxidizing compounds and should not contact organic materials, eg, oil, greases, etc. This is especially true while oxygen is being produced. Caustic residues that may remain after use of peroxides and superoxides require disposal appropriate to alkali metal hydroxides. Spent candles containing barium may require special disposal considerations.

Dusts associated with these oxidizing compounds produce caustic irritation of skin, eyes, and nasal membranes. Appropriate protection should be worn when handling. Skin contact should be treated as for any caustic material, ie, flush with water and neutralize. Toxicity is low to moderate and is the same as for the hydroxides. Toxicity of the chlorate is greater than for the peroxides and superoxides, and the chlorate material also causes local irritation.

Properly designed respiratory support equipment is configured to safely contain the oxygen-generating chemicals. These systems, when hermetically sealed, are stable to storage and safe to use.

Economic Aspects

Potassium superoxide is utilized primarily in respiratory support equipment. The material is produced in the United States (Callery Chemical Co.), France (Air Liquide), and China.

The large candles used by the U.S. Navy have been produced in the United States by three companies, Mine Safety Appliances Co., Puritan-Bennett Corp., and Scott Aviation. These sell for $50–$60. Production is less than 10,000/yr. Smaller candles incorporated in breathing apparatus are produced by equipment suppliers. Production quantities are tied to the number of complete units and the candles are a small percentage of the total price. Production for aircraft oxygen supply during a decompression incident is about 50,000 units per year. In the United States, Puritan-Bennett and Scott Aviation are the primary suppliers as is Draeger in Germany.

BIBLIOGRAPHY

"Oxygen Generation Systems" in *ECT* 2nd ed., Suppl. Vol., pp. 658–667, by J. W. Mausteller, MSA Research Corp.; " Oxygen-Generation Systems" in *ECT* 3rd ed., Vol. 16, pp. 673–683, by J. W. Mausteller, MSA Research Corp.

1. E. B. Thompson, Jr., *An Investigation of the Chemical Formulation of Solid Chemical Oxygen Generators for Aircraft Emergency Oxygen Supply*, AFFDL-TM-70-4-FEFE, Flight Dynamics Laboratory, Wright-Patterson Air Force Base, Ohio, Sept. 1970.
2. U.S. Pat. 3,174,936 (Mar. 23, 1965), P. R. Gustafson and R. R. Miller (to U.S. Government).
3. W. H. Schecter, R. R. Miller, R. M. Bovard, C. B. Jackson, and J. R. Pappenheimer, *Ind. Eng. Chem.* **42**, 2348 (1950).
4. V. Hwoschinsky, *Safe Eng.* **4**(2), 21 (1970).
5. Y. Zhang, G. Kshirsagur, and J. Cannon, *Ind. Eng. Chem. Res.* **32**, 966 (1993).
6. W. J. O'Reilly and co-workers, *Development of Sodium Chlorate Candles for the Storage and Supply of Oxygen for Space Exploration Applications*, Rept. No. 69-4695, Air Research Corp., Los Angeles, Calif., July 1969.
7. U.S. Pat. 5,338,316 (Aug. 16, 1994), Y. Zhang and J. Cannon (to Puritan-Bennett Corp.).
8. U.S. Pat. 5,049,306 (Sept. 17, 1991), J. Greer (to Mine Safety Appliances Co.).
9. H. Nagate, K. Kusomoto, and N. K. Makai, *J. Ind. Explos. Soc. Jpn.* **27**, 99 (1966).

10. P. R. Gustafson, S. H. Smith, Jr., and R. R. Miller, *Chlorate Candle Fabrication by Hot Pressing*, NRL Rept. 5732, AP No. 272580, Naval Research Laboratory, Washington, D.C., Jan. 1962.

11. U.S. Pat. 5,198,147 (May 30, 1993), Y. Zhang and J. Cannon (to Puritan-Bennett Corp.).

12. U.S. Pat. 3,861,880 (Jan. 21, 1975), T. Thompson (to Dow Chemical Investment and Finance Corp.).

13. J. W. Mausteller and M. J. McGoff, MSA Research Corp., unpublished data, 1968.

14. R. N. Prince, T. L. Illes, and W. J. O'Reilly, "Development of the Portable Environment Control System," *Conference on Portable Life Support Systems*, NASA-Ames Research Center, Moffett Field, Calif., Apr. 1969.

15. *Exploratory Study of Potassium and Sodium Superoxides for Oxygen Control in Manned Space Vehicles*, Contract No. NASW-90, MSA Research Corp., Evans City, Pa., Mar. 1962.

16. *New Sci.* **27**, 628 (1970).

17. M. M. Markowitz and E. W. Dezmelyk, *A Study of the Application of Lithium Chemicals to Air Regeneration Techniques in Manned Sealed Environments*, AMRL-TDR-64-1, Wright-Patterson Air Force Base, Ohio, Feb. 1964.

18. K. J. Dressler and R. N. Prince, "Lithium Peroxide for Portable Life Support System Atmospheric Regeneration," *Conference on Portable Life Support Systems*, NASA-Ames Research Center, Moffett Field, Calif., Apr. 1969.

19. E. V. Ballan, P. C. Wood, L. A. Spitze, and T. Wydeven, paper presented at *The Intersociety Conference on Environmental Systems*, ASME, San Diego, Calif., July 12–15, 1976.

20. P. Wood, L. Spitze, and T. Wydeven, *The Use of Superoxides for Hyperbaric Air Revitalization*, NASA Technical Briefs ARC-11695, 1986.

21. K. Kyriazi and J. Shubilla, RI 8876, *Performance of Oxygen Self-Rescuers* U.S. Department of the Interior, U.S. Bureau of Mines, Washington, D.C., 1984.

22. J. W. Mausteller, M. J. McGoff, D. A. Keating, and K. Weiswurn, paper presented at *36th Aerospace Medical Association Meeting*, New York, Apr. 1965.

23. J. Presti, H. Waalman, and A. Petrocelli, *Undersea Technol.* **8**(6), 20 (1967).

24. H. M. David, *Missiles Rockets* **9**(15), 30 (1961).

25. T. V. Bolles and H. Wallman, *Aerosp. Med.* **37**, 675 (1966).

26. A. J. Adduci, *Chemtech*, 575 (Sept. 1976).

27. R. Tallman, B. Seberg, and S. Budzynaki, *On-Board Oxygen Generating Systems*, Final Report on Contract N62269-72-C-0144 by Grumman Aerospace Corp. for Naval Air Development Center, Oct. 1972.

28. U.S. Pat. 4,545,790 (Oct. 8, 1985), G. Miller and S. Clark (to Bio-Care, Inc.).

29. U.S. Pat. 4,561,287 (Dec. 31, 1985), R. Rowland (to Hudson Oxygen Therapy Sales Co.).

30. R. E. Shearer, J. C. King, and J. M. Mausteller, *Aerosp. Med.* **33**, 213 (1962).

J. WILSON MAUSTELLER
Consultant

OXYGEN RADICAL SCAVENGERS.   See ANTIAGING AGENTS; MEMORY-ENHANCING DRUGS.

OZONE

Ozone [*10028-15-6*], O_3, is an allotropic form of oxygen first recognized as a unique substance in 1840 (1). Its pungent odor is detectable at ~0.01 ppm. It is thermally unstable and explosive in the gas, liquid, and solid phases. In addition to being an excellent disinfectant, ozone is a powerful oxidant not only thermodynamically, but also kinetically, and has many useful synthetic applications in research and industry. Its strong oxidizing and disinfecting properties and its innocuous by-product, oxygen, make it ideal for the treatment of water. Indeed, the most important application of ozone is in the treatment of drinking water, which began in Europe in 1903; in the mid-1990s, there are well over 2000 such water-treatment installations, primarily in Europe. The treatment of swimming pool water was also developed in Europe during the 1960s (see WATER, TREATMENT OF SWIMMING POOLS, SPAS, AND HOT TUBS). Another important ozone application is for odor control in industrial processes and municipal wastewater-treatment plants. Ozone also is used on a large scale for the treatment of municipal secondary effluents (see WATER, MUNICIPAL WATER TREATMENT). Industrial high quality water supplies are also treated with ozone (see WATER, INDUSTRIAL WATER TREATMENT). In addition, ozone has applications in the treatment of cooling-tower water and in pulp bleaching. Advanced oxidation processes employing ozone in combination with uv, H_2O_2, and/or solid catalysts such as TiO_2 greatly improve the reactivity of ozone toward organic contaminants.

Ozone, which occurs in the stratosphere (15–50 km) in concentrations of 1–10 ppm, is formed by the action of solar radiation on molecular oxygen. It absorbs biologically damaging ultraviolet radiation (200–300 nm), prevents the radiation from reaching the surface of the earth, and contributes to thermal equilibrium on earth.

Properties

Gas–Liquid. At ordinary temperatures, pure ozone is a pale blue gas ($d = 2.1415 \, g/L$ at 0°C and 101.3 kPa (1 atm)) that can be condensed to an indigo blue liquid ($d = 1.354 \, kg/L$ at −112°C; bp = −111.9°C) which freezes to a deep blue-violet solid ($d = 1.728 \, kg/L$ at −195.8°C; mp = −192.5°C) (2). At the critical point, $T_c = -12.1°C$, $P_c = 5532.8 \, kPa$, and $\rho_c = 0.436 \, kg/L$. The vapor pressure of the liquid is given by $\log P \, (kPa) = 7.378 - 814.94/T - 0.00197 \, T$, for $T = 90\text{–}243 \, K$ (2), and $\log P \, (kPa) = 7.6689 - 867.6/T$, for $T = 85\text{–}95 \, K$ (3). Additional property data on ozone and ozone–oxygen mixtures are given in Reference 2.

Solubility. The solubility of gaseous ozone at atmospheric pressure and 0°C is 1.1 g/L H_2O. Henry's law constant, $K_H(kPa \cdot L/mol)$, is given by $\ln K_H = -2297/T + 2.659 \, \mu - 688 \, \mu/T + 16.808$, where T is in kelvin and μ is the molar ionic strength (4). Data on the effect of pH and ionic strength are given in Reference 5; data on the solubility of ozone in organic solvents are in Reference 6.

Adsorptivity. Gaseous ozone can be adsorbed by porous solid substrates such as silica gel and is often used in this form in organic synthesis (7).

Flammability and Explosivity. Ozone is endothermic, thus it can burn or detonate by itself and represents the simplest combustible and explosive system. The concentration threshold for spark-initiated explosion of liquid ozone in oxygen at $-183°C$ is 18.6 mol % O_3; the concentration limit for shock wave-initiated detonation of gaseous ozone–oxygen at 25°C is 9.2 mol % O_3 (2). Explosions of gaseous ozone can be initiated by shock wave, electrical spark, heat, or sufficiently intense light flash; explosion of pure liquid ozone and concentrated solutions in oxygen can be initiated by impurities, sudden change in temperature or pressure, heat, electrical spark, or mechanical shock.

Thermodynamic Properties. Values for gaseous ozone (8,9) are $\Delta H_f^0 = 142.7$ kJ/mol (34.1 kcal/mol), $\Delta G_f^0 = 163.2$ kJ/mol (39.0 kcal/mol), $S^0 = 238.8$ J/(mol·K) (57.08 cal/mol·K), and $C_p = 39.2$ J/mol·K (9.37 kcal/mol·K); heat of solution in water is $\Delta H_{soln}^0 = -19.1$ kJ/mol (-4.57 kcal/mol) (4). The heat of vaporization of liquid ozone is $\Delta H_v^0 = 12.47$ kJ/mol (2.59 kcal/mol). The standard reduction potential of ozone is 2.07 V for the half-cell reaction $O_3 + 2\,H^+ + 2\,e^- \rightarrow O_2 + H_2O$.

Spectral Properties. Gaseous ozone exhibits three principal absorptions in the infrared at 710, 1043, and 2105 cm^{-1} (10). Ozone has a diffuse absorption band in the visible region centered at $\sim$600 nm and an absorption cross section, σ, of 489×10^{-20} cm^2. Molar absorptivity is 2.94 L mol^{-1} cm^{-1} (11). The uv spectrum shows a single broad absorption centered at $\sim$254 nm. At 253.7 nm and 273 K, the absorption cross section, σ, is equal to $(1147 \pm 20) \times 10^{-20}$ cm^2. Molar absorptivity is 3000 ± 52 L mol^{-1} cm^{-1} (11). The molar absorptivity of aqueous ozone at 260 nm is 2930 ± 32 L mol^{-1} cm^{-1} at 25°C (12).

Structure. Ozone is a triangular molecule; its bond angle (116.8°) was established by microwave spectroscopy (13). The bond length of the ozone molecule (0.1278 nm) is intermediate to that of a single and double oxygen bond, corresponding to a bond order of 1.7. Ozone is diamagnetic with C_{2v} symmetry and has a low dipole moment of 1.77×10^{-30} C·m (0.53 D) (14). Based on Pauling resonance concepts, the structure of ozone is a hybrid, principally of form (**1**), with a small contribution from (**2**) (13).

$$O{=}O^+{-}O^- \longleftrightarrow {}^-O{-}O^+{=}O \qquad {}^-O{-}O{-}O^+ \longleftrightarrow {}^+O{-}O{-}O^-$$

$$\text{(1)} \hspace{10em} \text{(2)}$$

The bonding in (**1**) consists of two σ-bonds and one three-center–four-electron π-bond. Molecular orbital and valence bond calculations indicate that the 1,3-diradical, $\cdot O{-}O{-}O\cdot$, having two σ-bonds and weak bonding between the singly occupied π-atomic orbitals on the terminal O atoms, may contribute significantly to the ground state structure of ozone (15). Other studies suggest that ozone may have a hypervalent structure, with bonding similar to (**1**), and negligible diradical character (16).

Thermal Decomposition

Gas Phase. The decomposition of gaseous ozone is sensitive not only to homogeneous catalysis by light, trace organic matter, nitrogen oxides, mercury vapor, and peroxides, but also to heterogeneous catalysis by metals and metal oxides. The mechanism of the homogeneous decomposition of gaseous ozone is represented by the following reactions, where M represents the weighted sum of all species present in the gas phase, including ozone:

$$O_3 + M \underset{2}{\overset{1}{\rightleftharpoons}} O + O_2 + M - 106.5 \text{ kJ } (-25.5 \text{ kcal})$$

$$O_2 + O \xrightarrow{3} 2 O_2 + 391.8 \text{ kJ } (93.7 \text{ kcal})$$

Reaction 1 is the rate-controlling step. The decomposition rate of pure ozone decreases markedly as oxygen builds up due to the effect of reaction 2, which reforms ozone from oxygen atoms. Temperature-dependent equations for the three rate constants obtained by measuring the decomposition of concentrated and dilute ozone have been given (17–19).

The ratio k_2/k_3 has been estimated by measuring the steady-state ozone concentration, formed by means of electric discharge (20), photochemically (21,22), and low temperature reaction of molecular oxygen and glow-discharge generated oxygen atoms (23). The reaction of ozone with glow discharge-generated oxygen atoms has been studied in rapid flow systems (24). The preferred temperature-dependent equation for k_3 determined by laser photolysis of ozone and monitoring the oxygen atom concentration by time-resolved resonance fluorescence is $4.8 \times 10^9 \exp(-2060/T)$ L/(mol·s) (11). For k_2 (M = O_2), the preferred equation is $2.2 \times 10^8 (T/300)^{-2.8}$ L^2/(mol^2·s) (11), which is based on the generation of oxygen atoms by flash photolysis and measurement of ozone by uv spectrometry.

The calculated half-life of 1 mol % (1.5 wt %) of pure gaseous ozone diluted with oxygen at 25, 100, and 250°C (based on rate constants from Ref. 19) is 19.3 yr, 5.2 h, and 0.1 s, respectively. Although pure ozone–oxygen mixtures are stable at ordinary temperatures in the absence of catalysts and light, ozone produced on an industrial scale by silent discharge is less stable due to the presence of impurities; however, ozone produced from oxygen is more stable than that from air. At 20°C, 1 mol % ozone produced from air is ~30% decomposed in 12 h.

Aqueous Phase. In pure water, the decomposition of ozone at 20°C involves a complex radical chain mechanism, initiated by OH$^-$ and propagated by O_2^- radical ions and HO radicals (25). O_3^- is a radical ion.

$$O_3 + OH^- \longrightarrow HO_2 + O_2^- \qquad k = 70 \text{ L/(mol·s)}$$
$$HO_2 \rightleftharpoons O_2^- + H^+$$
$$O_2^- + O_3 \longrightarrow O_3^- + O_2$$
$$O_3^- + H^+ \rightleftharpoons HO_3$$
$$HO_3 \longrightarrow HO + O_2$$
$$HO + O_3 \rightleftharpoons HO_4$$
$$HO_4 \longrightarrow HO_2 + O_2$$

The two main termination steps for neutral solutions are $HO_4 + HO_4 \rightarrow H_2O_2 + 2O_3$ and $HO_4 + HO_3 \rightarrow H_2O_2 + O_3 + O_2$. An alternative mechanism has been proposed that does not involve HO_3 and HO_4, but has a different initiation step (26). Three ozone molecules are destroyed for each primary event. In the presence of excess HO radical scavengers, ie, bicarbonate, the pseudo-first-order rate constant at 20°C for the initiation step is $175 \times 10^{pH-14}$ s^{-1}. This yields an ozone half-life of 66 min at pH 8. In distilled water ($[PO_4]_{tot} = 50$ mmol/L), the half-life is significantly lower, ie, 7 min. In natural water, the half-lives fall between these extremes. For example, the half-life of Lake Zürich water (pH 8, 1.5 meq/L HCO_3^-) is 10 min (27). The decomposition in natural water also can be initiated by trace metal ions, eg, Fe^{2+}, promoted by impurities such as organic matter, and inhibited by HO radical scavengers, eg, HCO_3^-, CO_3^{2-}, HPO_4^{2-} (25,28).

Hydrogen peroxide greatly accelerates the decomposition of ozone in alkaline solutions because of formation of HO_2^-, which reacts rapidly with ozone to form the radical ion O_2^- (25). When the concentration of H_2O_2 exceeds 10^{-7} M, the decomposition of ozone is initiated faster by HO_2^- than by OH^- for pH <11.6.

Photochemical Decomposition

Gas Phase. Gaseous ozone is decomposed to oxygen atoms and molecules by absorbing radiation in the visible and uv spectrum: $O_3 + h\nu \rightarrow O_2 + O$. When visible light is above 400 nm, only ground-state oxygen molecules ($^3\Sigma_g^-$) and oxygen atoms (^3P) are formed (22). The initiation step is followed by ozone formation and destruction reactions, $O + O_2 + M \rightarrow O_3 + M$ and $O_3 + O \rightarrow 2O_2$. At ~600 nm, the maximum quantum yield, Φ, is 2.0 mol/$h\nu$.

With 334-nm uv radiation, the decomposition mechanism is similar to that with red light, except that the maximum quantum yield is four (in the absence of O_2) due to the formation of singlet delta oxygen ($^1\Delta_g$), which decomposes ozone (29). At 313 and 254 nm, a chain decomposition occurs with a maximum Φ of 6 (30,31). The primary step involves photodissociation of ozone to produce an excited oxygen atom (^1D) and a singlet delta oxygen molecule ($^1\Delta_g$). Reaction of ^1D oxygen atoms with molecular oxygen produces singlet sigma oxygen molecules ($^1\Sigma_g^+$) as well as ground-state (^3P) oxygen atoms. The energetic, electronically excited singlet oxygen molecules can decompose ozone: $O_2(^1\Delta_g$ and $^1\Sigma_g^+) + O_3 \rightarrow 2O_2 + O$ (32). Reaction of ^1D oxygen atoms with ozone produces vibrationally excited ground-state oxygen molecules $O_2^*(^3\Sigma_g^-)$ which also can decompose ozone (33). Highly vibrationally excited oxygen molecules ($X^3\Sigma_g^-$) formed at 226 nm also can form ozone by reaction with normal oxygen molecules (34). Vibrationally excited ozone molecules (O_3^*) are formed by collisional transfer between vibrationally excited oxygen molecules and ozone, and reaction of $O(^3P)$ with oxygen molecules. Although O_3^* molecules react faster than ground-state O_3 with O atoms, they are largely collisionally quenched by O and O_2 (32), and are reportedly unimportant in the decomposition mechanism (31). The rate of ozone decomposition is much greater in the uv than in the visible region because of the greater absorptivity of ozone in the uv region.

In the presence of water vapor, ^1D oxygen atoms formed by uv radiation react to form hydroxyl radicals (35), which can destroy ozone catalytically.

$$O(^1D) + H_2O \longrightarrow 2\,HO \qquad k(25°C) = 6.6 \times 10^{10}\ L/(mol \cdot s)$$

Aqueous Phase. In contrast to photolysis of ozone in moist air, photolysis in the aqueous phase can produce hydrogen peroxide initially because the hydroxyl radicals do not escape the solvent cage in which they are formed (36).

$$O(^1D) + H_2O \longrightarrow H_2O_2$$

Hydrogen peroxide is photolyzed slowly to hydroxyl radicals, which decompose ozone.

Chemistry of Ozone

The inorganic chemistry of ozone is extensive, encompassing virtually every element except most noble metals, fluorine (qv), and the inert gases. Reported second-order rate constants ($L/(mol \cdot s)$) at 20–23°C refer to the disappearance of ozone unless otherwise stated.

Inorganic Reactions. Ozone reacts rapidly with various free radicals and radical ions such as O, O_2^-, H, HO, N, NO, Cl, and Br. Some of these radicals (HO, NO, Cl, and Br) can initiate the catalytic decomposition of ozone.

Protonated Ozone. Gaseous ozone can be protonated with strong Brønsted acids, AH^+, eg, H_3^+, KrH^+, XeH^+, and CH_5^+, according to the reaction $AH^+ + O_3 \rightleftharpoons A + O_3H^+$. Protonated ozone reacts exothermically with methane: $O_3H^+ + CH_4 \rightarrow CH_3^+ + H_2O + O_2$, followed by $CH_3^+ + CH_4 \rightarrow C_2H_5^+ + H_2$ (37).

Halogen Compounds. Fluorine is unreactive toward ozone at ordinary temperatures. Chlorine is oxidized to Cl_2O_6 and Cl_2O_7, bromine to Br_3O_8, and iodine to I_2O_3 and I_4O_9. Oxidation of halide ions by ozone increases with the atomic number of halide. Fluoride is unreactive; chloride reacts slowly, ultimately forming chlorate; and bromide is readily oxidized to hypobromite (38). Oxidation of iodide is extremely rapid, initially yielding hypoiodite; the estimated rate constant is $\sim 2 \times 10^9$ (39). Hypohalite ions are oxidized to halites; hypobromite reacts faster than hypochlorite (40).

$$O_3 + Br^- \longrightarrow O_2 + BrO^- \qquad k = 160$$

$$2\,O_3 + XO^- \longrightarrow 2\,O_2 + XO_3^- \qquad k(ClO^-) = 30 \ \text{ and } \ k(BrO^-) = 100$$

Formation of halide is a competing reaction.

$$O_3 + XO^- \longrightarrow 2\,O_2 + X^- \qquad k(ClO^-) = 100 \ \text{ and } \ k(BrO^-) = 330$$

In contrast to reaction of ozone with nucleophilic halide and hypohalite ions, reaction of ozone with electrophilic hypohalous acids is very slow.

$$O_3 + HOX \longrightarrow 2\,O_2 + H^+ + X^- \qquad k(HClO) = 0.002 \ \text{ and } \ k(HBrO) < 0.01$$

Chlorite ion is oxidized rapidly to chlorine dioxide by ozone at pH 4, yielding one mol ClO_2 per mol O_3 when chlorite is in excess ($k > 10^4$) (39). The oxidation of bromite to bromate by ozone is too rapid to measure. Chlorine dioxide is oxidized rapidly to chlorate. Chlorate, bromate, and iodate ions do not react with ozone.

$$O_3 + ClO_2 \longrightarrow O_2 + ClO_3 \quad k = 1100$$

Nitrogen Compounds. Ammonium ion is not oxidized because, like ozone, it is an electrophilic reagent. Raising the pH to the 7–9 range shifts the equilibrium toward free ammonia where slow rates of oxidation are observed (41).

$$4\,O_3 + NH_3 \longrightarrow H^+ + NO_3^- + H_2O + 4\,O_2 \quad k = {\sim}5$$

Above pH 9, decomposition of ozone to the reactive intermediate, HO, determines the kinetics of ammonia oxidation. Catalysts, such as WO_3, Pt, Pd, Ir, and Rh, promote the oxidation of dilute aqueous solutions of ammonia; at 25°C, only two of the three oxygen atoms of ozone can react, whereas at 75°C, all three atoms react (42). The oxidation of ammonia by ozone depends not only on the pH of the system but also on the presence of other oxidizable species (39, 43,44). Because the ozonation rate of organic materials in wastewater is much faster than that of ammonia, oxidation of ammonia does not occur in the presence of ozone-reactive organics.

Monochloramine is oxidized slightly faster than ammonia (40). Monobromamine reacts much faster, $k = 40$ (39). By contrast, dichloramine and dibromamine both react slower than monohalamines; their k values are 1.3 and 20, respectively.

$$3\,O_3 + NH_2Cl \longrightarrow 2\,H^+ + NO_3^- + Cl^- + 3\,O_2 \quad k = 8.7$$

Unprotonated hydroxylamine is oxidized rapidly by ozone, $k = 2.1 \times 10^4$ (39). The reaction of ozone with the lower oxides of nitrogen (NO and NO_2) is also rapid and quantitative; the end product is nitrogen pentoxide, which is also a catalyst for the decomposition of ozone (45). Nitrous oxide, however, reacts slowly ($k < 10^{-3}$) (39). Nitrogen-containing anions, eg, nitrite and cyanide, also are oxidized by ozone (39). Nitrite is oxidized to nitrate ($k = 3.7 \times 10^5$) and cyanide is oxidized rapidly to cyanate ($k = 2.6 \times 10^3$ (46) and 10^3–10^5 (39)). Cyanate, however, is oxidized slowly.

$$CNO^- + 2\,O_3 \longrightarrow CO_2 + NO_3^- + O_2 \quad k \le 10^{-2}$$

Cyanic acid (HOCN) reacts slowly, whereas undissociated HCN and HNO_2 do not react at all (39).

Oxygen Compounds. Although hydrogen peroxide is unreactive toward ozone at room temperature, hydroperoxyl ion reacts rapidly (39). The ozonide ion, after protonation, decomposes to hydroxyl radicals and oxygen. Hydroxyl ions react at a moderate rate with ozone ($k = 70$).

$$HO_2^- + O_3 \longrightarrow HO_2 + O_3^- \quad k = 5.5 \times 10^6$$

Sulfur Compounds. Aqueous sulfide and H_2S, an odiferous compound in some waters, are oxidized rapidly (initially to sulfite and sulfurous acid); the rate constants are 3×10^9 and $\sim 3 \times 10^4$, respectively. Thiocyanate is oxidized by ozone to cyanide and sulfate via the intermediate formation of sulfite (47).

$$CNS^- + 2\,O_3 + 2\,OH^- \longrightarrow CN^- + SO_3^{2-} + 2\,O_2 + H_2O$$

Sulfite is oxidized rapidly ($k = 1 \times 10^9$) to sulfate by ozone (39). Bisulfite ion and sulfurous acid also are oxidized rapidly (to bisulfate and sulfuric acid) with k values of 3.2×10^5 and 2×10^4, respectively.

Metals and Metallic Ions. Under appropriate conditions, ozone oxidizes most metals with the exception of gold and the platinum group. When oxidized by ozone, heavy metal ions, such as Fe^{2+} and Mn^{2+}, result in the precipitation of insoluble hydroxides or oxides upon hydrolysis (48–50). Excess ozone oxidizes ferric hydroxide in alkaline media to ferrate, and MnO_2 to MnO_4^-.

$$2\,Fe^{2+} + O_3 + 5\,H_2O \longrightarrow 2\,Fe(OH)_3 + O_2 + 4\,H^+$$
$$Mn^{2+} + O_3 + H_2O \longrightarrow MnO_2 + O_2 + 2\,H^+$$

Formation of Ozonides. Although the parent compound, HO_3, is too unstable to be isolated, metal and nonmetal ozonides have been prepared. All ozonides are paramagnetic and are colored due to absorption at 400–600 nm. Alkali metal and alkaline-earth metal ozonides are crystalline ionic solids soluble in ammonia, amines, and amides. Potassium ozonide [*12030-89-6*], KO_3, is prepared by reaction of ozone with powdered potassium hydroxide at -10 to $-15°C$, followed by extraction with liquid ammonia. It also can be prepared by ozonation of metallic potassium dissolved in liquid ammonia or the low temperature reaction of KO_2 with ozone in a fluorocarbon oil. Lithium ozonide can only be isolated as the ammonia adduct, $LiO_3 \cdot 4NH_3$. Ozonides are thermally unstable, decomposing to superoxides.

$$5\,O_3 + 2\,KOH \longrightarrow 5\,O_2 + 2\,KO_3 + H_2O$$
$$2\,MO_3 \longrightarrow 2\,MO_2 + O_2$$

The stability of the alkali metal ozonides increases from Li to Cs; alkaline-earth ozonides exhibit a similar stability pattern. Reaction of metal ozonides with water proceeds through the intermediate formation of hydroxyl radicals.

$$4\,KO_3 + 2\,H_2O \longrightarrow 4\,KOH + 5\,O_2$$

The unstable ammonium ozonide [*12161-20-5*], NH_4O_3, prepared at low temperatures by reaction of ozone with liquid ammonia, decomposes rapidly at room temperature to NH_4NO_3, oxygen, and water (51). Tetramethylammonium ozonide [*78657-29-1*] also has been prepared.

Formation of Hydrogen Trioxide. Formation of hydrogen trioxide, HOOOH, has been observed as a transient intermediate in the ozonation of various organic molecules. For example, ozonation of 2-ethylanthrahydroquinone at $-78°C$

produces an organic hydrotrioxide and hydrogen trioxide in ~40% yield (52). Hydrogen trioxide decomposes by first-order kinetics beginning at about $-40°C$, forming water and singlet delta oxygen; in its reactions, hydrogen trioxide is even more electrophilic than ozone.

Formation of Hydrogen Tetroxide. The reaction of hydrogen atoms with liquid ozone at $-196°C$ proceeds through the intermediate formation of hydroperoxyl radicals forming hydrogen tetroxide, H_2O_4, which decomposes on warming to produce equimolar amounts of H_2O_2 and O_2 (53).

Organic Reactions. The strong electrophilicity of ozone is manifested in its reaction with a wide variety of organic and organometallic functional groups, eg, olefins, acetylenes, aromatics (carbocyclic and heterocyclic), activated C–H bonds (acetals, alcohols, aldehydes, ethers, and glycosides), unactivated C–H bonds (alkanes, cycloalkanes, and alkyl aromatics), deactivated C–H bonds (carboxylic acids and ketones), $C=N$ and $N=N$ bonds, Si–H and Si–C bonds, organometallic bonds (eg, Grignard reagents), and nucleophiles (eg, ammonia, amines, amino acids, arsines, disulfides, hydroxylamines, nitriles, phosphites, selenides, sulfides, and thioethers) (54,55). Ozone also acts as a nucleophile, eg, in its reaction with carbocations. Numerous synthetic applications of ozone have been described (7).

Kinetics and Mechanism of Ozone Reactions. Ozone attacks nucleophilic centers, ie, points of high electron density, in organic substrates. Reactivity of potential reaction sites is enhanced by the presence of electron-donating groups such as CH_3, and decreased by electron-withdrawing groups such as $C=O, COOH, Cl$, and NO_2. Reaction products depend on solvent type (reactive or nonreactive) and ozonation conditions. Ozone does not totally mineralize, ie, convert to CO_2 and water, most organic compounds during water treatment. Except in rare cases, such as the oxidation of formate, only partial oxidation is achieved on account of the low reactivity of common intermediate oxidation products, eg, acetic and oxalic acids. Although ozone has a high thermodynamic oxidation potential, its effectiveness in water treatment depends on the kinetics of its reactions, which can vary widely; indeed, rate constants can vary over 14 orders of magnitude, from $~10^{-5}$ for acetic acid to $~10^9$ L/(mol·s) for phenolate ion (56).

Below pH 7 and normal temperature, ozone reacts selectively, ie, directly, with organic compounds. However, at pH >7, common in natural waters, ozone also can react nonselectively or indirectly through HO radicals formed by the hydroxyl ion-catalyzed decomposition of ozone. Hydroxyl radicals are nonselective oxidants that react rapidly, by radical addition, hydrogen abstraction, and electron transfer, with functional groups (eg, alkyl) that normally are resistant to ozone. Rate constants generally vary over the narrow range of 10^7–10^{10} L/(mol·s). However, the availability of hydroxyl radicals for oxidation of organics is decreased by radical scavengers such as bicarbonate and carbonate ions.

The rate of aqueous ozonation reactions is affected by various factors such as the pH, temperature, and concentration of ozone, substrate, and radical scavengers. Kinetic measurements have been carried out in dilute aqueous solution on a large number of organic compounds from different classes (56,57). Some of the chemistry discussed in the following sections occurs more readily at high ozone and high substrate concentrations.

Ozonation of Aliphatics. Although ozone shows little tendency to react with saturated hydrocarbons under water treatment conditions, it can react via radical and ionic mechanisms with neat hydrocarbons containing secondary or tertiary hydrogen atoms. Isobutane yields mainly *t*-butyl alcohol (major) and acetone (minor); cyclohexane gives cyclohexanol and cyclohexanone as the primary oxidation products. Isolated methylene groups are oxidized to carbonyl. For example, ozonation of malonic acid yields oxalic and hydroxymalonic acids; the latter is oxidized further to ketomalonic acid and H_2O_2. Primary alcohols are oxidized by ozone to carboxylic acids, H_2O_2 (major), and aldehydes (minor); secondary alcohols form ketones, which are cleaved by ozone, eg, methyl ethyl ketone yields two moles of acetic acid. With the exception of formic and malonic acids, saturated mono- and dicarboxylic acids are relatively unreactive toward ozone. Glyoxylic and malonic acids react at moderate rates and are observed as intermediate products of ozonation of many organics. Acetic and oxalic acids are relatively stable to ozone and often are the end products of ozonation in water treatment. Acetals are converted to esters in high yield. Simple alkyl monochloramines are readily oxidized by ozone. Ethers are cleaved by ozone yielding alcohols, aldehydes, ketones, and esters; ozone also cleaves other compounds such as olefins, acetylenes, disulfides, and those with carbon–nitrogen double and triple bonds.

Mercaptans, thioethers, and disulfides are readily oxidized by ozone.

$$RSH + 3\,O_3 \longrightarrow RSO_3H + 3\,O_2$$

$$RSR + O_3 \longrightarrow R_2SO + O_2 \xrightarrow{O_3} R_2SO_2 + 2\,O_2$$

$$RSSR + 3\,O_3 \xrightarrow{-2\,O_2} RSO_2OSO_2R \xrightarrow{+H_2O} 2\,RSO_3H$$

Primary amino groups are oxidized stepwise by ozone to hydroxylamine, nitroso, and nitro (54,58); tertiary amines are oxidized to amine oxides.

$$RNH_2 \xrightarrow{+O_3,\,-O_2} RNHOH \xrightarrow{+O_3,\,-O_2,\,-H_2O} RNO \xrightarrow{+O_3,\,-O_2} RNO_2$$

Ozone converts nitro compounds, where R' is alkyl or hydrogen, to aldehydes and ketones.

$$RCH(NO_2)R' \xrightarrow[\text{CH}_3\text{OH,CH}_3\text{O}^-]{1.\ O_3,\ 2.\ (CH_3)_2S} RC(O)R'$$

Carbon–nitrogen double bonds in imines, hydrazones, oximes, nitrones, azines, and substituted diazomethanes can be cleaved, yielding mainly ketones, aldehydes and/or carboxylic acids. Ozonation of acetylene gives primarily glyoxal. With substituted compounds, carboxylic acids and dicarbonyl compounds are obtained; for instance, stearolic acid yields mainly azelaic acid, and a small amount of 9,10-diketostearic acid.

Although aliphatic amino acids in protonated form are unreactive toward ozone (56), sulfur-containing protonated amino acids (cystine, cysteine, and methionine) can react rapidly; the point of attack is the sulfhydryl group, which initially gives sulfoxides and disulfides (59,60). Unprotonated aliphatic amino acids are readily ozonated, resulting in the cleavage of the carbon–nitrogen bond. With an α-amino acid such as alanine, the amine group is converted to NH_3 and then to NO_3^-; decarboxylation yields CO_2 and acetic acid.

$$RCH(NH_2)COOH + O_3 \longrightarrow RCHO + CO_2 + NH_3 + O_2$$

Ozonation of Alkenes.   The most common ozone reaction involves the cleavage of olefinic carbon–carbon double bonds. Electrophilic attack by ozone on carbon–carbon double bonds is concerted and stereospecific (54). The modified three-step Criegee mechanism involves a 1,3-dipolar cycloaddition of ozone to an olefinic double bond via a transitory π-complex (**3**) to form an initial unstable ozonide, a 1,2,3-trioxolane or molozonide (**4**), where R is hydrogen or alkyl. The molozonide rearranges via a 1,3-cycloreversion to a carbonyl fragment (**5**) and a peroxidic dipolar ion or zwitterion (**6**).

(3)   (4)   (5)   (6)   (7)

The dipolar ion (**6**) and its alternative resonance form $R_2C=O^+-O^-$, exist in both *syn-* and *anti-*configurations that are formed in different amounts.

The dipolar ion can react in several ways according to the solvent and the structure of the olefin. In inert solvents, if the carbonyl compound is highly reactive (eg, an aldehyde), the dipolar ion can be added to the carbonyl fragment to give the normal ozonide or 1,2,4-trioxolane (**7**); for example, 1,1- and 1,2-dialkylethylenes react in this manner. Tri- or tetraalkyl-substituted olefins produce a small, if any, yield of an ozonide when the ozonolysis is performed in nonreactive solvents (61). However, the ozonide can be prepared indirectly; eg, ozonation of acetone *O*-methyloxime in acetone gives 3,3,5,5-tetramethyl-1,2,4-trioxolane, which is the normal ozonide of tetramethylethylene (62). The same approach is used to prepare ozonides corresponding to those of vinyl ethers and derivatives of acrylonitrile. If the carbonyl fragment (**5**) is less reactive (eg, in a ketone), the dipolar ion (**6**) can dimerize to a cyclic bisperoxide (**8**) or polymerize to a linear peroxide (**9**). The undesirable reaction intermediates (**8**) and (**9**) can be avoided by performing the reaction in reactive solvents, such as methanol, followed by hydrolysis (63).

(8)   (9)

Steric hindrance can cause polymerization to predominate even if (**5**) is an aldehyde. With increasing bulk of substituents on one side of a double bond, epoxidation can compete with ozonolysis.

Commercially, pure ozonides generally are not isolated or handled because of the explosive nature of lower molecular weight species. Ozonides can be hydrolyzed or reduced (eg, by Zn/CH_3COOH) to aldehydes and/or ketones. Hydrolysis of the cyclic bisperoxide (**8**) gives similar products. Catalytic (Pt/excess H_2) or hydride (eg, $LiAlH_4$) reduction of (**7**) provides alcohols. Oxidation (O_2, H_2O_2, peracids) leads to ketones and/or carboxylic acids. Ozonides also can be catalytically converted to amines by NH_3 and H_2. Reaction with an alcohol and anhydrous HCl gives carboxylic esters.

The zwitterion (**6**) can react with protic solvents to produce a variety of products. Reaction with water yields a transient hydroperoxy alcohol (**10**) that can dehydrate to a carboxylic acid or split out H_2O_2 to form a carbonyl compound (aldehyde or ketone, R_2CO). In alcoholic media, the product is an isolable hydroperoxy ether (**11**) that can be hydrolyzed or reduced (with $(CH_3O)_3P$ or $(CH_3)_2S$) to a carbonyl compound. Reductive amination of (**11**) over Raney nickel produces amides and amines (64). Reaction of the zwitterion with a carboxylic acid to form a hydroperoxy ester (**12**) is commercially important because it can be oxidized to other acids, RCOOH and R'COOH. Reaction of zwitterion with HCN produces α-hydroxy nitriles that can be hydrolyzed to α-hydroxy carboxylic acids. Carboxylates are obtained with H_2O_2/OH^- (65). The zwitterion can be reduced during the course of the reaction by tetracyanoethylene to produce its epoxide (66).

$$R_2C\begin{matrix}OOH\\OH\end{matrix} \qquad R_2C\begin{matrix}OOH\\OR\end{matrix} \qquad R_2C\begin{matrix}OOH\\OC(O)R'\end{matrix}$$

$$\text{(10)} \qquad\qquad\qquad \text{(11)} \qquad\qquad\qquad \text{(12)}$$

$$(\mathbf{6}) + (NC)_2C{=}C(CN)_2 \longrightarrow R_2CO + (NC)_2C\overset{O}{\overline{\quad\quad}}C(CN)_2$$

Ozonation of Aromatics. Aromatic ring unsaturation is attacked much slower than olefinic double bonds, but behaves as if the double bonds in the classical Kekule structures really do exist. Thus, benzene yields three moles of glyoxal, which can be oxidized further to glyoxylic acid and then to oxalic acid. Substituted aromatics give mixtures of aliphatic acids. Ring substituents such as amino, nitro, and sulfonate are cleaved during ozonation. Partial oxidation of phenol with ozone produces dihydroxybenzenes: catechol, hydroquinone, and resorcinol. The last two compounds are further oxidized to *o*- and *p*-benzoquinone, respectively. Further oxidation produces a mixture of products; ie, aliphatic unsaturated diacids (muconic, fumaric, and maleic), hydroxylated saturated diacids (tartaric and mesotartaric), as well as glyoxal, glyoxylic acid, and CO_2. Chlorobenzene reacts more slowly with ozone than does phenol but gives the same ring-opening products. Intermediate oxidation products include *o*-, *m*, and *p*-chlorophenols, as well as chlorotartaric acid. Chlorocresols and thiophenols also give similar ozonation products as does phenol. 3,4-Dimethylpyridine yields glyoxal, biacetal, ammonia, as well as formic, acetic, pyruvic, oxalic, and glyoxylic acids.

Activation of Ozone Reactions.   Reactivity of ozone can be enhanced by uv, H_2O_2, and heterogeneous catalysts such as TiO_2, all of which increase the decomposition of ozone and generate highly reactive species such as hydroxyl radicals and singlet oxygen (67). The main parameters affecting the efficiency of O_3–uv are ozonation rate, uv intensity, pH, alkalinity, and the type of compound. The O_3–uv system is similar to the O_3–H_2O_2 system except for substrates that absorb uv. The effectiveness of the ozonation of organic compounds is also physically enhanced by ultrasound (Sonozone) .

The O_3–uv photooxidation system, initially investigated in 1975, was shown to be effective in the oxidation of complexed cyanides, chlorinated compounds, and pesticides; total mineralization had been achieved in most cases (68). Although acetic and oxalic acids are relatively stable to ozone, they are more readily oxidized by ozone to CO_2 in the presence of uv or H_2O_2; acetic acid is oxidized stepwise, at pH 7, to glycolate, oxalate, and then to CO_2 (48). Although degradation of halogenated aliphatics such as chloroform, trichloroethylene, tetrachlorethylene, and CCl_4 is too slow to be practical with ozone alone, it is greatly accelerated by using uv radiation (69,70) or H_2O_2 (67). In simulated natural water containing fulvic acid, protein, and a disaccharide, the order of effectiveness in reducing the levels of TOC is O_3–TiO_2 > O_3–H_2O_2 > O_3 (71).

Ozone Adducts.   At low (−78°C) temperatures ozone forms stable adducts with electron-rich compounds such as phosphites (72), phosphines, tertiary amines, sulfides, and sulfoxides. The adduct is a cyclic, four-membered, trioxygen-containing ring that can be regarded as a moloxide. On warming, it decomposes and yields an oxidized form of the substrate, eg, triphenylphosphite gives triphenylphosphate, and singlet oxygen, $O_2(^1\Delta_g)$. Decomposition of adducts such as triphenyl phosphite–ozone provides a convenient method for accomplishing chemical oxidations involving singlet oxygen and making it a useful oxygenating agent for synthetic and mechanistic applications.

Formation of Hydrotrioxides.   Hydrotrioxides are intermediates in the ozonation of various organic substrates, including hydrocarbons (with reactive secondary or tertiary hydrogen atoms), ethers, alcohols, aldehydes, ketones, acetals, hydroquinones, amines, diazo compounds, and silanes. Hydrotrioxides formed by low (−78°C) temperature ozonation have been isolated and characterized; they can be formed by a 1,3-dipolar insertion mechanism and decompose by radical and ionic pathways in the temperature range of −45 to −10°C, releasing singlet oxygen. For example, ozone inserted into the C–H bond of the carbonyl group of benzaldehyde at −78°C can produce the organic hydrotrioxide (**13**), which on decomposition yields singlet O_2 and benzoic acid (73). Ozonation of 2-ethylanthrahydroquinone at −78°C produces 2-ethylanthraquinone, an organic hydrotrioxide, and hydrogen trioxide (H_2O_3) in ~40% yield (52).

$$\underset{\text{(13)}}{C_6H_5-\overset{\overset{\displaystyle O}{\|}}{C}-OOOH \longrightarrow O_2(^1\Delta_g) + C_6H_5-\overset{\overset{\displaystyle O}{\|}}{C}-OH}$$

Atmospheric Ozone

STRATOSPHERE

Ozone is formed rapidly in the stratosphere (15–50 km) by the action of short-wave ultraviolet solar radiation (<240 nm) on molecular oxygen, $O_2 + h\nu \rightarrow 2\,O$. At wavelengths above 175 nm, only ground-state (^3P) atoms are formed; whereas at wavelengths below 175 nm, one ground-state and one excited (^1D) atom are formed. Ground-state atoms also can be formed by the predissociation of electronically excited O_2. The oxygen atoms can react with molecular oxygen to yield ozone: $O + O_2 + M \rightarrow O_3 + M$. Ozone can be destroyed photochemically: $O_3 + h\nu \rightarrow O + O_2$; at 226 nm, however, this reaction also can produce vibrationally excited O_2 capable of forming ozone: $O_2(X^3\textstyle\sum_g^-) + O_2 \rightarrow O_3 + O$ (34). In addition, ozone can be destroyed by reaction with oxygen atoms ($O + O_3 \rightarrow 2\,O_2$) as well as with excited O_2 molecules ($^1\Delta_g, ^1\textstyle\sum_g^+,$) and $^3\textstyle\sum_g^-$ and other free radicals (11). Since the early 1960s, it has been recognized that radicals such as NO, OH, Cl, and Br affect the abundance and distribution of ozone in the stratosphere (32). Earlier studies simulating stratospheric chemistry concluded that ozone formation is significantly less than its destruction, hence the ozone deficit problem (34). However, studies indicate that this may not be the case (74).

Most ozone is formed near the equator, where solar radiation is greatest, and transported toward the poles by normal circulation patterns in the stratosphere. Consequently, the concentration is minimum at the equator and maximum for most of the year at the north pole and about 60°S latitude. The equilibrium ozone concentration also varies with altitude; the maximum occurs at about 25 km at the equator and 15–20 km at or near the poles. It also varies seasonally, daily, as well as interannually. Absorption of solar radiation (200–300 nm) by ozone and heat liberated in ozone formation and destruction together create a warm layer in the upper atmosphere at 40–50 km, which helps to maintain thermal equilibrium on earth.

Effect of Nitric Oxide on Ozone Depletion.   Nitrous oxide is injected into the atmosphere from natural sources on earth; about 10% is converted to nitric oxide ($N_2O + O(^1D) \longrightarrow 2\,NO$), which in turn can catalyze the destruction of ozone (11,32,75). The two main cycles are 1 and 2. Rate constant data are given in Reference 11.

Cycle 1:

$$O_3 + NO \longrightarrow NO_2 + O_2$$
$$NO_2 + O \longrightarrow NO + O_2$$
$$\textit{Net}: O_3 + O \longrightarrow 2\,O_2$$

Cycle 2:

$$O_3 + NO \longrightarrow NO_2 + O_2$$
$$NO_2 + O_3 \longrightarrow NO_3 + O_2$$
$$NO_3 + h\nu \longrightarrow NO + O_2$$
$$\textit{Net}: 2\,O_3 + h\nu \longrightarrow 3\,O_2$$

These cycles account for less than 20% of the total ozone loss observed during spring in the lower stratosphere at mid-north latitudes (76). During nighttime,

NO_2 and NO_3 reform equilibrium amounts of N_2O_5. Nitrogen pentoxide can react on the surface of sulfuric acid aerosols releasing gaseous nitric acid. Reaction of NO_2 with HO radicals also forms HNO_3. Nitric acid and N_2O_5 photolyze slowly in the gas phase and regenerate NO_2 radicals.

Effect of Hydroxyl Radicals on Ozone Depletion. Hydroxyl radicals, formed by reaction of (^{1}D) oxygen atoms with water or CH_4, can destroy ozone catalytically (11,32); as shown in the following reactions.

Cycle 3:
$$O_3 + HO \longrightarrow O_2 + HO_2$$
$$HO_2 + O_3 \longrightarrow HO + 2\,O_2$$

This cycle accounts for 30–50% of the total photochemical ozone loss observed during spring in the lower stratosphere at mid-north latitudes (76).

Effect of Halofluorocarbons on Ozone Depletion. The release of chlorofluorocarbons (CFCs) from industrial production of aerosol sprays, coolants for refrigerators and air conditioners, blowing agents for foams, cleaners for electronic parts, and Halons (bromofluorocarbons) from fire-extinguishing agents, has created conditions for the depletion of O_3 in the stratosphere (77). In addition to the main CFCs, $CFCl_3$ (CFC-11) and CF_2Cl_2 (CFC-12), other sources of chlorine are $CHFCl_2$ (CFC-22), $C_2F_3Cl_3$ (CFC-113), CCl_4, CH_3CCl_3, and CH_3Cl from the burning of biomass (see FLUORINE COMPOUNDS, ORGANIC–FLUORINATED ALIPHATIC COMPOUNDS). Methyl bromide is another important source of bromine, released into the atmosphere by marine plankton, through the burning of biomass, or by its use as a soil fumigant. Hydrogen-containing molecules are degraded by HO radicals in the troposphere, limiting the fraction that reaches the stratosphere.

Although inert in the lower atmosphere (troposphere), the fully halogenated CFCs and Halons diffuse into the upper stratosphere where they are photodissociated, ie, photolyzed, by the intense ultraviolet radiation.

$$CFCl_3 + h\nu \longrightarrow CFCl_2 + Cl$$
$$CF_3Br + h\nu \longrightarrow CF_3 + Br$$

The C–F bonds also are cleaved, resulting ultimately in the formation of HF, a stable fluorine reservoir. The Cl (or Br) atoms formed can convert ozone to oxygen catalytically; indeed, one chlorine atom is capable of destroying thousands of ozone molecules. Although much less abundant in the stratosphere than chlorine, bromine is considerably more reactive in some reactions, thus accounting for a significant fraction of ozone destruction in certain catalytic cycles. Satellite and airborne observations have shown significant decreases in total column ozone since 1978, ranging from zero near the equator to 6–8% at high latitudes (78). The most dramatic ozone decrease occurs every year in spring at the Antarctic ozone hole. In 1992 and 1993, the ozone levels were about 50% less than they were in 1979.

Halogen radicals account for about one-third of photochemical ozone loss observed in the spring in the lower stratosphere (below 21 km) at 15–60°N latitude (76). The following three cycles (4–6) are the most important. Rate constant data are given in Reference 11.

Cycle 4:
$$O_3 + Cl \longrightarrow O_2 + ClO$$
$$ClO + O \longrightarrow Cl + O_2$$

This cycle is terminated by the reaction of chlorine atoms with methane: $Cl + CH_4 \rightarrow HCl + CH_3$. The importance of this cycle depends on the availability of oxygen atoms and varies with altitude as well as the time of year; it accounts for only 5% of the halogen-controlled loss at 15 km, but increases to 25% at 21 km.

The following mechanisms are more important for explaining ozone destruction at lower altitudes, where the availability of oxygen atoms is limited; X = Cl and Br:

Cycle 5:
$$HO_2 + XO \longrightarrow HOX + O_2$$
$$HOX + h\nu \longrightarrow HO + X$$
$$X + O_3 \longrightarrow XO + O_2$$
$$HO + O_3 \longrightarrow HO_2 + O_2$$

The cycle accounts for ~30 and 20–30% of the halogen-controlled loss for chlorine and bromine, respectively.

Cycle 6:
$$BrO + ClO \longrightarrow Br + ClOO \xrightarrow{+M} Br + Cl + O_2$$
$$Br + O_3 \longrightarrow BrO + O_2$$
$$Cl + O_3 \longrightarrow ClO + O_2$$

The cycle accounts for 20–25% of the halogen-controlled loss (M is a third body molecule). The possible contribution of naturally occurring iodine compounds to ozone destruction (via I and IO radicals) is being investigated (79).

Other reactions, such as the reaction of ClO with NO_2 to form chlorine nitrate ($ClO + NO_2 \longrightarrow ClONO_2$), are also important. Reaction of N_2O_5 on sulfate aerosols forms HNO_3, reducing the availability of NO_2. This allows higher free ClO concentrations, shifting ozone catalytic decomposition from NO_x to ClO_x and HO_x (80). Chlorine nitrate and HCl can form temporary inert chlorine reservoirs and retard ozone depletion; they are the most abundant chlorine species at lower latitudes before winter. However, $ClONO_2$ can slowly photolyze, yielding chlorine atoms capable of decomposing ozone; this accounts for 10–15% of halogen-controlled losses (76). With the exception of NO_2, all radicals are produced photolytically each day and their concentrations fall to zero during the night.

Although relatively unreactive in the gas phase, chlorine nitrate and HCl can react on the surface of frozen particles in polar stratospheric clouds (PSCs), which form at altitudes of 10–25 km when air trapped within the polar vortex (a circular wind pattern) that circles the Antarctic in winter is cooled to low temperatures during the polar night, thus preventing replenishment by fresh ozone (81). Type I PSCs form at 193 K and type II (ice) at 187 K. Type I PSCs, originally thought to consist of nitric acid trihydrate, may have a more complex structure (82). The heterogeneous reactions release gaseous chlorine and HOCl:

$$HCl + ClONO_2 \xrightarrow{\text{type I and II PSCs}} Cl_2 + HNO_3$$

$$H_2O + ClONO_2 \xrightarrow{\text{type II PSC}} HOCl + HNO_3$$

N_2O_5 is also converted to HNO_3 on PSCs. Since the nitric acid formed remains with the cloud particle, PSC chemistry produces an atmosphere rich in reactive chlorine species but depleted in NO_2. These reactions are the key to the formation of the Antarctic hole, and continue until the break-up of the polar vortex and warming of the stratosphere in the spring. PSC particles can grow sufficiently large to sediment out of the stratosphere. This denitrification process removes HNO_3 as a source of NO_2 (83). The volatilized gaseous chlorine and HOCl are photolyzed to chlorine atoms when sunlight returns in the spring. Chlorine atoms re-enter the ozone catalytic destruction cycle (6) and the following (84):

$$ClO + ClO \xrightarrow{+M} (ClO)_2 \xrightarrow{h\nu} Cl + ClOO$$

$$ClOO \xrightarrow{+M} Cl + O_2 \qquad 2\,Cl + 2\,O_3 \longrightarrow 2\,ClO + 2\,O_2$$

Similar heterogeneous reactions also can occur, but somewhat less efficiently, in the lower stratosphere on global sulfate clouds (ie, aerosols of sulfuric acid), which are formed by oxidation of SO_2 and COS from volcanic and biological activity, respectively (80). The effect is most pronounced in the colder regions of the stratosphere at high latitudes. Indeed, the sulfate aerosols resulting from eruptions of El Chicon in 1982 and Mt. Pinatubo in 1991 have been implicated in subsequent reduced ozone concentrations (85).

Arctic polar stratospheric clouds (mostly type I) form sporadically because only rarely are temperatures sufficiently low. Nevertheless, the northern polar vortex contains enhanced ClO levels, similar to the Antarctic, that contribute to significant ozone loss in late winter (86). This may indicate that crucial heterogeneous chemistry on sulfate aerosols is occurring, although no hole has been observed. This is attributed to the relatively short period of significant loss and to the fact that the warmer, less stable Arctic vortex dissipates by late winter. In addition, there is a lack of denitrification, ie, removal of nitric acid via sedimentation of PSCs, observed in the Antarctic (83). Thus, the presence of HNO_3 throughout the arctic winter moderates ozone destruction by providing a source of NO_2 to quench ClO. Nevertheless, record-low ozone levels, up to 40% below normal and approaching those in the recurring Antarctic ozone hole, were observed in the winter of 1995 due to exceptionally low temperatures in the Arctic vortex (87).

TROPOSPHERE

Ozone and nitric oxide are transported from the stratosphere to the troposphere, the region of the atmosphere below 15 km. Though only about 10% of the atmospheric ozone is present in the troposphere, this small fraction plays a fundamental role in atmospheric chemistry because it leads to the formation of hydroxyl radicals. Hydroxyl radicals initiate the oxidation and prevent the buildup of many organic and inorganic pollutants in the atmosphere. The radical-dominated chemistry of the troposphere is complex, involving intertwining cycles of gas-phase, condensed-phase, and multiple-phase reactions; only the more important reactions are shown. In unpolluted atmosphere, HO radicals react with naturally occurring CO and methane, resulting in a net increase in ozone concentration (32,75,88); rate constant data are given in Reference 11.

$$O_3 + h\nu \longrightarrow O_2 + O(^1D) \qquad\qquad O(^1D) + H_2O \longrightarrow 2\,HO$$

$$HO + CO \xrightarrow{+O_2} CO_2 + HO_2 \qquad\qquad HO_2 + NO \longrightarrow NO_2 + HO$$

$$HO + RCH_3 \xrightarrow{+O_2} RCH_2O_2 + H_2O \qquad RCH_2O_2 + NO \xrightarrow{+O_2} NO_2 + RCHO + HO_2$$

$$2\,NO_2 + 2\,h\nu \longrightarrow 2\,NO + 2\,O \qquad\qquad 2\,O + 2\,O_2 \longrightarrow 2\,O_3$$

R is hydrogen, alkenyl, or alkyne. In remote tropospheric air where NO concentrations are sometimes quite low, HO_2 radicals can react with ozone ($HO_2 + O_3 \rightarrow HO + 2\,O_2$) and result in net ozone destruction rather than formation. The ambient ozone concentration depends on cloud cover, time of day and year, and geographical location.

Although the naturally occurring concentration of ozone at the earth's surface is very low, this distribution has been altered by the emission of anthropogenic pollutants which increase the production of ozone via the above mechanism. Photochemical smog, an aerosol irritant gas mixture, occurs in urban industrialized areas where heavy motor vehicle traffic is common, especially those areas where temperature inversions are common (eg, Denver, Los Angeles, Washington, D.C., and Mexico City). It forms at low altitudes by photolytic reactions involving nonmethane hydrocarbons, NO, and CO, resulting in low but potentially harmful concentrations of ozone and other irritating substances, such as aldehydes, ketones, acids, H_2O_2, organic peroxides, and peroxyacetyl nitrate.

Although the background concentration of ozone in surface air is ~0.01–0.03 ppm, during severe smog days in the Los Angeles area, for example, it has often reached 0.5 ppm, and a maximum of 1 ppm in 1957. The main automobile exhaust pollutants (CO, NO, and hydrocarbons) peak at about 7 a.m. in Los Angeles, coinciding with the morning rush hour traffic (75); the concentrations of nitrogen dioxide and ozone reach a maximum about three and five hours later, respectively. The evening rush hour traffic does not produce much smog primarily because the inversion layer has undergone considerable thermal expansion and the sunlight intensity is weaker. In the early morning hours, NO is removed slowly by the oxygen atom chain, which is initiated by the photolysis of NO_2 and subsequently by the photolysis of ozone. Later in the day when the light intensity is higher, the hydroxyl chain causes the NO conversion to accelerate.

Ozone can react rapidly with NO to produce NO_2, which re-enters the ozone formation cycle: $O_3 + NO \rightarrow O_2 + NO_2$. This is the main ozone-depleting reaction in the absence of sunlight. Ozone also reacts with NO_2 (to form NO_3, which in turn reacts with NO_2 to form N_2O_5), C_2H_4, as well as HO and HO_2 radicals. Nitric acid formed by the reaction $HO + NO_2 \rightarrow HNO_3$ is removed from the atmosphere by rain-out.

Ozone Generation

Ozone can be generated by a variety of methods, the most common of which involves the dissociation of molecular oxygen electrically (silent discharge) or photochemically (uv). The short-lived oxygen atoms (lifetime $\sim 10^{-5}$ s) react rapidly with oxygen molecules to form ozone. The widely employed technique of electric discharge produces much higher concentrations than the ultraviolet technique and is more practical and efficient for production of large quantities. A less common method of ozone formation is electrochemical generation.

SILENT ELECTRIC DISCHARGE

Commercial production and utilization of ozone by silent electric discharge consists of five basic unit operations: gas preparation, electrical power supply, ozone generation, contacting (ie, ozone dissolution in water), and destruction of ozone in contactor off-gases (Fig. 1).

Discharge Characteristics. The energy for chemical reaction is transferred to oxygen molecules by energetic electrons producing atoms, excited

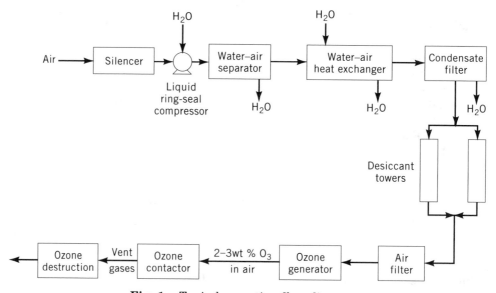

Fig. 1. Typical ozonation flow diagram.

molecules, and ions. In an ozone generator, the feed gas (oxygen or air) passes between two closely spaced electrodes (one of which is coated with a dielectric) under an applied nominal potential of $\sim$10 kV. A silent or dielectric barrier discharge occurs when the gas becomes partially ionized, resulting in a characteristic violet glow in air. Silent discharge consists of numerous randomly distributed, low current (but high current density) microdischarge pulses (89–92). The approximately columnar streamers or filaments (100–200 μm dia) emanating from the metal electrode discharge at the dielectric and extinguish within 10 ns.

Because ozone formation occurs only within these microdischarge channels, ozone-production efficiency for the most part depends on the strength of the microdischarges, which is influenced by a number of factors such as the gap width, pressure, properties of the dielectric and metal electrode, power supply, and moisture. In weak discharges, a significant fraction of the energy is consumed by ions, whereas in stronger discharges, almost all of the discharge energy is transferred to electrons responsible for the formation of ozone. The optimum is a compromise that avoids energy losses to ions but at the same time obtains a reasonable conversion efficiency of oxygen atoms to ozone.

Ozone Generation from Oxygen. Oxygen is dissociated into atoms by inelastic collisions with energetic electrons (6–7 eV) (89,90).

$$O_2 + e^- \longrightarrow 2\,O(^3P \text{ and } ^1D) + e^-$$

Whereas oxygen atoms are formed in nanoseconds, their subsequent reactions occur on a microsecond time scale. The highly reactive oxygen atoms can recombine in the gas phase and on the wall. For the gas phase, the reaction is $2\,O + M \rightarrow O_2 + M + 498.3\,kJ\,(119.1\,kcal)$, where $M = O_2$ and $k\,(25°C) = \sim 1 \times 10^{11}\,L^2/(mol^2 \cdot s)$ (93). Recombination of oxygen atoms is unimportant until $O/O_2 > 0.01$. The main reaction is with oxygen molecules in the presence of a third body, M, forming ozone: $O + O_2 + M \rightarrow O_3 + M + 106.5\,kJ\,(25.5\,kcal)$. The lower the relative oxygen atom concentration, O/O_2, the greater the fraction of oxygen atoms forming ozone.

Ozone formed in the above reaction is initially in a vibrationally excited state (O_3^*). Although most is quenched by collision with other molecules, a small fraction can react faster with oxygen atoms than ground-state ozone. Vibrationally excited ozone also can be formed by collision with (^1D) oxygen atoms and with vibrationally excited oxygen molecules. Vibrationally excited singlet oxygen molecules ($^1\Delta_g$ and $^1\Sigma_g^+$) can be formed by electron impact or by recombination of oxygen atoms.

Ozone can be destroyed thermally, by electron impact, by reaction with oxygen atoms, and by reaction with electronically and vibrationally excited oxygen molecules (90). Rate constants for these reactions are given in References 11 and 93. Processes involving ions such as O_2^-, O_2^+, O^-, O^+, and O_3^- are of minor importance. The reaction $O_3 + O(^3P) \rightarrow 2\,O_2$, is exothermic and can contribute significantly to heat evolution. Efficiently cooled ozone generators with typical short residence times (seconds) can operate near ambient temperature where thermal decomposition is small.

Experimental studies show that the ozone concentration increases with specific energy (eV/O_2) before reaching a steady state. The steady-state ozone concentration varies inversely with temperature but directly with pressure, reaching a maximum at about 101.3 kPa (1 atm). Above atmospheric pressure the steady-state ozone concentration decreases with pressure, apparently due to the pressure dependence of the rate constant ratio k_{O_2}/k_{O_3} for the reactions of O_2 and O_3 with energetic electrons. The preparation of ozone from oxygen presents fewer operational problems than that from air because significant amounts of moisture and large concentrations of nitrogen are absent. However, small amounts of nitrogen ($\sim$4%) actually increase the ozone concentration.

Ozone Generation from Air. Although the use of air for ozone generation has the advantage that air, unlike oxygen, is readily available, the concentration of ozone produced with air is lower than that produced with oxygen. In addition, the presence of moisture in air interferes with discharge formation and reaction kinetics and creates potential for corrosion that can adversely affect the performance of the ozone generator and increase the need for maintenance. The basic chemistry of ozone generation from oxygen is more complex when air is employed because of formation of nitrogen atoms, vibrationally excited nitrogen molecules, and nitrogen oxides (89,94). Nitrogen atoms are formed by the dissociation of nitrogen molecules by electron impact; they can generate oxygen atoms via the following reactions: $N + O_2 \rightarrow NO + O$ and $N + NO \rightarrow N_2 + O$. Oxygen atoms also can be formed by the dissociation of molecular oxygen by vibrationally excited nitrogen molecules. Thus, atomic nitrogen and excited nitrogen molecules enhance the formation of ozone by increasing the atomic oxygen concentration.

Nitric oxide can initiate catalytic decomposition of ozone, as previously discussed, by the reactions $NO + O_3 \rightarrow NO_2 + O_2$ and $NO_2 + O \rightarrow NO + O_2$. Nitrogen dioxide also can destroy ozone: $NO_2 + O_3 \rightarrow NO_3 + O_2$. NO_3 reacts with NO to form NO_2 as indicated by $NO_3 + NO \rightarrow 2\,NO_2$. Nitrogen dioxide and trioxide are in equilibrium with nitrogen pentoxide: $NO_2 + NO_3 \rightleftharpoons N_2O_5$. Decreasing the temperature and increasing the pressure shifts the equilibrium to the right, reducing the effect of lower oxides on the decomposition of ozone. Nitrogen pentoxide can react with oxygen atoms to form nitrogen dioxide. Only N_2O_5 and N_2O are present in the gases exiting the discharge. The nitrous oxide (N_2O) that is formed is inert toward ozone. Of less importance is the destruction of ozone by nitrogen atoms, $N + O_3 \rightarrow NO + O_2$, which is much slower than that by oxygen atoms.

Because of the formation of nitrogen oxides, a steady-state ozone concentration cannot be obtained; instead, due to the buildup of nitrogen oxides, an increase in residence time in the discharge results in a decrease in ozone concentration beyond the maximum value. Thus, there is an optimum residence time for maximum ozone production.

Suppression of Nitrogen Oxides. The concentration of nitrogen oxides during preparation of ozone from air increases linearly with the energy density in the discharge, causing a decrease in the formation rate of ozone. Most commercial ozone generators produce 0.5 kg of nitrogen oxides for every 100 kg of ozone generated. The formation of nitrogen oxides at a given energy density is minimized by decreasing the residence time and temperature, increasing the pressure, and reducing the dew point of air.

Feed Gas Preparation.   The use of oxygen for industrial ozone generation is significant and increasing. Oxygen provides a higher ozone concentration and more efficient ozone dissolution than air, and does not add nitrogen oxides to the water. It is prepared from dry, filtered air by liquefaction and fractional distillation. Liquid oxygen (LOX) can be prepared on-site or purchased from vendors. Oxygen is sometimes used to enrich air-fed systems. Although oxygen-rich off-gases from ozone contactors can be recycled, more often they are discarded to avoid redrying costs.

Air is widely used as the feed gas for commercial ozone generators. The air feed gas to the ozone generator should be dry and free of foreign matter. Filtered ambient air is drawn into the plant by vacuum, blower, or compressor. The pressure of the treated air can vary from subatmospheric to >400 kPa (4 atm). Since compression heats the air, cooling is necessary. The air is filtered again to remove oil droplets that can foul the desiccant dryers and interfere with ozone generation. Any hydrocarbons in the air can be removed with activated carbon. Moisture is removed by desiccant-drying or a combination of refrigerant- and desiccant-drying. Desiccant-drying is accomplished by using molecular sieves (qv), silica gel, or activated alumina, all of which are capable of regeneration. Liquid water droplets in refrigerant-dried air should be removed by filtration prior to contacting the desiccant dryers. A final filtration is necessary to remove desiccant dust particles down to 1 μm. The efficiency of ozone generation decreases with increasing moisture content in the air (95). At high dew points, nitrous and nitric acids are deposited within the ozone generator, decreasing performance and substantially increasing the maintenance frequency. The air feed to the ozone generator should have a dew point of at least $-60°C$, corresponding to a moisture content of ≤ 20 ppmv; some systems, however, operate at a dew point of $-80°C$. A sensor should be placed in the air stream entering the generator that can shut the system off and sound an alarm if the dew point increases above the desired level. In high pressure systems, the pressure of the compressed air prior to entering the ozone generator is reduced by means of a pressure-reducing valve. The pressure employed depends on the ozone generator type and can vary from 100 to 240 kPa (0–20 psig). The pressure of the ozone generator feed should be maintained at a constant level to avoid affecting power draw and applied voltage.

Cooling Requirements.   Since the majority of the electrical energy input to the electric discharge is dissipated as heat, cooling is necessary to minimize decomposition of ozone and extend dielectric life. Double-sided cooling is more effective than single-sided cooling in removing heat from the ozone generator. The gas exiting an efficiently cooled ozone generator normally is near ambient temperatures where the rate of decomposition is low.

Electrical Characteristics.   The basic features of an electric discharge cell are depicted in Figure 2. Electrical energy to the ozone generator is provided by a power supply, a frequency converter, and a transformer. Ozone formation is directly proportional to the power consumed in the discharge at constant O_3 concentration, temperature, and pressure. The average discharge power consumption P (W) is given by (96):

$$P = 4 C_d v_s f[v_o - (C_d + C_g)/C_d) v_s]$$

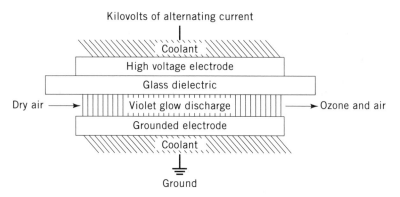

Fig. 2. Basic configuration of ozone generators.

where C_d and C_g are the dielectric and gap capacitances (F), v_s is the peak gap sparking potential (V), v_o is the peak driving potential (V), and f is the frequency (Hz). The sparking potential is given by $v_s(O_2) = 26.55\,pt_g + 1480$ and $v_s(\text{air}) = 29.64\,pt_g + 1350$, where p is the absolute pressure (kPa) and t_g is the gap thickness (mm). For a given geometry, gap, and pressure, the power consumed by the discharge can be increased by operating at higher driving voltages and frequencies, and employing thinner dielectrics having higher dielectric constants.

The potential and frequency employed in commercial ozone generators varies with the type of design and can range from 5–20 kV and 50–3000 Hz, respectively. Typically, high frequency ozone generators operate at lower voltages, where the expected lifetime of the high voltage electrode is virtually unlimited. Although lower voltage decreases the ozone production rate, when combined with high frequency it can produce more ozone per unit electrode area. Modern ozone generators operating at 10 kV rms and 600–1000 Hz employ power densities of 3–4 kW/m^2, resulting in production densities of 0.2–0.25 kg/hm^2 in air and 0.35–0.45 kg/hm^2 in oxygen (97).

Ozone Concentration and Yield. The output of an ozone generator can be increased by raising the power input at constant temperature and feed gas flow rate, but the increase in output is less than proportional unless the gas flow is increased to maintain a constant ozone concentration. Raising the flow rate at constant power input decreases the ozone concentration but increases the ozone and energy yields. At low flow rates, although the ozone concentration is high, the yield is low because the specific energy is high. At higher flow rates, the ozone concentration decreases; the yield approaches a limiting value because the specific energy does not change much at low ozone concentrations.

Most ozone generators currently in operation produce ozone in concentrations of 1–1.5 wt % (12–18 g/m^3) in air and 2–3 wt % (27–40 g/m^3) in oxygen. However, modern, well-cooled ozone generators can produce ozone efficiently at double these concentrations, ie, 2–3 wt % (24–37 g/m^3) in air and 4–6 wt % (54–81 g/m^3) in oxygen (97). The required ozone concentration depends on the application; concentrations as high as 16 wt % have been produced in oxygen. Commercial ozone generators are available that have different ozone production rates from air, ranging from 10 g/h to 90 kg/h; higher production capacities are obtained by combining multiple units.

Energy Requirements and Efficiency. The thermodynamics of ozone synthesis require the expenditure of 142.7 kJ/mol (34.1 kcal/mol); thus the formation of 1 kg of ozone requires 2.97 MJ (711 kcal) or 0.85 kWh/kg at 100% efficiency. The more concentrated the ozone, the higher the specific energy (kWh/kg) and the lower the efficiency. The specific energy for ozone production from dry oxygen varies from 7–14 kWh/kg over the 1–6 wt % range. For dry air, the specific energy (15–22 kWh/kg for 0.5–3.0 wt % ozone) is lower than expected due to the contribution of atomic nitrogen to ozone formation. The higher-than-theoretical specific energy requirements are due to the fact that most of the supplied energy is converted to heat resulting from ozone formation and decomposition reactions. These specific energy requirements correspond to ozone synthesis efficiencies from oxygen and air of 6–12% and 4–6%, respectively. Thus, the portion of the input synthesis energy dissipated as heat is 88–94% for oxygen and 94–96% for air. In addition to the power requirements for the ozone generator, the air-preparation unit requires 4.4–7.7 (kWh)/kg ozone, and the oxygen-recycle unit an additional 2–7 (kWh)/kg ozone.

Ozone Generator Design. A better understanding of discharge physics and the chemistry of ozone formation has led to improvements in power density, efficiency, and ozone concentration, initiating a trend toward downsizing.

The basic configuration of an electric discharge cell consists of two closely spaced electrodes (one of which is coated with a dielectric), supplied with high voltage ac and filled with a flowing oxygen-containing gas. The gap width varies from 1 to 3 mm, depending on whether oxygen or air is employed. The purpose of the dielectric, usually made of glass or ceramic, is to limit current flow, resulting in the formation of a relatively cold plasma. A thin dielectric with a high dielectric constant facilitates heat removal and improves ozone-generating efficiency. The higher the applied voltage, the thicker the electrode should be to prevent failure by electrical arcing. The dielectric must be strong enough to withstand mechanical shock and prevent puncturing by the applied voltage. High peak voltages induce dielectric failure, as do high power densities, the latter on account of its dielectric heating.

The electric discharge ozone generator is equivalent to a gas-phase reactor having internal heat generation; its design also bears some similarity to heat exchangers such as shell and tube. The gap width influences both the voltage requirement and the back pressure and should be uniform to avoid hot spots. Small gap widths typically are employed to facilitate heat removal; the smaller the discharge gap, the greater the power efficiency.

Types of Ozone Generators. Since 1906, a number of different ozone generators have been developed, including the plate-type (water- or air-cooled), the horizontal tube (water-cooled), and the vertical tube (water- or oil- and water-cooled) generators. Originally introduced in 1906, plate-type ozone generators have experienced operational problems and have been discontinued for use in some countries, even though many installations remain operational and the technology is still being promoted by some manufacturers.

The horizontal tube-type ozone generator (Fig. 3**a**) was a significant improvement over the plate-type ozone generators. A single unit consists of two

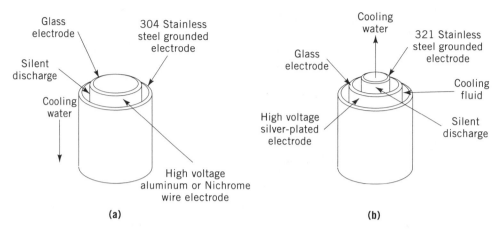

Fig. 3. Comparison of ozone-generator construction: (**a**) single-fluid-cooled; (**b**) double-fluid-cooled generator.

concentric tubes, an outer stainless steel tube that serves as the ground electrode, and an inner glass tube sealed at one end (which functions as the dielectric) with an inner conductive coating that acts as the high voltage electrode. Discharge occurs in the annular space between the two tubes through which the feed gas flows. A group of tubular units (eg, up to 1000) is arranged in parallel and enclosed in a cylindrical housing so that the ends protrude out of the cooling water jacket, which cools the outside stainless steel tube. Manifolds distribute the feed gas to the annular discharge spaces of the tubes at one end and collect the ozone-containing gas at the other. The tubular electrodes are supplied with low frequency power (~60 Hz). When higher production rates (>10 kg/h) are required, the use of solid-state variable medium frequency (600–1000 Hz) power supplies may be cost-effective.

The vertical tube (water-cooled) generator consists of two concentric tubes; the outer of which is cooled with water and acts as the ground electrode. Feed gas is introduced into the top of the inner stainless steel tube (which serves as the high voltage electrode), exits at the bottom of the outer tube, flows upward through the annular space (which contains the electric discharge), and emerges at the top of the outer tube into a product gas manifold.

Developed in the late 1960s, the double-fluid-cooled, high frequency (1000–3000 Hz) vertical tube generator (Fig. 3**b**) has a more complex design (three tubes) that advantageously cools both electrodes: the high voltage with a nonconducting oil (cooled with a heat exchanger) and the ground with water. It has higher production rates at higher ozone concentrations than other types of generators.

Production Costs and Capital Requirements. The production cost for 1.5 wt % O_3 from air, assuming 100% on-stream time, varies from ~\$8.25 to ~\$2.40/kg for production rates of 1–500 kg/h (98). The assumptions are power, 6.5¢/kWh; interest, 12.5%/yr; depreciation time, 10 yr; on-stream time, 8760 h/yr; and labor and maintenance, 1% of capital investment/yr. The production cost strongly depends on the size of the installation as well as the extent of instrumentation and controls, and varies inversely with on-stream time and ozone concentration. Production costs for manufacturing 5 wt % ozone from oxygen can vary from ~$10.45 to ~$2.50/kg for the same on-stream time and production rates. The cost of liquid oxygen is the main contributor to the total production cost. For large production rates, there is little difference between production costs for air and for oxygen; capital costs are in the ranges of ~$5.70 to ~$0.50/kg for air and ~$2.85 to ~$0.22/kg for oxygen.

ULTRAVIOLET LIGHT

The mechanism of the practical photochemical production of ozone is similar to that in the stratosphere; that is, oxygen atoms, formed by the photodissociation of oxygen by short-wavelength uv radiation ($\leq$240 nm), react with oxygen molecules to form ozone. At low conversions, the limiting quantum yield is ~2 (21). The steady-state ozone concentration (~3.5 mol % max) depends on temperature, pressure, and whether oxygen or air is employed; the time-to-steady-state depends on the light intensity. Efficiencies as high as 9% can be obtained, at low ozone concentration, by using narrow-band uv radiation produced by a xenon excimer laser operating at 172 nm (see LASERS) (99).

In practice, ozone concentrations obtained by commercial uv devices are low. This is because the low intensity, low pressure mercury lamps employed produce not only the 185-nm radiation responsible for ozone formation, but also the 254-nm radiation that destroys ozone, resulting in a quantum yield of ~0.5 compared to the theoretical yield of 2.0. Furthermore, the low efficiency (~1%) of these lamps results in a low ozone production rate of ~2 g/kWh (100).

Typical output of a commercial 40-watt uv lamp using air is ~0.5 g/h of ozone and a maximum concentration of 0.25 wt %. However, these maximum ozone yields and concentrations cannot be obtained simultaneously by the uv method. The low concentrations of ozone available from uv generators preclude their use for water treatment because the transfer efficiencies of ozone from air into water is low and large volumes of carrier gas must be handled. More than 44 kWh are required to generate one kg of ozone from dry air by uv radiation under high gas flow rates and low concentrations.

OTHER METHODS

High current density electrolysis of aqueous phosphate solutions at room temperature produces ozone and oxygen in the anodic gas. Electrolysis of 68 wt % sulfuric acid can produce 18–25 wt % ozone in oxygen when a well-cooled cell is used. Although electrolysis of water can produce high concentrations of ozone, the output is low, and the cost is several times more than that of electric discharge processes (101).

Ozone can be prepared radiochemically by irradiation of gaseous or liquid oxygen with β- and γ-rays from radioactive isotopes or a nuclear reactor (102). Although its energy efficiency is greater than that of ozone produced by silent discharge, this complex process has never been commercialized due to problems associated with recovery of ozone and separation of by-products and radioactive material.

In the laboratory, pure liquid ozone can be produced quantitatively by cooling a stream of atomic oxygen in oxygen at liquid nitrogen temperatures:

$$O + O_2 + M \xrightarrow{-196\,°C} O_3 + M$$

where M is the cold reactor wall (103). Pure stable gaseous ozone also can be prepared quantitatively by silent discharge through oxygen cooled with liquid nitrogen followed by warming to room temperature (18).

Although ozone can be formed in certain chemical reactions, eg, $F_2 + H_2O$ and $P + O_2$, and by rapid quenching of plasma-heated oxygen (>3000°C) with liquid oxygen, these methods have no commercial importance.

Transfer of Ozone into Water.   The solubility of ozone varies inversely with temperature and ionic strength and directly with pressure. The equilibrium partitioning of ozone between the gas and liquid phases is governed by Henry's law: $P = K_H C$, where P is the ozone partial pressure (kPa), C is the concentration of ozone in solution (mol/m^3), and K_H (kPa·m^3/mol) is Henry's law constant that should be independent of pH for weakly solvated and nonhydrolyzing gases such as ozone. It can be represented by the distribution constant K_D in the following equation: $K_D = [O_3]_{aq}/[O_3]_g = RT/K_H$, where $[O_3]_{aq}$ and $[O_3]_g$ are the aqueous and gas phase concentrations (in similar units, eg, mol/m^3 or mg/L), R is the gas constant (kPa·m^3/mol·K), and T is the kelvin temperature. The equilibrium fraction of ozone dissolved (α) at atmospheric pressure in single-stage absorbers depends only on K_D at the temperature of interest and the relative gas and water flow rates (V, L/min):

$$\alpha = (K_D V_{aq}/V_g)/(1 + K_D V_{aq}/V_g)$$

The applied dosage (D_a) in mg/L, assuming no decomposition or reaction, is $D_a = Y_i V_g/V_{aq}$, where Y_i is the ozone concentration (mg/L) in the inlet gas. The applied dose can be varied by changing Y_i or the V_g/V_{aq} ratio. The transferred ozone dose (D_t) in mg/L is $D_t = \alpha D_a = (Y_i - Y_o)V_g/V_{aq}$, where Y_o is the concentration of ozone in the vent gas. The transfer efficiency (TE) is $TE = 100\alpha = 100(D_t/D_a) = 100(Y_i - Y_o)/Y_i$.

The physical mass-transfer rate of ozone into water is affected by the gaseous ozone concentration, temperature, pressure, gas dispersion, turbulence, mixing, and composition of the solution, ie, pH, ionic strength, and the presence of reactive substances. Mass transfer of gaseous ozone into water is a diffusion process through the interface of gas and liquid films. Since ozone is only slightly soluble, Henry's law constant is large. This means that the resistance in the gas film is negligible compared to that in the liquid film, and diffusion through the liquid film controls the process. The instantaneous rate (mol/m^3/s) of O_3 mass transfer as a function of the concentration gradient ($C_L - C$) is given by the following: $dC/dt = (D/L)a(C_L - C) - r = k_L a(C_L - C) - r = k_L a(P/K_H - C) - r$, where D is the diffusion coefficient of ozone (m^2/s), L is the thickness of the liquid film (m), $k_L a$ is the experimentally determined liquid-phase volumetric mass-transfer coefficient (s^{-1}) that varies with the apparatus,

k_L is the liquid-phase mass-transfer coefficient (m/s), a is the effective interfacial mass-transfer area per unit volume of liquid phase (m^{-1}), C_L and C are the aqueous ozone concentrations in the liquid film and bulk solution (mol/m^3), and r is the rate of ozone decomposition and chemical reaction (mol/m^3·s).

Efficient transfer of ozone into solution requires the dispersion of gaseous ozone into small bubbles. This is accomplished in various types of ozone contactors such as porous diffuser bubble columns (co- or countercurrent flow), mechanically agitated vessels, turbine mixers, tubular reactors, in-line static mixers, as well as negative pressure (venturi) and positive pressure injectors. In turbines and injectors, ozone and water are forced or drawn cocurrently through a small opening. Under intense mixing, bubbles are sheared and mixed thoroughly, decreasing the liquid film thickness but increasing both interfacial area and contact time. Plate and packed columns and spray towers also can be employed to increase the gas–liquid contact. In plate and packed columns, ozone and water can flow co- or countercurrently. Single or multiple contact chambers (up to 5) can be used. Contact time varies from 2–20 minutes, depending on the application. Faster ozone-transfer rates result in faster disinfection rates. Because ozone is a strong oxidant, corrosion-resistant materials of construction such as stainless steel, glass, and Teflon, should be employed.

Off-Gas Treatment. Ozone-transfer efficiencies vary with the number of stages and are typically above 90%. However, since even a 95% ozone absorption efficiency can result in a contactor off-gas containing as much as 740 ppmw ozone (based on a 1.5 wt % O$_3$ feed gas), treatment is required to reduce the ozone concentration to an acceptable maximum level of 0.2 mg/m^3. Ozone in the vent gases from water-treatment ozone contact chambers is destroyed mainly by thermal (300–350°C for ≤5 s) and/or catalytic means, and sometimes by wet granular-activated carbon (GAC). Another option is recycling the off-gas to points in the water-treatment system having a high ozone demand. Dilution of ozone vent gases with air has been employed whenever practical. When oxygen is used as the feed gas, it can be recycled to the ozone-generation step; however, once-through operation is common in order to avoid redrying costs.

Uses

Ozone Treatment of Drinking Water. Ozonation and treatment strategy depends on the quality of the source water. Ozone is typically applied as a pre-disinfectant for the control of algae and inactivation of bacteria and viruses in direct filtration processes, and as a pre- and/or intermediate oxidant for inorganic and organic matter to eliminate taste, odor, and color compounds; remove turbidity, iron, and manganese; and reduce levels of trihalomethane (THM) and related organic precursors. Precipitated trace metals, microflocculated organics, and destabilized colloidal particles enhance coagulation and filtration. Ozone is frequently applied at an intermediate point in conventional treatment processes as an oxidant and for primary disinfection of viruses and cysts.

Ozone also can be used in conjunction with biological filtration through sand, mixed media, or GAC, for removing biodegradable organics produced during ozonation, especially aldehydes, ketones, acids, etc, which can be precursors to THMs, haloacetic acids (HAAs), and other halogenated by-products. For example, ozonation (1.5–4 mg/L) followed by GAC filtration produces biological activated carbon (BAC). Partial oxidation of organic matter by ozone can increase biodegradability, although decreasing adsorption efficiency. Oxygen in the carrier gas assists bacteria in oxididation of ammonia and in mineralization of adsorbed organics, providing an efficient process for removing low concentrations of dissolved organic compounds. Ozone plus BAC extends the life of the GAC by reducing the adsorptive load. Ozone can be used again to disinfect the resultant biologically stabilized water prior to chlorination and/or chloramination and distribution.

Advanced oxidation processes may be necessary in some cases for the destruction of ozone-refractory compounds. These processes employ various methods to cause ozone to react by free-radical mechanisms in order to increase the rate and efficiency of the oxidation of organic matter; such processes include, primarily, O_3 + uv radiation and O_3 + H_2O_2 (Peroxone), as well as O_3 + TiO_2 (Catazone). More than 100 plants employing these technologies are in operation (104); there are reportedly over 30 O_3–uv and O_3–H_2O_2 plants designed and/or operated by Ultrox, SolarChem, and other vendors in North America for treating a variety of organic pollutants. Many European landfills treat leachates with O_3–uv for the destruction of chemical oxygen demand (COD), and many French and U.K. plants use O_3–H_2O_2 to meet European Economic Community (EEC) pesticide standards.

Disinfection. Ozone is a more effective broad-spectrum disinfectant than chlorine-based compounds (105). Ozone is very effective against bacteria because even concentrations as low as 0.01 ppm are toxic to bacteria. Whereas disinfection of bacteria by chlorine involves the diffusion of HOCl through the cell membrane, disinfection by ozone occurs with the lysing (ie, rupture) of the cell wall. The disinfection rate depends on the type of organism and is affected by ozone concentration, temperature (106), pH, turbidity, clumping of organisms, oxidizable substances, and the type of contactor employed (107). The presence of oxidizable substances in ordinary water can retard disinfection until the initial ozone demand is satisfied, at which point rapid disinfection is observed.

There are three National Primary Drinking Water Regulations controlling disease-causing pathogens in public drinking water: the Total Coliform Rule (TCR) (108), the Surface Water Treatment Rule (SWTR) (109), and the Enhanced Surface Water Treatment Rule (ESWTR) (110); a fourth regulation for groundwater is under development. In the SWTR and ESWTR, the U.S. Environmental Protection Agency (EPA) redefines disinfection in terms of treatment systems to assure adequate removal and inactivation of *Giardia lamblia*, Cryptosporidium, enteric viruses, *Legionella*, and heterotrophic bacteria. Disinfection and filtration of most surface waters is required. Disinfection levels are determined by CT values, ie, the product of the residual disinfectant concentration (mg/L) and the effective contact time (min), which vary with primary disinfectant (chlorine, chlorine dioxide, and ozone), pH, and temperature. The SWTR requires all systems to achieve 99.9% and 99.99% removal and inactivation of *Giardia lamblia* cysts and enteric viruses, respectively. Proposed CT values for pH 6–9 over the 5–25°C temperature range are 1.9–0.46 for ozone, 26–11 for ClO_2, 179–45 for chlorine (at pH 7.5 and 1 ppm), and 2200–750 for NH_2Cl, which requires chlorination followed by ammoniation.

Ozone cannot be employed for the disinfection of distributed water because of its short lifetime. Unless removed from the treatment plant, the presence of low molecular weight, oxygenated, and biodegradable organics that ozone produces can promote biological growth.

Tastes and Odors. The origin of most tastes and odors in water supplies is synthetic organic compounds (eg, phenols) as well as naturally occurring inorganic (eg, Fe^{2+}, Mn^{2+}, and H_2S) and organic materials (biologically and chemically altered). The action of algae and actinomycetes on humic materials can produce distasteful water-soluble compounds such as geomycin and 2-methylisoborneol. Regrowth in the distribution system also can impart taste. Additionally, oxidation during water treatment can generate other odorous compounds (eg, aldehydes). Although many taste and odorous compounds are readily oxidized by ozone (typically 1.5–2.5 ppm), some compounds are more resistant and may require biofiltration or advanced oxidation processes. A pilot-scale study showed that the Peroxone process ($O_3 - H_2O_2$) was more effective in removing taste and odor compounds than ozone alone (111).

Color. Surface waters generally are colored by naturally occurring organic materials such as humic, fulvic, and tannic acids. These compounds result from the decay of vegetative materials and usually are related to condensation products of phenol-like compounds. Such color-causing compounds contain multiple conjugated double bonds, some of which are readily split by ozone. Cleavage of only one double bond generally destroys the chromophoric properties of the molecule. However, with insufficient oxidant, the color can return during storage or distribution. Ozone dosage levels for lake water decolorization usually are 2–4 mg/L.

Turbidity. Turbidity in water is removed by ozonation (0.5–2 ppm) through a combination of chemical oxidation and charge neutralization. Colloidal particles that cause turbidity are maintained in suspension by negatively charged particles which are neutralized by ozone. Ozone further alters the surface properties of colloidal materials by oxidizing the organic materials that occur on the surface of the colloidal spherical particles.

Iron and Manganese Removal. Soluble ferrous and manganous ions are oxidized by ozone (typically 2.5 ppm) to less soluble higher oxidation states, which can hydrolyze to $Fe(OH)_3$ and MnO_2 and be removed by settling, filtration, or both. Iron(II) is oxidized faster than manganese(II). The presence of organic matter retards the oxidation by consuming ozone and/or metal complex formation (112–113). Alkalinity (HCO_3^- and CO_3^{2-}) reduces the ozone dosage required for metal ion oxidation by scavenging hydroxyl radicals, which are formed by chain decomposition of ozone initiated by OH^- and organic matter (114). Excess ozone oxidizes manganese dioxide to permanganate, which can be reduced by organic matter to form MnO_2 scale in the distribution lines. As a result, permanganate must be avoided or allowed to dissipate or reduced by filtration through GAC prior to distribution.

Degradation of Readily Oxidizable Organics. Organic contaminants in water supplies consist of not only natural organic matter from various sources but also synthetic organic chemicals. More than 700 organic compounds have been identified in drinking water. Not all of these materials are oxidized by ozone at the same rate; some halogenated hydrocarbons are not oxidized at all. Materials that are oxidized readily by ozone include certain phenolics, detergents, pesticides, herbicides (qv), chemical-manufacturing wastes, humic acids, aromatic compounds, proteins, and most amino acids.

Although little or no reduction in total organic carbon occurs during ozonation, partially oxidized polar compounds (containing $>C=O$, COOH, and OH groups) are formed, which are all readily biodegradable. Some of these can chelate with polyvalent cations, yielding insoluble material that can be removed by coagulation with Al or Fe salts, polyelectrolytes, etc. This microflocculation phenomenon is responsible for the removal of humic substances (eg, humic and fulvic acids) having molecular weights of ~200 to several million, which account for the bulk of the dissolved organic carbon (DOC) in natural waters. Partial oxidation with ozone produces lower molecular weight by-products whose structures depend on the level of ozonation. Lower molecular weight, partially oxidized organics that are not removed by microflocculation or coagulation can be removed by biofiltration.

Removal of Refractory Organics. Ozone reacts slowly or insignificantly with certain micropollutants in some source waters such as carbon tetrachloride, trichlorethylene (TCE), and perchlorethylene (PCE), as well as in chlorinated waters, ie, trihalomethanes, THMs (eg, chloroform and bromoform), and haloacetic acids (HAAs) (eg, trichloroacetic acid). Some removal of these compounds occurs in the ozone contactor as a result of volatilization (115). Air-stripping in a packed column is effective for removing some THMs, but not $CHBr_3$. THMs can be adsorbed on granular activated carbon (GAC) but the adsorption efficiency is low.

Advanced oxidation processes can reduce the concentration of refractory organics and are partially effective in destroying TCE and PCE (104,116). The $O_3-H_2O_2$ process is used for removing TCE and PCE from groundwater (11 ML/d) in a plant operated by the city of Los Angeles. Similar technology is being employed in France and the U.K. for treatment of waters contaminated with herbicides and pesticides. The O_3-uv process has been installed for TCE and PCE removal (5 ML/d) in a plant in South Gate, California.

THM and HAA Formation Potential. Trace concentrations of organic materials, such as humic, tannic, and fulvic acids, and synthetic organic chemicals, in treated water react with HOCl and ClO^- to produce THMs and HAAs. In bromide-containing waters, both chlorinated and brominated THMs and HAAs are formed. Because some of these compounds are carcinogenic, the U.S. EPA has set the maximum contaminant level (MCL) for total THMs (TTHMs) at 0.1 mg/L (117). Ozone-treated waters generally have lower THM and HAA levels as a result of lower THM- and HAA-formation potentials (THMFP and HAAFP).

The main strategy for controlling THMs and HAAs is to lower the THMFP and HAAFP by reducing THM and HAA precursors. In preozonation, ozone is added in low dosage levels at the front of the plant to aid the coagulation and partial removal of THM and HAA precursors. Ozone also can be added as an intermediate (prefiltration) unit process to oxidize many remaining THM and HAA precursors; however, caution must be exercised when high concentrations of organics remain in the water. Partial oxidation of high concentrations of organics may alter their structure so that, if biofiltration is not practiced, they produce more THMs and HAAs on subsequent chlorination. In these isolated cases, THM and HAA formation can be controlled by lowering the ozone dosage level, although primary disinfection requirements may not be met. Low pH and high alkalinity enhance the oxidation of THM and HAA precursors by molecular ozone, whereas high pH and low alkalinity enhance precursor oxidation by HO radicals.

Some THM and HAA precursors also can be removed by biological filtration. The resulting biological stability of the effluent can reduce the residual disinfectant requirement, consequently reducing the formation of THMs and HAAs. More organics or THM and HAA precursors are removed in this manner than by using ozone or GAC alone. Formation of THMs and HAAs during disinfection also can be reduced by adding chlorine at the end of the treatment process or by following chlorination with ammoniation, which produces monochloramine, but not halogenated by-products. The EPA is proposing a two-stage lowering of the MCL on TTHMs and HAAs to 0.08 and 0.04 mg/L, respectively; the former will take effect in 1997, the latter in 2002 (118). These lower limits may require either advanced oxidation processes (119) or may be met by ozonation, biofiltration, and chloramination.

Control of Other Disinfection By-Products. The EPA is also proposing strict limits for bromate ion (118). Ozonation and biodegradation, which have shown promise in controlling TTHMs as well as THAAs in high bromide-containing water, do not result in excessive bromate formation (120). Aldehydes, which have been placed on the EPA's Drinking Water Priority List as candidates for future regulation, as well as other organic ozonation by-products, can be mineralized by biofiltration.

High Purity Water Systems. *Bottling and Canning Plants.* Breweries ozonate the brewing water to remove any residuals of taste and odor and to ensure the absence of microorganisms. The soft drink industry removes the ozone residual by vacuum-stripping in a degassing chamber before bottling (see CARBONATED BEVERAGES). The bottled water industry requires that an ozone residual be included with the water in the bottle. The ozone residual disinfects the inside of the bottle where contact is made with the water; some ozone, however, escapes into the gas phase where it also disinfects the inside of the cap and the container, which is not in contact with water. Finally, the ozone residual disappears as it decomposes to oxygen. In similar applications, the inside of bottles and cans is sprayed with water containing an ozone residual for disinfection prior to the introduction of food.

Pharmaceutical Industry. In the pharmaceutical industry, sterility of deionized water systems is maintained by using an ozone residual. The ozone residual concentration is maintained at >0.3 ppm in the water recirculation loop. Prior to product compounding, the ozone residual is removed by contact with uv irradiaton for <1 s. Ozone also is used to oxidize pyrogens from distilled water destined for intravenous solutions.

Electronics Industry. Organic materials on the surface of electronic components are oxidized by immersion in deionized water that contains 0.5–2 ppm ozone residual. In some cases, gaseous ozone is employed to oxidize organic films on the surface of electronic components. In this treatment process, 5–6 wt % ozone, made in an oxygen-fed generator, is required.

Industrial Wastewater Pollution Control. Nonbiodegradable industrial waste can be made degradable by treatment with ozone (121). The nonbiodegradable molecules are made biodegradable by the introduction of an oxygenated organic functional group at which site the metabolic process can begin. In a similar manner, oxygenated functional groups also are formed on the surface of furnace carbon black. Carbon black is suspended in a gas stream and is treated with ozone at 50–150°C. This process produces a long flow, high color ink containing good dispersive properties. One of the first industrial ozone wastewater applications involved the oxidation of phenol and cyanide in a solution that was used to strip paint from airplanes. Destruction of phenol and cyanide, together with the treatment of textile-dye wastewater, remain the three largest applications of ozone to industrial wastes.

Phenols. The first stable ozone oxidation product of phenol in water is *cis,cis*-muconic acid, which requires ~2 mol O_3/mol phenol. In practice, larger dosage levels of ozone are required because other ozone-reactive substances are present in most wastes. Ozone oxidation of phenolic effluents is employed in paper mills, coke mills, oil refineries, and thermoplastic resin manufacture, producing effluents that are safe to freshwater biota (122,123) (see LIGNIN; PULP).

Cyanide Wastes. Ozone is employed to oxidize cyanide-bearing wastes to yield the nontoxic cyanate ion, which can be further oxidized by ozone to carbon dioxide and nitrate ion (46,124). Ozone has been applied to cyanide-bearing plating wastes, photographic wastes, mining wastes, and textile dyeing plant wastewater; a concurrent reduction of heavy-metal concentrations occurs in these processes (125). The photographic industry treats their effluent with zone to remove the chemical oxygen demand (COD), heavy metal ions, chelating agents (qv)s, and cyanide ion. A combination of ozone and uv irradiation is used to destroy iron–cyanide complexes and other ozone refractory compounds, eg, glycine, palmitic acid, glycerol, ethanol, and acetic acid (126).

Color Bodies. Effluents possessing either natural color bodies, eg, tannins and lignins from pulp and paper operations, or synthetic color bodies, eg, organic dyes, can be decolorized by ozone (127,128). Ozone cleavage of the carbon–carbon double bonds in the chromophoric molecules shifts the absorption spectrum from the visible to the uv region.

Wastewater Disinfection. Ozone is used in disinfection of secondary or biologically treated wastewater and in situations where a high quality effluent is required, eg, when the effluent flows into a reservoir or when game fish are in the receiving stream. After ozone is used to disinfect a secondary municipal effluent, it is difficult visually to distinguish it from potable water (129). Ozone-disinfected water is saturated with oxygen and safe to the receiving stream, whereas chlorinated water is toxic to trout and other forms of aquatic life. In addition, ozone is a more effective disinfectant than chlorine since it inactivates all viruses. Concurrent with the disinfection process, ozone lowers the color, chemical oxygen demand, and turbidity. Ozone also removes suspended solids by an ozone-induced flotation process; this process is initiated by precipitating naturally occurring residual ferrous ion as ferric hydroxide, which functions as a coagulant (130) (see FLOTATION).

Odor Control. The largest number of ozone generators in the United States are employed for sewage odor control (see WATER, SEWAGE). More than 1000 ozone generators are used in wastewater treatment plants and in wastewater pumping stations. Ozone, applied at 2–10 ppm, oxidizes odor-forming compounds in the gas or aqueous phase. Odor properties usually are associated with functional groups of high electron density, eg, sulfides, amines, and olefins. Some industrial odors oxidized by ozone include those from dairy processing plants, compositing operations, fish processing, rubber compounding, rendering plants, commercial kitchens, paper mills, chemical plants, food processing, pharmaceutical fermentation, and phenols (131).

Process Water. Ozone is employed to a limited extent in the treatment of process water, such as cooling tower water, where ozone functions primarily as a biocide (132,133). In 1994, there were an estimated 300 installations employing ozone. Ozone has no direct effect on corrosion or scaling, since these are dependent on water chemistry and water quality. Of the various possible applications, including heat ventilation and air conditioning (HVAC), light manufacturing, oil refineries, chemical plants, utilities, and steel mills, only the first two areas, which involve moderate temperatures, have good applicability. In addition, use of ozone and bromide ion shows promise in industrial cooling towers.

Treatment of Swimming Pools and Spas. Ozone generated by silent discharge is employed primarily as an oxidant in treating swimming pool and spa water, mainly in Europe and to a small extent in North America. The largest application is in large municipal pools and spas, where chlorine typically is used as the disinfectant. Sometimes ozone is used in combination with NaBr for the generation of $HOBr-BrO^-$. Although inadequate for treatment of pools, uv ozone generators have some utility for small spas or hot tubs when employed at room temperature (see WATER, TREATMENT OF SWIMMING POOLS, SPAS, AND HOT TUBS).

Pulp Bleaching. Ozone is finding increasing use in pulp delignification and bleaching because of environmental concerns with conventional techniques. Lignin (qv), a high molecular weight phenolic material, is readily degraded by ozone. Since cellulose is a β-glycoside, it is also susceptible to attack by ozone. Research efforts are being directed toward the development of elemental chlorine-free, totally chlorine-free, and totally effluent-free technologies. There are numerous mills worldwide that use ozone bleaching on a commercial scale, and others are in the planning stage (134). Various bleaching sequences are employed, depending on the type of pulp and application. Ozone typically is employed in concentrations of 6–15%. For example, a plant producing 1000 t/d of a high consistency softwood kraft pulp employs an OZED sequence, where O is oxygen, Z ozone, E caustic, and D chlorine dioxide. A totally chlorine-free process for a medium consistency sulfite pulp in Säffle, Sweden uses a ZEP sequence, where P is hydrogen peroxide. Ozone also is used to treat effluents from pulp bleaching plants. Development of OZP bleaching has the potential to eliminate effluents (see BLEACHING AGENTS, PULP AND PAPER).

Organic Synthesis. Ozone is employed as a selective oxidant in laboratory-scale synthesis (7) and in commercial-scale production of specialty organic chemicals and intermediates such as fragrances, perfumes (qv), flavors, antibiotics (qv), hormones (qv), and vitamins (qv). In Japan, several metric tons per day (t/d) of piperonal [120-57-0] (3,4-methylenedioxybenzaldehyde) is manufactured in 87% yield via ozonolysis and reduction of isosafrole [93-16-3]. Piperonal (or heliotropine [120-57-0]) has a pleasant odor and is used in perfumery. Oleic acid [112-80-1], $CH_3(CH_2)_7CH = CH(CH_2)_7CO_2H$, from tall oil (qv) is ozonated on a t/d scale to produce pelargonic, $C_8H_{17}CO_2H$, and azelaic, $HO_2C(CH_2)_7CO_2H$, acids. Oleic acid also is ozonated in Japan to produce the azealic half-aldehyde that is reduced with hydrogen and ammonia to ω-aminononanoic acid. Polymerization of the latter acid yields nylon-9 or Azelon (135) (see POLYAMIDES). Ozonation of cyclohexene followed by oxidation yields adipic acid, a nylon-6 intermediate. Undesirable residual double bonds in cross-linked elastomeric polymers can be eliminated by ozonation to produce soluble elastomeric polymer latexes (136).

Ozone accelerates the autoxidation of acetaldehyde to peracetic acid at below 15°C. Acetaldehyde hemiacetal peracetate, an intermediate product, is decomposed by heating to peracetic acid. Reaction of peracids with olefins produces epoxides that are hydrolyzed to diols useful in automotive antifreeze mixtures (see ANTIFREEZES AND DEICING FLUIDS).

The pharmaceutical industry employs ozone in organic reactions to produce peroxides as germicides in skin lotions, for the oxidation of intermediates for bacteriostats, and in the synthesis of steroids (qv) such as cortisone (see DISINFECTANTS AND ANTISEPTICS). Vitamin E can be prepared by ozonation of trimethylhydroquinone.

Medical Applications. Among other uses, ozone therapy, employing O_3-O_2, is increasingly being employed and studied in dentistry, veterinary and sports medicine, and proctology; in the treatment of dermatoses, osteoporosis, tumors (137), ulcers, HIV and Hepatitus B viruses, ar_d sickle cell anemia; and for the healing of wounds and diseases of the extremities such as diabetic leg ulcers (138). It also has been used to treat certain cancers, to eliminate arterial blood clots and cholesterol deposits, and during brain surgery.

Miscellaneous Uses. Ozone is used as an aquatic oxidant and disinfectant in zoos, large aquariums, as well as fish and shrimp hatcheries. Ozone also is used for food preservation, in cold storage rooms, brewery cellars, hotel and hospital air ducts, and air conditioning systems. Ozone has also been used in textile bleaching and in the bleaching of esters, oils, fats, waxes (qv), starch, flour, ivory, etc. Oxidation of Ag^+ by ozone is employed commercially to produce high purity AgO. The use of ozone as a chemical agent decontaminant has been patented (139).

Analytical and Test Methods

Ozone can be analyzed by titrimetry, direct and colorimetric spectrometry, amperometry, oxidation–reduction potential (ORP), chemiluminescence, calorimetry, thermal conductivity, and isothermal pressure change on decomposition. The last three methods are not frequently employed. Proper measurement of ozone in water requires an awareness of its reactivity, instability, volatility, and the potential effect of interfering substances. To eliminate interferences, ozone sometimes is sparged out of solution by using an inert gas for analysis in the gas phase or on reabsorption in a clean solution. Historically, the most common analytical procedure has been the iodometric method in which gaseous ozone is absorbed by aqueous KI.

$$O_3 + 2\,I^- + H_2O \longrightarrow I_2 + O_2 + 2\,OH^-$$

The liberated iodine is measured spectrometrically or titrated with standard sodium thiosulfate solution ($I_2 + 2\,S_2O_3^{2-} \rightarrow 2\,I^- + S_4O_6^{2-}$) following acidification with sulfuric acid; buffers are sometimes employed. The method requires measurement of the total gas volume used in the procedure. The presence of other oxidants, such as H_2O_2 and NO_x, can interfere with the analysis. The analysis is also technique-sensitive, since it can be affected by a number of variables, including temperature, time, pH, iodide concentration, sampling techniques, etc (140). A detailed procedure is given in Reference 141.

Ozone in the gas phase can be determined by direct uv spectrometry at 254 nm via its strong absorption. The accuracy of this method depends on the molar absorptivity, which is known to $\pm1\%$; interference by CO, hydrocarbons, NO_x, or H_2O vapor is not significant. The method also can be employed to measure ozone in aqueous solution, but is subject to interference from turbidity as well as dissolved inorganics and organics. To eliminate interferences, ozone sometimes is sparged into the gas phase for measurement.

Various colorimetric methods have been employed for measuring ozone residuals, although most of these are susceptible to significant interferences (142). The indigo trisulfonate method (143), however, has been approved by the Standard Methods Committee of the American Public Health Association (141) and the International Ozone Association for ozone residual measurement.

Electrochemical methods can involve the amperometric titration of liberated iodine with phenylarsine oxide, using a rotating Au/Cu electrode or excess phenylarsine oxide, and measuring the excess by direct pulse polarography (144). Ozone also can be measured directly by reduction to O_2 on a rotating Au/Cu electrode or a Ni/Ag–AgCl electrode pair (145). To eliminate interferences, membrane cells have been developed that consist of a semipermeable membrane such as Teflon or dimethyl silicone and a suitable sensor (146). Ozone diffuses into the cell through the porous membrane and is measured polarographically by employing electrode combinations such as gold–calomel or Au/Ag–AgCl. ORP cells consist of an indicating and a reference electrode combination, which generates a potential proportional to the ozone concentration. They are typically employed as a monitor or controller in water treatment applications.

Chemiluminescent analyzers are based on the light (chemiluminescence) emitted in the gas-phase reaction of ozone with ethylene, which is measured with a photomultiplier tube. The resulting current is proportional to the ozone concentration (see LUMINESCENT MATERIALS, CHEMILUMINESCENCE).

Laboratory, portable, and rugged industrial ozone meters employing uv, amperometry, or chemiluminescence are available for continuous or semicontinuous analysis of either gaseous or aqueous ozone.

Health and Safety

Depletion of the Ozone Layer. As a constituent of the atmosphere, ozone forms a protective screen by absorbing radiation of wavelengths between 200 and 300 nm, which can damage DNA and be harmful to life. Consequently, a decrease in the stratospheric ozone concentration results in an increase in the uv radiation reaching the earth's surfaces, thus adversely affecting the climate as well as plant and animal life. For example, the incidence of skin cancer is related to the amount of exposure to uv radiation.

Laboratory studies in 1974 (76,147) indicated that chlorine radicals from photodegradation of CFCs (chlorofluorocarbons) can destroy ozone. Because of these studies and the fact that the two main varieties (CFC-11($CFCl_3$) and CFC-12(CF_2Cl_2)) can persist in the atmosphere for 75–100 years, the United States in 1978 banned their use in aerosols such as hair sprays and certain deodorants. However, pressure to eliminate CFCs slackened until 1985, when the Antarctic ozone hole was discovered (148). The ozone hole is a dramatic seasonal thinning of the normal ozone concentration over the continent of Antarctica, attributed to catalytic decomposition of ozone by halogen radicals ($Cl^{\bullet}$ and $Br^{\bullet}$), which are formed by photodecomposition of CFCs and Halons (bromofluorocarbons). A similar but less severe reduction in ozone levels in the Arctic also has been observed. Reduced ozone levels extend to the lower latitudes and can pose a threat to human, animal, marine, and plant life as a result of increased energetic uv–B radiation (280–320 nm) reaching the surface of the earth. Data from Antarctica show that uv radiation soars under the ozone hole, where fully half of the atmospheric ozone is destroyed each spring. The global drop in amphibian population such as frogs, toads, and salamanders has been ascribed to increased uv radiation from ozone thinning (149).

Confirmation of the destruction of ozone by chlorine and bromine from halofluorocarbons has led to international efforts to reduce emissions of ozone-destroying CFCs and Halons into the atmosphere. The 1987 Montreal Protocol on Substances That Deplete the Ozone Layer (150) (and its 1990 and 1992 revisions) calls for an end to the production of Halons in 1994 and CFCs, carbon tetrachloride, and methylchloroform by January 1, 1996. In 1993, worldwide production of CFCs was reduced to 50% of 1986 levels of 1.13×10^6 and decreases in growth rates of CFC-11 and CFC-12 have been observed (151).

In order to meet the established goals, the industry has accelerated efforts to find alternatives to CFCs (152). In 1988, the first International CFC & Halon Alternatives Conference was held for the purpose of sharing technology. The U.S. industry is switching to hydrofluorocarbons (HFCs), which do not contain chlorine or bromine that can decompose ozone, for use in compressors of home refrigerators and almost all new car and truck air-conditioning systems, and to hydrochlorofluorocarbons (HCFCs) for industrial refrigeration units and in foam-blowing applications. The Montreal Protocol classifies HCFCs as transitional, setting a limit on their production in 1996 and gradually phasing them out. There is some concern that HCFCs may slow the ozone layer recovery (153). Under the stratospheric ozone protection provision of the Clean Air Act, the EPA has issued its final rule regarding the evaluation and regulation of ozone-depleting substitutes (154) (see FLUORINE COMPOUNDS, ORGANIC–FLUORINATED ALIPHATIC COMPOUNDS).

A smaller factor in ozone depletion is the rising levels of N_2O in the atmosphere from combustion and the use of nitrogen-rich fertilizers, since they are the sources of NO in the stratosphere that can destroy ozone catalytically. Another concern in the depletion of ozone layer, under study by the National Aeronautics and Space Administration (NASA), is a proposed fleet of supersonic aircraft that can inject additional nitrogen oxides, as well as sulfur dioxide and moisture, into the stratosphere via their exhaust gases (155). Although sulfate aerosols can suppress the amount of nitrogen oxides in the stratosphere by converting N_2O_5 to HNO_3, the actual effect depends on where in the atmosphere the plane's exhaust gases finally accumulate.

Environmental Impact of Ambient Ozone. Ozone can be toxic to plants, animals, and fish. The lethal dose, LD_{50}, for albino mice is 3.8 ppmv for a 4-h exposure (156); the 96-h LC_{50} for striped bass, channel catfish, and rainbow trout is 80, 30, and 9.3 ppb, respectively. Small, natural, and anthropogenic atmospheric ozone concentrations can increase the weathering and aging of materials such as plastics, paint, textiles, and rubber. For example, rubber is degraded by reaction of ozone with carbon–carbon double bonds of the rubber polymer, requiring the addition of aromatic amines as ozone scavengers (see ANTIOXIDANTS; ANTIOZONANTS). An ozone decomposing polymer (noXon) has been developed that destroys ozone in air or water (157).

Although the naturally occurring concentration of ozone at earth's surface is low, the distribution has been altered by the emission of pollutants, primarily by automobiles but also from industrial sources, which lead to the formation of ozone. The strategy for controlling ambient ozone concentrations arising from automobile exhaust emissions is based on the control of hydrocarbons, CO, and NO via catalytic converters. As a result, peak ozone levels in Los Angeles, for instance, have decreased from 0.58 ppm in 1970 to 0.33 ppm in 1990, despite a 66% increase in the number of vehicles.

The EPA is reviewing and revising the Air Quality and Other Photochemical Oxidants (Criteria Document) and reevaluating the national ambient air quality standards (158). The EPA is also proposing that manufacturers ($\geq 10,000$ lb O_3/yr) and users ($\geq 25,000$ lb O_3/yr) submit data annually to the EPA on estimated ozone releases (159).

Human Exposure to Ozone. The toxicity of ozone is largely related to its powerful oxidizing properties. The odor threshold of ozone varies among individuals but most people can detect 0.01 ppm in air, which is well below the limit for general comfort. OSHA has established a time-weighted average permissible exposure level for workers for an eight-hour day of 0.10 ppm v/v (0.2 mg/m^3) and a short-term exposure limit of 0.30 ppm v/v (0.6 mg/m^3) for an exposure less than 15 minutes (160). The latter is based on observations showing that significant declines in pulmonary function can result from repeated intermittent exposures or from a single short-term exposure to ozone. The toxicity of gaseous ozone varies with concentration and exposure time (161). The symptoms experienced on exposure to 0.1–1 ppm ozone are headache, throat dryness, irritation of the respiratory passages, and burning of the eyes caused by the formation of aldehydes and peroxyacyl nitrates. Exposure to 1–100 ppm ozone can cause asthma-like symptoms such as tiredness and lack of appetite. Short-term exposure to higher concentrations can cause throat irritations, hemorrhaging, and pulmonary edema. Additional toxicity data is given in Reference 162.

The presence of naturally occurring ozone in the lower stratosphere creates a potential hazard for passengers and crew members of high flying aircraft (163,164). Ozone in the inlet air to the aircraft cabin, which can reach 1.2 ppm, is destroyed catalytically.

Ozone Disinfection By-Products. Ozonation of drinking water produces various by-products such as aldehydes, ketones, carboxylic acids, organic peroxides, epoxides, nitrosamines, N-oxy compounds, quinones, hydroxylated aromatic compounds, brominated organics, and bromate ion. Although some of these compounds are potentially toxic or carcinogenic, most bioassay-screening studies have shown that ozonated water induces substantially less mutagenicity than chlorinated water (165–167). However, further work is necessary to identify and screen (Ames test) ozonation by-products formed under typical water treatment conditions (168). Ozonation by-products are on the Drinking Water Priority List as candidates for future regulation (169). The Disinfection and Disinfection By-Products Rule proposed by the U.S. EPA will set limits for both disinfectants (excluding ozone) and disinfection by-products (eg, bromate) and require biofiltration following ozone use (118).

BIBLIOGRAPHY

"Ozone" in *ECT* 1st ed., Vol. 9, pp. 735–753, by V. A. Hann and T. C. Manley, The Welsbach Corp.; in *ECT* 2nd ed., Vol. 14, pp. 410–432, by T. C. Manley and S. J. Niegowski, The Welsbach Corp.; in *ECT* 3rd ed., Vol 16, pp. 683–713, by C. Nebel, PCI Ozone Corp.

1. C. Schönbein, *Compt. Rend. Hebd. Seances Acad. Sci.* **10**, 706 (1840).
2. A. G. Streng, *J. Chem. Eng. Data* **6**, 431 (1961).
3. D. Hanson and K. Mauersberger, *J. Chem. Phys.* **83**, 326 (1985).
4. L. F. Kosak-Channing and G. R. Helz, *Environ. Sci. Technol.* **17**, 145 (1983).
5. J. A. Roth and D. E. Sullivan, *Ind. Eng. Chem. Fundam.* **20**, 137, (1981); V. Tarinina and co-workers, *Zh. Obshch. Khim.* **53**, 1441 (1983); R. R. Munter, Deposited Doc. *VINITI*, 3996–3984 (1984).
6. R. Battino, ed., *Oxygen and Ozone, IUPAC Solubility Series*, Vol. 7, Pergamon, Oxford, U.K., 1981.
7. L. F. Fieser and M. Fieser, *Reagents for Organic Synthesis*, John Wiley & Sons, Inc., New York, Vols. 1–17.
8. JANAF Thermochemical Tables, *J. Phys. Chem. Ref. Data* **14**(Suppl. 1) 1695 (1985).
9. J. A. Dean, ed., *Lange's Handbook of Chemistry*, 13th ed., McGraw-Hill Book Co., Inc., New York, 1985.
10. G. Hettner, R. Pohlman, and H. J. Schumacher, *Z. Physik* **91**, 372 (1934).
11. R. Atkinson and co-workers, *J. Phys. Chem. Ref. Data* **21**, 1125 (1992).
12. M. L. Kilpatrick and co-workers, *J. Am. Chem. Soc.* **78**, 1784 (1956).
13. R. Tramborubo and co-workers, *J. Chem. Phys.* **21**, 851 (1953).
14. M. Mack and J. S. Muenter, *J. Chem. Phys.* **66**, 5279 (1977).
15. P. C. Hiberty, *Israel J. Chem.* **23**, 10 (1983); R. D. Harcourt and co-workers, *J. Chem. Soc. Faraday Trans.* **82**, 495 (1986); P. Borowski, and co-workers, *J. Chem Phys.* **97**, 5568 (1992); W. Wu and co-workers, *Chin. J. Chem.* **11**, 490 (1993).
16. D. L. Cooper and co-workers, *J. Chem. Soc. Perkin Trans.* **2**, 1187 (1989); D. V. Kostikova and co-workers, *Dokl. Akad. Nauk* **296**, 914 (1987).
17. A. Glissman and H. J. Schumacher, *Z. Physik. Chem.* **21B**, 323 (1933).
18. J. A. Zaslowsky and co-workers, *J. Am. Chem. Soc.* **82**, 2682 (1960).

19. S. W. Benson and A. E. Axworthy, *J. Chem. Phys.* **26**, 1718 (1957).
20. J. C. Devins, *J. Electrochem. Soc.* **103**, 460 (1956).
21. A. Eucken and F. Patat, *Z. Physik. Chem.* **B33** 459 (1936).
22. E. Castellano and H. J. Schumacher, *J. Chem. Phys.* **36**, 2238 (1962).
23. J. A. Wojtowicz and co-workers, paper No. 110, 136th American Chemical Society Meeting, Atlantic City, N.J., 1959.
24. J. A. Wojtowicz and co-workers, paper No. 109, in Ref. 23; J. A. Wojtowicz, M.S. dissertation, Niagara Univ., N.Y. 1965; L. F. Phillips and H. I. Schiff, *J. Chem. Phys.* **88**, 1 (1960).
25. J. Staehelin and J. Hoigné, *Environ. Sci. Technol.* **16**, 676 (1982); R. E. Bühler, J. Staehelin, and J. Hoigné, *J. Phys. Chem.* **88**, 1560 (1984); *Ibid.*, 5999.
26. K. Cheklkowska and co-workers, *Ozone Sci. Eng.* **14**, 33 (1992).
27. J. Hoigné, in R. G. Rice and A. Netzer, eds., *Hankbook of Ozone Technology and Applications*, Vol. 1, Ann Arbor Science Publishers, Inc., Ann Arbor, Mich., 1982, p. 341.
28. C. Yurteri and M. Gurol, *Ozone Sci. Eng.* **10**, 277 (1988).
29. E. Castellano and H. J. Schumacher, *Z. Phys. Chem.* **76**, 258 (1971).
30. E. Castellano and H. J. Schumacher, *Chem. Phys. Lett.* **13**, 625 (1972).
31. C. Cobos, E. Castellano, and H. J. Schumacher, *J. Photochem.* **21**, 291 (1983).
32. D. L. Baulch and co-workers, *J. Phys. Chem. Ref. Data* **11**, 327 (1982).
33. J. I. Steinfeld, S. M. Adler-Golden, and J. W. Gallagher, *J. Phys. Chem. Ref. Data* **16**, 911 (1987).
34. R. L. Miller and co-workers, *Science* **265**, 1831 (1994).
35. W. D. McGrath and R. G. W. Norrish, *Proc. Roy. Soc.* **A254**, 317 (1960).
36. H. Taube, *Trans. Faraday Soc.* **53**, 656 (1956).
37. F. Cocace and M. Speranza, *Science* **265**, 208 (1994).
38. W. R. Haag and J. Hoigné, *Environ. Sci. Technol.* **17**, 261 (1983).
39. J. Hoigné and co-workers, *Water Res.* **19**, 993 (1985).
40. W. R. Haag and J. Hoigné, *Water Res.* **17**, 1397 (1983); *Ozone Sci. Eng.* **6**, 103 (1984).
41. J. Hoigné and H. Bader, *Environ. Sci. Technol.* **12**, 79 (1978).
42. S. Palpko, *Zh. Prikl. Khim.* **30**, 1286 (1957).
43. C. Nebel and co-workers, *Water Sewage Works* **123**, 88 (1976).
44. P. Singer and W. Zilli, *Water Res.* **9**, 127 (1975).
45. D. Wulf and R. C. Tolman, *J. Am. Chem. Soc.* **49**, 1650 (1927).
46. M. D. Gurol and W. M. Bremen, *Environ. Sci. Technol.* **19**, 804 (1985).
47. J. N. Jensen and Y.-J. Tuan, *Ozone Sci. Eng.* **15**, 343 (1993).
48. D. J. Kjos, R. R. Furgason, and L. L. Edwards, in R. G. Rice and M. E. Browning, eds., *Proceedings of the First International Symposium on Ozone for Water and Wastewater Treatment*, International Ozone Association, Stamford, Conn., 1973, p. 194.
49. B. Kreingang and D. Chezhikov, *Izv. Akad. Nauk SSSR Otd. Tekh. Nauk* **4**, 141 (1955).
50. A. Netzer and A. Bowers, in Ref. 48, p. 731.
51. I. J. Solomon and co-workers, *J. Am. Chem. Soc.* **84**, 34 (1962).
52. J. Cerkovnik and B. Plesnicar, *J. Am. Chem. Soc.* **115**, 12,169 (1993).
53. J. A. Wojtowicz, F. Martinez, and J. A. Zaslowsky, *J. Phys. Chem.* **67**, 849 (1963); D. A. Csejka and co-workers, *J. Phys. Chem.* **68**, 3878 (1964).
54. P. S. Bailey, *Ozonation in Organic Chemistry*, Vols. 1 and 2, Academic Press, Inc., New York, 1978–1982.
55. J. Katz, ed., *Ozone and Chlorine Dioxide Technology for Disinfection of Drinking Water*, Noyes Data Corp., Park Ridge, N.J., 1980, p. 314.
56. J. Hoigné and H. Bader, *Water Res.* **17**, 185 (1983).
57. J. Hoigné and H. Bader, *Water Res.* **17**, 173 (1983).

58. M. Elmghari-Tabib and co-workers, *Water Res.* **16**, 223 (1982).

59. J. Jandik, Ph.D. dissertation, Technical Univ. of Münich, Münich, Germany, 1977.

60. J. Mudd and co-workers, *Atmos. Environ.* **3**, 669 (1969).

61. W. Mosher, in *Ozone Chemistry and Technology* (Advances in Chemistry Series, 21), American Chemical Society, Washington, D.C., 1959, p. 140.

62. K. Griesbaum and co-workers, *Proceedings of 209th American Chemical Society Meeting*, Anaheim, Calif., 1995.

63. K. Pollart and R. Miller, *J. Am. Chem. Soc.* **27**, 2392 (1962).

64. E. Fields, *Proceedings of 7th World Petroleum Congress*; *Chem. Eng. News*, 30 (Dec. 23, 1963).

65. C. Nebel, *Chem. Commun.* **2**, 101 (1968).

66. R. Criegee and P. Gunther, *Chem. Ber.* **96**, 1564 (1963).

67. W. H. Glaze, J. W. Kang, and D. H. Chapin, *Ozone Sci. Eng.* **9**, 335 (1987); H. Paillard, R. Brunet, and M. Doré, *Ozone Sci. Eng.*, 391; *Proceedings of First International EPRI/NSF Symposium on Advanced Oxidation*, San Francisco, Calif., 1993.

68. R. L. Garrison, C. E. Mauk, and H. W. Prengle, in Ref. 48, p. 551.

69. G. R. Peyton and co-workers, *Environ. Sci. Technol.* **16**, 448 (1982); H. W. Prengle, *ibid.*, **17**, 743 (1983).

70. P. D. Francis, *Ozone Sci. Eng.* **9**, 369 (1987).

71. H. Allemane and co-workers, *Ozone Sci. Eng.* **15**, 419 (1993).

72. R. W. Murray and M. L. Kaplan, *J. Am. Chem. Soc.* **91**, 5358 (1969).

73. F. E. Stary, D. E. Emge, and R. W. Murray, *J. Am. Chem. Soc.* **98**, 1880 (1976).

74. P. J. Crutzen and co-workers, *Science* **268**, 705 (1995).

75. J. Heicklen, *Atmospheric Chemistry*, Academic Press, Inc., New York, 1976.

76. P. O. Wennberg and co-workers, *Science* **266**, 398 (1994).

77. M. J. Molina and F. S. Rowland, *Nature* **249**, 810 (1974).

78. J. F. Gleason and co-workers, *Science* **260**, 523 (1993).

79. S. Solomon and co-workers, *J. Geophys. Res.* **99**, 20,491 (1994); *Ibid.*, 20,929.

80. D. W. Fahey and co-workers, *Nature* **363**, 509 (1993).

81. S. Solomon and co-workers, *Nature* **321**, 755 (1986).

82. O. B. Toon and M. A. Tolbert, *Nature* **375**, 218 (1995).

83. M. L. Santee and co-workers, *Science* **267**, 849 (1995).

84. R. J. Salawitch and co-workers, *Science* **261**, 1146 (1993).

85. J. C. Wilson and co-workers, *Science* **261**, 1140 (1993).

86. G. L. Manney and co-workers, *Nature* **370**, 429 (1994).

87. *Second European Stratospheric Arctic & Mid-Latitide Experiment*, U.K. Dept. of the Environment (London) News Release, Mar. 30, 1995.

88. A. M. Thompson and R. J. Cicerone, *J. Geophys. Res.* **91**, 10,853 (1986); R. W. Stewart, *Ibid.*, **98**, 20,601 (1993); A. M. Thompson, *Ozone Sci. Eng.* **12**, 195 (1990).

89. U. Kogelschatz, B. Eliasson, and M. Hirth, *Ozone Sci. Eng.* **10**, 367 (1988).

90. B. Eliasson, M. Hirth, and U. Kogelschatz, *J. Phys. D: Appl. Sci.* **20**, 1421 (1987).

91. B. Eliasson and U. Kogelschatz, *IEEE Trans. Plasma Sci.* **19**, 309, 1063 (1991).

92. D. Braun, U. Küchler, and G. Pietsch, *J. Phys. D: Appl. Sci.* **24**, 564 (1991).

93. B. Eliasson and U. Kogelschatz, *Basic Data For Modelling of Electrical Discharges In Gases: Oxygen*, Asea Brown Boveri, Baden, Switzerland, 1986.

94. S. Yagi and M. Tanaka, *J. Phys. D: Appl. Sci.* **12**, 1509 (1979).

95. K. Rakness and co-workers, paper presented at Water Pollution Control Federation Conference Workshop, Oct. 1979.

96. T. C. Manley, *Trans. Electrochem. Soc.* **84**, 83 (1943); J. J. Carlins and R. G. Clark, in Ref. 27, p. 41.

97. M. Fischer and co-workers, *Ozone Sci. Eng.* **9**, 93 (1987).

98. E. Maerz and F. Gaia, *Ozone Sci. Eng.* **12**, 401 (1990).

99. B. Elliasson and U. Kogelschatz, *Ozone Sci. Eng.* **13**, 365 (1991).

100. B. DuRon, in Ref. 27, p. 77; J. M. Dohan and W. J. Masschelein, *Ozone Sci. Eng.* **9**, 315 (1987).

101. C. Fabjan, *Proceedings of the Third Ozone World Congress*, Paris, France, 1977; P. C. Foller, in Ref. 27, p. 89.

102. M. Steinberg and M. Beller, *Chem. Eng. Prog. Symp. Series* **66**, 205 (1970).

103. J. A. Wojtowicz, R. B. Urbach, and J. A. Zaslowsky, *J. Phys. Chem.* **67**, 713 (1963).

104. W. H. Glaze, *Ozone News* **22**, 42 (1994).

105. J. C. Hoff, EPA Report 600/2-86/067, U.S. Environmental Protection Agency, Research Triangle Park, N.C., 1986.

106. G. B. Wickramanayake and O. J. Sproul, *Ozone Sci. Eng.* **10**, 123 (1988).

107. Q. Zhu, C. Liu, and Z. Xu, *Ozone Sci. Eng.* **11**, 169, 189 (1989).

108. *Fed. Reg.* **54**, 27544 (1989).

109. E. C. Nieminski, *Ozone Sci. Eng.* **12**, 133 (1990); *Fed. Reg.* **54**, 27486 (1989).

110. *Fed. Reg.* **59**, 38832 (1994).

111. D. W. Ferguson and co-workers, *J. Am. Water Works Assoc.* **82**, 181 (1990).

112. T. L. Theis and P. C. Singer, *Environ. Sci. Tech.* **8** 569 (1974).

113. H. Paillard and co-workers, *Ozone Sci. Eng.* **11**, 93 (1989).

114. W. R. Knocke, *J. Am. Water Works Assoc.* **79**, 75 (1987).

115. J. Symons and co-workers, *Removal of Organic Contaminants from Drinking Water Using Techniques Other Than Granular Activated Carbon Alone*, EPA Progress Report, U.S. Environmental Protection Agency, Research Triangle Park, N.C., 1979.

116. W. H. Glaze and J. W. Kang, *J. Am. Water Works Assoc.* **80**, 57 (1988); E. M. Aieta and co-workers, *ibid.*, 64.

117. *Fed. Reg.* **44**, 68624 (1979); **56**, 1470 (1991).

118. *Fed. Reg.* **59**, 38668 (1994).

119. J. L. Wallace and co-workers, *Ozone Sci. Eng.* **10**, 103 (1988).

120. J. M. Symons and co-workers, *J. Am. Water Works Assoc.* **86**, 48 (1994).

121. G. Davis, *Proceedings of the Second International Symposium on Ozone Technology*, International Ozone Association, Pan American Group, Stamford, Conn., 1975, p. 421.

122. C. Nebel, in Ref. 121, p. 374.

123. R. Patrick, *Report of Toxicity Tests*, Academy of Natural Sciences, Philadelphia, Pa., 1951.

124. L. J. Bollyky, *Ozone Treatment of Cyanide-Bearing Wastes*, EPA Report 600/2-77-104, U.S. Environmental Protection Agency, Research Triangle Park, N.C., 1977; M. D. Gurol, W. M. Bremen, and T. E. Holden, *Environ. Prog.* **4**, 46 (1985).

125. T. Hendrickson, in Ref. 48, p. 579.

126. H. Prengle and co-workers, in Ref. 121, p. 224.

127. C. Nebel and co-workers, *Pulp Paper* **48**, 142 (1974).

128. C. Nebel and L. Stuber, in Ref. 121, p. 336.

129. C. Nebel and co-workers, *J. Water Poll. Fed.* **45**, 2493 (1973).

130. C. Nebel and co-workers, *J. Boston Soc. Civil Eng.* **62**, 161 (1976).

131. C. Nebel, *Industrial Odor Technology Assessment*, Ann Arbor Science Publishers, Inc., Ann Arbor, Mich., 1975, Chapt. 23.

132. R. J. Strittmatter and co-workers, *Ozone Sci. Eng.* **15**, 47 (1993).

133. P. R. Puckorius, *Ozone Sci. Eng.* **15**, 81 (1993).

134. International Non-Chlorine Bleaching Conference, Amelia Island, Fla., Mar. 1994 and 1995; International Pulp Bleaching Conference, Vancouver, B.C., June 1994.

135. H. Otsuki and H. Funakashi, in Ref. 61, p. 205.

136. U.S. Pat. 5.039,737 (Aug. 23, 1991), D. K. Parker and J. R. Purdon (to Goodyear Tire and Rubber Co.).

137. J. Washüttl, R. Viebahn, and I. Steiner, *Ozone Sci. Eng.* **12**, 65 (1990).

138. J. S. Latino and co-workers, *Blood* **78**, 1882 (1991).

139. U.S. Pat. 4,784,699 (Nov. 15, 1988), D. R. Cowsar and co-workers (to U.S. Government).

140. G. Gordon and co-workers, *J. Am. Water Works Assoc.* **81**, 72 (1989).

141. *Standard Methods for the Examination of Water and Wastewater*, 18th ed., American Public Health Association, Washington, D.C., 1994, p. 4–105.

142. J. H. Stanley and J. D. Johnson, in Ref. 27, p. 255.

143. J. Hoigné and H. Bader, *Vom Wasser* **55**, 261 (1980).

144. R. B. Smart, J. H. Lowry, and K. H. Nancy, *Environ. Sci. Tech.* **13**, 89 (1979).

145. W. J. Masschelein and co-workers, *Analusis* **7**, 432 (1979).

146. J. H. Stanley and J. D. Johnson, *Anal. Chem.* **51**, 2144 (1979).

147. R. S. Stolarski, *Sci. Amer.* **258**, 30 (1988); P. S. Zurer, *Chem. Eng. News* 8 (May 24, 1993); *Ibid.*, (Aug. 15, 1994).

148. J. C. Farnam and co-workers, *Nature* **315**, 207 (1985).

149. A. R. Blaustein and co-workers, *Proc. Nat. Acad. Sci.* **91**, 1791 (1994).

150. *Montreal Protocol to Reduce Substances that Deplete the Ozone Layer, Final Report* U.N. Environmental Programme, New York, 1987.

151. J. W. Elkins and co-workers, *Nature* **364**, 780 (1993).

152. P. S. Zurer, *Chem. Eng. News.* 12 (Nov. 15, 1993).

153. Stratospheric Ozone Review Group, *Stratospheric Ozone 1993*, HMSO Books, London, 1993.

154. *Fed. Reg.* **59**, 13041 (1994).

155. P. S. Zurer, *Chem. Eng. News* 6 (Oct. 24, 1994); *Ibid.*, 10 (Apr. 24, 1995).

156. C. S. Weil, *Biometrica* **8**, 249 (1952).

157. P. Layman, *Chem. Eng. News* 9 (Aug. 22, 1994).

158. *Fed. Reg.* **59**, 5164 (1994).

159. *Fed. Reg.* **59**, 1787 (1994).

160. *Fed. Reg.* **40**, 47261 (1975).

161. J. M. Langerwerf, *Aerospace Medicine* **36**, June 1963.

162. R. J. Lewis, Jr., ed., *Sax's Dangerous Properties of Industrial Materials*, Van Nostrand Rienhold, Co., Inc., New York, 1993.

163. M. Letagola, C. Melton, and E. Higgins, *Aviat. Space Environ. Med.* **51**, 237 (1980).

164. W. Broad, *Science* **205**, 767 (1979).

165. R. G. Rice and J. A. Cotruvo, eds., *Ozone/Chlorine Dioxide Oxidation Products of Organic Materials*, International Ozone Association, Cleveland, Ohio, 1978.

166. W. H. Glaze, *Environ, Sci. Technol.* **21**, 224 (1987).

167. P. M. Huck and co-workers, *Ozone Sci. Eng.* **11**, 245 (1989).

168. W. H. Glaze and co-workers, *J. Am. Water Works Assoc.* **85**, 53 (1993).

169. *Fed. Reg.* **53**, 1892 (1988).

J. A. WOJTOWICZ
Consultant

PACKAGING

CONTAINERS FOR INDUSTRIAL MATERIALS

In any operation involving the manufacturing, distribution, and use of chemical substances, it is essential that consideration be given to packaging at an early stage of the manufacturing process. Container systems for industrial chemicals must fulfill several important functions: they must contain the product in order to be able to move it safely from point of manufacture to use; protect both the product from contamination and the immediate surroundings (plant, people, equipment) from the potential harm caused by the product itself; provide features that aid users in the effective utilization of the product; and communicate valuable information such as product identity, potential hazards, and handling information, to shippers and users.

The environmental impact of packages and packaging materials has come under increasingly vigorous scrutiny by all kinds of interests, ie, government and regulatory agencies, consumer groups, and environmentalists. It is becoming increasingly important that packaging for all kinds of products be developed in a rational manner, use less material, and have the ability to be recovered and reused whenever it is economically feasible and permitted by regulation.

Virtually any chemical can be stored and transported safely and effectively by using one of many package types. The choice of a container system, in general, is dictated by manufacturing, marketing, and economic considerations; for a chemical, however, the choice of packaging materials often is influenced primarily by safety and chemical compatibility factors. Both aspects can affect the cost of physical distribution, which often is comparable to the cost of the product being packaged.

Regulations

Regulations governing packaging and shipping of chemicals depend on the classification of the chemical as hazardous or nonhazardous. For nonhazardous chemicals, the packaging and shipping requirements are subject to the rules issued by the carrier. The most common of these rules are published in the *Uniform Freight Classification* for railroads and the *National Motor Freight Classification* for trucks. These rules are similar in that they both include sections listing the participating carriers, index to articles, article requirements, and packaging descriptions. The participating carriers have the right to collect a surcharge as well as to refuse handling or paying damage claims for articles not packaged according to the classification requirements (1,2).

There is also a procedure for trial shipments of new or improved packages. In such cases, the shipper must file an application for a test permit with the appropriate classification committee, supplying technical data to support the request. If the request is granted, a permit is issued. Test shipments must be marked as such, including reports filed on a timely basis during the test period. The shipper of the test package is responsible for all damage to the product, and must file reports with the commission that issued the permit. If, in the opinion of the commission, the package is good, a new package number will be issued.

Regulations controlling the packaging and shipping of hazardous materials in the United States are prepared by the Research and Special Programs Administration (RSPA) of the U.S. Department of Transportation (DOT). The primary document is Hazardous Materials Regulations (HMR) 49 CFR, parts 171–179. The *Code of Federal Regulations* (CFR) has been extensively changed to bring it into agreement with the international rules recommended by the United Nations Committee of Experts. Recent amendments to the CFR are contained in Docket Number HM-181, the most current of which, as of this writing, is HM-181F, with an effective date of October 1, 1993 (3). As a result of these changes, materials packaged for shipment in the United States are acceptable worldwide; although in Europe there can be some differences based on the mode of transportation (qv).

In the words of RSPA, the changes to the regulations will (*1*) simplify and reduce the volume of the HMR, (*2*) enhance safety through better classification and packaging, (*3*) promote flexibility and technological innovation in packaging, (*4*) reduce the need for exemptions from the HMR, and (*5*) facilitate international commerce (3).

The primary change in the HMR is in replacing the specific container requirements for products with performance orientated packaging (POP). In general, this means any package can be used as long as it passes certain rigid test requirements. The certification of a package is the responsibility of the shipper, who can self-certify the packaging or have the tests performed by an approved test laboratory. In the latter case, the test laboratory itself must be approved as a test facility by the DOT. Present DOT-specified packages may not be used after October 1, 1996. The other change in the HMR is that shippers must provide the telephone number of either a company representative or a service that can be reached 24 h/d to answer questions regarding the nature of the hazardous material.

The CFR is divided into the following numbered sections: 171 for general information, regulations, and definitions; 172 for hazardous materials table, special provisions, hazardous materials communications requirements, and emergency response information requirements; 173 for general shipment and packaging requirements; 174 for carriage by rail; 175 for carriage by aircraft; 176 for carriage by vessel; 177 for carriage by public highway; 178 for packaging specifications; and 179 for tank cars specifications.

The HMR classifies hazardous materials into nine classes, some with subdivisions, as shown in Table 1.

Materials with the exception of explosives, gases, or radioactive classifications are also placed in packing groups. These groups define the relative danger of a material and, in turn, the package test requirements. Packing Group I denotes materials of great danger; II, of medium danger; and III, of minor danger.

The proper method for determining the packaging requirements for a hazardous material is to find the material listing in the Hazardous Materials Table

Table 1. HMR Classification

Subdivision	Description
	Class 1: explosives
1.1	mass explosion hazard
1.2	projection hazard; no mass explosion hazard
1.3	fire hazard and minor projection or blast
1.4	no significant blast hazard
1.5	very insensitive mass explosion hazard
1.6	extremely insensitive detonating substances
	Class 2: compressed gases
2.1	flammable gas
2.2	nonflammable gas
2.3	poison gas
	Class 3: flammable liquids
	Class 4: flammable solids
4.1	flammable solid
4.2	spontaneously combustible
4.3	dangerous when wet
	Class 5: oxidizing substances and organic peroxides
5.1	oxidizer
5.2	organic peroxide
	Class 6: poisonous and infectious substances
6.1	poisonous substances
6.2	infectious substances
	Class 7: radioactive materials
	Class 8: corrosives
	Class 9: miscellaneous dangerous substances

of 49 CFR, section 172. From here the hazard class, identification number, packaging group, label requirements, special provisions, and packaging authorizations can all be ascertained. If any company has a tested package design for a similar type of product, that company may be allowed to use the design for the new product; otherwise, a new package must be designed and tested. Once the package has been approved, it must be marked with the U.N. package markings that specify the material, packing group for which the package has been approved, maximum gross weight or relative density of material for which the package was tested, whether the material is solid or under pressure, the year manufactured, the state or country of origin, and testing facility.

When ready for shipment, a package must have the identification number and proper shipping name marked in its upper left corner, the hazardous material labels centered in the panel, and the package marking in the lower left corner (Fig. 1). In addition, the shipping documents must show the hazardous material identification, hazard class, packing group, as well as a telephone number in case of emergency. Improper packaging, marking, documentation, or handling can result in civil and criminal liabilities against the shipper, package manufacturer, or carrier.

Transportation and Storage.   Three considerations apply to both transportation (qv) and storage: compliance with legal requirements; package compatibility with the product as well as manufacturing and physical distribution requirements, such as safety requirements; and selection of an optimal-cost pack-

Fig. 1. Hazardous materials labeling and marking. The codes are as follows: $\binom{u}{n}$ is the United Nations packaging symbol; 4G, a fiberboard box; Y, the container may be used for materials in packing groups II and III, ie, products that pose a medium or minor danger; 145, the maximum gross mass in kilograms that may be packed into the container; S, the product is a solid; 93, the year in which the container was made; USA, the country authorizing the allocation of the markings; and XYZ, the name of the manufacturer of the container.

aging system consistent with the preceding considerations. The interrelated nature of these factors, the variety of products to be shipped, the numerous packaging methods, and the large costs which are often associated with such decisions are all details that should be evaluated by a packaging specialist.

Bulk Handling of Products

Liquids. Approximately 170,000 railroad tank cars are used in the United States. The interior surfaces of these cars are tailored to carry a wide variety of products and are constructed of steel which is either unlined or lined with materials to enhance the chemical compatibility with a specific product; these lining materials include synthetic rubber, phenolic or modified epoxy resins, or corrosion-resistant materials such as aluminum, nickel-bearing steel, or stainless steel.

For commodities that solidify at temperatures commonly encountered during shipping, tank cars are equipped with internal or external heating coils. In some cases, cars are insulated with both sides of the insulation protected by thin steel shells. Approximately 15% of the tank cars in the United States are constructed for the transportation of pressurized commodities, such as anhydrous ammonia and propane.

If distribution requirements cannot be met by available tank cars, new car designs can be implemented; however, they are subject to the approvals of the Association of American Railroads and the DOT.

Tank cars have been constructed with capacities as great as 130,000 L (~34,000 gal) and weights as much as around 91 metric tons. The utilization of tank cars often represents an economical approach to the transportation of bulk commodities.

Solids. Increasing use of bulk cars, especially of covered hopper cars, has accompanied the expansion of the tank-car fleet. The principal drawback of bulk cars is the requirement for limited use, specialized cars, which necessitates a large investment. However, if such investment can be justified, the cost of transportation for dry bulk materials in hopper cars usually is less than those for goods in shipping containers. In many instances, such cars are used in closed-loop service; that is, they shuttle in unit trains between filling and discharge points. Similar equipment is also used in specialized highway vehicles whose truck bodies can incorporate dump hoppers and built-in conveyors.

Exotic materials are less likely to be used in the construction of hopper cars than in tank cars because of the lower chemical aggressiveness of solid chemicals compared to that of liquids and gases.

Semibulk Containers. Use of semibulk containers falls between bulk handling, eg, accomplished by tank cars and hopper cars, and individual package handling, which is often performed manually. Semibulk containers are also known as intermediate bulk containers (IBCs), the provisions and requirements for the construction and testing of which can be found in the U.N. recommendations (4).

Because volume capacity of a semibulk package can range from 400 L (110 gal) to 3000 L (790 gal), having a net weight of 225–2270 kg, cranes, lift trucks, and other powered materials-handling equipment are required for packaging and handling processes. Semibulk containers are either returnable or nonreturnable. Returnable models generally are made of metal and may be collapsible to minimize the cost of return shipment. Nonreturnable semibulk containers tend to be constructed of multiwall corrugated paperboard, typically with heavy gauge plastic liners, with pallets to facilitate moving and handling. The choice of either type of container is based on economics and industry practice, since functionally satisfactory versions can be produced for either type.

Note that Systéme Internationale (SI) units are employed in this article preceding English units. In U.S. trade practices and packaging specifications, however, English units are employed to the virtual exclusion of metric units.

Steel Drums and Pails

Drums are single-walled, cylindrical or bilged shipping containers with capacities of 49–416 L (12–110 gal). Although there are a small number of common-size steel drums, about 80% of all drums made have a capacity of 208 L (55 gal). Pails are containers having capacities of 45 L (12 gal) or less, usually characterized by cylindrical or truncated cone configurations, and often having a carrying handle (5).

When selecting a steel drum or pail of a given size, the most important factors to be considered are wall thickness and, if applicable, interior lining. Wall thicknesses, ie, general physical strength, usually are around 0.328–2.4 mm (28–12 U.S. standard gauge); linings are selected according to the product's chemical containment needs. Although drums can be reconditioned and reused under certain circumstances by specialized commercial operations, particular care must be taken in observing the associated regulatory and safety requirements. Steel drums are categorized as heavy, if the wall thicknesses are 1.09–2.40-mm thick (18–12 U.S. standard steel gauge), or light, if wall thicknesses are 0.328–0.960-mm (28–19 U.S. standard steel gauge). One common steel drum structure uses 20-gauge steel in the sides and 18-gauge steel for the ends (6). Although the use of steel drums in transportation of chemicals is in general wide-ranging and safe, their use for interstate transportation of hazardous materials is still subject to a set of minimum requirements.

Tight-head drums, ie, those having nonremovable ends, are usually equipped with two screw-type openings; typically the diameter of one is 19 mm (0.75 in.), and the other 51 mm (2 in.) (Fig. 2). The plug and gasket that comprise these closures may be metal or plastic. Other sizes of openings are also available; sometimes an opening can be located between the rolling hoops in the drum body.

The formulation and application of interior linings for steel drums have been the subject of intense development. Available in both clear and pigmented formulations, common varieties of linings are phenolic resin, polyethylene, or combinations of phenolic and epoxy resins. The lining formulation and thickness, which may be a function of its formulation, can be varied depending on the service needs (temperature range, handling abuse, corrosiveness, etc) of any particular lading. Most drum linings are applied by spray-painting.

Polyethylene-coated drums have been used for particularly aggressive products. Polyethylene (PE) coatings tend to be thicker than ordinary drum coatings;

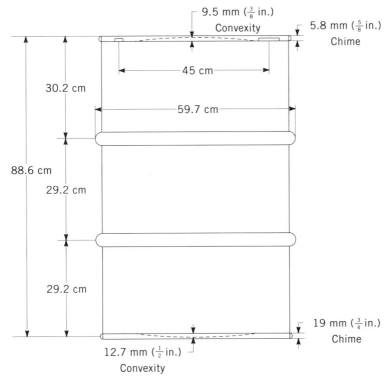

Fig. 2. Measurements of a tight-head drum; to convert cm to in., divide by 2.54.

PE-coated drums have been used for storage and transportation of organic and inorganic acids, food products, paints, janitorial supplies, surfactants (qv), and agricultural chemicals. In certain instances, a PE-lined steel drum can be more economical than both glass carboys and stainless steel or aluminum drums. However, the chemical resistance of a PE-lined drum may not be as high, and storage life not as long, as those of drums that might be considered more permanent containers. Drums characterized by a virtually permanent life, and which can be used in severe environments, can be justified economically; such drums are generally fabricated from aluminum, Monel, stainless steel, or nickel.

U.N. packaging codes for metal drums are given in Table 2; thickness requirements for reusable drums are shown in Table 3.

Table 2. U.N. Packaging Codes for Drums[a]

Metal drums		Plastic drums	
Code	Description	Code	Description
1A1	steel nonremovable head	1H1	nonremovable head
1A2	steel removable head	1H2	removable head
1B1	aluminum nonremovable head		
1B2	aluminum removable head		

[a]Maximum capacity 450 L (119 gal) and maximum net mass 400 kg (882 lb); Ref. 4.

Table 3. Thickness Requirements for Reusable Drums

Volume, L	Metal drum thickness, mm[a]	Plastic drum thickness, mm[a]
20	0.60	1.2
40	0.70	1.8
120	0.90	2.2
220	0.96	2.2
450	1.80	5.0

[a]For approximate conversion to in., multiply by 0.04.

Plastic Drums

Use of plastic drums is widespread in the chemical industry. These drums are used for packaging liquids, semisolids, powders, and granulated products. The shipper is responsible, though not solely, for verifying the suitability of containers, for checking DOT regulations and determining the suitability of plastic drums for a particular product, and for evaluating whether certain structural or material requirements are applicable. Two types of plastic drums are manufactured: free-standing plastic drums and plastic inserts for steel or fiber outer drums.

The principal advantage of plastic drums and liners is their resistance to corrosion. This aspect of their performance requires the lading to be investigated in terms of capacity for chemical attack on the drum. Stress-cracking tests should be performed in all instances where the compatibility of lading and drum material has not been established (6).

Plastic drums are made with either integral or removable heads, depending on the need of the customer. The chemical industry is more likely to use integral or tight-head drums, whereas the food industry is prone to use open-head drums to which separate covers can be fastened. The choice between these two styles often is determined by the requirement for cleaning the drum interior.

Capacities of plastic drums range from 9.5–208 L (2.5–55 gal) and the materials of construction generally are polyolefins. Small plastic containers are often placed and transported inside of corrugated boxes for convenience in handling. Large containers, especially the 208-L (55-gal) variety, are designed to be handled by a wide range of materials-handling equipments, including hand trucks, lift trucks, and cranes.

U.N. packaging codes for plastic drums are given in Table 2; thickness requirements for reusable plastic drums are shown in Table 3.

A composite drum is constructed of a plastic, semirigid inner liner (usually polyethylene) and either a steel or fiber outer container or overpack. The composite drum may be a useful compromise between a plastic drum and a conventional steel drum. It is strong enough to be handled like a steel drum, yet provides an enclosure for chemicals that cannot be shipped safely in steel or fiber drums. Capacities of composite drums are 19–208 L (5–55 gal). Such drums are used in chemical, pharmaceutical, and food industries.

Wooden Barrels

A barrel is a bilged, cylindrical container of greater length than width, having two flat ends (heads) of equal diameter. Because of their compound curves (the double-arch principle), wooden barrels can be extremely strong, capable of outperforming steel barrels in high stacking performance (6). A keg is a small barrel with a ≤ 38-L (≤ 10-gal) capacity. In general, capacities of barrels are not uniformly defined and can vary depending on the product to be carried. For example, the capacity of a U.S. standard cranberry barrel is 95 L (25 gal), a U.S. standard fruit or vegetable barrel is 178–189 L (47–50 gal), a beer barrel is 117 L (31 gal), and a barrel of oil is 159 L (42 gal). Between one and two million wooden barrels have been made in the United States in the mid-1990s; nearly all of these are 189-L (50-gal) charred-oak barrels used for long-term storage and flavor development of whiskey.

Wooden barrels are made of staves that are bound with hoops, which are usually metallic. A slack barrel is not watertight, hence it is only suitable for powders, semisolids, and solids. A tight barrel is watertight and suitable for storage and shipment of liquids, but may also be used for semisolids and heavy solids. Like steel drums, wooden barrels can be reconditioned and recoopered. Except in rare instances, wooden barrels are not the preferred form of packaging for the storage and shipment of chemicals. U.N. packaging codes for wooden barrels are as follows: 2C1 for bung-type and 2C2 for removable head having a maximum capacity of 250 L (66 gal) and a maximum net mass of 400 kg (882 lb) (4).

Fiber Drums

A fiber drum is a cylindrical shipping container made of convolutely wound layers of fiberboard adhered to each other to form a solid wall. Construction materials for the ends include metal, plastic, and fiberboard. Fiber drums are subject to DOT and, for international shipment, U.N. regulations when used for hazardous materials. Reference to these regulations is necessary before selecting a fiber drum.

Although fiber drums are mostly used for the shipment of dry granular and semisolid materials, they are also used in conjunction with liners and coatings for liquid products. Usual capacities are 4–380 L (1–100 gal), but larger-volume drums are also available; inside diameters range from 20–58 cm and heights from 7.6–107 cm. The net weight packing limit for fiber drums generally is 250 kg, although higher weights can also be authorized for certain types of commodities and shipments.

In addition, because of the extreme variation in the bulk density of products, use of the standard 19-L (5-gal), 38-L (10-gal), 61-L (16-gal), 114-L (30-gal), and 208-L (55-gal) metal drums often results in excessive outage. Fiber drums, in contrast, are available in a wide range of sizes, and can be sized to meet product volumes, thus allowing for little outage as well as saving storage and shipping space.

The most common style is a fiber shell with a metal chime at the top and bottom of the drum. The chime connects the bottom securely to the side wall, thereby preventing leakage, infestation, and sifting. The top of the container is grooved so that a cover and locking band can be applied. There are several types of closures, including friction or telescoping slip-on covers (which are secured by gummed or pressure-sensitive tape), lever-locking rings and other types of bands, expanding or crimped lids, and metal lugs or clips.

Fiber drums can be produced to meet a wide variety of requirements. They can be constructed with adhesives for water resistance, their interiors can be coated, their walls and ends can incorporate metal foil or asphalt-impregnated plies, and their exteriors can be decorated by painting, varnishing, and silk screening for both ornamental and functional purposes.

A basic drum is often supplemented by blow-molded plastic inserts that fit closely into the interior cavity. These inserts generally are provided with molded fittings for filling and sealing. The combination, which is referred to as a composite drum, enhances the fiber drum for additional applications, especially for liquid and chemically aggressive contents that cannot be contained safely in ordinary fiber drums. The U.N. packaging code for fiber drums is 1G for a fiber drum having a maximum capacity of 450 L (119 gal) and a maximum net mass of 400 kg (882 lb) (4).

Bags

Bags of various constructions are used in the storage and transportation of dry chemicals. The choice of which type of bag to use should be based on the needs of the product for adequate protection and the requirements of the distribution network. To a certain degree, bags can be custom-made for a particular product; indeed, almost any shipping requirement can be satisfied by one of many combinations of paper, plastic, and natural fibers incorporated in the design of bags.

Textile. Textile bags are made from natural fibers such as cotton and burlap (see FIBERS, VEGETABLE). Burlap or Hessian cloth is woven from jute fibers. Because the supply of jute and, consequently, its price have been uncertain for many years, textile bags gradually have been replaced by various combinations of textile components with plastic or paper, multiwall paper bags, or plastic bags (see TEXTILES).

Burlap bags, which can be made in various weights according to product requirements, are exceptionally strong and highly resistant to tear and snag. They are available with special finishes, and may be printed for brand identification; they are used largely for agricultural crops and animal feeds. Collection and renovation services buy used bags for resale.

Cotton fabrics are used for domestic and export shipments of flour, processed grain products, animal feeds, and seeds. Although cotton bags possess various advantages such as attractive appearance and good resale value, their high price has caused them to decline significantly as a packaging medium.

Bag sizes are specified by the width of material used for fabrication; for instance, "102-cm, 0.3-kg burlap, cut 107 cm" denotes that a single bag is made from a web of 0.3-kg burlap which is cut 102-cm wide (40 in.) and 107-cm long (42 in.).

Laminated Textile. Laminated-textile bags combine strength and flexibility. They usually are made from burlap or cotton fabric bonded to one or more plies of various combinations of creped kraft paper, plastic film, and aluminum foil. Barrier material is selected to provide the appropriate degree and type of protection. Bags can be constructed with burlap or other fabric strength component either on the outside or buried as an intermediate ply in the wall of the bag. Combinations of materials result in laminated-textile shipping bags that are exceptionally strong and tear- and puncture-resistant. These bags can be stacked higher than conventional multiwall bags and be made to resist extreme weather conditions as well as moisture, odor, acid, and grease contamination; they can also package materials of high intrinsic value and replace rigid drums, thus saving storage and shipping space. Products commonly packaged in them include petrochemicals, drugs, insecticides, fertilizers, seeds, etc.

Multiwall Paper. Multiwall paper bags represent by far the most common type of bag used to store and transport chemicals. The terms sack and bag are often used interchangeably, although sack usually refers to heavier-duty bags. Several types of paper shipping sacks are used commonly in industry, featuring such constructions as sewn or pasted ends, gussets or flat, and a variety of valves (see PAPER).

Multiwall paper bags can be comprised of 2–6 tubular plies that are nested within each other. Generally, the material used in individual plies is heavy-duty kraft paper having a basis weight of 18–32 kg (39.69–70.56 lb) per ply. Basis weight is the weight, in pounds, of 3000 sq ft (279 m^2) of paper, based on a standard ream of 500 sheets measuring 24×36 in. ($\sim 60 \times 90$ cm). Inner plies may be either plastic-coated or plastic films. Individual plies may also be treated to provide wet strength, acid or alkali resistance, moisture barriers, and nonskid surfaces (7).

The versatility of the modern paper shipping sack is enormous. Thousands of products are packaged in multiwall paper bags, particularly agricultural products and supplies, but also chemical, food, and rock products. A few attributes that can be provided for paper shipping sacks are the inhibition of moisture migration in both directions, resistance to insect and rodent infestation, water repellency, containment of aromatic ingredients, protection against abrasion and scuffing, nonskid properties on outer surfaces for more stable storage and handling, U.S. FDA approval for food packaging (qv), and a billboard appearance on the exterior for advertising purposes.

Protection capabilities of bags can be maximized by maintaining the moisture content of the paper at 6–8% by weight, relative to a shipping environment of 21°C at 60% in relative humidity. When the moisture content of shipping-sack paper drops below 6%, the sacks become excessively brittle; however, if short-term storage under such conditions is unavoidable, an attempt should be made to place them in an environment with adequate relative humidity for 24–48 hours prior to their use.

Multiwall shipping sacks are suitable for transport in all kinds of carriers from trucks to seafaring vessels. Particular care must be taken to prevent the storage of objects with sharp projections adjacent to the bags; in case of doubt, it is advisable to line the enclosure with kraft paper of a basis weight > 23 kg. Additional rules for the transportation of multiwall sacks are contained in applicable tariff specifications and should be consulted for rail, truck, and ocean-going shipments.

Plastic. A plastic bag usually consists of a single heavy wall of plastic film, woven sheets of plastic tape, or laminates. Principal materials of construction are polyethylene and polypropylene (see OLEFIN FIBERS). Both transparent and opaque sheeting are used, and printability usually is excellent. Plastic bags can be filled and closed with conventional equipment; heat-sealing is essential for open-mouthed bags to effect a moisture barrier.

Even though the use of woven plastic bags may be economically advantageous, their applications in the chemical industry are limited. For example, woven polyethylene bags made from film that has been slit into tapes is used for a number of agricultural and industrial applications. Such bags are more likely to compete with burlap bags than with the multiwall paper bag; in comparison with burlap, woven polyethylene bags are usually lighter in weight but equal in strength and superior in moisture and chemical resistance.

Filling. Almost all modern filling equipment is designed to insert a predetermined weight or volume of the product being shipped. Modern sack-filling equipment either delivers a predetermined weight of the product to the sack, or controls the weight that has been delivered by automatic shutoff after a predetermined level has been reached in the bag; weight accuracy is ca 0.25–1%. Most equipment used to fill open-mouthed bags operates on one of three principles: gravity feeding, auger feeding, or belt-conveyor feeding. Filling of valve-type sacks generally utilizes one of the following principles: centrifugal belt feeding, impeller feeding, auger feeding, or fluidized-bed feeding.

Closing. The closing of open-mouthed sacks can be accomplished by sewing, taping, wire tying, heat-sealing, or ultrasonic sealing. Where absolute tightness is required, thermal heat-sealing and, less frequently, ultrasonic sealing are preferred. Heat-sealing requires a heat-sealable coating on the inner ply. The closing of valve-type sacks is essentially automatic because of the construction of the valve.

U.N. packaging codes for bags are given in Table 4.

Table 4. U.N. Packaging Codes for Bags and Composite Packagings[a]

Bags[b]		Composite packagings[c]	
Code	Description	Code	Description
5H1	unlined or noncoated woven plastic bag	6PA1	within a steel drum
		6PA2	within a steel crate or box
5H2	sift-proof woven plastic bag		
		6PB1	within an aluminum drum
5H3	water-resistant woven plastic bag	6PB2	within an aluminum crate or box
5H4	plastic bag	6PC	within a wooden box
5L1	unlined or noncoated textile bag	6PD1	within a plywood drum
5L2	sift-proof textile bag	6PD2	within a wickerwork hamper
5L3	water-resistant textile bag	6PG1	within a fiber drum
5M1	multiwall bag of at least three plies, sift-proof	6PG2	within a fiberboard box
5M2	multiwall bag of at least three plies, having at least one waterproof ply, sift-proof	6PH1	within an expanded plastic package
		6PH2	within a solid plastic package

[a] Ref. 4.
[b] Maximum net mass 50 kg (110 lb).
[c] Composite packagings with inner glass, porcelain, or stoneware receptacles; maximum capacity 60 L (16 gal) and maximum net mass 75 kg (165 lb).

Carboys and Bottles

Carboys are straight-sided, cylindrical bottles with characteristic capacities of 10–50 L (ca 3–13 gal) (5). They generally are made of glass, but they can also be made with earthenware, porcelain, or plastic. Carboys typically are enclosed in a protective outer container or frame to preserve glass carboys from damage and sometimes also to provide stacking and handling features for shipment. Carboys are identified as composite packaging in the U.N. recommendations; packaging codes for a variety of composite packagings are given in Table 4.

The traditional carboy is used primarily for shipping liquids that can be safely held only in glass containers, or products of extremely high purity. Such contents must be protected from all possible sources of contamination, including extraction of contaminants from the walls of the container. Glass carboys are heavy for the volume of product carried, and can weigh as much as 0.6 kg/L (5 lb/gal). Available in one-way and returnable varieties, carboys are in widespread use for the local shipment of spring water, distilled water, or drinking water; the typical outer protective frame can be of simple construction, from which the carboy is easily removed for product dispensation (5).

Boxes

When used in an industrial environment, the term boxes usually refers to rectangular containers made of corrugated or occasionally solid fiberboard. The general category of boxes may also include crates, which are rigid shipping containers typically made of natural wood or plywood, although they are rarely used in the transportation of chemicals. The term boxes as used here does not include folding cartons, which are unlikely to be used as packaging for storage and transportation purposes.

The designs and varieties of boxes that can be constructed and used in chemical packaging are many. This flexibility in design is possible because boxes are manufactured with minimal tooling expense under specifications that are unique to each order, but which are subject to various material specifications for meeting freight and shipping requirements.

The walls of corrugated boxes are comprised of at least two flat outer facings (liners) and a fluted (corrugated) interior component. Solid fiber containers are made from layers of paperboard, usually three or four plies in thickness and 1.5–3.6 mm in caliper. Dimensions and style of the box are controlled by the product's needs as well as the shipper's and receiver's materials-handling requirements. The basic style of shipping box is the regular slotted container (RSC); other styles include folders and telescoping constructions as well as constructions having interior flaps meeting at the centerline, overlapping, or with a gap (8). The U.N. packaging code for fiberboard boxes is 4G for fiberboard boxes made with a water-resistant outer liner and having a maximum net mass of 400 kg (882 lb) (4).

BIBLIOGRAPHY

"Packaging and Packages" in *ECT* 1st ed., Vol. 9, pp. 754–762, by R. D. Minteer, Monsanto Chemical Co.; in *ECT* 2nd ed., Vol. 14, pp. 432–443, by G. T. Stewart, The Dow Chemical Co.; "Packaging Materials, Industrial" in *ECT* 3rd ed., Vol. 16, pp. 714–724, by S. J. Fraenkel, Container Corp. of America.

1. *Uniform Freight Classification 6000*, Uniform Classification Committee, Chicago, Ill., revised annually.
2. *National Motor Freight Classification 100-S*, National Classification Board, Washington, D.C., 1992.
3. DOT DOCKET HM-181, Performance Packaging, in *Fed. Reg.* **55**(246) (Dec. 21, 1990).
4. *U.N. Recommendations on the Transport of Dangerous Goods*, United Nations, New York, 6th ed., 1990.
5. *Glossary of Packaging Terms*, Packaging Institute International, Stamford, Conn., 1988.
6. M. Bakker, ed., *The Wiley Encyclopedia of Packaging Technology*, John Wiley & Sons, Inc., New York, 1986.
7. J. F. Hanlon, *Handbook of Package Engineering*, Technomic Publishing Co., Inc., Lancaster, Pa., 1992.
8. *Fibre Box Handbook*, Fiber Box Association, Rolling Meadows, Ill., 1992.

General References

Refs. 3–8 are also useful general references.
W. F. Friedman and J. J. Kipnees, *Distribution Packaging*, Krieger Publishing Co., Malabar, Fla., 1977.
R. C. Griffin, Jr., S. Sacharow, and A. L. Brody, *Principles of Package Development*, Van Nostrand Reinhold Co., Inc., New York, 1985.
G. G. Maltenfort, *Corrugated Shipping Containers: An Engineering Approach*, Jelmar Publishing Co., Inc., Plainview, N.J., 1990.
Modern Plastics Encyclopedia, McGraw-Hill Book Co., Inc., New York; published annually in November.
C. A. Taff, *Management of Physical Distribution and Transportation*, 7th Ed., Richard D. Irwin, Inc., Homewood, Ill., 1984.

David L. Olsson
A. Ray Chapman
Rochester Institute of Technology

CONVERTING

Well over 90% of consumer goods sold in the United States is shipped, stored, and purchased in some form of packaging. The total expenditure for packaging products in 1990 was $73.0 billion, of which 25.2% was for paperboard corrugated and solid fiber boxes, 21.3% for paper bags, molded pulp, and other paper-based packages, 20.5% for metal packages, 13.7% for plastic packages, 9.0% for paperboard folding cartons, 6.5% for glass packages, and 3.8% for wood and textile packages (1).

The selection of a specific package type by a product manufacturer is based on a number of factors, including the characteristics of the product being packaged; the nature of the shipping and storage environment to which it will be exposed; the physical strengths and properties of the package; regulatory requirements for packaging, eg, by the United Nations, U.S. Food and Drug Administration (FDA), U.S. Department of Agriculture, U.S. Military Regulations, U.S. Department of Transportation, U.S. Uniform Classification Committee, U.S. National Motor Freight Classification Committee, and ISO; consumer preference; and unit packaging cost. In many instances, different packaging types compete for the same packaging application. For example, glass jars, plastic jars, and metal cans may compete with each other, or corrugated boxes, solid fiber boxes, wooden crates, and paper or plastic bags may compete with each other. Assuming there are no regulatory restrictions or consumer preference differences, the choice of a packaging type is dependent on the relative direct costs of the competing packaging products, and the relative indirect costs of the package types throughout the entire filling, shipping, storage, and usage cycle, for example, package-filling line jams, product damage, warehouse stacking life, retail shelf life, moisture resistance, and pilferage.

In order to understand the various packaging options available for a given packaging application, there must be an understanding of the converting process and materials used to produce the various packages and the range of characteristics that can be achieved with each package type. This discussion deals with processes and chemical components involved with the production of the various types of paper-based packaging. Printing ink materials are not included herein (see INKS).

Paper and Paperboard Materials

The paper and paperboard materials, hereafter referred to as paper (qv), used to manufacture corrugated boxes, solid fiber boxes, folding cartons, paper bags, fiber drums, and fiber cans, are generally made from natural cellulosic fibers, such as wood fiber, and are typically supplied to the package manufacturing plant in roll form. Although there are many specific grades of paper available commercially, they can be divided into specific categories: coated or noncoated, and regular or wet strength. Many different types of coating are used by the paper and paperboard packaging industry and are discussed later (2).

The strength properties of regular grades of paper are dependent on the strength of the individual fibers and the natural bonding that occurs between the cellulosic fibers (see CELLULOSE). Paper fibers readily absorb moisture when exposed to humid or moist conditions. This absorbed moisture causes the fibers to swell, reducing the individual fiber strength and disrupting physical bonding between fibrils of adjacent fibers. Absorbed water molecules also disrupt hydrogen bonding between the cellulose molecules within the fibers and between the fibers at the bond sites. This bond disruption results in a significant loss of strength properties by the paper. The wet-strength grades of paper are made by adding resin to the paper pulp or the finished paper. The resins act by protecting and strengthening the interfiber bonding sites from the effect of the absorbed moisture. Typically, regular grades of paper lose 94% of their tensile strength when wet, however wet-strength grades of paper, made with 5 wt % resin, lose only 61% of their tensile strength when wet (2,3).

Resin additives that have been used to develop wet-strength characteristics in paper include urea–formaldehyde, melamine–formaldehyde, polyamide–polyamine–epichlorohydrin, polyethyleneimine, dialdehyde starches, insolubilized protein, vegetable gums and extracts, and silicates and silicones (see SILICON COMPOUNDS). The degree of wet-strength retention achieved in the paper is dependent on the type of additive, amount of additive application, amount of resin retained in the paper, method of additive application (pulp slurry or finished dry paper), and degree of resin cure achieved. Because of environmental concerns, several aspects must be considered: (1) the effect of any resin additive on the ability of paper mills to recycle fibers from used wet-strength packages (3); (2) the potential of having unacceptable odors transferred from the resin additive to the packaged product, particularly in the case of food or drug products (methods for measuring odor transfer potential, using gas chromatography–infrared–mass spectroscopy, have been developed (4)); (3) possible chemical adulteration of food and drug products by the resin additive, within the meaning of the U.S. Food, Drug and Cosmetic Act; and (4) possible release of chemical substances into the atmosphere during packaging manufacturing in terms of any OSHA or U.S. EPA restrictions.

Corrugated Paperboard Boxes

Corrugated paperboard is a sandwich structure formed by gluing a fluted corrugating medium ply to two linerboard facings. The first corrugated fiberboard shipping containers were used commercially in 1903 (5). Corrugated paperboard sheets are produced in a single, continuous manufacturing process consisting of five operations (Fig. 1). The paperboard from a roll of corrugating medium is shaped into the flute form by passing the paperboard between the nip formed by two steel gear-shaped rolls. An adhesive is applied to one side of the flute tips and a linerboard facing is brought into contact with the adhesive and bonded to the medium to produce a single-faced web. The single-faced web proceeds to the double-facer operation where adhesive is applied to the opposite side of the flute tips and a second linerboard facing is glued to the medium (6). This simplest form of the sandwich structure is called single-wall corrugated fiberboard. Double-wall corrugated fiberboard has two fluted mediums, produced using two single-facers, and three linerboard plies. Triple-wall corrugated fiberboard has three fluted mediums and four linerboard plies (1). Finally, the corrugated board web is scored and slit in a triplex operation to achieve the required box height and to produce the areas which will become the box top and bottom closure flaps. The scored and slit corrugated web is cut in the cut-off knife operation to the required sheet length to produce the specified box perimeter size. The finished box is then formed from the sheets in a discrete process operation where the corrugated sheets are printed, slotted to form the top and bottom flap closures, scored to provide the required box length and width dimension, and folded and fastened to form a continuous box perimeter and the finished box. This perimeter closure is known as the manufacturer's joint and is usually formed by gluing. A small percentage of boxes have stapled or taped manufacturer's joints.

The chemical materials required for this process include the corrugating adhesive, 1–1.2 MPa (150–175 psi) process steam to provide heat to set the

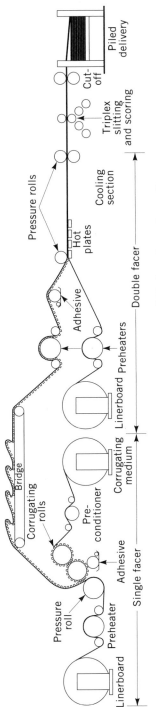

Fig. 1. Schematic of a corrugator (simplified and condensed).

1011

corrugator adhesive, possible coatings applied to the surface of the linerboard facings, possible chemicals impregnated into the linerboard and/or medium, the manufacturer's (glue) joint adhesive, and the printing ink (see INKS).

The formation of a uniformly strong and tough (not brittle) corrugator bond between the fluted medium and the linerboard facings is essential to corrugated paperboard package strength and performance. Without this bond to maintain the sandwich structure, the package would be a bag rather than a box. A silicate adhesive was used to form these bonds until the mid-1940s when a starch-base adhesive was developed for this application (see ADHESIVES).

On average, the corrugator starch adhesive in use in the 1990s contains approximately 78% water, 18% uncooked raw pearl starch, 3% cooked and gelled pearl starch, 0.5% borax, and 0.5% caustic. The raw pearl starch is the process adhesive, the cooked pearl starch holds the raw starch granules in suspension in the adhesive slurry, the borax serves as a buffering agent and tackifier by controlling the swelling of the raw starch granules at the bond sites when the adhesive is heated, and the caustic controls the gel point of the adhesive (7–11). The starch adhesive typically is characterized by its solids content, low shear viscosity, and gel point (11–14). A water-resistant corrugator adhesive is often used to make corrugated packages that may be exposed to moisture or high humidity conditions during their use cycle. The water-resistant properties are developed by adding a cross-linking agent to the starch adhesive slurry. The water-resistance agents typically act by cross-linking with the starch adhesive and paperboard cellulose molecules. The corrugating process heat serves to activate the resin. Typical slurry additives used to provide the starch and starch/paper cross-linking include urea–formaldehyde, phenol–formaldehyde, cyanamide–formaldehyde, resorcinol, resorcinol resins, ketone–aldehyde resins, high amylose adhesives, and ketone–acrylamide–formaldehyde copolymer resins (9,11). The selection of an additive to produce a water-resistant corrugator adhesive should include consideration of any U.S. EPA, OSHA, or FDA regulatory concerns, for example, the release of free formaldehyde.

A typical single-wall corrugated paperboard product has approximately 4.5 kg/m^2 (1.0 lb/msf) of regular starch adhesive solids applied in the single-facer operation and approximately 5.4 kg/m^2 (1.2 lb/msf) starch solids applied in the double-backer operation. Water-resistant adhesive application rates are approximately twice those of regular adhesives. This water-base corrugating adhesive adds considerable moisture to the linerboard facings and medium. Paperboard exhibits hygroexpansivity characteristics with changing moisture content. This dimensional instability can result in a warped (bowed) corrugated sheet if the two outer linerboard facings are at different moisture content levels at the time that corrugated bonds set. Warp is a serious quality defect (15).

The double-back bond is formed as the corrugated board web travels sandwiched between a weighted fabric belt and a steam heated cast-iron plate. Typical double-backers are 24.4 m in length and double-face bond sites are under the influence of both heat and pressure as the bond sets for ~4.8 s minimum (16,17). Typically, the single-facer bond is under heat and pressure only as it passes through a pressure nip formed by the lower corrugating roll steel flute tips and the steel pressure roll. The single-face bond must develop sufficient adhesive tack in this nip, referred to as a green-bond, to hold the medium and linerboard facing together until the final cured bond strength is developed (6,18,19).

Attempts have been made to use cold-set adhesives in the corrugating operation, such as poly(vinyl acetate) and modified, precooked starch formulations, but these have not achieved any appreciable degree of commercial acceptance (20). The use of a polyethylene film applied to the inside surface of the linerboard facing, which serves as a hot-melt corrugator adhesive, has achieved some commercial usage. However, its use is limited to the small, specialty product niche of fast-food hamburger cartons (see OLEFIN POLYMERS, POLYETHYLENE).

Chemical treatments commonly applied to corrugated paperboard packaging materials include additives that impart various degrees of water resistance, humidity resistance, oil and grease resistance, product abrasion resistance, product corrosion resistance, adhesion release properties, flame-retardant properties, nonskid properties, and static electricity control properties to the finished package (1,2).

Surface coatings and pulp slurry additives to impart linerboard and/or medium product abrasion resistance, adhesion release properties, corrosion resistance, flame-retardant properties, oil and grease resistance, and static electric control are typically applied by the paper mill during production of paperboard (2). Abrasion resistance and adhesion release properties are typically achieved by the use of silicone surface applications. The silicone may be applied directly to the paperboard surface or the paperboard surface may first be clay coated or polyethylene (PE) film laminated to achieve better silicone holdout properties on the paperboard surface. Maximum release levels of 20 g/cm (TAPPI UM-502 test method) have been reported for paperboard products consisting of silicone applied to a PE film that is laminated to the paperboard surface (21). Corrosion resistance may be achieved by controlling the finished paperboard extraction pH through the addition of buffering chemicals, and by minimizing the residual reducible sulfur content of the paperboard. Flame-retardant properties are typically achieved by the pulp slurry addition of chemicals such as antimony pentoxide (see FLAME RETARDANTS). Oil and grease resistance properties are typically achieved by a surface application of fluorocarbon or stearato chromic chloride materials which act by reducing the ability of the oil and grease to wet the treated paperboard surface. The control of static electricity is typically achieved by the pulp slurry or paperboard surface application of a grounding material, such as graphite.

Although the nonskid and moisture/humidity resistance properties of a corrugated fiberboard package can be achieved by the use of pretreated linerboard and medium materials, they are generally achieved by the application of treatment chemicals during the package-making process. Increased nonskid properties are achieved by a surface application to the double-face linerboard of a material which increases the coefficient of friction. The nonskid agent is applied in the double-face section of the corrugator by the use of an applicator roll. It may be applied to the total surface of the corrugated board or only to the surface areas that become the box closure flaps. The coefficient of friction of untreated linerboard is 0.2 to 0.5. The nonskid application increases the coefficient of friction to the range of 0.5 to 0.8. The nonskid materials applied to the linerboard consist of aqueous colloids of silica or alumina (22).

Moisture/humidity resistance is achieved in the corrugated board by the addition of a sizing agent to the linerboard and/or medium, or by the application of a surface coating to the linerboard which acts as a moisture barrier. The sizing material can be added either on the corrugator or in an off-line dipping and cascading process. The surface coating approach can be done on the corrugator by the use of a roll or a flooded nip applicator, or in an off-line curtain coating operation.

The most commonly used corrugator-applied sizing material is a lower melting point paraffin wax; the most commonly used off-line sizing material is a higher melting point paraffin wax. The most common corrugator-applied coating is an aqueous emulsion of a wax or acrylic polymer, and the most common off-line coating is a modified higher melting point/microcrystalline wax blend (1,10,23–25). Some of the properties of the various wax applications materials are shown in Table 1 (10,26) (see WAXES).

Poly(vinyl acetate) emulsions or hot-melt adhesives are typically used to form the manufacturer's or glue lap joint of the box. The main criteria for the adhesive is that it provide a strong and tough final bond and that it set up quickly enough to allow fast box production speeds. Production rates in excess of 240 boxes per minute are not uncommon in the industry.

Table 1. Wax Sizing Materials

Property	Corrugator wax	Dip/cascade wax	Off-line coating wax
congealing point, °C	50–57	63–68	66–71
application temp, °C	78–84	90–93	129–163
elongation, % at 23°C	1–3	1–3	50–85
viscosity, mPa·s(=cP) at 177°C	150–180	150–180	50
application rate, wt %	3–10	45–55	7–9

Solid Fiber Paperboard Boxes

A solid fiber paperboard package is a sandwich structure composed of inner plies called filler stock and outer plies called facings. Solid fiber differs from corrugated in that the inner plies are not fluted. Solid fiber packaging is typically used where toughness is important. Many solid fiber packages are reusable.

The solid fiber sheets are produced in a continuous process using a machine called a paster. Between two and six rolls of filler stock and facings are unwound onto the paster, coated on one side with an adhesive, and brought together in an unheated pressure roll nip to bond the plies together. The solid fiber web then passes through slitting and cut-off devices, similar to those described for the corrugated paperboard boxes, to form the proper sheet size needed to make the package. Die cutters are typically used to cut the specific package shape from the sheets because of the toughness of the solid paperboard material.

The paster is a nonheated operation. The most common paster adhesive formulation consists of poly(vinyl alcohol)–clay–starch blends (10). A 100% area adhesive coverage is used. The rate of bond strength development of the adhesive is an important commercial concern and rapid bond formation rates are desirable.

Various types of surface coating materials can be used with solid fiber paperboard to achieve desired package properties. The coatings and treatments described for corrugated paperboard apply to solid fiber paperboard as well. Various solvent and aqueous-based polymeric emulsion coatings are also commonly used (see EMULSIONS).

When selecting a particular paster adhesive, coating material, and coating process for solid fiber packaging, the fact that solid fiberboard lacks the open-flute structure (to facilitate exit of moisture from internal plies of paperboard) should be considered. It is generally desirable to minimize the amount of process water added with the adhesive or coating with solid fiber products. Moisture trapped in the filler plies can result in pin holes or blisters in the coating and/or warp of the sheet (15).

Paper Bags

Paper bag packaging includes multiwall bags, for consumer and commercial packaging applications, as well as grocery sacks and retail bags. Bags and sacks are manufactured in a continuous operation from rolls of paper. The paper web is folded and glued at a seam to form a continuous tube, which is cut to the required length for the specific bag size. The bag bottom is then folded and glued closed (10).

Grocery sacks typically are made from regular or wet-strength grades of paper material. Multiwall sacks are typically made from combinations of paper, plastic film, and composite material plies. The various plies used in multiwall bags can have special coatings or additives to achieve specialized and unique properties for any given packaging application. The types of properties sought include water and moisture resistance, oil and grease resistance, adhesion release properties, nonskid properties, air permeability, and gloss and color (1). Chemicals used to achieve these characteristics are the same as those used for corrugated paperboard boxes.

The adhesives (qv) used to form tube seams and bag bottoms include unborated dextrin, borated dextrin, casein, latex–casein, latex, poly(vinyl acetate), vinyl acetate copolymers, and hot-melt materials (10,27). Dextrin and casein adhesives are more commonly used in the production of grocery sacks; vinyl acetate-type adhesives are commonly used in all paper multiwall bags. The hot-melt adhesives are typically used to tack the plies of the multiwall bag together and to form the seam and bottom joints when polymer film plies or coated paper plies are used in bag construction.

Folding Cartons

Folding cartons are used extensively for point-of-purchase packaging and generally include high quality print graphics. The paperboard material typically used has smooth surface properties and often is clay coated. The paperboard roll stock is printed and/or coated, cut into sheets of an appropriate size, and die cut into individual package blanks, which are then folded and glued to form the finished package (10).

The over-coatings that are applied are used to achieve barrier and gloss properties. The exact chemical composition of the coatings varies greatly depending on specific packaging needs. Common coating types include paraffin wax, polyethylene and vinyl acetate copolymers, and organic solvent-based coatings and lacquers (10).

The seam closure on a folding carton is typically made using a latex, poly(vinyl acetate), vinyl acetate copolymer, or hot-melt adhesive (27). The choice of adhesive depends on a number of factors, including the nature of any coating used on the package and the production speeds required.

Fiber Drums and Cans

Fiber drum packaging and fiber can packaging are included together because of the similarity of their production processes. Both types of packaging are produced by wrapping plies of paperboard into a tube form. Two basic winding processes, convolute and spiral, are in use. Convolute winding is a discrete process where only one roll of paperboard is used at a time. The paper ply is wound so that it is layered upon itself and only one side seam, where the starting edge and ending edge of the paperboard meet, is produced. A 100% area adhesive coverage is used. The winding is done on a mandrel having the same diameter as that of the desired package. The finished tube must be removed from the mandrel before the next tube can be started. Spiral winding is a continuous process where multiple rolls of paperboard, plastic film, foil, and/or composite materials can be wound at the same time. Each ply is wound at an angle so that the side edges abut. Adjacent plies are slightly offset so that the seams for each of the plies is not at exactly the same position. Spiral winding therefore produces packages with multiple side seams which spiral around the tube perimeter. A 100% area adhesive coverage is used. Spiral winding is done on a mandrel having the same diameter as the desired package. The mandrel is lubricated so that the finished tube can move continuously down the mandrel to a rotary knife that cuts the tube to the proper length for the required package height. In general, fiber drums are produced using the convolute winding method, and fiber cans are produced using the spiral winding method.

The web materials used to form drums and cans can be customized to each packaging application. The materials are selected based on the package properties of moisture and humidity resistance, nonstick resistance, or barrier properties required. The adhesives used to bond the plies together include silicates, poly(vinyl alcohol), and poly(vinyl acetate) (10). Silicate adhesives are most commonly used in the manufacture of drum packages.

BIBLIOGRAPHY

1. *Fiber Box Handbook*, Fiber Box Association, Rolling Meadows, Ill., 1992.
2. C. E. Libby, ed., *Pulp and Paper Science and Technology*, Vol. II paper, McGraw-Hill Book Co., Inc., New York, 1962.

3. *Wet Strength in Paper and Paperboard*, TAPPI Monograph Series No. 29, Technical Association of the Pulp and Paper Industry, Atlanta, Ga., 1965.

4. P. A. Tice and C. P. Offen, *Tappi J.* **77**(12), 149–154 (Dec. 1994).

5. G. J. Daly, *The Corrugated Container Industry*, Harry J. Bettendorf, Oak Park, Ill., 1971.

6. J. J. Batelka, *Corrugating Medium—Its Influence on Box Plant Operations and Combined Board Properties and Package Performance*, Institute of Paper Science and Technology, Atlanta, Ga., 1993.

7. T. Pickens, *Southern Pulp Paper* **46**(3), 12–14 (Mar. 1983).

8. R. H. Williams, C. H. Leake, and M. A. Silano, *Tappi J.* **60**(4), 86–89 (Apr. 1977).

9. *Starch Based Corrugating Adhesives and their Application*, Harper/Love Adhesives Corp., Charlotte, N.C., 1980.

10. J. C. W. Evans, ed., *Trends in Paper and Paperboard Converting*, Lockwood Trade Journal Co., New York, 1965.

11. W. O. Kroeschell, ed., *Preparation of Corrugating Adhesives*, TAPPI Press, Atlanta, Ga., 1977.

12. Technical information sheets, Vol. 1: Series 0100-0300, *Coating and Graphic Arts Division*, TSI-0304-05, Technical Association of the Pulp and Paper Industry, Atlanta, Ga., 1989.

13. Ref. 12, TSI-0304-06.

14. Ref. 12, TSI-0304-08.

15. Ref. 12, TSI-0304-07.

16. *Compression and Converting Properties of Board*, Continuing Education Center, Institute of Paper Science and Technology, Atlanta, Ga., 1988.

17. Ref. 7, pp. 28–31.

18. J. J. Batelka, *Tappi J.* **75**(10), 94–101 (Oct. 1992).

19. J. J. Batelka, *Tappi J.* **77**(7), 71–73 (July 1994).

20. J. J. Becher, G. R. Hoffman, J. W. Swanson, and R. C. McKee, *Development of Cold-Set Adhesive*, Project 2696-11, Report One to the Fourdrinier Kraft Board Institute from the Institute of Paper Science and Technology, Atlanta, Ga., Nov. 1973.

21. *Release Paper and Unsupported Release Film*, H-P Smith, Chicago, Ill., 1981.

22. *Recommended Practices: General Notes on Anti-Skid Treatments for Fibreboard Shipping Containers*, Fiber Box Association, Rolling Meadows, Ill., 1992.

23. L. Roth and J. Weiner, *Waxes, Waxing and Wax Modifiers*, Bibliography Series No. 198, Institute of Paper Science and Technology, Atlanta, Ga., 1961.

24. Ref. 23, Suppl. I, 1967.

25. Ref. 23, Suppl. II, 1973.

26. Ref. 12, TSI-0206-01.

27. *Adhesive Selection Chart*, McGraw-Hill Book Co., Inc., New York, 1971.

JOSEPH J. BATELKA
Consultant

COSMETICS AND PHARMACEUTICALS

Cosmetics (qv) and pharmaceuticals (qv) each have their own special packaging requirements. The primary purpose of packaging for both broad classifications is to provide a means to store and distribute the product until the contents of the package are used. Each product must be analyzed for stability in the package being considered for use by the manufacturer; changes in container material, resin formulation, color, and closure system can all affect product stability. Although the distribution function of the packaging is always important, each product has other objectives that packaging components must achieve.

Cosmetic packaging, in addition to the above functions, is used to enhance the image of the product. This can be accomplished by frosting the container, graphics, proprietary design of the package, or use of metallized closures. The display package, or other secondary packaging, is also used to promote the image of the product. Principal products may have proprietary designs; smaller cosmetic manufacturers are able to distinguish their products through creative combinations of stock designs and graphics.

Although pharmaceutical packaging has the same basic objectives as cosmetic packaging, different parameters dictate product stability and safe packaging requirements. Both classes of products and their packaging are regulated by the U.S. Food and Drug Administration (FDA), but requirements for pharmaceutical packaging are more stringent because of product tampering prevention and child safety requirements of the FDA and the Consumer Product Safety Commission, respectively.

Child-Resistant Packaging

Under the Poison Prevention Packaging Act of 1970, any product that, if consumed by a child, could result in harm to the child must be packaged using components difficult for a child to open. This is referred to as child-resistant packaging.

The Consumer Product Safety Commission is responsible for administering the packaging rule under 16 CFR 1700, and the procedures for testing packages to assure compliance with the rule are included in the Code of Federal Regulations (CFR) (1). In 1995 the Commission concluded hearings on changing the protocol to require child-resistant packaging be user friendly, that is, easy to open by senior adults. The outcome of the change in this protocol is reflected in 16 CFR 1700 and the CFR should be reviewed for current testing requirements (1).

Product Tampering

In 1982, seven people died from consuming cyanide-laced Tylenol capsules. The incident resulted in a total product recall, massive negative publicity for the product, new requirements for safe packaging, and a federal statute making product tampering a crime (2). Since that time, the packaging industry has become visible to most consumers. This awareness has benefited the consumer by a reduction in loss of life due to consumption of adulterated products from tampering. Never before has an industry reacted so swiftly to resolve a problem.

There were incidents of product tampering prior to 1982, however, the exact number of incidents per year is unknown due to various methods of reporting. According to government figures, the problem peaked in the United States in 1986 when ~1800 claims of possible product tampering were reported. The number has decreased to around 500 per year. The decrease may be the result of better packaging or discouragement of potential violators by the penalties for violating product tampering laws. Most probably the decrease is caused by a change in the way claims are recorded.

Every developed nation has experienced product tampering incidents. The principal difference between domestic and foreign incidents is the motive of the tamperers. In the United States, typically random tampering without prior threat occurs; whereas outside the United States, extortion prior to injury occurs, with money appearing to be the primary motive. Most developed nations are either implementing or modifying their rules on the use of tamper-evident packaging. Some features as they are used in the United States would have to be modified or the use of a secondary feature required to meet the standards of various other countries.

In the late 1970s representatives from the U.S. FBI, Commerce Department, Defense Department, State Department, and CIA met to address the problem of state-sponsored terrorism in detail. One of the chief concerns was the threat of retail product tampering by a state-sponsored organization, ie, any group of terrorists supported financially, logistically, or with intelligence by the government body of any country. In certain countries that sponsor terrorist groups, training in retail product tampering, and how such acts can be used to further the cause, is being conducted. An example of the potential for disaster that exists if a tamperer has the resources available to build a complete packaging line and can print duplicate labels, occurred in South America when a drug organization bought a beverage plant in order to smuggle cocaine into the United States. At least one bottle in a specially marked case contained the drug in a liquid form, and when the contents of the bottle were distilled in the United States it yielded a powder that could be cut in strength and distributed to dealers. Unfortunately, one bottle was overlooked and sold to a consumer who died from a massive cocaine overdose.

The FDA Rule

The FDA has passed a rule (21 CFR 211.132) (3) requiring the use of tamper-evident packaging on all over-the-counter (OTC) drugs and some cosmetics (qv), while ignoring other products they regulate (2). Table 1 offers examples of such packaging forms.

Product tampering may be a foreseeable possibility and manufacturers have a responsibility to protect consumers against such possible acts. If a product in an adulterated form could harm a consumer, manufacturers have the responsibility of protecting the product and consumer against such acts, meaning the use of tamper-evident packaging transcends FDA regulations.

Table 1. FDA Examples of Tamper-Resistant Package Forms[a]

Type	Description
film wrappers	transparent film[b] with distinctive design wrapped securely around a product or product container
blister or strip packs	dosage units individually sealed[c] in clear plastic or foil
bubble packs	product and container sealed in plastic[d] and mounted in or on display card
shrink seals and bands	bands or wrappers with distinctive design are shrunk by heat or drying to seal[b] union of cap and container
foil, paper, or plastic pouches	product enclosed in individual pouches[c]
bottle seals	paper or foil with distinctive design sealed[e] to mouth of container under cap
tape seals	paper or foil with distinctive design sealed[e] over all carton flaps or bottle cap
breakable caps	container sealed by plastic or metal cap[f] that either breaks away completely when removed from container or leaves part of cap attached to container
sealed tubes	mouth of tube is sealed and seal must be punctured to obtain product
sealed carton	all flaps of carton securely sealed and carton must be visibly damaged when opened to remove product
aerosol containers	inherently tamper resistant

[a]Refs. 3 and 4.
[b]Must be cut or torn to open container and remove product.
[c]Must be torn or broken to obtain product.
[d]Must be torn or broken to remove product.
[e]Must be torn or broken to open container and remove product.
[f]Must be broken to open container and remove product.

The FDA Rule specifies which packaging features are acceptable for use in providing resistance to tampering; the list was modified in 1988. There is a proposal to modify the rule again and make the acceptability of a feature based on performance. The FDA has a procedure by which methods of providing protection that are not on the approved list may obtain approval on a case-specific basis. To obtain approval, samples of the complete package must be submitted to the FDA along with a written request for a waiver.

The inclusion of a specific form of protection does not warrant that the feature will deter violation, nor does it prevent legal action in the event of a claim of injury related to product tampering. There are variances in designs and tooling of the same design that affects each feature. Any evaluation of a package relates to the exact components used in the test. Material from a different manufacturer usually results in a different level of effectiveness, much the same as using a different resin or closure liner affects a stability study.

Consumer Preferences.   Studies into consumer preferences for tamper-evident (TE) packaging have consistently revealed that consumers prefer products that are resistant to tampering and have shelf-visible features. The same studies have indicated that a consumer is willing to pay slightly more than a competing brand that is not TE, indicating consumer awareness of packaging.

Cost to the Industry.   When compared to the potential expense for defending a single claim of tampering, the cost of effective tamper-evident packaging becomes insignificant. Many firms simply cannot afford the cost of responding to product tampering claims, especially if the firm is a small one with a limited or totally related product line where the reputation of the entire product line can be affected by adverse publicity on one item. Liability insurance cannot return lost customer confidence.

Tamper-Evident Features

Selection of features to use should be done by objective testing during the package development stage. During the design stage, the package engineer should consider the function of the product and how the consumer intends to use it. Next, each tamper-evident feature that is usable on the package should be tested to determine which feature offers the greatest protection to the consumer. The test used should be objective, consistent, and replicable. Records of the test results should be retained indefinitely. If a feature selected for use achieves a lower value than others that were rejected, reasons for the selection should be recorded and retained with the test results. Cost should not be a factor in selecting which feature to use. It would be a false economy to accept less effectiveness to save a few cents when compared to the cost of potential injury to a consumer.

One form of testing the effectiveness of tamper-evident packaging is the Rosette protocol which measures the degree of difficulty in violating a specific package and restoring it to a near original condition. The Rosette protocol also measures increases in effectiveness through the use of multiple features. The value for a specific combination of features is not equal to the sum of each feature. Some factors cover the combination rather than each feature separately. For example, the knowledge factor is applied once. Regardless of the number of features in a combined package, only one knowledge level is required. Time is cumulative; if it takes 20 minutes to violate each feature, the time required is not the value for 20 minutes times the number of features used on the package. In this example, the time factor is the value for one hour. Only one category of equipment may be required if all tools or equipment required to violate the different features in the combination are in the same class. The feature visibility values for all used on multiple feature packages are multiplied; even the use of multiple features not shelf-visible increases the effectiveness of the package. The feature material is added for each feature replaced or reused to determine the feature material value. The value of the feature, used with the specific package components, on the specific product and form of product tested, is the sum of all the factors.

The FDA and the Non-Prescription Drug Manufacturers Association have expressed concern that the industry may gravitate to the feature achieving the highest score in any testing procedure. The value for a specific feature varies depending on the exact product and all other packaging components used in a specific package. The same feature from different manufacturers may achieve different values, because there are slight variations in design and manufacture, even though the features may appear to be identical. If a value were to be established at 20 on a scale of 0–50 for a single feature, the value of 20 could be attained through the use of multiple features if necessary. This would preclude any single feature from becoming the industry mandated method to provide protection. The value of 20 is for illustration purposes only; prior testing has shown the minimum design for FDA acceptance to be around 11, with some minimum values for approved features scoring much higher. A standard requiring 20 as the minimum value would require most packages to be improved before they met the standard. The use of multiple features can result in a value higher than 50. Although the Rosette protocol has been tested and in use for several years, there are others that may be as objective and in use by other companies.

No single TE feature is best for all products. There are variations in effectiveness of similar features from different manufacturers, as well as variations in effectiveness where the product contributes to the effectiveness. An example is a metal can that is much more effective for a carbonated product than a noncarbonated product. The product can direct which feature provides the most protection, eg, products that can be adulterated effectively by penetration require a more rigid outer container than one that degrades visibly upon violation by penetration. The best feature for a product is the one that provides the greatest resistance to violation for the product in its current form and size. All features can be violated in some manner, but effective TE features provide greater difficulty in violating the product. In a particular instance a package was opened, the original product was replaced with a toxic substance, and no attempt was made to restore the package to its original appearance. The package worked as intended, ie, it showed it had been opened, but because there was no indication of violation to the actual product, the consumer still experienced injury.

Productiveness of Tamper-Evidence

Increased consumer awareness of packaging has led to an increase in the number of complaints of possible product tampering, although most are later dismissed as unfounded. Tamper-evident packaging prevents in-store tasting and violation, and if the feature is intact assures the consumer that the product is safe. Effective tamper-evident packaging acts as a deterrent to most persons who would commit such acts of violation and makes it difficult for others to violate the package and restore it to its original appearance. Effective tamper-evident packaging works, provided the consumer is aware of the feature and pays attention to what is being used. Most experts agree that consumers should be more aware of what to look for in tamper-evident packaging. Educating the consumer may include pictures of the feature and product on the label and in media ads.

To make tamper-evident packaging work as it should requires the effort of all involved, ie, manufacturers of packaging components should only promote effective packaging for use in providing protection, product manufacturers should use the most effective feature for their product, retailers should be aware of the potential for violated products being in their display, and consumers should maintain an awareness of what to look for in a secure package and refuse to buy suspicious packages. Ideally, law enforcement agencies should conduct professional investigations and prosecute all violations, government agencies should provide for legal relief to companies meeting or exceeding government standards, and the media should only report those incidents that are verified as representing a threat to the health and welfare of the public.

BIBLIOGRAPHY

"Packaging and Packages" in *ECT* 1st ed., Vol. 9, pp. 754–762, by R. D. Minteer, Monsanto Chemical Co.; in *ECT* 2nd ed., Vol. 14, pp. 432–443, by G. T. Stewart, The Dow Chemical Co.

1. *Code of Federal Regulations*, Title 16, part 1700, Consumer Product Safety Commission, U.S. Government Printing Office, Washington, D.C.
2. Ref. 1, Title 18, USC 1365.
3. *Fed. Reg.* **47**, 50442–50456 (Nov. 5, 1982).
4. M. Bakker, ed., *The Wiley Encyclopedia of Packaging Technology*, John Wiley & Sons, Inc., New York, 1986.

General References

J. L. Rosette, "Development of an Index For Rating the Effectiveness of Tamper-Evident Packaging Features," Masters Thesis, California Coast University, Santa Ana, Calif., 1985.
J. L. Rosette, *Improving Tamper-Evident Packaging*, Technomic Publishing Co., Inc., Lancaster, Pa., 1992.
J. L. Rosette, *Product Tampering Detection*, Forensic Packaging Concepts, Inc., Atlanta, Ga., 1993.
Food & Drug Packaging, published monthly.
Packaging, published monthly.
Packaging Digest, published monthly.
Packaging Technology & Engineering, published eight times per year.
D. Lowe, "Crisis Management Marketing, Tampering Strategies for International and Domestic Marketplace Terrorism," *San Fran. St. Univ. J.* **1**, (1990).
J. Rosette, *Defending Against Terroristic Tampering*, 1991.
J. Rosette, *Product Tampering As It Affects Consumers and Law Enforcement*, FBI Law Enforcement Bulletin, Sept. 1992.
Product Tampering in the Marketplace, Foundation for American Communications, 1993.
Product Tampering Problems, Foundation for American Communications, 1986.
J. Sneden, H. Lockhart, and M. Richmond, *Tamper-Resistant Packaging*, Michigan State University, East Lansing, 1983.
Trends in Product Tampering, FDA & FBI data on product tampering, 1990.

JACK L. ROSETTE
Forensic Packaging Concepts, Inc.

ELECTRONIC MATERIALS

Materials play an important role in the electronics industry. The effectiveness of the electrical performance of the system, its reliability, and its cost all depend on the packaging materials used, which are chosen for their properties and applications. As a result, the practicing engineer must have ready access to current information on the materials that can be used in product development. This article gives an overview of the various material choices for the elements of an electronic product.

Electronic packaging refers to the placement and connection of many, sometimes thousands, electronic and electromechanical components in an enclosure that protects the system from the environment and provides easy access for routine maintenance (1). The packaging process starts with a chip, which is fabricated from a wafer. A chip is electrically connected to a chip carrier or substrate. This is the first level of interconnection, for which various interconnection methods are available, including wire-bonding, tape automated bonding (TAB), and flip-chip (2). The substrate-chip assembly is packaged in a case; the chip carriers are then mounted to a common base, normally a printed circuit board. Various chip carriers are electrically connected by metallized conductor paths, forming the second level of interconnection. Next, the various printed circuit boards are mounted on a backplane, forming the third level of interconnection. Several racks may be interconnected to form a cabinet. The fourth level of interconnection involves the interconnection between various cabinets. Figure 1 illustrates the various levels of packaging, Figure 2 shows a ceramic package, and Figure 3 shows a plastic package.

Semiconductor Materials

Semiconductors (qv) are materials with resistivities between those of conductors and those of insulators (between 10^6 and 10^{-3} $\Omega \cdot$cm). The electrical properties of a semiconductor determine the functional performance of the device. Important electrical properties of semiconductors are resistivity and dielectric constant. The resistivity of a semiconductor can be varied by introducing small amounts of material impurities or dopants. Through proper material doping, electron movement can be precisely controlled, producing functions such as rectification, switching, detection, and modulation.

Silicon [7440-21-3] is by far the most commonly used die material in electronic packaging, employed by over 90% of the electronic industry. Among the other semiconductors popularly used are gallium arsenide [1303-00-0], indium phosphide [22398-80-7], germanium [7440-56-4], and silicon carbide [409-21-2]. A list of the materials used as semiconductors, along with their various properties, is given in Table 1. Silicon is similar to diamond and germanium, and is characterized by a negative temperature coefficient of resistance. Silicon, which has a gray metallic luster, is normally considered a brittle material, with a hardness intermediate between those of germanium and quartz. Silicon is most widely used in electronic applications as a material for dies, and also in the production of integrated circuits, rectifiers, diodes, transistors, and triacs.

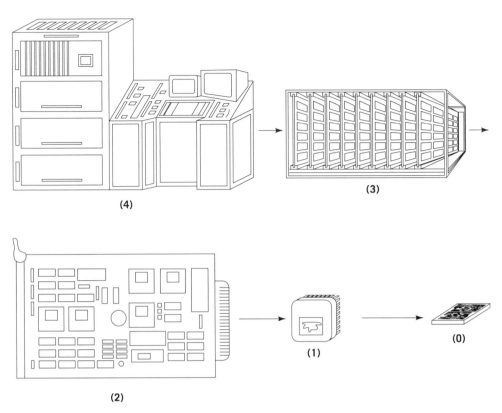

Fig. 1. The different levels of electronic packaging: (**0**) chip; (**1**) chip carrier; (**2**) printed circuit board (electronic card assembly); (**3**) rack; and (**4**) system (2).

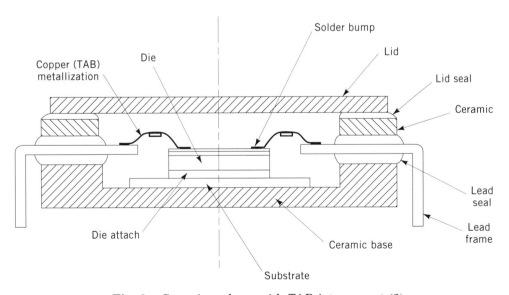

Fig. 2. Ceramic package with TAB interconnect (3).

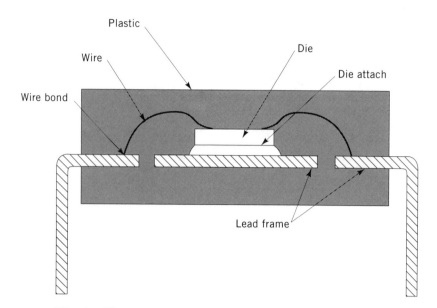

Fig. 3. Plastic package with wire bond interconnect (3).

Table 1. Properties[a] **of Semiconductor and Die Materials**

Material	Dielectric constant at high frequency	Density, kg/m³	Knoop hardness, kg/mm³	Thermal conductivity, W/(m·K)	Melting point, °C
Si	11.9	2330	850–1150	124–148	1685
Ge	16.0	5320–5360	750–780	2.0–3.4	958–1231
Te		6230–6240		990–2300	723
diamond	5.7	3500–3530	7000–8800	283	>3823
SiC	10.0	3210	2200–2950	67–80	3070
InP	12.4	4787	420–535	29	1344
InAs	14.6	5775	220–225	16	1216
InSb	17.7	3810	510	84	808
AlAs	8.5–10.9	2420–2850	430–560	92	>1873
AlP		4218	360–408	46–60	>1773
AlSb	11.6	5316	535–765	44–58	1323
GaAs	11.0–14.4	4130	950–964	75–79	1511
GaP	13.1	5619	450	27–33	1738
GaSb	11.1	4135–4820	55	40.1	985
CdS	15.7	5660–5810		31.6	1750
CdSe	5.4	6200	61	6	1512
CdTe	10.0	3980–4100	180	25–46	1365
ZnS	72–10.2	5420	138	14.0	2122
ZnSe	5.2–8.9	6340	92	10.8	1790
ZnTe	5.9–9.2	7500–7610		2.3	1511
PbS	9.0–10.4	8100–8150		1.7	1386
PbSe	17.0	8160–8164		2.3	1338
PbTe	24.0–161				1190

[a]Other properties, eg, strength, resistivity, heat capacity, and thermal expansion, are given in the reference (2,4) from which this table is compiled.

Gallium arsenide is a dark gray material that is becoming increasingly popular as a semiconductor material. It is a compound semiconductor; the crystalline structure consists of gallium atoms alternating with arsenic atoms. Gallium arsenide is used in microwave devices, varactor diodes, schotty barrier diodes, light-emitting diodes, injection lasers, as well as Gunn-mode oscillators.

Substrate Materials

A substrate is a robust element that provides mechanical support for the die. It can be mounted with more than one die; such packages are called multichip modules. Because parasitic capacitance effects are directly proportional to the dielectric constant, substrate material should have a low dielectric constant. To minimize electrical losses, especially at high frequencies, a low dissipation factor is required. High volume resistance provides good insulation to prevent electrical current leakage between the conductor tracks. Since the substrate provides mechanical support to the die, its material should have good mechanical strength; it should also be thermally conductive to dissipate heat produced by the active devices. The coefficient of thermal expansion (CTE) of the substrate should match closely to that of the die to avoid thermomechanical stresses due to a CTE mismatch. The CTE of silicon is 2.3–4.7 ppm/°C and for GaAs, 5.4–5.72 (2,4).

Materials used for substrates can be broadly classified into ceramics and metals. Commonly used ceramics, ie, alumina, aluminum nitride, and beryllia, can be easily incorporated into a hermetic package, ie, a package permanently sealed by fusion or soldering to prevent the transmission of moisture, air, and other gases.

Alumina, or aluminum oxide [1344-28-1], has a thermal conductivity 20 times higher than that of most oxides (5). The flexural strength of commercial high alumina ceramics is two to four times greater than those of most oxide ceramics. The drawbacks of alumina ceramics are their relatively high thermal expansion compared to the chip material (silicon) and their moderately high dielectric constant.

Although beryllium oxide [1304-56-9] is in many ways superior to most commonly used alumina-based ceramics, the principal drawback of beryllia-based ceramics is their toxicity; thus they should be handled with care. The thermal conductivity of beryllia is roughly about 10 times that of commonly used alumina-based materials (5). Beryllia [1304-56-9] has a lower dielectric constant, a lower coefficient of thermal expansion, and slightly less strength than alumina. Aluminum nitride materials have begun to appear as alternatives to beryllia. Aluminum nitride [24304-00-5] has a thermal conductivity comparable to that of beryllia, but deteriorates less with temperature; the thermal conductivity of aluminum nitride can, theoretically, be raised to over 300 W/(m·K) (6). The dielectric constant of aluminum nitride is comparable to that of alumina, but the coefficient of thermal expansion is lower.

Single-crystal silicon has also been employed as substrate material, particularly in multichip module (MCM)-Si applications. As a substrate, silicon offers good thermal conductivity and matches the CTE of the devices mounted on it; it does, however, have a relatively high dielectric constant and is very brittle.

In applications in which electrical conductivity is required, metals, copper, tungsten, molybdenum, and Kovar [*12606-16-5*] are the preferred chip-carrier materials. Metals have excellent thermal conductivities. Tables 2 and 3 list the various materials used for substrates, along with their mechanical, electrical, and thermal properties.

Attachment Materials

Die attach, the process by which the die is anchored to the substrate, can be accomplished by either (*1*) introducing an adhesive between the backside of the die and the substrate, or (*2*) using an electrical connection procedure, such as TAB or solder bumping. One of the most important properties of the attachment material is its bonding strength, which ensures that the die and the substrate stay in place when subjected to the stresses imposed during manufacture, storage, and operation. The other important properties for attachment materials are tensile strength, shear strength, and fatigue endurance. If the attachment material must transmit heat from the die to the substrate, its thermal conductivity is also a critical material property. The electrical properties of the attachment materials assume importance when the attachment material serves as an ohmic contact.

Adhesive Connection.   Adhesives can be classified as organic or inorganic. Inorganic materials, eg, soft solders, have the advantage of producing low contamination by-products, but require relatively high processing temperatures. Gold-based eutectics such as gold–tin (Au–Sn), gold–silicon (Au–Si), and gold–germanium (Au–Ge) have high strength and therefore offer excellent fatigue and creep resistance, but lack plastic flow which can lead to high stresses in the semiconductor due to the thermal expansion mismatch between the die and substrate (8). Their high cost also makes gold-based eutectics less popular. Organic compounds, on the other hand, have nearly the opposite characteristics. They can be effectively used on large dies since they can withstand large strains, but may be inappropriate for hermetic environments, due to out-gassing.

Epoxies are the most commonly used adhesives (qv). Silver and gold are sometimes added to an epoxy to improve its thermal conductivity. Polyimide, also used as an adhesive, has low shrinkage as well as low viscosity and can be cured at 180°C; its primary drawback is a tendency to absorb water, as much as 6% by weight.

Engineering rework is possible with eutectic and solder materials, but impossible with silver–glass. This constraint severely limits the usefulness of the material. Tables 4 and 5 give the electrical, mechanical, and thermal properties for various adhesives.

Wire Interconnect Materials.   Wire-bonding is accomplished by bringing the two conductors to be joined into such intimate contact that the atoms of the materials interdiffuse (2). Wire is a fundamental element of interconnection, providing electrical connection between first-level (ie, the chip or die) and second-level (ie, the chip carrier, or the leadframe in a single-chip carrier) packages.

Characteristics of the wire materials, which are crucial to the strength of the wire bond, include wire dimensions, tensile strength, elongation, and contamination. Wire dimensions are important because the quality of the bond

Table 2. Mechanical Properties of Substrates[a]

Material	Tensile strength, MPa[b]	Compressive strength, MPa[b]	Flexural strength, MPa[b]	Elastic modulus, GPa[c]	Density, kg/m³	Hardness[d]	Impact strength, J[e]
BeO	230		250–490	345	3000	100 K	
SiC	17.24	490	440–460	412	3160	2800 K	65
AlN		392–441	360–490	310–343	3260	1200 K	
Si			580	190	2330	850–950 K	
Si₃N₄	96.5–965		275–932	314	2400–3440		
SiO₂	96.5–386		30–100	69	2190–2320		
Al₂O₃ 85%	124.11	1620	290	221	3970	9 MH	8.5–8.8
Al₂O₃ 95%	127.4–193.0	2069–2413	310–338	296–317	3970	9 MH	8.8–10.3
Al₂O₃ 99.8%	206.9	2758	350–414	386	3970	93.5 RA	9.5
Al₂O₃ silicate	17.24	276	62	55.2	3970	6 MH	4.5
diamond					3500–3530	7000–8800 K	
quartz	48.3	1103	71.4	71.7	2200	5 MH	
steatite	55.2–69.0	448–896	110–165	90–103	2500–2700		0.4–0.5
forsterite	55.2–69.0	414–690	124–138	90–103	2700–2900		0.04–0.05
titanate	27.6–69.0	276–827	69–152	69–103	3500–5500		0.4–0.7
cordierite	55.2–69.0	138–310.3	10.3–48.0	13.8–34.0	1600–2100		0.3–0.34
mullite			125–275	175			
molybdenum	655			324	10,240		
10% Cu–90% W	489.5		1062	331	17,300	427 K	
tungsten	310–1517		28.3	345	19,300	485 K	
Kovar	522–552			138	8360	68 RB	

[a]Refs. 2, 4, and 7.
[b]To convert MPa to psi, multiply by 145.
[c]To convert GPa to psi, multiply by 145,000.
[d]K = Knoop; MH = Moh; RA = Rockwell A; B = Brinnell.
[e]To convert J to ft·lb, divide by 1.356.

Table 3. Thermal[a] and Electrical[b] Properties of Substrates[c]

Material	Thermal conductivity, W/(m·K)	CTE, ppm/°C	Heat capacity, J/(kg·K)	Dielectric constant at 1 MHz	Dielectric strength, kV/mm
BeO	150–300	6.3–7.5	1047–2093	20–42	0.04
SiC	120–270	3.5–4.6	675	8.5–10	0.55
AlN	82–320	4.3–4.7	745	11.9	
Si	125–148	2.33	712	6–10	196.8
Si_3N_4	25–35	2.8–3.2	691	3.5–4.0	196.8
SiO_2	1.5	0.6		4.5–10	0.33
Al_2O_3	15–33	4.3–7.4	765	5.7	
quartz	43	1.0–5.5	816–1193	4.6	
diamond	2000–2300	1.0–1.2	509	5.5–7.5	7.9–15.7
steatite	2.1–2.5	8.6–10.5		6.2	7.9–11.8
forsterite	2.1–4.2	11		15–12,000	2.0–11.8
titanate	3.3–4.2	7–10		4.5–5.5	1.6–9.8
cordierite	1.3–4.0	2.5–3.0	770	6.6–6.8	
mullite	5.0–6.7	4.0–4.2			
molybdenum	138	3.0–5.5	251		
10% Cu–90% W	209.3	6.0	209		
tungsten	174–177	4.5	132		
Kovar	15.5–17.0	5.87	439		

[a]Maximum use temperatures exceed 1000°C except for SiO_2 (>800°C); melting points range from 1450°C (Kovar) to 3660°C (tungsten).
[b]Resistivity for nonmetallic materials is 10^4–10^{14} Ω·m; for metals the order of magnitude is 10^{-8} Ω·m.
[c]Refs. 2, 4, and 7.

Table 4. Electrical and Thermal Properties of Attachment Materials[a]

Materials	Volume resistivity, Ω·m	Dielectric constant at 1 MHz	Dielectric strength, kV/mm	Dissipation factor at 1 MHz	Thermal conductivity, W/(m·K)	CTE, ppm/°C	Maximum use temperature, °C
silicone	10^{13}–10^{15}	2.9–4.0	15.8–27.6	0.001–0.002	6.4–7.5	262–300	260
polyurethane	0.3×10^9	5.9–8.5	12.9–27.6	0.05–0.06	1.9–4.6	90–450	65.6
acrylic	7×10^{11}						93.3
epoxy novolak	10^{13}–10^{16}	3.4–3.6		0.016			
epoxy phenolic	6.1×10^{14}	3.4	15.8	0.32	25.0–74.7	33	87.8
epoxy bisphenol A	10^{14}–10^{16}	3.2–3.8		0.013–0.024			
epoxy silicone					13–26	60–80	260
cyanoacrylate							82.2
polyimide					0.2	40–50	

[a]Refs. 2, 9, and 10.

Table 5. Mechanical Properties of Attachment Materials[a]

Material	Tensile strength, MPa[b]	Shear strength, MPa[b]	Elongation, %	Modulus of elasticity, GPa[c]	Specific gravity	Hardness[d]
silicone	10.3		100–800	0.06–2.21	1.02–1.2	20–90 A
urethane	5.5–55	15.5	250–800	0.172–34.5	1.1–1.6	10 A–80 D
acrylic	12.4–13.8		100–400	0.69–10.3	1.09	40–90 A
epoxy silicone		11.7				
epoxy novolak	55.0–82.7	26.2	2–5	2.76–3.45	1.2	
epoxy bisphenol A	43–85		4.40–11.0	2.7–3.3	1.15	106 RM
polyimide		16.5		3.0		
epoxy polyimide		41				
epoxy polyure-thane (50:50)	34		10			

[a]Ref. 2.
[b]To convert MPa to psi, multiply by 145.
[c]To convert GPa to psi, multiply by 145,000.
[d]A = Shore A; D = Shore D; RM = Rockwell M.

depends on the mass of the wire involved in making the bond. The tensile strength of the material is another significant factor for both ultrasonic and thermocompression bonds. Other important properties of wire and bond pad materials include shear and yield strength, elastic modulus, Poisson's ratio, hardness, CTE, elongation, fracture toughness, and fatigue endurance in tension and shear.

Gold and aluminum are the most commonly used wire materials, although copper and silver have also been used. Gold is mostly used in thermocompression bonding, aluminum in ultrasonic bonding. Both surface finish and cleanliness should be carefully controlled for gold-bonding wire to ensure a strong bond and prevent clogging of bonding capillaries. Pure gold can usually be drawn to an adequate breaking strength and proper elongation. Pure gold is very soft, small amounts of impurities can be added to make the gold wire workable. Beryllium-doped wire is stronger than copper-doped wire under most conditions by about 10 to 20%.

Because pure aluminum is typically too soft to be drawn into a fine wire, it is often alloyed with 1% silicon or 1% magnesium to provide a solid solution-strengthening mechanism. The resistance of Al-1% Mg wire to fatigue failure and to degradation of ultimate strength after exposure to elevated temperatures is superior to that of Al–1% Si wire.

Copper wires are used primarily because of their economy and resistance to sweep, ie, tendency of the wire to move in a plane perpendicular to its length, during plastic encapsulation (11–16). Because copper is harder than gold, more attention is needed during the bonding operation to prevent the chip from cratering. Table 6 lists the various properties for wire materials.

Tape-Automated Bonding (TAB). This is an approach to fine-pitch interconnections between the chip and the leadframe. The interconnections are patterned on a polymer tape; the chip to be bonded is positioned above the tape so that the metal tracks on the polymer tape correspond to the bonding sites on the

Table 6. Properties of Wire and Bond Pad Materials[a]

Property	Aluminum [7429-90-5]	Copper [7440-50-8]	Gold [7440–57-5]
specific heat, (W·s)/(g·°C)	0.9	0.385	0.129
thermal conductivity, W/(cm·°C)	2.37	4.03	3.19
specific gravity	2.7	9.0	19.3
melting point, °C	660	1083	1064
electrical resistivity, Ω·cm	2.65×10^{-6}	1.7×10^{-6}	2.2×10^{-6}
temperature coefficient of electrical resistivity, Ω·cm/°C	4.3×10^{-9}	6.8×10^{-9}	4×10^{-9}
elastic modulus, GPa[b]	34.5	1324	77.2
yield strength, MPa[c]	10.34	68.95	172.38
ultimate tensile strength, MPa[c]	44.82	220.64	206.85
coefficient of thermal expansion, ppm/°C	46.4	16.12	14.2
Poisson's ratio	0.346	0.339	0.291
hardness, Brinnell	17	37	18.5
elongation, %	50	51	4

[a]At 27°C (3).
[b]To convert GPa to psi, multiply by 145,000.
[c]To convert MPa to psi, multiply by 145.

chip. A thermocompression process is then used for bonding all joints in a single operation. Figure 4 shows an example of a TAB tape.

The basic elements of a tape-automated bond include metal plating, interface metallurgy, bumps, adhesive, and tape. Each of these elements requires materials containing unique properties. Metal plating should have excellent electrical conductivity for good performance and high thermal conductivity for heat dissipation. High toughness ensures that the material can stand more stress during thermal cycling in the bonding operation. In addition, metal plating should have good adherence to the plastic carrier and the capacity to be easily etched. Bump material should be soft and ductile.

Tape material, on the other hand, should have dimensional stability, good surface flatness, nonflammability, low shrinkage, low moisture absorption, a high tensile modulus, a low coefficient of thermal expansion, a high dielectric constant, and the capacities to withstand exposure to plating, etching, and soldering, as well as short-term elevated temperatures which are required during solder reflow and thermocompression bonding.

Copper is universally used as the metal plating for tape because it can be easily laminated with copper and the various plastic tapes. Copper is readily etched and has excellent electrical and thermal conductivity in both electrodeposited and rolled-annealed form. The tape metal plating is normally gold- or tin-plated to ensure good bondability during inner- and outer-lead bonding operations and to provide better shelf life and corrosion resistance.

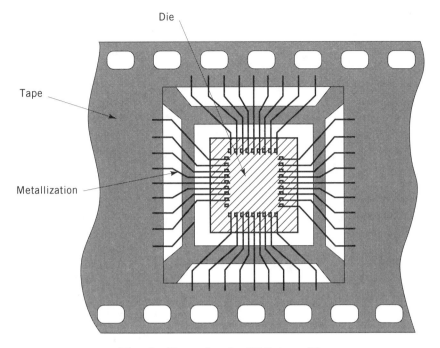

Fig. 4. Example of a TAB tape (3).

Common materials for interface metallurgy are titanium or chromium for the adhesion layer, copper, nickel, tungsten, palladium, or platinum for the barrier layer, and gold for the bonding layer. To accommodate thermocompression bonding, soft gold is preferred. The solder bump in solder-reflow TAB is usually 95% lead and 5% tin by weight. Such a high lead solder has a 315°C melting point, which is higher than most other temperatures expected in the packaging process (17).

Possible tape materials include polyimide, polyester, polyethersulfone (PES), and polyparabanic acid (PPA) (18). Of these, polyimide is the most widely used material because its high melting point allows it to survive at temperatures up to 365°C. Although polyester is much cheaper than other materials, its use is limited to temperatures less than 160°C. PES and PPA, on the other hand, are half as costly as polyimide, and can survive maximum short-term temperatures of 220 and 275°C, respectively. PES has better dimensional stability than polyimide, absorbs less moisture, and does not tear as easily; however, it is inflammable and can be attacked by certain common solvents. Table 7 lists various plastic tapes and their properties. Common bump materials are gold, copper, and 95% Pb/5% Sn solder (see Tables 6 and 8 for properties; see also References 2 and 21).

Flip-Chip Materials. Flip-chip bonding is an interconnection technique in which the die is turned upside down and connected to the substrate by a medium that includes solder bumps, conduction polymers, and z-axis adhesives. Flip-chip bonds provide a high density, low inductance direct electrical path between the die and the substrate. Figure 5 shows the various layers of the flip-chip bond.

Table 7. Properties of Plastic Films[a]

Plastic film	Melting point, °C	Moisture absorption, %	Tensile strength, MPa[b]	Ultimate elongation, %	Coefficient of thermal expansion, 10^{-6}/K	Dissipation factor at 1 kHz	Dielectric constant at 1 kHz	Dielectric strength, MV/m	Cost factor
polyimide (Kapton)	none	3.0	230	72	18–50	0.0018–0.0026	3.4–3.5	150–300	17.0
polyester (Mylar)	180	<0.8	170	120	31	0.005	3.25	300	1.0
polytetrafluoroethylene (Teflon)	328	<0.01	28	350	12.2	0.0002	2.0	17	9.2
fluorinated ethylene–propylene (Teflon)	280	<0.01	21	300	9.7	0.0002	2.0	255	16.0
polyamide (Nomex)	482	3.0	75	10	38–154	0.007	2.0	18	1.9
poly(vinyl chloride)	163	<0.5	35	130	63	0.009	3.0	40	0.36
poly(vinyl fluoride) (Tedlar)	299	<0.05	70–130	110–300	50	0.02	8.5–10.5	140	3.5
polyethylene	121	<0.01	21	>300	198	0.0003	2.2	20	0.17
polypropylene	204	<0.005	170	250	104–184	0.0003	2.1	160	0.48
polycarbonate	132	0.35	62	110	68	1.0000	3.2	16	1.32
polysulfone	190	0.22	68	95	56	0.001	3.1	295	3.33
polyparabanic acid (PPA)	299	1.8	97	10	51	0.004	3.8	235	8.0
polyether sulfone (PES)	203		84			0.001	3.5		4.5

[a]Ref. 19.
[b]To convert MPa to psi, multiply by 145.

1034

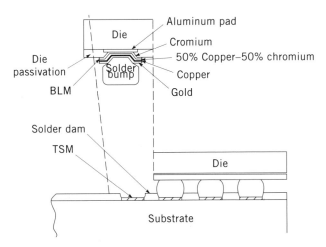

Fig. 5. Various layers in a flip-chip bond, where BLM = ball-limiting metallurgy and TSM = top-surface metallurgy (3).

Table 8. Mechanical and Electrical Properties of Solder Materials[a]

Materials	Density, g/mL	Hardness[b]	Electrical conductivity, % IACS	Electrical resistivity, $\mu\Omega\cdot$cm	Young's modulus, GN/m$^{2 c}$
99% Sn–1% Cu	7.31	9 HB			
97% Sn–3% Cu	7.34	11 HB			
96.5% Sn–3.5% Ag	10.38	14.8 HV	14	12.31	
95% Sn–5% Sb	7.25	15 HB	11.9	14.5	49.99
60% Sn–40% Pb	8.52	16 HV	11.5	14.99	29.99
50% Sn–50% Pb	8.89	14 HV	10.9	15.8	
42% Sn–58% Bi	8.72	22 HB	4.5	34.5	
40% Sn–60% Pb	9.28	12 HV	10.1	17.07	23.03
62% Sn–36% Pb–2% Ag	8.50				22.96
40% Sn– 58% Pb–2% Sb	9.23		9.2		
1% Sn–97.5% Pb–1.5% Ag	11.28	13 HB			

[a]Ref. 20.
[b]Brinnell (HB) or Vickers (HV).
[c]To convert GN/m^2 (GPa) to psi, multiply by 145,000.

Three classes of materials are used in making a flip-chip solder bump; they are solder, ball-limiting metallurgy (BLM), and top-surface metallurgy (TSM).

High melting point solders are used at the die level to permit the use of lower melting point solders for packaging at the board level without remelting the die-level interconnections. The strain-hardening exponent of the solder, which describes the strain induced in the solder by a given stress condition, is a material property that influences the thermal fatigue life.

As well as a barrier, functioning as the BLM adhesive layer must have excellent adhesion to both the metallization pad and the passivation layer. The solder-wetting layer of the BLM must have good solderability and the ability to prevent interdiffusion between the solder and the metallization pad. The oxidation-barrier layer provides an inert surface during bonding and protects the

BLM metal layers from oxidation during storage. In contrast, the upper layer of the TSM structure must have good wetting properties, because it contacts the solder, and should have an adequate shelf life prior to die attachment. This layer is usually a metal that increases the wettability and makes the surface suitable for solder reflow.

The most widely used high melting point solder for flip-chip bonding is 95% lead–5% tin solder (mp ~315°C); eutectic 60% lead–40% tin solder (mp ~183°C) is used to attach dies to organic substrates such as polyimide films and glass–epoxy printed circuit boards. Lead–indium alloys offer significant fatigue-life enhancement over lead–tin solder alloys (22), showing parabolic dependence between the indium concentration and fatigue life. Alloys of 3–5% indium provide a compromise between fatigue-life enhancement and resistance against corrosion (23).

The BLM layer uses a glue layer of chromium or titanium. These metals stick well to other metals and most dielectrics, but they are not solderable. Copper, nickel, and silver have been used as the solder-wetting layer for BLM in applications involving 95% lead–5% tin solders. Gold is commonly used as the oxidation layer on account of its resistance to oxidation and its excellent solderability.

Case Materials

A case provides mechanical support and protection for the devices, interconnects, and substrate mounted in it; it also helps to dissipate heat during component operation and offers protection to the contents of the package from environmental stresses, contaminants, and, in the case of hermetic packages, moisture.

Cases can be classified as either hermetic or nonhermetic, based on their permeability to moisture. Ceramics and metals are usually used for hermetic cases, whereas plastic materials are used for nonhermetic applications. Cases should have good electrical insulation properties. The coefficient of thermal expansion of a particular case should closely match those of the substrate, die, and sealing materials to avoid excessive residual stresses and fatigue damage under thermal cycling loads. Moreover, since cases must provide a path for heat dissipation, high thermal conductivity is also desirable.

Ceramics and Metals. Among ceramics, the most commonly used material is alumina, which has good electrical resistivity ($\sim 10^{12}$ $\Omega \cdot$cm). The coefficient of thermal expansion of alumina fairly matches that of commonly used sealing glasses. Alumina cannot be used in high power applications due to its low thermal conductivity; for these hybrid applications, other ceramics, such as beryllia, aluminum nitride, or silicon carbide, are used. Although beryllia has a high thermal conductivity, it is toxic and must be handled with care, a precaution that also increases its processing costs. A comparatively low cost alternative to beryllia is aluminum nitride, which has a high thermal conductivity and a coefficient of thermal expansion matching that of silicon.

Kovar, one of the most commonly used metals for hermetic cases, is composed of 54% iron, 29% nickel, and 17% cobalt. It has a good CTE (5.1–5.9 ppm/°C), matching that of the commonly used sealing glasses (5.25–6.96 ppm/°C). Its thermal conductivity, however, is lower than that of the copper alloys. As a result, Kovar is not effective as a heat dissipator in high power applications. Instead, copper tungsten, which has a high thermal conductivity, is used for such applications. However, copper alloys pose a problem when sealing a lid to a copper case with electrical resistance welding, because the high thermal conductivity of copper causes the heat generated during the welding process to dissipate quickly. Tables 2 and 3 give properties of many materials used for cases.

Plastic Encapsulant Materials. A PEM encapsulant is generally an electrically insulating plastic material formulation that protects an electronic device and die-leadframe assembly from the adverse effects of handling, storage, and operation (see EMBEDDING). Various molding compounds are used as encapsulants. The molding compound is a proprietary multicomponent mixture of an encapsulating resin with various types of additives. The principal active and passive (inert) components in a molding compound include curing agents or hardeners, accelerators, inert fillers, coupling agents, flame retardants, and stress-relief additives.

ASTM D883 defines a filler as "... a relatively inert material added to a plastic to modify its strength, permanence, working properties, or other qualities or to lower costs." Fillers (qv) that modify the properties and characteristics of epoxies are employed in epoxy resins for a variety of reasons. Their principal functions are to control viscosity, reduce shrinkage and the coefficient of thermal expansion, effect a cost reduction, and color the epoxy resins.

The binder system of a plastic encapsulant consists of an epoxy resin, a hardener or curing agent, and an accelerating catalyst system. The conversion of epoxies from the liquid (thermoplastic) state to tough, hard, thermoset solids is accomplished by the addition of chemically active compounds known as curing agents. Flame retardants (qv), usually in the form of halogens, are added to the epoxy resin backbone because epoxy resins are inherently flammable.

The polymeric encapsulating resin, modified by additives, must possess adequate mechanical strength, adhesion to package components, manufacturing and environmental chemical resistance, electrical resistance, CTE matching, as well as thermal and moisture resistance in the use-temperature range.

Epoxy resins cannot be used alone as encapsulants because of their high coefficient of thermal expansion and low thermal conductivity. Inert inorganic fillers are added to the molding compound to lower the coefficient of thermal expansion, increase thermal conductivity, raise the elastic modulus, prevent resin bleed at the molding-tool parting line, and reduce encapsulant shrinkage during cure, reducing residual thermomechanical stress. Curing agents and accelerators perform the primary function of setting the extent and rate of resin polymerization. The toughness and stress-relaxation response of the epoxy resins can be enhanced by adding flexibilizers and stress-relief agents.

Thermosetting-encapsulation compounds, based on epoxy resins (qv) or, in some niche applications, organosilicon polymers, are widely used to encase electronic devices. Polyurethanes, polyimides, and polyesters are used to encase modules and hybrids intended for use under low temperature, low humidity conditions. Modified polyimides have the advantages of thermal and moisture stability, low coefficients of thermal expansion, and high material purity. Thermoplastics are rarely used for PEMs, because they are low in purity, require unacceptably high temperature and pressure processing conditions, and lead to moisture-induced stresses.

All epoxy resins contain compounds from the epoxide, ethoxylene, or oxirane groups, in which an oxygen atom is bonded to two adjacent (end) bonded carbon atoms. In electrical and electronic applications, three types of epoxy resins are commonly used: diglycidyl ethers of bisphenol A (DGEBA) or bisphenol F (DGEBF), phenolic and cresol novolaks, and cycloaliphatic epoxides. DGEBF is less viscous than DGEBA. Liquid DGEBAs synthesized from petrochemical derivatives are most common and are readily adaptable for electrical and electronic device encapsulation. The epoxy novolaks, essentially synthesized in the same way as DGEBA, are primarily solids; because of their relatively superior performance at elevated temperatures, they are widely used as molding compounds. The cycloaliphatic epoxides or peracid epoxides, usually cured with dicarboxylic acid anhydrides, offer excellent electrical properties and resistance to environmental exposure. Reference 10 compiles the mechanical and electrical properties of the various epoxies (see also EPOXY RESINS).

Silicones (qv) have an advantage over organic resins in their superior thermal stability and low dielectric constants. Polyurethanes, when cured, are tough and possess outstanding abrasion and thermal shock resistance. They also have favorable electrical properties and good adhesion to most surfaces. However, polyurethanes are extremely sensitive to and can degrade after prolonged contact with moisture; as a result, they are not as commonly used as epoxies and silicones (see URETHANE POLYMERS).

Additives. Historically, the filler that exhibits the optimum combination of required properties has been crystalline silica or alpha quartz. A typical crystalline silica-filled molding compound, loaded to 73% by weight, offers a coefficient of thermal expansion of about 32 ppm/°C, a thermal conductivity of around 15 kW/m·K and good abrasion and moisture resistance, electrical properties, and dimensional stability. Crystalline silica has been replaced by ground-fused silica, which provides lower density and viscosity. Thermal conductivity can be increased by a factor of about five with the addition of fillers such as alumina and copper. Generally, an increase in filler concentration increases thermal conductivity, which can also be enhanced by the addition of other filler materials, such as aluminum nitride, silicon carbide, magnesium oxide, and silicon nitride (24). In low stress epoxy resins, spherical silica particles are usually blended with crushed silica, thus increasing the filler loading with a nonlinear increase of viscosity as well as further lowering the coefficient of thermal expansion.

For many electronic and electrical applications, electrically conductive resins are required. Most polymeric resins exhibit high levels of electrical resistivity. Conductivity can be improved, however, by the judicious use of fillers; eg, in epoxy, silver (in either flake or powdered form) is used as a filler. Sometimes other fillers such as copper are also used, but result in reduced efficiency. The popularity of silver is due to the absence of the oxide layer formation, which imparts electrical insulating characteristics. Consequently, metallic fibers such as aluminum are rarely considered for this application.

Epoxidized phenol novolak and cresol novolak are the most common curing agents. The composition of the resin and hardener system is optimized for each specific application; eg, incorporating phenol novolaks in the matrix resin can increase cure speed.

In epoxy molding compounds, the principal stress-relief agents used are silicones, acrylonitrile–butadiene rubbers, and poly(butyl acrylate) (PBA). Silicone elastomers, with their high purity and high temperature properties, are the preferred agents. Silicone elastomers, interface-modified with poly(methyl methacrylate) (PMMA), have uniform domain sizes (1–100 μm) and inhibit passivation layer cracking, aluminum line deformation, and package cracking.

One of the more important flame retardants is homologous brominated DGEBA (diglycidyl ethers of bisphenol A). When used with a normal nonhalogenated resin, they impart self-extinguishing properties and achieve flame retardancy via the bromine liberated at the decomposition temperature. Previously, halogen additives were used as hardeners and/or fillers, but these tended to lower the heat-distortion temperature. Halogenated epoxy resins can be used without sacrificing properties. Antimony trioxide used as a filler is another common flame retardant, but its cost is higher; homologous brominated DGEBA epoxy and heterogenous antimony oxides can also be used together (24).

Lead Materials

Leads serve as the input–output interconnections between the component package and the mounting platform. Sometimes leads also aid in the dissipation of heat generated in the package. In the case of plastic packages, leads are formed from the leadframe, which also acts as a heat-dissipation path and a mechanical support for the die.

Because they serve as a path for electrical signals, leads should have good electrical conductivity. They should also have good corrosion resistance, since corrosion products can change the electrical properties of the lead, and because leads are soldered to the board, they should be wettable by solder. The coefficient of thermal expansion should closely match that of the die material for leadframes, substrate, and sealing glass. Other important properties of leads are yield strength and fatigue properties. Good thermal conductivity is also desirable to enhance the path for heat dissipation.

Alloy 42 (42% Ni–58% Fe) and Kovar are lead materials commonly used in ceramic chip carriers. The coefficients of thermal expansion of both materials match moderately well with those of silicon and ceramic substrate materials (3.4–7.4 ppm/°C). In addition, both materials have high fatigue strength; Alloy 42 has a fatigue strength of 620 MPa, compared with only 380–550 MPa for most copper alloys (3). Copper alloys are particularly suitable for leadframes in plastic packages because they have higher electrical and thermal conductivities than either Kovar or Alloy 42. Copper alloys also exhibit better solderability characteristics than Kovar or Alloy 42. Table 9 gives the key properties for various lead and leadframe materials.

Solder Materials

Solders are alloys that have melting temperatures below 300°C, formed from elements such as tin, lead, antimony, bismuth, and cadmium. Tin–lead solders are commonly used for electronic applications, showing traces of other elements that can tailor the solder properties for specific applications.

Alloy selection depends on several factors, including electrical properties, alloy melting range, wetting characteristics, resistance to oxidation, mechanical and thermomechanical properties, formation of intermetallics, and ionic migra-

Table 9. Properties of Lead–Lead Frame Alloys[a]

Symbol	Nominal composition, %	Thermal conductivity, W/(m·K)	Electrical resistivity, $\mu\Omega\cdot$cm	Yield bend fatigue strength[b], MPa[c]
		Cu–Fe		
C19400	Fe 2.35; P 0.03; Zn 0.12	260	2.54	475
C19500	Fe 1.5; Co 0.8; P 0.05; Sn 0.6	200	3.44	
C19700	Fe 0.6; P 0.2; Mg 0.04	320	2.16	450
C19210	Fe 0.10; Mg 0.034	340	2.03	380
		Cu–Cr		
CCZ	Cr 0.55; Zr 0.25	340	2.03	430
EFTEC647	Cr 0.3; Sn 0.25; Zn 0.2			
		Cu–Ni–Si		
C7025	Ni 3.0; Si 0.65; Mg 0.15	160	4.31	620
KLF 125	Ni 3.2; Si 0.7; Sn 1.25; Zn 0.3	140	4.89	
C19010	Ni 1.0; Si 0.2; P 0.03	240	2.87	585
		Cu–Sn		
C50715	Sn 2.0; Fe 0.1; P 0.03	140	4.89	550
C50710	Sn 2.0; Ni 0.2; P 0.05	120	5.75	450
		Others		
C15100	Zr 1.0; Cu 99.0	380	1.81	380
C15500	Mg 0.11; P 0.06; Cu 99.83	344	1.99	
		Fe–Ni		
ASTM F30 (Alloy 42)	Ni 42; Fe 58	12	70	620
		Fe–Ni–Co		
ASTM F15 (Kovar)	Ni 29; Co 17; Fe 54	40	49	

[a]Ref. 25.
[b]Load bend fatigue strength of alloys capable of withstanding 4–5 cycles before failure in 0–90–0 degree cycles, which is above the three-cycles-to-failure minimum in MIL-STD-883; values pertain to a 0.25-mm thick strip that has been sheared to 0.45-mm width.
[c]To convert MPa to psi, multiply by 145.

tion characteristics (26). These properties determine whether a particular solder joint can meet the mechanical, thermal, chemical, and electrical demands placed on it.

Tin–lead solders are popular because of their melting range (180–230°C), good wetting characteristics, and affordable prices. Their thermal conductivity ranges from 43–50 W/(m·K). In general, lead takes little part in the reactions at the interface, but affects the solidus temperature of the alloy. Tin is more reactive and affects metallurgical reactions with metals. Solders having high land content possess higher solidus temperatures and are used for high temperature applications. Other metals are also added to tin–lead solders for specific applications. Silver is added (62% Sn–36% Pb–2% Ag) for soldering silver-coated surfaces because it reduces silver dissolution and creep. Antimony is added to improve solder strength. Table 8 lists the electrical and mechanical properties of various tin–lead solders. Tin–indium solders provide good solderability, ductility, and thermal fatigue properties; they are used for soldering glasses and ceramics.

Printed Circuit Board Materials

A printed circuit board (PCB) typically consists of a copper circuit pattern created on a copper-clad composite laminate by lithography and electoplating technologies (2). A typical printed circuit board serves the electronic system in three ways: mechanically, by providing support to the components and the conductors; thermally, by offering a thermal conduction path to dissipate heat generated by the devices mounted on it; and electrically, by serving as an insulator for the conductors. Chemical properties also need to be considered, since moisture absorption and chemical attack can both degrade the material properties.

Most of the laminates used for rigid printed circuit boards have been classified, by the National Electrical Manufacturers Association (NEMA), according to the combination of properties that determine the suitability of a laminate for a particular use. Fiber reinforcements make laminate-effective properties orthotropic.

E-glass–epoxy laminates, by far the most widely used circuit board materials, have been designated FR-4 (fire retardant epoxy-glass cloth) by NEMA. The glass-transition temperature, T_g, for FR-4 ranges from 120 to 135°C, which is acceptable for most commercial PCB applications. The advantages of FR-4 are ease of processing, punchability, machinability, low cost, and availability. E-glass–epoxy laminates provide an in-plane thermal expansion closely matched to that of inner-layer copper, with a fiber volume fraction in the range of 0.28–0.30 (27). E-glass–bismaleimide triazine (BT) resin laminates offer similar advantages as FR-4 laminates as well as sustained insulation resistance against moisture absorption (28). E- glass–polyimide laminate has a higher glass-transition temperature (260°C), and is used for military applications.

Ceramic laminates withstand a harsher environment than organic laminates. They also have good thermal conductivities, higher flexural strength, and a better match of coefficients of thermal expansion with the components mounted on them, ensuring good board interconnection reliability under cyclic thermal loading. The disadvantages of ceramic laminates are their relatively high cost, high dielectric constant, and high density. Table 10 gives the properties of various laminate materials.

Conformal Coating Materials

Conformal coatings are protective coatings applied to circuit board assemblies. They protect the interconnect conductors, solder joints, components, and the board itself; they reduce permeability to moisture, hostile chemical vapors, and solvents in the coating. Use of conformal coatings eliminates dendritic growth between conductors, conductor bridging from moisture condensation, and reduction in insulation resistance by water absorption.

The critical property for conformal coatings is resistance to chemicals, moisture, and abrasion. Other properties, such as the coefficient of thermal expansion, thermal conductivity, flexibility, and modulus of elasticity, are significant only in particular applications. The dielectric constant and loss tangent of the conformal coating are important for high speed applications.

Materials for coating must last as long as the product itself, be easy to apply and rework, and be cost-effective. Various materials are used for conformal coatings. Polyurethanes are the most widely used and offer good resistance to moisture, fungus, abrasion, solvent and chemical. In addition, they have good adhesion, low shrinkage, flexibility, and elasticity, and are particularly suited to applications requiring good humidity resistance. However, polyurethanes are difficult to rework due to their high chemical resistance, and are unsuitable for high frequency and high temperature applications due to the strong dependence of their dielectric constants and loss tangents on these variables.

Acrylic resins have excellent moisture resistance, dielectric properties, and reworkability, but poor abrasion resistance. Their dielectric constant, which decreases with increasing frequency, makes them attractive candidates for high frequency applications.

Epoxies provide excellent chemical, moisture, and abrasion resistance. However, because a solvent can sometimes dissolve the epoxy present in the laminate, rework on an epoxy coating requires heat or abrasion to remove the coating.

Silicone-based coatings are well suited for high temperature and high speed applications. They are flexible, tough, and resistant to thermal and oxidative deterioration. They have good surface resistance and are fungus- and flame resistant. However, they possess a high coefficient of thermal expansion and have poor adhesion.

Table 11 gives the properties for various commercial parylenes (*para*-xylylene). These materials have excellent moisture and chemical resistance as well as good electrical and mechanical properties. They are noted for their ability to penetrate small spaces by vapor deposition (see XYLYLENE POLYMERS).

Connector Materials

Connectors are third-level interconnections between daughter boards, between subassemblies, between systems, and to peripherals. The connector assembly normally consists of insert, pin, and contact materials. Platings and/or lubricants are often used to reduce wear and improve contact resistance between the pin and the contact. The insert material must provide support and insulation for the pin and contacts. The coefficient of thermal expansion of the material and the mold shrinkage should be low to ensure pin-to-pin dimensional stability in high density connectors. At higher pin densities, the volumetric and surface resistivity should be high for proper electrical performance. Flexibility is essential for a snap fit. Properties required of metals used for pin and contacts are low electrical resistivity, high strength, wear resistance, corrosion resistance, and a high modulus of elasticity. A low dielectric constant is required for high speed connectors.

Polymers are used as inserts for pins and contacts. Important properties of the commonly used insert materials have been compiled (31). Polysulfones are high temperature thermoplastics that have high rigidity, low creep, excellent thermal stability, flame resistance, low loss tangents, and low dielectric constants. The principal weakness of polysulfones is their low chemical resistance.

Table 10. Properties of Selected PWB Laminate Materials[a]

Laminate	T_g, °C	CTE below T_g, ppm/°C		Water uptake, MIL-P-13949F, mg	Dielectric constant at 1 MHz	Dissipation factor at 1 MHz	Tensile strength, MPa[b]	Modulus of elasticity, GPa[c]	Thermal conductivity, W/(m·K)
measurement direction		x, y	z		z	z	x, y	x, y	z
E-glass–epoxy	120	12–16	60–80	10	4.7	0.021	276	17.4	0.35
E-glass–polyimide	220–300	11–14	60–80	25	4.5	0.018	345	19.6	0.35
E-glass–PTFE	75	24	261		2.3	0.006	68–103	1.0	0.26
quartz–polyimide	260	6–12	34	25	3.6	0.010		27.6	0.13
quartz–Quatrex	185		62		3.5			18.6	
Kevlar-49–Quatrex	185	3–8	105	10	3.7	0.030	22–28	0.16
Kevlar-49–polyimide	180–200	3–8	83	25	3.6	0.008	20–27	0.12

[a]Refs. 2, 29, and 30.
[b]To convert MPa to psi, multiply by 145.
[c]To convert GPa to psi, multiply by 145,000.

Table 11. Typical Properties of Parylene Coatings[a]

Properties	Parylene-N	Parylene-C	Parylene-D
density, g/cm^3	1.11	1.289	1.418
tensile modulus, GPa[b]	2.4	3.2	2.8
tensile strength, MPa[c]	45	70	75
yield strength, MPa[c]	42	55	60
elongation to break, %	30	200	10
yield elongation, %	2.5	2.9	3
hardness, Rockwell R scale	85	80	
melting point, °C	420	290	380
CTE at 25°C, ppm/°C	69	35	
specific heat capacity at 25°C, J(g·K)	1.3	1.0	
thermal conductivity at 25°C, W(m·K)	0.12	0.082	
dielectric constant			
at 60 Hz	2.65	3.15	2.84
1 kHz	2.65	3.10	2.82
1 MHz	2.65	2.95	2.80
loss tangent			
at 60 Hz	0.0002	0.02	0.004
1 kHz	0.0002	0.019	0.003
1 MHz	0.0006	0.013	0.002
dielectric strength, 25-μm thick, MV/m			
sort time	275	220	215
step by step	235	186	
volume resistivity,[d] MΩ	14	8.8	2
surface resistivity,[d] Ω	10^{13}	10^{14}	10^{16}
water absorption, %	<0.1	<0.1	<0.1
gas permeability at 25°C, mol/(Pa·s·m)			
N_2	15.4	2.0	9.0
O_2	78.4	14.4	64.0
CO_2	429.0	15.4	26.0
H_2S	1590	26.0	2.9
SO_2	3790	22.0	9.53
Cl_2	148	0.7	1.1

[a]Ref. 9.
[b]To convert GPa to psi, multiply by 145,000.
[c]To convert MPa to psi, multiply by 145.
[d]At 23°C and 50% rh.

Commonly used materials for pins and contacts are brass, beryllium copper, phosphor–bronze, and copper–nickel. Beryllium copper, a popular contact material, has good spring properties, corrosion resistance, high strength, hardness, fatigue endurance, and wear resistance. Phosphor–bronze, made with a large percentage of tin, is resistant to fatigue, wear, and corrosion, and has a high tensile strength and a high capacity to absorb energy. Properties of commonly used pin and contact materials are given in Tables 12 and 13.

Cable and Flex Circuit Materials

A cable or flex circuit is often employed as an interconnection between electronic circuits that are not easily connected by other means. The interconnection density obtained with a cable or flex circuit is lower than that achieved with a backpanel.

Table 12. Copper-Based Metal Alloys[a]

Property[b]	Alloy 260[c] brass	Alloy 172[c] beryllium–copper[d]	Alloy 510, grade A, phosphor–bronze	Alloy 638	Alloy 725	Alloy 762, nickel–silver
nominal composition, %	Cu 70 Zn 30	Cu 98.1 Be 1.9	Cu 94.81 Sn 5.0	Cu 95 Al 2.8 Si 1.8 Co 0.4	Cu 88.2 Ni 9.5 Sn 2.3	Cu 59.25 Zn 28.75 Ni 12
electrical conductivity at 20°C, MΩ/cm	0.163	0.128	0.087	0.058	0.064	0.050
thermal conductivity at 20°C, W(m·K)	121	109–130	68.6	40.6	54.4	41.8
density at 20°C, g/cm³	8.54	8.260	8.86	8.29	8.89	8.70
modulus of elasticity, GPa[e]	112	130	112	117	135	127
yield strength, 0.2%						
offset, MPa[f]	70–220	109	150	410–470	180	200
annealed	290–410	123	330–480	530–630	400–510	410–580
half hard	460–530	127	520–620	640–720	520–560	580–680
hard	580–630		650–760	700–790	550–650	710–770
spring	600–690		690–770	750 min	630–720	720 min

[a]Ref. 32.
[b]At 0.020-mm grain size except in the case of Alloy 638, which is the standard annealed condition.
[c]Designation of Copper Development Association.
[d]All property data for beryllium–copper are for material after age-hardening heat treatment.
[e]To convert GPa to psi, multiply by 145,000.
[f]To convert MPa to psi, multiply by 145.

Table 13. Properties of Nickel-Based Alloys[a]

Alloy	Nominal composition, %	Tensile strength, MPa[b]	Yield strength, MPa[b]	Elongation, %	Electrical conductivity, MΩ·cm
nickel 200[c]	99 Ni	379–758	103–689	55–10	0.106
nickel 270	99.97 Ni	344–655	103–620	50–4	0.134
Duranickel 301[c]	93 Ni 4.5 Al	620–1448	209–1206	55–15	0.024
beryllium–nickel 440[d]	98 Ni 2 Be	655–1861	275–1586	30–8	0.027

[a]Ref. 32.
[b]To convert MPa to psi, multiply by 145.
[c]Modulus of elasticity = 206 GPa (29.9×10^6 psi).
[d]Modulus of elasticity = 186–207 GPa ($27–30 \times 10^6$ psi).

Properties desired in cable insulation and flexible circuit substrate materials include mechanical flexibility, fatigue endurance, and resistance to chemicals, water absorption, and abrasion. Both thermoplasts and thermosets are used as cable-insulating materials. Thermoplastic materials possess excellent electrical characteristics and are available at relatively low cost.

Table 14. Mechanical and Electrical[a] Properties of Cable Insulating Materials[b]

Insulation	Common designation	Tensile strength, MPa[c]	Elongation, %	Specific gravity	Dielectric strength, kV/mm	Dielectric constant at 1 kHz	Dissipation factor at 1 kHz
poly(vinyl chloride)	PVC	16.6	260	1.2–1.5	15.8	5–7	0.02
polyethylene	PE	9.6	300	0.92	18.9	2.3	0.005
polypropylene		41.4	25	1.4	29.5	2.54	0.006
cross-linked polyethylene	IMP	20.7	120	1.2	27.6	2.3	0.005
polytetrafluoroethylene	TFE	20.7	150	2.15	18.9	2.1	0.0003
fluorinated ethylene–propylene	FEP	20.7	150	2.15	19.9	2.1	0.0003
chlorotrifluoroethylene	Kel-F	34.5	120	2.13	17.0	2.45	0.025
poly(vinylidene fluoride)	Kynar	49.0	300	1.76	50.4 (8 mil)	7.7	0.02
silicone rubber	Silicone	5.5–12.6	100–800	1.15–1.38	22.6–27.6	3.0–3.6	0.003
polychloroprene	Neoprene	1.0–27.6	600–700	1.23	32.0	9.0	0.03
butyl rubber	Butyl	4.8–10.3	500–700	0.92	23.6	2.3	0.003
fluorocarbon rubber	Viton	16.6	350	1.4–1.95	19.7	4.2	0.14
polyurethane	Urethane	34.5–55.2	100–600	1.24–1.26	17.7–19.7	6.7–7.5	0.055
polyamide	nylon	27.6–48.3	300–600	1.10	15.2	4–10	0.02
polyimide film	Kapton	124.1	707	1.42	212.6 (8 mil)	3.5	0.003
polyester film	Mylar	89.6	185	1.39	102.4	3.1	0.15
polyalkene		13.8–48.3	200–300	1.76	73.6	3.5	0.028
polysulfone		69.0	50–100	1.24	16.73	3.13	0.0011
polyimide-coated TFE	TFE/ML	20.7	150	2.2	18.9	2.2	0.0003
polyimide-coated FEP	FEP/ML	20.7	150	2.2	18.9	2.2	0.0003

[a] Volume resistivity ranges from 10^{11}–10^{18} $\Omega\cdot$cm.
[b] Ref. 5.
[c] To convert MPa to psi, multiply by 145.

Copper is by far the most widely used conductor material. It has high electrical conductivity, thermal conductivity, solderability, and resistance to corrosion, wear, and fatigue. Annealed copper conductors can withstand flex and vibration stresses normally encountered in use.

Commonly used materials for cable insulation are poly(vinyl chloride) (PVC) compounds, polyamides, polyethylenes, polypropylenes, polyurethanes, and fluoropolymers. PVC compounds possess high dielectric and mechanical strength, flexibility, and resistance to flame, water, and abrasion. Polyethylene and polypropylene are used for high speed applications that require a low dielectric constant and low loss tangent. At low temperatures, these materials are stiff but bendable without breaking. They are also resistant to moisture, chemical attack, heat, and abrasion. Table 14 gives the mechanical and electrical properties of materials used for cable insulation.

BIBLIOGRAPHY

1. J. Dally, *Packaging of Electronic Systems: A Mechanical Engineering Approach*, McGraw-Hill Book Co., Inc., New York, 1990.
2. M. Pecht, *Handbook of Electronic Package Design*, Marcel Dekker, Inc., New York, 1991.
3. M. Pecht, *Integrated Circuit, Hybrid, and Multichip Module Package Design Guidelines: A Focus on Reliability*, John Wiley & Sons, Inc., New York, 1994.
4. D. R. Lide, *Handbook of Chemistry and Physics*, 7th ed., CRC Press, Inc., Boca Raton, Fla., 1990.
5. C. A. Harper, *Electronic Packaging and Interconnection Handbook*, McGraw-Hill Book Co., Inc., New York, 1991.
6. N. Iwase, K. Anzai, and K. Shinozaki, *Solid State Technol.* **29**(10), 135–138 (Oct. 1986).
7. A. H. Kumar and R. R. Tummala, *Int. J. Hybrid Microelectronics*, **14**(4), 137–150 (Dec. 1991).
8. R. K. Shukla and N. P. Mecinger, *Solid State Technol.* **29**(7), 67–74 (July 1985).
9. F. B. Beach and R. Olson, in M. L. Minges and co-workers, eds., *Electronic Materials Handbook*, Vol. 1, ASM International, Materials Park, Ohio, 1989.
10. M. L. Minges and co-workers, eds., *Electronic Materials Handbook*, Vol. 1, ASM International, Materials Park, Ohio, 1989.
11. K. Atsumi and co-workers, "Ball Bonding Technique for Copper Wire," *36th Proceedings of the IEEE Electronic Components Conference*, Seattle, Wash., May 5–7, 1986, pp. 312–317.
12. J. Hirota and co-workers, "The Development of Copper Wire Bonding for Plastic Molded Semiconductor Packages," *35th Electronic Component Conference Proceedings*, Washington, D.C. 1985, pp. 116–121.
13. J. Kurtz, D. Cousens, and M. Dufour, *34th IEEE Electronic Components Conference*, New Orleans, La., May 14–16, 1984, pp. 1–5.
14. L. Levine and M. Shaeffer, *Semiconductor Int.* **9**(8), 126–129 (Aug. 1986).
15. J. Onuki, M. Koizumi, and I. Araki, *IEEE Transactions on Components, Hybrids, and Manufacturing Technology*, Vol. 10, CHMT-10, Dec. 1987, pp. 550–555.
16. S. T. Riches and N. R. Stockham, *Proceedings of the 6th European Microelectronics Conference (ISHM)*, Bournemouth, U.K., June 3–5, 1987, pp. 27–33.
17. C. Speerschneider and J. Lee, *Solid State Technol.*, **32**(7) (July 1989).
18. J. Lyman, *Packag. Prod. Ed.*, 175–182 (Dec. 1975).

19. J. H. Lau, S. J. Erasmus, and D. W. Rice, in Ref. 9, pp. 274–296.
20. *Solder Alloy Data-Mechanical Properties of Solders and Soldered Joints*, I.T.R.I. Pub. No. 656, International Tin Research Institute.
21. *Metal Foil for Printed Wiring Applications*, IPC-MF-150, Institute for Interconnecting and Packaging Electronic Circuits, Lincolnwood, Ill., 1991.
22. L. S. Goldmann, *Proceeding of the 27th Electronic Components Conference*, Arlington, Va., 1977, pp. 25–29.
23. T. R. Howard, *IBM J. Res. Dev.* **3**, 372–378 (1982).
24. R. K. Rosler, in Ref. 9, pp. 810–816.
25. J. Crane and J. F. Breedis, in Ref. 9, pp. 483–492.
26. M. G. Pecht, *Soldering Processes and Equipment*, John Wiley & Sons, Inc., New York, 1993.
27. C. L. Guiles, *CTE Materials for PWBs*, IPC Technical Paper 914, IPC Seminar Meeting, San Diego, Calif., 1990.
28. G. Morio, N. Ikeguchi, and S. Ayano, *Electronic Packaging Prod.* **25**(12), 30–33 (Dec. 1985).
29. F. L. Gray, in Ref. 9, pp. 611–629.
30. S. J. Mumby, *J. Electronic Mater.* **18**(2), 241–250 (Mar. 1989).
31. W. T. Shugg, *Handbook of Electrical and Electronic Insulating Materials*, Van Nostrand Reinhold Co., Inc., New York, 1986.
32. G. L. Ginsberg, ed., *Connectors and Interconnection Handbook*, Vols. 1–3, The Electronic Connector Study Group, Inc., Camden, N.J., 1978.

General References

B. Ellis, *Chemistry and Technology of Epoxy Resins*, Chapman and Hall, Glasgow, U.K., 1993.
M. G. Pecht, L. Nyugen, and E. B. Hakim, *Plastic Encapsulated Microelectronics: Materials, Processes Quality, Reliability and Applications*, John Wiley & Sons, Inc., New York, 1995.

MICHAEL PECHT
University of Maryland at College Park

ASOHK PRABHU
Nitto Denko America Inc.

PACKING MATERIAL.   See PUMPS.

PAINT

ARCHITECTURAL

Paint is defined as "any liquid, liquefiable, or mastic composition designed for application to a substrate in a thin layer which is converted to an opaque solid film after application." Paint is "used for protection, decoration, or identification, or to serve some functional purpose such as the filling or concealing of surface irregularities, the modification of light and heat, etc" (1). This article focuses on the types of paints commonly referred to as architectural coatings, house paints, and trade sales paints. Although some of the information in this article is applicable to other types of paints, factory-applied finishes, automobile, industrial, and marine coatings are covered elsewhere (see COATINGS).

Paint is one of the most common and widely used materials in home and building construction and decoration (see BUILDING MATERIALS). Its broad use comes from its ability to provide not only improved appearance and decoration but also protection of a substrate to which it is applied. Evidence of the historical uses of paint goes back over 25,000 years to cave paintings found in Europe. The Bible describes pitch being used to coat and protect Noah's Ark. Over 10,000 years ago in the Middle East, various minerals and metals such as lime, silica, copper and iron oxides, and chalk were mixed and reacted to produce many colors. Resins from plant sap and casein were also used. Over 2000 years ago in Asia, resins refined from insect secretions and sap from trees were used to make clear lacquers and varnishes (2).

Paint provides outstanding value, considering that such a thin film (usually <0.1-mm thick), provides protection for materials that would have a much shorter life span without a protective coating. Paints are most often described according to the type of binder or solvent employed, eg, latex-, alkyd-, or oil-based paint describes the type of binder or resin system used. Paints are also described as water- or solvent-based, referring to the type of solvent used in the formulation. In the paint industry, the word solvent usually refers to organic hydrocarbon solvents and does not include water. The conditions for which a paint is developed is also an important part of its description, ie, a paint recommended for interior or exterior application. Properties such as color and appearance are also terms for describing paints. Flat, satin, semigloss, and gloss paints refer to how dull or shiny the dried paint film can be expected to appear. The terminology for paint color is so varied that color swatches or chips often accompany a paint purchase so the user can visualize how the color will appear.

Chemical and Physical Properties

Most paints are made up of four basic groups of chemical raw materials: binders or resins, pigments, solvents, and additives. When a paint is applied to a surface, the solvents begin to evaporate while the binder, pigments, and additives remain on the surface to form a hard, dry solid film. The paint formulator selects the proper type and concentration of raw material from each of these groups that will provide paint with the desired end use properties. The chemical and physical properties of paints are directly related to the choice and concentration of raw materials used to make the paints. Paint ingredients are selected to provide the desired appearance, performance, application, and rheological properties of the finished product. Other criteria for a good paint formulation include raw material cost, ease of manufacture and application, and package stability.

Examples of binders and resins used in paints include latex emulsions based on acrylic and vinyl acetate, as well as styrene polymer and copolymer systems. Alkyds, linseed oil, and oil-modified epoxy and polyurethane resins are common binders in solvent-based paints. Water-reducible alkyd and oil systems are also available. The types of pigments used include organic and inorganic colored pigments as well as inorganic extenders and filler pigments. Solvent choice is limited mainly to a solvent that is compatible with the binder and has the desired evaporation rate and toxicity profile. Additives include thickeners, biocides, driers, pigment dispersants, surfactants (qv), defoamers, and other specialty ingredients used at relatively low levels in a paint formulation.

Binders and Resins.   The choice of binder is the most important ingredient choice in the formulation process because the binder affects the performance properties of a paint more than any other single ingredient (3). The physical properties of binders required for paints include the ability to dry or cure under various ambient conditions, good adhesion to various substrates, abrasion resistance, washability, flexibility, water resistance, and ultraviolet light resistance. The balance of these required properties is mostly dependent on whether the paint is being developed for interior or exterior applications.

Up until the 1940s, the only types of paint broadly utilized were based on solvent-based oil or alkyd resins (qv). About the same time, the development of water-based paints, primarily based on latex emulsions, began; in the 1990s, these types of paints represent over 70% of the architectural coatings market (4). Latex emulsions for paints are available in a wide variety of compositions and particle size ranges (see LATEX TECHNOLOGY). Monomer and polymer compositions control the durability, hardness (qv), water resistance, and gloss potential of the latex emulsion. The size and distribution of the latex particles in the emulsion influence the rheology of the paint, its film-forming ability, and gloss potential. Carboxylic acid comonomers are an important part of the stabilization system of most latices. Their distribution in the interior or at the surface of the particles can influence the rheology of the latex and the paint made from it, and adhesion to metal surfaces. Surfactants are necessary for controlling particle size and stabilizing the latex. Their proper selection involves balancing these necessary features against negatives such as contributing to foaming and water sensitivity in the dry paint film. Adhesion promoters are used to improve adhesion to wood and aged alkyd paint surfaces.

All-acrylic (100%) latex emulsions are commonly recognized as the most durable paints for exterior use. Exterior grades are usually copolymers of methyl methacrylate with butyl acrylate or 2-ethylhexyl acrylate (see ACRYLIC ESTER POLYMERS). Interior grades are based on methyl methacrylate copolymerized with butyl acrylate or ethyl acrylate. Acrylic latex emulsions are not commonly used in interior flat paints because these paints typically do not require the kind of performance characteristics that acrylics offer. However, for interior semigloss or gloss paints, all-acrylic polymers and acrylic copolymers are used almost exclusively due to their excellent gloss potential, adhesion characteristics, as well as block and print resistance.

Vinyl acetate is another monomer used in latex manufacture for architectural coatings. When copolymerized with butyl acrylate, it provides a good balance of cost and performance. The interior flat latex paint market in North America is almost completely dominated by vinyl acetate–acrylic copolymers. Vinyl acetate copolymers are typically more hydrophilic than all-acrylic polymers and do not have the same ultraviolet light resistance as acrylics; as a result, their use in exterior applications is limited (see VINYL POLYMERS, POLY(VINYL ACETATE)).

Styrene–acrylic copolymers provide latices with good water resistance and gloss potential in both interior and exterior latex paints. However, they are typically regarded as having limited exterior durability compared to all-acrylic latex emulsions that are designed for exterior use.

Alkyd resins are produced by reaction of a polybasic acid, such as phthalic or maleic anhydride, with a polyhydric alcohol, such as glycerol, pentaerythritol, or glycol, in the presence of an oil or fatty acid. The resulting polymeric material can be further modified with other polymers and chemicals such as acrylics, silicones, and natural oils. On account of the broad selection of various polybasic acids, polyhydric alcohols, oils and fatty acids, and other modifying ingredients, many different types of alkyd resins can be produced that have a wide range of coating properties (see ALKYD RESINS).

Alkyd resins are described as long-oil, medium-oil, and short-oil alkyds. Such description is based on the amount of oils and/or fatty acids in the resin. Long-oil alkyds generally have an oil content of 60% or more; short-oil alkyds, less than 45%; and medium-oil alkyds have an oil content in between the two. The short- and medium-oil alkyds are based on semidrying and nondrying oils, whereas long-oil alkyds are based on semidrying and drying oils (qv). Common examples of nondrying oils are castor and coconut oil; of semidrying oil, soybean oil; and of drying oils, linseed and dehydrated castor oil. Long- and medium-oil alkyds cure and form a film by autoxidation. Short-oil alkyds cure mostly by heat (baking) (3).

Both regulatory limits on the amount of organic solvents allowed in paints and advancements in alkyd resin technology have resulted in the development of higher solids alkyd resins that require less solvent for dilution and viscosity reduction. In addition, developments of water-reducible alkyds and alkyd emulsions have resulted in alkyd-based paints that require less organic solvent in their formulations.

Pigments and Extenders.   Pigments are selected for use in house paints based on their appearance and performance qualities. Appearance includes color and opacifying ability. Performance qualities include ultraviolet light resistance, fade resistance, exterior weatherability, chemical resistance, as well as particle size and shape. Toxicity profiles and safety and health related properties are also important criteria in pigment selection.

Color pigments can be classified chemically as organic or inorganic. Inorganic color pigments are more commonly used than organic pigments in house paints because of their relatively low cost and better opacifying ability. Also, the exterior durability properties of inorganic color pigments are generally better than those of organics. As a result of the advantages of inorganic color pigments, organic color pigments use is limited mostly to situations where the more saturated and brilliant colors available from them are desired and not available with inorganic pigments (qv).

Color pigments are often regarded as falling into three general hue classifications: white, black, and other colors. The most popular pigment in use is titanium dioxide, TiO_2, because of its high brightness and opacifying nature; TiO_2 is also known for its good chemical and heat stability (5).

Titanium dioxide has the highest refractive index of any pigment used in paints and can be processed to produce particle sizes close to the wavelength of light for maximum light-scattering capability. These two properties, as well as the ability to reflect almost all wavelengths of visible light, make TiO_2 the most efficient light-scattering pigment in existence (5). TiO_2 is also characterized by high brightness, a result of its inability to absorb light and its superior ability to scatter it.

Most extender pigments have refractive indexes close to the binder and do not contribute opacity to the paint. Extenders are primarily cheap fillers, but choice of the extender type can dramatically affect film properties. Examples of some popular extender pigments are calcium carbonate, talc, clay, silica, and feldspar. Although calcium carbonates have such features as low cost flattening in interior paints and good tint retention and chalk resistance in exterior paints, it tends to cause frosting on the paint surface as a result of leaching from the paint film by moisture. Clays typically have relatively high binder demand that can result in good hiding at higher concentrations, but they are also soft and easily burnished. Silica extenders are hard, abrasion- and burnish-resistant, but they cause wear of milling and spray equipment.

Whereas semigloss paints can use only one extender, flat paints, especially interior, can use combinations of three or more. Gloss paints typically do not use extenders, which decrease gloss. As a result, gloss paints lack the hardening effect of extenders and must rely on harder binders for adequate durability. Other types of extenders used in paints can have functional properties such as corrosion resistance, mildew resistance, and film-hardening effects. Their functional properties result from their reactive nature in the paint film. Zinc oxide is an example of a functional extender that contributes to these properties in a paint film.

Solvents.   Solvents in house paints serve several essential purposes. They keep the binder dispersed or dissolved and the pigments dispersed in an easy-to-use state. Solvents allow the paint to be applied in the correct thickness and uniformity, and evaporate from the paint film after the paint is applied. Solvent choice is limited mainly to a solvent that is compatible with the binder system and that has the desired evaporation rate and toxicity profile. The volatility or evaporation rate of a solvent determines to a large extent the open-time and dry-time properties of a paint (6).

Mineral spirits, a type of petroleum distillate popular for use in solvent-based house paints, consist mainly of aliphatic hydrocarbons with a trace of aromatics. This type of solvent finds use in oil- and alkyd-based house paints because of its good solvency with typical house paint binders and its relatively slow evaporation rate which imparts good brushability, open-time, and leveling.

Other properties include lower odor, relatively lower cost, as well as safety and health hazard characteristics comparable to most other organic solvents.

Organic solvents are also used in water- or latex-based house paints to coalesce latex particles, prevent freezing, control drying rates, and help reduce the surface tension of water for improved pigment and substrate wetting. The use of organic solvents, such as glycol ethers, in latex paints to coalesce the latex particles is critical for ensuring proper film formation of the latex upon application and subsequent drying. When a latex paint is applied to a substrate, the water in the wet paint begins to evaporate. As the water evaporates, the polymer particles, along with the pigment particles in the paint, come closer to each other and eventually fuse together (coalesce) to form a cohesive film. Without coalescing agents, the development of a cohesive paint film can be impaired, especially if the paint is applied at low temperatures. This is because at lower ambient temperatures, polymer particles in a latex paint become harder and more difficult to coalesce and form a continuous film. Harder latex polymers generally require more coalescents for adequate film formation than softer latex polymers. There is also a trend in the architectural coatings industry to use new, softer, and specially designed latex emulsions that require fewer or no coalescents in the paint for film formation. This practice allows greater reduction in the solvent volume used in latex paints.

Water-immiscible coalescents usually partition into the polymer phase and, because they must diffuse out of the polymer in order to leave the film, stay in the film longer and promote better film formation. They also tend to have higher boiling points and slower evaporation rates. Water-soluble coalescents usually have lower boiling points and faster evaporation rates than water-immiscible coalescents; because they do not partition as strongly into the polymer, they leave more quickly from the latex film. Therefore, higher levels of water-soluble coalescents may be required for film formation, which may be worse under poor ambient application conditions. Water-soluble coalescents also can lower the viscosity of latex paints containing associative thickeners.

Glycol solvents are also commonly used in latex paint formulations and are multifunctional as both additives and cosolvents. A primary purpose of glycol solvents in latex paints is to act as an antifreeze in the presence of water. This enhances the long-term package stability properties of latex paints as the paints are better able to withstand a wider range of temperatures during storage. Another property of glycol solvents is that they help give water-based paints longer wet-edge, ie, the amount of time that a paint stays wet after application, and open-times during application.

Ethylene glycol [107-21-1] is the most efficient antifreeze and can slow latex paint dry-times considerably. Its performance is sensitive to ambient humidity, and high levels of ethylene glycol can slow high humidity drying to an unacceptable degree. However, because of its potential toxicity, ethylene glycol is falling out of favor with paint manufacturers. Propylene glycol [57-55-6] is also widely used as a cosolvent in latex paints. It is essentially nontoxic and odorless, and can act to improve the ease of incorporation of water-immiscible coalescents into the latex and promote antifreeze and wet-edge properties in latex paints.

Additives. The principal components of house paints are the binder, solvents, pigments, and extenders. However, also crucial to the performance of the paint are minor components known as additives, both by performing functions not covered by the principal components and by aiding the binder and pigments to fulfill their functions in the paint formulation. Additives collectively represent the smallest fraction of the paint in terms of weight and volume, but because of their specialized functions, they are usually sold as high performance chemicals and often command a high price. Care in the selection and use of additives is important for obtaining the desired paint performance at an affordable cost. Often, the higher the performance demanded of the paint, the greater the role the additives play. The kinds of additives used in house paints depend on the type of paint being formulated. Additives for latex paints typically include thickeners, pigment dispersants, surfactants, defoamers, biocides, and amines or other chemicals for pH adjustment. Additives for solvent-based paints typically include thickeners, mildewcides, driers, and antiskinning agents. Dispersants and surfactants are also used in solvent-based paints, but not with the same degree of necessity as in latex or other water-based paints.

Paint Formulations.   Tables 1 and 2 provide examples of generic water-based latex and solvent-based alkyd oil paints. These formulations exhibit typical proportions of the paint ingredients discussed.

Table 1. Exterior Acrylic Latex Flat Water-Based House Paint

Raw material ingredients	Weight, kg	Volume, L
Grind portion[a]		
water	144.1	144.1
propylene glycol	72.3	69.2
in-can preservative	2.0	2.0
cellulosic thickener, 100%	3.6	2.5
dispersant, 25%	14.7	13.3
surfactant	2.4	2.2
defoamer	2.4	2.6
titanium dioxide	210.8	52.5
zinc oxide	30.1	5.3
extenders	192.8	80.7
Let-down portion		
latex emulsion, 53.3%	391.0	365.8
polymeric opacifier	79.5	76.7
texanol	11.9	12.4
defoamer	2.4	2.6
mildewcide	2.4	2.3
polyurethane thickener, 25%	12.0	12.5
aqueous ammonia, 28%	2.7	2.9
water	150.4	150.4
Total	*1327.5*	*1000.0*
Properties		
pigment volume concentration, %	49	
volume solids, %	37.5	
weight solids, %	51.5	
density, g/L	1.3	
VOC, g/L	200	

[a]Pigment, dispersion, or millbase.

Table 2. Exterior Alkyd–Linseed Oil Flat Solvent-Based House Paint

Raw material ingredients	Weight, kg	Volume, L
Grind portion[a]		
mineral spirits	78.3	100.0
clay thickener	13.5	9.2
mildewcide	13.5	8.4
alkyd resin, 60%	186.7	200.0
surfactant	12.8	12.5
titanium dioxide	321.0	80.0
extenders	319.3	120.0
Let-down portion		
alkyd resin, 60%	166.3	177.0
linseed oil, 80%	119.9	130.0
mineral spirits	121.7	155.4
organometallic driers	6.0	5.5
antiskinning agent	1.8	2.0
Total	*1360.8*	*1000.0*
Properties		
pigment volume concentration, %	40	
volume solids, %	54	
weight solids, %	72	
density, g/L	1.4	
VOC, g/L	360	

[a]Pigment, dispersion, or millbase.

Application and Appearance Properties

A well-formulated paint that is properly manufactured and packaged should have good product uniformity in the container without either pigment settling, phase-separation of the liquids, or color float. The paint should also have an acceptable viscosity and rheological profile. The expectation for most paints are that they will have a degree of structure and viscosity in the can, and load heavily but easily on a brush or roller with minimum dripping, making it easy to apply heavy coats quickly and uniformly.

Whether applied by roller or brush, the paint should flow uniformly onto the surface being coated, and deposit a thick enough film that offers satisfactory hiding and coverage. High brushing and application viscosity forces the painter to apply more paint, increasing the likelihood that a satisfactory finish will be obtained with the minimum number of coats. Generally, it is not possible to guarantee that a paint will provide satisfactory coverage in only one coat; however, by using proper application technique, this can be achieved with high quality paints and a good brush or roller. When applied by roller, latex paints have traditionally tended to spatter much more than solvent-based paints. However, technological developments in additives for latex paints allow them to be formulated with little spatter. Once applied, the paint should level well, ie, brush marks or roller stipple should flow out and disappear. At the same time, the paint should not sag or run. Unfortunately, these two requirements tend to contradict one another, since leveling of brush marks requires low viscosity at low shear rates, but sag resistance requires high viscosity at low shear rates. Achieving an acceptable balance of leveling and sag resistance is one of the many compromises the paint formulator must make.

For the most part, additives control the application or rheological properties of a paint. These additives include materials for latex paints such as hydroxyethylcellulose, hydrophobically modified alkali-soluble emulsions, and hydrophobically modified ethylene oxide urethanes. Solvent-based alkyd paints typically use castor oil derivatives and attapulgite and bentonite clays. The volume solids of a paint is an equally important physical property affecting the application and rheological properties. Without adequate volume solids, the desired application and rheological properties may be impossible to achieve, no matter how much or many additives are incorporated into the paint.

The color and appearance of an object are among the first things that cause people to evaluate and judge it as acceptable or not. Thus, color and appearance of a paint are critical properties for analyzing product quality. To this end, paint researchers must study not only the ingredients that provide coatings with the desired color, opacity or hiding, and appearance, but also how they can best maintain these properties over a long period of time. Color (qv) is the result of the visual phenomenon wherein an object illuminated by light changes the physical properties and pathways of the light. In order to have color, three conditions are required: a light source, an object to be illuminated, and an observer to see the color (7). A change in any of these can result in a color change. When light interacts with a paint, the paint can transmit, reflect, scatter, and/or absorb the light to varying degrees, producing varying color and opacity. Paints contain raw materials with different refractive indexes. Light-scattering involves light falling on small particles that have a different refractive index from that of the material surrounding them, such as titanium dioxide particles in the paint binder (7).

Light reflection and absorption are the properties that give paints their particular color. For example, a red paint appears red because it absorbs all the colors of white light except red. The red light is reflected back to the observer and the result is that the paint appears red. The overall appearance of a paint is a combination of its surface qualities and its color. The surface qualities of a paint are described in terms of gloss and texture. Gloss is a function of light reflection off a surface and can be defined as the degree of approach to that of a mirror surface (8). Smooth paint surfaces have a high gloss that are mirror-like in appearance; dull or rough paint surfaces have a nonglossy or flat appearance. Paints of the same color but different surface qualities, eg, one gloss and one flat, appear different to an observer and are often mistakenly described as being slightly different in color.

Manufacture and Processing

Along with the choice of a binder for the paint formula, the selection of pigments and their volume relationship to the binder govern the performance properties of the resultant paint (3). When evaluating the properties of paint formulations, volume relationships of the pigments to the binder are used in preference to weight relationships because of the wide density range of various binders and pigments used in paints. The volume relationships between pigments and binder in a paint formula provide more precise correlation to paint performance than do weight relationships (3). Knowledge of this relationship is the key to accurate evaluations of test results on the performance properties of paint formulas having different compositions.

Pigment Volume Concentration.   The volume relationship of pigments to binders is known as the pigment volume concentration (*PVC*). The mathematical calculation of *PVC*, as shown in equation 1, is the total pigment volume divided by the total volume of pigment (V_p) and solid binder (V_b) in the formula.

$$PVC = \frac{V_p}{V_p + V_b} \tag{1}$$

PVC is usually expressed in percentage, but the % sign is often omitted. Although many additives in the paint formulation are nonvolatile, they are often omitted from this calculation, because they represent a small fraction of the volume of a newly formed paint film and, in exterior paints, are often water-soluble materials leached out by rainfall, and therefore will probably not factor into the long-term performance of the paint film.

An important aspect in the *PVC* determination of a paint formula is the knowledge of the critical *PVC* (*CPVC*). The *CPVC* of a paint is the pigment volume concentration at which there is just a sufficient amount of binder to fill all the voids between the pigment particles in a dry paint film. The *CPVC* depends on the binding efficiency of the binder or resin, and the binder demand of the pigmentation. A paint that has a *PVC* value below its *CPVC* is characterized by having an excess of binder where pigment particles are separated by the binder. In this case there is enough binder to coat the surface of the pigment particles and fill in the interstices between pigment particles. As the *PVC* of a paint formulation is increased, there is less binder available to hold the pigment particles together. When the *PVC* is greater than its *CPVC*, the paint consists of a dry film composed of pigment, binder, and air. In this case the pigment particles are closely packed, and both binder and air fill the voids and interstices between the pigment particles (9).

Knowledge of the *CPVC* of a paint formula is important because dramatic changes in the behavior and properties of the paint occur when the *PVC* is in the *CPVC* region. In a paint formula where the *PVC* is below the *CPVC*, the paint is characterized by having an airless film, higher gloss, lower porosity, good weather resistance, flexibility, and abrasion resistance. When the *PVC* of a paint is greater than its *CPVC*, the resultant properties are air voids in the dry film, greater porosity, poor weatherability, low gloss, and poor flexibility and abrasion resistance. As a result, house paints produced for exterior use must be formulated so that their *PVC* is below their *CPVC* (10). Because film porosity increases rapidly above the *CPVC*, light-scattering from air voids, whose refractive index is much lower than the binder, extenders, and hiding pigments, also increases rapidly. This is an inexpensive way to obtain opacity, but it comes at the cost of other performance properties. The ratio of *PVC* to *CPVC* is often used as a measure of the film quality, especially above the *CPVC*; when below *CPVC*, the impact of this ratio is less.

The *CPVC* of a paint film can be determined by mathematical calculation or through a variety of physical test methods performed on a dry paint film. Analysis of *CPVC* includes measurements of paint film density, tensile strength, porosity, and hiding power (9). Test procedures for the determination of *CPVC* in a paint require the evaluation of paint samples having different *PVC* values, eg, 40, 45, 50, 55, etc. Test results are plotted graphically and almost always show a distinct, abrupt change in properties at the *CPVC* value. *CPVC* can also be calculated if the oil absorption values of the pigments are known. Calculation of *CPVC* is based on the following equation, where OA_V = volume of linseed oil/volume of pigment.

$$CPVC = \frac{1}{1 + OA_V} \qquad (2)$$

The oil absorption of any pigment or combination of pigments is most commonly determined experimentally by titrating a fixed weight of the pigments(s) with linseed oil. After each aliquot of linseed oil added, the pigment–oil mixture is sheared with a spatula for a thorough blending. The end point is reached when the pigment–oil mixture changes from a putty-like to a fluid consistency (9). The oil absorption value is defined as the weight in grams of linseed oil needed to wet 100 grams of a particular pigment or pigments. Expressing the amounts of pigment and linseed oil at the end point in volumetric terms, the *CPVC* of the linseed oil–pigment combination is the volume of pigments divided by the total volume of pigment plus linseed oil. Factors that affect the oil absorption value of a pigment or extender include not only the chemical and mineralogical identities of the pigment but also its shape and particle size. Pigment particles that have relatively large surface areas due to nonspherical shapes or small particle sizes effectively increase the oil absorption value for a given pigment composition. Oil absorption values are often given by pigment and extender manufacturers and can be found in several coatings handbooks.

Calculation of *CPVC* for a latex paint cannot be made accurately from equation 2 because the pigment-binding efficiencies of latex emulsions differ from those of solution resins such as alkyds or linseed oil. The binding efficiency, or binding power index (BPI), of a latex is the ratio of linseed oil to latex volume solids needed to bind a given volume of pigment. It can be calculated from the experimentally determined *CPVC* of the latex having a given pigmentation, and the *CPVC* of linseed oil containing the same pigmentation calculated from the oil absorption. Calculation of the BPI of a latex emulsion is based on the following equation:

$$\text{BPI} = \frac{CPVC_b(1 - CPVC)}{CPVC(1 - CPVC_b)} \qquad (3)$$

where $CPVC_b$ is the *CPVC* of latex emulsion and pigments, and *CPVC*, the *CPVC* of linseed oil and pigments.

Since the BPI for a given latex is fairly consistent according to experiments with different pigments, it is most convenient to determine the $CPVC_b$ of a latex emulsion with an extender pigment that gives the emulsion and pigment little opacity below the *CPVC*, but provides dry hiding from air voids introduced into the film above the *CPVC*. *CPVC* can then be interpolated by plotting some measure of the hiding power for a series of paints varying in *PVC* that are either well above or below the expected critical point. This usually gives two straight lines having a clearly defined break intersecting at the *CPVC* point. The calculated binding efficiency or BPI of a latex is always less than 1.0 because of the particulate nature of the latex and its resistance to complete deformation prevents the latex from binding as well as a solution resin. Smaller particle-size latices generally have better binding efficiencies than larger-particle-size emulsions due to better packing of the smaller particles. However, this holds true only when comparing latices of identical compositions because the binding efficiency of a latex varies with composition. Experiments have shown that the easier film formation of softer latices also improves their pigment-binding efficiencies.

Pigment Dispersion and Paint Volume Solids. Another important aspect in pigment–binder relationships is the degree of dispersion of pigment particles in the binder. Pigment particles tend to agglomerate both in their dry state before manufacture and later when they are part of the paint formula. As a result, energy in the form of paint-mixing equipment is required to disperse the pigment agglomerates in the paint vehicle (binder and solvent). Along with proper mixing, chemical additives such as dispersants, surfactants, and thickeners are necessary ingredients of the paint formulation for controlling and inhibiting pigment agglomeration during and after the mixing process. If pigment agglomeration is not controlled in a paint formula and pigment particles flocculate together, the result is a lower *CPVC* for that particular formula and worse appearance properties (11). Good pigment dispersion is necessary for a paint to maximize its *CPVC* value.

A third criterion for the proper formulation of house paints is volume solids level. Paint with low volume solids are characterized by poor adhesion and poor exterior durability. A low volume solids paint film does not have the tensile strength properties necessary for good adhesion found in a higher volume solids paint (10). Also, any slight disruption or imperfection in the dried paint film or in the application of a low volume solids paint can result in poor durability properties.

Paint Processing. The manufacture of house paints involves mixing together the raw material ingredients in such a way that the finished paint is a homogeneous mixture of evenly distributed ingredients. The particular raw material ingredients used in house paints are chosen not only for their appearance and performance attributes, but also for their relative compatibility with each other and their ability to be mixed together to produce a relatively stable and homogenous paint product without significant chemical reactions during processing. There may be many interactions and chemical associations that do occur, but most paint manufacturing processes result in a 100% yield of final product, mostly due to the fact that no chemical reactions take place that can result in an unwanted chemical by-product. Therefore, the manufacture of house paints is concerned mostly with proper blending and mixing of the raw material ingredients in order to ensure that the ingredients are evenly dispersed in the finished paint.

Most house paints manufactured in the 1990s rely on high speed mixers or dispersators. Components of high speed mixers include a mixing vessel and a simple configuration of a long vertical rotating shaft attached to a high power electric motor at the top end, typically ranging from 7.46–149 kW (10–200 hp), and to an adapter for different mixing blades at the bottom end (see MIXING AND BLENDING). The pigment(s) in a paint formulation are the ingredients that require the most energy to disperse, wet out, and keep separated so that they do not agglomerate in the paint (2). The selection of the mixing blade used to disperse the pigments in the liquid vehicle is the key component choice for proper pigment-grinding when using this type of equipment. The most common design for mixing blades is a flat disk having metal teeth on the outer edges of the disk that point up and down perpendicular to the horizontal plane of the disk. The purpose of this design is that a spinning disk creates shear forces in thick liquids due to the inertial drag of one layer of material moving over another. The higher the viscosity of the liquid surrounding the disk, the higher are the shear forces being generated. The teeth, on the other hand, add circulation to the liquid in the mixing vessel. This circulation provides fresh liquid material for the spinning disk surface and ensures that all the liquid in the mixing vessel comes in contact with the spinning disk and is effectively sheared. This combination of strong shear forces developed by the spinning disk and good circulation provided by the toothed edges of the disk results in efficient pigment dispersion in the grind portion of the paint formulation.

Both water-based latex paint and solvent-based alkyd and oil paint have most manufacturing processes in common. The paint formulations in Tables 1 and 2 list the paint raw material ingredients according to the ingredients in both the grind portion of the paint and the let-down portion of the paint. The order of adding these ingredients into the paint mixing vessel essentially follows the order of ingredient listing. The grind portion of a paint formulation is typically made in the vessel by using a high speed disperser, whereas the let-down portion of the paint is typically made in another vessel that may have a smaller motor with a paddle-type mixing blade to keep the liquids stirring but not necessarily under shear. Liquids in the form of water, solvents, and/or resins are added to the grind mixing vessel first. Additives that assist pigment dispersion and wetting are added next. Pigments are then added to the liquid; the motor speed and the height of the blade in the mixing vessel are adjusted as more pigments are added. For efficient mixing it is important to have the mixing blade submerged approximately halfway in the grind volume of the liquid–pigment mixture. Once all the materials in the grind portion are added to the mixing vessel the high speed mixing should continue for 15–20 minutes to ensure complete dispersion of the pigment particles.

Grind temperature is an important factor to monitor during the paint manufacturing process, because a low grind temperature sometimes indicates that no friction is developing from shearing the ingredients but usually indicates that too much liquid has been added for the amount of pigment being dispersed. A grind temperature that is too high may indicate a high grind viscosity which can lead to an unsafe work environment and can even burn out the motor. In such a case more liquid must be added to the grind to reduce viscosity. Certain types of paint ingredients in the grind portion, such as castor-oil derivative thickeners, can also require temperature activation between a narrow range of temperatures so the grind viscosity must be balanced to accommodate these temperature ranges. The let-down portion of a paint formulation is typically made up of all-liquid ingredients that do not require high shear mixing to disperse the ingredients into a homogeneous mixture. A common paint manufacturing procedure is to make both the grind and let-down portions of a paint formulation at the same time but in different tanks and then pump the completed grind portion into the let-down portion to finish making the paint. When this is done it is best if the viscosities of the grind and let-down portions are similar so that pigment agglomeration or colloidal shock can be prevented (2). Other paint manufacturing procedures begin by adding the components of the let-down portion to the grind portion once the proper amount of mixing in the grind portion has been completed; this allows the paint formulation to be completed in a single vessel.

Although most house paints manufactured in the 1990s use high speed mixing equipment to disperse and deagglomerate pigments, other types of pigment grinding and milling equipment, such as ball mills, sand mills, and roller mills, are also used (2). A ball mill is a horizontal cylinder that contains the liquid vehicle and pigments to be dispersed, along with many ceramic or steel balls. As the horizontal cylinder rotates, the balls tumble and create shear forces on the pigment agglomerates that disperse the pigment particles. Although this process disperses pigments well, including relatively hard-to-disperse pigments, ball-milling is not popular because processing times of several hours to a few days are required for adequate mixing. Sand mills operate in a similar fashion to ball mills, but use much smaller glass or ceramic beads for grind media. Sand mills can have either horizontal or vertical configurations and generally represent smaller size grinding equipment and require less production space to operate than ball mills. Further, because of the smaller grind media used and the efficiency of the milling equipment, sand mills are able to deliver finished pigment grinds much faster than ball mills. Roller mills operate by feeding the grind portion of a paint formulation between two or more rollers closely spaced together and rotating at different speeds so that high shear forces are created to disperse the pigment particles. However, roller mills are not efficient for processing large quantities of grind mix, but are still used to make special high quality paints and inks (qv) (2).

Paint Types and End Uses

Because a paint's appearance is mostly a function of *PVC*, paints can be described in reference to their *PVC* as in Figure 1.

Gloss and semigloss paints are formulated at relatively low *PVC*s and often use only color pigments with no extenders. Gloss enamels can have *PVC*s ranging from 18–23, whereas semigloss paints can range up to 30 *PVC* or more. Terms such as semigloss, satin, and luster are used to market paints in this range, though there is no universal acceptance of the actual measured gloss range that constitutes a gloss or a semigloss paint. Gloss paints are used almost exclusively for interior and exterior doors and trim, whereas semigloss paint may be used for trim, bathroom and kitchen walls, or other surfaces that may be washed or subjected to high humidity. Satin paints are always extended and usually

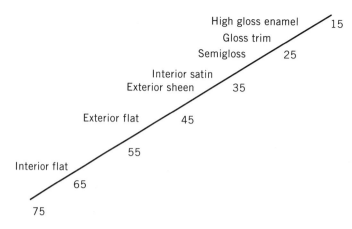

Fig. 1. Formulation types where each number represents the percent of pigment volume concentration for the paint types shown.

lower in gloss than semigloss paint made by the same manufacturer. Satin paints are mostly used for walls but can also be used for trim. They are usually more washable than flat paints since they cannot be penetrated as easily by stains, and they can better resist scrubbing due to their lower *PVC*s and greater film integrity (~30–38 *PVC*). Sheen paints are similar to satin, but use more extenders so that they give off a luster at low angles close to the surface, but appear flat in a head-on view. Sheen paints are popular as exterior paints for the body of a house, and can give a freshly painted look for many years.

Exterior House Paints. Quality exterior flat paints, which provide protection for the exterior surfaces of the house, are characterized by proper choice of binder, lower *PVC*s formulated below *CPVC*, and higher volume solids. Exterior paints are more likely to be applied at marginal temperature, or subjected to rainfall soon after the paint is applied, perhaps before the film is fully formed. It is also more likely to be applied to deteriorating surfaces, such as chalking paint, weathered wood, or corroded metal. In order to protect the substrate, the paint film must adhere to it well and have the flexibility to expand and contract as the substrate does. This is especially important in painting over wood, where the winter and summer grain expand at different rates in response to changing temperature and humidity. Grain cracking is a common mode of failure over wood, and is soon followed by flaking and peeling of the paint. In order to maintain the average water content of wood below the threshold for rotting, the paint film must prevent liquid water from entering the wood from outside the house, yet allow moisture vapor from the interior of the house to pass through. This is sometimes referred to as breathing. In order to fulfill this function, the paint film must have the proper permeability, maintain film integrity, and adhere well to the substrate. An exterior paint is subjected to ultraviolet light radiation, temperature and humidity cycling, and attack from acid rain. If any of the paint components are degraded by these conditions, colors may fade, binder may erode, chalking may occur, and the permeability of the film may increase unacceptably. Exterior paint must also resist dirt that settles on it from the atmosphere from embedding into its surface; thus is necessary not just to maintain appearance, but also to minimize the nutrients available for mildew to feed on. Mildew can cause defacement of paint, and eventually attack wood substrates. In addition to maintaining cleanliness, the paint should act as a barrier to microorganisms attacking the substrate.

Interior House Paints. Interior flat paints range in *PVC* from below to well above *CPVC*, depending on the quality and intended service. Ceiling paints, for example, are not usually scrubbed, and can be formulated well above *CPVC* to obtain good hiding at low cost. Trim paints such as semigloss and gloss paints are lower in *PVC* and usually receive the most wear and abuse on doors, windows, and surrounding trim. They must resist permanent staining, make removal of stains relatively easy, and not be marred by the washing or scrubbing necessary to remove dirt or stains. This is a principal reason that gloss and semigloss paints are used for trim. Freshly painted doors and windows should not block (painted surfaces sticking to each other) after a reasonable drying time. Objects placed on painted surfaces such as windowsills or shelves should not stick to the paint or mar the paint when removed. This requires the paint film to develop its hardness quickly. Interior paints must have good adhesion to the substrate, so that paint is not removed when washed or rubbed or if blocking occurs. Adhesion under wet conditions can present a special problem, eg, in a bathroom where a shower is used frequently. Paints that are water-sensitive can blister under such conditions. Freshly painted trim surfaces can be subject to wet conditions when wall coverings are applied to adjacent walls and overlapped onto the painted surface before trimming. The average home owner probably expects some paint odor after application but prefers lower odor paints, especially after the paint has dried for a few days. Increased consumer awareness of chemical hazards and the development of new paint technology for solvent-free water-based latex paints is likely to accelerate this trend, which is already underway in Europe.

Specialty House Paints and Finishes. Several other related types of house paint finishes not already mentioned are formulated for specific end uses or appearance characteristics. These products include primers, sealers, and opaque solid stains, and are available from most paint manufacturers in either water-based latex or solvent-based alkyd and oil. Primers are used for many reasons. First, they help improve the adhesion of the topcoat to the substrate underneath. Features of a primer include their ability to penetrate, seal, and help prepare a new or weathered substrate for topcoat application. Second, primers can help prevent stains, tannins, and moisture from wood or other building materials from passing through the finished coat of paint and ruining the finished paint job. For example, top quality primers can block out stains caused by water, grease, rust, smoke, and other substances, and they are particularly helpful when used over staining woods such as cedar and redwood. Third, the application of a primer can help give a more uniform surface, and hence a more attractive finish topcoat, to the paint job. Primers tend to be high quality paint formulations low in *PVC*s and high in volume solids. They may not provide superior hiding or coverage because these properties are not as important for primers as they are for topcoat paints. Primers may also contain specialty pigments and additives to help resist corrosion or prevent stain bleed-through.

Opaque solid stains are different from most paint formulations in that they are typically characterized by lower viscosities, lower hiding and volume solids, and lower *PVC*s. Further, they are almost always recommended for wood surfaces only. The effective life of a stain is usually less than that of a quality house paint. However, the advantage for using stains is mostly based on an appearance preference to fully cover a substrate with a color in a single coat without obscuring the texture of the surface. Also, the maintenance of a stain finish usually requires restaining only without the need for more extensive surface preparation as is required of paints. Solids stains are similar to paints in that they have only slightly lower viscosities and often the same volume solids and *PVC*s as found in a paint. Also, solids stains can be formulated to give special performance features such as water repellency and wood preservation (12).

Application Methods and Surface Preparation. For good durability and performance, proper surface preparation and correct application of house paints are as important as the formulation of high quality paint. Proper surface preparation prior to painting involves several considerations. For new construction, proper installation and protection of the substrate material are necessary. For previously painted surfaces, preparation involves mostly cleaning and removing any existing paint that is unstable. Once surface preparation is complete, the application process can begin.

The most common application tools for house painting are brushes, spray equipment, rollers, and pad applicators. The choice of application equipment is largely based on use preference and the paint manufacturers' recommendations. When applying paint it is most important to apply it uniformly over the whole surface of the substrate. If paint is applied excessively, sparingly, or unevenly, both durability and performance can be adversely affected and nonuniform or premature paint failures may result. A coat of paint applied too liberally results in slow drying, sagging, and film wrinkling. Conversely, a light application of paint does not provide the film thickness necessary for the paint to protect and cover the substrate and maintain good durability, appearance, and adhesion properties.

Analysis and Testing of Paints

Quality Control Testing. After a paint formulation is manufactured and processed, several quality control tests are commonly used to ensure that the paint complies with the desired formula specifications and that it was made properly. These tests include fineness of grind, viscosity, weight per gallon, percent weight solids, pH of water-based paints, color and appearance characteristics, and volatile organic compound (VOC) content. The fineness-of-grind test measures how well and to what degree the pigments are dispersed in the paint vehicle. This test, which ensures that the pigment particles have been properly dispersed in the paint, is commonly measured on the grind portion of the paint and not on the completed paint batch. In ASTM D1210, the fineness of grind is measured by applying a small sample of the grind portion of a paint to a special gauge, such as the popular Hegman gauge, which contains a grooved channel that goes from a depth of 0–101.6 μm (0–4 mils). The small sample of paint is placed in the deepest part of the channel and spread along it with a scraper until the channel ends at the gauge surface. A visual inspection is then made of the paint in the gauge channel to see at what point on the scale the pigment particles become visible. The results of the Hegman reading should correspond to the known particle sizes of the pigments used in the paint. Unfortunately, this method for determining the fineness of pigment dispersion shows only how well the coarsest particles are dispersed.

Viscosity of paint samples are most often measured with the use of a Stormer viscometer. The procedure for using this viscometer follows ASTM D652. The instrument is a paddle-type viscometer that measures viscosity by correlating a weighted load on a pulley apparatus that is required to rotate the paddle 200 rpm in a paint sample having a viscosity reading reported in Krebs units (KU). The viscosity of paints can change depending on the amount of shear applied to the paint, eg, some paints may seem thick when stirred gently (low shear) but become thinner when stirred vigorously (high shear). The Stormer viscometer produces moderate shear during sample analysis and thus is an effective tool for evaluating the handling and application properties of house paints. Other types of viscometers commonly used in the paint industry include the Brookfield viscometer for measuring very low shear viscosities of paints which correlate to the standing or storage viscosity of the paint. Another popular instrument is the ICI viscometer, used to measure high shear or brushed application viscosity of paints.

Historically, color and appearance measurements of paint were made by human visual inspection only. Although visual inspection is still important in the final analysis of a paint's color and appearance, modern paint laboratories rely on computer color-measuring instruments and gloss meters to provide greater accuracy and more objective quality control of these properties. The measurement and assessment of color are made with instruments known as spectrophotometers and colorimeters. The assessment of a paint color and its acceptability is usually compared to a color standard which is required for the paint to match. ASTM D2244 outlines a method for measuring the color difference between a paint sample and a color standard by using color-measuring instruments; such instruments basically consist of a view area to which a paint sample is mounted, a filtered light that illuminates the sample, various optical lenses of clear or colored glass, and photodetectors that measure the intensity of the light reflected off or transmitted through the sample. A color spectrophotometer measures the complete spectral range of visible light, whereas a colorimeter measures the light intensity that filters through a colored glass. Results from color-measuring instruments are automatically given in CIELAB and/or Tristimulus (X, Y, Z) values (see COLOR). If the resulting color measurement value of a paint differs from the color standard that it is required to match, color-matching software can be used with the color-measuring instruments to calculate the exact amounts of colorants that need to be added to the paint for an acceptable color match.

The measurement of gloss and sheen is also an important part of appearance measurement. Glossmeters work by sending an incident beam of light to the surface of the paint sample. An amplified photodetector opposite the light source receives a portion of the light reflected off by the sample. The amount of light received by the photodetector is the gloss measurement. The gloss reading is taken by placing the glossmeter over a flat surface of the paint sample to be measured. Glossmeters are available in different geometries such as 20°, 60°, and 85°, which represent the angle of incidence of the light beam and photodetector to the surface being measured. A 20° glossmeter is best for assessing the appearance characteristics of high gloss paints, a 60° glossmeter is best for semigloss and satin paints, and an 85° glossmeter is best for the sheen of flat paints. The ASTM test method for measurement of specular gloss is D523.

The volatile organic compound (VOC) content of a paint is an important analytical test for compliance to various state and federal regulations which limit the amount of volatile organic materials that can be used in a paint. The VOC content of a paint is reported as the weight of total volatile organics in a specific volume of solid coating, excluding any water present in the formulation. VOC content is normally expressed in terms of pounds per gallon or grams per liter. The most comprehensive guide to the methodology of testing the VOC content of paints can be found in the ASTM *Manual on the Determination of Volatile Organic Compounds in Paints, Inks, and Related Coating Products* (13). This manual includes not only all related ASTM test methods for measuring and calculating the VOC content of coatings but also specific ASTM test methods designated by the U.S. Environmental Protection Agency for determining and reporting VOC levels in paints. Also included in this manual is information related to the expected accuracy of the test methods for providing repeatable and reproducible results as well as common errors that can occur in the measurement and determination of VOC.

Paint Performance Testing. The types and methods of performance testing on house paints are related mostly to the recommended end use of the type of paint being developed, ie, interior or exterior use. Most tests that have been developed in the paint industry focus on simulating typical end use environment of paints. However, the paint industry is also finding greater use of chemical analytical instrumentation for both evaluating and predicting paint performance and other characteristics of paints (14). The common performance tests for an exterior paint include evaluating hiding and appearance, package stability, color retention or fade resistance, good adhesion to a substrate, resistance to chalking, flexibility to resist cracking and flaking, water and ultraviolet light resistance, mildew resistance, resistance to surfactant leaching, and alkali resistance when applied over masonry surfaces. These properties are best evaluated by exposing the paints outdoors in locations and environments that can produce the desired weather conditions for the assessment of paint performance. Sometimes this process can be time-consuming, lasting several years before the exterior durability properties of an exterior house paint are fully understood. To accelerate this process, artificial weathering test equipment is available to simulate the degradative effects of natural outdoor weathering conditions. The advantages of this testing method are the ability to control the exposure test conditions and to minimize the time necessary to get weathering results from exposure of paint samples. However, artificial weathering cannot exactly match natural outdoor weather conditions and does not simulate certain effects of weathering, such as air pollution, mildew growth, and wind.

The characteristics of interior paints that require testing and analysis include hiding and appearance, package stability, adhesion, spatter resistance, flow and leveling, color and sheen uniformity, touch-up, stain removal, burnish resistance, and block and print resistance. A popular test that assesses the wet abrasion resistance of an interior paint is to measure its scrub resistance. A mechanical device is used to scrub a paint film of a specified thickness with a standard brush and abrasive cleanser suspension. The number of scrub cycles (back and forth movements of the weighted brush) at various end points (first cut through, or 50% removal of the film) is then recorded. Scrub resistance usually holds steady or decreases slightly as *PVC* is increased, but drops quickly once the *CPVC* is exceeded in a paint formulation.

Economic Aspects

For 1993, the U.S. paint and coatings industry had total sales of $13 billion and total shipments of 1.1 billion gallons of product. Of these totals, the architectural coatings house paint segment had total sales of more than $5.5 billion and total shipments of 604 million gallons. According to these figures, architectural coatings account for 55% of all paint shipments but only 43% of total paint sales. Other analyses of this data show that the average producer cost per gallon of paint equals $11.82/gal, and the average producer cost of architectural house paint is $9.19/gal. Retail consumers, on average, pay $13.25/gal for architectural paint. Below are five-year industry figures for architectural coatings shipments in terms of total gallons, dollars, and average cost per gallon (13). To convert gal to L, multiply by 3.785; to convert $/gal to $/L, divide by 3.785.

Total shipments	1989	1990	1991	1992	1993
gal, in millions	540	560	540	575	600
$, in billions	4.55	4.9	4.85	5.25	5.55
$ per gallon	8.4	8.8	9.1	9.1	9.2

In the 1990s, the price of paint and raw material costs have been growing at a rate lower than that of inflation, and increases in the price of paints have been slim. To control costs, paint manufacturers have generally held industry employment at static or slightly lower levels. In 1993, U.S. paint industry employment stood at 58,200, compared to 61,200 in 1990; the total number of paint manufacturing companies is estimated to be between 800 and 900. Information shows that the industry trend for larger companies increasing in size as the result of mergers and acquisitions has continued. In the early 1980s, it was estimated that 1300 paint companies were producing paint in the United States (15).

Environmental, Health, and Safety Factors

The most significant environmental and health issues affecting the paint and coatings industry in the 1990s are regulations to lower the VOC content for virtually all types of paints and to restrict the use of certain solvents known as hazardous air pollutants (HAPs) under the federal Clean Air Act. Except for the water in a latex paint or in other water-based coatings, solvents used in house paints are mostly all VOCs. Several states, along with the U.S. EPA, have implemented environmental regulations to restrict the VOC content of paints, as mandated by the Clean Air Act. These regulations are aimed at minimizing the emission of organic compounds from paints that contribute to the formation of air pollution in the form of smog or ground-level ozone.

Health and safety issues affect both the professional painter and paint manufacturer who as part of the occupation can be exposed to high concentrations of organic solvent for extended periods of time. Environmental issues focus on the contribution of organic solvents to air pollution and other issues such as hazardous waste disposal. Paint companies are reducing their use of organic solvents in the manufacture and development of architectural coatings by offering more water-borne and higher solids alternatives to conventional solvent-borne paints. Some paint manufacturers are offering solvent-free latex paint alternatives to water-borne paints that contain organic solvents. A typical solvent-borne paint may contain 30–60% of organic solvent. By contrast, a water-based paint may only contain 5–10% of organic solvent. This is a significant reduction in solvent content, as water-borne paints are becoming more and more predominant in the architectural coatings market, and are part of the focus for reducing the use of organic solvents in paints.

Restriction on the use of certain types of solvents, listed as HAPs under the Clean Air Act, are forcing paint manufacturers not only to lower the limits on the amount of organic solvents in a paint, but also to eliminate certain types of solvents. Thus paint manufacturers are challenged to comply simultaneously with both VOC and HAP regulations. These Clean Air Act mandates are expected to affect most types of paints and paint manufacturers beginning in 1996.

Another issue affecting the architectural paint industry is the remediation of homes, buildings, and structures that contain lead-based paint. Lead poisoning in children has been linked to ingestion of paint dust or paint chips that contain lead pigments and this has resulted in U.S. government regulations to reduce the lead content in paint to no more than 0.06%. This restriction essentially bans the use of lead-containing materials in paints, with the exception of lead impurities in the water and minerals used in paints. Prior to these government regulations, most paint companies had already phased out their use of lead pigments. However, homes and other structures of older construction (pre-1960) may still contain lead paint (16). Efforts in some states have required the removal of lead-containing paints in homes and buildings inhabited by young children (see PAINT AND FINISH REMOVERS). However, lead paint removal is expensive and in some cases has exacerbated the problem as lead-containing dust or chemicals are generated during the removal process. As a result, remediation efforts have begun to allow the use of lead paint encapsulants to coat over lead paint.

BIBLIOGRAPHY

"Paint" in *ECT* 1st ed., Vol. 9, pp. 770–803, by H. E. Hillman, Eagle Paint & Varnish Corp., W. G. Vannoy, E. I. du Pont de Nemours & Co., Inc., and R. Carlisle and L. K. Doraiswamy, R. L. Carlisle Chemical & Manufacturing Co., Inc., in *ECT* 2nd ed., Vol. 14, pp. 426–485, by D. Cannell, The Sherwin-Williams Co.; in *ECT* 3rd ed., Vol. 16, pp. 742–761, by G. G. Schurr, The Sherwin-Williams Co.

1. S. LeSota, *Paint/Coatings Dictionary*, Federation of Societies for Coatings Technology, Blue Bell, Pa., 1978.

2. A. Brandau, *Introduction to Coatings Technology*, Federation of Societies for Coatings Technology, Blue Bell, Pa., 1990.

3. J. Boxall and J. VonFraunhofer, *Paint Formulation*, Industrial Press, New York, 1981.

4. *Am. Paint. Cont.* **72**, 1 (1995).

5. J. H. Braun, *Prog. Org. Coat.* **15**, 249 (1987).

6. W. H. Ellis, *Solvents*, Federation of Societies for Coatings Technology, Blue Bell, Pa., 1990.

7. F. W. Billmeyer and M. Saltzman, *Principles of Color Technology*, 2nd ed., John Wiley & Sons, Inc., New York, 1981.

8. D. Judd and G. Wyszecki, *Color in Business, Science, and Industry*, 3rd ed., John Wiley & Sons, Inc., New York, 1975.

9. T. C. Patton, *Paint Flow and Pigment Dispersion*, 3rd ed., John Wiley & Sons, Inc., New York, 1979.

10. M. B. Freedman, *Paint Coat. Ind.* **1**, 2 (Mar./Apr. 1985).

11. W. A. Asbeck, *J. Coat. Tech.* **49**, 635 (Dec. 1977).

12. D. L. Cassens, and W. C. Feist, *USDA Forest Service Handbook*, 647 (May 1986).

13. J. J. Brezinski, *Manual on the Determination of Volatile Organic Compounds in Paints, Inks, and Related Coating Products*, 2nd ed., ASTM Manual Series, MNL4, American Society for Testing and Materials, Philadelphia, Pa., 1992.

14. W. C. Golton, ed., *Analysis of Paints and Related Materials*, ASTM Publication STP 1119, American Society for Testing and Materials, ASTM Publication STP 1119, American Society for Testing and Materials, Philadelphia, Pa., 1992.

15. M. S. Reisch, *Chem. Eng. News* **72**, 40 (Oct. 1994).

16. *Dealing with Lead-Based Paint: A Practical Guide for Consumers*, National Paint and Coatings Association (NPCA), Washington, D.C.

General References

Ref. 1 is also a general reference.

Federation Series on Coatings Technology, Federation of Societies for Coatings Technology, Blue Bell, Pa.

Journal on Coatings Technology, Federation of Societies for Coatings Technology, Blue Bell, Pa.

Annual Book of ASTM Standards, Section 6: Paints and Related Coatings, American Society for Testing and Materials, Philadelphia, Pa.

ARTHUR A. LEMAN
Rohm and Haas Company

PAINT AND FINISH REMOVERS

The term finish denotes the final process of manufacturing. Finishing operations include such processes as clear coating (varnishes and lacquers), painting, plating, anodizing, phosphatizing, galvanizing, and blueing, all of which take place at the terminal point of manufacturing. Finishing is defined as the process of coating or treating a surface for the purpose of protecting and/or decorating the product (1). The useful life of most usable objects is greater than the finish. This results in a periodic need to remove and replace the finish. Many times the appearance of the item is outdated and the finish is removed to allow a more modern finish to be applied. An organic or inorganic chemical solution can be used to lift or remove the finish as well as a mechanical process such as media blasting.

Antiques, household furniture, kitchen cabinets, pianos, aircraft, and buildings can have their useful life extended by removing the old coating and applying new. Original equipment manufacturers (OEMs) remove coatings from rejected parts to avoid scrapping the items. Finish removers are used to remove lead paint from woodwork, windows, or entire buildings to abate the risk of lead exposure. There are over 104 different industries that use finish removers (2). The use of and need for finish removers will probably expand with the increasing importance of recycling (qv), refinishing, and the restoration of durable items.

Components of Finish Removers

The physical properties of finish removers vary considerably due to the diverse uses and requirements of the removers. Finish removers can be grouped by the principal ingredient of the formula, method of application, method of removal, chemical base, viscosity, or hazardous classification. Except for method of application, a paint remover formulation usually has one aspect of each group, by which it can be used for one or more applications. A list of the most common organic solvents used in finish removers has been compiled (3). Many are mentioned throughout this article; others include ethyl lactate [97-64-3], propylene carbonate [108-32-7], furfural alcohol [98-01-1], dimethyl formamide [68-12-2], tetrahydrofuran [109-99-9], methyl amyl ketone [110-43-0], dipropylene glycol methyl ether [34590-94-8], and Exxate 600, a trade name of Exxon Chemicals.

Finish removers are applied by brushing, spraying, troweling, flowing, or soaking. Removal is by water rinse, wipe and let dry, or solvent rinse. Removers may be neutral, basic, or acidic. The viscosity can vary from water thin, to a thick spray-on, to a paste trowel-on remover. The hazard classification, such as flammable or corrosive, is assigned by the U.S. Department of Transportation (DOT) for the hazardous materials contained in the remover.

Major Ingredients.   The major ingredient of a formula is the chemical of greatest volume. Current formulas include approximately five major ingredient formulas: methylene chloride [75-09-2], petroleum base, oxygenates, aqueous alkaline, and other organic blends. Pure methylene chloride is only a fair remover. When mixed with methanol [67-56-1] a synergistic effect is produced that results in a solvent blend with removal characteristics far superior to either of the solvents when used independently. The petroleum base removers include ingredients such as toluene [108-88-3] and monochlorotoluene [25168-05-2]. Acetone [67-64-1], an oxygenate, is an aggressive finish remover and may be the major ingredient. Aqueous alkaline removers are used in soak tanks or as a paste. The major ingredients are sodium hydroxide [1310-73-2] or potassium hydroxide [1310-58-3].Other remover blends include solvents such as glycol ethers, N-methylpyrrolidinone [872-50-4], and esters such as propylene glycol methyl ether [107-98-2], dimethyl adipate [627-93-0], and ethyl 3-ethoxypropionate [763-69-9].

Cosolvents and Diluents. Cosolvents and diluents are solvents that can be added to a remover to expand the stripability or to reduce the cost of raw materials without unacceptably compromising the performance of the product. Mineral spirits [8052-41-3] and other hydrocarbon solvents are often used as diluents. Cosolvents are solvents that may not be a significant remover by themselves, but when added to a major-ingredient solvent increase the stripping efficiency of the mixture. Small amounts of alcohol and toluene are commonly used in methylene chloride removers and act as both cosolvents and diluents. In N-methylpyrrolidinone removers, middle molecular weight alkyl aryl compounds are important as dilutants because of the extreme cost of the major ingredients. They are also significant as cosolvents, since the maximum stripping efficiency is reached when the mole percentage of N-methylpyrrolidinone is between 40 and 50% (4).

Evaporation Retardants. Small molecule solvents that make up the most effective paint removers also have high vapor pressure and evaporate easily, sometimes before the remover has time to penetrate the finish. Low vapor pressure cosolvents are added to help reduce evaporation. The best approach has been to add a low melting point paraffin wax (mp = 46–57°C) to the paint remover formulation. When evaporation occurs the solvent is chilled and the wax is shocked-out forming a film on the surface of the remover that acts as a barrier to evaporation (5,6). The addition of certain esters enhances the effectiveness of the wax film. It is important not to break the wax film with excessive brushing or scraping until the remover has penetrated and lifted the finish from the substrate. Likewise, it is important that the remover be used at warm temperatures, since at cool temperatures the wax film may not form, or if it does it will be brittle and fracture. Rapid evaporation occurs when the wax film is absent or broken.

Emulsifiers. Removing the remover is just as important as removing the finish. For water rinse removers, a detergent that is compatible with the remover formula must be selected. Many organic solvents used in removers are not water soluble, so emulsifiers are often added (see EMULSIONS). Anionic types such as alkyl aryl sulfonates or tolyl fatty acid salts are used. In other applications, nonionic surfactants are preferred and hydrophilic–lipophilic balance is an important consideration.

Activators. Activators are often added to removers to make them more efficient. Acids such as phenol [108-95-2], phosphoric acid [7664-38-2], acetic acid [64-19-7], formic acid [64-18-6], and citric acid [5949-29-1] are used to increase the cutting ability on epoxide-type paints and other modern finishes. Strongly alkaline activators are effective on enamel and latex paints. Other activators include ammonia [7664-41-7], monoethylamine [75-04-7], and N-phenyldiethanolamines. Acid and base activators shorten the shelf life of some removers.

Thickeners. Thickeners are added to remover formulas to increase the viscosity which allows the remover to cling to vertical surfaces. Natural and synthetic polymers are used as thickeners. They are generally dispersed and then caused to swell by the addition of a protic solvent or by adjusting the pH of the remover. When the polymer swells, it causes the viscosity of the mixture to increase. Viscosity is controlled by the amount of thickener added. Common thickeners used in organic removers include hydroxypropyl methylcellulose [9004-65-3], hydroxypropylcellulose [9004-64-2], hydroxyethyl cellulose, and poly(acrylic acid) [9003-01-4]. Thickeners used in aqueous removers include acrylic polymers and latex-type polymers. Some thickeners are not stable in very acidic or very basic environments, so careful selection is important.

Corrosion Inhibitors. Corrosion inhibitors are added to the formula to promote packaging stability. If the formula is acidic and will be used to remove finishes from nonferrous metals, inhibitors are incorporated to protect the substrate.

Organic Finish Removers

METHYLENE CHLORIDE FINISH REMOVERS

Methylene chloride formulas are the most common organic chemical removers. The low molar volume of methylene chloride allows it to rapidly penetrate the finish by entering the microvoids of the finish. When the solvent reaches the substrate, the remover releases the adhesive bond between the finish and the substrate and causes the finish to swell. The result is a blistering effect and an efficient rapid lifting action. Larger molecule solvents generally cannot cause this lifting action and must dissolve the finish. When methylene chloride is used in amounts of 78% or more, even with flammable cosolvents, the mixture is nonflammable. A typical methylene chloride base remover includes cosolvents, activators, evaporation retarders, corrosion inhibitors, thickeners, and wetting agents.

Typical cosolvents include methanol [67-56-1], ethanol [64-17-5], isopropyl alcohol [67-63-0], or toluene. The selection of cosolvents depends on the requirement of the formula and their interaction with other ingredients. Methanol is a common cosolvent in methylene chloride formulas since it has good solvency and is needed to swell cellulose-type thickening agents. A typical methylene chloride formula used to strip wood is as follows (7).

Ingredient	Wt %
methylene chloride	81.1
toluene	2.1
paraffin wax (ASTM 50–53°C mp)	1.6
methycellulose	1.2
methanol	7.8
mineral spirits	6.2

The rate of stripping or the stripability on catalyzed urethane and epoxy resin finishes can be increased by adding formic acid, acetic acid, and phenol. Sodium hydroxide, potassium hydroxide, and trisodium phosphate [10101-89-0] may be added to the formula to increase the stripability on enamel and latex paints. Other activators include oleic acid [112-80-1], trichloroacetic acid [76-83-9], ammonia, triethanolamine [102-71-6], and monoethylamine. Methylene chloride-type removers are unique in their ability to accept cosolvents and activators that allow the solution to be neutral, alkaline, or acidic. This ability greatly expands the number of coatings that can be removed with methylene chloride removers.

Paraffin wax vapor barriers are used in water rinse removers that can disperse the wax without coating the substrate. In soak tank applications, water is sometimes floated on top of an all-solvent, neutral pH, nonwater rinse remover to prevent evaporation. Flotation devices that cover the exposed surface area may be used with other formulas.

Corrosion inhibitors are used to protect both the container and the metal substrate being stripped. Acid activated removers use inhibitors to block corrosion on active metals. Typical inhibitors are propylene oxide [75-56-9], butylene oxide [9106-88-7], triethylammonium phosphates, and sodium benzoate [532-32-1] (see CORROSION AND CORROSION CONTROL).

A cellulose ether thickener such as hydroxypropyl methylcellulose can be added to thicken a methylene chloride remover. Removers that are acid or alkaline activated require thickeners that will not be hydrolyzed by the acid or the base.

A detergent that is compatible with the remover formula must be developed for water rinse removers. Anionic or nonionic surfactants should be selected, depending on the pH and intended application of the remover.

Health and Safety. Remover formulas that are nonflammable may be used in any area that provides adequate ventilation. Most manufacturers recommend a use environment of 50–100 parts per million (ppm) time weighted average (TWA). The environment can be monitored with passive detection badges or by active air sampling and charcoal absorption tube analysis. The vapor of methylene chloride produces hydrogen chloride and phosgene gas when burned. Methylene chloride-type removers should not be used in the presence of an open flame or other heat sources such as kerosene heaters (8).

Persons exposed to methylene chloride removers should wear protective clothing and eye protection. Glove selection varies with the components of the formula. Thirty mils (~0.75 mm) natural rubber gloves with laminated polyethylene and ethylene–vinyl alcohol glove liners are normally worn for industrial uses (9). Extended skin contact is usually limited during use because of the immediate discomfort when the remover touches the skin. The area should be adequately ventilated as recommended by the manufacturer. Canister respirators should not be used for methylene chloride protection because the charcoal in the canister is rapidly saturated. Because the permitted exposure level is lower than the level at which humans can detect the odor, the worker cannot tell when the canister is no longer effectively cleaning the air, resulting in exposure to unsafe levels. Fresh air-supplied respirators should be used in areas that cannot be properly ventilated. Tests indicate that long-term overexposure to methylene chloride vapors causes cancer in test animals; further tests are in progress.

Environmental Impact. Methylene chloride is nonphotochemically reactive and is not listed as an ozone (qv) depleter. Methylene chloride removers can easily be recovered from paint chips and other residue sedimentation, thus allowing recovery of remover and its continued use. This greatly increases the useful life of the remover and, when mixed with fresh remover, eliminates the need for disposing of the used remover. This process requires no special recovery equipment. The high volatility of methylene chloride allows the waste residue from the stripping process to be easily dried. The resulting waste is normally considered hazardous because of the amounts of heavy metals from old finishes.

PETROLEUM AND OXYGENATE FINISH REMOVERS

Many older finishes can be removed with single solvents or blends of petroleum solvents and oxygenates. Varnish can be removed with mineral spirits, shellac can be stripped with alcohols, and lacquers can be removed with blends of acetates and alcohols (lacquer thinners). The removal mechanism is one of dissolving the coating, then washing the surface or wiping away the finish. This method is often used to reamalgamate or liquefy old finishes on antique items of furniture.

In petroleum and oxygenate finish removers, the major ingredient is normally acetone, methyl ethyl ketone [78-93-3], or toluene. Cosolvents include methanol, n-butanol [71-36-3], sec-butyl alcohol [78-92-2], or xylene [1330-20-7]. Sodium hydroxide or amines are used to activate the remover. Paraffin wax is used as an evaporation retarder though its effectiveness is limited because it is highly soluble in the petroleum solvents. Cellulose thickeners are sometimes added to liquid formulas to assist in pulling the paraffin wax from the liquid to form a vapor barrier or to make a thick formula. Corrosion inhibitors are added to stabilize the formula for packaging (qv).

Wetting agents are used to make a water rinse remover. Water rinse removers are normally used for removing paint, where the surfactants help remove paint and remover from the substrate. Solvent rinse removers or wipe and dry formulas may be used for stripping clear finishes. A typical petroleum and oxygenate formula is as follows (10).

Ingredient	Wt %
toluene	21
acetone	19
alkyl acetate	31
methyl ethyl ketone	19
butyl alcohol	10

This is a liquid scrape-off remover for brush or soak applications. Clean up is with a solvent that is compatible with finish to be used, or wipe and dry.

Health and Safety. Petroleum and oxygenate formulas are either flammable or combustible. Flammables must be used in facilities that meet requirements for hazardous locations. Soak tanks and other equipment used in the removing process must meet Occupational Safety and Health Administration (OSHA) standards for use with flammable liquids. Adequate ventilation that meets the exposure level for the major ingredient must be attained. The work environment can be monitored by active air sampling and analysis of charcoal tubes.

Extreme caution must be taken to prevent the possibility of fire when using flammable removers. Extra care must be taken when stripping on location to secure the area of ignition sources. When used on lacquer finishes, the dissolved finish and remover combined are extremely flammable. Natural rubber, neoprene, or other gloves suitable for use with the remover formula must be worn. The effect of skin contact with the remover is limited because there is immediate irritation and discomfort. Canister respirators are available for most petroleum and oxygenate remover solvents. Symptoms of long-term overexposure should be compared to symptoms of the major ingredients in the formula.

Environmental Impact.   Most petroleum and oxygenate removers are photochemically reactive and classed as volatile organic compounds (VOCs). Air districts, such as California South Coast Air Quality Management District (Ca SCAQMD), restrict the use of these products. There are statewide regulations for VOCs in New Jersey and Rhode Island, and several states are developing their own statewide regulations. Recovery of the remover after use is difficult because the finish is resolubilized by the remover. Disposal of this type of remover is difficult because the dissolved finish cannot be separated from the spent remover and the whole mixture must be disposed as a liquid hazardous waste. Distillation to recover the solvents is dangerous because the nitrocellulose from lacquer finish may cause autoignition in the still (11).

OTHER ORGANIC FINISH REMOVERS

Concerns over the reported toxicity and carcinogenicity of methylene chloride have stimulated research for alternative solvents in remover formulas. N-Methylpyrrolidinone [872-50-4] and dibasic esters (dimethyl glutarate [1119-40-0] or dimethyl adipate [627-93-0]) have been used in removers. They remove single-component finishes but work much more slowly than methylene chloride, petroleum, and oxygenate group removers. They have little success on epoxy and catalyzed finishes. The area of greatest use appears to be as special-purpose removers for known finishes that are compatible with the remover formula. Cosolvents for N-methylpyrrolidinone removers include diacetone alcohol [123-42-2], γ-butyrolactone [96-48-0], and ethyl 3-ethoxypropionate [763-69-9]. Monoethanolamine [141-43-5], citric acid [5949-29-1], formic acid [64-18-6], and acetic acid [64-19-7] are used as activators. Strong bases such as sodium hydroxide [1310-73-2] or sodium methoxide [124-41-4] may not be used since they react with N-methylpyrrolidinone causing oxidation and polymerization. Alkyl aryl sulfonates, tolyl fatty acids, and varieties of nonionic surfactants are used in water rinse formulas. Co-solvents for dibasic ester finish removers include N-methylpyrrolidinone, aromatic 150, cyclohexanone [108-94-1], acetone, diacetone alcohol, and methanol. The same surfactants and activators are used with dibasic ester removers as with N-methylpyrrolidinone removers. A typical N-methylpyrrolidinone formula is as follows. It is a semipaste, water rinse finish remover.

Ingredient	Wt %
N-methylpyrrolidinone	27.5
ethoxylated alkylphenol	2.0
N-butanol	2.0
triethanolamine	4.0
dibasic ester	18.5
aromatic 100	18.0
cyclohexanone	17.0
xylene	8.0
masking agent	1.0
phosphate ester	1.0
thickener	1.0

Health and Safety. Both N-methylpyrrolidinone and dibasic esters have very low vapor pressure which limits worker exposure to vapors. Manufacturers recommend that the same safety precautions be taken as with other organic solvents. Hazardous location requirements must be considered if the formula is flammable. Ventilation that reduces vapors to manufacturer's recommended exposure levels should be used.

Protective clothing must be worn during use. Natural rubber or neoprene gloves should be worn with dibasic ester removers and ~0.75-mm (30 mils) natural rubber or neoprene with laminated polyethylene and ethylene–vinyl alcohol liners should be worn with N-methylpyrrolidinone. Since both dibasic ester and N-methylpyrrolidinone have low vapor pressures, the off-gassing period on porous substrates like wood is much longer. N-Methylpyrrolidinone is transdermal and can cause skin and nerve damage due to exposure from working on an item before the solvents have been released. Tests are also being conducted to determine if N-methylpyrrolidinone may be a subchronic toxin. Symptoms of long-term overexposure to the formulation should be compared to the symptoms caused from overexposure to the major ingredients of the formula.

Environmental Impact. The volume of waste remover from these products is remarkably increased when compared to methylene chloride, petroleum, and oxygenate removers, since both N-methylpyrrolidinone and dibasic esters have low vapor pressures. Recovery of the remover after use is difficult because the finish is resolubilized by the remover. A representative dibasic ester formula appears below for a thickened water rinse finish remover.

Ingredient	Wt %
dibasic ester	40
N-methylpyrrolidinone	15
aromatic 150 solvent	35–38
monoethanolamine	2
potassium oleate	4
thickener	0.8–4

Inorganic Finish Removers

LIQUID ALKALINE REMOVERS

This group consists of alkaline materials that are dissolved in water then heated to an appropriate temperature to remove finishes. In a typical application, a hot water bath large enough to submerge an item is used. The tank is heated from 43 to 74°C for stripping metals. Various alkaline materials may be used to provide the desired alkalinity. Of these, sodium compounds are preferred, such as sodium hydroxide [1310-73-2], sodium carbonate [5968-11-6], sodium silicates, mono-, di-, and trisodium phosphates, tetrasodium pyrophosphate [7722-88-5], and sodium tripolyphosphate [7758-29-4]. Compounds of other metals, such as potassium or lithium, may be used. Activators such as phenol, gluconic acid [526-95-4], and alkali metal gluconate are dissolved into a suitable solvent that itself can be dissolved in and added to the caustic solution. Surfactants are added to aid in removal of the solution and increase the penetration of the finish. Various suitable surfactants of the anionic, nonionic, and cationic types may be used, provided they are soluble and effective in the alkaline stripping solution. A typical aqueous alkaline remover formula is as follows.

Ingredient	Wt %
sodium hydroxide	87.0
sodium carbonate	6.0
sodium hydrogen phosphates	3.0
sodium gluconate	3.0
phosphate ester	0.5
pine oil	0.5

This aqueous alkaline remover is used for stripping the finish from wood or ferrous metals at a mix ratio of 30–600 g/L (0.25–5 lbs/gal).

PASTE-TYPE ALKALINE REMOVERS

Sodium hydroxide, potassium hydroxide, or other caustic compounds are blended to make these types of removers. Polymer-type thickeners are added to increase the viscosity that allows the remover to be applied with a brush, trowel, or spray. Some of these products use a paper or fabric covering to allow the remover finish mixture to be peeled away. The most common application for this group of removers is the removal of architectural finishes from the interior and exterior of buildings. The long dwell time allows for many layers of finish to be removed with one thick application of remover.

Fused State Baths. Sodium hydroxide and salt can be heated to a fused state in baths to allow the removal of finishes from ferrous metals. The most common use of this method is the removal of heavy concentrations of paint on conveyer parts and hangers used in production spray systems.

Health and Safety. Protective clothing that is compatible with the remover formula must be worn. Caustic soda baths should be ventilated to remove vapors from the work area. Most caustic removers are corrosive and cause severe burns with minimal contact to the skin. Canister respirators that are compatible with the remover should be worn.

The liquid from spent caustic soda baths must be disposed of or treated as a hazardous waste. The finish residue may contain heavy metals as well as caustic thus requiring treatment as a hazardous waste.

Manufacturing and Processing

Finish removers are manufactured in open or closed kettles. Closed kettles are preferred because they prevent solvent loss and exposure to personnel. To reduce air emissions from the solvents, condensers are employed on vent stacks. Mild steel or black iron kettles are used for neutral or basic removers; stainless steel (316 or 317) or reinforced polyethylene kettles are used for acidic removers. The kettles are heated to increase dispersion of paraffin waxes and aid in the mixing of other ingredients. Electric or air driven motors drive either sweeping blade or propeller mixers that give sufficient lift to rotate and mix the liquid. Dispenser-type mixers are used to manufacture thick and viscous removers. Kettle, fittings, mixer, and fill equipment must be fabricated with materials resistant to the chemicals in remover formulas.

Standard 0.25 or 0.50 lb (227 g) tin coated cans are used for packaging liquid with neutral and mildly alkaline base formulas; polypropylene is used for acid–base removers. Steel and polypropylene drums are used for industrial removers. Viscous removers are packaged in removable top containers. Dry caustic removers are packaged in bag-lined boxes or fiber drums.

The DOT has established standards for the packaging and labeling of hazardous materials offered for shipment by public transportation. The Consumer Product Safety Commission (CPSC) has set standards for retail labeling and packaging. OSHA and the U.S. EPA have labeling requirements. States such as California have adopted their own independent standards. Manufacturers should consult with regulatory agencies and/or trade associations before marketing finish removers.

Test Methods

Quality control in the production of organic solvent finish removers may be done by gas–liquid chromatography, which allows the manufacturer to determine the actual ratio of volatile solvent present in the finished product. If the product does not meet specifications, solvents can be added to bring the product to an acceptable composition. A less expensive approach is to use a hydrometer to determine the specific gravity of the product. The specific gravity indicates if the proper blend has been reached. Nonaqueous acid–base titration may be used to determine the amount of acid or alkaline activator present in a remover.

The solvent ratio of a semipaste remover may also be analyzed by gas–liquid chromatography by separating the solvents from the thickener. It is also useful to determine the viscosity and flow characteristics of the semipaste remover. A Brookfield viscometer is effective in determining the viscosity of most semipaste removers. Flow characteristics may be determined by a constantometer.

Economic Aspects

The demand for finish removers is increasing due to the profitability of recycling and refinishing durable goods. As the cost of new goods has steadily increased, many items that were previously disposed of are being recycled. The concern of lead poisoning from lead base paints will create further demand for finish removers in the housing industry.

Few data are available on the amounts of finish remover production except for methylene chloride removers. Methylene chloride consumption by finish removal groups is estimated in Table 1.

Table 1. Consumption of Methylene Chloride by Finish Removal Groups[a]

Application groups	Estimated CH_2Cl_2 consumed, 10^3 t	Fraction, %
aircraft refurbishing	16.75	25
durable goods recycling	6.03	9
furniture refinishing and restoration	11.40	17
military	11.40	17
original equipment manufacturers	2.70	4
retail products	16.75	25
transportation	2.00	3
Total	*67.03*	*100*

[a]Ref. 12.

There are approximately 114 paint remover manufacturers in the United States. Among them are companies that specialize in the manufacture of paint removers only, paint and coating manufacturers that produce a line of paint removers, specialty products manufacturers, and other manufacturers of finish removers for unique or special applications. Each group of paint remover manufacturer can be divided into retail or industrial products. A list of U.S. paint remover manufacturers is given in Table 2.

Table 2. Paint and Coating Remover Manufacturers[a]

Manufacturer	Location	Category[b] Paint removers	Category[b] Paint and coating	Category[b] Specialty products
	Industrial			
Certified Coatings Products	Los Angeles, Calif.		+	
Turco Products	Westminster, N.Y.	+		
Stripping Products, Inc.	Bethel, Conn.	+		
James B. Day	Carpentersville, Ill.		+	
Reliable Finishing Products	Elk Grove, Ill.		+	
Dober Chemical	Midlothian, Ill.			
Kwick Kleen Industrial Solvents, Inc.	Vincennes, Ind.	+		
Amax Industrial Products	Louisville, Ky.			+
Sterling-Clark-Lurton	Malden, Mass.	+		
Dell Marking Systems	Ferndale, Mich.			+
Gage Specialty Products	Ferndale, Mich.			+

Table 2. (*Continued*)

Manufacturer	Location	Category[b]		
		Paint removers	Paint and coating	Specialty products
Chem-Elast Coatings	St. Louis, Mo.			+
Sentry/Custom Services	Allamuchy, N.Y.			+
Oakite Products	Berkley Heights, N.J.			+
Bryn Mawr Concrete Maint.	Carlstadt, N.J.			+
Octagon Process, Inc.	Edgewater, N.J.	+		
Dynaloy, Inc.	Hanover, N.J.			+
Orelite Technical Coatings	Irvington, N.J.		+	
Penetone Corp.	Tenafly, N.J.	+		
AS Chemicals, Inc.	Bronx, N.Y.	+		
U.S. Chemical & Plastics	Canton, Ohio			+
Scot Laboratories	Chagrin Falls, Ohio			+
Texo Corp.	Cincinnati, Ohio	+		
Man-Gill Chemical	Cleveland, Ohio			
Watson-Standard	Harwick, Pa.		+	
Stuart-Ironsides, Inc.	Chicago, Ill.	+		
Benco Sales, Inc.	Crossville, Tenn.	+		
Besway Chemical Systems	Madison, Tenn.	+		
Ecco Chemicals	Dallas, Tex.	+		
Greater Southwest Chemicals	Dallas, Tex.	+		
Panther Industries	Fort Worth, Tex.	+		
CD Products	Appleton, Wis.	+		
Crown Paint Co.	Oklahoma City, Okla.		+	
Nutec	Seattle, Wash.			
B & B	Miami, Fla.	+		
Custom Chemical Eng.	Springfield, Ill.	+		
Diversey Corp.	Wyandotte, Mich.	+		
Retail				
Fuller-O'Brien Paints	San Francisco, Calif.		+	
Sunnyside Corp.	Wheeling, Ill.		+	
Savogran Co.	Norwood, Mass.	+		
Parks Co.	Sumerset, Mass.	+		
Thompson & Formby, Inc.	Olive Branch, Miss.	+		
Bix Mfg. Co., Inc.	Ashland City, Tenn.	+		
Sansher Corp.	Ft. Wayne, Ind.	+		
Dumond	New York, N.Y.	+		
Creative Technology	Charlotte, N.C.	+		
Industrial and retail				
Jasco Chemical	Mountain View, Calif.	+		
Rap Products, Inc.	Bay City, Mich.	+		
Reliable Remover & Lacquer	Irvington, N.J.	+		
Wilson-Imperial Co.	Newark, N.J.	+		
Red Devil Coatings	Mt. Vernon, N.Y.		+	
Star Bronze Co.	Alliance, Ohio	+		
Sherwin-Williams	Cleveland, Ohio		+	
National Solvents Corp.	Medina, Ohio	+		
W. M. Barr	Memphis, Tenn.	+		
Prillaman Chemical Corp.	Martinsville, Va.	+		
Zep Manufacturing	Atlanta, Ga.	+		
Green Chemical	Richmond, Calif.	+		

Table 2. (*Continued*)

Manufacturer	Location	Category[b] Paint removers	Paint and coating	Specialty products
	Unspecified			
National Aerosol Products	Los Angeles, Calif.		+	
Charles Crosbie Laboratories	Van Huy, Calif.			+
U.S. Technology	Danielson, Conn.			+
Composition Materials Co.	Fairfield, Conn.			
Special Machine & Tool	Forestville, Conn.			
Machinery Services	Hartford, Conn.			
Mitchell-Bradford	Milford, Conn.			+
Permatix Industries	Newington, Conn.			+
Du Pont	Wilmington, Del.			+
Lester Laboratories	Atlanta, Ga.			+
Pave-Mark Corp.	Atlanta, Ga.			+
Craftsman Chemical	Marietta, Ga.			+
Urban Chemical Co.	Deerfield, Ill.			
KCI Chemical	Matteson, Ill.			
Enterprise Companies	Wheeling, Ill.		+	
Brulin & Co.	Indianapolis, Ind.		+	
Maxi-Blast, Inc.	South Bend, Ind.			
Higley Chemical Co.	Dubuque, Iowa	+		
Alvin Products, Inc.	Worchester, Mass.	+		
Gibral Tar National Co.	Detroit, Mich.			+
Parker & Amchem	Madison Heights, Wis.			+
Hillyard Chemical Co.	St. Joseph, Mo.			+
Painter Design & Eng.	New Baltimore, Mich.			
H. F. Staples	Merrimack, N.H.			
Harley Chemicals	Camden, N.J.			+
Chem Power Mfg.	Cedar Knolls, N.J.			+
Bio Plex Environmental	Edgewood Cliffs, N.J.			+
PPS Industries	Manalapan, N.J.			+
Fidelity Chemical Products	Newark, N.J.			+
International Paint	Union, N.J.		+	
H. Behlen	Amsterdam, N.J.			
Rite-Off, Inc.	Bay Shore, N.J.			+
Nuvite Chemical Compounds	Brooklyn, N.Y.			+
Pyrock Chemical Corp.	Brooklyn, N.Y.			+
Pride Group	Farmingdale, N.Y.			+
Beck Chemicals	Cleveland, Ohio			+
Ensign Products	Cleveland, Ohio			
Excelsior Varnish & Chem.	Cleveland, Ohio		+	
Glidden Paint	Cleveland, Ohio		+	
Revere Products	Cleveland, Ohio			+
Ashland Chemicals	Columbus, Ohio			+
Pioneer Chemical Mfg.	West Point, Ohio	+		
Tower Chemical Corp.	Easton, Pa.			+
Delaware Valley Paint	Philadelphia, Pa.		+	
Haas Corp.	Philadelphia, Pa.			
International Chemical Co.	Philadelphia, Pa.			+
Beaver Alkali Products	Rochester, Pa.			
United Gilsonite Lab	Scranton, Pa.			+
CRC Industries	Warminster, Pa.			+

1081

Table 2. (*Continued*)

Manufacturer	Location	Category[b] Paint removers	Paint and coating	Specialty products
EZE Products, Inc.	Greenville, S.C.			+
Delta Foremost Chemical Corp.	Memphis, Tenn.			
Texas Refinery Corp.	Fort Worth, Tex.			+
Amity	Sun Prairie, Wis.	+		
Dumond	New York, N.Y.			
3-M	St. Paul, Minn.			+
American Building Restoration	Franklin, Wis.			
Eldorado Chemicals	San Antonio, Tex.	+		

[a]Ref. 2.

[b]If no + appears for a manufacturer, they produce some other category of finish remover.

BIBLIOGRAPHY

"Paint Removers" in *ECT* 1st ed., Vol. 9, pp. 803–804, by B. N. Allnutt, Chemical Products Co., Inc., and J. R. Holland, Wiley & Co., Inc.; "Paint and Varnish Removers" in *ECT* 2nd ed., Vol. 14, pp. 485–493, by R. S. Downing, Sherwin-Williams Co.; in *ECT* 3rd ed., Vol. 16, pp. 762–768, by W. R. Mallarnee, The Sherwin-Williams Co.

1. G. A. Soderberg, *Finishing Technol.* **3** (1969).
2. *Industrial Users of Paint and Finish Removers*, Paint Remover Manufacturer's Association, 1992.
3. *Solvents Used in Paint Removers*, Paint Remover Manufacturer's Association, Sept. 1991.
4. U.S. Pat. 4,120,810 (Oct. 17, 1978), D. A. Palmer.
5. U.S. Pat. 1,023,213 (Nov. 13, 1962), (to Omage Chemicals Corp.).
6. U.S. Pat. 4,645,617 (Feb. 24, 1987), T. A. Vivian (to Dow Chemical Co.).
7. *Dow Methylene Chloride—An Effective Solvent for Industry*, Dow Chemical Co., 1981, p. 3.
8. W. B. Gerritsen and C. H. Buschmann, *Brit. J. Indust. Med.* **17**, 187 (1960).
9. *Chemical Protective Clothing for Furniture Stripping*, Department of Health and Human Services, Washington, D.C., Mar. 11, 1991.
10. *Exxate Solvents are Setting the Pace in Paint Strippers*, Exxon Chemicals, 1989, p. 3.
11. K. Claunch, *Solvent Recovery From Waste Solvents Containing Nitrocellulose*, Aug. 1, 1985.
12. *Methylene Chloride Consumption By Paint and Coating Removal Groups*, Paint Remover Manufacturer's Association, 1992.

DAVID L. WHITE
Kwick Kleen Industrial Solvents, Inc.

JAY A. BARDOLE
Vincennes University

PALLADIUM.   See PLATINUM-GROUP METALS.